Periodic Table of the Elements

Group						
	13 IIIA	14 IVA	15 VA	16 VIA	17 VIIA	18 VIIIA

Period 1
- 2 | −269 / −272 / 0.179 | **He** helium | 4.00

Period 2
- 5 | 4000 / 2075 / — ; 2.0 ; 2.34 | **B** boron | 10.81
- 6 | 4027 / 3527 / 2.26 ; 2.6 | **C** carbon | 12.01
- 7 | −196 / −210 / 1.25 ; 3.0 ; 3− | **N** nitrogen | 14.01
- 8 | −183 / −219 / 1.43 ; 3.4 ; 2− | **O** oxygen | 16.00
- 9 | −188 / −220 / 1.70 ; 4.0 ; 1− | **F** fluorine | 19.00
- 10 | −246 / −249 / 0.900 | **Ne** neon | 20.18

Period 3
- 13 | 2519 / 660 / 2.70 ; 1.6 ; 3+ | **Al** aluminium | 26.98
- 14 | 3265 / 1414 / 2.33 ; 1.9 | **Si** silicon | 28.09
- 15 | 281 / 44 / 1.82 ; 2.2 ; 3− | **P** phosphorus | 30.97
- 16 | 445 / 115 / 2.07 ; 2.6 ; 2− | **S** sulfur | 32.07
- 17 | −34 / −101 / 3.21 ; 3.2 ; 1− | **Cl** chlorine | 35.45
- 18 | −186 / −189 / 1.78 ; 3.0 | **Ar** argon | 39.95

10	11 IB	12 IIB

Period 4
- 28 | 2913 / 1455 / 8.90 ; 1.9 ; 2+, 3+ | **Ni** nickel | 58.69
- 29 | 2562 / 1085 / 8.92 ; 1.9 ; 2+, 1+ | **Cu** copper | 63.55
- 30 | 907 / 420 / 7.14 ; 1.7 ; 2+ | **Zn** zinc | 65.41
- 31 | 2204 / 30 / 5.90 ; 1.8 ; 3+ | **Ga** gallium | 69.72
- 32 | 2833 / 938 / 5.35 ; 2.0 ; 4+ | **Ge** germanium | 72.64
- 33 | 614 / 817 / 5.73 ; 2.2 ; 3− | **As** arsenic | 74.92
- 34 | 685 / 221 / 4.81 ; 2.6 ; 2− | **Se** selenium | 78.96
- 35 | 59 / −7 / 3.12 ; 3.0 ; 1− | **Br** bromine | 79.90
- 36 | −153 / −157 / 3.74 ; 3.0 | **Kr** krypton | 83.80

Period 5
- 46 | 2963 / 1555 / 12.0 ; 2.2 ; 2+, 3+ | **Pd** palladium | 106.42
- 47 | 2162 / 962 / 10.5 ; 1.9 ; 1+ | **Ag** silver | 107.87
- 48 | 767 / 321 / 8.64 ; 1.7 ; 2+ | **Cd** cadmium | 112.41
- 49 | 2072 / 157 / 7.30 ; 1.8 ; 3+ | **In** indium | 114.82
- 50 | 2602 / 232 / 7.31 ; 2.0 ; 4+, 2+ | **Sn** tin | 118.71
- 51 | 1587 / 631 / 6.68 ; 2.1 ; 3+, 5+ | **Sb** antimony | 121.76
- 52 | 988 / 450 / 6.2 ; 2.1 ; 2− | **Te** tellurium | 127.60
- 53 | 184 / 114 / 4.93 ; 2.7 ; 1− | **I** iodine | 126.90
- 54 | −108 / −112 / 5.89 ; 2.6 | **Xe** xenon | 131.29

Period 6
- 78 | 3825 / 1768 / 21.5 ; 2.2 ; 4+, 2+ | **Pt** platinum | 195.08
- 79 | 2856 / 1064 / 19.3 ; 2.4 ; 3+, 1+ | **Au** gold | 196.97
- 80 | 357 / −39.0 / 13.5 ; 1.9 ; 2+, 1+ | **Hg** mercury | 200.59
- 81 | 1473 / 304 / 11.85 ; 1.8 ; 1+, 3+ | **Tl** thallium | 204.38
- 82 | 1749 / 327 / 11.3 ; 1.8 ; 2+, 4+ | **Pb** lead | 207.2
- 83 | 1564 / 271 / 9.80 ; 1.9 ; 3+, 5+ | **Bi** bismuth | 208.98
- 84 | 962 / 254 / 9.40 ; 2.0 ; 4+ | **Po** polonium | (209)
- 85 | — / 302 / — ; 2.2 ; 1− | **At** astatine | (210)
- 86 | −62 / −71 / 9.73 | **Rn** radon | (222)

Period 7
- 110 | **Ds** darmstadtium | (281)
- 111 | **Rg** roentgenium | (272)
- 112 | **Uub** ununbium
- 113 | **Uut** ununtrium
- 114 | **Uuq** ununquadium
- 115 | **Uup** ununpentium
- 116 | **Uuh** ununhexium
- 117 | **Uus** ununseptium
- 118 | **Uuo** ununoctium

Lanthanides
- 63 | 1529 / 822 / 5.24 ; — ; 3+, 2+ | **Eu** europium | 151.96
- 64 | 3273 / 1313 / 7.90 ; 1.2 ; 3+ | **Gd** gadolinium | 157.25
- 65 | 3230 / 1356 / 8.23 ; — ; 3+ | **Tb** terbium | 158.93
- 66 | 2567 / 1412 / 8.55 ; 1.2 ; 3+ | **Dy** dysprosium | 162.50
- 67 | 2700 / 1474 / 8.80 ; 1.2 ; 3+ | **Ho** holmium | 164.93
- 68 | 2868 / 1529 / 9.07 ; 1.2 ; 3+ | **Er** erbium | 167.26
- 69 | 1950 / 1545 / 9.32 ; 1.3 ; 3+ | **Tm** thulium | 168.93
- 70 | 1196 / 819 / 6.97 ; — ; 3+, 2+ | **Yb** ytterbium | 173.04
- 71 | 3402 / 1663 / 9.84 ; 1.3 ; 2+ | **Lu** lutetium | 174.97

Actinides
- 95 | 2011 / 1176 / 13.7 ; — ; 3+, 4+ | **Am** americium | (243)
- 96 | 3100 / 1345 / 13.5 ; 1.3 ; 3+ | **Cm** curium | (247)
- 97 | — / 1050 / 14 ; 1.3 ; 3+, 4+ | **Bk** berkelium | (247)
- 98 | — / 900 / — ; 1.3 ; 3+ | **Cf** californium | (251)
- 99 | — / 860 / — ; 1.3 ; 3+ | **Es** einsteinium | (252)
- 100 | — / 1527 / — ; 1.3 ; 3+ | **Fm** fermium | (257)
- 101 | — / 827 / — ; 1.3 ; 2+, 3+ | **Md** mendelevium | (258)
- 102 | — / 827 / — ; 1.3 ; 2+, 3+ | **No** nobelium | (259)
- 103 | — / 1627 / — ; — ; 3+ | **Lr** lawrencium | (262)

Measured values are subject to change as experimental techniques improve. Atomic molar mass values in this table are based on IUPAC Web site values (2005). Some melting points can only be measured at higher than standard pressures.

Chemistry

Alberta 20–30

Authors

Dr. Frank Jenkins

Hans van Kessel

Dick Tompkins

Dr. Oliver Lantz

THIS BOOK IS THE PROPERTY OF THE
ALBERTA DISTANCE LEARNING CENTRE
BOX 4000, BARRHEAD, AB   T7N 1P4

N.B. DO NOT LOSE DESTROY OR DEFACE
THIS BOOK. THE PERSON WHO RENTS
IT NEXT YEAR WILL WANT A BOOK WHICH
IS WELL KEPT

NELSON EDUCATION

NELSON EDUCATION

Chemistry Alberta 20-30

Authors
Dr. Frank Jenkins
Hans van Kessel
Dick Tompkins
Dr. Oliver Lantz

Director of Publishing
Beverley Buxton

General Manager, Mathematics, Science, and Technology
Lenore Brooks

Publisher, Science
John Yip-Chuck

Executive Managing Editor, Development
Cheryl Turner

Managing Editors, Development
Susan Ball
Lois Beauchamp

Product Manager
Paul Masson

Program Manager
Julia Lee

Educational Publishing Consultant
Trudy Rising

Developmental Editors
Julie Bedford
Jenna Dunlop
Julia Lee
Lina Mockus
Betty Robinson

Editorial Assistants
Jacquie Busby
Christina D'Alimonte

Aisha Hammah
Alisa Yampolsky

Executive Director, Content and Media Production
Renate McCloy

Director, Content and Media Production
Linh Vu

Senior Content Production Editors
Deborah Lonergan
Sheila Stephenson

Copy Editor
Ruth Peckover

Proofreaders
Christine Hobberlin
Gilda Mekler
Paul Pettitt-Townsend
Linda Szostak

Indexer
Noeline Bridge

Production Manager
Cathy Deak

Production Coordinators
Sharon Latta Paterson
Helen Locsin
Kathrine Pummell

Design Director
Ken Phipps

Art Management
Suzanne Peden

Illustrators
Andrew Breithaupt
Steven Corrigan
Deborah Crowle
Irma Ikonen
Dave Mazierski
Dave McKay
Peter Papayanakis
Ken Phipps
Marie Price
Katherine Strain

Interior Design
Kyle Gell
Allan Moon

Cover Design
Johanna Liburd

Cover Image
© Daryl Benson/Masterfile

Compositors
Zenaida Diores
Nelson Gonzalez

Photo/Permissions Researcher
Robyn Craig

Printer
Transcontinental

COPYRIGHT © 2007 by Nelson Education Ltd.

ISBN-13: 978-0-17-6289300
ISBN-10: 0-17-6289305

Printed and bound in Canada
8 9 10 16 15 14

For more information contact Nelson Education Ltd., 1120 Birchmount Road, Toronto, Ontario, M1K 5G4. Or you can visit our Internet site at http://www.nelson.com

ALL RIGHTS RESERVED. No part of this work covered by the copyright herein, except for any reproducible pages included in this work, may be reproduced, transcribed, or used in any form or by any means— graphic, electronic, or mechanical, including photocopying, recording, taping, Web distribution, or information storage and retrieval systems— without the written permission of the publisher.

For permission to use material from this text or product, submit all requests online at www.cengage.com/permissions. Further questions about permissions can be emailed to permissionrequest@cengage.com

Every effort has been made to trace ownership of all copyrighted material and to secure permission from copyright holders. In the event of any question arising as to the use of any material, we will be pleased to make the necessary corrections in future printings.

Note: Photo credits are found at the end of the book.

Reviewers and Advisors

Teacher Reviewers

Mike Busby
Chemistry Teacher, Hunting Hills High School, Red Deer School District No.104

Edie Ferris
Chemistry Teacher, Edmonton School District No.7

Len Huculak
(Formerly) District Principal, Edmonton Catholic Separate School District No.7

Kevin Klemmer
Central Memorial High School, Calgary School District No.19

Don Langer
Chemistry Teacher, (Formerly) Calgary Roman Catholic Separate School District No. 1

Ed Leong
Science Curriculum Leader, James Fowler High School, Calgary School District No.19

Narsh Ramrattan
Peace River High School, Peace River School Division No.10

Olof J. Sandblom
Calgary School District No.19

Terry Sliwkanich
B.Sc., B.Ed., Fort Saskatchewan High School, Elk Island Public Schools Regional Division No.14

Matthew Wiebe
Ponoka Composite High School, Wolf Creek School Division No.72

Aboriginal Education Consultants

Dean Cunningham
Consultant, Edmonton

Frank Elliott
Ph.D., Amiskwaciy Academy, Edmonton School District No.7

Accuracy Reviewers

Dr. Lucio Gelmini
Chair-Chemistry, Grant MacEwan College

Dr. Peter Mahaffy
Professor of Chemistry, The King's University College

Dr. Sieghard E. Wanke
Professor, Department of Chemical and Materials Engineering, University of Alberta

Web Quest Writers

Mike Busby
Chemistry Teacher, Hunting Hills High School, Red Deer School District No.104

Craig Frelich
Strathcona-Tweedsmuir School, Private

Milan Sanader
Department Head of Science, Holy Name of Mary Secondary School, Dufferin-Peel RCSSB

Audio Clip Writer

Jayni Caldwell
Foothills Composite High School, Foothills School Division No. 38

Assessment Consultants

Gary Glover
Strathcona-Tweedsmuir School, Private

Dan Leskiw
Edmonton School District No.7

Careers Consultant

Art Bauer
Science Coordinator, Living Waters Catholic Regional Division No. 42

Literacy Reviewer

Mary McDougall
(Formerly) Secondary Science Consultant, Calgary Roman Catholic Separate School District No. 1

Safety Reviewer

Art Bauer
Science Coordinator, Living Waters Catholic Regional Division No. 42

Technology Consultants

Gary Glover
Strathcona-Tweedsmuir School, Private

William Konrad
Boreal-Northwest

Dr. Norma Nocente
Associate Professor, Faculty of Education, University of Alberta

International Baccalaureate Advisors

Peggy Au
Winston Churchill High School, Lethbridge School District No.51

Natalie Deptuck
(Formerly) Chemistry Teacher, Harry Ainlay School, Edmonton School District No.7

Kay J. Jauch
McNally School, Edmonton School District No.7

Gary Glover
Strathcona-Tweedsmuir School, Private

Dr. Frank Jenkins
University of Alberta

Hans van Kessel
(Formerly) Bellerose Composite High School, St. Albert Protestant Separate School District No.6

Field Test Schools

Alberta Education Field Test Schools

Calgary Roman Catholic Separate School District No.1: Bishop Grandin High School, Father Lacombe High School

Calgary School District No.19: Lord Beaverbrook

Chinooks Edge School Division No.73: Olds Jr. Sr. High School

Edmonton Public School District No.7: Strathcona High School

Fort McMurray Roman Catholic Separate School District No.32: Father Mercredi Catholic High School

Grande Yellowhead Regional Division No.35: Harry Collinge High School

Grasslands Regional Division No.6: Brooks Composite School

Parkland School Division No.70: Memorial Composite

Prairie Rose Regional Division No.8: South Central High School

Private: Concordia High School

St. Albert Protestant Separate School District No.6: Bellerose Composite High School

Thomson Nelson Field Test Schools

Buffalo Trail Public Schools Regional Division No.28: Kitscoty Junior Senior High School

Calgary School District No.19: Centennial High School, James Fowler High School, John G. Diefenbaker

Edmonton Catholic Separate School District No.7: J.H. Picard,

Edmonton Public School District No.7: Harry Ainlay, Jasper Place, Victoria

Elk Island Catholic Separate Regional Division No.41: St. Mary's School

Elk Island Public Schools Regional Division No.14: Salisbury Composite

Foothills School Division No.38: Foothills Composite High School

Grande Prairie Roman Catholic Separate School Division No.28: St. Joseph Catholic High

Lethbridge School District No.51: Lethbridge Collegiate Institute

Northern Gateway Regional Division No.10: Onoway High School

Pembina Hills Regional Division No.7: Swan Hills School

Red Deer Catholic Regional Division: Notre Dame High School

Red Deer School District No.104: Hunting Hills High School

Rocky View School District No.41: Bert Church High School, Bow Valley High School

St. Thomas Aquinas Roman Catholic Separate Regional Division No.38: St. Augustine School

Sturgeon School Division No.24: Sturgeon Composite

Wolf Creek School Division No.72: Lacombe Composite High School

CONTENTS

Your Guide to this Textbook — viii

▶ Review Unit: Chemistry Review

Are You Ready? — 4

Chapter 1 Elements and Compounds — 6
- Exploration: Combustion of Magnesium (Demonstration) — 7
- 1.1 Introduction: Science and Technology — 8
- 1.2 Classifying Matter — 12
- 1.3 Classifying Elements — 14
  - Web Activity: Case Study–Groups of Elements — 15
- 1.4 Theories and Atomic Theories — 18
  - Web Activity: Simulation–The Rutherford Scattering Experiment — 20
  - Web Activity: Simulation–Emission and Absorption Spectra for Hydrogen — 23
  - Web Activity: Canadian Achievers–Harriet Brooks — 25
- 1.5 Classifying Compounds — 27
- 1.6 Molecular Elements and Compounds — 33

Chapter 1 Summary — 38
Chapter 1 Review — 39

Chapter 2 Chemical Reactions — 42
- Exploration: Molecules Making Magic (Demonstration) — 43
- 2.1 Science and Technology in Society — 44
- 2.2 Changes in Matter — 46
  - Web Activity: Case Study–States of Matter and Changes in Matter — 46
- 2.3 Balancing Chemical Reaction Equations — 51
- 2.4 Chemical Amount — 55
- 2.5 Classifying Chemical Reactions — 58
- 2.6 Chemical Reactions in Solution — 61

Chapter 2 Summary — 65
Chapter 2 Review — 66
Unit Review — 68

▶ Unit 1: Chemical Bonding—Explaining the Diversity of Matter

Are You Ready? — 74

Chapter 3 Understanding Chemical Compounds — 76
- Exploration: Properties and Forces — 77
- 3.1 Bonding Theory and Lewis Formulas — 78
  - Explore an Issue: Funding Research and Development — 79
- 3.2 Explaining Molecular Formulas — 85
- 3.3 Molecular Shapes and Dipoles — 91
  - Web Activity: Web Quest–Cleaning Up Dry Cleaning — 104
- 3.4 Intermolecular Forces — 105
  - Web Activity: Canadian Achievers–Gerhard Herzberg — 109
  - Lab Exercise 3.A: Boiling Points and Intermolecular Forces — 110
  - Web Activity: Web Quest–Cloud Seeding — 112
- Case Study: Current Research in Intermolecular Forces — 114
- Web Activity: Simulation–Modelling Molecules — 117
- 3.5 Structures and Physical Properties of Solids — 119
  - Web Activity: Canadian Achievers–Jillian Buriak — 123

Chapter 3 Investigations — 131
- Investigation 3.1: Molecular Models — 131
- Investigation 3.2: Evidence for Polar Molecules — 131
- Investigation 3.3: Molecular Compound Melting Points — 132
- Investigation 3.4: Hydrogen Bonding — 134
- Investigation 3.5: Classifying Unknown Solids — 134

Chapter 3 Summary — 135
Unit 1 Review — 137

▶ Unit 2: Gases as a Form of Matter

Are You Ready? — 144

Chapter 4 Gases — 146
- Exploration: Creating and Testing Ideas about Gases — 147
- 4.1 Empirical Properties of Gases — 148
  - Lab Exercise 4.A: A Thought Experiment about Gas Properties — 148
  - Web Activity: Simulation–The Combined Gas Law — 157
  - Web Activity: Canadian Achievers–Malcolm King — 158
  - Case Study: Compressed Gases — 159
- 4.2 Explaining the Properties of Gases — 163
  - Web Activity: Canadian Achievers–Elizabeth MacGill — 164
  - Web Activity: Web Quest–"Designer Air" for Tires — 166
  - Case Study: Weather Forecasts — 167
- 4.3 Molar Volume of Gases — 169
- 4.4 The Ideal Gas Law — 172
  - Web Activity: Simulation–The Ideal Gas Law — 173
  - Lab Exercise 4.B: Evaluating an Experimental Design — 175

Chapter 4 Investigations — 177
- Investigation 4.1: Pressure and Volume of a Gas — 177
- Investigation 4.2: Temperature and Volume of a Gas — 178
- Investigation 4.3: Using the Ideal Gas Law — 179

Chapter 4 Summary — 180
Unit 2 Review — 181

▶ Unit 3: Solutions, Acids, and Bases

Are You Ready? — 188

Chapter 5 The Nature and Properties of Solutions — 190
- Exploration: Substances in Water (Demonstration) — 191
- 5.1 Solutions and Mixtures — 192
  - Lab Exercise 5.A: Identifying Solutions — 195
- 5.2 Explaining Solutions — 197
  - Lab Exercise 5.B: Qualitative Analysis — 202
- 5.3 Solution Concentration — 203
  - Web Activity: Canadian Achievers–David Schindler — 206

- Case Study: Household Chemical Solutions — 212
- Web Activity: Web Quest–Hot Tub Safety — 213
- **5.4** Preparation of Solutions — 215
- **5.5** Solubility — 220
  - Lab Exercise 5.C: Solubility and Temperature — 223
  - Explore an Issue: Pesticides — 224
- **Chapter 5 Investigations** — 227
  - Investigation 5.1: Qualitative Chemical Analysis — 227
  - Investigation 5.2: A Standard Solution from a Solid — 227
  - Investigation 5.3: A Standard Solution by Dilution — 228
  - Investigation 5.4: The Iodine Clock Reaction — 228
  - Investigation 5.5: The Solubility of Sodium Chloride in Water — 229
- **Chapter 5 Summary** — 230
- **Chapter 5 Review** — 231

Chapter 6 Acids and Bases — 234
- Exploration: Consumer Products — 235
- **6.1** Properties of Acids and Bases — 236
- **6.2** pH and pOH Calculations — 238
  - Lab Exercise 6.A: The Relationship between pH and Hydronium Ion Concentration — 240
  - Web Activity: Web Quest–Bad Hair Day? — 243
- **6.3** Acid-Base Indicators — 245
  - Lab Exercise 6.B: Using Indicators to Determine pH — 247
- **6.4** Explaining Acids and Bases — 248
  - Case Study: Acid Deposition — 252
- **6.5** The Strength of Acids and Bases — 254
- **Chapter 6 Investigations** — 260
  - Investigation 6.1: Properties of Acids and Bases — 260
  - Investigation 6.2: Testing Arrhenius' Acid-Base Definitions — 260
  - Investigation 6.3: Comparing the Properties of Acids (Demonstration) — 261
- **Chapter 6 Summary** — 262
- **Chapter 6 Review** — 263
- **Unit 3 Review** — 265

▶ Unit 4: Quantitative Relationships in Chemical Changes

Are You Ready? — 272

Chapter 7 Stoichiometry — 274
- Exploration: The Problem Is What You Don't See! — 275
- **7.1** Interpreting Chemical Reaction Equations — 276
  - Web Activity: Canadian Achievers–Roberta Bondar — 278
- **7.2** Gravimetric Stoichiometry — 286
  - Lab Exercise 7.A: Testing the Stoichiometric Method — 291
  - Lab Exercise 7.B: Testing a Chemical Process — 293
- **7.3** Gas Stoichiometry — 294
  - Case Study: Producing Hydrogen for Fuel Cells — 297
- **7.4** Solution Stoichiometry — 300
  - Lab Exercise 7.C: Testing Solution Stoichiometry — 302
  - Lab Exercise 7.D: Determining a Solution Concentration — 303
- **Chapter 7 Investigations** — 304
  - Investigation 7.1: Decomposing Malachite — 304
  - Investigation 7.2: Gravimetric Stoichiometry — 305
  - Investigation 7.3: Producing Hydrogen — 305
  - Investigation 7.4: Analysis of Silver Nitrate (Demonstration) — 307
- **Chapter 7 Summary** — 308
- **Chapter 7 Review** — 309

Chapter 8 Chemical Analysis — 312
- Exploration: Test Your Drinking Water — 313
- **8.1** Introduction to Chemical Analysis — 314
  - Web Activity: Web Quest–Is Your Classroom Putting You to Sleep? — 316
- **8.2** Gravimetric Analysis — 317
  - Lab Exercise 8.A: Chemical Analysis Using a Graph — 317
- **8.3** Stoichiometry: Limiting and Excess Reagent Calculations — 320
  - Web Activity: Canadian Achievers–Ursula Franklin — 321
  - Case Study: The Haber Process — 325
- **8.4** Titration Analysis — 328
- **8.5** Acid-Base Titration Curves and Indicators — 333
  - Web Activity: Web Quest–Blood Alcohol Content — 333
  - Web Activity: Simulation–Titration Curves — 337
  - Case Study: Analytic Measurement Technology — 337
- **Chapter 8 Investigations** — 340
  - Investigation 8.1: Analysis of Sodium Carbonate — 340
  - Investigation 8.2: Percent Yield of Barium Sulfate — 341
  - Investigation 8.3: Standardization Analysis of NaOH(aq) (Demonstration) — 342
  - Investigation 8.4: Titration Analysis of Vinegar — 343
  - Investigation 8.5: pH Curves (Demonstration) — 344
  - Investigation 8.6: Titration Analysis of ASA — 345
- **Chapter 8 Summary** — 346
- **Chapter 8 Review** — 347
- **Unit 4 Review** — 349

▶ Unit 5: Organic Chemistry

Are You Ready? — 354

Chapter 9 Hydrocarbons from Petroleum — 356
- Exploration: Burning Fossil Fuels — 357
- **9.1** Fossil Fuels — 358
  - Case Study: Fossil Fuel Industries in Alberta — 359
  - Web Activity: Case Study–Coal in Alberta — 361
- **9.2** Alkanes from Natural Gas — 362
  - Explore an Issue: Coalbed Methane — 365
- **9.3** Alkenes and Alkynes–Cracking Natural Gas — 374
- **9.4** Aromatics — 381
  - Web Activity: Web Quest–West Nile Denial — 383
  - Lab Exercise 9.A: Chemical Properties of Aliphatics and Aromatics — 384
  - Lab Exercise 9.B: Boiling Points of Sample Aliphatics and Aromatics — 384
- **9.5** Crude Oil Refining — 386
  - Web Activity: Canadian Achievers–Karl Chuang — 390

- Case Study: Octane Number — 392
- Case Study: The Athabasca Oil Sands — 395

9.6 Complete and Incomplete Combustion Reactions — 398

Chapter 9 Investigations — 401
- Investigation 9.1: Classifying Fossil Fuels — 401
- Investigation 9.2: Structures and Properties of Isomers — 401
- Investigation 9.3: Fractional Distillation (Demonstration) — 402
- Investigation 9.4: Bitumen from Oil Sands — 403
- Investigation 9.5: Solvent Extraction — 404
- Investigation 9.6: Complete and Incomplete Combustion — 404

Chapter 9 Summary — 405
Chapter 9 Review — 406

Chapter 10 Hydrocarbon Derivatives, Organic Reactions, and Petrochemicals — 410
- Exploration: Burning Fossil Fuels — 411

10.1 Petrochemicals in Alberta — 412

10.2 Organic Halides and Addition and Substitution Reactions — 417
- Lab Exercise 10.A: Synthesis of an Organic Halide — 423

10.3 Alcohols and Elimination Reactions — 425
- Web Activity: Web Quest–Cellulosic Ethanol — 434

10.4 Carboxylic Acids, Esters, and Esterification Reactions — 436
- Lab Exercise 10.B: Explaining Physical Property Trends — 437

10.5 Polymerization Reactions–Monomers and Polymers — 445
- Web Activity: Web Quest–Teflon: Healthy or Hazardous — 448
- Web Activity: Case Study–Recycling Plastics — 449
- Web Activity: Simulation–Molecular Modelling — 452
- Explore an Issue: Natural or Artificial Polymers? — 455
- Web Activity: Case Study–Cellulose Acetate — 458

Chapter 10 Investigations — 461
- Investigation 10.1: Substitution and Addition Reactions — 461
- Investigation 10.2: Isomers of Butanol — 462
- Investigation 10.3: Synthesis of an Ester — 463
- Investigation 10.4: Testing with Models — 464
- Investigation 10.5: Preparing Nylon 6,10 (Demonstration) — 464

Chapter 10 Summary — 465
Chapter 10 Review — 466

Unit 6: Chemical Energy

Unit 5 Review — 468

Are You Ready? — 476

Chapter 11 Enthalpy Change — 478
- Exploration: Burning Oil — 479

11.1 Energy Demands and Sources — 480
- Case Study: Personal Use of Chemical Energy — 482

11.2 Calorimetry — 485
- Web Activity: Case Study–Thermal Insulation — 490
- Lab Exercise 11.A: Molar Enthalpy of Neutralization — 492
- Web Activity: Simulation–Calorimetry — 493

11.3 Communicating Enthalpy Changes — 495

11.4 Hess' Law — 502
- Web Activity: Simulation–Hess' Law — 506
- Lab Exercise 11.B: Testing Hess' Law — 506
- Lab Exercise 11.C: Analysis Using Hess' Law — 506
- Explore an Issue: Alternative Energy Sources and Technologies — 507

11.5 Molar Enthalpies of Formation — 510
- Lab Exercise 11.D: Testing $\Delta_r H°$ from Formation Data — 513
- Web Activity: Web Quest–Rocket Fuel Thermochemistry — 513

Chapter 11 Investigations — 516
- Investigation 11.1: Designing and Evaluating a Calorimeter — 516
- Investigation 11.2: Molar Enthalpy of Reaction — 516
- Investigation 11.3: Applying Hess' Law — 517

Chapter 11 Summary — 518
Chapter 11 Review — 519

Chapter 12 Explaining Chemical Changes — 522
- Exploration: Starting, Comparing, and Altering Reactions — 523

12.1 Activation Energy — 524
- Web Activity: Simulation–Collisions and Reactions — 526
- Web Activity: Simulation–Collision-Reaction Theory — 529
- Web Activity: Web Quest–Neurotransmitters and Nerve Agents — 530

12.2 Bond Energy and Reactions — 532
- Web Activity: Canadian Achievers–John Polanyi — 534

12.3 Catalysis and Reaction Rates — 535
- Web Activity: Simulation –A Catalyzed Reaction — 539

Chapter 12 Investigations — 543
- Investigation 12.1: Iodine Clock Reaction — 543
- Investigation 12.2: Evidence for an Activated Complex — 543

Chapter 12 Summary — 544
Chapter 12 Review — 545

Unit 7: Electrochemistry

Unit 6 Review — 547

Are You Ready? — 554

Chapter 13 Redox Reactions — 556
- Exploration: Cleaning Silver — 557

13.1 Oxidation and Reduction — 558
- Case Study: Early Metallurgy — 560
- Web Activity: Canadian Achievers–Henry Taube and Rudolph Marcus — 561
- Web Activity: Simulation–Redox Reaction — 563

13.2 Predicting Redox Reactions — 568
- Web Activity: Web Quest–Piercings: A Rash Decision — 570
- Lab Exercise 13.A: Building a Redox Table — 572

13.3 Oxidation States — 583
- Lab Exercise 13.B: Oxidation States of Vanadium — 586
- Web Activity: Case Study–Catalytic Converters — 588
- Case Study: Bleaching Wood Pulp — 594

13.4	Redox Stoichiometry	596
	• Lab Exercise 13.C: Analyzing for Tin	598
	• Web Activity: Canadian Achievers—Imants Lauks	598
	• Lab Exercise 13.D: Analyzing for Chromium in Steel	599
	Chapter 13 Investigations	**601**
	• Investigation 13.1: Single Replacement Reactions	601
	• Investigation 13.2: Spontaneity of Redox Reactions	602
	• Investigation 13.3: Predicting the Reaction of Sodium Metal (Demonstration)	602
	• Investigation 13.4: Analyzing a Hydrogen Peroxide Solution	603
	Chapter 13 Summary	**604**
	Chapter 13 Review	**605**

Chapter 14 Electrochemical Cells — 610

	• Exploration: A Simple Electric Cell	611
14.1	Technology of Cells and Batteries	612
	• Web Activity: Web Quest—Hydrogen: Wonderfuel or Hype?	618
	• Web Activity: Canadian Achievers—Lewis Urry	619
	• Case Study: The Ballard Fuel Cell	620
14.2	Voltaic Cells	622
	• Web Activity: Simulation—Voltaic Cells Under Standard Conditions	630
	• Lab Exercise 14.A: Developing a Redox Table	633
	• Web Activity: Case Study—Galvanizing Steel	637
14.3	Electrolytic Cells	639
14.4	Cell Stoichiometry	652
	• Web Activity: Simulation—Electrolytic Cell Stoichiometry	657
	Chapter 14 Investigations	**658**
	• Investigation 14.1: Designing an Electric Cell	658
	• Investigation 14.2: A Voltaic Cell (Demonstration)	659
	• Investigation 14.3: Testing Voltaic Cells	660
	• Investigation 14.4: A Potassium Iodide Electrolytic Cell	661
	• Investigation 14.5: Electrolysis (Demonstration)	662
	Chapter 14 Summary	**663**
	Chapter 14 Review	**664**
	Unit 7 Review	**666**

▶ Unit 8: Chemical Equilibrium Focusing on Acid-Base Systems

Are You Ready?		672

Chapter 15 Equilibrium Systems — 674

	• Exploration: Shakin' the Blues	675
15.1	Explaining Equilibrium Systems	676
	• Web Activity: Simulation—Equilibrium State	678
	• Lab Exercise 15.A: The Synthesis of an Equilibrium Law	683
	• Web Activity: Canadian Achievers—Paul Kebarle	683
	• Lab Exercise 15.B: Determining an Equilibrium Constant	686
	• Web Activity: Simulation—Writing Equilibrium Expressions	688
15.2	Qualitative Change in Equilibrium Systems	690
	• Case Study: Urea Production in Alberta	696
	• Lab Exercise 15.C: The Nitrogen Dioxide-Dinitrogen Tetroxide Equilibrium	698
	• Web Activity: Web Quest—Poison Afloat	698
	Chapter 15 Investigations	**700**
	• Investigation 15.1: The Extent of a Chemical Reaction	700
	• Investigation 15.2: Equilibrium Shifts (Demonstration)	700
	• Investigation 15.3: Testing Le Châtelier's Principle	701
	• Investigation 15.4: Studying a Chemical Equilibrium System	703
	Chapter 15 Summary	**704**
	Chapter 15 Review	**705**

Chapter 16 Equilibrium in Acid-Base Systems — 710

	• Exploration: Salty Acid or Acidic Salt?	711
16.1	Water Ionization and Acid-Base Strength	712
	• Lab Exercise 16.A: The Chromate-Dichromate Equilibrium	712
	• Web Activity: Canadian Achievers—Edgar Steacie	721
16.2	The Brønsted-Lowry Acid-Base Concept	722
	• Lab Exercise 16.B: Predicting Acid-Base Equilibrium	727
	• Web Activity: Web Quest—Pool Chemistry	727
	• Lab Exercise 16.C: Aqueous Bicarbonate Ion Acid-Base Reactions	732
	• Lab Exercise 16.D: Creating an Acid-Base Table	733
	• Case Study: Changing Ideas on Acids and Bases— The Evolution of a Scientific Theory	733
16.3	Acid-Base Strength and the Equilibrium Law	737
16.4	Acid-Base Equilibrium and pH Curves	751
	• Web Activity: Simulation—Titration of Polyprotic Acids and Bases	762
	• Web Activity: Simulation—Buffer Systems	765
	• Web Activity: Canadian Achievers—Maud Menten	766
	Chapter 16 Investigations	**768**
	• Investigation 16.1: Creating an Acid-Base Strength Table	768
	• Investigation 16.2: Testing Brønsted-Lowry Reaction Predictions	768
	• Investigation 16.3: Testing a Buffer Effect	769
	Chapter 16 Summary	**770**
	Chapter 16 Review	**772**
	Unit 8 Review	**774**

Appendices	**781**
Glossary	**831**
Index	**845**
Credits	**854**

Your Guide to this Textbook

Each unit begins with a two-page set of questions: **Are You Ready?** These questions will help you assess which concepts you should review before you begin the unit.

You can review prerequisite concepts and skills on the Nelson Web site and in the Appendices. The **Appendices** are useful resources for your reference throughout the course.

Each chapter begins with **Starting Points** questions, helping you assess what you already know about the concepts for that chapter. Continue to consider these as you go through the chapter.

Throughout the chapters are activities that have one of the following icons on the left-hand side of their banner. Icons indicate the curriculum emphasis to which the activity relates.

Icon	Curriculum emphasis of activity
	Nature of Science
	Science and Technology
	Social and Environmental Contexts

Investigations are labs in which you will make predictions, form hypotheses, gather and record evidence, then analyze, evaluate, and communicate your results. **Report Checklists** in Investigations show you the parts of lab reports you will need to complete.

You will see an **Investigation Introduction** at the point in the chapter where you will most likely perform each Investigation. The Evaluation numbers refer to the parts of the Evaluation you need to address. (See Appendix B.2.)

Lab Exercises are similar to investigations, but you will not perform the Procedure so the Evidence is provided for you. Again, the **Report Checklist** tells you which parts of the lab report you should complete.

A **Case Study** provides you with information or data, and then guides you in analyzing, decision making, or problem solving by a series of questions.

Case Study	
The Haber Process	With the help of Carl Bosch, BASF's chief chemical engineer, BASF built a giant industrial plant that was capable of and

In an **Explore an Issue** feature, you have the opportunity to define, research, analyze, and report on issues affecting our planet. The **Issue Checklist** shows you the parts of the decision-making process you will need to complete.

EXPLORE an issue	Issue Checklist		
Pesticides	○ Issue ○ Resolution	○ Design ● Evidence	● Analysis ● Evaluation

Sample Problems guide you step-by-step to a solution.

> **SAMPLE** problem 5.1

Communication Examples show you how to clearly present your solutions to word problems.

> **COMMUNICATION** example 8

You will have many opportunities for practice and review. Practice questions are provided throughout the chapter, and will help you to assess your understanding as you work.

1. Describe the three different systems of expressing the concentration of a solution.

You can review and demonstrate your understanding of the concepts and skills in each section using the **Summaries** and **Section Questions**.

The **Chapter Summary** lists the outcomes that you should have mastered as you worked through the chapter, as well as equations and key terms. There is also a box of questions—**Make a Summary**—to help you consolidate your understanding of the concepts addressed.

Chapter and **Unit Reviews** give you practice in answering questions similar to those on the Alberta Diploma Exam.

Appendix A contains numerical answers for questions in this textbook, and **Appendix H** provides Diploma Exam tips.

Other icons throughout the textbook will direct you to features to aid you in your learning.

Icon	Explanation	Icon	Explanation
GO	• an invitation to go to the **Nelson Web site** for research, additional information, or to reach an activity	+	**Extension** • reading material, an audio clip, or an activity related to concepts or skills beyond the Alberta Chemistry 20-30 curriculum
(lightbulb)	**Starting Points** • questions for you to check your knowledge of upcoming concepts, and to revisit at the end of the chapter to assess your learning	(DNA)	**Biology Connection** • online information about a related topic in your biology course
(hand)	**Caution** • a warning of particular safety concerns	(hand pointer)	**Web Activity** • activities on the Nelson Web site that are an integral part of your Chemistry textbook – Web Quests (investigations in which you gather, analyze, and use online information); – Case Studies (activities that provide information and then ask you to analyze it and draw conclusions); – Simulations (interactive online activities or investigations); and – Canadian Achievers (explorations of science-related careers of exceptional Canadians)
(arrows)	**Career Connection** • online information about a science-related career		
(speaker)	**Audio** • an audio file on the Student CD and the Nelson Web site that may be a walk-through of a Sample Problem, an Extension, or pronunciation of Key Terms		
(film)	**Video** • a video or animation on the Nelson Web site that demonstrates a technique or illustrates a concept	NR	**Diploma Exam numerical response** style questions
(play)	**Explorations and Mini Investigations** • brief activities that introduce new concepts or skills and help you to explore concepts being discussed	DE	**Diploma Exam written response** style questions

Preparation for Alberta Diploma Exams

We hope that your interest in science will grow and deepen as you work through your Chemistry course with the aid of this textbook. As your knowledge, skills, and attitudes develop, you will also be working toward the Chemistry Diploma Exam. This resource has been developed to help you achieve your best on the Alberta Chemistry 30 Diploma Exam. **Appendix H** provides specific tips on writing the exam. Part 1 of the Chapter and Unit Reviews contain multiple choice and numerical response questions like you will find on the Diploma Exam. The numerical response questions are marked with this icon NR. Your teacher can provide you with additional questions we have provided to her/him. The Case Studies provide practice in answering closed response written questions based on a scenario. In completing an Explore an Issue, you will develop skills for answering open response written questions on the Diploma Exam. Here you will find and read information about a science-related issue and then formulate and communicate your ideas, supported by your research. You will apply these skills to the written response (Part 2) questions in the Chapter Reviews and Unit Reviews, and in the **additional Diploma Exam-style Review questions** on the Nelson Web site. These questions are longer scenario-based questions, some using published articles. In the Chapter and Unit Reviews, this icon, DE, indicates a question that is in the format of a Diploma Exam written response question. (Note that other Review questions, in addition to these, cover the required content for the Diploma Exams.)

unit Review

Chemistry Review

Figure 1
Peggy Au

"As a high school chemistry teacher, I have had opportunities to be creative, tell stories, play, learn, and teach chemistry in everyday life. I have explored the chemistry of pottery, food and cooking, and silver-smithing, and toured local industrial plants in the coal, aluminium, iron, and oil industries. I have enjoyed the taste and chemistry of wine, beer, and chocolate with fellow colleagues, and learned about the properties and characteristics that make each of these products so special."

Peggy Au, Winston Churchill High School
Lethbridge, Alberta

Chemistry is everywhere, from the colourful Canada Day fireworks display to the ingredients in toothpaste and cosmetics at the local pharmacy and grocery store. "Chemical" is one of those words that people often associate with negative feelings or dangerous consequences. In fact, the comfortable lives we lead are due in large part to our understanding and application of chemistry. Some chemicals are harmful to people or the environment, but many are integral to life, such as the carbon dioxide, oxygen, water, and glucose in the cycle of photosynthesis and cellular respiration. The air we breathe and the food we eat are all based on the elements of the periodic table and their combinations. The water we drink must be processed chemically and physically before it is considered safe. Rocks and minerals form the foundation of the non-living environment. For example, magnesium carbonate from the Rocky Mountains becomes a main component of sidewalks in our bustling cities. Mined metal ores become pots, pans, building components, and jewellery. Copper pans cannot be used for cooking acidic foods such as tomatoes and lemon juice because they cause the copper to oxidize. Cookies baked without baking soda are as hard as rocks. Chemistry is discovery, and a sense of wonder about how and why elements interact and combine in the natural and human-made world. Chemistry surrounds us.

Review Unit

GENERAL OUTCOMES

In this unit, you will

- use atomic theory and the periodic table to classify, describe, explain, and predict the properties of the elements
- use atomic, ionic, and bonding theories to describe, explain, and predict the properties and chemical formulas for compounds
- use reaction generalizations to describe, explain, and predict simple chemical reactions
- describe the processes of science and the nature of scientific knowledge
- describe the differences between and interdependence of science and technology
- employ decision-making processes on science–technology–society issues

Chemistry Review

ARE YOU READY?

These questions will help you find out what you already know, and what you need to review, before you continue with this unit.

Knowledge

1. To facilitate the use of this textbook as a reference, answer the following questions. Use the periodic table and the Appendices to help you find the answers in the most efficient way. Include with your answer the textbook section where you found the answer.
   (a) What is the atomic number and melting point for the element copper?
   (b) From the Appendix, what is the chemical formula and recommended name for baking soda?
   (c) Sketch an Erlenmeyer flask.
   (d) Which variable should be listed on the vertical (y) axis of a graph?
   (e) Write the definition for "science" as found in the textbook.
   (f) What is the melting point of aluminium?
   (g) Describe the first step in the procedure for lighting a laboratory burner.
   (h) From the Appendix, list the headings for writing a laboratory report.
   (i) Describe the WHMIS symbol for a flammable chemical.

Skills

Refer to the Appendices to help you answer the following questions.

2. A student attempting to identify a pure substance from its density obtained the evidence shown in **Table 1**.
   (a) Construct and label a mass–volume graph from the evidence in **Table 1**.
   (b) From the graph, what mass of the substance has a volume of 12.7 mL?
   (c) From the graph, describe the relationship of mass and volume for this solid.

Table 1 Mass and Volume of a Solid

Mass (g)	Volume (mL)
1.2	3.6
1.8	5.5
2.3	6.9
3.1	9.2
6.9	20.7

3. Imagine that a vacuum cleaner salesperson comes to your home to demonstrate a new model. The salesperson cleans a part of your carpet with your vacuum cleaner, and then cleans the same area again using the new model. A special attachment on the new model lets you see the additional dirt that the new model picked up. Analysis seems to indicate that the new model does a better job. Evaluate the experimental design and provide your reasoning.

4. List the manipulated and responding variables and one controlled variable of this experimental problem: "How does altitude affect the boiling point of pure water?"

5. Write an experimental design to answer the problem in question 4.

6. How must you dispose of the following substances in the laboratory?
   (a) broken beaker
   (b) corrosive solutions
   (c) toxic compounds

▶ Prerequisites

Skills

- use a textbook, periodic table, and other references efficiently and effectively
- identify and know how to use basic chemistry laboratory equipment
- create and interpret a table of evidence and a graph of the evidence
- read and write laboratory reports
- create and critique experimental designs
- interpret WHMIS symbols and MSDS data sheets
- operate safely in a chemistry laboratory
- appropriately dispose of waste in a chemistry laboratory

You can review prerequisite concepts and skills on the Nelson Web site and in the Appendices.
  A Unit Pre-Test is also available online.

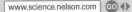

7. Draw a floor plan of the laboratory where you will be working. On your plan indicate the location of the following:
   (a) entrances (exits), including the fire exit
   (b) storage for aprons and eye protection
   (c) eyewash station
   (d) first-aid kit
   (e) fire extinguisher(s)
   (f) MSDS binders
   (g) container for broken glass

8. List the actions you should take if
   (a) your clothing catches fire
   (b) someone else's clothing catches fire

9. Examine **Figure 1**. What safety rules are the students breaking?

10. (a) Identify the WHMIS symbols in **Figure 2**.
    (b) What should you do immediately if any chemical comes in contact with your skin?

11. Describe the procedure for lighting a burner by giving the correct sequence for the photographs in **Figure 3**. You may use a photograph more than once.

Figure 1
What is unsafe in this picture?

Figure 2
The Workplace Hazardous Materials Information System (WHMIS) provides information regarding hazardous products.

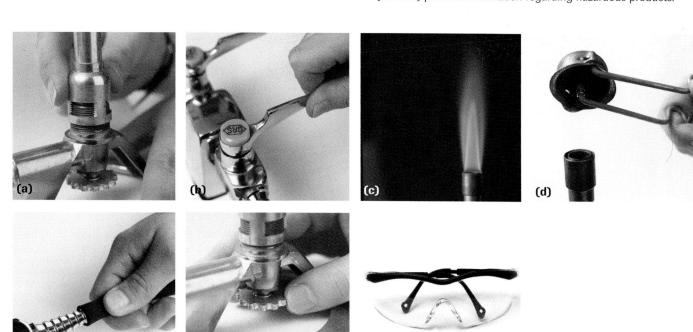

Figure 3
Lighting a laboratory burner

chapter 1

Elements and Compounds

▶ In this chapter

- ▶ Exploration: Combustion of Magnesium (Demonstration)
- Web Activity: Groups of Elements
- Web Activity: The Rutherford Scattering Experiment
- Web Activity: Emission and Absorption Spectra for Hydrogen
- Web Activity: Harriet Brooks

We could argue that chemistry is responsible for some of the hazards of modern life: environmental damage resulting from resource extraction; the toxic effects of some products; and the challenge of garbage disposal. However, that argument ignores the underlying truth: Chemistry has been fundamental to the development of society as we know it. We now have cleaner fuel, more durable and safer paints, easy-care clothing, inexpensive fertilizers, life-saving pharmaceuticals, corrosion-resistant tools and machinery, and unusual new materials that we are using in interesting new ways. Much of this innovation has made our lives better to some degree.

Chemistry is just another way to say "the understanding of the nature of matter." Chemists through the ages and around the world have relied upon scientific inquiry, carrying out investigations and making careful observations. The periodic table sums up the results of many of those investigations and presents information about the elements (**Figure 1**). The observations that went into the creation of the periodic table also helped to create modern atomic theory. In turn, we can explain many of the patterns in the properties of the elements in terms of atomic theory. In this chapter, we will discuss the patterns used to classify elements and compounds, and consider how these patterns are explained by atomic theory.

💡 STARTING Points

Answer these questions as best you can with your current knowledge. Then, using the concepts and skills you have learned, you will revise your answers at the end of the chapter.

1. Examine the periodic table on the inside front cover of this book. Identify and describe some similarities and differences between this table and the ones you used in previous grades.
2. Identify the parts of the atom and describe how they are arranged. According to this model, how do the atoms of the various elements differ from each other?
3. Identify and describe patterns in properties that you are aware of among the elements of the periodic table. Explain these patterns, using your model of the atom.
4. Why do elements form compounds? Use examples of compounds you are familiar with in your explanation.
5. Describe how the scientific community names and writes formulas for chemicals such as sodium, chlorine, table salt, sugar, and battery acid.

 Career Connections:
Chemistry Teacher; Careers with Chemistry

Figure 1
Since phosphorus (shown) spontaneously ignites in air, it must be stored in water. Other metals react with both oxygen and water and so must be stored in oil. Magnesium oxidizes more slowly in air, but can be ignited with a flame (see the Exploration below). Each element has its own set of physical and chemical properties.

▶ Exploration Combustion of Magnesium (Demonstration)

We know things in several different ways. We might see something happening with our own eyes, or we might take a measurement of some variable. These are qualitative and quantitative observations, respectively. Interpretations are statements that go beyond direct observation; for example, the magnesium reacted with oxygen.

In this demonstration you will watch a reaction and classify what you see as a qualitative or quantitative observation or an interpretation. The reaction is the burning of magnesium.

 Only observe the burning of the magnesium when it is within the glass beaker. Never look directly at burning magnesium. The bright flame emits ultraviolet radiation that could harm your eyes.

Due to possible reaction to bright light, persons known to have had seizures should not participate in the demonstration.

Because of its hazardous nature, this demonstration should never be carried out by students.

Materials: lab apron, eye protection, rubber gloves, magnesium ribbon (5 cm), steel wool, laboratory burner and striker, crucible tongs, large glass beaker

- Observe the magnesium before, during, and after it is burned in air. Record all your observations.
- Take safety precautions, then light the laboratory burner (Appendix C).
- Use rubber gloves when handling the steel wool.
- Clean the magnesium ribbon with steel wool. Record any observations.
- Use tongs to hold the magnesium ribbon.
- Light the magnesium ribbon in the burner flame and hold the burning magnesium inside the glass beaker to observe.

(a) Classify your observations as qualitative or quantitative observations.

(b) Indicate which, if any, of your written statements are interpretations.

Elements and Compounds

1.1 Introduction: Science and Technology

What Is Science?

Science involves describing, predicting, and explaining nature and its changes in the simplest way possible. Scientists refine the descriptions of the natural world so that these descriptions are as precise and complete as possible. In science, reliable and accurate descriptions of phenomena become scientific laws.

In scientific problem solving, descriptions, predictions, and explanations are developed and tested through experimentation. In the normal progress of science, scientists ask questions, make predictions based on scientific concepts, and design and conduct experiments to obtain experimental answers. As shown in **Figure 1**, scientists evaluate this process by comparing the results they predicted with their experimental results.

Scientists make predictions that can be tested by performing experiments. Experiments that verify predictions lend support to the concepts on which the predictions are based. We try to explain events in order to understand them. Scientists, like young children, try to understand and explain the world by constructing concepts. Scientific explanations are refined to be as logical, consistent, and simple as possible.

Every investigation has a purpose—a reason why the experimental work is done. The purpose of scientific work is usually to create, test, or use a scientific concept. This order is chronological; for example, one scientist creates a concept, others test the concept, and, if the test is passed, many scientists (plus teachers and students) use the concept.

Scientific research is often very complex and involves, by analogy, many roadblocks, road repairs, and detours along the way. Scientists record all of their work, but the formal report submitted for publication often does not reflect the difficulties and circling back that is part of the process. The order of the report headings does not reflect a specific scientific method. **Figure 1** illustrates the sections and sequence of a laboratory report. Depending on the purpose of the investigation, some sections (Hypothesis, Prediction, and Evaluation) may not be required. See Appendix B.

The Natures of Science and Technology

Science and technology are two different but parallel and intertwined human activities. **Science** is the study of the natural world with the goal of describing, explaining, and predicting substances and changes. The purpose of scientific investigations is to create, test, and/or use scientific concepts. **Technology** is the skills, processes, and equipment required to manufacture useful products or to perform useful tasks. Technology employs a systematic trial-and-error process whose goal is to get process or equipment to work. Technology generally does not seek scientific explanations for why it works.

Technology is an activity that runs parallel to science (**Figure 2**). Technology often leads science as in the development of processes for creating fire, cooking, farming, refining metal, and the invention of the battery. The use of fire, cooking, farming, and refining preceded the scientific understanding of these processes by thousands of years. The invention of the battery in 1800 was not understood scientifically until the early 1900s. Seldom does technology develop out of scientific research, although as the growth of scientific knowledge increases, the number of instances of technology as applied science is increasing.

The discovery of fire and the invention of the battery provided science with sources of energy with which to conduct experimental designs that would not otherwise have been

> **DID YOU KNOW?**
> **Scientific Values**
> Besides the processes of describing, predicting, and explaining, science also includes specialized methods, such as designing investigations, attitudes, such as open-mindedness, and values, such as honesty.

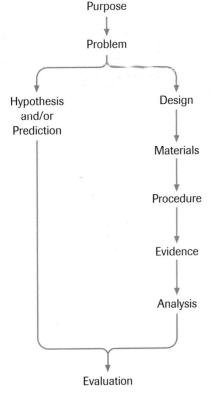

Figure 1
A model for reporting scientific problem solving. During laboratory investigations, the processes overlap and cycle back for modification. Although the reporting may look linear, the actual process of scientific problem solving is not linear and may involve many cycles. An hypothesis or prediction is not always present. In the evaluation section, the answer from the hypothesis or prediction is compared with the answer from the analysis. (See Appendix B.)

possible. Science would not progress very far without the increasingly advanced technologies available to scientists. Often scientific advances have to wait on the development of technologies for research to be done; for example, glassware, the battery, the laser, and the computer.

Often science is blamed for the effects of technology. Often people say, "Science did this," or "Science did that." Most often, though, it was a technological development that was responsible. Technologies and scientific concepts are created by people and used by people. We have to learn how to intelligently control and evaluate technological developments and scientific research, but we can't unless we are scientifically and technologically literate.

Science is an intellectual pursuit founded on research. Concepts start as hypotheses (tentative explanations) and end as accepted or discarded theories or laws. Science meets its goal of concept creation by continually testing concepts. Old concepts are restricted, revised, or replaced when they do not pass the testing process. Scientific knowledge is one of many kinds of knowledge that is trusted and relied upon around the world. Scientists employ a skeptical attitude that results in scientific knowledge being constantly tested worldwide against the best evidence and logic available. Attitudes such as open-mindedness, a respect for evidence, and a tolerance of reasonable uncertainty are qualities found in a scientist. These attitudes represent a predisposition to act in a certain way, without claiming absolute knowledge.

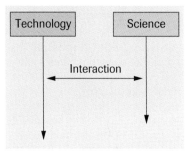

Figure 2
Science and technology are parallel streams of activity where, historically, the development of a technology has most often led the scientific explanation of the technology. Increasingly, technologies are being developed from an application of scientific knowledge. Certainly, science and technology feed off of one another.

What Is Chemistry?

Chemistry is the physical science that deals with the composition, properties, and changes in matter. Chemistry is everywhere around you, because you and your surroundings are composed of chemicals with a variety of properties. However, chemistry involves more than the study of chemicals. It also includes studying chemical reactions, chemical technologies, and their effects on the environment.

Chemistry is primarily the study of changes in matter. For example, coals burning, fireworks exploding, and iron rusting are all changes studied in chemistry. A *chemical change* or *chemical reaction* is a change in which one or more new substances with different properties are formed. Chemistry also includes the study of *physical changes*, such as water freezing to form ice crystals and boiling to form water vapour, during which no new substances are formed. (Physical changes are sometimes called phase changes or changes of state.)

Classifying Knowledge

Classification helps us to organize our knowledge. Classifying knowledge itself is an even more powerful tool. Here are some examples.

The Evidence section of an investigation report includes all observations related to a problem under investigation. An **observation** is a direct form of knowledge obtained by means of one of your five senses—seeing, smelling, tasting, hearing, or feeling. An observation might also be obtained with the aid of an instrument, such as a balance, a microscope, or a stopwatch.

Observations may also be classified as qualitative or quantitative. A *qualitative observation* describes qualities of matter or changes in matter; for example, a substance's colour, odour, or physical state. A *quantitative observation* involves the quantity of matter or the degree of change in matter; for example, a measurement of the length or mass of magnesium ribbon. All quantitative observations include a number; qualitative observations do not.

An **interpretation**, which is included in the Analysis section of an investigation report, is an indirect form of knowledge that builds on a concept or an experience to further

CAREER CONNECTION

Chemistry Teacher
Chemistry teachers, like all science teachers, are in great demand in high schools, colleges, and universities. In large schools, a chemistry teacher may teach only chemistry. However, in smaller schools, the chemistry teacher may also teach other sciences and even other subjects. What education is required to become a high school chemistry teacher?

www.science.nelson.com

DID YOU KNOW?

Scientific Attitudes
Scientific attitudes (also called habits of mind) represent a predisposition to act in a certain way when searching for an answer to a question. Examples of scientific attitudes include
- critical-mindedness
- suspended judgement
- respect for evidence
- honesty
- objectivity
- willingness to change
- open-mindedness
- questioning

CAREER CONNECTION

Careers with Chemistry
There are numerous careers and life experiences that require some knowledge of chemical composition and change. Some careers with chemistry are shown below. There are also many ways to find out about these careers.

pharmacist
forester
dental assistant
farmer
nurse
forensic chemist
nutritionist
veterinarian
cosmetologist
mechanic
miner
firefighter
pulp-mill worker
welder
toxicologist
baker
fertilizer salesperson
wine producer
petroleum engineer
environmental lawyer
chemistry teacher
science writer or reporter
environmental technologist

www.science.nelson.com

describe or explain an observation. For example, observing the light and the heat from burning magnesium might suggest, based on your experience, that a chemical reaction is taking place. A chemist's interpretation might be more detailed: The oxygen molecules collide with the magnesium atoms and remove electrons to form magnesium and oxide ions. Clearly, this statement is not an observation. The chemist did not observe the exchange of electrons.

Observable knowledge is called **empirical knowledge**. Observations are always empirical. **Theoretical knowledge**, on the other hand, explains and describes scientific observations in terms of ideas; theoretical knowledge is *not observable*. Interpretations may be either empirical or theoretical, and depend to a large extent on your previous experience of the subject. **Table 1** gives examples of both kinds of knowledge.

Table 1 Classification of Knowledge

Type of knowledge	Example
empirical	• observation of the colour and size of the flame when magnesium burns
theoretical	• the idea that "magnesium atoms lose electrons to form magnesium ions, while oxygen atoms gain electrons to form oxide ions"

Communicating Empirical Knowledge in Science

Communication is an important aspect of science. Scientists use several means of communicating knowledge in their reports or presentations. Some ways of communicating empirical knowledge are presented below:

- *Empirical descriptions* communicate a single item of empirical knowledge, that is, an observation. For example, you might communicate the simple description that magnesium burns in air to form a white, powdery solid.
- *Tables of evidence* report a number of observations. The manipulated (independent) variable is usually listed first followed by the responding (dependent) variable. **Table 2** shows results from a quantitative experiment that involved burning magnesium.

Table 2 Mass of Magnesium Burned and Mass of Ash Produced

Trial	Mass of magnesium (g)	Mass of ash produced (g)
1	3.6	6.0
2	6.0	9.9
3	9.1	15.1

- *Graphs* are visual presentations of observations. According to convention, the manipulated variable is labelled on the *x*-axis, and the responding variable is labelled on the *y*-axis (Appendix F.4). For example, the evidence reported in **Table 2** is shown as a graph in **Figure 3**. Graphs appear in the Analysis section of a lab report.
- **Empirical hypotheses** are preliminary generalizations that require further testing. Based on **Figure 3**, for example, you might tentatively suggest that the mass of the product of a reaction will always vary directly with the mass of a reacting substance.
- **Empirical definitions** are statements that define an object or a process in terms of observable properties. For example, a metal is a shiny, flexible solid.

- **Generalizations** are statements that summarize a limited number of empirical results. Generalizations are usually broader in scope than empirical definitions and often deal with a minor or sub-concept. For example, many metals slowly react with oxygen from the air in a process known as corrosion.
- **Scientific laws** are statements of major concepts based on a large body of empirical knowledge. Laws are more important and summarize more empirical knowledge than generalizations. For example, the burning of magnesium, when studied in greater detail (**Table 3**), illustrates the law of conservation of mass.

Table 3 Mass of Magnesium, Oxygen, and Product of Reaction

Trial	Mass of magnesium (g)	Mass of oxygen (g)	Mass of product (g)
1	3.2	2.1	5.3
2	5.8	3.8	9.6
3	8.5	5.6	14.1

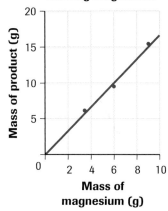

Figure 3
The relationship between the mass of magnesium that reacted and the mass of product obtained is a straight line (direct proportion).

According to the evidence in **Table 3**, the total mass of magnesium and oxygen is generally equal to the mass of the product. Similar studies of many different reactions reflect the **law of conservation of mass**: *In any physical or chemical change, the total initial mass of reactant(s) is equal to the total final mass of product(s).*

For a statement to become accepted as a scientific law, evidence must first be collected from many examples and replicated by many scientists. Even after the scientific community recognizes a new law, that law is subjected to continuous experimental tests based on the ability of the law to describe, explain, and predict nature. Laws must accurately describe and explain current observations and predict future events in a simple manner.

▶ Section 1.1 Questions

1. Classify the following statements about carbon as observations or interpretations:
   (a) Carbon burns with a yellow flame.
   (b) Carbon atoms react with oxygen molecules to produce carbon dioxide molecules.
   (c) Global warming is caused by carbon dioxide.

2. Classify each of the following statements as one of the forms of empirical knowledge:
   (a) Elements are defined as substances that cannot be decomposed by heat or electricity.
   (b) The mass of products in a chemical reaction is always equal to the mass of reactants.
   (c) Graphite is a black powdery substance.
   (d) I believe that, in this case, the temperature will affect the rate of reaction.
   (e) Based upon the limited evidence available, the metals are all bendable.

3. Classify the following statements about carbon as empirical or theoretical:
   (a) Carbon in the form of graphite conducts electricity, whereas carbon in the form of diamond does not.
   (b) Graphite contains some loosely held electrons, whereas the electrons in diamond are all tightly bound in the atoms.

4. Scientific knowledge can be classified as empirical or theoretical.
   (a) What is the key distinction between these two types of knowledge?
   (b) How would you classify the knowledge in the Analysis section of an investigation report?

5. According to research by historians and philosophers, science and technology are related but different disciplines. Classify each of the following activities by Canadians as science or technology:
   (a) Aboriginal Canadians built birch bark canoes.
   (b) Harriet Brooks is the only person known to have worked in the laboratories of Ernest Rutherford, J. J. Thomson, and Marie Curie.
   (c) The telephone was invented by Alexander Graham Bell.
   (d) In 1990, Richard Taylor, who grew up in Medicine Hat, Alberta, was awarded the Nobel Prize for empirically testing and verifying the existence of quarks.

6. List the four characteristics of scientific communication.

7. Describe how a generalization differs from a scientific law.

8. Classify the following activities as science or non-science.
   (a) predicting the weather
   (b) fortune telling
   (c) astronomy
   (d) astrology
   (e) studying animal behaviour
   (f) observing the Northern Lights
   (g) ESP (extrasensory perception)

1.2 Classifying Matter

Matter is anything that has mass and occupies space. Anything that does not have mass or that does not occupy space—energy, happiness, and philosophy are examples—is not matter. To organize their knowledge of substances, scientists classify matter (**Figure 1**). A common classification differentiates matter as **pure substances**, whose composition is constant and uniform, and **mixtures**, whose composition is variable and may or may not be uniform throughout the sample. Empirically, **heterogeneous mixtures** are non-uniform and may consist of more than one phase. By analogy, your bedroom, for example, is a heterogeneous mixture because it consists of solids such as furniture, gases such as air, and perhaps liquids such as soft drinks. **Homogeneous mixtures** are uniform and consist of only one phase. Examples are alloys, tap water, aqueous solutions, and air.

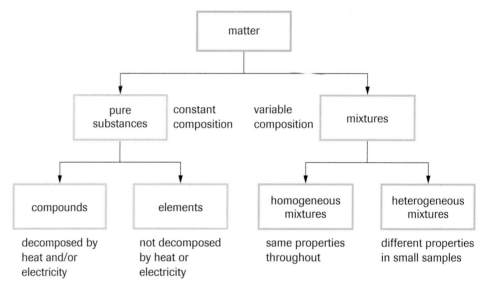

Figure 1
An empirical classification of matter

You can classify many substances as heterogeneous or homogeneous by making simple observations. However, some substances that appear homogeneous may, on closer inspection, prove to be heterogeneous (**Figure 2**). Introductory chemistry focuses on pure substances and homogeneous mixtures, commonly known as solutions.

This empirical (observable) classification system is based on the methods used to separate matter. The parts of both heterogeneous mixtures and solutions can be separated by physical means, such as filtration; distillation; chromatography; mechanically extracting one component from the mixture; allowing one component to settle; or using a magnet to separate certain metals. A pure substance cannot be separated by physical methods. A compound can be separated into more than one substance only by means of a chemical change involving heat or electricity, called chemical decomposition. **Elements** cannot be broken down into simpler chemical substances by any physical or chemical means.

An **entity** is a general term that includes particles (sub-atomic entities such as protons, electrons, and neutrons), atoms, ions, molecules, and formula units. In this textbook, we restrict the use of the term "particle" to sub-atomic entities. Although the classification of matter is based on experimental work, theory lends support to this system. According to theory, elements are composed entirely of only one kind of atom. An **atom**, according to theory, is the smallest entity of an element that is still characteristic of that element. According to this same theory, **compounds** contain atoms of more than one element combined in a definite fixed proportion. Both elements and

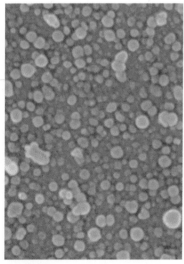

Figure 2
Although milk is called "homogenized," close examination through a microscope reveals solid and liquid phases. Milk is a heterogeneous mixture.

compounds may consist of molecules, distinct entities composed of two or more atoms. Solutions, unlike elements and compounds, contain entities of more than one substance, uniformly distributed throughout them.

A pure substance can be represented by a **chemical formula**, which consists of element symbols representing the entities and their proportions present in the substance. You can use chemical formulas to distinguish between elements, which are represented by a single symbol, and compounds, which are represented by two or more different symbols. Examples of formulas, along with empirical and theoretical definitions, are summarized in **Table 1**.

Table 1 Definitions of Elements and Compounds

Substance	Empirical definition	Theoretical definition	Examples
element	substance that cannot be broken down chemically into simpler units by heat or electricity	substance composed of only one kind of atom	Mg(s) (magnesium) O_2(g) (oxygen) C(s) (carbon)
compound	substance that can be decomposed chemically by heat or electricity	substance composed of two or more kinds of atoms	H_2O(l) (water) NaCl(s) (table salt) $C_{12}H_{22}O_{11}$(s) (sugar)

▶ Section 1.2 Questions

1. Describe how to distinguish experimentally between each of the following pairs of substances:
   (a) heterogeneous and homogeneous mixtures
   (b) solutions and pure substances
   (c) compounds and elements

2. The purpose of the investigation in this problem is to test the Design of decomposition by electricity to determine whether a pure substance is an element or a compound. Complete the Analysis and Evaluation of the investigation report. In your Evaluation, evaluate the Design only (Part 1; see Appendix B.2).
   Problem
   Are water and table salt classified as elements or compounds?
   Prediction
   According to current theoretical definitions of element and compound, as well as the given chemical formulas, water and table salt are classified as compounds. The chemical formulas indicate that water, H_2O(l), and table salt, NaCl(s), are composed of more than one kind of entity.

   Design
   Electricity is passed through water and through molten salt. Any apparent evidence of decomposition is noted.
   Evidence

   Table 2 Passing Electricity through Samples

Sample	Description	Observations after passing electricity through sample
water	colourless liquid	two colourless gases produced
molten table salt	colourless liquid	silvery solid and pale yellow-green gas formed (**Figure 3**)

3. Name two examples of each of the following:
   (a) pure substance
   (b) homogeneous mixture
   (c) heterogeneous mixture

4. Write a Design to determine whether a substance is an element or a compound. Make the Design extensive enough to provide a high degree of certainty in the answer.

5. John Dalton erroneously classified lime and several other compounds as elements because they would not decompose on heating. What does this indicate about the certainty of scientific knowledge?

Figure 3
(a) Sodium chloride (table salt)
(b) Sodium (dangerously reactive metal)
(c) Chlorine (poisonous, reactive gas)

1.3 Classifying Elements

Dmitri Mendeleev, a Russian chemist, created a periodic table in 1869. His periodic table communicated the **periodic law**: *chemical and physical properties of elements repeat themselves in regular intervals, when the elements are arranged in order of increasing atomic number.* (See the periodic table on the inside cover of this textbook.) A periodic table is a very useful tool upon which to base our chemical knowledge. The periodic table organizes the elements, for example, in groups and periods and as metals and nonmetals (**Figure 1**).

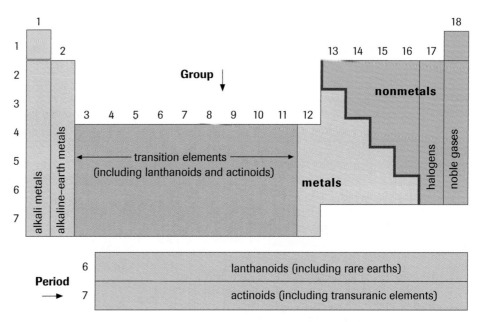

Figure 1
The periodic table is divided into sections, each with commonly used names. The main group, or representative, elements are those in Groups 1, 2, and 12 to 18.

- A **family** or **group** of elements has similar chemical properties and includes the elements in a vertical column in the main part of the table.
- A **period** is a horizontal row of elements whose properties gradually change from metallic to nonmetallic from left to right along the row.
- Metals are located to the left of the "staircase line" in the periodic table, and nonmetals are located to the right. **Semi-metals**—sometimes called metalloids—are a class of elements that are distributed along the staircase line (**Figure 2**).

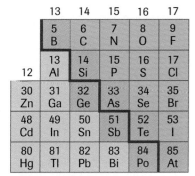

Figure 2
The properties of some elements have led to the creation of a class of elements called semi-metals.

When chemists investigate the properties of materials, they must specify the conditions under which the investigations were carried out. For example, water is a liquid under normal conditions indoors, but in subzero winter temperatures, it becomes a solid outdoors. Ordinarily, tin is a silvery-white metal, but at temperatures below 13 °C, it gradually turns grey and crumbles easily. For the sake of accuracy and consistency, the International Union of Pure and Applied Chemistry (IUPAC), a governing body for scientific communication, has defined a set of standard conditions. Unless other conditions are specified, descriptions of materials are assumed to be at **standard ambient temperature and pressure**. Under these ambient (surrounding) conditions, known as **SATP**, the materials are at a temperature of 25 °C and a pressure of 100 kPa.

From many observations of the properties of elements, scientists have found that **metals** are shiny, bendable, and good conductors of heat and electricity. The majority of the known elements are metals, and all metals except mercury are solids at SATP. The remaining known elements are mostly nonmetals. **Nonmetals** are not shiny, not bendable, and generally not good conductors of heat and electricity in their solid form. At SATP, most nonmetals are gases and a few are solids. Solid nonmetals are brittle and lack the lustre of metals. Most nonmetals exist in compounds rather than in element form.

In the definitions for metals and nonmetals, bendable, ductile, and malleable are often used interchangeably, although they are different properties. Ductile means that the metal can be drawn (stretched) into a wire or a tube, and malleable means that the metal can be hammered into a thin sheet. The word *lustrous* is often used in these definitions in place of shiny.

WEB Activity

Case Study—Groups of Elements

Chemists classify elements, based on their similar physical and chemical properties, into (vertical) groups in our periodic table. Choose a vertical group to investigate. Report on similarities and differences among the elements of the group that you chose.

Periodic tables usually include each element's symbol, atomic number, and atomic mass, along with other information that varies from table to table. The periodic table on the inside front cover features a box of data for each of the elements, and a key explaining the information in each box. The key is also shown in **Figure 3**. Note that theoretical data are listed in the column on the left, and empirically determined data are listed on the right.

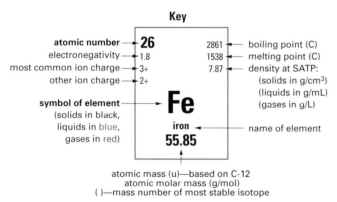

Figure 3
This key, which also appears on this book's inside front cover, helps you to determine the meaning of the numbers in the periodic table.

IUPAC specifies rules for chemical names and symbols. The IUPAC rules, which are summarized in many scientific references, are used all over the world.

SUMMARY — IUPAC Rules for Element Symbols and Names

- Element names should differ as little as possible among different languages. However, only the symbols are truly international.
- The first letter (only) of the symbol is always an uppercase letter (e.g., the symbol for cobalt is Co, not CO, co, or cO).

DID YOU KNOW?
Malleability of Gold
Gold is one of the most malleable metals. It is used when super-thin foil (~0.1 µm thick) is required in experiments and technologies. Gold foil is used for decorative gilding on cakes and books and for coating artificial satellites to reflect infrared radiation.

DID YOU KNOW?
Ida Noddack, Discoverer of Rhenium
Ida Noddack (**Figure 4**) and her husband, Walter Karl Friedrich, predicted the properties of undiscovered elements in Group 7 and searched for these elements.

Less well-known among Noddack's scientific achievements is the initial concept of nuclear fission. Her idea conflicted with theories of atomic structure at that time, so it was ignored for several years.

Figure 4
Ida Noddack (1896–1978)

DID YOU KNOW?
International Element Symbols
The Chinese character for hydrogen is unique to that language, but the symbol (on the right) is recognizable worldwide.

Table 1 IUPAC Element Roots

Number	Root
0	nil
1	un
2	bi
3	tri
4	quad
5	pent
6	hex
7	sept
8	oct
9	enn

DID YOU KNOW ?

Naming the Classes of Elements

IUPAC, by international agreement, sets conventions of communication for the scientific community. In some older textbooks, you may find different names for some classes of elements. For example, the main group elements have been called representative elements, and semi-metals have been called metalloids. Even the definitions might change; for example, in the past, alkaline-earth elements have omitted beryllium and magnesium, and transition elements have included Group 12. Also, lanthanoids and actinoids have, in the past, been called lanthanides and actinides (although with a slightly different membership).

- The name of any new metallic element after uranium should end in -ium.
- Artificial elements beyond atomic number 103 have IUPAC temporary names and symbols derived by combining (in order) the "element roots" for the atomic number and adding the suffix "-ium." The symbols of these elements consist of three letters, each being the first letter of the three element root names corresponding to the atomic number. (See **Table 1**.) For example, element 112 is called ununbium, Uub (one-one-two). The permanent names and symbols are determined by a vote of IUPAC representatives from each country.

Families and Series of Elements

Some groups of elements have family and series names that are commonly used in scientific communication. It is important to learn these family and series names (**Figure 1**).

- The **alkali metals** are Group 1 elements. They are soft, silver-coloured metals that react violently with water to form basic solutions. The most reactive alkali metals are cesium and francium.
- The **alkaline-earth metals** are the Group 2 elements. They are light, reactive metals that form oxide coatings when exposed to air.
- The **halogens** are the elements in Group 17. They are all extremely reactive, with fluorine being the most reactive.
- The **noble gases** are the elements in Group 18. They are special because of their extremely low chemical reactivity.
- The **main group elements** are the elements in Groups 1, 2, and 12 to 18. Of all the elements, the main group elements best follow the periodic law.
- The **transition elements** are the elements in Groups 3 to 11. These elements exhibit a wide range of chemical and physical properties.

In addition to the common classes of elements described above, the bottom two rows in the periodic table also have common names. The *lanthanoids* are the elements with atomic numbers 58 to 71. The *rare earth elements* include the lanthanoids, and yttrium and scandium. The *actinoids* are the elements with atomic numbers 90 to 103. The synthetic (not naturally occurring) elements that have atomic numbers of 93 or greater are referred to as *transuranic elements* (beyond uranium).

▶ Section 1.3 Questions

1. What does the acronym IUPAC stand for?
2. Define SATP and state the reasons why IUPAC defined a set of standard conditions.
3. Classify the following chemicals as metals, semi-metals, or nonmetals:
   (a) iron
   (b) sulfur
   (c) silicon
   (d) gallium
4. List two physical and two chemical properties of the alkali metals.
5. Describe how the reactivity varies within the alkali metal family compared with the halogen family.
6. Nitrogen and hydrogen form a well-known compound, $NH_3(g)$, ammonia. According to the position of phosphorus in the periodic table, predict the most likely chemical formula for a compound of phosphorus and hydrogen.
7. Complete the Prediction, Analysis, and Evaluation (Parts 2 and 3; see Appendix B.2) of the investigation report.

Purpose
The purpose of this investigation is to test the empirical definitions of metals and nonmetals.

Problem
What are the properties of the selected elements?

Design
Each element is observed at SATP, and the malleability and electrical conductivity are determined for the solid form.

Evidence

Table 2 Properties of Selected Metals and Nonmetals

Element	Appearance	Malleability of solid	Electrical conductivity of solid
bromine	red-brown liquid	no	no
cadmium	shiny solid	yes	yes
chlorine	yellow-green gas	no	no
chromium	shiny solid	yes	yes
nickel	shiny solid	yes	yes
oxygen	colourless gas	no	no
platinum	shiny solid	yes	yes
phosphorus	white solid	no	no

8. **Table 3** lists modern element symbols and ancient technological applications. Write the English IUPAC name for each element symbol.

Table 3 Ancient Technological Applications of Elements

International symbol	Technological application
Sn	part of bronze (Cu and Sn) cutting tools, weapons, and mirrors used by Mayan and Inca civilizations
Cu	primary component of bronze and brass (Cu and Zn) alloys
Pb	used by Romans to make water pipes
Hg	a liquid metal used as a laxative by the Romans
Fe	produced by Egyptian iron smelters in 3000 B.C.E.
S	burned for fumigation by Greeks in 1000 B.C.E.
Ag	used in gold-silver alloys made by Greeks in 800 B.C.E.
Sb	an element in ground ore used in early Egyptian cosmetics
Co	used in Egyptian blue-stained glass in 1500 B.C.E.
Al	part of alum used as a fire retardant in 500 B.C.E.
Zn	part of brass mentioned by Aristotle in 350 B.C.E.

9. Copy and complete **Table 4**. Note that the SATP states of matter are solid (s), liquid (l), and gas (g).

Table 4 Elements and Mineral Resources

Mineral resource or use	Element name	Atomic number	Element symbol	Group number	Period number	SATP state
(a) high-quality ores at Great Bear Lake, NT	radium					
(b) potash deposits in Saskatchewan		19				
(c) extracted from Alberta sour natural gas			S			
(d) radiation source for cancer treatment				9	4	
(e) fuel in CANDU nuclear reactors			U	—		
(f) mined in the Northwest Territories				14	2	

Extension

10. Search the Internet for the latest information on the discovery and naming of element 104 and beyond.

1.4 Theories and Atomic Theories

Empirical knowledge based on observation is the foundation for ideas in science. Usually, experimentation comes first and theoretical understanding follows. For example, the properties of some elements were known for thousands of years before a theoretical explanation was available.

So far in this chapter, you have encountered only empirical knowledge of elements, based on what has been observed. Curiosity leads scientists to try to explain nature in terms of what cannot be observed. This step—formulating ideas to explain observations—is the essence of theoretical knowledge in science. Albert Einstein (**Figure 1**) referred to theoretical knowledge as "free creations of the human mind."

Figure 1
"No amount of experimentation can ever prove me right; a single experiment can prove me wrong."
—Albert Einstein's view of the nature of science.

Communicating Theoretical Knowledge in Science
Scientists communicate theoretical knowledge in several ways:

- *Theoretical descriptions* are specific descriptive statements based on theories or models. For example, "a molecule of water is composed of two hydrogen atoms and one oxygen atom."
- **Theoretical hypotheses** are ideas that are untested or extremely tentative. For example, "protons are composed of quarks that may themselves be composed of smaller particles."
- **Theoretical definitions** are general statements that characterize the nature of a substance or a process in terms of a non-observable idea. For example, a solid is theoretically defined as "a closely packed arrangement of atoms, each atom vibrating about a fixed location in the substance."
- **Theories** are comprehensive sets of ideas based on general principles that explain a large number of observations. For example, the idea that materials are composed of atoms is one of the principles of atomic theory; atomic theory explains many of the properties of materials. Theories are dynamic; they continually undergo refinement and change.
- *Analogies* are comparisons that communicate an idea in more familiar or recognizable terms. For example, an atom may be conceived as behaving like a billiard ball. All analogies "break down" at some level; that is, they have limited usefulness.
- *Models* are physical, graphic, or mental representations used to communicate an abstract idea. For example, marbles in a vibrating box could be used to study and explain the three states of matter. Like analogies, models are always limited in their application.

Theories that are acceptable to the scientific community must describe observations in terms of non-observable ideas, explain observations by means of ideas, predict results in future experiments that have not yet been tried, and be as simple as possible in concept and application.

Dalton's Atomic Theory
Recall that by the use of logic, the Greeks (Democritus) in about 300 BCE hypothesized that matter cut into smaller and smaller pieces would eventually reach what they called the atom—literally meaning indivisible. This idea was reintroduced over two thousand years later in 1805 by English chemist/schoolteacher John Dalton. Dalton created the modern theory of atoms to explain three important scientific laws—the laws

+ EXTENSION

Analogies in Science
An analogy is a comparison of a familiar object or process with something that is unfamiliar. Hear more about analogies in the world of science.

www.science.nelson.com

DID YOU KNOW?

Earth, Air, Fire, and Water
"I do not find that anyone has doubted that there are four elements. The highest of these is supposed to be fire, and hence proceed the eyes of so many glittering stars. The next is that spirit, which both the Greeks and ourselves call by the same name, air. It is by the force of this vital principle, pervading all things and mingling with all, that the earth, together with the fourth element, water, is balanced in the middle of space."
– Pliny the Elder (Gaius Plinius Secundus), naturalist and historian (BCE 23–79)

of definite composition, multiple proportions, and conservation of mass (**Table 1**). Dalton's model of the atom was that of a featureless sphere—by analogy, a billiard ball (**Figure 2**). Dalton's atomic theory lasted for about a century, although it came under increasing criticism during the latter part of the 1800s.

SUMMARY: Creating Dalton's Atomic Theory (1805)

Table 1 Empirical Work Leading to Dalton's Atomic Theory

Key experimental work	Theoretical explanation	Atomic theory
Law of definite composition: Elements combine in a characteristic mass ratio.	Each atom has a particular combining capacity.	Matter is composed of indestructible, indivisible atoms, which are identical for one element, but different from other elements.
Law of multiple proportions: There may be more than one mass ratio.	Some atoms have more than one combining capacity.	
Law of conservation of mass: Total mass remains constant (the same).	Atoms are neither created nor destroyed in a chemical reaction.	

Figure 2
In Dalton's atomic model, an atom is a solid sphere, similar to a billiard ball. This simple model is still used today to represent the arrangement of atoms in molecules.

Thomson's Atomic Model

Thomson's model of the atom (1897) was a hypothesis that the atom was composed of electrons (negative particles) embedded in a positively charged sphere (**Figure 3(a)**). Thomson's model of the atom is often communicated by using the analogy of a raisin bun (**Figure 3(b)**).

Table 2 summarizes the key experimental work that led to the creation of Thomson's atomic theory, along with the theoretical explanations. Although you are not required to describe the experimental work, you do need to know that theories are created to explain evidence.

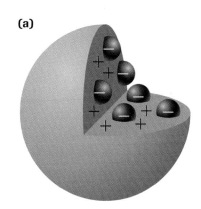

(a)

SUMMARY: Creating Thomson's Atomic Theory (1897)

Table 2 Empirical Work Leading to Thomson's Atomic Theory

Key experimental work	Theoretical explanation	Atomic theory
Arrhenius: the electrical nature of chemical solutions	Atoms may gain or lose electrons to form ions in solution.	Matter is composed of atoms that contain electrons (negatively charged particles) embedded in a positively charged material. The kind of element is characterized by the number of electrons in the atom.
Faraday: quantitative work with electricity and solutions	Particular atoms and ions gain or lose a specific number of electrons.	
Crookes: qualitative studies of cathode rays	Electricity is composed of negatively charged particles.	
Thomson: quantitative studies of cathode rays	Electrons are a component of all matter.	
Millikan: charged oil drop experiment	Electrons have a specific fixed electric charge.	

(b)

Figure 3
(a) In Thomson's atomic model, the atom is a positive sphere with embedded electrons.
(b) This model can be compared to a raisin bun, in which the raisins represent the negative electrons and the bun represents the region of positive charge.

Figure 4
Rutherford's work with radioactive materials at McGill helped prepare him for his challenge to Thomson's atomic theory.

Rutherford's Atomic Theory

One of Thomson's students, Ernest Rutherford (**Figure 4**), eventually showed that some parts of Thomson's atomic theory were incorrect. Rutherford developed an expertise with nuclear radiation during the nine years he spent at McGill University in Montreal. Working with his team of graduate students, at Manchester in England, he devised an experiment to test Thomson's model of the atom. The prediction, based on Thomson's model, was that alpha particles should be deflected little, if at all. When some of the alpha particles were deflected at large angles and even backwards from the foil, the prediction was shown to be false, and Thomson's model judged unacceptable (**Figure 5**). Rutherford created a nuclear model of the atom to explain the evidence gathered in this scattering experiment. The theoretical explanations for the evidence gathered are presented in **Figure 6** and in **Table 3**.

Prediction

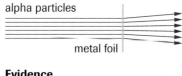

Evidence

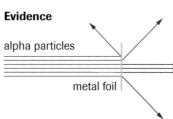

Figure 5
Rutherford's experimental observations were dramatically different from what he had expected based on Thomson's model.

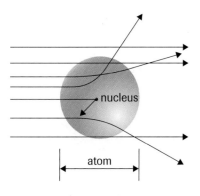

Figure 6
To explain his results, Rutherford suggested in his classic 1911 paper "that the atom consists of a central charge supposedly concentrated at a point."

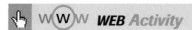

WEB Activity

Simulation—The Rutherford Scattering Experiment

Find out how Rutherford conducted his gold foil experiment.

DID YOU KNOW?

Gold
The Incas used the malleability of gold to create many functional works of art. Gold was not viewed as an economic commodity, but was valued partly because it did not rust (oxidize). Due to its malleability, gold was used as the super-thin foil in Rutherford's scattering experiment.

SUMMARY — Creating Rutherford's Atomic Theory (1911)

Table 3 Empirical Work Leading to Rutherford's Atomic Theory

Key experimental work	Theoretical explanation	Atomic theory
Rutherford: A few positive alpha particles are deflected at large angles when fired at a gold foil.	The positive charge in the atom must be concentrated in a very small volume of the atom.	An atom is composed of a very tiny nucleus, which contains positive charges and most of the mass of the atom. Very small negative electrons occupy most of the volume of the atom.
Most materials are very stable and do not fly apart (break down).	A very strong nuclear force holds the positive charges within the nucleus.	
Rutherford: Most alpha particles pass straight through gold foil.	Most of the atom is empty space.	

Further research by several scientists led to creating the concepts of protons, neutrons, and isotopes (**Table 4**).

SUMMARY: Creating the Concepts of Protons, Neutrons, and Isotopes

Table 4 Experimental Work Leading to Theories of New Particles

Key experimental work	Theoretical explanation	Atomic theory
Soddy (1913): Radioactive decay suggests different nuclei of the same element.	Isotopes of an element have a fixed number of protons, but varying stability and mass (**Figure 7**).	Atoms are composed of protons, neutrons, and electrons. Atoms of the same element have the same number of protons and electrons, but may have a varying number of neutrons (isotopes of the element).
Rutherford (1914): The lowest charge on an ionized gas particle is from the hydrogen ion.	The smallest particle of positive charge is the proton.	
Aston (1919): Mass spectrometer work indicates different masses for some atoms of the same element.	The nucleus contains neutral particles called neutrons.	
Radiation is produced by bombarding elements with alpha particles.		

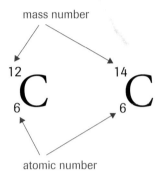

Figure 7
Two isotopes of carbon. Carbon-12 is stable, but carbon-14 is radioactive. Carbon-14 is used for carbon dating of ancient artifacts.

Table 5 Masses and Charges of Subatomic Entities

Particle	Relative mass	Relative charge
electron	1	1−
proton	1836.12	1+
neutron	1838.65	0

Still further empirical research with increasingly higher technologies allowed more precise determination of the relative masses of the subatomic particles. This more precise work confirmed that the electron and the proton had opposite charges, but that the charge was of the same magnitude (quantity) (**Table 5**).

The theoretical explanation for isotopes and the different masses of atoms of the same element is that atoms of elements can have a varying number of neutrons. Isotopes are designated by their **mass number**—the number of protons plus neutrons in their nucleus. The **atomic number** of an element could now be explained as the characteristic number of protons in the nucleus of atoms of that particular element. The number of neutrons can be calculated by subtracting the atomic number from the mass number. The atomic number and the mass number are shown in **Figure 7**.

Bohr's Atomic Theory

The genius of Niels Bohr lay in his ability to combine aspects of several theories and atomic models. He created a theory that, for the first time, could explain the periodic law. Bohr saw a relationship between the sudden end of a period in the periodic table and the quantum theory of energy proposed by German physicist Max Planck in 1900 and applied by Albert Einstein in 1905.

According to the Bohr atomic model, periods in the periodic table result from the filling of electron energy levels in the atom; for example, atoms in Period 3 have electrons in three energy levels. A period comes to an end when the maximum number of electrons is reached for the outer level. The maximum number of electrons in each energy level is given by the number of elements in each period of the periodic table; that is, 2, 8, 8, 18, etc. You may also recall that the last digit of the group number in the periodic table provides the number of electrons in the valence (outer) energy level. Although Bohr did his

DID YOU KNOW ?

Canadian Diamonds
Diamonds are the hardest mineral on Earth. Scientists indicate that diamonds from the Canadian north were formed 2.5 to 3.3 Ga (billion years) ago from carbon under high pressure and temperature at depths of 150 to 225 km. Volcanic action has brought diamonds closer to the surface in structures called kimberlite pipes. Once the pipes are discovered, they are drilled to test for the presence of diamonds. Buffalo Head Hills in Alberta is a diamond exploration area.

www.science.nelson.com

(a) An electron gains a quantum of energy and moves to a higher energy level.

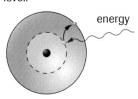

(b) An electron loses a quantum of energy and falls to a lower energy level.

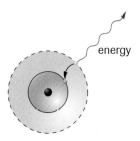

(c) A line spectrum indicates that several different electron jumps/transitions are possible.

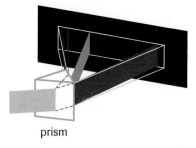

Figure 8
The theory of line spectra

calculations as if electrons were in circular orbits, the most important property of the electrons was their energy, not their motion. Energy-level diagrams for Bohr atoms are presented in the sample problem below. These diagrams have the same procedure and rationale as the orbit diagrams that you have drawn in past years. Since the emphasis here is on the energy of the electron, rather than the motion or position of the electron, orbits are not used.

▶ SAMPLE problem 1.1

Use the Bohr theory and the periodic table to draw energy-level diagrams for the phosphorus atom.

First, refer to the periodic table to find the position of phosphorus. Use your finger or eye to move through the periodic table from the top left along each period until you get to the element phosphorus. Starting with Period 1, your finger must pass through 2 elements, indicating that there is the maximum of 2 electrons in energy level 1. Moving on to Period 2, your finger moves through the full 8 elements, indicating 8 electrons in energy level 2. Finally, moving on to Period 3, your finger moves 5 positions to phosphorus, indicating 5 electrons in energy level 3 for this element.

The position of 2, 8, and 5 elements per period for phosphorus tells you that there are 2, 8, and 5 electrons per energy level for this atom. The information about phosphorus atoms in the periodic table can be interpreted as follows:

atomic number 15:	15 protons and 15 electrons (for the atom)	
period number 3:	electrons in 3 energy levels	
group number 15:	5 valence electrons (the last digit of the group number)	

To draw the energy-level diagram, work from the bottom up:

Sixth, the 3rd energy level:	5 e$^-$	(from group 15)
Fifth, the 2nd energy level:	8 e$^-$	(from eight elements in Period 2)
Fourth, the 1st energy level:	2 e$^-$	(from two elements in Period 1)
Third, the protons:	15 p$^+$	(from the atomic number)
Second, the symbol:	P	(uppercase symbol from the table)
First, the name of the atom:	phosphorus atom	(lowercase name)

Although the energy levels in this energy-level diagram are, for convenience, shown as equal distances apart, this is contrary to the evidence. Line spectra evidence indicates that higher energy levels are increasingly closer together (**Figure 8**).

▶ COMMUNICATION example

Use Bohr's theory and the periodic table to draw energy-level diagrams for hydrogen, carbon, and sulfur atoms.

Solution

$$\frac{1\ e^-}{1\ p^+}$$
H
hydrogen atom

$$\frac{4\ e^-}{\frac{2\ e^-}{6\ p^+}}$$
C
carbon atom

$$\frac{6\ e^-}{\frac{8\ e^-}{\frac{2\ e^-}{16\ p^+}}}$$
S
sulfur atom

▶ Practice

1. Draw an electron energy-level diagram for each of the following:
   (a) an atom of boron
   (b) an atom of aluminium
   (c) an atom of helium

Section **1.4**

Simulation—Emission and Absorption Spectra for Hydrogen

How do the transitions to and from the $n = 2$ energy level provide the emission and absorption spectra for the hydrogen atom? Try this simulation to find out.

SUMMARY   Creating Bohr's Atomic Theory (1913)

Table 6 Experimental Work Leading to Bohr's Atomic Theory

Key experimental evidence	Theoretical explanation	Bohr's atomic theory
Mendeleev (1869–1872): There is a periodicity of the physical and chemical properties of the elements.	A new period begins in the periodic table when a new energy level of electrons is started in the atom.	• Electrons travel in the atom in circular orbits with quantized energy—energy is restricted to only certain discrete quantities. • There is a maximum number of electrons allowed in each orbit. • Electrons "jump" to a higher level when a photon is absorbed. A photon is emitted when the electron "drops" to a lower level.
Mendeleev (1872): There are two elements in the first period and eight elements in the second period of the periodic table.	There are two electrons maximum in the first electron energy level and eight in the next level.	
Kirchhoff and Bunsen (1859), Johann Balmer (1885): Gaseous elements have line spectra for emission and absorption, not continuous spectra.	Since the energy of light absorbed and emitted is quantized, the energy of electrons in atoms is quantized.	

DID YOU KNOW ?

Richard Taylor and Quarks
Richard Taylor, born in Medicine Hat, Alberta in 1929, was jointly awarded the 1990 Nobel Prize in Physics for his 1969 work in verifying the existence of quarks. From that point forward, common matter could be described as being composed of electrons, up quarks (u), and down quarks (d). A proton is composed of two up quarks and one down quark (uud), while a neutron is composed of one up quark and two down quarks (udd).

Formation of Monatomic Ions

In the laboratory, sodium metal and chlorine gas can react violently to produce a white solid, sodium chloride, commonly known as table salt (**Figure 9**). Sodium chloride is very stable and unreactive compared with the elements sodium and chlorine. Bohr originally suggested that the stable, unreactive behaviour of the noble gases was explained by their full outer electron orbits. According to this theory, when the neutral atoms collide, an electron is transferred from one atom to the other, and both atoms become entities called **ions**, which have an electrical charge (**Figure 10**). Sodium ions and chloride ions are **monatomic ions**—single atoms that have gained or lost electrons.

The theory of monatomic ion formation can be used to predict the formation of ions by most representative elements. However, the theory is restricted to these elements. Predictions cannot be made about:

- transition metals. Information about the ions of these elements can be obtained from the data in the periodic table on the inside front cover of this book.
- boron, carbon, and silicon. Experimental evidence indicates that these elements rarely form ions.
- hydrogen. Hydrogen atoms usually form positive ions by losing an electron. Although unusual, a negative hydrogen ion can be formed.

Figure 9
A very reactive metal (sodium) reacts with a poisonous, reactive nonmetal (chlorine) to produce a relatively inert compound (sodium chloride).

> **Learning Tip**
>
> Note that sodium (atomic number 11) has one more electron than the nearest noble gas, neon (atomic number 10). Chlorine (atomic number 17) has one fewer electron than its nearest noble gas, argon (atomic number 18). A transfer of one electron from a sodium atom to a chlorine atom will result in both entities having filled energy levels.

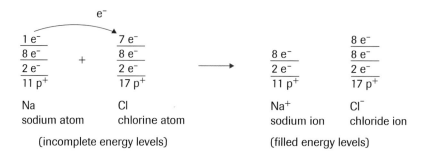

Figure 10
Energy-level models for the reaction of sodium and chlorine. The models are used to explain how two very reactive elements could react to form a relatively inert compound.

> **DID YOU KNOW ?**
>
> **Useful Isotopes**
>
> The following radioisotopes (radioactive isotopes) are produced artificially, for example, within the core of CANDU nuclear reactors. Most of the radioisotopes are used for medical diagnosis or therapy or for industrial or research work.
>
> Ir-192 — analysis of welds
> Co-60 — cancer treatment
> Tc-99 — monitoring blood flow
> I-131 — hyperthyroid treatment
> C-14 — archeological dating
> Hg-203 — dialysis monitoring
> P-32 — white cell reduction
> Co-57 — monitoring vitamin B_{12}
> Tl-201 — monitoring blood flow
> Sr-85 — bone scanning
> In-111 — brain tumor scanning
> Se-75 — pancreas tumor scanning

Positively charged ions are called **cations**. All of the monatomic cations are formed from the metallic and semi-metal elements when they lose electrons in an electron transfer reaction.

Negatively charged ions are called **anions**. All of the monatomic anions come from the nonmetallic and semi-metal elements. Names for monatomic anions use the stem of the English name of the element with the suffix "-ide" and the word "ion" (**Table 7**).

Table 7 Names and Symbols of Monatomic Anions

Group 15	Group 16	Group 17
nitride ion, N^{3-}	oxide ion, O^{2-}	fluoride ion, F^-
phosphide ion, P^{3-}	sulfide ion, S^{2-}	chloride ion, Cl^-
arsenide ion, As^{3-}	selenide ion, Se^{2-}	bromide ion, Br^-
	telluride ion, Te^{2-}	iodide ion, I^-

The symbols for monatomic ions include the element symbol with a superscript indicating the net charge. The symbols "+" and "−" represent the words "positive" and "negative." For example, the charge on a sodium ion Na^+ is "positive one," an aluminium ion Al^{3+} is "positive three," and an oxide ion O^{2-} is "negative two."

Evaluation of Scientific Theories

"Physical concepts are free creations of the human mind, and are not, however it may seem, uniquely determined by the external world. In our endeavor to understand reality we are somewhat like a man trying to understand the mechanism of a closed watch. He sees the face and the moving hands, even hears its ticking, but he has no way of opening the case. If he is ingenious he may form some picture of a mechanism which could be responsible for all the things he observes, but he may never be quite sure his picture is the only one which could explain his observations. He will never be able to compare his picture with the real mechanism and he cannot even imagine the possibility of the meaning of such a comparison."

(Albert Einstein and Leopold Infeld, *The Evolution of Physics*, New York: Simon and Schuster, 1938, page 31.)

> **DID YOU KNOW ?**
>
> **Analogy for Electron Transitions**
>
> In an automobile, the transmission shifts the gears from lower to higher gears, such as from first to second, or downshifts from higher to lower gears. The gears are fixed: first, second, third. You cannot shift to "$2\frac{1}{2}$." Similarly, electron energies in the Bohr model are fixed and electron transitions can only be up or down between specific energy levels.

It is never possible to *prove* theories in science. A theory is accepted if it logically describes, explains, and predicts observations. A major endeavour of science is to make predictions based on theories and then to test the predictions. Once the evidence is collected, a prediction may be

- *verified* if the evidence agrees within reasonable experimental error with the prediction. If this evidence can be replicated, the scientific theory used to make the prediction is judged to be acceptable, and the evidence adds further support and certainty to the theory;
- *falsified* if the evidence obviously contradicts the prediction. If this evidence can be replicated, then the scientific theory used to make the prediction is judged to be unacceptable.

The ultimate authority in scientific work is the evidence (empirical knowledge) gathered during valid experimental work.

An unacceptable theory requires further action; there are three possible strategies.

- *Restrict* the theory. Treat the conflicting evidence as an exception and use the existing theory within a restricted range of situations.
- *Revise* the theory. This option is the most common. The new evidence becomes part of an improved theory.
- *Replace* the existing theory with a totally new concept.

These choices are often referred to as the three Rs.

DID YOU KNOW?
"Seeing" Atoms
In the 1980s, two kinds of microscopes were developed that can produce images of atoms and molecules. These microscopes have magnifications up to 24 million times. The scanning tunnelling microscope (STM) produces images indirectly by using feedback to draw electrons from a material's surface at a constant rate; the atomic force microscope (AFM) probes the surface of a specimen with a diamond stylus whose tip may be only one atom wide. Nanoscience and nanotechnology have advanced substantially due to the invention of these microscopes.

DID YOU KNOW?
Falsification
Karl Popper (1902–1994) is known as the Father of Falsification. He developed the concept of falsification to change scientific work from a conservative attitude of trying to verify to a critical attitude of trying to refute. The Popper concept of scientific work is used extensively in this textbook.

SUMMARY: Theoretical Descriptions of Atoms and Ions

Table 8 Cation and Anion Formation

	Atoms	Cations formed by metals	Anions formed by nonmetals
Name	element name	element name	element root + -ide
Nucleus	#p$^+$ = atomic number	#p$^+$ = atomic number	#p$^+$ = atomic number
Electrons	#e$^-$ = #p$^+$	#e$^-$ < #p$^+$	#e$^-$ > #p$^+$

WEB Activity

Canadian Achievers—Harriet Brooks

Ontario-born Harriet Brooks (**Figure 11**) was Rutherford's first graduate student at McGill University, Montreal. She was also the first woman at McGill to receive a Master of Science degree in physics. She made many contributions to science, including the discovery of radon as a radioactive by-product of radium. She was also the first person to realize that one element can change into another in a series of transformations. Research Brooks's other research interests, and what life was like for a female scientist in the early 20th century.

www.science.nelson.com

Figure 11
Harriet Brooks (1876–1933)

Section 1.4 Questions

1. What is the key difference between empirical and theoretical knowledge?
2. List four characteristics a theory must have to be accepted by the scientific community.
3. Describe the difference between a theory and a law.
4. According to the Bohr theory, what is the significance of full, outer electron orbits of atoms? Which chemical family has this unique property?
5. The alkali metals have similar physical and chemical properties (**Figure 12**). According to the Bohr theory, what theoretical similarity of alkali metal atoms helps to explain their empirical properties?

Figure 12
The alkali metals, such as potassium, are soft metals that react vigorously with water.

6. What is the ultimate authority in scientific work (what kind of knowledge is most trusted)?
7. Use the periodic table and theoretical rules to predict the number of occupied energy levels and the number of valence electrons for each of the following neutral atoms: beryllium, chlorine, krypton, iodine, lead, arsenic, and cesium.
8. If a scientific theory or other scientific knowledge is found to be unacceptable as a result of falsified predictions, what three options are used by scientists?
9. Write a theoretical definition of cation and anion.
10. The alkali metals all react violently with halogens to produce stable white solids. Draw energy-level diagrams (like **Figure 10**, page 24) for each of the following reactions:
    (a) lithium + chlorine → lithium chloride
    (b) potassium + fluorine → potassium fluoride
11. List the ion charges of the monatomic ions for the following families: alkali metals, alkaline-earth metals, Group 13, Group 15, Group 16, and halogens.
12. Draw energy-level diagrams for the following reactions between Group 2 elements and oxygen.
    (a) magnesium with oxygen (**Figure 13**)
    (b) calcium with oxygen

Figure 13
Magnesium burns vigorously in air, giving off visible and ultraviolet light. The reaction should, therefore, not be observed directly.

13. All the electron energy-level diagrams drawn in the previous questions have complete or filled outer energy levels. Describe the experimental evidence for these filled outer energy levels.
14. Write the symbols for the following atoms and ions (e.g., the sodium atom is Na, while the chloride ion is Cl^-).
    (a) sulfur atom
    (b) oxide ion
    (c) lithium ion
    (d) phosphide ion
    (e) aluminium atom
    (f) gallium ion
    (g) rubidium ion
    (h) iodide ion
15. The purpose of this investigation is to test the theory of ions presented in this section. Complete the Prediction and Evaluation (Parts 2 and 3) of the investigation report.

Problem
What is the chemical formula of the compound formed by the reaction of aluminium and fluorine?

Prediction
According to the Bohr theory of atoms and ions, the chemical formula of the compound formed by the reaction of aluminium and fluorine is [your answer]. The reasoning behind this prediction is [your reasoning, including electron energy-level diagrams].

Design
Aluminium and fluorine react in a closed vessel, and the chemical formula is calculated from the masses of reactants and products.

Analysis
According to the evidence gathered in the laboratory, the chemical formula of the compound formed by the reaction of aluminium and fluorine is AlF_3.

Extension

16. Describe some contributions Canadian scientists and/or scientists working in Canadian laboratories made to the advancement of knowledge about the nature of matter.

www.science.nelson.com

Classifying Compounds 1.5

Before chemists could understand compounds, they had to devise ways to distinguish them from elements. Once they achieved this distinction, they could begin to organize their knowledge by classifying compounds. In this section, classification of compounds is approached in three different ways: by convention, empirically, and theoretically.

Classification of Compounds by Convention

Elements are commonly classified as metals or nonmetals. Given that compounds contain atoms of more than one kind of element, what combinations can result? Three classes of compounds are possible: *metal–nonmetal*, *nonmetal–nonmetal*, and *metal–metal* combinations (**Figure 1**).

- Metal–nonmetal combinations are called **ionic compounds**. An example is sodium chloride, NaCl (**Figures 1** and **2**).
- Nonmetal–nonmetal combinations are called **molecular compounds**. An example is sulfur dioxide, SO_2.
- Metal–metal combinations are called alloys and inter-metallic compounds. Alloys include common metal–metal solutions (silver–gold alloys in coins). Inter-metallic compounds include CuZn, Cu_5Zn_8, and $CuZn_3$ in brass at certain temperatures.

DID YOU KNOW ?
Alternative Terms
- Molecular compounds are sometimes called *covalent compounds*, but in this textbook the preferred term is *molecular compounds*.
- *Alloys*, such as brass and amalgams, are solutions of a metal with other metals and, sometimes, nonmetals. Alloys do not combine in definite proportions whereas inter-metallic compounds do.

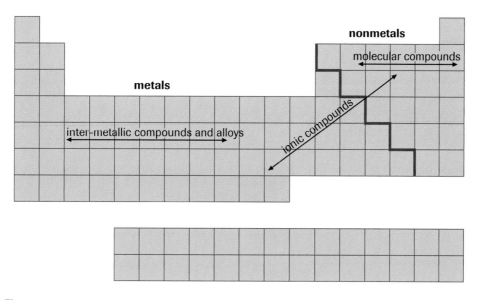

Figure 1
From two classes of elements, there can be three classes of compounds. This chapter covers ionic and molecular compounds.

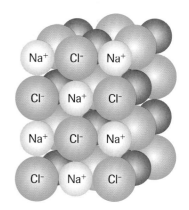

Figure 2
To agree with the explanation of the empirically determined formula for sodium chloride, the model of a sodium chloride crystal must represent both a 1:1 ratio of ions and the shape of the salt crystal. The chemical formula, NaCl(s), represents one formula unit of the crystal, representing a 1:1 ratio of ions.

Empirical Classification of Ionic and Molecular Compounds

The properties of compounds can be used to classify compounds as ionic or molecular. Many properties are common to each of these classes, but by restricting the properties of each empirical definition to those easiest to identify, we can design diagnostic tests for

Figure 3
Conductivity is used to distinguish between aqueous solutions of soluble ionic and molecular compounds. Solutions of ionic compounds conduct electricity, but solutions of molecular compounds do not.

Learning Tip

general properties
↓
defining properties
↓
empirical definitions
↓
diagnostic tests

---

ionic and molecular compounds. A *diagnostic test* is a laboratory procedure conducted to identify or classify chemicals. Some of the common diagnostic tests used in chemistry are described in Appendix C.3.

Empirical Definitions of Compounds

In a series of replicated investigations scientists have found that ionic compounds are all solids at SATP. When dissolved in water, these compounds form solutions that conduct electricity. Scientists have also discovered that molecular compounds at SATP are solids, liquids, or gases that, when dissolved in water, form solutions that generally do not conduct electricity. These *empirical definitions*—a list of empirical properties that define a class of chemicals—will prove helpful throughout your study of chemistry. For example, electrical conductivity of a solution is an efficient and effective diagnostic test that determines whether a compound is ionic or molecular (**Figure 3**).

Empirical Definition of Acids, Bases, and Neutral Compounds

In science, it is not uncommon for new evidence to conflict with widely known and accepted theories, laws, and generalizations. Rather than viewing this as a problem, it is best regarded as an opportunity to improve our understanding of nature. As a result of the new evidence, the scientific concept is either restricted, revised, or replaced.

Not all compounds are either ionic or molecular. For example, aqueous hydrogen citrate (citric acid)—whose chemical formula is $C_3H_4OH(COOH)_3$—is a compound composed of nonmetals. You might predict that this compound is molecular. However, a citric acid solution conducts electricity, which might lead you to predict that the compound is ionic (**Figure 4**). This conflicting evidence necessitates a revision of the classification system. A third class of compounds, called acids, has been identified, and the three classes together provide a more complete description of the chemical world.

Acids are solids, liquids, or gases as pure compounds at SATP that form conducting aqueous solutions that make blue litmus paper turn red. Acids exhibit their special properties only when dissolved in water. As pure substances, all acids, at this point in your chemistry education, have the properties of molecular compounds.

Experimental work has also shown that some substances make red litmus paper turn blue. This evidence has led to another class of substances: **bases** are empirically defined as compounds whose aqueous solutions make red litmus paper turn blue. Compounds whose aqueous solutions do not affect litmus paper are said to be **neutral**. These empirical definitions will be expanded in Chapter 6.

The properties of ionic compounds, molecular compounds, acids, and bases are summarized in **Tables 1** and **2**.

Table 1 Properties of Ionic and Molecular Compounds and Acids

	Ionic	Molecular	Acids
State at SATP (pure substance)	(s) only	(s), (l), or (g)	(s), (l), or (g)
Conductivity of aqueous solution	high to low	none	high to low

Table 2 Properties of Aqueous Solutions of Acids, Bases, and Neutral Compounds

	Acidic	Basic	Neutral
Effect on blue litmus paper	turns red	none	none
Effect on red litmus paper	none	turns blue	none

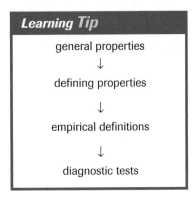

Names and Formulas of Ionic Compounds

Communication systems in chemistry are governed by IUPAC. This organization establishes rules of communication to facilitate the international exchange of chemical knowledge. However, even when a system of communication is international, logical, precise, and simple, it may not be generally accepted if people prefer to use old names and are reluctant to change. There are many examples of chemicals that have both traditional names and IUPAC names. Chemical *nomenclature* is the systematic method for naming substances. Although names of chemicals are language-specific, the rules for each language are governed by IUPAC.

States of Matter in Chemical Formulas

Chemical formulas include information about the numbers and kinds of atoms or ions in a compound. It is also common practice in a formula to specify the state of matter: (s) to indicate "solid"; (l) to indicate "liquid"; (g) to indicate "gas"; and (aq) to indicate "aqueous," which refers to solutions in water. Substances that readily form **aqueous solutions** are very soluble in water. Some examples of chemical formulas with states of matter are:

$NaCl(s)$	pure table salt
$CH_3OH(l)$	pure methanol
$O_2(g)$	pure oxygen
$C_{12}H_{22}O_{11}(aq)$	aqueous sugar solution

Predicting and Naming Ionic Compounds

Besides *explaining* empirically determined formulas, an acceptable theory must also be able to predict future empirical formulas correctly. Experimental evidence provides the test for a prediction made from a theory. A major purpose of scientific work is to test concepts by making predictions.

Binary Compounds

To predict an ionic formula from the name of a binary (two-element) compound, write the chemical symbol, with its charge, for each of the two ions. Then predict the simplest whole-number ratio of ions to obtain a net charge of zero. For example, for the compound aluminium chloride, the ions are Al^{3+} and Cl^-. For a net charge of 0, the ratio of aluminium ions to chloride ions must be 1:3. The formula for aluminium chloride is, therefore, $AlCl_3$. This prediction agrees with the chemical formula determined empirically in the laboratory.

A complete chemical formula should also include the state of matter at SATP. Recall the generalization that all ionic compounds are solids at SATP. The complete formula is therefore, $AlCl_3(s)$.

$$Al(s) + Cl_2(g) \rightarrow Al^{3+}Cl^- Cl^- Cl^- \text{ or } AlCl_3(s)$$
aluminium + chlorine → aluminium chloride

The name of a binary ionic compound is the name of the cation followed by the name of the anion. The name of the metal ion is stated in full and the name of the nonmetal ion has an *-ide* suffix, for example, magnesium oxide, sodium fluoride, and aluminium sulfide. Remember, name the two ions.

Multi-Valent Metals

Most transition metals and some main group metals can form more than one kind of ion, that is, they are multi-valent. For example, iron can form an Fe^{3+} ion or an Fe^{2+} ion, although Fe^{3+} is more common. In the reaction between iron and oxygen, two

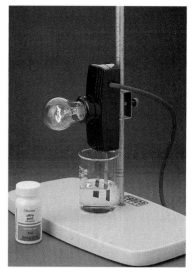

Figure 4
Aqueous hydrogen citrate (citric acid)—an acid in the juice of citrus fruits such as oranges and grapefruits—might be predicted to be molecular, but it forms a conducting solution. It is, therefore, classified as an acid.

DID YOU KNOW ?

Logical Consistency
Explanations are judged on their ability to explain an observation, generalization, or law in a logical, consistent, and simple fashion. For example, the inert character of noble gases and the inert chemical properties of ionic compounds (relative to elements) are explained logically and consistently by using the *same* atomic theory. This atomic theory suggests that there is a magic number of electrons to fill each energy level. When this maximum number is reached, the entity is inert—a nice, simple concept.

Learning *Tip*

Note that the focus here is on writing the correct chemical formulas for reactants and products, not on balancing the equation. The chemical equation is shown here along with the word equation.

Learning Tip

You can use the most common ion to help you predict the product of a chemical reaction; for example, iron(III) oxide in preference to iron(II) oxide. This indicates that iron(III) oxide is more commonly found in nature than is iron(II) oxide. Common occurrence, thermal stability, and electrochemical stability are considered when determining the most common ion.

Figure 5
This information from the periodic table indicates that the most stable ion formed from Fe atoms is Fe^{3+}. Some metals have more than two possible ion charges, but only the most common two are listed in the periodic table.

DID YOU KNOW ?

Classical System

An older (classical) system for naming ions of multi-valent metals uses the Latin name for the element with an *-ic* suffix for the larger charge and an *-ous* suffix for the smaller charge. In this system, iron(III) oxide is "ferric oxide" and iron(II) oxide is "ferrous oxide."

Fe^{3+}	ferric	Fe^{2+}	ferrous
Sn^{4+}	stannic	Sn^{2+}	stannous
Pb^{4+}	plumbic	Pb^{2+}	plumbous
Cu^{2+}	cupric	Cu^{+}	cuprous
Sb^{5+}	stibnic	Sb^{3+}	stibnous

Learning Tip

Recognize the difference between the chemical formula for the element oxygen, $O_2(g)$, and the oxide ion, O^{2-}. Oxide ions occur in compounds such as aluminium oxide, $Al_2O_3(s)$, but not in elemental oxygen, $O_2(g)$.

---

possible products form stable compounds. We can predict the chemical formulas for the possible ionic compounds formed by the reaction by examining ion charges and balancing charges—the total positive charge plus the total negative charge equals zero.

$$\overset{2(3+)}{Fe^{3+}}\overset{3(2-)}{{}_2O^{2-}{}_3} \qquad \overset{2+}{Fe^{2+}}\overset{2-}{O^{2-}}$$
$$Fe_2O_3(s) \qquad FeO(s)$$

In the periodic table on this book's inside front cover, selected ion charges are shown, with the most common (stable) charge listed first (**Figure 5**). If the ion of a multi-valent metal is not specified in a description or an exercise question, you can assume the charge on the ion is the most common one.

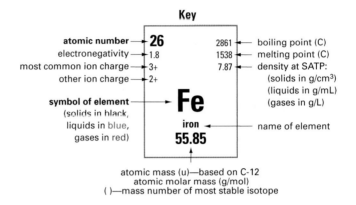

atomic mass (u)—based on C-12
atomic molar mass (g/mol)
( )—mass number of most stable isotope

To name the compounds, name the two ions. In the IUPAC system, the name of the multi-valent metal includes the ion charge. The ion charge is given in Roman numerals in brackets; for example, iron(III) is the name of the Fe^{3+} ion and iron(II) is the name of the Fe^{2+} ion. The Roman numerals indicate the charge on the ion, not the number of ions in the formula. The names of the previously mentioned compounds are

$Fe_2O_3(s)$, iron(III) oxide and $FeO(s)$, iron(II) oxide

Compounds with Polyatomic Ions

Charges on **polyatomic ions**—ions containing a group of atoms with a net positive or negative charge—can be found in a table of polyatomic ions (see the inside back cover). Predicting the formula of ionic compounds involving polyatomic ions is done in the same way as for binary ionic compounds. Write the ion charges and then use a ratio of ions that yields a net charge of zero. For example, to predict the formula of a compound containing copper ions and nitrate ions, write the following:

$$\overset{2+}{Cu^{2+}}\overset{2(1-)}{(NO_3^{-})_2} \qquad Cu(NO_3)_2(s) \qquad copper(II)\ nitrate$$

Two nitrate ions are required to balance the charge on one copper(II) ion (**Figure 6**). Note that parentheses are used in the formula to indicate the presence of more than one polyatomic ion. Do not use parentheses with one polyatomic ion or with simple ions. Do not write: $Ag_2(SO_4)(s)$ or $(Ag)_2SO_4(s)$.

A **formula unit** of an ionic compound is a representation of the simplest whole number ratio of ions; for example, NaCl is a formula unit of sodium chloride. There is no such thing as a molecule of NaCl, only a formula unit. (See Figure 2, page 27.) The simplest ratio formula is also referred to as the **empirical formula**. All ionic formulas are empirical formulas.

Ionic Hydrates

Empirical work indicates that some ionic compounds exist as hydrates; for example, white $CuSO_4(s)$ also exists as blue $CuSO_4 \cdot 5H_2O(s)$ (**Figure 7**). **Hydrates** are compounds that decompose at relatively low temperatures to produce water and an associated compound. You cannot predict the number of water molecules added to the ionic formula unit. You need to be given or to reference this information. The following examples illustrate the nomenclature of ionic hydrates that is recommended by IUPAC, with older nomenclature in parentheses.

$CuSO_4 \cdot 5H_2O(s)$ is copper(II) sulfate—water (1/5) (copper(II) sulfate pentahydrate)

$Na_2CO_3 \cdot 10H_2O(s)$ is sodium carbonate—water (1/10) (sodium carbonate decahydrate)

sodium sulfate—water (1/7) is $Na_2SO_4 \cdot 7H_2O(s)$ (sodium sulfate heptahydrate)

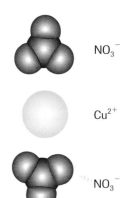

Figure 6
According to theory, two nitrate groups are required to balance the charge on one copper(II) ion. This theory agrees with observations.

Learning *Tip*

The convention for naming ionic hydrates has undergone a couple of changes in the last few decades. For example, copper(II) sulfate pentahydrate became copper(II) sulfate-5-water and now is copper(II) sulfate—water (1/5).

Figure 7
Heating bluestone crystals, $CuSO_4 \cdot 5H_2O(s)$, produces a white powder, $CuSO_4(s)$, according to the reaction $CuSO_4 \cdot 5H_2O(s) + \text{heat} \rightarrow CuSO_4(s) + 5\,H_2O(g)$
Adding water to the white powder produces bluestone.

DID YOU KNOW ?

Baking Soda

Baking soda (**Figure 8**) is one of the most versatile chemicals known. If you were to be deserted on an isolated island, baking soda would be a chemical of choice. Baking soda can be used for bathing, for brushing your teeth, for cleaning pots, for baking, and for extinguishing fires.

SUMMARY   Ionic Compounds

Laboratory investigations indicate that there are classes of ionic compounds:
- binary ionic compounds such as NaCl, $MgBr_2$, and Al_2S_3
- polyatomic ionic compounds such as Li_2CO_3 and $(NH_4)_2SO_4$
- compounds of multi-valent metals such as $CoCl_2$ and $CoCl_3$

The empirically determined formulas of these types of compounds can be explained theoretically in a logically consistent way, using two concepts:
- Ionic compounds are composed of two kinds of ions: cations and anions.
- The sum of the charges on all the ions is zero.

Naming ionic compounds and writing ionic formulas:
- To name an ionic compound, name the two ions: first the cation and then the anion.
- To write an ionic formula, determine the ratio of ions that yields a net charge of zero.

Figure 8
Baking soda

Section 1.5 Questions

1. Distinguish, empirically and theoretically, between an element and a compound.
2. Distinguish, empirically and theoretically, between a metal and a nonmetal.
3. Classify each of the following as an element or compound:
   (a) $C_{12}H_{22}O_{11}(s)$ (sugar)
   (b) Fe(s) (in steel)
   (c) CO(g) (poisonous)
   (d) oxygen (20% of air)
4. Classify each of the following as a metal or nonmetal:
   (a) lead (poisonous)
   (b) chlorine (poisonous)
   (c) Hg(l)
   (d) Br_2(l)
5. Classify each of the following as ionic or molecular:
   (a) $C_6H_{12}O_6(s)$ (glucose)
   (b) $Fe_2O_3 \cdot 3H_2O(s)$ (rust)
   (c) H_2O(l) (water)
   (d) potassium chloride (fertilizer)
6. Once empirical definitions of compounds are established, what kind of knowledge about compounds is likely to follow?
7. State the general names given to a positive ion and a negative ion.
8. Write the symbol, complete with charge, for each of the following ions:
   (a) chloride
   (b) chlorate
   (c) nitride
   (d) iron(III)
   (e) ammonium
   (f) hydroxide
9. Use IUPAC rules to name the following binary ionic compounds:
   (a) lime, CaO(s)
   (b) road salt, $CaCl_2$(s)
   (c) potash, KCl(s)
   (d) a hydride, CaH_2(s)
10. Write the chemical formulas and IUPAC names for the binary ionic products of the following chemical reactions. Do not get distracted by the formulas for the nonmetals or try to balance the equations. For example,
    Li(s) + Br_2(l) → Li^+Br^-(s) or LiBr(s) (lithium bromide)
    (a) Sr(s) + O_2(g) →
    (b) Ag(s) + S_8(s) →
11. Write the chemical formula and IUPAC name of the most common ionic product for each of the following chemical reactions. Do not get distracted by the formulas for the nonmetals or try to balance the equations. For example,
    Bi(s) + O_2(g) → Bi_2O_3(s) (bismuth(III) oxide)
    (a) Ni(s) + O_2(g) →
    (b) Pb(s) + S_8(s) →
    (c) Sn(s) + I_2(s) →
    (d) Fe(s) + O_2(g) →
12. Sketch diagrams of the sulfate and carbonate polyatomic ions.
13. Write empirical and theoretical definitions of an ionic compound.
14. For the IUPAC chemical names in each of the following word equations, write the corresponding chemical formulas (including the state at SATP) to form a chemical equation. (It is not necessary to balance the chemical equation.)
    (a) Sodium hypochlorite is a common disinfectant and bleaching agent. This compound is produced by the reaction of chlorine, Cl_2(g), with lye:
    chlorine gas + aqueous sodium hydroxide → aqueous sodium chloride + liquid water + aqueous sodium hypochlorite
    (b) Sodium hypochlorite solutions are unstable when heated and slowly decompose:
    aqueous sodium hypochlorite → aqueous sodium chloride + aqueous sodium chlorate
    (c) The calcium oxalate produced in the following reaction is used in a further reaction to produce oxalic acid, a common rust remover:
    aqueous sodium oxalate + solid calcium hydroxide → solid calcium oxalate + aqueous sodium hydroxide
15. Predict the international chemical formulas with states at SATP for the compounds formed from the following elements. Unless otherwise indicated, assume that the most stable metal ion is formed. (Write the full chemical equations, but do not balance the equations.) Also write the IUPAC name of the product.
    (a) Mg(s) + O_2(g) →
    (b) Ba(s) + S_8(s) →
    (c) Sc(s) + F_2(g) →
    (d) Fe(s) + O_2(g) →
    (e) Hg(l) + Cl_2(g) →
    (f) Pb(s) + Br_2(l) →
    (g) Co(s) + I_2(s) →
16. For the chemical formulas in each of the following equations, write the corresponding IUPAC names to form a word equation. (Refer to the table of polyatomic ions.)
    (a) The main product of the following reaction (besides table salt) is used as a food preservative.
    NH_4Cl(aq) + NaC_6H_5COO(aq) → $NH_4C_6H_5COO$(aq) + NaCl(aq)
    (b) Aluminium compounds, such as the one produced in the following reaction, are important constituents of cement:
    $Al(NO_3)_3$(aq) + Na_2SiO_3(aq) → $Al_2(SiO_3)_3$(s) + $NaNO_3$(aq)
    (c) Sulfides are foul-smelling compounds that can react with water to produce basic solutions:
    Na_2S(s) + H_2O(l) → NaHS(aq) + NaOH(aq)
    (d) Nickel(II) fluoride may be prepared by the reaction of nickel ore with hydrofluoric acid:
    NiO(s) + HF(aq) → NiF_2(aq) + H_2O(l)
17. Use current IUPAC rules to name the following hydrates. An older name is provided for reference, if possible.
    (a) $FeSO_4 \cdot 7H_2O$(s) (ferrous sulfate heptahydrate)
    (b) $NiNO_3 \cdot 6H_2O$(s) (nickelous nitrate hexahydrate)
    (c) $Al_2(SO_4)_3 \cdot 18H_2O$(s) (aluminium sulfate-18-water)
    (d) $3CdSO_4 \cdot 8H_2O$(s) (no name possible under older systems of nomenclature for hydrates)

Extension

18. Why do systems of nomenclature change over time?
19. Why is it important to have internationally accepted systems of communication?

Molecular Elements and Compounds 1.6

Many molecular formulas, such as H_2O, NH_3, and CH_4, had been determined empirically in the laboratory by the early 1800s, but chemists could not explain or predict molecular formulas using the same theory as for ionic compounds. The theory that was accepted for these compounds was the idea that nonmetal atoms share electrons and that the sharing holds the atoms together in a group called a **molecule**. The chemical formula of a molecular substance—called a **molecular formula**—indicates the number of atoms of each kind in a molecule (**Figure 1**).

water (H_2O)

Molecular Elements

As you have seen from the given chemical formulas for elements in the preceding examples and exercises, the chemical formula of all metals is shown as a single atom, whereas nonmetals frequently form **diatomic molecules** (i.e., molecules containing two atoms). Some useful rules are provided in **Table 1**. (Memorize the examples in this table.) An explanation of these rules is given in Unit 1. The diatomic elements end in *-gen*; for example, hydro*gen*, nitro*gen*, oxy*gen*, and the halo*gens*. $O_3(g)$ is a special unstable form of oxygen called ozone. $S_8(s)$ is called cycloöctasulfur, octasulfur, or usually just sulfur. **Figure 2** (on the next page) illustrates models of some of these molecules.

ammonia (NH_3)

Table 1 Chemical Formulas of Metallic and Molecular Elements

Class of elements	Chemical structure	Examples
metals	all are monatomic	Na(s), Hg(l), Zn(s), Pb(s)
molecular elements (nonmetals)	some are diatomic	$H_2(g)$, $N_2(g)$, $O_2(g)$, $F_2(g)$, $Cl_2(g)$, $Br_2(l)$, $I_2(s)$
	some have molecules containing more than two atoms	$O_3(g)$, $P_4(s)$, $S_8(s)$
	all noble gases are monatomic	He(g), Ne(g), Ar(g)
other elements (semi-metals)	the rest of the elements can be assumed to be monatomic	C(s), Si(s)

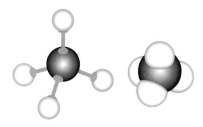

methane (CH_4)

Figure 1
Molecular models of H_2O, NH_3, and CH_4 help us understand the theoretical explanation for the empirically determined formulas of water, ammonia, and methane. Ball-and-stick and space-filling models, as above, are types of molecular models.

Molecular Compounds

The names of some compounds communicate the number of atoms in a molecule. IUPAC has assigned Greek numerical prefixes to the names of molecular compounds formed from two different elements (**Table 2**). Other naming systems are used when a molecule has more than two kinds of atoms.

The following are examples of names of *binary molecular compounds*. Recall that *binary* refers to compounds composed of only two kinds of atoms and that *molecular* refers to compounds composed only of nonmetals.

Reactants		Product	Name
$C(s) + S_8(s)$	→	$CS_2(l)$	carbon disulfide
$N_2(g) + I_2(s)$	→	$NI_3(s)$	nitrogen triiodide
$N_2(g) + O_2(g)$	→	$NO_2(g)$	nitrogen dioxide
$P_4(s) + O_2(g)$	→	$P_4O_{10}(s)$	tetraphosphorus decaoxide

You will predict the chemical formulas for molecular compounds in Unit 1.

Table 2 Prefixes Used in Chemical Names

Prefix	Number of atoms
mono	1
di	2
tri	3
tetra	4
penta	5
hexa	6
septa	7
octa	8
nona	9
deca	10

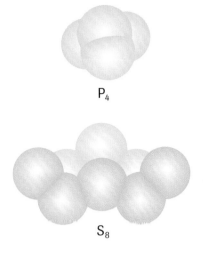

Naming Molecular Compounds

According to IUPAC rules, the prefix system is used only for naming binary molecular compounds—molecular compounds composed of two kinds of atoms.

For hydrogen compounds such as hydrogen sulfide, $H_2S(g)$, the common practice is *not* to use the prefix system. In other words, we do not call this compound dihydrogen sulfide.

The molecular formulas and names of many molecular compounds must be memorized, referenced, or given. Some common molecular compounds whose names and formulas should be memorized are given in **Table 3**.

SUMMARY Elements and Molecular Compounds

- Empirically, molecular elements and compounds as pure substances are solids, liquids, or gases at SATP. If they dissolve in water, their aqueous solutions do not conduct electricity.
- Theoretically, molecular elements and compounds are formed by nonmetal atoms bonding covalently to share electrons in an attempt to obtain the same number of electrons as the nearest noble gas.
- The chemical formulas for molecular elements should be memorized from **Table 1** on page 33. SATP states of matter are obtained from the periodic table.
- The chemical names and formulas for most binary molecular compounds are obtained from the memorized prefixes (see **Table 2** on page 33); for example, $N_2S_5(l)$ is dinitrogen pentasulfide and dinitrogen tetraoxide gas is $N_2O_4(g)$.
- Memorize the chemical formulas, names, and states of matter at SATP for common binary and ternary molecular compounds in **Table 3**. For other molecular compounds referred to in questions, you are given the states of matter.

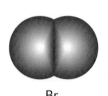

Figure 2
Models representing the molecular elements $P_4(s)$, $S_8(s)$, $H_2(g)$, and $Br_2(l)$.

Table 3 Common Molecular Compounds

IUPAC name	Molecular formula
water	$H_2O(l)$ or $HOH(l)$
hydrogen peroxide	$H_2O_2(l)$
ammonia	$NH_3(g)$
glucose	$C_6H_{12}O_6(s)$
sucrose	$C_{12}H_{22}O_{11}(s)$
methane	$CH_4(g)$
propane	$C_3H_8(g)$
octane	$C_8H_{18}(l)$
methanol	$CH_3OH(l)$
ethanol	$C_2H_5OH(l)$
hydrogen sulfide	$H_2S(g)$

Naming and Writing Formulas for Acids and Bases

In this chapter, acids and bases are given very restricted empirical and theoretical definitions. Aqueous hydrogen compounds that make blue litmus paper turn red are classified as acids and are written with the hydrogen appearing first in the formula. For example, $HCl(aq)$ and $H_2SO_4(aq)$ are acids. $CH_4(g)$ and $NH_3(g)$ are not acids, so hydrogen is written last in the formula. In some cases, hydrogen is written last if it is part of a group such as the COOH group; for example, $CH_3COOH(aq)$. These –COOH acids are organic acids and are described in more detail in Chapter 9.

Acids

Empirically, acids as pure substances are molecular compounds, as evident from their solid, liquid, and gas states of matter. Theoretically, they are composed of nonmetals that share electrons. However, the formulas of acids can be explained and predicted by assuming that they are ionic compounds. For example, the chemical formulas for the acids HCl(aq), H_2SO_4(aq), and CH_3COOH(aq) can be *explained* as follows:

H^+Cl^-(aq), $H^+_2SO_4^{2-}$(aq), $CH_3COO^-H^+$(aq)

The chemical formulas for acids can also be *predicted* by assuming that these aqueous molecular compounds of hydrogen are ionic:

aqueous hydrogen sulfide is $H^+_2S^{2-}$(aq), or H_2S(aq)
aqueous hydrogen sulfate is $H^+_2SO_4^{2-}$(aq), or H_2SO_4(aq)
aqueous hydrogen sulfite is $H^+_2SO_3^{2-}$(aq), or H_2SO_3(aq)

Acids are often named according to more than one system because they have been known for so long that the use of traditional names persists (**Figure 3**). IUPAC suggests that names of acids should be derived from the IUPAC name for the compound. In this system, sulfuric acid would be named aqueous hydrogen sulfate. However, the classical system of nomenclature is well entrenched, so it is necessary to know two or more names for many acids, especially the common ones.

The classical names for acids are based on anion names, according to three simple rules:

- If the anion name ends in "-ide," the corresponding acid is named as a "hydro —— ic" acid. Examples are hydrochloric acid, HCl(aq), hydrosulfuric acid, H_2S(aq), and hydrocyanic acid, HCN(aq).
- If the anion name ends in "-ate," the acid is named as a " —— ic" acid. Examples are nitric acid, HNO_3(aq), sulfuric acid, H_2SO_4(aq), and phosphoric acid, H_3PO_4(aq).
- If the anion name ends in "-ite," the acid is named as a " —— ous" acid. Sulfurous acid, H_2SO_3(aq), nitrous acid, HNO_2(aq), and chlorous acid, $HClO_2$(aq), are examples.

The classical system of acid nomenclature is part of a system for naming a series of related compounds. **Table 4** lists the acids formed from five different chlorine-based anions to illustrate this naming system.

Figure 3
In addition to classical and systematic scientific names, many acids also have common commercial names. For example, muriatic acid is the commercial name for HCl(aq).

Table 4 Classical Acid Nomenclature System of Chlorine Anion Acids

Ion Formula	IUPAC Ion name	Classical Acid name	Systematic IUPAC Acid name	Acid Formula
ClO_4^-	perchlorate ion	perchloric acid	aqueous hydrogen perchlorate	$HClO_4$(aq)
ClO_3^-	chlorate ion	chloric acid	aqueous hydrogen chlorate	$HClO_3$(aq)
ClO_2^-	chlorite ion	chlorous acid	aqueous hydrogen chlorite	$HClO_2$(aq)
ClO^-	hypochlorite ion	hypochlorous acid	aqueous hydrogen hypochlorite	HClO(aq)
Cl^-	chloride ion	hydrochloric acid	aqueous hydrogen chloride	HCl(aq)

Learning Tip

For oxy-ions such as ClO_4^-, the most common ion has an "... ate" suffix (e.g., chlorate). One extra oxygen is "per ... ate"; one less oxygen is "... ite"; and two less oxygens is "hypo ... ite".

Bases

Chemists have discovered that all aqueous solutions of ionic hydroxides make red litmus paper turn blue; that is, these compounds are *bases*. Other solutions have been classified as bases, but for the time being, restrict your definition of bases to aqueous ionic hydroxides such as NaOH(aq) and $Ba(OH)_2$(aq). The name of the base is the name of the ionic hydroxide; for example, aqueous sodium hydroxide and aqueous barium hydroxide.

Learning Tip

Scientists create classification systems to help them organize their knowledge. These classification systems also help you to organize the knowledge being learned.

SUMMARY: Acids and Bases

- Empirically, acids are aqueous molecular compounds of hydrogen that form electrically conductive solutions and turn blue litmus paper red.
- By convention, the formula for an empirically identified acid is written as H___(aq) or ___COOH(aq).
- As pure substances, acids are molecular compounds, and, thus, can be solids, liquids, or gases; $HCl(g)$, $CH_3COOH(l)$, and $C_3H_4OH(COOH)_3(s)$.
- The chemical formulas and electrical conductivity of aqueous solutions of acids can be explained and predicted by assuming that these molecular compounds are ionic; for example, $H^+_2SO_4^{2-}(aq)$ or $H_2SO_4(aq)$.
- The classical names for acids follow this pattern: hydrogen ___ide becomes a "hydro___ic" acid; hydrogen ___ate is a "___ic" acid; hydrogen ___ite is a "___ous" acid; and hydrogen hypo___ite is a "hypo___ous" acid.
- The IUPAC name for an acid is aqueous hydrogen ___; for example, aqueous hydrogen sulfate for $H_2SO_4(aq)$.
- Empirically, bases are aqueous ionic hydroxides that form electrically conductive solutions and turn red litmus paper blue.
- There is no special nomenclature system for bases. They are named as ionic hydroxides; for example, $KOH(aq)$ is aqueous potassium hydroxide.

+ EXTENSION

Naming Compounds
Improve your understanding of naming ionic and molecular compounds.

www.science.nelson.com

Section 1.6 Questions

1. Until a theoretical way of knowing molecular formulas is available, you must be given the formula or name and then rely on memory or use the prefix system to provide the name or formula as required. Provide the names or formulas (complete with the SATP states of matter) for the following substances:
   (a) chlorine (toxic)
   (b) phosphorus (reacts with air)
   (c) $C_2H_5OH(l)$ (alcohol)
   (d) methane (fuel)
   (e) helium (inert)
   (f) carbon (black)
   (g) $NH_3(g)$ (smelling salts)

2. Write the chemical formulas for the following molecular substances emitted as gases from the exhaust system of an automobile. Some of these substances may produce acid rain:
   (a) carbon dioxide
   (b) carbon monoxide
   (c) nitrogen dioxide
   (d) sulfur dioxide
   (e) nitrogen
   (f) octane
   (g) nitrogen monoxide
   (h) dinitrogen oxide
   (i) dinitrogen tetraoxide
   (j) water

3. Write unbalanced chemical equations to accompany the given statements or word equations, including the states at SATP. For example,
   nitrogen + oxygen → nitrogen dioxide
   $N_2(g) + O_2(g) \rightarrow NO_2(g)$
   (a) Solid silicon reacts with gaseous fluorine to produce gaseous silicon tetrafluoride.
   (b) Solid boron reacts with gaseous hydrogen to produce gaseous diboron tetrahydride.
   (c) Aqueous sucrose and water react to produce aqueous ethanol and carbon dioxide gas.
   (d) Methane gas reacts with oxygen gas to produce liquid methanol.
   (e) nitrogen + oxygen → nitrogen monoxide
   (f) nitrogen monoxide + oxygen → nitrogen dioxide
   (g) octane + oxygen → carbon dioxide + water vapour
   (h) octane + oxygen → carbon dioxide + carbon monoxide + carbon + water vapour

4. Classify the following as acidic, basic, neutral ionic, or neutral molecular:
   (a) $KCl(aq)$ (fertilizer component)
   (b) $HCl(aq)$ (in stomach)
   (c) sodium hydroxide (oven/drain cleaner)
   (d) ethanol (beverage alcohol)

5. Write the chemical formulas for the following acids:
   (a) aqueous hydrogen chloride (from a gas)
   (b) hydrochloric acid (stomach acid)
   (c) aqueous hydrogen acetate (from a liquid)
   (d) acetic acid (vinegar)
   (e) aqueous hydrogen sulfate (from a liquid)
   (f) sulfuric acid (car battery)
   (g) aqueous hydrogen nitrite (from a gas)
   (h) nitric acid (for making fertilizers)

6. Write accepted names for the following acids:
   (a) H_2SO_3(aq) (acid rain)
   (b) HF(aq) (used for etching glass)
   (c) H_2CO_3(aq) (carbonated beverages)
   (d) H_2S(aq) (rotten egg odour)
   (e) H_3PO_4(aq) (rust remover)
   (f) HCN(aq) (rat killer)
   (g) H_3BO_4(aq) (insecticide)
   (h) C_6H_5COOH(aq) (preservative)

7. Write chemical equations, including the states at SATP, for the following reactions involved in the manufacture and use of sulfuric acid:
   (a) Sulfur reacts with oxygen to produce sulfur dioxide gas.
   (b) Sulfur dioxide reacts with oxygen to produce sulfur trioxide gas.
   (c) Sulfur trioxide gas reacts with water to produce sulfuric acid.
   (d) Sulfuric acid reacts with ammonia gas to produce aqueous ammonium sulfate (a fertilizer).
   (e) Sulfuric acid reacts with rock phosphorus, $Ca_3(PO_4)_2$(s), to produce phosphoric acid and solid calcium sulfate (gypsum).

8. Write chemical equations, including states at SATP, for the following reactions involved in the destructive reactions of acid rain:
   (a) Sulfuric acid in rain reacts with limestone (see Appendix J), causes deterioration of buildings, statues, and gravestones, and produces aqueous hydrogen carbonate (carbonic acid) and solid calcium sulfate.
   (b) Sulfuric acid from rain reacts with solid aluminium silicate in the bottom of a lake and releases aqueous hydrogen silicate (silicic acid) and toxic (to the fish) aqueous aluminium sulfate.

9. Write chemical equations, including states at SATP, for each of the following reactions involved in the control of acid rain:
   (a) Sulfur dioxide emissions can be reduced in the exhaust stack of an oil sands refinery by reacting the sulfur dioxide gas with lime (see Appendix J) and oxygen to produce solid calcium sulfate (gypsum).
   (b) Sulfuric acid in an acid lake can be neutralized by adding slaked lime (see Appendix J) to produce water and solid calcium sulfate.

10. An investigation is planned to explore the conductivity of various categories of substances. Complete the Analysis and Evaluation sections (Parts 2 and 3) of the investigation report.

Purpose

The purpose of this investigation is to extend the previously determined empirical definitions of ionic and molecular compounds to include the electrical conductivity of the solid, liquid, and aqueous states of matter.

Problem

What are the empirical definitions of ionic and molecular compounds?

Prediction

According to the current definitions of ionic and molecular compounds, ionic compounds are all solids at SATP that form electrically conductive solutions, whereas molecular compounds are solids, liquids, or gases at SATP that form non-conductive solutions.

Design

Pure samples of water (H_2O), calcium chloride ($CaCl_2$), sucrose ($C_{12}H_{22}O_{11}$), methanol (CH_3OH), sodium hydroxide (NaOH), and potassium iodide (KI) are tested for electrical conductivity in the pure state at SATP, in the pure molten state, and in aqueous solution.

manipulated variable:	compound tested
responding variable:	electrical conductivity
controlled variables:	temperature
	quantity of chemical
	quantity of water
	conductivity apparatus
	conductivity of the water

Evidence

Table 5 Electrical Conductivity of Compounds in Different States

Chemical formula	Pure state at SATP	Conductivity		
		Pure at SATP	Molten	Aqueous
H_2O	liquid	none	none	n/a*
$CaCl_2$	solid	none	high	high
$C_{12}H_{22}O_{11}$	solid	none	none	none
CH_3OH	liquid	none	none	none
NaOH	solid	none	high	high
KI	solid	none	high	high

*not applicable

Chapter 1 SUMMARY

Outcomes

Knowledge
- classify matter as pure and mixtures as homogeneous and heterogeneous (1.2)
- interpret the periodic table of the elements (1.3)
- use atomic theory to explain the periodic table (1.4)
- classify elements and compounds and know the properties of each class (1.3, 1.4)
- explain and predict chemical formulas for and name ionic and molecular compounds, acids, and bases (1.5, 1.6)
- identify the state of matter of substances (1.5, 1.6)
- write chemical equations when given reactants and products (1.5, 1.6)
- classify scientific knowledge as qualitative and quantitative, as observations and interpretations, and as empirical and theoretical (1.1)

STS
- describe the natures of science and technology (1.1)
- describe the application of some common chemicals (1.3, 1.5, 1.6)

Skills
- use a textbook, a periodic table, and other references efficiently and effectively (1.1–1.6)
- interpret and write laboratory reports (1.1, 1.2, 1.3, 1.4, 1.6)
- select and use diagnostic tests (1.2, 1.3, 1.4, 1.5, 1.6)

Key Terms

1.1
science
technology
chemistry
observation
interpretation
empirical knowledge
theoretical knowledge
empirical hypothesis
empirical definition
generalization
scientific law
law of conservation of mass

1.2
matter
pure substance
mixture
heterogeneous mixture
homogeneous mixture
element
entity
atom
compound
chemical formula

1.3
periodic law
family
group
period
semi-metal
standard ambient temperature and pressure (SATP)
metal
nonmetal
alkali metal
alkaline-earth metal
halogen
noble gas
main group element
transition element

1.4
theoretical hypothesis
theoretical definition
theory
mass number
atomic number
ion
monatomic ion
cation
anion

1.5
ionic compound
molecular compound
acid
base
neutral
aqueous solution
polyatomic ion
formula unit
empirical formula
hydrate

1.6
molecule
molecular formula
diatomic molecule

MAKE a summary

1. Prepare a concept map that is centred on pure substances. Include classes of substances along with their properties and nomenclature. See the Key Terms list.
2. Revisit your answers to the Starting Points questions at the start of this chapter. How has your thinking changed?

Go To

The following components are available on the Nelson Web site. Follow the links for *Nelson Chemistry Alberta 20-30*.
- an interactive Self Quiz for Chapter 1
- additional Diploma Exam-style Review questions
- Illustrated Glossary
- additional IB-related material

There is more information on the Web site wherever you see the Go icon in the chapter.

EXTENSION

Lightning
This video looks into the mystery of lightning from a scientific perspective. An understanding of positive and negative charges, related to the atomic theory, explains the phenomenon.

www.science.nelson.com

Chapter 1 REVIEW

Many of these questions are in the style of the Diploma Exam. You will find guidance for writing Diploma Exams in Appendix H. Exam study tips and test-taking suggestions are on the Nelson Web site. Science Directing Words used in Diploma Exams are in bold type.

www.science.nelson.com

DO NOT WRITE IN THIS TEXTBOOK.

Part 1

Scientists not only classify natural objects and phenomena, but they also classify knowledge. In both cases, the process of classifying helps them (and us) to better organize knowledge.

1. The classification of knowledge includes
   NR 1. interpretations   2. observations
   Use the above classes of knowledge to classify the following statements, in order.
   • Carbon (as graphite) conducts electricity.
   • Aluminium can pass electrons from atom to atom.
   • Sodium is a metal.
   • Magnesium has two valence electrons.

   ___ ___ ___ ___

2. The classification of knowledge also includes
   NR 1. quantitative observation   2. qualitative observation
   Use the classes of observations to classify the following statements in order:
   • To form an ion, the chlorine atom gains one electron.
   • Gold is malleable.
   • A sodium carbonate solution conducts electricity.
   • The mass of magnesium burned is 2.0 g.

   ___ ___ ___ ___

3. Scientific knowledge can also be classified as
   NR 1. empirical   2. theoretical
   Use the above classes of knowledge to classify the following statements:
   • Molecular compounds form nonconducting aqueous solutions.
   • Ionic compounds dissolve as ions.
   • Acids form conducting aqueous solutions.
   • Bases dissolve to increase the hydroxide ion concentration.

   ___ ___ ___ ___

4. A substance that cannot be decomposed is _i_ definition for _ii_.
   The above sentence is completed by the information in which row?

Row	i	ii
A.	an empirical	a compound
B.	an empirical	an element
C.	a theoretical	a compound
D.	a theoretical	an element

Science and technology are influential disciplines in our developed country. Understanding the natures of science and technology and their relationship to each other is important to being a citizen of Canada.

5. The purpose of scientific investigations does **not** include
   A. creating a concept
   B. verifying a concept
   C. using a concept
   D. testing a concept

6. The criterion that is **not** used to evaluate a scientific concept is its ability to
   A. explain
   B. predict
   C. describe
   D. prove

7. The relationship of science and technology is best described as
   A. science leading technology
   B. parallel supporting activities
   C. science involving more trial and error
   D. adversarial (in conflict)

Use this information to answer questions 8 and 9.

The format of a laboratory report reflects the basic pattern of scientific research. The following sections of a laboratory report are listed in random order.

1. Analysis   6. Problem
2. Evaluation   7. Procedure
3. Evidence   8. Purpose
4. Design   9. Prediction
5. Materials

8. Once the Purpose and Problem have been chosen, **identify**, in order, the parts of an investigation report that are done before the work in a laboratory begins.

   ___ ___ ___ ___

9. **Identify**, in order, the parts of the investigation report that are completed during and after the work in a laboratory.

   ___ ___ ___ ___

10. The Evaluation of an investigation does **not** involve
    A. evaluating the evidence based upon whether the prediction is verified or falsified
    B. evaluating the evidence based upon whether the design, materials, and procedure are adequate
    C. evaluating the prediction by comparing the answer in the Analysis with the answer in the Prediction
    D. evaluating the hypothesis by comparing the answer in the Analysis with the answer in the Prediction

Use this information and the periodic table to answer questions 11 to 13.

Scientific research shows that heavy metals such as cadmium, lead, and mercury damage the human nervous and reproductive systems. This research has led to legislation restricting the use of these metals in Canada.

11. The name for an entity containing 48 electrons and 48 protons is
    A. cadmium atom
    B. cadmium ion
    C. mercury atom
    D. mercury ion

12. The symbol for an entity containing 80 electrons and 82 protons is
    A. Hg(l)
    B. Pb(s)
    C. Hg^{2+}(aq)
    D. Pb^{2+}(aq)

13. Chemical and Physical Properties of Elements
    1. brittleness
    2. form cations
    3. solid at SATP
    4. good insulators
    5. silvery-grey colour
    6. good conductors of heat
    7. good conductors of electricity

    The four properties shared by cadmium, lead, and mercury, listed in numerical order, are:

    ___ ___ ___ ___

Use this information to answer questions 14 to 16.

Some of the chemicals produced in Alberta are listed below.
1. ammonium sulfate, $(NH_4)_2SO_4$(s)
2. ammonia, NH_3(g)
3. caustic soda, NaOH(s)
4. chlorine gas, Cl_2(g)
5. ethene, C_2H_4(g)
6. methane, CH_4(g)
7. methanol, CH_3OH(l)
8. sodium chlorate, $NaClO_3$(s)
9. sulfur, S_8(s)

14. **Identify** the molecular compounds in numerical order.
    ___ ___ ___ ___

15. **Identify** the ionic compounds in numerical order.
    ___ ___ ___ ___

16. **Identify** the elements in numerical order.
    ___ ___ ___ ___

Use a periodic table to answer questions 17 to 19.

17. The element that has the **least** similar properties to the rest is
    A. oxygen
    B. sulfur
    C. bromine
    D. silver

18. The element that does **not** fit with the chemical properties of the rest is
    A. sodium
    B. potassium
    C. lithium
    D. cerium

19. The melting point of aluminium is °C.

Part 2

Scientific concepts can be defined in empirical terms or theoretical terms.

20. **Define** the following concepts empirically:
    (a) metals
    (b) nonmetals
    (c) molecular compounds
    (d) ionic compounds

21. **Define** the following concepts theoretically:
    (a) atomic number
    (b) mass number
    (c) isotopes
    (d) anion
    (e) cation

22. Draw energy-level diagrams for the following entities:
    (a) silicon atom
    (b) potassium ion
    (c) fluoride ion
    (d) calcium atom
    (e) sulfide ion

23. Table salt is often used in cooking. Many cook books recommend using salt in the water for cooking green beans. Some books suggest that the salt is necessary for maintaining the green colour of the beans.
    (a) **Plan** a simple experimental design to test the hypothesis that salt is necessary to maintain the colour of green beans. (Refer to Appendix B.)
    (b) If you know about single and double blind studies (see Appendix B.4), **plan** a more sophisticated experimental design for this research.

24. Copy and complete **Table 1** by classifying the compounds as ionic or molecular and writing the chemical formulas or IUPAC names.

Table 1 Ionic and Molecular Compounds

Use	IUPAC name	Ionic or molecular?	Formula
leavening agent	sodium hydrogen carbonate		
home heating fuel	methane		
bleach			NaClO(s)
masonry	calcium oxide		
dry ice			$CO_2(s)$
gas-line antifreeze	methanol		
in laundry detergent	sodium carbonate		
melts ice on sidewalks			$CaCl_2(s)$
sweetener			$C_{12}H_{22}O_{11}(s)$
fungicide	copper(II) sulfate		
prevents tooth decay			$SnF_2(s)$
car batteries	lead(IV) oxide		
food seasoning	sodium chloride		
solvent for oils and fats			$CCl_4(l)$
produces nitric acid	nitrogen dioxide		

25. A qualitative analysis of four compounds is carried out. Complete the Analysis of the following investigation report.

    Purpose
    The purpose of this investigation is to use evidence and empirical definitions to identify four solutions.

    Problem
    Which of the solutions labelled 1, 2, 3, and 4 is KCl(aq), $C_2H_5OH(aq)$, HCl(aq), and $Ba(OH)_2(aq)$?

    Design
    Each solution is tested with a conductivity apparatus and with litmus paper to determine its identity. A sample of the water used for preparing the solutions is tested for conductivity as a control. Taste tests are ruled out because they are unsafe.

Evidence

Table 2 Qualitative Analysis Results

Solution	Conductivity	Litmus paper
water	none	no change
1	high	no change
2	high	blue to red
3	none	no change
4	high	red to blue

26. Memorizing is often an initial way of knowing something, such as a chemical formula. What other ways of knowing are available to you?

27. Chemistry is one way of knowing about nature. What are some other ways of knowing about natural phenomena?

Extension

28. Use the Internet to investigate how the modern view of fluctuating electrons in an atom compares with an Aboriginal view of a fluctuating universe. Briefly **describe** your findings.

    www.science.nelson.com

29. Search for high-technology images, similar to **Figure 1**, of atoms in elements and molecules, and ions in ionic compounds.

    www.science.nelson.com

Figure 1
Image of a silver crystal taken with a scanning electron microscope (SEM).

30. Niels Bohr introduced his quantum model of the atom in his 1913 paper. In the last paragraph of the paper, Bohr wrote: "The foundation of the hypothesis has been sought entirely in its relation with Planck's theory of radiation; ... later it will be attempted to throw some further light on the foundation of it from another point of view."
    The other point of view that Bohr refers to in this paragraph is that of testing the explanatory power of his quantum-model hypothesis by explaining the periodicity of the elements as displayed in the periodic table.
    In your own words, use the Bohr theory of the atom to **explain** the periodic law. You can read Bohr's classic paper on the Nelson Web site.

    www.science.nelson.com

Elements and Compounds

chapter 2

Chemical Reactions

In this chapter

 Molecules Making Magic (Demonstration)

 Web Activity: States of Matter and Changes in Matter

Chemical reactions are not just something that scientists study in a laboratory. Chemical changes or reactions are an essential part of nature—past, present, and future. For example, lightning initiates many chemical reactions in the atmosphere including the formation of nitrogen compounds that provide nutrients to plants on Earth. Plants carry out many chemical reactions, the most important of which is photosynthesis. Animals eat plants and use the chemical reactions of metabolism to survive and grow. Eventually plants and animals die, and their decomposition is another set of chemical reactions. The cycle of life is a cycle of chemical reactions.

Although you can learn some things about chemicals by observing their physical properties, laboratory work involving chemical reactions reveals a great deal more about the compounds. You can make inferences about chemicals based on the changes that occur in chemical reactions. By studying chemical reactions, you can construct generalizations and laws and, eventually, infer the theoretical structure of the compounds involved.

Initially, theories of the structure of matter attempt to explain the known chemical properties of a substance. The validity of a theory is determined by its ability to both explain and predict changes in matter. How and why do chemicals react? What compounds will form as a result of a reaction? How do we explain the different properties of compounds? Chemical combination represents not only the compounds we know, but also the processes or chemical reactions by which we know them.

💡 STARTING Points

Answer these questions as best you can with your current knowledge. Then, using the concepts and skills you have learned, you will revise your answers at the end of the chapter.

1. List the kinds of evidence that indicate the occurrence of a chemical reaction.
2. Provide a theoretical explanation as to *how* and *why* chemical reactions occur.
3. Describe how an understanding of chemical reaction types can be used to predict the products of a chemical reaction.
4. Describe the process for balancing a chemical reaction equation.
5. The chemical amount of a substance is measured in moles. What is a mole?
6. List some perspectives on a science–technology–society (STS) issue.

 Career Connection:
Forensic Laboratory Analyst

Figure 1
What burns, when a server sets a dish of saganaki aflame?

▶ Exploration — Molecules Making Magic (Demonstration)

Have you ever wondered how magicians set fire to $100 bills without apparently damaging the paper? Or how a server can flambé a dish of food without burning it to a crisp (**Figure 1**)? Here is a teacher demonstration that might help you answer the puzzle of "how do they do that?"

Materials: eye protection, lab apron, protective gloves, a 13 × 100 mm test tube with stopper, a small beaker, propan-2-ol (rubbing alcohol), water, table salt, tongs, paper, scissors, safety lighter or match

 Propan-2-ol is highly flammable. Ensure that containers of the alcohol are sealed and stored far from any open flame.

- Add a few crystals of table salt to a small test tube.
- Add propan-2-ol to nearly half-fill the test tube.
- Add water to nearly fill the test tube.
- Stopper the test tube and invert it several times to dissolve the components.
- Remove the stopper and pour the contents into a small beaker.
- Cut and insert a strip of paper into the alcohol solution.
- Use tongs to remove the paper.
- Remove the beaker of solution from the work area.
- Use the matches or lighter to light the wet paper.
- Repeat.
- Dispose of any excess solution in a waste container.

2.1 Science and Technology in Society

Figure 1
Many familiar materials, such as plastics, metals, ceramics, and soap, are products of both science and technology.

The science of chemistry goes hand-in-hand with the technology of chemistry: the skills, processes, and equipment required to make useful products, such as water-resistant adhesives, or to perform useful tasks, such as water purification. Chemists make use of many technologies, from test tubes to computers. Chemists may or may not thoroughly understand a particular chemical technology. For example, the technologies of glass-making and soap-making existed long before scientists could explain these processes. We now use thousands of metals, plastics, ceramics, and composite materials developed by chemical engineers and technologists (**Figure 1**). However, chemists do not have a complete understanding of superconductors, ceramics, chrome-plating, and some metallurgical processes. Sometimes technology leads science—as in glass-making and soap-making—and sometimes science leads technology. Overall, science and technology complement one another.

Science, technology, and society (**STS**) are interrelated in complex ways (**Figure 2**). In this chapter and throughout this book, the nature of science, technology, and STS interrelationships will be introduced gradually, so that you can prepare for decision making about STS issues in the 21st century.

You can acquire specialized knowledge, skills, and attitudes for understanding STS issues by studying science. For example, a discussion of global warming becomes an informed debate when you have specific scientific knowledge about the topic; scientific skills to acquire and test new knowledge; and scientific attitudes and values to guide your thinking and your actions. You also need an understanding of the nature of science and of scientific knowledge.

Scientists have indicated that, at present, both the observations and the interpretations of global warming are inadequate for fully understanding the present phenomenon and for accurately predicting the situation in the future. However, scientists will always state qualifications such as these, even 100 years from now, no matter how much more evidence is available. In a science course, you learn that scientific knowledge is never completely certain or absolute. When scientists testify in courts of law, present reports to parliamentary committees, or publish scientific papers, they tend to avoid authoritarian, exact statements. Instead, they state their results with some degree of uncertainty. In studying science, you learn to look for evidence, to evaluate experiments, and to attach a degree of certainty to scientific statements. You learn to expect and to accept uncertainty, but to search for increasingly greater certainty. This is the nature of scientific inquiry.

Chemicals and chemical processes represent both a benefit and a risk for our planet and its inhabitants. Chemistry has enabled people to produce more food, to dwell more comfortably in homes insulated with fibreglass and polystyrene, and to live longer, thanks to

DID YOU KNOW ?

Imagination in Science
"We especially need imagination in science. It is not all mathematics, nor all logic, but it is somewhat beauty and poetry."

–Maria Mitchell, American astronomer (1818–1889)

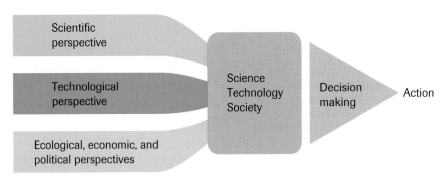

Figure 2
Many issues in society involve science and technology. Both the problems and the solutions involve complex interrelationships among these three categories. An example of an STS issue is the problem of climate change.

clean water supplies, more varied diets, and modern drugs. While enjoying these benefits, we also consciously and unconsciously assume certain risks. For example, when chemical wastes are dumped or oil spills into the environment, the effects can be disastrous. Assessing benefits and risks is a part of evaluating advances in science and technology.

The world increasingly depends on science and technology. Our society's affluence has led to countless technological applications of metal, paper, plastic, glass, wood, and other materials. Thousands of new scientific discoveries and technological advances are made each year. As our society embraces more and more sophisticated technology, we tend to seek technological "fixes" for problems, such as chemotherapy in treating cancer, and the use of fertilizers in agriculture. However, a strictly technological approach to problem solving overlooks the multidimensional nature of the problems confronting us.

Deciding how to use science and technology to benefit society is extremely complex. Most STS issues can be discussed from many different points of view, or **perspectives**. Even pure scientific research is complicated by economic and social perspectives. For example, should governments increase funding for scientific research when money is needed for social assistance programs? Environmental problems, such as discharge from pulp mills and air pollution, are controversial issues. For rational discussion and acceptable action on STS issues, a variety of perspectives must be taken into account. For example, five of many possible STS perspectives on air pollution are listed below:

- A **scientific perspective** leads to researching and explaining natural phenomena. Research into sources of air pollution and its effects involves a scientific perspective.
- A **technological perspective** is concerned with the development and use of machines, instruments, and processes that have a social purpose. The use of instruments to measure air pollution and the development of processes to prevent air pollution reflect a technological approach to the issue.
- An **ecological perspective** considers relationships between living organisms and the environment. Concern about the effect of a smelter's sulfur dioxide emissions on plants and animals, including humans, reflects an ecological perspective.
- An **economic perspective** focuses on the production, distribution, and consumption of wealth. The financial costs of preventing air pollution and the cost of repairing damage caused by pollution reflect an economic perspective.
- A **political perspective** involves governments, vote-getting actions or arguments based on an ideology. Debate over proposed legislation to control air pollution involves a political perspective.

DID YOU KNOW ?

Technological Fixes
Often our first thought when change is required is to ask for a technology to be invented and/or used to solve the problem. Some examples of technological fixes for societal problems (which themselves may create problems) are
- escalators and elevators
- cars and airplanes
- radios and televisions
- the Internet and cellphones
- pesticides and fertilizers
- gasoline and plastics

Learning Tip

A mnemonic that may be used to recall these five STS perspectives is STEEP.

▶ Section 2.1 Questions

1. Identify four or more current STS issues.
2. Classify each of the following statements about aluminium as representing a scientific, technological, ecological, economic, or political perspective:
   (a) Recycled aluminium costs less than one-tenth as much as aluminium produced from ore.
   (b) Aluminium ore mines in South America have destroyed the natural habitat of plants and animals.
   (c) Aluminium is refined in Canada using electricity from hydro-electric dams.
   (d) In Quebec, aluminium is refined using hydro-electric power that some politicians in Newfoundland have claimed belongs to their constituents.
   (e) In 1886, American chemist Charles Hall discovered through research that aluminium can be produced by using electricity to decompose aluminium oxide dissolved in molten cryolite.
3. Instead of changing their lifestyles, many people look to technology to solve problems caused by the use of technology! Suggest one technological fix and one lifestyle change that would help to solve each of the following problems:
   (a) Aluminium ore used to produce aluminium cans will be in short supply soon.
   (b) Aluminium cans are not magnetic and are, therefore, difficult to separate from the rest of the garbage.
   (c) People throw garbage into bins for recyclable aluminium cans.
4. Aluminium is used extensively for making beverage cans. List some benefits and risks of this practice. Are there any alternatives that might have equal or better benefits and fewer risks? Be prepared to argue your case.

www.science.nelson.com

2.2 Changes in Matter

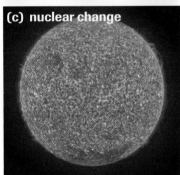

Figure 1
(a) Hydrogen is liquified at –253 °C.
$H_2(g) \rightarrow H_2(l)$
(b) Hydrogen is burning—as it does during the space shuttle launch and in hydrogen-fueled automobiles.
$2\,H_2(g) + O_2(g) \rightarrow 2\,H_2O(g)$
(c) Hydrogen is undergoing nuclear fusion on the Sun and is being converted into helium.
$H(g) + H(g) \rightarrow He(g)$

The explanation of natural events is one of the aims in science. Careful observation, leading to the formation of a concept or theory, and followed by testing and evaluating the ideas involved, defines the basic process scientists use to increase understanding of the changes going on in the world around us. A useful way to begin is to classify the types of changes that occur in matter. Changes in matter can be explained at three levels according to size. Modern scientists study and discuss matter at a *macroscopic* (naked eye observable) level, or at a *microscopic* (too small to see without a microscope) level, or at a *molecular* (smallest entities of a substance) level. To understand their observations at a molecular level, chemists usually start by basing their explanations on the atomic theory proposed by John Dalton in 1803.

Types of Changes in Matter

Chemists often describe changes in matter as a physical change, chemical change, or nuclear change (**Figure 1**), depending on whether they believe that a change has occurred in the molecules, electrons, or nuclei of the substance being changed. The quantity of energy associated with every change in matter can also help classify the type of change.

Physical changes are any changes where the fundamental entities remain unchanged at a molecular level, such as the phase changes of evaporation and melting. There is no change in the chemical formula of the substance involved (**Figure 1(a)**). Dissolving a chemical is usually classified as a physical change. Other examples include changes in physical structure that change only the shape and appearance. Physical changes in matter usually involve relatively small amounts of energy change.

A **chemical change** involves some kind of change in the chemical bonds within the fundamental entities (between atoms and/or ions) of a substance, and is represented by a change in the chemical formula (**Figure 1(b)**). At least one new substance is formed, with physical and chemical properties different from those of the original matter. Normally, chemical changes involve larger energy changes than physical changes.

Nuclear changes (changes within the nucleus) create entirely new atomic entities (**Figure 1(c)**). These entities are represented by formulas that show new atomic symbols, different from those of the original matter. Nuclear changes involve extremely large changes in energy, which allow them to be identified. In 1896, Henri Becquerel noticed the continuous production of energy from a piece of rock that showed no other changes at all. His observation led to the discovery of radioactivity—a nuclear change.

Physical, chemical, and nuclear changes can be described both empirically and theoretically. **Table 1** provides these descriptions, along with an example of each. In this course, we will focus our attention almost entirely on chemical change. We can use classification systems, such as Table 1, to help us to organize our knowledge.

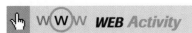 **WEB Activity**

Case Study—States of Matter and Changes in Matter

Watch the movie about properties of matter and physical changes.
(a) What properties of solids, liquids, and gases make them different from one another?
(b) What kinds of physical changes are not mentioned in the movie?
(c) What are some clues that a chemical reaction has taken place?

www.science.nelson.com

Table 1 Physical, Chemical, and Nuclear Change

Change	Empirical description	Theoretical description
physical	• state or energy change • solid ⇌ liquid ⇌ gas • no new substance • small energy change	• $H_2(g) \rightarrow H_2(l)$ + energy • $H_2O(s) \rightleftharpoons H_2O(l) \rightleftharpoons H_2O(g)$ • no new atoms/ions/molecules • intermolecular forces broken and made
chemical	• colour, odour, state, and/or energy change • new substance formed • new permanent properties • medium energy change	• $2 H_2(g) + O_2(g) \rightarrow 2 H_2O(g)$ + energy • old (reactants) → new (products) • atoms/ions/electrons rearranged • chemical bonds broken and made
nuclear	• often radiation emitted • new elements formed • enormous energy change	• $^{2}_{1}H + ^{1}_{1}H \rightarrow ^{3}_{2}He$ + energy • new atoms formed • nuclear bonds broken and made

Nuclear Change
There are two types of nuclear change. Read about both and then run the fission and fusion animations.

www.science.nelson.com

The Kinetic Molecular Theory

Scientists observed gas pressure, diffusion, and chemical reactions, and eventually explained their observations using the concept of molecular motion. The idea of molecular motion led to the kinetic molecular theory, which has become a cornerstone of modern science.

The central idea of the **kinetic molecular theory** is that the smallest entities of a substance are in continuous motion. These entities may be atoms, ions, or molecules. As they move about, the entities collide with each other and with objects in their path (**Figure 2**).

How and Why Chemical Reactions Occur

Chemical changes are also called chemical reactions. To explain chemical reactions, we can expand the kinetic molecular theory to create a theory of chemical reactions. According to the kinetic molecular theory, the entities of a substance are in continuous, random motion. This motion inevitably results in collisions between the entities. If different substances are present, all the different entities will collide randomly with each other. If the collision has a certain orientation and sufficient energy, the components of the entities will rearrange to form new entities. The rearrangement of entities that occurs is the chemical reaction. This general view of a chemical reaction is known as the *collision–reaction theory* (**Figure 3**).

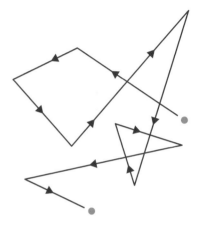

Figure 2
Observation of microscopic particles, such as pollen grains or specks of smoke, shows a continuous, random motion. This movement is known as Brownian motion, named for Scottish scientist Robert Brown, who first described it. Scientists' interpretations of this evidence led to the creation of the kinetic molecular theory.

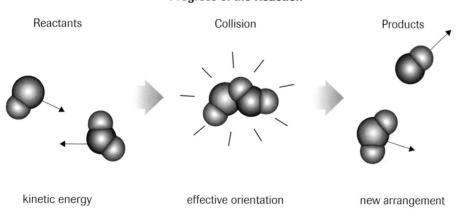

Figure 3
Collision–reaction theory explains that chemical entities must collide with the correct orientation to react.

CAREER CONNECTION

Forensic Laboratory Analyst
As vital members of an investigative team, forensic laboratory analysts examine and test physical evidence related to a criminal offence, and provide expert testimony. With their knowledge of biology and chemistry, they conduct experiments and tests on documents, firearms, and biological tissues.

Find out more about the personal characteristics of these valued law-enforcement scientists.

www.science.nelson.com

A theoretical explanation of why chemical reactions occur is partially covered in Chapter 1. Atoms often react in order to obtain a more stable electron arrangement (often an octet of electrons like the nearest noble gas). The how and why of chemical reactions can be summarized in a statement such as, "when chemical entities collide, they may exchange or share electrons to obtain a more favourable (stable) electron arrangement."

Chemical Reactions

Recall that chemical reactions produce new substances. How do you know if an unfamiliar change is a chemical reaction? Certain characteristic evidence is associated with chemical reactions (**Table 2**).

Table 2 Evidence of Chemical Reactions

Evidence	Description and example
colour change	The final product(s) may have a different colour than the colour(s) of the starting material(s). For example, the solution changes from colourless to blue.
odour change	The final material(s) may have a different odour than the odour(s) of the starting material(s). For example, mixing solutions of sodium acetate and hydrochloric acid produces a mixture that smells like vinegar.
state change	The final material(s) may include a substance in a state that differs from the starting material(s). Most commonly, either a gas or a solid (precipitate) is produced.
energy change	When a chemical reaction occurs, energy in the form of heat, light, sound, or electricity is absorbed or released. For most chemical reactions, the energy absorbed or released is in the form of heat. A common example of an energy change is the combustion, or burning, of a fuel. If energy is absorbed, the reaction is *endothermic*. If energy is released, the reaction is *exothermic*.

A **diagnostic test** is a short and specific laboratory procedure, with expected evidence and analysis, that is used as an empirical test for the presence of a substance. Diagnostic tests increase the certainty that a new substance has formed in a chemical reaction. Appendix C.4 describes diagnostic tests for chemicals such as hydrogen and oxygen. If the diagnostic test entails a single step for a specific chemical, you may find it convenient to summarize this test using the format, "If [procedure] and [evidence], then [analysis]." An example of a diagnostic test is shown in **Figure 4**.

Conservation of Mass in Chemical Changes

Experimenters have found that the total mass of matter present after a chemical change is always the same as the total mass present before the change, no matter how different the new substances appear. This finding, called the law of conservation of mass, was one of the compelling reasons why scientists accepted the atomic theory of matter. If a chemical change is thought of as a rearrangement of entities at the molecular level, then it is simple to argue that the mass must be constant. The individual entities do not change, except in the ways they are associated with each other.

Communicating Chemical Reactions

A **balanced chemical equation** is one in which the total number of each kind of atom or ion in the reactants is equal to the total number of the same kind of atom or ion in the products.

Figure 4
If an unidentified gas is bubbled through a limewater solution, and the mixture becomes cloudy, then the gas most likely contains carbon dioxide. The limewater diagnostic test provides evidence for carbon dioxide gas in the breath you exhale.

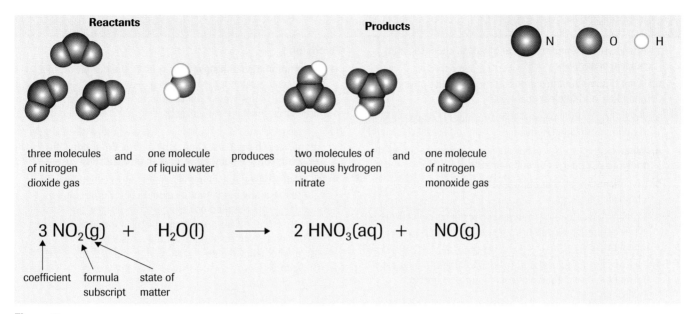

Figure 5
The balanced chemical equation and the molecular models for the reactants and products in the chemical reaction of nitrogen dioxide and water. Models such as these help us to visualize non-observable processes.

Figure 5 shows the balanced chemical equation and molecular models representing the reaction of nitrogen dioxide gas and water to produce nitric acid and nitrogen monoxide gas. By studying the molecular models, you can see there are three nitrogen atoms on the reactant side and three on the product side of the equation arrow. Likewise, there are seven oxygen atoms on both sides of the equation arrow, and two hydrogen atoms.

If more than one molecule is involved (for example, three molecules of nitrogen dioxide), then a number called a **coefficient** is placed in front of the chemical formula. In this example, *three* molecules of nitrogen dioxide and *one* molecule of water react to produce *two* molecules of nitric acid and *one* molecule of nitrogen monoxide. Coefficients are part of a balanced chemical equation and should not be confused with formula subscripts, which are part of the chemical formula for a substance.

A substance's state of matter is given in parentheses after the chemical formula. It is not part of the theoretical description given by the molecular models. Chemical formulas showing states of matter provide both a theoretical and an empirical description of a substance.

SUMMARY: Chemical Reaction Equations

- A chemical reaction is communicated by a balanced chemical equation in which the same number of each kind of atom or ion appears on the reactant and product sides of the equation.
- A *coefficient* in front of a chemical formula in a chemical equation communicates the number of molecules or formula units of a reactant or product that are involved in the reaction.
- Within formulas, a numerical subscript communicates the number of atoms or ions present in one molecule or formula unit of a substance.
- A state of matter in parentheses in a chemical equation communicates the physical state of the reactants and products at SATP.

DID YOU KNOW?

Language
Learning the language of chemistry is essential to communicating chemical knowledge.
"Language grows out of life, out of its needs and experiences.... Language and knowledge are indissolubly connected; they are interdependent. Good work in language presupposes and depends on a real knowledge of things."
—Annie Sullivan, American educator of persons who were visually and hearing impaired (1866–1936)

Section 2.2 Questions

1. Provide two examples each of physical, chemical, and nuclear changes.
2. What different entities are rearranged during physical, chemical, and nuclear changes?
3. According to the collision–reaction theory, identify the requirements for a chemical reaction to take place.
4. What is the purpose of classification systems, such as those for types of changes by substances?
5. List four changes that can be used as evidence for chemical reactions.
6. Provide two examples from everyday life of each of the four types of changes listed in your answer to the previous question.
7. Use an "If [procedure] and [evidence], then [analysis]" format to write diagnostic tests for an acid and for hydrogen (see Appendix C.4).
8. Identify one scientific law that led John Dalton to create the theory that atoms are conserved in a chemical reaction.
9. Distinguish between a formula subscript, such as H_2, and a coefficient, such as 2 H. How are they similar and how are they different?
10. An investigation is conducted to observe and classify evidence of chemical changes. The substances are mixed and the evidence obtained is recorded in **Table 3**.
    (a) Match as many mixtures as possible to each of the four categories of evidence of chemical reactions (see **Table 2**, page 48).
    (b) Which mixture did not appear to have a chemical reaction? How certain are you about this interpretation?
    (c) In general terms, what additional laboratory work could be done to improve the certainty that a chemical reaction has occurred?

Extension

11. A common media mistake is to refer to the dissolving of a chemical in water or the reaction of a metal with a solution as melting. How would you explain to the media that their concept of melting is incorrect?
12. There is debate among chemists as to whether dissolving is a physical change or a chemical change. What does the existence of a debate tell you about classification systems?
13. Evidence-based reasoning is a mainstay of scientific work. There are a few terms that a scientifically literate person needs to know to read newspapers, magazines, and scientific reports (see Appendix B.4.) What do the following terms mean with respect to scientific research?
    (a) anecdotal evidence
    (b) sample size
    (c) replication
    (d) placebo
14. Scholars publish their research in a refereed (peer-reviewed) journal. These journals are like scholarly magazines. Scientists who submit their articles for publication have their research reviewed by their peers (experts in their field of study). Many research reports submitted for publication are rejected because of some fault in the study. As a referee (peer-reviewer), critique the following experimental designs.
    (a) One group of 10 patients is given experimental medication for an illness while another group of 10 patients is not given the medication. The health of all twenty patients is monitored for six months.
    (b) One chemistry teacher completes a unit of study without doing any laboratory work; another teacher completes the same unit of study by doing four laboratory investigations. Student achievement is compared on the results of a unit test.

Table 3 Evidence for Chemical Reactions

Mixture	Procedure	Evidence
1	A zinc strip is briefly dipped into a hydrochloric acid solution.	Colourless gas bubbles formed on the strip.
2	A couple of drops of blue bromothymol blue solution are added to hydrochloric acid.	The blue bromothymol blue solution turned yellow.
3	A few drops of silver nitrate solution are added to hydrochloric acid.	A white solid (precipitate) formed.
4	Hydrochloric acid is added to a sodium acetate solution.	The odour of vinegar was evident.
5	Ammonium nitrate crystals are stirred into water.	The water (solution) felt cool.
6	Hydrochloric acid is added to a sodium bicarbonate solution.	Colourless gas bubbles formed.
7	A couple of drops of phenolphthalein solution are added to an ammonia solution.	The colourless phenolphthalein turned red.
8	Sodium hydroxide solution is added to a cobalt(II) chloride solution.	A blue and/or pink solid (precipitate) formed.
9	Sodium nitrate solution is added to a potassium chloride solution.	No change was observed.
10	A copper wire is placed into a silver nitrate solution.	Silvery crystals formed on the wire.

Balancing Chemical Reaction Equations 2.3

You are already familiar with some terms used to define convenient numbers (**Table 1**). For example, a dozen is a convenient number referring to such items as eggs or doughnuts. Since atoms, ions, and molecules are extremely small entities, a convenient number for them must be much greater than a dozen. A convenient *amount of substance*—also called the **chemical amount**—is the SI quantity for the number of entities in a substance. It is measured in units of moles (SI symbol, mol). Modern methods of estimating this number of entities have led to the value 6.02×10^{23}. This value is called **Avogadro's number**, named after the Italian chemist, Amedeo Avogadro (1776–1856). (Avogadro did not determine the number, but he created the research idea.) A **mole** is the unit of chemical amount of substance with the number of entities corresponding to Avogadro's number. For example:

- one mole of sodium is 6.02×10^{23} Na atoms
- one mole of chlorine is 6.02×10^{23} Cl_2 molecules
- one mole of sodium chloride is 6.02×10^{23} NaCl formula units

Essentially, a mole represents a number (6.02×10^{23}, Avogadro's number), just as a dozen represents the number 12.

Although the mole represents an extraordinarily large number, a mole of a substance is an observable quantity that is convenient to measure and handle. **Figure 1** shows a mole of each of three common substances: one element, one ionic compound, and one molecular compound. In each case, a mole of entities is a sample size that is convenient for lab work.

Translating Balanced Chemical Equations

A balanced chemical equation can be interpreted theoretically in terms of individual atoms, ions, or molecules, or groups of them. Consider the reaction equation for the industrial production of the fertilizer ammonia:

$N_2(g)$	+	$3\,H_2(g)$	→	$2\,NH_3(g)$
1 molecule		3 molecules		2 molecules
1 dozen molecules		3 dozen molecules		2 dozen molecules
1 mol nitrogen		3 mol hydrogen		2 mol ammonia
6.02×10^{23} molecules		$3(6.02 \times 10^{23})$ molecules		$2(6.02 \times 10^{23})$ molecules
6.02×10^{23} molecules		18.06×10^{23} molecules		12.04×10^{23} molecules

Note that the numbers in each row are *in the same ratio (1:3:2)* whether individual molecules, large numbers of molecules, or moles are considered. When moles are used to express the coefficients in the balanced equation, the ratio of reacting amounts is called the *mole ratio*.

A complete translation of the balanced chemical equation for the formation of ammonia is: "One mole of nitrogen gas and three moles of hydrogen gas react to form two moles of ammonia gas." This translation includes all the symbols in the equation, including coefficients and states of matter.

Table 1 Convenient Numbers

Quantity	Number	Example
pair	2	shoes
dozen	12	eggs
gross	144	pencils
ream	500	paper
mole	6.02×10^{23}	molecules

+ **EXTENSION**

How Big Is a Mole?
Extend your understanding of the magnitude of Avogadro's number using analogies.

www.science.nelson.com GO

Figure 1
These amounts of carbon, table salt, and sugar each contain about a mole of entities (atoms, formula units, molecules) of the substance. The mole represents a convenient and specific quantity of a chemical. Note that equal numbers of entities can have very different volumes and masses.

DID YOU KNOW ?

The Mole

It is difficult for us to get an idea of how large a number the mole is and how small atoms and molecules are. Consider that one mole of water molecules occupies 18 mL of $H_2O(l)$, whereas one mole of marbles would cover planet Earth with a layer 80 km thick.

An experimental method for determining the number of molecules in a mole is to place a one-molecule-thick layer of an insoluble substance on water and use the measurements of mass and volume to calculate Avogadro's number.

▶ COMMUNICATION example

Translate the following chemical equation into an English sentence.

$6\ CO_2(g) + 6\ H_2O(l) \rightarrow C_6H_{12}O_6(aq) + 6\ O_2(g)$

Solution

Six moles of carbon dioxide gas react with six moles of liquid water to produce one mole of aqueous glucose and six moles of oxygen gas.

Balancing Chemical Equations

A chemical equation is a simple, precise, logical, and international method of communicating the experimental evidence of a reaction. The evidence used when writing a chemical equation is often obtained in stages. First there are some general observations that suggest a chemical change has occurred. These are likely followed by a series of diagnostic tests to identify the products of the reaction. At this stage an unbalanced chemical equation can be written, and then the theory of conservation of atoms can be used to predict the coefficients necessary to balance the reaction equation. In most cases, trial and error, as well as intuition and experience, play an important role in successfully balancing chemical equations. The following summary outlines a systematic approach to balancing equations. Use it as a guide as you study Sample Problem 2.1.

SUMMARY | Balancing Chemical Equations

Step 1: Write the chemical formula for each reactant and product, including the state of matter for each one.

Step 2: Try balancing the atom or ion present in the greatest number. Find the lowest common multiple to obtain coefficients to balance this particular atom or ion.

Step 3: Repeat step 2 to balance each of the remaining atoms and ions.

Step 4: Check the final reaction equation to ensure that all atoms and ions are balanced.

▶ SAMPLE problem 2.1

A simple technology for recycling silver is to trickle waste solutions containing silver ions over scrap copper. Copper metal reacts with aqueous silver nitrate to produce silver metal and aqueous copper(II) nitrate (**Figure 2**). Write the balanced chemical equation.

Step 1: $?\ Cu(s) + ?\ AgNO_3(aq) \rightarrow ?\ Ag(s) + ?\ Cu(NO_3)_2(aq)$

Step 2:   Oxygen atoms are present in the greatest number as part of the nitrate ion, so balance this first. Balance the nitrate ion as a group.

$?\ Cu(s) + 2\ AgNO_3(aq) \rightarrow ?\ Ag(s) + 1\ Cu(NO_3)_2(aq)$

Step 3:   Balance Ag and Cu atoms. (Always balance elements last.)

$Cu(s) + 2\ AgNO_3(aq) \rightarrow 2\ Ag(s) + Cu(NO_3)_2(aq)$

Step 4:   The chemical amounts in moles of copper, silver, and nitrate are one, two, and two on both the reactant and product sides of the equation arrow. (This is a mental check; no statement is required.)

Translation: One mole of solid copper reacts with two moles of aqueous silver nitrate to produce two moles of solid silver and one mole of aqueous copper(II) nitrate.

Figure 2
The chemical equation must represent the evidence from a reaction between copper and aqueous silver nitrate.

Use the following techniques for balancing chemical equations:

- Persevere and realize that, like solving puzzles, several attempts may be necessary for more complicated chemical equations.
- The most common student error is to use incorrect chemical formulas to balance the chemical equation. *Always write correct chemical formulas first and then balance the equation as a separate step.*
- If polyatomic ions remain intact, balance them as a single unit.
- Delay balancing any atom that is present in more than two substances in the chemical equation until all other atoms or ions are balanced. (Oxygen atoms in several entities is a common example.)
- Balance elements (entities with only one kind of atom) last.
- If a fractional coefficient is required to balance an atom, multiply all coefficients by the denominator of the fraction to obtain integer values. (Balancing with fractions is correct, but not preferred.)

For example, in balancing the following reaction equation, hydrogen atoms are balanced first, then nitrogen, and oxygen is balanced last. This equation requires 7 mol of oxygen atoms:

$$2\,NH_3(g) + ?\,O_2(g) \rightarrow 3\,H_2O(g) + 2\,NO_2(g)$$

The only number that can balance the oxygen atoms is $\frac{7}{2}$. By doubling all coefficients, the reaction equation can then be balanced using only integers:

$$4\,NH_3(g) + 7\,O_2(g) \rightarrow 6\,H_2O(g) + 4\,NO_2(g)$$

> **DID YOU KNOW?**
> **Balancing Equations**
> There are two general purposes for balancing chemical equations:
> - to communicate that atoms/ions are neither created nor destroyed during a chemical reaction
> - to calculate the masses and volumes of reactants and products by using the mole ratios of reactants and products (see Unit 4)

> **Learning Tip**
> When water is a product of a reaction, it is often produced as a gas (vapour), $H_2O(g)$. The heat of combustion produces temperatures above the boiling point of water, 100 °C. If the temperature is low and/or the humidity is high, the water condenses to $H_2O(l)$ or freezes to $H_2O(s)$, as seen in vapour trails of cars and jets.

▶ Section 2.3 Questions

1. Translate the following English sentences into internationally understood balanced chemical equations:
   (a) Two moles of solid aluminium and three moles of aqueous copper(II) chloride react to form three moles of solid copper and two moles of aqueous aluminium chloride. (This reaction does not always produce the expected products listed here.)
   (b) One mole of solid copper reacts with two moles of hydrochloric acid to produce one mole of hydrogen gas and one mole of copper(II) chloride. (When tested in the laboratory, this prediction of products is falsified.)
   (c) Two moles of solid mercury(II) oxide decomposes to produce two moles of liquid mercury and one mole of oxygen gas. (This decomposition reaction is a historical but dangerous method of producing oxygen. Research an MSDS for mercury(II) oxide.)
   (d) Methanol (used in windshield washer antifreeze and as a fuel) is produced from natural gas in world-scale quantities in Medicine Hat, Alberta by the following reaction series:
   (i) One mole of methane gas reacts with one mole of steam to produce one mole of carbon monoxide gas and three moles of hydrogen gas.
   (ii) One mole of carbon monoxide gas reacts with two moles of hydrogen gas to produce one mole of liquid methanol.

2. Translate each of the following chemical equations into an English sentence including the chemical amounts and states of matter for all the substances involved:
   (a) Fire-starters for camp fires often involve the following reaction. Methanol is also a fondue fuel.
   $$2\,CH_3OH(l) + 3\,O_2(g) \rightarrow 2\,CO_2(g) + 4\,H_2O(g)$$
   (b) Phosphoric acid for fertilizer production is produced from rock phosphorus at Fort Saskatchewan, Alberta:
   $$Ca_3(PO_4)_2(s) + 3\,H_2SO_4(aq) \rightarrow 2\,H_3PO_4(aq) + 3\,CaSO_4(s)$$
   (c) The reaction of sodium with water is potentially dangerous.
   $$2\,Na(s) + 2\,H_2O(l) \rightarrow H_2(g) + 2\,NaOH(aq)$$
   (d) Sulfuric acid can be used as a catalyst to dehydrate sugar.
   $$C_{12}H_{22}O_{11}(s) \rightarrow 12\,C(s) + 11\,H_2O(g)$$

3. Which of the following chemical equations is balanced correctly?
   (a) $H_2(g) + O_2(g) \rightarrow H_2O(g)$
   (b) $2\,NaOH(aq) + Cu(ClO_3)_2(aq) \rightarrow$
       $Cu(OH)_2(s) + 2\,NaClO_3(aq)$
   (c) $Pb(s) + AgNO_3(aq) \rightarrow Ag(s) + Pb(NO_3)_2(aq)$
   (d) $2\,NaHCO_3(s) \rightarrow Na_2CO_3(s) + CO_2(g) + H_2O(l)$

4. Write a balanced chemical equation for each of the following reactions. Assume that substances are pure and at SATP unless the states of matter are given. Also classify the primary perspective presented in the accompanying statements: The possible perspectives are scientific, technological, ecological, economic, and political.
   (a) Research indicates that sulfur dioxide gas reacts with oxygen in the air to produce sulfur trioxide gas.
   (b) Sulfur trioxide gas travelling across international boundaries causes disagreements between governments.
       sulfur trioxide + water → sulfuric acid
   (c) The means exist for industry to reduce sulfur dioxide emissions; for example, by treatment with lime.
       calcium oxide + sulfur dioxide + oxygen → calcium sulfate
   (d) Restoring acidic lakes to normal pH (acid–base balance) is expensive; for example, adding lime to lakes using aircraft (**Figure 3**).
       calcium oxide + sulfurous acid → water + calcium sulfite
   (e) Fish in overly acidic lakes may die from mineral poisoning due to the leaching of, for example, aluminium ions from lake bottoms.
       solid aluminium silicate + sulfuric acid → aqueous hydrogen silicate + aqueous aluminium sulfate
   (f) Chemical engineers prepare sodium fluoride in large batches for city water or toothpaste distributors.
       aqueous sodium chloride + hydrofluoric acid → aqueous sodium fluoride + hydrochloric acid

5. Balance the following equations that communicate reactions that occur before, during, and after the formation of acid rain.
   (a) $C(s) + O_2(g) \rightarrow CO_2(g)$
   (b) $S_8(s) + O_2(g) \rightarrow SO_2(g)$
   (c) $CaSiO_3(s) + H_2SO_3(aq) \rightarrow H_2SiO_3(aq) + CaSO_3(s)$
   (d) $CaCO_3(s) + HNO_3(aq) \rightarrow H_2CO_3(aq) + Ca(NO_3)_2(aq)$
   (e) $Al(s) + H_2SO_4(aq) \rightarrow H_2(g) + Al_2(SO_4)_3(aq)$
   (f) $SO_2(g) + H_2O(l) \rightarrow H_2SO_3(aq)$
   (g) $Fe(s) + H_2SO_3(aq) \rightarrow H_2(g) + Fe_2(SO_3)_3(s)$
   (h) $CO_2(g) + H_2O(l) \rightarrow H_2CO_3(aq)$
   (i) $CH_4(g) + O_2(g) \rightarrow CO_2(g) + H_2O(g)$
   (j) $FeS(s) + O_2(g) \rightarrow FeO(s) + SO_2(g)$
   (k) $H_2S(g) + O_2(g) \rightarrow H_2O(g) + SO_2(g)$
   (l) $CaCO_3(s) + SO_2(g) + O_2(g) \rightarrow CaSO_4(s) + CO_2(g)$

6. Balance the following chemical reaction equations and draw molecular models to represent the number and kind of atoms in each molecule.
   (a) Hydrogen is used as fuel for the space shuttle.
       $H_2(g) + O_2(g) \rightarrow H_2O(g)$
   (b) Ammonia fertilizer is produced for agricultural use.
       $N_2(g) + H_2(g) \rightarrow NH_3(g)$
   (c) Hydrogen chloride gas is produced and then dissolved to make hydrochloric acid to etch concrete.
       $H_2(g) + Cl_2(g) \rightarrow HCl(g)$

Extension

7. Scientists have a high standard for accepting knowledge as valid/acceptable. Write a short procedure for a double blind experimental design to test the effectiveness of using lime to neutralize acid lakes. Reference Appendix B.4.

Figure 3
Helicopters are used to add lime to acidic lakes.

Chemical Amount 2.4

Chemists created the concept of the mole. They needed a convenient unit for determining the quantities of chemicals that react and/or are produced in a chemical reaction. You have seen that the mole is a useful unit for communicating reacting amounts in a balanced chemical equation. Now you will learn how to measure chemical amount (in moles) indirectly, by measuring mass directly.

Molar Mass

The **molar mass**, M, of a substance is the mass of one mole of the substance. The unit for molar mass is grams per mole (g/mol) (**Figure 1**). Each substance has a different molar mass, which can be calculated as follows:

1. Write the correct chemical formula for the substance.
2. Determine the chemical amount (in moles) of each atom (or monatomic ion) in one mole of the chemical.
3. Use the atomic molar masses from the periodic table and the chemical amounts (in moles) to determine the mass of one mole of the chemical.
4. Communicate the mass of one mole as the molar mass in units of grams per mole, precise to two decimal places; for example, 78.50 g/mol.

You may also think of the molar mass as a ratio of the mass of a particular chemical to the amount of the chemical in moles. Molar mass is a convenient factor to use when converting between mass and chemical amount. In Sample Problems 2.2 and 2.3, M represents the molar mass, and the numbers and atomic molar masses are written out.

When doing calculations involving molar masses, remember that the relevant SI quantities and units are:

- mass, m, in grams, g
- chemical amount, n, in moles, mol
- molar mass, M, in grams per mole, g/mol

Figure 1
The molar mass of water is 18.02 g/mol. One mole of water has a mass of 18.02 g and contains 6.02×10^{23} molecules of H_2O. One mole of liquid water occupies about 18.0 mL.

Learning Tip

When adding and subtracting measured values, the answer should have the same precision (number of decimal places) as the value with the least precision (number of decimal places). See Appendix F.3.

> ### SAMPLE problem 2.2
>
> Determine the molar mass of water.
>
> Step 1: Write the correct chemical formula for water: $H_2O(l)$.
>
> Step 2: Determine the chemical amount of each atom in one mole of $H_2O(l)$: one mole of H_2O is composed of 2 mol_H + 1 mol_O
>
> Step 3: Find the mass of one mole of water using the number of atoms and their atomic molar masses from the periodic table.
>
> $$m_{H_2O} = 2\,M_H + 1\,M_O$$
> $$= (2\text{ mol} \times 1.01\text{ g/mol}) + (1\text{ mol} \times 16.00\text{ g/mol})$$
> $$= 18.02\text{ g}$$
>
> Step 4: Since this quantity is the mass of one mole of water, the molar mass of water is 18.02 g/mol; that is,
>
> $$M_{H_2O} = 18.02\text{ g/mol}$$

DID YOU KNOW ?

SI

SI is the Système International d'Unités or International System of Units. SI is used in all languages to denote this system of units. SI was created by international agreement in 1960. Canada officially became an SI metric country in the early 1970s.

> **SAMPLE** problem 2.3

What is the molar mass of ammonium phosphate, a fertilizer?

Step 1: Write the correct chemical formula for ammonium phosphate: $(NH_4)_3PO_4(s)$.

Step 2: Determine the chemical amount of each atom in one mole of $(NH_4)_3PO_4(s)$:
One mole of $(NH_4)_3PO_4$ is composed of $3\ mol_N + 12\ mol_H + 1\ mol_P + 4\ mol_O$

Step 3: Find the mass of one mole of ammonium phosphate.

$$m_{(NH_4)_3PO_4} = 3\ M_N + 12\ M_H + 1\ M_P + 4\ M_O$$
$$= (3\ mol \times 14.01\ g/mol) + (12\ mol \times 1.01\ g/mol) +$$
$$(1\ mol \times 30.97\ g/mol) + (4\ mol \times 16.00\ g/mol)$$
$$= 149.12\ g$$

Step 4: Since this quantity is the mass of one mole of ammonium phosphate, the molar mass of this chemical is 149.12 g/mol.

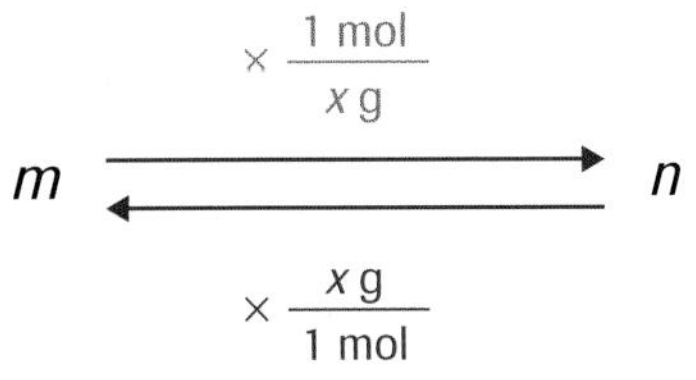

Figure 2
Chemists created the concept of molar mass to convert between mass and chemical amount.

Mass–Amount Conversions

In order to use the mole ratio from the balanced equation to determine the masses of reactants and products in chemical reactions, you must be able to convert a mass to a chemical amount (an amount in moles) and vice versa. To do this, you use either the molar mass as a conversion factor in grams per mole (g/mol), or the reciprocal of the molar mass, in moles per gram (mol/g), and cancel the units (**Figure 2**). Examples of each conversion follow; n represents chemical amount in moles and m represents mass in grams.

For these Communication Examples, note that the molar mass is calculated separately (e.g., in your calculator) and inserted where needed. Your teacher may tell you whether you need to state it separately.

Learning Tip

The certainty rule is presented in Appendix F.3. A value calculated by multiplying or dividing is no more certain than the least certain of the initial values. (A chain is no stronger than its weakest link.) Certainty is measured in significant digits. All digits except the zeros in front in a correctly measured or calculated value are counted; for example, 0.06 g has a certainty of one significant digit.

> **COMMUNICATION** example 1

Calcium carbonate helps to neutralize acidic soil under spruce trees. Convert a mass of 1500 g of calcium carbonate to a chemical amount.

Solution

$$M_{CaCO_3} = 100.09\ g/mol$$

$$n_{CaCO_3} = 1500\ g \times \frac{1\ mol}{100.09\ g}$$

$$= 14.99\ mol$$

In Communication Example 1, the appropriate conversion factor is the reciprocal of the molar mass. The certainty of the value for chemical amount (14.99 mol) is four significant digits, resulting from the least certain of four significant digits for 1500 g and five significant digits for 100.09 g/mol.

COMMUNICATION example 2

Sodium sulfate is mined from lakes and deposits along the Alberta–Saskatchewan border. Convert a reacting amount of 3.46 mmol of sodium sulfate into mass in grams.

Solution

$M_{Na_2SO_4} = 142.05$ g/mol

$m_{Na_2SO_4} = 3.46 \text{ mmol} \times \dfrac{142.05 \text{ g}}{1 \text{ mol}}$

$= 491$ mg

or

$m_{Na_2SO_4} = 3.46 \text{ mmol} \times \dfrac{1 \text{ mol}}{1000 \text{ mmol}} \times \dfrac{142.05 \text{ g}}{1 \text{ mol}}$

$= 0.491$ g

DID YOU KNOW ?

Sodium Sulfate
Sodium sulfate as a decahydrate is called Glauber's salt, $Na_2SO_4 \cdot 10 H_2O(s)$. The hydrate is used in solar energy storage units. Solar energy as heat is stored when the hydrate changes to an aqueous solution. Sodium sulfate is sometimes used by Alberta pulp and paper plants in the pulp bleaching process.

In Communication Example 2, the certainty of the value for mass (491 mg or 0.491 g) is three significant digits, resulting from the least certain of 3.46 mmol and 142.04 g/mol.

Section 2.4 Questions

It helps to memorize the chemical formula or name for each technological or natural substance marked with an asterisk (*).

1. Calculate the molar mass of each of the following substances. The molar masses of water and of carbon dioxide should be memorized for efficient work:
   (a) $H_2O(l)$ (water)*
   (b) $CO_2(g)$ (respiration product)*
   (c) $NaCl(s)$ (pickling salt, sodium chloride)*
   (d) $C_{12}H_{22}O_{11}(s)$ (table sugar, sucrose)*
   (e) $(NH_4)_2Cr_2O_7(s)$ (ammonium dichromate)

2. Communicate the certainty of the following measured or calculated values as a number of significant digits.
   (a) 16.05 g   (d) 0.563 kg
   (b) 7.0 mL   (e) 0.000 5 L
   (c) 10 cm^2 (f) 90.00 g/mol

3. Perform the following calculations and express the answer to the correct certainty (number of significant digits).
   (a) $n_{Cu} = 7.46 \text{ g} \times \dfrac{1 \text{ mol}}{63.55 \text{ g}} =$
   (b) $m_C = 2.0 \text{ mol} \times \dfrac{12.01 \text{ g}}{1 \text{ mol}} =$
   (c) $n_{CuSO_4} = 100.0 \text{ mL} \times \dfrac{0.500 \text{ mol}}{1 \text{ L}} =$
   (d) $m_{CuSO_4 \cdot 5H_2O} = 0.05000 \text{ mol} \times \dfrac{249.71 \text{ g}}{1 \text{ mol}} =$

4. Perform the following more advanced calculations by using the precision and/or uncertainty rule where appropriate.
   (a) $m_{NH_3} = 101 \text{ mol} \times \dfrac{17.04 \text{ g}}{1 \text{ mol}} =$
   (b) $V_{n_{CuSO_4}} = 250.0 \text{ mol} \times \dfrac{1 \text{ L}}{5.00 \text{ mol}} =$
   (c) $c_{NaCl} = \dfrac{15.5 \text{ mmol}}{10.00 \text{ mL}} =$
   (d) $V_{avg} = \dfrac{13.6 \text{ mL} + 13.5 \text{ mL} + 13.6 \text{ mL}}{3} =$
   (e) % difference $= \dfrac{|3.67 \text{ g} - 3.61 \text{ g}|}{3.61 \text{ g}} \times 100 =$
   (f) $Q = 50.0 \text{ g} \times 4.19 \text{ J/(g} \cdot °C) \times (34.2 - 15.4)°C$

5. Calculate the chemical amount of pure substance present in each of the following samples:
   (a) 40.0 g of propane, $C_3H_8(l)$, in a camp stove cylinder
   (b) A 500 g box of pickling salt*
   (c) A 10.00 kg bag of table sugar*
   (d) 325 mg of acetylsalicylic acid (ASA), $C_6H_4COOCH_3COOH(s)$, in a headache relief tablet
   (e) 150 g of isopropanol (rubbing alcohol), $CH_3CH_2OHCH_3(l)$, from a pharmacy

6. Calculate the mass of each of the following specified chemical amounts of pure substances:
   (a) 4.22 mol of ammonia in a window-cleaning solution*
   (b) 0.224 mol of sodium hydroxide (lye) in a drain-cleaning solution*
   (c) 57.3 mmol of water vapour produced by burning methane in a laboratory burner
   (d) 9.44 kmol of potassium permanganate fungicide
   (e) 0.77 mol of ammonium sulfate fertilizer

7. Calculate the mass of each reactant and product from the chemical amount shown in the following equations, and show how your calculations agree with the law of conservation of mass:
   (a) $H_2(g) + Cl_2(g) \rightarrow 2 HCl(g)$
   (b) $2 CH_3OH(l) + 3 O_2(g) \rightarrow 2 CO_2(g) + 4 H_2O(g)$

2.5 Classifying Chemical Reactions

By analyzing the evidence obtained from many chemical reactions, it is possible to distinguish patterns. On the basis of these patterns, certain generalizations about reactions can be formulated. The generalizations in **Table 1** are based on extensive evidence and provide an empirical classification of most, but not all, common chemical reactions. The five types of reactions are described in the sections that follow. (For now, any reactions that do not fit these categories are classified as "other.")

Table 1 Chemical Reactions

Reaction type	Generalization
formation	elements → compound
simple decomposition	compound → elements
complete combustion	substance + oxygen → most common oxides
single replacement	element + compound → element + compound (metal + compound → metal + compound or nonmetal + compound → nonmetal + compound)
double replacement	compound + compound → compound + compound

Figure 1
A burning sparkler demonstrates the formation of magnesium oxide by the combustion of magnesium and oxygen. This reaction can be classified as a formation reaction and as a combustion reaction.

Formation Reactions

A **formation reaction** is the reaction of two or more elements to form either an ionic compound (from a metal and a nonmetal) or a molecular compound (from two or more nonmetals). An example of a reaction forming on ionic compound is the reaction of magnesium and oxygen shown in **Figure 1**.

word equation: magnesium + oxygen → magnesium oxide

chemical equation: $2 Mg(s) + O_2(g) \rightarrow 2 MgO(s)$

For chemical reactions producing molecular substances, the only products that you will be able to predict at this time are those whose formulas you memorized from **Table 3** in Section 1.6 (page 34); for example, H_2O.

Simple Decomposition Reactions

A **simple decomposition reaction** is the breakdown of a compound into its component elements, that is, the reverse of a formation reaction. Simple decomposition reactions are important historically since they were used to determine chemical formulas. They remain important today in the industrial production of some elements from compounds available in the natural environment. A well-known example that is easy to demonstrate is the simple decomposition of water (**Figure 2**).

word equation: water → hydrogen + oxygen

chemical equation: $2 H_2O(l) \rightarrow 2 H_2(g) + O_2(g)$

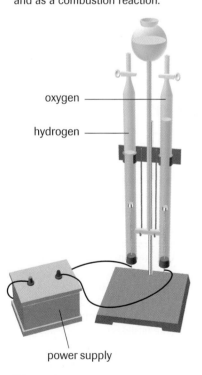

Figure 2
The simple decomposition of water is accomplished using a Hoffman apparatus.

Combustion Reactions

A **complete combustion reaction** is the burning of a substance with sufficient oxygen available to produce the most common oxides of the elements making up the substance that is burned. Some combustions, like those in a burning candle or an untuned automobile engine, are incomplete, and also produce the less common oxides such as carbon monoxide. Combustion reactions (**Figure 3**) are exothermic; these reactions provide the major source of energy for technological use in our society.

To successfully predict the products of a complete combustion reaction, you must know the composition of the most common oxides. If the substance being burned contains

- carbon, then $CO_2(g)$ is produced
- hydrogen, then $H_2O(g)$ is produced
- sulfur, then $SO_2(g)$ is produced
- nitrogen, then assume $NO_2(g)$ is produced
- a metal, then the oxide of the metal with the most common ion charge is produced (**Figure 3**)

A typical example of a complete combustion reaction is the burning of butane, $C_4H_{10}(g)$:

word equation: butane + oxygen → carbon dioxide + water

chemical equation: $2 C_4H_{10}(g) + 13 O_2(g) \rightarrow 8 CO_2(g) + 10 H_2O(g)$

You will learn more about combustion reactions in Section 9.6.

Figure 3
A spectacular combustion of a metal is the burning of steel wool in pure oxygen. This reaction is used in fireworks; note that it is also a formation reaction for iron(III) oxide.
$4 Fe(s) + 3 O_2(g) \rightarrow 2 Fe_2O_3(s)$

▶ Section 2.5 Questions

1. Rewrite each of the following reactions as a word or balanced chemical equation, and classify each reaction as formation, simple decomposition, or complete combustion. Assume the SATP states of matter unless otherwise indicated. For example,
   $2 Na(s) + Cl_2(g) \rightarrow 2 NaCl(s)$ (formation)
   sodium + chlorine → sodium chloride

   (a) lithium oxide → lithium + oxygen
   (b) $2 KBr(s) \rightarrow 2 K(s) + Br_2(l)$
   (c) $6 K(s) + N_2(g) \rightarrow 2 K_3N(s)$
   (d) magnesium oxide → magnesium + oxygen
   (e) $16 Al(s) + 3 S_8(s) \rightarrow 8 Al_2S_3(s)$
   (f) methane + oxygen → carbon dioxide + water (vapour)

2. For each of the following reactions,
   - classify the reaction type as formation or simple decomposition,
   - predict the product(s) of the reaction,
   - complete and balance the chemical equation
   - complete the word equation.
   Assume the most common ion charges and that the products are at SATP.
   (a) Since the Bronze Age (about 3000 B.C.E.), copper has been produced by heating the ore that contains CuO(s).

   copper(II) oxide →

   (b) When aluminium reacts with air, a tough protective coating forms. This coating helps prevent acidic

Figure 4
Aluminium does not corrode in air because of a strongly adhering oxide coating.

substances, such as soft drinks (**Figure 4**), from reacting with the acids and thereby corroding the aluminium.

$Al(s) + O_2(g) \rightarrow$

(c) Sodium hydroxide can be decomposed into its elements by melting it and passing electricity through it.

$NaOH(l) \rightarrow$

(d) Very reactive sodium metal reacts with the poisonous gas chlorine to produce an inert, edible chemical.

$Na(s) + Cl_2(g) \rightarrow$

(e) A frequent technological problem associated with the operation of swimming pools is that copper pipes react with aqueous chlorine.

$Cu(s) + Cl_2(aq) \rightarrow$

(f) A major scientific breakthrough occurred in 1807 when Sir Humphry Davy isolated potassium by passing electricity through molten (melted) potassium oxide.

$K_2O(l) \rightarrow$

(g) When zinc is exposed to oxygen, a protective coating forms on the surface of the metal. This reaction makes zinc coating of metals (galvanizing) a desirable process for resisting corrosion.

zinc + oxygen →

(h) Translate the last equation above into an English sentence. Include the chemical amounts in moles.

3. State the names and chemical formulas for the most common oxides of carbon, hydrogen, sulfur, nitrogen, and iron.

4. For each of the following complete combustion reactions, complete and balance the chemical equation or complete the word equation. Assume the pure state of matter at SATP unless otherwise indicated.

(a) In Canada, many homes are heated by the combustion of natural gas (assume methane).

$CH_4(g) + O_2(g) \rightarrow$

(b) Nitromethane, $CH_3NO_2(l)$, is a fuel commonly burned in drag-racing vehicles.

nitromethane + oxygen →

(c) Mercaptans (assume $C_4H_9SH(g)$) are added to natural gas to give it a distinct odour. The mercaptan burns with the natural gas.

$C_4H_9SH(g) + O_2(g) \rightarrow$

(d) Ethanol from grain can be added to gasoline as a fuel and antifreeze. It burns along with the gasoline.

ethanol + oxygen →

(e) Write a balanced equation for (d), and then translate the equation into an English sentence. Include the chemical amounts and states of matter.

(f) Most automobiles currently burn gasoline (assume octane) as a fuel.

octane + oxygen →

5. Rewrite each of the following reactions as a word equation or a balanced chemical equation, and classify each reaction as formation, simple decomposition, or complete combustion. (Some reactions may have two classifications.)

(a) Electricity is used to produce elements from molten potassium bromide at a high temperature.

$2\ KBr(l) \rightarrow 2\ K(l) + Br_2(g)$

(b) Coal burns in a power plant to produce heat for generating electrical energy.

carbon + oxygen → carbon dioxide

(c) Gasoline antifreeze burns in an automobile engine.

methanol + oxygen → carbon dioxide + water

(d) Poisonous hydrogen sulfide from natural gas is eventually converted to elemental sulfur using this reaction as a first step in about 50 gas plants in Alberta.

$2\ H_2S(g) + 3\ O_2(g) \rightarrow 2\ SO_2(g) + 2\ H_2O(g)$

(e) Hydrogen gas may be the automobile fuel of the future.

hydrogen + oxygen → water

(f) Toxic hydrogen cyanide gas can be destroyed in a waste treatment plant, such as the one at Swan Hills, Alberta.

Four moles of hydrogen cyanide gas react with nine moles of oxygen gas to produce four moles of carbon dioxide gas, two moles of water vapour, and four moles of nitrogen dioxide gas.

6. Classify the following reactions as formation, simple decomposition, or complete combustion. Predict the products of the reactions, write the formulas and states of matter, and balance the reaction equations:

(a) $Al(s) + F_2(g) \rightarrow$

(b) $NaCl(s) \rightarrow$

(c) $S_8(s) + O_2(g) \rightarrow$

(d) methane + oxygen →

(e) aluminium oxide →

(f) propane burns

(g) $C_4H_{10}(g) + O_2(g) \rightarrow$

7. Describe a technological application for two of the chemical reactions in question 6.

8. Write a brief empirical description of reactants and products for two of the chemical reactions in question 6.

9. List two benefits and two risks of using combustion reactions. Use examples of a fuel used in your area. Try to use a variety of perspectives in your answer (see Section 2.1, page 45).

Extension

10. Hydrogen-burning cars may become common in the future. Write perspective statements, pro and/or con, to the resolution that most cars should be burning hydrogen in twenty years. Provide at least one statement from each of scientific, technological, ecological, economic, and political perspectives.

www.science.nelson.com

11. Dr. John Polanyi is a Canadian Nobel laureate. One reaction that he studied was that of atomic hydrogen gas, H(g), with chlorine gas. Read the story on the Web site and write the chemical equation for this reaction.

www.science.nelson.com

Chemical Reactions in Solution 2.6

The reactions reviewed so far involve pure substances. The remaining two reaction types, single and double replacements, usually occur in aqueous solutions. As you know, substances dissolved in water are indicated by (aq).

A **solution** is a homogeneous mixture (page 12) of a **solute** (the substance dissolved) and a **solvent** (the substance, usually a liquid, that does the dissolving). **Figure 1** shows a common example involving table salt and water. The **solubility** of a substance, which is covered in more detail in Chapter 5, is the maximum quantity of the substance that will dissolve in a solvent at a given temperature. For substances like sodium chloride (in table salt), the maximum quantity that dissolves in certain solvents is large compared with other solutes. Such solutes are said to be very soluble. When very soluble substances are formed as products in a single or double replacement reaction, the maximum quantity of solute that can dissolve is rarely reached; thus, the new substance remains in solution, and an (aq) notation is appropriate. Other substances, such as calcium carbonate (in limestone and chalk), are only slightly soluble. When these substances are formed in a chemical reaction, the maximum quantity that can dissolve is usually reached and most of this substance settles to the bottom as a solid. Solid substances formed from reactions in solution are known as **precipitates** (**Figure 2**, on the next page). They are indicated in a chemical reaction equation by (s).

A *solubility chart* outlines solubility generalizations for a large number of ionic compounds; see **Table 1**. A major purpose of this chart is to predict the state of matter for ionic compounds formed as products of chemical reactions in solution. This summary of solubility evidence is listed in two categories—*very soluble (aq)* (for example, sodium chloride) and *slightly soluble (s)* (for example, calcium carbonate). The solubility of ionic compounds in water can be predicted from the solubility chart.

At this point, you will not be expected to predict the solubility of molecular compounds in water, but you should memorize the examples in **Table 2**. Some elements, like the alkali metals, react with water, but most elements do not react or dissolve in water to any noticeable extent. In general, if an element is a reactant or product in a chemical reaction in an aqueous solution then assume its pure state of matter unless otherwise indicated.

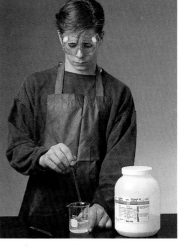

Figure 1
Table salt (the solute) is being dissolved in water (the solvent) to make a solution.

Table 2 Solubility of Selected Molecular Compounds in Water

Solubility	Examples
very soluble	$NH_3(aq)$, $H_2S(aq)$, $H_2O_2(aq)$, $CH_3OH(aq)$, $C_2H_5OH(aq)$, $C_{12}H_{22}O_{11}(aq)$, $C_6H_{12}O_6(aq)$
slightly soluble	$CH_4(g)$, $C_3H_8(g)$, $C_8H_{18}(l)$

Table 1 Solubility of Ionic Compounds at SATP–Generalizations*

Ion	Cl^- Br^- I^-	S^{2-}	OH^-	SO_4^{2-}	CO_3^{2-} PO_4^{3-} SO_3^{2-}	CH_3COO^-	NO_3^- ClO_3^- ClO_4^-	Group 1 NH_4^+ H_3O^+ (H^+)
very soluble (aq) $\geq$ 0.1 mol/L	most	Group 1, NH_4^+, Group 2	Group 1, NH_4^+, Sr^{2+}, Ba^{2+}, Tl^+	most	Group 1, NH_4^+	most	all	all
slightly soluble (s) < 0.1 mol/L (at SATP)	Ag^+, Pb^{2+}, Tl^+, Hg_2^{2+}, Cu^+	most	most	Ag^+, Pb^{2+}, Ca^{2+}, Ba^{2+}, Sr^{2+}, Ra^{2+}	most	Ag^+, Hg_2^{2+}	none	none

*Although these are particularly reliable, all generalizations have exceptions. This textbook specifically identifies any reference to an ionic compound solubility that is an exception to these generalizations.

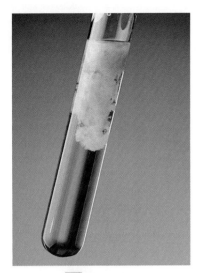

Figure 2
When an iron(III) nitrate solution is added to a sodium phosphate solution, a yellow precipitate forms immediately. Diagnostic tests indicate that the low-solubility product is iron(III) phosphate, as predicted.

> ▶ **SAMPLE** problem 2.4
>
> Iron(III) phosphate is predicted as a product in a chemical reaction. What is the solubility of FePO$_4$?
>
> 1. In the top row of the solubility chart, locate the column containing the negative ion PO$_4^{3-}$.
> 2. Look at the two boxes below this anion to determine in which category the positive ion Fe^{3+} belongs. (A process of elimination may be necessary.)
> 3. Since Fe^{3+} is not in Group 1, it must belong in the *slightly soluble* category.
>
> We can predict that a precipitate of FePO$_4$(s) forms.

> ▶ **COMMUNICATION** example
>
> Predict the solubility of each of the following ionic compounds by writing their chemical formulas with an (aq) or an (s), indicating very soluble or slightly soluble, respectively.
>
> (a) copper(II) sulfate (b) magnesium carbonate (c) silver acetate (d) potassium iodide (e) ammonium chloride
>
> **Solution**
> (a) CuSO$_4$(aq) (b) MgCO$_3$(s) (c) AgCH$_3$COO(s) (d) KI(aq) (e) NH$_4$Cl(aq)

Single Replacement Reactions

A **single replacement reaction** is the reaction of an element with a compound to produce a new element and an ionic compound. This reaction usually occurs in aqueous solutions. For example, silver can be produced from copper and a solution of silver ions (**Figure 3**):

$$Cu(s) + 2\,AgNO_3(aq) \rightarrow 2\,Ag(s) + Cu(NO_3)_2(aq)$$

copper + silver nitrate → silver + copper(II) nitrate
(metal) (compound)   (metal) (compound)

Iodine can be produced from chlorine and aqueous sodium iodide:

$$Cl_2(g) + 2\,NaI(aq) \rightarrow I_2(s) + 2\,NaCl(aq)$$

chlorine + sodium iodide → iodine + sodium chloride
(nonmetal) (compound)   (nonmetal) (compound)

We can predict the very soluble (aq) states for the two ionic products, copper(II) nitrate and sodium chloride, from the solubility chart (**Table 1**, page 61). Evidence shows that *a metal replaces a metal ion to liberate a different metal as a product* (as in the first preceding example) and *a nonmetal replaces a nonmetal ion to liberate a different nonmetal as a product* (as in the second example). Reactive metals, such as those in Groups 1 and 2, react with water to replace the hydrogen, forming hydrogen gas and a hydroxide compound. (In these reactions, hydrogen acts like a metal.)

Double Replacement Reactions

A **double replacement reaction** can occur between two ionic compounds in solution. In the reaction, the ions, by analogy, "change partners" to form the products. If one of the products is slightly soluble, it may form a precipitate, as shown in **Figure 4**. As the

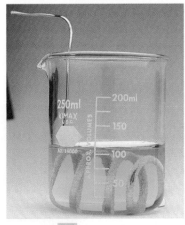

Figure 3
The blue solution verifies the prediction that the most common ion of copper, Cu^{2+}(aq), is formed.

term implies, **precipitation** is a double replacement reaction in which a solid substance forms. For example:

CaCl$_2$(aq)   +   Na$_2$CO$_3$(aq)   →   CaCO$_3$(s)   +   2 NaCl(aq)
compound   +   compound   →   compound   +   compound

Remember that chemists have created solubility generalizations to help us organize our knowledge. The generalizations serve our purpose and are only approximations of what is found in nature. There are always exceptions to generalizations.

In another kind of double replacement reaction, an acid reacts with a base, producing water and an ionic compound. This kind of double replacement reaction is known as **neutralization**. The reaction between hydrochloric acid and potassium hydroxide is an example:

HCl(aq)   +   KOH(aq)   →   H$_2$O(l)   +   KCl(aq)
acid   +   base   →   water   +   ionic compound (a salt)

When writing chemical equations for both precipitation and neutralization reactions, consult the solubility chart (**Table 1**, page 61) to determine the state of matter of the ionic products. For neutralization reactions, it may be easier to balance the equation if you temporarily write the chemical formula for water as HOH(l) rather than H$_2$O(l).

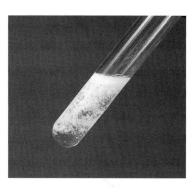

Figure 4
A white precipitate forms when colourless aqueous solutions of calcium chloride and sodium carbonate are mixed.

SUMMARY   Predicting Chemical Reactions

Table 3 Reaction Generalizations

Type of reaction	Generalization	Notes
formation	element + element → compound	metal + nonmetal → ionic compound nonmetal + nonmetal → molecular compound
simple decomposition	compound → element + element	
complete combustion	element or compound + oxygen → oxide	assume most common oxides are formed
single replacement	element + compound → element + compound	metal + compound → metal + compound nonmetal + compound → nonmetal + compound metal + acid → hydrogen + compound
double replacement	compound + compound → compound + compound	precipitation reaction: solution + solution → precipitate + solution neutralization reaction: acid + base → water + aqueous ionic compound

To write the correct chemical equation for a reaction, follow these steps:
- Step 1: Use the reaction generalizations to classify the reaction.
- Step 2: Use the reaction generalizations to predict the products of the chemical reaction and write the chemical equation.
    (a) Predict, from theory, the chemical formulas for ionic compounds, and write the formulas from memory for molecular compounds and elements.
    (b) Include states of matter, using the rules and generalizations.
- Step 3: Balance the equation without changing the chemical formulas.

+ EXTENSION

Hard and Soft Water
Precipitation reactions affect our water supply. Extend your understanding of precipitation by hearing about the everyday chemistry of hard and soft water.

www.science.nelson.com GO

Section 2.6 Questions

1. The following chemical reactions occur in a water environment (e.g., in water in a beaker). Write the balanced chemical equation, including the states of matter, (s) or (aq), for each reaction:
   (a) lead(II) nitrate + lithium chloride → lead(II) chloride + lithium nitrate
   (b) ammonium iodide + silver nitrate → silver iodide + ammonium nitrate
   (c) The net reaction during the discharge cycle of a car battery is one mole of lead and one mole of solid lead(IV) oxide reacting with two moles of sulfuric acid to produce two moles of water and two moles of lead(II) sulfate.

2. The following reactions occur in a water environment. Write the balanced chemical equation, including SATP states of matter, for each chemical.
   (a) Copper can be extracted from solution by reusing cans that contain iron.
   iron + copper(II) sulfate → copper + iron(III) sulfate
   (b) Water can be clarified by producing a gelatinous precipitate.
   aluminium sulfate + calcium hydroxide → aluminium hydroxide + calcium sulfate
   (c) Chlorine is used to extract bromine from sea water.
   $Cl_2(g) + NaBr(aq) \rightarrow Br_2(l) + NaCl(aq)$
   (d) During photosynthesis in a plant, carbon dioxide reacts with water to produce glucose and oxygen.
   carbon dioxide + water → glucose + oxygen

3. Use the solubility table (**Table 1**, page 61), the generalization that all elements (except chlorine) are slightly soluble in water, and the molecular solubility in **Table 2** to predict the solubility of the following chemicals in water. Classify the chemical and then write the chemical formula with (aq) to indicate that the chemical is very soluble and with the pure state of matter, (s), (l), or (g), to indicate that the chemical is slightly soluble.
   (a) Zn (dry cell container and reactant)
   (b) P_4 (white phosphorus)
   (c) $C_{12}H_{22}O_{11}$ (sugar)
   (d) methanol (windshield and gasoline antifreeze)
   (e) octane (gasoline component)
   (f) barium sulfate (gastric X-rays)
   (g) sodium hydroxide (drain cleaner)
   (h) ammonia (window and general cleaner)
   (i) hydrogen fluoride (used to etch glass)

4. Communication systems are very important in chemistry. Describe the difference in what is being communicated in the following sets of symbols.
   (a) P_4 and 4 P
   (b) $Cl_2(g)$ and $Cl(g)$
   (c) $Hg(l)$ and $Hg(g)$
   (d) H_2O_2 and $H_2 + O_2$
   (e) $NaCl(l)$ and $NaCl(aq)$
   (f) Mg and Mg^{2+}

5. For each of the following reactions, classify the reaction, predict the products of the reaction, and complete and balance the chemical equation. (Assume the most common ion charge and state at SATP if not indicated otherwise.)
   (a) Expensive silver metal is recovered in a lab by placing inexpensive aluminium foil in aqueous silver nitrate.
   $Al(s) + AgNO_3(aq) \rightarrow$
   (b) When aqueous potassium hydroxide is added to a well-water sample, the formation of a rusty-brown precipitate indicates the presence of an iron(III) compound in the water.
   $KOH(aq) + FeCl_3(aq) \rightarrow$
   (c) A chemist in a consumer-protection laboratory adds aqueous sodium hydroxide to determine the concentration of acetic acid, $CH_3COOH(aq)$, in a vinegar sample.
   (d) A dishonest 16th-century alchemist, who tried to fool people into believing that iron could be changed into gold, dipped an iron strip into aqueous copper(II) sulfate.
   (e) Translate equation (d) into an English sentence. Include the chemical amounts and states of matter.

6. Complete the Prediction and diagnostic tests of the investigation report. Write up the diagnostic tests (Appendix C.4), including any controls, as part of the Design.

Problem

What are the products of the reaction of sodium metal and water?

Design

A very small piece of sodium metal is placed in distilled water and some diagnostic tests are carried out to identify the products....

7. State diagnostic tests for each of the product(s) in the following chemical reactions. Use the "If (procedure) and (evidence), then (analysis)" format.
   (a) $2 H_2O(l) \rightarrow 2 H_2(g) + O_2(g)$
   (b) $H_2(g) + Cl_2(aq) \rightarrow 2 HCl(aq)$
   (c) $NH_3(g) + H_2O(l) \rightarrow NH_4OH(aq)$

Extension

8. Critiquing and creating experimental designs are important skills for scientific literacy. Create experimental designs to test at least two of the claims made by the following individuals.
   (a) A salesperson claims that wrapping magnets around water pipes will reduce the amount of hard-water scaling that accumulates on the inside of the pipes.
   (b) A psychic claims that he can see halos over the heads of some identified individuals in the audience and not over others.
   (c) A psychic claims to be able to bend spoons with his mind—a feat of psychokinesis.
   (d) A salesperson claims that a copper bracelet relieves pain in the wrist.

Chapter 2 SUMMARY

Outcomes

Knowledge
- use kinetic molecular theory and collision theory to explain how chemical reactions occur (2.2)
- write balanced chemical equations (2.2, 2.3)
- interpret balanced chemical equations in terms of chemical amount (in moles) (2.3)
- convert between chemical amount and mass (2.4)
- classify chemical reactions (2.5, 2.6)
- predict the solubility of elements and ionic and molecular compounds in water (2.6)
- predict products for chemical reactions (2.5, 2.6)

STS
- state the technological application of important chemicals and chemical reactions (2.1, 2.3, 2.4, 2.5, 2.6)
- identify risks and benefits of some important chemical reactions (2.1, 2.3, 2.5)

Skills
- read and write laboratory reports (2.6)
- create and critique experimental designs (2.6)

Key Terms

2.1
STS
perspective
scientific
technological
ecological
economic
political

2.2
physical change
chemical change
nuclear change
kinetic molecular theory
diagnostic test
balanced chemical equation
coefficient

2.3
chemical amount
Avogadro's number
mole

2.4
molar mass

2.5
formation reaction
simple decomposition reaction
complete combustion reaction

2.6
solution
solute
solvent
solubility
precipitate
single replacement reaction
double replacement reaction
precipitation
neutralization

▶ MAKE a summary

1. Use the Key Terms to prepare concept maps that are centred on chemical reactions. Include, along with an example of each:
   (a) evidence for the occurrence of chemical reactions
   (b) classes of chemical reactions
   (c) solubility of elements, ionic compounds, and molecular compounds
   (d) perspectives on an STS issue
2. Search for ways to link the concept maps prepared in the previous question.
3. Refer back to your answers to the Starting Points questions at the beginning of this chapter. How has your thinking changed?

The following components are available on the Nelson Web site. Follow the links for *Nelson Chemistry Alberta 20-30*.
- an interactive Self Quiz for Chapter 2
- additional Diploma Exam-style Review questions
- Illustrated Glossary
- additional IB-related material

There is more information on the Web site wherever you see the Go icon in the chapter.

➕ EXTENSION

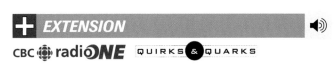

Science and the Courts
Scientific evidence is increasingly being introduced into legal cases: DNA, blood alcohol levels, psychological profiles—all are intended to support either the prosecution's or the defence's case. But are the "facts" beyond doubt? Three experts give their opinions about interpreting scientific evidence for the courts.

www.science.nelson.com

Chapter 2 REVIEW

Many of these questions are in the style of the Diploma Exam. You will find guidance for writing Diploma Exams in Appendix H. Exam study tips and test-taking suggestions are on the Nelson Web site. Science Directing Words used in Diploma Exams are in bold type.

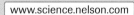

DO NOT WRITE IN THIS TEXTBOOK.

Part 1

1. Pentane, $C_5H_{12}(g)$, is a major component of naptha, the fuel manufactured by chemical engineers for camp stoves (**Figure 1**). The complete combustion of pentane is represented by the chemical equation
   $\_\_ C_5H_{12}(g) + \_\_ O_2(g) \rightarrow \_\_ CO_2(g) + \_\_ H_2O(g)$
   When the equation is balanced, the coefficients are
   ___, ___, ___, ___.

Figure 1
Naphtha is a popular camp fuel because it is easy to transport. It does, however, require careful handling, as the HHP symbols indicate.

2. Environmental chemists have found that nitrogen oxides, NO_x, can cause acid rain, as shown by the chemical equation
   $\_\_ NO_2(g) + \_\_ H_2O(l) \rightarrow \_\_ HNO_3(aq) + \_\_ NO(g)$
   When the equation is balanced the coefficients are
   ___, ___, ___, ___.

3. Classification systems help us organize our knowledge. A reaction in which a precipitate is formed when two solutions of ionic compounds are mixed is classified as a
   A. formation reaction
   B. single replacement reaction
   C. double replacement reaction
   D. simple decomposition reaction

4. Research indicates that manufactured chemicals are among the most persistent pollutants on Earth. Scientists have found some chemicals in the tissues of polar bears, seals, tropical birds, dolphins, and humans. The primary perspective of this research is
   A. technological
   B. ecological
   C. economic
   D. scientific

5. The concept that the smallest entities of a substance are in continuous motion is central to
   A. Dalton's atomic theory
   B. Rutherford's atomic theory
   C. the kinetic molecular theory
   D. the law of conservation of mass

6. Chemists classify an acid–base neutralization reaction as a
   A. formation reaction
   B. single replacement reaction
   C. double replacement reaction
   D. simple decomposition reaction

Part 2

7. Use the solubility table (**Table 1** on page 61 or inside back cover), the generalization that all elements (except chlorine) are only slightly soluble in water, and the molecular solubilities in **Table 2** (page 61) to **predict** the solubility of the following chemicals in water. Classify the chemical and then write the chemical formula with (aq) if chemical is very soluble and with the pure state of matter, (s), (l), or (g), if the chemical is slightly soluble in a water environment.
   (a) sucrose (table sugar)
   (b) methane (natural gas)
   (c) calcium sulfate (gypsum)
   (d) carbon (charcoal/graphite)
   (e) sulfuric acid (car batteries)
   (f) sodium carbonate (water softener)
   (g) ammonium nitrate (fertilizer)
   (h) sulfur (from H_2S in sour gas)
   (i) silver bromide (photographic film)
   (j) magnesium hydroxide (milk of magnesia)

8. For each of the following word equations, write a balanced chemical equation, and classify the reaction.
   (a) aqueous sodium hydroxide + sulfuric acid $\rightarrow$ water + sodium sulfate
   (b) propane + oxygen $\rightarrow$ carbon dioxide + water
   (c) aluminium + aqueous copper(II) chloride $\rightarrow$ copper + aluminium chloride
   (d) molten sodium hydroxide $\rightarrow$ sodium + oxygen + hydrogen
   (e) calcium + chlorine $\rightarrow$ calcium chloride
   (f) aqueous lead(II) nitrate + aqueous sodium chloride $\rightarrow$ lead(II) chloride + sodium nitrate

9. Classify each of the following reactions as formation, simple decomposition, complete combustion, single replacement, or double replacement. **Predict** the products of the reactions, write the chemical formulas and states of matter, and balance the reaction equations.
   (a) $KCl(s) \rightarrow$
   (b) $Cu(s) + Cl_2(g) \rightarrow$
   (c) $C_5H_{12}(g) + O_2(g) \rightarrow$
   (d) $AgNO_3(aq) + NaCl(aq) \rightarrow$
   (e) $Al(s) + Cu(NO_3)_2(aq) \rightarrow$
   (f) $C_8H_{18}(l) + O_2(g) \rightarrow$
   (g) $Al_2O_3(s) \rightarrow$
   (h) $Fe(s) + Br_2(l) \rightarrow$
   (i) $Cu(NO_3)_2(aq) + NaOH(aq) \rightarrow$
   (j) $H_3PO_4(aq) + Ca(OH)_2(aq) \rightarrow$

10. Translate each of the following balanced equations into an English sentence. Include the chemical amounts and the states of matter for all the substances involved.
    (a) $4\ NH_3(g) + 7\ O_2(g) \rightarrow 4\ NO_2(g) + 6\ H_2O(g)$
    (b) $3\ CaCl_2(aq) + 2\ Na_3PO_4(aq) \rightarrow Ca_3(PO_4)_2(s) + 6\ NaCl(aq)$
    (c) $2\ NaCl(l) \rightarrow 2\ Na(l) + Cl_2(g)$

11. **Describe** diagnostic tests for each of the product(s) in the following chemical reactions. Use the "If (procedure) and (evidence), then (analysis)" format (Appendix C.4).
    (a) $2\ K(s) + 2\ H_2O(l) \rightarrow 2\ KOH(aq) + H_2(g)$
    (b) $SO_3(g) + H_2O(l) \rightarrow H_2SO_4(aq)$
    (c) $Cl_2(g) + 2\ NaI(aq) \rightarrow 2\ NaCl(aq) + I_2(s)$

12. Complete the Prediction and diagnostic tests of the investigation report. **Describe** the diagnostic tests, including any controls, as part of the Design. Use the "If (procedure) and (evidence), then (analysis)" format for the diagnostic tests.

    ### Problem
    What are the products of the reaction of iron metal and hydrochloric acid?

    ### Design
    A short piece of iron wire is placed in a dilute solution of hydrochloric acid and some diagnostic tests are carried out to identify the products....

Unit REVIEW

Many of these questions are in the style of the Diploma Exam. You will find guidance for writing Diploma Exams in Appendix H. Exam study tips and test-taking suggestions are on the Nelson Web site. Science Directing Words used in Diploma Exams are in **bold** type.

www.science.nelson.com

DO NOT WRITE IN THIS TEXTBOOK.

Part 1

1. The modern periodic table was developed from evidence of periodicity in chemical and physical properties. The periodic table is an efficient means of organizing a vast body of empirical knowledge of the elements.

    The scientist who created the first periodic table was
    A. Niels Bohr
    B. John Dalton
    C. Albert Einstein
    D. Dmitri Mendeleev

    Use a periodic table to answer questions 2 to 6.

2. The family of elements listed in Group 2 of the periodic table is known as the
    A. actinoids
    B. alkali metals
    C. alkaline-earth metals
    D. transition elements

3. All the members of a certain family of elements are soft, silver-coloured conductors of electricity that react violently with water and form ions with a 1+ charge. The family name for this group of elements is
    A. halogens
    B. noble gases
    C. alkali metals
    D. alkaline-earth metals

4. All the members of a certain family of elements are very reactive and form ions with a 1− charge. The family name for this group of elements is
    A. actinides
    B. halogens
    C. alkali metals
    D. lanthanoids

5. The number of protons, electrons, and neutrons, respectively in an ion of lithium-7 are
    __ __ __

6. The number of protons, electrons, and neutrons, respectively in an atom of carbon-14 are
    __ __ __

7. Chemists often classify changes in matter as physical change, chemical change, or nuclear change. An example of a physical change is
    A. formation
    B. combustion
    C. evaporation
    D. neutralization

8. An equation that represents a physical change is
    A. $H_2O(g) \rightarrow H_2O(l)$
    B. $2 H_2(g) + O_2(g) \rightarrow 2 H_2O(g)$
    C. $2 H_2(g) + O_2(g) \rightarrow 2 H_2O(l)$
    D. $2 H_2O(l) \rightarrow 2 H_2(g) + O_2(g)$

Use this information to answer questions 9 and 10. Some descriptors may be used more than once.

Physical and chemical changes can be described empirically. The empirical descriptions include:

1. odour change
2. state change
3. colour change

List, in numerical order, the description(s) that applies/apply to each of the following reactions.

9. $2 C_8H_{18}(l) + 25 O_2(g) \rightarrow 16 CO_2(g) + 18 H_2O(g)$
   NR __ __ __ __

10. $Zn(s) + CuSO_4(aq) \rightarrow Cu(s) + ZnSO_4(aq)$
    NR __ __ __ __

11. Knowledge can be classified as
    NR
    1. empirical
    2. theoretical

    Use the above classes of knowledge (i.e., 1 and 2) to classify the following statements, in order:
    - A yellow precipitate formed.
    - Atoms/ions were rearranged.
    - Chemical bonds were broken and formed.
    - The solution changed from purple to colourless.

    __ __ __ __

12. The multiple-step industrial process for producing sodium carbonate uses common salt (sodium chloride) and limestone (calcium carbonate) as raw material.
    __$NaCl(s)$ + __$CaCO_3(s) \rightarrow$ __$CaCl_2(s)$ + __$Na_2CO_3(s)$

    The coefficients for balancing the overall reaction equation are, in order, __ __ __ __

13. Propane is widely used as fuel for heating and cooking.
    __$C_3H_8(g)$ + __$O_2(g) \rightarrow$ __$CO_2(g)$ __$H_2O(g)$

    The coefficients for balancing the reaction equation for the complete combustion of propane are, in order,

    __ __ __ __

14. A spontaneous chemical reaction occurs when a crumpled piece of aluminium foil is dropped into a beaker of aqueous copper(II) sulfate.
    __$Al(s)$ + __$CuSO_4(aq) \rightarrow$ __$Cu(s)$ + __$Al_2(SO_4)_3(aq)$

    The coefficients for the balanced reaction equation are, in order, __ __ __ __

15. An early industrial process for producing sodium metal involved the electrolysis of molten sodium hydroxide.
    __$NaOH(l) \rightarrow$ __$Na(l)$ + __$H_2(g)$ + __$O_2(g)$

    The coefficients for the balanced reaction equation are, in order, __ __ __ __

16. The basic SI unit for chemical amount is the
    A. gram
    B. litre
    C. metre
    D. mole

17. The units for the molar mass of a chemical are
    A. g/mol
    B. g/L
    C. mol/L
    D. mol/g

18. An alcohol lamp uses 2.50 mol of methanol, $CH_3OH(l)$, while providing emergency lighting. The mass of methanol burned is _____ g.

19. A student obtains 150 g of ammonium sulfate, $(NH_4)_2SO_4(s)$, to prepare a fertilizer solution. The chemical amount of ammonium sulfate obtained is _____ mol.

20. A laboratory technician needs 7.50 mmol of potassium permanganate, $KMnO_4(s)$, to prepare a solution for use in a chemical analysis. The mass of potassium permanganate required is _____ g.

21. The lead(II) ions in a waste laboratory solution precipitate as lead(II) carbonate, $PbCO_3(s)$. If 285 g of dry lead(II) carbonate is produced, the chemical amount of lead(II) ions precipitated is _____ mol.

Part 2

Use a periodic table to answer questions 22 to 24.

22. The periodic table summarizes and organizes a wealth of empirical and theoretical knowledge about chemical elements. **Define** the following terms associated with the periodic table:
    (a) group
    (b) period
    (c) staircase line

23. Copy and complete the following table of atoms and ions.

Table 1 Atoms and Ions

Symbol	Name	# protons	# electrons	Net charge
	sulfide ion			
		35	36	
Ca^{2+}				
		23	23	
		26		3+
			18	0

24. Many radioisotopes are produced artificially for use in medical diagnosis or therapy. Copy and complete the following table of radioisotopes assuming they are used as atoms.

Table 2 Radioisotopes

Name	Use	# protons	# electrons	# neutrons
cobalt-60	cancer treatment			
	hyperthyroid treatment	53		78
	reduces white cell count		15	17
strontium-85	bone scanning			

25. For each of the following word equations, write a balanced chemical equation and classify the reaction.
    (a) Butane, $C_4H_{10}(g)$, is a convenient fuel for camping (**Figure 1**).
        butane + oxygen → carbon dioxide + water vapour

Figure 1
Many campers cook with a butane camp stove.

   (b) Beautiful crystals form when copper objects are immersed in silver nitrate solutions.
        copper + silver nitrate → copper(II) nitrate + silver
   (c) Toxic cadmium ions can be removed from industrial effluent by sodium carbonate.
        cadmium nitrate + sodium carbonate →
            sodium nitrate + cadmium carbonate
   (d) Sir Humphry Davy discovered potassium by using electricity to decompose molten potassium hydroxide.
        potassium hydroxide →
            potassium + oxygen + hydrogen
   (e) Very pure or freshly cleaned aluminium forms a protective oxide coating when it is exposed to air.
        aluminium + oxygen → aluminium oxide

26. Balance the following reaction equations, and then write the equation in words, including the states of matter and chemical amounts.
    (a) Methanol burners keep food warm at a buffet.
    $CH_3OH(l) + O_2(g) \rightarrow CO_2(g) + H_2O(g)$
    (b) Phosphate ions can be removed from solution by adding calcium chloride.
    $Na_3PO_4(aq) + CaCl_2(aq) \rightarrow NaCl(aq) + Ca_3(PO_4)_2(s)$

27. Complete and then balance the following chemical equations.
    (a) Magnesium burns with a brilliant white light.
    $Mg(s) + O_2(g) \rightarrow$
    (b) The industrial production of sodium metal involves using electricity to decompose molten sodium chloride.
    $NaCl(l) \rightarrow$
    (c) Aluminium sulfate and sodium hydroxide solutions react to form a gelatinous precipitate.
    $Al_2(SO_4)_3(aq) + NaOH(aq) \rightarrow$
    (d) When its protective coating is removed, aluminium reacts vigorously with hydrochloric acid.
    $Al(s) + HCl(aq) \rightarrow$
    (e) Candles are included in emergency kits because they can produce both heat and light.
    $C_{25}H_{52}(s) + O_2(g) \rightarrow$

28. Complete the Prediction and Design of the investigation report. Include three diagnostic tests in the Design to determine whether the predicted reaction has taken place and the predicted products have formed.

    Problem
    What are the products of the reaction of aqueous copper(II) chloride and sodium hydroxide solution?

29. For each of the following reactions, translate the information into a balanced reaction equation. Then classify the main perspective—scientific, technological, ecological, economic, or political—suggested by the introductory statement.
    (a) Oxyacetylene torches are used to produce high temperatures for cutting and welding metals such as steel (**Figure 2**). This process involves burning acetylene, $C_2H_2(g)$, in pure oxygen.
    (b) During chemical research conducted in 1808, Sir Humphry Davy produced magnesium metal by decomposing molten magnesium chloride using electricity.
    (c) An inexpensive application of single replacement reactions uses scrap iron to produce copper metal from waste copper(II) sulfate solutions.
    (d) The emission of sulfur dioxide into the atmosphere creates problems between different levels of government, both nationally and internationally. Sulfur dioxide is produced when zinc sulfide is roasted in a combustion-like reaction in a zinc smelter.
    (e) Burning leaded gasoline added toxic lead compounds to the environment, which damaged both plants and animals. Leaded gasoline contained tetraethyl lead, $Pb(C_2H_5)_4(l)$, which undergoes a complete combustion reaction in a car engine.

Figure 2
The flame of an oxyacetylene torch is hot enough to melt most metals.

30. Classify the perspective being communicated by each of the following statements about global warming.
    (a) An invention is needed to remove carbon dioxide from gases emitted from oil refineries and power plants.
    (b) More research is needed to confirm or refute the causes of global warming.
    (c) The cost for stopping global warming is enormous.
    (d) Votes can be won or lost over global warming.
    (e) Profits will be reduced if greenhouse gas emissions are to be reduced.

31. Critiquing and creating experimental designs are important skills for scientific literacy. Create experimental designs to test at least two of the claims made by the following individuals. (See Appendix B.4.)
    (a) A believer in the power of magnets claims that sleeping with flexible and padded magnets in your pillow provides a more restful sleep.
    (b) A group of disbelievers claims that the photos and video from the 1969 Apollo 11 landing on the moon are fake because the shadows created by the Sun are not parallel in the video and photos.
    (c) A psychic claims to be able to reproduce simple drawings made by a person who comes up from the audience.
    (d) An alternative medical care provider claims to be able to cure a disease that standard medical practices cannot.
    (e) A commercial for a shampoo claims that your hair will have more body if you use this shampoo.
    (f) A commercial claims that a particular detergent removes grass stains from clothes better than other leading detergents.

32. Complete the Prediction (including possible diagnostic tests), Analysis (including reaction types), and Evaluation (Parts 2 and 3) for the following investigation report.

Purpose
To test the single and double replacement reaction generalizations.

Problem
What reaction products are formed when the following substances are mixed?
(a) aqueous chlorine and potassium iodide solution
(b) solutions of magnesium chloride and sodium hydroxide
(c) solutions of aluminium nitrate and sodium phosphate
(d) magnesium metal and hydrochloric acid
(e) sodium hydroxide solution and chromium(III) chloride solution
(f) lithium metal and water
(g) a clean cobalt strip and a silver nitrate solution
(h) nitric acid and an ammonium acetate solution

Design
Diagnostic test information such as evidence of chemical reactions (**Table 2**, page 48), ion colours and solubilities (reference tables, inside back cover), and specific tests for products (Appendix C.4), are predicted, for convenience, along with the balanced chemical equations in the Prediction. The general plan is to observe the substances before and after mixing, and conduct the appropriate diagnostic tests.

Evidence

Table 3 Single and Double Replacement Reactions

Reaction	Observations
(a)	• The colourless solutions produced a yellow-brown colour when mixed (**Figure 3**). • A violet-purple colour appeared in the chlorinated layer when a hydrocarbon was added.
(b)	• The colourless solutions produced a white precipitate when mixed.
(c)	• The colourless solutions produced a white precipitate when mixed.
(d)	• The silvery solid added to the colourless solution produced gas bubbles and a green solution. • The gas produced a pop sound when ignited.
(e)	• The colourless sodium hydroxide and green chromium(III) chloride solutions produced a dark precipitate and a colourless solution.
(f)	• The soft, silvery solid and colourless liquid produced gas bubbles and a colourless solution (**Figure 4**). • The gas produced a pop sound when ignited. • Red litmus turned blue in the final solution. • The final solution produced a bright red flame colour.
(g)	• The silvery solid and colourless solution produced a pink solution and silvery needles.
(h)	• The colourless solutions remained colourless when mixed. • A vinegar odour was produced.

Figure 3
Chlorine reacting with potassium iodide.

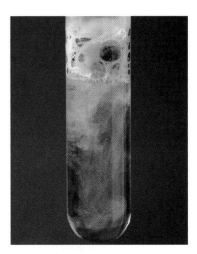

Figure 4
The reaction of lithium with water.

unit 1
Chemical Bonding—Explaining the Diversity of Matter

Winter in Alberta! While the snowboard, the clothing, and the energy all come from chemical reactions, the fun depends largely on the way water behaves when solid. This most common of substances has properties unlike any other; for example, water expands upon freezing, and ice liquefies momentarily under pressure. These properties help make skiing, snowboarding, skating, and sledding possible.

Describing, explaining, and predicting the nature and behaviour of substances is fundamental to chemistry. Inquiring scientists assumed there must be a reason why the complex structures of snowflakes always have hexagonal symmetries, and they applied scientific inquiry and techniques to find a plausible theory. When their initial theory also worked to describe many other properties of water, scientists considered the theory more valid. When scientists found the theory to be useful in predicting the properties of many other substances, it became generally accepted within the scientific community. All theories develop in this way: we constantly test them against what we know of the real world. When such testing eventually produces conflicting evidence, a theory is *restricted*, or *revised*, or perhaps even *replaced*, resulting in a newer theory that is better able to *describe*, *explain*, and *predict* what we observe. The knowledge-gathering process of science really has only two fundamental rules: theories must be based on reliable evidence, and theories must be tested for accuracy whenever possible.

Bonding theories assume that an understanding of how substances behave depends on understanding how the tiniest units of matter are held together. In addition, understanding chemical change always depends on the concept of breaking bonds and forming new ones. In this unit, you will study chemical bonding—the concept of forces that hold atoms, ions, and molecules together—and the relationship of these bonds to the properties and structure of matter.

As you progress through the unit, think about these focusing questions:
- Why do some substances dissolve easily while others do not?
- Why do different substances have widely different melting and boiling points?
- How can models increase our understanding of invisible things such as bonds?

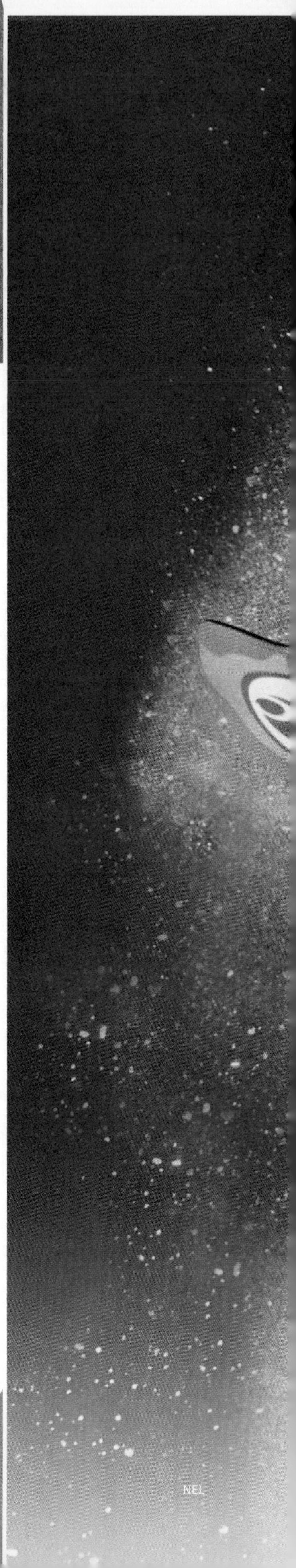

GENERAL OUTCOME

In this unit, you will

- describe the roles of modelling, evidence, and theory used in explaining and predicting the structure, chemical bonding, and properties of ionic and molecular substances

Chemical Bonding—Explaining the Diversity of Matter

Unit 1
Chemical Bonding—Explaining the Diversity of Matter

▶ **Prerequisites**

Concepts
- ionic and molecular compounds
- atomic structure
- formation of ions
- nature and strength of electrostatic forces
- relationship between laws and theories

Skills
- laboratory safety rules

You can review prerequisite concepts and skills in the Chemistry Review unit and in the Appendices.
   A Unit Pre-Test is also available online.

www.science.nelson.com

ARE YOU READY?

These questions will help you find out what you already know, and what you need to review, before you continue with this unit.

Knowledge

1. According to a chemistry classification system, each of the following statements applies only to elements or only to compounds:
   i. cannot be decomposed into simpler substances by chemical means
   ii. composed of two or more kinds of atoms
   iii. can be decomposed into simpler substances using heat or electricity
   iv. composed of only one kind of atom

   (a) Which statements apply only to elements?
   (b) Which statements apply only to compounds?
   (c) Which statements are empirical?
   (d) Which statements are theoretical?

2. Describe the atomic models presented by each of the following:
   (a) J. J. Thomson
   (b) Ernest Rutherford
   (c) John Dalton

3. Provide the labels, (a) to (e), for **Figure 1** to describe an atom. In addition to the name, provide the international symbol for the three subatomic particles.

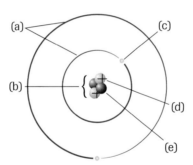

Figure 1
An atom model, as used to represent an atom (very simplistically) in the 1930s

4. Copy **Table 1**, and complete it using the periodic table and atomic theory.

Table 1 Components of Atoms and Ions

Entity	Number of protons	Number of electrons	Net charge
hydrogen atom			
sodium atom			
chlorine atom			
hydrogen ion			
sodium ion			
chloride ion			

5. Niels Bohr explained the periodic law and line spectra by creating the Bohr model of the atom. Draw energy-level diagrams to describe the following atoms:
   (a) nitrogen, N
   (b) calcium, Ca
   (c) chlorine, Cl

6. Compounds can often be classified, based upon empirical definitions, as ionic or molecular. Copy and complete **Table 2**, indicating the properties of these compounds.

Table 2 Properties of Ionic and Molecular Compounds

Class of compound	Classes of elements involved	Properties		
		Melting point (high/low)	State at SATP (s, l, g)	Electrolytes (yes/no)
ionic				
molecular				

7. Classify compounds according to the following empirical properties as ionic, molecular, or either:
   (a) high solubility in water; aqueous solution conducts electricity
   (b) solid at SATP; low solubility in water
   (c) solid at SATP; low melting point; aqueous solution does not conduct electricity

8. According to atomic and bonding theories, atoms react by rearranging their electrons.
   (a) Silicon tetrafluoride gas is used to produce ultra-pure silicon wafers (**Figure 2**). Draw an energy-level diagram for each atom or molecule in the following word equation for the formation of silicon tetrafluoride:
   silicon + fluorine → silicon tetrafluoride
   (b) Calcium fluoride occurs naturally as fluorite (pure compound) and as fluorspar (mineral, **Figure 3**). Calcium fluoride is the principal source of the element fluorine, and is also used in a wide variety of applications such as metal smelting, certain cements, and paint pigments. Draw an energy-level diagram for each atom or ion in the following word equation for the formation of calcium fluoride:
   calcium + fluorine → calcium fluoride
   (c) What is the difference in the electron rearrangement in (a) compared with (b)?

9. What is the difference between a scientific law and a scientific theory?
10. In the progress of chemistry, what generally comes first, laws or theories?

Figure 2
In 1972, Intel's 8008 computer processor had 3500 transistors. By 2000, the Pentium 4 processor had 42 million transistors on a silicon chip.

Figure 3
Fluorspar is a mineral that can be found in many countries. It takes a variety of colours and has different properties, depending on contaminants.

Skills

11. What document provides information about safe handling and safe disposal of a chemical?
12. If a corrosive chemical comes in contact with your skin, what is the normal procedure to follow?
13. List two items common to school labs that should routinely be worn for personal protection against chemical hazards.
14. List three types of general safety equipment common to school chemistry labs.
15. When evaluating a scientific theory or concept, what three basic criteria are used?

chapter 3
Understanding Chemical Compounds

In this chapter

- Exploration: Properties and Forces
- Explore an Issue: Funding Research and Development
- Investigation 3.1: Molecular Models
- Mini Investigation: Electrostatic Repulsion Model
- Investigation 3.2: Evidence for Polar Molecules
- Web Activity: Cleaning Up Dry Cleaning
- Web Activity: Gerhard Herzberg
- Lab Exercise 3.A: Boiling Points and Intermolecular Forces
- Investigation 3.3: Molecular Compound Melting Points
- Web Activity: Cloud Seeding
- Investigation 3.4: Hydrogen Bonding
- Case Study: Current Research in Intermolecular Forces
- Mini Investigation: Floating Pins
- Web Activity: Modelling Molecules
- Mini Investigation: Building an Ionic Crystal Model
- Web Activity: Jillian Buriak
- Investigation 3.5: Classifying Unknown Solids

When enough atoms or molecules bond to each other, they must eventually form a quantity of substance that is big enough for a human to detect and measure. The compound shown crystallizing from solution in **Figure 1** illustrates this process.

As you study this chapter, you will follow much the same sequence of ideas as chemists did in the early 1900s. As some chemists refined atomic theory, others developed ideas about how atoms bond to each other. Initial bonding theories were restricted to simple ionic and molecular compounds, but expanded over time through constant questioning and testing. All theories are considered to be only as good as their ability to accurately describe, explain, and predict phenomena. Eventually, bonding theory became highly sophisticated, allowing structural descriptions of very complex substances. Bonding theory development is a good example of the cyclical nature of science. Curiosity about initial observations leads to creation of theory, which is then used to describe, explain, and predict new observations. The new observations in turn may require better theories for a more complete explanation. The net result of this cycle is a constant accumulation of knowledge.

Chemists use ideas about bonding to explain physical properties such as hardness and melting point, as well as chemical properties such as reactivity and acidity. Chemical engineers study bonding and molecular structure to synthesize useful new substances such as semiconductors and detergents. Ultimately, chemistry is about reactions. When a reaction is thought of as the breaking of bonds holding the original substances together, followed by the formation of different bonds to form new substances, it becomes obvious why a study of bonding is essential to any study of theoretical chemistry.

STARTING Points

Answer these questions as best you can with your current knowledge. Then, using the concepts and skills you have learned, you will revise your answers at the end of the chapter.

1. When atoms collide, what rearrangement of their structures can take place?
2. What properties of a substance are affected by the shape of its molecules?
3. How can the behaviour and properties of a pure substance or a composite material be explained by theories of bonding between its basic entities?
4. How can differences in physical properties of molecular liquid substances, such as $C_2H_5OH(l)$ and $CH_3OCH_3(l)$, and of solid substances, such as $SiO_2(s)$ and $CO_2(s)$, be predicted by understanding current bonding theories?

Career Connections:
Biochemist; Geologist/Gemmologist

Figure 1
These delicate structures are solid crystals being formed by the bonding together of enormous numbers of individual entities from the surrounding solution.

▶ Exploration   *Properties and Forces*

Every property of a substance should be predictable once we have a complete understanding of the interactions between atoms and molecules. To obtain this understanding, we need to observe carefully and develop explanations. Record your observations for the following activity, and see what explanations flow from them.

Materials: 2 small, flat-bottom drinking glasses or beakers; 2 small ceramic bread plates (china, stoneware, or glass); some dishwashing liquid; some canola oil

- Place the drinking glass on the bread plate, and press down firmly while trying to move the glass in a small horizontal circle.
(a) Both the glass and the plate are very smooth. Does this mean they slide over each other easily?
- Add dishwashing liquid to the plate until it is about 2 mm deep, and try the first step again with a clean glass.
(b) Dishwashing liquid makes your fingers feel slippery. Does dishwashing liquid actually make the contact surface between the glass and plate slippery?
- Add canola oil to the second plate until it is about 2 mm deep, and try the first step again with a clean glass.
(c) The canola oil makes your fingers feel slippery. Does oil actually make the contact surface between the glass and plate slippery?
- Lift the glass on both plates vertically, and note the tendency of the plate to stick.
(d) Which liquid seems to be a more effective adhesive?
- In one small glass (or beaker) add about 1 cm of detergent to 2 cm of water, and stir. In the other glass, add about 1 cm of canola oil to 2 cm of water, and stir. Let each container sit undisturbed for approximately one minute.
(e) What evidence do you have for these liquids about the strengths of the attractive forces between their molecules (cohesion), and between their molecules and the water molecules (adhesion)?
- Dispose of the liquids down the sink with lots of water.

 Wash your hands thoroughly after completing this activity.

Understanding Chemical Compounds

3.1 Bonding Theory and Lewis Formulas

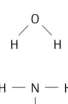

Figure 1
Structural formulas (also called line formulas and structural diagrams) for water, ammonia, and methane. Structural formulas are more descriptive than molecular formulas, since they show which atoms are bonded to each other.

Figure 2
Light is polarized by passing through polarized lenses like those in sunglasses. When the polarized light passes through most transparent substances, we notice nothing unusual. However, certain substances dramatically change the light to produce some beautiful effects. This effect can be explained only by 3-D versions of structural formulas.

Bonding is one of the most theoretical concepts in chemistry. While we can now detect single atoms (just barely) with the most advanced microscopes, we have no direct visible evidence for bonds between atoms. Our concepts of bonding must be created from, and based on, indirect experimental evidence and logic.

The development of bonding theory starts before the Mendeleyev periodic table with Edward Frankland (1852) stating that each element has a fixed bonding capacity. Friedrich Kekulé (1858) extended the idea to illustrating a bond as a dash between bonding atoms, that is, what we now call a **structural formula** (Figure 1).

Sixteen years after Kekulé created these diagrams, Jacobus van't Hoff and Joseph Le Bel independently extended them to three dimensions (3-D). They revised existing theory in order to explain the ability of certain substances to change light as it passes through a sample of the substance (optical activity, **Figure 2**). Note that all of this work was done by working only with the Dalton atom, before any concept of a nucleus or bonding electrons. As yet, there was no explanation for the bonds that were being represented in the diagrams.

That explanation started with Richard Abegg, a German chemist, in 1904. Abegg was the first to suggest that bonding capacity must somehow be associated with an atom's electron structure. Almost immediately, Rutherford's evidence for the nuclear atom made Abegg's theory seem probable, since it became obvious that any interaction between atoms had to involve their surrounding electrons. Abegg suggested that the stability of the "inert" (noble) gases was due to the number of electrons in the atom. He looked at the periodic table and noted that a chlorine atom had one less electron than the stable electron structure of argon. He theorized that, in a reaction with another atom, a chlorine atom was likely to gain one electron to form a stable, unreactive chloride ion. Likewise, he suggested that a sodium atom had one more electron than needed for stability. A reacting sodium atom should, according to his theory, lose one electron to form a stable sodium ion. Such a transfer of electrons, between a sodium and a chlorine atom, is indicated below. The newly formed ions would, in turn, be held together by electrostatic forces. In a reaction between sodium metal and chlorine gas, enormous numbers of cations and anions can form and bond this way to form visible crystals of table salt. The attraction force holding the ions together is called ionic bonding.

$$Na + Cl \rightarrow Na^+ + Cl^- \rightarrow NaCl$$

sodium + chlorine → sodium chloride

In 1916, Gilbert Lewis, an American chemist, combined the evidence of many known chemical formulas, the concept of valence, and the concept of stable electron numbers for atoms. He proposed that atoms could achieve stable electron arrangements by sharing electrons as well as by transferring them. This would only work if the atoms stayed close together, which would have to result in an attraction force. This force became known as the covalent bond. The electrons involved were called **valence electrons**. These are the electrons in the highest energy level of the atom. Immediately, many molecular compounds became more understandable. An ionic bond, then, is explained as the simultaneous attraction between positive and negative ions resulting from the transfer of one or more valence electrons. A covalent bond is explained as the simultaneous attraction of the nuclei of two atoms for valence electrons that they share between them. These types of bonding are formally defined and discussed in more detail later in this section.

The culmination of all this work was the brilliant synthesis of all these concepts with the complex ideas of quantum mechanics. *Quantum mechanics*, which developed from Einstein's revolutionary theories of matter, energy, space, and time, is a mathematical model. In it, electrons are described in terms of their energy content, and orbitals in terms of the calculated probability of an electron being (at any instant) at any given point relative to the atomic nucleus. In 1939, Dr. Linus Pauling published a book called *The Nature of the Chemical Bond*, which he dedicated to Gilbert Lewis. Pauling's work explained, for the first time, why certain electron arrangements are stable; it also showed that electron sharing must cover a complete range from equal attraction to total transfer. The theory was so complete and explained both known and unknown situations so well, that in 1954 it won Dr. Pauling his first Nobel Prize. Just like all other scientific theories, Pauling's theory was valued for how well it was able to describe, explain, and predict.

DID YOU KNOW?

Simplicity

Scientists generally agree that simplicity is a characteristic of an acceptable theory. Einstein never fully accepted the theory of quantum mechanics, partly because of its complexity. In spite of his suspicions, he admired the ability of quantum mechanics to describe, explain, and predict observations.

EXPLORE an issue

Funding Research and Development

Issue Checklist
- ○ Issue
- ○ Resolution
- ○ Design
- ● Evidence
- ● Analysis
- ● Evaluation

Science, technology, and society are interdependent: Society depends on science to provide explanations for natural phenomena and on technology to provide beneficial materials and processes. Science and technology both depend on society for funding their research. The goal of research and development (R&D) is to gain and use knowledge. Research and development can be classified into three broad categories:

- pure research: to advance knowledge for its own sake, such as researching the empirical properties of semi-metals (e.g., silicon) to develop theoretical explanations for their crystal structure and semiconductivity
- applied research: to advance technology, for example, researching how the semiconductivity of semi-metals can be enhanced to make even smaller integrated circuits in electronic devices
- development: to transform technological knowledge into concrete operational hardware, such as a microscopic electronic circuit to be used in a medical diagnostic device. The end product is judged by the criteria of cost, efficiency, reliability, and simplicity of use.

A simple view of R&D is that applied research uses the ideas generated by pure research in making inventions, which, in turn, are made commercially viable through development. In reality, science and technology are mutually dependent, and developments in one field prompt developments in the other. Technology provides the tools for scientific research, while science provides the theoretical background that guides research in technology.

Society supports R&D through grants from both federal and provincial governments. The Natural Sciences and Engineering Research Council of Canada (NSERC) is the primary federal agency investing in university research and training in the natural sciences and engineering. NSERC supports R&D by supporting more than 20 000 university students and post-doctoral fellows in their advanced studies and funding more than 10 000 university professors every year. NSERC also promotes innovation by encouraging Canadian companies to invest in university research. In 2004–2005, NSERC invested $850 million in university-based research and training in all the natural sciences and engineering.

Provincial governments also provide funding for R&D. For example, the Alberta Science and Research Authority (ASRA) provides strategic direction to research funding priorities for the provincial government, which centre on agriculture, energy, forestry, and information/communication technology. In addition, the Alberta Research Council receives government funds to conduct applied R&D focusing on energy, life sciences, engineered products and services, and integrated resource management.

While many Canadians believe that federal and provincial governments should increase funding for R&D, there are many others who question why any public money should be used for scientific research. The second group maintains that taxpayer dollars would be better spent on improving health care and social services than on improving theories of solid-state physics. In their view, the research required to develop new technologies should be paid for by the private industries that benefit from selling the new products. The issue for Canadian society is to find the best long-term strategy for using public money to improve the quality of life for all its citizens.

Issue
What is the most effective use of government funding for research in science and technology?

Resolution
The provincial and federal governments should direct all their research funding to applied research.

Design
Within small groups, research the pros and cons of using public money to fund each of the three categories of research. Gather information from a wide variety of perspectives.

www.science.nelson.com

Bonding Theory: Valence Electrons and Orbitals

Based upon evidence and logic, chemists believe that bonding changes for main group atoms in chemical reactions involve only the valence electrons occupying the highest energy level. Atomic theory says that electrons in lower energy levels are held so strongly by their (positively charged) nucleus that, during a reaction, they remain essentially unchanged.

To describe where electrons exist in the atom, chemists created the concept of an orbital. According to quantum mechanics, an **orbital** is a specific volume of space in which an electron of certain energy is likely to be found. Each orbital may contain two, one, or no electrons. The word "orbital" was an unfortunate choice because it tends to make you think of little particles in an orbit, and we now know that electrons are nothing like that. An orbital may be thought of as a sort of 3-D space that defines where an electron may be. For bonding study, we are only concerned with an atom's **valence orbitals**, the volumes of space that can be occupied by electrons in an atom's highest energy level.

According to this theory, valence electrons are classified in terms of orbital occupancy. An atom with a valence orbital that is occupied by a single electron can theoretically share that electron with another atom. Such an electron is, therefore, called a **bonding electron**. A full valence orbital, occupied by two electrons, has a repelling effect on electrons in any nearby orbitals. Two electrons occupying the same orbital are called a **lone pair**.

According to quantum mechanics, the number and occupancy of valence orbitals in the atoms of the main group elements are determined by the following theoretical rules plus the descriptions in **Table 1**, and the energy-level diagrams in **Figure 3**:

- The first energy level has room for only one orbital with a maximum of two electrons. Hydrogen, the smallest reactive atom with the simplest structure, has only one energy level. This gives hydrogen unique properties; it is an exception to most rules and generalizations that apply to other atoms.

- Energy levels above the first contain four orbitals, that is, eight electrons maximum. The noble gases have this valence electronic structure; their lack of reactivity indicates that a structure with eight electrons filling a valence level is very stable. This is known as the **octet rule** and is usually obeyed by main group atoms. However, only C, N, O, and F atoms *always* obey the octet rule when bonding.

- An orbital may be unoccupied, or it may contain one or two electrons. This means that two (but never more than two) electrons may share the same region of space at the same time. (This is called the Pauli exclusion principle, first stated in 1925 by Austrian scientist Wolfgang Pauli.)

- Electrons "spread out" to occupy any empty valence orbitals before forming electron pairs.

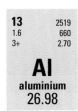

3 e⁻ → 1 e⁻ 1 e⁻ 1 e⁻ 0 e⁻
8 e⁻
2 e⁻
13 p⁺
Al
aluminium atom

6 e⁻ → 2 e⁻ 2 e⁻ 1 e⁻ 1 e⁻
8 e⁻
2 e⁻
16 p⁺
S
sulfur atom

Figure 3
To explain bonding, the valence (highest) energy level is considered to have four orbitals. Each orbital may be unoccupied, or it may contain one or two electrons. In these diagrams for aluminium and sulfur, the distribution of the valence electrons in the four valence orbitals is shown to the right of the arrow. Aluminium has three half-filled valence orbitals and one vacant valence orbital. Sulfur has two full valence orbitals and two half-filled valence orbitals.

Table 1 Theoretical Definitions of Orbitals

Orbital	Number of electrons in the orbital	Description of electrons	Type of electrons
empty	0	—	—
half-filled	1	unpaired	bonding
filled	2	lone pair	nonbonding

Atomic Models: Lewis Symbols

Gilbert Lewis also created a simple model of the arrangement of electrons in atoms that explains and predicts empirical formulas. He used the symbol for the element to represent the nucleus plus all the electrons *except* the valence electrons. He reasoned that in reactions, this collection of subatomic particles, with its net positive charge, did not change. The valence electrons, which do change, are shown as dots around the central symbol, with the entire diagram showing (for an atom) a net charge of zero. This is called a **Lewis symbol** and is an extremely useful model to help describe how atoms bond. **Figure 4** shows the relationship of a Lewis symbol to an energy-level diagram for the oxygen atom. In **Figure 4(b)**, the symbol O represents the oxygen nucleus and the two electrons in the first energy level, with a net charge of 6+. The oxygen atom is thought to have two lone pairs and two bonding electrons in its four valence orbitals.

To draw Lewis symbols for main group atoms:

- Write the element symbol to represent the nucleus and any filled energy levels of the atom.
- Add a dot to represent each valence electron.
- Start by placing valence electrons singly into each of four valence orbitals (represented by the four sides of the element symbol).
- If additional locations are required for electrons once each orbital is half-filled, start filling each of the four orbitals with a second electron until up to eight valence electrons have been represented by dots.

The Lewis symbols for atoms of the Period 2 elements show the dot arrangement for main group atoms.

(a) Electron energy-level diagram of oxygen atom
(b) Lewis symbol for oxygen atom

Figure 4
An energy-level diagram and a Lewis symbol are two (of many) models that can be used to represent the structure of an atom.

Li·   ·Be·   ·B·   ·C·   ·N·   :O·   :F·   :Ne:

It is important to understand that Lewis symbols do not mean that electrons are dots or that they are stationary. The four sides of the atomic symbol just represent the four valence level space regions (orbitals) that may be occupied by electrons. The dots simply keep count of how many electrons are in each orbital (or not); that's all the information that Lewis symbols provide. They are simplistic 2-D diagrams of complex 3-D structures that help us visualize and account for each electron.

Electronegativity

On theoretical grounds, chemists believe that atoms have different abilities to attract valence electrons. For example, the farther away from the nucleus that electrons are, the weaker their attraction to the nucleus. In addition, inner electrons (those closer to the nucleus) shield the valence electrons from the attraction of the positive nucleus. Finally, the greater the number of protons in the nucleus, the greater the attraction for electrons must be. Combining these three points, it is possible to assign a value to any atom, describing how well it attracts electrons shared in a covalent bond with another atom.

Chemists use the term **electronegativity** to describe the relative ability of an atom to attract a pair of bonding electrons in its valence level. Electronegativity is usually assigned on a scale developed by Linus Pauling (**Figure 5**). Empirically, Pauling based his scale on energy changes in chemical reactions. According to his scale, fluorine has the highest electronegativity, 4.0. Of the nonradioactive metal atoms, cesium has the lowest electronegativity, 0.8. Note that these are the most reactive nonmetal and metal, respectively. Metals tend to have low electronegativities, and nonmetals tend to have high electronegativities. See the inside front cover of this book for data on electronegativities of atoms.

Figure 5
Linus Pauling (1901–1994) was a dual winner of the Nobel Prize. In 1954, he won the prize in chemistry for his work on molecular structure, and in 1962, he received the Nobel Peace Prize for campaigning against the nuclear bomb.

DID YOU KNOW ?

Crystallography
Louis Pasteur (1848) studied crystals (as well as germs) under a microscope; Friedrich Kekulé (1858) established a method of communicating the bonding within crystals; Jacobus van't Hoff (1874) described crystal structures with different abilities to polarize light in terms of orientations of atoms (**Figure 6**); and Linus Pauling used X-ray diffraction of crystals to further his understanding of chemical bonding. Crystallography forced the theory of chemical bonding to greater and greater refinements.

Figure 6
A photomicrograph of a solid's crystal structure—taken with polarized light.

▶ Practice

1. Place the following chemistry concepts in the order in which they were created:
   (a) Lewis symbols (d) Kekulé structures
   (b) empirical formulas (e) quantum mechanics
   (c) Dalton atom

2. Write the Lewis formula, the electron energy-level diagram, and the electronegativitiy value for each of the following atoms:
   (a) aluminium (c) calcium
   (b) chlorine (d) argon

3. (a) Draw the Lewis symbol for a calcium atom, but omit the two dots for valence electrons. Show that the remaining structure has a double positive charge by enclosing the Lewis structure in large square brackets and writing the overall charge to the upper right side, outside the brackets.
   (b) What structure is represented by your symbol? What is it called?

4. In a Lewis symbol of a potassium atom, describe what entities are assumed to be included in its element symbol.

5. List the requirements (criteria) for a new scientific concept, such as the Lewis theory, before it becomes accepted by the scientific community.

Bonding

Imagine that two atoms, each with an orbital containing one bonding electron, collide in such a way that these half-filled orbitals overlap. As the two atoms collide, the nucleus of each atom attracts and attempts to "capture" the bonding electron of the other atom. A "tug-of-war" over the bonding electrons occurs. Which atom wins? Comparing the electronegativities of the two atoms can predict the result of the contest.

Covalent Bonding

If the electronegativities of both atoms are relatively high, neither atom will "win," and the pair of bonding electrons will be shared between the two atoms. The simultaneous attraction of two nuclei for a shared pair of bonding electrons is known as a **covalent bond**. This kind of bond normally forms between two nonmetal atoms. When this kind of bond forms during chemical reactions, often the products are molecular substances (**Figure 7**). The electron sharing may be equal (between two carbon atoms) or unequal (between a carbon atom and an oxygen atom or between a carbon atom and a hydrogen atom).

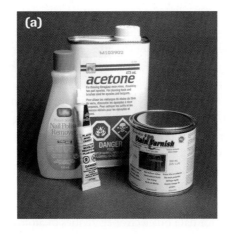
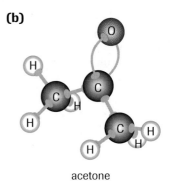

Figure 7
In the molecular substance acetone (a), each molecule is described as containing three carbon atoms, six hydrogen atoms, and one oxygen atom; theory suggests that all 10 atoms are held together by 9 covalent bonds (b). The bond between the carbon and oxygen atoms can only be explained if it involves four electrons, which is called a double bond, a type discussed in more detail in Section 3.2. This is a type of stereochemical formula, which is a formula that shows the molecule in three dimensions.

82 Chapter 3

Ionic Bonding

If the electronegativities of two colliding atoms are quite different, the atom with the stronger attraction for electrons may succeed in removing the bonding electron from the other atom. An *electron transfer* then occurs, and positive and negative ions are formed. The term **ionic bond** refers generally only to the attraction between any specific cation and any specific anion. However, the bonding in ionic compounds necessarily involves enormous numbers of both kinds of ions, and is not nearly as simple as an ionic formula makes it seem. After electron transfer, the ions arrange themselves in positions where the maximum total attraction between positive and negative charges occurs. This determines the numerical ratio of ions in a compound, for example, 1:2 for the compound calcium chloride, $CaCl_2(s)$. This happens because ions are always attracted such that the total net charge becomes zero. If there were to be any extra negative charge, more positive ions would be attracted and vice versa. Ions in an ionic compound always "pack together" to arrange themselves in some regular, repeating 3-D pattern. This 3-D arrangement is called the crystal lattice because, when enough ions assemble to form a visible amount of the compound, we see that the structure is always crystalline in form, with a particular predictable crystal shape for each compound (**Figure 8**). Ordinary table salt, for example, always forms crystals in a cubic pattern. Within such a crystal, we use the term ionic bonding to describe the overall force holding it together. (Crystals are discussed in more detail in Section 3.5.)

Figure 8
The regular geometric shape of ionic crystals is evidence for an orderly array of positive and negative ions.

Metallic Bonding

If both types of colliding atoms have relatively low electronegativities, the atoms can share valence electrons, but no actual chemical reaction takes place between the substances. Mixing melted lead and tin, for example, followed by cooling the mixture until it solidifies, produces a solid, called solder, that is shiny, flexible, and conducts well. The bonding in such substances is called a *metallic bond*. In metallic bonding, the valence electrons are not held very strongly by their atoms, and the atoms have vacant valence orbitals. The result is that the valence electrons are free to move about between the atoms. Wherever the electrons move, they are acting to hold atoms together because positive nuclei on either side will both be attracting the electrons between them. The bonding in metallic substances has been described as a great number of positive ions surrounded by a "sea" of mobile electrons (**Figure 9**). The valence electrons act like a glue that holds the whole structure together.

Due to this mobility of valence electrons, the attractive force around a metal atom acts in every direction, somewhat like the attraction force around any ion in an ionic compound. But the attractive force is very unlike the strong attractions between covalently bonded atoms, which act only in one location and, therefore, in only one specific direction. In addition, unlike the ions in ionic solids, metal atoms do not have to be in any particular arrangement to attract each other, which results in the most useful property observed in metals. The atoms in solid metals may be moved around each other without moving them farther apart from each other, so the bonds between them are not weakened or broken by changing the shape of the solid. This allows the physical formation of metals into any convenient shape, creating the properties we call flexibility, malleability, and ductility. Making objects from metal has defined the progress of human technology since the Bronze Age, when metal alloys first came into use to make edged weapons and armour that did not break upon impact.

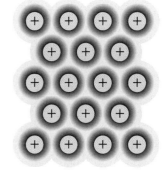

Figure 9
In this model of metallic bonding, each circled positive charge represents the nucleus and inner electrons of a metal atom, surrounded by a mobile "sea" or "cloud" of valence electrons (occupying the grey area).

SUMMARY Bonding Theory

- The formation of a chemical bond involves competition for bonding electrons occupying valence orbitals.
- If the competing atoms have equal electronegativities, the bonding electrons are shared equally.
- Electron sharing between atoms with high electronegativity results in covalent bonding.
- Electron sharing between atoms with low electronegativity often results in metallic bonding.
- If the competing atoms have unequal electronegativities, the result may be unequal covalent bonding; if unequal enough that electrons transfer, the result will be ionic bonding.
- Bonding theory was created by chemists to describe, explain, and predict natural events and observed properties.

Section 3.1 Questions

1. Use bonding theory to describe the following in terms of electrons and orbitals: bonding electron, lone pair.
2. Write a theoretical definition of electronegativity, covalent bond, and ionic bond, and describe the bonds in terms of a difference in electron rearrangement.
3. Copy **Table 2** using computer software, if available, and fill it in for each of the main group elements of Period 3.
4. How do Lewis symbols that represent metal atoms differ from those of nonmetals?
5. Using the electronegativity data in the periodic table, describe the variation in electronegativities within a group and a period.
6. Which element is an exception to almost every rule or generalization about elements? What is unique about this element compared with all other elements in the periodic table?
7. What category of substances would have perfectly equal sharing of valence electrons?
8. What characteristic of valence electrons makes most metals very good conductors of electricity?
9. Potassium and calcium both have valence electrons in the fourth energy level, presumably about the same distance from their nuclei, yet calcium has higher electronegativity (**Figure 10**). Why?

(a) $1\ e^-$
$8\ e^-$
$8\ e^-$
$2\ e^-$
$19\ p^+$
K
potassium atom

(b) $2\ e^-$
$8\ e^-$
$8\ e^-$
$2\ e^-$
$20\ p^+$
Ca
calcium atom

Figure 10
Energy-level diagrams for (a) potassium and (b) calcium

10. Electronegativities are an indication of how strongly atoms attract electron pairs shared in bonds. Why are no electronegativities listed for the two smallest Group 18 atoms?
11. Briefly outline how bonding theory has developed and has been improved through successive contributions by scientists from 1904 to the present day.
12. Dr. Linus Pauling's bonding theory gained rapid acceptance by the scientific community. What characteristics must any scientific theory have in order to be successful?

Table 2 Theoretical Descriptions of Main Group Elements of Period 3

Element symbol	Electronegativity	Group number	Number of valence electrons	Lewis symbol	Number of bonding electrons	Number of lone pairs of electrons

Explaining Molecular Formulas 3.2

The theory of chemical bonding developed in Section 3.1 paves the way for a more complete understanding of molecular elements and compounds. In this section, molecular substances are discussed in terms of their chemical bonding.

Molecular Elements

Using evidence that gases react in simple ratios of whole numbers and Avogadro's theory that equal volumes of gases contain equal numbers of molecules, early scientists were able to determine that the most common forms of hydrogen, oxygen, nitrogen, and the halogens are diatomic molecules. Modern evidence shows that phosphorus and sulfur commonly occur as $P_4(s)$ and $S_8(s)$. An acceptable theory of molecular elements must provide an explanation for evidence such as these empirical formulas. Recall that, according to atomic theory, an atom such as chlorine, with seven valence electrons, requires one electron to complete the stable octet. This electron may be obtained from a metal by electron transfer or by sharing a valence electron with another atom. Two chlorine atoms could each obtain a stable octet of electrons if they shared a pair of electrons with each other. Bonding theory suggests that a covalent bond between the atoms results from the simultaneous attraction of two nuclei for a shared pair of electrons, explaining why chlorine molecules are diatomic (**Figure 1**). Lewis formulas are particularly useful for communicating electron sharing and octet formation:

$$:\ddot{Cl}\cdot + \cdot\ddot{Cl}: \rightarrow :\ddot{Cl}:\ddot{Cl}:$$

According to atomic theory, oxygen atoms have six valence electrons, and evidence shows that the element is diatomic. Sharing a pair of bonding electrons would leave both oxygen atoms with less than a stable octet. Initially, molecular theory could not explain the diatomic character of oxygen. Instead of replacing the theory, scientists revised it by introducing the idea of a double bond. If two oxygen atoms can share a pair of electrons, chemists hypothesized that perhaps they can share two pairs of electrons at once to form a double covalent bond:

$$:\ddot{O}\cdot + \cdot\ddot{O}: \rightarrow :\ddot{O}::\ddot{O}:$$

This arrangement is consistent with accepted theory, since stable octets of electrons result. This idea explains many empirically known molecular formulas, in addition to $O_2(g)$, in a simple way, without the necessity of changing most of the previous assumptions. Chemists created the concept of a triple bond to explain the empirically determined chemical formulas for nitrogen (N_2) and hydrogen cyanide (HCN). This triple bond results from two atoms sharing three pairs of electrons.

Lewis formulas are a form of electron "bookkeeping" to account for valence electrons. They do not show what orbitals look like or where electrons may actually be at any instant—they simply keep track of which electrons are involved in bonds. Once this is understood, the simple and efficient structural formula can be used to represent bonding in molecules (**Figure 2**). In structural formulas, lone pairs of electrons are not indicated, and each shared pair of bonding electrons is represented by a line. Although we indicate a double bond with two lines, and a triple bond with three lines, each type is only one bond. The lines just tell us how many electrons are shared by the bonded atoms.

Learning Tip

Many molecular elements are diatomic, and some are polyatomic. It is very useful to memorize the formulas of the nine molecular elements:

hydrogen	$H_2(g)$
nitrogen	$N_2(g)$
oxygen	$O_2(g)$
fluorine	$F_2(g)$
chlorine	$Cl_2(g)$
iodine	$I_2(s)$
bromine	$Br_2(l)$
phosphorus	$P_4(s)$
sulfur	$S_8(s)$

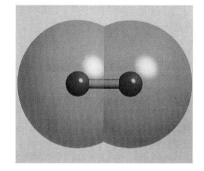

Figure 1
The chlorine, Cl_2, molecule is represented here by a computer-generated combination of space-filling and ball-and-stick models.

$$Cl-Cl \qquad O=O$$

Figure 2
To help visualize bonding, chemists use structural formulas. A single line represents a single covalent bond (as shown in the chlorine molecule), and a double line represents a double covalent bond (as shown in the oxygen molecule).

> **Practice**
>
> 1. Use bonding theory to draw Lewis formulas for the elements in the halogen family. How are these diagrams consistent with the concept of a chemical family?
> 2. The Lewis formula of a hydrogen molecule is automatically an exception to the octet rule. Considering the positions of hydrogen and helium in the periodic table, how is the Lewis formula for hydrogen a good explanation of its empirical formula?
> 3. Use a Lewis formula to explain the molecular formula for nitrogen, $N_2(g)$. Recall that N atoms always obey the octet rule.
> 4. According to atomic theory, a sulfur atom has six valence electrons, including two bonding electrons. Therefore, any molecule containing only sulfur would have either two single covalent bonds or one double covalent bond from each sulfur atom.
>   (a) Use a Lewis formula to predict the simplest chemical formula for sulfur.
>   (b) Assuming two single covalent bonds from each sulfur atom, draw a structural formula that might explain the molecular formula for sulfur, $S_8(s)$.

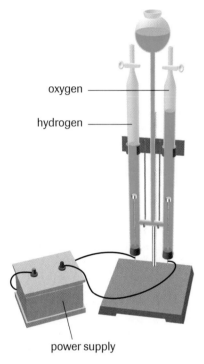

Figure 3
When water is decomposed by electricity, the volume of hydrogen gas produced is twice the volume of oxygen gas produced. This evidence led to the acceptance of H_2O as the empirical formula for water.

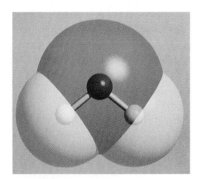

Figure 4
The empirical formula for water is described theoretically by this computer-generated model.

Molecular Compounds

Molecular compounds cannot usually be represented by a simplest ratio formula in the way that ionic compounds can. Simplest ratio formulas indicate only the relative numbers of atoms or ions in a compound; they provide no evidence for the actual number or arrangement. For example, the simplest ratio formula CH represents a compound composed of molecules containing equal numbers of carbon and hydrogen atoms. Empirical work indicates that several very different common compounds, for example, acetylene, $C_2H_2(g)$, and benzene, $C_6H_6(l)$, can be described by means of the simplest ratio formula CH. To distinguish between these compounds, it is necessary to represent them with molecular formulas. Molecular formulas also accurately represent the actual composition of the smallest units of molecular compounds, which is very different from ionic compound structure, where the entire crystal lattice is really all one continuously bonded unit.

Explanations of molecular formulas are based on the same concepts used to explain molecular elements. The rules of the atomic theory, the idea of overlapping half-filled orbitals, and a consideration of differences in electronegativity all work together to produce a logical, consistent, simple explanation of experimentally determined molecular formulas.

A covalent bond in molecular compounds, just as in molecular elements, is a strong, directional force within a complete structural unit: the molecule. The theoretical interpretation of the empirical formula for water, $H_2O(l)$, is that a single molecule contains two hydrogen atoms and one oxygen atom held together by covalent bonds (**Figures 3** and **4**). The purposes of explaining a molecular formula are to show the arrangement of the atoms that are bonded together and to test the explanatory power of the bonding theory. As shown below, an oxygen atom requires two electrons to complete a stable octet. These two electrons are thought to be supplied by the bonding electrons of two hydrogen atoms. In this way, oxygen achieves a stable octet, and all atoms complete their unfilled energy levels.

$$:\!\ddot{O}\!: \;+\; \begin{matrix}\cdot H\\ \cdot H\end{matrix} \;\rightarrow\; :\!\ddot{O}\!:\!H \quad \text{or} \quad \begin{matrix} O-H \\ | \\ H \end{matrix}$$
$$H$$

The concept of a double covalent bond explains empirically known molecular formulas such as $O_2(g)$ and $C_2H_4(g)$. The atoms involved must share more than one bonding electron. A double bond involves the sharing of two pairs of electrons between two atoms.

Similarly, a triple covalent bond involves two atoms sharing three pairs of electrons. In common molecular compounds, there is no empirical evidence for the formation of a bond involving more than three pairs of electrons. According to accepted rules and models, there are many atoms that can form more than one kind of covalent bond. For example, carbon, nitrogen, and oxygen, which are three of the most important elements in molecules in living organisms, can all form more than one kind of covalent bond. The maximum number of single covalent bonds that an atom can form is known as its **bonding capacity**, which is determined by its number of bonding electrons (**Table 1**). For example, nitrogen, with a covalent bonding capacity of three, can form three single bonds, one single bond and one double bond, or one triple bond. The theory suggests that carbon, with a bonding capacity of four, can form four single bonds, two double bonds, one single and one triple, or one double and two single bonds.

Table 1 Bonding Capacities of Some Common Atoms

Atom	Number of valence electrons	Number of bonding electrons	Bonding capacity
carbon	4	4	4
nitrogen	5	3	3
oxygen	6	2	2
halogens	7	1	1
hydrogen	1	1	1

Evidence for the reaction of ammonia with boron trihydride required a revision to bonding theory. A covalent bond in which one of the atoms donates both electrons is called a *coordinate covalent bond*. This concept is useful in explaining the structure of many molecules and polyatomic ions. The following equation shows the formation of a coordinate covalent bond, where a nitrogen atom overlaps a full valence orbital (its lone pair) with the empty (unoccupied) valence orbital of a boron atom:

$$\begin{array}{c} H \quad H \\ H:\!\ddot{N}\!: \\ H \end{array} + \begin{array}{c} H \quad H \\ \ddot{B}\!:\!H \\ H \end{array} \rightarrow \begin{array}{c} H \quad H \\ H:\!\ddot{N}\!:\!\ddot{B}\!:\!H \\ H \quad H \end{array}$$

The properties of coordinate covalent bonds do not differ from those of a normal covalent bond because all electrons are alike, regardless of their source. For purposes of writing Lewis formulas, it is irrelevant which atom the electrons "come from."

In the discussion presented in this textbook, the sharing of electrons has been restricted to valence electrons. This restricted theory also requires that all valence electrons in molecules be paired. Of course, no theory in science is absolute, and there are exceptions to both parts of this theory. Some molecules, such as nitrogen monoxide, appear to have unpaired electrons. Other molecules, like the boron trihydride shown above, appear not to follow the octet rule. Rather than developing a more detailed theory, this textbook specifically notes such cases as exceptions.

Types of Formulas

Over time, different groups of consumers, technologists, and scientists have developed different terms for the same thing. For accuracy in communication, it is necessary that terms be mutually understood, so a convention is normally established. Some conventions, like those for symbols for elements and SI units, are international and universal. Others are more regional or specialized, for example, the "barrel" volume unit (151 L) used only by the oil industry.

DID YOU KNOW ?

Creativity in Science

Many people consider science to be very structured, with no room for human creativity. Nothing could be further from the truth. Forming new concepts and thinking up new areas to explore require scientists to act like artists. Consider the following quote from Dr. John Polanyi (**Figure 5**), a Canadian Nobel Prize winner for chemistry:

"To obtain an answer of note, one must ask a question of note; a question that is exquisitely phrased. It must be one that matters, on a topic that is new, to which an answer can be found. So, yes, we do 'turn over stones', but the art is to pick the right stone."

Figure 5
John Polanyi (1929–)

Ionic Compound Formulas

All ionic compounds have empirical formulas, representing the crystal lattice with a formula unit that shows only the simplest number ratio of cations to anions. The ion charges are omitted, so the formula must be memorized or referenced. The molar mass of an ionic compound is the mass of a mole of formula units.

Molecular Compound Formulas

Table 2 defines and gives examples of the terminology used in this book to refer to the various kinds of molecular compound formulas. Acetic acid is used as a comparative example.

Table 2 Names of Types of Formulas for Molecular Compounds

Empirical formula	CH_2O	An empirical formula shows the simplest whole-number ratio of atoms in the compound. Empirical formulas are rarely useful for molecular compounds.
Molecular formula	$C_2H_4O_2$	A molecular formula shows the actual number of atoms that are covalently bonded to make up each molecule.
	CH_3COOH	A molecular formula often has the atom symbols written in a sequence that helps you determine which atoms are bonded to which.
Lewis formula	(Lewis structure of acetic acid)	*A Lewis formula is also commonly called a Lewis diagram, or an electron-dot diagram.* It uses Lewis symbols to show electron sharing in covalent bonds, and the formation of stable valence octets of electrons in molecules and ions.
Structural formula	(structural formula of acetic acid)	*A structural formula is also commonly called a structural diagram.* As well as showing which atoms are bonded, the type of covalent bond is represented by the number of lines drawn between atomic symbols.
Stereochemical formula	(stereochemical formula of acetic acid)	A stereochemical formula is a structural formula drawn to try to represent the three-dimensional molecular shape. For much larger molecules, this style of representation often becomes too complex to be practical.

* It is strongly recommended that you memorize this terminology information.

Several kinds of physical models (or visual depictions of such models) can also help us understand the shape, relative size, and structure of molecules. Pictures and diagrams of various kinds of models are used throughout this textbook (**Figure 6**).

Learning Tip

The word "empirical" has two common meanings, both of which are used in this unit. To this point, "empirical" has been used to mean "as obtained by direct observation, such as in a laboratory." Thus, any formula determined from evidence in a laboratory can be said to be empirically determined.

When a formula is called an empirical formula, however, the meaning is quite different. In this case, the term means that the formula shows the simplest whole-number ratio of atoms or ions present in a compound.

+ EXTENSION

Computer-Generated Models

Scientists have been collecting empirical evidence of the shapes of molecules for decades. Recently, as computing power and speed increase, software engineers have converted scientific findings into interactive models that appear to be three dimensional. There are many Web sites where you can explore these models.

www.science.nelson.com

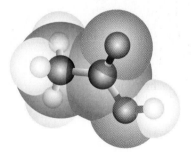

Figure 6
A computer-generated image, combining two models (ball-and-stick and space-filling) for an acetic acid molecule

Determining Lewis Formulas

Chemists make use of bonding theory to predict valence electron distribution for molecules and polyatomic ions. A Lewis formula allows further predictions of entity structure, polarity, and shape. In this textbook, Lewis formula predictions are limited to entities with only one central atom, unless extra information is included with the question. The **central atom** is the atom to which all the other atoms—**peripheral atoms**—are bonded.

Simple Lewis formulas can be predicted with a series of five steps.

Learning Tip

You will find that structural formulas can be written directly for most molecules (even large ones) that have central C atoms, just by using atom bonding capacity. This works because C, N, O, and F atoms always obey the octet rule, and so do other nonmetals when they are peripheral atoms.

▶ SAMPLE problem 3.1

Determine the Lewis formula and the structural formula for sulfur trioxide, $SO_3(g)$.

1. Count the total valence electrons in the entity by adding the valence electrons of each atom. If the entity is a polyatomic ion, add (usually) or subtract valence electrons to account for the net charge, one for each unit of charge.

   3 oxygen atoms and 1 sulfur atom and no net charge
   $3(6e^-) + 1(6e^-) + 0e^- = 24e^-$

It is usually obvious from relative numbers which atom is central; if not, the central atom will usually be the one with the highest bonding capacity. In this molecule, the single S atom must be central.

2. Arrange the peripheral atom symbols around the central atom symbol, and place one pair of valence electrons between each peripheral atom and the central atom (bond pairs).

3. Place more pairs of valence electrons (lone pairs) on all the peripheral atoms, to complete their octets. Recall that a hydrogen atom's energy level is completed with only two valence electrons.

4. Place any remaining valence electrons on the central atom as lone pairs. For this example, all 24 valence electrons have already been assigned.

   Sulfur atom has an incomplete octet.

5. If the central atom's octet is not complete, move a lone pair from a peripheral atom to a new position between that peripheral atom and the central atom. Repeat until the central atom has a complete octet. For a molecule, this completes the Lewis formula.

If the entity is a polyatomic ion, place square brackets around the entire Lewis formula, and then write the net charge outside the bracket on the upper right.

To show the structural formula, omit all lone pairs and replace every bond pair with a line.

+ EXTENSION

Resonance

What if empirical evidence does not support the theoretical Lewis structure? Or if more than one Lewis structure can be drawn? Find out about resonance, the apparent existence of multiple structures at the same time.

www.science.nelson.com

Learning Tip

When molecules have more than one possible arrangement of atoms, the molecular formula may be written differently to indicate groups of atoms that are bonded together. For example, the molecular formula for ethanol, C_2H_5OH, clearly shows that one hydrogen atom is bonded to the oxygen atom. Dimethyl ether, which has exactly the same number and kind of atoms but very different properties, is written as CH_3OCH_3 or $(CH_3)_2O$ to distinguish it from ethanol. Structural formulas make the difference much easier to see.

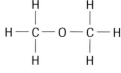

▶ Practice

5. Use both Lewis formulas and structural formulas to represent molecules of the following compounds:
   (a) $CS_2(l)$, carbon disulfide
   (b) $PH_3(g)$, phosphine
   (c) $H_2S(g)$, hydrogen sulfide
   (d) $H_2Se(g)$, hydrogen selenide
   (e) $CH_3SH(g)$, methanethiol
   (f) $SF_6(g)$, sulfur hexafluoride*

   *This molecule's central atom does not follow the octet rule.

6. Use both Lewis formulas and structural formulas to represent the following ions:
   (a) hydrogen ion
   (b) hydroxide ion
   (c) hydrogen sulfide ion
   (d) nitrate ion
   (e) carbonate ion
   (f) hydrogen carbonate ion*

   *Hint: A hydrogen ion (proton) bonds to any lone pair from answer (e).

INVESTIGATION 3.1 Introduction

Molecular Models

Chemists use molecular models to explain and predict molecular structure, relating structure to the properties and reactions of substances.

Report Checklist
- ○ Purpose
- ○ Problem
- ○ Hypothesis
- ○ Prediction
- ○ Design
- ○ Materials
- ○ Procedure
- ● Evidence
- ● Analysis
- ○ Evaluation

Purpose
The scientific purpose of this investigation is to test the ability of bonding theory to explain some known chemical reactions by using molecular models.

Problem
How can theory, represented by molecular models, explain certain chemical reactions?

Design
Chemical reactions are simulated with model kits (**Figure 7**) to test the ability of bonding theory to explain reaction evidence.

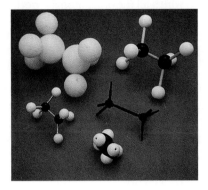

Figure 7
You can use molecular model kits to test the explanatory power of bonding theories.

To perform this investigation, turn to page 131.

Section 3.2 Questions

1. Why is it incorrect to write the structural formula of the H_2S molecule as H–H–S?

2. Why is the molecular formula for the methanol molecule usually written as CH_3OH instead of CH_4O?

3. For each of the following molecular compounds, name the compound, and explain the empirically determined formula by drawing a Lewis formula and a structural formula:
   (a) HCl
   (b) NH_3
   (c) H_2S
   (d) CO_2

4. Use the bonding capacities listed in **Table 1** (page 87) to draw a structural formula of each of the following entities. In each case (for these particular molecules), every C must connect to 4 lines, every N to 3 lines, every O to 2 lines, and every H to one line because C, N, and O atoms always obey the octet rule. Hint: Use the sequence of atoms in some of these molecular formulas to guide you.
   (a) H_2O_2
   (b) C_2H_4
   (c) HCN
   (d) C_2H_5OH
   (e) CH_3OCH_3
   (f) CH_3NH_2

5. Using Lewis symbols, predict the simplest binary molecular compound and write the chemical name for a product of each of the following reactions. Include a structural formula for a molecule of the product.
   (a) $I_2(s) + Br_2(l) \rightarrow$
   (b) $P_4(s) + Cl_2(g) \rightarrow$
   (c) $O_2(g) + Cl_2(g) \rightarrow$
   (d) $C(s) + S_8(s) \rightarrow$
   (e) $S_8(s) + O_2(g) \rightarrow$

6. Draw Lewis formulas for the following common polyatomic anions:
   (a) hypochlorite ion, OCl^-
   (b) bromite ion, BrO_2^-
   (c) iodate ion, IO_3^-
   (d) perchlorate ion, ClO_4^-
   (e) ammonium ion, NH_4^+

Extension

7. Compare your predictions from question 5 with empirical evidence from a reference such as *The CRC Handbook of Chemistry and Physics*.

Molecular Shapes and Dipoles 3.3

The shape of molecules has long been investigated through crystallography, using microscopes and polarimeters in the late 1800s, and X-ray and other spectrographic techniques since the early 1900s. One of the most important applications of molecular shape research is the study of enzymes. Enzymes are large proteins that, because of their shape, will react only with specific molecules, much like a key's shape will fit only one specific lock. There are about three thousand enzymes in an average living cell, and each one carries out (catalyzes) a specific reaction. There is no room for error without affecting the normal functioning of the cell; different molecular shapes help to ensure that all processes occur properly. Despite extensive knowledge of existing enzymes, the structure of these proteins is so complex that it is still effectively impossible to predict the shape an enzyme will take, even though the sequence of its constituent amino acids is known. The study of molecular shapes, particularly of complex biological molecules, is still a dynamic field.

VSEPR Theory

The valence bond theory created and popularized by Linus Pauling in the late 1930s successfully explained many of the atomic orientations in molecules and ions. Pauling's main empirical work was with the X-ray analysis of crystals. The valence bond theory of bonding was created to explain what he observed in the laboratory. Pauling extended the work of his friend and colleague, Gilbert Lewis, who, as you saw earlier, is famous for creating Lewis symbols (also called electron-dot diagrams).

However, it was not until 1957 that Ronald Nyholm from Australia and Ron Gillespie (**Figure 1**) from England created a much simpler theory for describing, explaining, and predicting the stereochemistry of chemical elements and compounds. **Stereochemistry** is the study of the 3-D spatial configuration of molecules and how this affects their reactions. The theory that Nyholm and Gillespie created is highly effective for predicting the shape of molecules.

The name of the Nyholm-Gillespie theory is the valence-shell-electron-pair-repulsion theory, or **VSEPR** (pronounced "vesper") **theory**. The theory is based on the electrical repulsion of bonded and unbonded electron pairs in a molecule: pairs of electrons in the valence shell of an atom stay as far apart as possible because of the repulsion of their negative charges. The number of electron pairs can be counted by adding the number of bonded atoms plus the number of lone pairs of electrons (**Figure 2**). Once the counting is done, the 3-D distribution about the central atom can be predicted by arranging all pairs of electrons as far apart as possible. The type, number, and direction of bonds to the central atom determine the shape of the resulting molecule or polyatomic ion.

According to VSEPR theory:

- Only the valence electrons of the central atom(s) are important for molecular shape.
- Valence electrons are paired in a molecule or polyatomic ion.
- Bonded pairs of electrons and lone pairs of electrons are treated approximately equally.
- Valence electron pairs repel each other electrostatically.
- The molecular shape is determined by the positions of the electron pairs when they are a maximum distance apart.

CAREER CONNECTION

Biochemist
Biochemists study the chemistry of living matter, including enzymes and proteins. They conduct research in a variety of areas such as medicine, agriculture, and environmental science.
  Find a university that offers a program in biochemistry, and find out what other disciplines are part of the program.

www.science.nelson.com GO

Figure 1
Dr. Ronald Gillespie co-created VSEPR theory in 1957. He moved from England to McMaster University in Hamilton, Ontario, the following year. His work in molecular geometry and in chemistry education is renowned.

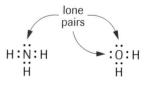

Figure 2
Both the ammonia molecule and the water molecule have four pairs of electrons surrounding the central atom. Some of these are bonding pairs, and some are lone pairs.

mini Investigation

Electrostatic Repulsion Model

The electrostatic repulsion of electron pairs around a central atom in a molecule can be modelled using balloons.

Materials: eye protection; four balloons; string; one wooden dowel, 1 cm in diameter and 50 cm long, with a small "eye" screw (metal loop) in one end.

- Inflate the balloons and tie them off, leaving long (50 cm) pieces of string attached to each.
- Thread the strings of two balloons through the metal ring on the end of the dowel, and pull the strings tight, to pull the balloons as close to each other as possible.

(a) What is the orientation (e.g., angle of a line through the centres) of the two balloons?
- Repeat with three and then four balloons.

(b) What is the orientation of the three balloons?

(c) What is the orientation of the four balloons?
- Reuse or recycle the balloons as directed.

(d) What are the pros and cons of using balloons for a physical model of the repulsion of electron pairs about a central atom in a molecule?

Using VSEPR Theory to Predict Molecular Shape

Chemists use VSEPR theory to predict molecular shape. What is the shape of the hydrogen compounds of Period 2: $BeH_2(s)$, $BH_3(g)$, $CH_4(g)$, $NH_3(g)$, $H_2O(l)$, and $HF(g)$?

First, we draw Lewis formulas of each of the molecules and then consider the arrangement of all pairs of valence electrons. *The key concept is that all pairs of valence electrons repel each other and try to get as far from each other as possible.*

Beryllium Dihydride

Table 1 Geometry of Beryllium Dihydride

Lewis formula	Bond pairs	Lone pairs	Total pairs	General formula	Electron pair arrangement	Stereochemical formula
H:Be:H	2	0	2	AX_2	linear	H —— Be —— H linear

The Lewis formula indicates that BeH_2 has two bonds and no lone pairs of electrons. The number of bond pairs of electrons around the central atom (Be) is two. This is represented by X_2 in the general formula; the A represents the central atom. VSEPR theory suggests that these electron pairs repel each other. According to VSEPR theory, the farthest the electron pairs can get away from each other is to move to opposite sides of the Be atom. This gives the molecule a linear orientation, with the two bonds at an angle of 180°.

Boron Trihydride

Table 2 Geometry of Boron Trihydride

Lewis formula	Bond pairs	Lone pairs	Total pairs	General formula	Electron pair arrangement	Stereochemical formula
H:B:H with H above	3	0	3	AX_3	trigonal planar	H—B with H, H trigonal planar

As indicated in **Table 2**, BH_3 has three bonds and three pairs of electrons around the central atom, B. The three pairs of electrons repel one another to form a plane of bonds at 120° to each other. According to VSEPR theory, this arrangement or geometry is called trigonal planar.

Learning Tip

Tetrahedral, trigonal planar, and linear orientations of atoms are common in molecules. To represent 3-D shapes on paper, we use the following conventions: a solid line —— is a bond in the plane of the page; a dashed line --- is a bond to an atom behind the plane of the page (away from the viewer); and a wedged line ◢ is a bond to an atom ahead of the plane of the page (toward the viewer). See the stereochemical formula for acetic acid in Table 2, page 88, for a typical example.

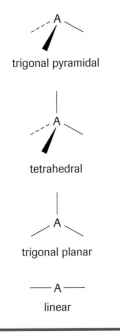

trigonal pyramidal

tetrahedral

trigonal planar

linear

Methane

Table 3 Geometry of Methane

Lewis formula	Bond pairs	Lone pairs	Total pairs	General formula	Electron pair arrangement	Stereochemical formula
H:C:H (with H above and below)	4	0	4	AX_4	tetrahedral	H—C(—H)(—H)—H tetrahedral

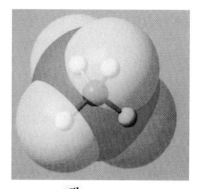

Figure 3
This image shows the ball-and-stick model of methane superimposed on its space-filling model. This model most accurately represents scientists' empirical and theoretical knowledge of the CH_4 molecule. The methane molecule has a classic tetrahedral shape.

Lewis theory indicates that CH_4 has four bonds or four pairs of electrons repelling each other around the central atom, C. Experimental work and VSEPR theory both agree that a tetrahedral arrangement minimizes the repulsion. Tetrahedral bonds, if identical, orient in three dimensions so that every bond makes an angle of 109.5° with each of the others (**Figure 3**).

Ammonia

Table 4 Geometry of Ammonia

Lewis formula	Bond pairs	Lone pairs	Total pairs	General formula	Electron pair arrangement	Stereochemical formula
H:N:H (with H below)	3	1	4	AX_3E	tetrahedral	N(—H)(—H)—H trigonal pyramidal

DID YOU KNOW ?

Ronald Gillespie
Professor Gillespie is best known for developing a theory of molecular shapes and angles known as VSEPR theory. Gillespie has also done ground-breaking work on acids, including superacids, and has identified previously unknown polyatomic nonmetal cations.

The Lewis formula shows that NH_3 has three bonding pairs and one lone pair (represented by E in the general formula) of electrons. VSEPR theory indicates that the four groups of electrons should repel each other to form a tetrahedral arrangement of the electron pairs just like methane, CH_4. The molecular geometry is always based on the atoms present; therefore, if we ignore the lone pair, the shape of the ammonia molecule is like a three-sided (triangular) pyramid (called trigonal pyramidal). We would expect the angle between the atoms H—N—H to be 109.5°, which is the angle for an ideal tetrahedral arrangement of electron pairs. However, in ammonia, the atoms form a trigonal pyramidal arrangement with an angle of 107.3° (**Figure 4**). Chemists hypothesize that this occurs because there is slightly stronger repulsion between the lone pair of electrons and the bonding pairs than between the bonding pairs. According to VSEPR theory, this causes the bonding pairs to be pushed closer together.

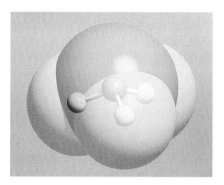

Figure 4
Evidence indicates that the shape of the ammonia molecule, NH_3, is trigonal pyramidal.

Water

Table 5 Geometry of Water

Lewis formula	Bond pairs	Lone pairs	Total pairs	General formula	Electron pair arrangement	Stereochemical formula
:Ö:H H	2	2	4	AX_2E_2	tetrahedral	H—O—H angular (V-shaped)

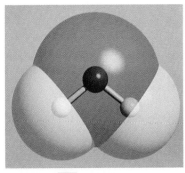

Figure 5
Empirical evidence shows that in water molecules, the two hydrogens bond to the central oxygen at an angle. Such molecules are said to be angular, bent, or V-shaped. The most appropriate and correctly descriptive term is probably Dr. Gillespie's original word, "angular."

According to the Lewis formula, the water molecule has two bonding pairs and two lone pairs of electrons. Based upon VSEPR theory, the four pairs of electrons repel each other to produce a tetrahedral orientation. The geometry of the water molecule is called angular with an angle of 104.5° (**Figure 5**). Notice that this angle is again less than the ideal angle of 109.5° for a tetrahedral arrangement of electron pairs. Again, the slightly stronger repulsion exerted by lone pairs is thought to force the bonding electron pairs a bit closer together.

Hydrogen Fluoride

Table 6 Geometry of Hydrogen Fluoride

Lewis formula	Bond pairs	Lone pairs	Total pairs	General formula	Electron pair arrangement	Stereochemical formula
H:F:	1	3	4	AXE_3	tetrahedral	H—F linear

Based upon the Lewis theory of bonding, the hydrogen fluoride molecule has one bonding pair and three lone pairs of electrons. VSEPR theory indicates that the four electron pairs repel to create a tetrahedral arrangement for the electrons. Since there are only two atoms with one covalent bond holding them together, by definition, the shape of HF is linear, as is the shape of every other diatomic molecule.

Learning Tip

Although beryllium is classified as a metal, this very small atom forms some molecular compounds when reacted with nonmetals. Such a molecule obviously cannot obey the octet rule, because beryllium has only two bonding electrons. Similarly, boron (classified as a semi-metal, or metalloid) forms several molecular compounds. With only three bonding electrons, boron cannot form an octet in a simple molecule where one boron is the central atom.

SUMMARY   Shapes of Molecules

VSEPR theory describes, explains, and predicts the geometry of molecules by counting pairs of electrons that repel each other to minimize repulsion. The process for predicting the shape of a molecule is summarized below:

Step 1: Draw the Lewis formula for the molecule, including the electron pairs around the central atom.
Step 2: Count the total number of bonding pairs (bonded atoms) and lone pairs of electrons around the central atom.
Step 3: Refer to **Table 7**, and use the number of pairs of electrons to predict the shape of the molecule.

Table 7 Using VSEPR Theory to Predict Molecular Shape

General formula*	Bond pairs	Lone pairs	Total pairs	Molecular shape Geometry**	Stereochemical formula	Examples
AX_2	2	0	2	linear (linear)	X — A — X	CO_2, CS_2
AX_3	3	0	3	trigonal planar (trigonal planar)		CH_2O, BH_3
AX_4	4	0	4	tetrahedral (tetrahedral)		CH_4, SiH_4
AX_3E	3	1	4	trigonal pyramidal (tetrahedral)		NH_3, PCl_3
AX_2E_2	2	2	4	angular (V-shaped)		H_2O, OCl_2
AXE_3	1	3	4	linear (tetrahedral)	A — X	HCl, BrF

*A is the central atom; X is another atom; E is a lone pair of electrons.
**The electron pair arrangement is in parentheses.

DID YOU KNOW?

Confusing Mosquitoes
Research has shown that molecules with a round shape are better in mosquito repellents than long, thin molecules. It seems round molecules are better able to block the sensory nerves in the mosquito's antennae. This makes it difficult for mosquitoes to detect carbon dioxide, moisture, and heat from humans and animals.

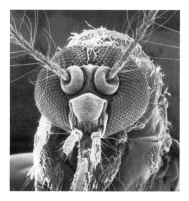

Figure 6
Mosquitoes can be dangerous when they transmit diseases such as malaria and West Nile fever.

SAMPLE problem 3.2

Use the Lewis formula and VSEPR theory to predict the shape of a sulfate ion, SO_4^{2-}.

Determining the shape of a polyatomic ion is no different than determining the shape of a molecule. Again, you first draw the Lewis formula of the ion. For the sulfate ion, the central sulfur atom is surrounded by four oxygen atoms. Add (or subtract) valence electrons to get the net ion charge. In this case, add 2 electrons for a total of $32e^-$. See question 6, page 89, for a hint about predicting this Lewis formula.

$$1(6e^-) + 4(6e^-) + 2e^- = 32e^-$$

Notice that you have four pairs of electrons around the central sulfur atom. This corresponds to the VSEPR theory AX_4 category in **Table 7**; therefore, the ion has a tetrahedral shape.

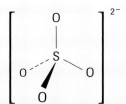

DID YOU KNOW ?

Predictive Power

Concepts are tested by their power to explain and to predict. In this section, VSEPR theory is used to predict the shape of molecules with one central atom. Of course, these predictions must be tested in the laboratory. Some of the evidence is provided herein, but more is needed. If a concept, such as VSEPR theory, predicts accurately in a reliable fashion, then the concept is said to have predictive power.

▶ COMMUNICATION example 1

Draw Lewis and stereochemical formulas for a chlorate ion, ClO_3^-, and predict the shape.

Solution

According to VSEPR theory, the chlorate ion has a trigonal pyramidal shape.

▶ Practice

1. Explain how the words that the VSEPR acronym represents communicate the main ideas of this theory.
2. Use VSEPR theory to predict the geometry of a molecule of each of the following substances. Draw a stereochemical formula for each molecule, and state whether the central atom obeys the octet rule.
   (a) $Cl_2O(g)$ (c) $H_2S(g)$ (e) $SiBr_4(l)$
   (b) $PF_3(g)$ (d) $BBr_3(l)$ (f) $HCl(g)$
3. Use VSEPR theory to predict the shape of each of the following polyatomic ions:
   (a) PO_4^{3-} (b) BrO_3^- (c) NH_4^+
4. Cubane is a hydrocarbon with the formula $C_8H_8(s)$. It has a cubic shape, as its name implies, with a carbon atom at each corner of the cube (**Figure 7**). This molecule is very unstable, and some researchers have been seriously injured when crystals of the compound exploded while being scooped out of a bottle. Not surprisingly, cubane has been the subject of some research as an explosive.
   (a) According to VSEPR theory, what should be the shape around each carbon atom? Why?
   (b) If we assume an ideal cubic shape, what are the actual bond angles around each carbon?
   (c) Explain how your answers to (a) and (b) suggest why this molecule is so unstable.
5. Where did the evidence come from that led to the creation of VSEPR theory?

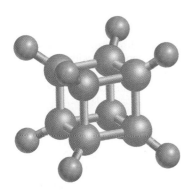

Figure 7
Cubane

Learning Tip

It is important to always remember that a double bond or a triple bond is one bond, and to treat it as such, when using VSEPR theory to predict shapes of molecules containing these bonds.

The Multiple Bond in VSEPR Models

By observing the rate at which hydrocarbon substances react with bromine, a highly reactive reagent, chemists have determined that molecules can contain a double covalent bond or a triple covalent bond. Hydrocarbons with molecules that contain a multiple bond react rapidly, quickly removing the orange-brown colour of the reagent. Hydrocarbon molecules with only single bonds react extremely slowly. Further evidence indicates that a double bond or a triple bond is always shorter and stronger than a single bond between the same kind of atoms. Evidence from crystallography (such as the X-ray analysis of crystals) indicates that any multiple bond is treated just like any single bond for describing, explaining, or predicting the shape of a molecule. This has implications for using VSEPR theory for molecules containing a multiple bond. Let's look at some examples.

SAMPLE problem 3.3

Ethene (ethylene, $C_2H_4(g)$) is the most common hydrocarbon with a multiple covalent bond. Crystallography indicates that the orientation around the central carbon atoms is trigonal planar. Is VSEPR theory able to explain the empirically determined shape of this molecule?

The first step in testing the ability of VSEPR theory to explain the shape of ethene is to draw a Lewis formula for the molecule.

The second step is to count the number of "pairs" of electrons around the central atoms (the carbon atoms). If a multiple bond is present, there are more than two shared electrons involved in that bond. For VSEPR theory to explain the trigonal planar shape around each carbon, it must assume that any "group" of electrons shared in a bond behaves in much the same way, regardless of how many electrons (2, 4, or 6) are being shared. In other words, you count the numbers of lone pairs and bonded atoms around a central atom to determine the shape. This is a typical revision of a theory, one that increases its power to explain. Each carbon is seen to have three bonds (two single and one double) and no lone pairs. This is an AX_3 configuration, and it explains a trigonal planar shape around each carbon atom, which agrees with the evidence from crystallography.

Figure 8
Ethyne (acetylene), when mixed with oxygen, burns with a very high-temperature flame. Operating at up to 2500 °C, an oxy-acetylene torch will melt most metals easily.

VSEPR theory passes the test by being able to explain the trigonal planar shape of ethene. Now let's see if VSEPR theory can pass another test by predicting the stereochemistry of the ethyne molecule. Ethyne is the IUPAC name for the substance commonly called acetylene (**Figure 8**). Acetylene is widely used in high-temperature torches for cutting and welding metals.

COMMUNICATION example 2

Draw a stereochemical formula of ethyne, $C_2H_2(g)$, to predict the shape of the molecule.

Solution

H : C :: : C : H

H — C ≡ C — H

According to VSEPR theory, the shape of the ethyne molecule is linear.

COMMUNICATION example 3

Draw a stereochemical formula for a nitrite ion, NO_2^-, to predict the shape of the ion.

Solution

[:Ö:N::Ö:]⁻   [O—N=O]⁻

The central N atom has one lone pair and forms two bonds. Therefore, according to VSEPR theory, a nitrite ion has an angular shape.

DID YOU KNOW ?

Prediction Accuracy
To show octets for all atoms, bonding theory suggests that ethene has one double bond and ethyne has one triple bond. This can be tested empirically, using the evidence that a multiple covalent bond is shorter than a single bond. For both molecules, X-ray evidence agrees that shapes and bond types are both predicted accurately.

The nitrite ion, however, is predicted to have one single and one double bond. In this case, the evidence does not agree; it indicates that both bonds are identical in length and strength. The shape prediction works very well, but the bond type prediction does not.

As previously stated, electrons are not really small particles, and bonding theory that treats them as such is necessarily limited, a typical case of restriction of a scientific theory.

DID YOU KNOW ?

Theories

Just as scientists have special definitions for words that apply in the context of science (e.g., the definition of "work" in physics), philosophers who study the nature of chemistry have special definitions for terms such as "theory." To them, a theory is not a hypothesis. To philosophers of chemistry a theory uses the unobservable (such as electrons and bonds) to explain observables (such as chemical and physical properties).

Practice

6. In order to make the rules of VSEPR theory work, how must a multiple (double or triple) bond be treated?

7. Use Lewis formulas and VSEPR theory to predict the shapes around each central atom of the following molecules:
   (a) $CO_2(g)$, carbon dioxide (in "carbonated" beverages)
   (b) $HCN(g)$, hydrogen cyanide (odour of bitter almonds, extremely toxic)
   (c) $CH_2CHCH_3(g)$, propene (monomer for polypropylene)
   (d) $CHCCH_3$, propyne (in specialty fuels, such as MAPP gas, for welding)
   (e) $H_2CO(g)$, methanal (formaldehyde)
   (f) $CO(g)$, carbon monoxide (highly toxic gas)

8. Is VSEPR a successful scientific theory? Defend your answer.

Molecular Polarity: Dipole Theory

Chemists believe that molecules are made up of charged particles (electrons and nuclei). A **polar molecule** is one in which the negative (electron) charge is not distributed symmetrically among the atoms making up the molecule. Thus, it will have partial positive and negative charges on opposite sides of the molecule. A molecule with symmetrical electron distribution is a **nonpolar molecule**. The existence of polar molecules can be demonstrated by running a stream of water past a charged object (**Figure 9**). When repeated with a large number of pure liquids, this experiment produces a set of empirical rules for predicting whether a molecule is polar (**Table 8**).

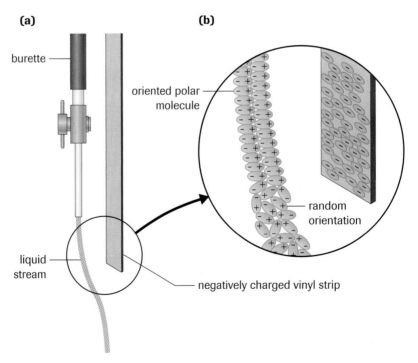

Figure 9
(a) Testing a liquid with a charged object, in this case a vinyl strip, provides evidence for the existence of polar molecules in a substance.
(b) In a liquid, molecules are able to rotate freely. Polar molecules in a liquid will rotate so that their positive sides are closer to a negatively charged material. Near a positively charged material they become oriented in the opposite direction. Logically, then, we can hypothesize that polar molecules will be attracted by either kind of charge.

Table 8 Empirical Rules for Polar and Nonpolar Molecules

Type		Description of molecule	Examples
Polar	AB	diatomic with different atoms	HCl(g), CO(g)
	N_xA_y	containing nitrogen and other atoms	NH_3(g), NF_3(g)
	O_xA_y	containing oxygen and other atoms	H_2O(l), OCl_2(g)
	$C_xA_yB_z$	containing carbon and two other kinds of atoms	$CHCl_3$(l), C_2H_5OH(l)
Nonpolar	A_x	all elements	Cl_2(g), N_2(g)
	C_xA_y	containing carbon and only one other kind of atom	CO_2(g), CH_4(g)

INVESTIGATION 3.2 Introduction

Evidence for Polar Molecules

Use the empirical rules (**Table 8**) to predict the molecular polarity of the liquids provided.

Report Checklist
- ○ Purpose
- ○ Problem
- ○ Hypothesis
- ● Prediction
- ● Design
- ○ Materials
- ○ Procedure
- ● Evidence
- ● Analysis
- ● Evaluation (1, 2, 3)

Purpose
The purpose of this investigation is to test the empirical rules for predicting molecular polarity.

Problem
Which of the liquids provided have polar molecules?

To perform this investigation, turn to page 131.

Electronegativity and Bond Polarity

Linus Pauling saw the need for a theory to explain and predict the observed polarity of some molecular substances. He combined valence bond theory and bond energy theory, as well as several empirical measures, to create the concept of an atomic property that he called electronegativity (Section 3.1). Electronegativity is found to be a periodic property of atoms, increasing as an atom's position on the periodic table is located farther to the right in a row or higher up in a column (**Figure 10**).

Pauling explained the polarity of a covalent bond as the difference in electronegativity of the bonded atoms. If the bonded atoms have the same electronegativity, they will attract any shared electrons equally and form a **nonpolar covalent bond**. If the atoms have different electronegativities, they will form a **polar covalent bond**. The greater the electronegativity difference, the more polar the bond will be. For a very large electronegativity difference (combined with other factors), the difference in attraction may transfer one or more electrons from one atom to the other, forming cations and anions. The resulting ions will group to form an ionic compound, held together by ionic bonding.

Pauling explained that polar covalent bonds form when two different atoms share electrons. Because the attraction for those electrons is unequal, *the electrons spend more of their time closer to one atomic nucleus than the other*. The side of the bond where electrons spend more time is labelled as being partially negative ($\delta-$); we use the Greek lowercase delta to represent partial charge. The side of the bond that is partially positive is labelled $\delta+$ (**Figure 11**).

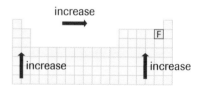

Figure 10
Electronegativity of the elements increases as you move up the periodic table and to the right.

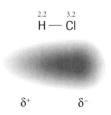

Figure 11
Unequal attraction of the shared electrons gives one side of the covalent bond a partial negative charge ($\delta-$) and the other side a partial positive charge ($\delta+$).

Pauling liked to think of chemical bonds as being different in degree rather than different in kind. According to him, all chemical bonds involve a sharing of electrons, with ionic bonds and nonpolar covalent bonds being just the two extreme cases (**Figure 12**). The bonding in substances therefore ranges anywhere along a continuum from nonpolar covalent to polar covalent to ionic. For polar covalent bonds, the greater the electronegativity difference of the atoms, the more polar the bond.

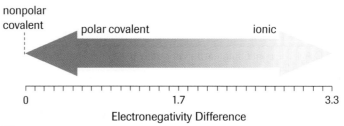

Figure 12
This model of the bonding type continuum shows the relationship of bonded atom electronegativity to bond polarity.

Learning Tip

The gradual variation in bond character is shown in beryllium compounds. $BeF_2(s)$ behaves more like an ionic substance, while $BeCl_2(s)$ behaves more like a molecular substance. Empirical behaviour is also partly due to size of atoms and shape of molecules, so predicting bond type from electronegativity only is not very reliable, except at the far ends of the scale shown.

DID YOU KNOW ?

Evidence for Polar Bonds
The energy required to break a bond can be determined experimentally. The energy required to break the H−F bond is considerably greater than the energy required to break either H−H or F−F bonds. Pauling realized that an unequal sharing of the electron pair produced a polar bond that enhanced the bonding between the atoms.

SAMPLE problem 3.4

Consider the bond that forms between the entities in the following pairs. Label each entity and bond with its electronegativity and bond polarity, and classify the bond:
(a) H and H
(b) P and Cl
(c) Na and Br

From the periodic table, assign each atom an electronegativity.
(a) H − H
   2.2 2.2
   The electronegativity difference is 0.0, indicating a nonpolar covalent bond.
(b) Chlorine is more electronegative, so it is assigned the δ− charge.
   δ+   δ−
   P — Cl
   2.2 3.2
   The electronegativity difference is 1.0, indicating a polar covalent bond.
(c) +   −
    Na   Br
    0.9 3.0
   The electronegativity difference is 2.1. These atoms are a metal and a nonmetal, and the electronegativity difference is large. In such a case we assume that an electron transfers completely, forming a cation and an anion. The electrostatic attraction between the newly formed ions is called an ionic bond.

Practice

9. Draw the following bonds, label the electronegativities, and label the charges (if any) on the ends of the bond. Classify the bond as ionic, polar covalent, or nonpolar covalent:
   (a) H and Cl
   (b) C and H
   (c) N and O
   (d) I and Br
   (e) Mg and S
   (f) P and H

10. Using electronegativity as a guide, classify the following bonds as ionic, polar covalent, or nonpolar covalent:
    (a) the bond in HBr(g)
    (b) the bond in LiF(s)
    (c) a C−C bond in propane, C_3H_8(g)

11. List and order the bonds in the following substances according to increasing bond polarity. Provide your reasoning.
    (a) H₂O(l), H₂(g), CH₄(g), HF(g), NH₃(g), LiH(s), BeH₂(s)
    (b) PCl₃(l), LiI(s), I₂(s), ICl(s), RbF(s), AlCl₃(s)
    (c) CH₃OH(l)
    (d) CHFCl₂(g)

Bond Polarity and Molecular Polarity

Chemists have found that, logically and experimentally, the existence of polar bonds in a molecule does not necessarily mean that you have a polar molecule. For example, carbon dioxide is found to be a nonpolar molecule, although each of the C=O bonds is a polar bond. To resolve this apparent contradiction, we need to look at this molecule more closely. Based on the Lewis formula and the rules of VSEPR, carbon dioxide is a linear molecule (**Figure 13(a)**). Using electronegativities, we can predict the polarity of each of the bonds. It is customary to show the bond polarity as an arrow, pointing from the positive (δ+) to the negative (δ−) side of the bond (**Figure 13(b)**). This arrow represents the **bond dipole**. The bond dipole is the charge separation that occurs when the electronegativity difference of two bonded atoms shifts the shared electrons, making one end of the bond partially positive and the other partially negative. The arrow representing a bond dipole points from lower to higher electronegativity.

These arrows are vectors and when added together, produce a zero total. In other words, the equal and opposite bond dipoles balance each other, and the result is that the molecule is nonpolar. A nonpolar molecule is one where the bond dipoles balance each other, producing a molecular dipole (vector sum) of zero.

Let's try this procedure again with another small molecule. As you know from Investigation 3.2, water is a polar substance and the O−H bonds in water are polar bonds. The Lewis formula and VSEPR rules predict an angular molecule, shown in **Figure 14** together with its bond dipoles.

In this case, logic indicates that the bond dipoles do not balance. Instead, they add together to produce a nonzero molecular dipole (shown in red). Whenever bond dipoles add to produce an overall dipole for the molecule, the result is a polar molecule. Notice that the water molecule has a partial negative charge on the side by the oxygen and a partial positive charge on the side by the two hydrogens. This explains why a stream of water is attracted to a positively charged strip or rod. This example shows how we can use theory to explain our empirical observations.

From the two examples, carbon dioxide and water, you can see that the shape of the molecule is as important as the bond polarity.

Evidence indicates that methane is a nonpolar substance, although its C−H bonds are polar. Does the same explanation we used for carbon dioxide apply to methane? The stereochemical formula with bond dipoles (**Figure 15**) shows four equal bond dipoles at tetrahedral (3-D) angles. The bond dipoles balance, the vector sum is zero, and the molecule has no overall dipole. This is true because the CH₄ molecule is symmetrical. In fact, all symmetrical molecules with one central atom, such as CCl₄ and BF₃, are nonpolar for the same reason.

The theory created by combining the concepts of covalent bonds, electronegativity, bond polarity, and VSEPR logically and consistently explains the polar or nonpolar nature of molecules. We are now ready to put this combination of concepts to a further test—to *predict* the polarity of a molecule.

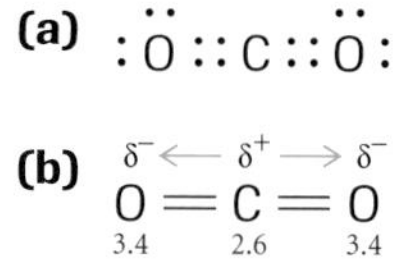

Figure 13
(a) The central carbon atom has no lone pairs and two groups of electrons (remember that a double bond counts as one group of electrons). According to VSEPR theory, the least repulsion of two groups of electrons is a linear arrangement. (b) Bond polarity for carbon dioxide

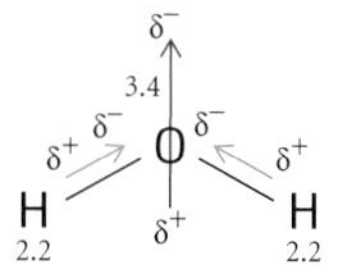

Figure 14
In water, the two bond dipoles are at an angle and do not balance. This creates a resultant molecular dipole (in red) equal to the sum of the bond dipoles.

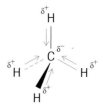

Figure 15
Notice how all the bond dipoles point into the central carbon atom. There are no positive and negative areas on the outer part of the methane molecule. A tetrahedral molecule is symmetrical in three dimensions, and four equal tetrahedral bond dipoles always sum to zero.

Learning Tip

A vector quantity is a quantity that has a size and a direction and is often represented by a vector, which is an arrow pointing in a particular direction. When we want to add vector quantities, we usually add the vectors (arrows) "head to tail" to obtain the final (resultant) vector. Bond dipoles are vector quantities because they have a size (difference in electronegativities) and a direction, defined as pointing toward the negative end of the bond.

1. head-to-tail addition (resultant vectors in red)

2.

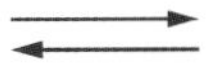

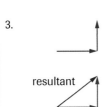

resultant: none

3.

Adding vectors in three dimensions follows the same rules but it is mathematically more difficult. For symmetrical molecules, like CH_4, it is much easier to use the symmetry of the molecule to reach the conclusion that the molecule is nonpolar.

SAMPLE problem 3.5

Predict the polarity of the ammonia, NH_3, molecule. Include your reasoning.

First, draw the Lewis formula.

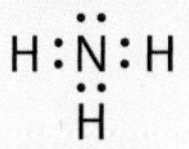

Based on VSEPR theory, draw the stereochemical formula.

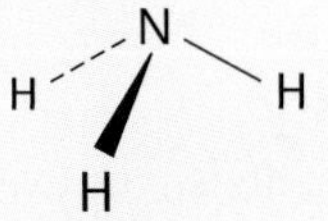

Add the electronegativities of the atoms, from the periodic table, and assign $\delta+$ and $\delta-$ to the bonds.

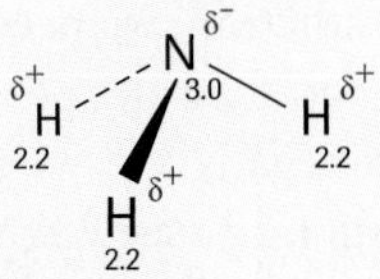

Draw in the bond dipoles.

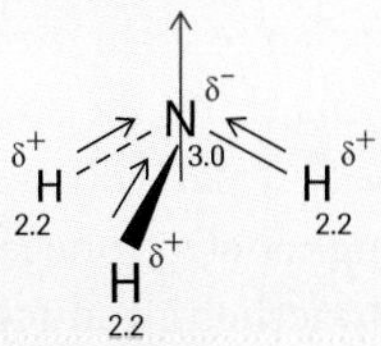

The ammonia molecule is polar because it has bond dipoles that do not balance (cancel). Each of the bond dipoles shown is angled upward as well as sideways. The molecule is symmetrical sideways (viewed from above), but not vertically (viewed from the side), so there is a resulting molecular dipole, oriented vertically. An ammonia molecule is polar, slightly positive on the side with the hydrogens, and slightly negative on the opposite side.

SUMMARY — Theoretical Prediction of Molecular Polarity

To use molecular shape and bond polarity to predict the polarity of a molecule, complete these steps:

Step 1: Draw a Lewis formula for the molecule.
Step 2: Use the number of electron pairs and VSEPR rules to determine the shape around each central atom.
Step 3: Use electronegativities to determine the polarity of each bond.
Step 4: Add the bond dipole vectors to determine whether the final result is zero (nonpolar molecule) or nonzero (polar molecule).

Practice

12. Predict the shape of the following molecules. Provide Lewis formulas and stereochemical formulas.
    (a) silicon tetrabromide, $SiBr_4(l)$
    (b) nitrogen trichloride, $NCl_3(l)$
    (c) boron trifluoride, $BF_3(g)$
    (d) sulfur dichloride, $SCl_2(l)$

13. Predict the bond polarity for the following bonds. Use a diagram that includes the partial negative and positive charges and direction of the bond dipole:
    (a) C≡N in hydrogen cyanide
    (b) N=O in nitrogen dioxide
    (c) P–S in $P(SCN)_3(s)$
    (d) C–C in $C_8H_{18}(l)$

14. Predict the polarity of the following molecules. Include a stereochemical formula, bond dipoles, and the final resultant dipole (if nonzero) of the molecule.
    (a) boron trifluoride, $BF_3(g)$
    (b) oxygen difluoride, $OF_2(g)$
    (c) carbon disulfide, $CS_2(l)$
    (d) phosphorus trichloride, $PCl_3(l)$

15. Use the empirical rules from **Table 8**, page 99, to predict the polarity of an octane, $C_8H_{18}(l)$, molecule. Explain your answer without drawing the molecule.

16. Use bonding theory to predict the polarity of hydrogen sulfide, $H_2S(g)$, a toxic gas with a rotten-egg odour. Design an experiment to test your prediction.

Learning Tip

Parallel empirical and theoretical ways of knowing are common in chemistry. For example, the polarity of the simplest molecules can be predicted from either empirical or theoretical rules.

Polar Substances

Whether or not a substance has polar molecules will necessarily affect how it interacts with other substances. For example, solubility evidence consistently shows that "like dissolves like." We find experimentally that polar compounds are soluble in polar solvents, and nonpolar compounds are soluble in nonpolar solvents. Research indicates that greases and oils are nonpolar substances, and water is an extremely polar liquid. Therefore, water is not very good at dissolving oil or grease. This can be a difficult problem when trying to remove an oily stain from clothing (**Figure 16**). On the other hand, we often use greases and oils as protective coatings *because* they repel water and therefore keep metal parts from rusting. Aboriginal peoples knew that waterproofing leather clothing was easily done by rubbing animal fat into the surface. Modern silicone water repellent sprays use synthetic molecules to accomplish the same end.

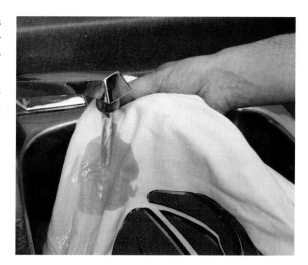

Figure 16
Polar water is useless for removing a nonpolar oil stain.

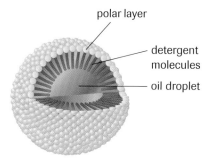

Figure 17
Long, nonpolar sections of detergent molecules are attracted to (dissolve in) a tiny oil droplet. The polar end of each of these detergent molecules helps form a polar "layer" around the droplet, which attracts polar water molecules. This allows them to pull the oil droplet away from a stained area of fabric and hold it suspended in the wash water.

Cleaning clothing is a matter of concern to everyone. Soap has been made (from lye and animal fats) and used for all of recorded history, because it helps make oily dirt dissolve in wash water. It does not work well in cold or hard water, though. By learning the molecular structure of soap, scientists were able to create new molecules that would do this job better. Detergents are artificially created molecules with structures that are very long chains of atoms. Most of the length of a detergent molecule is nonpolar and will mix easily with the oil in a stain. At one end of a detergent molecule, however, are many highly polar bonds; that end is attracted strongly by water. The net result is a process that pulls the oil away from the fabric and suspends it in the wash water as tiny droplets (**Figure 17**), another science research and development success story.

In the other cleaning process, called dry cleaning, clothing is washed with a nonpolar liquid solvent that dissolves the oil directly. (Dry cleaning is not really dry; it just doesn't use water.) This is necessarily a commercial process because it involves toxic solvents and specialized equipment.

Mixing nonpolar and polar liquids results in them forming layers, with the least dense liquid on top. This occurs because the polar molecules attract each other more strongly; thus they stay close together, excluding the nonpolar molecules (**Figure 18**).

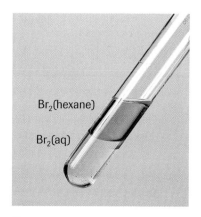

Figure 18
Two clear liquids formed layers in this tube: nonpolar hexane, $C_6H_{14}(l)$, on top, and polar water, $H_2O(l)$, below. Nonpolar dark orange liquid bromine, $Br_2(l)$, was then added. The bromine dissolves much more readily in the nonpolar hexane.

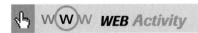

Web Quest—Cleaning Up Dry Cleaning

An organic solvent called perchloroethylene, or PERC, is widely used by the dry-cleaning industry because both PERC and most stains are nonpolar. Like so many other organic solvents, however, PERC can be harmful to both the environment and human health. This Web Quest requires you to look into the chemistry, health, and legislation of the dry-cleaning industry.

www.science.nelson.com

Section 3.3 Questions

You can predict the structural formula for any molecule that has more than one central atom by assuming that all atoms (except hydrogen) follow the octet rule and by using the bonding capacity of those atoms.

1. Use Lewis formulas and VSEPR theory to predict the molecular shape of the following molecules. Include a stereochemical formula for each molecule.
   (a) $H_2S(g)$, hydrogen sulfide (poisonous gas)
   (b) $BBr_3(l)$, boron tribromide (density of 2.7 g/mL)
   (c) $PCl_3(l)$, phosphorus trichloride
   (d) $SiBr_4(l)$, silicon tetrabromide
   (e) $BeI_2(s)$, beryllium iodide (soluble in $CS_2(l)$)

2. Use Lewis formulas and VSEPR theory to predict the molecular shape around the central atom(s) of each of the following molecules. Provide stereochemical formulas.
   (a) $CS_2(l)$, carbon disulfide (solvent)
   (b) $HCOOH(g)$, formic acid (lacquer finishes)
   (c) $N_2H_4(l)$, hydrazine (toxic; explosive)
   (d) $H_2O_2(l)$, hydrogen peroxide (disinfectant)
   (e) $CH_3CCCH_3(l)$, 2-butyne (reacts rapidly with bromine)

3. Draw the Lewis formula and describe the shape of each of the following ions:
   (a) IO_4^- (b) SO_3^{2-} (c) ClO_2^-

4. Scientific concepts are tested by their ability to explain current observations and predict future observations. To this end, explain why the following molecules are polar or nonpolar, as indicated by the results of diagnostic tests:
   (a) beryllium bromide, $BeBr_2(s)$; nonpolar
   (b) nitrogen trifluoride, $NF_3(g)$; polar
   (c) methanol, $CH_3OH(l)$; polar
   (d) hydrogen peroxide, $H_2O_2(l)$; polar
   (e) ethylene glycol, $C_2H_4(OH)_2(l)$; polar

5. Predict the polarity of the following molecules:
   (a) dichlorofluoromethane, $CHFCl_2(g)$
   (b) ethene, $C_2H_4(g)$
   (c) chloroethane, $C_2H_5Cl(g)$
   (d) methylamine, $CH_3NH_2(g)$
   (e) ethanol, $C_2H_5OH(l)$

6. Polar substances may be used in a capacitor, a device for storing electrical energy. For example, a capacitor may store enough electrical energy to allow you to change the battery in your calculator without losing data in the memory. Based upon polarity alone, is water or pentane, $C_5H_{12}(l)$, better for use in a capacitor? Provide your reasoning.

7. Search the Internet for information on the current workplace and position of Dr. Ronald Gillespie, the co-creator of VSEPR theory. What degrees does he hold? What are some of the awards that he has won? What is his major topic of research?

   www.science.nelson.com

8. Some scientists argue that taste has developed as a protective mechanism. Many poisonous molecules taste bitter, and ones that are useful to us have a more pleasant, often sweet, taste. On the Internet, research the relationship between taste and molecular structure, and write a brief summary.

   www.science.nelson.com

9. Various consumer products and books exist to help people remove greasy stains from clothing, carpets, etc. Discuss how a knowledge of polar and nonpolar substances is related to the removal of these kinds of stains.

10. For each of the following pairs of atoms, label the electronegativity and bond polarity. In addition, classify the bond as nonpolar covalent, polar covalent, or ionic.
    (a) Cl and Cl   (d) O and H
    (b) K and I   (e) Mg and O
    (c) P and Cl   (f) Xe and F

Extension

11. Astronomers have detected an amazing variety of molecules in interstellar space.
    (a) One interesting molecule is cyanodiacetylene, HC_5N. Draw a structural formula and predict its shape.
    (b) How do astronomers detect molecules in space?

    www.science.nelson.com

12. Values for the Pauling electronegativities have changed over time. Why would these values change? Are the new values the "true" values?

Intermolecular Forces 3.4

There are many physical properties that demonstrate the existence of **intermolecular forces**, the forces of attraction and repulsion *between molecules*. (This term is not to be confused with *intramolecular*, meaning *within a molecule*, a term that is sometimes used to refer to the covalent bonds that hold a molecule together.) For example, chemists suggest that a simple property like "wetting" depends to a large extent on intermolecular forces. Consider cotton: The molecules that make up cotton can form many intermolecular attractions with water molecules. Therefore, cotton is very good at absorbing water and easily gets wet. You would not want a raincoat made from cotton. On the other hand, rubber and plastic materials do not absorb water because there is little intermolecular attraction between water molecules and the molecules of the rubber or plastic (**Figure 1**). Rubber and plastics may not be comfortable materials to wear. Alternatively, you can use cotton if it is treated with a water repellent, a coating that has little attraction to water molecules. As you can see, the development of water repellents requires a good knowledge of intermolecular forces.

Figure 1
Cotton (left) will absorb a lot more water than polyester, because the molecules in cotton are better able to attract and hold water molecules.

Another property that demonstrates the existence of an intermolecular force is capillary action. Capillary action allows trees to move water from the ground to the leaves (**Figure 2**). We will look at this and other properties in terms of intermolecular forces.

Van der Waals Forces

In 1873, Johannes van der Waals suggested that there must be a reason that all gases will condense if cooled. Although molecules in gases act as if they do not affect each other, when the molecular motion is slowed enough, the molecules collect together and form a liquid. Van der Waals assumed that the molecules of a gas must have a small but definite volume, and the molecules must exert weak attractive forces on each other. These forces are often simply referred to as **van der Waals forces**. It is now believed that in many substances, van der Waals forces are actually a combination of many types of intermolecular forces. In general, intermolecular forces vary over a much wider range and are always considerably weaker than the covalent bonds inside a molecule.

The evidence for this comparison comes primarily from experiments that measure bond energies. For example, it takes much less energy to boil water (overcoming intermolecular forces) than it does to decompose water (breaking covalent bonds), even though the intermolecular forces in water are some of the strongest known.

$$H_2O(l) \rightarrow H_2O(g) \quad\quad 41 \text{ kJ/mol}$$
$$H_2O(l) \rightarrow H_2(g) + \frac{1}{2}O_2(g) \quad 286 \text{ kJ/mol}$$

Figure 2
Research shows that water is transported in thin, hollow tubes in the trunk and branches of a tree. This is accomplished by several processes, including capillary action. Intermolecular bonding theory suggests that it is necessary that water molecules attract each other to maintain a continuous column of water. They must also attract the molecules in the walls of the tiny tubes in the wood.

Figure 3
Sucrose, $C_{12}H_{22}O_{11}(s)$

Chemists theorize that intermolecular forces are responsible for what we physically observe about molecular substances. For example, to have a sample of sucrose (table sugar) (**Figure 3**) that is big enough to see, there must be trillions of trillions of molecules present, all attracting each other to hold the sample together. These trillions of forces are strong enough to create a crystalline solid structure from the molecules.

Intermolecular forces also appear to control the physical behaviour of molecular substances. In the case of sugar, water dissolves it extremely well but alcohol does not. How do we explain this? We can heat sugar to try to break the bonds between its molecules, but that doesn't work. When heated to a temperature of about 170 °C, long before the molecules start to separate completely from each other (called vaporization, or boiling), the covalent bonds in the molecules start to break, and we don't have sugar any more. However, there is a change—a chemical reaction (decomposition)—that creates totally new substances.

Now we continue the study of intermolecular forces and their effect on several other physical properties of substances, such as boiling point, rate of evaporation, and surface tension. We will interpret these properties using intermolecular forces, but there are other factors that also affect these properties. To minimize this problem, we will try to compare simple, similar substances and do only qualitative comparisons. Before we try to tackle this problem, we need to look at various kinds of intermolecular forces.

Dipole–Dipole Force

In Investigation 3.2, you learned how to test a stream of liquid to see whether the molecules of the liquid are polar. You also learned, in Section 3.3, how to predict whether a molecule was polar or nonpolar. (Recall that polar molecules have dipoles—oppositely charged sides.) Attraction between dipoles is called the **dipole–dipole force** and is thought to be due to a simultaneous attraction between any dipole and surrounding dipoles (**Figure 4**). Dipole–dipole forces are among the weakest intermolecular forces; nonetheless, they can still control important properties. For instance, they have a pronounced effect on the ability of solvents to dissolve some solutes and not others.

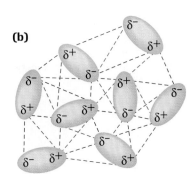

Figure 4
(a) Oppositely charged sides of polar molecules attract.
(b) In a liquid, polar molecules can move and rotate to maximize attractions and minimize repulsions. The net effect is greater overall attraction.

London Force

After repeated failures to find any pattern in physical properties like the boiling points of polar substances, Fritz London (**Figure 5**) suggested that the van der Waals force was actually two forces: the dipole–dipole force and what we now call the **London force** (or dispersion force). The evidence for this was that substances with nonpolar molecules would still liquefy upon cooling; something other than dipole–dipole attraction had to be pulling such molecules together. London knew that any forces must be due to the electrostatic forces acting between positive nuclei and negative electrons. He reasoned that, although electron distribution in a molecule might *average* to a zero dipole, any electron movement *within* the molecule could still produce a *momentary dipole*, one that would last for just the instant that the electrons were not distributed perfectly evenly. London also showed that momentary dipoles occurring in adjacent molecules would result in an overall attraction.

A full treatment of London's theory is too involved for the scope of this book. The key point is that the more electrons a molecule has, the more easily momentary dipoles will form, and the greater the effect of the London force will be. Note that London force is necessarily present between *all* molecules, whether or not any other types of attractions are present. Because weak momentary dipoles cause effective attraction only over very short distances, the shape of larger molecules also has an effect on London force.

Figure 5
German-American physicist Fritz London (1900–1954). Linus Pauling indicated that London made "the greatest single contribution to the chemists conception of valence" since Lewis created the concept of the shared pair of electrons. Pauling was referring to the use of quantum mechanics by London to describe the covalent bond in the hydrogen molecule.

SUMMARY: Dipole–Dipole and London Forces

- The dipole–dipole force is due to the simultaneous attraction between any one dipole and all surrounding dipoles.
- The strength of the dipole–dipole force is dependent on the overall polarity of the molecule.
- The London force is due to the simultaneous attraction between a momentary dipole in a molecule and the momentary dipoles in the surrounding molecules.
- The strength of the London force is directly related to the number of electrons in the molecule, and inversely related to the distance between the molecules.

DID YOU KNOW?
A Synthesis
Scientists are always seeking a synthesis of different explanations. For example, all bonding can be explained by referring to simultaneous electrostatic attractions. The covalent bond involves a simultaneous attraction of a pair of negative electrons by two positive nuclei. An ionic bond is the simultaneous attraction of any two oppositely charged ions. Note, also, the theoretical definitions of dipole–dipole force and London force in the Summary.

Using Dipole–Dipole and London Forces to Predict Boiling Points

Let's take a look at the boiling points of Group 14 hydrogen compounds (**Table 1**). Based on VSEPR and molecular polarity theories, we would expect these molecules to be nonpolar, based on their four equivalent bonds and their tetrahedral shape.

Table 1 Boiling Points of Group 14 Hydrogen Compounds

Compound (at SATP)	Electrons	Boiling point (°C)
$CH_4(g)$	10	−164
$SiH_4(g)$	18	−112
$GeH_4(g)$	36	−89
$SnH_4(g)$	54	−52

In **Table 1**, you can see that as the number of electrons (and protons) in the molecule increases (from 10 to 54), the boiling point increases (from −164 °C to −52 °C). *Since boiling point measures how difficult it is to separate the molecules completely from each other, we assume it is a good relative measure of intermolecular force.* The evidence presented in **Table 1** supports London's theory, and provides a generalization for describing, explaining, and predicting the relative strength of London forces between molecules. For simplicity, we just count the electrons in a molecule as an indicator of relative London force strength.

Table 2 Boiling Points of Hydrocarbons

Alkane	Boiling point (°C)
methane	−162
ethane	−89
propane	−42
butane	−0.5

> ### SAMPLE problem 3.6
>
> Use London force theory to predict which of these hydrocarbons has the highest boiling point: methane (CH_4), ethane (C_2H_6), propane (C_3H_8), or butane (C_4H_{10}).
>
> According to intermolecular force theory, butane should have the highest boiling point. The reasoning behind this prediction is that, according to empirical rules, all of these molecules are nonpolar, but butane has the most attractive London force, because it has the greatest number of electrons in its molecules. This prediction is verified by the evidence shown in **Table 2**.

Isoelectronic molecules, that is, molecules with the same number of electrons, are predicted to have the same or nearly the same strengths for the London force of intermolecular attraction, provided the shapes of the molecules are similar. Isoelectronic molecules help us study intermolecular forces. For example, if one of two isoelectronic substances is polar and the other is nonpolar, then the polar molecule should have a higher boiling point, as shown in Sample Problem 3.7.

Table 3 Isoelectronic Substances

Substance	Electrons	Boiling point (°C)
$Br_2(l)$	70	59
$ICl(l)$	70	97

> ### SAMPLE problem 3.7
>
> Consider two isoelectronic molecules, bromine (Br_2) and iodine monochloride (ICl). Based upon your knowledge of intermolecular forces, explain the difference in the boiling points of the two substances: bromine, 59 °C, and iodine monochloride, 97 °C.
>
> Both bromine and iodine monochloride have 70 electrons per molecule (**Table 3**). Therefore, the strength of the London forces between molecules of each should be about the same. Bromine is nonpolar and, therefore, has only London forces between its molecules. Iodine monochloride is polar, which means it also has dipole–dipole forces between its molecules, in addition to London forces. This extra attraction between ICl molecules produces a higher boiling point. Both molecules are diatomic, with similar shape; so shape should not be a factor affecting London force. All atoms in each molecule have complete octets.

SUMMARY: Predicting with Dipole–Dipole and London Forces

- Isoelectronic molecules of similar shape have approximately the same strength of London force between them.
- If all other factors are equal, then
  — the more polar the molecule, the stronger the dipole–dipole force and, therefore, the higher the boiling point
  — the greater the number of electrons per molecule, the stronger the London force and, therefore, the higher the boiling point
- You can explain and predict the relative boiling points of two substances if
  — the London force is the same, but the dipole–dipole force is different
  — the dipole–dipole force is the same, but the London force is different
  — both the London force and the dipole–dipole force are greater for one substance
- You cannot explain and predict the relative boiling points of two substances if
  — one of the substances has a stronger dipole–dipole force, and the other substance has a stronger London force
  — the molecules of the two substances differ significantly in shape
  — the central atom of either molecule has an incomplete octet

Section **3.4**

Canadian Achievers—Gerhard Herzberg

In 1971, Gerhard Herzberg (**Figure 6**) became the first Canadian to win the Nobel Prize in Chemistry. In announcing the prize, the Swedish Academy stated that Herzberg's ideas and discoveries stimulated "the whole modern development of chemistry from chemical kinetics to cosmochemistry."

1. What technological device was central to Herzberg's research?
2. What is a free radical?
3. Herzberg was a physicist; why was he awarded the Nobel Prize in Chemistry?

www.science.nelson.com

Figure 6
Gerhard Herzberg (1904–1999)

DID YOU KNOW

Canadian Nobel Laureates in Chemistry
Gerhard Herzberg (1971)
Henry Taube (1983)
John Polanyi (1986)
Sidney Altman (1989)
Rudolph Marcus (1992)
Michael Smith (1993)

www.science.nelson.com

▶ Practice

1. Using London forces and dipole–dipole forces, state the kind of intermolecular force(s) present between molecules of the following substances in their liquid or solid state:
   (a) water (solvent)
   (b) carbon dioxide (dry ice)
   (c) methane (in natural gas)
   (d) ethanol (beverage alcohol)
   (e) ammonia (cleaning agent)
   (f) iodine (disinfectant)

2. Which compound in each of the following pairs has stronger dipole–dipole forces than the other in their liquid or solid state? Provide your reasoning.
   (a) hydrogen chloride or hydrogen fluoride
   (b) chloromethane, $CH_3Cl(g)$, or iodomethane, $CH_3I(l)$
   (c) nitrogen tribromide or ammonia
   (d) water or hydrogen sulfide

3. Based upon London force theory, which of the following pure substances has the stronger London forces in their liquid or solid state? Provide your reasoning.
   (a) methane, CH_4, or ethane, C_2H_6
   (b) oxygen or nitrogen
   (c) sulfur dioxide or nitrogen dioxide
   (d) methane or ammonia

4. Based upon dipole–dipole and London forces, predict, where possible, which molecular substance in the following pairs has the higher boiling point. Provide your reasoning.
   (a) boron trifluoride or nitrogen trifluoride
   (b) chloromethane or ethane

5. Why is it difficult to predict whether NF_3 or Cl_2O has the higher boiling point?

6. Using a chemical reference, look up the boiling points for the substances in questions 4 and 5. Evaluate your predictions for question 4.

Learning Tip

In the liquid state, molecules are held close together by intermolecular forces. Boiling (or vaporizing) a liquid means adding enough energy to overcome these forces and to separate the molecules to positions far from each other in the gas state. The temperature at which a liquid boils reflects the strength of the intermolecular forces present among the molecules. A higher boiling point temperature means more energy has to be added, and thus the intermolecular forces must be stronger.

LAB EXERCISE 3.A

Boiling Points and Intermolecular Forces

Report Checklist
- ○ Purpose
- ○ Problem
- ● Hypothesis
- ○ Prediction
- ○ Design
- ○ Materials
- ○ Procedure
- ○ Evidence
- ● Analysis
- ● Evaluation (2, 3)

In this investigation, you will state a hypothesis about the trend in boiling points of compounds containing elements within and between Groups 14 to 17. Your Hypothesis should refer to dipole–dipole and London forces as part of the general reasoning. It could include a general sketch of a predicted graph of boiling point versus number of electrons per molecule. For Analysis of the Evidence, plot the data for each group as a graph of boiling point versus number of electrons per molecule, using the same axes for all four groups. Evaluate your hypothesis as well as the concept of intermolecular forces used to make the hypothesis. Note, and suggest a revised hypothesis to explain, any anomalies (unexpected results).

Purpose
The scientific purpose of this lab exercise is to test a new hypothesis based on the theory and rules for London and dipole–dipole forces.

Problem
What is the trend in boiling points of the hydrogen compounds of the elements of each of the periodic table Groups 14 to 17 (**Table 4**)?

Evidence
See **Table 4**.

Table 4 Boiling Points of the Hydrogen Compounds of Elements in Groups 14–17

Group	Hydrogen compound	Boiling point (°C)
14	$CH_4(g)$	−162
	$SiH_4(g)$	−112
	$GeH_4(g)$	−89
	$SnH_4(g)$	−52
15	$NH_3(g)$	−33
	$PH_3(g)$	−87
	$AsH_3(g)$	−55
	$SbH_3(g)$	−17
16	$H_2O(l)$	100
	$H_2S(g)$	−61
	$H_2Se(g)$	−42
	$H_2Te(g)$	−2
17	$HF(g)$	20
	$HCl(g)$	−85
	$HBr(g)$	−67
	$HI(g)$	−36

INVESTIGATION 3.3 Introduction

Molecular Compound Melting Points

Report Checklist
- ○ Purpose
- ○ Problem
- ○ Hypothesis
- ○ Prediction
- ○ Design
- ○ Materials
- ○ Procedure
- ● Evidence
- ● Analysis
- ● Evaluation (1, 2, 3)

In this investigation, you will accurately measure the melting point of a molecular compound. You will then combine your laboratory evidence with that of other research groups to see if a hypothesized relationship between melting point and molecular size is supported.

Purpose
The purpose of this investigation is to test the hypothesis of a relationship between melting point and molecular size.

Hypothesis
The melting point of similar molecular compounds increases with increased molecular size, as described by the total number of electrons in the molecule.

Problem
What is the relationship between molecular size and melting point?

Design
Hot and cold water baths are used to melt a solid compound and then to freeze and remelt it, with the temperature measured at regular intervals. The phase change (freezing point/melting point) temperature is found from superimposed temperature-versus-time graphs of this evidence. This is combined with evidence from other research groups for a second similar compound and with information from other references to test the Hypothesis.

To perform this investigation, turn to page 132.

Hydrogen Bonding

The graph you drew for Lab Exercise 3.A makes it very evident that ammonia, water, and hydrogen fluoride have anomalously high boiling points, far higher than would be predicted from their size or polarity. Some significant attraction effect other than dipole–dipole and London forces (and stronger than either) must be acting between the molecules of these compounds.

In 1920, Maurice Huggins (a graduate student of Lewis), Wendell Latimer, and Worth Rodebush devised the concept of a **hydrogen bond**, where a hydrogen nucleus (a proton) could be shared between pairs of electrons on adjacent molecules. A hydrogen bond can be thought of as a very special type of polarity that results in an unusually strong intermolecular force. For hydrogen bonding to occur, two things must be simultaneously true about a molecular structure.

- First, a hydrogen atom must be covalently bonded to another very electronegative atom, one that can pull the hydrogen's electron strongly away from the proton that is its nucleus. The proton is then said to be "unshielded," meaning that another molecule's lone pair can approach this proton much more closely than before, on the side opposite its covalent bond.

- Second, there must also be at least one lone pair of electrons on the atom bonded to the hydrogen. This sets up a condition where the lone pair on one molecule can attract the unshielded proton on the next molecule (**Figure 7**). The proton has a pair of electrons on one side shared in a covalent bond, and a more distant pair on the other side shared in the much weaker (intermolecular) hydrogen bond.

In practice, only three possible structures show hydrogen bonding, and one case is trivial because it only includes one possible compound: HF(g). The other two cases are extremely important, however. *Any molecule with an –OH bond or an –NH bond in any part of its structure will show hydrogen bonding.* That includes most of the millions of different molecules found in living things. It is important to note that hydrogen bonding is not a "bond" in the sense that a covalent bond is. For substances in liquid state, hydrogen bonds are momentary attractive forces between passing mobile molecules. Hydrogen bonds only act as continuous bonds between molecules in solids, where the molecules are moving slowly enough to be locked into position. Hydrogen force would have been a better name for this intermolecular force.

For a common example of some effects of hydrogen bonding, consider table sugar again (**Figure 8**). The formula is often written $C_{12}H_{22}O_{11}(s)$ for brevity, but that doesn't give any hint about the structure. Suppose we now write the formula in a less condensed form, as $C_{12}H_{14}O_3(OH)_8(s)$, to show that this molecule has no fewer than eight –OH groups in its structure. Now the question of what causes the strong intermolecular attractions (that make this substance a crystalline solid) is easily explained: Every sucrose molecule has multiple hydrogen bonds holding it firmly to the surrounding sucrose molecules. In addition, the extreme solubility of sugar in water is explained as being due to the formation of many new hydrogen bonds to water molecules.

Additional evidence for hydrogen bonding can be obtained by looking at energy changes associated with the formation of hydrogen bonds. Chemists theorize that endothermic and exothermic reactions are explained by the difference between the energy absorbed to break bonds in the reactants and the energy released when new bonds in the products are formed. For example, in the exothermic formation of water from its elements, more energy is released in forming the new O—H bonds than was required to break the old H—H and O=O bonds. (See equations on page 105.) In a sample of glycerol, you would expect some hydrogen bonding between glycerol molecules. However, these molecules are rather bulky, and their shape limits the number of

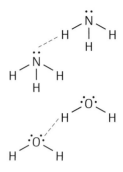

Figure 7
A hydrogen bond (----) occurs when a hydrogen atom bonded to a strongly electronegative atom is attracted to a lone pair of electrons in an adjacent molecule. Another way of thinking of this is that a hydrogen nucleus (proton) is simultaneously attracted to two pairs of electrons; one closer (in the same molecule) and one further away (on the next molecule).

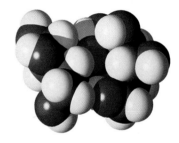

Figure 8
Every white-to-red part of this sucrose model represents an –OH group. Theory suggests that each of these is a site where as many as three hydrogen bonds to this molecule could form.

Learning *Tip*

Don't be confused about the difference between intermolecular and intramolecular (covalent) bonds. Intermolecular bonds act *between* molecules, and are weaker than the covalent bonds that act *within* molecules (or polyatomic ions). The hydrogen nucleus in any hydrogen bond has an intramolecular (covalent) bond on one side and an intermolecular (hydrogen) bond on the other.

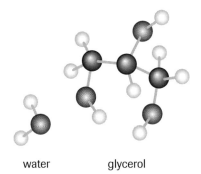

water   glycerol

Figure 9
Water and glycerol molecules. Glycerol is commonly called glycerine; it is an ingredient in many soaps and cosmetics.

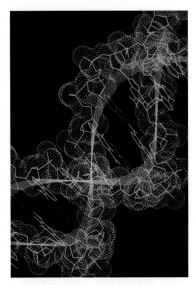

Figure 10
According to Francis Crick, co-discoverer with James D. Watson of the DNA structure, "If you want to understand function, study structure." Hydrogen bonding (blue dashes) explains the shape and function of the DNA molecule. The interior of the double helix is cross-linked by hydrogen bonds.

EXTENSION

Water Transportation in Plants

Why are intermolecular forces so important to plants? Without the forces that result in water's physical properties, plants as we know them could not exist. Find out more about plants and intermolecular forces.

possible hydrogen bonds that can form. If water is mixed with glycerol (**Figure 9**), theory suggests that additional hydrogen bonds should be possible. The small size of water molecules should allow water to form many new hydrogen bonds to each glycerol molecule. Experimentally, you find that mixing water with each glycerol is an exothermic process, indicating the formation of new bonds. The temperature increase upon mixing is evidence that verifies the prediction from theory.

The theory of hydrogen bonding is necessary to explain the functions of biologically important molecules. Proteins are chains of amino acids, and amino acids have $-NH_2$ and $-COOH$ structural groups, both of which fulfill the conditions for hydrogen bonding. Similarly, the double helix of the DNA molecule owes its unique structure largely to hydrogen bonding. The central bonds that hold the double helix together are hydrogen bonds (**Figure 10**). If the helix were held together by covalent bonds, the DNA molecule would not be able to unravel and replicate because the bonds would be too strong to break and reform (react) at body temperature.

Unlike nearly all other substances on Earth, water expands rather than contracts when it freezes. This means that the solid form (ice) is less dense than the liquid and will float. Thus, a body of water freezes from the surface down, rather than from the bottom up. When an ice layer forms on the surface, it insulates the liquid beneath, so that "frozen" lakes, rivers, and oceans remain mostly liquid—a very good thing for everything living in them! The reason ice is less dense than water has to do both with hydrogen bonding and the angular shape of water molecules (**Figure 11**). The hydrogen bonds hold water molecules in a hexagonal lattice with open space in the centre, explaining the low density. This also explains why individual snowflakes are hexagonal in structure: the hexagonal lattice repeats throughout each one.

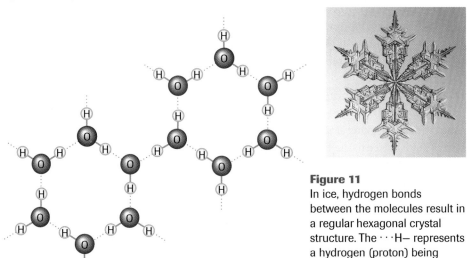

Figure 11
In ice, hydrogen bonds between the molecules result in a regular hexagonal crystal structure. The ···H— represents a hydrogen (proton) being shared unequally between two pairs of electrons.

WEB Activity

Web Quest—Cloud Seeding

Attempting to control natural forces is a goal for many researchers because the possible benefits are tremendous. Seeding clouds, for example, uses properties of chemical bonding to change how much, where, and when precipitation will fall. Unfortunately, changing a natural system may have unforeseen consequences. This Web Quest will lead you to explore cloud seeding and its results, both expected and unexpected!

INVESTIGATION 3.4 *Introduction*

Hydrogen Bonding

Breaking bonds absorbs energy, and forming bonds releases energy. Forming new bonds within a liquid should produce a temperature effect. For this investigation, write a Prediction that includes a bonding concept. Add to the design by identifying the manipulated and responding variables as well as the major controlled variable.

Purpose
The purpose of this investigation is to test the concept of hydrogen bonding.

Report Checklist
- ○ Purpose
- ○ Problem
- ○ Hypothesis
- ● Prediction
- ○ Design
- ○ Materials
- ● Procedure
- ● Evidence
- ● Analysis
- ● Evaluation (1, 2, 3)

Problem
How does the temperature change upon mixing ethanol, $C_2H_5OH(l)$, with water compare with the temperature change upon mixing glycerol (glycerin), $C_3H_5(OH)_3(l)$, with water?

Design
Equal volumes of ethanol and water, and then of glycerol and water, are mixed. The change in temperature is recorded in each case.

To perform this investigation, turn to page 134.

Physical Properties of Liquids

Liquids have a variety of physical properties that can be explained by intermolecular forces. As you have seen, comparing boiling points provides a relatively simple comparison of intermolecular forces in liquids, if we assume that the gases produced have essentially no intermolecular forces between their molecules. What about some other properties of liquids, such as surface tension, shape of a meniscus, volatility, and capillary action? Surface tension is pretty important for water insects (**Figure 12**). The surface tension on a liquid seems to act like an elastic skin. Molecules within a liquid are attracted by other molecules in all directions equally, but molecules right at the surface are only attracted downward and sideways (**Figure 13**). This means that the surface tends to stay intact. Consistent with this theory, substances containing molecules with stronger intermolecular forces have higher surface tensions. Water is a good example—it has one of the highest surface tensions of all liquids.

The shape of the meniscus of a liquid and capillary action in a narrow tube are both thought to be due to intermolecular forces. In both cases, two intermolecular attractions need to be considered—attraction between like molecules (called cohesion) and attraction between unlike molecules (called adhesion). Both cohesion and adhesion are intermolecular attractions. For example, compare water and mercury, two very different liquids: water rises in a narrow tube, but mercury does not (**Figure 14**). The adhesion between the water and the glass is thought to be greater than the cohesion between the water molecules. In a sense, the water is pulled up the tube by the intermolecular forces between the water molecules and the glass.

Another property connected to intermolecular forces is the volatility of a liquid, that is, how rapidly it tends to evaporate. Gasoline is a mix of nonpolar hydrocarbon liquids that evaporate very readily because their intermolecular attractions are not very strong. For gasoline, high volatility is a necessary property; it has to vaporize almost instantly when injected into the cylinder of a car engine. Liquids that are both volatile and flammable are always dangerous because the mix of their vapour with air can be explosive.

Figure 12
The weight of the water strider is not enough to overcome the intermolecular forces between the water molecules. This would be like you walking on a trampoline. The fabric of the trampoline is strong enough to support your weight.

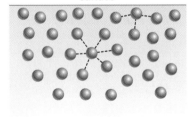

Figure 13
The model suggests that the intermolecular forces on a molecule inside a liquid are relatively balanced. The forces on a molecule right at the surface are not balanced; the net pull is downward, creating surface tension.

Figure 14
Capillary action is the movement of a liquid up a narrow tube. For water, capillary action is very noticeable; for mercury, it is insignificant. (The water is coloured red with dye to make it more visible.)

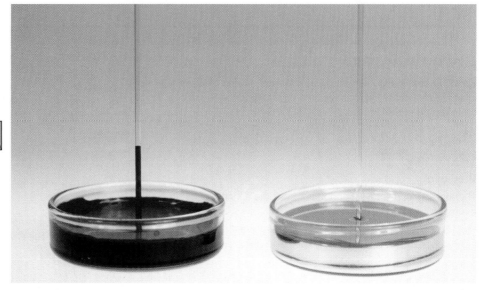

BIOLOGY CONNECTION

Water-Dependent Processes
Water plays a vital role in many biological and geological cycles. If you are studying biology, you will find out about many processes that depend on the chemical and physical properties of water.

www.science.nelson.com

Case Study

Current Research in Intermolecular Forces

In almost any area of science today, experimental work runs parallel to the theoretical work, and there is constant interplay between the two areas. In Canada, there are several theorists whose research teams examine the forces between atoms and molecules, to increase our understanding of the properties of substances. One such individual is Dr. Robert J. Le Roy (**Figure 15**), currently working in theoretical chemical physics at the University of Waterloo, in Ontario.

Figure 15
Dr. Robert J. Le Roy and his research team study intermolecular forces and the behaviour of small molecules and molecular clusters; develop methods to simulate and analyze the decomposition of small molecules; and create computer models to simulate and predict molecular properties.

Dr. Le Roy's field of interest is intermolecular forces. He uses quantum mechanics and computer modelling to define and analyze the basic forces between atoms and molecules. Early in his career, Dr. Le Roy developed a criterion for defining a radius for a small molecule, now known as the Le Roy radius. This defines a theoretical boundary around the molecule. For atomic separations inside this boundary, chemical bonding and electron exchange between atoms are most important; outside this boundary, the classical "physical" interactions between charge distributions predominate. In his work, the study of atomic and molecular spectra (called spectroscopy) plays a crucial role. Measurements from spectroscopy help theoreticians develop better models and theories for explaining molecular structure. Computer programs that Dr. Le Roy developed for converting experimental data information on intermolecular forces, shape, and structure (and vice versa) are freely available on his Web site, and are widely used in laboratories around the world.

It is important not to assume that theories of molecular forces and structures are well established. Our knowledge of bonding and structure becomes scantier and more unreliable for larger molecular structures. A huge amount of research remains to be done if we are ever to be able to describe bonding and structure very accurately for complex substances. As Dr. Le Roy states on his Web site, "… except for the simplest systems, our knowledge of (interactions between molecules) is fairly primitive." The classic example is our poor understanding of the forms and activity of proteins—the stuff of life. Proteins are extremely large molecules with highly complex shapes. We know the atomic composition of many proteins quite precisely, but protein reactions depend largely on how bonding has folded and shaped the protein's structure. How a protein *behaves* depends on its precise bonding, structure, and shape, which is something scientists describe with the common scientific phrase "not well understood."

Living things build proteins in multi-step chemical reactions that bond together long chains of small molecules called amino acids. There are about 25 different amino acids that make up human proteins, of which only about half are synthesized in the body. One fascinating thing about bonding between amino acids is that almost all of them are chiral. This means they can exist in two different structures that are identical, except that the two forms are mirror images of each other, like your hands. In fact, such molecules are often called left- or right-handed to identify which form is being discussed. Understanding how this difference in form can affect intermolecular forces is critical, because in living things only one form of such a molecule will "work" in a given chemical reaction. For example, all the amino acids in the human body are "left-handed," and all the simple sugars (such as glucose) are "right-handed" (**Figure 16**). Only left-handed amino acids can bond together to build proteins, and only right-handed sugars can be digested for energy. The two forms of any chiral substance will rotate beams of polarized light in opposite directions, so polarized light microscopes can be useful for determining molecular and crystal structures (see **Figure 2** in Section 3.1, page 78).

Current research in this area is being done by Dr. Yunjie Xu, working in the Chemistry Department of the University of Alberta (**Figure 17**). She is studying the precise mechanisms, at a molecular level, that enable one molecule to accept or reject bonding to another molecule, based on its specific chiral structure. This requires learning exactly where on the molecules various intermolecular forces come into play and how a molecule's orientation in space plays a part in this process. Dr. Xu uses very sophisticated equipment and techniques to work at the molecular level.

Figure 16
This computer-generated model of a glucose molecule is shown in its "right-handed" form, as produced by living plants. An old name for this simple sugar was dextrose, from the Latin word for right. A left-handed form of glucose can be produced in a laboratory but is not found naturally in life on Earth.

Figure 17
Dr. Xu and her research team study intermolecular forces between different forms (isomers) of molecules to better understand how only specific molecules are selected for reaction by living things, and to develop systems to better identify such molecules in chemical samples.

Case Study Questions

1. Explain briefly what the Le Roy radius of a molecule represents, and discuss how it might be useful to someone studying interactions between small molecules.
2. The simplest (smallest) amino acid in your body is glycine, H_2NCH_2COOH. Draw a structural formula for this molecule. Discuss the probable solubility of glycine in terms of intermolecular bonding.

Extension

3. The smallest chiral amino acid is alanine, H_2NCHCH_3COOH. It may be thought of as a glycine where one of the hydrogens attached to the central carbon has been replaced by a carbon (which also has three hydrogens attached to it; a $-CH_3$ group). Draw its structural formula, and note that the central carbon has four *different* things attached to it. *Any* such molecule is chiral, including all the other amino acids. To examine this feature, you can use any molecular model kit that shows 3-D (tetrahedral) bonding to carbon. Make two exact copies of a model by attaching four spheres (all of different colours) to a central sphere representing carbon. Then, on one model, switch any two of the coloured spheres attached to the central carbon. You now have models representing a chiral substance. All the covalent bonds are identical and the molecular formulas are exactly the same, but the orientation in space is different. Draw stereochemical formulas for your models to show more clearly what this difference is. To better appreciate why this might be important to the way molecules fit together, just try to put your left glove on your right hand.

▶ mini *Investigation*   *Floating Pins*

Materials: beaker or glass, water, several other different liquids, dishwashing detergent, straight pin, tweezers, toothpick

- Make sure the straight pin is clean and dry.
- Using clean tweezers, carefully place the pin in a horizontal position on the surface of each liquid, one at a time. Wash and dry the pin between tests.
(a) What happens for each liquid? Why?

- Using tweezers, carefully place the pin vertically into the surface of the water. Try both ends of the pin.
(b) What happens this time? Why do you think the result is different than before?
- Place the pin horizontally onto the surface of water. Using a toothpick, add a small quantity of dish detergent to the water surface away from the pin.
(c) Describe and explain what happens.

Figure 18
Modern spray "protectants" have molecules that join together (polymerize) as the liquid dries to form a super thin layer that bonds very strongly to the surface. The "beading" you see is explained by differences in the attraction of water molecules for each other and the attraction of water molecules for molecules of the surface protectant layer.

Figure 19
A lava lamp

Figure 20
Setup for question 13

Practice

7. For each of the following compounds, evidence indicates that hydrogen bonds contribute to the attraction between molecules. Draw a Lewis formula using a dashed line to represent any hydrogen bond between two molecules of the substance.
   (a) hydrogen peroxide, $H_2O_2(l)$ (disinfectant)
   (b) hydrogen fluoride, $HF(l)$ (aqueous solution etches glass)
   (c) methanol, $CH_3OH(l)$ (wood alcohol)
   (d) ammonia, $NH_3(l)$ (anhydrous ammonia for fertilizer)

8. (a) Refer to or construct a graph of the evidence from Lab Exercise 3.A. Extrapolate the Group 15 and 16 lines to estimate the boiling points of water and ammonia if they followed the trend of the rest of their family members.
   (b) Approximately how many degrees higher are the actual boiling points for water and ammonia compared to your estimate in (a)?
   (c) Explain why the actual boiling points are significantly higher for both water and ammonia.
   (d) Propose an explanation why the difference from (b) is much greater for water than for ammonia.

9. Water beads on the surface of a freshly waxed car hood. Use your knowledge of intermolecular forces to explain "beading" (**Figure 18**).

10. A lava lamp is a mixture of two liquids with a light bulb at the bottom to provide heat and light (**Figure 19**). What interpretations can you make about the liquids, intermolecular forces, and the operation of the lamp?

11. To gather evidence for the existence of hydrogen bonding in a series of chemicals, what variables must be controlled?

12. (a) Design an experiment to determine the volatility (rate of evaporation) of several liquids. Be sure to include a list of variables.
    (b) Suggest some liquids to be used in this experiment. Predict the results. Provide your reasoning.
    (c) What safety precautions should be taken with volatile liquids?

13. Theories are valued by their ability to describe, explain, and predict. Intermolecular bond theory may be tested by investigating the solubility of substances in the laboratory. Complete the Prediction, Analysis, and Evaluation of this investigation report. Evaluate the Prediction as well as the reasoning used to make the Prediction.

    Purpose
    The purpose of this investigation is to test the ability of intermolecular bond theory to predict.

    Problem
    Does ammonia gas have high or low solubility in water?

    Design
    A Florence flask full of ammonia is inverted above a beaker of coloured water. The ammonia flask has a medicine dropper of water inserted through one hole of its rubber stopper. A glass tube through the other hole in the stopper connects the flask to coloured water in the beaker below. The medicine dropper is squeezed to add a small amount of water to the ammonia gas in the flask.

    Evidence
    The coloured water in the beaker is drawn up the tube and sprays into the flask like a fountain (**Figure 20**); this flow continues until the beaker is completely emptied.

14. Natural substances were used for their adhesive and cohesive properties long before scientific knowledge allowed us to synthesize glues and sealants. Aboriginal peoples of North America used spruce and pine tree secretions for a wide variety of purposes, including the "leakproofing" of birchbark canoes. Does this application illustrate adhesion, cohesion, or both? Explain your answer in terms of intermolecular forces.

Section **3.4**

SUMMARY *Intermolecular Forces*

- Intermolecular forces, like all forces involving atoms, are electrostatic—they involve the attraction of positive and negative charges.
- All molecules attract each other through the London force, the simultaneous attraction of momentary dipoles in adjacent molecules.
- Dipole−dipole force exists between polar molecules, the simultaneous attraction of permanent dipoles in adjacent molecules.
- Hydrogen bonding exists when hydrogen atoms are bonded to the highly electronegative atoms N, O, and F—the proton is simultaneously attracted to a lone pair of electrons on the N, O, or F atom of an adjacent molecule.
- Intermolecular forces affect many physical properties, for example, the melting point, boiling point, capillary action, surface tension, volatility, and solubility of substances.

WEB Activity

Simulation—Modelling Molecules

Many computer programs that model a wide variety of molecular shapes (and other features) are available on the Internet. Examine the programs on the links provided to see how well the models correspond to the kinds of predictions you have been making in this chapter. You can explore VSEPR and Lewis theory and examine molecular structures and shapes that are not predictable within the scope of this chapter.

www.science.nelson.com

▶ Section 3.4 *Questions*

1. All molecular compounds may have London, dipole–dipole, and hydrogen-bonding intermolecular forces affecting their physical and chemical properties. Using the rules and definitions for intermolecular forces and **Table 8** (page 99), indicate which intermolecular forces contribute to the attraction between molecules in each of the following classes of organic compounds (many of which are described in detail in Chapter 10). It should not be necessary to draw Lewis formulas. Look carefully at the sequence of atoms in the molecular formula.
   (a) hydrocarbon; e.g., pentane, $C_5H_{12}(l)$ (in gasoline)
   (b) alcohol; e.g., propan-2-ol, $CH_3CHOHCH_3(l)$ (rubbing alcohol)
   (c) ether; e.g., dimethylether, $CH_3OCH_3(g)$ (polymerization catalyst)
   (d) carboxylic acid; e.g., acetic acid, $CH_3COOH(l)$ (in vinegar)
   (e) ester; e.g., ethylbenzoate, $C_6H_5COOC_2H_5(l)$ (cherry flavour)
   (f) amine; e.g., dimethylamine, $CH_3NHCH_3(g)$ (depilatory agent)
   (g) amide; e.g., ethanamide, $CH_3CONH_2(s)$ (lacquers)
   (h) aldehyde; e.g., methanal, $HCHO(g)$ (corrosion inhibitor)
   (i) ketone; e.g., acetone, $(CH_3)_2CO(l)$ (varnish solvent)

2. Use structural formulas and hydrogen bonding theory to explain the very high solubility of methanol, $CH_3OH(l)$, in water.

3. Predict the solubility of the following organic compounds in water as low (negligible), medium, or high. Provide your reasoning.
   (a) 2-chloropropane, $C_3H_7Cl(l)$ (solvent)
   (b) propan-1-ol, $C_3H_7OH(l)$ (brake fluids)
   (c) propanone, $(CH_3)_2CO(l)$ (cleaning precision equipment)
   (d) propane, $C_3H_8(g)$ (gas barbecue fuel)

4. For each of the following pairs of chemicals, which one is predicted to have the stronger intermolecular attraction? Provide your reasoning.
   (a) chlorine or bromine
   (b) fluorine or hydrogen chloride
   (c) methane or ammonia
   (d) water or hydrogen sulfide
   (e) silicon tetrahydride or methane
   (f) chloromethane, $CH_3Cl(g)$, or ethanol

5. Which liquid, propane (C_3H_8) or ethanol (C_2H_5OH), would have the greater surface tension? Justify your answer.

6. In cold climates, outside water pipes, such as those used in underground sprinkler systems, need to have the water removed before it freezes. What might happen if water freezes in the pipes? Explain your answer.

7. A glass can be filled slightly above the brim with water without the water running down the outside. Explain why the water does not overflow even though some of it is above the glass rim.

8. Design an experiment to determine whether hydrogen bonding has an effect on the surface tension of a liquid. Clearly indicate the variables in this experiment.

9. Critique the following experimental design:
The relative strength of intermolecular forces in a variety of liquids is determined by measuring the height to which the liquids rise in a variety of capillary tubes.

10. A student is intent upon finding a way to reduce the effect of "hot" spices in spicy food. The student finds that water does not help alleviate the effect of spices but sour cream does. Use your knowledge of chemistry to explain this finding.

11. Volatile organic compounds (VOCs) are compounds that vaporize and are distributed in Earth's atmosphere. VOCs can increase smog formation in the lower atmosphere and promote global warming in the upper atmosphere. Predict the order of the following compounds in terms of decreasing volatility. Provide your reasoning.
    - hexane, C_6H_{14}
    - hexan-1-ol $C_6H_{13}OH$
    - 1-chlorohexane, $C_6H_{13}Cl$

    [Note: The "1" in the names tells you that the OH and the Cl are bonded to the end of a carbon chain.]

12. (a) Draw a bar graph with temperature on the vertical axis and the three isoelectronic compounds listed below on the horizontal axis. Mark the temperature scale from absolute zero, -273 °C (the lowest temperature possible) to $+100$ °C. For each compound, draw a vertical bar from absolute zero to its boiling point: propane, $CH_3CH_2CH_3(g)$ (-42 °C); fluoroethane, $CH_3CH_2F(g)$ (-38 °C); and ethanol, $CH_3CH_2OH(l)$ (78 °C).

    (b) Divide each of the three bar graphs into the approximate component for the intermolecular force involved. (Assume that the London force is the same for each chemical and that the dipole–dipole force is the same for the two polar molecules.)

    (c) Based upon the proportional components for the three possible intermolecular forces, infer the relative strength of these forces.

13. In the far North, groundwater is permanently in the solid state (called permafrost), because the ground does not warm enough in the short, cool summers for this water to thaw. In some areas of Siberia, the permafrost layer is over a kilometre deep! In northern Canadian communities, warm structures such as heated buildings, sewer systems, and oil pipelines must be built over the permafrost. The unique intermolecular bonding of water creates a huge problem if the permafrost melts under a building. State which physical property of water (other than its physical state) changes significantly when it thaws, and explain how this change causes the problem depicted in **Figure 21**.

Figure 21
These two buildings, in Dawson City, Yukon, have been nick-named the "kissing buildings"; they date back to the Gold Rush era.

Extension

14. Some vitamins are water soluble (e.g., B series and C), while others are fat soluble (e.g., vitamins A, D, E, and K).
    (a) What can you infer about the polarity of these chemicals?
    (b) Find and draw the structure of at least one of the water-soluble and one of the fat-soluble vitamins.
    (c) When taking vitamins naturally or as supplements, what dietary requirements are necessary to make sure that the vitamins are used by the body?
    (d) More of a vitamin is not necessarily better. Explain why you can take a fairly large quantity of vitamin C with no harm (other than the cost), but an excess of vitamin E can be dangerous.

    www.science.nelson.com

15. Plastic cling wrap is widely used in North America. Why does it cling well to smooth glass and ceramics but not to metals? Describe the controversial social issue associated with the use of this plastic wrap. How are intermolecular forces involved in starting the process that leads to this controversy?

    www.science.nelson.com

Structures and Physical Properties of Solids 3.5

All solids, including elements and compounds, have a definite shape and volume, are virtually incompressible, and do not flow readily. However, research shows that there are many specific properties, such as hardness, melting point, mechanical characteristics, and conductivity, that vary considerably between different solids. For example, if you hit a piece of copper with a hammer, you can easily change its shape. If you do the same thing to a lump of sulfur, you crush it. A block of paraffin wax when hit with a hammer will deform and may break (**Figure 1**). Why do these solids behave differently?

The way solids break is called fracture, a term familiar to anyone who has ever broken a bone. Glass is a special case, because it has a special kind of bonding (discussed in detail later in this section). Glass has conchoidal fracture; this means it breaks along a curve. Wherever two fracture curves meet, the edge is extremely sharp. The natural minerals obsidian (volcanic glass), flint, and chert all have a bonding structure like that of glass; that is, they break with conchoidal fracture. Aboriginal cultures depended on this property for thousands of years. A piece of flint can be chipped carefully to produce a stone knife (or spearhead or arrowhead) that is much superior to any point or edge that can be made on a wooden object. An entire historic period based on this technique is called the Stone Age. Archaeologists believe Aboriginal peoples moved into North America as the last Ice Age retreated. Much of the evidence for this belief comes from the spread throughout the continent of what are called Clovis points (**Figure 2**). A testimony to the efficiency of hunters using such spearheads may be the extinction of large mammals like the mammoth and the short-faced bear at about this time. Later techniques allowed small sharp points to be made that could be used to tip small throwing spears and arrows, creating even more formidable weapons.

By observing properties of chemical substances, we were able to classify them empirically (**Table 1**). To now explain the properties of each category, we use our knowledge of chemical bonding. In both elements and compounds, the structure and properties of the solid are related to the forces between the particles. Although all forces are electrostatic in nature, they vary in strength.

Figure 1
Different solids—copper, sulfur, and paraffin wax—behave very differently under mechanical stress.

Figure 2
Clovis points very similar to these are found in hundreds of archaeological sites throughout North America, dating back well over 10 000 years. Considerable skill is required to chip a stone point this way; even experts break more than they finish. Consequently, to the delight of later archaeologists, Aboriginal point makers always left behind many fragments and substandard points. "Clovis" is from the name of the town in New Mexico near the site where these points were first found.

Table 1 Classifying Solids

Class of substance	Elements combined	Examples
ionic	metal + nonmetal	$NaCl(s)$, $CaCO_3(s)$
metallic	metal(s)	$Cu(s)$, $CuZn_3(s)$
molecular	nonmetal(s)	$I_2(s)$, $H_2O(s)$, $CO_2(s)$
covalent network	semi-metals/nonmetals	$C(s)$, $SiC(s)$, $SiO_2(s)$

Ionic Crystals

Ionic compounds are abundant in nature. Soluble ionic compounds are present in both fresh water and salt water. Ionic compounds with low solubility make up most rocks and minerals. Relatively pure deposits of sodium chloride (table salt) occur in Alberta, and Saskatchewan has the world's largest deposits of potassium chloride (potash fertilizer) and sodium sulfate (used for manufacturing paper).

You can also find ionic compounds at home. Iodized table salt consists of sodium chloride with a little potassium iodide added. Antacids contain a variety of compounds, such as magnesium hydroxide and calcium carbonate. Many home cleaning products

Figure 3
Rusting is a common and expensive problem.

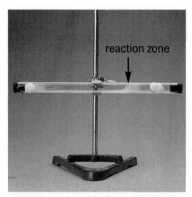

Figure 4
The solid ionic compound ammonium chloride forms from the reaction of ammonia and hydrogen chloride gases when small quantities of concentrated solutions are placed into the ends of a tube such as this.

Learning Tip

Some confusion may arise from use of the word "particle" in a chemistry textbook. In general use, it often means a tiny bit of a solid, as when we refer to the "smoke" particles of $NH_4Cl(s)$ that are forming in **Figure 4**. In bonding theory, it more often refers to subatomic particles, such as electrons and protons, and is usually used to distinguish these from chemical entities (molecules, atoms, or ions).

contain sodium hydroxide. Other examples of ionic compounds are the rust (iron(III) hydroxide) that forms on the steel bodies of cars and the tarnish (silver sulfide) that forms on silver. Ionic compounds have low chemical reactivity compared with the elements from which they are formed. Some ionic compounds, such as lime (calcium oxide) and the mineral sylvite (potassium chloride), are so stable that they were classified as elements until the early 1800s. (At that time, elements were defined as pure substances that could not be decomposed by strong heating.)

Reactive metals, such as sodium, magnesium, and calcium, are not found as elements in nature but occur instead in ionic compounds. The least active metals, such as silver, gold, platinum, and mercury, do not react readily to form compounds and may, therefore, be found uncombined in nature.

For thousands of years, metallurgists have used ionic compounds to extract metals from naturally occurring compounds. In these processes, an ionic compound such as hematite (Fe_2O_3) in iron ore is reduced to a pure metal, which can then be used to make tools, weapons, and machines. Iron, the main constituent of steel, is the most widely used metal. Unfortunately, iron is reactive and readily corrodes, or reacts with substances in the environment, re-forming to an ionic compound (**Figure 3**). A lot of time and money is spent trying to prevent or slow the corrosion of iron and other metals, for example, by having cars rust-proofed at automotive centres.

Formation of Ionic Compounds

Chemical research shows that ionic compounds are formed in many ways. Binary ionic compounds are the simplest ionic compounds; they may be formed by the reaction of a metal with a nonmetal. For example:

$$2\,Na(s) + Cl_2(g) \rightarrow 2\,NaCl(s)$$

The reaction of ammonia and hydrogen chloride gases produces the ionic compound ammonium chloride, which appears as a white smoke of tiny solid particles (**Figure 4**):

$$NH_3(g) + HCl(g) \rightarrow NH_4Cl(s)$$

The conductivity of molten ionic compounds and aqueous solutions of soluble ionic compounds suggests that charged entities are present. According to atomic theory, the stability of ionic compounds suggests that their electronic structure is similar to that of the noble gases, which have filled energy levels. By tying in these concepts with the collision–reaction theory, scientists explain the formation of an ionic compound as involving collisions between metal and nonmetal atoms that result in a transfer of electrons; the transfer of electrons forms cations and anions that have filled valence energy levels. Scientists believe that the electron transfer is encouraged by the large difference in electronegativity between metal and nonmetal atoms. Lewis formulas represent these concepts. For example, a theoretical description of the formation of sodium fluoride shows how a stable octet structure is formed:

$$Na \cdot + \cdot \ddot{\underset{\cdot\cdot}{F}}: \;\rightarrow\; Na^+[:\ddot{\underset{\cdot\cdot}{F}}:]^-$$

Sodium fluoride is added to many toothpastes; research shows that it acts to harden enamel in teeth, so that cavities are less likely to form.

A Model for Ionic Compounds

To be acceptable, a theory of bonding must be able to explain the properties of ionic compounds. All ionic compounds are solids at SATP, so the ions must be held together or bonded strongly in a rigid structure. In the model for ionic compounds, ions are shown as spheres arranged in a regular pattern. Depending on the sizes, shapes, and charges of the ions, different arrangements are possible, but whatever the pattern, it will allow the greatest number of oppositely charged ions to approach each other closely while preventing the close approach of ions with the same charge. In all cases, the model describes any ion as being surrounded by ions of opposite charge. This creates strong attractions and explains why ionic compounds are always solids with high melting and boiling points. The arrangement of ions for a given compound is called its **crystal lattice**. The model also explains why ionic compounds are brittle—the ions cannot be rearranged without breaking the ordered structure of the crystal lattice apart.

Chemists represent ionic crystals with models, such as the diagram of sodium chloride in **Figure 5(b)**. The diagram shows each ion surrounded by six ions of opposite charge, held firmly within the crystal by strong electrostatic attractions. The observable cubic shape of sodium chloride crystals supports this model. All binary ionic compounds have brittle crystalline forms and medium to high melting and boiling points. These properties vary greatly in degree, however, depending on the nature of the ions forming the compound. Binary ionic compounds have the simplest structures. They are often quite hard solids with very high melting points. Sodium chloride, for example, melts at 800 °C. If you fracture a crystal of salt, it will break along surfaces at 90° to each other, called cubic fracture. These fracture planes follow layers of ions in the crystal lattice.

We have only examined binary ionic compounds here. However, there are also many ionic compounds of polyatomic ions, such as sodium carbonate, $Na_2CO_3(s)$. Chemists explain that these compounds have covalent bonds within each polyatomic ion. The bonds hold the polyatomic ion together as a group and cause the whole group to act much like a monatomic ion. Crystal lattices for such ionic compounds are more complex, and the ranges of hardnesses and melting and boiling points vary much more widely. Some of these compounds will decompose before they get hot enough to melt. The theoretical explanation is that the covalent bonds in the polyatomic ions begin to break before the ionic bonding forces in the crystal are overcome.

According to laboratory evidence and the ion model, ion attractions are nondirectional—all positive ions attract all nearby negative ions. There are no distinct neutral molecules in ionic compounds. The chemical formula shows only a formula unit (empirical formula) expressing the simplest whole-number ratio of ions. For example, any crystal of sodium chloride, NaCl(s), contains equal numbers of sodium ions and

> **DID YOU KNOW ?**
> **X-Rays and 3-D Models**
> Three-dimensional models of crystals are determined experimentally by X-ray diffraction (**Figure 6**). When a beam of X-rays is reflected from the top layers of an ionic crystal, a regular pattern is obtained. The wavelength of the X-rays determines the way they are reflected from the spaces between ions. (This is similar to the way in which the wavelength of light determines how it is reflected from the grooves on the surface of a compact disc.) Scientists can use the pattern of either reflected or transmitted X-rays and their knowledge of the wavelength to infer the arrangement, size, and separation of the ions.

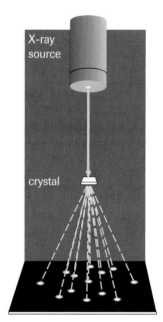

Figure 6
Diagram of X-ray diffraction apparatus

 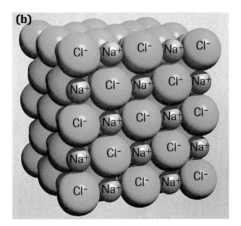

Figure 5
From a cubic crystal of table salt **(a)** and from X-ray analysis, scientists infer the 3-D arrangement for sodium chloride **(b)**. In this cubic crystal, each ion is surrounded by six ions of opposite charge. The sodium and chloride ions are shown to the correct relative scale in this model.

chloride ions, and any crystal of calcium fluoride, $CaF_2(s)$, contains one calcium ion for every two fluoride ions. Any ionic crystal is explained as being all one structure. You don't know how many positive or negative ions it might contain, but you know exactly how their numbers compare. Logically, ions must always collect to form a crystal lattice in a such way that the total charge balances to zero. In reading chemical formulas, it is very important to remember two things: First, ionic compound formulas do not represent molecules. Second, ion charges are never shown, so they must be referenced or memorized.

mini Investigation: Building an Ionic Crystal Model

The crystal lattice arrangement for an ionic substance can be quite complex, depending on the sizes, shapes, and charges of the cations and anions that make up the compound. It can get even more complicated when (highly polar) water molecules make up part of the lattice, as happens in hydrated ionic compounds like $CuSO_4 \cdot 5H_2O(s)$. Interestingly, only about nine common crystal shapes result from all these possible combinations. Of these, the most familiar is the cubic structure seen in common table salt, $NaCl(s)$, which is fairly easy to model.

Materials: 18 large Styrofoam spheres, 18 small Styrofoam spheres (half the diameter of the large spheres), toothpicks

- Assemble two separate three-by-three ion layers for the lattice by attaching spheres to each other using toothpicks. Use large spheres for the chloride ions, and small spheres for the sodium ions. Place the chloride ions at the corners and centre of each of the two layers. Spheres representing oppositely charged ions should be as close as possible, and spheres representing like charged ions should not touch each other.
- Assemble two more ion layers, but place the sodium ions at the corners.
- Stack the layers alternately, noting how they fit together.

(a) In the assembled lattice, what are the nearest ions to each chloride? How many of these ions are near any chloride ion that is located completely inside the crystal lattice?

(b) In the assembled lattice, what are the nearest ions to each sodium? How many of these ions are near any sodium ion that is completely inside the crystal lattice?

The formula for a crystal of sodium chloride could be written as $Na^+_n Cl^-_n$.

(c) Why are the ion charges not written in the chemical formula for an ionic compound?

(d) What number does n represent in the above formula?

(e) Why is this numbering symbolism not normally used?

(f) What, precisely, do the subscript numbers in an ionic compound formula refer to?

(g) Observe other models constructed by other groups in your class. Consider and discuss their ion spacing and structural differences and similarities. What theoretical arrangement maximizes attractions of unlike charges and minimizes repulsions of like charges to produce a maximum net attraction?

Practice

1. Draw Lewis formulas for each of the following ions. In each case, begin by first adjusting the number of valence electrons of the central atom to account for the ion charge. Bracket the final formula, and show the net ion charge outside the brackets.
   (a) CN^-
   (b) Se^{2-}
   (c) OH^-
   (d) NH_4^+
   (e) SO_4^{2-}
   (f) Mg^{2+}

2. Hydrogen chloride and ammonia gases are mixed in a flask to form a white solid product.
   (a) Write a balanced chemical equation for the reaction of ammonia and hydrogen chloride gases to form ammonium chloride.
   (b) Rewrite the equation using Lewis formulas.

3. Where are ionic compounds abundant in nature?

4. Write a brief explanation for the formation of a binary ionic compound from its constituent elements.

5. What evidence suggests that ionic bonds are strong?

6. Potassium chloride is a substitute for table salt for people who need to reduce their intake of sodium ions. Use Lewis formulas to represent the formation of potassium chloride from its elements. Show the electronegativities of the reactant atoms.

7. Use Lewis formulas to represent the reaction of calcium and oxygen atoms. Name the ionic product.

8. The empirically determined chemical formula for magnesium chloride is $MgCl_2$. Create Lewis formulas to explain the empirical formula of magnesium chloride.

9. Create Lewis formulas to predict the chemical formula of the product of the reaction of aluminium and oxygen.

10. Based only on differences in electronegativity, what compound would you expect to be the most strongly ionic of all binary compounds?

11. What is the difference between the information communicated by the empirical formula of an ionic compound such as NaCl and the molecular formula of a substance such as H_2O?

Metallic Crystals

Metals are shiny, silvery, flexible solids with good electrical and thermal conductivity. The hardness varies from soft to hard (e.g., lead to chromium) and the melting points from low to high (e.g., mercury at -39 °C to tungsten at 3410 °C). Further evidence from the analysis of X-ray diffraction patterns shows that all metals have a continuous and very compact crystalline structure (**Figure 7**). With few exceptions, all metals have closely packed structures.

Figure 7
The crystal structure of metals normally can be seen only under a microscope. On the surface of zinc-plated (galvanized) objects, however, large flat crystals of zinc are plainly visible.

WEB Activity

Canadian Achievers—Jillian Buriak

Jillian Buriak is a chemistry professor at the University of Alberta and a senior research officer at the NRC's National Institute for Nanotechnology (**Figure 8**). Dr. Buriak is a leading expert in semiconductor surface chemistry, and she leads a team in basic research essential to developing new technologies such as microelectronics, nano-electromechanical systems, sensors and diagnostics which communicate directly with cells, viruses, and bacteria, and a host of other applications.

1. What is nanotechnology?
2. What major award did Buriak win in 2005?
3. List five recent projects of Buriak's research group.

www.science.nelson.com

Figure 8
Dr. Jillian Buriak

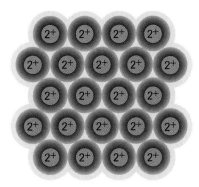

Figure 9
This model of bonding in copper shows a continuous (grey) area occupied by mobile valence electrons, attracting all the positive metal ions at once, to hold them together in a regular structure with a net (overall) charge of zero.

An acceptable theory for bonding in metals must describe and explain characteristic metallic properties, provide testable predictions, and be as simple as possible. According to current theory, the properties of metals are the result of metallic bonding. Recall from Section 3.1 that with metallic bonding, the valence electrons are not held strongly by their atoms, so they can easily become mobile. This attraction is not localized or directed between specific entities, as occurs with ionic crystals. Instead, the electrons act like a negative "glue" surrounding the positive metal ions, acting to hold them firmly together. As illustrated in **Figure 9**, valence electrons are believed to occupy the spaces between the positive metal ions. This model incorporates the ideas of

- low electronegativity of metal atoms to explain loosely held electrons
- empty valence orbitals to explain electron mobility
- electrostatic attractions of positive ions and the negatively charged mobile valence electrons to explain the strong, nondirectional bonding

The model is used to explain the empirical properties of metals (**Table 2**).

Table 2 Explaining the Properties of Metals

Property	Explanation
shiny, silvery	valence electrons absorb and re-emit the energy from all wavelengths of visible and near-visible light
flexible	nondirectional bonds mean that the planes of ions can slide over each other while remaining bonded
electrical conductivity	valence electrons can freely move throughout the metal
crystalline	electrons provide the "electrostatic glue" holding the metal ions together producing structures that are continuous and closely packed

The next time you have a rice crisp square, look at it carefully and play with it. The marshmallow is the glue that binds the rice crisps together. If you push on the square, you can easily deform it, without breaking it. The marshmallow acts like the valence electrons in a metal; the crisps represent the metal cations. The mechanical properties are somewhat similar to those of a metal.

Molecular Crystals

Molecular solids can be elements such as iodine and sulfur, or compounds such as ice and carbon dioxide. Substances with small molecules form solid crystals that have relatively low melting points, are not very hard, and are nonconductors of electricity in their pure form as well as in solution. From X-ray analysis, chemists find that these crystals have molecules arranged in a regular lattice (**Figure 10**). In general, the molecules are packed as close together as their size and shape allow (**Figure 11**).

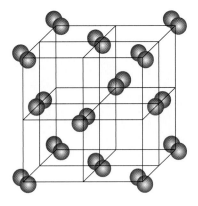

Figure 10
A model of an iodine crystal based on X-ray analysis shows a regular arrangement of iodine molecules.

The properties of molecular crystals can be explained by their structure and the intermolecular forces that hold them together. Chemists understand that London, dipole–dipole, and hydrogen bonding forces are not very strong compared with ionic and covalent bonds. This would explain why molecular crystals have relatively low melting points and a general lack of hardness. Because individual entities are neutral molecules, they cannot conduct an electric current even when the molecules are free to move in the liquid state.

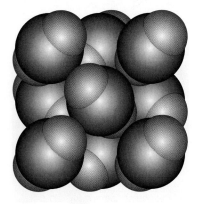

Figure 11
Solid carbon dioxide, or dry ice, also has a crystal structure containing individual carbon dioxide molecules.

Section **3.5**

Covalent Network Crystals

Two commonly recognized crystal substances (because they are used in jewellery) are diamond and quartz (**Figure 12**). These substances are among the hardest materials on Earth and belong to a group known as covalent network crystals. Experiments show that such substances are very hard and brittle, have very high melting points, are insoluble, and are nonconductors of electricity. Covalent network crystals are usually much harder and have much higher melting points than ionic and molecular crystals. They are described as brittle because they do not bend under pressure, but they are so hard they seldom break. One example is silicon carbide (carborundum), SiC(s), used for sandpaper and for grinding stones to sharpen metal tools. Carbide-tipped saw blades (**Figure 13**) have steel teeth tipped with much harder tungsten carbide, WC(s). However, diamond, C(s), is the classic example of a covalent crystal. It is so hard that it can be used to make drill bits for drilling through the hardest rock on Earth.

CAREER CONNECTION

Geologist/Gemmologist
Geologists explore the Earth's crust for minerals and hydrocarbon formations, such as oil and gas. Gemmologists work specifically with natural gemstones such as diamonds. These scientists work outdoors and in laboratories conducting geological surveys and field research.

Find out what these careers offer and how to train for them.

www.science.nelson.com

Figure 12
Amethyst **(a)**, rose quartz **(b)**, and citrine **(c)** are all variations of quartz, which is SiO_2(s). The different colours are due to tiny amounts of impurities; pure quartz is colourless.

Figure 13
Circular saw blades with tungsten carbide-tipped teeth stay sharp much longer.

The shape and X-ray diffraction analysis of diamond show that the carbon atoms are in a large tetrahedral network with each carbon covalently bonded to four other carbon atoms (**Figure 14**). Each diamond is a crystal and can be described theoretically as a single macromolecule with the chemical formula C(s). The network of covalent bonds leads to a common name for these covalent crystals: **covalent network**, a 3-D arrangement of atoms continuously linked throughout the crystal by strong covalent bonds. This term helps differentiate between the covalent bonds within molecules and polyatomic ions and the covalent bonds within covalent network crystals. Like ionic compounds, covalent network compounds have continuous bonding holding every entity in the crystal together. However, covalent network compounds consist of atoms held together by covalent bonds rather than ions held together by ionic bonding. Like ionic compounds, the formulas for covalent network compounds are empirical formulas, showing the simplest whole-number ratio of atoms in the crystalline solid. Most covalent networks involve carbon and/or silicon atoms, which bond strongly (and often) to themselves and to many other atoms, as well.

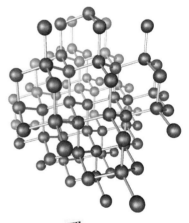

Figure 14
In a model of diamond, each carbon atom has four single covalent bonds to each of four other carbon atoms. As you know from VSEPR theory, four pairs of electrons predict a tetrahedral shape around each carbon atom. This bonding forms a continuous network out to the boundary of the crystal surface.

Understanding Chemical Compounds

DID YOU KNOW ?

Glass, An Ancient Technology
Glass is one of the oldest, most useful, and versatile materials used by humans. It has been produced for at least four thousand years. Ordinary glass is made from sand (silicon dioxide), limestone (calcium carbonate), and soda ash (sodium carbonate), all very common materials.

Crystalline quartz is described theoretically as a covalent network of $SiO_2(s)$. The structure can be determined by X-ray analysis (**Figure 15(a)**). When heated above its melting point of 1600 °C and then cooled rapidly, it forms a quartz glass, in which the atoms are quite disordered (**Figure 15(b)**). The properties become more like those of a very stiff liquid rather than a crystalline solid. Common commercial glass has many other compounds mixed with quartz and melts at a much lower temperature. Its properties can be controlled by changing the kinds and amounts of the additives. Purposely, glass is cooled to a rigid state in such a way that it will not crystallize as a regular lattice, so it can be worked into various shapes while hot.

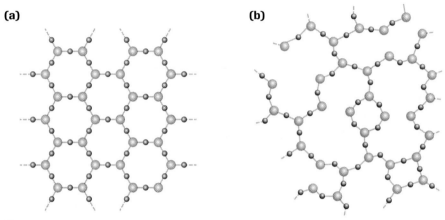

Figure 15
(a) X-ray analysis indicates that quartz in its crystalline form has a regular, repeating 3-D network of covalently bonded silicon and oxygen atoms.
(b) Quartz glass is not crystalline because it does not have an extended order; at the atomic level, it is quite disordered and irregular in structure.

DID YOU KNOW ?

Traffic Control
Semiconductor "sandwiches" that convert electricity directly to light are called LEDs (light-emitting diodes). LEDs are being used increasingly in brake and tail lights of automobiles, and in traffic control signs (**Figure 16**). Small LEDs are used in large groups, so if one fails, the sign/signal is not lost. Although more expensive to produce than conventional lights, LEDs use less energy and last much longer.

Figure 16
The city of Edmonton saves power costs of $800 000 per year because it converted all red and green traffic signal lights to LEDs.

The properties of hardness and high melting point provide the evidence that the overall bonding in the large macromolecule of a covalent network is very strong—stronger than most ionic bonding and all intermolecular bonding. Although an individual carbon–carbon bond in diamond is not much different in strength from any other single carbon–carbon covalent bond, it is the interlocking structure that is thought to be responsible for the strength of the material. The final structure is stronger than any individual component. This means that individual atoms are not easily displaced, and that is why the sample is very hard. In order to melt a covalent network crystal, many covalent bonds need to be broken, which requires considerable energy, so the melting points are very high. Electrons in covalent network crystals are held either within the atoms or in the covalent bonds. In either case, they are not free to move through the network. This explains why these substances are nonconductors of electricity.

Other Covalent Networks of Carbon

Carbon is an extremely versatile atom in terms of its bonding and structures. More than any other atom, carbon can bond to itself to form a variety of pure carbon substances. It can form 3-D tetrahedral arrangements (diamond), layers of sheets (graphite), 60-atom spherical molecules (buckyballs), and long, thin tubes (carbon nanotubes) (**Figure 17**). Graphite is unlike most covalent crystals in that it readily conducts electricity, but it still has a high melting point. Graphite also acts as a lubricant. All of these properties, plus the X-ray diffraction of the crystals, indicate that the structure for graphite is hexagonal sheets of carbon atoms. Within these planar sheets, the bonding is a covalent network and, therefore, strong, but between the sheets, the bonding is relatively

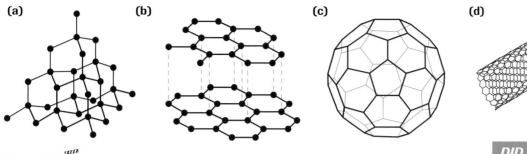

Figure 17
Models of the many forms of pure carbon:
(a) diamond
(b) graphite
(c) buckyball
(d) carbon nanotube

weak—due to London forces. The lubricating property of graphite can be explained as the covalent network planes sliding over one another easily, like individual sheets of paper in a stack. Powdered graphite makes an excellent lubricant anywhere a dry lubricant is desirable, since it will not attract and hold dust and dirt the way oily lubricants do. The electrical conductivity of graphite is explained by the concept that each carbon atom has only three covalent bonds. The unbonded fourth valence electron of each atom is free to move through the space between the 2-D sheets of atoms. The conductivity of graphite, coupled with a very high melting point, makes it very useful in high-temperature electrical applications, such as contact brushes in electric motors. Another form of carbon, known as buckminsterfullerene (named for the American architect, Buckminster Fuller), is also a very useful dry lubricant. The smallest fullerene is a 60-carbon sphere, affectionately known as a buckyball because of its resemblance to a soccer ball (**Figure 17(c)**).

Semiconductors

In recent decades we have seen an electronic technological revolution driven by the discovery of the transistor—a solid-state "sandwich" of crystalline semiconductors. Semiconductor material used in transistors is usually pure crystalline silicon or germanium with a tiny quantity (e.g., 5 ppm) of either a Group 13 or 15 element added to the crystal in a process called doping. The purpose of doping is to control the electrical properties of the covalent crystal to produce the conductive properties desired. Transistors are the working components of almost everything electronic (**Figure 18**).

In an atom of a semiconductor, the highest energy levels may be thought of as being full of electrons that are unable to move from atom to atom. Normally, this would make the substance a nonconductor, like glass or quartz. In a semiconductor, however, electrons require only a small amount of energy to jump to the next higher energy level, which is empty. Once in this level, they may move to another atom easily. Semiconductors are an example of a chemical curiosity where research into atomic structure has turned out to be amazingly useful. Power supplies for many satellites and for the International Space Station (**Figure 19**) come from solar cells that are semiconductors arranged to convert sunlight directly to electricity. Other arrangements convert heat to electricity, or electricity to heat, or electricity to light—all without moving parts in a small, solid device. Obviously, improving the understanding of semiconductor structure was of great value to society (**Figure 16**).

DID YOU KNOW?

Diamond Cutting and Polishing
Rough diamonds are mined northeast of Yellowknife, Northwest Territories. A copper saw blade pasted with diamond powder and oil is used to cut along the crystal planes. Facets are added by polishing the diamond with a disk impregnated with diamond powder. Only diamond powder is hard enough to cut and polish diamonds.

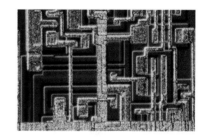

Figure 18
Semiconductors in transistors are covalent crystals that have been manipulated by doping with atoms that have more or fewer electrons than the atoms in the main crystal.

Figure 19
The solar panels that power the International Space Station use semiconductors to change light energy to electric current.

DID YOU KNOW?

Canadian Gems

In recent years, Canada's Northwest Territories have produced excellent quality diamonds, emeralds, and sapphires. Diamonds are a crystallized form of pure carbon and are the hardest natural substance known. Emeralds are a type of beryl ($Be_3Al_2Si_6O_{18}$). Ordinary beryl is colourless, but emeralds are green because some of the aluminium has been replaced by either chromium or vanadium. Sapphires are a form of crystallized aluminium oxide (Al_2O_3). Trace amounts of titanium in the crystals give sapphires their blue colour. Rubies are another type of corundum that owes its red colour to tiny quantities of chromium.

DID YOU KNOW?

The Mohs Hardness Scale

Geologists often identify minerals by relative hardness. They do a "scratch test" to see whether one substance will scratch another. This is really comparing the relative strengths of bonds in the crystals of the minerals. The most common relative hardness scale is called the Mohs Hardness Scale. Some of the crystal lattices of minerals are very simple, and some are not. Note the range of strengths of ionic and covalent bonds in the crystals, from very soft talc (magnesium silicate) to diamond, the hardest known:

1	talc	$Mg_3Si_4O_{10}\cdot H_2O$
2	gypsum	$CaSO_4\cdot 2H_2O$
3	calcite	$CaCO_3$
4	fluorite	CaF_2
5	apatite	$CaF_2\cdot 3Ca_3(PO_4)_2$
6	orthoclase	$K_2O\cdot Al_2O_3\cdot 6SiO_2$
7	quartz	SiO_2
8	topaz	$(AlF)_2SiO_4$
9	corundum	Al_2O_3
10	diamond	C

For comparison purposes, using this scale, the hardness of your fingernail is 2.5, that of a copper penny is 3, and that of a steel knife blade is 5.5. The Mohs scale is not linear: the numbers (1–10) are not proportional to the absolute hardness of these substances.

Practice

12. In terms of chemical bonds, what are some factors that determine the hardness of a solid?
13. Identify the main type of bonding and the type of solid for each of the following:
    (a) SiO_2
    (b) Na_2S
    (c) CH_4
    (d) C
    (e) Cr
    (f) CaO
14. How does the melting point of a solid relate to the type of entities and forces present?
15. Explain why metals are generally malleable, ductile, and flexible.
16. State the similarities and differences in the properties of each of the following pairs of substances. In terms of the entities and forces present, briefly defend each answer.
    (a) Al(s) and Al_2O_3(s)
    (b) CO_2(s) and SiC(s)
17. To cleave or split a crystal, you tap a sharp knife on the crystal surface with a small hammer.
    (a) Why is the angle of the blade on the crystal important to cleanly split the crystal?
    (b) If you wanted to cleave a sodium chloride crystal, at what angle to the surface would you place the knife blade?
    (c) Speculate about what would happen if you tried to cleave a crystal in the wrong location or at the wrong angle.
    (d) State one technological application of this technique.
18. Match the solids NaBr(s), V(s), P_2O_5(s), and SiO_2(s) to the property listed below:
    (a) high melting point, conducts electricity
    (b) low melting point, soft
    (c) high melting point, soluble in water
    (d) very high melting point, nonconductor
19. Metals are generally good conductors of heat and electricity. Is there a relationship between a metal's ability to conduct heat and its ability to conduct electricity?
    (a) Create a hypothesis to answer this question. Include your reasoning.
    (b) Design an experiment to test your hypothesis and reasoning using common examples of metals.
20. Suggest some reasons why graphite may be better than oil in lubricating moving parts of a machine.
21. Flint is a form of rock, primarily composed of SiO_2. Light-coloured forms are called chert, and forms with different colour bands are called agate. Flint can be shaped to produce an extremely sharp edge in a procedure called flintknapping. This technique has been used for millennia by Aboriginal peoples and has now become a popular hobby for many people (**Figure 20**). Given that flint is very hard and fractures much the same way glass or obsidian does, describe its general bonding in terms of bond type and structural regularity.
22. Write a design for an experiment to measure some physical properties of the ionic compounds NaCl(s), NaBr(s), NaI(s), and NaF(s). Write a prediction for a trend you might expect to see in one of the physical properties you will measure for these substances.

Figure 20
Flintknapping is a traditional Aboriginal stoneworking technique that has been taught in North America for thousands of years.

Section **3.5**

SUMMARY: Structures and Properties of Crystals

Table 3 Properties of Ionic, Metallic, Molecular, and Covalent Network Crystals

Crystal	Entities	Force/Bond	Properties	Examples
ionic	cations anions	ionic	hard; brittle; medium to high melting point; liquid and solution conducts	$NaCl(s)$, $Na_3PO_4(s)$, $CuSO_4 \cdot 5H_2O(s)$
metallic	cations electrons	metallic	soft to very hard; solid and liquid conducts; ductile; malleable; lustrous	$Pb(s)$, $Fe(s)$, $Cu(s)$, $Al(s)$
molecular	molecules	London dipole–dipole hydrogen	soft; low melting point; nonconducting solid, liquid, and solution	$H_2O(s)$ (ice), $CO_2(s)$ (dry ice), $I_2(s)$
covalent network	atoms	covalent	very hard; very high melting point; nonconducting	$C(s)$, $SiC(s)$, $SiO_2(s)$

+ EXTENSION

How do Molecules Interact?
This short quiz will help you figure out whether you have understood the concepts in this unit.

www.science.nelson.com GO

INVESTIGATION 3.5 Introduction

Classifying Unknown Solids

Design an investigation that enables you to use physical properties, such as conductivity, hardness, solubility, and melting point, to classify four unknown solids as ionic, metallic, molecular, or covalent network.

Purpose
The purpose of this investigation is to use empirical definitions to classify solid substances.

To perform this investigation, turn to page 134.

Report Checklist
- ○ Purpose
- ○ Problem
- ○ Hypothesis
- ○ Prediction
- ● Design
- ● Materials
- ● Procedure
- ● Evidence
- ● Analysis
- ● Evaluation (1, 3)

Problem
To what class of solids do the four mystery solids belong?

▶ Section 3.5 Questions

1. Describe and explain the different electrical conductivity properties of ionic substances under different conditions.

2. In terms of entities and forces present, what determines whether a solid substance conducts electricity?

3. Calcium oxide (m.p. 2700 °C) and sodium chloride (m.p. 801 °C) have the same crystal structure, and the ions are about the same distance apart in each crystal. Explain the significant difference in their melting points.

4. Compare and contrast the bonding forces in carbon dioxide (dry ice) and silicon dioxide (quartz). How does this explain the difference in their properties?

5. Why do most metals have a relatively high density?

6. State the general order of strength of intermolecular, ionic, covalent network, and metallic bonding. Defend your answer.

7. If the zipper on your jacket does not slide easily, how could using your pencil help? Describe what you would do, and explain why this would work.

8. Compare diamond and graphite using the following categories: appearance, hardness, electrical conductivity, crystal structure, and uses.

Understanding Chemical Compounds   **129**

9. Use the evidence in **Table 4** to classify the type of substance and type of bonding for each unknown chloride listed.

Table 4 Physical Properties of Two Chlorides

Unknown chloride	XCl_a	XCl_b
Melting point (°C)	750	−25
Boiling point (°C)	1250	92
Solubility in water	high	very low
Solubility in $C_6H_6(l)$	very low	high

10. Describe two areas where research into the structure of molecules has caused a dramatic improvement in materials available to society in general.

Extension

11. Ceramics are products manufactured from earth materials in which silicon oxides and silicate compounds are predominant components. They include products like brick and tile, chinaware, glass (**Figure 21**), porcelain, and abrasives. How do the structure and bonding change when clay is fired (heated very strongly) to create a ceramic pot or vase? Present your findings, with illustrations.

www.science.nelson.com

12. Research and report on the properties, applications, structure, and bonding in boron nitride, BN(s).

www.science.nelson.com

13. Composite materials combine materials with two or more different types of bonds to produce something with desired properties. For example, concrete embeds rock and sand in a surrounding mix (matrix) of cement to make a building material as strong as stone but easier to work with. Fibreglass embeds glass strands for strength in a plastic matrix for flexibility. New composite materials combine high strength with very light weight, so now they show up everywhere, from graphite golf club shafts to auto engine parts. Research composite materials. Write a brief summary on how the strength and weight of composite materials have dramatically influenced the design of commercial airliners for the 21st century.

www.science.nelson.com

14. Hemoglobin is a very large molecule (more than 9500 atoms) found in red blood cells. It carries oxygen from the lungs to body tissues (**Figure 22**), and carbon dioxide from body tissues to the lungs. Research hemoglobin on the Internet and answer the following questions:
    (a) How many oxygen molecules can one hemoglobin molecule transport?
    (b) How does hemoglobin bind to oxygen and to carbon dioxide?
    (c) How strong are these "transport" bonds compared to each other?

www.science.nelson.com

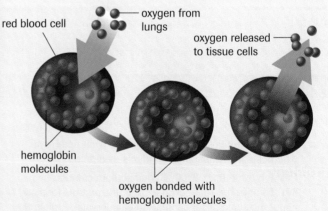

Figure 22
Hemoglobin transports oxygen to body tissue.

Figure 21
Glass is one special type of ceramic material.

Chapter 3 INVESTIGATIONS

INVESTIGATION 3.1

Molecular Models

Chemists use molecular models to explain and predict molecular structure, relating structure to the properties and reactions of substances.

Report Checklist

- ○ Purpose   ○ Design   ● Analysis
- ○ Problem   ○ Materials   ○ Evaluation
- ○ Hypothesis   ○ Procedure
- ○ Prediction   ● Evidence

Purpose
The purpose of this investigation is to test the ability of bonding theory to explain some known chemical reactions by using molecular models.

Problem
How can theory, represented by molecular models, explain the following series of chemical reactions that have been studied in a laboratory?

(a) $CH_4 + Cl_2 \rightarrow CH_3Cl + HCl$
(b) $C_2H_4 + Cl_2 \rightarrow C_2H_4Cl_2$
(c) $N_2H_4 + O_2 \rightarrow N_2 + 2 H_2O$
(d) $CH_3CH_2CH_2OH \rightarrow CH_3CHCH_2 + H_2O$
(e) $HCOOH + CH_3OH \rightarrow HCOOCH_3 + H_2O$

Design
Chemical reactions are simulated with model kits to test the ability of bonding theory to explain reaction evidence.

Materials
molecular model kit

Procedure
1. For reaction (a), construct a model for each reactant; record the structures in your notebook.
2. Rearrange the reactant models to produce structural models of the products, recording the new structures in your notebook.
3. Repeat steps 1 and 2 for each chemical reaction.

INVESTIGATION 3.2

Evidence for Polar Molecules

Use the empirical rules (**Table 8**, page 99) to predict the molecular polarity of the liquids provided.

Report Checklist

- ○ Purpose   ● Design   ● Analysis
- ○ Problem   ● Materials   ● Evaluation (1, 2, 3)
- ○ Hypothesis   ○ Procedure
- ● Prediction   ● Evidence

Purpose
The purpose of this investigation is to test the empirical rules for predicting molecular polarity.

Problem
Which of the liquids provided have polar molecules?

⚠ Check the MSDS for all liquids used, and follow appropriate safety precautions.

Materials
lab apron
eye protection
50 mL burette
clamp and stand
burette funnel
400 mL beaker
acetate strip (marked $+$)
vinyl strip (marked $-$)
paper towel
various liquids

Procedure
1. Charge an acetate plastic strip by rubbing with a paper towel.
2. Allow a thin stream of liquid to flow into a catch beaker.
3. Hold the charged strip close to the stream and observe the effect, if any.
4. Repeat steps 1 to 3 with the vinyl plastic strip.
5. Move to the next station and repeat steps 1 to 4.

Understanding Chemical Compounds 131

INVESTIGATION 3.3

Molecular Compound Melting Points

Report Checklist
- ○ Purpose
- ○ Problem
- ○ Hypothesis
- ○ Prediction
- ○ Design
- ○ Materials
- ○ Procedure
- ● Evidence
- ● Analysis
- ● Evaluation (1, 2, 3)

Intermolecular bonding can be expected to have some effect on the melting (freezing) point of a molecular substance. London force theory suggests that, if other factors are equal, substances composed of larger molecules should have higher melting points. In this investigation, you accurately measure the melting point of one of two molecular compounds. You will then combine your laboratory evidence with that of other research groups to see if a hypothesized relationship between melting point and molecular size is supported.

The molecular compounds used are both saturated fatty acids. One is commonly called lauric acid, and has the molecular formula $C_{12}H_{24}O_2(s)$. A molecule of this compound has its 12 carbons joined by single covalent bonds in a continuous "chain." Its IUPAC chemical name is dodecanoic acid ("dodec" is the prefix for 12), and its molecular formula is

$CH_3(CH_2)_{10}COOH(s)$

The other compound used for this investigation is stearic (octadecanoic) acid, $CH_3(CH_2)_{16}COOH(s)$. As you can see from its formula, a stearic acid molecule is very similar to a molecule of lauric acid, but the carbon chain is longer.

For the Analysis, plot both melting and freezing temperature–time curves on the same axes, and interpret the graph shapes for the most accurate determination of the melting point temperature. The class results for both compounds will then be combined with researched melting points for

- capric acid, $CH_3(CH_2)_8COOH(s)$
- myristic acid, $CH_3(CH_2)_{12}COOH(s)$
- palmitic acid, $CH_3(CH_2)_{14}COOH(s)$, and
- arachidic acid, $CH_3(CH_2)_{18}COOH(s)$

The melting points for these compounds may be found in chemistry reference books and online.

www.science.nelson.com

A plot of melting point versus electron count for the molecules of all six saturated fatty acids can then be used as evidence for a statement of any obvious relationship.

Purpose
The purpose of this investigation is to test the hypothesis of a relationship between melting point and molecular size.

Problem
What is the relationship between molecular size and melting point?

Hypothesis
The melting point of similar molecular compounds increases with an increase in molecular size, as described by the total number of electrons in the molecule.

Design
Hot and cold water baths are used to melt a solid compound and then to freeze and remelt it, with the temperature measured at regular intervals during both the freezing and remelting processes. The phase change (freezing point/melting point) temperature is found from superimposed temperature-versus-time graphs of this evidence. This is combined with evidence from other groups for a second similar compound and with information from other references to test the Hypothesis.

Materials
lab apron
eye protection
sample of lauric acid *or* stearic acid in a
   15 × 150 mm test tube
laboratory stand
test tube clamp
two thermometers or temperature probes
hotplate
two 400 mL beakers

Procedure

 Be careful not to splash the hot water. Avoid touching the hotplate surface with your hands. You may find the compound's odour objectionable; use a fume hood.
The glass of a thermometer bulb is very thin and fragile. Be careful not to bump, pull, or twist the thermometer.

1. On the hotplate, heat approximately 300 mL of water in a 400 mL beaker until the temperature is about 80 °C. Control the hotplate (turn it down, or switch it off and on) to keep the water hot but not boiling.
2. Set (or hold) the test tube containing the solid sample into the beaker of hot water on the hotplate until the sample is completely melted. Place the second thermometer or probe in the melted compound in the test tube.
3. Fill the second 400 mL beaker with cold tap water and set it on the base of a laboratory stand. Adjust a test tube clamp on the stand so that it will hold the test tube with the sample suspended in the water in the beaker (**Figure 1**).
4. Remove the test tube from the hot water, and place it in the clamp. Hold the thermometer against the side of the test tube so that the bulb is located about halfway between the bottom and the top of the liquid compound in the test tube. The thermometer should be positioned so that it will be held in this position when the liquid compound solidifies. Do *not* attempt to move the thermometer once it is immobilized by the solidified compound.
5. Measure and record the temperature as precisely as possible every 15 s as the compound cools. (See Appendix F.3 for a note on precision when reading a thermometer.)
6. When the compound has obviously been completely solidified for a few minutes, stop recording measurements.
7. Lift the test tube assembly by raising the clamp out of the cold water beaker, and lower the test tube assembly into the beaker of hot water on the hotplate, and begin recording warming measurements by repeating step 5.
8. When the compound has obviously been completely liquified for a few minutes, stop recording measurements.
9. Remove the thermometer from the melted sample, and wipe it clean with a paper towel. Place the test tube back in the cold water beaker to solidify the sample before returning it.

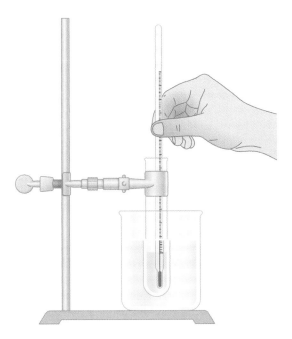

Figure 1
Setup for step 3

INVESTIGATION 3.4

Hydrogen Bonding

Report Checklist
- ○ Purpose
- ○ Problem
- ○ Hypothesis
- ● Prediction
- ○ Design
- ○ Materials
- ● Procedure
- ● Evidence
- ● Analysis
- ● Evaluation (1, 2, 3)

Breaking bonds absorbs energy, and forming bonds releases energy. Forming new bonds within a liquid should produce a temperature effect. For this investigation, write a Prediction that includes a bonding concept. The Prediction should include reference to a measurable (testable) result of mixing the liquids. Add to the design by identifying the manipulated and responding variables, and the major controlled variable.

Purpose
The purpose of this investigation is to test the concept of hydrogen bonding.

Problem
How does the temperature change upon mixing ethanol, $C_2H_5OH(l)$, with water compare with the temperature change upon mixing glycerol (glycerin), $C_3H_5(OH)_3(l)$, with water?

Design
Equal volumes of ethanol and water, and then of glycerol and water, are mixed. The temperature change is recorded in each case (**Figure 2**).

Materials
- lab apron
- eye protection
- distilled water
- ethanol
- glycerol
- two nested polystyrene cups
- two 10 mL graduated cylinders
- cup lid with centre hole
- two thermometers or temperature probes
- 250 mL beaker (to support the cups)

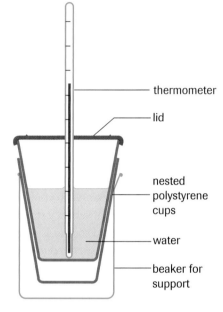

Figure 2
Suggested set-up of apparatus

INVESTIGATION 3.5

Classifying Unknown Solids

Report Checklist
- ○ Purpose
- ○ Problem
- ○ Hypothesis
- ○ Prediction
- ● Design
- ● Materials
- ● Procedure
- ● Evidence
- ● Analysis
- ● Evaluation (1, 3)

Design an investigation that enables you to use physical properties, such as conductivity, hardness, solubility, and melting point, to classify four unknown solids as ionic, metallic, molecular, or covalent network.

Purpose
The purpose of this investigation is to use empirical definitions to classify solid substances.

Problem
To what class of solids do the four mystery solids belong?

Chapter 3 SUMMARY

Outcomes

Knowledge

- explain why formulas for ionic compounds refer to the simplest whole-number ratio of ions that result in a net charge of zero (3.1)
- define valence electron, electronegativity, and ionic bond (3.1, 3.3)
- use the periodic table and Lewis structures to support and explain ionic bonding theory (3.1)
- explain how an ionic bond results from the simultaneous attraction of oppositely charged ions (3.1)
- draw or build models of common ionic lattices and relate structures and properties (3.5)
- explain why the formulas for molecular substances refer to the number of atoms of each constituent element (3.2)
- relate electron pairing to covalent bonds (3.1, 3.2)
- build models depicting the structure of simple covalent molecules, including selected organic compounds (3.2)
- draw electron-dot diagrams of atoms and molecules, writing structural formulas for molecular substances and using Lewis structures to predict bonding in simple molecules (3.2)
- apply VSEPR theory to predict molecular shapes (3.3)
- illustrate, by drawing or building models, the structure of simple molecular substances (3.2)
- explain intermolecular forces, London (dispersion) forces, dipole–dipole attractions, and hydrogen bonding (3.4)
- relate properties of substances to the predicted intermolecular bonding in the substance (3.4, 3.5)
- determine the polarity of a molecule based on simple structural shapes and unequal charge distribution (3.3)
- describe bonding as a continuum ranging from complete electron transfer to equal sharing of electrons. (3.3, 3.4)

STS

- state that the goal of science is knowledge about the natural world (3.1, 3.3, 3.5)
- list the characteristics of empirical and theoretical knowledge (3.1)
- evaluate scientific knowledge and restrict, revise, or replace it where necessary (3.1, 3.4, 3.5)
- state examples of science leading technology and technology leading science (3.1, 3.5)

Skills

- initiating and planning: design an investigation to determine the properties of ionic compounds (3.5); describe procedures for safe handling, storage, and disposal of laboratory materials (3.3, 3.4, 3.5); state a hypothesis and make a prediction about the properties of molecular substances based on attractive forces (3.3, 3.4)
- performing and recording: draw Lewis formulas and build models of ionic solids (3.5); build models depicting the structure of simple covalent molecules (3.2, 3.4); carry out an investigation to determine the melting points of molecular substances (3.4)
- analyzing and interpreting: identify trends and patterns in the melting points of a related series of molecular substances (3.4); determine the properties of ionic compounds (3.5)
- communication and teamwork: working cooperatively, critically analyze and evaluate models and graphs constructed by others (3.2, 3.3, 3.5)

Key Terms

3.1
structural formula
valence electron
orbital
valence orbital
bonding electron
lone pair
octet rule
Lewis symbol
electronegativity
covalent bond
ionic bond

3.2
bonding capacity
empirical formula
molecular formula
Lewis formula
structural formula
stereochemical formula

3.3
stereochemistry
VSEPR theory
polar molecule
nonpolar molecule
nonpolar covalent bond
polar covalent bond
bond dipole

3.4
intermolecular force
van der Waals force
dipole–dipole force
London force
isoelectronic molecules
hydrogen bond

3.5
crystal lattice
covalent network

EXTENSION

CBC radioONE QUIRKS & QUARKS

The Big Rip
The Universe will end not with a bang, or a whimper, but with a rip, according to cosmologist Dr. Robert Caldwell. He has been thinking about a mysterious repulsive force often called the "dark energy." Dr. Caldwell's calculations suggest that the dark energy could actually be a different kind of force, previously unknown, and could grow even more powerful as the universe gets older—possibly resulting in the ripping apart of all matter.

www.science.nelson.com

MAKE a summary

1. Covalent bonds and intermolecular forces can be explained in a unified way by describing the central entity that is simultaneously attracted (electrostatically) to the surrounding entities. Copy and complete **Table 1**.

Table 1 Forces Acting Between Entities

Force or bond	Central entity	Surrounding entities
covalent		
covalent network		
dipole–dipole		
hydrogen		
ionic		
London		
metallic		

2. Each class of substance has a characteristic set of properties. Copy and complete **Table 2** using relative descriptions such as low, high, and variable. (Indicate n/a if not applicable.)

Table 2 Summary of Properties of Substances

Substance	Hardness	Melting point	Electrical conductivity		
			Solid	Liquid	Solution
molecular					
ionic					
covalent network					
metallic					

3. Refer back to your answers to the Starting Points questions at the beginning of this chapter. How has your thinking changed?

Go To

The following components are available on the Nelson Web site. Follow the links for *Nelson Chemistry Alberta 20-30*.
- an interactive Self Quiz for Chapter 3
- additional Diploma Exam-style Review questions
- Illustrated Glossary
- additional IB-related material

There is more information on the Web site wherever you see the Go icon in this chapter.

Unit 1 REVIEW

Many of these questions are in the style of the Diploma Exam. You will find guidance for writing Diploma Exams in Appendix H. Exam study tips and test-taking suggestions are on the Nelson Web site. Science Directing Words used in Diploma Exams are in bold type.

DO NOT WRITE IN THIS TEXTBOOK.

Part 1

1. The following scientists made major contributions to bonding theory.
   NR
   1. Abegg
   2. Kekulé
   3. Pauling
   4. Lewis

   When arranged in chronological order, the order is ___, ___, ___, and ___.

2. The most significant new concept of Lewis theory was
   A. the octet rule for atom stability
   B. electron gain or loss in ion formation
   C. electron sharing in covalent bonds
   D. different electron energy levels

3. The Lewis symbol for a single nitrogen atom would show
   A. 1 bonding electron and 3 lone pairs
   B. 2 bonding electrons and 2 lone pairs
   C. 1 bonding electron and 2 lone pairs
   D. 3 bonding electrons and 1 lone pair

4. A Lewis formula for the molecule PCl_3 would show
   A. 13 electron pairs
   B. 10 electron pairs
   C. 8 electron pairs
   D. 4 electron pairs

5. X-ray diffraction evidence led to development of
   A. Lewis formulas
   B. crystal lattice models
   C. VSEPR theory
   D. the octet rule

6. According to VSEPR theory, which molecule has a trigonal pyramidal shape?
   A. BF_3
   B. CO_2
   C. NCl_3
   D. H_2S

7. When the following covalent bonds are arranged in order
   NR from least polar to most polar, the order is ___, ___, ___, and ___.
   1. N–O
   2. C–I
   3. O–H
   4. H–Cl

8. Intermolecular force theory best explains
   A. surface tension of a liquid
   B. electrical conductivity of a metal
   C. hardness of a covalent network solid
   D. melting point of an ionic solid

9. Theory indicates that metallic bonding depends on
   A. high electronegativity
   B. mobile valence electrons
   C. polar covalent bonds
   D. electrical conductivity

10. A molecule of a substance with physical properties primarily explained by London forces is
    A. SiC
    B. KCl
    C. Na_3P
    D. PH_3

11. Hydrazine, $N_2H_4(l)$, is used in rocket fuels. The shape of the bonding around each nitrogen atom in a molecule of hydrazine is predicted to be
    A. tetrahedral
    B. linear
    C. trigonal pyramidal
    D. trigonal planar

12. Which of the following molecules is polar?
    A. C_2H_6
    B. CH_3OH
    C. CO_2
    D. CCl_4

13. The *correct* statement about covalent bonding is:
    A. Almost all covalent bonds involve equally shared electrons and so are normally nonpolar.
    B. Different covalent bonds can range from equal to extremely unequal electron sharing.
    C. The total number of electrons shared in covalent bonds by a central atom is always eight.
    D. Each atom always "contributes" one bonding electron when a covalent bond forms.

14. The compound strontium nitrate, $Sr(NO_3)_2(s)$, has which of the following structures?
    A. covalent bonding and ionic bonding
    B. covalent bonding only
    C. ionic bonding only
    D. metallic and ionic bonding

15. The shape of permanganate ion, MnO_4^-, cannot be predicted from the level of VSEPR theory covered in this textbook. However, if you know that the ion's Lewis diagram shows no lone pairs on the central atom, you can conclude that the shape must be
    A. trigonal pyramidal
    B. linear
    C. trigonal planar
    D. tetrahedral

Chemical Bonding—Explaining the Diversity of Matter 137

Part 2

Use the information below, **Figure 1**, and **Table 1** to answer questions 16 to 19.

When dry chlorine gas is passed over heated phosphorus, it ignites, and the reaction produces two different products: mostly phosphorus trichloride, with some phosphorus pentachloride mixed in.

Figure 1
Lewis formulas for phosphorus trichloride and phosphorus pentachloride.

Table 1 Physical Properties of Phosphorus Chlorides

Compound	Melting point (°C)	Boiling point (°C)	Colour/State (SATP)
PCl_3	−112	76	clear, colourless liquid
PCl_5	148 (under pressure)	sublimes (s → g) at 165	pale yellow crystals

16. Which of the compounds has molecular structure that does not follow the octet rule?

17. **Explain**, using intermolecular bonding theory, why only one of the compounds is a solid at SATP. Your explanation should state which type of intermolecular bonding is stronger, and why.

18. Both compounds are corrosive to skin and tissue, and both react with water. Since human tissue is largely water, the compounds are doubly hazardous because reaction with water produces a very common acid that is also extremely corrosive to tissue. Using each compound's atomic composition, **predict** which common acid is produced.

19. State which compound has a molecular shape that is easily predictable using VSEPR theory, and what shape those molecules will have.

20. Empirically, how does the chemical reactivity vary
    (a) among the elements in Groups 1 and 2 of the periodic table?
    (b) among the elements in Groups 16 and 17?
    (c) within Period 3?
    (d) within Group 18?

21. **Identify** two generalizations that describe how the trends in atomic electronegativity vary with atom position in the periodic table.

22. How are the positions of two reacting elements in the periodic table related to the type of compound and bond formed? State two generalizations.

23. What is the maximum number of electrons in the valence level of an atom of a representative element?

24. **Compare** the electronegativities of main group metals with those of main group nonmetals.

25. Draw a Lewis formula for each of the following atoms. Then **determine** the number of bonding electrons and lone pairs for each atom.
    (a) Ne
    (b) Al
    (c) Ge
    (d) N
    (e) Br

26. Draw Lewis formulas for atoms of the following elements and **predict** their bonding capacity:
    (a) calcium
    (b) chlorine
    (c) phosphorus
    (d) silicon
    (e) sulfur

27. **Describe** the requirements for valence electrons and orbital occupancy in order for a covalent bond to form between two approaching atoms.

28. According to atomic theory, how many lone electron pairs are on the central atom in molecules of the following substances?
    (a) $SCl_2(l)$
    (b) $NH_3(g)$
    (c) $H_2O(l)$
    (d) $CCl_4(l)$
    (e) $PCl_3(l)$

29. Compounds of metals and carbon are used in engineering because of their extreme hardness and strength (**Figure 2**). The carbon in these metallic carbides behaves as a C^{4-} ion, for example, in tungsten carbide, WC(s). Draw the Lewis formula for a carbide ion.

Figure 2
Steel drill bits, when tipped with tungsten carbide, become hard enough to drill holes in solid rock, concrete, and ceramic tile.

30. The American chemist Gilbert Lewis suggested that when atoms react, they achieve a more stable electron configuration. **Describe** the electron arrangement that gives an atom maximum stability.

31. Draw a Lewis formula and a structural formula to describe each molecule shown in these unbalanced equations.
    (a) $N_2(g) + I_2(s) \rightarrow NI_3(s)$
    (b) $H_2O_2(aq) \rightarrow H_2O(l) + O_2(g)$

32. **Why** did scientists create the concept of double and triple covalent bonds?

33. What empirical evidence is there for double and triple bonds?

34. What information does a boiling point provide about intermolecular forces?

35. List three types of intermolecular forces, and give an example of a substance having each type of force.

36. Write an empirical definition (based on observable properties) for an ionic compound.

37. **Summarize** the theoretical structure (based on concepts and theory) of ionic compounds.

38. Use Lewis formulas to **describe** the electron rearrangement in the following chemical reactions:
    (a) magnesium atoms + sulfur atoms → magnesium sulfide
    (b) aluminium atoms + chlorine atoms → aluminium chloride

39. The VSEPR model includes several concepts related to atomic theory. **Describe** the following concepts:
    (a) valence shell
    (b) bonding pair
    (c) lone pair
    (d) electron pair repulsion

40. **Outline** the steps involved in predicting the shape of a molecule using VSEPR theory.

41. Using VSEPR theory, **predict** the shape around the central atom of each of the following molecules:
    (a) $SCl_2(l)$
    (b) $BCl_3(g)$
    (c) $SiH_4(g)$
    (d) $CCl_4(l)$
    (e) $HCN(g)$
    (f) $OCl_2(g)$
    (g) $NCl_3(g)$
    (h) $H_2O_2(l)$

42. What is the difference in the meaning of the subscripted numbers in a molecular formula and in an ionic formula?

43. List the types of bonds and forces that, according to our current theories, are believed to be present in
    (a) $C_6H_{14}(l)$
    (b) $Fe(s)$
    (c) $CaCl_2(s)$
    (d) $C_2H_3Cl_3(l)$
    (e) $C(s)$ (diamond)
    (f) $Fe_2O_3(s)$
    (g) $C_2H_5OH(l)$ (**Figure 3**)
    (h) $C_{12}H_{22}O_{11}(s)$

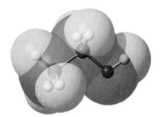

Figure 3
The structure of ethanol molecules results in intermolecular bonding that gives the liquid some very useful properties.

44. Draw diagrams of models to describe the following types of bonding:
    (a) ionic bonding
    (b) dipole–dipole forces
    (c) hydrogen bonding
    (d) metallic bonding

45. Metals are shiny, malleable (bendable) conductors of heat and electricity (**Figure 4**).
    (a) Write full electron energy-level diagrams for these metals: magnesium, potassium, lithium, and aluminium.
    (b) **How** do the valence electron arrangements of metals explain their conductivity and malleability?

46. Carbon dioxide is used by green plants in the process of photosynthesis and is also a greenhouse gas produced by fossil fuel combustion.
    (a) Draw a Lewis formula for carbon dioxide.
    (b) **Predict** its shape and bond angle.
    (c) Using appropriate bonding theories, **predict** and **explain** the polarity of carbon dioxide.

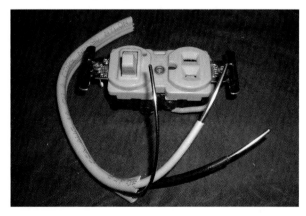

Figure 4
Copper is strong, flexible, unreactive, and an excellent conductor of electricity, making it the material of choice for wiring household electric circuits.

47. The polarity of a molecule is determined by theory from bond polarity and molecular shape.
    (a) **Determine** the polarity of the bonds N–Cl and C–Cl.
    (b) **Predict** whether the molecules NCl_3 and CCl_4 are polar or nonpolar. Justify your predictions.

48. Use appropriate bonding theory to **explain** the following:
    (a) BeH_2 is nonpolar; H_2S is polar.
    (b) BH_3 is trigonal planar; NH_3 is trigonal pyramidal.
    (c) LiH has a melting point of 688 °C; that of HF is −83 °C.

49. Use a theory of intermolecular bonding to **explain** the sequence of boiling points in the following compounds, from a family called alkyl bromides:
    - $CH_3Br(g)$ (4 °C)
    - $C_2H_5Br(l)$ (38 °C)
    - $C_3H_7Br(l)$ (71 °C)

50. Name the intermolecular forces present in the following compounds, and account for the difference in their boiling points:
    (a) $CH_4(g)$ (−164 °C)
    (b) $NH_3(g)$ (−33 °C)
    (c) $BF_3(g)$ (−100 °C)

51. All chemical bonds are ultimately thought to be the result of simultaneous attractions between opposite electrostatic charges. For each chemical bond listed below, indicate which types of charged entities and/or particles are involved:
    (a) covalent bond
    (b) London forces
    (c) dipole–dipole forces
    (d) hydrogen bonds
    (e) ionic bond

52. Ionic compounds and metals have different physical properties because of the different forces involved. For example, sodium chloride and nickel have nearly identical molar masses; however, their melting points, conductivity, and solubility in water are quite different.
    (a) **Explain** the large difference in melting point between sodium chloride (801 °C) and nickel metal (1453 °C).
    (b) **Predict** the electrical conductivity of each of these substances in the solid state, and provide a theoretical explanation for your prediction.
    (c) **Predict** the solubility in water of each substance, and provide a theoretical explanation for your prediction.

53. Name all the bonding forces acting in each of the following substances:
    (a) hexane, $C_6H_{14}(l)$
    (b) 1-butanol, $C_4H_9OH(l)$
    (c) ethylamine, $C_2H_5NH_2(l)$
    (d) chloroethane, $C_2H_5Cl(l)$
    (e) calcium carbonate, $CaCO_3(s)$
    (f) diamond, $C(s)$

54. **Compare** entities and bonding forces in the following pairs of solids:
    (a) metallic and covalent network
    (b) molecular and ionic

55. Given the very wide range of boiling points of metal elements, what can be hypothesized about the strength of metallic bonding?

56. **How** can the boiling of water be used as an example to illustrate the relative strengths of hydrogen bonds and covalent bonds?

57. Some people find the odour from cooking fish to be unpleasant. Methylamine, $CH_3NH_2(g)$, is one of the compounds responsible for the odour (**Figure 5**).
    (a) Draw Lewis and structural formulas to describe methylamine.
    (b) Use VSEPR theory to **predict** the shape around the carbon and nitrogen atoms in methylamine.
    (c) Methylamine and ethane, $C_2H_6(g)$, have similar molar masses. **Explain** why the boiling point of methylamine is −6 °C while that of ethane is much lower, −89 °C.
    (d) Use structural formulas to rewrite the following equation for the reaction of methylamine with acetic acid:
    $CH_3NH_2(aq) + CH_3COOH(aq) \rightarrow CH_3NH_3^+(aq) + CH_3COO^-(aq)$
    (e) Give a practical reason why vinegar (5% acetic acid) and lemon juice (dilute citric acid) are often served along with cooked fish.

Figure 5
A "fishy" smell is partly due to the presence of the volatile substance methylamine.

58. The most common oxides of Period 2 elements are as follows:
 $Na_2O, MgO, Al_2O_3, SiO_2, P_2O_5, SO_2, Cl_2O$
 (a) Which oxides are classified as ionic, and which are classified as molecular when using the simple metal-nonmetal or nonmetal-nonmetal combination rule?
 (b) Which oxide is observed to be neither ionic nor molecular, and how does bonding theory explain this?
 (c) Calculate the difference in electronegativity between the two elements in each oxide.
 (d) **How** is the difference in electronegativity related to the properties of the compound?

59. **Predict** the structural formula, molecular formula, and name for the simplest product in each of the following chemical reactions. Which product does *not* follow the octet rule of covalent bonding?
 (a) $H_2(g) + P_4(s) \rightarrow$
 (b) $Si(s) + Cl_2(g) \rightarrow$
 (c) $C(s) + O_2(g) \rightarrow$
 (d) $B(s) + F_2(g) \rightarrow$

60. Chlorine is a very reactive element that forms stable compounds with most other elements. For each of the following chlorine compounds, draw Lewis and structural diagrams, and then **predict** the polarity of the molecules that obey the octet rule. Use the concept of symmetry to speculate about the polarity of the molecules that do *not* obey the octet rule.
 (a) NCl_3
 (b) $SiCl_4$
 (c) PCl_5
 (d) SCl_6

61. Review the focusing questions on page 72. Using the knowledge you have gained from this unit, briefly **outline** a response to each of these questions.

Extension

62. Modern technologies like spectroscopy and X-ray diffraction allow scientists to measure indirectly the length of chemical bonds. Use the evidence in **Table 2** to create a generalization about the effect of bond type on bond length.

Table 2

Typical compound	Covalent bond type	Bond length (nm)
CH_4	C–H	0.109
C_2H_6	C–C	0.154
C_2H_4	C=C	0.134
C_2H_2	C≡C	0.120
CH_3OH	C–O	0.143
CH_3COCH_3	C=O	0.123
CH_3NH_2	C–N	0.147
CH_3CN	C≡N	0.116

63. Medical professionals are concerned about the level of saturated and unsaturated fats in the foods we eat. Of even greater concern is the consumption of "trans fats," which are produced by chemically reacting hydrogen with natural oils. Research "saturated fats," "unsaturated fats," and "trans fats," and find out how these terms are related to the concept of single and double bonds. Locate three products such as margarine and snack foods at home or in a grocery store, and list any information printed on the labels or packaging that describes the products' fat content (**Figure 6**). Briefly **summarize** the medical concerns about consumption of trans fats.

www.science.nelson.com

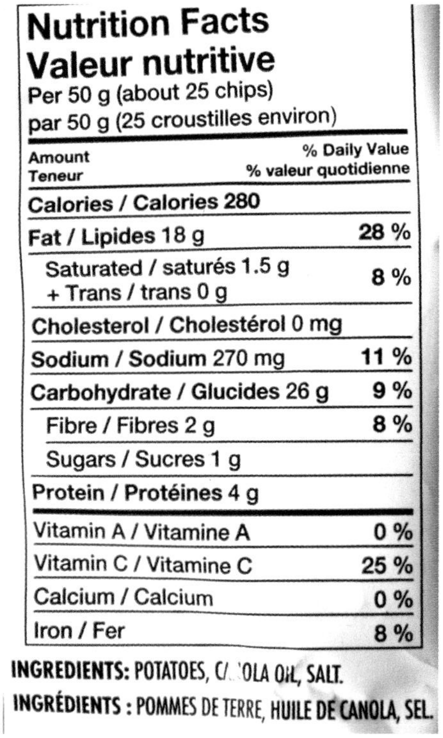

Figure 6
Potato chip nutritional data

unit 2
Gases as a Form of Matter

Modern science has a relatively short history of just a few hundred years. An important goal of modern science is to acquire empirical and theoretical knowledge about aspects of the natural world such as the air. This is usually accomplished by systematically reducing nature into categories that are studied in ever-increasing detail. This method has not been the only way of knowing about our world. For example, Aboriginal peoples developed an indigenous knowledge of nature as evidenced by tens of thousands of years of survival. From an Aboriginal perspective, air is one of the four inter-connected "elements" of nature—earth, air, fire, and water. Aboriginal peoples traditionally had a knowledge of and respect for the air in which we live. They recognized that air plays an important role in maintaining the balance of nature. Their way of knowing did not include experimenting with air; systematic experimentation is more in the tradition of modern science.

According to modern scientific knowledge, air is a mixture of gases. The most important gas for humans is oxygen, which is produced by photosynthesis. This now common-place knowledge required some time to develop. Even at the heyday of alchemy, no one had conceived of the idea that a gas might be produced from a chemical reaction.

The study of gases was essential to the development of many chemical theories and quantitative descriptions. Advancements in the study of gases progressed hand-in-hand with the development of technologies that enabled more precise measurements of temperature and pressure. Today, the study of gases is crucial to our understanding of natural and technological phenomena. We need to understand gases to study natural phenomena such as weather patterns. Scientific knowledge of gases helps us understand modern technologies such as air bags, gas turbines, and lasers.

As you progress through the unit, think about these focusing questions:
- How do familiar observations of gases relate to modern scientific models describing the behaviour of gases?
- What is the relationship between the pressure, temperature, volume, and chemical amount of a gas?

Unit **2**

GENERAL OUTCOME

In this unit, you will
- explain molecular behaviour using models of the gaseous state of matter

Gases as a Form of Matter

Unit 2
Gases as a Form of Matter

ARE YOU READY?

These questions will help you find out what you already know, and what you need to review, before you continue with this unit.

Knowledge

1. This unit is about gases. Evidence shows that the state of a substance (solid, liquid, or gas) is determined by temperature and pressure. Copy and complete **Table 1** by indicating the state of each substance for the given temperature at standard ambient pressure. The melting and boiling points of the elements can be referenced on the periodic table.

Table 1 States of Matter of Substances at Different Temperatures

Substance	State at −150 °C	State at 25 °C	State at 150 °C
argon			
bromine			
chlorine			
sulfur			
water			

2. Copy **Table 2** and, using your own experience, complete the Empirical properties column by describing the shape and volume, compressibility, and ability to flow for each state of matter.

Table 2 Empirical Properties of States of Matter

State	Empirical properties	Example
solid		 **Figure 1** A diamond ring
liquid		 **Figure 2** Coloured liquids
gas		 **Figure 3** A helium-filled balloon

▶ Prerequisites

Concepts
- states of matter
- empirical and theoretical knowledge
- chemical names and formulas
- mole concept and calculations
- balanced chemical reaction equations

Skills
- scientific problem solving
- WHMIS symbols
- laboratory safety rules
- graphing

You can review prerequisite concepts and skills in the **Chemistry Review Unit and in the Appendices.**
A Unit Pre-Test is also available online.

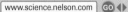

3. Scientists have empirically determined that the production of sulfur dioxide is a significant factor in the formation of acid rain. Sulfur dioxide is oxidized in air (oxygen) to form sulfur trioxide. The sulfur trioxide readily combines with water to form sulfuric acid.
   (a) Write the balanced chemical equations to communicate the conversion of sulfur dioxide gas to sulfur trioxide gas, and sulfur trioxide gas to sulfuric acid.
   (b) Translate each chemical reaction equation in (a) into an English sentence. Include chemical names and chemical amounts.

STS Connections

4. Distinguish between the two main types of scientific knowledge. Which type usually comes first?

5. Research a local weather forecast, then copy and complete **Table 3**.

   Table 3 Meteorological Information

Information	Value/Description	Technology (instrument)
date		
location		
high/low temperatures (°C)		
local atmospheric pressure (kPa)		
local relative humidity (%)		

Skills

6. Scientists frequently communicate empirical knowledge with tables and graphs. In many jobs, the salary (without deductions) is directly related to the time spent on the job.
   (a) Using axes like those in **Figure 4**, sketch a line graph for the direct variation between these two variables. (No numbers are required.)
   (b) Like many relationships in science, the slope of the graph has a specific meaning. What does the slope of your graph in (a) represent?
   (c) Suppose that the salary varied inversely with the time spent. Using axes like those in (a), draw the graph for this inverse relationship. Would you want a job that pays this way?

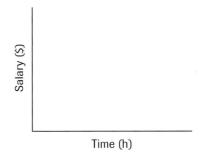
Figure 4
Salary–time axes

7. Flammable and combustible materials require special attention in a laboratory.
   (a) What do the WHMIS symbols in **Figure 5** represent?
   (b) Copy and complete **Table 4** by using check marks to indicate the types of fire extinguishers that are suitable for the various classes of fire.

   Table 4 Classes of Fire and Extinguishers

Class of fire	Water	Carbon dioxide	Dry chemical
Class A (wood, paper, cloth)			
Class B (flammable liquids)			
Class C (live electrical equipment)			

Figure 5
WHMIS symbols (Appendix E.2)

chapter 4

Gases

In this chapter

- Exploration: Creating and Testing Ideas about Gases
- Lab Exercise 4.A: A Thought Experiment about Gas Properties
- Investigation 4.1: Pressure and Volume of a Gas
- Investigation 4.2: Temperature and Volume of a Gas
- Biology Connection: Gas-Dependent Processes
- Web Activity: The Combined Gas Law
- Web Activity: Malcolm King
- Case Study: Compressed Gases
- Web Activity: Elizabeth MacGill
- Web Activity: "Designer Air" for Tires
- Case Study: Weather Forecasts
- Web Activity: The Ideal Gas Law
- Lab Exercise 4.B: Evaluating an Experimental Design
- Investigation 4.3: Using the Ideal Gas Law

Career Connections:
Respiratory Therapist; Meteorologist

The scientific study of the properties of gases has led not only to a better understanding of the natural world, but also to many modern technologies that involve gases. The photograph in **Figure 1** is a dramatic depiction of how a gas can save a human life. In a car crash, an air bag, especially in combination with a seat belt, can protect an adult occupant from serious injury. Upon collision, sensors in the steering column and in the bumper initiate the decomposition of sodium azide into sodium metal and nitrogen gas. This reaction is extremely fast. Nitrogen gas is produced and expands into the bag in less than 0.04 s. After cushioning the impact, the air bag quickly deflates as the nitrogen gas escapes through the permeable fabric. Instead of taking a trip to the hospital, the driver takes a trip to the automobile body shop to have the air bag mechanism recharged and the triggering devices reset.

Air bags are not the only use of gases in the operation of automobiles. Tires and shock absorbers are inflated with pressurized air to provide a safe and comfortable ride. Air enters through the car's vents and is either cooled by the air conditioner to keep us comfortable on hot summer days or heated by the car engine to keep us warm in winter. Inside the combustion cylinder of the engine, a gasoline and oxygen explosion produces a gas at a high temperature, which moves a piston. This is an example of converting chemical energy into kinetic energy (motion). Finally, the gases emitted by the exhaust, such as carbon oxides and nitrogen oxides, diffuse into the atmosphere as pollutants.

As you can see, gases play an important role in both technology and our environment. The study of gases illustrates how empirical descriptions and advances in technology have led to a better understanding of gases in both natural and technological phenomena.

STARTING Points

Answer these questions as best you can with your current knowledge. Then, using the concepts and skills you have learned, you will revise your answers at the end of the chapter.

1. Since many gases are invisible, how do you think we can study them?
2. When you are solving problems, the gas laws are like tools. How do you know which gas law to use?
3. Do all gases have the same physical properties? Why or why not?

Figure 1
Empirical and theoretical knowledge of gas properties have useful applications. Air bags are a good example of how gas concepts are used in a life-saving technology.

▶ Exploration Creating and Testing Ideas about Gases

Coming up with ideas to explain the results of experiments with gases requires some imagination, because most gases are invisible. Try to figure out what is happening to the aluminium cans and the water vapour (a gas in our atmosphere) in this activity.

Materials: water, 5 aluminium pop cans, 5 large beakers or containers, hot plate, beaker tongs, ice cubes, eye protection, heat-proof gloves or mitts

 Use care when handling hot items. Steam can scald skin. Switch off the hot plate immediately after use.

- Place about 20 mL of water in an empty aluminium pop can.
- Heat the can on a hot plate until steam rises steadily out of the top for a couple of minutes.
- Fill a large beaker to near the top with cold water.
- Using the tongs, lift the can off the hot plate.
- Invert the can, and dip the top rim of the can just under the surface of the water in the beaker.
- Record your observations.

(a) Create a Hypothesis for what happens.

- Repeat the Procedure without placing any water in the can. (Heat the can for a few minutes.)

(b) What happened? Does this support or refute your Hypothesis?

(c) Using your original or revised Hypothesis, predict the results if you repeat the Procedure inverting the steaming can into ice water and warm water.

(d) Try each of these tests, then judge your Predictions and Hypotheses.

- Recycle the cans.

4.1 Empirical Properties of Gases

CAREER CONNECTION

Respiratory Therapist
As vital members of a medical team, respiratory therapists help diagnose and treat patients with breathing problems.
Discover more about the responsibilities and working conditions of respiratory therapists.

www.science.nelson.com

Gases have always been important to us—we need them to breathe, after all. Aboriginal peoples recognize that air provides a physical interconnection, linking us with everything else in the world. The wind is important for survival, bringing information about weather and game, so it is highly valued. As Western society has advanced technologically, the importance of gases has been expanding. We use gases in our daily lives—natural gas as fuel, gases as refrigerants, and anesthetic gases for surgery. We also generate gases for special uses. For example, we create artificial atmospheres for deep-sea diving and for the exploration of outer space. It is not surprising, then, that the study of gases has quite a long history in chemistry. People studied many empirical properties of gases long before the development of our modern understanding of the composition and molecular motion of substances. In fact, it was the large body of empirical knowledge about gases that made possible the development of some important ideas, such as atomic theory, kinetic molecular theory, and the mole concept. Let's now look more closely at the empirical properties of gases. Keep in mind that any empirical knowledge in science can be communicated in several ways: simple descriptions like observations, tables of evidence, graphs, empirical hypotheses, empirical definitions, generalizations, and scientific laws.

LAB EXERCISE 4.A

A Thought Experiment about Gas Properties

Report Checklist

- ○ Purpose
- ○ Problem
- ○ Hypothesis
- ○ Prediction
- ○ Design
- ○ Materials
- ○ Procedure
- ○ Evidence
- ● Analysis
- ● Evaluation (1, 3)

Imagine that you have five gas cylinders that have the volumes and temperatures listed in **Table 1** and contain the same mass of nitrogen gas. Complete the Analysis by proposing a hypothesis (with reasoning) to answer the Problem. In the Evaluation, identify any difficulties you had in answering the question, state how certain you are about your answer, and list some additional information you think you need to improve the certainty.

Purpose
The purpose of this exercise is to create a hypothesis based on differing gas properties.

Problem
What is the order of gas cylinders, from most likely to explode to least likely to explode?

Evidence

Table 1 Comparison of Nitrogen Gas Cylinders

Cylinder number	Volume (L)	Temperature (°C)
1	1.0	800
2	2.0	200
3	2.0	300
4	4.0	200
5	4.0	800

Pressure

We now understand that we live at the bottom of an ocean of air. Air has many properties that can be altered experimentally in a laboratory, including temperature and pressure (familiar to us from weather reports), volume, and chemical amount of gas.

In any controlled experiment, the plan is to manipulate one variable and observe its effect on another variable while keeping all other properties constant. We begin our study of gases by looking at the relationship between pressure and volume with temperature and chemical amount of gas kept constant.

Your weight, due to Earth's gravity, is the force you exert on the ground, and the ground pushes back with an equal but opposite force. However, the force you exert can be distributed over a larger or smaller area. The area is larger when you lie down and smaller when you stand on the tips of your toes. The greater the area, the lower the **pressure**, or force per unit area. This observation was put to practical use by many Aboriginal groups as they developed snowshoes (**Figure 1**). The pressure of a gas is also a measure of force per unit area, but in this case, according to modern scientific theories, the force is exerted by the moving molecules as they collide with objects in their path, particularly the walls of a container.

Scientists have agreed, internationally, on units, symbols, and standard values for pressure. The SI unit for pressure, the pascal (Pa), represents a force of 1 N (newton) on an area of 1 m^2; 1 Pa = 1 N/m^2. Atmospheric pressure and the pressure of many gases are conveniently measured in kilopascals (kPa); 1 kPa = 1000 Pa = 1 kN/m^2 (exactly).

Atmospheric pressure is the force per unit area exerted by air on all objects. At sea level, average atmospheric pressure is about 101 kPa. Scientists used this value as a basis to define one standard atmosphere (1 atm), or *standard pressure,* as exactly 101.325 kPa. For many years, standard conditions for work with gases were a temperature of 0 °C and a pressure of 1 atm (101.325 kPa); these conditions are known as **standard temperature and pressure (STP)**. However, 0 °C is not a convenient temperature, because laboratory temperatures are not close to 0 °C. Scientists have since agreed to use another set of standard conditions, not only for gases but also for reporting the properties of other substances. The new standard is called **standard ambient temperature and pressure (SATP)**, defined as 25 °C and 100 kPa. The new standard is much closer to laboratory conditions and therefore more convenient.

Since the empirical properties of gases were measured long before the development of SI, the pressure of gas has been expressed in a bewildering variety of units over the years. In 1643, Evangelista Torricelli (1608–1647), following up on a suggestion from Galileo, accidentally invented a way of measuring atmospheric pressure. He was investigating Aristotle's notion that a vacuum cannot exist in nature. Torricelli's experimental design involved inverting a glass tube filled with mercury and placing it into a tub also containing mercury (**Figure 2**). Noticing that the mercury level changed from day to day, he realized that his device, which came to be called a mercury barometer, was a means of measuring atmospheric pressure. In Torricelli's honour, standard pressure was at one time defined as 760 torr, or 760 mm Hg. (Mercury vapour is toxic. In modern mercury barometers, a thin film of water or oil is added to prevent the evaporation of mercury from the open reservoir, as shown in Figure 2.)

Many areas of study that use gases, such as medicine and meteorology, and several technological applications, such as deep-sea diving, still use non-SI units (**Table 2**). Using the definitions in **Table 2**, we can convert between SI and non-SI units.

Figure 1
When you wear snowshoes, the force is distributed over the surface area of the snowshoes, so you exert less pressure on the ground than you would if you were wearing regular shoes. This allows you to walk over snow instead of sinking into it.

Figure 2
When a tube filled with mercury is inverted, the weight of the column of mercury pulls it toward Earth. However, the weight of the air directly above the open dish pushes down on the surface of the mercury and prevents all of the mercury from falling out of the tube. The two opposing forces balance each other when the height of mercury is about 760 mm. If the vertical mercury-filled tube is longer than 760 mm, the mercury drops to 760 mm. This leaves a vacuum above the liquid.

Table 2 SI and Non-SI Units of Gas Pressure

Unit name	Unit symbol	Definition/Conversion
pascal (SI unit)	Pa	1 Pa = 1 N/m^2
atmosphere	atm	1 atm = 101.325 kPa (exactly)
millimetres of mercury	mm Hg	760 mm Hg = 1 atm = 101.325 kPa
torricelli	torr	1 torr = 1 mm Hg

Learning Tip

When converting units, multiply the given quantity by a ratio obtained from the definition in **Table 2**. Make sure the ratio is oriented so that the unit of the given quantity is cancelled out, as shown in the example.

COMMUNICATION example 1

Standard ambient pressure is defined as 100 kPa. Convert this value to the corresponding values in atmospheres and millimetres of mercury.

Solution

$$100 \text{ kPa} \times \frac{1 \text{ atm}}{101.325 \text{ kPa}} = 0.987 \text{ atm}$$

$$100 \text{ kPa} \times \frac{760 \text{ mm Hg}}{101.325 \text{ kPa}} = 750 \text{ mm Hg}$$

+ EXTENSION

Mars Dead or Alive

You may think of parachutes and airbags as useful emergency technologies here on Earth, but they are also being put to work in the exploration of Mars. These two video segments show you some applications of gases that are real-world—but not *this* world!

www.science.nelson.com GO

Practice

1. Define STP and SATP. What is the advantage of using SATP over STP?
2. Copy and convert the measured values of pressure in **Table 3**. Show your work using appropriate conversion factors from **Table 2**.

Table 3 Converting between Pressure Units

	Pressure (kPa)	Pressure (atm)	Pressure (mm Hg)
(a)	96.5		
(b)			825
(c)		2.50	

3. What are the advantages of having only one SI unit for pressure?
4. When using a medicine dropper or a meat baster, you squeeze the rubber bulb and insert the end of the tube into a liquid. Why does the liquid rise inside the dropper or baster when you release the bulb?

The Relationship between Pressure and Volume

An important goal of science is to obtain knowledge about the natural world. One common method for accomplishing this goal is a controlled experiment involving known variables. Investigation 4.1 is such an experiment.

INVESTIGATION 4.1 Introduction

Pressure and Volume of a Gas

This investigation is a replication of a famous experiment first done by Robert Boyle in 1662. Write your Design using the information given and remember to include a plan and identify the variables. Include a graph and a word statement describing the relationship as part of your Analysis.

Purpose

The purpose of this investigation is to create a general relationship between the pressure and volume of a gas.

Report Checklist

- ○ Purpose
- ○ Problem
- ○ Hypothesis
- ○ Prediction
- ● Design
- ○ Materials
- ○ Procedure
- ● Evidence
- ● Analysis
- ○ Evaluation

Problem

What effect does increasing the pressure have on the volume of a gas?

To perform this investigation, turn to page 177.

Analysis of the evidence produced in an investigation similar to Investigation 4.1 suggests an inverse variation between the pressure and volume of a gas; that is, as the pressure increases, the volume decreases (**Figure 3**). Using the evidence given in SI units in **Table 4**, you can see that, when the pressure is doubled (100 kPa to 200 kPa), the volume is reduced to about one-half (3.00 L to 1.52 L). If the pressure is tripled, the volume is reduced to one-third. Test this hypothesis by checking the other values in Table 4.

Boyle's Law

If P_1 and V_1 represent the initial conditions of pressure and volume of a gas, the other values of pressure and volume from Table 4 may be stated as follows:

$$(P_1, V_1) \quad \left(2P_1, \tfrac{1}{2}V_1\right) \quad \left(3P_1, \tfrac{1}{3}V_1\right) \quad \left(4P_1, \tfrac{1}{4}V_1\right) \quad \left(5P_1, \tfrac{1}{5}V_1\right)$$

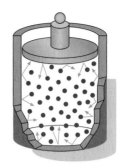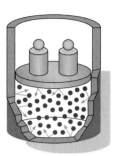

lower pressure greater volume   higher pressure smaller volume

Figure 3
As the pressure on a gas increases, the volume decreases, as illustrated by this combined empirical–theoretical model.

For all the conditions listed above, the product of the pressure and volume is equal to P_1V_1. Mathematically, the relationship is represented as $PV = k$, where k is a constant. This simple quantitative relationship was first determined by Robert Boyle in 1662 (**Figure 4**). Boyle's hypothesis was tested many times and, after successful replication by others, is now accepted as a scientific law. **Boyle's law** states that *as the pressure on a gas increases, the volume of the gas decreases proportionally, provided that the temperature and chemical amount of gas remain constant.* In other words, the volume of a gas is inversely proportional to the pressure of the gas, providing that the temperature and chemical amount of gas are held constant. Boyle's law can be conveniently written comparing any two sets of pressure and volume measurements:

$$P_1V_1 = P_2V_2 \quad \text{(Boyle's law)}$$

Table 4 Pressure and Volume of Gas Samples

Pressure (kPa)	Volume (L)	PV (kPa·L)
100	3.00	300
200	1.52	304
300	1.01	303
400	0.74	296
500	0.60	300

► COMMUNICATION example 2

A 2.0 L party balloon at 98 kPa is taken to the top of a mountain where the pressure is 75 kPa. Assume that the temperature and chemical amount of the gas remain the same. Use Boyle's law to determine the new volume of the balloon.

Solution

$$P_1V_1 = P_2V_2$$

$$V_2 = \frac{P_1V_1}{P_2}$$

$$= \frac{98 \text{ kPa} \times 2.0 \text{ L}}{75 \text{ kPa}}$$

$$= 2.6 \text{ L}$$

or $\quad V_2 = 2.0 \text{ L} \times \dfrac{98 \text{ kPa}}{75 \text{ kPa}} = 2.6 \text{ L}$

According to Boyle's law, the balloon would have a new volume of 2.6 L.

Figure 4
Anglo-Irish chemist Robert Boyle (1627–1691) was a founding member of the Royal Society of London. He is reported to have coined its anti-Aristotelian motto: "Nothing by Authority."

Practice

5. Write an empirical definition of atmospheric pressure.
6. A bicycle pump contains 0.650 L of air at 101 kPa. If the pump is closed, what pressure is required to change the volume to 0.250 L?
7. A weather balloon containing 35.0 L of helium at 98.0 kPa is released and rises. Assuming that the temperature is constant, find the volume of the balloon when the atmospheric pressure is 25.0 kPa at a height of about 25 km.
8. A small oxygen canister contains 110 mL of oxygen gas at a pressure of 3.0 atm. All of the oxygen is released into a balloon with a final pressure of 2.0 atm.
   (a) Predict whether the final volume will be smaller, greater, or the same. Justify your answer.
   (b) What is the final volume of the balloon?
9. A diving bell contains 32 kL of air at a pressure of 98 kPa at the surface. About 5 m below the surface, the volume of air trapped inside the bell is 21 kL (**Figure 5**). What is the pressure of the air in the bell, if you assume that the temperature remains constant?
10. Why does atmospheric pressure depend on your location or vary over time at your location?

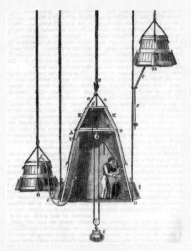

Figure 5
Before underwater diving apparatus became common, divers used a diving bell to explore underwater.

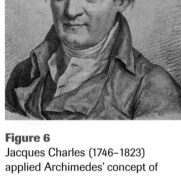

Figure 6
Jacques Charles (1746–1823) applied Archimedes' concept of buoyancy, Henry Cavendish's calculations for the density of hydrogen, and his own observations to invent the hydrogen balloon. His first flight was in 1783. His experiences and experiments led to the formulation of Charles' law in 1787.

The Relationship between Temperature and Volume

More than a century after Boyle had determined the relationship between the pressure and volume of a gas, French physicist Jacques Charles (**Figure 6**) determined the relationship between the temperature and volume of a gas. Charles became interested in the effect of temperature on gas volume after observing the hot air balloons that had become popular as flying machines. This is another example in which a technological development (hot air balloons) led to advances in science.

INVESTIGATION 4.2 Introduction

Temperature and Volume of a Gas

This investigation, like Investigation 4.1, is a controlled experiment in which all variables are kept constant except the two variables being investigated, in this case temperature and volume. Include a graph and a word statement describing the relationship as part of your Analysis. In your Evaluation, pay particular attention to the sources of experimental uncertainties (see Appendix B.2).

Purpose
The purpose of this investigation is to create a general relationship between the temperature and volume of a gas.

Problem
What effect does increasing the temperature have on the volume of a gas?

Report Checklist
- ○ Purpose
- ○ Problem
- ○ Hypothesis
- ○ Prediction
- ○ Design
- ○ Materials
- ○ Procedure
- ● Evidence
- ● Analysis
- ● Evaluation (1, 3)

Design
A volume of air is sealed inside a syringe, which is then placed in a water bath. As the temperature of the water is manipulated, the volume of the air is measured as the responding variable. Two controlled variables are the chemical amount of gas inside the syringe and the pressure on the gas.

To perform this investigation, turn to page 178.

152 Chapter 4

Absolute Temperature Scale

The mathematical equation describing the relationship between volume and temperature may not be apparent from the graph you created in Investigation 4.2; however, if you draw a graph of the two variables, as in **Figure 7(a)**, you get a straight line. This is evidence that a simple relationship does exist. When the line is extrapolated backwards, it crosses the horizontal axis at –273 °C. It appears that, if the gas did not liquefy, its volume would become zero at –273 °C. If this experiment is repeated with different quantities of gas or with samples of different gases, straight-line relationships between temperature and volume are also observed. When the lines are extrapolated, they all meet at –273 °C, as shown in **Figure 7(b)**. This temperature, called **absolute zero**, is the lowest possible temperature. Scientists with sophisticated technology are coming within an increasingly smaller fraction of a degree from absolute zero.

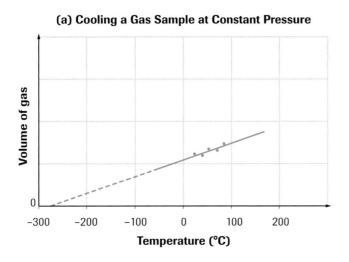
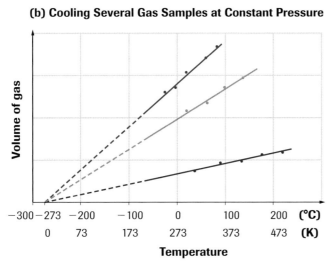

Figure 7
When the graphs from several careful volume–temperature experiments are extrapolated, all the lines meet at what scientists define as absolute zero, −273 °C or 0 K.

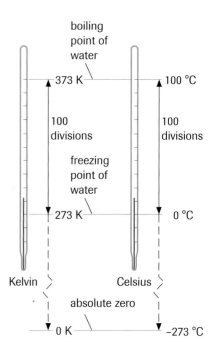

Figure 8
Jacques Charles predicted the absolute zero temperature and Lord Kelvin explained it as the temperature at which the kinetic energy of all entities of solids, liquids, or gases would become zero.

Absolute zero is the basis of another temperature scale, called the **absolute** or Kelvin **temperature scale**. On the absolute temperature scale, absolute zero (−273 °C) is zero kelvin (0 K), as shown in **Figure 7(b)**. (Note that no degree symbol is used for kelvin.) The absolute temperature scale has the same size divisions as the Celsius temperature scale. To convert degrees Celsius to kelvin, you add 273 (**Figure 8**). STP and SATP are each defined by two exact values with infinite significant digits. STP is 273.15 K and 101.325 kPa; SATP is 298.15 K and 100 kPa. For convenience, however, use STP as 273 K and 101 kPa and SATP as 298 K and 100 kPa.

> **DID YOU KNOW ?**
>
> **Lord Kelvin**
> Sir William Thomson (1824–1907), also known as Lord Kelvin, was a Scottish engineer, mathematician, and physicist. His work, during his very long and productive life, profoundly influenced the scientific thought of his generation. Thomson started attending Glasgow University when he was only eleven years old. His brilliant mind ranged from the constitution of matter to the age of Earth and the challenges of laying the Atlantic Cable. It was this last challenge that earned him his peerage and the title "Lord Kelvin."

▶ **Practice**

11. What is the approximate temperature for absolute zero in degrees Celsius and kelvin?
12. Convert the following Celsius temperatures (t) to absolute temperatures (T).
    (a) 0 °C
    (b) 100 °C
    (c) −30 °C
    (d) 25 °C
13. Convert the following absolute temperatures (T) to Celsius temperatures (t).
    (a) 0 K
    (b) 100 K
    (c) 300 K
    (d) 373 K

Charles' Law

The relationship between the volume and absolute temperature of a gas is shown in **Figure 7(b)**. This relationship is described as a direct variation; that is, as the temperature increases, the volume increases. The simplest mathematical relationship is obtained when the absolute temperature is used. Mathematically, this relationship is represented as:

$$V = kT$$

where T represents the absolute temperature in kelvin. This means that the quotient of the two variables $\left(\frac{V}{T}\right)$ has a constant value (k), which is the slope of the straight-line graph (Figure 7(b) on the previous page). A constant value is clearly shown by the analysis in **Table 5**.

Table 5 Analysis of Temperature and Volume of a Gas Sample

Temperature, t (°C)	Temperature, T (K)	Volume, V (L)	Constant, V/T (L/K)
25	298	5.00	0.0168
50	323	5.42	0.0168
75	348	5.84	0.0168
100	373	6.26	0.0168
125	398	6.68	0.0168

> **DID YOU KNOW ?**
>
> **Development of Scientific Laws**
> Boyle's or Charles' initial experiments led to an empirical hypothesis in their first result and then a generalization after investigating several gases. Only after replication by many other scientists with many examples does the result attain the status of a scientific law.
>
> empirical hypothesis
> ↓
> generalization
> ↓
> scientific law

The relationship between volume and absolute temperature, after verification by other scientists, became known as **Charles' law**. Charles' law states that, *as the temperature of a gas increases, the volume increases proportionally, provided that the pressure and chemical amount of gas remain constant* (**Figure 9**). Charles' law can be conveniently written to compare any two sets of volume and temperature measurements at constant pressure and chemical amount:

$$\frac{V_1}{T_1} = k \text{ and } \frac{V_2}{T_2} = k$$

Therefore, $\qquad \dfrac{V_1}{T_1} = \dfrac{V_2}{T_2}$ (**Charles' law**)

Section **4.1**

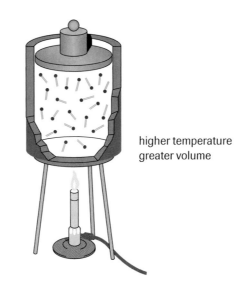

lower temperature
smaller volume

higher temperature
greater volume

Figure 9
This diagram of cylinders with movable pistons illustrates a combined empirical–theoretical model. The model shows that the volume of a gas increases as the temperature of the gas increases. The pressure, equal to the pressure exerted by the mass, the piston, and the atmosphere, remains constant.

▶ **COMMUNICATION** example 3

In a test of Charles' law, a gas inside a cylinder with a movable piston (**Figure 9**) is heated to 315 °C. The initial volume of gas in the cylinder is 0.30 L at 25 °C. What will be the final volume when the temperature is 315 °C?

Solution

$T_1 = (25 + 273)\text{ K} = 298\text{ K}$

$T_2 = (315 + 273)\text{ K} = 588\text{ K}$

$$\frac{V_1}{T_1} = \frac{V_2}{T_2}$$

$$V_2 = \frac{V_1 T_2}{T_1}$$

$$= \frac{0.30\text{ L} \times 588\text{ K}}{298\text{ K}}$$

$$= 0.59\text{ L}$$

or $\quad V_2 = 0.30\text{ L} \times \dfrac{588\text{ K}}{298\text{ K}}$

$\quad\quad\quad = 0.59\text{ L}$

According to Charles' law, the final volume will be 0.59 L.

Learning Tip

- Charles' law only works if you use absolute temperatures (in kelvin). Check the example to see that you get a totally different (and incorrect) answer using Celsius temperatures.
- Some calculators allow you to store (STO) numbers and recall (RCL) them in later calculations. Check your calculator to see if you can calculate and then store the values of T_1 and T_2. Alternatively, you can include the calculation of T_1 and T_2 in the final calculation if you use parentheses around the addition operation.

Temperature Measurement Technologies

The first thermometer in the early 1600s was an air-filled glass bulb with a long stem inverted in a bowl of coloured water. This device was not very accurate or precise. The next technological development in the mid-1600s was the invention of sealed, alcohol-in-glass devices such as the common lab thermometer. Initially this thermometer had only fifty equal divisions but no zero reference point, so was not very useful for communicating actual temperatures. About 1725, the invention of the mercury-filled thermometer and the identification of two reproducible reference points (freezing point and boiling point of pure water) finally allowed different experimenters to measure temperatures with reasonable accuracy and reproducibility. These developments were essential for Charles to be able to conduct his experiments in the late 1700s.

More recently, advances in science and technology have produced a variety of electrical temperature sensors such as thermocouples (temperature-sensitive voltage of dissimilar

Figure 10
The exothermic combustion of propane provides the heat to warm the air in this hot-air balloon.

BIOLOGY CONNECTION

Gas-Dependent Processes

There are many biological processes that depend on the properties of gases. You might come across some if you are taking a biology course.

www.science.nelson.com

Learning Tip

Recall from previous work (Chapter 2, Section 2.4) that n represents the chemical amount of the gas, in moles.

metals) and thermistors (temperature-sensitive resistance). Thermocouples are best suited for high temperatures (such as ovens) and thermistors are best for lower temperatures (such as the digital home thermometers).

> ### Practice

14. Butane is a gas that can be used in simple lighters.
    (a) If 15 mL of butane gas at 0 °C is warmed to 25 °C, calculate its final volume.
    (b) What assumptions did you make in your calculation?
    (c) How does this calculation illustrate the need to use absolute temperatures?

15. An open, "empty" 2 L plastic pop container, which has an actual inside volume of 2.05 L, is removed from a refrigerator at 5 °C and allowed to warm up to 21 °C on a kitchen counter. What volume of air, measured at 21 °C, will leave the container as it warms?

16. Cooking pots have loose-fitting lids to allow air to escape while food is being heated. If a 1.5 L saucepan is heated from 22 °C to 100 °C, by what percentage will any gas in the pan increase in volume?

17. Jacques Charles became interested in temperature–volume relationships for gases because of his curiosity about hot-air balloon flight. Hot-air balloons are open containers that maintain the air inside at (very nearly) atmospheric pressure. Imagine that a modern balloon is being prepared for a flight when the air temperature is 20 °C. The propane burner (**Figure 10**) warms the air, causing it to expand by 20% (so every 1.00 L of air becomes 1.20 L). Calculate the final temperature, in degrees Celsius, of the air in the balloon.

18. A student decides to make a gas expansion thermometer by trapping some air (about 50–70 mL) inside an inverted 100 mL graduated cylinder, the open end of which is submerged in a beaker of water. The student reasons that she should be able to calculate the temperature of the surrounding air by measuring the volume of air inside the cylinder using the graduated scale on the cylinder walls.
    (a) Draw a diagram of the apparatus and describe how it can function as a thermometer.
    (b) Evaluate the design of this technology, using your knowledge of gas behaviour, and predict whether this design would provide accurate values. Suggest possible improvements.

The Combined Gas Law

One of the goals of science is the synthesis or putting together of different concepts. When Boyle's and Charles' laws are combined, the resulting **combined gas law** produces a relationship among the volume, temperature, and pressure of any fixed chemical amount of gas: *the product of the pressure and volume of a gas sample is proportional to its absolute temperature in kelvin: $PV = kT$.*

Boyle's law: $PV =$ a constant (T and n are controlled variables.)

Charles' law: $\dfrac{V}{T} =$ a constant (P and n are controlled variables.)

If the product PV is constant at a fixed temperature, then $P\left(\dfrac{V}{T}\right)$ should also be a constant because V divided by a constant temperature is also constant. If the temperature changes, then Charles' law tells us that the ratio $\dfrac{V}{T}$ is constant at a fixed pressure. Therefore, multiplying a constant pressure by a constant ratio of volume to temperature certainly produces a number that is a constant. Using this reasoning (a mathematical method of joint variation), we can conclude that the product of the pressure and volume of a gas divided by its absolute temperature is a constant as long as the chemical amount of gas does not change:

$$\dfrac{PV}{T} = k$$

The relationship can be expressed in a convenient form for calculations involving changes in volume, temperature, or pressure for a particular gas sample:

$$\frac{P_1 V_1}{T_1} = \frac{P_2 V_2}{T_2} \quad \text{(combined gas law)}$$

 **WEB** Activity

Simulation—The Combined Gas Law

This gas simulation program provides a visual representation of gas molecules and the effect of changing gas properties.

www.science.nelson.com GO

The combined gas law is a useful starting point for all cases involving pressure, volume, and temperature, even if one of these variables is a constant (as in Boyle's and Charles' laws). A variable that is constant can easily be eliminated from the combined gas law equation, reducing it to an equation for either Boyle's or Charles' law as shown by Sample Problem 4.1.

▶ SAMPLE problem 4.1

A steel cylinder with a fixed volume contains a gas at a pressure of 652 kPa and a temperature of 25 °C. If the cylinder is heated to 150 °C, use the combined gas law to calculate the new pressure.

Because the volume is constant, we can cancel V_1 and V_2 from the combined gas law equation because $V_1 = V_2$:

$$\frac{P_1 \cancel{V_1}}{T_1} = \frac{P_2 \cancel{V_2}}{T_2}$$

We can now solve for P_2 and then substitute the pressures and temperatures (after converting to kelvin):

$T_1 = (25 + 273)\text{ K} = 298\text{ K}$

$T_2 = (150 + 273)\text{ K} = 423\text{ K}$

$$P_2 = \frac{P_1 T_2}{T_1}$$

$$= \frac{652\text{ kPa} \times 423\text{ K}}{298\text{ K}}$$

$$= 925\text{ kPa}$$

Alternatively, this can be expressed as a calculation of a new pressure directly related to the ratio of the temperatures and inversely related to the ratio of the volumes (which cancel to 1 in this case):

$$P_2 = P_1 \frac{T_2}{T_1} \frac{\cancel{V_1}}{\cancel{V_2}}$$

$$= 652\text{ kPa} \times \frac{423\text{ K}}{298\text{ K}}$$

$$= 925\text{ kPa}$$

If we assume that the steel walls are sufficiently strong, the gas will have a pressure of 925 kPa inside the cylinder.

DID YOU KNOW?

Gas Laws and Breathing
Breathing is something we all do. Experiments show that, on average, you inhale and exhale about 0.5 L of air, roughly 12 times per minute. How? Your lungs are like elastic balloons inside a flexible chest cavity. By activating muscles that raise your rib cage and lower your diaphragm, you increase the volume of your chest cavity (**Figure 11a**). Exhaling reverses this process (**Figure 11b**).

(a) Inhaling

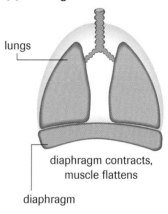

(b) Exhaling

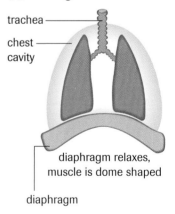

Figure 11
The diaphragm can be compared to a piston.

▶ **COMMUNICATION** example 4

A balloon containing helium gas at 20 °C and a pressure of 100 kPa has a volume of 7.50 L (**Figure 12**). Calculate the volume of the balloon after it rises 10 km into the upper atmosphere, where the temperature is −36 °C and the outside air pressure is 28 kPa. Assume that no gas escapes and that the balloon is free to expand so that the gas pressure within it remains equal to the air pressure outside.

Solution

$T_1 = (20 + 273)\ K = 293\ K$

$T_2 = (-36 + 273)\ K = 237\ K$

$$\frac{P_1 V_1}{T_1} = \frac{P_2 V_2}{T_2}$$

$$V_2 = \frac{100\ kPa \times 7.50\ L \times 237\ K}{28\ kPa \times 293\ K}$$

$$= 22\ L$$

or $\quad V_2 = 7.50\ L \times \dfrac{100\ kPa}{28\ kPa} \times \dfrac{237\ K}{293\ K}$

$$= 22\ L$$

According to the combined gas law, the volume of the balloon in the upper atmosphere will be 22 L.

Figure 12
Various shapes and sizes of weather balloons are launched several times a day at more than 1000 sites in North America. (See Weather Forecasts, page 165.) Some of these weather balloons are reported as UFOs.

SUMMARY Gas Properties and Laws

STP: 0 °C and 101.325 kPa (exact values)
SATP: 25 °C and 100 kPa (exact values)
101.325 kPa = 1 atm = 760 mm Hg (exact values)
absolute zero = 0 K or −273.15 °C
$T\ (K) = t\ (°C) + 273$ (for calculation)

Boyle's law: $\quad P_1 V_1 = P_2 V_2 \quad$ (for constant temperature and chemical amount of gas)

Charles' law: $\quad \dfrac{V_1}{T_1} = \dfrac{V_2}{T_2} \quad$ (for constant pressure and chemical amount of gas)

combined gas law: $\quad \dfrac{P_1 V_1}{T_1} = \dfrac{P_2 V_2}{T_2} \quad$ (for constant chemical amount of gas)

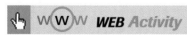 **WEB Activity**

Canadian Achievers—Malcolm King

Dr. Malcolm King switched from his training in polymer chemistry to a career searching for therapies for lung diseases. Now at the University of Alberta, Dr. King is drawing on his heritage as he researches the effectiveness of traditional Aboriginal remedies. What is Dr. King's theory on why so much of the world's research into lung function is concentrated in Canada? Why did he become interested in this field? What has King done to improve the health of Aboriginal Canadians?

www.science.nelson.com

Practice

19. Express the combined gas law in your own words.
20. A large party balloon has a volume of 5.00 L at 20 °C and 100 kPa. Calculate the pressure for a volume of 6.00 L at 35 °C.
21. A cylinder of helium gas has a volume of 1.0 L. The gas in the cylinder exerts a pressure of 800 kPa at 30 °C. What volume would this gas occupy at SATP?
22. For any of the calculations in the previous questions, does the result depend on the identity of the gas? Explain briefly.
23. A 2.0 mL bubble of gas is released at the bottom of a lake where the pressure is 6.5 atm and the temperature is 10 °C. Predict the Celsius temperature of the gas bubble at the surface, where the pressure is 0.95 atm and the volume becomes 14.4 mL.
24. What assumption was made in all of the previous calculations?
25. Use Boyle's law to describe what happens during inhaling and exhaling (**Figure 11**).
26. Popcorn is a favourite snack food for many people (**Figure 13**). The corn kernel is heated, and some of the moisture inside the kernel vaporizes, starting a chain of events that leads to the tasty popped corn.
    (a) If we assume a constant volume kernel (before popping), what happens to the pressure inside the kernel as the temperature increases? Justify your answer using appropriate mathematical equations or relationships.
    (b) The pressure inside the kernel forces some superheated water and steam to penetrate into the starch granules, making them soft and gelatinous. When the hull of the kernel breaks at about 900 kPa, what happens to the volume of water vapour when the pressure quickly drops to about 100 kPa? Justify your answer using appropriate mathematical equations or relationships.

Figure 13
Popcorn was invented by Aboriginal peoples in North America long before the arrival of the Europeans. The popping method used very hot clay pots, which is a method similar to today's hot-air poppers.

Case Study

Compressed Gases

Not only are gases a major part of our lives, but compressed gases—that is, gases at pressures above atmospheric pressure—are particularly useful:
- The tires of vehicles contain pressurized air.
- Many people use gas barbecues with a pressurized propane fuel tank.
- Aerosol cans contain a propellant that carries the contents of the can out the nozzle; the propellant is a pressurized gas.
- Major surgery usually involves oxygen administered from a pressurized oxygen tank. It is often accompanied by an anesthetic, which may also be a pressurized gas, such as dinitrogen monoxide.

Certain occupations require some work with pressurized gases. In the medical field, paramedics and doctors use oxygen tanks. Firefighters use compressed air tanks like those used by underwater divers. Some welders use oxyacetylene torches (**Figure 14**). This form of welding requires both a pressurized oxygen tank and a pressurized acetylene tank. Many scientists and their graduate students routinely use a pressurized gas for research because the gas is part of the reaction system or because it provides an inert (nonreactive) environment. Noble gases, such as argon, are also used to provide an inert environment in the computer chip industry, where oxygen would cause undesirable reactions.

Figure 14
The use of a controlled mixture of oxygen and acetylene provides the best combustion and very high temperatures necessary for cutting or welding metal.

The chemical safety hazards of some gases are similar to those of many other chemicals, which may be corrosive, toxic, flammable, dangerously reactive, or oxidizing agents. What makes compressed gases much more dangerous is the physical hazard of it becoming a "rocket": gas pressures can be as high as 15 MPa (about 150 atm). The hole in the tank, to which the valve stem and valve are connected, is the diameter of a pencil. If the gas is suddenly released through such a small opening, the very great pressure propels the tank, making it a formidable projectile. A damaged tank can go through a solid brick wall and cause considerable damage.

Professional and recreational underwater divers face risks associated with the use of a pressurized air tank or scuba (self-contained underwater breathing apparatus). The tank containing compressed air is attached to a regulator that releases the air at the same pressure as the underwater surroundings (**Figure 15**). The pressure underwater can be quite substantial. Breathing pressurized air is necessary to balance the internal and external pressures on the chest to allow divers to inflate their lungs. However, this creates problems if divers ascend to normal pressure too quickly or while holding their breath. According to Boyle's law, if the pressure is decreased, the volume of air increases. When the volume of air is contained in the lungs, the lungs would expand to accommodate the increased volume. This is very dangerous because the lungs could rupture. This is one reason why a person needs an understanding of gases and gas laws in order to obtain a scuba-diving licence and to dive safely.

Figure 15
Every 10 m of depth adds about 1 atm (100 kPa) of pressure to the normal air pressure. At a depth of 20 m, the total pressure is about 300 kPa. In order to breathe, the air pressure must be about 300 kPa.

A European company recently developed an automobile powered by compressed gas. The Air Car (**Figure 16**) uses an electric pump to compress air into a tank. This compressed air, in turn, pumps the pistons to rotate the wheels and make the car move without the need for gasoline. The only exhaust that comes out of the tail pipe is cold air. The air pump that puts air into the tank plugs into an ordinary household outlet and takes four hours to refill. On a full tank of compressed air the vehicle can travel 80 km at a speed of 110 km/h, and farther at lower speeds. The manufacturer, Luxembourg-based Moteur Developpement International, describes the car as safe, nonpolluting, and inexpensive. Environmental scientists, however, are leery of the Air Car's claimed benefits. They point out that converting energy from electricity to compressed air is inefficient and uses more energy from the power plant than it delivers on the road. While there may not be any pollution from the car itself, the vehicle merely transfers the environmental burden to another place.

Case Study Questions

1. Identify two consumer or commercial products that use or contain compressed air and two that involve other compressed gases.
2. Identify several careers that involve work with pressurized gases. How does knowledge of gas properties help the people in those careers?
3. Another problem of breathing air under pressure while scuba diving is that it forces more air to dissolve in the diver's bloodstream. Using gas laws and other information, describe why this can be dangerous to the diver. What can be done to prevent the problem or to solve the problem after it has been created?

www.science.nelson.com

Extension

4. Helium has many uses (**Figure 17**) but one that is familiar to many people is its use in party balloons. Sometimes people inhale helium because it produces an unusual change in a person's voice. In a paragraph based on your Internet research, describe this effect as well as the dangers of inhaling helium.

www.science.nelson.com

Figure 17
Modern airships are filled with the light noble gas, helium. Helium is also used in party balloons.

Figure 16
The Air Car is very quiet and produces no pollution during its operation.

Section 4.1 Questions

1. Copy and complete **Table 6**. Show your work using appropriate conversion factors.

   Table 6 Converting between Pressure Units

	Pressure (kPa)	Pressure (atm)	Pressure (mm Hg)
(a)		0.0875	
(b)	25.0		
(c)			842

2. Copy and complete **Table 7**.

   Table 7 Converting between Celsius and Kelvin

	t (°C)	T (K)
(a)	25	
(b)	−35	
(c)		312
(d)		208

3. A syringe contains 50.0 mL of a gas at a pressure of 96.0 mm Hg. The end is sealed, and the plunger is pushed to compress the gas to 12.5 mL. What is the new pressure of the gas inside the syringe, assuming constant temperature?

4. Carbon dioxide produced by yeast in bread dough causes the dough to rise, even before it is baked (**Figure 18**). During baking, the carbon dioxide gas expands. Predict the final volume of 0.10 L of carbon dioxide in bread dough that is heated from 25 °C to 190 °C at a constant pressure.

5. An automobile tire has an internal volume of 27 L at 225 kPa and 18 °C.
   (a) What volume would the air inside the tire occupy if it escaped? (Atmospheric pressure at the time is 98 kPa and the temperature remains the same.)
   (b) How many times larger is the new volume compared with the original volume? How does this compare with the change in pressure?

6. In a cylinder of a diesel engine, 500 mL of air at 40.0 °C and 1.00 atm is powerfully compressed just before the diesel fuel is injected. The resulting pressure is 35.0 atm. If the final volume is 23.0 mL, what is the final temperature in the cylinder?

7. The gas laws described in this section involve the properties of volume, pressure, and temperature. Some of these variables have a direct relationship (as one increases, so does the other), and some have an inverse relationship (as one increases, the other decreases).
   (a) For each pair of the following variables, state whether the relationship is direct or inverse and sketch a graph:
      (i) pressure and volume at a constant temperature
      (ii) temperature and volume at a constant pressure
   (b) What other property of a gas must also be constant for all of the above?

Figure 18
The lightness of baked goods, such as bread and cakes, is a result of gas bubbles trapped in the dough or batter when it is heated. The leavening, or production of gas bubbles, can be due to vaporization of water, expansion of gases already in the dough or batter, or leavening agents such as yeast and baking powder. In contrast, bannock made by Aboriginal peoples does not require a leavening agent and is therefore easier to use when surviving on the land.

8. List the seven ways that empirical knowledge can be communicated. Where possible, provide a specific example from this section for each of these ways.

9. The air exhaled by a scuba diver (**Figure 15**) rises to the surface.
   (a) What happens to the bubbles as they rise? Describe your reasoning.
   (b) What assumptions did you make in your answer to (a)?

10. Why do aerosol cans have a warning not to incinerate the container?

11. Jet aircraft engines use energy from burning fuel to power the process of taking in cold air and releasing hot gases. Most of the intake air is nitrogen, which reacts only in negligible quantities. The expanding gas mixture escapes backward, and the reaction of this force drives the engine forward.
    (a) Assuming the $N_2(g)$ in the air is heated from −60 °C to 540 °C in the engine, express the volume increase as a ratio of final volume to initial volume, to a certainty of three significant digits.
    (b) Describe what other work the expanding gases in a jet engine must do, besides providing forward thrust. (*Hint:* An older term is "turbojet" engine.)

12. Read the following partial lab report. Complete the Prediction, Design (based on the apparatus shown in **Figure 19**), Analysis (of the Evidence in **Table 8**), and Evaluation (2, 3) portions of a lab report for this investigation. Include a graph in your Analysis.

    Purpose
    The purpose of this investigation is to test the combined gas law for the relationship between the pressure and the temperature of a gas.

    Problem
    What effect does the temperature of nitrogen gas have on the pressure it exerts?

Figure 19
This apparatus consists of a hollow metal sphere to which a pressure gauge is attached. Because the gas inside the sphere cannot expand, the relationship between temperature and pressure of a gas can be determined.

Table 8 Evidence: Variation of Pressure with Temperature

Temperature (°C)	Pressure (kPa)
0	100
20	106
40	115
60	123
80	129
100	135

13. Scientific laws are the most important type of empirical knowledge in science. Using an example from this section, write a brief summary describing how a scientific law becomes established.

14. The traditional or indigenous knowledge of the Aboriginal peoples is very important to their culture and survival. In a sentence or two, describe the characteristics of indigenous knowledge (IK). Why is this type of knowledge still important today? Identify some similarities and differences between indigenous knowledge and Western scientific knowledge.

    www.science.nelson.com

15. For a typical geyser (**Figure 20**), underground water seeps into a deep narrow shaft in the ground and is heated from below. Because of the depth, the pressure on the water is high, so the water at the bottom of the shaft boils at a much higher temperature than normal.
    (a) What happens to the volume of a 1.0 L bubble of water vapour at 130 °C and 3.05 atm when it reaches the surface, where the conditions are 100 °C and 1.01 atm?
    (b) Why is a narrow shaft necessary to produce the geyser effect?

Figure 20
Geysers are unusual and dramatic examples of geothermal energy used to heat water in a confined space.

16. Barometers, manometers, and Bourdon gauges all measure gas pressure. How are they similar and how are they different? How did the invention of these devices advance the science of gases? Working in a group, use the Internet to answer these questions and present your findings as a poster or multimedia presentation.

    www.science.nelson.com

Extension

17. Search the Internet for research reports on how close scientists have come to reaching absolute zero. What do the reports say about whether the kinetic energy of all entities is zero at absolute zero?

    www.science.nelson.com

18. Why is mercury used in most barometers, and not other liquids such as water, which is plentiful and nontoxic?

    www.science.nelson.com

Explaining the Properties of Gases 4.2

The early study of gases was strictly empirical. Boyle and Charles conducted their initial experiments that, after independent replication, led to the laws named in their honour. Boyle's and Charles' work occurred well before Dalton's atomic theory was published in 1803. The kinetic molecular theory (Chapter 2, Section 2.2) was developed by about 1860, thanks to the use of mathematical tools such as statistical analysis.

Before any law or theory is accepted by the scientific community, supporting evidence must be available. An acceptable theory must describe observations in terms of non-observable ideas, explain existing evidence, predict results of future experiments, and be as simple as possible in concept and application. The easiest thing for a theory to do is to explain existing evidence; the hardest thing is to successfully predict new evidence. Over the years the kinetic molecular theory (**Figure 1**) has been strongly supported by experimental evidence and is very useful in explaining the properties of gases. For the restricted theory presented in this textbook, we will limit predictions to qualitative properties and comparisons.

- The kinetic molecular theory explains why gases, unlike solids and liquids, are compressible. In a molecular solid, the distance between molecules is about the same size as the molecules themselves; in a liquid, it is generally slightly greater; and in a gas, the distance between molecules is about 20 to 30 times the size of the molecules. If most of the volume of a gas sample is empty space, it should be possible to force the molecules closer together.

- The kinetic molecular theory explains the concept of gas pressure. Pressure is considered to be the result of gas molecules colliding with objects—particularly the walls of a container. Based on this theory, the pressure exerted by a gas sample is the total force of these collisions distributed over an area of the container wall; in other words, force per unit area.

- The kinetic molecular theory explains Boyle's law. If the volume of a container is reduced, gas molecules will move a shorter distance before colliding with the walls of the container. The kinetic molecular theory suggests that they will collide with the walls more frequently, resulting in increased pressure on the container.

- The kinetic molecular theory explains Charles' law. An increase in temperature represents an increase in the average kinetic energy and therefore the average speed of the entities' motion. In a container in which the pressure can be kept constant (for example, in a cylinder with a piston or in a flexible-walled container such as a

DID YOU KNOW ?

Analogy for States of Matter
To picture yourself as an entity in a solid, imagine yourself seated in a regular classroom. For a liquid, picture a school dance. For a gas, imagine yourself and three friends skating randomly in a large ice-hockey arena.

DID YOU KNOW ?

Scientific versus Indigenous Knowledge
The usual development of scientific knowledge is from empirical to theoretical and an important goal is to explain natural substances and processes. In contrast, the indigenous knowledge of Aboriginal peoples is largely empirical and its goal is to use natural substances and processes to live in harmony with nature in order to survive on the land.

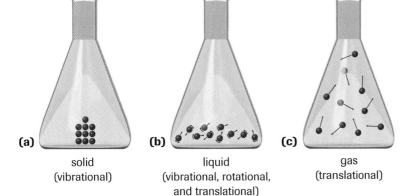

(a) solid (vibrational)
(b) liquid (vibrational, rotational, and translational)
(c) gas (translational)

Figure 1
According to the kinetic molecular theory, the motion of molecules is different in solids, liquids, and gases.
(a) Entities in solids have primarily vibrational motion.
(b) Entities in liquids have vibrational, rotational, and some translational motion.
(c) The most important form of motion in gases is translational.

weather balloon), faster-moving molecules will collide more frequently with the container walls. They will also collide with more force, causing the walls to move outward. Thus, the volume of a gas sample increases with increasing temperature.

> ### Practice
>
> 1. Use the kinetic molecular theory to explain the following observed properties of gases.
>   (a) Gas pressure increases when the volume of the gas is kept constant and the temperature increases.
>   (b) Gas pressure increases when the temperature is kept constant and the volume of the gas decreases.
>   (c) Gases mix much more quickly than do liquids.
>   (d) Oil, not air, is used in hydraulic systems.
>   (e) At SATP, the average speed of air (oxygen and nitrogen) molecules is about 450 m/s, which is approximately the speed of a bullet fired from a rifle. Nevertheless, it takes several minutes for the odour of a perfume to diffuse throughout a room.
> 2. Describe at least three examples that show how kinetic molecular theory provides an explanation for natural or technological products or processes.
> 3. For a theory, such as the kinetic molecular theory, to become acceptable to the scientific community it must have satisfied some criteria. List four characteristics of an accepted scientific theory.

WEB Activity

Canadian Achievers—Elizabeth MacGill

Elsie MacGill had a distinguished career in aeronautics—the science and technology of objects moving through the air. In her lifetime, she accomplished many "firsts" for women.

1. Identify some of these accomplishments.
2. List some gas properties that are important in aeronautics.

www.science.nelson.com

Figure 2
Elizabeth ("Elsie") MacGill, (1905–1980)

Explaining the Law of Combining Volumes

The kinetic molecular theory explains many physical properties of gases. But what about their chemical properties? In other words, what happens during chemical reactions? Many chemical reactions involve gases as reactants and/or products.

In 1809, Joseph Gay-Lussac, a French scientist and a colleague of Jacques Charles, measured the relative volumes of gases involved in chemical reactions. For example, when he combined hydrogen and chlorine gases at the same temperature and pressure, he noticed that every one litre of hydrogen gas, $H_2(g)$, reacted with one litre of chlorine gas, $Cl_2(g)$, to produce two litres of hydrogen chloride gas, $HCl(g)$.

	hydrogen	+	chlorine	→	hydrogen chloride
	1.0 L		1.0 L		2.0 L
ratio	1		1		2

His observations of several gas reactions led, after independent replication, to the **law of combining volumes**, which states that *when measured at the same temperature and pressure, volumes of gaseous reactants and products of chemical reactions are always in simple ratios of whole numbers.* This law is also known as Gay-Lussac's law of combining volumes. Not all reactants and products need be gases, but the law deals only with the gases consumed or produced. An example of this is the simple decomposition of liquid water, in which the volumes of hydrogen gas and oxygen gas produced are always in the ratio of 2:1 (**Figure 3**).

Two years after this law was formulated, Amedeo Avogadro proposed an explanation in terms of numbers of molecules. Unfortunately, Avogadro's idea was largely ignored for about half a century. When Gay-Lussac published his work about combining volumes, Avogadro was intrigued by the fact that reacting volumes of gases were in whole-number ratios, just like the coefficients in a balanced equation. Suggesting an explanation for the relationship between the volume ratios and coefficient ratios, Avogadro proposed that *equal volumes of gases at the same temperature and pressure contain equal numbers of molecules*, a statement that is now best called **Avogadro's theory**. Avogadro's initial idea was a hypothesis. Although it is still sometimes referred to as a hypothesis, the idea is firmly established. Therefore, Avogadro's idea now has the status of a theory.

This theoretical concept explains the law of combining volumes. For example, if a reaction occurs between two volumes of one gas and one volume of another gas at the same temperature and pressure, the theory says that two molecules of the first gas react with one molecule of the second gas. Another example is the reaction of nitrogen and hydrogen, in which ammonia is produced (**Figure 4**).

When all gases are at the same temperature and pressure, the law of combining volumes provides an efficient way of predicting the volumes of gases involved in a chemical reaction. As explained by Avogadro's theory, the mole ratios provided by the balanced equation are also the volume ratios. For example,

coefficients	2 $C_4H_{10}(g)$	+	13 $O_2(g)$	→	8 $CO_2(g)$	+	10 $H_2O(g)$
chemical amounts	2 mol		13 mol		8 mol		10 mol
volumes	2 L		13 L		8 L		10 L
example	4 mL		26 mL		16 mL		20 mL

Figure 3
Water decomposes to hydrogen and oxygen gases in a 2:1 volume ratio.

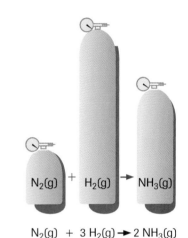

$N_2(g)$ + 3 $H_2(g)$ → 2 $NH_3(g)$

coefficients	1		3		2
mole ratio	1	:	3	:	2
volume ratio	1	:	3	:	2

Figure 4
One volume of nitrogen reacts with three volumes of hydrogen, producing two volumes of ammonia.

▶ **SAMPLE** problem **4.2**

Use the law of combining volumes to predict the volume of oxygen required for the complete combustion of 120 mL of butane gas from a lighter.

The first step is to write the balanced chemical equation, including what you are given and what you need to find:

2 $C_4H_{10}(g)$ + 13 $O_2(g)$ → 8 $CO_2(g)$ + 10 $H_2O(g)$
120 mL   V

From this chemical equation you can see that 13 mol of oxygen is required for every 2 mol of butane. Therefore, the volume of oxygen has to be greater than 120 mL by a factor of $\frac{13}{2}$:

$$V_{O_2} = 120 \text{ mL} \times \frac{13}{2}$$
$$= 780 \text{ mL}$$

To make sure that the ratio is used in the correct order, you could include the chemical formula with each quantity as shown below:

$$V_{O_2} = 120 \text{ mL } C_4H_{10} \times \frac{13 \text{ mL } O_2}{2 \text{ mL } C_4H_{10}} = 780 \text{ mL } O_2$$

Note the cancellation of the units and chemical formulas.

Learning Tip

This equivalence between the chemical amounts (coefficients) and the volumes only works for gases, and only if they are at the same temperature and pressure.

▶ COMMUNICATION example

A catalytic converter in the exhaust system of a car uses oxygen (from the air) to convert carbon monoxide to carbon dioxide, which is released through the tailpipe. If we assume the same temperature and pressure, what volume of oxygen is required to react with 125 L of carbon monoxide produced during a 100 km trip?

Solution

$$2\,CO(g) + O_2(g) \rightarrow 2\,CO_2(g)$$
$$125\,L \qquad V$$

$$V_{O_2} = 125\,L \times \frac{1}{2} = 62.5\,L$$

or $\quad V_{O_2} = 125\,\cancel{L\,CO} \times \dfrac{1\,L\,O_2}{2\,\cancel{L\,CO}} = 62.5\,L\,O_2$

According to the law of combining volumes, 62.5 L of oxygen is required.

▶ Practice

4. Gay-Lussac was the first to notice and publish evidence of simple volume ratios of reacting gases.
   (a) State the law that was created from this evidence.
   (b) State the theory that can be used to explain this law.
   (c) Compare the characteristics of scientific laws and theories.

5. Gas barbecues burn propane using oxygen from the air. If 5.00 L of propane is burned, predict the volume of oxygen, at the same temperature and pressure, required for complete combustion.

6. In modern automobile catalytic converters, nitrogen monoxide (a pollutant) reacts with hydrogen to produce nitrogen and water vapour (part of the exhaust). The catalytic converter of a car meeting current emission standards removes about 1.2 L of nitrogen monoxide at SATP for every kilometre of driving. What volume of nitrogen gas is formed from 1.2 L of nitrogen monoxide at the same temperature and pressure?

7. Ammonia is produced from its elements in huge quantities at many facilities in Alberta.
   (a) Predict the volume of hydrogen (from natural gas) that is required to produce 1.0 ML of ammonia.
   (b) State the important assumption that must be made in this calculation.

CAREER CONNECTION

Meteorologist

Meteorologists use advanced knowledge of the physical and chemical properties of gases to study the atmosphere and how it interacts with the rest of the planet and everything on it. Meteorologists can work as weather forecasters or they can consult and provide specific analyses or weather warnings to governments and businesses.

Find out more about the possibilities of being a meteorologist, the work of meteorologists, and the training required.

www.science.nelson.com

🖱 WWW WEB Activity

Web Quest—"Designer Air" for Tires

Most tires contain plain old air, but one of the latest trends is to fill car tires with pure nitrogen to increase safety and decrease gas consumption. This Web Quest lets you find out more about using pure nitrogen in tires. Your consulting group will be hired to make recommendations about this trend and answer this question: How is pure nitrogen gas different from air, and would those differences change how your vehicle performs?

www.science.nelson.com

Section **4.2**

Case Study

Weather Forecasts

Meteorology is the study of the atmosphere and weather forecasting. This study is a good example of collecting information with various technologies, and then using this information to explain and predict events.

In Canada, human weather watchers and automated meteorological stations take readings to determine atmospheric pressure, temperature, humidity, wind, type and extent of cloud cover, and precipitation (rain, snow, and hail). Helium weather balloons are also used to collect data. Some weather balloons may go as high as 30 km, where the air pressure is 1.2 kPa and the air temperature is -47 °C. The designers of these balloons must therefore know about gas properties. Satellites take pictures of cloud development that indicate wind patterns, as well as snow and ice cover.

All the data, collected from these different sources, are then transmitted via a communications network to the computers at a regional weather office. Meteorologists use sophisticated computer models to study and simulate weather changes based on vast quantities of data collected around the world.

In Canada, the supercomputer at the Canadian Meteorological Centre in Montreal generates forecasts of atmospheric conditions that are used to develop regional predictions. The meteorologists at about 14 weather offices across Canada adapt these forecasts for the general public and the aviation, agricultural, energy, forestry, and shipping sectors. Every day, Environment Canada issues about 1300 public weather predictions for more than 200 geographic areas, and about 1000 aviation forecasts for 175 airports. The primary responsibility of Environment Canada is to warn Canadians about severe weather conditions that could threaten lives and property.

Weather maps summarize predictions for that day's weather, based on the conditions observed the previous day. These maps typically show the predicted temperatures for selected cities, warm fronts and cold fronts, and areas of precipitation. You have probably seen weather reports citing high- and low-pressure systems (**Figure 5**). Lows are associated with overcast skies and precipitation, while highs generally bring clear skies. The interaction between high- and low-pressure systems drives active weather.

Long before modern weather analysis and prediction, Aboriginal Canadians were able to anticipate the local weather. People living on the land were keen observers of natural phenomena, such as wind direction and cloud patterns, and could predict weather changes based on this traditional knowledge. For example, sun-dogs (bright spots usually occurring in pairs on either side of the Sun) signalled a drastic change in the weather. On the western Prairies, a bow-shaped cloud formation bordering an expanse of clear sky indicated that a Chinook wind was coming. Air rushing out of caves warned of a fast-moving low-pressure system bringing stormy weather. The reaction of animals to oncoming weather was also important forecast information.

Case Study Questions

1. Use the gas laws to describe how a fast-moving low-pressure system could cause air to rush out of a cave.
2. State some examples of observations used by Aboriginal peoples to predict weather changes.
3. Historically, measurements of pressure, temperature, and relative humidity were made using a mercury barometer, thermometer, and sling psychrometer (wet/dry bulb thermometer) respectively. How has the technology used to measure these three important atmospheric variables changed? Briefly describe each modern example, including the type of sensor and typical precision values.

www.science.nelson.com

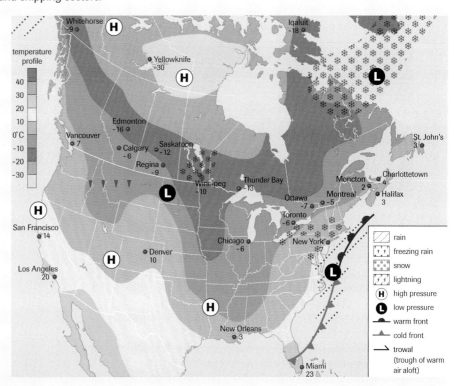

Figure 5

Other than temperature, the most commonly used gas property in weather reports is pressure, often referred to as "highs" and "lows."

Section 4.2 Questions

1. Use the kinetic molecular theory to explain in words or diagrams each of the following observations.
   (a) A drop of food colouring is carefully added to a glass of water. After sitting without stirring for a short period of time, the colour is evenly distributed throughout the water.
   (b) At the same pressure, cooler air has a higher density than warmer air.
   (c) A layer of pure zinc is deposited onto the surface of a piece of copper. After heating briefly in a flame, a brass (solid solution) layer forms on the surface.

2. In the development of the gas laws in the previous section, it was necessary to always keep the chemical amount of the gas constant. Suppose a container with a fixed volume contains a certain chemical amount of gas. If the chemical amount of gas is doubled and the temperature kept constant, predict the change in pressure. Justify your prediction using the kinetic molecular theory.

3. (a) State the law of combining volumes.
   (b) Using the formation reaction for nitrogen dioxide gas as an example, predict the ratio of combining volumes.
   (c) Explain your prediction using simple molecular model diagrams.

4. Some sulfur compounds are undesirable components of much of Alberta's fossil fuels. For example, sour natural gas contains hydrogen sulfide in varying proportions. In some cases, hydrogen sulfide is burned in gas flares often seen around Alberta. In other cases, the pure sulfur is extracted from the gas as a useful byproduct (**Figure 6**).

Figure 6
Pure elemental sulfur is a useful byproduct of the natural gas processing industry. Here at the Shell Waterton gas plant near Pincher Creek, Alberta, pure sulfur extracted from the sour gas is stored in blocks before being marketed to the fertilizer, pharmaceutical, and manufacturing industries. Sulfur is one of the world's most common elements.

   (a) One technology for removing hydrogen sulfide from sour natural gas involves converting part of the hydrogen sulfide to sulfur dioxide by burning in air. Predict the volume of oxygen required to burn 124 kL of hydrogen sulfide measured at the same temperature and pressure.
   (b) The remaining hydrogen sulfide then reacts with the sulfur dioxide as shown in the reaction equation below. Calculate the volume of sulfur dioxide a chemical engineer would predict to react completely with 248 kL of hydrogen sulfide. The gases are measured at 350 °C and 250 kPa.

   $$16\ H_2S(g) + 8\ SO_2(g) \rightarrow 3\ S_8(s) + 16\ H_2O(g)$$

5. The production of nitric acid is important to the fertilizer and explosives industries. Chemical engineers routinely use gas laws to design and control processes such as the Ostwald process.
   (a) The production of nitric acid by the Ostwald process begins with the combustion of ammonia:

   $$4\ NH_3(g) + 5\ O_2(g) \rightarrow 4\ NO(g) + 6\ H_2O(g)$$

   Predict the volume of oxygen required to react with 100 L of ammonia as well as the volumes of nitrogen oxide and water vapour produced. All gases are measured at 800 °C and 200 kPa.
   (b) In another step of the Ostwald process, nitrogen monoxide reacts with oxygen to form nitrogen dioxide. Predict the volume of oxygen at 800 °C and 200 kPa required to produce 750 L of nitrogen dioxide at the same temperature and pressure.
   (c) Nitric acid is produced by reacting nitrogen dioxide with water:

   $$3\ NO_2(g) + H_2O(l) \rightarrow 2\ HNO_3(aq) + NO(g)$$

   Predict the volume of nitrogen monoxide produced by the reaction of 100 L of nitrogen dioxide with excess water. Both gases are measured at the same temperature and pressure as in (b).
   (d) A high-nitrogen fertilizer is made by reacting ammonia gas with nitric acid to produce aqueous ammonium nitrate. Can the law of combining volumes be used to predict the volume of ammonia gas required to react with 100 L of nitric acid? Justify your answer.

6. Science provides the theoretical basis for interpreting and explaining natural and technological products and processes. Provide two natural and two technological examples that can be explained using the kinetic molecular theory. Give a brief explanation for each example.

Extension

7. Explanations from scientific theories are judged on their ability to explain empirical knowledge in a logical, consistent, and simple fashion. How well does Avogadro's theory fit these criteria? Consider and evaluate whether this theory makes sense (is logical) and agrees with other theories (is consistent).

Molar Volume of Gases 4.3

The evolution of scientific knowledge often involves integrating two or more concepts. An example of this is the combination of Boyle's and Charles' laws to create the combined gas law. Avogadro's idea and the mole concept (Chapter 2) can also be combined. According to Avogadro's theory, equal volumes of any gases at the same temperature and pressure contain an equal number of entities (usually molecules). The mole concept indicates that a mole is a specific number of entities—Avogadro's number of entities. Therefore, for all gases at a specific temperature and pressure, there must be a certain volume—the molar volume—that contains one mole of entities. **Molar volume** is the volume that one mole of a gas occupies at a specified temperature and pressure. Logically, molar volume should be the same for all gases at the same temperature and pressure. For scientific work, the most useful temperature and pressure conditions are either SATP or STP. It has been determined empirically that the molar volume of a gas at SATP is approximately 24.8 L/mol. The molar volume of a gas at STP is approximately 22.4 L/mol (**Figure 1**). Molar volume is often given the symbol V_m.

Figure 1
At STP, one mole of gas has a volume of 22.4 L, which is approximately the volume of 11 "empty" 2 L pop bottles.

Knowing the molar volume of gases allows scientists to work with easily measured volumes of gases when specific chemical amounts of gases are needed. Measuring the volume of a gas is much more convenient than measuring its mass. Imagine trapping a gas in a container and trying to measure its mass on a balance—and then making corrections for the buoyant force of the surrounding air. Also, working with gas volumes is more precise because the process involves measuring relatively large volumes rather than relatively small masses. Molar volume can be used as a conversion factor to convert chemical amount to volume, as shown in the following example.

▸ COMMUNICATION example 1

Calculate the volume occupied by 0.024 mol of carbon dioxide gas at SATP.

Solution

$$V_{CO_2} = 0.024 \text{ mol} \times \frac{24.8 \text{ L}}{1 \text{ mol}}$$

$$= 0.60 \text{ L}$$

Using molar volume at SATP, the gas will occupy 0.60 L.

Notice how the units cancel in the example above. A molar volume can also be used to convert from a volume to a chemical amount. In this case, the molar volume ratio must be inverted to allow for the correct cancellation of units.

Learning Tip

In SI symbols, the relationship of chemical amount (n), volume (V), and molar volume (V_m) is expressed as

$$n = \frac{V}{V_m} \quad \text{or} \quad V = nV_m$$

▸ COMMUNICATION example 2

What chemical amount of oxygen is available for a combustion reaction in a volume of 5.6 L at STP?

Solution

$$n_{O_2} = 5.6 \text{ L} \times \frac{1 \text{ mol}}{22.4 \text{ L}}$$

$$= 0.25 \text{ mol}$$

Using molar volume at STP, 0.25 mol of oxygen is available.

Gases such as oxygen and nitrogen are often liquefied for storage and transportation, and then allowed to vaporize for use in a technological application. Helium is stored and transported as a compressed gas. Both liquefied and compressed gases are sold by mass. Molar volume and molar mass can be combined to calculate the volume of gas that is available from a known mass of a substance.

SAMPLE problem 4.3

Helium-filled balloons (**Figure 2**), often used for party decorations, are less dense than air, so they stay aloft and will rise unless tied down by a string. What volume does 3.50 g of helium gas occupy at SATP?

To answer this question, we first need to convert the mass into the chemical amount:

$$n_{He} = 3.50 \text{ g} \times \frac{1 \text{ mol}}{4.00 \text{ g}}$$

$$= 0.875 \text{ mol}$$

Now we can convert this chemical amount into a volume at SATP, using the molar volume:

$$V_{He} = 0.875 \text{ mol} \times \frac{24.8 \text{ L}}{1 \text{ mol}}$$

$$= 21.7 \text{ L}$$

Once these two steps are clearly understood, they can be combined into a single calculation, as shown below. Notice how you can plan your use of the conversion factors by planning the cancellation of the units. All units except the final unit cancel.

$$V_{He} = 3.50 \text{ g} \times \frac{1 \text{ mol}}{4.00 \text{ g}} \times \frac{24.8 \text{ L}}{1 \text{ mol}}$$

$$= 21.7 \text{ L}$$

Figure 2
Helium-filled balloons are popular items for parties and store promotions.

COMMUNICATION example 3

A propane tank for a barbecue contains liquefied propane. If the tank mass drops by 9.1 kg after a month's use, what volume of propane gas at SATP was used for cooking?

Solution

$$V_{C_3H_8} = 9.1 \text{ kg} \times \frac{1 \text{ mol}}{44.11 \text{ g}} \times \frac{24.8 \text{ L}}{1 \text{ mol}}$$

$$= 5.2 \text{ kL}$$

Using molar volume at SATP, 5.2 kL of gas was used.

SUMMARY   Molar Volumes

molar volume: the volume that one mole of a gas occupies at a specified temperature and pressure

$$V_m = 22.4 \text{ L/mol at STP} \qquad V_m = 24.8 \text{ L/mol at SATP}$$

Section 4.3 Questions

1. Describe the concept of molar volume in your own words.
2. Justify the fact that molar volume requires a specified temperature and pressure.
3. Relate the molar volume at SATP to some familiar volume.
4. Describe the similarities and differences between calculations of chemical amounts using molar masses and molar volumes.
5. Weather balloons filled with hydrogen gas are occasionally reported as UFOs. They can reach altitudes of about 40 km. What volume does 7.50 mol of hydrogen gas in a weather balloon occupy at SATP?
6. Sulfur dioxide gas is emitted from marshes, volcanoes, and refineries that process crude oil and natural gas. Calculate the chemical amount of sulfur dioxide contained in 50 mL at SATP.
7. Neon gas under low pressure and high voltage emits the red light that glows in advertising signs (**Figure 3**). Determine the volume occupied by 2.25 mol of neon gas at STP before the gas is added to neon tubes in a sign.

Figure 3
When an electric current is passed through a glass tube containing the noble gas neon, the gas glows a characteristic red colour.

8. Oxygen is released by plants during photosynthesis and is used by plants and animals during respiration. Predict the chemical amount of oxygen present in 20.0 L of air at STP. Assume that air is 20% oxygen by volume.
9. One gram of baking powder produces about 0.13 g of carbon dioxide. Predict the volume occupied by 0.13 g of carbon dioxide gas at SATP.
10. Volatile liquids vaporize rapidly from opened containers or spills. Some vapours, such as those from gasoline, contribute to the formation of smog. Calculate the volume at STP occupied by vapours from 50.0 g of spilled gasoline (assume complete vaporization of octane, $C_8H_{18}(l)$).
11. Millions of tonnes of nitrogen dioxide is dumped into the atmosphere each year by automobiles. This is a major cause of smog formation. Calculate the volume of 1.00 t (1.00 Mg) of nitrogen dioxide at SATP.
12. Water vapour plays an important role in the weather patterns on Earth. What mass of water must vaporize to produce 1.00 L of water vapour at SATP?
13. Natural gas (assume pure methane) reacts with oxygen in the air to heat most homes in Alberta. Predict the mass of oxygen gas required to react completely with 1.00 L of methane, with all gases measured at SATP.
14. Carbon dioxide is commonly used in fire extinguishers.
    (a) Determine the density (in grams per litre) of carbon dioxide at SATP (to a certainty of two significant digits).
    (b) If the density of air at SATP is 1.2 g/L, use this value and your answer to (a) to suggest one reason for the use of carbon dioxide as a fire-extinguishing agent.
    (c) State another important characteristic of carbon dioxide that makes it suitable for use in a fire extinguisher.

Extension

15. When performing calculations, we assume in this textbook that all gases behave exactly the same. You can make your own judgment on how valid this assumption is by considering the empirical values for molar volumes of selected gases (**Table 1**).

Table 1 Molar Volumes of Five Gases

Gas	Molar volume (L/mol at STP)
O_2	22.397
N_2	22.402
CO_2	22.260
H_2	22.433
NH_3	22.097

(a) Which gas has a molar volume that differs the most from the given (rounded) value of 22.4 L/mol?
(b) Using your knowledge of molecular properties, suggest a reason why this gas is different from the rest.

4.4 The Ideal Gas Law

Up to this point in your study of gases you have studied various laws and generalizations such as Boyle's law, Charles' law, the law of combining volumes, and molar volumes. All of these empirical properties are assumed to apply perfectly to all gases; in other words, all gases behave like an ideal gas. An **ideal gas** is a hypothetical gas that obeys all the gas laws perfectly under all conditions; that is, it does not condense into a liquid when cooled, and graphs of its volume and temperature and of pressure and temperature are perfectly straight lines (see **Figure 7** in Section 4.1).

The kinetic molecular theory (KMT) provides a good explanation of gas pressure, temperature, and the gas laws for an ideal gas. What about for real gases? **Table 1** shows the comparison between ideal and real gases in terms of the kinetic molecular theory.

Table 1 Comparison of Ideal and Real Gases

KMT assumption for ideal gases	Interpretation for real gases
Gas molecules are very far apart compared to their size. In other words, the molecules' size is negligible.	For high pressures, the molecules are forced much closer together and their size becomes significant. In other words, the empty space available is less than the size of the container.
Gas molecules are in constant, random, straight-line motion because no forces exist between them.	As the temperature decreases, the molecules slow down. At some point, the intermolecular attractions may cause the molecules to stick together and the gas becomes a liquid.
Gas molecules undergo perfectly elastic collisions in which no energy is lost and collisions (and rebounds) occur very quickly (**Figure 1(a)**).	Molecules of a real gas are more like "soft" spheres (**Figure 1(b)**). Shape change during collision and rebound makes this process occur a little more slowly. This means that the pressure of the gas is actually a little less than ideal.

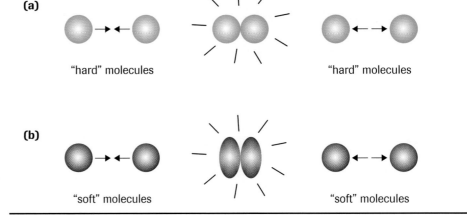

Figure 1
(a) In an ideal gas, the molecules collide like perfectly hard spheres and rebound very quickly after collision.
(b) In a real gas, the molecules are "soft" (can be deformed) and intermolecular attractions are important. The process of collision takes a slightly longer time, as a result.

There is considerable evidence to suggest that, for relatively low pressures and high temperatures such as STP and SATP conditions, real gases behave very nearly like ideal gases. It is only when the pressures become very large (>1 MPa) and the temperatures become very low (approaching the condensation point of the gas) that any differences between real and ideal become significant. In this textbook, all gases are dealt with as if they are ideal.

DID YOU KNOW?
Ideal Concepts
The discovery that the gas laws are not perfect should come as no surprise. No scientific knowledge can perfectly describe or explain natural phenomena (the real world). Scientific knowledge is always an ideal simplification. Some scientists claim that science is *reductionist* and only works by viewing the world as something much simpler than it really is. Other scientists respond that, although this may be true, the continuing challenge in science is to get closer and closer to the truth—whatever that might be. Science in this sense is open-ended—there are many discoveries yet to be made. Some scientists are beginning to appreciate Aboriginal concepts of interconnectedness and finding relevant applications in science.

Section **4.4**

A single, ideal-gas equation describes the interrelationship of pressure, temperature, volume, and chemical amount of matter—the four variables that define a gaseous system.

- According to Boyle's law, the volume of a gas is inversely proportional to the pressure: $V \propto \frac{1}{P}$.
- According to Charles' law, the volume of a gas is directly proportional to the absolute temperature: $V \propto T$.
- According to Avogadro's theory, the volume of a gas is directly proportional to the chemical amount of matter: $V \propto n$.

Combining these three statements produces the following relationship:

$$V \propto \frac{1}{P} \times T \times n$$

Another way of stating this is:

$$V = (\text{a constant}, R) \times \frac{1}{P} \times T \times n$$

$$V = \frac{nRT}{P}$$

$$PV = nRT \quad (\text{ideal gas law})$$

Synthesis is a goal of science. This equation is another example in which several concepts are combined (synthesized) into one broader concept. Scientists call this concept the **ideal gas law**. The constant, R, is known as the **universal gas constant**. The value for the universal gas constant can be obtained by substituting STP (or SATP) conditions for one mole of an ideal gas into the ideal gas law and solving for R. Using more certain values for STP and the molar volume at STP,

$$R = \frac{PV}{nT}$$

$$= \frac{101.325 \text{ kPa} \times 22.414 \text{ L}}{1.000 \text{ mol} \times 273.15 \text{ K}}$$

$$= \frac{8.314 \text{ kPa} \cdot \text{L}}{\text{mol} \cdot \text{K}}$$

The value of the universal gas constant depends on the units chosen to measure volume, pressure, and temperature.

If any three of the four variables in the ideal gas law are known, the fourth can be calculated by means of the ideal gas law equation. Often, however, the mass of a gas, rather than the chemical amount, is the known quantity. In this case, a two-step calculation is required.

DID YOU KNOW?

van der Waals Forces
In 1873, Johannes van der Waals hypothesized the existence of attraction between gas molecules to explain deviations from the ideal gas law. The general forces of attraction, called van der Waals forces, include London forces and dipole–dipole forces.

Figure 2
Johannes van der Waals (1937–1923)

 **WEB** Activity

Simulation—The Ideal Gas Law

In this simulation you can choose from a variety of different gases. By manipulating one variable—pressure, mass, or temperature—you can create graphs and observe mathematical relationships between any of these variables and volume. You can also use the values obtained to test the ideal gas law.

SAMPLE problem 4.4

A common use of the ideal gas law is to predict volumes or masses of gases under specified conditions of temperature and pressure. For example, predict the volume occupied by 0.78 g of hydrogen at 22 °C and 125 kPa.

To use the ideal gas law, you first need to convert the mass into the chemical amount of hydrogen.

$$n_{H_2} = 0.78 \text{ g} \times \frac{1 \text{ mol}}{2.02 \text{ g}} = 0.39 \text{ mol}$$

Retain the unrounded value in your calculator for the calculation.
Now you can use the ideal gas law to determine the volume of hydrogen at the conditions specified.

$$PV = nRT$$

$$V_{H_2} = \frac{nRT}{P}$$

$$= \frac{0.39 \text{ mol} \times 8.314 \text{ kPa·L} \times 295 \text{ K}}{125 \text{ kPa} \times 1 \text{ mol·K}} = 7.6 \text{ L}$$

or $V_{H_2} = 0.78 \text{ g} \times \dfrac{1 \text{ mol}}{2.02 \text{ g}} \times \dfrac{8.314 \text{ kPa·L}}{1 \text{ mol·K}} \times \dfrac{295 \text{ K}}{125 \text{ kPa}} = 7.6 \text{ L}$

Notice how the units cancel. This is a good check of your calculation.

COMMUNICATION example

What mass of argon gas (**Figure 3**) should be introduced into an evacuated 0.88 L tube to produce a pressure of 90 kPa at 30 °C?

Solution

$$PV = nRT$$

$$n_{Ar} = \frac{PV}{RT}$$

$$= \frac{90 \text{ kPa} \times 0.88 \text{ L·mol·K}}{8.314 \text{ kPa·L} \times 303 \text{ K}} = 0.031 \text{ mol}$$

$$m_{Ar} = 0.31 \text{ mol} \times \frac{39.95 \text{ g}}{1 \text{ mol}} = 1.3 \text{ g}$$

or $m_{Ar} = 0.88 \text{ L} \times \dfrac{1 \text{ mol·K}}{8.314 \text{ kPa·L}} \times \dfrac{90 \text{ kPa}}{303 \text{ K}} \times \dfrac{39.95 \text{ g}}{1 \text{ mol}} = 1.3 \text{ g}$

According to the ideal gas law, 1.3 g of argon should be used.

Figure 3
Argon has many practical uses, such as in welding, surgical lasers, and light bulbs. Argon and other noble gases are also used to produce "neon art." Argon produces a lavender colour when subjected to a high voltage. What colour does neon produce?

Practice

1. In your own words, state the three assumptions of the kinetic molecular theory for ideal gases.
2. List the previously known concepts that have been synthesized into the ideal gas law.
3. Determine the pressure in a 50 L compressed air cylinder containing 30 mol of air at a temperature of 40 °C.

Section **4.4**

4. Predict the chemical amount of methane gas present in a sample that has a volume of 500 mL at 35.0 °C and 210 kPa.

5. What volume does 50 kg of oxygen gas occupy at a pressure of 150 kPa and a temperature of 125 °C?

SUMMARY Properties of an Ideal Gas

Empirical
- V–T and P–T graphs are straight lines
- gas does not condense to a liquid when cooled
- $PV = nRT$

Theoretical
- volume (size) of molecules is negligible
- there are no forces of attraction between molecules
- collisions are elastic (no energy is lost)

+ EXTENSION

Collecting Gases Over Water
Extend your understanding of the gas laws with a refinement of Investigation 4.3.

www.science.nelson.com

LAB EXERCISE 4.B

Evaluating an Experimental Design

A good test of the design of an experiment is to use it to find the value of a well-known constant. In the Evaluation, determine the percent difference in Part 2 and evaluate the Design in Part 1.

Report Checklist
- ○ Purpose
- ○ Problem
- ○ Hypothesis
- ● Prediction
- ○ Design
- ○ Materials
- ○ Procedure
- ○ Evidence
- ● Analysis
- ● Evaluation (2, 1, 3)

Purpose
The purpose of this experiment is to evaluate the Design by determining the value of the universal gas constant.

Problem
What is the value of the universal gas constant, R?

Design
A measured mass of oxygen gas is collected by water displacement, and the volume, temperature, and pressure of the gas are measured.

Evidence
$m = 1.27$ g
$P = 99.0$ kPa
$V = 1.00$ L
$t = 21$ °C

INVESTIGATION 4.3 Introduction

Using the Ideal Gas Law

The molar mass of a compound is an important characteristic that, in some cases, can help to identify a substance. Molar masses of known compounds are obtained using accepted atomic molar masses from the periodic table.

Report Checklist
- ○ Purpose
- ○ Problem
- ○ Hypothesis
- ○ Prediction
- ○ Design
- ○ Materials
- ○ Procedure
- ● Evidence
- ● Analysis
- ● Evaluation (1, 3)

Purpose
The purpose of this investigation is to use the ideal gas law to determine an important characteristic of an unknown substance.

Problem
What is the molar mass of an unknown gas?

Design
A sample of gas is collected in a graduated cylinder by downward displacement of water. The volume, temperature, and pressure of the gas are measured, along with the change in mass of the original container.

To perform this investigation, turn to page 179.

Section 4.4 Questions

1. Using the kinetic molecular theory, describe the three main ways that an ideal gas differs from a real gas.
2. Under what conditions does a real gas most resemble an ideal gas? Explain, using the kinetic molecular theory and intermolecular forces.
3. When sweetgrass is burned or a leaf of fresh sage is rubbed between your hands, the fragrance is immediately obvious. Write a theoretical explanation for these observations.
4. Unlike an ideal gas, a real gas condenses to a liquid when the temperature is low enough. What does this indicate about the interaction between the molecules and why the gas is real versus ideal?
5. When an air bag is activated in a collision, sodium azide rapidly decomposes to produce nitrogen gas. Chemical engineers carefully choose the quantity of sodium azide to produce the required chemical amount of nitrogen gas. Use the ideal gas law to predict the chemical amount of nitrogen gas required to fill a 60 L air bag at a pressure of 233 kPa and a temperature of 25 °C.
6. At what temperature does 10.5 g of ammonia gas exert a pressure of 85.0 kPa in a 30.0 L container?
7. Use the ideal gas law to determine three ways to reduce the volume of gas in the shock absorber (cylinder and piston) of an automobile.
8. Use the ideal gas law and the molar volume at STP to calculate a value for the universal gas constant using units of atmospheres (atm) instead of kilopascals (kPa).
9. A 1.49 g sample of a pure gas occupies a volume of 981 mL at 42.0 °C and 117 kPa.
   (a) Determine the molar mass of the compound.
   (b) If the chemical formula is known to be XH_3, identify the element "X."
10. What is the volume of an ideal gas at STP and at SATP? What is the main reason that these values differ? Justify your answer with suitable calculations.
11. The density of a gas is the mass per unit volume of the gas in units of, for example, grams per litre. By finding the mass of one litre (assume 1.00 L) of gas, you can then calculate the density of the gas. Knowledge of the densities of gases compared with the density of air (at 1.2 g/L) can save your life.
    (a) What is the density of carbon monoxide gas at 20 °C and 98 kPa in a home?
    (b) Using your answer to (a), where should a carbon monoxide detector be located, close to the floor or close to the ceiling?
    (c) If potentially lethal carbon dioxide comes from a fire and carbon monoxide comes from a furnace, what other variable might affect the densities of these gases released within a home?
    (d) What is the density of propane, $C_3H_8(g)$, at 22 °C and 96.7 kPa?
    (e) If the density of air at 20 °C is 1.2 g/L, what happens to propane gas that may leak from a propane cylinder in a basement or from the tank of an automobile in an underground parkade? Why is this a problem?
12. A hot-air balloon rises up through the air because the density of the air inside the balloon is less than the density of the outside air. Using the ideal gas law, describe how this occurs.
13. Skiing in the back country can be very dangerous because of the possibility of being buried alive in an avalanche. Some enterprising people have come up with an avalanche air bag (**Figure 4**) to save lives. How does it work? Describe, in steps, how you would calculate how much nitrogen gas is required in the canister.

www.science.nelson.com

Figure 4
This avalanche air bag system is normally contained in a lightweight backpack while skiing. It only inflates when the skier is caught in an avalanche and pulls the emergency cord.

Extension

14. Alberta produces about 80% of the natural gas produced in Canada and exports about three quarters of its production, mostly to the United States. Scientists and engineers are developing the technology to obtain maximum benefit from this natural resource. Their work requires that they understand the science of gases.
    (a) Natural gas, from gas wells drilled into pockets of gas far underground, is a mixture of gases. List the main gases present.
    (b) Which non-hydrocarbon impurities are removed first?
    (c) Which hydrocarbon gases are typically found along with methane? Why is it important to understand real and ideal gases in order to remove these gases?
    (d) What are NGLs? List some important products made from NGLs.

www.science.nelson.com

Chapter 4 INVESTIGATIONS

INVESTIGATION 4.1

Pressure and Volume of a Gas

Report Checklist
- ○ Purpose
- ○ Problem
- ○ Hypothesis
- ○ Prediction
- ● Design
- ○ Materials
- ○ Procedure
- ● Evidence
- ● Analysis
- ○ Evaluation

This investigation is a replication of a famous experiment first done by Robert Boyle in 1662. Write your Design using the information given, and remember to include a plan and identify the variables. Include a graph and a word statement describing the relationship as part of your Analysis.

Purpose
The purpose of this investigation is to create a general relationship between the pressure and volume of a gas.

Problem
What effect does increasing the pressure have on the volume of a gas?

Materials
eye protection
Boyle's law apparatus or 35 mL plastic syringe
large rubber stopper or support block
cork borer
5 textbooks or equal masses (1 kg)

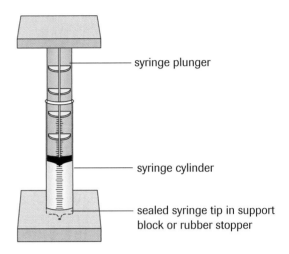

Figure 1
Setup of apparatus for Investigation 4.1

Procedure

1. Pull out the syringe plunger so that 30 mL of air is inside the cylinder.
2. If a syringe cap is not provided, bore a small hole in the rubber stopper deep enough so that the tip of the syringe fits tightly inside the stopper. Make sure that the tip of the syringe does not leak.
3. Hold the syringe barrel vertical, and measure the initial volume of air in the syringe.
4. While holding the syringe securely, carefully place one textbook or mass on the end of the plunger (**Figure 1**). (Your partner should balance the mass and be prepared to catch it if it starts to tilt.)
5. Record the mass placed on the syringe and the new volume of air.
6. Repeat steps 4 and 5 for a total of 4 or 5 books or masses.

INVESTIGATION 4.2

Temperature and Volume of a Gas

Report Checklist
- ○ Purpose
- ○ Problem
- ○ Hypothesis
- ○ Prediction
- ○ Design
- ○ Materials
- ○ Procedure
- ● Evidence
- ● Analysis
- ● Evaluation (1, 3)

This investigation, like Investigation 4.1, is a controlled experiment in which all variables are kept constant except the two variables being investigated, in this case temperature and volume. Include a graph and a word statement describing the relationship as part of your Analysis. In your Evaluation, pay particular attention to the sources of experimental uncertainties.

Purpose
The purpose of this investigation is to create a general relationship between the temperature and volume of a gas.

Problem
What effect does increasing the temperature have on the volume of a gas?

Design
A volume of air is sealed inside a syringe, which is then placed in a water bath. As the temperature of the air is manipulated, the volume of air inside the syringe is measured as the responding variable. Two controlled variables are the chemical amount of gas inside the syringe and the pressure of the gas.

Materials
lab apron
eye protection
plastic syringe (35–60 mL)
cap or stopper for the syringe tip
burette clamp
thermometer or temperature probe and clamp
600 mL beaker
ring stand
hot plate
stirring rod
boiling stones or chips

 Heat the water slowly, and make sure that the tested gas in the syringe does not eject the plunger. Wear eye protection.

Procedure
1. Set the syringe plunger to about 15–20 mL of air.
2. Seal the tip of the syringe with a cap or stopper.
3. Set up the ring stand with the 600 mL beaker on the hot plate (**Figure 2**).
4. Use the burette clamp to hold the syringe as far as possible into the beaker without touching the sides or bottom.
5. Clamp the thermometer or probe so that the bulb is close to, but not touching, the syringe.
6. Add water at room temperature to about 1 cm from the top of the beaker. Drop in a few boiling stones.
7. After a few minutes, record the temperature and volume of air. (See Appendix F.3 for a note on precision when reading a thermometer.)
8. Turn on the hot plate. Heat the water slowly, stirring occasionally.
9. Record the gas volume and temperature about every 10 °C until about 90 °C. (It may be necessary to tap or twist the plunger occasionally to make sure that it is not stuck.)

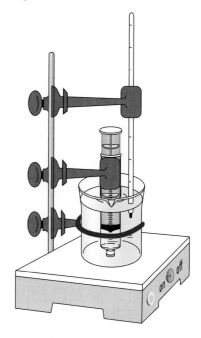

Figure 2
Setup of apparatus for Investigation 4.2

INVESTIGATION 4.3

Using the Ideal Gas Law

Report Checklist
- ○ Purpose
- ○ Problem
- ○ Hypothesis
- ○ Prediction
- ○ Design
- ○ Materials
- ○ Procedure
- ● Evidence
- ● Analysis
- ● Evaluation (1, 3)

The molar mass of a pure substance is an important characteristic that, in some cases, can help to identify the substance. Molar masses of known compounds are obtained using accepted atomic molar masses from the periodic table.

Purpose
The purpose of this investigation is to use the ideal gas law to determine an important characteristic of an unknown substance.

Problem
What is the molar mass of an unknown gas?

Design
A sample of gas is collected in a graduated cylinder by downward displacement of water (**Figure 3**). The volume, temperature, and pressure of the gas are measured, along with the change in mass of the original container.

Materials
lab apron
eye protection
lighter (with flint removed) or cylinder of unknown gas, with tubing
plastic bucket (approx. 4 L)
500 mL graduated cylinder or 600 mL graduated beaker
balance
thermometer or temperature probe
barometer

 The gas used may be flammable. Work in a well-ventilated area. Keep away from sparks and flames.

Procedure
1. Determine the initial mass of the lighter or gas canister.
2. Pour water into the bucket until it is two-thirds full. Completely fill the graduated cylinder with water and invert it in the bucket (**Figure 3**). Ensure that no air has been trapped in the cylinder.

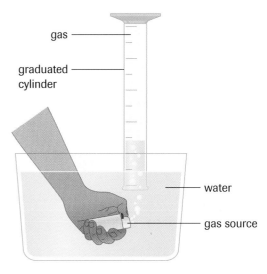

Figure 3
The gas can be collected by downward displacement of water.

3. Hold the lighter or tubing from the gas canister in the water and under the cylinder. Release the gas until you have collected 400 to 500 mL of gas. Make sure that all the bubbles enter the cylinder.
4. Equalize the pressures inside and outside the cylinder by adjusting the position of the cylinder until the water levels inside and outside the cylinder are the same.
5. Read the measurement on the cylinder, and record the volume of gas collected.
6. Record the ambient (room) temperature and pressure.
7. Dry the lighter or gas canister (if necessary), and determine its final mass.
8. Release the gas from the cylinder in a fume hood or outdoors.

Chapter 4 SUMMARY

Outcomes

Knowledge
- express atmospheric pressure in a variety of ways, including units of mm Hg, atm, and kPa (4.1)
- convert between the Celsius and absolute (kelvin) temperature scales (4.1, 4.4)
- describe and compare the behaviour of real and ideal gases in terms of kinetic molecular theory (4.2, 4.4)
- explain the law of combining volumes (4.2)
- illustrate how Boyle's, Charles', and combined gas laws are related to the ideal gas law (4.4)
- perform calculations based on the ideal gas law under STP, SATP, and other conditions (4.4)

STS
- identify and use a scientific problem-solving model (all sections)
- state that the goal of science is knowledge about the natural world (all sections)

Skills
- initiating and planning: state hypotheses and make predictions related to the pressure, temperature, and volume of a gas (4.1, 4.4); describe procedures for safe use and disposal of laboratory materials (4.1, 4.4)
- performing and recording: perform laboratory and simulated experiments to illustrate the gas laws, identifying and controlling variables (4.1, 4.4); use thermometers, balances, and other measuring devices to collect data on gases (4.1, 4.4); use research tools to collect information about real and ideal gases and applications of gases (all sections); perform an investigation to determine the molar mass from gaseous volume (4.4)
- analyzing and interpreting: draw and interpret graphs of experimental evidence that relate pressure and temperature to gas volume (4.1); identify the limitations of measurement (4.1, 4.4); identify a gas based on an analysis of experimental evidence (4.4)
- communication and teamwork: use appropriate SI notation and certainty in significant digits (all sections); work collaboratively and communicate effectively (all sections)

Key Terms 🔊

4.1
pressure
atmospheric pressure
STP
SATP
Boyle's law
absolute zero
absolute temperature scale
Charles' law
combined gas law

4.2
law of combining volumes
Avogadro's theory

4.3
molar volume

4.4
ideal gas
ideal gas law
universal gas constant

Key Equations

Boyle's law: $P_1V_1 = P_2V_2$ (4.1)

Charles' law: $\dfrac{V_1}{T_1} = \dfrac{V_2}{T_2}$ (4.1)

Combined gas law: $\dfrac{P_1V_1}{T_1} = \dfrac{P_2V_2}{T_2}$ (4.1)

$T(K) = t(°C) + 273$ (4.1)

Ideas gas law: $PV = nRT$ (4.4)

▶ MAKE a summary

1. Prepare a concept map for this topic. One suggested organization is:
   1st level: Gas
   2nd level: empirical, theoretical
   3rd level: qualitative properties; quantitative properties (with graph sketches); real and ideal gases (with an explanation); law of combining volumes (with an explanation)
   additional levels: (complete details)

2. Revisit your answers to the Starting Points questions at the beginnng of this chapter. How would you answer the questions differently now? Why?

▶ Go To

The following components are available on the Nelson Web site. Follow the links for *Nelson Chemistry Alberta 20-30*.
- an interactive Self Quiz for Chapter 4
- additional Diploma Exam-style Review questions
- Illustrated Glossary
- additional IB-related material

There is more information on the Web site wherever you see the Go icon in this chapter.

➕ EXTENSION 🔊

Kettle Call
Why does the noise in a kettle increase in volume until just before the boiling point and then go silent for a brief time? Dr. Ron Kydd, from the University of Calgary, suggests that collapsing bubbles may be the cause.

Unit 2 REVIEW

Many of these questions are in the style of the Diploma Exam. You will find guidance for writing Diploma Exams in Appendix H. Exam study tips and test-taking suggestions are on the Nelson Web site. Science Directing Words used in Diploma Exams are in bold type.

DO NOT WRITE IN THIS TEXTBOOK.

Part 1

1. Which one of the following is **not** a physical property of a gas?
   A. is incompressible
   B. flows easily
   C. assumes the shape of the container
   D. assumes the volume of the container

Use this information to answer questions 2 and 3.

Atmospheric pressure is an important variable for describing and explaining weather patterns and for predicting future weather. At sea level, the average atmospheric pressure is about 101 kPa.

2. Atmospheric pressure is defined as
   A. the weight of air
   B. the force exerted by air on all objects
   C. the force per unit area exerted by air on all objects
   D. the height of a column of mercury in a long, narrow tube

3. Average atmospheric pressure at sea level is approximately equal to
   A. 0.1 Pa
   B. 1 atm
   C. 76 mm Hg
   D. 101 atm

4. Based on many measurements, the normal body temperature of a human is about 37 °C. This corresponds to an absolute temperature of _____ K.

5. A bicycle pump contains 250 mL of air at a pressure of 102 kPa. If the pump is closed and the pressure increases to 210 kPa, the new volume becomes _____ mL.

6. A party balloon containing 3.5 L of air is taken from inside a house at 22 °C to the outdoors where the temperature is −15 °C. The new volume of the balloon is _____ L.

7. Under what conditions is a real gas most similar to an ideal gas?
   A. low temperature and low pressure
   B. low temperature and high pressure
   C. high temperature and high pressure
   D. high temperature and low pressure

8. According to the kinetic molecular theory, an ideal gas has all of the following characteristics *except*
   A. molecules of zero or insignificant size
   B. molecules that attract each other
   C. molecules that move randomly
   D. molecules that are widely separated

Use this information to answer questions 9 to 11.

The empirical study of gases provided a number of laws that formed the basis for important developments in chemistry such as atomic theory and the mole concept.

Statements
1. The volume of a gas varies inversely with the pressure on the gas.
2. Volumes of reacting gases are always in simple, whole number ratios.
3. The volume of a gas varies directly with the absolute temperature of the gas.
4. The volume of a gas varies directly with the absolute temperature and inversely with the pressure.

9. The statements that correspond to Boyle's law, Charles' law, the combined gas law, and the law of combining volumes are, respectively, __, __, __, and __.

10. Which statements require that the temperature be a controlled variable?
    A. 1, 2, 3, and 4
    B. 1, 3, and 4 only
    C. 1 and 2 only
    D. 3 and 4 only

11. Identify the statement that is best explained by Avogadro's theory.
    A. 1
    B. 2
    C. 3
    D. 4

12. A Chinook is a warm, dry winter wind that causes a rapid change in the weather in southern Alberta. The final volume of a cubic metre (1.00 kL) of air at −23 °C and 102 kPa when the conditions change to 12 °C and 96 kPa is predicted to be _____ kL.

Use this information to answer questions 13 to 15.

Disastrous explosions have resulted from the unsafe storage and handling of the fertilizer, ammonium nitrate, which can decompose rapidly according to the following unbalanced chemical equation.

__ $NH_4NO_3(s)$ → __ $N_2(g)$ + __ $H_2O(g)$ + __ $O_2(g)$

13. **NR** The simplest, whole-number coefficients to balance this chemical equation are, in order, __, __, __, and __.

14. According to Avogadro's theory, the ratio of the coefficients of the gaseous products represents the relative quantities of
    A. molecules
    B. masses
    C. volumes
    D. pressures

15. **NR** If 49.6 L of water vapour is produced, the volume of oxygen gas produced at the same temperature and pressure is _____ L.

16. Argon gas is an inert gas that moves other gases through a research or industrial system. The volume occupied by 4.2 kg of argon gas at SATP is
    A. 4.2 L
    B. 1.3 kL
    C. 2.4 kL
    D. 2.6 kL

17. Large quantities of chlorine gas are produced from salt to make bleach and treat water. The chemical amount of chlorine in 170 L of gas at 35 °C and 400 kPa is
    A. 13.3 mol
    B. 26.6 mol
    C. 37.6 mol
    D. 234 mol

Part 2

18. Convert each of the following temperatures into absolute temperatures in kelvin.
    (a) freezing point of water
    (b) 21 °C (room temperature)
    (c) absolute zero

19. Convert the following gas pressures to units of pascals:
    (a) A bike tire is pumped up to a pressure of 4.00 atm.
    (b) During manufacture, light bulbs are filled with argon to a pressure of 763 mm Hg.
    (c) Ammonia is produced in a reaction vessel at a pressure of 450 atm.

20. Convert each of the following gas volumes into a chemical amount.
    (a) 5.1 L of carbon monoxide gas at SATP
    (b) 20.7 mL of fluorine gas at STP

21. Calculate the volume at SATP, of each of the following chemical amounts of gas.
    (a) 500 mol of hydrogen (most common element in the universe)
    (b) 56 kmol of hydrogen sulfide (a toxin found in sour natural gas)

22. State each of the following laws in a sentence beginning "The volume of a gas sample ..."
    (a) Boyle's law
    (b) Charles' law
    (c) the ideal gas law

23. **Compare** a law and a theory. Use Charles' law and the kinetic molecular theory to support your answer.

24. Avogadro's idea is sometimes called a principle, a hypothesis, a law, or a theory. Is Avogadro's idea empirical or theoretical? **Explain** your answer.

25. **Describe** and **compare** the behaviour of real and ideal gases in terms of the kinetic molecular theory and intermolecular forces at
    (a) low temperature and high pressure
    (b) high temperature and low pressure

26. Pressurized hydrogen gas is used to fuel some prototype automobiles. **Determine** the new volume of a 28.8 L sample of hydrogen for which the pressure is increased from 100 kPa to 350 kPa? State the assumptions that you need to make in order to answer this question.

27. A student buys a 4.0 L party balloon in a store where the inside temperature is 23 °C. Outside, the balloon shrinks to 3.5 L. **Predict** the outside temperature in Celsius.

28. In an industrial process, bromine is produced by reacting chlorine with bromide ions in seawater. **Determine** what chemical amount of bromine is present in an 18.8 L sample of bromine gas at 60 kPa and 140 °C.

29. A 5.00 L balloon contains helium at SATP at ground level. **Predict** the balloon's volume when it floats to an altitude where the temperature is −15 °C and the atmospheric pressure is 91.5 kPa.

30. Before Boyle's and Charles' initial hypotheses became laws, their experiments had to be replicated by other scientists. For each of Boyle's and Charles' experiments, write a Problem statement in terms of variables and write a general Design including the identification of manipulated, responding, and important controlled variables.

31. Improper automobile tire inflation is one of the leading causes of tire failure. Manufacturers recommend that tire pressure be checked once a month. The recommended tire pressure for a particular vehicle is printed on a sticker (**Figure 1**) that is often attached to the driver-side door jamb or in the glove compartment. You should never use the pressure printed on the outside of the tire. This pressure is the maximum tire pressure from the tire manufacturer and is not the pressure recommended for a particular vehicle.

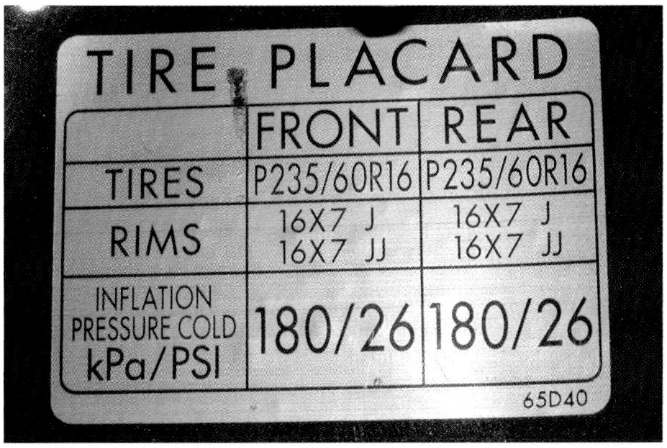

Figure 1
An example of a tire pressure sticker found on a vehicle

(a) Convert the recommended cold tire pressure (Figure 1) from units of kilopascals to atmospheres.
(b) Assuming a "cold" temperature of 15 °C and a constant volume, **predict** the tire pressure if the temperature is 40 °C.
(c) **Explain** your answer to (b) using the kinetic molecular theory.
(d) Manufacturers recommend that tire pressure be checked when the tires are cold (i.e., not driven for several hours). Suggest a problem that may occur if the tire pressure is adjusted to the recommended value when the tires are very hot after driving for a long time.

32. A particular glass container can hold an internal pressure of only 195 kPa before breaking. The container is filled with a gas at 19.5 °C and 96.7 kPa and then heated. **Predict** the temperature at which the container will break.

33. Electrical power plants commonly use steam to drive turbines, producing mechanical energy from the pressure of the steam. The rotating turbine is connected to a generator that produces electricity. Steam enters a turbine at a high temperature and pressure and exits, still a gas, at a lower temperature and pressure.
    (a) **Determine** the final pressure of steam that is converted from 10.0 kL at 600 kPa and 150 °C to 18.0 kL at 110 °C.
    (b) **Determine** the mass of steam that has gone through the turbine.

34. On summer afternoons, warm air masses often rise rapidly through the atmosphere, creating cumulus clouds or, sometimes, cumulonimbus (thunderstorm) clouds (**Figure 2**). Use the kinetic molecular theory to **explain** this rising of the warm air mass.

Figure 2
Cumulus clouds

35. During the production of nitric acid, ammonia reacts with oxygen to produce nitrogen monoxide:

    $4 NH_3(g) + 5 O_2(g) \rightarrow 4 NO(g) + 6 H_2O(g)$

    (a) Calculate the volumes of ammonia and oxygen required to produce 1.00 L of nitrogen monoxide. All gases are measured at the same temperature and pressure.
    (b) Use Avogadro's theory to **explain** the relationship used to calculate the volumes in (a).

36. Yeast cells in bread dough convert glucose into either carbon dioxide and water or carbon dioxide and ethanol, as shown in the following chemical equations:

    $C_6H_{12}O_6(aq) + 6 O_2(g) \rightarrow 6 CO_2(g) + 6 H_2O(l)$

    $C_6H_{12}O_6(aq) \rightarrow 2 CO_2(g) + 2 C_2H_5OH(l)$

    (a) Use the law of combining volumes to **predict** the volume of carbon dioxide produced when 50 mL of oxygen gas reacts with excess glucose.
    (b) For equal chemical amounts of glucose reacted, which of the two reactions will produce the greater degree of leavening (rising due to gas formation)? **Justify** your answer.

37. Suppose that you were trapped in a room in which there was a slow natural gas leak (assume pure methane). In order to breathe as little natural gas as possible, should you be near the ceiling or the floor? Use the molar volume at SATP to calculate the densities of methane and nitrogen to **justify** your answer.

38. One of the most common uses of carbon dioxide gas is carbonating beverages, such as soft drinks.
    (a) Squeezing a plastic bottle increases the pressure inside the bottle. What is the new volume of a 300 mL sample of carbon dioxide gas when the pressure doubles?
    (b) A carbonated drink contains pressurized carbon dioxide gas above the liquid. In addition to the usual factors such as temperature and pressure that affect the volume of a gas, what other variable is important?
    (c) **Explain** why a can of carbonated pop sometimes overflows when opened.

39. An investigation was conducted to determine the relationship between the pressure and solubility of nitrogen in water.

    Purpose
    The purpose of this investigation is to create a possible relationship between two variables.

    Problem
    What effect does the pressure of nitrogen gas have on its solubility in water at a fixed temperature?

    Evidence

    Table 1 Solubility of Nitrogen Gas in Water at 25 °C

Pressure (kPa)	Solubility (mmol/L)
50	0.33
100	0.67
150	1.04
200	1.35
250	1.61
300	1.98

    (a) Complete the Analysis, including a graph.
    (b) From the graph, **infer** the chemical amount of nitrogen gas that could dissolve at 225 kPa in 5.00 L of blood (assume mostly water) in a scuba diver.
    (c) Calculate the volume of nitrogen gas that would come out of 5.00 L of solution at 100 kPa if the diver had been submerged at 300 kPa and surfaced too quickly.
    (d) **Describe** briefly how knowledge of gas properties is important for scuba diving.

40. Hydrochlorofluorocarbons (HCFCs) are being used to replace chlorofluorocarbons (CFCs) because HCFCs are believed to do less damage to the ozone layer. The purpose of the following investigation is to use molar mass to identify an HCFC gas. Complete the Analysis section of the following report.

Purpose
The purpose of this investigation is to use gas concepts in a chemical analysis.

Problem
Is the HCFC sample tested $CHF_2Cl(g)$, $C_2H_3FCl_2(g)$, or $C_2H_3F_2Cl(g)$?

Design
A sample of an HCFC from a canister of the compressed gas is collected in a graduated cylinder by the downward displacement of water (**Figure 3**). The volume, temperature, and pressure of the gas are measured, along with the change in mass of the gas canister. Assume that the HCFC is not soluble in water.

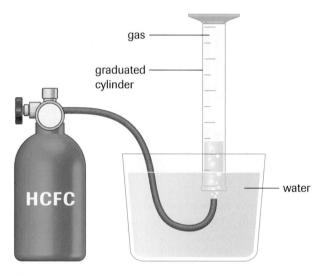

Figure 3
Collection of a known mass of HCFC

Evidence
initial mass of canister = 457.64 g
atmospheric pressure = 100.1 kPa
final mass of canister = 454.26 g
ambient temperature = 22.0 °C
volume = 840 mL

41. **Design** an experiment to test one of the gas laws. Assume that you have only everyday materials available to you, such as a pump, a pressure gauge, a balloon, a pail, hot and cold water, a measuring cup, a tape measure, and an outdoor alcohol thermometer. Complete the following categories: Problem, Prediction, Design, Materials (including sketch of the apparatus), Procedure, and outline the steps required for the Analysis.

42. A temperature inversion is a weather pattern that can trap polluted air near ground level. **Describe** the circumstances and the process by which the polluted air becomes trapped.

www.science.nelson.com

43. Review the focusing questions on page 142. Using the knowledge you have gained from this unit, briefly **outline** a response to each of these questions.

Extension

44. Research how gases are used in both medical anaesthetics and undersea exploration. Choose one application for each, and prepare a two-paragraph description of this application.

    www.science.nelson.com GO

45. There are many examples in science where new ideas are immediately rejected (without any of the usual testing) because they do not fit within the existing theories and beliefs of the scientific community. One example is the explanation of gas pressure by Daniel Bernoulli in 1738—about sixty years before Dalton's atomic theory and over a hundred years before emergence of the kinetic molecular theory. **Summarize** Bernoulli's hypothesis. Why was this idea not widely accepted? What does this example illustrate about the nature of science?

    www.science.nelson.com GO

46. The Bermuda Triangle has claimed many boats and planes (**Figure 4**). One hypothesis is that large volumes of natural gas released from the ocean floor may cause the boats to sink and the planes to drop.
    (a) **Explain** why boats might sink.
    (b) **Explain** why planes might drop.
    (c) Use the Internet to research the Bermuda Triangle. Which explanation for the losses seems most reasonable to you? Defend your choice in a brief report.

    www.science.nelson.com GO

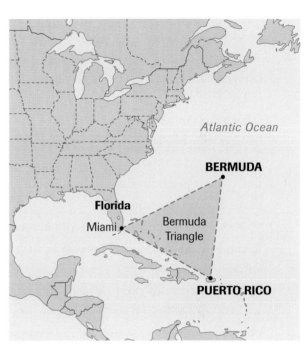

Figure 4
The Bermuda Triangle

unit 3
Solutions, Acids, and Bases

Solutions, especially of the liquid variety, are everywhere. All fresh water in streams, rivers, and lakes, salt water in the oceans, and even the rain that falls from the sky are examples of solutions. In general, what we call "water" is a solution that is essential to life. The characteristics of natural water can be quite complicated. Various physical, chemical, and biological factors need to be considered. For example, the colour of lakes may be due to dissolved minerals, decomposition of plant materials, and light reflected from suspended solids. In the glacial lake shown in the photograph, the colour is primarily due to the reflection of light from finely ground rock washed down in glacial streams.

Once you include manufactured solutions, there are literally countless examples to consider. Many consumer products, such as liquid cleaners, various drinks, and antiseptics, to name a few types, are solutions. Most industries use solutions in the cleaning, preparation, or treatment of the products they produce. Generally, when we have used or finished with a manufactured solution, we either recycle or dispose of it. Disposal usually means putting the unwanted solution into the environment, with or without some sort of treatment, and assuming that the environment will take care of our wastes. Our view of the environment needs to change from some external entity that can be exploited, to a more holistic view in which we are part of the system. In other words, our attitudes toward the environment need to become more like the traditional attitudes of Aboriginal peoples.

If you want to better understand our natural environment, how we produce solutions from the water in this environment, and the effects of our disposal of wastes, then an empirical and theoretical knowledge of solutions is essential.

As you progress through the unit, think about these focusing questions:

- How can we describe and explain matter as solutions, acids, and bases using empirical and theoretical descriptions?
- Why is an understanding of acid–base and solution chemistry important in our daily lives and in the environment?

Unit 3

GENERAL OUTCOMES

In this unit, you will

- investigate solutions, describing their physical and chemical properties
- describe acid and base solutions qualitatively and quantitatively

Solutions, Acids, and Bases

Unit 3
Solutions, Acids, and Bases

ARE YOU READY?

These questions will help you find out what you already know, and what you need to review, before you continue with this unit.

Prerequisites

Concepts
- classes of matter
- states of matter
- chemical and physical properties
- periodic table
- elements and compounds
- atomic theory
- ions
- chemical reactions
- chemical formulas and equations
- acids and bases

Skills
- WHMIS symbols
- laboratory safety rules
- scientific problem solving
- graphing, and solving linear and exponential equations

You can review prerequisite concepts and skills on the Nelson Web site, in the Chemistry Review unit, and in the Appendices.
A Unit Pre-Test is also available online.

www.science.nelson.com

Knowledge

1. Copy and complete the classification scheme in **Figure 1**.

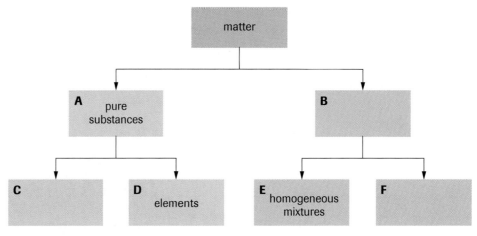

Figure 1
A classification of matter

2. Match each of the substances in **Table 1** to the classification categories illustrated in **Figure 1**.

3. Distinguish between ionic and molecular compounds based on their
   (a) chemical name or formula
   (b) empirical (observable) properties

Table 1 Classification of Substances

Substance	A or B	C or D or E or F
(a) vinegar		
(b) pure water		
(c) sulfur		
(d) air		
(e) milk		

4. In **Table 2**, match each term in column I with its corresponding description in column II.

Table 2 Definitions of Types of Matter

I	II
(a) compound	A. Cannot be broken down into simpler substances
(b) solution	B. Contains two or more visible components
(c) element	C. Can be identified by a single chemical formula
(d) heterogeneous mixture	D. A mixture of two or more pure substances with a single visible component

5. Write the missing words from the following statement in your notebook:

According to modern atomic theory, an atom contains a number of positively charged _____, determined by the _____ _____ of the element, and an equal number of negatively charged _____.

188 Unit 3

6. Atoms of the main group (representative) elements generally form predictable ions. Using calcium and fluorine as examples, draw Lewis symbols showing atoms and ions.

7. Draw a diagram to illustrate the model of a small sample (a few particles) of
   (a) sodium chloride
   (b) bromine

8. What is the type of bond between the atoms of a water molecule? What are the types of bonds between the molecules of water in a sample (**Figure 2**)?

9. Refer to the list of substances in **Table 3** to answer the following questions.
   (a) Which substances have London (dispersion) forces present between the molecules?
   (b) Classify each substance as polar or nonpolar.
   (c) Which substance would be expected to have hydrogen bonding as part of the intermolecular forces between the molecules?
   (d) Distinguish between intermolecular and intramolecular forces.

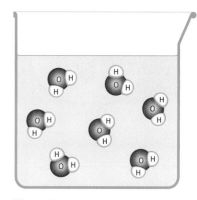

Figure 2
A model of a sample of liquid water

Table 3 Substances and Their Uses

Substance	Chemical formula	Use
propane	$C_3H_8(g)$	propane barbecues
ethanol	$C_2H_5OH(l)$	in gasohol (gasoline-alcohol fuel)
dichloromethane	$CH_2Cl_2(l)$	paint stripper

10. For each of the following pairs of reactants:
    - write a balanced chemical equation, including states of matter at SATP, for the expected reaction
    - translate the balanced chemical equation into an English sentence, including the coefficients and states of matter
    - state one diagnostic test that could be used to test the predicted reaction

    (a) aqueous iron(III) chloride and aqueous sodium hydroxide
    (b) aqueous silver nitrate and copper metal
    (c) sulfuric acid and aqueous potassium hydroxide
    (d) aqueous chlorine and aqueous sodium bromide

11. Many reactions in solution are single or double replacement reactions.
    (a) Write the word generalization for these two reaction types.
    (b) Classify each of the reactions in question 10 as a single or double replacement reaction.

12. Chemical reactions can also be classified as endothermic or exothermic. What do these terms mean? What diagnostic test would be used for this classification?

Skills

13. In this unit, you will work with many different solutions. What should you do immediately if some solution is spilled on your hand?

14. State the hazard communicated by each of the following WHMIS symbols.

(a)   (b)   (c)

chapter 5

The Nature and Properties of Solutions

In this chapter

- Exploration: Substances in Water (Demonstration)
- Mini Investigation: Solutions and Reactions
- Investigation 5.1: Qualitative Chemical Analysis
- Lab Exercise 5.A: Identifying Solutions
- Mini Investigation: Hot and Cold Solutions
- Biology Connection: Ions in Blood
- Lab Exercise 5.B: Qualitative Analysis
- Web Activity: David Schindler
- Biology Connection: Pollutants
- Case Study: Household Chemical Solutions
- Web Activity: Hot Tub Safety
- Investigation 5.2: A Standard Solution from a Solid
- Investigation 5.3: A Standard Solution by Dilution
- Investigation 5.4: The Iodine Clock Reaction
- Mini Investigation: Measuring the Dissolving Process
- Investigation 5.5: The Solubility of Sodium Chloride in Water
- Lab Exercise 5.C: Solubility and Temperature
- Explore an Issue: Pesticides

Is there such a thing as pure, natural water? Certainly it can't be found in the oceans. Drinking the water of the sea, which is rich in dissolved solutes, can be fatal. Today, seagoing ships carry distillation equipment to convert salt water into drinking water by removing most of those solutes.

Water from lakes and rivers (**Figure 1**), which we depend on for drinking, cooking, irrigation, electric power generation, and recreation, is also impure. Even direct from a spring, fresh water is a solution that contains dissolved minerals and gases. So many substances dissolve in water that it has been called "the universal solvent." Many household products, including soft drinks, fruit juices, vinegar, cleaners, and medicines, are aqueous (water) solutions. ("Aqueous" comes from the Latin *aqua* for "water," as in aquatics.) Our blood plasma is mostly water, and many substances essential to life are dissolved in it, including glucose.

The ability of so many materials to dissolve in water also has some negative implications. Human activities have introduced thousands of unwanted substances into water supplies. These substances include paints, cleaners, industrial waste, insecticides, fertilizers, salt from highways, and other contaminants. Even the atmosphere is contaminated with gases produced when fossil fuels are burned. Rain, falling through these contaminants, may become acidic. From an Aboriginal perspective, we are all connected to water. It flows through us and does not stay in us. If water is contaminated, the contaminates will also flow through us. The effects of water contamination in Walkerton, Ontario were a tragic illustration of this connection and the importance of water to our survival. Learning about aqueous solutions and the limits to purity will help you understand science-related social issues forming around the quality of our water.

STARTING Points

Answer these questions as best you can with your current knowledge. Then, using the concepts and skills you have learned, you will revise your answers at the end of the chapter.

1. What happens when a pure substance dissolves in water? How is dissolving related to any subsequent chemical reactions of the solution?
2. List the different ways you can express the concentration of a solution. Are some ways more useful than others?
3. Is there a limit to how much of a substance dissolves to make a solution? Explain.

Career Connections:
Water and Waste Water Treatment Plant Operator; Toxicologist

Figure 1
The water in this waterfall is clear and clean, but is it pure?

▶ Exploration Substances in Water (Demonstration)

Even the "universal solvent" does not dissolve all solutes equally well. See what happens when you add different substances to water.

Materials: overhead projector; 5 petri dishes; water; 5 substances, such as fruit drink crystals, a marble chip (calcium carbonate), a sugar cube, a few drops of alcohol (ethanol), a few drops of vegetable oil

- Pour a few millilitres of water into each of five petri dishes, and then carefully add one of the substances to each dish, without stirring. Record your observations.

(a) Which substances dissolved in water?
(b) How certain are you about each substance in (a)? Give your reasons.
(c) Which substances do not appear to dissolve?
(d) How certain are you about your answer in (c)?
(e) Do the mixtures all have the same properties? Other than visible differences, hypothesize how they might differ.
(f) Design some tests for your hypotheses.
- Dispose of any solids into the waste paper basket and pour the liquids down the drain.

The Nature and Properties of Solutions

5.1 Solutions and Mixtures

Many of the substances that we use every day come dissolved in water. We buy other substances with little or no water, but then mix water with them before use. For example, we may purchase syrup, household ammonia, and pop with water already added, but we mix baking soda, salt, sugar, and powdered drinks with water. Most of the chemical reactions that you see in high school occur in a water environment. Indeed, most of the chemical reactions necessary for life on our planet occur in water.

Science and technology provide us with many useful products and processes that involve substances dissolved in water, such as cleaning solutions and pharmaceuticals. However, as with all technologies, there are risks and benefits arising from the use, misuse, or disposal of these products. A key to understanding the risks and benefits starts with understanding solutions, and in particular, solutions containing water.

Solutions

Solutions are homogeneous mixtures of substances composed of at least one **solute**—a substance that is dissolved, such as salt, NaCl—and one **solvent**—the medium in which a solute is dissolved, such as water. Most liquid-state and gas-state solutions are clear (transparent)—you can see through them; they are not cloudy or murky in appearance. Solutions may be coloured or colourless. Opaque or translucent (cloudy) mixtures, such as milk (Figure 1), contain undissolved particles large enough to block or scatter light waves. These mixtures are considered to be heterogeneous.

It is not immediately obvious whether a clear substance is pure or a mixture, but it is certainly homogeneous. If you were to do a chemical analysis of a sample of a homogeneous mixture (i.e., a solution), you would find that the proportion of each chemical in the sample remains the same, regardless of how small the sample is. This is explained by the idea that there is a uniform mixture of entities (atoms, ions, and/or molecules) in a solution. Empirically, a solution is homogeneous; theoretically, it is uniform at the atomic and molecular levels.

Both solutes and solvents may be gases, liquids, or solids, producing a number of different combinations (**Table 1**). In metal alloys, such as bronze, the dissolving has taken place in liquid form before the solution is used in solid form. Common liquid solutions that have a solvent other than water include varnish, spray furniture polish, and gasoline. Gasoline, for example, is a mixture of many different hydrocarbons and other compounds. These substances form a solution—a uniform mixture at the molecular level. There are many such hydrocarbon solutions, including kerosene (a Canadian-invented fuel for lamps and stoves), and turpentine (used for cleaning paintbrushes). Most greases and oils dissolve in hydrocarbon solvents.

DID YOU KNOW?

Alternative Meanings
Milk is sometimes labelled as "homogenized," meaning that the cream is equally distributed throughout the milk. This use of the word does not match the chemistry definition of homogeneous. Using the strict chemistry definition, milk is not a homogeneous mixture, but a heterogeneous mixture (**Figure 1**).

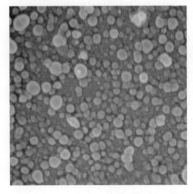

Figure 1
Milk is not a solution. This is quite obvious under magnification.

Table 1 Classification of Solutions

Solute in solvent	Example of solution	Source or use
gas in gas	oxygen in nitrogen	air
gas in liquid	oxygen in water	water
gas in solid	oxygen in solid water	ice
liquid in liquid	methanol in water	antifreeze
solid in liquid	sugar in water	syrup
solid in solid	tin in copper	bronze alloy

Other examples of liquids and solids dissolving in solvents other than water include the many chemicals that dissolve in alcohols. For example, solid iodine dissolved in ethanol (an alcohol) is used as an antiseptic (**Figure 2**). Some glues and sealants make use of other solvents: acetic acid is used as a solvent of the components of silicone sealants. You can smell the vinegar odour of acetic acid when sealing around tubs and fish tanks.

The chemical formula representing a solution specifies the solute by using its chemical formula and shows the solvent in parentheses. For example:

$NH_3(aq)$ ammonia gas (solute) dissolved in water (solvent)

$NaCl(aq)$ solid sodium chloride (solute) dissolved in water (solvent)

$I_2(alc)$ solid iodine (solute) dissolved in alcohol (solvent)

$C_2H_5OH(aq)$ liquid ethanol (solute) dissolved in water (solvent)

By far the most numerous and versatile solutions are those in which water is the solvent. Water can dissolve many substances, forming many unique solutions. All *aqueous solutions* have water as the solvent. They may be either coloured or colourless. Although water solutions are all different, they have some similarities and can be classified or described in a number of ways. This chapter deals primarily with the characteristics of aqueous solutions.

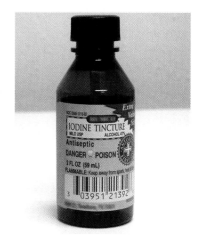

Figure 2
Tincture of iodine is a solution of the element iodine and the compound potassium or sodium iodide dissolved in ethanol. It is often found in first aid kits and is used to prevent the infection of minor cuts and scrapes.

▶ mini **Investigation**   *Solutions and Reactions*

For a chemical reaction to occur, does it matter if the reactants are in their pure form or dissolved in a solution when mixed?

Materials: 3 small test tubes with stoppers, vials of $Pb(NO_3)_2(s)$ and $NaI(s)$, wash bottle with pure water, laboratory scoop, test tube rack or small beaker, lead waste container

(a) Describe the appearance of each solid.
- Place a few crystals of $Pb(NO_3)_2(s)$ in one clean, dry test tube.
- Add an equal quantity of $NaI(s)$ to the same test tube. Stopper and shake.

(b) Describe the appearance of the solid mixture.

(c) Is there any evidence of a chemical reaction? Justify your answer.
- Set up two clean, dry test tubes with separate, small quantities (a few crystals) of $Pb(NO_3)_2(s)$ and $NaI(s)$.
- Add pure water to each test tube to a depth of about one-quarter of the test tube. Stopper each test tube and shake to dissolve the solids.

(d) Describe the appearance of each solution.
- Remove the stoppers and pour the contents of one test tube into the other. Stopper and invert to mix.

(e) Is there any evidence of a chemical reaction? Justify your answer.

(f) Compare your answers to (c) and (e). What conclusion can be made?

(g) Suggest a hypothesis to explain your answer to (f).
- Dispose of all materials into the lead waste container.

Properties of Aqueous Solutions

Compounds can be classified as either electrolytes or nonelectrolytes. At this point we will restrict ourselves to compounds in aqueous solutions. Compounds are **electrolytes** if their aqueous solutions conduct electricity. Compounds are **nonelectrolytes** if their aqueous solutions do not conduct electricity. Most household aqueous solutions, such as fruit juices and cleaning solutions, contain electrolytes. The conductivity of a solution

Figure 3
The bulb in this conductivity apparatus lights up if the solute is an electrolyte.

is easily tested with a simple conductivity apparatus (**Figure 3**) or an ohmmeter. This evidence also provides a diagnostic test to determine the class of a solute—electrolyte or nonelectrolyte. This very broad classification of compounds into electrolyte and nonelectrolyte categories can be related to the main types of compounds classified in Chapter 2. Electrolytes are mostly highly soluble ionic compounds (such as KBr(aq)), including bases such as ionic hydroxides (for example, sodium hydroxide, NaOH(aq)). Most molecular compounds (such as ethanol, C_2H_5OH(aq)) are nonelectrolytes, with the exception of acids. Acids (such as nitric acid, HNO_3(aq)) are molecular compounds in their pure form but in aqueous solution, conduct electricity.

Another empirical method of classifying solutions uses litmus paper as a test to classify solutes as *acids*, *bases*, or *neutral* substances. Acids form acidic solutions, bases form basic solutions, and most other ionic and molecular compounds form neutral solutions (**Table 2**). These definitions of acids, bases, and neutral substances are empirical, based on the results of the litmus and conductivity tests. Later in this unit, you will encounter theoretical definitions.

EXTENSION

Fastest Glacier
Water, life's universal solvent, has a very different impact on the environment depending on whether it is in its liquid or solid form. This simple difference is one of the major concerns of environmentalists today. This video shows how the rapid melting of glaciers could have a drastic impact on the environment worldwide.

www.science.nelson.com

Table 2 Properties of Solutes and Their Solutions

Type of solute	Conductivity test
electrolyte	light on conductivity apparatus glows; needle on ohmmeter moves compared to the control
nonelectrolyte	light on conductivity apparatus does not glow; needle on ohmmeter does not move compared to the control
Type of solution	**Litmus test**
acidic	blue litmus turns red
basic	red litmus turns blue
neutral	no change in colour of litmus paper

INVESTIGATION 5.1 *Introduction*

Qualitative Chemical Analysis

Solutions have properties determined by the solute that is present. Diagnostic tests based on characteristic properties can be used to identify substances in a qualitative analysis.

Report Checklist
○ Purpose ● Design ● Analysis
○ Problem ● Materials ○ Evaluation
○ Hypothesis ● Procedure
○ Prediction ● Evidence

Purpose
The purpose of this investigation is to use known diagnostic tests to distinguish among several pure substances.

Problem
Which of the white solids labelled 1, 2, 3, and 4 is calcium chloride, citric acid, glucose, and calcium hydroxide?

To perform this investigation, turn to page 227.

LAB EXERCISE 5.A

Identifying Solutions

For this investigation, assume that the labels on the four containers have been removed (perhaps washed off). Your task as a laboratory technician is to match the labels to the containers to identify the solutions.

Purpose
The purpose of this investigation is to use diagnostic tests to identify some solutions.

Problem
Which of the solutions labelled 1, 2, 3, and 4 is hydrobromic acid, sodium nitrate, lithium hydroxide, and methanol?

Design
Each solution is tested with both red and blue litmus paper and with conductivity apparatus. The temperature and concentration of the solutions are controlled variables.

Report Checklist
- ○ Purpose
- ○ Problem
- ○ Hypothesis
- ○ Prediction
- ○ Design
- ○ Materials
- ○ Procedure
- ○ Evidence
- ● Analysis
- ○ Evaluation

Evidence

Table 3 Properties of the Unidentified Solutions

Solution	Red litmus	Blue litmus	Conductivity
1	red	blue	none
2	red	red	high
3	red	blue	high
4	blue	blue	high

Section 5.1 Questions

1. Classify the following mixtures as heterogeneous or homogeneous. Justify your answers.
   (a) fresh-squeezed orange juice
   (b) white vinegar
   (c) an old lead water pipe
   (d) humid air
   (e) a cloud
   (f) a dirty puddle

2. Which of the following substances are solutions?
   (a) milk
   (b) pop
   (c) pure water
   (d) smoke-filled air
   (e) silt-filled water
   (f) rainwater

3. State at least three ways of classifying solutions.

4. (a) Describe an aqueous solution.
   (b) Give at least five examples of aqueous solutions that you can find at home.

5. (a) What types of solutes are electrolytes?
   (b) Write a definition of an electrolyte.

6. Classify each compound as an electrolyte or a nonelectrolyte:
   (a) sodium fluoride (in toothpaste)
   (b) sucrose (table sugar)
   (c) calcium chloride (a road salt)
   (d) ethanol (in wine)

7. Based upon your current knowledge, classify each of the following compounds (**Figure 4**) as forming an acidic, basic, or neutral aqueous solution, and predict the colour of litmus in each solution.
   (a) HCl(aq) (muriatic acid for concrete etching)
   (b) NaOH(aq) (oven and drain cleaner)
   (c) methanol (windshield washer antifreeze)
   (d) sodium hydrogen carbonate (baking soda)

Figure 4
Everyday chemicals form acidic, basic, or neutral solutions.

8. When assessing the risks and benefits of solutions, it is useful to consider multiple perspectives such as scientific, technological, ecological, economic, and political. In a few words, describe the main focus of each of these perspectives.

9. The importance of water can be described from several perspectives. Write a brief statement illustrating the importance of water for each of the following perspectives: technological, economic, ecological, and political.

10. Since grease dissolves in gasoline, some amateur mechanics use gasoline to clean car, bicycle, or motorcycle parts in their basements. Why is this practice unsafe? What precautions would make the use of gasoline for this purpose safer?

11. Electrolytes are lost during physical activity and in hot weather through sweating. The body sweats in order to keep cool—cooling by evaporation of water. Sweating removes water and the substances dissolved in the water, such as salts and other electrolytes. We replace lost electrolytes by eating and drinking. By law, the ingredients of a food item are required to be placed on the label in decreasing order of quantity, as they are in sports drinks.
    (a) Classify the ingredients of the sports drink in **Figure 5** as electrolytes or nonelectrolytes. How does the number and quantity of electrolytes and nonelectrolytes compare?
    (b) Which ingredients contain sodium ions? Which contain potassium ions? Are there more sodium or potassium ions in the drink? Justify your answer.
    (c) Does the most energy in the drink come from proteins, carbohydrates, or fats (oils)?
    (d) What three chemical needs does the drink attempt to satisfy?

Figure 5
Manufacturers recommend sports drinks to athletes, to restore electrolytes to the body.

Extension

12. Aboriginal peoples in North America use many solutions as medicine, both internal and external. Often, the solutions are like a tea, made by placing parts of plants in hot water. Traditional knowledge of the type of plant to use for different ailments also includes where to find the plant, what parts to use, and what procedure to follow.
    (a) Why were rosehips important to Aboriginal peoples of Canada's northwest like the Dene Tha' as well as to the first Europeans who came in contact with them (**Figure 6**)?
    (b) Given that vitamin C is a polar compound while vitamin A is nonpolar, which vitamin would be present in the greatest amount in a solution (tea) of rose hips?

www.science.nelson.com

Figure 6
Aboriginal peoples have traditionally used rosehips, as well as a wide variety of other natural remedies, for their well-being and survival.

Explaining Solutions 5.2

Water is the most important solvent on Earth. The oceans, lakes, rivers, and rain are aqueous solutions containing many different ionic compounds and a few molecular solutes. As you know, there are some ionic compounds that dissolve only very slightly in water, such as limestone (calcium carbonate) and various other rocks and minerals. Nevertheless, many more ionic compounds dissolve in water than in any other known solvent.

Why are ionic compounds so soluble in water? The key to the explanation came from the study of electrolytes. Electrolytes were first explained by Svante Arrhenius who was born in Wijk, Sweden, in 1859. While attending the University of Uppsala, he became intrigued by the problem of how and why some aqueous solutions conduct electricity, but others do not. This problem had puzzled chemists ever since Sir Humphry Davy and Michael Faraday performed experiments over half a century earlier, passing electric currents through chemical substances.

Faraday believed that an electric current produces new charged particles in a solution. He called these electric particles *ions* (a form of the Greek word for "to go"). He could not explain what ions were, or why they did not form in some solutions such as sugar or alcohol dissolved in water.

In 1887, Arrhenius proposed that when a substance dissolves, particles of the substance separate from each other and disperse into the solution. Nonelectrolytes disperse electrically neutral particles throughout the solution. As **Figure 1** shows, molecules of sucrose (a nonelectrolyte) separate from each other and disperse in an aqueous solution as individual molecules of sucrose surrounded by water molecules.

But what about the conductivity of solutions of electrolytes? Arrhenius' explanation for this observation was quite radical. He agreed with the accepted theory that electric current involves the movement of electric charge. Ionic compounds form conducting solutions. Therefore, according to Arrhenius, electrically charged particles must be present in their solutions. We now believe that, when a compound such as table salt dissolves, existing ions from the solid crystal lattice are separated as individual aqueous ions (**Figure 2**). The positive ions are surrounded by the negative ends of the polar water molecules, while the negative ions are surrounded by the positive ends of the polar water molecules (Chapter 3, page 101). **Dissociation** describes the separation of ions that occurs when an ionic compound dissolves in water. Dissociation equations, such as the following examples, show this separation of ions:

$$NaCl(s) \rightarrow Na^+(aq) + Cl^-(aq)$$

$$(NH_4)_2SO_4(s) \rightarrow 2\,NH_4^+(aq) + SO_4^{2-}(aq)$$

Notice that the formula for the solvent, $H_2O(l)$, does not appear as a reactant in the equation. Although water is necessary for the process of dissociation, it is not consumed and hence is not a reactant. The presence of water molecules surrounding the ions is indicated by (aq).

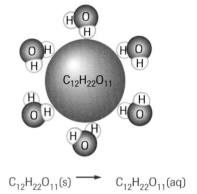

$$C_{12}H_{22}O_{11}(s) \rightarrow C_{12}H_{22}O_{11}(aq)$$

Figure 1
This model illustrates sucrose dissolved in water. The model, showing electrically neutral particles in solution, agrees with the evidence that a sucrose solution does not conduct electricity.

DID YOU KNOW ?

Restricting Terminology
Dissociation, strictly speaking, is the separation and dispersal of *any* entities that were initially bonded together. For our purposes, we restrict this term to refer to the separation and dispersal of ions as ionic compounds dissolve in water.

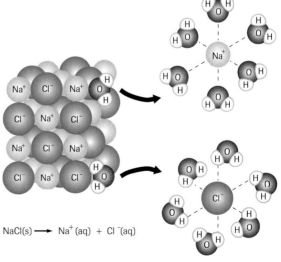

Figure 2
This model represents the dissociation of sodium chloride into positive and negative ions.

$$NaCl(s) \rightarrow Na^+(aq) + Cl^-(aq)$$

The Nature and Properties of Solutions 197

> ## Practice
>
> 1. Suppose you place a sugar cube (sucrose) and a lump of salt (sodium chloride) into separate glasses of water.
>   (a) Predict as many observations as you can about each mixture.
>   (b) According to theory, how is the dissolving process similar for both solutes?
>   (c) According to theory, how are the final solutions different?
>   (d) What theoretical properties of a water molecule help to explain the dissolving of both solutes?
> 2. Write equations to represent the dissociation of the following ionic compounds when they are placed in water:
>   (a) sodium fluoride
>   (b) sodium phosphate
>   (c) potassium nitrate
>   (d) cobalt(II) chloride
>   (e) aluminium sulfate
>   (f) ammonium hydrogen phosphate

Acids and Bases

Arrhenius eventually extended his theory to explain some of the properties of acids and bases (**Table 1**). According to Arrhenius, bases are ionic hydroxide compounds that dissociate into individual positive ions and negative hydroxide ions in solution. He believed the hydroxide ion was responsible for the properties of basic solutions; for example, turning red litmus paper blue. The dissociation of bases is similar to that of any other ionic compound, as shown in the following dissociation equation for barium hydroxide:

$$Ba(OH)_2(s) \rightarrow Ba^{2+}(aq) + 2\,OH^-(aq)$$

Acids, however, are a different story. The properties of acids appear only when compounds containing hydrogen, such as HCl(g) and H_2SO_4(l), dissolve in water. Since acids are electrolytes, the accepted theory is that acidic solutions must contain ions. However, the pure compounds that become acids in solution are usually molecular, so only neutral molecules are initially present. This unique behaviour requires an explanation other than dissociation. According to Arrhenius, acids ionize into positive hydrogen ions and negative ions when dissolved in water.

Ionization is the process by which a neutral atom or molecule is converted to an ion. In the case of acids, Arrhenius assumed that the water solvent somehow causes the acid molecules to ionize, but he did not propose an explanation for this process. The aqueous hydrogen ions are believed to be responsible for changing the colour of litmus in an acidic solution. Hydrogen chloride gas dissolving in water to form hydrochloric acid is a typical example of an acid. An ionization equation, shown below, is used to communicate this process:

$$\underset{\text{neutral molecules}}{HCl(g)} \xrightarrow{\text{ionization}} \underset{\text{ions}}{H^+(aq) + Cl^-(aq)}$$

Arrhenius' theory was a major advance in understanding chemical substances and solutions. Arrhenius also provided the first comprehensive theory of acids and bases. The empirical and theoretical definitions of acids and bases are summarized in Table 1.

DID YOU KNOW

When is an acid not an acid?
Pure acetic acid, CH_3COOH(l), or a solution of acetic acid in a nonpolar organic solvent such as gasoline, does not conduct electricity or change the colour of litmus. This behaviour is perfectly consistent with molecular substances. As soon as acetic acid is dissolved in water, however, the solution conducts electricity and changes the colour of litmus from blue to red.

EXTENSION

Dissociation vs. Ionization
What is the difference between dissociation and ionization? Both may produce aqueous ions. Dissociation, however, is the separation of ions that already exist before dissolving in water.

$$M^+X^-(s) \rightarrow M^+(aq) + X^-(aq)$$

Ionization involves the production of new ions, specifically hydrogen ions, in the case of acid solutions.

$$HX(aq) \rightarrow H^+(aq) + X^-(aq)$$

You can listen to a detailed discussion of when it is appropriate to use each of these theoretical terms.

www.science.nelson.com

Table 1 Acids, Bases, and Neutral Substances

Type of substance	Empirical definition	Arrhenius' theory
acids	Acids form solutions that • turn blue litmus red and are electrolytes • neutralize bases	• some hydrogen compounds ionize to produce H^+(aq) ions • H^+(aq) ions react with OH^-(aq) ions to produce water
bases	Bases form solutions that • turn red litmus blue and are electrolytes • neutralize acids	• ionic hydroxides dissociate to produce OH^-(aq) ions • OH^-(aq) ions react with H^+(aq) ions to produce water
neutral substances	Neutral substances form solutions that • do not affect litmus • some are electrolytes • some are nonelectrolytes	• no H^+(aq) or OH^-(aq) ions are formed • some are ions in solution • some are molecules in solution

Energy Changes

You already know that chemical reactions can be endothermic (absorb energy from the surroundings) or exothermic (release energy to the surroundings). What about the formation of solutions? Is energy absorbed or released when a substance dissolves in water?

▶ mini Investigation   Hot and Cold Solutions

In this short investigation, you will determine if changes are endothermic or exothermic simply by feeling the changes in the palm of your hand.

Materials: small plastic bag, plastic spoon, calcium chloride solid, sodium nitrate solid, 10 mL graduated cylinder

- Place about one teaspoon of calcium chloride in a bottom corner of a dry plastic bag.
- Measure about 5 mL of tap water and place into the other corner of the bag, keeping it separate from the solid.
- Lift the bag by the top, shake gently to mix, and immediately place into the palm of your hand.

(a) Is the dissolving of calcium chloride endothermic or exothermic? Justify your answer.
- Dispose of contents into the sink. Rinse and dry the inside of the bag.
- Repeat this procedure using sodium nitrate and dispose of the contents into a waste container.

(b) Is the dissolving of sodium nitrate endothermic or exothermic? Justify your answer.
(c) Suggest some practical applications of this investigation.

 Solids may cause irritation to the eyes, skin, and respiratory tract, and are harmful if swallowed. Wear eye protection and avoid direct contact.

At this stage in your chemical education, it is difficult to predict whether the dissolving of a particular solute will be endothermic or exothermic. More extensive investigations have clearly shown that all solution formation involves some energy change. The accepted explanation depends upon two general theoretical principles:

- Breaking existing bonds uses energy.
- Forming new bonds releases energy.

In the dissolving of an ionic solid such as sodium chloride (see Figure 2, page 197), the ionic bond between the positive and negative ions needs to be broken to separate the ions. Energy is also required to overcome the intermolecular forces among the water molecules.

Simultaneously, new bonds between the individual ions and polar water molecules need to form, as shown in Figure 2.

$$NaCl(s) \rightarrow Na^+ + Cl^- \rightarrow Na^+(aq) + Cl^-(aq)$$
$$\text{uses energy} \qquad \text{releases energy}$$

$$NaCl(s) \xrightarrow{\text{in } H_2O(l)} Na^+(aq) + Cl^-(aq)$$

Bonds broken
- ionic bonds in solid
- intermolecular forces between water molecules

Energy absorbed

Bonds formed
- electrostatic forces between ions and water molecules

Energy released

Net energy change

We can only observe the effects of the net energy change, usually as a change in temperature. Based on the idea of energy changes in bond breaking and forming, we can explain that a dissolving process is either *endothermic* because more energy is absorbed than released or *exothermic* because more energy is released than absorbed. Typical of a simple theory, it is able to explain this process once you know the experimental answer, but is unable to predict the outcome.

When calcium chloride dissolves in water, for example, the evidence (increase in temperature) shows that the overall dissolving process is exothermic. Therefore, more energy must be released when the individual ions bond to water molecules than is used to break the bonds between the ions in the solid calcium chloride and the intermolecular forces between water molecules. For endothermic dissolving, the opposite applies: When sodium nitrate dissolves in water, the temperature of the solution decreases. Therefore, heat is absorbed from the surroundings as the sodium nitrate dissolves. Theoretically, this means that less energy is released when the solute particles bond to water molecules than is used to break the bonds between the particles in the pure substances. You will study endothermic and exothermic changes in more detail in Unit 6.

Substances in Water

The focus in this unit is on substances that dissolve in water. However, many naturally occurring substances, such as limestone (mainly calcium carbonate), do not dissolve in water to an appreciable extent. This property is fortunate for us because limestone is a common building material that would not be very useful if it dissolved when it rained. On the other hand, sodium chloride is a naturally occurring compound found in large underground deposits near Fort Saskatchewan. As you know, sodium chloride easily dissolves in water, and chemical engineers make good use of this property in a process called solution mining—extracting substances by dissolving them in water. Knowledge of the properties of materials is useful for meeting societal needs such as building materials, salt for flavouring, or salt as a raw material to make other useful products. The solubility of ionic compounds (including bases such as ionic hydroxides) can be predicted from a solubility chart (inside back cover of this textbook).

Acids also occur naturally. For example, you have hydrochloric acid in your stomach, carbonic acid is in natural rain, and acetic acid forms during the fermentation process. Acidic solutions vary in their electrical conductivity. Acids that are extremely good conductors are called *strong acids*. Sulfuric acid, nitric acid, and hydrochloric acid are examples of strong acids that are almost completely ionized when in solution. (These strong

CAREER CONNECTION

Water and Waste Water Treatment Plant Operator

These operators are essential to everyone's health and safety. They monitor and control our water purification and waste-water treatment facilities. As stewards of our drinking water and freshwater ecosystems, these water specialists use a variety of skills to manage public safety and health. Learn more about the variety of duties and certification requirements for this career direction.

www.science.nelson.com GO

BIOLOGY CONNECTION

Ions in Blood

Human blood plasma contains the following ions: $Na^+(aq)$, $K^+(aq)$, $Ca^{2+}(aq)$, $Mg^{2+}(aq)$, $HCO_3^-(aq)$, $Cl^-(aq)$, $HPO_4^{2-}(aq)$, and $SO_4^{2-}(aq)$, as well as many complex acid and protein molecules. A physician can gain information about the state of your health by testing for the quantity of these substances present in a blood sample.

www.science.nelson.com GO

acids are listed on the inside back cover of this textbook under "Concentrated Reagents.") Most other common acids, such as acetic acid, are *weak acids*. The conductivity of acidic solutions varies a great deal. The accepted explanation, presented in more detail in Chapter 6, is that the degree of ionization of acids varies.

Molecular compounds abound in nature. They include most of the compounds that make up living things, as well as fossil fuels such as oil and natural gas. Like ionic compounds, the solubility of molecular compounds in water varies tremendously, but unlike ionic compounds, there is no simple solubility chart to make specific predictions. According to theories of intermolecular bonding (Chapter 3), however, we can make some general predictions about solubility in water—a polar liquid with hydrogen bonding. Nonpolar molecular compounds generally do not dissolve in water; polar compounds may be slightly soluble in water; and polar compounds with hydrogen bonding are the most likely to be very soluble in water. For efficiency in studying the examples in this textbook, you should memorize the examples in **Table 2**.

To understand the properties of aqueous solutions and the reactions that take place in solutions, it is necessary to know the major entities present when any substance is in a water environment. **Table 3** summarizes this information. The information is based on the solubility and electrical conductivity of substances as determined in the laboratory. Your initial work in chemistry will deal mainly with strong acids and other highly soluble compounds.

Table 2 Solubility of Selected Molecular Compounds

Solubility	Examples
high	ammonia, $NH_3(g)$
	hydrogen peroxide, $H_2O_2(l)$
	methanol, $CH_3OH(l)$
	ethanol, $C_2H_5OH(l)$
	glucose, $C_6H_{12}O_6(s)$
	sucrose, $C_{12}H_{22}O_{11}(s)$
low	methane, $CH_4(g)$
	propane, $C_3H_8(g)$
	octane, $C_8H_{18}(l)$

Table 3 Major Entities Present in a Water Environment

Type of substance	Solubility in water	Typical pure substance	Major entities present when substance is placed in water
ionic compounds	high	$NaCl(s)$	$Na^+(aq)$, $Cl^-(aq)$, $H_2O(l)$
	low	$CaCO_3(s)$	$CaCO_3(s)$, $H_2O(l)$
bases	high	$NaOH(s)$	$Na^+(aq)$, $OH^-(aq)$, $H_2O(l)$
	low	$Ca(OH)_2(s)$	$Ca(OH)_2(s)$, $H_2O(l)$
molecular substances	high	$C_{12}H_{22}O_{11}(s)$	$C_{12}H_{22}O_{11}(aq)$, $H_2O(l)$
	low	$C_8H_{18}(l)$	$C_8H_{18}(l)$, $H_2O(l)$
strong acids	high	$HCl(g)$	$H^+(aq)$, $Cl^-(aq)$, $H_2O(l)$
weak acids	high	$CH_3COOH(l)$	$CH_3COOH(aq)$, $H_2O(l)$
elements	low	$Cu(s)$	$Cu(s)$, $H_2O(l)$
		$N_2(g)$	$N_2(g)$, $H_2O(l)$

SUMMARY Explaining Solutions

Table 4 Arrhenius' Theory of Solutions

Substance	Process	General equation
molecular	disperse as individual molecules	$XY(s/l/g) \rightarrow XY(aq)$
ionic	dissociate as individual cations and anions	$MX(s) \rightarrow M^+(aq) + X^-(aq)$
base (ionic hydroxide)	dissociate as cations and hydroxide ions	$MOH(s) \rightarrow M^+(aq) + OH^-(aq)$
acid	ionize to form new hydrogen ions and anions	$HX(s/l/g) \rightarrow H^+(aq) + X^-(aq)$

LAB EXERCISE 5.B

Qualitative Analysis

In your evaluation, suggest improvements to the Design, using your knowledge of chemicals and **Table 5**.

Report Checklist
- ○ Purpose
- ○ Problem
- ○ Hypothesis
- ○ Prediction
- ○ Design
- ○ Materials
- ○ Procedure
- ○ Evidence
- ● Analysis
- ● Evaluation (1, 3)

Table 5 Electrical Conductivity

Class	Solid	Liquid	Aqueous
metal	✓	✓	–
nonmetal	X	X	–
ionic	X	✓	✓
molecular	X	X	X
acid	X	X	✓

Purpose
The purpose of this investigation is to test the diagnostic tests for different classes of substances as a means of qualitative analysis.

Problem
Which of the chemicals numbered 1 to 7 is $KCl(s)$, $Ba(OH)_2(s)$, $Zn(s)$, $C_6H_5COOH(s)$, $Ca_3(PO_4)_2(s)$, $C_{25}H_{52}(s)$ (paraffin wax), and $C_{12}H_{22}O_{11}(s)$?

Design
The chemicals are tested for solubility, conductivity, and effect on litmus paper. Equal amounts of each chemical are added to equal volumes of water.

Evidence

Table 6 Solubility, Conductivity, and Litmus Test Results

Chemical	Solubility in water	Conductivity of solution	Effect of solution on litmus paper
1	high	none	no change
2	high	high	no change
3	none	none	no change
4	high	high	red to blue
5	none	none	no change
6	none	none	no change
7	low	low	blue to red

▶ Section 5.2 Questions

1. In your own words, describe what is believed to happen when an ionic compound dissolves in water.

2. Write dissociation or ionization equations for each of the following pure substances when they dissolve in water:
   (a) $CaCl_2(s)$ (road salt)
   (b) $HF(g)$ (etching glass)
   (c) $(NH_4)_2HPO_4(s)$ (fertilizer)
   (d) $Al_2(SO_4)_3(s)$ (making pickles)

3. Compare dissociation and ionization by listing similarities and differences.

4. According to Arrhenius' theory, what is the explanation for
   (a) an acid turning blue litmus red?
   (b) a base turning red litmus blue?
   (c) neutralization of an acid and a base?

5. List three examples of solutions in consumer products.

6. A key characteristic of science is the goal of explaining natural products and processes. Do we need a theoretical explanation of solutions in order to use them? Answer from consumer and Aboriginal perspectives.

7. Many substances dissolve in water because water is such a polar solvent.
   (a) Are energy changes always involved when substances dissolve in water? Justify your answer.
   (b) Describe a brief experimental design to test your answer to (a).
   (c) What are some limitations that might be encountered if you were to perform this experiment?

8. List the chemical formulas for the major entities present in water for each of the following:
   (a) zinc
   (b) sodium bromide
   (c) oxygen
   (d) nitric acid
   (e) calcium phosphate
   (f) methanol
   (g) aluminium sulfate
   (h) potassium dichromate
   (i) acetic acid
   (j) sulfur
   (k) copper(II) sulfate
   (l) silver chloride
   (m) paraffin wax, $C_{25}H_{52}(s)$

9. Why is water such a good solvent for dissolving many ionic and molecular compounds? How is this property an advantage and a disadvantage?

10. The dissolving of calcium chloride in water is very exothermic compared with dissolving sodium chloride in water. Would calcium chloride be an appropriate substitute for a sidewalk deicer? Identify some positive and negative aspects, including several perspectives.

Extension

11. List some personal values demonstrated by Svante Arrhenius while developing his ideas about solutions.

www.science.nelson.com

12. List some vitamins that are water soluble and some that are fat soluble. Using one example of each, draw a structural formula and explain its solubility. How does the solubility in water and fat relate to how quickly a vitamin is excreted?

www.science.nelson.com

Solution Concentration 5.3

Most aqueous solutions are colourless, so there is no way of knowing, by looking at them, how much of the solute is present in the solution. As we often need to know the quantity of solute in the solution, it is important that solutions be labelled with this information. We use a ratio that compares the quantity of solute to the quantity of the solution. This ratio is called the solution's **concentration**. Chemists describe a solution of a given substance as *dilute* if it has a relatively small quantity of solute per unit volume of solution (**Figure 1**). A *concentrated* solution, on the other hand, has a relatively large quantity of solute per unit volume of solution.

In general, the concentration, c, of any solution is expressed by the ratio

$$\text{concentration} = \frac{\text{quantity of solute}}{\text{quantity of solution}}$$

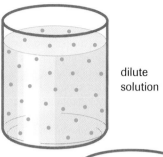

Figure 1
The theoretical model of the dilute solution shows fewer solute entities (particles) per unit volume compared with the model of the concentrated solution.

Percentage Concentration

Many consumer products, such as vinegar (acetic acid), are conveniently labelled with their concentration ratios expressed as percentages (**Figure 2**). A vinegar label listing "5% acetic acid (by volume)" means that there is 5 mL of pure acetic acid dissolved in every 100 mL of the vinegar solution. This type of concentration is often designated as % V/V, percentage volume by volume, or percentage by volume.

$$c_{CH_3COOH} = \frac{5 \text{ mL}}{100 \text{ mL}} = 5\% \text{ V/V}$$

In general, a percentage by volume concentration may be defined as

$$c = \frac{V_{solute}}{V_{solution}} \times 100\%$$

> ### ▶ COMMUNICATION example 1
>
> A photographic "stop bath" contains 140 mL of pure acetic acid in a 500 mL bottle of solution. What is the percentage by volume concentration of acetic acid?
>
> **Solution**
>
> $$c_{CH_3COOH} = \frac{140 \text{ mL}}{500 \text{ mL}} \times 100\%$$
>
> $$= 28.0\% \text{ V/V}$$
>
> The percentage by volume concentration of acetic acid is 28.0%, or the concentration of acetic acid is 28.0% V/V.

Another common concentration ratio used for consumer products is "percentage weight by volume" or % W/V. (In consumer and commercial applications, "weight" is used instead of "mass," which explains the W in the W/V label.) For example, a hydrogen peroxide topical solution used as an antiseptic is 3% W/V (**Figure 2**). This means that 3 g of hydrogen peroxide is dissolved in every 100 mL of solution.

$$c_{H_2O_2} = \frac{3 \text{ g}}{100 \text{ mL}}$$

$$= 3\% \text{ W/V}$$

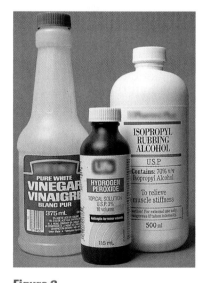

Figure 2
The concentrations of different consumer products are usually expressed as a percentage because percentages are generally easy for consumers to understand.

In general, we write a percentage weight by volume concentration as

$$c = \frac{m_{solute}}{V_{solution}} \times 100\%$$

A third concentration ratio is the "percentage weight by weight," or % W/W:

$$c = \frac{m_{solute}}{m_{solution}} \times 100\%$$

CAREER CONNECTION

Toxicologist
Toxicologists are investigators who specialize in detecting poisons and other harmful substances. Toxicologists can specialize in many areas, including analyzing natural substances such as venoms, testing new industrial products, or, in forensic science, testing for poisons in suspicious deaths. Learn more about possibilities of working as a toxicologist, including different specializations, education, and salary.

www.science.nelson.com GO

▶ COMMUNICATION example 2

A sterling silver ring has a mass of 12.0 g and contains 11.1 g of pure silver. What is the percentage weight by weight concentration of silver in the metal?

Solution

$$c_{Ag} = \frac{11.1 \text{ g}}{12.0 \text{ g}} \times 100\%$$
$$= 92.5\% \text{ W/W}$$

The percentage weight by weight of silver is 92.5%, or the concentration of silver is 92.5% W/W.

DID YOU KNOW ?

Understanding ppm
- A concentration of 1% W/W is equivalent to 10 000 ppm.
- 1 ppm is approximately the concentration obtained by dissolving one grain of salt (about 0.1 mg) in a 100 mL glass of water.

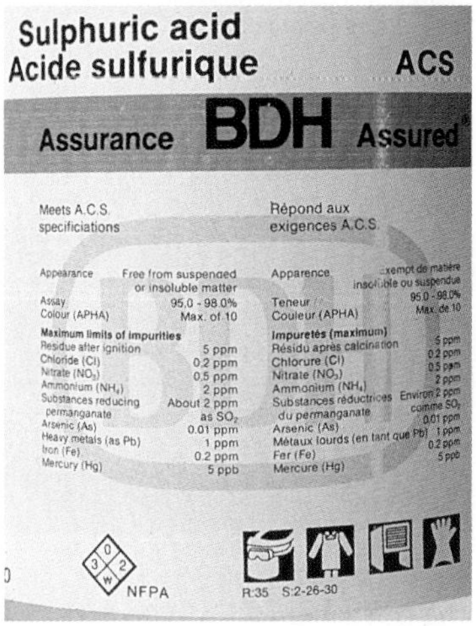

Figure 3
This concentrated sulfuric acid has an amount concentration of 17.8 mol/L. Note the impurities listed in units of parts per million.

Parts per Million Concentration

In studies of solutions in the environment, we often encounter very low concentrations. For very dilute solutions, we choose a concentration unit to give reasonable numbers for very small quantities of solute. For example, the concentration of toxic substances in the environment, of chlorine in a swimming pool, and of impurities in laboratory chemicals (**Figure 3**) is usually expressed as parts per million (ppm, $1:10^6$) or even smaller ratios, such as parts per billion (ppb, $1:10^9$) or parts per trillion (ppt, $1:10^{12}$). These ratios are a special case of the weight by weight (W/W) ratio. By definition, parts per million means

$$\frac{m_{solute}(\text{g})}{m_{solution}(\text{g})} \times 10^6$$

However, this calculation can be simplified by altering the units to incorporate the factor of 10^6. Therefore, a concentration in parts per million (ppm) can also be expressed as

$$c = \frac{m_{solute}(\text{mg})}{m_{solution}(\text{kg})}$$

which means that 1 ppm = 1 mg/kg.

Because very dilute aqueous solutions are similar to pure water, their densities are considered to be the same: 1 g/mL. Therefore, 1 ppm of chlorine is 1 g in 10^6 g or 10^6 mL (1000 L) of pool water, which is equivalent to 1 mg of chlorine per litre of water. For dilute aqueous solutions only,

1 ppm = 1 g/10^6 mL = 1 mg/L = 1 mg/kg

Small concentrations such as ppm, ppb, and ppt are difficult to imagine, but are very important in environmental studies and in the reporting of toxic effects (**Table 1**).

> **COMMUNICATION** *example 3*

Dissolved oxygen in natural waters is an important measure of the health of the ecosystem. In a chemical analysis of 250 mL of water at SATP, 2.2 mg of oxygen was measured. What is the concentration of oxygen in parts per million?

Solution

$$c_{O_2} = \frac{2.2 \text{ mg}}{0.250 \text{ L}}$$
$$= 8.8 \text{ mg/L}$$
$$= 8.8 \text{ ppm}$$

The concentration of dissolved oxygen is 8.8 ppm.

Table 1 Maximum Acceptable Concentration (MAC) of Chemicals in Canadian Drinking Water

Substance	Typical source	MAC (ppm)
cadmium	batteries in landfills	0.005
lead	old plumbing	0.010
nitrates	fertilizers	45.0
cyanides	mining waste	0.2

Amount Concentration

Chemistry is primarily the study of chemical reactions, which we communicate using balanced chemical equations. The coefficients in these equations represent chemical amounts in units of moles. Concentration is therefore communicated using amount concentration. **Amount concentration**, c, is the chemical amount of solute dissolved in one litre of solution.

$$\text{amount concentration} = \frac{\text{chemical amount of solute (in moles)}}{\text{volume of solution (in litres)}}$$

$$c = \frac{n}{V}$$

The units of amount concentration (mol/L) come directly from this ratio.

Amount concentration can also be indicated by the use of square brackets. For example, the amount concentration of sodium hydroxide in water could be represented by [NaOH(aq)].

DID YOU KNOW

Amount Concentration

According to IUPAC, the molar concentration of X is now officially called the "amount concentration of X" and is denoted by c_X or [X]. Amount concentration replaces the older terms, molar concentration and molarity. Remember that in chemistry, the "amount" of a substance is always a quantity in units of moles.

> **COMMUNICATION** *example 4*

In a quantitative analysis, a stoichiometry calculation produced 0.186 mol of sodium hydroxide in 0.250 L of solution. Calculate the amount concentration of sodium hydroxide.

Solution

$$c_{NaOH} = \frac{0.186 \text{ mol}}{0.250 \text{ L}}$$
$$= 0.744 \text{ mol/L}$$

The amount concentration of sodium hydroxide is 0.744 mol/L.

> **Practice**

1. Describe the three different systems of expressing the concentration of a solution.
2. Gasohol, a solution of ethanol and gasoline, is considered to be a cleaner fuel than gasoline alone. A typical gasohol mixture available across Canada contains 4.1 L of ethanol in a 55 L tank of fuel. Calculate the percentage by volume concentration of ethanol.
3. Solder flux, available at hardware and craft stores, contains 16 g of zinc chloride in 50 mL of solution. The solvent is aqueous hydrochloric acid. What is the percentage weight by volume of zinc chloride in the solution?

4. Brass is a copper–zinc alloy. If the concentration of zinc is relatively low, the brass has a golden colour and is often used for inexpensive jewellery. If a 35.0 g pendant contains 1.7 g of zinc, what is the percentage weight by weight of zinc in this brass?
5. Formaldehyde, $CH_2O(g)$, an indoor air pollutant that is found in synthetic materials and cigarette smoke, is a carcinogen. If an indoor air sample with a mass of 0.59 kg contained 3.2 mg of formaldehyde, this level would be considered dangerous. What would be the concentration of formaldehyde in parts per million?
6. A plastic dropper bottle for a chemical analysis contains 0.11 mol of calcium chloride in 60 mL of solution. Calculate the amount concentration of calcium chloride.

WEB Activity

Canadian Achievers—David Schindler

Dr. David Schindler (**Figure 4**) is a professor of ecology at the University of Alberta, where he specializes in researching land-water interactions. State three specific examples of studies Dr. Schindler has conducted that would involve concentration measurements. Identify some personal values and attitudes that make him both a renowned and a controversial scientist.

www.science.nelson.com

Figure 4
David Schindler

Calculations Involving Concentrations

Solutions are so commonly used in chemistry that calculating concentrations might be the primary reason why chemists pull out their calculators. Chemists and chemical technicians also frequently need to calculate a quantity of solute or solution. Any of these calculations may involve percentage concentrations, ppm concentrations, or amount concentrations. When we know two of these values—quantity of solute, quantity of solution, and concentration of solution—we can calculate the third quantity. Because concentration is a ratio, a simple procedure is to use the concentration ratio (quantity of solute/quantity of solution) as a conversion factor. This approach parallels the one you followed when using molar mass as a conversion factor.

Suppose you are a nurse who needs to calculate the mass of dextrose, $C_6H_{12}O_6(s)$, present in a 1000 mL intravenous feeding of D5W, which is a solution of 5.0% W/V dextrose in water. The conversion factor you need to use is the mass/volume ratio:

$$m_{C_6H_{12}O_6} = 1000 \text{ mL} \times \frac{5.0 \text{ g}}{100 \text{ mL}} = 50 \text{ g}$$

In some calculations, you may want to find the quantity of solution, in which case you will have to invert the ratio to quantity of solution/quantity of solute. This is then the appropriate conversion factor.

For example, what volume of 30.0% W/V hydrogen peroxide solution can be made from 125 g of pure hydrogen peroxide? You know that the answer must be greater than 100 mL because 125 g is greater than 30.0 g (the quantity in 100 mL). Notice how the units cancel to produce the expected volume unit, millilitres, when we use the volume/mass ratio:

$$V_{H_2O_2} = 125 \text{ g} \times \frac{100 \text{ mL}}{30.0 \text{ g}} = 417 \text{ mL}$$

Learning Tip

You can set up calculations involving concentrations as a proportion. For example,

$$\frac{m_{C_6H_{12}O_6}}{1000 \text{ mL}} = \frac{5.0 \text{ g}}{100 \text{ mL}}$$

You can invert the ratios, if it is more convenient. For example,

$$\frac{V_{H_2O_2}}{125 \text{ g}} = \frac{100 \text{ mL}}{30.0 \text{ g}}$$

In both cases, make sure the quantities in the numerator and the denominator on both sides of the equation match.

Section **5.3**

Thinking about the quantity given and the concentration ratio helps to ensure you are calculating correctly. This method also works for other concentration ratios.

▸ COMMUNICATION example 5

A box of apple juice has a fructose (sugar) concentration of 12 g/100 mL (12% W/V) (**Figure 5**). What mass of fructose is present in a 175 mL glass of juice? (The chemical formula for fructose is $C_6H_{12}O_6$.)

Solution

$$m_{C_6H_{12}O_6} = 175 \text{ mL} \times \frac{12 \text{ g}}{100 \text{ mL}}$$

$$= 21 \text{ g}$$

The mass of fructose present in 175 mL of apple juice is 21 g.

▸ COMMUNICATION example 6

People with diabetes have to monitor and restrict their sugar intake. What volume of apple juice could a diabetic person drink, if the person's sugar allowance for that beverage was 9.0 g? Assume that the apple juice has a sugar concentration of 12 g/100 mL (12% W/V), and that the sugar in apple juice is fructose.

Solution

$$V_{C_6H_{12}O_6} = 9.0 \text{ g} \times \frac{100 \text{ mL}}{12 \text{ g}}$$

$$= 75 \text{ mL}$$

The volume of apple juice allowed is 75 mL.

Figure 5
The label on a box of apple juice gives the ingredients and some nutritional information, but not the concentration of the various solutes.

When you are given a concentration in parts per million (ppm), it is usually easier to convert the parts per million into units of milligrams per kilogram (1 ppm = 1 mg/kg) before doing your calculation. Remember that 1 ppm = 1 mg/L only applies to aqueous solutions. If, for example, you are given a value of 99 ppm of DDT in a 2 kg gull, what mass of DDT is present? The concentration ratio is 99 mg/kg. Note the cancellation of kilograms.

$$m_{DDT} = 2 \text{ kg} \times \frac{99 \text{ mg}}{1 \text{ kg}}$$

$$= 0.2 \text{ g (rounded from 198 mg)}$$

▸ COMMUNICATION example 7

A sample of well water contains 0.24 ppm of iron(III) sulfate dissolved from the surrounding rocks. What mass of iron(III) sulfate is present in 1.2 L of water in a kettle?

Solution

$$m_{Fe_2(SO_4)_3} = 1.2 \text{ L} \times \frac{0.24 \text{ mg}}{1 \text{ L}}$$

$$= 0.29 \text{ mg}$$

The mass of iron(III) sulfate in 1.2 L is 0.29 mg.

BIOLOGY CONNECTION

Pollutants
The effects of pollutants in the environment, such as toxicity, are an important topic in biology. You will see a much more detailed discussion of this topic if you are taking a biology course.

www.science.nelson.com

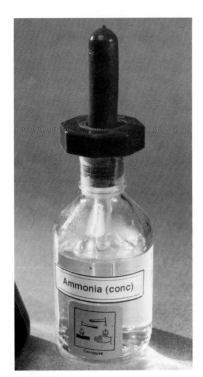

Figure 6
Aqueous ammonia is purchased for science laboratories as a concentrated solution.

> ### COMMUNICATION example 8
>
> A sample of laboratory ammonia solution has an amount concentration of 14.8 mol/L (**Figure 6**). What chemical amount of ammonia is present in a 2.5 L bottle?
>
> **Solution**
>
> $$n_{NH_3} = 2.5 \text{ L} \times \frac{14.8 \text{ mol}}{1 \text{ L}}$$
>
> $$= 37 \text{ mol}$$
>
> The chemical amount of ammonia in 2.5 L is 37 mol.

You should always check that your answer makes sense. For example, in Communication Example 8, 14.8 mol/L means that there is 14.8 mol of ammonia in 1 L of solution. Therefore, 2.5 L, which is greater than 1 L, must contain a chemical amount greater than 14.8 mol.

In some situations, you may know the amount concentration and need to find either the volume of solution or amount (in moles) of solute. In these situations, use either the volume/amount or amount/volume ratio. Notice that *the units of the quantity you want to find should be the units in the numerator of the conversion factor ratio.*

> ### COMMUNICATION example 9
>
> What volume of a 0.25 mol/L salt solution in a laboratory contains 0.10 mol of sodium chloride?
>
> **Solution**
>
> $$V_{NaCl} = 0.10 \text{ mol} \times \frac{1 \text{ L}}{0.25 \text{ mol}}$$
>
> $$= 0.40 \text{ L}$$
>
> The volume of salt solution is 0.40 L.

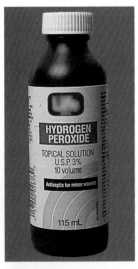

Figure 7
Hydrogen peroxide solutions are stored in dark bottles to keep out light, which promotes the decomposition of hydrogen peroxide to water and oxygen.

> ### Practice
>
> 7. Rubbing alcohol, $C_3H_7OH(l)$, is sold as a 70.0% V/V solution for external use only. What volume of pure $C_3H_7OH(l)$ is present in a 500 mL bottle?
> 8. Suppose your company makes hydrogen peroxide solution with a generic label for drugstores in your area (**Figure 7**). Calculate the mass of pure hydrogen peroxide needed to make 1000 bottles, each containing 250 mL of 3.0% W/V $H_2O_2(aq)$.
> 9. Seawater contains approximately 0.055 mol/L of magnesium chloride. Determine the chemical amount of magnesium chloride present in 75 L of seawater.
> 10. A bottle of 5.0 mol/L hydrochloric acid is opened in the laboratory, and 50 mL of it is poured into a beaker. What chemical amount of acid is in the beaker?
> 11. A household ammonia solution (e.g., a window-cleaning solution) has an amount concentration of 1.24 mol/L. What volume of this solution would contain 0.500 mol of $NH_3(aq)$?
> 12. A student needs 0.14 mol of $Na_2SO_4(aq)$ to do a quantitative analysis. The amount concentration of the student's solution is 2.6 mol/L $Na_2SO_4(aq)$. What volume of solution does the student need to measure?

Mass, Volume, and Concentration Calculations

Even though the mole is a very important unit, measurements in a chemistry laboratory are usually of mass (in grams) and of volume (in millilitres). A common chemistry calculation involves the mass of a substance, the volume of a solution, and the amount concentration of that solution. This type of calculation requires the use of two conversion factors—one for molar mass and one for amount concentration. Calculations using molar mass are just like the ones you did in previous units.

▸ SAMPLE problem 5.1

A chemical analysis requires 2.00 L of 0.150 mol/L $AgNO_3$(aq). What mass of solid silver nitrate is required to prepare this solution?

First determine the chemical amount of silver nitrate needed.

$$n_{AgNO_3} = 2.00 \; \cancel{L} \times \frac{0.150 \; mol}{1 \; \cancel{L}}$$

$$= 0.300 \; mol$$

Then convert this amount into a mass of silver nitrate by using its molar mass, M. The molar mass of silver nitrate is 169.88 g/mol.

$$m_{AgNO_3} = 0.300 \; \cancel{mol} \times \frac{169.88 \; g}{1 \; \cancel{mol}}$$

$$= 51.0 \; g$$

If you clearly understand these two steps, you could combine them into one calculation.

$$m_{AgNO_3} = 2.00 \; \cancel{L} \times \frac{0.150 \; mol}{1 \; \cancel{L}} \times \frac{169.88 \; g}{1 \; \cancel{mol}}$$

$$= 51.0 \; g$$

Learning Tip

If you prefer to use mathematical formulas, for this sample problem you may use

$n = Vc$
$m = nM$

In order to successfully combine the steps into one operation, as shown above, you need to pay particular attention to the units in the calculation. Cancelling the units will help you to check your procedure.

▸ COMMUNICATION example 10

To study part of the water treatment process in a laboratory, a student requires 1.50 L of 0.12 mol/L aluminium sulfate solution. What mass of aluminium sulfate must she measure for this solution?

Solution

$$n_{Al_2(SO_4)_3} = 1.50 \; \cancel{L} \times \frac{0.12 \; mol}{1 \; \cancel{L}} \quad \text{or} \quad m_{Al_2(SO_4)_3} = 1.50 \; \cancel{L} \times \frac{0.12 \; \cancel{mol}}{1 \; \cancel{L}} \times \frac{342.14 \; g}{1 \; \cancel{mol}}$$

$$= 0.180 \; mol \qquad\qquad\qquad\qquad\qquad = 61.6 \; g$$

$$m_{Al_2(SO_4)_3} = 0.180 \; \cancel{mol} \times \frac{342.14 \; g}{1 \; \cancel{mol}}$$

$$= 61.6 \; g$$

The mass of aluminium sulfate required is 61.6 g.

Another similar calculation involves the use of a known mass and volume to calculate the amount concentration of a solution. This calculation is similar to the examples given above, using the same conversion factors.

> **COMMUNICATION** example 11

Sodium carbonate is a water softener that is an important part of the detergent used in a washing machine. A student dissolves 5.00 g of solid sodium carbonate to make 250 mL of a solution to study the properties of this component of detergent. What is the amount concentration of the solution?

Solution

$$n_{Na_2CO_3} = 5.00 \text{ g} \times \frac{1 \text{ mol}}{105.99 \text{ g}}$$
$$= 0.0472 \text{ mol}$$

$$c_{Na_2CO_3} = \frac{0.0472 \text{ mol}}{0.250 \text{ L}}$$
$$= 0.189 \text{ mol/L}$$

or

$$c_{Na_2CO_3} = 5.00 \text{ g} \times \frac{1 \text{ mol}}{105.99 \text{ g}} \times \frac{1}{0.250 \text{ L}}$$
$$= 0.189 \text{ mol/L}$$

The amount concentration of sodium carbonate is 0.189 mol/L.

> **Practice**

13. A chemical technician needs 3.00 L of 0.125 mol/L sodium hydroxide solution. What mass of solid sodium hydroxide must be measured?
14. Seawater is mostly a solution of sodium chloride in water. The concentration varies, but marine biologists took a sample with an amount concentration of 0.56 mol/L. Calculate the mass of sodium chloride in the biologists' 5.0 L sample.
15. Acid rain may have 355 ppm of dissolved carbon dioxide.
    (a) What mass of carbon dioxide is present in 1.00 L of acid rain?
    (b) Calculate the amount concentration of carbon dioxide in the acid rain sample.
16. A brine (sodium chloride) solution used in pickling contains 235 g of pure sodium chloride dissolved in 3.00 L of solution.
    (a) Determine the percent concentration (% W/V) of sodium chloride.
    (b) What is the amount concentration of sodium chloride?

Concentration of Ions

In solutions of ionic compounds and strong acids, the electrical conductivity suggests the presence of ions in the solution. When these solutes produce aqueous ions, expressing the concentration of individual ions in moles per litre (mol/L) is important. The amount concentrations of the ions in a solution depend on the relative numbers of ions making up the compound: for example, Cl^- ions in NaCl(aq) and $CaCl_2$(aq).

The dissociation or ionization equations for ionic compounds or strong acids allow you to determine the amount concentration of either the ions or the compounds in solution. The ion concentration is always equal to a whole number multiple of the compound concentration. For convenience, square brackets are commonly placed around formulas to indicate the amount concentration of the substance within the brackets. For example, $[NH_3(aq)]$ and $[H^+(aq)]$ indicate the amount concentrations of aqueous ammonia and hydrogen ions respectively.

When sodium chloride dissolves in water, each mole of sodium chloride produces one mole of sodium ions and one mole of chloride ions.

$$\text{NaCl(aq)} \rightarrow \text{Na}^+\text{(aq)} + \text{Cl}^-\text{(aq)}$$
1 mol   1 mol   1 mol

If $[NaCl(aq)] = 1$ mol/L, then $[Na^+(aq)] = 1$ mol/L and $[Cl^-(aq)] = 1$ mol/L because the mole ratio from the dissociation equation is 1:1:1.

Figure 8
Two moles per litre

Calcium chloride dissociates in water to produce individual calcium and chloride ions. Each mole of calcium chloride produces one mole of calcium ions and two moles of chloride ions.

$$CaCl_2(aq) \rightarrow Ca^{2+}(aq) + 2\,Cl^-(aq)$$
$$1\text{ mol} \quad\quad 1\text{ mol} \quad\quad 2\text{ mol}$$

If $[CaCl_2(aq)] = 1$ mol/L, then $[Ca^{2+}(aq)] = 1$ mol/L. The $[Cl^-(aq)] = 2$ mol/L because the dissociation equation shows that 2 mol of chloride ions is produced from 1 mol of calcium chloride—a 2:1 mole ratio. Notice that you can easily predict the individual ion concentrations from the concentration of the compound and the subscripts of the ions in the formula of the compound. Even so, it is good practice to write the dissociation or ionization equation prior to calculating concentrations. This practice will help you avoid errors and is good preparation for your later study of stoichiometry in Unit 4.

Learning Tip

Initially it might appear that there is a "lack of conservation" because the dissociation equation shows a greater amount on the product side compared to the reactant side. Here is a simple model for the dissolving of calcium chloride that may help you to understand that no rules are being broken.

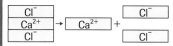

1 sheet of paper → cut into 3 smaller pieces

Although the number of pieces of paper has increased by cutting one sheet into smaller pieces, the total mass of paper has not changed. If you started with five sheets of paper ($CaCl_2$), how many pieces of paper, labelled chloride ion, would you obtain?

▶ COMMUNICATION *example* 12

What is the amount concentration of aluminium ions and sulfate ions in a 0.40 mol/L solution of $Al_2(SO_4)_3(aq)$?

Solution

$$Al_2(SO_4)_3(aq) \rightarrow 2\,Al^{3+}(aq) + 3\,SO_4^{2-}(aq)$$

$$[Al^{3+}(aq)] = 0.40 \text{ mol/L } Al_2(SO_4)_3(aq) \times \frac{2 \text{ mol } Al^{3+}(aq)}{1 \text{ mol } Al_2(SO_4)_3(aq)} = 0.80 \text{ mol/L}$$

$$[SO_4^{2-}(aq)] = 0.40 \text{ mol/L } Al_2(SO_4)_3(aq) \times \frac{3 \text{ mol } SO_4^{2-}(aq)}{1 \text{ mol } Al_2(SO_4)_3(aq)} = 1.20 \text{ mol/L}$$

The amount concentration of aluminium ions is 0.80 mol/L and of sulfate ions is 1.20 mol/L.

Note that, in Communication Example 12, the chemical formula of the entity you wish to find appears in the numerator of the ratio. The formula of the known entity is in the denominator and cancels with the chemical formula of the known solution concentration. The chemical formulas are omitted from the ratios in the following examples.

▶ COMMUNICATION *example* 13

Determine the amount concentration of barium and hydroxide ions in a solution made by dissolving 5.48 g of barium hydroxide to make a volume of 250 mL.

Solution

$$Ba(OH)_2(aq) \rightarrow Ba^{2+}(aq) + 2\,OH^-(aq)$$

$$n_{Ba(OH)_2} = 5.48 \text{ g} \times \frac{1 \text{ mol}}{171.35 \text{ g}} = 0.0320 \text{ mol}$$

$$[Ba(OH)_2(aq)] = \frac{0.0320 \text{ mol}}{0.250 \text{ L}} = 0.128 \text{ mol/L}$$

$$[Ba^{2+}(aq)] = 0.128 \text{ mol/L} \times \frac{1}{1} = 0.128 \text{ mol/L}$$

$$[OH^-(aq)] = 0.128 \text{ mol/L} \times \frac{2}{1} = 0.256 \text{ mol/L}$$

The amount concentration of barium ions is 0.128 mol/L and of hydroxide ions is 0.256 mol/L.

Learning Tip

Notice that in this example, we are now "going backwards" from the ion concentration to the solute concentration. The first calculation step means that for every two moles of bromide ions, we need only one mole of magnesium bromide.

COMMUNICATION example 14

What mass of magnesium bromide must be dissolved to make 1.50 L of solution with a bromide ion concentration of 0.30 mol/L?

Solution

$$MgBr_2(aq) \rightarrow Mg^{2+}(aq) + 2\,Br^-(aq)$$

$$[MgBr_2(aq)] = 0.30 \text{ mol/L} \times \frac{1}{2} = 0.15 \text{ mol/L}$$

$$n_{MgBr_2} = 1.50 \text{ L} \times \frac{0.15 \text{ mol}}{1 \text{ L}} = 0.23 \text{ mol}$$

$$m_{MgBr_2} = 0.23 \text{ mol} \times \frac{184.11 \text{ g}}{1 \text{ mol}} = 41 \text{ g}$$

The mass of magnesium bromide required is 41 g.

Practice

17. Find the amount concentration of each ion in the following solutions:
    (a) 0.41 mol/L $Na_2S(aq)$
    (b) 1.2 mol/L $Sr(NO_3)_2(aq)$
    (c) 0.13 mol/L $(NH_4)_3PO_4(aq)$

18. A 250 mL solution is prepared by dissolving 2.01 g of iron(III) chloride in water. What is the amount concentration of each ion in the solution?

19. In order to prepare for a chemical analysis, a lab technician requires 500 mL of each of the following solutions. Calculate the mass of solid required for each solution:
    (a) $[Cl^-(aq)] = 0.400$ mol/L from $CaCl_2(s)$
    (b) $[CO_3^{2-}(aq)] = 0.35$ mol/L from $Na_2CO_3(s)$

Case Study

Household Chemical Solutions

An amazing number of solutions is available for household use at your local drugstore, hardware store, and supermarket in the form of food products, household cleaners, and health and personal care products (**Figure 9**). They come with a bewildering array of names, instructions, warnings, and concentration labels. For consumer convenience and safety, it is important that household chemical solutions be labelled accurately and honestly. Unfortunately, manufacturers and distributors of household cleaning products are not required by law to list ingredients on their labels. In some cases, you can find the Material Data Safety Sheet (MSDS) for the product on the manufacturer's Web site or phone to request it. Societal concerns about safety and disposal of chemicals have resulted in efforts by chemical manufacturers to promote safe and environmentally sound practices through their Responsible Care® program.

Being able to read information on household product labels is important for personal safety and proper disposal. Hazard symbols and safety warnings on labels are pointless if they go

Figure 9
Corrosive substances, such as acids and bases, are found in many household products, including cleaning solutions.

unnoticed, or are not understood. Every year people are injured because they are unaware that bleach (sodium hypochlorite solution) should never be mixed with acids such as vinegar. Although both solutions are effective cleaners for certain stains, when they are combined, they react to produce a highly toxic gas, chlorine. Trying to use both at once—for example, in cleaning a toilet—has been known to transform a bathroom into a deathtrap.

Concentration is another factor to consider when buying solutions. The labels on consumer products usually give concentration as a percentage, which is easier for the general public to understand than moles per litre. Some consumer products, such as insect repellant, are sold in different concentrations, so it is important to know which one is most suitable for your needs. Another household chemical, isopropyl alcohol, is sold in pure form as a disinfectant, and in 70% concentration as rubbing alcohol. Knowing the key ingredients and their concentrations is useful for safe and proper use and disposal.

Case Study Questions

1. How and why are concentrations for household chemical solutions expressed differently than for laboratory work?
2. According to Health Canada, it is the manufacturer's responsibility to assess and report the hazards associated with a chemical product. What personal or social values would you expect a manufacturer to demonstrate? To what extent do you think these values are demonstrated?
3. Survey your home and list any household solutions you have that are commonly brought to the Household Hazardous Waste Round-Up. When is your local Round-Up? If your location is not listed, phone the local government office to ask why you are not listed and what you should do to dispose of household hazardous materials.

www.science.nelson.com

4. The Canadian Chemical Producers' Association (CCPA) represents over 65 chemical manufacturing industries with over 200 plants across Canada. These industries collectively produce over 90% of all chemicals in Canada. CCPA is the driving force behind the Responsible Care® initiative—a global effort aimed at addressing public concerns about the manufacture, distribution, use, and disposal of chemicals. State their Ethic and list their six Codes of Practice. Do you feel that this program is a suitable replacement for government regulation?

www.science.nelson.com

5. If household products are brought into a workplace, they become restricted products and a MSDS is required. Choose one household product and record the name of one key ingredient. Find the MSDS and identify three pieces of information supplied on this sheet that you think should be on the label.

www.science.nelson.com

 WEB Activity

Web Quest—Hot Tub Safety

Hot tubs are very popular in private homes, public recreation centres, and commercial hotels (**Figure 10**). How are you protected from infectious diseases transmitted via hot tubs?

www.science.nelson.com

Figure 10
How safe are hot tubs?

SUMMARY Concentration of a Solution

Type	Definition	Units	
percentage by volume	$c = \dfrac{V_{solute}}{V_{solution}} \times 100\%$	% V/V	(or mL/100 mL)
mass by volume	$c = \dfrac{m_{solute}}{V_{solution}} \times 100\%$	% W/V	(or g/100 mL)
by mass	$c = \dfrac{m_{solute}}{m_{solution}} \times 100\%$	% W/W	(or g/100 g)
parts per million	$c = \dfrac{m_{solute}}{m_{solution}}$	ppm	(or mg/kg)
amount	$c = \dfrac{n_{solute}}{V_{solution}}$	mol/L	

Section 5.3 Questions

1. What concentration ratio is often found on the labels of consumer products? Why is this unit used?

2. What concentration unit is most useful in the study of chemistry? Briefly describe why this unit is useful.

3. Bags of a D5W intravenous sugar solution used in hospitals contain 50 g of dextrose (glucose) in a 1.00 L bag.
   (a) Calculate the percentage weight by volume concentration of dextrose.
   (b) Suggest a reason why the bags are labelled D5W.

4. An Olympic-bound athlete tested positive for the anabolic steroid nandrolone. The athlete's urine test results showed 0.20 mg of nandrolone in a 10.0 mL urine sample. Convert the test result concentration to parts per million.

5. A 15 mL dose of a cough syrup contains 4.8 mmol of ammonium carbonate, $(NH_4)_2CO_3(aq)$.
   (a) Determine the amount concentration of ammonium carbonate.
   (b) What is the amount concentration of each ion in this solution?

6. The maximum concentration of salt in water at 0 °C is 31.6 g/100 mL. What mass of salt can be dissolved in 250 mL of solution?

7. Bald eagle chicks raised in northern Alberta were found to contain PCBs (polychlorinated biphenyls) at an average concentration of 18.9 ppm. If a chick had a mass of 0.60 kg, predict the mass of PCBs it would contain.

8. An experiment is planned to study the chemistry of a home water-softening process. The brine (sodium chloride solution) used in this process has a concentration of 25 g in every 100 mL of solution. Calculate the amount concentration of this solution.

9. To prepare for an experiment using flame tests, a school lab technician requires 100 mL of 0.10 mol/L solutions of each of the following substances. Calculate the required mass of each solid.
   (a) NaCl(s)
   (b) KCl(s)
   (c) $CaCl_2(s)$

10. What volume of 0.055 mol/L glucose solution found in a plant contains 2.0 g of glucose, $C_6H_{12}O_6(aq)$?

11. In an experiment, 28.6 g of aluminium chloride is dissolved in 1.50 L of solution.
    (a) Calculate the amount concentration of aluminium chloride.
    (b) Determine the amount concentration of each ion in the final solution.

12. As part of a chemical analysis, a technician requires a 0.25 mol/L bromide ion solution. What mass of magnesium bromide is required to prepare 100 mL of the required solution?

13. How is your report card mark in a subject similar to a concentration? What other ratios have you used that are similar to concentration ratios?

14. List several examples of how solutions and solution concentration are applied in products and processes we use in daily life.

15. Identify the implications of selling medicines in much more concentrated solutions. Present points both in favour and against.

16. Science and technology have both intended and unintended consequences. Illustrate this statement using DDT as your example. Include the role of biomagnification of DDT in the environment.

www.science.nelson.com

17. Very low concentrations of toxic substances sometimes require the use of the parts per billion (ppb) concentration.
    (a) Express parts per billion as a ratio, including the appropriate power of ten.
    (b) How much smaller is 1 ppb than 1 ppm?
    (c) Convert your answer in (a) to a concentration ratio using the appropriate SI prefixes to obtain a mass of solute per kilogram of solution.
    (d) Copper is an essential trace element for animal life. An average adult human requires the equivalent of a litre of water containing 30 ppb of copper a day. What is the mass of copper per kilogram of solution?

Extension

18. Toxicity of substances for animals is usually expressed by a quantity designated as "LD50." Use the Internet to research the use of this quantity. What does LD50 mean? What is the concentration in ppm for a substance considered "extremely toxic" and one considered "slightly toxic"?

www.science.nelson.com

19. Many chemicals that are potentially toxic or harmful to the environment and humans have maximum permissible concentration levels set by government legislation. Nevertheless, some people question the levels that are set and some suggest that the only safe level is zero.
    (a) To what extent should we trust our government agencies to set appropriate levels?
    (b) Outline some risks and benefits, from several perspectives, associated with the use of controversial chemicals such as pesticides.
    (c) What is chemical hormesis? Why might this effect have major implications for government regulatory agencies?
    (d) What does LC50 mean? List some advantages and disadvantages of this method of measuring toxicity.

www.science.nelson.com

Preparation of Solutions 5.4

When you prepare a jug of iced tea using a package of crystals and water, you are preparing a solution from a solid solute (actually, from several solid solutes). However, when you prepare the tea from a container of frozen concentrate, you are preparing a solution by dilution. Scientists use both of these methods to prepare solutions. In this course you will be preparing only aqueous solutions. The knowledge and skills for preparing solutions are necessary to complete some of the more complex laboratory investigations that come later in this course.

Preparation of Standard Solutions from a Solid

Solutions of accurate concentration, called **standard solutions**, are routinely prepared for use in both scientific research laboratories and industrial processes. They are used in chemical analysis as well as for the control of chemical reactions. To prepare a standard solution, good-quality equipment is required to measure the mass of solute and volume of solution. Electronic balances are used for precise and efficient measurement of mass (**Figure 1**). For measuring a precise volume of the final solution, a container called a volumetric flask is used (**Figure 2**).

Figure 1
An electronic balance is simpler to operate and more efficient than the older mechanical balance. Electronic balances also provide the convenience of taring (Appendix C.3).

Figure 2
Volumetric glassware comes in a variety of shapes and sizes. The Erlenmeyer flask on the far left has only approximate volume markings, as does the beaker. The graduated cylinders have much better precision, but for high precision, a volumetric flask (on the right) is used. The volumetric flask shown here, when filled to the line, contains 100.0 mL ± 0.16 mL at 20 °C. This means that a volume measured in this flask is uncertain by less than 0.2 mL at the specified temperature.

INVESTIGATION 5.2 Introduction

A Standard Solution from a Solid

In this investigation, you will practise the skills required to prepare a standard solution from a pure solid (Appendix C.4). You will need these skills in many investigations in this course.

Report Checklist
- ○ Purpose
- ○ Problem
- ○ Hypothesis
- ○ Prediction
- ○ Design
- ○ Materials
- ○ Procedure
- ○ Evidence
- ○ Analysis
- ○ Evaluation

Purpose
The purpose of this investigation is to acquire the skills required to prepare a standard solution starting with a pure solid.

To perform this investigation, turn to page 227.

Figure 3
Hard-water deposits such as calcium carbonate can seriously affect water flow in a pipe.

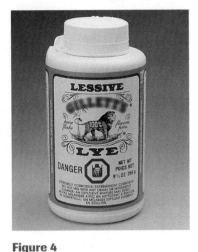

Figure 4
Solutions of sodium hydroxide in very high concentration are sold as cleaners for clogged drains. The same solution can be made less expensively by dissolving solid lye (a commercial name for sodium hydroxide) in water. The pure chemical is very caustic and the label on the lye container recommends rubber gloves and eye protection.

▶ Practice

1. To test the hardness of water (**Figure 3**), an industrial chemist performs an analysis using 100.0 mL of a 0.250 mol/L standard solution of ammonium oxalate. What mass of ammonium oxalate, $(NH_4)_2C_2O_4(s)$, is needed to make the standard solution?
2. Calculate the mass of solid lye (sodium hydroxide) (**Figure 4**) needed to make 500 mL of a 10.0 mol/L strong cleaning solution.
3. List several examples of solutions that you prepared from solids in the last week.
4. You have been asked to prepare 2.00 L of a 0.100 mol/L aqueous solution of cobalt(II) chloride for an experiment starting with $CoCl_2 \cdot 2H_2O(s)$.
   (a) Show your work for the pre-lab calculation.
   (b) Write a complete specific procedure for preparing this solution, as in Investigation 5.2. Be sure to include all necessary precautions.
5. (a) A technician prepares 500.0 mL of a 0.0750 mol/L solution of potassium permanganate as part of a quality-control analysis in the manufacture of hydrogen peroxide. Calculate the mass of potassium permanganate required to prepare the solution.
   (b) Write a laboratory procedure for preparing the potassium permanganate solution. Follow the conventions of communication for a procedure in a laboratory report.

Preparation of Standard Solutions by Dilution

A second method of preparing solutions is by dilution of an existing solution. You use this process when you add water to concentrated fruit juice, fabric softener, or a cleaning product. Many consumer and commercial products are purchased in concentrated form and then diluted before use. You can save money and help save the environment by diluting concentrated products. Doing so saves on shipping charges and reduces the size of the container, making the product less expensive and more environmentally friendly. Citizens who are comfortable with dilution techniques can live more lightly on Earth.

Because dilution is a simple, quick procedure, it is common scientific practice to begin with a stock solution and to add solvent (usually water) to decrease the concentration to the desired level. A **stock solution** is an initial, usually concentrated, solution from which samples are taken for a dilution. For the most accurate results, the stock solution should be a standard solution.

Even though there are no firm rules, we often describe solutions with an amount concentration of less than 0.1 mol/L as dilute, whereas solutions with a concentration of greater than 1 mol/L may be referred to as concentrated.

Calculating the new concentration after a dilution is straightforward because the quantity of solute is not changed by adding more solvent. Therefore, the mass (or chemical amount) of solute before dilution is the same as the mass (or chemical amount) of solute after dilution.

$$m_i = m_f$$
or
$$n_i = n_f$$

m_i = initial mass of solute
m_f = final mass of solute
n_i = initial chemical amount of solute
n_f = final chemical amount of solute

Using the definitions of solution concentration ($m = Vc$ or $n = Vc$), we can express the constant quantity of solute in terms of the volume and concentration of solution.

$$V_i c_i = V_f c_f$$

This equation means that the concentration is inversely related to the solution's volume. For example, if water is added to 6% hydrogen peroxide disinfectant until the total volume is doubled, the concentration becomes one-half the original value, or 3%.

Any one of the variables in this dilution equation may be calculated for the dilution of a solution, provided the other three values are known. (Note that the dilution calculation for percentage weight by weight (%W/W) will be slightly different because the mass of solution is used: $m_{solute} = m_{solution}c$.)

▶ COMMUNICATION example 1

Water is added to 0.200 L of 2.40 mol/L NH_3(aq) cleaning solution, until the final volume is 1.000 L. Find the amount concentration of the final, diluted solution.

Solution

$$V_i c_i = V_f c_f$$

$$c_f = \frac{V_i c_i}{V_f}$$

$$= \frac{0.200 \text{ L} \times 2.40 \text{ mol/L}}{1.000 \text{ L}}$$

$$= 0.480 \frac{\text{mol}}{\text{L}}$$

The amount concentration of the final, diluted ammonia solution is 0.480 mol/L.

Learning Tip

Alternatively, this problem can be solved another way:

$$n_{NH_3} = 0.200 \text{ L} \times \frac{2.40 \text{ mol}}{1.00 \text{ L}}$$

$$= 0.480 \text{ mol}$$

$$c_{NH_3} = \frac{0.480 \text{ mol}}{1.000 \text{ L}}$$

$$= 0.480 \text{ mol/L}$$

When diluting all concentrated reagents, especially acids, always add the concentrated reagent, with stirring, to less than the final required quantity of water (**Figure 5**), and then add the rest of the water.

▶ COMMUNICATION example 2

A student is instructed to dilute some concentrated HCl(aq) (36%) to make 4.00 L of 10% solution. What volume of hydrochloric acid solution should the student initially measure?

Solution

$$V_i c_i = V_f c_f$$

$$V_i = \frac{V_f c_f}{c_i}$$

$$= \frac{4.00 \text{ L} \times 10\%}{36\%}$$

$$= 1.1 \text{ L}$$

The volume of concentrated hydrochloric acid required is 1.1 L.

Figure 5
Handling concentrated reagents, especially acids, requires great care. Wear eye protection, a lab apron, and gloves, as shown in the photograph.

You can predict answers to dilution calculations if you understand the dilution process: As the volume increases, the concentration decreases. Use this principle as a useful check on your work. In Communication Example 1, the final concentration must be less than the initial concentration because the solution is being diluted. In Communication Example 2, the initial volume of acid required must be less than the final volume after the dilution.

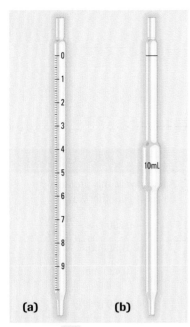

Figure 6
A graduated pipette **(a)** measures a range of volumes, whereas a volumetric pipette **(b)** is calibrated to deliver (TD) a fixed volume.

The dilution technique is especially useful when you need to decrease the concentration of a solution. For example, when doing scientific or technological research, you may want to slow down a reaction that proceeds too rapidly or too violently with a concentrated solution. You could slow down the reaction by lowering the concentration of the solution. In the medical and pharmaceutical industries, prescriptions require not only minute quantities, but also very accurate measurement. If the solutions are diluted before being sold, it is much easier for a patient to take the correct dose. For example, it's easier to accurately measure out 10 mL (two teaspoons) of a cough medicine than it is to measure one-fifth of a teaspoon, which the patient would have to do if the medicine were ten times more concentrated.

The preparation of standard solutions by dilution requires a means of transferring precise and accurate volumes of solution. You know how to use graduated cylinders to measure volumes of solution, but graduated cylinders are not precise enough when working with small volumes. To deliver a precise and accurate, small volume of solution, a laboratory device called a pipette is used. A 10 mL graduated pipette has graduation marks every tenth of a millilitre (**Figure 6**). This type of pipette can transfer any volume from 0.1 mL to 10.0 mL. A volumetric pipette transfers only one specific volume, but has a very high precision and accuracy. For example, a 10 mL volumetric (or delivery) pipette is designed to transfer 10.00 mL of solution with a precision of ± 0.02 mL. The volumetric pipette is often inscribed with TD to indicate that it is calibrated *to deliver* a particular volume with a specified precision. Both kinds of pipettes come in a range of sizes and are used with a pipette bulb. (See Appendix C.4.)

INVESTIGATION 5.3 *Introduction*

A Standard Solution by Dilution

In this investigation, you will practise the skills required to prepare a standard solution from a more concentrated or stock solution. This laboratory procedure is very common for preparing solutions (Appendix C.4).

Report Checklist
- ○ Purpose   ○ Design   ○ Analysis
- ○ Problem   ○ Materials   ○ Evaluation
- ○ Hypothesis   ○ Procedure
- ○ Prediction   ○ Evidence

Purpose
The purpose of this investigation is to acquire the skills required to prepare a standard solution by diluting a stock solution.

To perform this investigation, turn to page 228.

> ### Practice
>
> **6.** Radiator antifreeze (ethylene glycol) is diluted with an appropriate quantity of water to prevent freezing of the mixture in the radiator. A 4.00 L container of 94% V/V antifreeze is diluted to 9.00 L. Calculate the concentration of the final solution.
>
> **7.** Many solutions are prepared in the laboratory from purchased concentrated solutions. Calculate the volume of concentrated 17.8 mol/L stock solution of sulfuric acid a laboratory technician would need to make 2.00 L of 0.200 mol/L solution by dilution of the original concentrated solution.
>
> **8.** In a study of reaction rates, you need to dilute the copper(II) sulfate solution prepared in Investigation 5.3. You take 5.00 mL of 0.005000 mol/L $CuSO_4$(aq) and dilute it to a final volume of 100.0 mL.
>   (a) Determine the final concentration of the dilute solution.
>   (b) What mass of $CuSO_4$(s) is present in 10.0 mL of the final dilute solution?
>   (c) Can this final dilute solution be prepared directly using the pure solid? Defend your answer.

9. A student tries a reaction and finds that the volume of solution that reacts is too small to be measured with any available equipment. The student takes a 10.00 mL volume of the solution with a pipette, transfers it into a clean 250 mL volumetric flask containing some pure water, adds enough pure water to increase the volume to 250.0 mL, and mixes the solution thoroughly.
   (a) Compare the concentration of the dilute solution to that of the original solution.
   (b) Compare the volume that will react now to the volume that reacted initially.
   (c) Predict the speed or rate of the reaction using the diluted solution compared with the rate using the original solution. Explain your answer.

INVESTIGATION 5.4 Introduction

The Iodine Clock Reaction

Technological problem solving often involves a systematic trial-and-error approach that is guided by knowledge and experience. Usually one variable at a time is manipulated, while all other variables are controlled. Variables that may be manipulated include concentration, volume, and temperature. In this investigation, you will compete to see which team is the first to solve the Problem using a reliable process. Create a design to guide your work. Using this design, try several procedures to solve the Problem. The final Analysis will be the materials and procedure that best answer the Problem.

Report Checklist

- ○ Purpose
- ○ Problem
- ○ Hypothesis
- ○ Prediction
- ● Design
- ● Materials
- ● Procedure
- ● Evidence
- ● Analysis
- ○ Evaluation

Purpose
The purpose of this investigation is to find a method for getting a reaction to occur in a specified time period.

Problem
What technological process can be employed to have solution A react with solution B in a reliable time of 20 ± 1 s?

To perform this investigation, turn to page 228.

Section 5.4 Questions

1. List several reasons why scientists make solutions in the course of their work.

2. (a) Briefly describe two different ways of making a solution.
   (b) When should you use each method?

3. In an analysis for sulfate ions in a water treatment plant, a technician needs 100 mL of 0.125 mol/L barium nitrate solution. What mass of pure barium nitrate is required?

4. A 1.00 L bottle of purchased acetic acid is labelled with a concentration of 17.4 mol/L. A technician dilutes this entire bottle of concentrated acid to prepare a 0.400 mol/L solution. Calculate the volume of diluted solution prepared.

5. A 10.00 mL sample of a test solution is diluted in a laboratory to a final volume of 250.0 mL. The concentration of the diluted solution is 0.274 g/L. Determine the concentration of the original test solution.

6. A chemical analysis of silver uses 100 mL of a 0.155 mol/L solution of potassium thiocyanate, KSCN(aq). Write a complete, specific procedure for preparing the solution from the solid. Include all necessary calculations and precautions.

7. A laboratory technician needs 1.00 L of 0.125 mol/L sulfuric acid solution for a quantitative analysis experiment. A commercial 5.00 mol/L sulfuric acid solution is available from a chemical supply company. Write a complete, specific procedure for preparing the solution. Include all necessary calculations and safety precautions.

8. As part of a study of rates of reaction, you are to prepare two aqueous solutions of nickel(II) chloride.
   (a) Calculate the mass of solid nickel(II) chloride that you will need to prepare 100.0 mL of a 0.100 mol/L nickel(II) chloride solution.
   (b) Calculate how to dilute this solution to make 100.0 mL of a 0.0100 mol/L nickel(II) chloride solution.
   (c) Write a list of Materials, and a Procedure for the preparation of the two solutions. Be sure to include all necessary safety precautions and disposal steps.

9. It has been suggested that it is more environmentally friendly to transport chemicals in a highly concentrated state. List arguments for and against this position, including possible intended and unintended consequences.

10. For many years the adage "The solution to pollution is dilution" described the views of some individuals, industries, and governments. They did not realize at that time that chemicals, diluted by water or air, could be concentrated in another system later. What is biomagnification? Describe briefly using a specific chemical as an example. What implications does this effect have for the introduction of new technologies?

www.science.nelson.com

5.5 Solubility

It is easier to handle a great many chemicals when they are in solution, particularly those that are toxic, corrosive, or gaseous. Both in homes and at worksites, transporting, loading, and storing chemicals is more convenient and efficient when the chemicals are in solution rather than in solid or gaseous states. Also, performing a reaction in solution can change the rate (speed), the extent (completeness), and the type (kind of product) of the chemical reaction.

Solutions make it easy to
- handle chemicals—a solid or gas is dissolved in water for ease of use or transportation
- complete reactions—some chemicals do not react until in a solution where there is increased contact between the reacting entities
- control reactions—the rate, extent, and type of reactions are much more easily controlled when one or more reactants are in solution

These three points all apply to the liquid cleaning solution in **Figure 1**. First, the cleaning solution is easy to handle, and the fact that it is sold in a spray bottle adds to its convenience. Spraying a solution is an effective way of handling a chemical that is dissolved in water. Second, the solution allows a reaction to occur between the cleaning chemicals and the dirty deposit, whereas a pure gas or solid would not react well with a solid. Third, the manufacturer can control the rate of the reaction (and thus the safety) by choosing the ideal concentration of the cleaning solution. Having the chemical in solution rather than in its pure state increases our ability to handle and control its use.

Because solutions are very useful at home, in industry, and in scientific research, it is important to consider which substances dissolve easily in solvents such as water and how much of a substance you can dissolve.

Figure 1
The low solubility of the soap deposit is overcome by a chemical reaction—a good example of how science and technology provide products useful for daily life.

▶ mini Investigation   *Measuring the Dissolving Process*

Are there different kinds of salt? How much salt can you dissolve in a given volume of water? What happens to the volume of a solution when a solute is added to it? This quick mini investigation will help you to think about the answers to these questions.

Materials: distilled or deionized water, table salt, coarse pickling salt (pure NaCl(s)), a measuring teaspoon (5 mL), two 125 mL Erlenmeyer flasks with stoppers, one 50 mL or 100 mL graduated cylinder

- Place a level teaspoonful of table salt into 25 mL of pure water at room temperature in a 125 mL Erlenmeyer flask. Swirl the flask's contents thoroughly for a minute or two. Record your observations.
- Repeat with pickling salt, again recording your observations.
(a) What does the result, with common table salt as a solute, show about the nature of the substance being used? Compare it with the solution in the second flask.
(b) List the ingredients in common table salt, according to the package label, and explain your observations of the contents of the first flask.
- Add another teaspoon of pickling salt to the second flask, and swirl until the solid is again completely dissolved. Keeping track of how much pickling salt you add, continue to dissolve level teaspoons of salt until no amount of swirling will make all of the solid crystals disappear.
(c) How many level teaspoons of pickling salt (pure NaCl(s)) could you get to dissolve in 25 mL of $H_2O(l)$ in the second flask?
(d) What is the final volume of your NaCl(aq) solution in the second flask?
(e) If you dissolve 20.0 mL of NaCl(s) in 100.0 mL of liquid water, what do you suppose the volume of the solution would be? Describe a way to test your supposition. The answer is very interesting.

220 Chapter 5

Solubility of Solids

When you add a small amount of pickling salt (pure sodium chloride) to a jar of water and shake the jar, the salt dissolves and disappears completely. What happens if you continue adding salt and shaking? Eventually, some visible solid salt crystals will remain at the bottom of the jar, despite your efforts to make them dissolve. You have formed a **saturated solution**—a solution in which no more solute will dissolve at a specified temperature. We say it is at maximum solute concentration. If the container is sealed, and the temperature stays the same, no further changes will ever occur in the concentration of this solution. The quantity (mass) of solute that remains undissolved will also stay the same. **Solubility** is the concentration of a saturated solution. The units for solubility are simply units of concentration, such as % W/V or mol/L. You will learn in this section that solubility depends on the temperature, so it is a particular maximum concentration value. Every solubility value must be accompanied by a temperature value. When calculating and using solubility values, we have to make one assumption: The solute is not reacting with the solvent.

Every pure substance has its own unique solubility. Some references provide solubility data for substances in water using units of grams per hundred millilitres of water, not of solution. These units may be convenient for comparing solubilities, but they are not very convenient for calculations. For example, we can find from a reference source, such as the *CRC Handbook of Chemistry and Physics*, that the solubility of sodium sulfate in water at 0 °C is 4.76 g/100 mL H_2O. This means 4.76 g of solute can be dissolved in 100 mL of water—not that you will have 100 mL of solution after dissolving 4.76 g of solute. If more than 4.76 g of this solute is added to 100 mL of water in the container, the excess will not dissolve under the specified conditions (**Figure 2**). The quickest way to see whether you have a saturated solution is to look for the presence of undissolved solids in the solution. There are several experimental designs that can be used to determine the solubility of a solid. For example, the solvent from a measured volume of saturated solution might be removed by evaporation, leaving the crystallized solid solute behind, which can then be collected and measured.

Figure 2
The excess of solid solute in the mixture is visible evidence of a saturated solution.

INVESTIGATION 5.5 Introduction

The Solubility of Sodium Chloride in Water

A significant part of the work of science is to test existing theories, laws, and generalizations. You will create a graph from the solubility data provided and use this graph to predict the solubility of sodium chloride in water at a particular temperature. You will then compare the predicted value with a value that you determine experimentally—by crystallization of sodium chloride from a saturated solution.

Purpose
The purpose of this investigation is to test the known solubility data for a solid in water.

Report Checklist
- ○ Purpose
- ○ Problem
- ○ Hypothesis
- ● Prediction
- ○ Design
- ○ Materials
- ○ Procedure
- ● Evidence
- ● Analysis
- ● Evaluation (1, 2, 3)

Problem
What is the solubility of sodium chloride, in grams per 100 mL of solution, at room temperature?

Design
A precisely measured volume of a saturated NaCl(aq) solution at room temperature is heated to evaporate the solvent and crystallize the solute. The mass of the dry solute is measured and the concentration of the saturated solution is calculated.

To perform this investigation, turn to page 229.

> **DID YOU KNOW ?**
>
> **Generalizations and Indigenous Knowledge**
>
> In Western science, a generalization is a statement that summarizes a pattern of empirical properties or trends; for example, solids have a higher solubility in water at higher temperatures. Scientists rely on these generalizations to organize their knowledge and to make predictions. Even though you can find exceptions to all generalizations, they are still very useful.
>
> Aboriginal peoples developed a traditional or indigenous knowledge of their environment. The foundation of this knowledge is empirical and many Aboriginal traditions correspond to generalizations about properties and trends in the natural world. These generalizations allowed Aboriginal peoples to make predictions such as which plants to use for which ailments and weather forecasting.

> **DID YOU KNOW ?**
>
> **"The Bends"**
>
> When diving underwater using air tanks, a diver breathes air at the same pressure as the surroundings. The increased pressure underwater forces more air to dissolve in the diver's bloodstream. If a diver comes up too quickly, the solubility of air (mostly nitrogen) decreases as the pressure decreases, and nitrogen bubbles form in the blood vessels. These nitrogen bubbles are the cause of a diving danger known as "the bends" (so named because divers typically bend over in agony as they try to relieve the pain). Nitrogen bubbles are especially dangerous if they form in the brain or spinal cord. The bends may be avoided by ascending very slowly or corrected by using a decompression chamber.

Solubility in Water Generalizations

Scientists have carried out a very large number of experiments as they have investigated the effects of temperature on the solubility of various solutes. From the results of their experiments, they have developed several useful generalizations about the solubility of solids, liquids, and gases in water. In all cases, we assume that the solid, liquid, or gas does not react with the solvent, water. The following list outlines how the solubility of various solutes varies with temperature.

Solids

- Solids usually have higher solubility in water at higher temperatures. For example, sucrose has a solubility of about 180 g/100 mL at 0 °C and 487 g/100 mL at 100 °C.

Gases

- Gases always have higher solubility in water at lower temperatures. The solubility of gases decreases as the temperature increases. This inverse relationship is approximately linear.
- Gases always have higher solubility in water at higher pressures.

Liquids

- It is difficult to generalize about the effect of temperature on the solubility of liquids in water. However, for polar liquids in water, the solubility usually increases with temperature. A prediction of the solubility of liquids with temperature will not be as reliable as a prediction for solids and gases.
- Some liquids (mostly nonpolar liquids) do not dissolve in water to any appreciable extent, but form a separate layer. Liquids that behave in this way are said to be *immiscible* with water. For example, benzene, gasoline, and carbon disulfide (which is used in the process of turning wood pulp into rayon or cellophane) are all virtually insoluble in water.
- Some liquids (such as those containing small polar molecules with hydrogen bonding) dissolve completely in water in any proportion. Liquids that behave in this way are said to be *miscible* with water. For example, ethanol (in alcoholic beverages), acetic acid (in vinegar), and ethylene glycol (in antifreeze) all dissolve completely in water, regardless of the quantities mixed.

Elements

- Elements generally have low solubility in water. For example, carbon is used in many water filtration systems to remove organic compounds that cause odours. The carbon does not dissolve in the water passing through it.
- Although the halogens and oxygen dissolve in water to only a very tiny extent, they are so reactive that, even in tiny concentrations, they are often very important in solution reactions.

Solubility Table

A solubility table of ionic compounds (see the inside back cover of this textbook) is best understood by assuming that most substances dissolve in water to some extent. The solubilities of various ionic compounds range from very soluble, like table salt, to slightly soluble, like silver chloride. The classification of compounds into very soluble and slightly soluble categories allows you to predict the state of a compound formed in a reaction in aqueous solution. The cutoff point between very soluble and slightly soluble is arbitrary. A solubility of 0.1 mol/L is commonly used in chemistry as this cutoff point because most ionic compounds have solubilities significantly greater or less than this value, which is a typical concentration for laboratory work. Of course, some compounds seem to be exceptions to the rule. Calcium sulfate, for example, has a solubility close to our arbitrary cutoff point and enough of it will dissolve in water that the solution noticeably conducts electricity.

▶ Practice

1. List three reasons why solutions are useful in a chemistry laboratory or industry.
2. Distinguish between solubility and a saturated solution.
3. Describe in general terms how you would make a saturated solution of a solid in water. How would you know whether the solution is saturated or whether the solute is just very slow in dissolving?
4. For any solute, what important condition must be stated in order to report the solubility?
5. State why you think clothes might be easier to clean in hot water.
6. Sketch a solubility versus temperature graph showing two lines labelled "solids" and "gases." Assume a straight-line relationship and show the generalization for the change in solubility of each type of substance with increasing temperature.
7. Give examples of two liquids that are immiscible and two that are miscible with water.
8. Why do carbonated beverages go "flat" when opened and left at room temperature and pressure?
9. Can more oxygen dissolve in a litre of water in a cold stream or a litre of water in a warm lake? Include your reasoning, according to the kinetic molecular theory.
10. (a) The solubility of oxygen in blood is much greater than its solubility in pure water. Suggest a reason for this observation.
    (b) If the concentration of oxygen in blood were the same as in pure water, how would your life be different?
    (c) Is there an advantage for animals that are cold blooded? Explain briefly.

LAB EXERCISE 5.C

Solubility and Temperature

Report Checklist

○ Purpose ○ Design ● Analysis
○ Problem ○ Materials ● Evaluation (2, 3)
○ Hypothesis ○ Procedure
● Prediction ○ Evidence

Purpose
The purpose of this investigation is to test the generalization about the effect of temperature on the solubility of an ionic compound.

Problem
How does temperature affect the solubility of potassium nitrate?

Design
Solid potassium nitrate is added to four flasks of pure water until no more potassium nitrate will dissolve and there is excess solid in each beaker. Each mixture is sealed and stirred at a different temperature until no further changes occur. The same volume of each solution is removed and evaporated to crystallize the solid. The specific relationship of temperature to the solubility of potassium nitrate is determined by graphical analysis. The temperature is the manipulated variable and the solubility is the responding variable.

Evidence

Table 1 Solubility of Potassium Nitrate at Various Temperatures

Temperature (°C)	Volume of solution (mL)	Mass of empty beaker (g)	Mass of beaker plus solid (g)
0.0	10.0	92.74	93.99
12.5	10.0	91.75	93.95
23.0	10.0	98.43	101.71
41.5	10.0	93.37	100.15

EXPLORE an issue

Pesticides

Issue Checklist
- ○ Issue ○ Design ● Analysis
- ○ Resolution ● Evidence ● Evaluation

The tremendous advances made by science and technology have both intended and unintended consequences for humans and the environment. For example, the development of pesticides has greatly improved crop yields and human health by controlling insect populations. Pesticides are chemicals used to kill pests, including insects and plants. The downside of the use of many pesticides is that they are highly toxic and remain in the environment for years. Some of these pesticides and their residues are part of a group of chemicals known as POPs—persistent organic pollutants that have the potential to harm human health and damage the ecological system on which life depends.

On a more local level, many pesticides are used, particularly in urban areas, to maintain lush green lawns (**Figure 3**). Although these pesticides and their residues contribute slightly to the global problem, they are of more concern locally. The ability of municipalities to ban pesticide use was greatly enhanced by the June 2001 Supreme Court of Canada Ruling that upheld the 1991 pesticide ban in Hudson, Quebec. Since this decision, many other provinces, including Alberta, have passed legislation permitting municipalities to enact by-laws to regulate public health and safety.

Issue
The use of toxic chemicals for the cosmetic appearance of lawns may endanger human health.

Figure 3
Is this your lawn?

Resolution
All municipalities in Alberta should enact a complete ban on lawn pesticides.

Design
Within small groups, research the pros and cons of pesticide use on lawns. Gather information from a wide variety of perspectives.

www.science.nelson.com

Figure 4
In a saturated solution of iodine, the concentration of the dissolved solute is constant. According to the theory of dynamic equilibrium, the rate of the dissolving process is equal to the rate of the crystallizing process.

$$I_2(s) \rightleftarrows I_2(aq)$$

Explaining Saturated Solutions

Most substances dissolve in a solvent to a certain extent, and then dissolving appears to stop. If the solution is in a closed system, one in which no substance can enter or leave, then observable properties become constant, or are *in equilibrium* (**Figure 4**).

According to the kinetic molecular theory, particles are always moving and collisions are always occurring in a system, even if no changes are observed. The initial dissolving of sodium chloride in water is thought to be the result of collisions between water molecules and ions that make up the crystals. At equilibrium, water molecules still collide with the ions at the crystal surface. Chemists assume that dissolving of the solid sodium chloride is still occurring at equilibrium. Some of the dissolved sodium and chloride ions must, therefore, be colliding and crystallizing out of the solution to maintain a balance. If both dissolving and crystallizing take place at the same rate, no observable changes would occur in either the concentration of the solution or in the quantity of solid present. The balance that exists when two opposing processes occur at the same rate is known as **dynamic equilibrium** (**Figure 5**).

Testing the Theory of Dynamic Equilibrium

You can try a simple experiment to illustrate dynamic equilibrium. Dissolve pickling (coarse) salt to make a saturated solution with excess solid in a small jar. Ensure that the lid is firmly in place, and then shake the jar and record the time it takes for the contents to settle so that the solution is clear. Repeat this process once a day for two weeks. Although the same quantity of undissolved salt is present each day, the settling becomes much faster over time because the solid particles in the jar become fewer in number, but larger in size. Chemists usually allow precipitates to digest for a while before filtering them, because larger particles filter more quickly. This evidence supports the idea that both dissolving and crystallizing are occurring simultaneously.

The theory of dynamic equilibrium can be tested by using a saturated solution of iodine in water. Radioactive iodine is used as a marker to follow the movements of some of the molecules in the mixture. To one sample of a saturated solution containing an excess of solid normal iodine, a few crystals of radioactive iodine are added. To a similar second sample, a few millilitres of a saturated solution of radioactive iodine are added (**Figure 6**). The radioactive iodine emits radiation that can be detected by a Geiger counter to show the location of the radioactive iodine. After a few hours, the solution and the solid in both samples clearly show increased radioactivity over the average background readings. Assuming the radioactive iodine molecules are chemically identical to normal iodine, the experimental evidence supports the idea of simultaneous dissolving and crystallizing of iodine molecules in a saturated system.

Figure 5
In a saturated solution such as this one, with excess solute present, dissolving and crystallizing occur at the same rate. This situation is known as dynamic equilibrium.

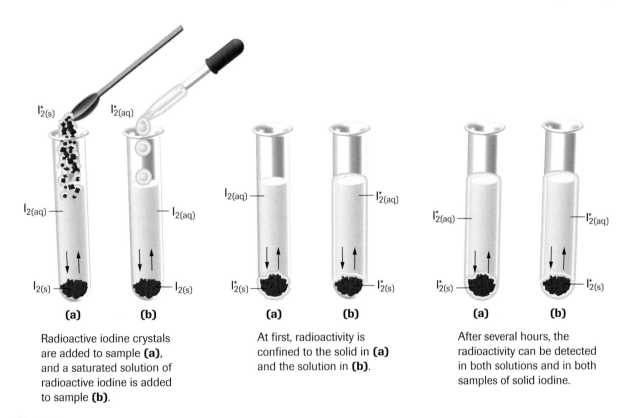

Figure 6
Radioactive iodine (indicated with an asterisk), added to a saturated solution of normal iodine (I_2), is eventually distributed throughout the mixture. A yellow outline indicates radioactivity.

A solubility equilibrium must contain both dissolved and undissolved solute at the same time. This state can be established by starting with a solute and adding it to a solvent. Consider adding calcium sulfate to water in a large enough quantity that not all will dissolve. We say we have added excess solute. A dissociation equation can be written for a saturated solution established this way:

$$CaSO_4(s) \rightleftarrows Ca^{2+}(aq) + SO_4^{2-}(aq)$$

Now consider a situation where two solutions, containing very high concentrations of calcium and sulfate ions respectively, are mixed. In this situation, the initial rate at which ions combine to form solid crystals is much greater than the rate at which those crystals dissolve, so we observe precipitation until the rates become equal and equilibrium is established.

$$Ca^{2+}(aq) + SO_4^{2-}(aq) \rightleftarrows CaSO_4(s)$$

How the equilibrium is established is not a factor. Viewed this way, most ionic compound precipitation reactions are examples of a dynamic equilibrium just like the equilibrium in a saturated solution.

▶ Section 5.5 Questions

1. Define solubility and state the main factors that affect the solubility of a substance in water.
2. Describe how the solubilities of solids and gases in water depend on temperature.

Use this information to answer questions 3 to 6.

In a chemical analysis experiment, a student notices that a precipitate has formed, and separates this precipitate by filtration. The collected liquid filtrate, which contains aqueous sodium bromide, is set aside in an open beaker. Several days later, some white solid is visible along the top edges of the liquid and at the bottom of the beaker.

3. What does the presence of the solid indicate about the nature of the solution?
4. What interpretation can be made about the concentration of the sodium bromide in the remaining solution? What is the term used for this concentration?
5. Write a brief theoretical explanation for this equilibrium mixture.
6. State two different ways to convert the mixture of the solid and solution into a homogeneous mixture.

7. Burping after drinking pop is common. What gas causes you to burp? Suggest a reason why burping occurs.
8. The purpose of the following investigation is to test the generalization about the effect of temperature on the solubility of an ionic compound known to be slightly soluble. Complete the Prediction and Design sections of the investigation report.

   Problem
   What is the relationship between temperature and the solubility of barium sulfate?

9. Different species of fish are adapted to live in different habitats. Some, such as carp, thrive in relatively warm, still water. Others, such as brook trout, need cold, fast-flowing streams, and will die if moved to the carp's habitat.
   (a) Describe and explain the oxygen conditions in the two habitats.
   (b) Hypothesize about the oxygen requirements of the two species of fish.
   (c) Thermal pollution is the large input of heated water into a lake or slow-moving stream from an industrial plant such as an electric generating station. Predict the effect of thermal pollution on trout in their lakes and streams.

10. Solubility also plays a role in cooking foods. Beans and broccoli should be cooked in water to retain their flavour but asparagus should be cooked in oil (or butter) and not water to best keep its flavour (**Figure 7**).
    (a) Based on this information, classify the solubility in water of the flavour molecules in these foods.
    (b) What interpretations can you make about the nature of the flavour molecules in beans and broccoli versus those in asparagus?

Figure 7
Fat retains the flavour of asparagus better than water does.

Chapter 5 INVESTIGATIONS

Unit 3

INVESTIGATION 5.1

Qualitative Chemical Analysis

Solutions have properties determined by the solute that is present. Diagnostic tests based on characteristic properties can be used to identify substances in a qualitative analysis.

Purpose
The purpose of this investigation is to use known diagnostic tests to distinguish among several pure substances.

Report Checklist
- ○ Purpose   ● Design   ● Analysis
- ○ Problem   ● Materials   ○ Evaluation
- ○ Hypothesis   ● Procedure
- ○ Prediction   ● Evidence

Problem
Which of the white solids labelled 1, 2, 3, and 4 is calcium chloride, citric acid, glucose, and calcium hydroxide?

 Calcium hydroxide is corrosive. Do not touch any of the solids. Wear eye protection, gloves, and an apron.

INVESTIGATION 5.2

A Standard Solution from a Solid

In this investigation, you will practise the skills required to prepare a standard solution from a pure solid (Appendix C.4). You will need these skills in many investigations in this course.

Purpose
The purpose of this investigation is to acquire the skills required to prepare a standard solution starting with a pure solid.

Materials
lab apron
eye protection
CuSO$_4$·5H$_2$O(s), copper(II) sulfate–water (1/5) or
   copper(II) sulfate pentahydrate
150 mL beaker
centigram balance
laboratory scoop
stirring rod
wash bottle of pure water (distilled or deionized)
100 mL volumetric flask with stopper
small funnel
medicine dropper
meniscus finder

 Copper(II) sulfate is harmful if swallowed.

Report Checklist
- ○ Purpose   ○ Design   ○ Analysis
- ○ Problem   ○ Materials   ○ Evaluation
- ○ Hypothesis   ○ Procedure
- ○ Prediction   ○ Evidence

Procedure

1. (Pre-lab) Calculate the mass of CuSO$_4$·5H$_2$O(s) needed to prepare 100.00 mL of a 0.05000 mol/L solution.

2. Measure the calculated mass of CuSO$_4$·5H$_2$O(s) in a clean, dry 150 mL beaker. (See Appendix C.3 for tips on using a laboratory balance. See Appendix C.4 and the Nelson Web site for tips on preparing a standard solution from a solid reagent.)

 www.science.nelson.com

3. Dissolve the solid in 40 mL to 50 mL of pure water. Use a stirring rod to help dissolve the solid. Be sure to rinse the stirring rod over your beaker of solution.

4. Transfer the solution into a 100 mL volumetric flask. Rinse the beaker two or three times with small quantities of pure water, transferring the rinsings into the volumetric flask.

5. Add pure water to the volumetric flask until the volume is 100.00 mL. Use the dropper and meniscus finder for the final few millilitres to set the bottom of the meniscus on the calibration line.

6. Stopper the flask and mix the contents thoroughly by repeatedly inverting the flask.

Note: Store your solution for the next investigation.

INVESTIGATION 5.3

A Standard Solution by Dilution

Report Checklist
- ○ Purpose   ○ Design   ○ Analysis
- ○ Problem   ○ Materials   ○ Evaluation
- ○ Hypothesis   ○ Procedure
- ○ Prediction   ○ Evidence

In this investigation, you will practise a very common laboratory procedure: preparing a standard solution from a more concentrated or stock solution.

Purpose
The purpose of this investigation is to acquire the skills required to prepare a standard solution by diluting a stock solution.

Materials
lab apron
eye protection
0.05000 mol/L $CuSO_4(aq)$ stock solution
150 mL beaker
10 mL volumetric pipette
pipette bulb
wash bottle of pure water
100 mL volumetric flask with stopper
small funnel
medicine dropper
meniscus finder

☠ Copper(II) sulfate is harmful if swallowed. Wear eye protection and a laboratory apron.

Use a pipette bulb. Do not pipette by mouth.

Procedure

1. (Pre-lab) Calculate the volume of a 0.05000 mol/L stock solution of $CuSO_4(aq)$ required to prepare 100.0 mL of a 0.005000 mol/L solution.
2. Measure the required volume of the stock solution using a 10 mL pipette. (See Appendix C.3, Appendix C.4, and the Nelson Web site for tips on pipetting and preparing a standard solution by dilution.)
3. Transfer the required volume of the stock solution into the 100 mL volumetric flask (**Figure 1(a)**).
4. Add pure water until the final volume is reached (**Figure 1(b)**). Use the dropper and meniscus finder for the final few millilitres to set the bottom of the meniscus on the calibration line.
5. Stopper the flask and mix the solution thoroughly.

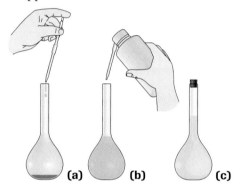

Figure 1
(a) The appropriate volume of $CuSO_4(aq)$ is transferred to a volumetric flask.
(b) Water is added to the flask.
(c) In the final dilute solution, the initial amount of copper(II) sulfate is still present, but it is diluted.

www.science.nelson.com GO ◀▶

INVESTIGATION 5.4

The Iodine Clock Reaction

Report Checklist
- ○ Purpose   ● Design   ● Analysis
- ○ Problem   ● Materials   ○ Evaluation
- ○ Hypothesis   ● Procedure
- ○ Prediction   ● Evidence

Technological problem solving often involves a systematic trial-and-error approach that is guided by knowledge and experience. Usually one variable at a time is manipulated, while all other variables are controlled. Variables that may be manipulated include concentration, volume, and temperature. In this investigation, you will compete to see which team is the first to solve the Problem using a reliable process. Create a design to guide your work. Using this design, try several procedures to solve the Problem. The final Analysis will be the materials and procedure that best answer the Problem.

Purpose
The purpose of this investigation is to find a method for getting a reaction to occur in a specified time period.

Problem
What technological process can be employed to have solution A react with solution B in a reliable time of 20 ± 1 s?

INVESTIGATION 5.5

The Solubility of Sodium Chloride in Water

Report Checklist
- ○ Purpose
- ○ Problem
- ○ Hypothesis
- ● Prediction
- ○ Design
- ○ Materials
- ○ Procedure
- ● Evidence
- ● Analysis
- ● Evaluation (1, 2, 3)

A significant part of the work of science is to test existing theories, laws, and generalizations. You will create a graph from the solubility data (**Table 1**) and use this graph to predict the solubility of sodium chloride in water at a particular temperature. You will then compare the predicted value with a value that you determine experimentally—by crystallization of sodium chloride from a saturated solution.

Table 1 Solubility of Sodium Chloride in Water

Temperature (°C)	Solubility (g /100 mL solution)
0	31.6
40	32.4
70	33.0
100	33.6

Purpose
The purpose of this investigation is to test the known solubility data for a solid in water.

Problem
What is the solubility of sodium chloride, in grams per 100 mL of solution, at room temperature?

Design
A precisely measured volume of a saturated NaCl(aq) solution at room temperature is heated to evaporate the solvent and crystallize the solute. The mass of the dry solute is measured and the concentration of the saturated solution is calculated.

Materials
lab apron
eye protection
oven mitts or heatproof gloves
saturated NaCl(aq) solution
laboratory burner with matches or striker, or hot plate
centigram balance
thermometer or temperature probe
laboratory stand
ring clamp
wire gauze
250 mL beaker
100 mL beaker
10 mL pipette with pipette bulb

 When using a laboratory burner, keep long hair tied back and loose clothing secured. If using a hot plate, take all necessary precautions.

Use oven mitts or heatproof gloves to handle hot apparatus.

Procedure

1. Measure and record the mass of a clean, dry 250 mL beaker. (See Appendix C.3 for tips on using a laboratory balance.)
2. Obtain about 40 mL to 50 mL of saturated NaCl(aq) in a 100 mL beaker.
3. Measure and record the temperature of the saturated solution to a precision of 0.2 °C. (See Appendix F.3 for a note on precision of readings.)
4. Pipette a 10.00 mL sample of the saturated solution into the 250 mL beaker. (See Appendix C.3 and the Nelson Web site for tips on pipetting.)

     www.science.nelson.com

5. Using a laboratory burner or hot plate, heat the solution evenly in the beaker until all the water boils away, and dry, crystalline NaCl(s) remains. (See Appendix C.3 for tips on using a laboratory burner. Also, see the video on the Nelson Web site.)

     www.science.nelson.com

6. Shut off the burner or hot plate, and allow the beaker and contents to cool for at least 5 min.
7. Measure and record the total mass of the beaker and contents.
8. Reheat the beaker and the residue and repeat steps 6 and 7 until two consecutive measurements of the mass give the same value. Record the final mass. (If the mass remains constant, this confirms that the sample is dry.)
9. Dispose of the salt as regular solid waste.

> **Learning Tip**
>
> You will be able to improve the precision of your prediction if you start the vertical axis of your graph at 31 g/100 mL instead of the usual zero value.

Chapter 5 SUMMARY

Outcomes

Knowledge
- explain the nature of solutions and the dissolving process (5.1, 5.2)
- illustrate how dissolving substances in water is often a prerequisite for chemical change (5.1, 5.2)
- differentiate between electrolytes and nonelectrolytes (5.1, 5.2)
- explain dissolving as an endothermic or an exothermic process with regard to breaking and forming of bonds (5.2)
- express concentration in various ways (5.3)
- perform calculations involving concentration, chemical amount, volume, and/or mass (5.3)
- use dissociation equations to calculate ion concentration (5.3)
- describe the procedures and calculations required for preparing solutions from a pure solid and by dilution (5.4)
- define solubility and identify the factors that affect it (5.5)
- explain a saturated solution in terms of equilibrium (5.5)

STS
- illustrate how science and technology are developed to meet societal needs and expand human capabilities (5.1)
- describe interactions of science, technology, and society (5.3, 5.5)
- relate scientific and technological work to personal and social values such as honesty, perseverance, tolerance, open-mindedness, critical-mindedness, creativity, and curiosity (5.1, 5.3, 5.4, 5.5)
- illustrate how science and technology have both intended and unintended consequences (5.3, 5.5)
- evaluate technologies from a variety of perspectives (5.4, 5.5)

Skills
- initiating and planning: design a procedure to identify the type of solution (5.1); design a procedure for determining the concentration of a solution containing a solid solute (5.4); describe procedures for safe handling, storing, and disposal of material used in the laboratory, with reference to WHMIS and consumer product labelling information (5.1, 5.4, 5.5)
- performing and recording: use a conductivity apparatus to classify solutions (5.1); perform an experiment to determine the concentration of a solution (5.4, 5.5); use a balance and volumetric glassware to prepare solutions of specified concentration (5.4); perform an investigation to determine the solubility of a solute in a saturated solution (5.5)
- analyzing and interpreting: use experimental data to determine the concentration of a solution (5.5)
- communication and teamwork: compare personal concentration data with the data of other groups (5.4, 5.5)

Key Terms

5.1
solution
solute
solvent
electrolyte
nonelectrolyte

5.2
dissociation
ionization

5.3
concentration
amount concentration

5.4
standard solution
stock solution

5.5
saturated solution
solubility
dynamic equilibrium

Key Equations

Concentration Types

percentage by volume $\quad c = \dfrac{V_{solute}}{V_{solution}} \times 100\% \quad$ % V/V (or mL/100 mL)

mass by volume $\quad c = \dfrac{m_{solute}}{V_{solution}} \times 100\% \quad$ % W/V (or g/100 mL)

by mass $\quad c = \dfrac{m_{solute}}{m_{solution}} \times 100\% \quad$ % W/W (or g/100 g)

parts per million $\quad c = \dfrac{m_{solute}}{m_{solution}} \quad$ ppm (typically mg/kg)

amount $\quad c = \dfrac{n_{solute}}{V_{solution}} \quad$ mol/L

Dilution

$$V_i c_i = V_f c_f$$

▶ MAKE a summary

1. Devise a concept map built around the subject "Solutions" and include all of the Key Terms listed above.
2. Refer back to your answers to the Starting Points questions at the beginning of this chapter. How has your thinking changed?

▶ Go To

The following components are available on the Nelson Web site. Follow the links for *Nelson Chemistry Alberta 20–30*.
- an interactive Self Quiz for Chapter 5
- additional Diploma Exam-style Review questions
- Illustrated Glossary
- additional IB-related material

There is more information on the Web site wherever you see the Go icon in this chapter.

Chapter 5 REVIEW

Many of these questions are in the style of the Diploma Exam. You will find guidance for writing Diploma Exams in Appendix H. Exam study tips and test-taking suggestions are on the Nelson Web site. Science Directing Words used in Diploma Exams are in **bold** type.

www.science.nelson.com

DO NOT WRITE IN THIS TEXTBOOK.

Part 1

1. Rusting of iron occurs extremely slowly in very dry climates. A likely reason for this observation is that
   A. iron is an inert material
   B. there is a lower concentration of oxygen in very dry climates
   C. dissolving substances in water is usually necessary for chemical change
   D. the higher temperatures prevent rusting because it is an exothermic reaction

2. Cold packs (**Figure 1**) contain an ionic compound such as ammonium nitrate and a separate pouch of water that is broken when the cold pack is needed. Which of the following rows indicates the type of change and the process that produces this change?

Row	Type of change	Process
A.	endothermic	ionization
B.	endothermic	dissociation
C.	exothermic	ionization
D.	exothermic	dissociation

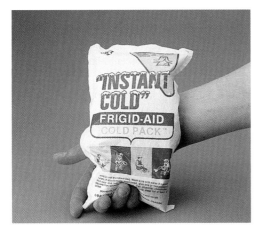

Figure 1
A cold pack

3. The maximum acceptable concentration of fluoride ions in municipal water supplies corresponds to 0.375 mg of fluoride in a 250 mL glass of water. The concentration of fluoride ions, in ppm, is _____.

Use this information to answer questions 4 to 7.

Hard water contains metal ions, most commonly calcium and magnesium ions. Some moderately hard water is found to contain 200 ppm of calcium hydrogen carbonate.

4. The dissociation equation for calcium hydrogen carbonate in water is
   A. $CaHCO_3(s) \rightarrow CaH^{2+}(aq) + CO_3^{2-}(aq)$
   B. $CaHCO_3(s) \rightarrow Ca^{2+}(aq) + HCO_3^-(aq)$
   C. $Ca(HCO_3)_2(s) \rightarrow Ca^{2+}(aq) + 2 H^+(aq) + 2 CO_3^{2-}(aq)$
   D. $Ca(HCO_3)_2(s) \rightarrow Ca^{2+}(aq) + 2 HCO_3^-(aq)$

5. The ppm concentration of hydrogen carbonate ions in the hard water is
   A. 400 ppm
   B. 300 ppm
   C. 200 ppm
   D. 100 ppm

6. The mass of calcium hydrogen carbonate that would be found in 52.0 L of hard water in a bathtub is _____ g.

7. The amount concentration of calcium hydrogen carbonate in the hard water is _____ mmol/L.

8. A Web site promoting eco-friendly alternatives to commercial cleaners suggests mixing 125 mL of vinegar with enough water to make 1.0 L of cleaning solution. If the vinegar used contains 5.0% acetic acid (by volume), what is the percentage concentration of acetic acid in the cleaning solution?
   A. 0.025%
   B. 0.13%
   C. 0.63%
   D. 8.0%

9. A 500 mL bottle of fireplace window cleaner contains 2.50 mol/L of potassium hydroxide. The mass of KOH(s) contained in the bottle is _____ g.

10. The main piece of laboratory equipment that is used in both procedures for the preparation of a standard solution, from a solid and by dilution, is a/an
    A. Erlenmeyer flask
    B. volumetric pipette
    C. volumetric flask
    D. graduated cylinder

Use this information to answer questions 11 to 13.

The salt tank attached to a water softener contains excess sodium chloride solid, sitting in a fixed quantity of water. This mixture remains for a long period of time before it is used to regenerate the resin in the water softener.

11. The best description of the salt solution is that it is
    A. dilute
    B. saturated
    C. miscible
    D. concentrated

12. There are no observable changes in any properties of the mixture because
    A. the rate of dissolving equals the rate of crystallizing
    B. no change is occurring at the molecular level
    C. the rates of dissolving and dissociating are equal
    D. there is no more space for any more salt to dissolve

13. The concentration of the salt solution can be increased by
    A. stirring vigorously
    B. adding more water
    C. removing some solution
    D. increasing the temperature

Part 2

14. **How** is a homogeneous mixture different from a heterogeneous mixture? Give one example of each.

15. A chemistry student was given the task of identifying four colourless solutions. Complete the **Analysis** of the investigation report.

    Problem
    Which of the solutions, labelled A, B, C, and D, is calcium hydroxide, glucose, potassium chloride, and sulfuric acid?

    Evidence

    Table 1 Litmus and Conductivity Tests

Solution*	Red litmus	Blue litmus	Conductivity
A	stays red	blue to red	high
B	stays red	stays blue	none
C	red to blue	stays blue	high
D	stays red	stays blue	high

    *same concentration and temperature

16. Scientists have developed a classification system to help organize the study of matter. **Describe** an empirical test that can be used to distinguish between the following classes of matter:
    (a) electrolytes and nonelectrolytes
    (b) acids, bases, and neutral compounds

17. What is a standard solution, and **why** is such a solution necessary?

18. **Describe** two methods used to prepare standard solutions.

19. Much of the food you eat is converted to glucose in your digestive tract. The glucose dissolves in the blood and circulates throughout your body. Cells use the glucose to produce energy in the process of cellular respiration. State two reasons why it is important for the glucose to be dissolved in a solution rather than remain as a solid.

20. From Mini Investigation: Hot and Cold Solutions (page 199), you know that the dissolving of sodium nitrate is endothermic. What does this mean, empirically and theoretically?

21. **Describe** the ways in which concentrations of solutions are expressed in chemistry laboratories, household products, and environmental studies.

22. A shopper has a choice of yogurt with three different concentrations (% W/W) of milk fat: 5.9%, 2.0%, and 1.2%. If the shopper wants to limit his or her milk fat intake to 3.0 g per serving, **determine** the mass of the largest serving the shopper could have for each type of yogurt.

23. What volume of vinegar contains 15 mL of pure acetic acid (**Figure 2**)?

Figure 2
The label tells us the concentration of acetic acid in vinegar.

24. **Determine** the amount concentration of the following solutions:
    (a) 0.35 mol copper(II) nitrate is dissolved in water to make 500 mL of solution.
    (b) 10.0 g of sodium hydroxide is dissolved in water to make 2.00 L of solution.
    (c) 25 mL of 11.6 mol/L HCl(aq) is diluted to a volume of 145 mL.
    (d) A sample of tap water contains 16 ppm of magnesium ions.

25. Standard solutions of sodium oxalate, $Na_2C_2O_4$(aq), are used in a variety of chemical analyses. **Determine** the mass of sodium oxalate required to prepare 250.0 mL of a 0.375 mol/L solution.

26. Phosphoric acid is the active ingredient in many commercial rust-removing solutions. **Determine** the volume of concentrated phosphoric acid (14.6 mol/L) that must be diluted to prepare 500 mL of a 1.25 mol/L solution.

Use this information to answer questions 27 to 29.

For people with diabetes, monitoring blood glucose levels is essential. There are many products available (**Figure 3**) that typically provide the concentration of glucose in units of millimoles per litre and use as little as 1 μL of blood.

Figure 3
A glucose meter

27. A glucose meter shows a normal reading of 7.8 mmol/L for an average adult, two hours after a meal. What mass of glucose is present in 4.7 L of blood of an average adult?

28. Glucose meters need to be checked periodically for accuracy. Checking is done using a standard glucose solution, such as one with a concentration of 3.1 mmol/L.
    (a) If you were to prepare 100.0 mL of this standard solution, what mass of solid glucose is required?
    (b) List the materials required to prepare this standard solution. Specify sizes and quantities.
    (c) Write a complete procedure for the preparation of the standard solution.

29. **How** does the glucose meter illustrate the interaction of science, technology, and society?

Use this information to answer questions 30 to 33.

Acids are usually purchased in their pure or concentrated form and then diluted to the concentration required for a particular use. Concentrated 17.8 mol/L sulfuric acid mixed with water can generate localized temperatures in excess of 100 °C. Sulfuric acid is a common example, but you need to be careful when diluting any acid.

30. When concentrated sulfuric acid dissolves in water, is this process endothermic or exothermic? State the evidence.

31. **Describe** the correct procedure for diluting concentrated reagents such as sulfuric acid. **Why** is it recommended that you always follow this procedure?

32. What volume of concentrated sulfuric acid would a technician require to prepare 2.00 L of 0.250 mol/L solution?

33. Write a dissociation equation to explain the electrical conductivity of each of the following chemicals:
    (a) potash: potassium chloride
    (b) Glauber's salt: sodium sulfate
    (c) TSP: trisodium phosphate

34. **Determine** the amount concentration of the cation and the anion in a 0.14 mol/L solution of each of the following chemicals.
    (a) saltpetre: KNO_3
    (b) road salt: calcium chloride
    (c) fertilizer: ammonium phosphate

Extension

35. The oil industry is an increasingly important component of Alberta's economy. Part of this industry involves transporting oil products in rail cars. The spill of oil products near Lake Wabamun in August 2005 is a dramatic example of the risks involved in getting products to market. Prepare a fact sheet on the transportation of oil products. Your response should include:
    - the nature of two spilled substances, including names, uses, and properties such as solubility, density, and toxicity
    - the risks and benefits of transporting these products, including several perspectives

www.science.nelson.com

Figure 4
An oil-covered Western Grebe

chapter 6
Acids and Bases

In this chapter

- Exploration: Consumer Products
- Investigation 6.1: Properties of Acids and Bases
- Lab Exercise 6.A: The Relationship between pH and Hydronium Ion Concentration
- Biology Connection: pH Dependence
- Web Activity: Bad Hair Day?
- Mini Investigation: pH of a Solution
- Lab Exercise 6.B: Using Indicators to Determine pH
- Investigation 6.2: Testing Arrhenius' Acid–Base Definitions
- Case Study: Acid Deposition
- Investigation 6.3: Comparing the Properties of Acids (Demonstration)

To survive in nature using only natural substances requires a very practical form of knowledge. The Cree and other Aboriginal peoples knew the best locations to harvest and plant manomin (wild rice), and various Aboriginal peoples used mosses as diaper materials and animal organs for tanning hides. From a scientific perspective, these examples all depend in some way on the presence of acids and bases. Antacid remedies for indigestion, pH-balanced shampoos, and acetylsalicylic acid (ASA) are just a few of the many acidic or basic products found in any drugstore.

Many acids and bases are sold under common or traditional names. Concentrated hydrochloric acid is sometimes sold as muriatic acid. Sodium hydroxide, called lye as a pure solid, has a variety of brand names under which it is sold as a concentrated solution for cleaning plugged drains. Generic or "no-name" products often contain the same kind and quantity of active ingredients as brand-name products. You can save time, trouble, and money by knowing that, in some cases, the chemical names of compounds used in home products are stated on the label.

Since so many technological applications involve acids and bases, it is important to understand these substances. References in the popular media offer little insight into what these substances are or what they do. Such references often emphasize one perspective, such as environmental damage caused by acid deposition or an acid spill. This can be confusing. An amateur gardener who has just read an article attributing the destruction of conifer forests to acid deposition (**Figure 1**) may be puzzled by instructions on a package of evergreen fertilizer stating that evergreens are acid-loving plants. Understanding acids and bases gives us the background to assess the beneficial and detrimental effects of acids and bases on the environment, and to evaluate technologies that are used to combat environmental problems.

STARTING Points

Answer these questions as best you can with your current knowledge. Then, using the concepts and skills you have learned, you will revise your answers at the end of the chapter.

1. What is pH? How is it determined, and how does pH relate to solutions encountered in everyday life?
2. How is scientific knowledge of acids and bases applied in industrial, environmental, and consumer contexts?
3. What is the distinction between strong and weak acids or bases?

Career Connections:
Ecologist; Medical Laboratory Technologist

Figure 1
If conifers like acidic soil, why is acid deposition believed to cause environmental problems?

▶ Exploration   Consumer Products

Look around at home or in a store and read the labels on a variety of consumer products, such as shampoos, soaps, and other cleaners. It is not necessary to open any containers. Find one type of product that mentions pH, and investigate other brands of this product. For each brand:

(a) Record the brand name and any pH information (none, qualitative, or quantitative).

(b) What claims are made about the product's performance, and the relationship between pH and the product's performance?

(c) Working within a group, consider what you would need to know to test the suggested relationship between the pH and the performance of the product. List a series of questions that could be answered after doing some research.

6.1 Properties of Acids and Bases

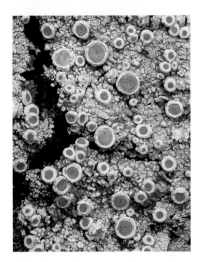

Figure 1
Lichen like this is used to make litmus.

Historically, only a few simple properties of acids and bases, such as their effect on the colour of litmus, were known by the middle of the 17th century (**Figure 1**). By the early 20th century, many other properties, such as pH, were discovered and new theories about acid–base reactions were created. Acids and bases are classes of compounds that have historically been distinguished by their empirical properties, primarily the behaviour of their aqueous solutions.

In general, science evolves from empirical knowledge to theoretical knowledge. In other words, to understand and explain acids and bases, scientists first need to have reliable empirical information about acids and bases before theories can be developed to explain them.

Technology that uses acids and bases depends on both empirical and theoretical knowledge of acids and bases for solving practical problems. The scientific knowledge of acids and bases is applied in several technological contexts: *consumer* (individuals), *commercial* (companies that use manufactured products to produce other products and processes), and *industrial* (large-scale companies that usually deal with raw materials and produce starting materials). Throughout this chapter, you will see examples of acids and bases used in each of these contexts.

INVESTIGATION 6.1 Introduction

Properties of Acids and Bases

In this investigation, you will use your previous knowledge of properties of substances, and practise your problem-solving skills. You are provided with solutions of approximately equal concentrations, at the same temperature, of the following pure substances: $CaCl_2(s)$, $C_3H_4OH(COOH)_3(l)$, $C_6H_{12}O_6(s)$, $Ca(OH)_2(s)$, $NH_3(g)$, $NaHSO_4(s)$, $CH_3OH(l)$, $H_2SO_4(l)$, $Na_2CO_3(s)$. Remember to include variables in your Design, and safety and disposal instructions in your Procedure.

Report Checklist
- ○ Purpose
- ○ Problem
- ○ Hypothesis
- ● Prediction
- ● Design
- ● Materials
- ● Procedure
- ● Evidence
- ● Analysis
- ● Evaluation (1, 2, 3)

Purpose
The purpose of this investigation is to test previous knowledge about the properties of acids and bases.

Problem
What properties are most useful for distinguishing acids and bases from other classes of compounds?

To perform this investigation, turn to page 260.

The Nature of Acidic and Basic Solutions

As shown in **Table 1**, acidic and basic solutions can be readily distinguished from each other by testing with an indicator (litmus) or pH paper (or meter). If you want to distinguish all four categories of solutes, a conductivity test would also be required.

Learning Tip

Note that the terms "acid" and "base" refer to chemical substances, whereas the terms "acidic" and "basic" refer to the properties of a solution.

Table 1 Diagnostic Tests for Various Types of Solutions

Type of solute	Type of solution	Conductivity	Litmus	pH
most molecular compounds	neutral	no	no effect	7
most ionic compounds	neutral	yes	no effect	7
acids	acidic	yes	blue to red	<7
bases	basic	yes	red to blue	>7

What is it about acids and bases that causes the change in the colour of an indicator or in values on a pH scale? Early attempts at a theory of acids and bases tended to focus on acids and ignore bases. Over time, several theories followed a cycle of creation, testing, acceptance, further testing, and eventual rejection. In this chapter, you will look briefly at a few early suggestions and at a more modern theory in the next chemistry course.

The idea that the presence of hydrogen gave a compound acidic properties started with Sir Humphry Davy in the early 1800s. A few decades later, Justus von Liebig expanded this theory to include the idea that acids are salts of hydrogen. According to his theory, acids could be thought of as *ionic compounds in which hydrogen had replaced the metal ion*. Liebig's theory, however, did not explain why many compounds containing hydrogen have neutral properties (such as CH_4) or basic properties (such as NH_3). In the late 1800s, Svante Arrhenius provided the first useful theoretical definition of acids and bases. *Acids are substances that ionize in aqueous solution to form hydrogen ions, and bases are substances that dissociate to form hydroxide ions in aqueous solution.* Although this theory has some drawbacks, as you will see later, it is still a widely used theory in many applications that require a simpler understanding of acids and bases.

For a long time, chemists have believed that acidic properties of a solution are related to the presence of hydrogen ions, whereas basic properties are related to the presence of hydroxide ions. The specific nature of the hydrogen ion in a solution has been the subject of much debate within the scientific community. Theoretical chemists thought that it was very unlikely that a hydrogen ion, which is a tiny proton with a very high charge-to-size ratio, could exist on its own in an aqueous solution. It is likely to bond strongly to polar water molecules. The first empirical evidence for this bonding was provided in 1957 by Paul Giguère at the Université Laval, Quebec, with his discovery of the existence of hydrated protons. The simplest representation of a hydrated proton is $H_3O^+(aq)$, commonly called the **hydronium ion** (**Figure 2**). A modern view of the nature of acidic and basic solutions is that hydronium ions are responsible for acidic properties and hydroxide ions are responsible for basic properties. The acidic or basic properties of a solution are most conveniently measured using paper test strips that have absorbed an indicator (such as litmus paper), but are most precisely measured using a pH meter.

Learning Tip

Chemists have known about some acids for hundreds of years. Because these acids are commonly known by familiar names, their IUPAC names are not often used:

Familiar name	IUPAC name
hydrochloric acid	aqueous hydrogen chloride
nitric acid	aqueous hydrogen nitrate
sulfuric acid	aqueous hydrogen sulfide
acetic acid	ethanoic acid

You should know these four acids by both names.

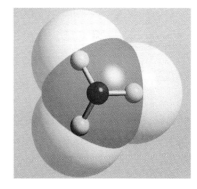

Figure 2
The hydronium ion has a trigonal pyramidal structure. The oxygen atom is the pyramid's apex; the hydrogen atoms form its base.

Section 6.1 Questions

1. In the usual progress of science, which comes first, empirical or theoretical knowledge?
2. List the three contexts that can be used to classify technological products or processes.
3. What is the most useful empirical property that can be used to distinguish acids, bases, and neutral compounds? Justify your answer.
4. How has the explanation of acidic properties changed from the early 1800s to the present day?
5. How is a hydronium ion different from a hydrogen ion? How is it similar?
6. Find and examine the label of one consumer product that contains an acid and one that contains a base. For each, identify any cautions noted for handling, storing, and disposing. Include the meaning(s) of any Household Hazardous Product symbols that indicate the primary hazard and the degree of hazard.
7. Environmental scientists and technicians often determine the acidity of aquatic environments.

Figure 3
Like other electronic devices, new pH meters are much smaller than earlier models. This one is very easy to take anywhere.

(a) Why is measuring pH with a meter, like the one in **Figure 3**, better than using an indicator such as litmus for this task?
(b) Which is the more common problem: acidic or basic aquatic environments? Briefly state your reasons for your answer.

6.2 pH and pOH Calculations

Acidic and basic solutions are an important part of many technological products, such as cleaning solutions, and of manufacturing processes, such as in the pulp and paper industry. In order to develop these technologies and deal with unintended consequences such as acid or base spills, a more detailed knowledge of these solutions is necessary.

At one time, the explanations of acidic and basic solutions were considered to be independent of one another—acidic solutions were explained by the presence of aqueous hydrogen ions (or the more modern concept of hydronium ions), whereas basic solutions were explained by the presence of aqueous hydroxide ions. Furthermore, the greater the concentration of hydronium ions is, the more acidic a solution is. Similarly, the higher the concentration of hydroxide ions is, the more basic a solution is.

You might expect that a neutral solution of pure water does not contain any hydronium or hydroxide ions. Careful testing, however, yields evidence that neutral water (pH 7) always contains trace amounts of both hydronium and hydroxide ions, due to a very slight ionization. In a sample of pure water, about two out of every billion molecular collisions are successful in forming hydronium and hydroxide ions.

$$2\ H_2O(l) \xrightarrow{\text{very slight}} H_3O^+(aq) + OH^-(aq)$$

In pure water at SATP, the hydronium ion concentration is very low: about 1×10^{-7} mol/L. This concentration is often negligible; for example, a conductivity test will show no conductivity for pure water unless the equipment is very sensitive (**Figure 1**).

Aqueous solutions exhibit a phenomenally wide range of hydronium ion concentrations—from more than 10 mol/L for a concentrated hydrochloric acid solution, to less than 10^{-15} mol/L for a concentrated sodium hydroxide solution. Any aqueous solution can be classified as acidic, neutral, or basic, using a scale based on the hydronium ion concentration.

At 25 °C, aqueous solutions can be classified as follows:
- neutral solution: $[H_3O^+(aq)] = 1 \times 10^{-7}$ mol/L
- acidic solution: $[H_3O^+(aq)] > 1 \times 10^{-7}$ mol/L
- basic solution: $[H_3O^+(aq)] < 1 \times 10^{-7}$ mol/L

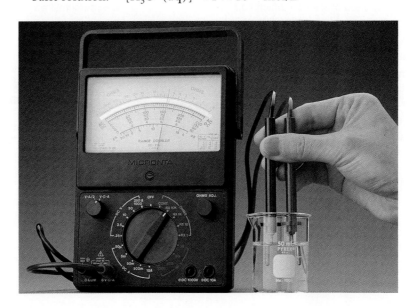

Figure 1
A sensitive multimeter shows the electrical conductivity of distilled water in a laboratory. Successive distillations to increase purity will lower, but never eliminate, the conductivity of water as measured by increasingly sensitive instruments. (See Appendix C.3 for instructions on using a multimeter.)

The extremely wide range of hydronium ion concentration led to a convenient shorthand method of communicating these concentrations. In 1909, Danish chemist Søren Sørensen introduced the term pH or "power of hydrogen." The **pH** of a solution is defined as the negative of the exponent to the base ten of the hydronium ion concentration (expressed as moles per litre). This definition is not as complicated as it sounds. For example, a concentration of 10^{-7} mol/L has a pH of 7 (neutral), and a pH of 2 corresponds to a much greater hydronium ion concentration of 10^{-2} mol/L (acidic). Notice from these two examples that pH has no units, and that the definition of pH can be used to create the following equation:

$$[H_3O^+(aq)] = 10^{-pH}$$

The pH is often mentioned on the labels of consumer products such as shampoos (**Figure 2**); in water-quality tests for pools and aquariums; in environmental studies of acid rain; and in laboratory investigations of acids and bases. Since each pH unit corresponds to a factor of 10 in the concentration, the huge $[H_3O^+(aq)]$ range can now be communicated by a much simpler set of positive numbers (**Figure 3**). Changes in pH can be deceptive. For example, if you add some vinegar to pure water, the pH might change from 7 to 4. This change of 3 pH units does not appear very significant, but the change in hydronium ion concentration is 10^3 or 1000 times larger.

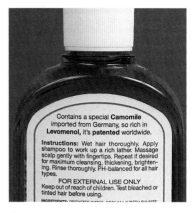

Figure 2
Some shampoos are pH balanced; that is, they have a pH similar to the natural pH of your scalp and hair (about 6.5). Do pH-balanced shampoos clean your hair as well as basic shampoos do?

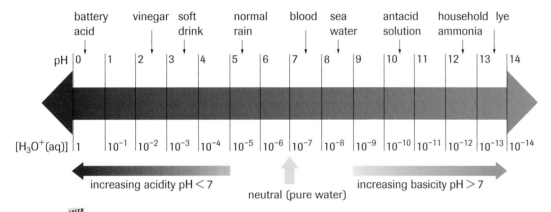

Figure 3
The pH scale is used to communicate a broad range of hydronium ion concentrations, in a wide variety of solutions. Most common acids and bases have pH values between 0 and 14.

Practice

1. Measurements of pH can be made using pH paper to provide a quick estimate of the hydronium ion concentration in an aqueous solution. What is the estimated hydronium ion concentration in each of the following solutions?
   (a) pure water: pH = 7
   (b) household ammonia: pH = 11
   (c) vinegar: pH = 2
   (d) soda pop: pH = 4
   (e) drain cleaner: pH = 14

2. Hydronium ion concentration is a theoretical concept used to explain the properties of acids. Express each of the following concentrations as pH values:
   (a) grapefruit juice: $[H_3O^+(aq)] = 10^{-3}$ mol/L
   (b) rainwater: $[H_3O^+(aq)] = 10^{-5}$ mol/L
   (c) milk: $[H_3O^+(aq)] = 10^{-7}$ mol/L
   (d) soap: $[H_3O^+(aq)] = 10^{-10}$ mol/L

3. If one water sample test shows a pH of 3 and another sample shows a pH of 5, by what factor do their hydronium ion concentrations differ? Justify your answer.

DID YOU KNOW?

The "p" in pH
pH was developed only about 100 years ago, but already the origin of the term has become blurred. Some scientists associate pH with **p**ower of **h**ydrogen, H; others with **p**otential of **h**ydrogen. Sørensen was Danish, so perhaps the "p" in pH comes from the Danish word "potenz," meaning "strength," or the French word "potentiel." It is strange that we have so quickly lost the origin of such a familiar term.

LAB EXERCISE 6.A

The Relationship between pH and Hydronium Ion Concentration

Report Checklist

- ○ Purpose
- ○ Problem
- ○ Hypothesis
- ○ Prediction
- ○ Design
- ○ Materials
- ○ Procedure
- ○ Evidence
- ● Analysis
- ○ Evaluation

Understanding pH and hydronium ion concentration is important in chemistry and also in the application of pH to other topics, such as environmental effects of pollution and consumer products. In the Analysis, use a spreadsheet program to produce the graphs.

Purpose
The purpose of this investigation is to create a relationship between the magnitude changes in pH and the changes in hydronium ion concentration of a solution.

Problem
What is the relationship between pH and hydronium ion concentration?

Design
A stock solution of hydrochloric acid is successively diluted and the pH of each diluted solution is measured using a calibrated pH meter. The manipulated variable is the concentration of hydrochloric acid (which equals the hydronium ion concentration). The responding variable is the pH. Some key controlled variables are temperature and type of acid used.

Evidence

Table 1 pH Readings of HCl(aq) Dilutions

$[H_3O^+(aq)]$ (mol/L)	pH
1.0×10^{-1}	0.9
1.0×10^{-2}	1.8
1.0×10^{-3}	2.8
1.0×10^{-4}	3.7
1.0×10^{-5}	4.9

Analysis

(a) Enter the values from **Table 1** into a spreadsheet. To enter scientific notation, use "E" to represent "× 10". For example, enter 1.0×10^{-1} as 1E-1. Plot a graph of pH versus $[H_3O^+(aq)]$.
(b) In words, describe the relationship shown by this graph.
(c) In a blank section of the spreadsheet, create a new table by converting $[H_3O^+(aq)]$ to a logarithm of $[H_3O^+(aq)]$ and keeping pH the same. To enter the logarithm, use log10( ). For example, enter the data in the first row of Table 1 as log10(1E-1) and 0.9. Plot a graph of pH versus log ($[H_3O^+(aq)]$).
(d) In words, describe the relationship shown by this graph. What is unusual about this graph?
(e) Alter your table of values by placing a negative sign in front of each log term. For example, enter the data in the first row of Table 1 as -log10(1E-1) and 0.9. The graph will change as you alter each number.
(f) In words, describe this new graph. Why is this graph simpler and better than any of the previous graphs?

CAREER CONNECTION

Ecologist

Ecologists conduct field research to study land, water, plant, and animal interactions, and make recommendations for environmental impact assessments. Some ecologists specialize in marine biology, microbiology, or ecosystem management, among other things. Most of us care about environmental impacts, so find out more about becoming an ecologist.

www.science.nelson.com

pH Calculations

Previously, pH was defined as the negative of the exponent to the base 10 of the hydronium ion concentration (expressed as moles per litre). This simplified definition is useful only if the pH values are integers and the concentration values are simply powers of 10. This situation is rarely the case. When environmental scientists or ecologists measure the pH of aquatic environments, they often require pH measurements to one or more decimal places.

The definition needs to be improved to be able to convert pH values such as 5.3 to a hydronium ion concentration, and hydronium ion concentrations such as 6.7×10^{-8} mol/L to a pH. Expressed as a numerical value without units, the pH of a solution is best defined as the negative of the logarithm to the base 10 of the hydronium ion concentration:

$$pH = -\log[H_3O^+(aq)]$$

SAMPLE problem 6.1

Given a hydronium ion concentration of 6.7×10^{-8} mol/L, calculate the pH.

According to the definition, $pH = -\log[H_3O^+(aq)]$, the pH is calculated as

$$pH = -\log(6.7 \times 10^{-8})$$

Notice that the units have been dropped because a logarithm has no units. When you enter the concentration value into your calculator and press "log," the number -7.1739252 appears (some calculators may display extra digits), which means

$$pH = -(-7.1739252)$$

The purpose of the negative sign in the pH definition is to change the initial "log" value from a negative to a positive value. Now, how many of the digits displayed on the calculator are significant? The answer to this question lies in the certainty (number of significant digits) of the concentration; 6.7×10^{-8} mol/L shows two significant digits. For the same reason that the exponent "8" in the concentration does not count as a significant digit, the integer part of the pH is also not counted. The correct answer is a pH of 7.17. *The number of digits following the decimal point in the pH value is equal to the number of significant digits in the hydronium ion concentration.* If you write the concentration in a nonstandard exponential notation, this rule may be clearer:

$[H_3O^+(aq)] = 0.67 \times 10^{-7}$ mol/L   [two significant digits]

$pH = 7.17$ [two decimal places]

COMMUNICATION example 1

Communicate a hydronium ion concentration of 4.7×10^{-11} mol/L as a pH value.

Solution

$pH = -\log[H_3O^+(aq)]$
$= -\log(4.7 \times 10^{-11})$
$= 10.33$

The pH of the solution is 10.33.

Learning Tip

Numbers in scientific notation are best entered on your calculator using the exponent (EE or EXP) key; for example, 8.7 EE -9. The calculator is programmed to treat this entry as one value. The 10^x key is not recommended because you may obtain incorrect results in some situations and on some calculators.

Learning Tip

Use your calculator to do the calculation shown in Communication Example 1. For common graphing calculators, the sequence of keystrokes is likely as follows:

For scientific calculators, the number (4.7 EE -11) is entered first, followed by the function (log) and then the sign change (+/−).

When pH is measured using a pH meter or pH paper in a chemical analysis, you may need to convert from pH to the amount concentration of hydronium ions. This conversion is based on the mathematical concept that a base 10 logarithm represents an exponent:

$$[H_3O^+(aq)] = 10^{-pH}$$

Notice that this definition is the same as that initially given for pH, but now it is more general and applies to all pH values, not only integers. Using this relationship, you can convert from a pH to a hydronium ion concentration. Because a pH (logarithm) has no units and the definition of pH includes the requirement that the concentration be in moles per litre, you will *always* need to add the units, mol/L, to your answer.

BIOLOGY CONNECTION

pH Dependence

Many biological processes, such as enzyme reactions, depend on pH. If you are taking a biology course, you will see several examples of pH dependent processes.

www.science.nelson.com

Figure 4
A pH meter measures a tiny voltage produced by a solution containing hydronium ions and converts this electrical measurement into a pH reading.

Learning Tip

Use your calculator to do the calculation shown in Communication Example 2. For common graphing calculators, the sequence of keystrokes is likely as follows:

For scientific calculators, the number (-10.33) is entered first, followed by the function (10^x).

SAMPLE problem 6.2

A solution has a pH of 5.3 (**Figure 4**). Calculate its hydronium ion concentration.

According to the definition, $[H_3O^+(aq)] = 10^{-pH}$, the hydronium ion concentration is calculated as

$$[H_3O^+(aq)] = 10^{-5.3} \text{ mol/L}$$

Note that the units have been added. Strictly speaking, this answer is mathematically correct, but it is not communicated in the usual or expected format. Your calculator will change the format of this answer into a properly expressed number. When you enter $10^{-5.3}$ on a calculator, you will obtain

5.01187 E -6 or 5.01187 -06 (which is 5.01187×10^{-06}).

Recall from Sample Problem 6.1 that the number of digits following the decimal point in the pH value is equal to the number of significant digits in the hydronium ion concentration. In the case of a pH equal to 5.3, one digit following the decimal point means one significant digit in the answer, 5×10^{-6} mol/L. If we write the concentration in a nonstandard exponential notation, this rule may be clearer:

$[H_3O^+(aq)] = 0.5 \times 10^{-5}$ mol/L   [one significant digit]

pH $= 5.3$ [one decimal place]

COMMUNICATION example 2

Communicate a pH of 10.33 as a hydronium ion concentration.

Solution

$$[H_3O^+(aq)] = 10^{-pH}$$
$$= 10^{-10.33} \text{ mol/L}$$
$$= 4.7 \times 10^{-11} \text{ mol/L}$$

The hydronium concentration is 4.7×10^{-11} mol/L.

Practice

4. State the rule for relating the certainty (number of significant digits) of a hydronium ion concentration to a pH measurement.

5. Knowing the pH of common substances is useful to a wide variety of people such as nutritionists, medical personnel, environmentalists, and consumers. Express each of the following concentrations as pH values.
   (a) grapefruit juice: $[H_3O^+(aq)] = 2.1 \times 10^{-3}$ mol/L
   (b) rainwater: $[H_3O^+(aq)] = 1 \times 10^{-5}$ mol/L
   (c) milk: $[H_3O^+(aq)] = 2.50 \times 10^{-7}$ mol/L
   (d) soap: $[H_3O^+(aq)] = 7.3 \times 10^{-9}$ mol/L

6. The technology of a pH meter provides an efficient way to obtain a pH that various people, such as chemists and biologists, often convert to a concentration. Calculate the hydronium ion concentration in each of the following household mixtures.
   (a) ammonia cleaner: pH = 11.3   (c) carbonated drink: pH = 4.2
   (b) salad dressing: pH = 2.65   (d) oven cleaner: pH = 13.755

7. As a result of an industrial accident, the concentration of hydronium ions in a small lake decreased one thousand times. How did the pH change? Did it increase or decrease?

8. What class of hazard and safety symbol is used on bottles of acids and bases? State the class letter and name, and describe or draw the symbol.

pOH and Hydroxide Ion Concentration

Although pH is used more commonly, in some applications, it may be more practical or convenient to describe hydroxide ion concentrations in a similar way. The definition of pOH follows the same format and the same certainty rule as pH. The **pOH** of a solution is the negative of the logarithm to the base 10 of the hydroxide ion concentration. Similarly, the hydroxide ion concentration is the negative of the exponent to the base 10 of the pOH:

$$\text{pOH} = -\log[\text{OH}^-(\text{aq})] \quad \text{and} \quad [\text{OH}^-(\text{aq})] = 10^{-\text{pOH}}$$

▸ COMMUNICATION example 3

The hydroxide ion concentration of a cleaning solution is determined to be 0.27 mol/L. What is its pOH?

Solution

$$\text{pOH} = -\log[\text{OH}^-(\text{aq})]$$
$$= -\log(0.27)$$
$$= 0.57$$

The pOH of the cleaning solution is 0.57.

▸ COMMUNICATION example 4

A sample of tap water has a pOH of 6.3. Calculate the hydroxide ion concentration.

Solution

$$[\text{OH}^-(\text{aq})] = 10^{-\text{pOH}}$$
$$= 10^{-6.3} \text{ mol/L}$$
$$= 5 \times 10^{-7} \text{ mol/L}$$

The hydroxide ion concentration of the tap water is 5×10^{-7} mol/L.

▸ Practice

9. A 4.5 mol/L solution of sodium hydroxide is prepared to clean a clogged drain. Calculate the pOH of this solution.
10. Soybean curd or tofu is one of the few foods that are basic. What is the hydroxide ion concentration in tofu if its pOH is 6.80?
11. The term "pH" is used in various technological contexts as well as in scientific fields of study. The term "pOH" seems to be used primarily in chemistry courses and materials. Suggest a reason for this practice.

DID YOU KNOW ?

Other Log Scales

The Richter scale for earthquakes is a log scale. An earthquake of magnitude 6 releases ten times as much energy as one of magnitude 5.

Another common example of a log scale is the decibel scale for sound intensity. Every change of 10 dB (decibels) corresponds to a change of 100 in the sound intensity (power per square metre), but for the average human, this difference is perceived as a change of a factor of two in the "loudness." So, a rock concert measured at 120 dB has a million times the intensity and sixty-four times the loudness of a normal conversation level of 64 dB.

▸ WEB Activity

Web Quest—Bad Hair Day?

Scientific research and technological developments are important in the development of hair products. Does pH matter? Many companies would like you to believe that it does.

www.science.nelson.com

SUMMARY: pH and pOH

$$pH = -\log[H_3O^+(aq)] \qquad pOH = -\log[OH^-(aq)]$$
$$[H_3O^+(aq)] = 10^{-pH} \qquad [OH^-(aq)] = 10^{-pOH}$$

- The number of digits following the decimal point in a pH or pOH value is equal to the number of significant digits in the corresponding hydronium or hydroxide ion concentration.
- For both pH and pOH, an inverse relationship exists between the ion concentration and the pH or pOH. For example, the greater the hydronium ion concentration, the lower the pH.

Section 6.2 Questions

1. Describe the main evidence that pure water contains hydronium and hydroxide ions.
2. What are some advantages of using pH compared to using the hydronium ion concentration?
3. Food scientists and dieticians measure the pH of foods when they devise recipes and special diets.
   (a) Copy and complete **Table 2**.
   (b) Based on pH only, predict which of the foods would taste most sour.

Table 2 Acidity of Foods

Food	$[H_3O^+(aq)]$ (mol/L)	pH
oranges	5.5×10^{-3}	
asparagus		8.4
olives		3.34
blackberries	4×10^{-4}	

4. If the pH is measured to be 0.0, is the hydronium ion concentration zero? Justify your answer.
5. The usefulness of many cleaning products depends, in part, on their basic nature.
   (a) Copy and complete **Table 3**.
   (b) Rank the products from least to most basic.
   (c) What safety precautions should you follow in handling acids and bases?

Table 3 Cleaning Products

Cleaning Solution	$[OH^-(aq)]$ (mol/L)	pOH
stain remover	6.7×10^{-2}	
baking soda		5.0
bleach	2.5×10^{-3}	
drain cleaner		0.1

6. An HCl(aq) solution with $[H_3O^+(aq)]$ of 1.0×10^{-2} mol/L is diluted by a factor of 1000. Determine the new pH.

7. Sketch a pOH scale similar to the pH scale (Figure 3, page 239). Label hydroxide ion concentrations in powers of 10, pOH, neutral point, and regions of acidic and basic solutions. (Examples are not required.)
8. If an acidic solution with a pH of 4 is completely neutralized, by what factor will the hydronium ion concentration change? Will it increase or decrease?
9. Most common acids and bases have a pH that falls in the range of 0 to 14. This range does not mean that other values do not exist. Calculate the pH of each of the following solutions.
   (a) a concentrated acid that has a hydronium ion concentration of 10 mol/L
   (b) a concentrated base that has a hydronium ion concentration of 1.6×10^{-15} mol/L
10. Design an experiment to determine if the vinegar from a fast-food outlet is normal household vinegar or diluted vinegar. State any controlled variables.
11. A scientist wants to determine the pH of several toothpastes, but must add water to the pastes in order to measure their pH with a pH meter. Critique this experimental design by supporting and defending the design, and by suggesting an alternative design.
12. Provide several examples of consumer products that have a relatively low pH and some that have a relatively high pH.

Extension

13. Research the effect of soil pH on deciduous versus evergreen plants. Report this information in a short margin note for a gardening magazine.

www.science.nelson.com

14. What is the normal pH range of human blood? If blood becomes too acidic, what symptoms appear? Can blood become too basic? If so, what happens? In both cases (when blood is too acidic or too basic), what medical technology is available to treat the problem?

www.science.nelson.com

Acid–Base Indicators 6.3

A number of synthetic dyes and compounds found in plants change colour when mixed with solutions of an acid or a base (**Figure 1**). Substances that change colour when the acidity of the solution changes are known as **acid–base indicators**. A very common indicator used in school laboratories is litmus, which is obtained from a lichen (see Figure 1, page 236). Litmus paper is prepared by soaking absorbent paper with litmus solution and then drying it. As you know, red and blue are the two colours of the litmus dye—red for acidic solutions and blue for basic solutions. Bromothymol blue and phenolphthalein are two other commonly used indicators (**Figure 2**).

Acid–base indicators are unique chemicals because they can exist in at least two forms, each with a distinctly different colour. The form of the chemical depends on the acidity of the solution. Acid–base indicators have very complicated molecular structures and formulas. Since it is inconvenient to write the actual formula, simple abbreviations are used instead. For example, litmus is abbreviated "Lt," bromothymol blue is "Bb," and "In" is a generic symbol for any indicator. The two forms of any indicator depend on whether a particular hydrogen atom is present in the indicator's molecule. In general, the lower pH form of the indicator is designated as "HIn(aq)" and the higher pH form as "In$^-$(aq)." Because the chemical structure of each indicator is different, the pH at which the indicator changes from the HIn(aq) form, with one colour, to the In$^-$(aq) form, with another colour, is different for each indicator.

Let's look more closely at litmus as an example. According to the table of acid–base indicators (see the inside back cover of this textbook), litmus changes colour between a pH of 6.0 and 8.0, which means that litmus is red at any pH less than or equal to 6.0 and is blue at any pH greater than or equal to 8.0 (**Figure 3**). Between 6.0 and 8.0, litmus is in the process of changing colour and you will see mixtures of red and blue. These intermediate colours are generally not useful, with the common exception of bromothymol blue, where an equal mixture of yellow and blue produces a distinct green in the middle of the pH range for this indicator (**Figure 2(a)**). The colours and pH of many other indicators have been measured and are reported in the table of acid–base indicators on the inside back cover of this textbook.

Acid–base indicators have two primary uses. As you will see in Chapter 8, Chemical Analysis, indicators are commonly used to mark the end of a titration. The other use that you will investigate below is to estimate the pH of a solution by using a number of different indicators. In this method, acid–base indicators are used to replace the more expensive pH meter, even though indicators are not as accurate.

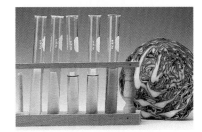

Figure 1
The test tubes show the colour of purple cabbage juice in solutions with different pHs, in order from left to right: 14, 9, 7, 4, and 1.

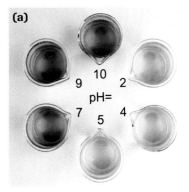

(a)

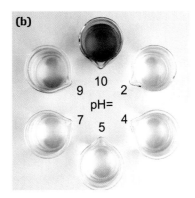

(b)

Figure 3
The colour changes of litmus are a little more complicated than what you have learned previously. There is a "fuzzy" region around the neutral point (pH = 7) where the colour is not easily distinguished. It is outside of this region where litmus is either definitely red (pH ≤ 6 in acidic solutions) or definitely blue (pH ≥ 8 in basic solutions).

Figure 2
Colour changes of common acid–base indicators
(a) bromothymol blue
(b) phenolphthalein

CAREER CONNECTION

Medical Laboratory Technologist

Medical laboratory technologists assist physicians in diagnosing and treating patients. They examine body fluids and tissues for the presence of disease. They can also specialize in transfusion medicine, or the microscopic examination of other types of cells and tissues. These diagnostic specialists are essential medical professionals.

Explore further to find out details on medical laboratory technologists, including where they are employed in your community and province.

www.science.nelson.com [GO]

▶ SAMPLE problem 6.3

Separate samples of an unknown solution were individually tested with several indicators. The solution containing methyl red turned red; the solution with thymol blue turned yellow; and the solution with methyl orange turned red. What is the most likely pH of the solution?

Refer to the table of acid–base indicators (inside back cover of the textbook).
- If methyl red turned red, then the pH must be ≤ 4.8.
- If thymol blue turned yellow, then the pH must be between 2.8 and 8.0. Notice that thymol blue is unusual because it has two colour changes over the normal pH range: yellow would mean a pH ≥ 2.8, but ≤ 8.0.
- If methyl orange turned red, then the pH must be ≤ 3.2.

Combining these results, the pH must be between 2.8 and 3.2. The hydronium ion concentration can then be calculated from your estimated pH, if required. This example shows that it is relatively simple and possible to determine the pH down to a small range, but this method is generally not as accurate as using a pH meter.

▶ COMMUNICATION example

In chemical analysis of separate samples of an unknown solution, phenolphthalein was colourless, bromothymol blue was blue, and phenol red was red. What is the estimated pH and hydronium ion concentration?

Solution

Indicator	Colour	pH
phenolphthalein	colourless	≤ 8.2
bromothymol blue	blue	≥ 7.6
phenol red	red	≥ 8.0

The pH is likely 8.1.

$[H_3O^+(aq)] = 10^{-8.1}$ mol/L
$= 8 \times 10^{-9}$ mol/L

The hydronium ion concentration is likely 8×10^{-9} mol/L.

pH Test Strips

Litmus is not the only indicator paper available. Bromothymol blue paper is sold to test aquarium water for pH values between 6.0 and 7.6, to a precision of about 0.1 pH unit. This precision is possible because subtle differences in colour over the range in which bromothymol blue changes can be matched to a colour comparison chart. Other test strips contain several different indicators and show different colours at different pH values. These test strips give a composite colour that can measure pH from 0 to 14 to within one to two pH units (**Figure 4**).

Figure 4
Comparing the colours of the test strip after dipping it into solution with the colours of the scale on the container gives an approximate pH. The inexpensive pH paper on the right bleeds (colour runs down the paper).

▶ mini Investigation pH of a Solution

In this activity, you will design an experiment to challenge your fellow students. Imagine that you have a solution with a pH known only to you. How can you give clues about its pH?

Materials: index cards or paper

- On one side of the card, write the pH of your solution.
- On the other side, write the names and colours (at that pH) of three or four indicators.
- Hand your card, "indicator" side up, to another student. See how close he or she can come to determining the pH of your solution without looking at the answer.

LAB EXERCISE 6.B

Using Indicators to Determine pH

One Design for determining the pH of a solution is testing the solution with indicators. Include a table of indicators and pH as part of your Analysis.

Report Checklist
- ○ Purpose
- ○ Problem
- ○ Hypothesis
- ○ Prediction
- ○ Design
- ○ Materials
- ○ Procedure
- ○ Evidence
- ● Analysis
- ○ Evaluation

Purpose
The purpose of this lab exercise is to use the concept of acid–base indicators and the reference table of indicator colours to determine the pH of three different solutions.

Problem
What is the approximate pH of three solutions?

Design
The solutions were labelled A, B, and C. Samples of each solution were tested with different indicators.

Evidence
Solution A: After addition to samples of the solution, methyl violet was blue, methyl orange was yellow, methyl red was red, and phenolphthalein was colourless.

Solution B: After addition to samples of the solution, indigo carmine was blue, phenol red was yellow, bromocresol green was blue, and methyl red was yellow.

Solution C: After addition to samples of the solution, phenolphthalein was colourless, thymol blue was yellow, bromocresol green was yellow, and methyl orange was orange.

SUMMARY Acid–Base Indicators

- Acid–base indicators are substances that change colour when the acidity of the solution changes.
- Acid–base indicators are unique chemicals because they can exist in at least two forms, each with a distinctly different colour. In general, the lower pH form of the indicator is "HIn(aq)" and the higher pH form is "In$^-$(aq)".
- The pH of a solution can be determined by comparing the resulting colours of several indicators in the solution with an indicator chart.

Section 6.3 Questions

1. According to the table of acid–base indicators on the inside back cover, what is the colour of each of the following indicators in the solutions of given pH?
   (a) phenolphthalein in a solution with a pH of 11.7
   (b) bromothymol blue in a solution with a pH of 2.8
   (c) litmus in a solution with a pH of 8.2
   (d) methyl orange in a solution with a pH of 3.9

2. Complete the Analysis for each of the following diagnostic tests. If the specified indicator is added to a solution, and the solution turns the given colour, then the solution's pH is __.
   (a) methyl red (red)
   (b) alizarin yellow (red)
   (c) bromocresol green (blue)
   (d) bromothymol blue (green)

3. Separate samples of a solution turned methyl orange and bromothymol blue indicators yellow, and a bromocresol green indicator blue.
   (a) Estimate the pH of the solution.
   (b) Calculate the approximate hydronium ion concentration of the solution.

4. Solving puzzles is a common feature of the scientific enterprise. Science and technology olympics often assign puzzles similar to this one: Design an experiment that uses indicators to identify which of three solutions, labelled X, Y, and Z, have pH values of 3.5, 5.8, and 7.8. There are several acceptable designs for this problem.

Extension

5. Measurement of pH is used in consumer, commercial, and industrial contexts. Make a list of different examples in which pH is likely used for each context. What different technologies exist for measuring pH, and how would the technology change with the context?

6. Using some of the concepts from Appendix B.4, design an experiment to test the role of pH in hair shampoos.

6.4 Explaining Acids and Bases

Learning Tip
Acids can have a variety of names, including classical and IUPAC names. For a review of acid nomenclature, see Section 1.6, in the Chemistry Review Unit.

In 1887, Svante Arrhenius introduced the first comprehensive theory of aqueous solutions. You have already seen his ideas about dissociation and ionization (Chapter 5, page 197). According to Arrhenius, ionic compounds dissociate into separate cations and anions when they dissolve to form a solution:

$$\text{ionic compound} \rightarrow \text{cation} + \text{anion}$$
$$\text{e.g., LiCl(s)} \rightarrow \text{Li}^+(aq) + \text{Cl}^-(aq)$$

Arrhenius proposed that bases are soluble ionic compounds that dissociate into a cation and the hydroxide ion, an anion:

$$\text{base} \rightarrow \text{cation} + \text{OH}^-(aq)$$
$$\text{e.g., LiOH(s)} \rightarrow \text{Li}^+(aq) + \text{OH}^-(aq)$$

Acids, according to Arrhenius, ionize in water to produce hydrogen ions plus an anion.

$$\text{acid} \rightarrow \text{H}^+(aq) + \text{anion}$$
$$\text{e.g., HCl(aq)} \rightarrow \text{H}^+(aq) + \text{Cl}^-(aq)$$

Arrhenius did not know that a hydrogen ion is better described as a hydronium ion—a hydrogen ion bonded to a water molecule. This discovery came much later. Nevertheless, it does not change his theory substantially because the "(aq)" part of $H^+(aq)$ can easily be interpreted to refer to some water molecule in the solvent.

Scientific knowledge is always subject to change. New evidence may arise unexpectedly or from the specific testing of generalizations, laws, and theories. If the new evidence supports the generalization, law, or theory, it strengthens existing support for it. If the new evidence contradicts the generalization, law, or theory, then changes are usually required.

INVESTIGATION 6.2 Introduction

Testing Arrhenius' Acid–Base Definitions

In this investigation, you will use Arrhenius' acid–base theory to make predictions, test these predictions using diagnostic tests, and finally, evaluate Arrhenius' theory. For simplicity, assume that Arrhenius' theory restricts dissociation and ionization to only two ions. In your Design, be sure to identify all variables, including any controls.

Purpose
The purpose of this investigation is to test Arrhenius' definitions of an acid and a base.

To perform this investigation, turn to page 260.

Report Checklist
- ○ Purpose
- ○ Problem
- ○ Hypothesis
- ● Prediction
- ● Design
- ○ Materials
- ● Procedure
- ● Evidence
- ● Analysis
- ● Evaluation (1, 2, 3)

Problem
Which of the substances tested may be classified as an acid, a base, or neutral, using Arrhenius' definitions?

A Revision of Arrhenius' Definitions

Evidence from Investigation 6.2 clearly indicates the limited ability of Arrhenius' definitions to predict acidic or basic properties of a substance in aqueous solution. Only five predictions that can reasonably be made using Arrhenius' definitions are verified: the acids are HCl(aq) and $CH_3COOH(aq)$, the bases are NaOH(aq) and $Ca(OH)_2(aq)$, and the neutral solution is NaNO$_3$(aq). Seven predictions were falsified. There were problems predicting the properties of solutions of compounds of polyatomic anions containing hydrogen, such as NaHCO$_3$(aq) and NaHSO$_4$(aq); solutions of oxides of metals and nonmetals, such as CaO(aq) or CO$_2$(aq); basic solutions that contain neither oxides nor hydroxides, such as NH$_3$(aq) and Na$_2$CO$_3$(aq); and acidic solutions such as Al(NO$_3$)$_3$(aq). Each of these substances fails to produce a neutral solution, as Arrhenius' definition would predict. Therefore, the theoretical definitions of acid and base need to be revised or replaced.

The ability of a theoretical concept to explain evidence is not valued as much as its ability to predict the results of new experiments. The best theories not only explain what is known, but enable correct predictions about new observations. Revising Arrhenius' acid–base definitions to explain the results of Investigation 6.2 involves two key ideas: (1) collisions of dissolved substances with water molecules and (2) the nature of the hydrogen ion. Because all substances tested are in aqueous solution, then particles will constantly be colliding with, and may also react with, the water molecules present.

The formation of acidic solutions by HCl may now be explained as a reaction with water molecules, resulting in hydronium ions being formed (**Figure 1**):

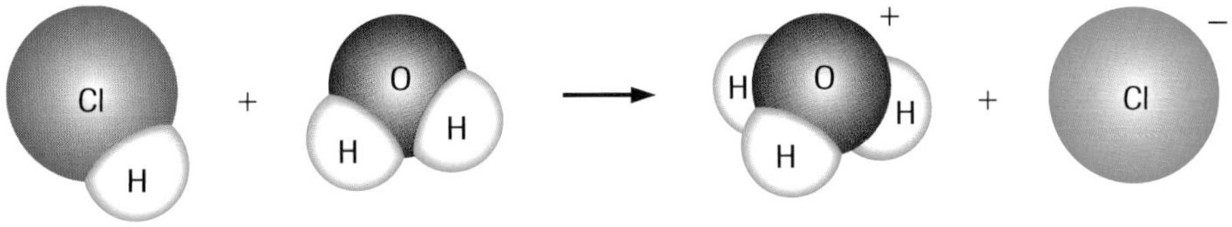

$$HCl(aq) + H_2O(l) \rightarrow H_3O^+(aq) + Cl^-(aq)$$

Figure 1
When gaseous hydrogen chloride dissolves in water, the HCl molecules are thought to collide and react with water molecules to form hydronium ions and chloride ions.

The idea of particles reacting with water produces some immediate improvements to Arrhenius' theory. First, Arrhenius could not explain how or why an acid molecule "falls apart" when it ionizes. Now, this reaction becomes like other chemical reactions—a collision between particles. Second, this idea includes the modern evidence for a hydronium ion (hydrated hydrogen ion) in a logical, balanced chemical equation.

Other substances whose solutions were found to be acidic can be explained in a similar way using this modified Arrhenius theory. For example, a solution of the hydrogen sulfate ion was shown to be acidic, and can be explained as a simple dissociation of sodium hydrogen sulfate into individual ions and then a reaction of the hydrogen sulfate ion with water:

$$NaHSO_4(s) \rightarrow Na^+(aq) + HSO_4^-(aq)$$

$$HSO_4^-(aq) + H_2O(l) \rightarrow H_3O^+(aq) + SO_4^{2-}(aq)$$

According to this modified Arrhenius theory, **acids** are substances that react with water to produce hydronium ions.

Reaction with water is also necessary to explain the behaviours of most bases. The main characteristic of bases is the production of hydroxide ions in solution. According to Arrhenius' original theory, bases produce hydroxide ions in solution by simple dissociation. For example:

$$Ca(OH)_2(s) \rightarrow Ca^{2+}(aq) + 2\,OH^-(aq)$$

In this example, there is no need to show a reaction with water because hydroxide ions are present as a result of the dissociation. As you have seen in Investigation 6.2, however, there are many other substances that are considered bases. For example, according to the evidence, ammonia and sodium carbonate form basic aqueous solutions. Arrhenius' theory did not help you to predict this result; nor does it explain this evidence. The

Learning Tip

The sodium ion is eliminated as a possible factor in acidic or basic solutions because many sodium compounds (e.g., NaCl(aq))—like all Group 1 ions—form neutral (pH 7) solutions.

EXTENSION

Oceans and CO$_2$
What is the effect of the increasing concentration of atmospheric carbon dioxide on the oceans? Researchers are investigating how the changing pH appears to be interfering with the growth of the shells of marine snails.

www.science.nelson.com

DID YOU KNOW?
Use of Theories
Chemists do not always use the most modern theories. As long as the essential idea is the same, different levels of theories are used in different applications. For example, the behaviour of acids in solution can be explained as either an ionization to produce hydrogen ions or a reaction with water to produce hydronium ions, depending on the context and level of sophistication required.

modified Arrhenius theory—the reaction-with-water theory—can explain the evidence, as illustrated in the chemical equations below:

$$NH_3(aq) + H_2O(l) \rightarrow OH^-(aq) + NH_4^+(aq)$$

Ammonia is considered a base because it produces hydroxide ions in solution. Sodium carbonate is an ionic compound with high solubility that dissociates in water to provide aqueous ions of sodium and carbonate. The basic character of carbonate ion solutions can also be explained as a reaction with water to produce hydroxide ions:

$$Na_2CO_3(s) \rightarrow 2\,Na^+(aq) + CO_3^{2-}(aq)$$

$$CO_3^{2-}(aq) + H_2O(l) \rightarrow OH^-(aq) + HCO_3^-(aq)$$

According to the modified Arrhenius theory, most **bases** are substances that react with water to produce hydroxide ions. Of course, if the base already contains hydroxide ions, a simple dissociation produces the hydroxide ions directly, as in calcium hydroxide.

SUMMARY: Writing Chemical Equations Using the Modified Arrhenius Theory

1. Write the chemical formulas for the reactants: molecule or polyatomic ion + water.
   - Ignore any Group 1 and 2 cations, and Group 7 anions in the compound formula. Evidence from many compounds shows that these ions do not produce acidic or basic solutions.
   - If the substance is a nonmetal oxide (e.g., CO$_2$(g), SO$_2$(g)), use two moles of water for every mole of the substance in the reactants.
2. Note the evidence provided. If the final solution is acidic, write hydronium ions as the first product. If the final solution is basic, then hydroxide ions are the first product.
3. Complete the other product by determining the combination of atoms and charge required to balance the chemical equation.
   - The other product should be a recognizable chemical formula—usually a polyatomic ion on your polyatomic ion chart (see the inside back cover).

COMMUNICATION example 1

In a test of the modified Arrhenius theory, a student tested the pH of a solution made by dissolving solid sodium cyanide in water, and found it to have a pH greater than 7. Can the modified Arrhenius theory explain this evidence? Provide your reasoning.

Solution

$$NaCN(s) \rightarrow Na^+(aq) + CN^-(aq)$$

$$CN^-(aq) + H_2O(l) \rightarrow OH^-(aq) + HCN(aq)$$

The modified Arrhenius theory can explain the basic character of a sodium cyanide solution because it is possible to write a balanced chemical equation with valid products, including the hydroxide ion.

COMMUNICATION example 2

Carbon dioxide is a major air pollutant from the combustion of fossil fuels. Suggest a possible chemical reaction that explains the acidity of a carbon dioxide solution.

Solution

$$CO_2(g) + 2 H_2O(l) \rightarrow H_3O^+(aq) + HCO_3^-(aq)$$

Practice

1. Use the modified Arrhenius theory to suggest a chemical reaction equation to explain the acidic properties of each of the following solutions:
   (a) HI(aq)   (b) HOCl(aq)   (c) H_3PO_4(aq)
2. Where possible, use the modified Arrhenius theory to explain the basic properties of each of the following solutions. Include appropriate chemical reaction equations:
   (a) Na_2SO_4(aq)   (b) $NaCH_3COO$(aq)   (c) $Sr(OH)_2$(aq)

DID YOU KNOW ?

Nonmetal Oxides in Water
Another interpretation for the production of an acidic solution by a nonmetal oxide is to consider a two-step reaction with water. For example,
$CO_2(g) + H_2O(l) \rightarrow H_2CO_3(aq)$
$H_2CO_3(aq) + H_2O(l) \rightarrow$
$\qquad H_3O^+(aq) + HCO_3^-(aq)$
The solution in Communication Example 2 is the sum of these two steps.

Scientists regard a theory as acceptable only if it can correctly predict results in new situations. The revisions made to Arrhenius' theoretical definitions do not offer reasons for these reactions, nor do they supply predictions in many situations. In other words, once you know the answer, you can usually explain it using the modified Arrhenius theory. Predicting the correct answer is always more difficult for any theory. Consider a solution formed by dissolving sodium hydrogen phosphate in water.

$$Na_2HPO_4(s) \rightarrow 2 Na^+(aq) + HPO_4^{2-}(aq)$$

Will the solution be acidic, basic, or neutral? We can write valid equations to predict that either hydronium ions or hydroxide ions will form when hydrogen phosphate ions react with water:

$$HPO_4^{2-}(aq) + H_2O(l) \rightarrow H_3O^+(aq) + PO_4^{3-}(aq)$$
or $\quad HPO_4^{2-}(aq) + H_2O(l) \rightarrow OH^-(aq) + H_2PO_4^-(aq)$

Nothing that you have studied so far in this textbook enables you to *predict* which of these equations is correct. If you know that the solution turns red litmus blue, however, then you can select one of the equations to *explain* the evidence. Clearly, the modified Arrhenius theory still needs some improvements.

Neutralization Reaction

You are familiar with the double replacement reaction generalization, which includes mostly precipitation reactions (Chapter 2, page 63). As you know, neutralization of an acid with a base is also a type of double replacement reaction. For example,

$$\underset{\text{acid}}{HCl(aq)} + \underset{\text{base}}{NaOH(aq)} \rightarrow \underset{\text{salt}}{NaCl(aq)} + \underset{\text{water}}{H_2O(l)}$$

According to the modified Arrhenius theory, acids produce hydronium ions in solution and bases produce hydroxide ions in solution. If we mix a solution of an acid with a solution of a base, we must be mixing solutions of hydronium and hydroxide ions. Therefore, when an acid neutralizes a base, the modified Arrhenius theory predicts

$$H_3O^+(aq) + OH^-(aq) \rightarrow 2 H_2O(l)$$

Neutralization can now be defined as the reaction between hydronium and hydroxide ions to produce water.

+ EXTENSION

Soil Acidity and Plant Growth
A fairly neutral soil pH is important for plant growth. At lower pH, metal ions can be released from the soil minerals. These ions can damage growing plants. Why is some soil acidic, and what can be done to reduce the ill effects? This audio clip tells you more.

Acids and Bases 251

Case Study

Acid Deposition

What exactly is acid deposition, or the more common term, acid rain? Like so many other issues related to science and technology, there are many perspectives and many technological solutions with various consequences. Here is a brief outline of different perspectives to help you organize your knowledge of acid deposition.

A Scientific Perspective

Normal precipitation, without any pollutants from human activity or catastrophic natural events like volcanic eruptions, is slightly acidic, having a pH of 5.6 or above. The acidity is the result of natural carbon dioxide dissolving in atmospheric moisture to form carbonic acid. Nitrogen oxides from lightning strikes and plant decay are other natural sources of acids in rain and other forms of precipitation.

Acid deposition is caused by nonmetal oxides. The formula NO_x represents several oxides of nitrogen, including $N_2O(g)$, $NO_2(g)$, and $N_2O_4(g)$. Likewise, SO_x represents sulfur oxides including $SO_2(g)$ and $SO_3(g)$, and CO_x represents gases such as $CO_2(g)$ and $CO(g)$.

Acid deposition includes any form of acid precipitation (rain, snow, or hail) and condensation from acid fog, as well as acid dust from dry air. A pH of less than 5.6 in the water from acid deposition is usually considered to be a potential problem. Empirical work indicates that the main causes of acid deposition in North America are sulfur dioxide and various nitrogen oxides, NO_x. These compounds react with water in the atmosphere to produce various acids. Oxides of sulfur and nitrogen are combustion products from sources such as automobiles, coal-burning power plants, and smelters. According to Statistics Canada, sulfur dioxide emissions have been declining across Canada, but nitrogen oxide emissions have remained relatively unchanged in the last decade.

An Ecological Perspective

Scientific research has repeatedly shown that acid deposition has increased the acidity of some lakes and streams to the point where aquatic life is depleted and waterfowl populations are threatened. Hundreds of lakes in Eastern Canada are now devoid of aquatic plant and animal life due to the lakes' high acidity from acid deposition. Some organisms are more susceptible than others to changes in acidity, but, eventually, the increased acidity (lower pH) leads to the "death" of a lake. Acid deposition has also been linked to forest decline in Eastern Canada and other parts of the world, but there is some dispute as to whether the major cause of the problem is particulates or acid precipitation (**Figure 2**).

If we see ourselves as an integral part of the environment, we may view environmental problems a little differently. These considerations are a feature of the Aboriginal worldview.

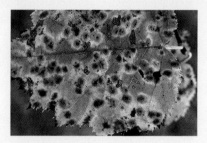

Figure 2
The effect of acid deposition on deciduous trees

An Economic Perspective

The environmental problems described above create economic problems. Acid deposition is endangering the fishing, tourism, agriculture, and forestry industries. Use of alternative energy sources or implementation of pollution-reducing technologies means spending money that consumers and industries may feel they cannot afford. If an industry shuts down, people lose jobs, and the cost of social assistance escalates. From the same perspective, future costs of doing nothing are likely to be staggering.

A Political Perspective

Political pressures and opinions have resulted in legislation limiting the production of sulfur oxides and nitrogen oxides. Some people argue that there should be even stiffer legislation to regulate industries and vehicle emissions, but others argue that there should be less government interference and voluntary industry compliance.

A Technological Perspective on Reducing Acid Deposition

Technologies are now available for the development and use of alternative fuels, the removal of sulfur from fossil fuels, and the recovery of oxides from exhaust gases.

Case Study Questions

Set up a research group to study technologies for reducing the sources of acid deposition.

In your group:

1. Prepare a list of possible technologies.
2. Assign one specific technological solution to each member or smaller sub-group. Using the library and the Internet, collect information about each technological solution, including its purpose, materials, processes, intended results, any information about how well it works, and any unintended consequences.

www.science.nelson.com

3. Compare the technologies. Are some technologies better than others? Why?
4. Assemble and communicate the information as directed by your teacher.

Section 6.4

The Modified Arrhenius Theoretical Definitions

- Acids are substances that react with water to produce hydronium ions.
- Most bases are substances that react with water to produce hydroxide ions.
- Neutralization can be explained as the reaction between hydronium ions and hydroxide ions to produce water.

Section 6.4 Questions

1. Identify two differences between the original Arrhenius definition of an acid and the modified Arrhenius definition.

2. Compare the original Arrhenius definition of a base with the modified Arrhenius definition. What is the same and what is different?

3. Evaluate the modified Arrhenius theory in terms of its ability to explain the results of Investigation 6.2.
   (a) Where possible, write a valid chemical equation to explain the evidence you collected.
   (b) Which substances can the modified Arrhenius theory not explain? Identify some questions that you would need to have answered in order to explain the evidence for these substances.
   (c) You know that Arrhenius' original theory did not work very well. In your opinion, does the modified Arrhenius theory get a "passing grade"? Why?

4. Test the explanatory power of the modified Arrhenius definitions by explaining the following evidence. For each of the following compounds, write a dissociation equation where appropriate, and then write a chemical equation showing reactions with water to produce either hydronium or hydroxide ions (consistent with the evidence):
   (a) HBr(g) in solution shows a pH of 2 on pH paper.
   (b) Na_3PO_4(s) forms a solution with a pH of 8.
   (c) $NaHSO_3$(s) in solution turns blue litmus red.
   (d) Na_2HPO_4(s) in solution turns red litmus blue.
   (e) Na_2O(s) in solution turns red litmus blue.
   (f) SO_3(g) in solution turns blue litmus red.
   (g) KOH(s) yields a solution with a pH of 12.

5. According to the modified Arrhenius theory, what is the balanced chemical equation for the reaction of aqueous nitric acid and aqueous potassium hydroxide?

6. Describe how the modified Arrhenius theory simplifies the many different examples that you have previously seen of the neutralization of an acid with a base.

7. What test do you put a theory through after testing its explanatory power?

8. Technological problems often have multiple solutions that involve different products and processes. All technological solutions or "fixes" have consequences. A decision to use one approach or another involves looking at the risks and benefits of each.

   Suppose a tanker truck, carrying a concentrated solution of sodium hydroxide destined for a pulp and paper company, crashes on the highway. Should this dangerous spill of a base be neutralized or diluted? Provide a brief list of pro and con arguments for each proposed solution.

Extension

9. Compare the animal skin tanning process traditionally used by Aboriginal peoples and a current process that uses synthetic chemicals. Outline the use of acids and bases in both processes. Evaluate the two technologies in terms of quality of the product and environmental impacts.

   www.science.nelson.com

6.5 The Strength of Acids and Bases

Acids and bases can be distinguished by means of a variety of properties (**Table 1**). Some properties of acids and bases are more useful than others to a chemist, especially those that can be used as diagnostic tests, such as the litmus or pH tests.

Table 1 Empirical Properties of Acids and Bases in Aqueous Solution

Acids	Bases
taste sour*	taste bitter and feel slippery*
turn blue litmus red	turn red litmus blue
have pH less than 7	have pH greater than 7
neutralize bases	neutralize acids
react with active metals to produce hydrogen gas	
react with carbonates to produce carbon dioxide	

*For reasons of safety, it is not appropriate to use taste or touch as diagnostic tests in the laboratory.

Acids are common substances. They occur naturally in many plants and animals. Technological products and processes employ acids in a wide variety of situations, from commonplace consumer products like cleaners, to large industrial processes like the fertilizer industry. Do all acids have the same properties and to the same degree?

DID YOU KNOW?
The Origin of "Acid"
The word "acid" comes from the Latin term "*acere*," which means "to make sour." A sour taste was likely associated with citrus fruits (citric acid) and sour wine (acetic acid).

INVESTIGATION 6.3 Introduction

Comparing the Properties of Acids (Demonstration)

Complete the Design by specifying the controlled variables. In the Evaluation, suggest several improvements to the Design.

Purpose
The purpose of this demonstration is to create the concept of strengths of acids.

Problem
How do the properties of two common acids compare?

To perform this investigation, turn to page 261.

Report Checklist
- ○ Purpose ● Design ● Analysis
- ○ Problem ○ Materials ● Evaluation (1, 3)
- ○ Hypothesis ○ Procedure
- ○ Prediction ● Evidence

Design
The properties of aqueous solutions of two acids, hydrochloric acid and acetic acid, are observed and compared. Diagnostic tests include the litmus test, conductivity test, and reaction with an active metal. Two important controlled variables are ...

Strong and Weak Acids

The evidence from Investigation 6.3 clearly shows that acids with the same initial concentration can have different degrees of acidic properties. The difference in degree of acidity was shown by the conductivity measurements and the differences in the rates of reaction. Controlling variables is very important when comparing properties such as conductivity and pH. For example, a very dilute solution of a strong acid could have a higher pH (lower hydronium ion concentration) than a more concentrated solution of a weak acid. Therefore, when comparing acid strengths, it is important that the concentration and temperature of the acids be controlled.

In terms of its empirical properties, an acid is described as a *weak acid* if its characteristic properties under the same conditions are less than those of a common strong acid, such as hydrochloric acid. There are relatively few strong acids; hydrochloric, sulfuric, and nitric acids are the most common strong acids. Most common acids are weak acids. The concept of strong and weak acids was developed to describe and explain the differences in properties of acids.

You can explain the differences in properties between strong and weak acids using the modified Arrhenius theory and the extent of reaction with water. For example, hydrochloric acid is a strong acid with a high conductivity, a high rate of reaction with active metals and carbonates, and a relatively low pH. These empirical properties suggest that there are many ions present in the solution and, specifically, a high concentration of hydronium ions. This evidence would be consistent with the idea that a **strong acid** reacts completely (more than 99%) with water to form hydronium ions. The high concentration of $H_3O^+(aq)$ ions gives the solution strong acid properties and a low pH.

$$HCl(aq) + H_2O(l) \xrightarrow{>99\%} H_3O^+(aq) + Cl^-(aq)$$

According to this chemical equation for the reaction of hydrochloric acid with water, for each mole of hydrochloric acid that reacts with water, about one mole of hydronium ions is produced.

A weak acid is an acid that has a relatively low conductivity, a lower rate of reaction with active metals and carbonates, and a relatively high pH compared with strong acids. This evidence suggests the presence of fewer hydronium ions in the acidic solution. Based on this evidence, a **weak acid** reacts incompletely with water to form relatively few hydronium ions. Measurements of pH indicate that most weak acids react much less than 50%. Acetic acid, a common weak acid, is a typical example. The relatively low concentration of hydronium ions gives the solution weaker acid properties and a pH closer to 7:

$$CH_3COOH(aq) + H_2O(l) \xrightarrow{<50\%} H_3O^+(aq) + CH_3COO^-(aq)$$

Under the same conditions, weak acids have fewer hydronium ions and are, therefore, less acidic. For this and other reasons, weak acids are generally much safer to handle than strong acids. You can even eat and drink many of them (**Figure 2**)! Lactic acid is found in dairy products, butanoic acid is found in rancid butter, citric acid is found in citrus fruits, oxalic acid is found in rhubarb, and long-chain fatty acids, such as stearic acid, are found in animal fats. Most of the acids you are likely to encounter are classed as weak acids.

DID YOU KNOW?

Aboriginal "Herbal Aspirin"—A Weak Acid
Aboriginal peoples of North America have a rich history of herbal medicine. In fact, most of the top 10 herbs used medically today were first used by Aboriginal peoples. Herbal medicine is part of their traditional knowledge—empirical knowledge developed over thousands of years. For example, Aboriginal peoples traditionally used the white willow (**Figure 1**) to obtain a substance that relieves pain and lowers fevers.

From a Western science perspective, this substance is called salicin. Salicin is converted to salicylic acid (closely related to acetylsalicylic acid, known as Aspirin) in the body. Salicylic acid is a weak acid from the same family of acids as the simpler acetic acid.

Figure 1
A white willow

Figure 2
Many naturally occurring acids are weak acids.

▶ Practice

1. State two diagnostic tests that can be used to distinguish between strong and weak acids. Word each test in the format, "*If* [procedure], *and* [evidence], *then* [analysis]." (See Appendix C.4.)
2. What is the key idea used to explain the difference between strong and weak acids?
3. State three common examples of strong acids.
4. A laboratory solution of hydrochloric acid has a concentration of 0.15 mol/L.
   (a) Calculate the concentration of hydronium ions in this solution.
   (b) Calculate the pH of this solution.
5. A laboratory solution of acetic acid has a concentration of 0.15 mol/L. Will the pH of this solution be lower or higher than the pH of the hydrochloric acid solution in the previous question? Briefly explain your answer.

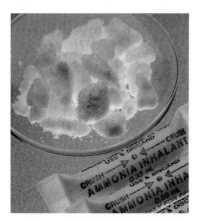

Figure 3
Ammonia gas is released from smelling salts such as ammonium carbonate. Ammonia irritates the nasal passages and stimulates a sharp intake of breath, rousing someone who feels faint or has fainted.

Strong and Weak Bases

Evidence indicates that there are both strong bases (such as sodium hydroxide) and weak bases (such as ammonia). For equal concentrations of solutions, strong bases have high electrical conductivity and very high pH (>>7); whereas weak bases have low electrical conductivity and pH closer to (but greater than) 7. We can explain the behaviour of strong bases as a dissociation in an aqueous solution to increase the hydroxide ion concentration. For example, if you dissolve some solid lye (sodium hydroxide) to make a solution for cleaning a drain, the sodium hydroxide dissociates completely as it dissolves.

$$NaOH(s) \rightarrow Na^+(aq) + OH^-(aq)$$

From this equation, for every one mole of sodium hydroxide that dissolves, one mole of hydroxide ion is produced in the solution. This mole ratio explains why bases like sodium hydroxide have such strong basic properties. Further evidence indicates that all soluble ionic hydroxides are **strong bases**: 100% of dissolved ionic hydroxides dissociate to release hydroxide ions. In general,

$$MOH(aq) \rightarrow M^+(aq) + OH^-(aq)$$

where M represents a metal ion.

What about weak bases? How can we explain their properties? The pure compounds (such as ammonia gas, $NH_3(g)$, **Figure 3**) do not contain hydroxide ions, so they cannot dissociate to release hydroxide ions. Nevertheless, pH measurements above 7 indicate that solutions of weak bases contain hydroxide ions in a higher concentration than in pure water. The explanation is found in the modified Arrhenius theory: **a weak base** reacts partially (usually much less than 50%) with water to produce relatively few hydroxide ions compared with a similar amount of a strong base, which accounts for the weaker basic properties of weak bases. Most bases, other than soluble ionic hydroxides, are weak bases. Weak bases may be either ionic or molecular compounds in their pure state. In general,

$$B(aq) + H_2O(l) \xrightarrow{<50\%} OH^-(aq) + \text{balancing entity}$$

where B represents a base (molecule or ion).

+ **EXTENSION**

Strong/Weak and Concentrated/Dilute
"Concentrated" and "dilute" are terms used to describe solutions qualitatively. Learn more about the distinction between these terms and "strong" and "weak."

www.science.nelson.com

> ▶ **SAMPLE problem 6.4**
>
> Explain the evidence that an ammonia solution (window cleaner) is a basic aqueous solution, as demonstrated by a litmus paper test (Investigation 6.2).
>
> According to the modified Arrhenius theory, ammonia reacts with water to produce hydroxide ions, which explains the evidence that the solution is basic:
>
> $$NH_3(aq) + H_2O(l) \rightarrow OH^-(aq) + ?$$
>
> Determine the second product by balancing the atoms and charge in the chemical reaction equation.
> Using the generalization that only soluble ionic hydroxides are strong bases, ammonia must be a weak base. According to the modified Arrhenius theory, this means less than 50% reaction with water.
> The final chemical equation for the reaction of ammonia with water illustrates the theory to explain the evidence:
>
> $$NH_3(aq) + H_2O(l) \xrightarrow{<50\%} OH^-(aq) + NH_4^+(aq)$$
>
> The presence of hydroxide ions explains why the solution is basic, and the less-than-complete reaction explains the weak base properties. (Note that both atoms and charge are conserved in the balanced equation.)

SAMPLE problem 6.5

Explain the weak base properties of baking soda.

Sodium hydrogen carbonate (baking soda, **Figure 4**) is an ionic compound with high solubility that dissociates in water to produce aqueous ions of sodium and hydrogen carbonate:

$$NaHCO_3(s) \rightarrow Na^+(aq) + HCO_3^-(aq)$$

The sodium ion cannot be responsible for the basic properties of the solution, because many sodium compounds (such as NaCl(aq)) form neutral solutions. The basic character of hydrogen carbonate ion solutions can be explained as resulting from their reaction with water:

$$HCO_3^-(aq) + H_2O(l) \xrightarrow{<50\%} OH^-(aq) + H_2CO_3(aq)$$

The presence of hydroxide ions explains the basic properties of sodium hydrogen carbonate, and the incomplete reaction explains the weak base properties.

Figure 4
Baking soda is a common household chemical, but it requires an uncommonly sophisticated theory to explain or predict its properties.

COMMUNICATION example

Solid sodium acetate is dissolved in water. The final solution is tested and found to have a pH of about 8. Explain this evidence by writing balanced chemical equations.

Solution

$$NaCH_3COO(s) \rightarrow Na^+(aq) + CH_3COO^-(aq)$$

$$CH_3COO^-(aq) + H_2O(l) \xrightarrow{<50\%} OH^-(aq) + CH_3COOH(aq)$$

We now have explanations for the production of hydroxide ions by bases: Strong bases (commonly ionic hydroxides) dissociate completely in solution to produce hydroxide ions; and weak bases partially react with water to increase the hydroxide ion concentration of the solution (modified Arrhenius theory).

Practice

6. Suppose you had two aqueous solutions with the same solute concentration, one containing a strong base and the other a weak base. Describe several ways to distinguish between these two solutions using diagnostic tests.

7. How would strong and weak bases be distinguished using pOH values? What variables would need to be controlled?

8. Describe how you can distinguish a strong base from a weak base using
   (a) the formula of the solute in the solution
   (b) the modified Arrhenius theory

9. Write balanced chemical equations to show all changes that occur when solid sodium sulfate dissolves in water to produce a solution that turns red litmus paper blue.

Polyprotic Substances

Some acids have only one acidic hydrogen atom in their compound formula (HA) and can react only once with water to produce hydronium ions. These acids are called **monoprotic acids**. Most strong acids, such as HCl(aq), as well as many weak acids, such as HCN(aq), are monoprotic. There are, however, some acids that contain more than one acidic hydrogen in their compound formulas (H_xA) and can react more than once with water. These acids are called **polyprotic acids**. For example, phosphoric acid, H_3PO_4(aq), found in cola soft drinks (**Figure 5**) and rust removers, initially reacts with water to produce hydronium ions:

$$H_3PO_4(aq) + H_2O(l) \xrightarrow{<50\%} H_3O^+(aq) + H_2PO_4^-(aq)$$

This reaction is less than 50% complete; therefore, phosphoric acid is a weak acid. The dihydrogen phosphate ion, a product of this first reaction with water, can react with an additional water molecule to produce more hydronium ions, but this reaction is even less complete (<1%):

$$H_2PO_4^-(aq) + H_2O(l) \xrightarrow{<1\%} H_3O^+(aq) + HPO_4^{2-}(aq)$$

Similarly, the hydrogen phosphate ion produced in this second reaction with water can in principle react again, but this reaction does not occur to any appreciable extent:

$$HPO_4^{2-}(aq) + H_2O(l) \xrightarrow{\sim 0\%} H_3O^+(aq) + PO_4^{3-}(aq)$$

With all of these possible reactions, you might think that phosphoric acid solutions would be very acidic, but only the first reaction is significant. Phosphoric acid is a weak acid: The pH of its solution is noticeably higher than the pH of a strong acid at the same concentration. Based on pH and other evidence, the first reaction is already incomplete and the subsequent reactions become even less complete. *In general, polyprotic acids are weak acids whose reaction with water decreases with each successive step.*

Acid	Concentration	pH
HCl(aq)	0.1 mol/L	1.0
H_3PO_4(aq)	0.1 mol/L	1.7

An important exception to the rule that polyprotic acids are weak acids is sulfuric acid. It is a strong acid because the first reaction with water is essentially complete. The second reaction, however, is much less than 50% complete.

The same concepts also apply to weak bases. Some bases, like CH_3COO^-(aq), are **monoprotic bases**, meaning they can react with water only once to produce hydroxide ions. Other bases, like CO_3^{2-}(aq), are **polyprotic bases** because they can react more than once with water, but all reactions with water are much less than 50%:

$$CO_3^{2-}(aq) + H_2O(l) \xrightarrow{<50\%} OH^-(aq) + HCO_3^-(aq)$$

$$HCO_3^-(aq) + H_2O(l) \xrightarrow{<1\%} OH^-(aq) + H_2CO_3(aq)$$

Evidence shows that a carbonate ion solution is less basic than a strong base at the same concentration. *In general, polyprotic bases are weak bases whose reaction with water decreases with each successive step.*

Base	Concentration	pH
NaOH(aq)	0.1 mol/L	13.0
Na_2CO_3(aq)	0.1 mol/L	11.4

Polyprotic substances are generally weak acids and weak bases that undergo multiple reactions with water to produce hydronium ions and hydroxide ions, respectively. In all cases, the successive reactions with water are increasingly incomplete.

Figure 5
A common ingredient in a solution used to remove rust is phosphoric acid, which surprisingly is also a common ingredient in cola soft drinks.

Learning Tip

Monoprotic bases are sometimes called monobasic bases; polyprotic bases are sometimes called polybasic bases, which could mean dibasic, tribasic, etc. This nomenclature is often used on consumer or commercial products.

Learning Tip

A base like barium hydroxide, $Ba(OH)_2$(aq), is not a polyprotic base because there is no possibility of successive reactions with water. Barium hydroxide dissociates to produce hydroxide ions. It does not matter that there are two moles of hydroxide ions produced in the dissociation of one mole of barium hydroxide. When barium hydroxide reacts with a strong acid, there is only one reaction that can take place: between hydroxide and hydronium ions.

SUMMARY: Strong and Weak Acids and Bases

Table 2 Strong and Weak Acids and Bases in Aqueous Solution

	Strong acids	Weak acids	Strong bases	Weak bases
empirical properties (same c and t)	very low pH ($\ll 7$)	medium to low pH (<7)	very high pH ($\gg 7$)	medium to high pH (>7)
	high conductivity	low conductivity	high conductivity	low conductivity*
	fast reaction rate	slow reaction rate	fast reaction rate	slow reaction rate
modified Arrhenius theory	completely react with water to form H_3O^+(aq)	partially react with water to form H_3O^+(aq)	completely dissociate to form OH^-(aq)	partially react with water to form OH^-(aq)

* applies only to weak bases that are molecular

Section 6.5 Questions

1. List three common examples of a strong acid and one common example of a strong base.
2. In a sentence, summarize the empirical evidence for strong and weak acids and bases.
3. According to the modified Arrhenius theory, explain the difference between
   (a) strong and weak acids   (b) strong and weak bases
4. Acids and bases can be concentrated or dilute, and strong or weak. Illustrate the meaning of these terms by sketching a diagram for each of the following solutions. Use simple symbols to represent molecules and ions, and dots to represent water molecules.
   (a) a concentrated strong acid
   (b) a dilute strong acid
   (c) a concentrated weak acid
   (d) a dilute weak acid
5. The pH values of two acids with the same concentration and temperature are 2 and 6. Which is likely the strong acid and which is likely the weak acid? Justify your answer.
6. Solid sodium fluoride dissolves in water to produce a solution with a pH of 8. Write chemical equations to show the changes that have occurred.
7. Define the term "polyprotic." State some examples of polyprotic acids and bases.
8. How does the completeness of the reaction change in successive reactions of a polyprotic substance with water?
9. Write at least three diagnostic tests to determine which of two unlabelled solutions is acetic acid or hydrochloric acid. Word each test in the format, "*If* [procedure], *and* [evidence], *then* [analysis]." (See Appendix C.4.)
10. Describe the precautions that should be taken for safe handling and disposal of acids and bases.
11. Write the Design (including variables) and the complete Analysis of the report for the following investigation:

    Purpose
    The purpose of this investigation is to compare the strengths of several acids.

    Problem
    What is the order of several common acids in terms of decreasing strength?

    Evidence
    Table 3 Acidity of 0.10 mol/L Acids

Acid solution	Formula	pH
hydrochloric acid	HCl(aq)	1.00
acetic (ethanoic) acid	CH_3COOH(aq)	2.89
hydrofluoric acid	HF(aq)	2.11
nitric acid	HNO_3(aq)	1.00
hydrocyanic acid	HCN(aq)	5.15

12. Identify the strong acids in **Table 3**. Justify your answer with appropriate chemical equations and calculations.
13. Many products found in our homes contain acids and bases. Find examples of these products and make a list of the acids and bases they contain. Where possible, identify the acids and bases as either strong or weak.

    Extension

14. Many cleaning products are often available in concentrated or dilute form. They may also contain different types of acids or bases. Using this general example, describe the importance of understanding the terms "strength" and "concentration." Include at least two perspectives.

Chapter 6 INVESTIGATIONS

INVESTIGATION 6.1

Properties of Acids and Bases

Report Checklist
- ○ Purpose
- ○ Problem
- ○ Hypothesis
- ● Prediction
- ● Design
- ● Materials
- ● Procedure
- ● Evidence
- ● Analysis
- ● Evaluation (1, 2, 3)

In this investigation, you will use your previous knowledge of properties of substances, and practise your problem-solving skills. You are provided with solutions of approximately equal concentrations, at the same temperature, of the following pure substances: $CaCl_2(s)$, $C_3H_4OH(COOH)_3(s)$, $C_6H_{12}O_6(s)$, $Ca(OH)_2(s)$, $NH_3(g)$, $NaHSO_4(s)$, $CH_3OH(l)$, $H_2SO_4(l)$, $Na_2CO_3(s)$. Remember to include variables in your Design, and safety and disposal instructions in your Procedure.

 Solutions may be toxic, irritant, or corrosive; avoid eye and skin contact.

Purpose
The purpose of this investigation is to test previous knowledge about the properties of acids and bases.

Problem
What properties are most useful for distinguishing acids and bases from other classes of compounds?

INVESTIGATION 6.2

Testing Arrhenius' Acid–Base Definitions

Report Checklist
- ○ Purpose
- ○ Problem
- ○ Hypothesis
- ● Prediction
- ● Design
- ○ Materials
- ● Procedure
- ● Evidence
- ● Analysis
- ● Evaluation (1, 2, 3)

In this investigation, you will use Arrhenius' acid–base theory to make predictions, test these predictions using diagnostic tests, and finally, evaluate Arrhenius' theory. For simplicity, assume that Arrhenius' theory restricts dissociation and ionization to only two ions. In your Design, be sure to identify all variables, including any controls.

Purpose
The purpose of this investigation is to test Arrhenius' definitions of an acid and a base.

Problem
Which of the substances tested may be classified as an acid, a base, or neutral, using Arrhenius' definitions?

Materials
lab apron
eye protection
conductivity apparatus
blue litmus paper
red litmus paper
any other materials necessary for diagnostic tests

aqueous 0.10 mol/L solutions of:
  hydrogen chloride (a gas in solution)
  acetic acid (vinegar)
  sodium hydroxide (lye, caustic soda)
  calcium hydroxide (slaked lime)
  ammonia (cleaning agent)
  sodium carbonate (washing soda, soda ash)
  sodium hydrogen carbonate (baking soda)
  sodium hydrogen sulfate (toilet bowl cleaner)
  calcium oxide (lime)
  carbon dioxide (carbonated beverages)
  aluminium nitrate (salt solution)
  sodium nitrate (fertilizer)

 Solutions may be toxic, irritant, or corrosive; avoid eye and skin contact.

INVESTIGATION 6.3

Comparing the Properties of Acids (Demonstration)

Report Checklist
○ Purpose ● Design ● Analysis
○ Problem ○ Materials ● Evaluation (1, 3)
○ Hypothesis ○ Procedure
○ Prediction ● Evidence

Complete the Design by specifying the controlled variables. In the Evaluation, suggest several improvements to the Design.

Purpose
The purpose of this demonstration is to create the concept of strengths of acids.

Problem
How do the properties of two common acids compare?

Design
The properties of aqueous solutions of two acids, hydrochloric acid and acetic acid, are observed and compared. Diagnostic tests include the litmus test, conductivity test, and reaction with an active metal. Two important controlled variables are …

Materials
lab apron
eye protection
2 150 mL beakers
2 petri dishes
conductivity apparatus or probe
steel wool
scissors
overhead projector
red and blue litmus paper
magnesium ribbon (about 5 cm)
100 mL of 1.0 mol/L $HCl(aq)$
100 mL of 1.0 mol/L $CH_3COOH(aq)$
distilled water bottle
baking soda

 Solutions may be corrosive; avoid eye and skin contact.

Procedure

1. Pour equal volumes of each acid into separate, clean beakers.
2. Dip a piece of red and a piece of blue litmus paper into each acid and record colour changes.
3. Rinse the conductivity probe with distilled water and insert into one acid. Record evidence of conductivity.
4. Repeat step 3 for the other acid.
5. Place two clean petri dishes on the stage of the overhead projector.
6. Carefully fill each dish about half full with each acid.
7. Clean a strip of magnesium ribbon with steel wool and cut into two equal lengths (about 2 cm each).
8. Add a piece of magnesium simultaneously to each petri dish.
9. Note and record evidence of the rate of reaction.
10. Neutralize the acids with sodium hydrogen carbonate (baking soda) and dispose of solutions into the sink with lots of running water.

Chapter 6 SUMMARY

Outcomes

Knowledge
- recall the empirical definitions of acidic, basic, and neutral solutions determined by using indicators, pH, and electrical conductivity (6.1)
- calculate H_3O^+(aq) and OH^-(aq) concentrations, pH, and pOH of acid and base solutions based on logarithmic expressions (6.2)
- use appropriate SI units to communicate the concentration of solutions and express pH and concentration to the correct number of significant digits (6.2)
- compare magnitude changes in pH and pOH with changes in concentration for acids and bases (6.2)
- explain how the use of indicators, pH meters, or pH paper can be used to measure [H_3O^+(aq)] (6.3)
- use the modified Arrhenius theory to define acids as substances that produce H_3O^+(aq) in aqueous solutions and bases as substances that produce OH^-(aq) in aqueous solutions and recognize that the definitions are limited (6.4)
- define neutralization as a reaction between hydronium and hydroxide ions (6.4)
- differentiate between strong acids and bases and weak acids and bases, qualitatively, using the modified Arrhenius (reaction with water) theory and dissociation (6.5)
- compare the reaction with water (ionization) of monoprotic with that of polyprotic acids and bases (6.5)

STS
- state that the goal of technology is to provide solutions to practical problems (all sections)
- recognize that solutions to technological problems may have both intended and unintended consequences (all sections)

Skills
- initiating and planning: design a procedure to determine the properties of acids and bases (6.1, 6.5); design an experiment to differentiate between weak and strong acids, and between weak and strong bases (6.1, 6.3, 6.4); describe procedures for safe handling, storing, and disposal of materials (6.1, 6.3, 6.4, 6.5)
- performing and recording: construct and analyze a table or graph comparing pH and hydronium ion concentration (6.2)
- analyzing and interpreting: use a pH meter (or paper) and indicators to determine acidity and pH (6.1, 6.3, 6.4, 6.5)
- communication and teamwork: work collaboratively to assess technologies (6.4)

Key Terms

6.1
hydronium ion

6.2
pH
pOH

6.3
acid–base indicator

6.4
acid (modified Arrhenius)
base (modified Arrhenius)
neutralization

6.5
strong acid
weak acid
strong base
weak base
monoprotic acid
polyprotic acid
monoprotic base
polyprotic base

Key Equations

$$pH = -\log[H_3O^+(aq)] \qquad [H_3O^+(aq)] = 10^{-pH} \qquad (6.2)$$

$$pOH = -\log[OH^-(aq)] \qquad [OH^-(aq)] = 10^{-pOH} \qquad (6.2)$$

▶ MAKE a summary

1. Create a concept map starting with "Acids and Bases" in the centre of a page. Include empirical and theoretical classification, all of the key terms and equations, as well as any other information you think may be useful in studying for a test on this chapter.

2. Refer back to your answers to the Starting Points questions at the beginning of this chapter. How has your thinking changed?

▶ Go To

The following components are available on the Nelson Web site. Follow the links for *Nelson Chemistry Alberta 20–30*.
- an interactive Self Quiz for Chapter 6
- additional Diploma Exam-style Review questions
- Illustrated Glossary
- additional IB-related material

There is more information on the Web site wherever you see the Go icon in this chapter.

Chapter 6 REVIEW

Unit 3

Many of these questions are in the style of the Diploma Exam. You will find guidance for writing Diploma Exams in Appendix H. Exam study tips and test-taking suggestions are on the Nelson Web site. Science Directing Words used in Diploma Exams are in bold type.

DO NOT WRITE IN THIS TEXTBOOK.

Part 1

1. Which set of properties best describes an acidic solution?

	conductivity	litmus	pH
A.	yes	blue	8
B.	yes	red	4
C.	no	red	5
D.	no	blue	7

2. What is the modern replacement for the original concept of a hydrogen ion in an acidic solution?
   A. $H^+_{(aq)}$
   B. $H_2^+_{(aq)}$
   C. $H_2O^+_{(aq)}$
   D. $H_3O^+_{(aq)}$

3. A window cleaning solution has a hydroxide ion concentration of 2.4×10^{-4} mol/L. The pOH of this solution is _____.

4. An acidic fruit juice has a hydronium ion concentration of 1.45×10^{-4} mol/L. The correctly reported pH is
   A. 3.8
   B. 3.84
   C. 3.839
   D. 3.8386

5. In a chemical analysis the hydronium ion concentration of a solution can be determined by all of the following *except*
   A. pH paper
   B. a pH meter
   C. indicators
   D. conductivity apparatus

6. A solution with a pH of 11.83 has a hydronium ion concentration of $a.b \times 10^{-cd}$ mol/L. The values of *a*, *b*, *c*, and *d* are ___ ___ ___ ___.

7. If a water sample test shows a pH of 5, by what factor would the hydronium ion concentration have to increase or decrease to completely neutralize the sample?
   A. increase by a factor of 2
   B. decrease by a factor 2
   C. increase by a factor of 100
   D. decrease by a factor of 100

Use the following information to answer questions 8 and 9.

As part of a chemical analysis, an unknown solution was tested and found to have a pH of 6.8.

8. What is the colour, in order, of the following indicators in this solution?
   methyl red
   chlorophenol red
   bromothymol blue
   methyl orange

   Using the colour codes: 1 = green, 2 = red, 3 = yellow, 4 = other colour, the list of numbered colour codes, in order of indicators, is ___ ___ ___ ___.

9. The hydronium ion concentration in this solution is
   A. 2×10^{-7} mol/L
   B. 8×10^{-6} mol/L
   C. 6.3×10^{6} mol/L
   D. 1.58×10^{-7} mol/L

10. The theoretical property that all strong acids have in common is that they all
    A. form very concentrated solutions
    B. react partially with water to form hydronium ions
    C. react completely with water to form hydronium ions
    D. are polyprotic acids forming multiple hydronium ions

11. Solutions of four different acids with the same concentration at the same temperature have the following characteristics:
    Acid 1 pH = 4
    Acid 2 pH = 1
    Acid 3 $[H_3O^+_{(aq)}] = 10^{-3}$ mol/L
    Acid 4 $[H_3O^+_{(aq)}] = 10^{-6}$ mol/L

    When the acids are arranged, in order, from strongest acid to weakest acid, the order is ___ ___ ___ ___. (Record all four digits.)

Part 2

12. Complete the Analysis and Evaluation (design only) for the following investigation:

    Problem
    Which of the chemicals, numbered 1 to 7, is $KCl_{(s)}$, $Ba(OH)_{2(s)}$, $Zn_{(s)}$, $C_6H_5COOH_{(s)}$, $Ca_3(PO_4)_{2(s)}$, $C_{25}H_{52(s)}$ (paraffin wax), and $C_{12}H_{22}O_{11(s)}$?

    Design
    Equal amounts of each chemical are added to equal volumes of water. The mixtures are tested for their conductivity and effect on litmus paper.

Evidence

Table 1 States of Matter of Elements

Chemical	Solubility in water	Conductivity of solution	Litmus paper test
1	high	none	no change
2	high	high	no change
3	none	none	no change
4	high	high	red to blue
5	none	none	no change
6	none	none	no change
7	low	low	blue to red

13. If you are given a measurement of the pH of a solution, what determines the certainty (number of significant digits) of the calculated hydronium ion concentration?

14. **Determine** the hydronium ion concentration in each of the following solutions.
    (a) cleaning solution with pH = 11.562
    (b) fruit juice with pH = 3.5

15. One sample of rainwater has a pH of 5, while another sample has a pH of 6. **Compare** the hydronium ion concentrations in the two samples.

16. Can a pOH equal zero? **Justify** your answer mathematically.

17. Separate samples of an unknown solution turned both methyl orange and bromothymol blue indicators to yellow, and turned bromocresol green indicator to blue.
    (a) Estimate the pH of the unknown solution.
    (b) **Determine** its approximate hydronium ion concentration.

18. Acids and bases may be either strong or weak.
    (a) **Design** an experiment to distinguish between strong and weak acids, and strong and weak bases. Include appropriate diagnostic tests and controlled variables.
    (b) Use the modified Arrhenius theory to **explain** how strong and weak acids differ, and how strong and weak bases differ.

19. Write appropriate chemical equations to **explain** the acidic or basic properties of each of the following substances when added to water:
    (a) sodium sulfide (basic)
    (b) hydrogen bromide (acidic)
    (c) potassium hydroxide (basic)
    (d) benzoic acid, $C_6H_5COOH_{(aq)}$ (acidic)

20. There are many factors that determine the effectiveness of a cleaner. Considering that most food is acidic and most cleaners are basic, what is the type of possible chemical reaction? **Illustrate** your answer with a chemical equation.

21. Citrus fruits such as oranges, lemons and grapefruit contain citric acid, $H_3C_6H_5O_7(aq)$ which is a weak polyprotic acid.
    (a) **Why** is citric acid called a polyprotic acid?
    (b) Write chemical equations for all possible reactions when pure citric acid dissolves in water.
    (c) In general, state how the reaction with water changes with each successive step.

22. Determining the acidity of a solution is an important task in many applications such as scientific research, industrial processes, and environmental studies. What technologies are available to determine the acidity of a solution? Use at least three different criteria to **compare** the technologies you identified.

Extension

23. Many chemicals that are potentially toxic or harmful to the environment have maximum permissible concentration levels set by government legislation.
    (a) If the chemical is dangerous, should the limit be zero?
    (b) Is a zero level theoretically possible?
    (c) Is a zero level empirically measurable?
    (d) If a non-zero limit is set, what groups should have input into this decision and how do you think this limit should be determined?

24. What is acid deposition? Your report should include
    • the names of the acids that are typically present
    • whether the acids are strong or weak
    • whether it is possible to predict which acids have a greater effect in the environment, with reasons

www.science.nelson.com

Unit 3 REVIEW

Many of these questions are in the style of the Diploma Exam. You will find guidance for writing Diploma Exams in Appendix H. Exam study tips and test-taking suggestions are on the Nelson Web site. Science Directing Words used in Diploma Exams are in bold type.

www.science.nelson.com GO

DO NOT WRITE IN THIS TEXTBOOK.

Part 1

1. A solution such as tap water is a
   A. pure substance
   B. compound
   C. homogeneous mixture
   D. heterogeneous mixture

Use this information to answer questions 2 and 3.

Iodine dissolved in alcohol is used as an external medication for its disinfecting and antibacterial properties. To study these effects, 4.52 g of pure iodine is dissolved in 150 mL of alcohol.

2. Evidence for the dissolving of iodine can be obtained from all of the following **except** the
   A. change in conductivity of the solution
   B. change in colour of the solution
   C. change in density of the solution
   D. heat transferred during the dissolving process

3. The amount concentration of the iodine solution is
   NR _____ mmol/L.

4. A sample of lake water contains 405 mg of dissolved minerals in 2.50 L of lake water. The concentration of dissolved minerals in parts per million is _____ ppm.

5. Dilution of stock solutions is an essential laboratory skill. If a lab technician needs to prepare 2.00 L of 2.50 mol/L ammonia solution, what volume of 14.8 mol/L concentrated ammonia will he require?
   A. 11.8 mL
   B. 33.8 mL
   C. 169 mL
   D. 338 mL

6. In the salt tank of a water softening system, excess sodium chloride forms a layer of solid below the salt solution. According to equilibrium theory, the explanation for the constant properties of the salt solution is that
   A. the concentration of the salt solution has become constant
   B. the rate of dissolving has become equal to the rate of crystallizing
   C. sodium chloride can no longer dissociate because the solution is saturated
   D. all of the water has been used up in dissolving the salt already present in the solution

Use this information to answer questions 7 and 8.

In a chemical analysis, a solution was found to have a hydronium ion concentration of 2×10^{-11} mol/L.

7. When this concentration is converted to a pH, the pH of the solution is _____.

8. Based on the pH, this solution may be classified as
   A. acidic
   B. basic
   C. neutral
   D. ionic

9. A cleaning solution with a pOH of 12.17 has a hydroxide ion concentration of $a.b \times 10^{-cd}$ mol/L. The values of *a, b, c,* and *d* are ___ ___ ___ ___.

10. As part of a chemical analysis of a window cleaning solution, the following evidence was obtained by testing samples of the cleaning solution with different indicators.
    - Phenol red turned red.
    - Both phenolphthalein and thymolphthalein were colourless.
    - Bromothymol blue was blue.

    The most likely pH of the solution is
    A. 10.0
    B. 9.4
    C. 8.1
    D. 7.6

11. If a solution of a strong acid is diluted by a factor of 10, the pH of the solution
    A. increases by one pH unit
    B. decreases by one pH unit
    C. increases by a factor of 10
    D. decreases by a factor of 10

12. One goal of chemical technology is to
    A. test current scientific laws and theories
    B. produce new chemical theories
    C. explain the nature of solutions
    D. solve practical problems

13. According to the modified Arrhenius theory, acids
    A. ionize into hydrogen ions
    B. react with water to produce hydronium ions
    C. react with water to produce hydrogen ions
    D. cause water to dissociate into hydrogen ions

14. A solution of sodium methanoate, NaHCOO(aq), has a pH of 8.3. Based on the modified Arrhenius theory, the chemical equation that explains this evidence is
    A. $HCOO^-(aq) \rightarrow H^+(aq) + CO_2(g)$
    B. $HCOO^-(aq) \rightarrow OH^-(aq) + CO(g)$
    C. $HCOO^-(aq) + H_2O(l) \rightarrow H_3O^+(aq) + COO^{2-}(aq)$
    D. $HCOO^-(aq) + H_2O(l) \rightarrow OH^-(aq) + HCOOH(aq)$

15. The following evidence was collected in an experiment to identify solutions at the same concentration and temperature.

    Table 1 Experimental Evidence

Solution	Conductivity	pH
1	low	2.9
2	low	11.1
3	none	7.0
4	high	13.0
5	high	1.0
6	high	7.0

    Identify, in order, the number of the solution that likely represents a
    - strong acid
    - weak acid
    - strong base
    - weak base

    ___ ___ ___ ___

16. According to the modified Arrhenius theory, a neutralization is a reaction between
    A. an acid and a carbonate
    B. an acid and an active metal
    C. hydrogen ions and hydroxide ions
    D. hydronium ions and hydroxide ions

17. When a polyprotic acid such as boric acid, $H_3BO_3(aq)$, reacts with water, the completeness of each successive reaction step
    A. increases
    B. decreases
    C. stays the same
    D. has no particular pattern

Part 2

18. **Distinguish** between the following terms:
    (a) homogeneous and heterogeneous mixtures
    (b) solute and solvent
    (c) electrolytes and nonelectrolytes
    (d) endothermic and exothermic dissolving

19. **Define** each of the following substances empirically:
    (a) acid
    (b) base
    (c) neutral ionic substance
    (d) neutral molecular substance

20. **How** is a hydronium ion different from an aqueous hydrogen ion? **How** is it similar?

21. Provide some examples from home or from your chemistry lab experience that illustrate the need to dissolve substances before they can react.

22. One brand of bottled water contains 150 mg of calcium in a 2.00 L bottle. **Determine** the concentration of calcium in
    (a) parts per million
    (b) moles per litre

23. A bottle of household vinegar is labelled 5% V/V acetic acid. **Determine** the minimum volume of vinegar that contains 60 mL of acetic acid.

24. Convert the following:
    (a) 0.35 mol of NaCl(aq) in 1.5 L of solution into an amount concentration
    (b) 25 mL of 0.80 mol/L $Mg(NO_3)_2$(aq) into a chemical amount
    (c) 0.246 mol of NH_3(aq) in a 2.40 mol/L solution into a volume of solution
    (d) 25.00 g of $CuCl_2$(s) in 1.20 L of solution into an amount concentration
    (e) 50.0 mL of 0.228 mol/L Na_2CO_3(aq) into a mass of Na_2CO_3(s)

25. If water is added to a 25.0 mL sample of 2.70 g/L NaOH(aq) until the volume becomes 4.00 L, **determine** the concentration of the final solution.

26. Cement floors, especially in workshops, are often protected with an epoxy coating. Before applying this coating, the cement floor needs to be cleaned thoroughly. One part of this procedure usually involves an acid etching of the cement using hydrochloric acid (**Figure 1**).
    (a) **Determine** the volume of 11.6 mol/L hydrochloric acid required to prepare 45 L of 3.5 mol/L solution.
    (b) The preparation of this solution is potentially dangerous. **Plan** a procedure, including appropriate safety instructions, for preparing this solution.

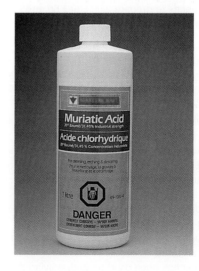

Figure 1
Hydrochloric acid is sold as muriatic acid.

27. Write dissociation equations and calculate the amount concentration of the cations and anions in each of the following solutions:
    (a) 2.24 mol/L Na₂S(aq)
    (b) 0.44 mol/L Fe(NO₃)₂(aq)
    (c) 0.175 mol/L K₃PO₄(aq)
    (d) 8.75 g of cobalt(III) sulfate in 0.500 L of solution

28. **Compare** the ways in which concentrations of solutions are expressed in chemistry laboratories, consumer products, and environmental studies. Provide at least one example for each situation and include a possible reason for this choice.

29. In your own words, **describe** the rule relating the number of digits of a pH to the certainty (number of significant digits) of the hydronium ion concentration.

30. **Determine** the pH of each of the following solutions:
    (a) lemon juice with $[H_3O^+(aq)] = 7.5 \times 10^{-3}$ mol/L
    (b) $[HNO_3(aq)] = 2.5 \times 10^{-3}$ mol/L

31. **Determine** the hydronium ion concentration in each of the following solutions:
    (a) cleaning solution with pH = 11.562
    (b) fruit juice with pH = 3.5

32. **Sketch** a graph showing the relationship between pH and the hydronium ion concentration. **Describe** the relationship in a sentence.

33. State two main ways in which a theory or theoretical definition may be tested. Which method is the more stringent test?

34. **How** is the knowledge of pH useful in aquatic environments? **How** does this knowledge relate to both natural and manufactured products? Briefly **describe** two examples where knowledge of pH is used to solve practical problems.

35. A household cleaning solution has a pH of 12 and some fruit juice has a pH of 3.
    (a) What is the hydronium ion concentration in each solution?
    (b) **Compare** the concentration of hydronium ions in the fruit juice to the concentration of hydronium ions in the cleaning solution. How many times more concentrated is the hydronium ion in the juice than in the cleaning solution?

36. **Design** an experiment to determine which of six acids are strong acids and which are weak acids. Assume that you have solutions of known concentration available and all common laboratory equipment.

37. Hydrangeas (**Figure 2**) are garden shrubs that may produce blue, purple, or pink flowers. Research has indicated that their colour depends on the pH of the soil: blue at pH 5.0–5.5, purple at pH 5.5–6.0, and pink at pH 6.0–6.5. What colour of flower would be produced for each of the following soil acidities?
    (a) $[H_3O^+(aq)] = 5 \times 10^{-7}$ mol/L
    (b) $[H_3O^+(aq)] = 7.9 \times 10^{-6}$ mol/L

Figure 2
Hydrangea flowers

38. The term "weak" is sometimes used in non-scientific applications to mean dilute solutions. Why do we have to be careful not to use "weak" in this context when referring to acids and bases?

Use this information to answer questions 39 to 41.

Standard solutions of potassium hydrogen tartrate, $KHC_4H_4O_6(s)$, are used in chemical analysis to precisely determine the concentration of bases, such as sodium hydroxide. In one particular analysis, 100.0 mL of a 0.150 mol/L solution is required.

39. **Determine** the mass of potassium hydrogen tartrate that is required.

40. **Plan** a complete procedure for the preparation of the standard solution, including specific quantities and equipment.

41. Consult the MSDS for potassium hydrogen tartrate and sodium hydroxide and note precautions for handling these substances.

42. In a neutralization reaction of a strong acid and a strong base, how do the pH and pOH values change from their initial to their final values? **Justify** your answer by including a balanced chemical equation.

43. Laboratory safety rules require students to wear eye protection when handling acids, such as hydrochloric acid and sulfuric acid, yet dilute boric acid, $H_3BO_3(aq)$, is an ingredient in eye drops sold in drugstores (**Figure 3**). Although borates have been used for thousands of years in China, boric acid was not synthesized until 1702 in France. Soon after its synthesis, its mild antiseptic properties were discovered. **Explain** why some acids are more harmful than others. Your response should include
    - at least two named examples of acids
    - balanced chemical equations
    - typical values of percent reaction with water

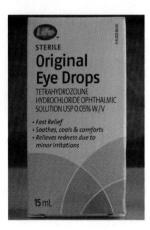

Figure 3
Boric acid is used in eye drops as a preservative and a buffer.

Use this information to answer questions 44 and 45.

Each of the following substances was dissolved in water to form a 0.1 mol/L solution, and then the pH of each solution was measured:
(a) $HCN(aq)$; pH = 5
(b) $HNO_3(aq)$; pH = 1
(c) $NaNO_2(aq)$; pH = 8
(d) $Sr(OH)_2(aq)$; pH = 13

44. Using the modified Arrhenius theory, write chemical equations to explain the pH evidence for each substance.

45. **Identify** the strong and weak acids and bases.

Use this information to answer questions 46 and 47.

Acids, such as nitric acid and nitrous acid, formed from the reaction of nitrogen oxides with water in the atmosphere, are components of acid deposition (such as acid rain).

nitric acid: $HNO_3(aq)$ nitrous acid: $HNO_2(aq)$

46. **Plan** an experiment to compare the acidity of nitric and nitrous acids. Your response should include
    - an experimental design, with variables
    - a list of materials required for the experiment
    - a procedure indicating all steps, plus safety and disposal instructions

47. What are some negative effects of acid deposition? Provide several brief statements from a number of perspectives.

Use this information to answer questions 48 to 52.

Glycolic acid (hydroxyacetic acid, $CH_2OHCOOH(s)$) has many different uses, such as in household cleaning solutions (such as CLR®, **Figure 4**), and personal care products (such as skin creams). Aboriginal peoples used natural glycolic acids found in animal parts, such as brains, to tan hides. Glycolic acid is often used where a common strong acid would not be suitable for safety reasons. A 1.00 mol/L solution of glycolic acid has a pH of 1.92.

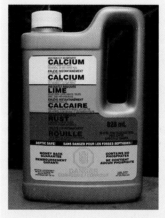

Figure 4
Glycolic acid is used in many household products, such as this bottle of cleaning solution.

48. Write the procedure and list the materials required to prepare glycolic acid solution and measure its pH as precisely as possible.

49. **Determine** the hydronium ion concentration of the solution.

50. Using the modified Arrhenius theory, write a chemical reaction equation to **explain** the acidity of glycolic acid.

51. If glycolic acid were a strong acid, what would its pH be? **Justify** your answer.

52. Suggest several other reasons why glycolic acid might be preferred in its many uses to a strong acid such as hydrochloric acid.

Use this information to answer questions 53 and 54.

Baking soda is one of the most versatile chemicals known. If you were to be stranded on an isolated island, baking soda would be one useful chemical to have available.

53. The acid–base properties of a baking soda solution are not easy to predict but can be explained using the modified Arrhenius theory. Knowing that a solution of baking soda turns litmus from red to blue, write chemical reaction equations, starting with solid baking soda, to explain the litmus evidence.

54. In a small group and using references, brainstorm a list of uses for baking soda. **Identify** any uses that involve acid–base reactions.

55. What are some benefits and risks of using acidic and basic solutions in your home? Try to be specific about the risks by using WHMIS and Household Hazardous Product information. **Illustrate** your answer with some examples where you consider the benefits to exceed the risks, and other examples where you consider the risks to exceed the benefits.

56. Vinegar, a dilute solution of acetic acid, has a long history in human civilization. The word "vinegar" originates from a Latin word meaning "sour wine," which also indicates how it was made.
    (a) Acetic acid is a common weak acid. Using the modified Arrhenius theory, write a chemical equation that explains the acidity of vinegar.
    (b) If the pH of a sample of vinegar is 2.4, **predict** the hydronium ion concentration.
    (c) If the sample of vinegar in (b) had a solute concentration of 1 mol/L, how does your answer to (b) show that acetic acid is a weak acid?
    (d) Research the history of the uses of vinegar. **Enumerate** as many different uses as possible.

57. The goal of technology is to provide solutions to practical problems. Most people associate technology with modern times, but technology, as practical products and processes, is as old as human civilization. Use the library or the Internet to research the modern-day precursor to Aspirin—the most widely used drug in the world—and how it was extracted and used by Aboriginal peoples in North America.

58. Review the focusing questions on page 186. Using the knowledge you have gained from this unit, briefly **outline** a response to each of these questions.

Extension

59. Sulfuric acid, a strong acid commonly found in car batteries, is the largest-volume industrial chemical produced worldwide. Research and report, in pairs or small groups, on sulfuric acid production in industrialized countries. Your response should include
    - why the volume of sulfuric acid consumed used as a measure of a country's degree of industrialization
    - the names of some of the processes and products that involve sulfuric acid

60. In the media, especially movies, acids are often portrayed as dangerous, with the ability to "burn through" or "eat away" almost anything. **Evaluate** this portrayal. Your response should include
    - the accuracy of the typical media portrayal of acids
    - justification of your evaluation with personal experience, examples, and explanations
    - a list of the most dangerous acids, with reasons
    - suggestions regarding how the media could more accurately portray the reactivity of acids

61. Many acids occur in nature. One common example is formic acid. Prepare a profile of formic acid. Your response should include
    - natural occurrence in plants and animals
    - general function in nature
    - chemical formula
    - safety and handling
    - some technological applications

62. The development of new pharmaceuticals usually involves acute toxicity tests on animals.
    (a) **Define** acute toxicity. **Describe** the general testing procedure used.
    (b) Animal testing is a controversial issue. **Identify** some positive and negative perspectives on this issue, including Aboriginal perspectives.

unit 4
Quantitative Relationships in Chemical Changes

Chemical substances are part of our lives at every level—medicines, fuels, plastics, fertilizers to grow our food; the list seems endless. This unit focuses on how chemistry and technology interact to allow us to identify, measure, produce, and use chemical substances.

Quantitative relationships in chemical processes have always been important to humankind. One of the earliest written records we have (preserved on clay tablets) is a pharmacopeia, on which a Sumerian physician recorded the proper amounts of ingredients for making prescriptions—more than 4000 years ago! Interestingly, one ingredient listed often is the ionic compound, potassium nitrate. This shows that the Sumerians had some knowledge of basic chemistry at that time. Similarly, food recipes emphasize the critical importance of using proper quantities. Oral traditions ensured that Canada's Aboriginal peoples retained knowledge of proper proportions of ingredients for making the essential high-energy and long-lasting food staple called pemmican.

As science explored the quantities and proportions of substances that were involved in chemical reactions, technology developed to create, monitor, and control processes using them. This technology is applied by industry, commerce, and consumers to improve quality of life and to solve problems.

When science and technology develop, there are always more problems and questions created by the new information, processes, and skills. Understanding is the key to being able to evaluate the risks and benefits of chemical substances, and this begins with knowledge of how to identify and measure quantities of substances in chemical reactions.

As you progress through the unit, think about these focusing questions:

- How do scientists, engineers, and technologists use mathematics to analyze chemical changes?
- How are balanced chemical equations used to predict yields in chemical reactions?

Unit **4**

GENERAL OUTCOMES

In this unit, you will
- explain how balanced chemical equations indicate the quantitative relationships between reactants and products involved in chemical changes
- use stoichiometry in quantitative analysis

Quantitative Relationships in Chemical Changes

Unit 4
Quantitative Relationships in Chemical Changes

ARE YOU READY?

These questions will help you find out what you already know, and what you need to review, before you continue with this unit.

Knowledge

1. For each of the following combinations of reagents, predict the products, and write a complete, balanced chemical reaction equation:
   (a) Calcium chloride and sodium carbonate solutions are mixed (**Figure 1**).
   (b) Zinc metal reacts in hydrochloric acid (**Figure 2**).
   (c) Water undergoes simple decomposition.
   (d) Iron burns in pure oxygen (**Figure 3**).
   (e) Copper metal is placed in an aqueous solution of silver nitrate (**Figure 4**).

Figure 1   Figure 2   Figure 3   Figure 4

Prerequisites

Concepts
- chemical formulas
- balanced reaction equations
- dissociation and ionization
- molar mass
- molar volume
- amount concentration
- ideal gas law
- acids and bases
- pH
- indicators

Skills
- WHMIS
- SI notation
- diagnostic tests

You can review prerequisite concepts and skills on the Nelson Web site, in the Chemistry Review unit, and in the Appendices.
   A Unit Pre-Test is also available online.

2. Many reactions will only occur in aqueous solution. Assume that each of the substances below is placed in water. Rewrite the formula for each substance, including the physical state, to indicate whether it has high or low solubility in water at SATP. Where appropriate, write a dissociation or ionization equation.
   (a) $Ca(NO_3)_2$
   (b) $PbCl_2$
   (c) HCl
   (d) $NaOH$

3. As part of an environmental analysis, the pH of a sample of lake water was measured to be 4.8.
   (a) Calculate the hydronium ion amount concentration in the lake water.
   (b) What would be the colours of the following indicators if placed in this lake water: methyl orange, bromothymol blue, phenolphthalein?

4. The molar mass of any substance is a useful conversion factor because it allows us to understand laboratory measurements of mass in terms of the chemical amount and vice versa. Copy and complete **Table 1**.

Table 1 Mass and Amount Conversions for Solids or Liquids

Formula	Mass (g)	Chemical amount (mol)
$(NH_4)_3PO_4(s)$	44.00	
$CH_3COOH(l)$		0.058

5. For reactions that occur in solution, the most useful conversion factor is the concentration, expressed as an amount concentration. Copy and complete **Table 2**.

Table 2 Volume, Concentration, and Amount Conversions for Solutions

Formula	Volume (L)	Amount concentration (mol/L)	Chemical amount (mol)
$NaOH(aq)$	2.20	0.500	
$HCl(aq)$		11.6	0.0400
$Na_2SO_4(aq)$	0.655		0.740

6. Calculating chemical amounts for gases often involves a number of variables. Sometimes a molar volume at STP or SATP can be used, but more often the ideal gas law is required. Copy and complete **Table 3**.

Table 3 Pressure, Temperature, Volume, and Amount Conversions for Gases

Formula	Pressure (kPa)	Temperature (K)	Volume (L)	Amount (mol)
$CH_4(g)$	101.325 (exact)	273.15 (exact)	13.7	
$UF_6(g)$	400	400	1.00	
$CO_2(g)$	100	298		2.0
$Ar(g)$	100	298	4.00	

Table 4 Matching WHMIS Symbols

Symbol	Class: Category
(a)	Class B: Flammable and Combustible Materials
(b)	Class C: Oxidizing Materials
(c)	Class D: Toxic Materials Immediate and Severe
(d)	Class F: Dangerously Reactive Materials

STS Connections

7. Provide one example of a chemical product or process that shows how chemical technology is used to solve practical problems.

Skills

8. Write a brief experimental design, including diagnostic tests, to distinguish between neutral ionic, neutral molecular, acidic, and basic solutions.

9. Match each WHMIS symbol in **Table 4** to what it represents in the second column.

10. Identify the substance in each test shown in **Figure 5**.

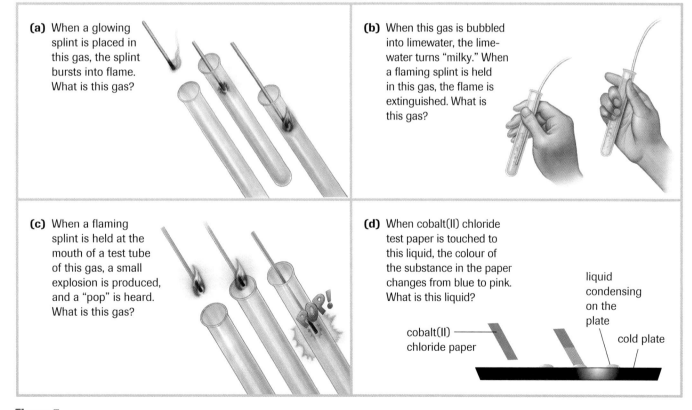

(a) When a glowing splint is placed in this gas, the splint bursts into flame. What is this gas?

(b) When this gas is bubbled into limewater, the limewater turns "milky." When a flaming splint is held in this gas, the flame is extinguished. What is this gas?

(c) When a flaming splint is held at the mouth of a test tube of this gas, a small explosion is produced, and a "pop" is heard. What is this gas?

(d) When cobalt(II) chloride test paper is touched to this liquid, the colour of the substance in the paper changes from blue to pink. What is this liquid?

Figure 5
Diagnostic tests

chapter 7

Stoichiometry

In this chapter

- Exploration: The Problem Is What You Don't See!
- Web Activity: Roberta Bondar
- Investigation 7.1: Decomposing Malachite
- Lab Exercise 7.A: Testing the Stoichiometric Method
- Investigation 7.2: Gravimetric Stoichiometry
- Lab Exercise 7.B: Testing a Chemical Process
- Investigation 7.3: Producing Hydrogen
- Case Study: Producing Hydrogen for Fuel Cells
- Investigation 7.4: Analysis of Silver Nitrate (Demonstration)
- Lab Exercise 7.C: Testing Solution Stoichiometry
- Lab Exercise 7.D: Determining a Solution Concentratian

Understanding quantitative relationships of reactants and products in chemical reactions is very important in chemical technology. The industrial production of fertilizer (**Figure 1**), the combustion of fuels, the treatment of water, and even our personal use of antacids are just a few examples. For each of these examples it is necessary to understand reacting quantities in order to understand the technology behind these products and processes.

In a general sense, food preparation is also chemical technology—products are produced from raw materials, using processes that are often based on chemical reactions. A recipe is a procedure that includes specific quantities and steps to be used to obtain the desired product. Characteristic of all technologies, you do not have to understand the science behind the process to be successful. However, if you want to explain the technology, the chemistry concepts become very important.

The laboratory study of chemical reactions requires simple technology: Chemists need to be able to identify the products when known substances react, often using the diagnostic tests that you already know. The study of chemical quantities used and/or produced in a reaction requires slightly more sophisticated technology and the skill to use it to make accurate measurements. A prediction of quantities also depends on scientific knowledge, such as balanced chemical reaction equations, chemical amounts, and their relationship to the chemical equation.

In this chapter, you will build on your previous knowledge of chemical formulas, chemical equations, and amount calculations. You will study reactions to interpret reaction equations and to calculate and predict the effects of controlling quantities of chemicals involved in a reaction.

💡 STARTING Points

Answer these questions as best you can with your current knowledge. Then, using the concepts and skills you have learned, you will revise your answers at the end of the chapter.

1. Are all individual entities in the reactants important when a chemical reaction occurs?
2. What is the simple relationship between the quantities of any two reactants and products in a chemical reaction?
3. Compare the procedures used for measuring and calculating the quantities of solids, liquids, gases, and solutions involved in chemical reactions.

Career Connections:
Aerospace Engineer; Chemical Engineer; Chemical Technologist; Soil Scientist

Figure 1
This Agrium™ industrial plant site near Redwater, Alberta, combines science and technology to produce nitrogen-based fertilizers using natural gas. Agrium™ operates other Alberta plants at Carseland, Joffre, and Fort Saskatchewan. These plants produce a gross combined total of nearly two and a half million tonnes of ammonia per year, as well as many other highly soluble nitrogen-rich compounds.

▶ Exploration   The Problem Is What You Don't See!

For many chemical reactions, the substances involved are invisible. This presents unique challenges in detecting, measuring, and calculating amounts of chemicals that you might not even notice. A balanced equation written for the combustion reaction of a typical hydrocarbon compound found in candle wax, $C_{17}H_{36}$, involves three substances that are invisible.

Materials: wax candle (about 2 cm in diameter), 250 mL Erlenmeyer flask, stopper to fit the flask, 600 mL beaker, cobalt(II) chloride test paper, limewater solution, knife

- Write and balance a reaction equation that describes the reaction of "burning" the candle.
- Cut the candle bottom level, so that when the candle is sitting on a flat surface, the candle flame height will be about 3 cm lower than the height of the 600 mL beaker.
- Light the candle. When it is burning well, invert the beaker over it and allow the flame to go out.
- Wipe the mist that forms in the beaker with a strip of cobalt(II) chloride test paper.
- Repeat the lighting and extinguishing procedure using a 500 mL Erlenmeyer flask.

 Take precautions when working near a flame: tie back long hair and keep clothing away from the flame.

- When the flame goes out, lift the flask and turn it upright. Add 10 to 20 mL of limewater to the flask, stopper the flask, and swirl and shake the solution.
- Relight the candle. Holding the beaker upright, lower it until the bottom is about halfway down the flame height. Move the beaker back and forth in the flame for a few seconds, and then remove it.

(a) What product can you detect by condensing it to visible liquid droplets?
(b) What is observed in the cobalt(II) chloride test, and how does this evidence verify your answer to (a)?
(c) What predicted product is not detectable by condensation?
(d) What is observed in the limewater test, and how does this evidence verify the answer to (c)?
(e) What invisible reactant is assumed to be involved in this reaction? What evidence supports this assumption?
(f) What is the black substance on the bottom of the beaker? Why isn't this substance predicted by the reaction equation?
(g) Is the black product a major or minor product? What would you need to know to answer this question?

7.1 Interpreting Chemical Reaction Equations

Science and technology are different activities, but they are mutually interdependent. Either may lead the other in a cycle of expanded knowledge and new abilities to do things. Problems solved by new technology naturally make scientists curious to know and explain why the technology works. The scientific activity that results in new knowledge often results in that knowledge being turned to practical uses by engineers.

Often in our society, and especially in the media, there is a tendency to confuse technology with gadgetry and to assume that the term "technology" just refers to the manufactured devices we use. But there is much more to it than that. Technology also includes the organized processes and skills we develop for doing things. For example, using alphabetical order in a filing system is a technology we use to make it easy to locate the files. In fact, using any alphabet is an incredibly important technology process, just as communicating by human speech is a critical technological skill.

The goal of science is to understand and explain the natural world. In addition, science

- is an *international* discipline
- is involved with *natural* products and processes
- is more *theoretical* in its approach (often based on pure imagination)
- emphasizes *ideas* and *concepts* over practical applications

Scientific concepts and theories are evaluated by how well they *describe, explain,* and *predict* natural phenomena. They do this on the basis of whether the concepts and theories are logical, supported by evidence, consistent with other theories, simple, and testable. The last point is critical, for a concept that cannot be tested, by definition, lies outside the realm of science.

The goal of technology is to provide solutions to practical problems. In addition, technology

- is often more *localized* in use
- is involved with *humanly developed* processes and products
- is more *empirical* in its approach (often based on pure trial-and-error experience)
- emphasizes *methods* and *materials* over understanding

Technological skills, products, and processes are evaluated by how well they *work* to solve practical problems. They do this on the basis of whether the skills, products, and processes are simple, reliable, efficient, sustainable, and economical. A trial-and-error approach can be an effective way to get an unknown system working. It is basically how we learn to walk. The trial-and-error approach is also the system we use to become proficient at a new video game. Anything people use or do to try to make their lives proceed the way they want may properly be termed a technology.

Technologies can be conveniently organized, by scale, into three approximate classes:

- *Industrial technologies* usually involve the very large-scale production of substances from natural raw materials. Examples include mining, oil refining, and the large-scale production of chemicals, such as ammonia and sulfuric acid.
- *Commercial technologies* are medium-scale processes involved in the production of goods at the level of individual business. Examples include the factory production of computers, home appliances, cleaning compounds, and radiator antifreeze.
- *Consumer technologies* involve the use, by individuals on a personal level, of products and processes such as mobile phones, shampoo, shrink-wrap packaging, remote car starters, online banking, and debit cards (**Figure 1**).

> **DID YOU KNOW?**
>
> **Technology—Human Nature?**
> Technological products and processes seem to have always been a part of all human societies. Anthropologists credit the mastery of stone chipping and fire making with starting humans on the road to better controlling and then better understanding their environment. All Aboriginal cultures have developed specialized technology based on solving specific problems. The processes use available materials, and the product addresses a particular need. One example is the birchbark canoe. The need was a light, sturdy craft for an area of many waterways, where land travel is often circuitous and difficult. The birchbark canoe was (and sometimes still is) an elegant solution.
>
> A second universal human factor is imagination—the ability to wonder about things that do not already exist. Wondering "why" and "how" is the key beginning to the knowledge-gathering system we call science.

Figure 1
Debit cards use consumer-scale technology.

Technologies are created and developed to solve social problems. In the process, they often help bring about significant social change. For example, buildings could be no higher than about six stories until the elevator came into use. Shopping malls would be impossible without common access to motor vehicles.

A serious practical question about all new technology is whether the technology will be *sustainable*. There are several reasons why a new technology may not last.

- A technology may become obsolete as a *better technology* replaces it. Computer technology is the classic example: most discarded computers still work perfectly well; they have just been replaced by faster, more powerful models.
- A technology may be based on *a nonrenewable resource*. Over some time period it will either become unavailable or prohibitively expensive. The petroleum industry (the main energy supply for our society) is based on technology that will become useless when we run out of accessible oil (**Figure 2**).
- A technology may produce by-products that *damage the environment*. Unintended harmful effects may well outweigh the usefulness of the technology. Even though the technology still works well for solving the original problem, it will be discontinued. For example, the fluids first used in refrigeration technology were discontinued because of damage to the atmosphere's ozone layer.

Figure 2
Gasoline as an auto fuel must inevitably run out someday when crude oil reserves are eventually used up.

Of course, nothing lasts indefinitely. If science has taught us anything, it has taught us that everything changes on Earth—it always has, including the shape and position of the continents. The only variation is the time scale involved. We strongly relate time to personal perception. Anything that changes very little over a century, for example, is often regarded as "unchanging," while changes (upgrades) to computer software may seem to happen every week. The point is that we cannot discuss the sustainability of any technology without specifying what we mean and how long a time is involved; otherwise, the discussion necessarily becomes meaningless.

Implementing new technology nearly always has unforeseen effects. Such unintended consequences may be unimportant, or even useful. However, sometimes they create a huge new problem. Use of the pesticide DDT in the mid-1900s decimated many raptor populations, including the Peregrine falcon (**Figure 3**). The subsequent banning of this pesticide has led to a recovery in the numbers of these birds. When society introduces a new technology, we must take extra care and make careful observations.

Figure 3
If a female Peregrine falcon ingests DDT pesticide residue, her eggs would form with shells that are far too thin. The population of these beautiful birds is on the rise today, but they were an endangered species in Canada just a few years ago due to this effect. An international ban on DDT has dramatically increased the survival rate of Peregrine falcon chicks.

Practice

1. List four significant differences between science and technology.
2. Give a clear example of a technology at each of the three levels of scale. Your example should not be one that has already been used in the above text.
3. Classify each of the following questions as to whether it would more likely require a scientific or a technological activity to find an answer. Do not actually answer the questions.
   (a) What coating on a nail will reduce corrosion?
   (b) Which chemical reactions are involved in the corrosion of iron?
   (c) What is the accepted explanation for the chemical formula of water?
   (d) What process produces a continuous thread of nylon?
   (e) Why is a copper(II) sulfate solution blue?
   (f) How can automobiles be designed to make them safer to operate?
4. What does it mean to say a technology must be sustainable? Give an example of any current technology that is likely to be sustainable and another that is certain not to be sustainable. State your reasoning.
5. We often call our society "technological" because our lifestyle depends so heavily on manmade things. Can the lifestyle of Canada's original Aboriginal peoples before 1500 AD be considered nontechnological? Explain your answer, giving examples.

Figure 4
Roberta Bondar (1945–)

CAREER CONNECTION

Aerospace Engineer
Aerospace engineers specialize in all aspects of developing structures and materials related to flight and space travel, including designing aircraft, surveillance systems, satellites, and rockets. Some aerospace engineers test aircraft and spacecraft prototypes experimentally by constructing models to test their performance.

Find out more about aerospace engineering in Canada and the educational requirements for Canada's space program.

www.science.nelson.com GO

Figure 5
(a) Copper wire and a beaker with aqueous silver nitrate solution
(b) A few moments after the wire is immersed
(c) The beaker contents after 24 h

 **WEB** *Activity*

Canadian Achievers—Roberta Bondar

In 1992, Roberta Lynn Bondar (**Figure 4**) became the first Canadian woman in space when she flew on the space shuttle *Discovery* as a payload specialist. Her mission was the first international microgravity laboratory mission.

1. What academic studies did Bondar complete before becoming an astronaut?
2. List three research projects conducted by Bondar.

www.science.nelson.com GO

Chemical Reaction Equations

Industrial chemists and chemical engineers must always be concerned about the conditions within and surrounding chemical reactions. After all, a main goal of technology is to develop products and processes that solve practical problems based on criteria such as efficiency, reliability, and cost. Students of chemistry must learn to think of reaction conditions. Chemistry students must also develop the ability to describe, explain, and predict practical outcomes from the scientific knowledge they have acquired. The first question to be asked is, Precisely what do reaction equations tell us? And the next and perhaps even more important question is, What do they not tell us?

Consider the single replacement reaction equation

$$Cu(s) + 2\ AgNO_3(aq) \rightarrow 2\ Ag(s) + Cu(NO_3)_2(aq)$$

How well does this balanced equation describe and explain the reaction? If you were asked to do the reaction, given that you now "know" the balanced chemical equation, you would have to consider several practical questions: What does pure copper look like? What does an aqueous solution of silver nitrate look like? What kind of apparatus should be used to contain this reaction? Does it matter if the silver nitrate is dissolved in water? How much copper would be reasonable to use? How much silver nitrate would be reasonable to use? How much water would be reasonable to use? Some of these questions can be answered by looking at **Figure 5**, which shows photographs taken before, during, and after this particular reaction. However, if you had never seen the reaction or pictures of it, consider how difficult finding the answers might be. For some of the questions, your memory could help; for others, the chemistry concepts you have learned to this point would indicate an answer, or at least a partial answer.

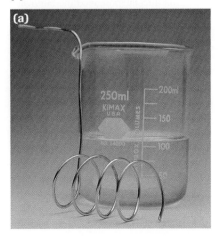
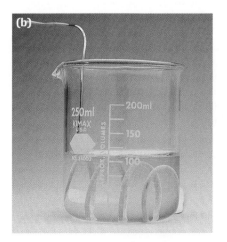
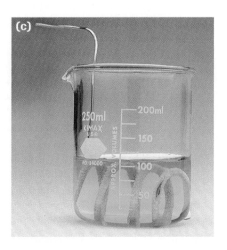

Limitations of Reaction Equations

If you performed the Exploration activity at the beginning of this chapter, then you detected evidence of an invisible substance. Consider the following similar reaction:

$$C_{25}H_{52}(s) + 38\, O_2(g) \rightarrow 25\, CO_2(g) + 26\, H_2O(g) + \text{heat energy}$$

This reaction equation can represent a typical substance found in candle wax reacting in a burning candle. In this reaction, the only thing you normally see is some light in a special region, the flame. The chemical products are invisible gases, and they are free to mix with other invisible gases (the air). Evidence suggests that the flame is visible for this reason: at one point in the reaction, carbon atoms have broken free of the wax molecule, but they have not yet combined with oxygen. These atoms are very hot, and they glow for an instant, emitting the light you see (**Figure 6**).

During winter months in Canada's far north, periods of darkness can last for a very long time. Aboriginal peoples learned long ago how to apply a very simple and effective technology to provide long-burning dependable light from readily available materials. Fat from animals (such as seals, bears, and whales) was rendered (clarified by cooking) to produce a clear oil. A bowl made of some noncombustible material (stone or shell) held the liquid (**Figure 7**), while a wick made of moss was placed at one end of the bowl. With the lower end of the wick submerged in the liquid, and the upper end above the liquid, the moss wick acted to draw the oil up to the flame zone by capillary action.

Note that the candle equation above gives "solid" as the state of the wax before it reacts. But wax must be in a gaseous state to burn—just try lighting a candle at the bottom, where there is no wick. The function of a wick is to draw melted wax up by capillary action, moving it into a very hot zone where the wax will vaporize. Once vaporized and mixed with air, it will react (burn) fast enough to set up a continuous reaction.

Chemical reaction equations also do not describe or explain the following:

- A reaction equation usually communicates little about the *pressure and temperature conditions* under which a reaction might occur or might actually be done. This necessary information is sometimes written above the arrow in an equation. To this point, you assume SATP conditions unless told otherwise, but many reactions occur at conditions that are not SATP. For instance, most of the complex reactions that happen inside your body occur at about 37 °C.
- A reaction equation communicates nothing about the *progress and process* of a reaction. It describes what is present before anything happens and what is present after any changes have stopped. It does not describe what actually happens during the reaction or anything about how long the reaction process might take.
- Most importantly, a reaction equation communicates nothing about measurable *quantities* of reactants in any form that you can use directly. An equation is "equal" in chemical symbols because a reaction is "equal" in chemical entities; that is, we believe that entities rearrange and become bonded differently in reactions, but the numbers and kinds of entities do not change. If the numbers of entities do not change, their total mass cannot change. This is the explanation for the law of conservation of mass, which describes a property that can be used directly. However, for most of the questions that arise about quantities of chemicals in reactions, this is not enough.

Equations may be read in terms of single entities (atoms/molecules/ions) or chemical amounts (in moles) of these entities, neither of which describes any property you can see, count, or measure directly in a laboratory or classroom. For instance, if you wanted to make 10.0 g of silver with the copper–silver nitrate reaction, a new set of questions would arise. This is the primary subject of this chapter.

Figure 6
A candle flame is easily visible because an intermediate stage of the combustion reaction involves the emission of light.

Figure 7
This stone bowl was formed by patiently rubbing with a harder type of stone. When filled with rendered seal oil and with a moss wick added at one end, it makes a good smoke-free light source.

Reaction Assumptions

Making assumptions is common, and not just in chemistry. We do it all the time; it's the way that the human mind works. The danger is that you might be unaware that an assumption is being made. This is like crossing the street without looking, assuming subconsciously that all the drivers on the street will obey crosswalk rules. It's a good idea to look anyway! In chemistry, you must be aware of the major assumptions normally made for reactions, so you will notice when an exception occurs. A reaction equation always carries with it many assumptions, untested statements considered to be correct without proof or demonstration. These must be known and understood to make practical predictions in many cases, especially in this unit. To this point in your study of chemistry, you assume the following:

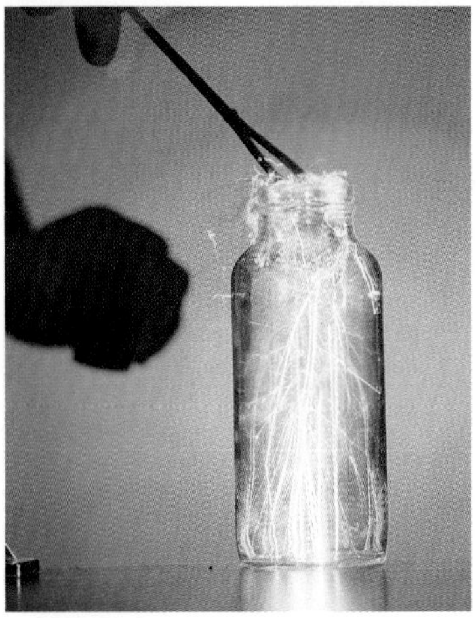

- Reactions are *spontaneous*. Reactions will occur when the reactants are mixed for all examples you are given. However, you have not yet studied any generalization, law, or theory that will let you predict whether given substances will react.
- Reactions are *fast*. For a reaction to be useful, either in a laboratory or in industry, the reaction must occur within a reasonable time. Consider the following simple balanced reaction equation, which gives no indication about the rate of the reaction:

$$4\ Fe(s)\ +\ 3\ O_2(g)\ \rightarrow\ 2\ Fe_2O_3(s)\ +\ heat\ energy$$

If this reaction represents fine strands of steel wool ignited in pure oxygen, it is over in a few seconds (**Figure 8**). If it represents the fine iron powder in a hand warmer pouch reacting in air, it will take several hours (**Figure 9**). If the reaction represents the eventual corrosion of an automobile, it could take many decades

Figure 8
A spectacular example of combustion of a metal is the burning of steel wool (iron) in oxygen.

Figure 9
A hand warmer pouch is sold sealed in an airtight plastic wrap. When the plastic outer wrap is removed, air filters in slowly through the porous pouch, and the reaction of the powdered iron (and other solids) inside the pouch can occur. The pouch is a carefully designed technological device, made to get the reaction to release heat at a rate that will warm your hand but not burn a hole in your coat pocket. One pouch has been cut open to show the powdered solid reagents inside.

(**Figure 10**). Controlling the rate of chemical reactions is extremely important; it concerns every aspect of society, from the speed at which fuel burns in a car, to the time dental epoxy takes to set and bond a tooth, to the speed of all the reactions that make up human metabolism.
- Reactions are *quantitative*. A **quantitative reaction** is one that is more than 99% complete; in other words, at least one reactant is essentially completely used up. Another way of saying this is that the reaction goes to completion.
- Reactions are *stoichiometric*. A **stoichiometric reaction** means that there is a simple, whole-number ratio of chemical amounts of reactants and products. In other words, the coefficients that you predict for a balanced chemical equation do not change when the reaction is repeated several times, even under different conditions.

Figure 10
When solid sheet metal is protected by paint, it can take many years to reach this level of oxidation (rusting). Spread over this time span, the heat energy released is completely undetectable.

These four key assumptions—that chemical reactions are spontaneous, fast, quantitative, and stoichiometric—will be tested and evaluated later in your chemistry education. These assumptions are particularly important when you do quantitative studies of chemical reactions in the rest of this unit.

> ### Practice
> 6. What does a balanced reaction equation directly communicate?
> 7. State three important aspects of a chemical reaction that are not communicated by the balanced chemical equation.
> 8. List the four major assumptions usually made about chemical reactions.
> 9. List three criteria that are often used to evaluate a technology.

Net Ionic Equations
Now we return to the original reaction equation example:

$$Cu(s) + 2\,AgNO_3(aq) \rightarrow 2\,Ag(s) + Cu(NO_3)_2(aq)$$

Collision–reaction theory is useful to consider when a solution is involved in a reaction. Many reactions will *only* occur in solution. Dissolving a reactant is often the only easy way to get its entities separated from each other so they can collide with entities of another reactant. To do this reaction, we could consider vaporizing the copper to separate its atoms. However, the temperature required is over 2500 °C, and the vapour is highly toxic and highly reactive. On the other hand, all the silver and nitrate ions in solid silver nitrate can be separated just by placing a sample in water and stirring it; an ordinary open glass beaker will work perfectly well as the reaction container. As a side benefit, since higher concentrations of substances react faster (more collisions), we can often easily control the rate of a reaction in solution.

One drawback, of course, is that often there is no direct visible way to tell a solution apart from the pure solvent. Silver nitrate solution looks exactly like pure water. In **Figure 5(a)** (page 278), the solid silver nitrate was already dissolved before the picture was taken, so you could not tell what the pure compound looked like. A few solutes produce visibly coloured solutions (like copper(II) compounds), and these sometimes allow visual identification of substances (**Figure 11**). However, usually chemical analysis techniques are needed to find out what, and how much, solute is in an aqueous solution. It is tempting to say that because the solution turns blue and copper(II) nitrate is produced in solution, that obviously the copper(II) nitrate is blue, but there is more to it than that. To really understand how dissolved ionic compounds react, it is necessary to write the equations in a form that more correctly represents the actual state of the entities present.

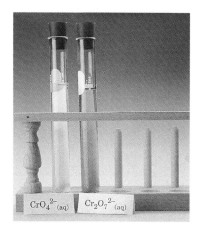

Figure 11
The yellow and orange colours seen here are characteristic of aqueous chromate and dichromate ions, respectively.

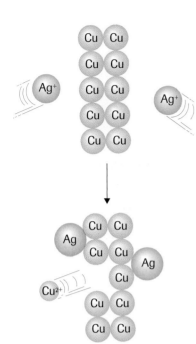

Figure 12
A model of the reaction of copper metal and silver nitrate solution illustrates aqueous silver ions reacting at the surface of a solid copper strip.

For writing net ionic equations, it is useful to refer to Table 3 in Section 5.2. The table indicates that a highly soluble ionic compound dissociates into individual ions as it dissolves in solution. If we rewrite the chemical equation to show that entities of the dissolved ionic compounds are actually present as separate (dissociated) aqueous ions, it then looks like the following:

$$Cu(s) + 2\,Ag^+(aq) + 2\,NO_3^-(aq) \rightarrow 2\,Ag(s) + Cu^{2+}(aq) + 2\,NO_3^-(aq)$$

Notice that when written this way, the equation makes it obvious that the nitrate ions do not change at all in the course of the reaction. They are like the beaker and the water in that they are not part of the reaction itself; they just help create an environment where the reaction can occur. If we write the equation again, leaving out any entities that do not change in the reaction, the result is the **net ionic equation**:

$$Cu(s) + 2\,Ag^+(aq) \rightarrow 2\,Ag(s) + Cu^{2+}(aq)$$

The equation can now be interpreted as follows: "If solid copper is placed in an aqueous solution of silver ions, solid silver will form, and copper(II) ions will form in solution" (**Figure 12**). This reaction statement looks different because it does not specify *what* silver compound was used to make the silver ion solution. How important is this point?

> ### SAMPLE problem 7.1
>
> A student mixed solutions of lead(II) nitrate and sodium iodide and observed the formation of a bright yellow precipitate. Another student recorded the same observation after mixing solutions of lead(II) acetate and magnesium iodide. Are these different reactions?
>
> The balanced chemical equations for these two double replacement reactions show some similarities and some differences.
>
> $Pb(NO_3)_2(aq) + 2\,NaI(aq) \rightarrow PbI_2(s) + 2\,NaNO_3(aq)$ (1)
>
> $Pb(CH_3COO)_2(aq) + MgI_2(aq) \rightarrow PbI_2(s) + Mg(CH_3COO)_2(aq)$ (2)
>
> Using Arrhenius' theory of dissociation, these reactions can be described more precisely. Each of the highly soluble ionic compounds is believed to exist in aqueous solution as separate ions. For reaction (1),
>
> $Pb^{2+}(aq) + 2\,NO_3^-(aq) + 2\,Na^+(aq) + 2\,I^-(aq) \rightarrow PbI_2(s) + 2\,Na^+(aq) + 2\,NO_3^-(aq)$
>
> It is apparent that some reactant ions—sodium and nitrate ions—are unchanged in this reaction. Ignoring these ions, you can write a net ionic equation, which shows only the entities that change in a chemical reaction:
>
> $Pb^{2+}(aq) + \cancel{2\,NO_3^-(aq)} + \cancel{2\,Na^+(aq)} + 2\,I^-(aq) \rightarrow PbI_2(s) + \cancel{2\,Na^+(aq)} + \cancel{2\,NO_3^-(aq)}$
>
> $Pb^{2+}(aq) + 2\,I^-(aq) \rightarrow PbI_2(s)$ (net ionic equation)
>
> Applying the same procedure to reaction (2):
>
> $Pb^{2+}(aq) + \cancel{2\,CH_3COO^-(aq)} + \cancel{Mg^{2+}(aq)} + 2\,I^-(aq) \rightarrow PbI_2(s) + \cancel{Mg^{2+}(aq)} + \cancel{2\,CH_3COO^-(aq)}$
>
> $Pb^{2+}(aq) + 2\,I^-(aq) \rightarrow PbI_2(s)$ (net ionic equation)
>
> The net ionic equations are identical for reactions (1) and (2), as are the observations (**Figure 13**). We can therefore say that the reactions are the same.

Figure 13
Solid, yellow lead(II) iodide precipitates when any lead(II) ion solution is mixed with any iodide ion solution. Reaction mixtures (1) and (2) from Sample Problem 7.1 produce exactly the same visible evidence.

Ions that are present but do not take part in (change during) a reaction are called **spectator ions**. These ions can be likened to spectators at a sports event; the spectators are present but are not part of the game. The conclusion to be drawn from the net ionic equations in Sample Problem 7.1 is that there were not two different reactions, just two different sets of substances used to make the same reaction occur.

▸ COMMUNICATION example 1

Write the net ionic equation for the reaction of aqueous barium chloride and aqueous sodium sulfate. Refer to Section 5.5 and to the solubility table on the inside back cover of this textbook.

Solution

$$BaCl_2(aq) + Na_2SO_4(aq) \rightarrow BaSO_4(s) + 2\,NaCl(aq)$$

$$Ba^{2+}(aq) + \cancel{2\,Cl^-(aq)} + \cancel{2\,Na^+(aq)} + SO_4^{2-}(aq) \rightarrow BaSO_4(s) + \cancel{2\,Na^+(aq)} + \cancel{2\,Cl^-(aq)}$$

$$Ba^{2+}(aq) + SO_4^{2-}(aq) \rightarrow BaSO_4(s)$$

> **Learning Tip**
>
> When eliminating or cancelling spectator ions, they must be identical in every way: chemical amount, form (atom/ion/molecule), and state of matter. Occasionally, the amount may be different while the form and state are identical. In this case, you may only cancel equal amounts.

Net ionic equations are useful in communicating reactions other than double replacement reactions. Communication Example 2 is a good illustration.

▸ COMMUNICATION example 2

Write the net ionic equation for the reaction of zinc metal and aqueous copper(II) sulfate.

Solution

$$Zn(s) + CuSO_4(aq) \rightarrow Cu(s) + ZnSO_4(aq)$$

$$Zn(s) + Cu^{2+}(aq) + \cancel{SO_4^{2-}(aq)} \rightarrow Cu(s) + Zn^{2+}(aq) + \cancel{SO_4^{2-}(aq)}$$

$$Zn(s) + Cu^{2+}(aq) \rightarrow Cu(s) + Zn^{2+}(aq)$$

▸ COMMUNICATION example 3

Write the net ionic equation for the reaction of hydrochloric acid and barium hydroxide solution.

Solution

$$2\,HCl(aq) + Ba(OH)_2(aq) \rightarrow BaCl_2(aq) + 2\,H_2O(l)$$

$$2\,H^+(aq) + \cancel{2\,Cl^-(aq)} + \cancel{Ba^{2+}(aq)} + 2\,OH^-(aq) \rightarrow \cancel{Ba^{2+}(aq)} + \cancel{2\,Cl^-(aq)} + 2\,H_2O(l)$$

$$H^+(aq) + OH^-(aq) \rightarrow H_2O(l) \text{ (coefficients reduced to 1)}$$

> **Learning Tip**
>
> Evidence indicates that protons in solution really exist attached to one or more water molecules. However, for writing ordinary net ionic equations for reactions involving aqueous strong acids, the entity symbol used is usually $H^+(aq)$, and not $H_3O^+(aq)$. This is just a matter of convenience; the H^+ symbol is quicker and easier to write. (Note this usage in Communication Example 3.)
> In Unit 8 you will study situations where use of the hydronium ion symbolism is necessary and more useful.

SUMMARY — Writing Net Ionic Equations

Step 1: Write a complete balanced chemical equation.
Step 2: Dissociate all high-solubility ionic compounds and ionize all strong acids to show the complete ionic equation.
Step 3: Cancel identical entities that appear on both reactant and product sides.
Step 4: Write the net ionic equation, reducing coefficients if necessary.

Figure 14
A water treatment facility

> ### Practice
>
> 10. In a laboratory test of the metal activity series, a student places a strip of lead metal into aqueous silver nitrate. Write the net ionic equation for the reaction that occurs.
> 11. (a) In a water treatment facility (**Figure 14**), sodium phosphate is added to remove calcium ions from the water. Write the net ionic equation for the reaction of aqueous calcium chloride and aqueous sodium phosphate.
>   (b) Identify the spectator ions in this reaction.
> 12. Some natural waters contain iron ions that affect the taste of the water and cause rust stains. Aeration converts any iron(II) ions into iron(III) ions. A basic solution (containing hydroxide ions) is added to produce a precipitate. Write the net ionic equation for the reaction of aqueous iron(III) ions and aqueous hydroxide ions.
> 13. A nitric acid spill is quickly neutralized by pouring a sodium hydrogen carbonate (baking soda) solution on it. Write the chemical equation and the net ionic equation for this neutralization reaction. Identify the spectator ions by name.
> 14. When you open a can of pop, the pressure inside the can is released. This allows the aqueous carbonic acid to decompose, forming carbon dioxide gas and water.
>   (a) Write the net ionic equation for this reaction.
>   (b) Write a statement about the dual role of water molecules in this particular reaction.

DID YOU KNOW?

Diagnostic Tests for Ions
You can easily create diagnostic tests for many specific ions using the solubility chart to find an oppositely charged ion that would produce a low solubility product. For example, here is a method to test for silver ions in a solution: Add a few drops of aqueous sodium chloride to the solution. If a precipitate forms, then silver ions are still present in the solution.

CAREER CONNECTION

Chemical Engineer
Chemical engineers are employed in a wide range of manufacturing and processing industries, consulting firms, government, research, and educational institutions. Among other duties, chemical engineers develop chemical processes in which reactions must be known.
  What education is required to become a chemical engineer?

Limiting and Excess Reagents

Revisiting the questions at the beginning of this section raises more points of interest. Consider the reaction demonstration in **Figure 5** (page 278) again. What is in the container after the reaction is finished? You can directly observe evidence for the silver, you know that the blue colour is likely from the copper(II) nitrate, and you can see that there is still a lot of solid copper left over. Of course, there is also a lot of water present. When no further changes appear to be occurring, we assume that all the silver nitrate that was initially present has now been completely reacted.

Is the silver nitrate really all gone? It is invisible in this system, so how can you tell? For reactions in which we care about quantities of substances involved, making sure that a measured reagent reacts completely becomes critically important. The standard method for this is to ensure that the measured reactant is a limiting reagent. A **limiting reagent** is the reactant whose entities are completely consumed in a reaction, meaning the reaction stops when—and because—all of this reactant is used up and none remains. To make sure that this happens, more of the other reactant must be present than is required for the reaction; otherwise, you would run out of it first. A greater quantity of this reactant than is necessary is deliberately added to the reaction system, and it is described as an excess reagent. An **excess reagent** is the reactant whose entities are present in surplus amounts, so that some remain after the reaction ends. In our reaction example, much more copper was used than was needed, as evidenced by the unreacted copper, so copper is the excess reagent. We assume that the reaction ended when there were no more silver ions left to react, so silver nitrate was the limiting reagent.

Most of our unanswered original questions are about "how much." After all, people do chemical reactions for specific reasons, and the activity nearly always involves knowing, measuring, or predicting quantities of something. The rest of this chapter is about combining many of the concepts you have learned so far to identify, calculate, and predict quantities of chemicals involved in reactions.

Section 7.1 Questions

1. Most Albertans use natural gas to heat their homes and to produce hot water. After refining, natural gas is composed almost entirely of methane.
   (a) Write the balanced chemical equation for the complete combustion of methane.
   (b) What specific information is given directly by this chemical equation?
   (c) What are some things that you do not know about this reaction?

2. At this stage in your chemistry education, you need to assume that chemical equations represent reactions that are spontaneous, fast, quantitative, and stoichiometric.
   (a) In your own words, explain what each of these assumptions means.
   (b) In general terms, explain how you would check these assumptions for a particular reaction.

3. An acceptable method for the treatment of soluble lead waste is to precipitate the lead as a low solubility lead(II) silicate.
   (a) Write the net ionic equation for the reaction of aqueous lead(II) nitrate and aqueous sodium silicate.
   (b) What can we assume about the ambient conditions and the container that likely could be used?
   (c) Identify the spectator ions in this reaction.

4. Bromine is a disinfectant commonly used in swimming pools. One industrial method of producing bromine is to react sea water, containing sodium bromide, with chlorine gas. Write the net ionic equation for this reaction.

5. (a) Strontium compounds are often used in flares because their flame colour is bright red (**Figure 15**). One commercial example of the production of strontium compounds is the reaction of aqueous solutions of strontium nitrate and sodium carbonate. Write the net ionic equation for this reaction.
   (b) Suggest another compound in solution (other than sodium carbonate) that would react with strontium nitrate solution to produce a reaction with the same net ionic equation as the reaction in (a).

Figure 15
The bright red of the flare is easily visible to passing motorists.

6. In a hard water analysis, sodium oxalate solution reacts with calcium hydrogen carbonate present in the hard water to precipitate a calcium compound. Write the net ionic equation for this reaction.

7. Write a net ionic equation for the reaction of vinegar (acetic acid solution) with a scale deposit in a kettle (assume solid calcium hydroxide).

8. State why it is desirable, in a quantitative chemical analysis of a substance, to use an excess of one reactant.

9. For a particular reaction, how are the interests of a research chemist different from the interests of an industrial chemist or engineer? How are they similar?

10. Introduction of new technology often has unintended consequences. It is not unusual for these consequences to be beneficial. One example is the discovery of using lasers to transfer information to and from DVDs. Unintended consequences can be very undesirable, however. Write statements including the terms "obsolete" and "recycle" to describe the current social problem that is an undesirable consequence of the rapid development of personal computer technology. Use an ecological point of view.

11. Mobile (cell) phone cameras (**Figure 16**) are a technology that is rapidly expanding in use worldwide. Write one statement about some aspect of this technology from each of the following perspectives:
    (a) economic
    (b) societal
    (c) ethical
    (d) environmental

Figure 16
Many cellphones now come with a camera, which is raising new issues regarding privacy.

12. State what measurements would normally have to be taken in a lab to allow you to calculate the chemical amount in a sample of each of the following substances:
    (a) $CH_4(g)$
    (b) $NaCl(s)$
    (c) $C_6H_6(l)$
    (d) 6.0 mol/L $HCl(aq)$

7.2 Gravimetric Stoichiometry

CAREER CONNECTION

Chemical Technologist
Chemical technologists and technicians find employment in the private sector as well as in government. They perform chemical tests and help create procedures in the laboratory, from routine processes to more intricate procedures needed in complex research projects.
   What level of education is required to become a chemical technologist?

www.science.nelson.com GO

Chemical engineers and technologists design and control chemical technology in a processing plant. Like all technology, the goal is to solve practical problems, such as producing fertilizers for the agricultural industry, improving the combustion qualities and environmental impacts of fuels, and creating better and safer water treatment processes. Typically, society provides the practical problem to be solved, science provides some or all of the understanding, and technology is developed to come up with the solution. The water purification unit used by Canada's Disaster Assistance Relief Team (DART) is a good example of this process (**Figure 1(a)**). The problem is the need for pure water in remote locations where only contaminated water is available; various sciences such as chemistry, biology, and physics provide the empirical and theoretical knowledge; and engineers and technologists create the technology to provide the solution to the original problem.

Typically, any such situation actually involves a whole series of problems, each requiring a specific technology to achieve the desired results. For example, the DART system produces a huge volume of water, which is temporarily stored in containers that must be very big, yet easily portable. As a result of polymer science, we now have high-strength synthetic plastics. Technologists used this knowledge to create collapsible water storage bladders for the DART team (**Figure 1(b)**). Then, of course, the water has to be pumped out of the storage bladders somehow, and that means energy must be provided to run the pumps, and on and on the process goes.

Most chemical technologies require quantitative predictions of raw materials and products. Quantitative predictions made to ensure that a commercial or industrial process works well are based largely on an understanding of the relative quantities of

(a)

(b)

Figure 1
(a) This rugged water treatment unit was used by the Canadian Armed Forces DART unit in the Asian tsunami relief effort in Sri Lanka in December 2004. The unit can produce 120 kL (120 m^3) of water per day from any water source, including chemically contaminated water. A large diesel engine runs powerful pumps that push water through large bundles of tubular membranes. Each individual tiny tube has many submicroscopic holes; the holes are so small that only water molecules can get through, so pure water emerges.
(b) Sergeant Shane Stachnick, from Rosetta, Alberta, distributes water from the collapsible storage bladder, which can hold 10 000 L (10 t), to the local people.

reactants consumed and products produced in a chemical system. This understanding can be entirely empirical, determined by trial and error, but more often, it is related to a knowledge of the balanced chemical reaction equation. For all chemical reactions where quantities are important—whether in industry, commerce, research, or analysis—a balanced equation is necessary because it describes the reaction stoichiometry. The **stoichiometry** of a reaction is the description of the relative quantities of the reactants and products by chemical amount, that is, in moles. Any prediction or calculation from any measured quantity of any substance in a reaction must necessarily be based on the stoichiometry of the reaction. Therefore, as you recall from the previous section, reactions must be stoichiometric, but we also assume that the reactions will be spontaneous, fast, and quantitative.

DID YOU KNOW ?

Meaning of Stoichiometry
Stoichiometry (stoy-kee-**ah**-meh-tree) is derived from the Greek words *stoicheion* (element) and *metron* (measure).

INVESTIGATION 7.1 Introduction

Decomposing Malachite

Copper(II) hydroxide carbonate, commonly called basic copper carbonate and also known as malachite, is a double salt with the chemical formula $Cu(OH)_2 \cdot CuCO_3(s)$. This double salt decomposes completely when heated to 200 °C, forming copper(II) oxide, carbon dioxide, and water vapour. Complete the Prediction using the balanced chemical equation. Include safety and disposal steps in your Procedure. Organize the data and create the graph using suitable software.

Purpose

The purpose of this investigation is to test the assumption that a chemical reaction is stoichiometric.

Report Checklist

- ○ Purpose
- ○ Problem
- ○ Hypothesis
- ● Prediction
- ○ Design
- ○ Materials
- ● Procedure
- ● Evidence
- ● Analysis
- ● Evaluation (1, 2, 3)

Problem

How is the chemical amount of copper(II) oxide produced related to the chemical amount of malachite reacted in the decomposition of malachite?

Design

A known mass of malachite (manipulated variable) is heated strongly until the colour changes completely from green to black. The mass of black copper(II) oxide (responding variable) is determined. The results from several laboratory groups are combined in a graph to determine the ratio of chemical amounts.

To perform this investigation, turn to page 304.

▶ Practice

1. What is the main goal of technology? Illustrate this with one example.
2. Most modern automobiles have improved fuel economy and produce less pollution compared to those built a number of years ago.
   (a) What role did science likely play in these technological developments?
   (b) What role did society likely play?
3. Technologies can be classified according to the scale of the technology. Identify the three scales, and write a brief description of each.
4. In your own words, explain what stoichiometry means. How have you been using this concept almost since you started studying chemistry?
5. Which of the four assumptions about chemical reactions is tested in Investigation 7.1? Was this assumption shown to be valid?
6. Using the balanced chemical equation for the decomposition of malachite, how does the chemical amount of copper(II) oxide product compare with the chemical amount of the carbon dioxide product?
7. What evidence do you have that the reaction in Investigation 7.1 was likely quantitative, that is, went to completion?

Calculating Masses Involved in Chemical Reactions

Analysis of the evidence from Investigation 7.1 indicates that when malachite is decomposed, the ratio of the chemical amounts of copper(II) oxide and malachite is a simple mole ratio of 2:1. This is the same ratio given by the coefficients of these substances in the balanced chemical equation. Two moles of copper(II) oxide are produced for each (one) mole of malachite that reacts.

$$Cu(OH)_2 \cdot CuCO_3(s) \rightarrow 2\,CuO(s) + CO_2(g) + H_2O(g)$$

Unfortunately, there is no instrument that measures amounts in moles directly. A measurable quantity such as mass is required, from which we can predict and analyze the quantities of reactants and products in a chemical reaction. However, the relationship between two substances in a chemical reaction is represented by the mole ratio from the balanced chemical equation.

The procedure for calculating the masses of reactants or products in a chemical reaction is called **gravimetric stoichiometry**. Gravimetric stoichiometry is restricted to chemical amount calculations from mass (gravity) measurement, so the measured substance has to be a pure solid or liquid. However, the calculated mass can be for any other substance in the reaction.

Gas stoichiometry, which you will study in Section 7.3, requires that volume, temperature, and pressure all be considered to calculate the chemical amount. This is because the entities are widely separated from each other and must be held in a sealed container.

Solution stoichiometry, which is covered in Section 7.4, also involves entities that are widely separated, but only the amount concentration and solution volume usually need to be measured to calculate the chemical amount.

▶ SAMPLE problem 7.2

If you decompose 1.00 g of malachite, what mass of copper(II) oxide would be formed?

First, write the balanced chemical equation. Underneath the balanced equation, write the mass that is given (measured) and the symbol m for the mass to be calculated, along with the conversion factors. In this example, one mass is given, and the conversion factors (the molar masses) are calculated from the chemical formulas and the information in the periodic table:

$$Cu(OH)_2 \cdot CuCO_3(s) \rightarrow 2\,CuO(s) + CO_2(g) + H_2O(g)$$
1.00 g m
221.13 g/mol   79.55 g/mol

Second, convert the measured mass of malachite to its chemical amount:

$$n_{Cu(OH)_2 \cdot CuCO_3} = 1.00\ \text{g} \times \frac{1\ \text{mol}}{221.13\ \text{g}}$$
$$= 0.004\,52\ \text{mol}$$

Third, calculate, using the mole ratio from the balanced equation, the amount of copper(II) oxide that will be produced:

$$\frac{n_{CuO}}{n_{Cu(OH)_2 \cdot CuCO_3}} = \frac{2}{1}$$

$$\frac{n_{CuO}}{0.004\,52\ \text{mol}} = \frac{2}{1}$$

> **Learning Tip**
>
> In all stoichiometric calculations, the third step is the same. The mole ratio is always used with the coefficient for the unknown (or required) substance as the numerator and the coefficient for the measured (or given) substance as the denominator. Where the single calculation method is shown in this textbook, formulas are written in with the mole ratio and the molar masses and then cancelled to make it clear how the calculation must be correctly set up.

$$n_{CuO} = 0.004\,52 \text{ mol} \times \frac{2}{1}$$
$$= 0.009\,04 \text{ mol}$$

Fourth, calculate the mass represented by this amount of CuO:

$$m_{CuO} = 0.009\,04 \text{ mol} \times \frac{79.55 \text{ g}}{1 \text{ mol}}$$
$$= 0.719 \text{ g}$$

Alternatively, all three steps of such a calculation can be expressed as a single "chained" calculation. When using this method, it is customary to label each quantity and conversion factor and to cancel quantities and labels carefully. The purpose is to keep track of the substances involved.

$$m_{CuO} = 1.00 \text{ g } Cu(OH)_2 \cdot CuCO_3 \times \frac{1 \text{ mol } Cu(OH)_2 \cdot CuCO_3}{221.13 \text{ g } Cu(OH)_2 \cdot CuCO_3} \times \frac{2 \text{ mol CuO}}{1 \text{ mol } Cu(OH)_2 \cdot CuCO_3} \times \frac{79.55 \text{ g CuO}}{1 \text{ mol CuO}}$$
$$= 0.719 \text{ g CuO}$$

The certainty of the answer, three significant digits, is determined by the least certain value used in the calculation, 1.00 g. Note that the mass of copper(II) oxide has been obtained by knowing only the balanced chemical equation and the molar masses. No actual experiment was necessary. The following example illustrates how to communicate the stoichiometric method.

Learning Tip

Remember to keep the unrounded values in your calculator for further calculation until the final answer is reported. The values for intermediate calculation are rounded when written down. Follow the calculation process for the Sample Problems on your calculator to review how to do this.

▶ COMMUNICATION *example*

Iron is the most widely used metal in North America (**Figure 2**). It may be produced by the reaction of iron(III) oxide, from iron ore, with carbon monoxide to produce iron metal and carbon dioxide. What mass of iron(III) oxide is required to produce 100.0 g of iron?

Solution

$$Fe_2O_3(s) + 3\,CO(g) \rightarrow 2\,Fe(s) + 3\,CO_2(g)$$

m			100.0 g	
159.70 g/mol			55.85 g/mol	

$$n_{Fe} = 100.0 \text{ g} \times \frac{1 \text{ mol}}{55.85 \text{ g}}$$
$$= 1.791 \text{ mol}$$

$$n_{Fe_2O_3} = 1.791 \text{ mol} \times \frac{1}{2}$$
$$= 0.8953 \text{ mol}$$

$$m_{Fe_2O_3} = 0.8953 \text{ mol} \times \frac{159.70 \text{ g}}{1 \text{ mol}}$$
$$= 143.0 \text{ g}$$

or

$$m_{Fe_2O_3} = 100.0 \text{ g Fe} \times \frac{1 \text{ mol Fe}}{55.85 \text{ g Fe}} \times \frac{1 \text{ mol } Fe_2O_3}{2 \text{ mol Fe}} \times \frac{159.70 \text{ g } Fe_2O_3}{1 \text{ mol } Fe_2O_3}$$
$$= 143.0 \text{ g } Fe_2O_3$$

According to gravimetric stoichiometry, 143.0 g of iron(III) oxide is needed to produce 100.0 g of iron.

Figure 2
Wrought iron is a very pure form of iron. The ornate gates on Parliament Hill in Ottawa are made of wrought iron. The metal is relatively soft and easily bent into decorative shapes. Wrought iron is also quite corrosion-resistant. When carbon is present in iron in small quantities, the metal becomes much harder and is called steel.

SUMMARY: Gravimetric Stoichiometry

Stoichiometry Calculations

measured quantity
solids/liquids $m \longrightarrow n$

required quantity   mole ratio
solids/liquids $m \longleftarrow n$

Step 1: Write a balanced chemical reaction equation, and list the measured mass, the unknown quantity (mass) symbol m, and conversion factors (the molar masses).
Step 2: Convert the mass of measured substance to its chemical amount.
Step 3: Calculate the chemical amount of required substance using the mole ratio from the balanced chemical equation.
Step 4: Convert the chemical amount of required substance to its mass.

DID YOU KNOW?

Refining Aluminium
Aluminium is the most abundant metal in Earth's crust, but it occurs only in chemical compounds such as aluminium oxide, Al_2O_3, the principal constituent of bauxite. Canada has little bauxite but has abundant hydroelectric power. Aluminium oxide imported principally from Jamaica and Australia is refined at Alcan's aluminium refinery in Kitimat, British Columbia, and then exported worldwide. The refinery uses just over 500 kt of aluminium oxide annually to produce about 272 kt of aluminium. To produce this amount of aluminium requires 896 MW of electricity. The B.C. refinery came at a social cost though, with many Aboriginal groups displaced from their homes to allow the flooding necessary for the power dams.

Practice

8. Why is a balanced chemical equation necessary when doing a stoichiometry calculation?
9. Powdered zinc metal reacts violently with sulfur (S_8) when heated to produce zinc sulfide (**Figure 3**). Predict the mass of sulfur required to react with 25 g of zinc.
10. Bauxite ore contains aluminium oxide, which is decomposed using electricity to produce aluminium metal (**Figure 4**). What mass of aluminium metal can be produced from 125 g of aluminium oxide?
11. Determine the mass of oxygen required to completely burn 10.0 g of propane.
12. Calculate the mass of lead(II) chloride precipitate produced when 2.57 g of sodium chloride in solution reacts in a double replacement reaction with excess aqueous lead(II) nitrate.
13. Predict the mass of hydrogen gas produced when 2.73 g of aluminium reacts in a single replacement reaction with excess sulfuric acid.
14. What mass of copper(II) hydroxide precipitate is produced by the reaction in solution of 2.67 g of potassium hydroxide with excess aqueous copper(II) nitrate?

Figure 3
The reaction of powdered zinc and sulfur is rapid and highly exothermic. Because of the numerous safety precautions that would be necessary, the reaction is not usually carried out in school laboratories.

Figure 4
An aluminium refinery

Testing the Stoichiometric Method

The most rigorous test of any scientific concept is whether it can be used to make predictions. If the prediction is shown to be valid, then the concept is judged to be acceptable. The prediction is falsified if the percent difference between the actual and the predicted values is considered to be too great, for example, more than 10%. The concept may then be judged unacceptable. (See "Evaluation" in Appendix B.2.) Percent difference between an experimental value and a predicted value is the primary criterion for the evaluation of an accepted value (such as a constant) or an accepted method (such as stoichiometry). It is assumed that reagents are pure and skills are adequate for the experiment that is conducted.

Filtration (see Appendix C.4) is a common technique used in experimental designs for testing stoichiometric predictions. Stoichiometry is used to predict the mass of precipitate that will be produced, and filtration is used to separate the mass of precipitate actually produced in a reaction (**Figure 5**). Lab Exercise 7.A and Investigation 7.2 test the validity of the stoichiometric method. In all examples, an excess of one reactant is used to ensure complete reaction of the limiting (measured or tested) reagent.

(a)

(b)

Figure 5
(a) A dissolved substance or ion can often be precipitated out of solution.
(b) A precipitate is filtered and dried, and its mass is measured to determine the amount of substance that was dissolved in the original solution.

The technique of filtration is explained in Appendix C.4, and demonstrated in the video.

www.science.nelson.com

LAB EXERCISE 7.A

Testing the Stoichiometric Method

Report Checklist
○ Purpose   ○ Design   ● Analysis
○ Problem   ○ Materials   ● Evaluation (2, 3)
○ Hypothesis   ○ Procedure
● Prediction   ○ Evidence

Purpose
The purpose of this investigation is to test the stoichiometric method. In your evaluation, assume the experiment was valid and that suitable quality evidence was obtained.

Problem
What mass of lead is produced by the reaction of 2.13 g of zinc with an excess of lead(II) nitrate in solution (**Figure 6**)?

Design
A known mass of zinc is placed in a beaker with an excess of lead(II) nitrate solution. The lead produced in the reaction is separated by filtration and dried. The mass of the lead is determined.

Evidence
In the beaker, crystals of a shiny black solid were produced, and all the zinc disappeared.
mass of filter paper = 0.92 g
mass of dried filter paper plus lead = 7.60 g

Figure 6
Zinc reacts with a solution of lead(II) nitrate.

INVESTIGATION 7.2 Introduction

Gravimetric Stoichiometry

In this investigation, you will use gravimetric stoichiometry to investigate the reaction of strontium nitrate with excess copper(II) sulfate in an aqueous solution. Use 2.00 g of strontium nitrate and about 3.5 g of copper(II) sulfate–water (1/5), initially dissolving each chemical in about 75 mL of water. Be sure to include safety and disposal instructions in your Procedure.

Purpose

The purpose of this investigation is to test the stoichiometric method.

To perform this investigation, turn to page 305.

Report Checklist

- ○ Purpose
- ○ Problem
- ○ Hypothesis
- ● Prediction
- ● Design
- ● Materials
- ● Procedure
- ● Evidence
- ● Analysis
- ● Evaluation (1, 2, 3)

Problem

What mass of precipitate is produced by the complete reaction of 2.00 g of strontium nitrate in solution with an excess of aqueous copper(II) sulfate?

Applications of Stoichiometry

Having tested the stoichiometric method several times, you now have evidence that it can be accepted as valid and used with confidence to answer questions.

Calculating Percent Yield for Reactions

We can use stoichiometry to test experimental designs, technological skills, purity of chemicals (**Figure 7**), and the quantitative nature of a particular reaction. For each of these situations, we can evaluate the overall experiment by calculating a **percent yield**. This is the ratio of the actual or experimental quantity of product obtained (actual yield) to the maximum quantity of product (**theoretical yield** or predicted yield) obtained from a stoichiometry calculation:

$$\text{percent yield} = \frac{\text{actual yield}}{\text{predicted yield}} \times 100$$

In laboratory work, many factors, called experimental uncertainties, can affect the percent yield of a chemical reaction. The only process you assume to be exact is stoichiometry itself. Some common sources of *experimental uncertainty* are the following:

- All measurements. Even assuming the experimenter makes careful measurements with the correct technique, inherent limitations in equipment always create some uncertainty. This also applies to initial values that are derived from measurements, such as molar masses and amount concentrations. For these values, the uncertainty is usually very small but it is not zero.
- The purity of the grade of chemical used (**Figure 7**). Where possible, consult the container label.
- Washing a precipitate. Very fine particles may be lost through the filter paper, or a very small amount of precipitate may be dissolved by repeated washings and lost that way.
- Any qualitative judgments that affect measurements. Estimation of colour or colour changes and estimation of reaction completion are two common examples.

It is not a simple matter to convert this list of experimental uncertainties into a numerical value, such as a percentage. *For school laboratories, investigations usually involve a total of all experimental uncertainties in the range of 5% to 10%.* This means that a percent yield as low as 90% could be considered quite acceptable for a particular experiment, depending on the equipment and chemicals used.

See Appendix B.2 for further tips on calculating and reporting percent yield.

Figure 7
Chemicals come in a wide variety of grades (purities). Some low-purity or technical grades may only be 80% to 90% pure, whereas high-purity or reagent grades may be better than 99.9% pure. The purity of a chemical can significantly affect experimental results when studying chemical reactions.

Learning Tip

Scientists and technicians recognize and accept that there are many sources of experimental uncertainty, sometimes called sources of error. However, "human error" is not an acceptable category. If an experimenter makes a mistake, then the trial or experiment is repeated.

LAB EXERCISE 7.B

Testing a Chemical Process

Report Checklist
- ○ Purpose
- ○ Problem
- ○ Hypothesis
- ○ Prediction
- ○ Design
- ○ Materials
- ○ Procedure
- ○ Evidence
- ● Analysis
- ● Evaluation (1, 2)

Some technological problem solving involves quality control tests. These are physical and/or chemical tests performed during or at the end of a chemical process. The tests make sure that the process is working within parameters determined by the person in charge of quality control. In the Evaluation, evaluate the Design, list sources of experimental uncertainty, and then evaluate the Prediction.

Purpose
The purpose of this exercise is to perform a quality control test on a chemical process.

Problem
What is the mass of sodium silicate in a 25.0 mL sample of the solution used in a chemical process?

Prediction
If the process is operating as expected, the mass of sodium silicate in a 25.0 mL sample should always be between 6.40 g and 6.49 g.

Design
An excess quantity of iron(III) nitrate is added to the sample of sodium silicate. The resulting precipitate is separated by filtration. After the precipitate has dried, its mass is determined.

Evidence
mass of filter paper = 0.98 g
mass of dried filter paper plus precipitate = 9.45 g
The colour of the filtrate was yellow-orange.

Section 7.2 Questions

1. A balanced chemical equation includes simple coefficients in front of the chemical formulas.
   (a) What do these coefficients represent?
   (b) What is the term for the overall relationship of chemical amounts of all reactants and products?

2. List four assumptions about chemical reactions. Which two assumptions cannot be tested simply by observing a reaction?

3. In your own words, explain gravimetric stoichiometry.

4. How is a scientific concept such as stoichiometry tested? Provide a specific example.

5. For automobiles powered by hydrogen fuel cells to become successful, a source of hydrogen is required. Hydrogen can easily be produced by the electrolysis (simple decomposition) of water, but this process is very expensive.
   (a) What perspective is being used to evaluate the production of hydrogen?
   (b) Write the balanced chemical equation for the simple decomposition equation of water.
   (c) Based on the coefficients in this chemical equation, if 100 g of hydrogen is produced, does this mean 50 g of oxygen will be formed? Justify your answer.

6. A chemical laboratory technician plans to react 3.50 g of lead(II) nitrate with excess potassium bromide in solution. Predict the mass of precipitate expected.

7. When calculating a percent yield for a reaction, where do the values for the actual yield and for the predicted yield come from?

8. In a chemical analysis, 3.00 g of silver nitrate in solution was reacted with excess sodium chromate to produce 2.81 g of filtered, dried precipitate.
   (a) Using stoichiometry, predict the mass of precipitate expected in this reaction.
   (b) Calculate the percent yield.

9. List four different sources of experimental uncertainty.

10. A solution made by dissolving 9.8 g of barium chloride is to be completely reacted with a second solution containing dissolved sodium sulfate.
    (a) Predict the mass of precipitate expected.
    (b) If 10.0 g of precipitate actually formed, calculate the percent yield.
    (c) Does the percent yield result indicate the reaction went as expected?

11. Air-bag technology in automobiles has saved many lives. Research air-bag technology on the Internet and using other sources, such as newspapers and periodicals. Working in a group, prepare a presentation to explain how air-bags work. Your presentation should include the following:
    - a list of the main chemicals and the main reactions
    - an evaluation of air-bag technology
    - an outline of the roles played by science and society in the development of this technology
    - communication technology (in making your presentation)

    www.science.nelson.com

Extension

12. If you have access to the software, develop a spreadsheet that will predict the mass of a reagent required to yield various masses of product for a given reaction.

7.3 Gas Stoichiometry

Many chemical reactions involve gases. One common consumer example is the combustion of propane in a home gas barbecue (**Figure 1**). The reaction of chlorine in a water treatment plant is a commercial example. An important industrial application of a chemical reaction involving gases is the production of the fertilizer ammonia from nitrogen and hydrogen gases. These technological examples feature gases as either valuable products, such as ammonia, or as part of an essential process, such as water treatment.

Studies of chemical reactions involving gases (e.g., the law of combining volumes, Chapter 4) have helped scientists develop concepts about molecules and explanations for chemical reactions, such as the collision–reaction theory (Chapter 4). In both technological applications and scientific studies of gases, it is necessary to accurately calculate quantities of gaseous reactants and products.

The method of stoichiometry applies to all chemical reactions. This section extends stoichiometry to gases—**gas stoichiometry**—using gas volume, pressure and temperature, molar volume, and the ideal gas law.

Figure 1
Propane gas barbecues have become very popular. Charcoal barbecues are now banned in parts of California because they produce five times as much pollution (nitrogen oxides, hydrocarbons, and particulates) as gas barbecues.

▶ SAMPLE problem 7.3

If 275 g of propane burns in a gas barbecue, what volume of oxygen measured at STP is required for the reaction?

First, write a balanced chemical equation to relate the amount of propane to the amount of oxygen. List the given and required values and the conversion factors for each chemical, just as you did in previous stoichiometry questions.

$$C_3H_8(g) + 5\,O_2(g) \rightarrow 3\,CO_2(g) + 4\,H_2O(g)$$

 275 g V (STP)
 44.11 g/mol 22.4 L/mol

Since propane and oxygen are related by their mole ratio, you must convert the mass of propane to its chemical amount:

$$n_{C_3H_8} = 275 \text{ g} \times \frac{1 \text{ mol}}{44.11 \text{ g}}$$
$$= 6.23 \text{ mol}$$

The balanced equation indicates that 1 mol of propane reacts with 5 mol of oxygen. Use this mole ratio to calculate the amount of oxygen required, in moles. (This step is common to all stoichiometry calculations.)

$$n_{O_2} = 6.23 \text{ mol} \times \frac{5}{1}$$
$$= 31.2 \text{ mol}$$

Finally, convert the amount of oxygen to the required quantity, in this case, volume:

$$V_{O_2} = 31.2 \text{ mol} \times \frac{22.4 \text{ L}}{1 \text{ mol}}$$
$$= 698 \text{ L}$$

Learning Tip

Symbols can be modified to clarify what they refer to by adding a subscript. For example, $n_{C_3H_8}$ represents the chemical amount of propane, and V_{O_2} represents the volume of oxygen.

Note that the final step used the molar volume at STP as a conversion factor, in the same way that molar mass is used in gravimetric stoichiometry.

As in gravimetric stoichiometry, all steps may be combined as a single calculation:

$$V_{O_2} = 275 \text{ g C}_3\text{H}_8 \times \frac{1 \text{ mol C}_3\text{H}_8}{44.11 \text{ g C}_3\text{H}_8} \times \frac{5 \text{ mol O}_2}{1 \text{ mol C}_3\text{H}_8} \times \frac{22.4 \text{ L O}_2}{1 \text{ mol O}_2}$$

$$= 698 \text{ L O}_2$$

▶ COMMUNICATION *example 1*

Hydrogen gas is produced when sodium metal is added to water. What mass of sodium is necessary to produce 20.0 L of hydrogen at SATP?

Solution

$2 \text{ Na(s)} + 2 \text{ H}_2\text{O(l)} \rightarrow \text{H}_2\text{(g)} + 2 \text{ NaOH(aq)}$

| m | 20.0 L |
| 22.99 g/mol | 24.8 L/mol |

$$n_{H_2} = 20.0 \text{ L} \times \frac{1 \text{ mol}}{24.8 \text{ L}}$$

$$= 0.806 \text{ mol}$$

$$n_{Na} = 0.806 \text{ mol} \times \frac{2}{1}$$

$$= 1.61 \text{ mol}$$

$$m_{Na} = 1.61 \text{ mol} \times \frac{22.99 \text{ g}}{1 \text{ mol}}$$

$$= 37.1 \text{ g}$$

or $\quad m_{Na} = 20.0 \text{ L H}_2 \times \frac{1 \text{ mol H}_2}{24.8 \text{ L H}_2} \times \frac{2 \text{ mol Na}}{1 \text{ mol H}_2} \times \frac{22.99 \text{ g Na}}{1 \text{ mol Na}}$

$$= 37.1 \text{ g Na}$$

According to gas stoichiometry, 37.1 g of sodium is needed to produce 20.0 L of hydrogen at SATP.

> **Learning Tip**
>
> Recall that when working with gas measurement, two sets of standard pressure and temperature conditions have been defined. These should be memorized. STP is a temperature of 0 °C and a pressure of 101.325 kPa (1 atm). SATP defines conditions closer to normal lab conditions: a pressure of 100 kPa and a temperature of 25 °C. These are exact values because they are definitions. Under conditions normal for laboratory work with gases, we assume the molar volume at STP is 22.4 L/mol and that the molar volume at SATP is 24.8 L/mol.

Note that the general steps of a stoichiometry calculation are the same for both solids and gases. Changes from mass to chemical amount or from volume to chemical amount, or vice versa, are done using the molar mass or the molar volume, respectively, of the substance. Although the molar mass depends on the chemical involved, the molar volume of a gas depends only on temperature and pressure. If the conditions are not standard (i.e., STP or SATP), then the ideal gas law ($PV = nRT$), rather than the molar volume, is used to find the amount or volume of a gas, as in the following example.

▶ COMMUNICATION *example 2*

Ammonia, which is widely used as a fertilizer, is produced from the reaction of nitrogen and hydrogen. What volume of ammonia at 450 kPa pressure and 80 °C can be obtained from the complete reaction of 7.5 kg of hydrogen?

Solution

$\text{N}_2\text{(g)} + 3 \text{ H}_2\text{(g)} \rightarrow 2 \text{ NH}_3\text{(g)}$

7.5 kg	V, 450 kPa, 353 K
2.02 g/mol	8.314 kPa·L/(mol·K)

> **Learning Tip**
>
> Recall that a Celsius temperature scale is not useful for gas quantity calculations because it is not an absolute scale. In other words, it does not start from zero temperature. Gas law calculations use the (absolute) Kelvin scale of temperature. (Recall Section 4.1.) The exact conversion is as follows:
> $T \text{ (K)} = t \text{ (°C)} + 273.15$
> For most purposes, this conversion value may be rounded off to
> $T \text{ (K)} = t \text{ (°C)} + 273$
> which is accurate enough for all calculation questions in this textbook.

Learning Tip

The ideal gas law relationship is necessarily complex, because three variables, P, V, and T, must all be known to define a specific amount, n, of any gas. For gas stoichiometry questions that involve the constant R, it is often useful to write down the variation of the memorized formula $PV = nRT$ that is being applied.

Note in Communication Example 2 that care must be taken in any cancellation of unit values written with prefixes. The unit g cancels (from the known mass of H_2), but the prefix k does not.

$$n_{H_2} = 7.5 \text{ kg} \times \frac{1 \text{ mol}}{2.02 \text{ g}}$$
$$= 3.7 \text{ kmol}$$

$$n_{NH_3} = 3.7 \text{ kmol} \times \frac{2}{3}$$
$$= 2.5 \text{ kmol}$$

$$V_{NH_3} = \frac{nRT}{P}$$

$$= 2.5 \text{ kmol} \times \frac{\frac{8.314 \text{ kPa} \cdot \text{L}}{1 \text{ mol} \cdot \text{K}} \times 353 \text{ K}}{450 \text{ kPa}}$$

$$= 16 \text{ kL}$$

or $V_{NH_3} = 7.5 \text{ kg H}_2 \times \dfrac{1 \text{ mol H}_2}{2.02 \text{ g H}_2} \times \dfrac{2 \text{ mol NH}_3}{3 \text{ mol H}_2} \times \dfrac{8.314 \text{ kPa NH}_3 \cdot \text{L NH}_3}{1 \text{ mol NH}_3 \cdot \text{K NH}_3}$
$\times \dfrac{353 \text{ K NH}_3}{450 \text{ kPa NH}_3}$

$= 16 \text{ kL NH}_3$

According to gas stoichiometry, from the complete reaction of 7.5 kg of hydrogen one can obtain 16 kL of ammonia.

SUMMARY: Gravimetric and Gas Stoichiometry

Step 1: Write a balanced chemical equation and list the measurements, unknown quantity symbol, and conversion factors for the measured and required substances.

Step 2: Convert the measured quantity to a chemical amount using the appropriate conversion factor.

Step 3: Calculate the chemical amount of the required substance using the mole ratio from the balanced equation.

Step 4: Convert the calculated chemical amount to the final quantity requested using the appropriate conversion factor.

Stoichiometry Calculations

measured quantity
solids/liquids m
gases V, T, P $\rightarrow n$

required quantity
solids/liquids m
gases V, T, P $\leftarrow n$

mole ratio

EXTENSION

Family Farming and Future Fuels

Ethanol created from agricultural waste can be used to produce hydrogen for fuel cells. This might give farmers a new source of income while providing a renewable and sustainable source of fuel. A research team at the University of Minnesota is investigating this process.

Practice

1. What volume of oxygen at STP is needed to completely burn 15 g of methanol in a fondue burner?

2. A Down's Cell is used in the industrial production of sodium from the decomposition of molten sodium chloride. A major advantage of this process compared with earlier technologies is the production of the valuable byproduct chlorine. What volume of chlorine gas at 30 °C and 95.7 kPa is produced, along with 105 kg of sodium metal, from the decomposition of sodium chloride?

3. Hydrogen gas is the fuel used in "pollution-free" vehicles in which hydrogen and oxygen gases react to produce water vapour and energy. Ballard Power Systems Inc. is a Canadian company pioneering the use of hydrogen fuel cells as power sources. Ballard heavy-duty fuel cells are currently being used to power a fleet of 30 Mercedes buses in trials in 10 European countries. What volume of oxygen at 40 °C and 1.50 atm is necessary to react with 300 L of hydrogen gas measured at the same conditions? (Recall the law of combining volumes.)

Section **7.3**

INVESTIGATION 7.3 Introduction

Producing Hydrogen

There are several possible methods that can be used in the Design and Analysis. The suggested method is to predict the volume of gas at STP and, in your Analysis, convert the measured volume to STP conditions using the combined gas law.

Purpose

The purpose of this investigation is to test the stoichiometric method applied to reactions that involve gases.

Problem

What is the volume at STP of hydrogen gas from the reaction of magnesium with excess hydrochloric acid?

To perform this investigation, turn to page 305.

Report Checklist

- ○ Purpose
- ○ Problem
- ○ Hypothesis
- ● Prediction
- ○ Design
- ○ Materials
- ○ Procedure
- ● Evidence
- ● Analysis
- ● Evaluation (1, 2, 3)

Design

A known mass of magnesium ribbon reacts with excess hydrochloric acid. The temperature, pressure, and volume of the hydrogen gas produced are measured.

Case Study

Producing Hydrogen for Fuel Cells

Hydrogen fuel cells are promoted as being environmentally friendly because their only product is water vapour. This claim is true only if the hydrogen used in the cells is produced with the minimum of environmental impact. Hydrogen is found in many compounds that occur in nature, but the element is very difficult to isolate in a reliable, efficient, and economic way.

At present, the hydrocarbon molecules in fossil fuels, primarily natural gas, are the main source of hydrogen. The industrial process of reforming fossil fuels to make hydrogen is a method called *steam reforming* (**Figure 2**), in which vaporized fossil fuels react with steam at high pressures and temperatures in the presence of a nickel-based catalyst:

$$CH_4(g) + 2 H_2O(g) \rightarrow CO_2(g) + 4 H_2(g)$$

Figure 2
Most hydrogen is produced industrially by steam reforming.

The steam reforming process is well established, and it is currently the most economic way to produce hydrogen. Some disadvantages of this process are that it consumes energy and dwindling fossil fuels and produces carbon dioxide, the primary greenhouse gas.

Another process for isolating hydrogen gas uses electricity. Electrolysis produces hydrogen by using an electrical current to separate water into hydrogen and oxygen:

$$2 H_2O(l) \rightarrow 2 H_2(g) + O_2(g)$$

Unfortunately, the energy for electrolysis usually comes from burning fossil fuels, so again, carbon dioxide production is a problem, as well as using up the limited supply of fossil fuels. Producing hydrogen by electrolysis is much more environmentally friendly when solar or wind power is used as the energy source (**Figure 3**). The development of hydrogen generators powered by low-voltage sources from renewable-power technology is currently an area of active research.

Biomass, such as carbohydrate-rich agricultural rubbish and wastewater from food processing, is also being examined as a source of hydrogen (**Figure 4**). Hydrogen-producing bacteria, which occur naturally in soil, can be used to produce commercial quantities of hydrogen gas, for example,

$$C_6H_{12}O_6(aq) + 2 H_2O(l) \rightarrow$$
$$2 CH_3COOH(aq) + 2 CO_2(g) + 4 H_2(g)$$

Proponents of producing hydrogen from biomass point out that the carbon dioxide produced can be absorbed by planting more agricultural crops. Any industrial scale process for hydrogen generation must be as reliable, economic, and efficient as the reforming of fossil fuels. The success of the search could well determine whether hydrogen's promise as the clean fuel of the future will be fully realized.

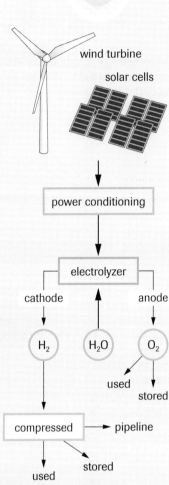

Figure 3
Solar-powered and wind-powered electrolysis is more environmentally friendly than electrolysis using fossil fuels.

Figure 4
Hydrogen-producing bacteria can produce commercial quantities of hydrogen.

Case Study Questions

1. Based on the stoichiometry of the reactions given on the previous page, what are the ratios of hydrogen to starting material for each of the three processes? To evaluate hydrogen sources, is it better to compare chemical amount ratios or mass ratios?

2. For the production of hydrogen or any other industrial technological process, what are the three main criteria that are used to judge the process?

3. A source of hydrogen is an important issue if hydrogen fuel cells are going to become useful for automobiles. What perspectives are mentioned in the Case Study? What important perspective is missing?

Extension

4. Search the Internet and other sources for reports of recent research on a process that extracts hydrogen from water or biomass. Evaluate the potential for this technology from a variety of perspectives. Include an analysis of the risks and benefits expected, and discuss the long-term sustainability of the process in the presentation of your findings.

www.science.nelson.com

Section 7.3 Questions

1. How does gravimetric stoichiometry compare with gas stoichiometry? Identify the similarities and differences in the procedures.

2. The first recorded observation of hydrogen gas was made by the famous alchemist Paracelsus (1493–1541) when he added iron to sulfuric acid. Predict the volume of hydrogen gas at STP produced by adding 10 g of iron to an excess of sulfuric acid.

3. A typical Alberta home heated with natural gas (assume methane, $CH_4(g)$) consumes 2.00 ML of natural gas during the month of December. What volume of oxygen at SATP is required to burn 2.00 ML of methane measured at 0 °C and 120 kPa?

4. Ammonia reacts with sulfuric acid to form the important fertilizer ammonium sulfate. What mass of ammonium sulfate can be produced from 75.0 kL of ammonia at 10 °C and 110 kPa?

5. Methane hydrate, a possible energy resource, looks like ice but is an unusual substance with the approximate chemical formula $CH_4 \cdot 6H_2O(s)$. It occurs in permafrost regions and in large quantities on the ocean floor (**Figure 5(a)**). Current, rough estimates of the quantity of methane hydrate suggest that it is at least twice the total known reserves of coal, oil, and natural gas combined. Considerable research is now underway to find ways to tap this huge energy resource. If 1.0 kg of solid methane hydrate decomposes to methane gas and water, what volume of methane is produced at 20 °C and 95 kPa (**Figure 5(b)**)?

(a)

(b)

Figure 5
(a) In the frigid ocean depths extreme pressure forms this mound of methane hydrate. When the camera's light warms the mound, bubbles of methane can be seen dissociating from the ice.
(b) The methane escaping from this block of methane hydrate has been ignited, and it burns while the ice melts to water.

6. As recently as the early 20th century, pinches of sulfur were sometimes burned in sickrooms. The pungent choking fumes produced were supposed to be effective against the "evil humours" of the disease. In fact, the sulfur dioxide gas produced is toxic and extremely irritating to lung tissue, where it dissolves to form sulfurous acid. Even today, a surprising number of people still believe that medicines are more likely to be effective if they have unpleasant tastes or odours. What volume of $SO_2(g)$ at SATP will be produced from the burning of 1.0 g of sulfur?

Figure 6
Alberta has large supplies of sulfur as a byproduct of natural gas production. Sulfur is used to make sulfuric acid and ammonium sulfate.

7. Alberta's natural gas often has hydrogen sulfide gas, $H_2S(g)$, mixed with it (among other things) when it comes out of a well. Hydrogen sulfide is highly toxic and must be removed from the gas stream. In the second step of this removal process, hydrogen sulfide reacts with sulfur dioxide gas at high temperatures to produce water vapour and sulfur vapour. Upon cooling, the sulfur condenses to a solid, which is then stockpiled (**Figure 6**). If 1000 L of $H_2S(g)$ at SATP reacts in this way, what mass of solid sulfur would be formed?

8. In this test, aqueous hydrogen peroxide is decomposed to water and oxygen gas. Complete the Prediction and Evaluation (Part 2 only) sections of the following report.

Purpose
The purpose of this investigation is to test the stoichiometric method for gas reactions.

Problem
What volume of oxygen at room conditions can be obtained from the decomposition of 50.0 mL of 0.88 mol/L aqueous hydrogen peroxide?

Design
A measured volume of a hydrogen peroxide solution (3%, 0.88 mol/L) is decomposed using manganese dioxide as a catalyst. The oxygen produced is collected by water displacement, just like the hydrogen in Investigation 7.3.

Evidence
volume of 0.88 mol/L $H_2O_2(aq)$ = 50.0 mL
volume of $O_2(g)$ = 556 mL
temperature = 21 °C
atmospheric pressure = 94.6 kPa

9. Describe briefly one consumer, one industrial, and one laboratory application of gases that involve a chemical reaction that uses or produces gases. For each example, include a complete balanced chemical equation.

7.4 Solution Stoichiometry

You have already seen the usefulness of gravimetric stoichiometry and gas stoichiometry for both predictions and analyses. However, the majority of stoichiometric work in research and in industry involves solutions, particularly aqueous solutions. Solutions are easy to handle and transport, and reactions in solution are relatively easy to control.

Solution stoichiometry is the application of stoichiometric calculation principles to substances in solution. The general stoichiometric method remains the same. The major difference is that the amount concentration and volume of a solution are used as conversion factors to convert to or from the chemical amount of substance.

INVESTIGATION 7.4 Introduction

Analysis of Silver Nitrate (Demonstration)

It is more financially viable to recycle metals if they are in fairly concentrated solutions, so recycling companies will pay more for these solutions than for dilute solutions. How do companies find out how much silver, for example, is in a solution? Technicians carry out a reaction that involves removing all the silver from a known volume of the solution, drying it, and measuring its mass.

Purpose
The purpose of this investigation is to use the stoichiometric method to find an unknown amount concentration.

Problem
What is the amount concentration of silver nitrate in solution?

To perform this investigation, turn to page 307.

Report Checklist
- ○ Purpose
- ○ Problem
- ○ Hypothesis
- ○ Prediction
- ○ Design
- ○ Materials
- ○ Procedure
- ● Evidence
- ● Analysis
- ● Evaluation (1, 2, 3)

Design
A precisely measured volume of aqueous silver nitrate solution, $AgNO_3(aq)$, reacts completely with excess copper metal, $Cu(s)$. The silver metal product, $Ag(s)$, is separated by filtration and dried, and the mass of silver is measured to the precision of the balance. The amount concentration of the initial solution is calculated from the mass of product by the stoichiometric method.

Figure 1
Fertilizers can have a dramatic effect on plant growth. The plants on the left were fertilized with an ammonium hydrogen phosphate fertilizer.

SAMPLE problem 7.4

Solutions of ammonia and phosphoric acid are used to produce ammonium hydrogen phosphate fertilizer (**Figure 1**). What volume of 14.8 mol/L $NH_3(aq)$ is needed for the ammonia to react completely with 1.00 kL of 12.9 mol/L $H_3PO_4(aq)$ to produce fertilizer?

First, write a balanced chemical equation so that the stoichiometry can be established. Beneath the equation, list both the given and the required measurements and the conversion factors:

$$2\,NH_3(aq) + H_3PO_4(aq) \rightarrow (NH_4)_2HPO_4(aq)$$
V 1.00 kL
14.8 mol/L   12.9 mol/L

Second, convert the information given for phosphoric acid to its chemical amount:

$$n_{H_3PO_4} = 1.00 \text{ kL} \times 12.9\,\frac{\text{mol}}{1\,\text{L}}$$

$$= 12.9 \text{ kmol}$$

300 Chapter 7

Third, use the mole ratio to calculate the amount of the required substance, ammonia. According to the balanced chemical equation, 2 mol of ammonia reacts for every 1 mol of phosphoric acid:

$$n_{NH_3} = 12.9 \text{ kmol} \times \frac{2}{1}$$
$$= 25.8 \text{ kmol}$$

Fourth, convert the amount of ammonia to the quantity requested in the question. The amount concentration is used to convert the chemical amount to the solution volume:

$$V_{NH_3} = 25.8 \text{ kmol} \times \frac{1 \text{ L}}{14.8 \text{ mol}}$$
$$= 1.74 \text{ kL}$$

As before, all steps may be combined as a single calculation:

$$V_{NH_3} = 1.00 \text{ kL } H_3PO_4 \times \frac{12.9 \text{ mol } H_3PO_4}{1 \text{ L } H_3PO_4} \times \frac{2 \text{ mol } NH_3}{1 \text{ mol } H_3PO_4} \times \frac{1 \text{ L } NH_3}{14.8 \text{ mol } NH_3}$$
$$= 1.74 \text{ kL } NH_3$$

▶ COMMUNICATION example

A technician determines the amount concentration, c, of a sulfuric acid solution. In the experiment, a 10.00 mL sample of sulfuric acid reacts completely with 15.9 mL of 0.150 mol/L potassium hydroxide solution. Calculate the amount concentration of the sulfuric acid.

Solution

$$H_2SO_4(aq) + 2 KOH(aq) \rightarrow 2 H_2O(l) + K_2SO_4(aq)$$

10.00 mL	15.9 mL
c	0.150 mol/L

$$n_{KOH} = 15.9 \text{ mL} \times \frac{0.150 \text{ mol}}{1 \text{ L}}$$
$$= 2.39 \text{ mmol}$$

$$n_{H_2SO_4} = 2.39 \text{ mmol} \times \frac{1}{2}$$
$$= 1.19 \text{ mmol}$$

$$[H_2SO_4(aq)] = \frac{1.19 \text{ mmol}}{10.00 \text{ mL}}$$
$$= 0.119 \text{ mol/L}$$

or

$$[H_2SO_4(aq)] = 15.9 \text{ mL KOH} \times \frac{0.150 \text{ mol KOH}}{1 \text{ L KOH}} \times \frac{1 \text{ mol } H_2SO_4}{2 \text{ mol KOH}} \times \frac{1}{10.00 \text{ mL } H_2SO_4}$$
$$= \frac{0.119 \text{ mol } H_2SO_4}{1 \text{ L } H_2SO_4}$$
$$= 0.119 \text{ mol/L } H_2SO_4$$

According to the stoichiometric method for solutions, the amount concentration of the sulfuric acid is 0.119 mol/L.

DID YOU KNOW

Plant Food
Nitrogen is the primary nutrient for plant growth. It promotes protein formation in crops and is a major component of chlorophyll, which helps promote healthy growth, producing high yields.

CAREER CONNECTION

Soil Scientist
Soil scientists use chemistry to investigate the composition of soil and how it behaves. They study how people and industry affect soil, and they create and monitor plans to remediate contaminated soils. Soil scientists also work closely with the agricultural and forestry industries to study and predict interactions of soil with other organisms.

Soil scientists are critical to understanding our environment. Find out whether this could be the career for you.

www.science.nelson.com

SUMMARY: Gravimetric, Gas, and Solution Stoichiometry

Stoichiometry Calculations

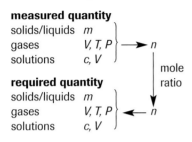

Step 1: Write a balanced chemical equation, and list the quantities and conversion factors for the given substance and the one to be calculated.
Step 2: Convert the given measurement to its chemical amount using the appropriate conversion factor.
Step 3: Calculate the amount of the other substance using the mole ratio from the balanced equation.
Step 4: Convert the calculated amount to the final quantity requested using the appropriate conversion factor.

Practice

1. Ammonium sulfate fertilizer is manufactured by having sulfuric acid react with ammonia. In a laboratory study of this process, 50.0 mL of sulfuric acid reacts with 24.4 mL of a 2.20 mol/L ammonia solution to produce the ammonium sulfate solution. From this evidence, calculate the amount concentration of the sulfuric acid at this stage in the process.

2. Slaked lime can be added to an aluminium sulfate solution in a water treatment plant to clarify the water. Fine particles in the water stick to the precipitate produced. Calculate the volume of 0.0250 mol/L calcium hydroxide solution required to react completely with 25.0 mL of 0.125 mol/L aluminium sulfate solution.

3. A chemistry teacher wants 75.0 mL of 0.200 mol/L iron(III) chloride solution to react completely with an excess quantity of 0.250 mol/L sodium carbonate solution. What is the minimum volume of sodium carbonate solution needed?

LAB EXERCISE 7.C

Testing Solution Stoichiometry

Report Checklist
- ○ Purpose
- ○ Problem
- ○ Hypothesis
- ● Prediction
- ○ Design
- ○ Materials
- ○ Procedure
- ○ Evidence
- ● Analysis
- ● Evaluation (1, 2, 3)

You have already tested the stoichiometric method for gravimetric and gas stoichiometry, but the testing of a scientific concept is never finished. Scientists keep looking for new experimental designs and new ways of testing a scientific concept. When completing the investigation report, pay particular attention to the evaluation of the Design.

Purpose
The purpose of this exercise is to test the stoichiometric method using solutions.

Problem
What mass of precipitate is produced by the reaction of 20.0 mL of 0.210 mol/L sodium sulfide with an excess quantity of aluminium nitrate solution?

Design
The two solutions provided react with each other, and the resulting precipitate is separated by filtration and dried. The mass of the dried precipitate is determined.

Evidence
A yellow precipitate resembling aluminium sulfide was formed.
mass of filter paper = 0.97 g
mass of dried filter paper plus precipitate = 1.17 g
A few additional drops of the sodium sulfide solution added to the filtrate produced a precipitate. *Hint*: What compound do you expect to be present in the filtrate solution?

LAB EXERCISE 7.D

Determining a Solution Concentration

Report Checklist
- ○ Purpose
- ○ Problem
- ○ Hypothesis
- ○ Prediction
- ● Design
- ○ Materials
- ○ Procedure
- ○ Evidence
- ● Analysis
- ○ Evaluation

Once a scientific concept has passed several tests, it can be used in industry. In your career as an industrial technician, you need to determine the amount concentration of a silver nitrate solution that, due to its cost, is being recycled.

Purpose
The purpose of this exercise is to use the stoichiometric method with solutions.

Problem
What is the amount concentration of silver nitrate in the solution to be recycled?

Evidence
A white precipitate was formed in the reaction with aqueous sodium sulfate.
volume of silver nitrate solution = 100 mL
mass of filter paper = 1.27 g
mass of dried filter paper plus precipitate = 6.74 g

Section 7.4 Questions

1. Some antacid products contain aluminium hydroxide to neutralize excess stomach acid. Determine the volume of 0.10 mol/L stomach acid (assumed to be HCl(aq)) that can be neutralized by 912 mg of aluminium hydroxide in an antacid tablet.

2. Sulfuric acid is produced on a large scale from readily available raw materials. One step in the industrial production of sulfuric acid is the reaction of sulfur trioxide with water. Calculate the amount concentration of sulfuric acid produced by the reaction of 10.0 Mg of sulfur trioxide with an excess quantity of water to produce 7.00 kL of acid.

3. Analysis shows that 9.44 mL of 50.6 mmol/L KOH(aq) is needed to completely react with 10.00 mL of water from an acidic lake. Determine the amount concentration of acid in the lake water, assuming that the acid is sulfuric acid.

4. Silver nitrate solution is used by electroplating businesses to replate silver tableware for their customers (**Figure 2**). To test the purity of the solution, a technician adds 10.00 mL of 0.500 mol/L silver nitrate to an excess quantity of 0.480 mol/L NaOH solution. From the reaction, 0.612 g of precipitate is obtained.
   (a) State a specific diagnostic test that could be done to verify that an excess had been added.
   (b) Calculate the predicted yield of precipitate.
   (c) What is the percent yield? What does this tell you about the purity of the solution?

5. Some commercial hydrochloric acid mixed with water produces 20.0 L of a 1.20 mol/L solution to be used to remove rust from car parts in a wrecking yard. What mass of rust can be reacted before the acid is used up? Assume solid Fe_2O_3 as a formula for rust.

6. Design an experiment to determine the amount concentration of a sodium sulfate solution. Include the Problem, Design, and Materials.

Figure 2
A small-scale electroplating business

Extension

7. In the late 1800s, two chemical processes, LeBlanc and Solvay, competed as methods for producing soda ash. The Solvay process clearly won and remains a major chemical industry today.
   (a) Why is soda ash important? List some uses.
   (b) What are the two major raw materials and two final products of the Solvay process?
   (c) Is the Solvay process a consumer-, commercial-, or industrial-scale technology?
   (d) Why was the Solvay process so successful? Working in a team, evaluate this technology compared to the LeBlanc process. Include a variety of perspectives.

www.science.nelson.com

Chapter 7 INVESTIGATIONS

INVESTIGATION 7.1

Decomposing Malachite

Copper(II) hydroxide carbonate, commonly called basic copper carbonate and also known as malachite, is a double salt with the chemical formula $Cu(OH)_2 \cdot CuCO_3(s)$ (**Figure 1**). This double salt decomposes completely when heated to 200 °C, forming copper(II) oxide, carbon dioxide, and water vapour. Complete the Prediction using the balanced chemical equation. Include safety and disposal steps in your Procedure. Organize the data and create the graph using suitable software.

Report Checklist
- ○ Purpose
- ○ Problem
- ○ Hypothesis
- ● Prediction
- ○ Design
- ○ Materials
- ● Procedure
- ● Evidence
- ● Analysis
- ● Evaluation (1, 2, 3)

Purpose
The purpose of this investigation is to test the assumption that a chemical reaction is stoichiometric.

Problem
How is the chemical amount of copper(II) oxide produced related to the chemical amount of malachite reacted in the decomposition of malachite?

Design
A known mass of malachite (manipulated variable) is heated strongly until the colour changes completely from green to black (**Figure 2**). The mass of black copper(II) oxide (responding variable) is determined. The results from several laboratory groups are combined in a graph to determine the ratio of chemical amounts.

Materials
lab apron
eye protection
porcelain dish (or crucible and clay triangle)
small ring stand
hot plate
glass stirring rod
sample of malachite (1 g to 3 g)
centigram balance
laboratory scoop or plastic spoon

  Malachite is toxic.
Do not touch the surface of the hot plate.

Figure 1
(a) The green mineral malachite is an important copper ore.
(b) When polished, it is also used as a semi-precious stone in jewellery. When prepared as a pure chemical substance, it is usually ground to a light green powder.

Figure 2
Use the glass stirring rod to break up lumps of powdered malachite and to mix the contents of the dish while they are being heated. Large lumps may decompose on the outside but not on the inside.

Chapter 7

INVESTIGATION 7.2

Gravimetric Stoichiometry

In this investigation, you will use gravimetric stoichiometry to investigate the reaction of strontium nitrate with excess copper(II) sulfate in an aqueous solution. Use 2.00 g of strontium nitrate and about 3.5 g of copper(II) sulfate–water (1/5), initially dissolving each chemical in about 75 mL of water. Be sure to include safety and disposal instructions in your Procedure. Refer to Appendix C.4 and the Nelson Web site for guidance on various lab techniques.

www.science.nelson.com

Purpose
The purpose of this investigation is to test the stoichiometric method.

Report Checklist
- ○ Purpose
- ○ Problem
- ○ Hypothesis
- ● Prediction
- ● Design
- ● Materials
- ● Procedure
- ● Evidence
- ● Analysis
- ● Evaluation (1, 2, 3)

Problem
What mass of precipitate is produced by the complete reaction of 2.00 g of strontium nitrate in solution with an excess of aqueous copper(II) sulfate?

 Strontium nitrate is moderately toxic; there is risk of fire when it is in contact with organic chemicals, and it may explode when bumped or heated. Copper(II) sulfate is a strong irritant and is toxic if ingested.

INVESTIGATION 7.3

Producing Hydrogen

There are several possible methods that can be used in the Design and Analysis. The suggested method is to predict the volume of gas at STP and, in your Analysis, convert the measured volume to STP conditions using the combined gas law.

Purpose
The purpose of this investigation is to test the stoichiometric method applied to reactions that involve gases.

Problem
What is the volume at STP of hydrogen gas from the reaction of magnesium with excess hydrochloric acid?

Design
A known mass of magnesium ribbon reacts with excess hydrochloric acid. The temperature, pressure, and volume of the hydrogen gas produced are measured.

Materials
lab apron
eye protection
disposable plastic gloves
magnesium ribbon, 60 mm to 70 mm
centigram or analytical balance
piece of fine copper wire, 100 mm to 150 mm

Report Checklist
- ○ Purpose
- ○ Problem
- ○ Hypothesis
- ○ Prediction
- ○ Design
- ○ Materials
- ○ Procedure
- ● Evidence
- ● Analysis
- ● Evaluation (1, 2, 3)

 Eye protection, a lab apron, and disposable gloves must be worn.

 Hydrochloric acid in 6 mol/L concentration is very corrosive. If acid is splashed into your eyes, rinse them immediately with water for 15 to 20 min. Acid splashed onto the skin should be rinsed immediately with plenty of water. Notify your teacher immediately. If acid is splashed onto your clothes, neutralize with baking soda, then wash thoroughly with plenty of water.

Rinse your hands well after step 8 in case any dilute acid got on your skin.

 Hydrogen gas, produced in the reaction of hydrochloric acid and magnesium, is flammable. Ensure that there is adequate ventilation and that there are no open flames in the classroom.

100 mL graduated cylinder
15 mL hydrochloric acid (6 mol/L)
250 mL beaker
water
large beaker (600 mL or 1000 mL)

INVESTIGATION 7.3 continued

two-hole stopper to fit cylinder
thermometer or temperature probe
barometer

Procedure

1. Measure and record the mass of the strip of magnesium.
2. Fold the magnesium ribbon to make a small compact bundle that can be held by a copper cage (**Figure 3**).

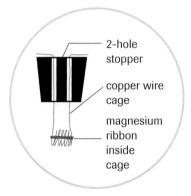

Figure 3
The magnesium should be small enough to fit into a copper cage (steps 2 and 3). Fasten the copper wire handle to the stopper (step 7).

3. Wrap the fine copper wire all around the magnesium, making a cage to hold it, but leaving 30 mm to 50 mm at each end of the wire free for a handle.
4. Carefully pour 10 mL to 15 mL of the hydrochloric acid into the graduated cylinder.
5. Slowly fill the graduated cylinder to the brim with water from a beaker. As you fill the cylinder, pour slowly down the side of the cylinder to minimize mixing of the water with the acid at the bottom. In this way, the liquid at the top of the cylinder is relatively pure water and the acid remains at the bottom.
6. Half-fill the large beaker with water.
7. Bend the copper wire handle through the holes in the stopper so that the cage holding the magnesium is positioned about 10 mm below the bottom of the stopper (**Figure 3**).
8. Insert the stopper into the graduated cylinder; the liquid in the cylinder will overflow a little. Cover the holes in the stopper with your finger. Working quickly, invert the cylinder, and immediately lower it into the large beaker so that the stopper is below the surface of the water before you remove your finger from the stopper holes (**Figure 4**).

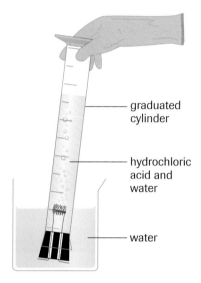

Figure 4
While holding the cylinder so it does not tip, rest it on the bottom of the beaker. The acid, which is denser than water, will flow down toward the stopper and react with the magnesium. The hydrogen produced should remain trapped in the graduated cylinder.

9. Observe the reaction, then wait about 5 min after the bubbling stops to allow the contents of the graduated cylinder to reach room temperature.
10. Raise or lower the graduated cylinder so that the level of liquid inside the beaker is the same as the level of liquid inside the graduated cylinder. (This equalizes the gas pressure in the cylinder with the pressure of the air in the room.)
11. Measure and record the volume of gas in the graduated cylinder.
12. Record the laboratory (ambient) temperature and pressure.
13. The liquids in this investigation may be poured down the sink, but rinse the sink with lots of water.

INVESTIGATION 7.4

Analysis of Silver Nitrate (Demonstration)

Report Checklist
- ○ Purpose
- ○ Problem
- ○ Hypothesis
- ○ Prediction
- ○ Design
- ○ Materials
- ○ Procedure
- ● Evidence
- ● Analysis
- ● Evaluation (1, 2, 3)

It is more financially viable to recycle metals if they are in fairly concentrated solutions, so recycling companies will pay more for these solutions than for more dilute solutions. How do companies find out how much silver, for example, is in a solution? Technicians carry out a reaction that involves removing all the silver from a known volume of the solution, drying it, and measuring its mass.

Purpose
The purpose of this investigation is to use the stoichiometric method to find an unknown amount concentration.

Problem
What is the amount concentration of silver nitrate in solution?

Design
A precisely measured volume of aqueous silver nitrate solution, $AgNO_3(aq)$, is completely reacted with excess copper metal, $Cu(s)$. The silver metal product, $Ag(s)$, is separated by filtration and dried, and the mass of silver measured to the precision of the balance. The amount concentration of the initial solution is calculated from the mass of product by the stoichiometric method.

Materials
lab apron
eye protection
>100 mL $AgNO_3(aq)$ of unknown amount concentration
centigram balance
#16–#20 gauge solid (not braided) copper wire
fine steel wool
wash bottle of pure water
wash bottle of pure acetone, $CH_3COCH_3(l)$
filtration apparatus
filter paper
250 mL beaker, with watch glass to fit
400 mL waste beaker for acetone
400 mL waste beaker for filtrate
100 mL graduated cylinder
stirring rod

 Acetone is volatile and flammable. Use only in a well-ventilated area. Keep away from any source of flame or sparks.

Procedure

Many of the skills and techniques required for this investigation are described in Appendix C.

Day 1

1. Using a graduated cylinder, measure 100 mL of silver nitrate solution and pour it into a 250 mL beaker.
2. Clean about 30 cm of solid copper wire with fine steel wool, and form about 20 cm of it into a coil with a 10 cm handle, so the coiled section will be submerged when placed in the silver nitrate solution.
3. Record any immediate evidence of chemical reaction, cover with a watch glass, and set aside.

Day 2

4. Check for completeness of reaction. If the coil is intact, with unreacted (excess) copper remaining, the reaction is complete; proceed to step 6. If all the copper has reacted, proceed to step 5.
5. Add another coil of copper wire, cover with a watch glass, and set aside until the next day.
6. Remove the wire coil. Shake the coil to ensure that all silver crystals remain in the beaker.
7. Measure and record the mass of a piece of filter paper.
8. Filter the beaker contents to separate the solid silver from the filtrate. (This technique is described in Appendix C.4, and demonstrated in a video on the Nelson Web site.)

www.science.nelson.com

9. Do the final three washes of the solid silver and filter paper with acetone from a wash bottle. Catch the rinsing acetone in a waste beaker.
10. Place the unfolded filter paper and contents on a paper towel to dry for a few minutes.
11. Measure and record the mass of the dry silver plus filter paper.
12. Dispose of solids in the garbage and the aqueous solutions (not acetone) down the drain with plenty of water. Transfer the acetone to a flammables disposal container.

Chapter 7 SUMMARY

Outcomes

Knowledge

- identify limitations and assumptions about chemical reactions (7.1)
- write balanced ionic and net ionic equations, including identification of spectator ions, for reactions taking place in aqueous solutions (7.1)
- recognize limiting and excess reagents in chemical reactions (7.1, 7.2, 7.3, 7.4)
- calculate quantities of reactants and/or products involved in chemical reactions using gravimetric, solution, or gas stoichiometry (7.2, 7.3, 7.4)
- define predicted (theoretical) and experimental (actual) yields, and explain the discrepancy between them (7.2, 7.3)
- identify sources of experimental uncertainty in experiments (7.2, 7.3, 7.4)

STS

- state that a goal of technology is to solve practical problems (7.2, 7.3, 7.4)
- recognize that technological problem solving may incorporate knowledge from various fields (7.2, 7.3)
- classify and evaluate technologies (7.2, 7.3, 7.4)
- explain how the appropriateness and the risks and benefits of technologies need to be assessed for each potential application from a variety of perspectives, including sustainability (7.3)

Skills

- initiating and planning: plan and predict states, products, and theoretical yields for chemical reactions (7.2); describe procedures for safe handling, storing, and disposal of materials used in the laboratory, with reference to WHMIS and consumer product labelling information (7.2, 7.4)
- performing and recording: translate word equations for chemical reactions into chemical equations, including states of matter for the products and reactants (7.2); balance chemical equations for chemical reactions, using lowest whole-number coefficients (7.2)
- analyzing and interpreting: interpret stoichiometric ratios from chemical reaction equations (7.2, 7.3, 7.4); perform calculations to determine theoretical yields and actual yields, percent yield, and error (7.2); use appropriate SI notation, fundamental and derived units, and significant digits when performing stoichiometry calculations (7.2, 7.3, 7.4)
- communication and teamwork: work collaboratively in addressing problems and applying the skills and conventions of science in communicating information and ideas and in assessing results (7.2)

Key Terms

7.1
quantitative reaction
stoichiometric reaction
net ionic equation
spectator ion
limiting reagent
excess reagent

7.2
stoichiometry

theoretical yield
gravimetric stoichiometry
percent yield

7.3
gas stoichiometry

7.4
solution stoichiometry

Key Equations

$$\text{percent yield} = \frac{\text{actual yield}}{\text{predicted yield}} \times 100 \quad (7.2)$$

▶ MAKE a summary

1. Expand the margin summary graphic on page 302 to clearly show the following:
   (a) the three systems used for initially calculating the amount of a measured substance and how many separate measurements are required for each
   (b) the six possible required final quantities from a stoichiometric calculation and the system used to calculate each of them
2. Refer back to your answers to the Starting Points questions at the beginning of this chapter. How has your thinking changed?

▶ Go To [www.science.nelson.com]

The following components are available on the Nelson Web site. Follow the links for *Nelson Chemistry Alberta 20–30*.
- an interactive Self Quiz for Chapter 7
- additional Diploma Exam-style Review questions
- Illustrated Glossary
- additional IB-related material

There is more information on the Web site wherever you see the Go icon in this chapter.

+ EXTENSION

CBC radioONE QUIRKS & QUARKS

Touchy-Feely Robots
One of the obstacles stopping robots from performing delicate manipulations is their lack of touch sensitivity. Researchers are working on a new material, incorporating nanoparticles and electrodes, that will give robots the sense of touch.

[www.science.nelson.com]

Chapter 7 REVIEW

Many of these questions are in the style of the Diploma Exam. You will find guidance for writing Diploma Exams in Appendix H. Exam study tips and test-taking suggestions are on the Nelson Web site. Science Directing Words used in Diploma Exams are in bold type.

www.science.nelson.com

DO NOT WRITE IN THIS TEXTBOOK.

Part 1

1. A main goal of technology is to
   A. advance science
   B. identify problems
   C. explain natural processes
   D. solve practical problems

2. In the reaction of aqueous solutions of sodium sulfide and zinc nitrate in a chemical analysis, the spectator ions are
   A. sodium and nitrate ions
   B. sulfide and zinc ions
   C. sodium and zinc ions
   D. sulfide and nitrate ions

3. In which sections of an investigation report are stoichiometric calculations most likely to be found?
   A. Problem and/or Procedure
   B. Prediction and/or Analysis
   C. Purpose and/or Materials
   D. Hypothesis and/or Evaluation

4. The four general steps of any stoichiometry calculation are given in the following numbered list.
   1. converting a chemical amount to another quantity
   2. writing a balanced equation and listing information
   3. converting another quantity (or more than one) to a chemical amount
   4. determining one chemical amount from another chemical amount

   List the order in which these steps occur:
   ___, ___, ___, and ___.

Use this information to answer questions 5 to 9.

The mineral malachite is mined for use as a copper ore (**Figure 1**). After malachite has been roasted (decomposed by heat), the next step in the production of copper metal is a single replacement reaction. Copper(II) oxide reacts with hot carbon to produce copper metal and carbon dioxide. Assume that a 1.00 kg sample of pure copper(II) oxide is reacted.

Figure 1
If copper ore is close to the surface, it is recovered using huge digging and transport machinery in an open pit mine.

5. The chemical amount of the copper(II) oxide to be reacted is _____ mol.

6. The mass of carbon that will be required to completely react with all the copper(II) oxide is _____ g.

7. This reaction situation suggests the use of an excess reagent. The substance that should be deliberately supplied in excess quantity is
   A. $CO_2(g)$
   B. $CuO(s)$
   C. $C(s)$
   D. $Cu(s)$

8. The mass of copper that should be formed by completely reacting all the copper(II) oxide is _____ g.

9. The carbon dioxide produced is vented to the atmosphere. What volume would this amount of carbon dioxide occupy at SATP?
   A. 156 L
   B. 77.9 L
   C. 39.0 L
   D. 24.8 L

Use this information to answer questions 10 to 13.

The balanced equation for the combustion of sulfur is
$S_8(s) + 8\,O_2(g) \rightarrow 8\,SO_2(g)$. Assume a sample of pure sulfur is burned in air, which is about 20% oxygen.

10. The balanced equation provides you with clear and direct information about
    A. the temperature and pressure at which the reaction will be spontaneous
    B. the likelihood that the reaction will be quantitative (complete)
    C. the initial rate of reaction and the time required for the reaction to finish
    D. the ratio of chemical amounts of reactants and products

11. At a glance, without any calculation, it is possible to confidently say that
    A. the mass of oxygen that reacts will be eight times the mass of sulfur that reacts
    B. the chemical amount of sulfur dioxide formed is greater than the amount of sulfur reacting
    C. oxygen will be the limiting reagent
    D. sulfur dioxide gas will form at SATP conditions

12. The mass of toxic sulfur dioxide gas produced by quantitative reaction of 4.00 g of sulfur is _____ g.

13. If this reaction were to be done inside a sealed container, with the only change in conditions being the use of 100% pure oxygen, the reaction should happen much faster because the rate of collisions between oxygen molecules and sulfur molecules should become much greater. What volume of oxygen, measured at SATP, would be required to burn each 1.00 kg of sulfur under these conditions?
    A. 773 L
    B. 544 L
    C. 224 L
    D. 24.8 L

Part 2

14. List the four basic assumptions made for chemical reactions when doing stoichiometric calculations.

15. What is meant by the percent yield of a reaction, for which a stoichiometric calculation was used to predict the quantity of a product?

16. List the common sources of experimental uncertainty that may account for some of the difference between predicted and experimental quantities.

17. **Explain** why, in all stoichiometric calculations, you always have to convert to or convert from chemical amounts.

18. Technology has always been a part of any society, even going back to the Stone Age. As knowledge, societal needs, and problems increase, more sophisticated technology develops.
    (a) Technologies may be classified according to their scale and use. What are three contexts used to classify technology?
    (b) For a particular technology, what are the main criteria used to judge the product or process?
    (c) List at least five perspectives that may be used when evaluating technologies.

19. A sodium phosphate solution is used to test tap water for the presence of calcium ions (**Figure 2**). A sample of tap water reacts with sodium phosphate solution to produce a precipitate.
    (a) Write the net ionic equation for the reaction.
    (b) **Identify** the spectator ion(s).
    (c) Based on the given design, **identify** the limiting and excess reagents.
    (d) **Identify** a possible, significant flaw in the design of this experiment.

Figure 2
The precipitate, formed when clear, colourless sodium phosphate solution is added to tap water, indicates that the tap water contains dissolved water-hardening "impurities."

20. When heated, baking soda (**Figure 3**) decomposes into solid sodium carbonate, carbon dioxide, and water vapour.
    (a) If 2.4 mol of baking soda decomposes, what chemical amount of each of the products is formed?
    (b) What mass of solid product will remain after complete decomposition of a 1.00 kg box of baking soda?
    (c) Suggest a reason why baking soda can be used as a fire extinguisher.

Figure 3
Baking soda is pure sodium bicarbonate. An open box should always be in any kitchen because it is an excellent extinguisher for small cooking fires.

21. A convenient source of oxygen in a laboratory is the decomposition of aqueous hydrogen peroxide to produce water and oxygen. What volume of 0.88 mol/L hydrogen peroxide solution (**Figure 4**) is required to produce 500 mL of oxygen at SATP?

Figure 4
When some coloured liquid soap is added, hydrogen peroxide decomposes rapidly to produce water and oxygen. The soap foams as it traps bubbles of the oxygen gas being formed.

Use this information to answer questions 22 to 25.

In a chemical reaction done to test the stoichiometric method, 3.00 g of silver nitrate in aqueous solution reacts with a large excess of sodium chromate in solution to produce 2.81 g of dry precipitate.

22. **Determine** the predicted mass of precipitate.

23. (a) What is the percent yield?
    (b) **Predict** some reasons that might account for the difference between the predicted and the actual yield.

24. Write the net ionic equation for this analysis.

25. **Identify** the spectator ions.

26. In plants, the process called photosynthesis involves a reaction to produce glucose and oxygen from carbon dioxide and water (**Figure 5**). This endothermic reaction is powered by light energy from the sun and is catalyzed by chlorophyll:

$$6\ CO_2(g) + 6\ H_2O(l) + energy \rightarrow C_6H_{12}O_6(aq) + 6\ O_2(g)$$

(a) **Predict** the mass of carbon dioxide consumed when a plant makes 10.0 g of glucose.
(b) **Predict** the mass of oxygen produced when a plant makes 10.0 g of glucose.

Figure 5
The production of carbohydrates by plants is the fundamental source of energy for almost all living things on Earth.

Use this information to answer questions 27 to 32.

Gasohol is a general term that refers to automobile fuel that has 10% ethanol, $C_2H_5OH(l)$, blended with unleaded gasoline (**Figure 6**). Using this fuel reduces some noxious exhaust emissions. Assume that 1.00 kg of ethanol reacts completely in a car engine.

Figure 6
Blending gasoline with ethanol not only produces a fuel that burns more cleanly, and also helps conserve petroleum resources.

27. Write the balanced equation for the complete combustion of ethanol.

28. **Predict** the mass of oxygen is required?

29. **Predict** the mass of carbon dioxide produced.

30. **Predict** the mass of water produced.

31. Show that your calculated answers to the previous questions agree with the law of conservation of mass.

32. **Illustrate**, using examples involving automobile transportation, polluting emissions, and catalytic converters, how science and technology have both intended and unintended consequences and also how science often leads technology, and technology often leads science.

33. A metal refinery that uses a hydrometallurgical (aqueous solution) process to produce pure metals uses a stoichiometric procedure to determine the cobalt(II) ion amount concentration in the process solution. Complete the Analysis of the investigation report.

Purpose
The purpose of this investigation is to use stoichiometric calculations to determine an unknown solution amount concentration.

Problem
What is the amount concentration of cobalt(II) sulfate in a 100.0 mL sample of process solution?

Design
Solid sodium carbonate is added to the hot aqueous sample and dissolved, with stirring. When adding more sodium carbonate causes no more precipitate to form, the precipitate is allowed to settle and is then filtered. The mass of the dried precipitate is determined.

Evidence
A red crystalline precipitate was formed in the reaction.

volume of cobalt(II) sulfate solution	100.0 mL
mass of filter paper	1.04 g
mass of dry precipitate plus paper	8.98 g

34. Make a list of theories, laws, generalizations, and rules that you must know in order to be able to solve a stoichiometry problem.

35. Chemical technicians in water treatment plants perform several routine reactions daily on a very large scale. Research and report on the use of stoichiometry for ensuring the quality of a municipal water supply. Your report should include
    - descriptions of two or three water-quality tests involving stoichiometry
    - graphics
    - properly referenced data

www.science.nelson.com

chapter 8

Chemical Analysis

In this chapter

- Exploration: Test Your Drinking Water
- Web Activity: Is Your Classroom Putting You to Sleep?
- Biology Connection: Quantitative Analysis
- Lab Exercise 8.A: Chemical Analysis Using a Graph
- Investigation 8.1: Analysis of Sodium Carbonate
- Web Activity: Ursula Franklin
- Investigation 8.2: Percent Yield of Barium Sulfate
- Case Study: The Haber Process
- Investigation 8.3: Standardization Analysis of NaOH(aq) (Demonstration)
- Investigation 8.4: Titration Analysis of Vinegar
- Investigation 8.5: pH Curves (Demonstration)
- Web Activity: Blood Alcohol Content
- Web Activity: Titration Curves
- Case Study: Analytic Measurement Technology
- Investigation 8.6: Titration Analysis of ASA

One of the earliest chemical technologies was the control of fermentation—the production of ethanol (ethyl alcohol) from plant sugars. In the human body, this alcohol induces chemical reactions that affect the coordination and judgment of the drinker. In turn, these changes can contribute to serious car accidents. This is why Canada has laws stipulating limits to the concentration of ethanol allowed in the blood of a motorist.

When a driver is asked to breathe into a breathalyzer, the device measures the alcohol content in the exhaled air. The reading from the device indicates the result as a concentration of alcohol in the blood (**Figure 1**). For example, a reading of 0.08 on a breathalyzer means that the blood alcohol content is 0.08%, or 80 mg of alcohol in 100 mL of blood. A police officer takes a breath sample for on-the-spot analysis. Because of the possibility of challenges in court, the officer must be prepared to defend the reliability and accuracy of the reading. If the breathalyzer test indicates an alcohol concentration above the legal limit (in Canada, the legal limit is 0.08), a blood sample may be analyzed more precisely in a laboratory. This test uses a technique called titration, which you will learn about in this chapter.

Chemical analysis involves knowledge of chemical reactions, an understanding of diverse experimental designs, and practical skills to apply this knowledge and understanding. In this chapter, you will have opportunities to develop all of these using the skills you developed in Chapter 7.

Chemical analysis is also closely tied to technology, using specialized equipment and techniques to detect and measure substances with ever-increasing accuracy and precision. Understanding the function and handling of such equipment is also part of your experience in this chapter.

STARTING Points

Answer these questions as best you can with your current knowledge. Then, using the concepts and skills you have learned, you will revise your answers at the end of the chapter.

1. Does the colour of a solution indicate which solutes it contains?
2. Do calculated quantities of a product necessarily predict how much product actually forms in a reaction?
3. Can the amount of a second reactant required for a reaction be determined without a stoichiometric calculation?
4. When an acidic solution is slowly poured into a sample of basic solution, how does the solution's pH change?

Career Connection:
Hydrologist

Figure 1
A breath sample is quickly analyzed for ethanol concentration by this compact technological device.

▶ Exploration — Test Your Drinking Water

Canadian government agencies routinely analyze our drinking water. The increasing sales of bottled water purified by reverse osmosis or distillation testify to public awareness and concern in this area. Government labs, of course, have access to high technology and highly trained and skilled personnel. The chemistry knowledge you have acquired, however, will let you make some informed statements about tap water with just a couple of simple tests.

Materials: tap water sample, pure water sample (distilled or filtered by reverse osmosis), dropper bottles of sodium carbonate solution, dropper bottles of silver nitrate solution, four small test tubes

- To each of the two water samples in separate test tubes, add a few drops of Na_2CO_3(aq).
- Carefully observe each sample for signs of precipitate formation.
- To each of the two water samples in new separate test tubes, add a few drops of $AgNO_3$(aq).
- Carefully observe each sample for signs of precipitate formation.

(a) Referring to a table of ionic compound solubility generalizations, state which ions are *not* present in significant concentration in the tap water sample.
(b) If you have evidence of precipitation, state which ions, according to the evidence, *may* be present in the tap water sample.
(c) List some ions that cannot be analyzed and identified by these two tests.
(d) Water purification usually involves some degree of chlorination to kill microorganisms. Given this fact, what common ion might you expect to be present in most municipally treated tap water?

Chemical Analysis

8.1 Introduction to Chemical Analysis

Analysis of an unknown chemical sample can include both *qualitative* analysis—the identification of a specific substance present—and *quantitative* analysis—the determination of the quantity of a substance present.

While there are many analytical technologies and procedures used, three methods predominate:

- **colorimetry**, or analysis by colour, which uses light emitted, absorbed, or transmitted by the chemical (Section 8.1)
- **gravimetric analysis**, which uses stoichiometric calculations from a measured mass of a reagent (Section 8.2)
- **titration analysis**, which uses stoichiometric calculations from a measured solution volume of a reagent (Section 8.4)

Colorimetry

First, consider the colours of aqueous solutions. Observation shows that most aqueous solutions are colourless. As **Table 1** shows, ions of elements in Groups 1, 2, and 17 impart no colour at all to solutions. Some other ions, not listed in the table, are also colourless. However, many solutions containing monatomic and polyatomic ions of the transition elements do have a visible colour.

Aqueous ion colour is due to the ion's interference with visible light. Ions absorb specific wavelengths, which makes analysis possible. A specific colour identifies a particular ion. The percentage of light that is absorbed depends on how many ions are in the light path, that is, on the concentration of that ion.

For example, in the reaction of copper with silver nitrate solution that you studied in Chapter 7, you observed that silver and nitrate ions are both colourless (and, thus, invisible) in solution. Copper(II) ions, however, are a characteristic blue colour, making it easy to tell when they are forming in the reaction. The blue colour becomes more intense as the reaction proceeds because the concentration of the copper(II) ions is increasing.

Table 1 Colours of Solutions

Ion	Solution colour
Groups 1, 2, 17	colourless
Cr^{2+}(aq)	blue
Cr^{3+}(aq)	green
Co^{2+}(aq)	pink
Cu^+(aq)	green
Cu^{2+}(aq)	blue
Fe^{2+}(aq)	pale green
Fe^{3+}(aq)	yellow-brown
Mn^{2+}(aq)	pale pink
Ni^{2+}(aq)	green
CrO_4^{2-}(aq)	yellow
$Cr_2O_7^{2-}$(aq)	orange
MnO_4^-(aq)	purple

▶ COMMUNICATION *example*

According to the evidence in **Figure 1**, which is organized in **Table 2**, which solution is potassium dichromate, sodium chloride, sodium chromate, potassium permanganate, nickel(II) nitrate, and copper(II) sulfate? (Refer to **Table 1**.)

Table 2 Colours of the Unknown Solutions

Solution	1	2	3	4	5	6
Colour	purple	colourless	green	yellow	blue	orange

Solution

According to the evidence and **Table 1**, the solutions are (1) potassium permanganate, (2) sodium chloride, (3) nickel(II) nitrate, (4) sodium chromate, (5) copper(II) sulfate, and (6) potassium dichromate.

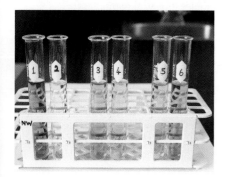

Figure 1
Which solution is which?

Section **8.1**

Aqueous ions can sometimes be identified qualitatively by eye, but for more precise identification or for quantitative measurement, technology must be used. A standard spectrophotometer is a device that measures the quantity of light absorbed at any desired visible wavelength when a light beam is passed through a solution sample. A spectrophotometer, like the one in **Figure 2**, can measure the concentration of any desired coloured ion, even in a solution that has several different mixed colours, because it can be adjusted to "see" only the precise colour (wavelength) selected.

We can also use flame tests to detect the presence of several metal ions, such as copper(II), calcium, and sodium (see the Selected Ion Colours table on the inside back cover). In a flame test, a clean platinum or nichrome wire is dipped into a test solution and then held in a nearly colourless flame (**Figure 3**). There are other ways to conduct flame tests: You could dip a wood splint in the aqueous solution and then hold it close to a flame; you could hold a tiny solid sample of a substance in the flame; or you could spray the aqueous solution into the flame.

Robert Wilhelm Bunsen and Gustav Robert Kirchhoff took the idea of the flame test and developed it into a technique called spectroscopy. Bunsen had previously invented an efficient gas laboratory burner that produced an easily adjustable, hot, nearly colourless flame. Bunsen's burner made better research possible—a classic example of technology leading science—and made his name famous. Bunsen and Kirchhoff soon discovered two new elements, cesium and rubidium, by examining the spectra produced by passing the light from flame tests through a prism.

Flame tests are still used for identification today. Of course, the technology has become much more sophisticated. An atomic absorption spectrophotometer (**Figure 4**) analyzes the light absorbed by samples vaporized in a flame. It can even detect wavelengths not visible to humans, which means that it can "see" and measure the concentrations of ions that are invisible to us, such as silver ions. This type of spectrophotometer can detect minute quantities of substances, in concentrations as tiny as parts per billion. By measuring the quantity of light absorbed, this device can also do quantitative analysis—measuring the concentrations of various elements precisely and accurately. Similar technology is used in a completely different branch of science: astronomy. Astronomers study the light spectra from distant stars to find out what elements make up the stars.

The Northern Lights (**Figure 5**) often create beautiful displays of moving colour in the sky in the Canadian North. Spectroscopy indicates that the various colours are due to high-energy charged particles from the Sun colliding with different molecules in the atmosphere at different altitudes. For example, a reddish colour is emitted by oxygen atoms at altitudes over 300 km, and greenish yellow (the most common) is from oxygen atoms at altitudes from 100 to 300 km.

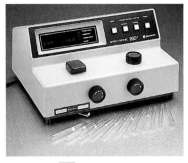

Figure 2
Solutions of different colour intensity absorb light to different degrees, so the concentration of a coloured product can be measured by light absorbence.

Figure 3
Copper(II) ions impart a green colour to this flame. This green flame and the characteristic blue colour in aqueous solution can be used as diagnostic tests for copper(II) ions.

Figure 5
Inuit nearly all share a common legend about the aurora, that the moving streamers of light are the spirits of their ancestors playing a game involving kicking a walrus skull around the sky, just as they did to while away the long winter darkness when they were living. The Innu word "aqsalijaat" can be loosely translated as "the trail of those playing soccer."

Figure 4
The atomic absorption spectrophotometer is a valuable tool, essential for precise qualitative and quantitative analyses in many areas of science.

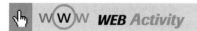

WEB Activity

Web Quest: Is Your Classroom Putting You to Sleep?

The air we breathe is invisible, so we cannot always tell if it is pure and healthy for us. Sometimes, pollutants can build up in the air and affect our health and our performance. How can you tell if the air you breathe has poisons in it? What are the signs of bad air? What is the impact on the individuals who breathe it? This Web Quest explores the dangers of toxins in the air, how they impact the people who breathe them, and what can be done to improve the situation.

www.science.nelson.com

Section 8.1 Questions

1. Describe the fundamental difference between quantitative and qualitative analysis.

2. What is the expected colour of solutions that contain the following? (Refer to **Table 1**.)
   (a) Na^+(aq)
   (b) Cu^{2+}(aq)
   (c) Fe^{3+}(aq)
   (d) $Cr_2O_7^{2-}$(aq)
   (e) Cl^-(aq)
   (f) Ni^{2+}(aq)

3. What colour is imparted to a flame by the following ions? Refer to the Selected Ion Colours table on the inside back cover.
   (a) calcium
   (b) copper(II)
   (c) Na^+
   (d) K^+
   (e) H^+

4. Flame tests on solids produce the same results as flame tests on solutions. These tests may be used as additional evidence to support the identification of precipitates. What colour would the following precipitates give to a flame?
   (a) $CaCO_3$(s)
   (b) $PbCl_2$(s)
   (c) $SrSO_4$(s)
   (d) $Cu(OH)_2$(s)

5. Complete the Analysis and Evaluation (Design only) of the following report.
   Problem
   What ions are present in the solutions provided?
   Design
   The solution colour is noted and a flame test is conducted on each solution.
   Evidence
   Table 3 Solution and Flame Colours

Solution	Solution colour	Flame colour
A	colourless	violet
B	blue	green
C	colourless	yellow
D	colourless	yellow-red
E	colourless	bright red

6. Artificial fire logs for home fireplaces are commonly available in supermarkets and hardware stores. Along with the combustible ingredients, the fire logs often have chemicals deliberately added to colour the flames. If such a fire log has copper(II) chloride near its core, sodium nitrate in layers farther out from the centre, and strontium chloride near the surface, describe how the flames will look over the normal three-hour burning period.

Extension

7. Identifying ions in an aqueous solution can be very important. Nitrate ions in well water, for example, must be identified because they may be harmful to health (especially for children) if the concentration is too high. To explore the use of chemical solubilities and flame colour in ion identification, assume that you have a solution containing several common cations and anions, which may or may not contain strontium ions. Write an experimental design for an analysis to determine whether strontium ions are present. Use two precipitation reactions followed by filtrations, and then a flame test. Use your Solubility of Ionic Compounds table (inside back cover) to decide what solutions you might use for the precipitation reactions. You may assume that no ions are present that are not listed in this table. *Hint:* Plan your first precipitation to remove most cations that are *not* strontium ions from the solution. Explain the logic you apply to each step of your design, in particular, why a flame test is required as a final step.

8. Forensic chemists with the RCMP and other forensic laboratories use flame emission spectroscopy to analyze glass and paint chips found at crime scenes.
   (a) Describe a scenario in which such an analysis would be useful.
   (b) What roles do you think science, society, and technology played in the development of this technology?

Gravimetric Analysis 8.2

Quantitative Analysis

The chemistry and the technology of quantitative analysis are closely related; knowledge and skills in both areas are essential for chemical technologists in medicine, agriculture, and industry.

In one type of chemical analysis, precipitation is part of the experimental design. As you know, precipitation occurs when a reaction forms a slightly soluble product. In a quantitative analysis involving precipitation, the sample under investigation is combined with an excess quantity of another reactant to ensure that all the sample reacts.

BIOLOGY CONNECTION

Quantitative Analysis
Chemistry is not the only science in which quantitative analysis is important. If you are studying biology, you may quantitatively analyze oxygen consumption of germinating seeds.

www.science.nelson.com

LAB EXERCISE 8.A

Chemical Analysis Using a Graph

Report Checklist
- ○ Purpose
- ○ Problem
- ○ Hypothesis
- ○ Prediction
- ○ Design
- ○ Materials
- ○ Procedure
- ○ Evidence
- ● Analysis
- ○ Evaluation

Lab technicians sometimes perform the same chemical analysis on hundreds of samples every day. For example, in a medical laboratory, blood and urine samples are routinely analyzed for specific chemicals such as cholesterol and sugar. In many industrial and commercial laboratories, technicians read the required quantity of a chemical from a graph that has been prepared in advance. This saves the time and trouble of doing a separate stoichiometric calculation for each analysis performed. By completing the Analysis of the investigation report, you will be illustrating this practice. Use graphics or spreadsheet software, if available, to create your graph.

Purpose
The purpose of this lab exercise is to use a graph of a precipitation reaction's stoichiometric relationship to determine the mass of lead(II) nitrate present in a sample solution.

Problem
What mass of lead(II) nitrate is in 20.0 mL of a solution?

Design
Samples of two different lead(II) nitrate solutions are used. Each sample is reacted with an excess quantity of a potassium iodide solution, producing lead(II) iodide, which has a low solubility and settles to the bottom of the beaker (**Figure 1**). After the contents of the beaker are filtered and dried, the mass of lead(II) iodide is determined. The reference data supplied in **Table 1**, relating the mass of $Pb(NO_3)_2$ to the mass of PbI_2 for this reaction, are graphed. The analysis is completed by reading the mass of lead(II) nitrate present in each solution from the graph.

Table 1 Reaction of Lead(II) Nitrate and Potassium Iodide

Mass of PbI_2 produced (g)	Mass of $Pb(NO_3)_2$ reacting (g)
1.39	1.00
2.78	2.00
4.18	3.00
5.57	4.00
6.96	5.00

Evidence

Table 2 Two Different $Pb(NO_3)_2$ Solutions

	Solution 1	Solution 2
Volume used (mL)	20.0	20.0
Mass of filter paper (g)	0.99	1.02
Mass of dried paper plus precipitate (g)	5.39	8.57

Figure 1
When lead(II) nitrate reacts with potassium iodide, a bright yellow precipitate forms.

Precipitation Completeness

In Lab Exercise 8.A, the Design states that an excess of potassium iodide solution is used, but no mention is made about how to determine that the quantity you choose to add is, in fact, an excess. In a gravimetric analysis where a precipitation reaction is used, it is not possible to predict the quantity of excess reagent required, because you do not initially *know* the amount of the limiting reagent; that is why you are doing an analysis. For such reactions, use the following *trial-and-error* procedure to verify that a sample of limiting reagent has completely reacted:

1. Precisely measure a sample volume of the solution containing the limiting reagent.
2. Add (while stirring) an approximately equal volume of the excess reagent solution.
3. Allow the precipitate that forms to settle, until the top layer of solution is clear.
4. With a medicine dropper, add a few more drops of excess reagent solution. Allow the drops to run down the side of the container, and watch for any cloudiness that may appear when the drops mix with the clear surface layer (**Figure 2(a)**).
5. If any new cloudiness is visible, the reaction of the limiting reagent sample is not yet complete. Repeat steps 2 to 4 of this procedure as many times as necessary, until no new precipitate forms during the test in step 4 (**Figure 2(b)**).
6. When no new cloudiness is visible (the test does not form any further precipitate), the reaction of the sample of limiting reagent is complete.

Recall from Chapter 7 that for stoichiometric calculations to provide useful information about any reaction done for purposes of chemical analysis, we must assume that the reaction will be spontaneous, rapid, quantitative, and stoichiometric. The precipitation reaction you will use in Investigation 8.1 (to provide evidence for the analysis of sodium carbonate) is a good example of a reaction that meets all four of these requirements.

Figure 2
When the precipitate has settled enough so that the top layer of solution is clear, you can test for completeness of reaction.
(a) If cloudiness forms, more of the excess reagent will have to be added.
(b) If no cloudiness forms, the reaction of limiting reagent is complete.

Practice

1. Explain, in terms of cations, anions, and collision–reaction theory, what completeness of reaction actually means for the reaction in Lab Exercise 8.A. Write a net ionic equation to illustrate your answer.
2. Is the trial-and-error procedure a scientific or a technological procedure?
3. Write a procedure for how you could make sure that a precipitation reaction "goes to completion" if the sample is filtered after the first precipitation, *without* testing for completeness before filtering. Explain what solution would be added, where, and why.

INVESTIGATION 8.1 Introduction

Analysis of Sodium Carbonate

Sodium carbonate has been used for all of recorded history in a variety of applications, from glassmaking to detergent manufacture to water treatment. A common name for sodium carbonate is soda ash, an appropriate name because it can easily be extracted from wood ashes. Sodium carbonate is one of the ten highest-volume chemicals produced in North America.

In this investigation, you will use techniques and equipment common to gravimetric analysis to analyze a sodium carbonate solution.

Purpose

The purpose of this investigation is to use the stoichiometric method as part of a gravimetric analysis.

To perform this investigation, turn to page 340.

Report Checklist
- ○ Purpose   ○ Design   ● Analysis
- ○ Problem   ○ Materials   ● Evaluation (1, 3)
- ○ Hypothesis   ○ Procedure
- ○ Prediction   ● Evidence

Problem
What is the mass of solute in a 50.0 mL sample of sodium carbonate solution?

Design
The mass of sodium carbonate present in the sample solution is determined by having it react with an excess quantity of a calcium chloride solution. The mass of calcium carbonate precipitate formed is used in stoichiometric calculation to determine the mass of sodium carbonate that reacted.

Section 8.2 Questions

1. A student wants to precipitate all the toxic lead(II) ions from 2.0 L of solution containing 0.34 mol/L $Pb(NO_3)_2$(aq). The purpose of this reaction is to make the filtrate solution nontoxic. If the student intends to precipitate lead(II) sulfate, suggest an appropriate solute, and calculate the minimum required mass of this solute.

2. A chemical analyst wants to determine the concentration of a solution of copper(II) sulfate that is used for treating wood, to prevent decay. A large strip of zinc metal is placed in a 200 mL sample of this solution. When the reaction shows no further change, much of the zinc strip remains. The originally blue solution is now colourless. A brownish layer of fine copper particles has formed, which when filtered and dried, has a mass of 1.72 g. What is the amount concentration of the sample solution?

3. Only quantitative reactions are suitable for use in a chemical analysis. Complete the Prediction, Analysis, and Evaluation (of the Design and Prediction) of the investigation report.

    Purpose
    The purpose of this investigation is to use the stoichiometric method to determine whether a reaction is quantitative.

    Problem
    What mass of precipitate is produced by the reaction of 20.0 mL of 0.210 mol/L sodium sulfide with an excess quantity of aluminium nitrate solution?

    Design
    The two solutions provided react with each other, and the resulting precipitate is separated by filtration and then dried. The mass of the dried precipitate is determined.

    Evidence
    A precipitate formed very rapidly when the solutions were mixed.

    A few additional drops of the aluminium nitrate solution added to the clear layer above the settled precipitate produced no additional cloudiness.

    mass of filter paper = 0.97 g

    mass of dried filter paper plus precipitate = 1.17 g

4. Many industries recycle valuable byproducts, such as silver nitrate solution. You are an industry technician who needs to determine the amount concentration of a solution. Complete the Analysis of the investigation report.

    Purpose
    The purpose of this investigation is to use the stoichiometric method to analyze a solution for its amount concentration.

    Problem
    What is the amount concentration of silver nitrate in the solution to be recycled?

    Design
    A sample of the silver nitrate solution reacts with an excess of sodium sulfate in solution. The precipitate is filtered and the mass of dried precipitate is measured.

    Evidence
    A white precipitate formed in the reaction.

    No further precipitate formed when a few extra drops of sodium sulfate were added to the clear solution layer above the settled precipitate.

    volume of silver nitrate solution = 100 mL

    mass of filter paper = 1.27 g

    mass of dried filter paper plus precipitate = 6.74 g

8.3 Stoichiometry: Limiting and Excess Reagent Calculations

Calculating Mass of Excess Reagents

For reaction situations other than analysis, you usually know (have measured values for) the quantities of one or more reagents. In these situations, it is often desirable to know in advance how much excess reagent will be required to ensure that the reaction goes to completion. When you know the quantity of more than one reagent, you also need to know which of those reagents will limit the reaction. Stoichiometric calculation can provide this kind of useful information.

For reacting a precisely measured quantity of one reagent with an excess of another, we use a "rule of thumb," a general but inexact guideline that works for most situations. For questions in this textbook, assume that a *reasonable quantity of excess reagent to use is 10% more than the quantity required for complete reaction*, as determined in a stoichiometric calculation. There are exceptions to this rule in practice, especially in industrial and commercial chemistry. When the excess reagent is inexpensive or free, using a larger excess is normal. For example, the burner on a propane barbecue is designed to supply a huge excess of oxygen to the reaction zone (the flame). The amount of propane is limited by the size of the hole in the burner supply pipe. In this case, not only is the oxygen free, but it is also extremely important that the reaction be complete, to minimize production of highly toxic carbon monoxide.

Learning Tip

Remember to keep the unrounded values in your calculator for further calculation until the final answer is reported. The values for intermediate calculation are rounded when written down. Follow the calculation process for the Sample Problems on your calculator to review how to do this.

> ### ▶ SAMPLE problem 8.1
>
> You decide to test the method of stoichiometry using the reaction of 2.00 g of copper(II) sulfate in solution with an excess of sodium hydroxide in solution. What would be a reasonable mass of sodium hydroxide to use?
>
> To answer this question, you need to calculate the minimum mass required and then add 10%. The first part of this plan follows the usual steps of stoichiometry:
>
> $$CuSO_4(aq) + 2\,NaOH(aq) \rightarrow Cu(OH)_2(s) + Na_2SO_4(aq)$$
>
> 2.00 g m
> 159.62 g/mol   40.00 g/mol
>
> $$n_{CuSO_4} = 2.00\ \cancel{g} \times \frac{1\ mol}{159.62\ \cancel{g}}$$
>
> $$= 0.0125\ mol$$
>
> $$n_{NaOH} = 0.0125\ mol \times \frac{2}{1}$$
>
> $$= 0.0251\ mol$$
>
> $$m_{NaOH} = 0.0251\ \cancel{mol} \times \frac{40.00\ g}{1\ \cancel{mol}}$$
>
> $$= 1.00\ g$$
>
> or $m_{NaOH} = 2.00\ \cancel{g\ CuSO_4} \times \dfrac{1\ \cancel{mol\ CuSO_4}}{159.62\ \cancel{g\ CuSO_4}} \times \dfrac{2\ \cancel{mol\ NaOH}}{1\ \cancel{mol\ CuSO_4}} \times \dfrac{40.00\ g\ NaOH}{1\ \cancel{mol\ NaOH}}$
>
> $$= 1.00\ g\ NaOH$$
>
> Now add 10% to this value: 1.00 g + 0.10 g = 1.10 g

Section **8.3**

 **WEB** *Activity*

Canadian Achievers—Ursula Franklin

In addition to a distinguished career in research, Ursula Franklin (**Figure 1**) has been a tireless advocate for the responsible use of scientific knowledge and an active member of Science for Peace.

1. Describe two fields of scientific research where Franklin did pioneering work.
2. Summarize briefly the objectives of the Science for Peace organization.

www.science.nelson.com

Figure 1
Ursula Franklin (1921–)

▶ *Practice*

1. A chemistry teacher wishes to have students perform a precipitation reaction, to practise filtration techniques and to test a stoichiometric prediction. Vials containing precisely measured 1.50 g samples of barium chloride are given to each student group. A stock supply of pure solid sodium sulfate is available in the laboratory. Both reagents are colourless in aqueous solution.
   (a) What would be a reasonable mass of sodium sulfate for each group to use to ensure complete reaction of their barium chloride sample?
   (b) When the precipitate is filtered, which aqueous ion should not be present in the filtrate?
   (c) Describe a procedure that students could use to test the filtrate to see whether the limiting reagent has all reacted.
   (d) What should the students do if the test shows the limiting reagent has not all reacted?

2. In a laboratory gas generator, zinc and an aqueous solution of hydrogen chloride are combined to produce hydrogen. If a 2.00 g sample of zinc is to react completely with an excess of 2.00 mol/L HCl(aq), what would be a reasonable volume of the acid to use?

Identifying Limiting and Excess Reagents

Another application of stoichiometry is the identification of limiting and excess reagents in a chemical reaction, when two known quantities of chemicals react. Which one is the limiting reagent? This is determined using the same stoichiometric principles as before. Like all stoichiometry problems, the mole ratio from the balanced chemical equation is the key part of the solution.

▶ SAMPLE problem 8.2

If 10.0 g of copper is placed in a solution of 20.0 g of silver nitrate, which reagent will be the limiting reagent?

$$Cu(s) + 2\,AgNO_3(aq) \rightarrow 2\,Ag(s) + Cu(NO_3)_2(aq)$$
10.0 g   20.0 g
63.55 g/mol   169.88 g/mol

According to the balanced equation, 1 mol of copper reacts completely with 2 mol of silver nitrate. To determine which reagent is limiting (and therefore which is in excess), convert the reactant quantities given into chemical amounts:

$$n_{Cu} = 10.0 \text{ g} \times \frac{1 \text{ mol}}{63.55 \text{ g}}$$
$$= 0.157 \text{ mol}$$

$$n_{AgNO_3} = 20.0 \text{ g} \times \frac{1 \text{ mol}}{169.88 \text{ g}}$$

$$= 0.118 \text{ mol}$$

You now need to test one of these values using the mole ratio from the chemical equation. In other words, assume that one chemical is completely used up and see if a sufficient amount of the second chemical is available. If copper is the limiting reagent, then the amount of silver nitrate required is calculated as follows:

$$n_{AgNO_3} = 0.157 \text{ mol} \times \frac{2}{1}$$

$$= 0.315 \text{ mol}$$

Obviously, this value (0.315 mol) is much greater than the amount we actually have available (0.118 mol). Therefore, the assumption is incorrect—copper cannot be the limiting reagent; it must be present in excess. Silver nitrate must be the limiting reagent.

Notice that it does not matter which chemical you initially assume to be limiting. You will be able to identify both the limiting and excess reagents no matter which chemical you first choose to assume as the limiting reagent.

Once you have identified the limiting and excess reagents, you can immediately answer a number of other questions. How much of the excess reagent will remain after the reaction? How much product will be obtained? *It is important to note that all predictions made from a balanced chemical equation must be based on the limiting reagent.*

Learning Tip

For any stoichiometric reaction:

$aA + bB \rightarrow$ product(s)

you can select either reactant for a calculation check to determine which reagent is actually limiting. If you assume that A is limiting, you use the mole ratio b/a to calculate the amount of B required to react with the amount of A that is present.
- If the amount of B present is enough, then A is limiting, and B is in excess.
- If the amount of B is not enough, then B is limiting, and A is in excess.

▶ SAMPLE problem 8.3

In the reaction of a 10.0 g sample of copper with 20.0 g of silver nitrate in solution in Sample Problem 8.2, what mass of copper will be in excess (left over when the reaction is complete)? What mass of silver will be produced?

Write and balance the reaction equation.

$$Cu(s) + 2 AgNO_3(aq) \rightarrow 2 Ag(s) + Cu(NO_3)_2(aq)$$

To find the excess quantity of copper, find the mass of copper required to react with the 20.0 g of silver nitrate and compare that with the starting mass of copper. From Sample Problem 8.2, you know that the 20.0 g of silver nitrate is equivalent to 0.118 mol, and the chemical equation shows that the mole ratio of copper to silver nitrate is 1:2. Therefore, the chemical amount of copper that reacts can be calculated as follows:

$$n_{Cu} = 0.118 \text{ mol} \times \frac{1}{2}$$

$$= 0.0589 \text{ mol}$$

This can now be converted, using the molar mass of copper, to a mass of copper:

$$m_{Cu} = 0.059 \text{ mol} \times \frac{63.55 \text{ g}}{1 \text{ mol}}$$

$$= 3.74 \text{ g}$$

The excess quantity of copper is 10.0 g − 3.74 g, or 6.3 g.

To find the yield of silver product expected, use the chemical amount of the limiting reagent, AgNO$_3$, to predict the mass of this product:

$$n_{Ag} = 0.118 \text{ mol} \times \frac{2}{2}$$
$$= 0.118 \text{ mol}$$

$$m_{Ag} = 0.118 \text{ mol} \times \frac{107.87 \text{ g}}{1 \text{ mol}}$$
$$= 12.7 \text{ g}$$

According to the stoichiometric method, the predicted (theoretical) yield of silver product is 12.7 g.

▶ COMMUNICATION example

In an experiment, 26.8 g of iron(III) chloride in solution is combined with 21.5 g of sodium hydroxide in solution. Which reactant is in excess, and by how much? What mass of each product will be obtained?

Solution

$$\text{FeCl}_3\text{(aq)} + 3 \text{ NaOH(aq)} \rightarrow \text{Fe(OH)}_3\text{(s)} + 3 \text{ NaCl(aq)}$$

| 26.8 g | 21.5 g | m | m |
| 162.20 g/mol | 40.00 g/mol | 106.88 g/mol | 58.44 g/mol |

$$n_{FeCl_3} = 26.8 \text{ g} \times \frac{1 \text{ mol}}{162.20 \text{ g}}$$
$$= 0.165 \text{ mol}$$

$$n_{NaOH} = 21.5 \text{ g} \times \frac{1 \text{ mol}}{40.00 \text{ g}}$$
$$= 0.538 \text{ mol}$$

If FeCl$_3$ is the limiting reagent, the amount of NaOH required is

$$n_{NaOH} = 0.165 \text{ mol} \times \frac{3}{1}$$
$$= 0.496 \text{ mol}$$

The sodium hydroxide is in excess. The excess amount and mass are

$$n_{NaOH} = 0.538 \text{ mol} - 0.496 \text{ mol}$$
$$= 0.042 \text{ mol}$$

$$m_{NaOH} = 0.042 \text{ mol} \times \frac{40.00 \text{ g}}{1 \text{ mol}}$$
$$= 1.7 \text{ g}$$

The mass of the two products is

$$m_{Fe(OH)_3} = 0.165 \text{ mol FeCl}_3 \times \frac{1 \text{ mol Fe(OH)}_3}{1 \text{ mol FeCl}_3} \times \frac{106.88 \text{ g Fe(OH)}_3}{1 \text{ mol Fe(OH)}_3}$$
$$= 17.7 \text{ g Fe(OH)}_3$$

$$m_{NaCl} = 0.165 \text{ mol FeCl}_3 \times \frac{3 \text{ mol NaCl}}{1 \text{ mol FeCl}_3} \times \frac{58.44 \text{ g NaCl}}{1 \text{ mol NaCl}}$$
$$= 29.0 \text{ g NaCl}$$

According to the stoichiometric method, sodium hydroxide is in excess by 1.7 g, the mass of iron(III) hydroxide produced is 17.7 g, and the mass of sodium chloride produced is 29.0 g.

Figure 2
If tap water contains iron ions, and is also slightly basic, precipitation of iron inside water pipes becomes a real problem. Ideally, tap water should be very slightly acidic, so it will keep any iron in solution. Neither copper nor the plastics normally used for water supply pipes are corroded by acids.

Figure 6
This huge production facility, located 50 km southeast of Calgary at Carseland, can synthesize 535 000 t of ammonia per year. Agrium Corporation, which operates this plant, has three other plants like it in Alberta, with a total ammonia production capacity of nearly 2.5 Mt per year. Most of this ammonia is not used directly as fertilizer; it is further processed (reacted) to make, among other chemicals, fertilizers such as urea, ammonium nitrate, and ammonium phosphate.

Figure 7
Ammonia fertilizer can be added directly to the soil.

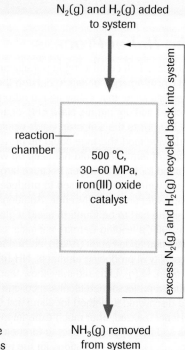

Figure 8
Outline of the Haber process

Today, the Haber process is used to produce ammonia from its elements in over 335 active synthetic ammonia plants worldwide (**Figure 6**). Much of the ammonia that is produced is used in agriculture (**Figure 7**). As a fertilizer, the ammonia dissolves in moisture that is present in the soil. If the soil is slightly acidic, the ammonia is converted to nitrate ions by soil bacteria. Nitrate ions are absorbed by the roots of plants and used in the synthesis of proteins, chlorophyll, and nucleic acids. Without a source of nitrogen, plants do not grow; they produce yellow leaves and die prematurely.

Case Study Questions

1. Recall that reactions for which stoichiometric calculations can be used for analysis are usually required to be spontaneous, fast, quantitative, and stoichiometric. Consider the information provided in this Case Study about the conditions under which the Haber process reaction is done industrially.
   (a) The Haber process reaction is not spontaneous. At SATP, nitrogen and hydrogen do not react at all. How is this problem overcome in the production of ammonia? Use the collision–reaction theory in your explanation.
   (b) When the reaction is set up under conditions in which it is spontaneous, it is not fast. This problem can be helped by adding finely powdered $Fe_2O_3(s)$ to act as a *catalyst*. Explain what effect this has on the reaction, and what role the iron(III) oxide plays.
   (c) Once the reaction rate is increased, the reaction is not quantitative. Even at the usual industrial conditions, ammonia seems to stop forming when less than 40% of the reactants have reacted. If increasing pressure increases percent yield, why is the process not just done at extremely high pressures? *Hint:* Think about the warnings on aerosol spray can labels.
   (d) Percent yield is improved by lowering the temperature. Why do industrial chemists not use low temperatures?
   (e) The reaction is stoichiometric. The amounts of nitrogen and hydrogen that react do so in an exact 1:3 proportion. Considering the composition of Earth's atmosphere, which of these reagents do you think would concern the company accounting department?

2. The process to remove the ammonia from the nitrogen and hydrogen remaining in the reaction pressure vessel is quite simple. When the three mixed gases are cooled at fairly high pressure, the ammonia condenses and can be drained out of the bottom of the vessel, leaving the hydrogen and nitrogen behind. More of each reagent is then added, and the temperature is raised to make the gases react again. This process is repeated continuously (**Figure 8**).
   (a) Explain why, in a chemical plant that runs continuously, the rate of reaction is much more important to industrial chemists than the percent reaction.
   (b) Ammonia is actually a smaller molecule (by mass, or electron count) than nitrogen. Explain why ammonia's condensation temperature should be so much higher than nitrogen's.

Extension

3. Working in a small group, research the Internet and other sources to find out where and how Alberta ammonia producers get the hydrogen they use for this reaction. Prepare a brief report or presentation, including graphics. Include a discussion of the likelihood of long-term (centuries) sustainability of Haber process technology as applied in Alberta, and its value from economic and environmental perspectives.

www.science.nelson.com

Section 8.3

SUMMARY Limiting/Excess Reagent Calculations

- Identify the limiting reagent by choosing either reagent amount, and use the reaction mole ratio to compare the required amount of the other reagent with the amount actually present.
- The quantity in excess is the difference between the amount of excess reagent present and the amount required for complete reaction.
- A reasonable reagent excess to use to ensure complete reaction is 10%.

+ EXTENSION

Limiting Reagents
Try this exercise to help consolidate your understanding of limiting and excess reagents.

www.science.nelson.com

Section 8.3 Questions

1. List the necessary assumptions about a reaction done for chemical analysis.

2. When testing gravimetric stoichiometry using an experiment, should you combine the two reactants in the same ratio as they appear in the balanced chemical equation? Justify your answer.

3. When calculating a percent yield, where does the value of the actual yield come from? Where does the predicted yield come from?

4. A quick, inexpensive source of hydrogen gas is the reaction of zinc with hydrochloric acid (**Figure 9**). If 0.35 mol of zinc is placed in 0.60 mol of hydrochloric acid,
   (a) which reactant will be completely consumed?
   (b) what mass of the other reactant will remain after the reaction is complete?

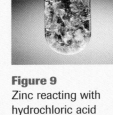

Figure 9
Zinc reacting with hydrochloric acid

5. A chemical technician is planning to react 3.50 g of lead(II) nitrate with excess potassium bromide in solution.
   (a) What would be a reasonable mass of potassium bromide to use in this reaction?
   (b) Predict the mass of precipitate expected.

6. In a chemical analysis, 3.40 g of silver nitrate in solution reacted with excess sodium chloride to produce 2.81 g of precipitate. What is the percent yield?

7. A solution containing 9.8 g of barium chloride is mixed with a solution containing 5.1 g of sodium sulfate.
   (a) Which reactant is in excess?
   (b) Determine the excess mass.
   (c) Predict the mass of precipitate.

8. A solution containing 18.6 g of chromium(III) chloride reacts with a 15.0 g piece of zinc to produce chromium metal (**Figure 10**).
   (a) Which reactant is in excess?
   (b) Determine the excess mass.
   (c) If 5.1 g of chromium metal is formed, what is the percent yield?

Figure 10
Electroplating produces a thin metal coating on objects such as the car door handles shown in the photograph. Chromium plating is used for esthetic as well as technical reasons because it creates a shiny surface and also prevents corrosion. Chromium ions are toxic, however, and environmental damage may result if the chromium solutions are dumped as waste. Treating toxic wastes to transform them into safe materials is sometimes prohibitively expensive.

9. A technical college instructor wishes a first-year chemistry group to perform an investigation to practise precipitation and filtration techniques and to calculate a percent yield. The class will react 50.00 mL pipetted samples of 0.200 mol/L potassium phosphate solution with an excess of 0.120 mol/L lead(II) nitrate solution.
   (a) Which reagent is intended to be the limiting reagent?
   (b) What is the minimum volume of lead(II) nitrate solution required?
   (c) What volume of lead(II) nitrate solution should the instructor tell the students to use?
   (d) Describe how the students can test for completeness of reaction of the limiting reagent.

Chemical Analysis 327

8.4 Titration Analysis

Titration is a common experimental design used to determine the amount concentration of substances in solution (see Appendix C.4). **Titration** is the process of carefully measuring and controlling the addition of a solution, called the **titrant**, from a burette into a measured fixed volume of another solution, called the **sample**, usually in an Erlenmeyer flask (**Figure 1**) until the reaction is judged to be complete. A *burette* is a precisely marked glass cylinder with a stopcock at one end. It allows precise, accurate measurement and control of the volume of reacting solution. This technique is a good example of a chemical technology that is reliable, efficient, economical, and simple to use.

When doing a titration, there will be a point at which the reaction is complete. In other words, chemically equivalent amounts of reactants, as determined by the mole ratio, have been combined. The point at which the exact theoretical amount of titrant has been added to completely react with the sample is called the **equivalence point**. To measure this point experimentally, we look for a sudden change in some observable property of the solution, such as colour, pH, or conductivity. The point during a titration when this sudden change is observed is called the **endpoint**. At the endpoint, the titration is stopped and the volume of titrant is determined. Ideally, the volume at the empirical endpoint and the volume corresponding to theoretical equivalence point should coincide.

A titration analysis should involve several trials, to improve the reliability of the answer. A typical requirement is to repeat titrations until three trials result in volumes within a range of 0.2 mL. These three results are then averaged before carrying out the solution stoichiometry calculation, disregarding any trial volumes that do not fall within this range.

> **Learning Tip**
>
> A titration analysis of an unknown amount concentration requires that the chemical reaction be spontaneous, fast, quantitative, and stoichiometric. The amount concentration of one reactant used must be accurately known. The solution of known amount concentration may be used as either the titrant or the sample; it makes no difference to the analysis.

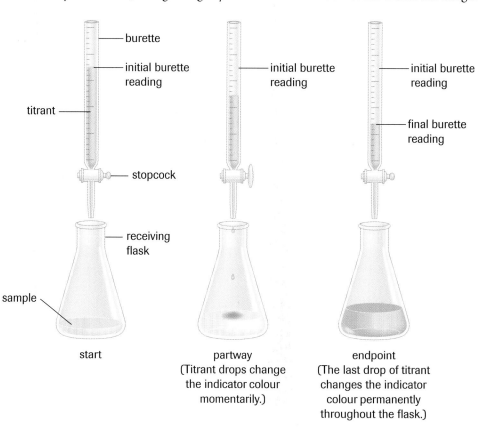

Figure 1
An initial reading of volume is made on the burette before any titrant is added to the sample solution. Then titrant is added until the reaction is complete; that is, when a final drop of titrant permanently changes the colour of the sample. The final burette reading is then taken. The difference in burette readings is the volume of titrant added. Appendix C.4 describes the titration process in detail.

SAMPLE problem 8.4

Determine the concentration of hydrochloric acid in a commercial solution that is used to treat concrete prior to painting.

A 1.59 g mass of sodium carbonate, $Na_2CO_3(s)$, was dissolved to make 100.0 mL of solution. Samples (10.00 mL) of this standard solution were then taken and titrated with solution, which was prepared by diluting the original commercial solution, HCl(aq), by a factor of 10. The titration evidence collected is shown in **Table 1**. Methyl orange indicator was used.

Table 1 Titration of 10.00 mL of $Na_2CO_3(aq)$ with Diluted HCl(aq)

Trial	1	2	3	4
final burette reading (mL)	13.3	26.0	38.8	13.4
initial burette reading (mL)	0.2	13.3	26.0	0.6
volume of HCl(aq) added (mL)	13.1	12.7	12.8	12.8
indicator colour	red	orange	orange	orange

First, calculate the amount concentration of the sodium carbonate solution:

$$n_{Na_2CO_3} = 1.59 \text{ g} \times \frac{1 \text{ mol}}{105.99 \text{ g}}$$

$$= 0.0150 \text{ mol}$$

$$[Na_2CO_3(aq)] = \frac{0.0150 \text{ mol}}{0.1000 \text{ L}}$$

$$= 0.150 \text{ mol/L}$$

Now write the balanced chemical equation:

$$2 \text{ HCl(aq)} + Na_2CO_3(aq) \rightarrow H_2CO_3(aq) + 2 \text{ NaCl(aq)}$$

| 12.8 mL* | 10.00 mL |
| c | 0.150 mol/L |

* The volume of HCl(aq) used is an average of trials 2, 3, and 4.

$$n_{Na_2CO_3} = 10.00 \text{ mL} \times \frac{0.150 \text{ mol}}{1 \text{ L}}$$

$$= 1.50 \text{ mmol}$$

$$n_{HCl} = 1.50 \text{ mmol} \times \frac{2}{1}$$

$$= 3.00 \text{ mmol}$$

$$[HCl(aq)] = \frac{3.00 \text{ mmol}}{12.8 \text{ mL}}$$

$$= 0.235 \text{ mol/L}$$

or

$$[HCl(aq)] = 10.00 \text{ mL Na}_2\text{CO}_3 \times \frac{0.150 \text{ mol Na}_2\text{CO}_3}{1 \text{ L Na}_2\text{CO}_3} \times \frac{2 \text{ mol HCl}}{2 \text{ mol Na}_2\text{CO}_3} \times \frac{1}{10.00 \text{ mL HCl}}$$

$$= 0.235 \text{ mol/L HCl (the diluted analysis solution)}$$

Since the sample of concrete cleaner had been diluted by a factor of 10, the original concentration of the commercial hydrochloric acid must be 10 times greater, or 2.35 mol/L.

Learning Tip

Notice in **Table 1** that four trials were done, and the volume added in the first trial is significantly higher than in the others. This value is thus disregarded when calculating an average volume of HCl(aq) that reacts. In titration analysis, the first trial is typically done very quickly. It is just for practice, to learn what the endpoint looks like and also to learn the approximate volume of titrant required to get to the endpoint. Then greater care is taken with subsequent trials.

Learning Tip

Any property of a solution, such as colour, conductivity, or pH, that changes abruptly can be used as an endpoint. However, some changes may not be very sharp or may be difficult to measure accurately. This may introduce error into the experiment. Any difference between the titrant volumes at the empirical (observed) endpoint and the theoretical equivalence point is known as the titration error.

SUMMARY: Titration Analysis

- Titration is the technique of carefully controlling the addition of a volume of solution (the titrant) from a burette into a measured fixed volume of a sample solution until the reaction is complete.
- The concentration of one reactant must be accurately known.
- The equivalence point is the point at which the exact theoretical (stoichiometric) reacting amount of titrant has been added to the sample.
- The endpoint is the point during the titration at which the sudden change of an observable property indicates that the reaction is complete.
- Several trials must be completed. When at least three trials result in values that are all within a range of 0.2 mL, those values are averaged. The average value is used for the stoichiometry calculation.

Standardizing Titrant Solutions

Before any titration is performed to analyze a solution, it is absolutely necessary that you know the amount concentration of one of the solutions to a high degree of certainty, because that value will be used to calculate your answer. Chemists call a solution of highly certain concentration a **standard solution**. In Chapter 5, you learned techniques for preparing standard solutions of accurately known amount concentration by carefully measuring both the mass of the solute and the volume of the solution. Sometimes, however, that process does not work because the solute you need is not what chemists call a primary standard.

A **primary standard** is a chemical that can be obtained at high purity, with mass that can be measured to high accuracy and precision. Some chemicals cannot be obtained at high purity, and measuring the mass of a chemical that is not pure is pointless—you have no way of knowing how much of that mass is made up of the impurities. Some chemicals, although pure, cannot be measured accurately with a balance because their mass will not remain constant. For example, sodium hydroxide in solution is a very common and useful strong base, but you cannot prepare a standard solution from solid sodium hydroxide. The pure solid compound attracts water so strongly that it will absorb water vapour rapidly from the air. If you take some NaOH(s) out of a closed container and place it on a balance, it will absorb water and increase in mass while you are trying to measure it. Once dissolved in pure water, however, the concentration of a dilute solution of sodium hydroxide will remain constant. Therefore, what is required is a way to analyze the concentration after the solution is prepared.

If a solute is a gas, it cannot be a primary standard. In anything other than a very dilute solution, some gas will escape any time the container is opened, decreasing the amount concentration. Ammonia and hydrochloric acid are common examples.

Standardizing a solution means finding the concentration of a solution *after* it is prepared, by reacting it with another solution that has been prepared from a primary standard. A stoichiometric calculation is then used to find the unknown concentration. This procedure is really just a titration analysis like any other. A primary standard often used to make a solution for standardizing basic solutions is potassium hydrogen phthalate, $KC_7H_4O_2COOH(s)$. Investigation 8.3 illustrates the total process of first making a standard solution, and then using it to analyze (and thus standardize) another solution. In this case, the sodium hydroxide solution to be standardized is also the titrant solution that will be used for Investigations 8.4 and 8.6.

CAREER CONNECTION

Hydrologist

Water is one of the most necessary substances for life. Hydrologists study water and how it flows around the atmosphere and the planet, looking at water's movement through rivers, glaciers, and geologic formations. These specialists examine data to make sure that water supply and quality meet public and industrial demands. They also make recommendations for environmental impact assessments.

Research this career and associated specialties online.

www.science.nelson.com

Section 8.4

INVESTIGATION 8.3 Introduction
Standardization Analysis of NaOH(aq) (Demonstration)

Report Checklist
- ○ Purpose
- ○ Problem
- ○ Hypothesis
- ○ Prediction
- ○ Design
- ○ Materials
- ○ Procedure
- ● Evidence
- ● Analysis
- ○ Evaluation

A large (stock) volume of sodium hydroxide solution is prepared for use in Investigations 8.4 and 8.6 by dissolving 6 g of solid for each litre of solution required. Because this solute is not a primary standard, this solution must be standardized to accurately determine its concentration. This will be accomplished by titration against a standard potassium hydrogen phthalate (KHP) solution.

Purpose
The purpose of this investigation is to use a titration design to standardize a solution for future chemical analysis.

Problem
What is the concentration of a stock NaOH(aq) solution?

Design
A standard solution of KHP is prepared, and it is then used to standardize a stock solution of sodium hydroxide. Samples of KHP are titrated with sodium hydroxide titrant, using phenolphthalein as an indicator.

To perform this investigation, turn to page 342.

Practice

1. When adding titrant to a burette, it is critical that the concentration of the solution remain constant. Ideally, the burette should be cleaned and dried just before use to make sure that no impurities change the titrant concentration. It is extremely difficult, however, to quickly dry a burette that has just been cleaned. What technique is used to solve this problem?

2. When a titration analysis is performed, multiple trials are normally run. This means that successive equal volumes (called aliquots) of the sample solution must be taken. What technology is used to ensure that the volumes of sample for each trial are as identical as possible?

3. One purpose of doing multiple trials for an analysis is to immediately identify any mistakes in procedure because these will cause discrepant results. What is the other reason for doing multiple trials?

4. Acid–base titrations, like Investigation 8.3, typically do not produce any visible product, which presents a problem. There is no direct way of knowing when such a reaction is complete. Explain how this problem is overcome by using another substance that is not part of the reaction, what characteristic of this substance is useful, and what characteristic of the reaction solution is detected by this substance. (*Hint:* Review Section 6.3.)

INVESTIGATION 8.4 Introduction
Titration Analysis of Vinegar

Report Checklist
- ○ Purpose
- ○ Problem
- ○ Hypothesis
- ○ Prediction
- ○ Design
- ● Materials
- ● Procedure
- ● Evidence
- ● Analysis
- ● Evaluation (1, 2, 3)

Some consumer food products are required by law to have the minimum quantity of the active ingredient listed on the product label. According to the label, a vinegar manufacturer states that the vinegar contains 5% acetic acid by volume, which translates to a minimum amount concentration of 0.83 mol/L.

Purpose
The purpose of this investigation is to test the manufacturer's claim of the concentration of acetic acid in a consumer sample of vinegar.

Problem
What is the amount concentration of acetic acid in a sample of vinegar?

Design
A sample of commercial white vinegar is diluted by a factor of 5 to make a 100.0 mL final solution. Samples of this diluted solution are titrated with a standardized sodium hydroxide solution (from Investigation 8.3) using phenolphthalein as the indicator.

To perform this investigation, turn to page 343.

Section 8.4 Questions

1. Ammonia is a very useful chemical; our society consumes it in huge quantities. Farmers use the pure substance in liquid form as a fertilizer. Pure liquid ammonia is called anhydrous, which means "without water," to distinguish it from aqueous solutions. In solution, ammonia has an outstanding ability to loosen dirt, oil, and grease, so it is commonly used in premixed home cleaners such as window cleaning sprays, along with other ingredients. Aqueous ammonia is also sold in most stores for household use, to be diluted at home to make solutions for cleaning and wax stripping (**Figure 2**). Such solutions can legally be anywhere from 5% to 30% ammonia by weight.

   A student wishing to find the concentration of ammonia in a commercial solution decides to do an analysis, titrating 10.00 mL samples of NH_3(aq) with a standardized solution of 1.48 mol/L HCl(aq). Her first trials use more than 50 mL (a burette full) of the acid, so she throws out the results and prepares a new ammonia sample solution by diluting the original commercial solution 10:1, that is, increasing a volume tenfold to reduce the concentration to precisely one-tenth of the original value. Using **Table 2**, complete the Analysis of her investigation report.

   Purpose
   The purpose of this investigation is to use a titration design to analyze a solution of ammonia.

   Problem
   What is the amount concentration of the original ammonia solution?

   Design
   The original ammonia solution is diluted tenfold. Samples of diluted solution are titrated with a standard 1.48 mol/L solution of hydrochloric acid. The colour change of bromocresol green indicator from blue to yellow is used as the endpoint.

   Evidence

   Table 2 Titration of 10.00 mL of NH_3(aq) with 1.48 mol/L HCl(aq)

Trial	1	2	3	4
final burette reading (mL)	15.0	29.1	43.0	14.4
initial burette reading (mL)	0.3	15.0	29.1	0.4
volume of HCl(aq) added (mL)				
colour at endpoint	yellow	green	green	green

2. Assume a hydrochloric acid solution is prepared by diluting commercial lab reagent solution (approximately 12 mol/L) by a factor of 20:1. For concentrated HCl(aq) solutions, complete the following:
   (a) Explain why the label concentration is necessarily uncertain for concentrated solutions of gases dissolved in water. *Hint:* Think about opening carbonated beverages.
   (b) Explain how the concentration changes each time the stock bottle is opened.

Figure 2
Household ammonia is sold as a fairly concentrated aqueous solution, making it very convenient to store, transport, dilute, and dispense.

3. Describe a design for precisely determining the concentration of (standardizing) a diluted hydrochloric acid solution, assuming you know that it will react quantitatively with the base sodium carbonate, which is a primary standard solid, and that methyl orange indicator's endpoint will accurately indicate the reaction equivalence point.

Extension

4. Sulfur impurities in fuels produce SO_2(g) when the fuel is burned. This is a pollutant that contributes to acid deposition and is a serious respiratory irritant (**Figure 3**). To analyze the sulfur content in a fuel, the sample may be burned, and the SO_2(g) may then be "dissolved" in water, which really means that it reacts with water to become sulfurous acid, H_2SO_3(aq). The sulfurous acid can then be analyzed by titration with a standardized solution of NaOH(aq). If, on average, 12.0 mL of 0.110 mol/L NaOH(aq) reacts with 100 mL samples of H_2SO_3(aq), what chemical amount of sulfur atoms was present in the 100 mL acid sample?

Figure 3
Acid rain is responsible for the damage to this sculpture. Sulfur dioxide is one of the two primary causes of acid rain. The other primary cause is nitrogen oxides.

Acid–Base Titration Curves and Indicators 8.5

Investigations 8.3 and 8.4 are titration analyses involving acid–base reactions. (You might find it helpful to review Sections 6.2 and 6.3.) While other types of reactions are sometimes useful for titration analysis, acid–base reactions predominate for several reasons. Perhaps most important, acidic and basic substances are very common, and a great number of reactions cause a change in the pH of a solution. Just as important is the fact that it is easy to find substances to act as indicators for acid–base titrations. Acid–base reactions are normally invisible in solution, so direct observation cannot tell you when a reaction is complete. Technological devices, such as pH meters, can be used to detect the equivalence point, but, as you have already experienced, there is an easier way. Because many organic substances (such as litmus) change colour depending on whether they are in an acidic or basic solution, such indicators (Chapter 6) make it easy to titrate acid–base reactions accurately to an observable endpoint. During titration, the indicator shows a momentary colour change where the titrant stream contacts the sample in the flask. Closer to the endpoint, the colour change lingers longer, allowing you to add titrant more slowly, drop by drop, until at the endpoint one final drop of titrant changes the colour of the flask contents permanently (**Figure 1**). The key to accurate titration analysis is making sure that the observed endpoint really occurs as close as possible to the reaction's equivalence point, as discussed in Section 8.4. To explore the connection of these two points for acid–base reactions, it is necessary to know exactly how the pH changes during a titration. This is demonstrated in Investigation 8.5.

Figure 1
Accurate titration analysis depends on applying a combination of specialized knowledge and specific skills. (See Appendix C.4.)

INVESTIGATION 8.5 Introduction

pH Curves (Demonstration)

When titrating a basic sample with an acidic titrant, you would expect the pH to be high initially, then to decrease as acid is progressively added, and finally to be low when a large excess of acid has been added. This expectation turns out to be correct. However, what is interesting and important is the *way* that the pH decreases. A titration pH curve is very useful evidence, providing valuable information about any acid–base reaction. Your Analysis involves plotting graphs of pH against volume of acid added. Alternatively, a computer program using a pH sensor probe may be used (if available) to plot the graph on screen as the titrations occur.

Purpose
The purpose of this demonstration is to create pH curves and observe the function of an indicator in an acid–base reaction.

To perform this investigation, turn to page 344.

Report Checklist
- ○ Purpose
- ○ Problem
- ○ Hypothesis
- ● Prediction
- ○ Design
- ○ Materials
- ○ Procedure
- ● Evidence
- ● Analysis
- ○ Evaluation

Problem
What are the shapes of the pH curves for the continuous addition of hydrochloric acid to a sample of a sodium hydroxide solution and to a sample of a sodium carbonate solution?

Design
Small volumes of hydrochloric acid are added continuously to a measured volume of a base. After each addition, the pH of the mixture is measured. The volume of hydrochloric acid is the manipulated variable, and the pH of the mixture is the responding variable.

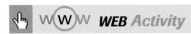

Web Quest—Blood Alcohol Content

This Web Quest will lead you to research the chemistry of blood alcohol analysis. You will be responsible for the defense of someone charged with impaired driving. How accurate are roadside alcohol tests? Learn the issues, and make a case to present in criminal court.

www.science.nelson.com

Learning Tip

As in many other areas, in chemistry the simplest system that "works" is usually preferred.

When chemists write net ionic acid–base equations for titration reactions, they often write the hydrogen ions as $H^+(aq)$ rather than $H_3O^+(aq)$, even though you learned in Chapter 6 that evidence indicates that the latter representation is more correct. The more complex notation is normally only used when the water *must* be considered as a reactant to understand the reaction.

Interpreting Titration pH Curves

When a titration is done to create a pH curve, the addition of titrant is not stopped at the endpoint, but is continued until a large excess has been added. This emphasizes the significance of such a curve: the very rapid change in pH passing the equivalence point. For the NaOH(aq)–HCl(aq) pH curve plotted in Investigation 8.5, the initial sample is a strong base, and the pH is high. As the titration proceeds and acid is added, some of the base is reacted with the added acid, but anywhere before the equivalence point some excess base will remain, so the pH stays relatively high. Very near the equivalence point, a small excess of base becomes a small excess of acid with the addition of just a few more drops of HCl(aq), and the pH abruptly changes from high to low. This rapid pH change is what makes an acid–base equivalence point easy to detect. The equivalence point is at the centre of the change, where the curve is most nearly vertical. Note that for strong monoprotic acid–strong monoprotic base reactions (Chapter 6), the net ionic equation will always be the same, because ions other than hydrogen and hydroxide ions are always spectator ions for these substances:

$$NaOH(aq) + HCl(aq) \rightarrow NaCl(aq) + H_2O(l)$$

$$\cancel{Na^+(aq)} + OH^-(aq) + H^+(aq) + \cancel{Cl^-(aq)} \rightarrow \cancel{Na^+(aq)} + \cancel{Cl^-(aq)} + H_2O(l)$$

$$OH^-(aq) + H^+(aq) \rightarrow H_2O(l) \text{ (net ionic equation)}$$

For convenience, aqueous hydrogen ions are written here in their simplest (Arrhenius) form. Now, consider this titration done in reverse, by titrating a strong acid sample with a large excess of strong base titrant so that the pH value will start low and end high. The net ionic equation is the same, and so is the equivalence point; the pH curve (**Figure 2**) is just a mirror image of the example in Investigation 8.5. When a strong monoprotic acid completely reacts with a strong monoprotic base, the products are always water and neutral spectator ions, so you can predict what the pH must be at the equivalence point. Recall from Chapter 6 that water has a (neutral) pH of 7, so a strong monoprotic acid–strong monoprotic base titration must have a pH of 7 at the equivalence point.

It is important to note that the equivalence point pH is 7 only *for this one specific type* of acid–base reaction. For every other acid–base reaction, the solution at the equivalence point will contain ions and/or molecules that are *not* spectators, and the pH will vary depending on which entities are present as well as on their concentration. This

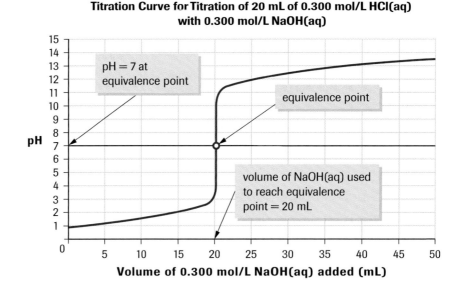

Figure 2
This curve is typical of curves depicting the titration of a strong acid with a strong base. Notice that the curve sweeps up and to the right as NaOH(aq) is added, beginning at a pH below 7 and ending at a pH above 7. After adding 20 mL of titrant, the pH is 7; the equivalence point has been reached.

means that a pH titration curve must always be done empirically, to determine the pH at the equivalence point before any acid–base reaction using an indicator can be used for titration analysis.

An equivalence point is read from a pH curve by estimating the inflection point position in the part of the curve where the slope steepens. An *inflection point* is the point where the direction of curvature changes, like the centre point of the letter S.

Recall the pH titration curve from Investigation 8.5 for a sodium carbonate solution sample titrated with hydrochloric acid. This titration produces a very different pH curve from the curve you just examined. A similar pH curve is shown in **Figure 3**. Observing this curve, you can see that the centre of the most rapid pH change (steepest slope) corresponds to a pH value of about 3.6. The balanced reaction equation follows:

$$Na_2CO_3(aq) + 2\,HCl(aq) \rightarrow 2\,NaCl(aq) + H_2CO_3(aq)$$

which can also be written as

$$\cancel{2\,Na^+(aq)} + CO_3^{2-}(aq) + 2\,H^+(aq) + \cancel{2\,Cl^-(aq)} \rightarrow$$
$$\cancel{2\,Na^+(aq)} + \cancel{2\,Cl^-(aq)} + H_2CO_3(aq) \text{ (total ionic equation)}$$

$$CO_3^{2-}(aq) + 2\,H^+(aq) \rightarrow H_2CO_3(aq) \text{ (net ionic equation)}$$

In this reaction, the base is diprotic, meaning that it will react with two hydrogen ions. If you observe the curve closely, you see that there are two places where the curve steepens as the titration proceeds. This happens because the two hydrogen ions attach to the carbonate ion one at a time. We use the second reaction equivalence point, because we want the pH value when the reaction is complete.

When this reaction is stoichiometrically complete (at the equivalence point), the only substances present are water, sodium ions, chloride ions, carbonic acid molecules, and a very small amount of methyl orange indicator. It seems logical to find experimentally that the solution pH is acidic at this reaction's equivalence point, since it contains water, spectator ions, and the weak acid $H_2CO_3(aq)$. Experimentally plotting a pH titration curve is essential for selecting the right indicator for acid–base titration analysis reactions. It is also critical that the amount of indicator used be extremely small. In theory, some titrant volume must be used to react with the indicator to make it change colour, but if the indicator amount is kept very small (a drop or two of solution), then the volume of titrant used in this way will be too small to be measurable, and the accuracy of the titration will not be affected.

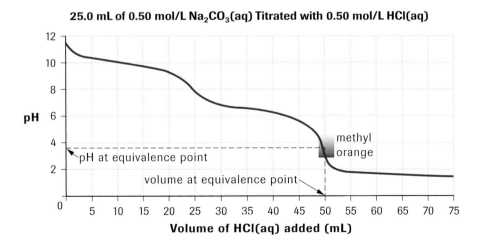

Figure 3
A pH curve for the addition of hydrochloric acid to a sample of sodium carbonate

> **Learning Tip**
>
> Acid–base reaction pH curves provide a wealth of information:
> - initial pH levels
> - volume of titrant at the equivalence point
> - pH (for indicator selection) at the equivalence point
> - number of reaction steps

Choosing Acid–Base Indicators for Titration

Experimentally plotting a pH titration curve is essential for selecting the right indicator for acid–base titration analysis reactions, so the endpoint observed for the indicator chosen will closely match the equivalence point of the reaction. For the equivalence point of the reaction in Figure 3, the pH was 3.6. To accurately show when this reaction is complete, an indicator must be chosen that changes colour across a pH range that has a central value close to 3.6. According to the Acid–Base Indicators table on the inside back cover, the indicator methyl orange is yellow above pH 4.4 and red below pH 3.1. This makes it a good choice for this reaction. The colour change pH range for methyl orange is superimposed on the pH curve in Figure 3 to show how an indicator is chosen to match observed endpoint to a reaction equivalence point. Showing three indicator choices for an HCl–NaOH titration is another way to illustrate this point (**Figure 4**).

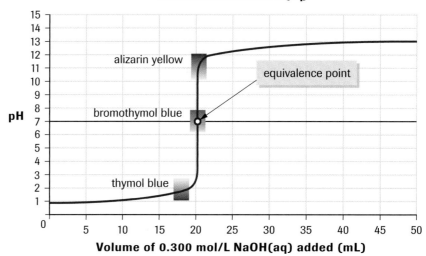

Figure 4
Thymol blue is an unsuitable indicator for this titration because it changes colour before the equivalence point (pH 7). Alizarin yellow is also unsuitable because it changes colour after the equivalence point. Bromothymol blue is suitable because its endpoint pH of 6.8 (assume the middle of its pH range) closely matches the reaction equivalence point pH of 7, and the colour change is completely on the vertical portion of the pH curve.

Indicator Choice
So, you are planning an investigation involving an acid–base titration. How do you know which indicator to use? This audio clip will help you understand how to choose an appropriate indicator for an acid–base titration.

www.science.nelson.com

Practice

1. What is the difference in meaning between endpoint and equivalence point?
2. (a) Sketch a pH curve for the titration of $HNO_3(aq)$ with $KOH(aq)$.
   (b) What will be the equivalence point pH? Why?
3. Which of the following indicators would show an "intermediate" endpoint colour of orange: bromocresol green, methyl red, phenolphthalein, or methyl violet?
4. According to the Acid–Base Indicators table on the inside back cover, what is an appropriate indicator for a titration with an equivalence point pH of 4.4?

WEB Activity

Simulation—Titration Curves

This computer simulation enables you to select from a variety of acids and bases, and to choose an indicator. The program automatically plots a pH curve as you add the titrant,

www.science.nelson.com GO

SUMMARY: Acid–Base Indicators, Endpoints, and Equivalence Points

- An indicator for an acid–base titration analysis must be chosen to have an endpoint (change of colour) at very nearly the same pH as the pH at the equivalence point of the reaction solution.
- The pH of the solution at the equivalence point for a strong monoprotic acid–strong monoprotic base reaction will be 7.
- The pH of the solution at the equivalence point for any other acid–base reaction must be determined experimentally, by plotting a titration pH curve.

Case Study

Analytic Measurement Technology

Developments in analytic measurement technology are giving scientists the ability to detect increasingly tiny amounts of substances. These technological developments are driven in part by the need to trace the path of toxic substances in ecosystems and in the human body. Contamination of the environment by heavy metals and semi-metals is a serious problem worldwide. Although the concentrations of most toxic substances in the environment are far below the lethal level, they may still cause serious damage to life processes when they are ultimately incorporated into drinking water and become concentrated by moving up the food chain. Top predators like swordfish and tuna have far more mercury in their tissue than the bait fish they eat, for example, so these fish should be consumed less frequently by humans. Since metals are biologically nondegradable, they tend to accumulate in vital organs, so that prolonged exposure to trace concentrations of metals and semi-metals can sometimes lead to long-term health effects.

Metals and semi-metals can be divided into three groups based on their toxicity:

- toxic at very low concentrations and with no known biological function, for example, lead, cadmium, and mercury
- toxic above trace amounts and with no known biological function, for example, arsenic, indium, antimony, and thallium
- toxic above certain concentrations and required for various biochemical processes, for example, copper, zinc, cobalt, selenium, potassium, and iron

One widely used technology for measuring very low concentrations of metals and semi-metals is *voltammetry*, which uses electrodes to apply a voltage to an aqueous sample and then measures the current produced (**Figure 5**). The magnitude of the current is proportional to the concentration of metal ions in the sample. Voltammetric methods have detection limits as low as a few picograms (10^{-12} g) and have been used to detect lead, cadmium, zinc, and copper in single raindrops. The main weakness of this technology is that its ability to distinguish between one element and another is poor at very low concentrations.

In the 1970s, a reliable method of counting individual atoms was developed to detect trace impurities in the materials used

Figure 5
Stationary voltammetry electrodes

to make computer chips (**Figure 6**). Some modern electronic components are so small that a few foreign atoms can cause them to malfunction. The technology for counting atoms uses photons from a laser to knock one electron out of the outer shell of each atom of a specific element; the electrons released are then counted. The energy of electrons in atoms is quantized into specific levels, rather like the rungs on a ladder that the electron has to climb to escape from the atom. The spacing of these energy rungs is different for every element, so by supplying the precisely correct amount of energy using the laser, electrons can be knocked off the atoms of a selected element while leaving the electrons in other atoms undisturbed. Counting these electrons provides an accurate measure of the number of atoms of a specific element in a sample.

In addition to being able to detect extremely small traces of various elements, atom counting technology dramatically reduces the size of samples required for laboratory tests. This technology is ideal for studying the effects of extremely small traces of various elements on the human body. For example, by using atom counting techniques, researchers have found that traces of metals like chromium, copper, and zinc, which are required for normal development, are transferred from the mother to the fetus late in pregnancy. Consequently, a very premature baby may lack these elements and suffer various ailments and birth defects unless supplied with the proper amounts of these elements.

On a smaller scale, University of Alberta Professor Jed Harrison (**Figure 7**) is researching the application of microfabrication technology to create tiny analytical instruments. Employing micromachining techniques, he uses semiconductors to create sensors that can detect tiny amounts of chemicals. These sensors will be useful in a wide range of situations, including blood tests for hormones and drugs, and soil testing (**Figure 8**).

Figure 7
Jed Harrison (1954-)

Figure 8
This 2 cm × 2 cm glass and silicon microchip uses electrolysis of water in the two lower chambers to create a gas pressure that pumps a blood sample and test mixture together, allowing the isolation of rare or diseased cells for clinical diagnostic assays.

Figure 6
Technology has reduced the limit of detectability of an element to a single atom. There is no longer a concentration that is too small to be measured. The atom-counting instrument shown is able to count atoms of specific elements in tiny samples, such as water, air, blood, and microchips.

Case Study Questions

1. What societal needs drive the development of increasingly sensitive measurement technology?
2. In the fictional television series, *Star Trek*, the crew of the space ship carried devices called "tricorders." These devices could perform a complete analysis of whatever organism was before them. State an argument for, and one against, the possibility of society ever developing analysis technology to the level imagined for the "tricorder."

Section **8.5**

INVESTIGATION 8.6 *Introduction*

Titration Analysis of ASA

Report Checklist
- ○ Purpose   ○ Design   ● Analysis
- ○ Problem   ○ Materials   ● Evaluation (1, 3)
- ○ Hypothesis ○ Procedure
- ○ Prediction   ● Evidence

Acetylsalicylic acid, known as ASA, is the most commonly used pharmaceutical drug, with over 10 000 t manufactured in North America every year. ASA, $C_8H_7O_2COOH(s)$, is an organic acid like acetic acid and reacts with strong bases such as sodium hydroxide in the same way that acetic acid does.

Purpose
The purpose of this investigation is to use titration analysis techniques to accurately determine the ASA content of a consumer product: a standard pain-relief tablet.

Problem
What is the mass of ASA in a consumer tablet?

Design
An ASA tablet is dissolved in methanol and then titrated with standardized sodium hydroxide solution (from Investigation 8.3) using phenolphthalein as the indicator.

To perform this investigation, turn to page 345.

▶ Section 8.5 *Questions*

For the following questions, use the Acid–Base Indicators table on the inside back cover where appropriate.

1. In the titration of dilute ammonia with dilute hydrochloric acid, a trial pH curve titration found the equivalence point pH of the solution to be 4.8. Explain why bromocresol green is a better indicator choice than alizarin yellow for this titration.

2. Why must only a very small amount of indicator be used in a titration analysis?

3. If congo red indicator is used in the titration of dilute nitric acid, $HNO_3(aq)$, with dilute sodium hydroxide, $NaOH(aq)$, will the indicator endpoint of the titration correspond to the equivalence point? Explain, using a sketch of the pH titration curve to illustrate your reasoning.

4. For a titration analysis to determine the concentration of an oxalic acid solution, complete the following:
   (a) What information must you have in order to select an indicator for this reaction?
   (b) What equipment and procedure would be required to get this information?

5. Why is it necessary to start a titration analysis with at least one standard solution?

6. Define the following terms:
   (a) titration
   (b) titrant
   (c) endpoint
   (d) equivalence point

7. In a titration analysis, state the function of
   (a) an Erlenmeyer flask
   (b) a volumetric pipette
   (c) a burette
   (d) a meniscus finder

8. A chemistry student is given the task of accurately and precisely determining the amount concentration of a hydrochloric acid solution so it can be used as a standard solution. She chooses sodium carbonate to prepare her initial primary standard solution.
   (a) What mass of pure dry sodium carbonate will she require to prepare 100.0 mL of 0.120 mol/L solution?
   (b) Write the steps for a complete procedure for her titration, including waste disposal.
   (c) What should she do to ensure her safety while performing this titration?

9. Copy and complete **Table 1** in Evidence, and complete the Analysis of the following investigation report.

   Purpose
   The purpose of this investigation is to use titration design to standardize a solution of hydrochloric acid.

   Problem
   What is the amount concentration of the hydrochloric acid solution?

   Design
   A standard sodium carbonate solution is prepared. Samples of this standard solution are titrated with the unknown solution of hydrochloric acid. The colour change of methyl orange is used to indicate the endpoint.

   Evidence

   Table 1 Titration of 10.00 mL of 0.120 mol/L $Na_2CO_3(aq)$ with $HCl(aq)$

Trial	1	2	3	4
final burette reading (mL)	17.9	35.0	22.9	40.1
initial burette reading (mL)	0.3	17.9	5.9	22.9
volume of HCl(aq) added (mL)				
colour at endpoint	red	orange	orange	orange

Chemical Analysis **339**

Chapter 8 INVESTIGATIONS

INVESTIGATION 8.1

Analysis of Sodium Carbonate

Report Checklist
- ○ Purpose
- ○ Problem
- ○ Hypothesis
- ○ Prediction
- ○ Design
- ○ Materials
- ○ Procedure
- ● Evidence
- ● Analysis
- ● Evaluation (1, 3)

Sodium carbonate has been used for all of recorded history in a variety of applications, from glassmaking to detergent manufacture to water treatment. A common name for sodium carbonate is soda ash, an appropriate name because it can easily be extracted from wood ashes. Sodium carbonate is one of the ten highest-volume chemicals produced in North America.

In this investigation, you will use techniques and equipment common to gravimetric analysis to analyze a sodium carbonate solution. For a description of a method for filtering a precipitate, see Appendix C.4.

Purpose
The purpose of this investigation is to use the stoichiometric method as part of a gravimetric analysis.

Problem
What is the mass of solute in a 50.0 mL sample of sodium carbonate solution?

Design
The mass of sodium carbonate present in the sample solution is determined by having it react with an excess quantity of a calcium chloride solution. The mass of calcium carbonate precipitate formed is used in a stoichiometric calculation to determine the mass of sodium carbonate that reacted.

 Sodium carbonate and calcium chloride solutions can irritate skin. As always, wash your hands before leaving the laboratory.

Materials
lab apron
eye protection
$Na_2CO_3(aq)$
$CaCl_2(aq)$
wash bottle of pure water
50 mL or 100 mL graduated cylinder
100 mL beaker
250 mL beaker
400 mL beaker
stirring rod
medicine dropper
filter paper
filter funnel, rack, and stand
centigram balance

Procedure
1. Measure 50.0 mL of $Na_2CO_3(aq)$ in the graduated cylinder, and transfer this sample into a clean 250 mL beaker. (See Appendix C.3.)
2. Measure 60 mL of $CaCl_2(aq)$ in a clean 100 mL beaker.
3. Slowly add about 50 mL of $CaCl_2(aq)$, with stirring, to the $Na_2CO_3(aq)$.
4. Allow the mixture to settle. When the top layer of the mixture becomes clear, use the dropper to add a few extra drops of $CaCl_2(aq)$.
5. If any new cloudiness is visible (**Figure 1(a)**), repeat steps 2 to 4, adding as much $CaCl_2(aq)$ as necessary, until this test indicates the reaction is complete (**Figure 1(b)**).
6. Measure the mass of a piece of filter paper.
7. Filter the mixture, and discard the filtrate in the sink. (See Appendix C.4.)
8. Dry the precipitate and filter paper overnight on a folded paper towel.
9. Measure the mass of the dried filter paper plus precipitate.
10. Dispose of the precipitate in the trash (solid waste).

Figure 1
(a) Visible evidence of incomplete reaction
(b) Visible evidence of complete reaction

INVESTIGATION 8.2

Percent Yield of Barium Sulfate

Report Checklist
- ○ Purpose
- ○ Problem
- ○ Hypothesis
- ● Prediction
- ○ Design
- ● Materials
- ● Procedure
- ● Evidence
- ● Analysis
- ● Evaluation (1, 2, 3)

Barium sulfate is a white, odourless, tasteless powder that has a variety of different uses: as a weighting mud in oil drilling; in the manufacture of paper, paints, and inks; and taken internally for gastrointestinal X-ray analysis. It is so insoluble that it is nontoxic, and is therefore safe to handle.

The reaction studied in this investigation is similar to the one used in the industrial manufacture of barium sulfate. To determine the predicted (theoretical) yield in the Prediction, you will need to identify the limiting and excess reagents before using the stoichiometric method to predict the expected mass of product.

Purpose
The purpose of this investigation is to use the stoichiometric method to evaluate a commercial procedure for producing barium sulfate.

Problem
What is the percent yield of barium sulfate in the reaction of aqueous solutions of barium chloride and sodium sulfate?

Design
A 40.0 mL sample of 0.15 mol/L sodium sulfate solution is mixed with 50.0 mL of 0.100 mol/L barium chloride solution. A diagnostic test is performed to check for completeness of precipitation of the limiting reagent (**Figure 2**). The mass of the filtered, dried precipitate is measured. The experimental mass of the precipitate is compared to the predicted mass.

 Soluble barium compounds, such as barium chloride, are toxic and must not be swallowed. Wear gloves and wash hands thoroughly after handling the barium ion solution.

Wear eye protection and a laboratory apron.

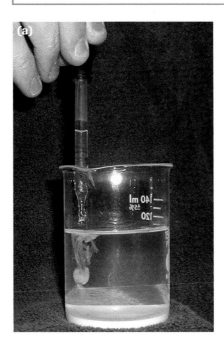

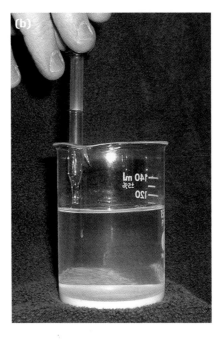

Figure 2
(a) Visible evidence of incomplete reaction
(b) Visible evidence of complete reaction

INVESTIGATION 8.3

Standardization Analysis of NaOH(aq) (Demonstration)

Report Checklist
- ○ Purpose
- ○ Problem
- ○ Hypothesis
- ○ Prediction
- ○ Design
- ○ Materials
- ○ Procedure
- ● Evidence
- ● Analysis
- ○ Evaluation

A large (stock) volume of sodium hydroxide solution is prepared for use in Investigations 8.4 and 8.6 by dissolving 6 g of solid for each litre of solution required. Because this solute is not a primary standard, this solution must be standardized to accurately determine its concentration. This will be accomplished by titration against a standard potassium hydrogen phthalate (KHP) solution. The balanced reaction equation is

$$KC_7H_4O_2COOH(aq) + NaOH(aq) \rightarrow$$
$$KNaC_7H_4O_2COO(aq) + H_2O(l)$$

Purpose
The purpose of this investigation is to use a titration design to standardize a solution for future chemical analysis.

Problem
What is the concentration of a stock NaOH(aq) solution?

Design
A standard solution of KHP is prepared, and is then used to standardize a stock solution of sodium hydroxide. Samples of KHP are titrated with sodium hydroxide titrant, using phenolphthalein as an indicator.

Materials
lab apron	50 mL burette
eye protection	10 mL volumetric pipette
$KC_7H_4O_2COOH(s)$	pipette bulb
NaOH(aq)	ring stand
phenolphthalein	centigram balance
wash bottle of pure water	laboratory scoop
150 mL beaker	stirring rod
250 mL beaker	small funnel
250 mL Erlenmeyer flask	meniscus finder
100 mL volumetric flask with stopper	

Procedure

1. Prepare a 100.0 mL standard solution of 0.150 mol/L KHP. (See Appendix C.4.)
2. Place approximately 70 mL of NaOH(aq) in a clean, dry, labelled 150 mL beaker.
3. Set up the burette with NaOH(aq), following the accepted procedure for rinsing and for clearing the air bubble. (See Appendix C.4 and the Nelson Web site.)

4. Pipette a 10.00 mL sample of KHP into a clean Erlenmeyer flask, and add 2 drops of phenolphthalein indicator.
5. Record the initial burette reading to the nearest 0.1 mL.
6. Titrate the KHP sample with NaOH(aq) until a single drop produces a permanent colour change, from colourless to pink.
7. Record the final burette reading to the nearest 0.1 mL.
8. Repeat steps 4 to 7 until three consistent results are obtained.
9. Dispose of all solutions in the sink, and flush with lots of water.

INVESTIGATION 8.4

Titration Analysis of Vinegar

Report Checklist
- ○ Purpose
- ○ Problem
- ○ Hypothesis
- ○ Prediction
- ○ Design
- ● Materials
- ● Procedure
- ● Evidence
- ● Analysis
- ● Evaluation (1, 2, 3)

Some consumer food products are required by law to have the minimum quantity of the active ingredient listed on the product label. Companies that produce chemical products usually employ analytical chemists and technicians to monitor the final product in a process known as quality control. Nevertheless, government consumer affairs departments also use chemists and technicians to check products, particularly in response to consumer complaints. According to the label, a vinegar manufacturer states that the vinegar contains 5% acetic acid by volume, which translates to a minimum amount concentration of 0.83 mol/L (**Figure 3**).

In the Evaluation section of your report, collect and average analysis values from all groups performing this investigation. Explain what should be done with discrepant results. Include safety and disposal instructions with the procedure.

Purpose
The purpose of this investigation is to test the manufacturer's claim of the concentration of acetic acid in a consumer sample of vinegar.

Problem
What is the amount concentration of acetic acid in a sample of vinegar?

  Chemicals used may be flammable or corrosive.

Design
A sample of commercial white vinegar is diluted by a factor of 5 to make a 100.0 mL final solution. Samples of this diluted solution are titrated with a standardized sodium hydroxide solution (from Investigation 8.3) using phenolphthalein as the indicator.

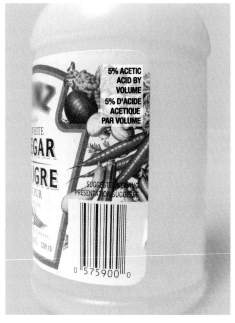

Figure 3
Is there really 5% acetic acid by volume, in this bottle of vinegar?

INVESTIGATION 8.5

pH Curves (Demonstration)

Report Checklist
- ○ Purpose
- ○ Problem
- ○ Hypothesis
- ● Prediction
- ○ Design
- ○ Materials
- ○ Procedure
- ● Evidence
- ● Analysis
- ○ Evaluation

When titrating a basic sample with an acidic titrant, you would expect the pH to be high initially, then to decrease as acid is progressively added, and finally to be low when a large excess of acid has been added. This expectation turns out to be correct. However, what is interesting and important is the *way* that the pH decreases. A titration pH curve is very useful evidence, providing valuable information about any acid–base reaction. Your Analysis involves plotting graphs of pH against volume of acid added. Alternatively, a computer program using a pH sensor probe may be used (if available) to plot the graph on screen as the titrations occur.

Purpose
The purpose of this demonstration is to create pH curves and observe the function of an indicator in an acid–base reaction.

Problem
What are the shapes of the pH curves for the continuous addition of hydrochloric acid to a sample of a sodium hydroxide solution and to a sample of a sodium carbonate solution?

Design
Small volumes of hydrochloric acid are added continuously to a measured volume of a base. After each addition, the pH of the mixture is measured. The volume of hydrochloric acid is the manipulated variable, and the pH of the mixture is the responding variable. **Figure 4** shows a stirring system that can be used for this lab, if available.

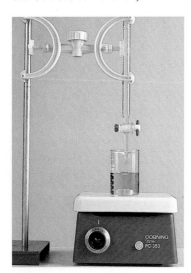

Figure 4
A burette and magnetic stirrer provide for a very efficient procedure.

Materials
lab apron
0.10 mol/L HCl(aq)
0.10 mol/L Na_2CO_3(aq)
methyl orange indicator
magnetic stirrer (optional)
150 mL beaker
50 mL graduated cylinders (2)
pH 7 buffer solution for calibration of pH meter
pH meter or pH probe with computer interface
eye protection
0.10 mol/L NaOH(aq)
bromothymol blue indicator
distilled water
50 mL burette and funnel
250 mL beakers (2)

 Acids and bases are corrosive and toxic. Avoid skin and eye contact. If you spill any of the chemical solutions on your skin, immediately rinse the area with lots of cool water. In the unlikely situation of getting some of the chemicals in your eye, immediately rinse your eye for at least 15 min and inform your teacher.

Procedure

1. Set the temperature on the pH meter and calibrate it by adjusting it to indicate the pH of the known pH 7 buffer solution.

2. Place 50 mL of 0.10 mol/L sodium hydroxide solution in a 150 mL beaker, and add a few drops of bromothymol blue indicator.

3. Measure and record the pH of the sodium hydroxide solution.

4. Successively add small quantities of HCl(aq), measuring the pH and noting any colour changes after each addition, until about 80 mL of acid has been added.

5. Repeat steps 1 to 4 for 50 mL of 0.10 mol/L sodium carbonate in a 250 mL beaker with the methyl orange indicator. Continue titrating until 130 mL of HCl(aq) has been added.

6. Dispose of all solutions in the sink, and flush with lots of water.

INVESTIGATION 8.6

Titration Analysis of ASA

Report Checklist
- ○ Purpose
- ○ Problem
- ○ Hypothesis
- ○ Prediction
- ○ Design
- ○ Materials
- ○ Procedure
- ● Evidence
- ● Analysis
- ● Evaluation (1, 3)

Acetylsalicylic acid, known as ASA, is the most commonly used pharmaceutical drug, with over 10 000 t manufactured in North America every year (**Figure 5**). ASA, $C_8H_7O_2COOH(s)$, is an organic acid like acetic acid and reacts with strong bases such as sodium hydroxide in the same way that acetic acid does.

As part of your evaluation, find the percent difference between your experimental averaged value for the mass of ASA in a sample tablet, and the value for the mass (in mg) that is listed on the product label.

Purpose
The purpose of this investigation is to use titration analysis techniques to accurately determine the ASA content of a consumer product: a standard pain-relief tablet.

Problem
What is the mass of ASA in a consumer tablet?

Design
An ASA tablet is dissolved in methanol and then titrated with standardized sodium hydroxide solution (from Investigation 8.3) using phenolphthalein as the indicator.

Materials
lab apron	50 mL burette
eye protection	10 mL volumetric pipette
standardized NaOH(aq)	pipette bulb
ASA tablets	ring stand
phenolphthalein	burette clamp
30 mL methanol	stirring rod
wash bottle of pure water	small funnel
50 mL beaker	meniscus finder
250 mL beaker	250 mL Erlenmeyer flask

 Methanol is toxic by ingestion. It may cause permanent blindness if swallowed.

 Methanol is volatile (it evaporates easily) and is very flammable. Do not use it near any ignition source or open flame.

Procedure

1. Add about 30 mL of methanol to one ASA tablet in a clean Erlenmeyer flask (**Figure 5**).

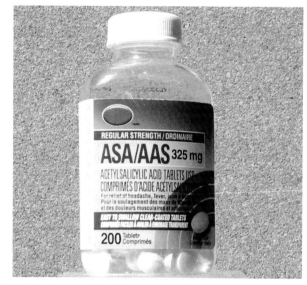

Figure 5
ASA, or acetylsalicylic acid

2. Stir and crush the tablet until the solid has mostly dissolved. (The final mixture will probably be slightly cloudy because of the presence of inert ingredients.)

3. Add 1 or 2 drops of phenolphthalein indicator.

4. Set up the burette with standardized NaOH(aq) titrant, using accepted techniques for rinsing and for eliminating the air bubbles (Appendix C.4).

5. Record the initial burette reading to the nearest 0.1 mL.

6. Titrate the ASA sample with NaOH(aq) until a single drop produces a permanent change from colourless to pink.

7. Record the final burette reading to the nearest 0.1 mL.

8. Repeat steps 1 to 7 until three consistent results are obtained.

9. Dispose of all solutions into the sink, and flush with lots of water.

12. Sources of experimental uncertainty are always present in an experiment. Which one of the following is not an acceptable example of an experimental uncertainty?
    A. all measurements
    B. empirical constants
    C. purity of chemicals
    D. human error

13. A primary standard for solution preparation must be
    A. stable when its container is opened
    B. an ionic compound
    C. an acidic substance
    D. colourless in solution

14. The main benefit of preparing a graph of relative proportions of substances in a chemical reaction, when that reaction is constantly used for repetitive analysis, is that the graph
    A. improves accuracy
    B. saves calculation time
    C. illustrates the proportionality visually
    D. allows percent yields to be found

15. A student analyzes the concentration of some propanoic acid, $C_2H_5COOH(aq)$, by titrating 10.00 mL samples with a standardized solution of 0.186 mol/L sodium hydroxide, NaOH(aq), using phenolphthalein as the indicator. If the average volume of sodium hydroxide used to reach endpoint is 14.0 mL, the amount concentration of the propanoic acid solution is _____ mmol/L.

16. Pure chromium metal can be obtained in a laboratory by heating chromium(III) oxide mixed with powdered aluminium. This reaction releases a great deal of heat. Assume that a 20% excess of Al(s) must be mixed with a 20.0 g sample of $Cr_2O_3(s)$ to ensure complete reaction. The mass of aluminium required will be _____ g.

17. Which indicator from **Table 1** would be most suitable to use for a titration of HCl(aq) with NaOH(aq)?

Table 1 For question 17

	Indicator	pH change range
A.	p-naphtholbenzene	8.2–10.0
B.	brilliant yellow	6.6–7.8
C.	propyl red	4.8–6.6
D.	2,4-dinitrophenol	2.8–4.0

Part 2

18. Briefly **describe** each of the four main assumptions usually made about chemical reactions done in school classrooms and laboratories.

19. In both scientific studies and technological applications of chemical reactions, one reactant is usually a limiting reagent, and the other reactant is an excess reagent. **Distinguish** between these two terms, and suggest a general reason for using one reactant in excess.

20. In a gravimetric analysis, **explain** how you know that the reaction is complete.

21. (a) In a titration analysis, explain how you know that the reaction is complete.
    (b) **Sketch** a pH curve of a titration reaction in which a strong monoprotic acid is titrated with a strong monoprotic base. On the curve, indicate the pH at the equivalence point.
    (c) Name three specific chemicals that would be appropriate materials for this titration.

Use this information to answer questions 22 to 24.

The common mineral apatite (**Figure 2**) has a characteristic hexagonal crystal structure. Apatite is an impure form of calcium phosphate, which is widely used as a supplement for all types of animal feed. Calcium phosphate supplies phosphorus and calcium, two essential nutrients for the growth of all animals. In addition, this compound is the chief component of the hard parts of teeth and the chief cementing material in the solid parts of bones. In a laboratory, pure calcium phosphate can be produced from the reaction of aqueous solutions of calcium chloride and sodium phosphate, because the product compound has extremely low solubility.

Figure 2
Apatite

22. Write the net ionic equation for this reaction.

23. **Identify** the spectator ions, and **explain** why only some of the ions present are described as spectator ions.

24. In a quantitative study of the laboratory reaction, a student predicted that 2.34 g of calcium phosphate product should form. In the experiment, 2.47 g of product was obtained.
    (a) **Determine** the percent yield.
    (b) What is unusual about your answer? Suggest several sources of experimental uncertainty that might account for this result.

25. In the Canadian steel industry, iron(III) oxide, from a previous roasting step, is reduced by reacting with carbon to produce iron metal and carbon dioxide. **Predict** what mass of iron metal could be produced from 1.00 t of iron(III) oxide.

26. **Identify** the limiting and excess reagents for each of the following pairs of reactants, and **determine** the chemical amount in excess in each case.
    (a) Zn(s)   + CuSO₄(aq) →
        0.42 mol   0.22 mol
    (b) Cl₂(aq)   + NaI(aq) →
        10 mmol   10 mmol
    (c) AlCl₃(aq) + NaOH(aq) →
        20 g   20 g

Use this information to answer questions 27 to 30.

Air bag technology has saved many lives (**Figure 3**). The reaction that causes the air bag to inflate (in less than 30 ms) is the simple decomposition of sodium azide, NaN₃(s). This reaction is initiated by an electrical signal from a sensor.

Figure 3
Crash test dummies testing air bags

27. If 70 L of gas at 20 °C and 98.5 kPa is required to completely fill a typical air bag, **determine** what mass of sodium azide must be stored in the inner reaction container.

28. This reaction is very fast and spontaneous when the electrical signal is sent. What other assumptions did you make when answering question 27?

29. If this were the only reaction to occur, what safety concerns would you have? Use your chemical knowledge and WHMIS to describe the possible dangers.

30. Research answers to the following questions.
    (a) What other substances are included along with the primary reactant, sodium azide?
    (b) What is the purpose of each of these substances?
    (c) The use of air bags is somewhat controversial. Comment on the pros and cons of air bags from four perspectives.

 www.science.nelson.com

31. In a study of rust-removing solutions, 27.8 mL of 0.115 mol/L phosphoric acid reacts completely with 0.245 mol/L sodium hydroxide. **Predict** the minimum volume of sodium hydroxide required for this reaction.

Use this information to answer questions 32 to 35.

Aqueous solutions of ammonia and hydrochloric acid produce strong, irritating fumes as the dissolved gas escapes from solution. When beakers of concentrated solutions of aqueous ammonia and hydrochloric acid are placed side by side, a white "smoke" forms in the air between them, and a solid powder deposits on the beakers (**Figure 4**). This white solid is ammonium chloride. Suppose that 2.00 g of hydrogen chloride gas were to be mixed with 2.00 g of ammonia gas, in a closed container.

Figure 4
Aqueous solutions of ammonia and hydrochloric acid

32. Which gas would be a limiting reagent?

33. What mass of solid product would form?

34. (a) Which reactant would remain when the reaction is complete?
    (b) **Predict** the mass that would remain.

35. **Explain** how your answers to questions 33 and 34 support the law of conservation of mass.

36. Scientific concepts need to be tested in many different situations. In the following investigation report, complete the Prediction, Analysis, and Evaluation.

    Purpose
    The purpose of this experiment is to test the concept of stoichiometry.

    Problem
    What mass of magnesium metal will react completely with 100.0 mL of 1.00 mol/L HCl(aq)?

    Evidence
    volume of 1.00 mol/L HCl(aq) = 100.0 mL
    initial mass of Mg(s) = 3.72 g
    final mass of Mg(s) = 2.45 g

37. Review the focusing questions on page 270. Using the knowledge you have gained from this unit, briefly **outline** a response to each of these questions.

unit 5
Organic Chemistry

"Syncrude Canada was just about to begin operating its oil sands plant when I was graduating. I had attended several talks about the oil sands, so my primary reason for interviewing with the company was to see their pilot plants. When they offered me a job, however, I accepted. One of the advantages of working in such a large company is that I have had the opportunity to carry out research in numerous areas. I like the variety of the work.

When I joined the work force, I was concerned whether I had the correct qualifications for the job. I have since learned that my formal training was really just the beginning of my learning. What I learned in school is how to approach a problem. On the job, I've expanded my knowledge and learned how to work in diverse teams to solve our problems."

Jean Cooley,
Research Associate,
Syncrude Canada

As you progress through the unit, think about these focusing questions:

- What are the common organic compounds and what is the system for naming them?
- How does society use the reactions of organic compounds?
- How can society ensure that the technological applications of organic chemistry are assessed to ensure future quality of life and a sustainable environment?

Unit **5**

GENERAL OUTCOMES

In this unit, you will
- explore organic compounds as a common form of matter
- describe chemical reactions of organic compounds

Unit 5 Organic Chemistry

ARE YOU READY?

Prerequisites

Concepts
- atomic structure
- physical and chemical properties of molecular compounds
- covalent bonding
- polar and nonpolar molecules
- intermolecular forces

Skills
- drawing diagrams and writing formulas of molecules
- building molecular models
- conducting and analyzing diagnostic tests
- describing safety and disposal

You can review prerequisite concepts and skills in the Chemistry Review unit, on the Nelson Web site, and in the Appendices.
A Unit Pre-Test is also available online.

Knowledge

1. According to Lewis' theory, the number of bonding electrons in an atom determines the number of bonds the atom will form. Complete **Table 1** for five elements that are often involved in organic compounds.

Table 1 Atomic Structures of Some Common Nonmetals

Element	Atomic number	Number of electrons	Number of valence electrons	Lewis symbol	Number of bonding electrons
carbon					
hydrogen					
oxygen					
nitrogen					
sulfur					

2. Different formulas for molecules can be used to predict the types of intermolecular bonding that exist in a substance. Complete **Table 2** for some common compounds of carbon.

Table 2 Some Common Carbon Compounds

IUPAC name	Molecular formula	Lewis formula	Structural formula	Types of intermolecular forces
	$CH_4(g)$			
dichloromethane	$CH_2Cl_2(l)$			
methanol				
methyl amine	$CH_3NH_2(l)$			
ethane				
ethene	$C_2H_4(g)$			
ethyne		H:C:::C:H	H—C≡C—H	
	$C_2H_5OH(l)$			
ethyl mercaptan	$C_2H_5SH(l)$			

3. Use VSEPR theory to predict the shapes of each of the following molecules. Predict whether each molecule is polar or nonpolar.
   (a) chloromethane, $CH_3Cl(g)$, used as a refrigerant
   (b) carbon dioxide, $CO_2(g)$, produced by hydrocarbon combustion
   (c) formaldehyde, $HCHO(g)$, used as a disinfectant
   (d) ethyne, $C_2H_2(g)$, used as a fuel for welding
   (e) ethene, $C_2H_4(g)$, used to make plastics

4. The octane used in automobile fuel is considered nonrenewable because it is produced from petroleum, a fossil fuel. Ethanol is considered a renewable fuel since it can be produced from biomass, such as grain.
   (a) Write a balanced chemical equation for the combustion of octane, $C_8H_{18}(l)$.
   (b) Write a balanced chemical equation for the combustion of ethanol, $C_2H_5OH(l)$.
   (c) Describe some diagnostic tests for the reaction products in (a) and (b).

5. Photosynthesis is the formation of carbohydrates and oxygen from carbon dioxide, water, and sunlight, catalyzed by chlorophyll in the green parts of a plant (**Figure 1**).
   (a) Write a balanced chemical equation for photosynthesis, using $C_6H_{12}O_6(aq)$ for the carbohydrate.
   (b) During summer in the Northern Hemisphere, the level of carbon dioxide in the atmosphere decreases slightly. Use the chemical equation in (a) to explain this phenomenon.

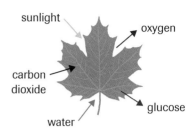

Figure 1
Photosynthesis consumes carbon dioxide and produces oxygen.

Skills

6. Boiling point is a measure of the strength of the intermolecular forces in a substance. There is evidence to suggest that London forces exist among all molecules in the liquid or solid state. Chemists explain the strength of London forces as being dependent on the number of electrons in the molecule. Polar molecules exert relatively weak dipole–dipole forces in addition to London forces. Molecules with OH, NH, or FH groups have relatively strong hydrogen bonds in addition to London forces. The purpose of this activity is to test the ability of the theoretical concepts of London forces, dipole–dipole forces, and hydrogen bonding to predict the relative boiling points of three families of compounds. Complete the Prediction, Analysis, and Evaluation (2, 3) of the investigation report.

Problem
What are the trends in boiling point for three families of carbon compounds?

Design
The boiling points of twelve organic compounds (from three chemical families) are measured experimentally. The boiling point evidence is used to test the Prediction. The polarity of the molecules and the hydrogen bonding are controlled variables within each family of compounds.

Evidence

Table 3 Boiling Points of Three Families of Carbon Compounds

Hydrocarbon	Boiling point (°C)	Organic halide	Boiling point (°C)	Alcohol	Boiling point (°C)
$CH_4(g)$	−164	$CH_3Cl(g)$	−24	$CH_3OH(l)$	65
$C_2H_6(g)$	−89	$C_2H_5Cl(g)$	12	$C_2H_5OH(l)$	78
$C_3H_8(g)$	−42	$C_3H_7Cl(l)$	47	$C_3H_7OH(l)$	97
$C_4H_{10}(g)$	−0.5	$C_4H_9Cl(l)$	78	$C_4H_9OH(l)$	117

7. Liquid hydrocarbons are readily flammable and often evaporate easily to form explosive mixtures with air.
   (a) What WHMIS symbol is found on a container of a liquid hydrocarbon?
   (b) Outline the precautions that must be taken when working with liquid hydrocarbons.
   (c) Describe an acceptable method for handling waste flammable liquids.

chapter 9

Hydrocarbons from Petroleum

In this chapter

- Exploration: Burning Fossil Fuels
- Investigation 9.1: Classifying Fossil Fuels
- Case Study: Fossil Fuel Industries in Alberta
- Web Activity: Coal in Alberta
- Explore an Issue: Coalbed Methane
- Investigation 9.2: Structures and Properties of Isomers
- Web Activity: West Nile Denial
- Lab Exercise 9.A: Chemical Properties of Aliphatics and Aromatics
- Lab Exercise 9.B: Boiling Points of Sample Aliphatics and Aromatics
- Investigation 9.3: Fractional Distillation (Demonstration)
- Web Activity: Karl Chuang
- Case Study: Octane Number
- Investigation 9.4: Bitumen from Oil Sands
- Investigation 9.5: Solvent Extraction
- Case Study: The Athabasca Oil Sands
- Investigation 9.6: Complete and Incomplete Combustion

The most widely accepted explanation of the origin of fossil fuels is that they come from decaying plant and animal material. The carbon cycle is a model connecting the organic reactions of photosynthesis, digestion, and respiration (**Figure 1**). The formation of fossil fuels can be seen as the end of the natural processes in the carbon cycle and the beginning of the technological processes described in this chapter.

Humans have invented methods to extract fossil fuels from below the surface of Earth. We mostly burn fossil fuels to produce heat and useful energy. These combustion reactions produce carbon dioxide, which feeds back into the carbon cycle.

Recently, we have developed many other technological uses of fossil fuels. We now produce vast quantities of petrochemicals (chemicals created from fossil fuels), most of which are not meant to be burned. Petrochemicals include methanol (windshield antifreeze), ethylene glycol (radiator antifreeze), chlorofluorocarbons (CFCs: refrigerator and air conditioner coolants), plastics (polyethylene and polyvinylchloride (PVC)), and pesticides. In this chapter, you will learn about the extraction and refining of fossil fuels. In the next chapter, you will learn about the use of fossil fuels to create petrochemicals. For both uses, fossil fuels are a very valuable nonrenewable resource. The impact of fossil fuel use on our lives and our planet is everywhere around us.

Historically, natural resources were seen as assets to be developed, and it was usual to imagine technological solutions to problems. In recent times, however, we have come to realize that technology is limited in its ability to cope with resource depletion and pollution. Many people, including scientists, believe we need to move toward an Aboriginal viewpoint that sees natural resources as gifts to be treasured for future generations. In order to save our planet from irreparable harm, we have a responsibility to live in harmony with nature for the mutual benefit of nature and humanity.

STARTING Points

Answer these questions as best you can with your current knowledge. Then, using the concepts and skills you have learned, you will revise your answers at the end of the chapter.

1. What are the origins and sources of fossil fuels?
2. How are fossil fuels extracted and refined?
3. What are the main hydrocarbon families, and how do they differ?
4. What are some uses of fossil fuels and their components?

Career Connections:
Petroleum Engineer; The Petroleum Industry

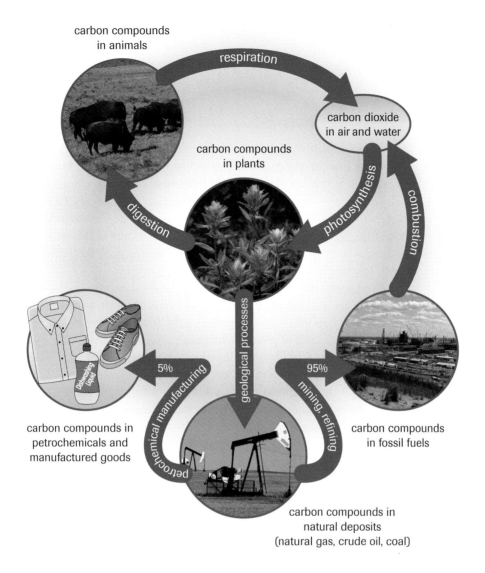

Figure 1
The carbon cycle is an illustration of the interrelationship of all living things with the environment and with technologies that refine and use fossil fuels.

▶ Exploration — Burning Fossil Fuels

Issue Checklist

○ Issue	● Design	● Analysis
○ Resolution	● Evidence	● Evaluation

Scientists agree that fossil fuels are a nonrenewable, finite resource. Fossil fuels in Alberta include coal, natural gas, crude oil, oil sands, and heavy oil. Studies have indicated that about 95% of the fossil fuels used in Alberta is used for energy production, including natural gas and propane for home heating, and gasoline and diesel fuel for cars, trucks, and buses. The evidence indicates that about 5% of fossil fuels is converted into petrochemicals (such as plastics) through technologies based on chemical reactions. Two issues related to burning fossil fuels for energy production are

- whether we should be burning a very valuable resource at the current rate
- whether we should be saving more of this resource for petrochemical use in the future

Investigate the first of these issues as you work through Chapter 9, and the second issue in Chapter 10.

Resolution: The burning of fossil fuels for heat and transportation should be significantly reduced.

While progressing through Chapter 9, complete the Design, Evidence, Analysis, and Evaluation components of the first issue report.

Governments, industry, and citizens alike speak of sustainable development. The United Nations Commission on Environment and Development defines sustainable development as development that meets the needs of the present without compromising the ability of future generations to meet their own needs. Keep this concept in mind while completing this unit-long exploration.

9.1 Fossil Fuels

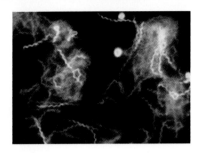

Figure 1
Extremophiles are microorganisms that thrive in harsh environments, such as boiling water, acid, rocks, the coolant of a nuclear reactor, and toxic wastes. Identification of these carbon-based organisms raises the hopes of astrobiologists that basic life forms may exist on moons and planets in the universe. On Earth, the study of organic chemistry began with living things. Beyond Earth, scientists look for evidence of living things.

Life as we know it is based on carbon chemistry (**Figure 1**). Therefore, it is not surprising that the early definition of organic chemistry was related to compounds obtained only from living things. Today, **organic chemistry** is a major branch of chemistry that deals with compounds of carbon, excluding oxides (such as $CO(g)$) and ionic compounds of carbon-based ions such as carbonate, cyanide, and carbide ions, for example, $Na_2CO_3(s)$, $NaCN(s)$, and $CaC_2(s)$, respectively. In spite of this broader definition, the major source of carbon compounds is still living or previously living things, such as plants, animals, and all types of fossil fuels.

Coal, oil sands, heavy oil, crude oil, and natural gas are nonrenewable sources of fossil fuels. They are also the primary sources of **hydrocarbons**—compounds containing carbon atoms bonded to hydrogen atoms. Hydrocarbons are the starting points in the synthesis of thousands of products, including specific fuels, plastics, and synthetic fibres. Some hydrocarbons are obtained directly by physically refining oil and natural gas (both called petroleum), whereas others come from further (chemical) refining (**Table 1**).

Refining is the technology that includes physical and chemical processes for separating complex mixtures into simpler mixtures or near-pure components. The refining of coal and natural gas involves physical processes; for example, coal may be crushed. Components of natural gas are separated either by solvent extraction or by condensation and distillation. Oil sands refining involves a chemically enhanced physical process followed by the complex refining of the bitumen/tar. Crude oil refining is more complex than coal or gas refining, but many more products are obtained from crude oil.

Table 1 Refining of Fossil Fuels—a Preview

Fossil fuel[1]	Extraction	Physical processing[2]	Chemical processing	Sample uses
Natural gas[3]	natural pressure underground	condensation and distillation	removal of hydrogen sulfide and carbon dioxide at a gas plant	heating buildings; source of ethane, propane, and butane
Coalbed methane	water removed from underground, if necessary	removal and disposal of saline water, if necessary	removal of non-combustibles at a gas plant	same as natural gas, including production of hydrogen and methanol
Crude oil	water or gas injection underground	water, sand, and salt removal, and fractional distillation in a tower	hydrocracking and catalytic reforming	gasoline, jet fuel, and asphalt
Heavy oil	steam injection underground	separation from water and solids; fractional distillation	heavy oil hydrocracking and catalytic reforming	same as crude oil
Oil sands (bitumen)	physical mining and in situ steam or hot water injection	hot water extraction and floatation; centrifugation; fractional distillation	coking, hydrocracking, hydrotreating	synthetic crude used as crude oil
Coal	surface and underground mines	sorting, crushing, and/or grinding	none or gasification for alternative delivery	energy for producing electricity

1. These Alberta fossil fuels are listed in order of increasing density. 2. In many cases, physical processing also includes removing water and solid contaminants from the raw material. 3. For natural gas, the chemical processing precedes the physical processing.

Section 9.1

INVESTIGATION 9.1 Introduction

Classifying Fossil Fuels

Alberta has many different fossil fuels. Samples of fossil fuels can be described and classified in terms of their physical properties of colour, transparency, viscosity, and density. In this investigation, the focus is on density.

Purpose
The purpose of this investigation is to use the known properties of fossil fuels to analyze the provided samples.

Problem
What are the classes of the fossil fuel samples provided?

To perform this investigation, turn to page 401.

Report Checklist
- ○ Purpose
- ○ Problem
- ○ Hypothesis
- ○ Prediction
- ○ Design
- ○ Materials
- ● Procedure
- ● Evidence
- ● Analysis
- ● Evaluation (1, 3)

Design
Equal-volume samples of fossil fuels are provided in sealed containers of equal mass. The mass of each sample, including the container, is determined. The fossil fuel samples are classified based on their densities.

Case Study

Fossil Fuel Industries in Alberta

Alberta is very rich in fossil fuels. The most widely accepted hypothesis for the origin of these fossil fuels is that they formed from sand, silt, and plant and animal remains starting about 500 Ma ago. The majority of hydrocarbons found in Canada exist in the Western Canada Sedimentary Basin, which covers all but the northeast corner of Alberta (**Figure 2**).

Coal, oil sands, heavy oil, crude oil, natural gas, and coalbed methane are fossil fuels found in significant quantities in Alberta. From a technological perspective, each of these fuels has historically had to go through several cycles of discovery, research, and development.

Discovery, Research, and Development
Discovery has occurred throughout history, from Aboriginal peoples discovering oil sands along the banks of the Athabasca River and farmers digging fence-post holes and finding coal, to today's sonar equipment used for conducting underground seismic surveys. Research and development (R & D) has historically been conducted separately and sequentially: research by universities and development by industry. However, this model of R & D has changed and continues to change, with more overlap of the roles of universities and industries. For example, on a small scale, government (the Alberta Research Council) scientists did research and development on the hot-water process for extracting oil from the oil sands. Industry and engineers then ran larger-scale tests and developed the sophisticated technologies required for economical world-scale plants. Researchers continue the R & D cycle in their attempts to extract oil from deep oil sands and depleted oil wells.

Petroleum Discoveries
Natural gas was first discovered by chance near Medicine Hat, Alberta, in 1883. The drillers were looking for water, and found a methane-rich flammable gas. The discovery, and the many subsequent fires, prompted author Rudyard Kipling to refer to

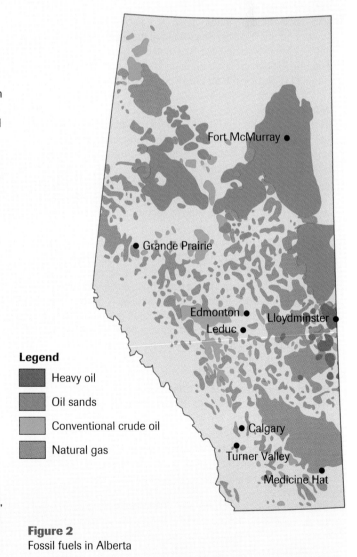

Figure 2
Fossil fuels in Alberta

NEL

Hydrocarbons from Petroleum **359**

Medicine Hat as Hell's Kitchen. The plentiful gas made Medicine Hat famous for natural gas street lights and pottery furnaces. Natural gas was then discovered in Cessford, Turner Valley, Suffield, Bow Island, and Viking, Alberta.

Hell's Half Acre was the nickname for Turner Valley, Alberta, where oil was discovered in 1914 and in 1936—beneath the natural gas. The next important discovery of crude oil was Leduc 1 on February 13, 1947 (**Figure 3**). This gusher near Devon, south of Edmonton, started an oil rush. Oil was struck in the Devonian Formation after Imperial Oil had drilled 133 unsuccessful wells. Perseverance, it appears, is just as important for technologists as it is for scientists (and students).

Figure 3
Leduc 1

The largest oil field in Canada is the Pembina field, near Drayton Valley, Alberta. Geologists continue to search for oil and gas in the sedimentary basin to Rainbow Lake, Norman Wells, and the Beaufort Sea. Much of the exploration in Alberta today is to find coalbed methane gas and oil sands (see the Explore an Issue on page 365 and the Case Study on page 395). The search for crude oil also continues.

Infrastructure

With the discovery and production of oil and natural gas came the need for infrastructure: oil and gas pipelines, gas plants, and oil refineries. Pipelines are needed to transport the oil or gas to a refinery, to local consumers, and to national and international consumers. These pipelines criss-cross Alberta—as evidenced by cut-lines and road-crossing signs. Pipelines were, and continue to be, built to eastern Canada, the west coast, and the United States.

Oil refineries were first built in 1923 and 1939 in Calgary. As the discovery of oil moved northward, so did the refineries. As the size of the discoveries grew, so did the size of the refineries. Refineries continue to be built, especially for the refining of bitumen (from oil sands) and heavy oil. The refining of natural gas in Alberta requires over 800 gas plants to remove impurities (such as hydrogen sulfide) and to extract the valuable components from the gas (including ethane, propane, and butane).

Social and Environmental Challenges

The advantages to Albertans of the large quantities of fossil fuels found in Alberta are numerous. Major examples include using gasoline (from crude oil) to propel our cars and trucks, using natural gas to heat our homes, and using coal to produce our electricity. In general, the fossil fuel industry in Alberta has important effects on our economy and our lifestyles.

Our fortune of fossil fuels also presents challenges. Some of these challenges include the effects on communities and infrastructure. For example, in fast-growing communities close to fossil fuel extraction and refining plants, there is a higher probability of a shortage of housing and social services. Rapid and extensive development of resources also presents environmental challenges. For example, the extraction and refining of many fossil fuels requires water. The extraction process disturbs surface land and air, and creates water pollution to varying degrees.

There are nearly always trade-offs (a negative for a positive result) when we decide to develop a natural resource. Research and technological fixes can only solve so many problems. The demands (needs and wants) of society and of individuals may be justified in our minds, but we also need to consider the potential negative consequences of our everyday actions.

Case Study Questions

1. (a) Based upon Figure 2 on the previous page, which are the most widespread fossil fuel deposits in Alberta?
   (b) What fossil fuels found in Alberta are not depicted in Figure 2?
2. (a) What is the classical relationship between research and development?
   (b) Which group of people performed each of the R & D?
3. What technological infrastructure is necessary for processing and transporting fossil fuels?
4. Give an example of a trade-off between a positive and a negative consequence of fossil fuel production.

Extension

5. Research and report alternative hypotheses to the currently accepted hypothesis for the formation of fossil fuels, such as the organic and the inorganic hypotheses.

   www.science.nelson.com

6. In what general regions of Alberta have we found
   (a) oil sands, heavy oil, crude oil, and natural gas?
   (b) coal and coalbed methane?
   (c) significant wind, solar, biomass, and geothermal energy production?

   www.science.nelson.com

7. Research when and where natural gas was discovered in Canada's Northwest Territories.

   www.science.nelson.com

8. Continue your research on the Resolution in the Exploration on page 357: The burning of fossil fuels for heat and transportation should be significantly reduced. Gather statements, pro and con, from a variety of perspectives. Focus on the effects of the discovery and extraction of fossil fuels in Alberta.

   www.science.nelson.com

Section 9.1

Case Study—Coal in Alberta

Research indicates that the most abundant fossil fuel reserve in Alberta and Canada is coal. To help describe coal, scientists have classified it into four broad classes based upon the percentage of carbon. From highest carbon content to lowest, the classes of coal are anthracite, bituminous, sub-bituminous, and lignite. You will find out more about coal as you pursue this Case Study.

 www.science.nelson.com

Could the Athabasca Tar Sands Catch Fire?
The tar sands, or oil sands, are rich in fossil fuels. So could a huge fire spread through the vast deposits? Listen to a University of Calgary professor's answer to find out.

 www.science.nelson.com

Section 9.1 Questions

1. Classify the following chemicals as organic or inorganic.
   (a) $CaCO_3(s)$
   (b) $C_{25}H_{52}(s)$
   (c) $CaC_2(s)$
   (d) $CCl_4(l)$
   (e) $CH_3COOH(l)$
   (f) $CO_2(g)$
   (g) $KCN(s)$
   (h) $C_{12}H_{22}O_{11}(s)$

2. List six different types of fossil fuels found in Alberta.

3. All fossil fuels require some refining after being extracted from the ground. State the two major kinds of refining and describe how, in general, they are different.

4. Write molecular and structural formula equations for the combustion of the following hydrocarbon components of natural gas. Assume complete combustion.
   (a) lighting a butane lighter, $C_4H_{10}(g)$
   (b) burning pentane, $C_5H_{12}(l)$, in winter gasoline

5. The carbon dioxide gas produced by combustion of hydrocarbons enters the atmosphere.
   (a) What change can the carbon dioxide induce on the climate? Explain briefly.
   (b) Write an equation to represent the production of an acid component of acid rain from carbon dioxide.

6. Classify the following statements as pro or con to the Resolution in the Exploration on page 357. Also classify the perspective from which the statement is made.
   (a) Research is being done to reduce or reuse carbon dioxide emissions.
   (b) Alternative energy sources, such as wind and solar power, may reduce energy-use effects on the environment.
   (c) Industries producing ethanol from grain should receive the same tax breaks as industries producing synthetic crude oil from oil sands.
   (d) The yields of crude oil from wells and of synthetic crude from oil sands are increasing due to better and better equipment and techniques.
   (e) Votes can be gained by accelerating the extraction of crude oil and of coalbed methane from available deposits.
   (f) The laws governing the extraction of fossil fuels by energy companies need to be more stringently enforced.
   (g) Developing our fossil fuel resources is the morally right thing to do—it helps people.
   (h) Communities depend on the energy industries for supporting their families and way of life.

7. What kinds of energy sources do you use in your home and from what fossil fuels do they come?

8. List three renewable sources of energy naturally available in Alberta.

9. What kinds of jobs are available working on oil and gas drilling rigs? Look at the options and choose an occupation on or around the rigs. Describe the job, its requirements, and the recommended salary. Where would you go to train for this job?

 www.science.nelson.com

9.2 Alkanes from Natural Gas

Natural gas is removed from underground by drilling a deep hole into the ground where geologists have predicted that natural gas and/or oil is to be found. Nature makes the geologists' task of predicting the presence of oil and gas very challenging, as illustrated in **Figure 1**, but modern technologies provide increased certainty to their predictions. From a chemistry perspective, the separation of the rock and water from the oil and gas can be explained by solubility theory. For example, nonpolar substances (such as oil) do not dissolve in polar substances (such as water). Some ionic substances (such as limestone) are only slightly soluble in water, and no ionic substances dissolve in oil.

Some common types of traps for oil and gas
1. Limestone reef trap of the type found in Leduc and Redwater.
2. Reservoirs in folded and faulted strata. Turner Valley's oil was found in traps like this.
3. Stratigraphic trap of the type found in southeast Saskatchewan.
4. Displacement of rock layers along a fault. This type of structure is found in some areas of the Peace River district.

Legend
 Sandstone
Limestone
Shale
 Gas
 Oil
 Water

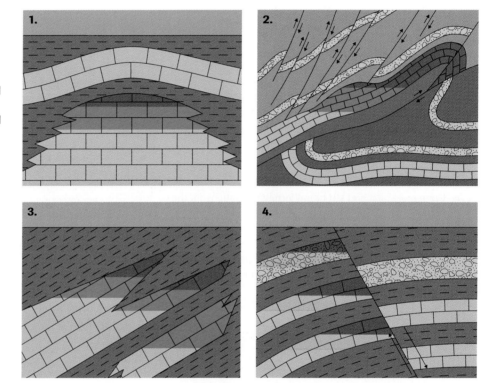

Figure 1
Trapping of natural gas and crude oil in geological formations (limestone and sandstone)

Natural gas has varying composition but it is primarily methane, $CH_4(g)$ (**Table 1**). Methane is the smallest of the alkane molecules. An *alkane* is a class of hydrocarbons that contains only single bonds and only carbon and hydrogen atoms. In this section, first you will learn about the natural occurrence and the technological extraction and refining of natural gas. Then you will learn about the chemical and physical properties of alkanes. Finally, you will investigate the theoretical structure and nomenclature of this important class of hydrocarbons.

Raw natural gas often contains impurities that make it "sour." This means that it contains the very toxic gas hydrogen sulfide, which forms acidic solutions when mixed with water. The hydrogen sulfide gas (also called rotten-egg gas) presents a major safety problem for gas-field workers. Exposure to only a small quantity of this gas is deadly. Fortunately, the human nose is extremely sensitive to hydrogen sulfide gas—an early warning system.

DID YOU KNOW ?

Smelly Additives
Mercaptans are added to natural gas and to propane. Natural gas and propane are odourless, which makes them difficult to detect and a potential explosion hazard. Mercaptans, however, have a very strong odour. The addition makes gas leaks much more noticeable, and, therefore, much safer.

Section **9.2**

Natural Gas Refining

Raw natural gas is piped from the well site to one of more than 800 gas treatment plants in Alberta. At the plant, water and liquid hydrocarbons are removed from the gas before it is chemically refined in an absorber tower. During this technological process, the sour natural gas reacts with an amine such as aqueous diethanolamine, $(C_2H_4OH)_2NH(aq)$, under relatively high pressure and low temperature. The amine "sweetens" the gas by removing any hydrogen sulfide (H_2S) and carbon dioxide gases, which are then released from the amine in a regenerator tower. The hydrogen sulfide then partially reacts with oxygen to produce sulfur dioxide. The mixture is then sent to a sulfur recovery unit containing Claus converters (**Figure 2**). The Claus converters convert the H_2S and SO_2 mixture to elemental sulfur and water vapour. You may have seen large yellow blocks of sulfur beside gas plants in Alberta (**Figure 3**). Natural gas may range from 0% to 80% hydrogen sulfide. Sweet natural gas directly from a well contains no hydrogen sulfide but still needs to have water, liquid hydrocarbons, and carbon dioxide removed before being pumped into a natural gas pipeline.

Figure 2
An upgraded sulfur recovery unit was installed at Shell's Burnt Timber sour gas processing plant, 117 km northwest of Calgary, in the early 1990s. This facility processes natural gas from seven different gas fields that contain 0–28% hydrogen sulfide.

Sweet natural gas (natural or refined) is a mixture of hydrocarbons (**Table 1**). The composition of natural gas depends on the region of the province and the specific well within the region from which it comes. Some industries may preferentially purchase gas with the composition that best suits them, such as high percentages of propane to sell to customers.

Table 1 Typical Composition of Sweet Natural Gas

Component	Chemical formula	Mole fraction (~%)	Boiling point (°C)
nitrogen	$N_2(g)$	2.7	−196
carbon dioxide	$CO_2(g)$	0.3	−78
methane	$CH_4(g)$	83.6	−162
ethane	$C_2H_6(g)$	11.6	−89
propane	$C_3H_8(g)$	1.75	−42
butane	$C_4H_{10}(g)$	0.15	−0.5
pentanes*	$C_5H_{12}(l/g)$	0.08	vary (e.g., ~36)
hexanes*	$C_6H_{14}(l/g)$	0.07	vary (e.g., ~69)

* Pentanes and hexanes are liquids under normal room conditions, but are vapour components of natural gas (just like water is a vapour component of air).

Figure 3
This photo shows a block of sulfur at the Caroline-Shantz sulfur-forming facility near Sundre, Alberta. Only about 60 out of 800 gas plants recover sulfur. Most gas plants process sweet (non-sulfur) natural gas.

Although some of the natural gas may be burned directly or exported as a mixture of gases, most of the natural gas consumed in Alberta is further refined into separate components. At these refining (fractional distillation) plants, a simplified description is that the natural gas is cooled under high pressure to condense all the components except the methane gas. The condensed (liquid) portion is then slowly distilled to separate out the ethane, propane, butane, and pentane fractions (**Figure 4**). Some of the uses of these fractions are presented in **Table 2**.

Table 2 Uses of Natural Gas Components

Component	Some uses for the component
methane, $CH_4(g)$	heating homes, fueling taxis, producing hydrogen and methanol
ethane, $C_2H_6(g)$	cracked into ethene for producing many petrochemicals
propane, $C_3H_8(g)$	heating homes and vacation trailers; fueling barbecues and stoves
pentane, $C_5H_{12}(l/g)$	winter additive to gasoline for greater vaporization

Figure 4
Fractionation of natural gas provides a supply of propane.

DID YOU KNOW ?

Helium From Natural Gas

Air ships used before and during the First World War were filled with explosive hydrogen gas. Looking for a safer alternative, Canadian John McLennan invented a method for extracting the tiny quantities of helium from natural gas. McLennan designed and built a plant in Calgary during the First World War to provide helium for safer air ships for the Allies.

DID YOU KNOW ?

Pipelines and Gas Processing Plants

Crude oil, synthetic crude, diluted bitumen, natural gas, ethane, and other pipelines criss-cross Alberta and Canada. Many farm fields in Alberta have one or more pipeline(s) under the ground. Is there a pipeline or a gas processing plant near you?

www.science.nelson.com

+ EXTENSION

CBC radiONE
QUIRKS & QUARKS

Mining Methane

In a fuel-hungry world, geologists and researchers are searching everywhere for new resources. The Mallik 2002 drilling program explored buried deposits of methane in Canada's Arctic, and has raised hopes that these resources might be financially viable. There could be as much fuel here as in the entire world's reserves of traditional fossil fuels.

www.science.nelson.com

▶ Practice

1. Describe the two main stages in the refining of most natural gas. Identify which changes are physical and which are chemical.

2. Draw structural formulas for the first four members of the alkanes. Show all atoms and bonds.

3. Chemists have determined the boiling point of the components of natural gas (see Table 1 on the previous page).
   (a) What is the trend in boiling point as the number of carbon atoms increases for the alkane family?
   (b) Explain any trend that you identify that relates the size of an alkane molecule to its boiling point.

4. Geologists are able to use various technologies to empirically determine the geological features of oil- and gas-containing formations (Figure 1). Use intermolecular force and solubility theories to help explain the separation of
   (a) water and crude oil
   (b) limestone (assume $CaCO_3(s)$) and crude oil
   (c) crude oil and natural gas

5. Chemical engineers indicate that hydrocarbon gases, such as methane ($CH_4(g)$), evaporate from or are easily boiled off from crude oil.
   (a) Use what you know about solubility to explain the presence of methane in crude oil.
   (b) Use what you know about crude-oil-containing geological formations to create a hypothesis as to how the methane may have gotten into the crude oil.

6. There is an increasing use of natural gas and a decreasing use of petroleum in Canada (**Figure 5**).

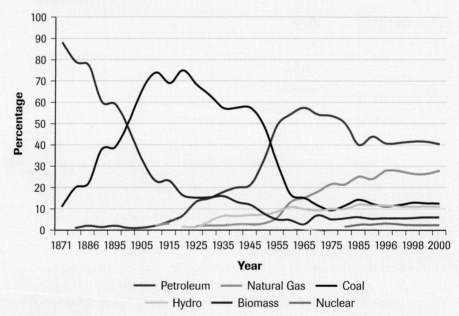

Figure 5
Energy use in Canada since 1871

   (a) Suggest some reasons for the changes in use of natural gas and petroleum in Canada.
   (b) Revisit your work on the Exploration: Burning Fossil Fuels at the beginning of this chapter. Add some of the statements from this section of study to your perspectives sheet relating to the resolution that the burning of fossil fuels should be decreased.

Section **9.2**

EXPLORE an issue

Coalbed Methane

Issue Checklist

○ Issue   ○ Design   ● Analysis
○ Resolution   ● Evidence   ● Evaluation

Coal miners have known about coalbed methane for centuries. Sudden releases of methane underground have killed hundreds of miners by suffocation or explosions. Similar gas releases have even been the subject of hypotheses to explain strange happenings in the Bermuda Triangle (see Chapter 4, Figure 2, page 185). Productively controlling the release of methane from coal beds is a developing technology in Alberta. Coalbed methane has been described as the oil sands of natural gas. Alberta geologists have mapped a potentially huge resource for coalbed methane (**Figure 6**). The methane is held underground by the surrounding pressure of trapped water. Releasing that pressure releases the methane. The method of extraction involves drilling a well to the methane-rich coal seam and, in many cases, pumping the water out (**Figure 7**). Removing the water releases the pressure on the methane and the low-pressure methane flows to the surface. A (potentially noisy) compressor at the surface pumps the methane along a pipeline to a natural gas refining site. Recall that natural gas is mainly methane. Coalbed methane is a natural gas that comes from underground coal beds. This coalbed natural gas has a very high percentage of methane, hence its name. The water removed from the coalbed is typically saline (salty) and is placed in holding ponds until it can be pumped back down the well.

Unlike natural gas wells, where only a couple of wells are needed to recover the gas on an owner's land, up to eight coalbed methane wells may be needed on a section of land. The landowners are paid for leasing the land to the natural gas company. Landowners do not, typically, own the mineral rights to the methane underground. They only own the surface rights.

Issue
Accessing coalbed methane may present challenges with land, noise, and water.

Resolution
Coalbed methane extraction in Alberta should be stopped until further research has been conducted.

Design
Research, from a variety of perspectives, the pros and cons of coalbed methane extraction in Alberta.

www.science.nelson.com

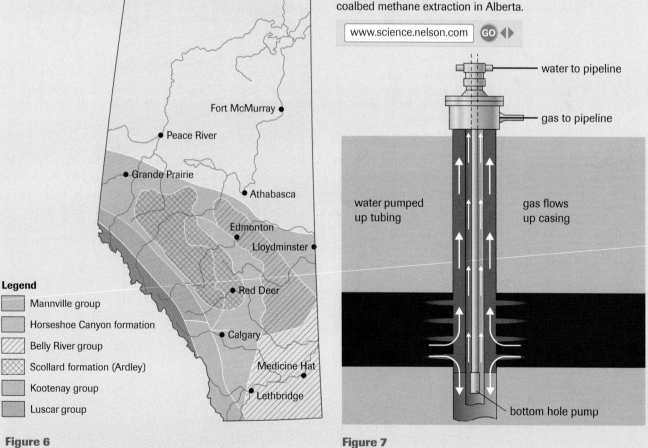

Figure 6
Zones with coalbed methane potential in Alberta

Legend:
- Mannville group
- Horseshoe Canyon formation
- Belly River group
- Scollard formation (Ardley)
- Kootenay group
- Luscar group

Figure 7
Coalbed methane extraction

Hydrocarbons from Petroleum

Table 3 The Alkane Family of Organic Compounds

IUPAC Name	Formula
methane	$CH_4(g)$
ethane	$C_2H_6(g)$
propane	$C_3H_8(g)$
butane	$C_4H_{10}(g)$
pentane	$C_5H_{12}(l)$
hexane	$C_6H_{14}(l)$
heptane	$C_7H_{16}(l)$
octane	$C_8H_{18}(l)$
nonane	$C_9H_{20}(l)$
decane	$C_{10}H_{22}(l)$
-ane	C_nH_{2n+2}

Learning Tip

Memorize the prefixes indicating one to ten carbon atoms. See Table 3.

Naming Alkanes

Although hydrocarbons can be classified according to empirical properties, a more common classification is based upon the chemical formulas of the compounds. Hydrocarbons whose empirically determined molecular formulas indicate that the carbon-to-carbon bonds are only single bonds are called **alkanes**. The simplest member of the alkane series is methane, $CH_4(g)$, which is the main constituent of the natural gas sold for home heating. The molecular formulas of the smallest alkanes (the first six of which are all typical constituents of natural gas) are shown in **Table 3**. Each formula in the series has one more CH_2 group than the one preceding it. Derived from empirically determined formulas and from bonding capacity, the general molecular formula for all alkanes is C_nH_{2n+2}; that is, a series of CH_2 units plus two terminal hydrogen atoms.

Alkanes are an example of what chemists call a **homologous series**—a sequence of molecules with similar structure and differing only the number of repeating units; e.g., CH_2. Chemists also classify alkanes as **saturated hydrocarbons**—compounds of carbon and hydrogen containing only carbon–carbon single bonds with the maximum number of hydrogen atoms bound to each carbon.

The first syllable in the name of an alkane is a prefix that indicates the number of carbon atoms in the molecule (**Figure 8**). The prefixes shown in **Table 3** are used in naming all organic compounds.

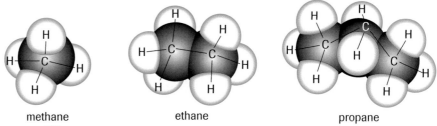

methane   ethane   propane

Figure 8

The prefix *meth–* indicates one carbon atom, *eth–* signifies two carbon atoms, and *prop–* signifies three carbon atoms. The ending *–ane* indicates a chain of carbon atoms with single bonds only.

Structural Isomers

Chemical formulas, such as the ones in Table 3, tell you the total number of each kind of atom in a molecule. Except for the three smallest molecules (**Figure 8**), there are several structures that can have the same molecular formula. For example, C_4H_{10} has two different structural formulas, both satisfying the rules for chemical bonding.

butane, $C_4H_{10}(g)$

methylpropane, $C_4H_{10}(g)$

So, while we can say that butane always has the formula, C_4H_{10}, we cannot say that a compound with the formula C_4H_{10} is always butane.

Compounds with the same molecular formula, but with different structures, are called **structural isomers**. Chemists created this theoretical concept to explain why isomers (different compounds with the same molecular formula) could have different physical and chemical properties. Compounds with the same molecular formula but different properties must be differentiated in some way. For this reason, the two structures for C_4H_{10} shown above must have different names to avoid confusion. (See the next subsection.)

As the number of carbon atoms in a hydrocarbon increases, the number of possible isomers increases dramatically. For example, $C_{10}H_{22}$ has 75 possible isomers, $C_{20}H_{42}$ has 366 319, and $C_{30}H_{62}$ has 4 111 846 763 possible isomers. No wonder there are more compounds of carbon than compounds of all other elements combined!

Names and Structures of Branched Alkanes

The name of the compound indicates whether there are branches on a carbon chain. Prefixes (as in **Table 3**) identify groups of atoms that form branches on the structures of larger molecules. A *branch* is any group of atoms that is not part of the main structure of the molecule. For example, a branch consisting of only singly bonded carbon and hydrogen atoms is called an **alkyl branch**. In the names of alkyl branches, the prefixes are followed by a *-yl* suffix (**Table 4**). For example, consider the three isomers of C_5H_{12} shown in **Figure 9**. The unbranched isomer on the left is named pentane.

Table 4 Examples of Alkyl Branches

Branch	Name
–CH_3	methyl
–C_2H_5 (–CH_2CH_3)	ethyl
–C_3H_7 (–$CH_2CH_2CH_3$)	propyl

Figure 9
Each of the three isomers of C_5H_{12} has different physical and chemical properties.

The "straight" chain description of C_5H_{12} is shown below. Of course, there is a tetrahedral arrangement of bonds around each carbon atom, which produces the "saw-tooth" chain of carbon atoms modelled in Figure 9.

```
    H   H   H   H   H
    |   |   |   |   |
H — C — C — C — C — C — H
    |   |   |   |   |
    H   H   H   H   H
          pentane
```

In the second isomer, there is a continuous chain of four carbon atoms with a methyl group on the second carbon atom. To name this structure, identify the *parent chain* — the longest continuous chain of carbon atoms. Here, the four carbons indicate that the parent chain is butane. Since a methyl group is added as a branch to the longest continuous chain of four carbon atoms, this isomer is called methylbutane. The same name applies when the methyl group replaces any of the four hydrogen atoms on the middle two carbon atoms of butane.

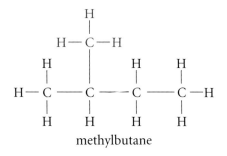

methylbutane

Learning Tip

Computer models of molecules can be displayed as
- wireframe (line structural)
- sticks (structural)
- ball and stick (structural)
- space filling

Interact with some of the alkane models at the following Web link.

www.science.nelson.com

Learning Tip

To decide whether two different-looking structural formulas are the same molecule, rotate the structure top to bottom or left to right. If, by rotating the molecule, you can make it the same as another molecule, then it is the same molecule and has the same name (and is not a different isomer).

In the third isomer of C_5H_{12} in Figure 9, two methyl groups are attached to a three-carbon (propane) parent chain. This third pentane isomer is named dimethylpropane. There is no other isomer of dimethylpropane. Attaching one of the two methyl groups to an end carbon makes the molecule methylbutane. Attaching a methyl group to each end of propane makes the molecule pentane. See Figure 9.

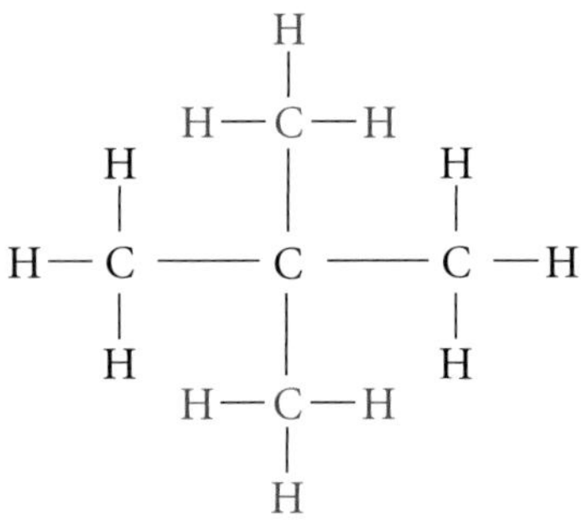

dimethylpropane

Since there is only one methylbutane isomer and only one dimethylpropane isomer, we usually do not use numbers in the names of these molecules: 2-methylbutane and 2,2-dimethylpropane are acceptable but not necessary. In most cases, however, we need to number the carbon atoms to indicate the position(s) of branch(es) (substitution group(s)). In these cases, follow the steps below to name the branched hydrocarbon.

SUMMARY Naming Branched Alkanes

Step 1: Identify the longest continuous chain of carbon atoms—the parent chain—in the structural formula. Number the carbon atoms, starting from the end closest to the branch(es), so that the numbers are the lowest possible.
Step 2: Identify any branches and their location number on the parent chain.
Step 3: Write the complete IUPAC name, following this format:
(number of location) – (branch name)(parent chain).

When writing the name of the alkane, list the branches in alphabetical order, ignoring, for example, the di- and tri- prefixes.

▶ SAMPLE problem 9.1

Write the IUPAC name corresponding to the following condensed structural formula for a component of gasoline:

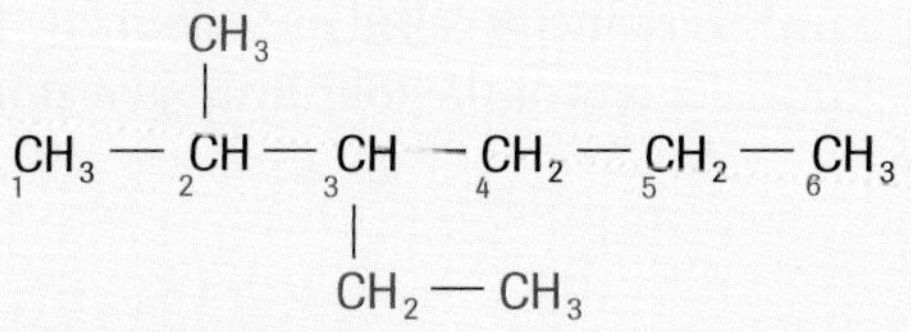

$$CH_3-\underset{2}{CH}(CH_3)-\underset{3}{CH}(CH_2CH_3)-\underset{4}{CH_2}-\underset{5}{CH_2}-\underset{6}{CH_3}$$

Step 1: The longest continuous chain has six carbon atoms. Therefore, the name of the parent chain is hexane.
Step 2: There is a methyl group branch at the second carbon atom, and an ethyl group branch at the third carbon atom of the parent chain.
Step 3: With the branches named in alphabetical order, the compound is 3-ethyl-2-methylhexane.

Learning Tip

Although IUPAC allows for alkyl groups (or any other atoms or groups of atoms substituted for hydrogen) to be placed in alphabetical order *or* in order of complexity (least to greatest), this textbook uses the rule of alphabetical order, for example, 3-ethyl-2,2-dimethylhexane. (Ignore the "di" and pay attention only to the "e" of ethyl and the "m" of methyl.)

Learning Tip

Structural formulas show all bonds between atoms. This system is not efficient, especially for C–H bonds. Thus, the diagram is often simplified by not showing the bonds with H atoms, as shown in Sample Problem 9.1.

Long-chain hydrocarbons such as

$CH_3-CH_2-CH_2-CH_2-CH_2-CH_3$

can be represented as

$CH_3-(CH_2)_4-CH_3$

Both of these formulas and the one in Sample Problem 9.1 are *condensed structural formulas*.

COMMUNICATION example 1

Write the names from the condensed structural formulas for the following components of gasoline. The second component (b) is an isomer of octane.

(a) 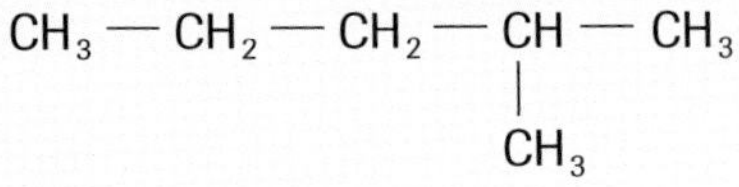 $CH_3 — CH_2 — CH_2 — CH — CH_3$
                              $|$
                              CH_3

(b) 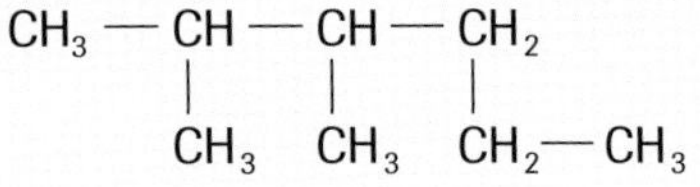 $CH_3 — CH — CH — CH_2$
           $|$ $|$ $|$
           CH_3 CH_3 $CH_2 — CH_3$

Solution

(a) 2-methylpentane

(b) 2,3-dimethylhexane

Learning Tip

Be careful when choosing the longest continuous chain (parent chain) because the structure may twist and turn. If your name ends up with an alkyl branch on the number 1 position, then you have made a mistake in identifying the parent chain. For example, 1,1-dimethylpropane is incorrect and should be 2-methylbutane.

A structural formula can illustrate an IUPAC name. For example, in the Communication Example above, 2,3-dimethylhexane has a hexane parent chain consisting of six carbon atoms joined by single covalent bonds. "2,3-dimethyl" tells us that there is a methyl group attached to the second carbon and another attached to the third carbon.

Above you have learned to name a branch-chained hydrocarbon when given the structural formula. The reverse process is to draw a structural formula when given the name. Both of these processes emphasize the conventions of communication agreed upon worldwide. These conventions are necessary for chemists to communicate their research in internationally distributed scientific journals and for health providers, for example, to administer the correct medication.

SUMMARY — Drawing Branched Alkane Structural Formulas

Step 1: Draw a straight chain containing the number of carbon atoms represented by the name of the parent chain, and number the carbon atoms from left to right.

Step 2: Attach all branches to their numbered locations on the parent chain.

Step 3: Add enough hydrogen atoms to show that each carbon has four single bonds.

SAMPLE problem 9.2

Draw a structural formula for 3-ethyl-2,4-dimethylpentane—possible component of gasoline.

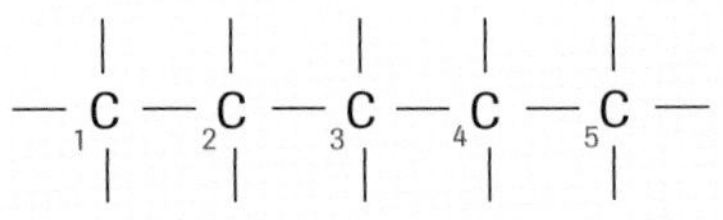

Numbering this straight chain from left to right establishes the location of the branches. An ethyl branch is attached to the third carbon atom and a methyl branch is attached to each of the second and fourth carbon atoms.

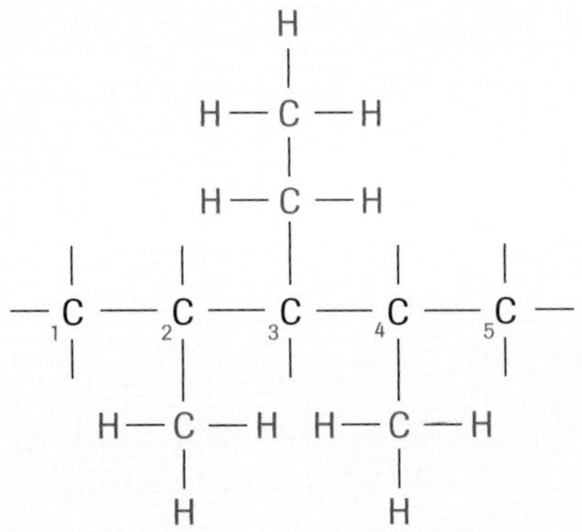

Learning Tip

A molecular formula of an organic compound is not very useful because many isomers are often possible. To more clearly specify a particular compound, the structure needs to be indicated either by the IUPAC name or by a diagram. Chemists have invented many different kinds of diagrams to show the structure of a molecule. For our purposes, the structural formula (either full or condensed) is the most useful.

Because structure is so important to the physical and chemical properties of organic molecules, chemists make extensive use of molecular models such as space-filling models (Figure 8) and ball-and-stick models (Figure 9). Just as there are many other kinds of diagrams, there is also a variety of molecular models.

For a little more practice naming alkanes, follow the electronic link.

www.science.nelson.com

In the following structural formula, hydrogen atoms are shown at any of the four bonds around each carbon atom that are left after the branches have been located.

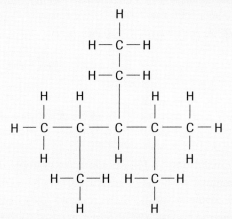

▶ COMMUNICATION example 2

There are many different hydrocarbon molecules in gasoline, including many isomers. Draw the structural formula for the first isomer and the condensed structural formula for the second isomer of octane.

(a) 2,3,4-trimethylpentane   (b) tetramethylbutane

Solution

(a)

(b)
$$CH_3-\underset{\underset{CH_3}{|}}{\overset{\overset{CH_3}{|}}{C}}-\underset{\underset{CH_3}{|}}{\overset{\overset{CH_3}{|}}{C}}-CH_3$$

Figure 10
There are many different hydrocarbon mixtures that can be found at hardware stores.

▶ Practice

7. Draw the structural formula of each alkane found in crude oil.
   (a) 2-methylpentane
   (b) 3-methylpentane
   (c) 2-methylhexane
   (d) 3-methylhexane
   (e) 3,4-dimethylhexane
   (f) 3-ethyl-2-methylhexane
   (g) 3,3-dimethylpentane
   (h) 2,5-dimethyl-4-propyloctane

8. There are many different naphtha mixtures. Petroleum naphtha, commonly used as a camping fuel, is a mixture of alkanes with 5 or 6 carbon atoms per molecule (**Figure 10**).
   (a) Draw structural formulas and write the IUPAC names for all the isomers of C_6H_{14}.
   (b) Why are these isomers highly soluble in each other?

9. Automotive gasoline is a mixture that is largely composed of alkanes containing 5 to 12 carbon atoms per molecule. Write IUPAC names for the following components of gasoline:

(a) CH₃—CH—CH₂—CH₂—CH₂—CH₂—CH₃
         |
         CH₃

(b) CH₃—CH—CH₂—CH₂—CH₂—CH—CH₃
         | |
         CH₂—CH₃   CH₂—CH₃

(c) CH₃—CH—CH₂—CH—CH₂—CH—CH₃
         | | |
         CH₃   CH₃   CH₃

(d) CH₃—CH—CH₂—CH₂—CH—CH₂—CH₃
         | |
         CH₂—CH₃   CH₂—CH₃

10. Kerosene, used as a fuel for diesel and jet engines, contains alkanes with 12 to 16 carbon atoms per molecule. Draw structural formulas for the following components of kerosene:
(a) 4-ethyl-3,5-dimethyloctane
(b) 3,3-diethyl-2-methylnonane
(c) 3,5-diethyl-4-propylheptane
(d) 5-butyl-4,6-dimethyldecane

11. IUPAC tries to standardize chemical communication throughout the world. For each of the following names, determine if it is a correct name for an organic compound. Give reasons for your answer and include a correct name.
(a) 2-dimethylhexane
(b) 3,3-ethylpentane
(c) 3,4-dimethylhexane
(d) 2-ethyl-2-methylpropane
(e) 1-methyl-2-ethylpentane
(f) 3,3-dimethylpropane

DID YOU KNOW?
Chewing Gum
American Aboriginal peoples introduced Europeans to chewing spruce gum, which the Europeans quickly commercialized. In the 1950s, paraffin wax, a solid alkane, became the base for chewing gum. Today, more than one thousand varieties of gum are manufactured using various synthetic materials to replace natural gum ingredients.

Learning Tip
When naming hydrocarbons, not only must the branches be put in alphabetical order but also the set of numbers must be the lowest set. The lowest set of numbers is defined by IUPAC as being the set that starts with the lowest number; e.g., 2, 4 and 5 are lower than 3, 3 and 4. Start numbering the longest continuous chain from the end of the molecule with the closest branch.

Cycloalkanes

On the evidence of empirical formulas and chemical properties, chemists believe that organic carbon compounds sometimes take the form of cyclic hydrocarbons—hydrocarbons with a closed ring. When all the carbon–carbon bonds in a cyclic hydrocarbon are single bonds, the compound is called a **cycloalkane** (C_nH_{2n}).

Cycloalkanes are named by placing the prefix *cyclo* in front of the alkane name, as in cyclopropane and cyclobutane (**Figure 11**). If branches are present, treat the cycloalkane as the parent chain and identify the branches. Since there is no end at which to start the numbering for the location of the branches, use the lowest possible numbers.

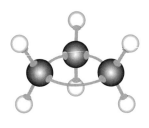

cyclopropane

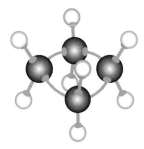

cyclobutane

Figure 11
Cycloalkanes such as cyclopropane and cyclobutane are similar to alkanes, except that the two ends of the molecule are joined to form a ring of atoms. This figure shows computer-generated, condensed structural, and line structural models of these two cycloalkanes. These models help to provide a logically consistent theoretical explanation for empirically determined formulas and help chemists to visualize the molecules.

(a)

(b)

(c)

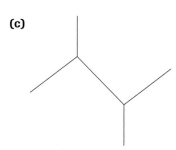

Figure 12
2,3-dimethylbutane can be represented by
(a) a structural formula,
(b) a condensed structural formula, or
(c) a line structural formula.

Start with the location of one of the branches. Omit the "1" if only one branch is present. For example, methylcyclopropane is a cyclopropane ring with a methyl branch on one of three equivalent carbon atoms of the ring.

In addition to structural and condensed structural formulas, we can use *line structural formulas*. Line structural formulas of hydrocarbons show the position of the carbon atoms as the intersections and ends of bonding lines; they do not show hydrogen atoms (**Figure 12(c)**).

▶ COMMUNICATION example 3

Provide structural, condensed structural, and line structural formulas for cyclohexane.

Solution

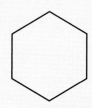

▶ COMMUNICATION example 4

Write the IUPAC name for this hydrocarbon, communicated with a structural, a condensed structural, and a line structural formula.

Solution
1,2-dimethylcyclopentane

Note that rotating the 1,2-dimethylcyclopentane molecule in Communication Example 4 does not change its name—all of the formulas represent 1,2-dimethylcyclopentane. We count from the carbon atom that gives the lowest possible first number.

▶ Section 9.2 Questions

1. List the alkanes that are typically present in natural gas before it is refined. How does this composition compare with the natural gas you use in your homes for heating?
2. What is the difference between sweet and sour natural gas? Why is it important to know the difference?
3. Describe a process used to remove alkanes, other than methane, from natural gas.
4. Natural gas burns readily with oxygen from the air. The primary product is energy.
   (a) List two consumer, two commercial, and two industrial uses of the energy obtained from natural gas.
   (b) Look around the room you are in. List some items in the room that likely required energy from natural gas for their production.

5. Draw structural formulas for all reactants and products in the following organic reactions:
   (a) methane + butane → pentane + hydrogen
   (b) propane + pentane → octane + hydrogen
   (c) decane + hydrogen → heptane + propane
   (d) 3-ethyl-5-methylheptane + hydrogen →
       ethane + propane + methylbutane
   (e) cyclohexane + ethane → ethylcyclohexane + hydrogen

6. Write the IUPAC names for each of the following possible components of gasoline:

   (a)
   $$CH_3-CH_2-\underset{\underset{CH_3}{|}}{CH}-CH_2-CH_3$$

   (b)
   $$CH_3-\underset{\underset{CH_3}{|}}{\overset{\overset{CH_3}{|}}{C}}-\overset{\overset{CH_3}{|}}{CH}-CH_3$$

   (c)
   $$CH_3-\overset{\overset{CH_3}{|}}{CH}-CH_2-\underset{\underset{CH_2-CH_3}{|}}{\overset{\overset{CH_2-CH_3}{|}}{C}}-CH_3$$

   (d)
   $$CH_3-\overset{\overset{CH_2-CH_3}{|}}{CH}-CH_2-\underset{\underset{CH_2-CH_2-CH_3}{|}}{CH}-CH_2-CH_2-CH_3$$

7. For each of the following chemical equations, write the IUPAC names for all reactants and products.
   (a) $CH_3-CH_2-CH_3 + CH_3-(CH_2)_3-CH_3 \rightarrow$
       $CH_3-(CH_2)_6-CH_3 + H-H$
   (b) $CH_3-(CH_2)_4-CH_3 \rightarrow CH_3-\underset{\underset{CH_3}{|}}{\overset{\overset{CH_3}{|}}{C}}-CH_2-CH_3$
   (c) Draw condensed structural and line structural formulas, and write the IUPAC names for two other isomers of the product given in (b).

8. Complete the following equations for complete combustion, using molecular formulas to write a balanced chemical equation. Include the structural formula or IUPAC name for each hydrocarbon.
   (a) 2,2,4-trimethylpentane + oxygen →
   (b) $CH_3-\overset{\overset{CH_3}{|}}{CH}-CH_2-CH_3 + O=O \rightarrow$
   (c) ⬠ + O=O →

9. Why do companies find and develop petroleum resources? Suggest an answer from at least two perspectives.

10. (a) Use the information in **Table 5** to plot a graph showing the relationship between the number of carbon atoms and the boiling points of the alkanes. Describe the relationship and propose an explanation for it.
    (b) Research a use for each of the first 10 alkanes, and suggest why each is appropriate for this use.

    www.science.nelson.com

Table 5 Boiling Points of the First 10 Straight-Chain Alkanes

Formula	Name	B.p. (°C)
$CH_4(g)$	methane	−161
$C_2H_6(g)$	ethane	−89
$C_3H_8(g)$	propane	−42
$C_4H_{10}(g)$	butane	−1
$C_5H_{12}(l)$	pentane	36
$C_6H_{14}(l)$	hexane	68
$C_7H_{16}(l)$	heptane	98
$C_8H_{18}(l)$	octane	125
$C_9H_{20}(l)$	nonane	151
$C_{10}H_{22}(l)$	decane	174

11. Use the information in this section and from your own research to continue gathering perspective statements concerning the statement that the burning of fossil fuels should be reduced.

Extension

12. Oil spills in Alberta or near water, like the 2005 rail-car spill near Wabamum, cause a wide variety of environmental problems. Use your knowledge of alkanes to describe and explain what happens physically and chemically when oil is spilled from a tanker car into water.

13. Alberta has extensive natural gas fields that supply gas for use by provincial consumers and industries. The Alberta government collects royalties on the production of all fossil fuels, including natural gas.
    (a) Describe or sketch a map of the distribution of natural gas fields in Alberta.
    (b) Approximately how much natural gas is produced in one year and how much of it is used within Alberta?
    (c) How much royalty has the Alberta government collected recently for natural gas? How does this sum compare with the total royalties the Alberta government collects for non-renewable resources?

    www.science.nelson.com

14. Aboriginal peoples used waxes to waterproof footwear, bags, and other clothing. Technologically, paraffins are waxes and chemically, paraffins are alkanes with twenty or more carbons; for example, $C_{25}H_{52}(s)$. Explain why waxes, such as paraffin, can be used as a waterproofing agent.

    www.science.nelson.com

9.3 Alkenes and Alkynes—Cracking Natural Gas

Alkenes and Alkynes

Analysis reveals that hydrocarbons containing double or triple covalent bonds make up a small percentage of petroleum (such as crude oil and natural gas). These compounds are also often formed during chemical refining and are valuable components of gasoline. Hydrocarbons containing double or triple bonds are important in the petrochemical industry because they are the starting materials for the manufacture of many derivatives, including plastics.

A double or a triple bond between two carbon atoms in a molecule affects the chemical and physical properties of the molecule. Organic compounds with carbon–carbon double or triple bonds are said to be **unsaturated**, because they have fewer hydrogen atoms than compounds with carbon–carbon single bonds. Unsaturated hydrocarbons can react readily (usually in the presence of a catalyst) with small diatomic molecules, such as hydrogen. This type of reaction is an addition reaction. Addition of a sufficient quantity of hydrogen, called **hydrogenation**, converts unsaturated hydrocarbons to saturated ones. Saturated hydrocarbons (alkanes) have no double or triple bonds. An excess of hydrogen is needed to convert an alkyne to an alkane, otherwise some of the alkyne is only converted to an alkene.

unsaturated hydrocarbon (e.g., an alkene) + hydrogen → saturated hydrocarbon

unsaturated hydrocarbon (e.g., an alkyne) + excess hydrogen → saturated hydrocarbon

Hydrocarbons with carbon–carbon double bonds are members of the **alkene** family (**Figure 1**). They all have the general formula of C_nH_{2n}. The names of alkenes with only one double bond have the same prefixes as the names of alkanes, together with the ending *-ene* (**Table 1**). (Ethene is the starting material for a huge variety of consumer, commercial, and industrial products, some of which are listed in Chapter 10, page 415.) Isomers exist for all alkenes larger than propene. Cycloalkanes are isomers of alkenes.

DID YOU KNOW ?

Margarine
Margarine containing vegetable oils whose molecules have many double bonds is said to be *polyunsaturated*. The molecules of *saturated* fats, in animal products such as butter, are fully hydrogenated (have only single bonds).

Table 1 The Alkene Family of Organic Compounds

IUPAC name (common name)	Molecular formula
ethene (ethylene)	$C_2H_4(g)$
propene (propylene)	$C_3H_6(g)$
but-1-ene (butylene)	$C_4H_8(g)$
pent-1-ene	$C_5H_{10}(l)$
hex-1-ene	$C_6H_{12}(l)$
–ene	C_nH_{2n}

Note: See Naming Alkenes and Alkynes on the next page for an explanation of the numbering system.

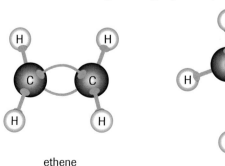

Figure 1
The two simplest and most commonly used alkenes are ethene (known as ethylene) and propene (propylene).

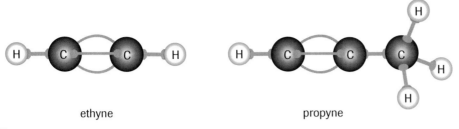

Figure 2
Sophisticated laboratory research indicates that the triple covalent bonds of ethyne (acetylene) and propyne are the shortest, strongest, and most reactive of all carbon–carbon bonds.

Table 2 The Alkyne Family of Organic Compounds

IUPAC name (common name)	Molecular Formula
ethyne (acetylene)	$C_2H_2(g)$
propyne	$C_3H_4(g)$
but-1-yne	$C_4H_6(g)$
pent-1-yne	$C_5H_8(l)$
hex-1-yne	$C_6H_{10}(l)$
-yne	C_nH_{2n-2}

The **alkyne** family has chemical properties that can be explained only by the presence of a triple bond between carbon atoms (**Figure 2**). Like alkenes, alkynes are unsaturated and can react with small molecules such as hydrogen or bromine in an addition reaction. Alkynes are named like alkenes, except for the *-yne* ending. The simplest alkyne, ethyne (acetylene), is commonly used in an oxyacetylene torch for cutting and welding metals. **Table 2** lists the first five members of the alkyne family. Empirical studies indicate that all alkynes with one triple bond have the general formula C_nH_{2n-2}. Isomers exist for all alkynes larger than propyne. Cycloalkenes are isomers of alkynes.

Naming Alkenes and Alkynes

Since the location of a multiple bond affects the chemical and physical properties of a compound, an effective naming system should specify the location of the multiple bond. Alkenes and alkynes are named much like alkanes, with two additional points to consider.

- The longest or parent chain of carbon atoms must contain the multiple bond, and the chain is numbered from the end closest to the multiple bond.
- The number that indicates the position of the multiple bond on the parent chain precedes the ending (-ene or -yne) of the parent chain.

For example, there are two possible butene isomers, but-1-ene and but-2-ene.

$$\underset{1}{CH_2}=\underset{2}{CH}-\underset{3}{CH_2}-\underset{4}{CH_3} \qquad \underset{1}{CH_3}-\underset{2}{CH}=\underset{3}{CH}-\underset{4}{CH_3}$$
$$\text{but-1-ene} \qquad\qquad \text{but-2-ene}$$

Note that the number in front of the ending specifies the location of the multiple bond, just as the number in front of the branch indicates its location.

▶ COMMUNICATION *example 1*

Name the hydrocarbons communicated by the following condensed structural and line structural formulas.

(a)
$$CH_3-\underset{\underset{CH_3}{|}}{CH}-CH=CH_2$$

(b)
$$CH_3-CH=CH-CH_2-\underset{\underset{CH_3}{|}}{CH}-CH_3$$

Solution

The IUPAC name for (a) is 3-methylbut-1-ene.

The IUPAC name for (b) is 5-methylhex-2-ene.

DID YOU KNOW ?

Steroids
Steroids are unsaturated compounds based on a structure of four rings of carbon atoms. The best known and most abundant steroid is cholesterol, which is an essential constituent of cell walls, but which has also been associated with diseases of the cardiovascular system. Cholesterol that coats the interior surfaces of arteries contributes to health problems such as high blood pressure. Other steroids include the male and female sex hormones, and anti-inflammatory agents such as cortisone. Oral contraceptives include two synthetic steroids. Some athletes have used anabolic steroids to enhance muscle development and physical performance, but such use may cause permanent damage.

You can view a computer-generated 3-D molecular model of cholesterol by visiting the Nelson Web site.

www.science.nelson.com

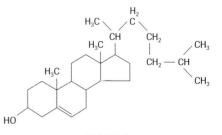

cholesterol

Learning Tip
Note that the numbering of carbons for alkenes and alkynes always starts from the end of the molecule closest to the double or triple bond(s). This method yields the lowest possible number for the position of the multiple bond.

In the following branched alkyne structure, the parent chain is a pentyne.

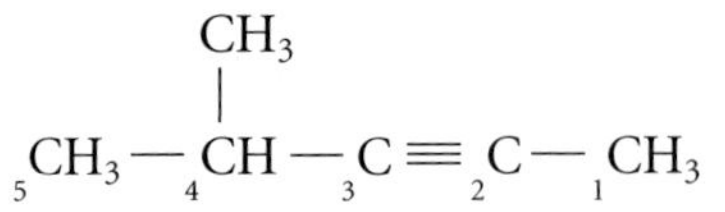

$$\underset{5}{CH_3}-\underset{4}{CH}-\underset{3}{C}\equiv\underset{2}{C}-\underset{1}{CH_3}$$
$$\quad\quad\quad|$$
$$\quad\quad CH_3$$

4-methylpent-2-yne

When numbering the carbon atoms of an alkene or alkyne parent chain, chemists give more importance to the location of the multiple bond(s) than to the location of any branches (alkyl groups). The IUPAC name, 4-methylpent-2-yne, follows the same format as that used for alkanes.

Drawing Structural Formulas of Alkenes and Alkynes
Whenever you need to draw a structural formula for any hydrocarbon, first look at the end of the name to find the parent chain. Draw the parent alkene or alkyne first, and then add the branches listed in the name. Be sure to finish the formula with sufficient hydrogen atoms to complete four bonds for each carbon atom. The following Communication Example shows some typical examples of alkenes and alkynes.

Learning Tip
The two general rules used throughout organic nomenclature are:
- use the longest continuous chain containing the primary characteristic group of the molecule to determine the prefix and ending of the parent name

Prefix +**Ending**
(number of C atoms (characteristic
in longest chain) group)
e.g., meth (1) ane (C—C)
    eth (2) ene (C=C)
    prop (3) yne (C≡C)

For example,
2-methylhexane
2-methylhex-3-ene
2-methylhex-3-yne

- use the lowest set of numbers to indicate the first characteristic feature and then the branches or substituents in the molecule

> ### COMMUNICATION example 2
> Draw condensed structural formulas for the following alkyne petrochemicals:
> (a) 4-methylpent-1-yne
> (b) dimethylbut-1-yne
>
> **Solution**
> (a) 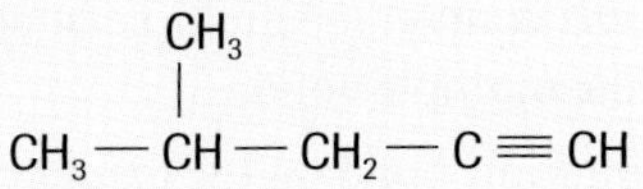
> $$CH_3-CH-CH_2-C\equiv CH$$
> $$\quad\quad\;|$$
> $$\quad\;CH_3$$
>
> (b)
> $$\quad\quad\quad CH_3$$
> $$\quad\quad\quad\;|$$
> $$CH\equiv C-C-CH_3$$
> $$\quad\quad\quad\;|$$
> $$\quad\quad\quad CH_3$$

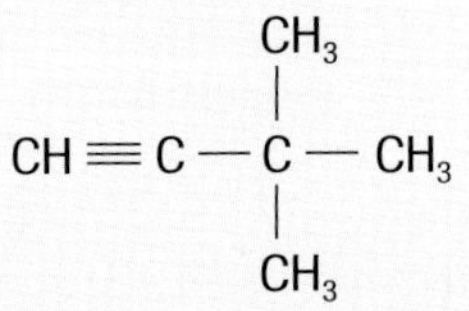

Cycloalkenes and Cycloalkynes
In addition to alkanes, cycloalkanes, alkenes, and alkynes, chemists discovered hydrocarbons whose properties can only be explained by creating a new class of hydrocarbons, called **cycloalkenes**. As the name of this class of hydrocarbons suggests, its molecules have a cycle of carbon atoms with at least one double bond. The structural formulas and names of cycloalkenes follow the same IUPAC rules as those for cycloalkanes. Here are some examples.

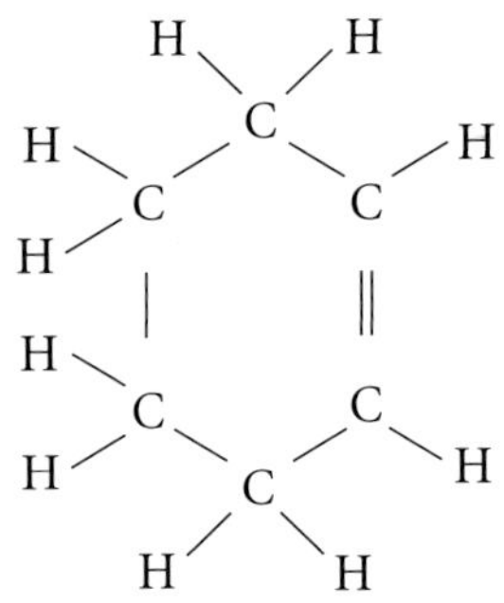

cylcohexene
(structural formula)

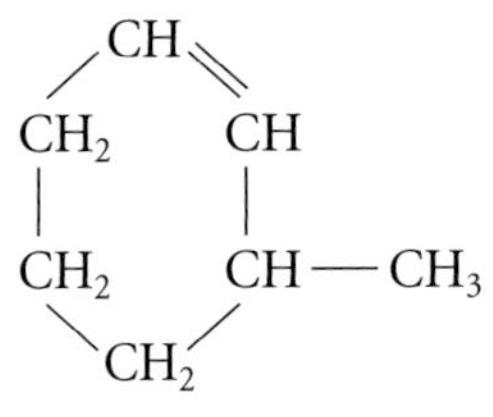

3-methylcyclohexene
(condensed structural formula)

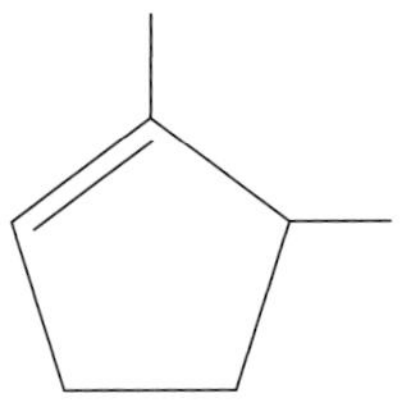

2,3-dimethylcyclopentene
(line structural formula)

Note that, when naming cycloalkenes with side branches, the numbering for the carbon atoms begins with the double bond: the carbons of the double bond are carbons 1 and 2. The carbons are numbered so as to provide *the lowest numbers possible*, as for all other hydrocarbons. The system for naming all hydrocarbons is logically consistent—the same rules apply to all classes of hydrocarbons.

There are few known cycloalkenes and cycloalkynes. Chemists explain this low membership by considering the bond-angle stress put on the double and triple bonds when a cyclic hydrocarbon is created.

Cycloalkanes are isomers of alkenes with the same number of carbon atoms, both with the general formula C_nH_{2n}. Similarly, cycloalkenes are isomers of alkynes, both with the general formula C_nH_{2n-2}. Isomers of alkenes and alkynes exist with different locations of the double or triple bond and by changing a straight-chain hydrocarbon into a branched hydrocarbon or into a cyclic hydrocarbon. If you find that several structures have the same formula but different names, then the structures are isomers. For example, cyclohexene is an isomer of a simple hexyne. Both have the same molecular formula, $C_6H_{10}(l)$ (**Figure 3**).

hex-1-yne

Figure 3
Hex-1-yne is an isomer of cyclohexene. Cycloalkenes are isomers of alkynes and vice versa.

▶ ***Practice***

1. Classify each of the following hydrocarbons as an alkane, alkene, or alkyne, and/or as a cycloalkane or cycloalkene:
   (a) $C_2H_4(g)$
   (b) $C_3H_8(g)$
   (c) $C_4H_6(g)$
   (d) $C_5H_{10}(l)$

2. Draw a structural formula and write a molecular formula for each of the following:
   (a) propane
   (b) propene
   (c) propyne
   (d) cyclopropane

3. Draw a condensed structural formula for each of the following petrochemicals.
   (a) propene
   (b) but-2-ene
   (c) 2,4-dimethylpent-2-ene
   (d) but-1-yne

4. Why are no numbers required for the location of the multiple bond in propene or propyne?

5. Draw line structural formulas and write IUPAC names for each of the following structures:

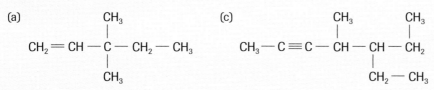

6. Draw line structural formulas and write the IUPAC names for the four structural isomers of $C_4H_8(g)$. (Remember cyclo compounds.)

7. Alkenes and alkynes are the starting materials in the manufacture of a wide variety of organic compounds. Draw structural formulas for the following starting materials that are used to make the products named in parentheses.
   (a) propene (polypropylene)
   (b) methylpropene (synthetic rubber)

✚ EXTENSION

Interactive Models

There are more and more computer-generated 3-D molecular models being created on Web sites. These models help chemists, teachers, and students to visualize the structures of molecules. (Crystallographic research from, for example, X-ray analysis, provides the data points for the atoms.) Most of the computer models allow you to select line, ball-and-stick, and space-filling models. You can also rotate the molecule by using your computer mouse. Many examples are provided at the following link. Using Internet resources such as this will help to increase your knowledge and feel for organic molecules.

www.science.nelson.com

INVESTIGATION 9.2 Introduction

Structures and Properties of Isomers

Molecular models are visual representations of molecules that help us to understand some of the properties and structures of molecules. In addition to building models, this investigation asks you to consult references to obtain data gathered by chemists over many years of laboratory work.

Purpose
The purpose of this investigation is to use models and a reference, respectively, to examine the structures and physical properties of some isomers of unsaturated hydrocarbons.

Report Checklist
- ○ Purpose
- ○ Problem
- ○ Hypothesis
- ○ Prediction
- ○ Design
- ○ Materials
- ○ Procedure
- ● Evidence
- ● Analysis
- ● Evaluation (3)

Problem
What are the structures and physical properties of the isomers of C_4H_8 and C_4H_6?

Design
Structures of possible isomers are determined by means of a molecular model kit. Once each structure is named, the boiling and melting points are obtained from a reference such as the *CRC Handbook of Chemistry and Physics* or *The Merck Index*.

To perform this investigation, turn to page 401.

Table 3 Boiling Points of Simple Alkanes in Natural Gas

Alkane	Boiling point (°C)
methane	−162
ethane	−89
propane	−42
methylpropane	−12
butane	−0.5
dimethylpropane	10
methylbutane	28
pentane	36

Learning Tip

The use of the term "cracking" in this chapter has two meanings. The first meaning, presented here, is that of removing (cracking) hydrogen from ethane to produce ethene. Technically this is not "real" cracking, but *ethane cracking* is a widely used term in industry.

The meaning presented in Section 9.5 is that of breaking or cracking a molecule between two carbon atoms to make smaller hydrocarbon molecules. This is the more common usage of the term cracking.

Ethane Cracking

The low boiling points in natural gas make it difficult to separate these components (**Table 3**). Alberta has several specialized gas plants where the various components of natural gas are separated. (See the list in the margin: Ethane Extraction Plants.) The condensation of the natural gas components is achieved with temperatures down to −100 °C and pressures up to 1500 to 5000 kPa (15 to 50 atm). The components are also known as LPGs (liquid petroleum gases) and are sold separately (or as a mixture) once they are removed from natural gas. The most familiar LPG to us is propane, used for combustion in barbecues, automobiles, and homes. From a chemical industry perspective, the most important LPG is ethane, used for manufacturing ethene. The uncondensed methane is returned to the natural gas pipeline for consumer, commercial, or industrial combustion or petrochemical use. Methane is the refined natural gas that may heat your home and your water.

Ethene (ethylene) is produced world wide by cracking either ethane or naphtha (a mixture of C_5–C_7 hydrocarbons). **Cracking** is an industrial process in which larger hydrocarbon molecules are broken down at high temperatures, with or without catalysts, to produce smaller hydrocarbon molecules. The large hydrocarbon molecules are known as cracking stock. Since there is a large quantity of ethane available from natural gas in Alberta, ethene is manufactured in western Canada by cracking ethane. First the ethane must be extracted from the natural gas. According to **Table 1** on page 363, a typical mole fraction of ethane in natural gas is 11.6% (although this value can vary considerably).

The ethane that is separated from the natural gas is put in a pipeline and piped to an ethane cracking plant in Alberta (such as the one in Joffre) or in eastern Canada or the United States. Sometimes the LPGs are stored in large salt caverns 1 to 2 km below the surface (**Figure 4**).

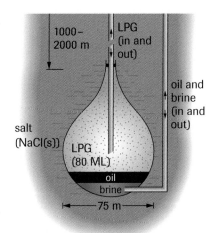

Figure 4
The salt caverns near Fort Saskatchewan, Alberta are enormous: about 80 ML or 80 km³. Situated at least 1 km below the surface, a typical salt cavern holds a volume equivalent to about 32 Olympic-size swimming pools.

Ethane cracking is also called dehydrogenation. The term cracking usually refers to breaking a large molecule down into a smaller molecule. In the case of ethane cracking, two hydrogen atoms are removed ("cracked") from an ethane molecule to convert it into an ethene molecule. A catalyst is used to increase the rate of the reaction and the hydrogen product is used in the plant.

$$H-\underset{\underset{H}{|}}{\overset{\overset{H}{|}}{C}}-\underset{\underset{H}{|}}{\overset{\overset{H}{|}}{C}}-H + \text{heat} \rightarrow \underset{H}{\overset{H}{}}C=C\underset{H}{\overset{H}{}} + H-H$$

$$\text{ethane} \rightarrow \text{ethene}$$

$$C_2H_6(g) + \text{heat} \rightarrow C_2H_4(g) + H_2(g)$$

You will find out much more about the reactions of ethane and ethene in Chapter 10. As with most chemical processes, the process is not as simple as the equation suggests. Temperature and pressure must be optimized, and a suitable catalyst must be found through experimentation. The incoming gas must be very pure: impurities such as carbon dioxide must be removed before cracking. During cracking, many side reactions produce hydrocarbons such as methane, ethyne (acetylene), propane, and butane. For economic and environmental reasons, some of the by-products are burnt rather than released into the atmosphere. The propene made at Joffre is sold as feedstock to produce other products, and the hydrogen is used to make ammonia (used in fertilizer production).

Aliphatic Hydrocarbons

So far you have encountered many different kinds of hydrocarbons: some with straight chains, some with branches, and some that are cyclic. You have also learned about saturated and unsaturated hydrocarbons. All of these compounds are classed together as **aliphatic hydrocarbons**. In the next section, you will encounter another class of hydrocarbons: the aromatics.

DID YOU KNOW ?

Ethane Extraction Plants
Alberta has over a dozen plants that extract (fractionate) ethane from natural gas. The ethane is gathered and distributed by pipelines.
The major extraction plants include
- Cochrane
- Empress
- Joffre

Other locations include
- Harmattan
- Judy Creek (near Swan Hills and Barrhead)
- Jumping Pound (near Sibbald Flats)
- Bonnie Glen (near Wetaskiwin)
- Waterton (near Pincher Creek)
- Elmworth and Wapiti (near Grande Prairie)
- Edmonton (23 Avenue)

Is there an ethane extraction plant near you?

SUMMARY Alkenes and Alkynes

- Alkenes are hydrocarbons that contain at least one carbon–carbon double bond, usually C_nH_{2n}; e.g., propene, $C_3H_6(g)$ or $CH_2=CH-CH_3$.
- Alkynes are hydrocarbons that contain at least one carbon–carbon triple bond, usually C_nH_{2n-2}; e.g., propyne, $C_3H_4(g)$ or $CH\equiv C-CH_3$.
- Alkenes and alkynes are unsaturated compounds that are easily converted to saturated (alkane) compounds by the addition of hydrogen (called hydrogenation).
  alkene/alkyne + $H_2(g)$ → alkane
  e.g., but-2-ene + hydrogen → butane
  $CH_3-CH=CH-CH_3 + H-H \rightarrow CH_3-CH_2-CH_2-CH_3$
- Rules for Naming
  1. Number the longest chain containing the multiple bond from the end closer to the multiple bond.
  2. Identify the type and location of each branch.
  3. Write the IUPAC name using the format:
     (branch location)-(branch name)(prefix)-(multiple bond location)-(ene/yne);
     e.g., 4-methylpent-2-ene
- Alkenes are isomers of cycloalkanes and alkynes are isomers of cycloalkenes.
- The cracking of ethane to ethene is a very important chemical reaction in Alberta.

Learning Tip

Classroom research indicates that the kind of learning required here is assisted by memorizing the rules, practising on many examples, and learning from your initial mistakes.

Initially there is a lot of memorizing, but eventually the logic of the nomenclature systems and of the structural formulas becomes easy when you persevere.

Section 9.3 Questions

1. Alkanes, alkenes, and alkynes are the three main families of aliphatic hydrocarbons.
   (a) What is the general molecular formula for each family?
   (b) What is the main structural feature of each family?
   (c) Why does the number of hydrogen atoms in the molecular formula decrease by two as you go from alkanes to alkenes, and then to alkynes?

2. Why are there more possible isomers of an alkene than an alkane with the same number of carbon atoms?

3. State one major use of the first member of the alkene and alkyne families.

4. Explain what is meant by the term "unsaturated" as applied to a hydrocarbon.

5. Draw structural formulas and write IUPAC names for the five aliphatic structural isomers of C_5H_{10}.

6. Write IUPAC names for the following hydrocarbons that are communicated as condensed structural formulas.
   (a) $CH_3 - CH_2 - CH_2 - CH_2 - CH_3$
   (b) $CH_3 - CH = CH - CH_2 - CH_3$
   (c) $CH \equiv C - CH_2 - CH_2 - CH_3$
   (d) $CH_2 = CH - CH_2 - CH_3$
   (e) $CH_3 - CH - CH = CH - CH_3$
                $|$
                CH_3

7. Write IUPAC names for the following hydrocarbons that are communicated as line structural formulas.
   (a)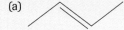
   (b)
   (c)
   (d)
   (e)
   (f)

8. Use structural or condensed structural formulas for some simple cycloalkanes to determine the general molecular formula for cycloalkanes.

9. Draw a line structural formula and write the IUPAC name for a cyclic hydrocarbon that is a structural isomer of but-1-yne.

10. Draw condensed structural formulas, labelled with IUPAC names, for all the straight-chain isomers of $C_4H_6(g)$.

11. Draw a line structural formula for each of the following hydrocarbons:
    (a) 3-ethyl-4-methylpent-2-ene
    (b) 5-ethyl-2,2,6-trimethylhept-3-yne

12. Use the information in this section and from your own research to continue gathering pro and con perspective statements concerning the resolution that the burning of fossil fuels should be reduced.

Extension

13. Ethyne (acetylene) is used in large quantities by industrial processes. Normally, gaseous substances are liquefied under high pressure and stored in steel cylinders in order to provide a reasonably large quantity for use; cylinders of propane are a typical example. Research why it is not advisable to highly compress acetylene, and how solubility is used to store $C_2H_2(g)$ in cylinders. Present your findings.

    www.science.nelson.com

14. Use a map of Alberta to find where the ethane extraction and ethane cracking plants are located. Which one is closest to where you live? In what ways does the presence of this plant affect you or your community?

    www.science.nelson.com

15. Use the Internet to find the home page of one of the companies involved with ethane production in Alberta. Access their career or human resources page and find information on the education required to work there.

    www.science.nelson.com

16. Computer-generated 3-D molecular models communicate the research done by chemists and help everyone visualize the shape of molecules. Find some alkenes, alkynes, and cycloalkenes at the following Web link. How has your personal view of these molecules changed when moving from viewing 2-D static models in this textbook or elsewhere to viewing 3-D (interactive) models on the Internet?

    www.science.nelson.com

Aromatics 9.4

Historically, organic compounds with an aroma or odour were called *aromatic compounds*. Today, chemists define **aromatics** as benzene, $C_6H_6(l)$, and all other carbon compounds that contain benzene-like structures and properties. Aromatic hydrocarbons are found naturally in petroleum (such as crude oil and natural gas) and are most often burned. Research has identified benzene as a potential carcinogen (cancer-producing substance), however. As a result, government legislation and industry innovations have reduced the benzene content of, for example, gasoline to no more than 1%. At the same time, refinery emissions of benzene have been reduced by about 90%. Research on urban air quality has shown that these reductions have, in fact, improved the air breathed by urbanites.

The molecular structure of benzene intrigued chemists for many years because they could not explain the properties of this compound (listed below), using accepted theories of bonding and reactivity.

- The molecular formula of benzene, based on its percentage composition and molar mass, is C_6H_6.
- The melting point of benzene is 5.5 °C, the boiling point is 80.1 °C, and tests show that the molecules are non-polar.
- There is no empirical support for the idea that there are double or triple bonds in benzene. For example, it is very unreactive with hydrogen.
- X-ray diffraction indicates that all the carbon–carbon bonds in benzene are the same length.
- Evidence from chemical reactions indicates that all carbons in benzene are identical and that each carbon is bonded to one hydrogen.

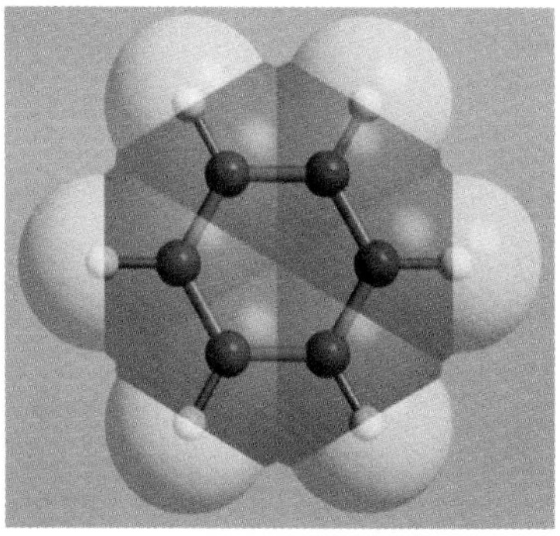

The empirical formula for benzene, $C_6H_6(l)$, was determined in 1825 by Michael Faraday (1791–1867). Visualizing a model of the benzene molecule that followed accepted bonding rules proved difficult. Finally, in 1865, August Kekulé (1829–1896) proposed a cyclic structure for benzene. The combination of the theoretical work on structural formulas by Kekulé and the empirical work of Josef Loschmidt (1821–1895) proved fruitful. Since evidence indicates that all bonds between the carbon atoms in benzene are identical in length and in strength, an acceptable model requires the even distribution of the extra six unbonded valence electrons around the entire molecule (**Figures 1 and 2**). Consider this molecule as having six unbonded valence electrons distributed around a 6-carbon-atom ring, forming a strong hexagonal structure.

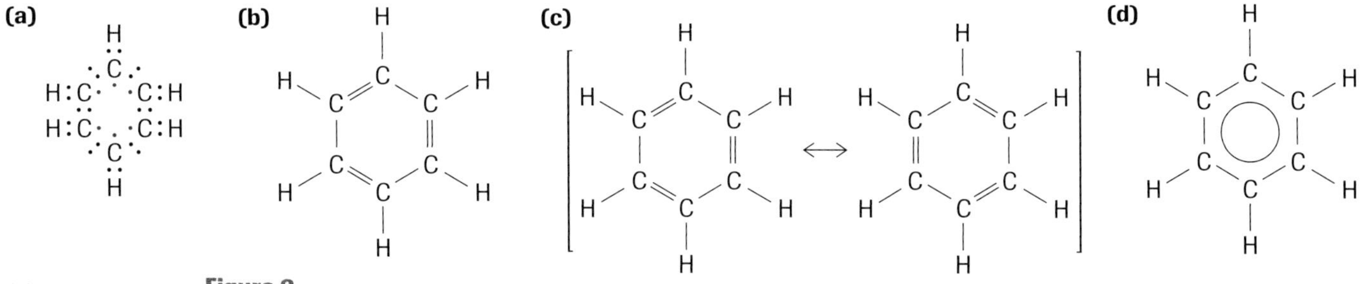

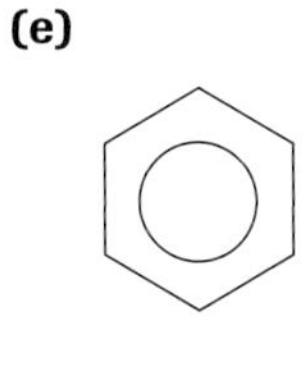

Figure 1
Although Loschmidt and others had proposed several hypotheses for the structure of benzene (C_6H_6), Kekulé championed a hexagon model. The evidence from structural isomers and chemical reactions and the logical consistency of this model with the bonding capacities of carbon and hydrogen led to the acceptance of the hexagon hypothesis (model) by the chemistry community.

Figure 2
A progressive series of models of benzene: **(a)** agrees with the evidence but is not consistent with the theory that carbon forms four bonds with other atoms; **(b)** disagrees with the evidence but is consistent with the four-bond theory for carbon; **(c)** is closer to the evidence and is consistent with the four-bond theory; **(d)** represents a new theory that agrees with the evidence; and **(e)** is a line structural formula that represents the new theory—the sharing of six valence electrons among six carbon atoms.

Hydrocarbons from Petroleum

Aspirin®
(ASA, acetylsalicylic acid)

vanillin

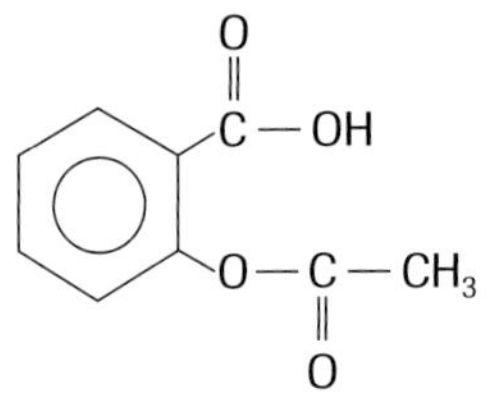

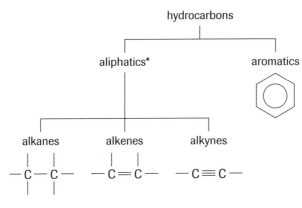

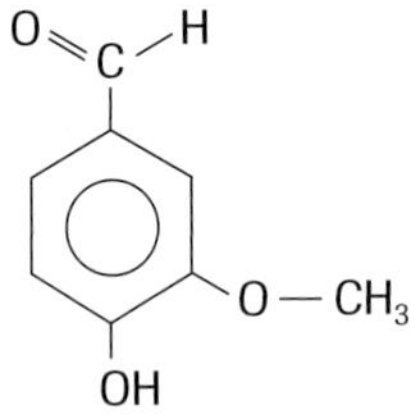

Figure 4
Common aromatic compounds include Aspirin® and vanillin. Vanillin is one of the flavour molecules in vanilla. You will notice that many aromatic molecules are often depicted using a condensed structural formula except for the benzene ring, which is shown as a line structural formula. This combination is commonly used by chemists.

*includes cyclo compounds of each family

Figure 3
Aliphatic and aromatic hydrocarbons

Based on the evidence of the properties of benzene, this structure must be particularly stable. The reactions of benzene are similar to those of alkanes—benzene behaves chemically like a saturated hydrocarbon (an alkane). Since 1865, hydrocarbons were classified as aliphatic or aromatic (**Figure 3**). For our purposes, aromatic compounds contain the benzene ring, which is represented by a hexagon with an inscribed circle. Structures of all aromatic compounds include bonding similar to that in the benzene ring (**Figure 4**).

Naming Aromatics

Simple aromatics are usually named as relatives of benzene. If an alkyl group is bonded to a benzene ring, it is named as an alkylbenzene (**Figure 5**). The alkyl group is considered a substitute for a hydrogen atom. Since all of the carbon atoms of benzene are equivalent to each other, no number is required in the names of compounds of benzene that contain only one branch.

When two hydrogen atoms of the benzene ring have been substituted, three isomers are possible. These isomers are named as alkylbenzenes, using the lowest possible pair of numbers to indicate the locations of the two alkyl groups on the benzene ring. The numbering starts at one of the branches and goes clockwise or counterclockwise to obtain the lowest possible pair of numbers (**Figure 6**).

For some larger molecules, it is more convenient to consider the benzene ring as a branch. In such molecules, the benzene ring is called a **phenyl group**, $-C_6H_5$. For example, the compound in **Figure 7** is named 2-phenylpentane, according to the naming system for branched alkanes (page 368). Either naming system for aromatic compounds is acceptable; choose the more convenient method for the compound in question.

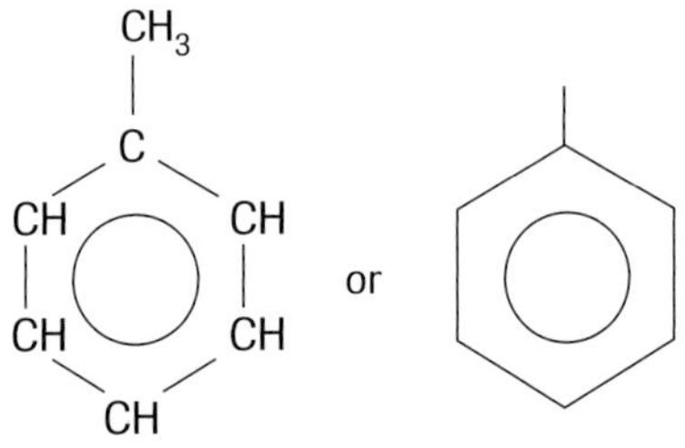

Figure 5
Methylbenzene ($C_6H_5CH_3(l)$), commonly known as toluene, is a solvent used in glues and lacquers. It is toxic to humans but is preferred to benzene as a solvent because benzene is both toxic and carcinogenic.

1,2-diethylbenzene (o-diethylbenzene) | 1,3-diethylbenzene (m-diethylbenzene) | 1,4-diethylbenzene (p-diethylbenzene)

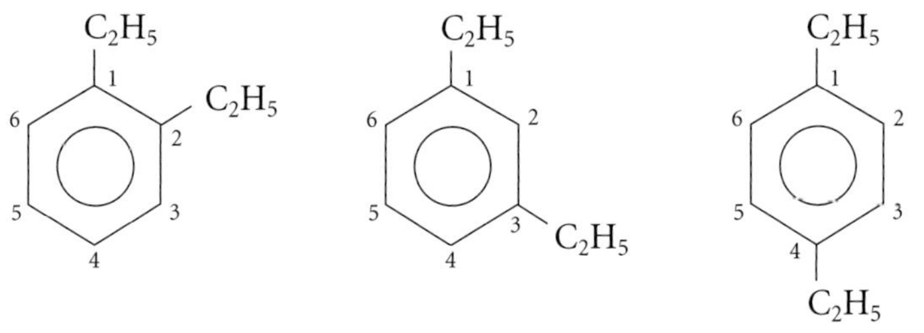

Figure 6
Three isomers of diethylbenzene. The *o*-, *m*-, and *p*- classical nomenclature is described in the Learning Tip on the next page.

When drawing a structural formula from a name, carbon 1 can be designated anywhere on the benzene ring. The numbering 1 to 6 can be done clockwise or counterclockwise: The molecules do not know which way is up.

COMMUNICATION example

Write IUPAC names for the following aromatic hydrocarbons:

(a) 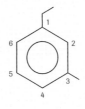 (b)

Solution

(a) 1-ethyl-2,4-dimethylbenzene   (b) 3-phenyl-4-propyloctane

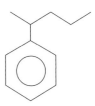

Figure 7
A line structural formula for 2-phenylpentane

When drawing a structural formula for an aromatic compound, always look at the end of the given name to identify the parent chain. Is the parent chain benzene or an aliphatic compound? Draw the parent chain first, then consider the placement of the branches.

SAMPLE problem 9.3

Draw the line structural formula for 1-ethyl-3-methylbenzene.

First, draw the benzene ring, and then add an ethyl group to any C atom of the ring; this C atom automatically becomes C 1 in the numbering system. Finally, add a methyl group to C 3, which can be clockwise or counterclockwise from C 1.

Note that, for line structural formulas, no hydrogen atoms are shown and the end of any line segment denotes a carbon atom.

Learning Tip

You may encounter a naming system—known as the classical system—for benzene rings with two substituted groups. The 1,2-, 1,3-, and 1,4- arrangements are denoted by the prefixes ortho- (*o*), meta- (*m*), and para- (*p*), respectively. These names, such as *o*-dimethylbenzene, are still used in industry, and you may encounter them in other references.

SUMMARY Naming Aromatic Hydrocarbons

1. If an alkyl branch is attached to a benzene ring, the compound is named as an alkylbenzene. Alternatively, the benzene ring may be considered as a branch of a large molecule; in this case, the benzene ring is called a phenyl branch.

2. If more than one alkyl branch is attached to a benzene ring, the branches are numbered using the lowest numbers possible, starting with one of the branches. Given a choice between two sets of lowest numbers, choose the set that is in both numerical and alphabetical order. See Sample Problem 9.3.

DID YOU KNOW ?

VOCs
Volatile organic compounds (VOCs) are gases and vapours, such as benzene, that are released throughout the petroleum refining processes. Petrochemical plants (see Chapter 10), plastics manufacturing, and the distribution and burning of gasoline all release VOCs into the atmosphere. Some VOCs, like benzene, are carcinogens. Other VOCs react in sunlight with nitrogen oxides from automobile exhaust to form ground-level ozone, a component of smog.

  WEB Activity

Web Quest—West Nile Denial

In this Web Quest, investigate traditional and modern mosquito repellents, and create a brochure to help people in the prairie provinces reduce the risk of contracting West Nile Virus.

www.science.nelson.com

LAB EXERCISE 9.A

Chemical Properties of Aliphatics and Aromatics

Report Checklist
- ○ Purpose
- ○ Problem
- ○ Hypothesis
- ● Prediction
- ○ Design
- ○ Materials
- ○ Procedure
- ○ Evidence
- ● Analysis
- ● Evaluation (2, 3)

Comparing chemical properties of different classes of hydrocarbons can be difficult in a high school chemistry laboratory. Many hydrocarbons are volatile organic compounds (VOCs), some are carcinogenic (such as benzene), and many are flammable. In this investigation, the properties of aliphatics are compared to those of aromatics by reporting the evidence for the reaction of a cycloalkane, a cycloalkene, and an aromatic with potassium permanganate, a strong oxidizing agent. When completing the Prediction, provide a theoretical justification.

Purpose
The purpose of this investigation is to test the generalization that aromatic hydrocarbons react like saturated rather than unsaturated hydrocarbons.

Problem
What is the order of reaction rate for cyclohexane, cyclohexene, and benzene?

Design
Each of cyclohexane, cyclohexene, and benzene is mixed vigorously with aqueous potassium permanganate in a fumehood. Evidence for a reaction is a change in the purple colour of the aqueous potassium permanganate.

Evidence
Cyclohexane
- Initially, no changes were observed in the mixture.
- After 5 min, the colour changed from purple to grey-purple.

Cyclohexene
- The colour of the mixture immediately changed to green.
- After 5 min, a brown precipitate formed.

Benzene
- Initially, no changes were observed in the liquid.
- After 5 min, the colour was still the same.

LAB EXERCISE 9.B

Boiling Points of Sample Aliphatics and Aromatics

Report Checklist
- ○ Purpose
- ○ Problem
- ○ Hypothesis
- ● Prediction
- ○ Design
- ○ Materials
- ○ Procedure
- ○ Evidence
- ● Analysis
- ● Evaluation (2, 3)

Most of the hydrocarbons with smaller molecules, such as methane, ethane, propane, and butane, are gases at SATP. Hydrocarbons with medium-sized molecules (C_5–C_{18}) are found to be liquids. We often fail to appreciate how easy it is to transport and dispense these liquid hydrocarbons. We can explain the relative boiling points of hydrocarbons theoretically with the concept of London forces. Evaluate the ability of the concept of London forces to predict the relative boiling points of different classes of hydrocarbons. (According to philosophers of science, in this lab exercise, you are testing the *predictive* power of a scientific concept.)

Purpose
The purpose of this lab exercise is to test the ability of the concept of London forces to predict the relative boiling points of aliphatic and aromatic hydrocarbons.

Problem
What is the relative order of boiling points of simple C_6 alkane, alkene, cycloalkane, cycloalkene, and aromatic hydrocarbons?

Design
The boiling points of hexane, hex-1-ene, cyclohexane, cyclohexene, and benzene are obtained from *Lange's Handbook of Chemistry, 13th Edition*.

Evidence

Table 1 Boiling Points of C_6 Aliphatics and Aromatics

Hydrocarbon	Hexane	Hex-1-ene	Cyclohexane	Cyclohexene	Benzene
Boiling point (°C)	68.7	63.5	80.5	83.0	80.10

Section 9.4 Questions

1. State the name and chemical formula of the simplest aromatic compound. In your own words, describe what is unusual about its structure.

2. There are over 4000 dehydrators at natural gas well sites throughout Canada. The older technology allowed for emissions of benzene, methylbenzene (toluene), ethylbenzene, and the three dimethylbenzenes (xylenes). Newer technologies have been invented to reduce these emissions. Draw line structural formulas for each of these six aromatic compounds.

3. Draw a combination of line and condensed structural formulas for the following aromatics:
   (a) 1,2,4-trimethylbenzene (c) 3-phenylpentane
   (b) 1-ethyl-2-methylbenzene (d) 3-methyl-1-phenylbutane

4. The isomers of dimethylbenzene, commonly called xylenes, are used as solvents for adhesives (**Figure 8**). Write alternative IUPAC names and draw structural formulas for the three xylene isomers.
   (a) 1,4-dimethylbenzene (c) 1,2-dimethylbenzene
   (b) 1,3-dimethylbenzene

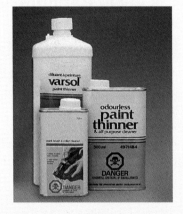

Figure 8
Isomers of dimethylbenzene, known as xylenes, are used as solvents.

5. Write IUPAC names for the following aromatic hydrocarbons:

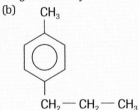

6. Aromatics can also take part in various chemical refining reactions. For each of the following reactions, draw a structural formula equation where names are given and provide names where structures are given.
   (a) propylbenzene → methylbenzene + ethene
   (b) [benzene with CH₃ → benzene with CH₃ on different position]
   (c) [benzene + CH₂=CH₂ → ethylbenzene]

7. Classify each of the following hydrocarbons as aromatic or aliphatic.
   (a) 2,3-dimethylhexane (d) 4-methylpent-2-yne
   (b) $CH_3-CH_2-CH(C_6H_5)-CH_3$ (e) 1,3-dimethylbenzene
   (c) 1,2-dimethylcyclohexane (f) 2-phenyl-3-hexene

8. Classify each of the following hydrocarbons as saturated or unsaturated.
   (a) $C_2H_4(g)$ (d) $C_3H_4(g)$
   (b) $CH_3-(CH_2)_4-CH_3$ (e) $CH_3-(CH)_2-CH_3$
   (c) $C_6H_5CH_3(l)$ (f) $C_6H_6(l)$

9. Why is benzene not called cyclohex-1,3,5-triene, to communicate three double bonds in a hexagon?

10. In Lab Exercise 9.B, you compared the boiling points of hexane, hex-1-ene, cyclohexane, cyclohexene, and benzene. Test the ability of the concept of London forces to explain the relative order of the boiling points of these C_6 hydrocarbons. Did the London-force concept or your understanding of it need to be revised to provide an acceptable explanation? (According to philosophers of science, in this question, you are testing the explanatory power of a scientific concept.)

11. Even though the use of benzene as a solvent was banned in Canada several decades ago, people's exposure to benzene has not stopped. A Materials Safety Data Sheet (MSDS) is the source of chemical safety information. List some common sources of benzene exposure and identify some short-term and long-term health effects.

www.science.nelson.com

9.5 Crude Oil Refining

Crude oil is pumped from the ground by thousands of pump jacks throughout Alberta (**Figure 1**). The crude oil then enters a pipeline through which it is shipped to an oil refinery. If the oil has to travel a long distance, pumping stations are situated along the way to assist the movement of the oil. Much of the oil in Alberta is exported to other provinces and other countries for refining. Some of the oil is refined in Alberta.

Crude oil throughout the world is classified on the basis of viscosity, hydrocarbon composition, and sulfur content. (Viscosity is a measure of how well a liquid pours: low-viscosity liquids pour like water, whereas high-viscosity liquids pour like molasses or liquid honey.) For example, light crude oil has a lower viscosity and requires less refining than heavy or high-viscosity crude oil. While some types of crude oil are best suited for gasoline production, other types of crude oil may be better suited for jet fuel, diesel fuel, home heating, motor oil, and asphalt end products (**Figure 2**). Most crude oil is separated into a variety of components that differ according to the percentage of each component obtained. This separation can involve both physical and chemical processes.

Figure 1
Pump jacks are found over oil wells in many parts of Alberta. This one is near Cochrane, Alberta.

Physical Processes in Oil Refining

Crude oil is a complex mixture of hundreds of thousands of compounds. Some of these compounds boil at temperatures as low as 20 °C. The least volatile components of crude oil, however, boil at temperatures above 400 °C. Chemical engineers take advantage of the differences in boiling points to physically separate the components. This technological process is called *fractional distillation*, or **fractionation**.

Any substance with a boiling point above 25 °C is a liquid or solid at SATP. Chemical engineers have found that, when crude oil is heated to 500 °C in the absence of air, most of its constituent compounds vaporize. The compounds with boiling points higher than 500 °C remain as mixtures called asphalts and tars. The vaporized components of the crude oil rise and gradually cool in a metal tower (**Figure 2**). The technological design is very ingenious: To get from one level to the next, the vapours have to force their way through the liquid in the next tray. Bubble caps, which were used in the past, illustrate the contact between liquid and vapour (**Figure 3**). Currently, valve trays, which are more efficient, are used in distillation columns. When the temperature of the liquids in the trays in the higher parts of the tower is below the boiling points of the vaporized compounds, the substances in the vapour begin to condense. Those substances with high boiling points condense in the lower, hotter trays of the tower, whereas those with lower boiling points condense in the cooler trays near the top of the tower. Side streams are withdrawn from various locations along the column. These various streams are called **fractions**. Some typical fractions and their properties are listed in **Table 1**.

Chemists have an explanation for the boiling points of these fractions. They have found that the fractions with the lowest boiling points generally contain the smallest molecules. Chemists explain that the low boiling points are due to small molecules,

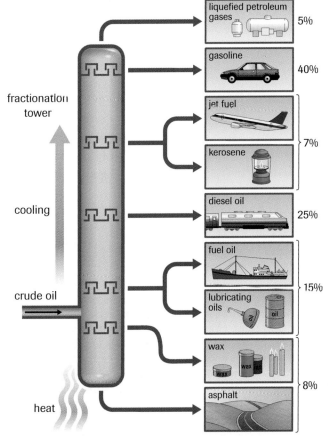

Figure 2
A fractional distillation (fractionation) tower contains trays positioned at various levels. Heated crude oil enters near the bottom of the tower. The bottom of the tower is kept hot, and the temperature gradually decreases toward the top of the tower. The concentration of components with lower boiling points increases from the bottom to the top of the tower. The percentage distributions shown vary with the type of crude oil and with seasonal demands.

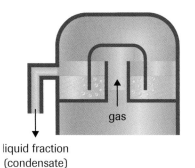

Figure 3
The vaporized components are forced to bubble through the liquids condensed in each tray. Any gas with its condensation point (boiling point) greater than the temperature of the liquid in the tray will condense. The liquid in each tray continuously overflows into the draining tube and the liquid mixture is piped from the fractionation tower.

Table 1 Fractional Distillation of Crude Oil

Boiling point range of fraction (°C)	Number of carbon atoms per molecule	Fraction (intermediate product)	Products
below 30	1 to 5	gases; LPGs	gaseous fuels for cooking and for heating homes
30 to 90	5 to 6	naphthas	dry-cleaning solvents, naphtha gas, camping fuel
30 to 200	5 to 12	straight-run gasoline	automotive gasoline
175 to 275	12 to 16	kerosene	diesel, jet, and kerosene fuels; cracking stock* (raw materials for fuel and petro-chemical industries)
250 to 375	15 to 18	light gas or fuel oil	furnace oil; cracking stock
over 350	16 to 22	heavy gas oil	lubricating oils; cracking stock
over 400	18 and up	greases	lubricating greases; cracking stock
over 450	20 and up	paraffin waxes	candles, waxed paper, cosmetics, polishes; cracking stock
over 500	26 and up	unvaporized residues	asphalts and tars for roofing and paving

* Cracking stock is feedstock (raw material) that is, for example, hydrocracked to make smaller molecules that are more in demand; e.g., gasoline molecules.

which have fewer electrons and weaker London forces compared with large molecules. The fractions with higher boiling points are found to contain much larger molecules. Some typical fractions are shown in **Table 1**. Recall from Chapter 3 that London forces are also dependent on the shape of the molecules; that is, on whether the molecule is branched. Since these molecules are hydrocarbons, chemists explain that dipole–dipole and hydrogen bonding are not involved.

Other Physical Processes in Oil Refining

In addition to fractional distillation, there are several other physical processes that are used to treat fractions before and/or after chemical processing. Several of these physical processes are **solvent extractions**—a solvent is added to selectively dissolve and remove an impurity or to separate some useful products from a mixture. Dewaxing is a simple process of cooling a mixture to precipitate the larger wax fraction.

DID YOU KNOW?

Naphthas
Like all of the fractions listed in Table 1, naphtha is a mixture with varying composition (including aliphatic and aromatic hydrocarbons). The mixture depends on the source of the hydrocarbon (e.g., light or heavy crude, bitumen, or coal) or customer demand. Naphtha, for example, may be petroleum naphtha with more aliphatic character, or bitumen naphtha with more aromatic character. What defines it as naphtha is the boiling point range over which it is collected in the fractionator.

CAREER CONNECTION

Petroleum Engineer
Petroleum engineers apply their knowledge of chemistry to the discovery, recovery, and processing of oil and gas deposits.
   Rob Manuel is a petroleum engineer. He combines his Western scientific world view with his Aboriginal world view (he grew up on the Upper Nicola Indian Band Reserve) to bring a "different way of thinking to the table." Find out about the education petroleum engineers require for the kind of work they do.

www.science.nelson.com

INVESTIGATION 9.3 Introduction

Fractional Distillation (Demonstration)

Fractional distillation is a process used commercially to separate components (fractions) of petroleum using boiling point differences.

Report Checklist
- ○ Purpose
- ○ Problem
- ○ Hypothesis
- ○ Prediction
- ○ Design
- ○ Materials
- ○ Procedure
- ● Evidence
- ● Analysis
- ○ Evaluation

Purpose
The purpose of this investigation is to use the concept and technique of fractional distillation to separate two hydrocarbon components from a liquid mixture.

Problem
What is the percent by volume of each of the two hydrocarbon components?

Design
A mixture of the two liquids is heated in a fractional distillation apparatus while the temperature is measured at regular intervals in a fume hood.

To perform this investigation, turn to page 402.

DID YOU KNOW ?

Oil Recovery
The recovery of oil can be enhanced by the use, for example, of water or of carbon dioxide. Recent research has shown that carbon dioxide can be captured from large industries and transported by pipeline to oil wells. The recovery of the oil from underground increases from about 25% to about 40%. The carbon dioxide used in this application helps reduce emissions and helps meet Kyoto targets.

Practice

1. State some similarities and differences between the operation of a fractionation tower and a laboratory-scale distillation apparatus.
2. (a) Why is crude oil heated in the absence of air in the fractionation tower?
   (b) Write chemical equations to represent the vaporization of pentane and of octane at the bottom of a fractionation tower.
   (c) Write chemical equations to represent the condensation of pentane and of octane in a bubble-cap tray of a fractionation tower.
   (d) Which of pentane and octane has a higher boiling point, and why?
   (e) Which of pentane and octane is removed in a higher bubble tray of the fractionation tower?
3. Feedstock is raw or semi-processed material that is fed into a chemical process to produce a more valuable product. Cracking stock is feedstock that is cracked into smaller or more branched hydrocarbons. List the different fractions that are used as cracking stock. Suggest some reasons why so many fractions are used for this purpose.
4. What percentage of petroleum is used as fuel? What is the major type of fuel that makes up this category?
5. Straight-run gasoline is gasoline that comes straight from a crude-oil, fractional distillation tower. There are various kinds of crude oil, including light, heavy, and synthetic. Some kinds of oil provide a greater percentage of straight-run gasoline than others.
   (a) Describe the straight-run gasoline fraction compared to other crude oil fractions.
   (b) Write balanced chemical equations for the complete combustion of the pentane and octane components of straight-run gasoline while driving a hybrid (gasoline and electric) automobile.
6. Crude oil is a solution of many, many hydrocarbons, including hydrocarbons that are solids, liquids, and gases as pure substances at SATP.
   (a) Explain how this mixture can be a solution.
   (b) Water is part of the mixture when extracting crude oil from oil wells. Use intermolecular forces concepts to explain why water would or would not be a solute in the crude oil solution.
7. Use the information in this section to add to your collection of statements, pro and con, from a variety of perspectives on the resolution that the burning of fossil fuels should be reduced.

Section **9.5**

Chemical Processes in Oil Refining

The refining of crude oil can be divided into two main types of processes: physical processes, such as fractionation and solvent extraction, and chemical processes, such as cracking and reforming (**Figure 4**). These chemical processes are necessary because the fractional distillation of crude oil does not produce enough of the hydrocarbons that are in demand (particularly the gasoline fraction) and produces too much of the heavier fractions. Compared to gasoline and diesel, there is limited market demand for fuel oil, lubricating oils, greases, and waxes. Notice that all of these fractions (**Table 1**, page 387) contain larger molecules than those typically found in gasoline (C_5 to C_{12}).

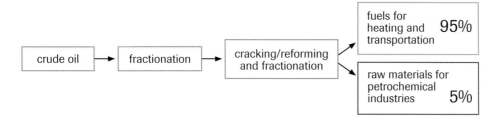

Figure 4
Only 5% of the original mass of crude oil is used as starting chemicals (called feedstock) in the manufacture of solvents, greases, plastics, synthetic fibres, and pharmaceuticals.

www.science.nelson.com

Cracking

Cracking is a chemical process in which larger molecules are broken down with heat and/or catalysts to produce smaller molecules. In the early 1900s, cracking was accomplished using only high temperatures and pressures in a process called *thermal cracking*. This process is fairly effective, but is messy and wasteful because it produces large quantities of solid coke (carbon). Thermal cracking is still used today, but to a limited extent. By 1937, basic research and technological development produced a new improved cracking technique called *catalytic cracking* (**Figure 5**). This process breaks apart larger molecules, but the presence of a catalyst, along with less severe reaction conditions, produces more desirable fractions and less residual materials such as tar, asphalt, and coke. (You will learn more about the uses of catalysts in chemical processes in Chapter 12.) A typical reaction equation might be

$$C_{17}H_{36}(l) \rightarrow C_9H_{20}(l) + C_7H_{16}(l) + C(s)$$

CAREER CONNECTION

The Petroleum Industry
The Petroleum Human Resources Council of Canada has been set up to handle the huge demand for employees in the areas of exploration, development, production, service industries, pipeline transmission, gas processing, and the mining, extracting, and upgrading of heavy oil and bitumen. Find out more about the Petroleum HR Council, the seven human resource issues it has identified, and the strategies it has developed to address any one of these issues. If you were interested in working in this field, how might the Petroleum HR Council's work benefit you?

www.science.nelson.com

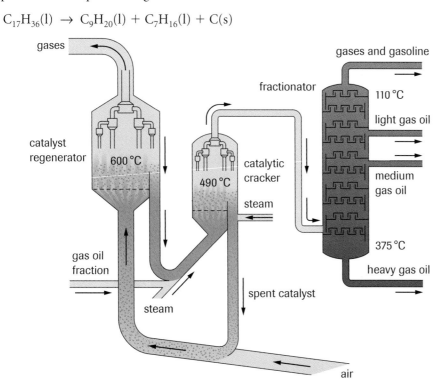

Figure 5
For catalytic cracking, the gas oil fraction, for example, enters the central vessel, is catalytically cracked, and then fractionated again in the column on the right. The catalyst is recycled into the vessel on the left. The catalyst is regenerated when the coke is burned off of it in the left vessel.

Hydrocarbons from Petroleum

DID YOU KNOW ?

Ammolite Gems and Ammonite Fossils

Ammolite (**Figure 6(a)**) is one of only three organic gemstones, besides pearl and amber, but contains traces of up to a dozen metals. The highly supported hypothesis is that ammolite comes from the shells of ammonites—squid-like creatures that lived about 75 Ma ago in the Bearpaw Sea that ran north to south through the middle of Alberta (**Figure 6(b)**). Aboriginal peoples, ranchers, and prospectors have found ammolite along the banks of the St. Mary River, south of Lethbridge, Alberta. Ammolite is the only uniquely Canadian gem, and has been mined and refined in Alberta since the 1980s.

Ammolite is also known as aapoak (Blackfoot for small, crawling stone) and buffalo stone (also named by the Blackfoot of southern Alberta). The latter name comes from the Blackfoot legend about a princess who found an ammolite gemstone that brought prosperity (in the form of a buffalo herd) back to the Blackfoot. Since then, the Blackfoot have regarded ammolite as a sign of prosperity.

(a)

(b)
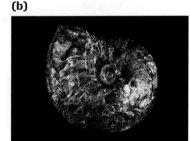

Figure 6
(a) An ammolite gem and (b) an ammonite fossil

The history of oil refining is a story of continuous technological development to meet societal needs for various petroleum products, in particular, gasoline. In 1960, another improvement, called hydrocracking, helped meet this demand. **Hydrocracking** is a combination of catalytic cracking and hydrogenation and is used for heavier feedstocks, particularly those containing complex aromatic compounds. During the hydrogenation process, no coke (carbon) is produced. A simplified hydrocracking reaction equation would be

$$C_{17}H_{36}(l) + H_2(g) \rightarrow C_9H_{20}(l) + C_8H_{18}(l)$$

Catalytic Reforming and Alkylation

Cracking reactions are like decomposition reactions because the main objective is to break larger molecules into smaller ones. Both the fractionation process and the cracking process produce large quantities of light fractions such as gases and naphthas (**Table 1**, page 387), as well as molecules whose structures are not suitable for our needs. For example, there might be too few branches on the molecules to produce a sufficiently high-quality gasoline. The chemical process of reforming and alkylation rearranges the bonding in molecules to improve the quality of the gasoline.

Catalytic reforming is the chemical process involved in converting molecules in a naphtha (gasoline) fraction into aromatic gasoline molecules. These aromatic molecules have better burning properties than the original aliphatic (non-aromatic) molecules. For example, heptane does not burn well in internal combustion (car) motors. If the heptane is converted into methylbenzene, the gasoline burns better in car motors. All reforming is now done with the use of catalysts to increase the rate of the reaction, and is, therefore, often called catalytic reforming. Hydrogen, a by-product of catalytic reforming, is usually recycled in other processes, such as hydrocracking.

$$CH_3-(CH_2)_5-CH_3 \rightarrow C_6H_5-CH_3 + 4\,H_2$$

heptane → methylbenzene

Alkylation

Another way of improving the quality of gasoline is to increase the branching of the molecules in a process called *alkylation*. This process is also called isomerization because it converts a molecule into a branched isomer. For example, heptane can be converted into 2,4-dimethylpentane, a molecule that burns better in an internal combustion motor.

$$CH_3-(CH_2)_5-CH_3 \rightarrow CH_3-CH(CH_3)-CH_2-CH(CH_3)-CH_3$$

heptane → 2,4-dimethylpentane

WEB Activity

Canadian Achievers—Karl Chuang

Karl Chuang studied chemical engineering at the University of Alberta in the 1970s. Today, he is a top distillation expert in Canada and one of the best in the world. He has developed distillation technologies that are used in refineries around the world.
What are Dr. Chuang's current research interests?

www.science.nelson.com

SUMMARY

Gasoline and diesel fuel are high-demand crude oil fractions. More of each of these fractions can be made by cracking heavier fractions. Also, molecules that burn better in internal combustion motors can be made by chemical processes (e.g., catalytic reforming and alkylation) controlled by chemical engineers.

Catalytic cracking: larger molecules → smaller molecules + carbon
Hydrocracking: larger molecule + hydrogen → smaller molecules
Catalytic reforming: aliphatic molecule → aromatic molecule + hydrogen
Alkylation (isomerization): aliphatic molecule → more branched isomer

Practice

8. Why are chemical processes necessary in oil refining, in addition to the physical process of fractionation?

9. Write a condensed structural formula equation for each of the following catalytic cracking reactions. Assume that unbranched alkanes are produced, in addition to carbon (coke). There are many possible product combinations. Predict one combination.
   (a) $CH_3—(CH_2)_{16}—CH_3$ →
   (b) decane →
   (c)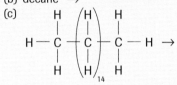

10. Compare hydrocracking and catalytic reforming. In what way are they complementary?

11. High-quality gasoline requires chains of C_5 to C_{12} molecules, branching within the molecules, and some aromatic molecules. For each of these three requirements, identify the physical and/or chemical process that accomplishes the requirement.

12. Since crude oil contains many alkanes, unwanted hydrocracking reactions are common in the first stage of oil refining during fractionation. The hydrogen is a product of thermal cracking. For each of the following word equations representing hydrocracking reactions, write a balanced condensed structural formula equation:
    (a) hexane + hydrogen → ethane + butane
    (b) 2-methylpentane + hydrogen → propane + propane
    (c) 2,2-dimethylbutane + hydrogen → ethane + methylpropane

13. After initial fractional distillation, catalytic reforming and alkylation reactions increase the yield of desirable products, such as compounds whose molecules have aromatic character and more branches. Complete each of the following equations. Draw condensed structural formulas when IUPAC names are given, and write IUPAC names when condensed structural formulas are given.
    (a) $CH_3—CH_2—CH_3 + CH_3—CH_2—CH_2—CH_2—CH_3$ →
    $CH_3—C(CH_3)_2—CH_2—CH(CH_3)—CH_3$
    (b) octane → 1,3-dimethylbenzene + hydrogen
    (c) $CH_3—(CH_2)_3—CH_3$ → $C_3H_5(C_2H_5) + H_2$
    (d) decane → 3-ethyl-2-methylheptane

DID YOU KNOW ?

Transportation Fuels
About 50% of crude oil processed in Canadian refineries is used as transportation fuels—gasoline, diesel fuel, jet fuel, and fuel oil for ships and locomotives. The demand keeps increasing, although, for example, the decreasing demand for fuel oil for heating homes (in Eastern Canada) has allowed more of the fuel-oil supply to be used as a transportation fuel.

DID YOU KNOW?

Oil Sands Discovery

In 1719, a Cree named Wa-pa-su took a sample of Athabasca oil sands to the Hudson Bay Post in Fort Churchill. Aboriginal peoples used the tarry substance, mixed with tree sap, to waterproof their canoes and other items.

Extension

14. Benzene and other aromatic hydrocarbons have a low H:C ratio (close to 1:1). Chemical engineers have invented two processes to increase this ratio to about 2:1.
    (a) Explain how the 1:1 hydrogen-to-carbon ratio is derived for aromatics.
    (b) What classes of hydrocarbons have H:C ratios close to 2:1?
    (c) Describe how hydrocracking works to improve the H:C ratio of aromatics from approximately 1:1 to approximately 2:1.

15. For economic and efficiency reasons, crude oil refining technologies have moved from thermal cracking to catalytic cracking. Both catalytic cracking and hydrocracking employ catalysis. Chemical engineers who are designing modern refineries are increasingly moving toward hydrocracking over straight catalytic cracking. What is an advantage of hydrocracking over catalytic cracking?

Case Study

Octane Number

Hydrocracking, alkylation, and catalytic reforming reactions produce not only more gasoline, but also convert low-grade gasoline into higher-grade gasoline. At gas stations, you see the grades displayed on the gas pumps, for example, as Regular and Super or as a specific number, such as 88 and 98. The numbers, 88 and 98, represent the octane rating (number) of the gasoline.

Low-octane gasoline, with long straight-chain molecules, burns quickly. Fuel is supposed to ignite in the cylinder as a result of a spark. The spark is timed to ignite the fuel when the piston is close to the top of its "up stroke" (see **Figure 7**). Sometimes the mixture auto-ignites as a result of the pressure in the cylinder, rather than burning when the spark is produced. If the fuel ignites earlier, high-performance vehicle engines lose some of their efficiency. They sometimes produce a knocking or pinging sound. High-performance gasoline engines require high-octane fuel, which contains more branched-chained and aromatic hydrocarbons. Higher-octane gasoline resists spontaneous (high-pressure, no-spark) auto-ignition of excess fuel. The engine runs more smoothly, and is less likely to be damaged or lose efficiency.

Russell Marker, a chemist, assigned 2,2,4-trimethylpentane (an isomer of octane) an octane number of 100, and heptane (a straight-chain alkane) an octane number of zero. These chemicals were chosen for comparison due to their similar boiling points and enthalpies of vaporization (**Table 2**).

In addition to using high-octane gasoline to reduce knocking in high-performance engines, an additive (octane booster) can be used. From the 1920s until the 1970s, the most common additive was tetraethyl lead, $Pb(C_2H_5)_4(l)$. Because lead deactivated the catalysts in the catalytic converters installed in cars starting in 1972, tetraethyl lead was removed from gasoline by the 1990s.

Several ingredients have replaced the tetraethyl lead, such as dibromoethene (ethylene dibromide, EDB), dichloroethene (ethylene chloride, EDC), and 2-methyl-2-methoxypropane (methyl ditertiary butyl ether, MTBE). Each substitute has had its own problems.

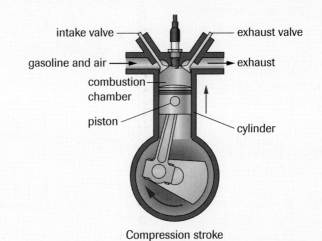

Figure 7
Engine knock in an internal combustion engine is caused by auto-ignition of excess fuel occurring between carefully timed sparks.

Table 2 Octane Rating Reference Scale

Chemical	Boiling point (°C)	Octane rating
heptane $CH_3 — (CH_2)_5 — CH_3$	98.4	0
2,2,4-trimethylpentane $CH_3-\underset{\underset{CH_3}{\vert}}{\overset{\overset{CH_3}{\vert}}{C}}-CH_2-\underset{}{\overset{\overset{CH_3}{\vert}}{CH}}-CH_3$	99.3	100

Case Study Questions

1. As they relate to octane rating of gasoline, why are alkylation and catalytic reforming reactions carried out?
2. Predict the reaction products in a word equation, and then write a structural formula equation for each of the following gasoline reactions. Recall that complete combustion involves a reaction with oxygen to produce the most common oxides.
   (a) octane + oxygen →
   (b) 2,2,4-trimethylpentane + oxygen →
   (c) heptane + oxygen →
   (d) methylbenzene (toluene) + oxygen →
3. Knocking occurs when compression ignites fuel during the power stroke that did not ignite initially from the spark provided by the spark plug (**Figure 7**). Write line structural formulas for the following hydrocarbons in the internal combustion engine.
   (a) 1,3-dimethylbenzene enters the cylinder through the intake port of the cylinder.
   (b) 2,3-dimethylbutane is ignited by a spark from the spark plug at the top of the compression stroke.
   (c) Hexane is ignited by heat and high pressure before completion of the compression stroke.
   (d) Decane (an excess/unreacted hydrocarbon) exits through the exhaust port of the cylinder.

Sulfur in Gasoline

Many organic compounds incorporate sulfur in their molecules. Sulfur in gasoline is a pollution problem. When the gasoline is burned, sulfur emissions reduce air quality and can also decrease the pH of rain, resulting in acid deposition. **Table 3** presents typical emissions found in the exhaust of a well-tuned automobile with a catalytic converter.

Besides contributing to the problem of acid deposition, sulfur dioxide causes problems even before leaving the exhaust pipe of an automobile. Sulfur dioxide and sulfur atoms in unburned fuel tend to reduce the effectiveness of the catalytic converter. (You will learn more about the catalytic converter in Chapter 12.) This reduced effectiveness increases the quantity of other pollutants, such as carbon monoxide and nitrogen oxides, that make their way through the exhaust system and into the air.

Research indicates that refined gasoline has a typical average sulfur content of 340 ppm, although the range is from 30 ppm to 1000 ppm. Some jurisdictions have legislated a sulfur content as low as 30 ppm.

The technology that scientists and engineers have created to reduce the sulfur content in gasoline is a process called hydrogenation or hydrotreating. Hydrogen gas (likely obtained by catalytic reforming of crude oil fractions) reacts with sulfur atoms in gasoline molecules to produce hydrogen sulfide gas. The $H_2S(g)$ is then converted to sulfur and water in a Claus converter (**Figure 8**).

Table 3 Typical Gasoline Automobile Exhaust

Gas	Volume (%)
carbon dioxide	9
oxygen	4
hydrogen	2
carbon monoxide	4–9
hydrocarbons	< 0.2
aldehydes	0.004
nitrogen oxides	0.05–0.4
sulfur dioxide	0.006
ammonia	0.0006
nitrogen, etc.	≐75*

*The approximate 75% includes, for example, water vapour that condenses into a vapour trail on a cold winter day.

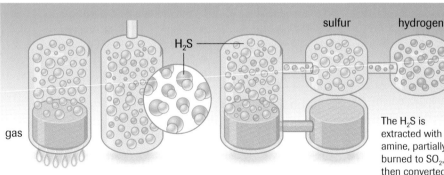

In a continuous process, gasoline is vaporized by heating it to more than 500 °C.

Hydrogen gas is injected into the vaporized gasoline. Hydrogen catalytically reacts with sulfur to form $H_2S(g)$.

The gasoline is cooled and it condenses. The hydrogen sulfide remains in vapour form and is extracted. Some hydrogen atoms remain in the gasoline, replacing the extracted sulfur.

The H_2S is extracted with an amine, partially burned to SO_2, and then converted to sulfur in a Claus converter. The sulfur is used, for example, in fertilizers and asphalt.

Figure 8
To achieve the federally mandated goal of sulfur reduction in gasoline, suppliers needed to invest in new or additional desulfurization equipment, another step in the refining process. Refineries in Alberta have upgraded to reduce sulfur content in their products. Alberta oil refineries include
- Ft. Saskatchewan (Shell)
- Edmonton (PetroCanada)
- Edmonton (Imperial Oil)

DID YOU KNOW ?

Sulfur Compounds

There are many organic compounds that contain sulfur. They have traditionally been identified by the use of suffixes:
thiol indicates R—S—H
thio represents R=S
sulfide communicates R—S—R
(R represents a hydrocarbon chain.)

Practice

16. There are many possible sulfur-bearing organic compounds in gasoline. Write a condensed structural formula equation to communicate the hydrogenation (hydrotreating) of sulfur to hydrogen sulfide for the following sulfur compounds. (You are not required to name organic compounds of sulfur.) For example, octanethiol:
$CH_3—(CH_2)_7—SH(l) + H_2(g) \rightarrow H_2S(g) + CH_3—(CH_2)_6—CH_3(l)$
    (a) $S=CH—CH_2—CH(C_6H_5)—CH_3$
    (b) methylphenylsulfide, $CH_3SC_6H_5$

17. Researchers have predicted that the price of gasoline must increase by 1–3 ¢/L in order to finance the further refining of gasoline to decrease the sulfur content to 30 ppm.
    (a) Would you be willing to pay the extra price? Why or why not?
    (b) What alternatives do we have to using sulfur-containing gasoline for fueling our transportation needs?

INVESTIGATION 9.4 Introduction

Bitumen from Oil Sands

In 1920, Canadian Karl Clark invented the hot water and caustic soda, NaOH(s), recovery process for extracting bitumen from oil sands. In 1923, Clark and associates at the Alberta Research Council set up a pilot project in the basement of the power plant at the University of Alberta. After decades of work, the first economically feasible world-scale plant for the extraction of bitumen from oil sands was opened in 1967 near Fort McMurray by what is now Suncor Energy.

Recent advances in bitumen extraction have called into question the need for caustic soda. This investigation involves testing the old process.

Purpose

The purpose of this investigation is to test the need for caustic soda in the extraction of bitumen from oil sands.

Report Checklist

- ○ Purpose
- ○ Problem
- ○ Hypothesis
- ○ Prediction
- ○ Design
- ○ Materials
- ○ Procedure
- ● Evidence
- ● Analysis
- ● Evaluation (3)

Problem

What is the percentage, by mass, of bitumen in an oil sands sample?

Design

The volume of caustic soda solution added to a hot water and oil sands mixture is systematically varied. Each research group uses a different volume of caustic soda and reports the percentage mass of bitumen obtained to the other groups.

To perform this investigation, turn to page 403.

INVESTIGATION 9.5 Introduction

Solvent Extraction (Demonstration)

Solvent extraction of bitumen from oil sands has significant importance for the development of the oil sands chemical industry in Alberta. Some in-situ bitumen extraction processes use a solvent. Most, if not all, of the surface mining bitumen extractions use a solvent to assist with bitumen extraction during the hot-water extraction process. This investigation asks you to predict which of water, propan-2-ol (an alcohol), or naphtha (a distillation fraction) is the best solvent for bitumen.

Purpose

The purpose of this investigation is to test solubility theory to predict which solvents can be used to dissolve bitumen.

Report Checklist

- ○ Purpose
- ○ Problem
- ○ Hypothesis
- ● Prediction
- ● Design
- ● Materials
- ● Procedure
- ● Evidence
- ● Analysis
- ● Evaluation (1, 2, 3)

Problem 1

Which of water, propan-2-ol, or naphtha is the best solvent for extracting bitumen from oil sands?

Problem 2

Which of water, propan-2-ol, or naphtha is the best solvent for extracting bitumen from water?

To perform this investigation, turn to page 404.

Section **9.5**

Case Study

The Athabasca Oil Sands

World technology at present is utterly dependent on energy from burning fuels derived from petroleum. We know that oil resources are finite, and that oil production must, inevitably at some point, reach a peak and then decline. In the continental United States (the world's biggest oil consumer), production peaked in 1970. The world's production peak is predicted at various times by different specialists in this area, but few predict it later than the mid-21st century, and some argue that it is already here. Alberta has a huge and growing interest in this area because Alberta's oil reserves are very large.

The special case in Alberta is that the reserves are mostly not liquid crude oil, but rather bitumen—a tarry residue coating the sand grains of the immense Athabasca and other oil sand deposits (**Figure 9**). Mining oil sand and then separating bitumen from the sand (**Figure 10**) is a huge operation, involving equipment and processes on a massive scale (**Figure 11**).

Bitumen Extraction

As indicated in Investigation 9.4, Karl Clark used controlled experiments to develop the caustic hot water process for bitumen extraction from oil sands. He and others found that oil sand is typically 84% sand, 12% bitumen, and 4% water. Further research has shown that the sand particles are surrounded by a thin layer of water and then a thicker layer of bitumen (**Figure 12**).

Experience shows that if the oil sand dries out, the hot water process for extracting bitumen from the sand does not work: The water layer disappears and the bitumen sticks very tightly to the sand.

A current technology is to have the initial separation occur in a hydrotransport pipeline that simultaneously carries the oil sands and caustic hot water, and separates the bitumen from the sand. At the plant, the mixture is physically separated by density in a large conical vessel into bitumen froth, water, and sand. The bitumen froth, created by blowing air into the mixture, floats to the top and is skimmed off.

Figure 9
Bitumen permeates this sample of "oil sand."

Figure 10
Chemical engineering is required to extract bitumen from oil sands (left) and then convert bitumen (right) into synthetic crude oil.

Figure 11
The complex processes in this refinery eventually produce a valuable "synthetic" crude oil product.

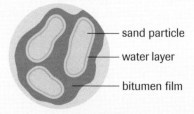

Figure 12
Close examination of oil sand shows a sand particle surrounded by water and bitumen.

Hydrocarbons from Petroleum

Many technological advances for bitumen extraction have occurred through research over the last couple of decades, and many more creative advances for bitumen extraction and upgrading are likely to occur in the next couple of decades.

The transforming of bitumen into useful liquid synthetic crude oil, however, is where the chemistry of organic reactions comes into play, in a reaction system that has great importance for Alberta, Canada, and the world. Bitumen is about 20% alkanes and cycloalkanes in the form of very complex branched chains; and about 80% aromatic compounds with multiple rings, forming many complex structures, from rather small to extremely large molecular sizes. Converting the bitumen into a useful product begins either with a process called coking or with hydrocracking. Both of these processes increase the hydrogen-to-carbon (H/C) ratio of bitumen. Coking removes carbon and hydrocracking adds hydrogen.

Bitumen Upgrading: Coking

Coking involves spraying bitumen onto a bed of hot (500 °C) coke particles. Coke is a material like charcoal, with a very high carbon content. Coking vaporizes the smaller molecular substances in bitumen, and causes thermal cracking in the larger molecular substances. (No catalyst is present, so the cracking is caused by high-temperature collisions.) The cracking produces many new substances that then vaporize. The process also causes some of the substances in the original bitumen to convert to solid, granular coke. When the hot vapour leaves the coking vessel and is cooled, much of it condenses to liquid that can be separated into fractions, just as is done with crude oil.

Bitumen Upgrading: Hydrocracking

Molecules in a sample of bitumen have a molar mass of 500–800 g/mol and an H/C ratio of less than one. Similar to coking, hydrocracking breaks the large aromatic molecules in bitumen into smaller aromatic and aliphatic molecules (increasing the H/C ratio). The aromatic character of bitumen results in a greater aromatic character to synthetic crude oil compared to conventional crude oil. Hydrocracking uses a catalyst to increase the rate of the cracking reaction at a lower temperature. As the name, hydrocracking, suggests, hydrogen is also a reactant—to increase the H/C ratio, convert unsaturated hydrocarbons to saturated hydrocarbons, and to remove, for example, nitrogen and sulfur impurities.

Bitumen Upgrading: Hydrotreating

Hydrotreating is the process of causing reactions of the organic substances in partially upgraded (cracked) bitumen with hydrogen, at high temperature and pressure. The reactions remove impurities, particularly nitrogen and sulfur, which cause problems with the uses for synthetic crude oil. The hydrogenation reactions also convert any double or triple carbon–carbon bonds in the liquids to single bonds, to make the molecules more stable (less reactive).

Case Study Questions

1. Use solubility theory to explain
    (a) the natural existence of oil sands as separate layers of bitumen and then water next to the sand
    (b) the reason for choosing water as the liquid for extracting bitumen from oil sands
    (c) the difficulty of separating bitumen from sand when oil sand is left standing to dry out
    (d) the need for a hydrocarbon solvent for extracting bitumen from oil sand that has been dried by our sun

2. What is the H/C ratio of the following compounds?
    (a) naphthalene in bitumen (**Figure 13**)
    (b) methylbenzene in gasoline
    (c) heptane in naphtha (camping fuel)

Figure 13
The molecular structure of naphthalene, a small molecule in bitumen

3. During coking, a naphthalene, $C_{10}H_8(l)$, molecule (Figure 13) in a bitumen sample is thermally cracked into a hexane molecule and a but-2-ene molecule as products.
    (a) Write and balance an equation for this reaction, using any structural formula to represent each of the substances involved.
    (b) What is the H/C ratio for each of naphthalene, hexane, and but-2-ene?

4. Consider the hydrotreating of this organic substance:
    $H_2NCH_2CSCH_2COOH + \_\_\_ H_2(g) \rightarrow$
    (a) Complete a balanced chemical equation to produce the most common hydrogen compounds of N, S, and O and a saturated hydrocarbon.
    (b) Why is more hydrotreating needed after coking than after hydrocracking?

5. For each of the following reactions of one mole of reactant, predict the product, including its structural formula and name, and state the chemical amount (in moles) of hydrogen required for complete reaction.
    (a) $CH_2CHCHCH_2 + \_\_\_ H_2(g) \rightarrow$
    (b) $HCCCH_2CH_2C_6H_5 + \_\_\_ H_2(g) \rightarrow$

6. The oil sands are a huge source of oil for Alberta, Canada, and the world. The development of this resource will cost billions of dollars and cause environmental damage. Why are we developing the oil sands? Answer, pro and con, from a variety of perspectives and add these statements to your Exploration issue concerning the resolution that the burning of fossil fuels for heat and transportation should be significantly reduced.

7. Fly through the whole process, from the initial mining of the oil sands to the final products, in this video. Prepare a flowchart to describe the oil sands process.

www.science.nelson.com

Section 9.5 Questions

1. (a) What physical property of hydrocarbons is used for fractionation?
   (b) What chemical theory explains the difference in the physical property used for fractionation?

2. Using Table 1 (page 387), identify all products that you have used and classify them as essential, useful, convenient, or a luxury.

3. (a) Why does crude oil have to be chemically processed, in addition to being fractionated?
   (b) How would our technologies and/or our society be different without the invention of these processes?

4. The conceptual understanding of any phenomenon in science usually starts with the creation of a classification system. Classify and write condensed structural formula equations for each of the following organic reactions. Catalysts are involved in all reactions other than combustion.
   (a) ethane → ethene + hydrogen
   (b) hexane → benzene + hydrogen
   (c) 4,4-dimethylpent-2-yne + hydrogen → propene + methylpropane
   (d) methylbenzene + oxygen →
   (e) nonane → 1-ethyl-3-methylbenzene + hydrogen
   (f) $CH_3(CH_2)_{16}CH_3$ + hydrogen → nonane + octane + methane

5. Predict a possible structural equation for the following reaction classes and reactants. There are many correct answers.
   (a) combustion of decane
   (b) hydrocracking of decane
   (c) catalytic reforming of decane

6. What chemical processes are used for the upgrading of bitumen?

7. What is the chemical difference between
   (a) bitumen and crude oil?
   (b) synthetic and conventional crude oil?

8. Evidence gathered in the laboratory by analytical chemists indicates that the molecules in bitumen have a hydrogen-to-carbon ratio of 1:1 or less. The interpretation placed on this evidence is that the molecules are composed of multiple benzene rings (**Figure 14**).

(a)

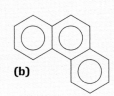
(b)

Figure 14
Anthracene (a) and phenanthrene (b) are a couple of the smaller aromatic compounds found in bitumen and coal.

   (a) Write the molecular formula for each molecule in Figure 14.
   (b) What is the H/C ratio for each of these aromatic compounds?
   (c) Propose a chemical reaction equation for the hydrocracking of anthracene.

9. In terms of molecules, compare low-octane and high-octane gasoline. What technological processes did engineers invent to convert low-octane gasoline into high-octane gasoline?

10. Only 10–25% of the crude oil in an oil well is initially recovered using current technology. 95% of the recovered oil is burned, and only 5% is used to make useful products, such as pharmaceuticals and plastics. Some people argue that it is morally wrong for our generation to be burning this valuable chemical resource rather than giving future generations the opportunity to extract more of it using more advanced technology. Use this information (and other information from this section) to add pro and con perspective statements to the Evidence section of your Exploration (STS issue) report from page 357.

Extension

11. Compare the following processes:
    (a) thermal cracking and catalytic cracking
    (b) hydrocracking and catalytic reforming
    (c) hydrocracking and hydrotreating
    (d) catalytic reforming and alkylation

12. In addition to alkanes, cracking reactions may also involve alkenes, alkynes, and aromatics. For each of the following reactions, draw a structural formula equation. Include all reactants and products.
    (a) but-1-ene → ethyne + ethane
    (b) 3-methylheptane → but-2-ene + butane
    (c) 3-methylheptane → propene + 2-methylbut-1-ene + hydrogen
    (d) propylbenzene → benzene + propene

13. For each of the following alkylation reactions in gasoline refining (octane boosting), draw a structural formula or write the IUPAC name for each reactant and product.
    (a) 2-methylpent-1-ene → 2,3-dimethylbut-1-ene
    (b)

9.6 Complete and Incomplete Combustion Reactions

When organic compounds undergo combustion reactions (burning), they release large quantities of energy (mainly heat and light) and chemical products, including carbon dioxide, carbon monoxide, particles of solid carbon (soot), and water vapour. In general, organic compounds may undergo two forms of combustion: complete combustion and incomplete combustion. Both of these reactions are exothermic.

Complete Combustion

In complete combustion, a hydrocarbon (fuel) reacts with oxygen to produce carbon dioxide and water vapour as the only chemical products. For example, the complete combustion of 2,2,4-trimethylpentane, a component of gasoline, may be communicated in the form of a word equation, a balanced chemical equation, and a structural formula equation as follows:

2,2,4-trimethylpentane + oxygen → carbon dioxide + water vapour + energy

$2\ C_8H_{18}(l) + 25\ O_2(g) \rightarrow 16\ CO_2(g) + 18\ H_2O(g)$

DID YOU KNOW?
Natural Gas Consumption
Chemical industries consume about half of the natural gas in Alberta. The natural gas serves as an energy source or as a feedstock (raw material) for the industries.

Figure 1
A yellow flame is evidence of incomplete combustion—in a candle, camp fire, laboratory burner, barbecue, or furnace. Soot and/or poisonous carbon monoxide gas are likely products, as evidenced by soot deposits on objects that are passed through the flame.

Incomplete Combustion

Unlike complete combustion, the incomplete combustion of an organic compound may include reactions that produce carbon monoxide and soot or any combination of carbon dioxide, carbon monoxide, and carbon (soot), in addition to water and energy (**Figure 1**). Using 2,2,4-trimethylpentane as an example, we can communicate two possible incomplete combustion reactions in the form of word equations and balanced chemical equations as follows:

2,2,4-trimethylpentane + oxygen → carbon monoxide + water vapour
$2\ C_8H_{18}(l) + 17\ O_2(g) \rightarrow 16\ CO(g) + 18\ H_2O(g)$

2,2,4-trimethylpentane + oxygen → carbon + water vapour
$2\ C_8H_{18}(l) + 9\ O_2(g) \rightarrow 16\ C(s) + 18\ H_2O(g)$

▶ COMMUNICATION example

For dimethylpropane, write balanced chemical equations for complete combustion, and for incomplete combustion to produce carbon monoxide.

Solution
Complete combustion: $C_5H_{12}(l) + 8\ O_2(g) \rightarrow 5\ CO_2(g) + 6\ H_2O(g)$

Incomplete combustion: $2\ C_5H_{12}(l) + 11\ O_2(g) \rightarrow 10\ CO(g) + 12\ H_2O(g)$

Practice

1. Write chemical and structural formula equations for complete and one possible incomplete combustion reaction of butane (in a lighter).
2. When automobiles burn gasoline, combustion is incomplete. Complete combustion produces carbon dioxide, a greenhouse gas. Incomplete combustion produces carbon monoxide, a toxic gas that becomes part of the airborne "chemical soup." Write a balanced chemical equation to represent one possible incomplete combustion of each of the following gasoline components to carbon monoxide, carbon dioxide, and water vapour.
   (a) 3-methylhexane (b) benzene (c) cyclohexane

Evidence from combustion studies indicates that all three types of combustion reactions occur simultaneously when an organic compound is burned, albeit not in equal proportions. Comparing the equations for complete and incomplete combustion of 2,2,4-trimethylpentane, you will notice that the fuel-to-oxygen mole ratio decreases from a high of 2:25 in complete combustion, to 2:17 in the reaction producing carbon monoxide, to 2:9 in the reaction producing soot. Thus, increasing the amount of oxygen available during combustion may increase the proportion of complete combustion reactions that occur. Even an excess of oxygen does not guarantee that only complete combustion will take place. The larger the excess of oxygen available during combustion, however, the smaller the amount of carbon monoxide and soot that is produced.

Carbon monoxide is a toxic gas that can reach dangerous levels in a confined area (such as in a house, a car, and even in a city). Reducing carbon monoxide emissions from oil and gas furnaces, gas stoves, and vehicles helps keep atmospheric concentrations of carbon monoxide low. In some places, alcohol is added to gasoline to help reduce carbon monoxide emissions. When added to gasoline, alcohols are often called *oxygenators* because they provide additional oxygen to the combustion reaction. Oxygenators make the combustion more complete, as evidenced by the reduced quantity of carbon monoxide and soot in the exhaust of automobiles that have been tested. In fact, some jurisdictions legislate the percentage of oxygenator that must be included in gasoline.

The following two equations allow you to compare the combustion of ethane and ethanol. Note the smaller chemical amount of oxygen required to burn ethanol:

$$C_2H_6(g) + \tfrac{7}{2} O_2(g) \rightarrow 2\,CO_2(g) + 3\,H_2O(g)$$

$$C_2H_5OH(l) + 3\,O_2(g) \rightarrow 2\,CO_2(g) + 3\,H_2O(g)$$

DID YOU KNOW?

Ethanol in Gasoline

In the 1920s and 1930s, cars could be converted to run on gasoline, ethanol (from corn), or a 25% blend of ethanol with gasoline. In 2002, about 620 000 m³ of corn were used to produce ethanol for blending with gasoline in hundreds of gas stations in Canada. The importance of ethanol has returned! It takes approximately 1.5 L of ethanol to deliver the same energy as 1.0 L of regular gasoline. The percentage of carbon monoxide emissions from burning ethanol or ethanol-blended gasoline, however, is significantly reduced. Despite this drawback, some provinces now have legislation that calls for mandatory blending of ethanol in gasoline.

INVESTIGATION 9.6 Introduction

Complete and Incomplete Combustion

Complete combustion has several benefits over incomplete combustion. Even complete combustion can create problems for the environment, however. In this investigation, you have an opportunity to start a research program (a major experimental design) on improving combustion. This is an open entry investigation. You create it and then possibly carry it out.

Purpose

The purpose of this investigation is to create a research program to improve combustion of, for example, candle wax.

To perform this investigation, turn to page 404.

Report Checklist

- ○ Purpose
- ○ Problem
- ○ Hypothesis
- ○ Prediction
- ● Design
- ● Materials
- ● Procedure
- ● Evidence
- ● Analysis
- ● Evaluation (1, 3)

Problem

How can the combustion of a hydrocarbon be improved?

SUMMARY: Complete and Incomplete Combustion

Complete Combustion

hydrocarbon + (excess) $O_2(g) \rightarrow CO_2(g) + H_2O(g)$

Incomplete Combustion

hydrocarbon + (insufficient) $O_2(g) \rightarrow xC(s) + yCO(g) + zCO_2(g) + H_2O(g)$
(The ratio of $x:y:z$ largely depends on the proportion of oxygen available.)

Section 9.6 Questions

1. The primary reaction of the components of natural gas is combustion with oxygen from the air. The primary product of these combustion reactions is energy. List three chemical by-products of this energy-producing reaction.

2. Write molecular and condensed structural formula equations for the combustion of pentane (in winter gasoline). Assume complete combustion.

3. Evaluate, with reasoning, the statement: "Fossil fuels must contain significant quantities of nitrogen because nitrogen oxides are found in the exhaust of cars and trucks."

4. Combustion is not always complete, usually because of an insufficient supply of oxygen. Write a hypothesis, including a chemical equation, to explain the following events.
   (a) When ethyne (acetylene) is burned in air, black soot appears.
   (b) Deaths have occurred when charcoal briquettes (assume pure carbon) are burned inside a tent to keep campers warm on a cold night.
   (c) Deaths have occurred when people fall asleep in the back seat of a car that has a leaky muffler.

5. Gasoline is not a mixture of pure hydrocarbons. Some of the hydrocarbon molecules are contaminated with nitrogen, sulfur, and oxygen atoms. Air pollutants, such as nitrogen dioxide and sulfur dioxide, are produced from the combustion of these molecules. Write unbalanced chemical equations to represent the complete combustion of
   (a) $C_6H_5NO(l)$ (b) $C_4H_8S(l)$ (c) $NHC_3H_4S(l)$

6. Canadians are among the highest producers of greenhouse gases in the world; for example, the average Albertan produces five tonnes per annum of greenhouse gases. The One-Tonne Challenge (OTC) to each Canadian is to reduce his or her greenhouse gas production by the equivalent of one tonne of carbon dioxide per year. Write pro and/or con statements from at least four perspectives concerning the resolution that GHGs should be reduced.

7. Use the information in this section to add to your list of pro and con perspective statements regarding your resolution on reducing fossil fuel consumption.

Extension

8. Evidence-based reasoning and scientific concept-based reasoning are goals of science and of science education. There are many examples of claims that gather a following even though they are based on anecdotal evidence and concepts that have not gained acceptance by the scientific community. Use the links provided to start an investigation into claims of spontaneous human combustion. What is your evaluation, based on the evidence and on accepted scientific concepts?

   www.science.nelson.com

9. Chemists communicate their research through peer-reviewed (refereed) journals. (See Appendix B.4.) Other chemists who do research on the same topic search for and read the abstract to see if they want to read the full research report. Read the abstract provided by the link below. What fuel was added to the natural gas and what effect did it have on the carbon dioxide emission?

   www.science.nelson.com

Chapter 9 INVESTIGATIONS

INVESTIGATION 9.1

Classifying Fossil Fuels

Report Checklist

- ○ Purpose
- ○ Problem
- ○ Hypothesis
- ○ Prediction
- ○ Design
- ○ Materials
- ● Procedure
- ● Evidence
- ● Analysis
- ● Evaluation (1, 3)

Alberta has many different fossil fuels. Samples of fossil fuels can be described and classified in terms of their physical properties of colour, transparency, viscosity, and density. In this investigation, the focus is on density.

Purpose
The purpose of this investigation is to use the known properties of fossil fuels to analyze the provided samples.

Problem
What are the classes of the fossil fuel samples provided?

Design
Equal-volume samples of fossil fuels are provided in sealed containers of equal mass. The mass of each sample, including the container, is determined. The fossil fuel samples are classified based upon their densities.

Materials
lab apron eye protection
equal volume samples, labelled A to G, of
  heavy oil natural gas
  light oil medium oil
  coalbed methane bitumen
  coal

Table 1 Densities of Fossil Fuels

Fossil fuel class	Density (g/cm³)
methane	0.0007165
light crude oil	0.800–0.850
medium crude oil	0.875–0.900
heavy crude oil	0.935–1.000
bitumen	>1.000–1.075
coal	1.3–>2.0

INVESTIGATION 9.2

Structures and Properties of Isomers

Report Checklist

- ○ Purpose
- ○ Problem
- ○ Hypothesis
- ○ Prediction
- ○ Design
- ○ Materials
- ○ Procedure
- ● Evidence
- ● Analysis
- ● Evaluation (3)

Molecular models are visual representations of molecules that help us to understand some of the properties and structures of molecules. In addition to building models, this investigation asks you to consult references to obtain data gathered by chemists over many years of laboratory work.

Purpose
The purpose of this investigation is to use models and a reference, respectively, to examine the structures and physical properties of some isomers of unsaturated hydrocarbons.

Problem
What are the structures and physical properties of the isomers of C_4H_8 and C_4H_6?

Design
Structures of possible isomers are determined by means of a molecular model kit. Once each structure is named, the boiling and melting points are obtained from a reference such as the *CRC Handbook of Chemistry and Physics* or *The Merck Index*.

 www.science.nelson.com GO

Materials
molecular model kits chemical reference

Procedure

1. Use the required "atoms" to make a model of C_4H_8.
2. Draw a structural formula of the model and write the IUPAC name for the structure.
3. Rearrange bonds to produce models for all other isomers of C_4H_8, including cyclic structures. Draw structural formulas and write the IUPAC name for each structure before disassembling the models.

Hydrocarbons from Petroleum 401

INVESTIGATION 9.2 continued

4. If you construct a model that contains a double C–C bond, test the restricted rotation of groups about the bond axis and the bond length.

5. Repeat steps 1 to 4 for C_4H_6.
6. In a reference, find the melting point and the boiling point of each of the compounds you identified.

INVESTIGATION 9.3

Fractional Distillation (Demonstration)

Report Checklist
- ○ Purpose
- ○ Problem
- ○ Hypothesis
- ○ Prediction
- ○ Design
- ○ Materials
- ○ Procedure
- ● Evidence
- ● Analysis
- ○ Evaluation

Fractional distillation is a process used commercially to separate molecular components of petroleum using boiling point differences. The same technique can be used to separate components of most liquid mixtures, provided a significant boiling point difference exists. In this demonstration, the vapours of the boiling mixture rise into a fractionating column, where initially they all condense and fall back into a flask. When the temperature in the column rises sufficiently, vapours of the component with the lower boiling point pass through the column and enter a condenser, which cools and condenses the component back to a liquid for collection. The component with the higher boiling point still condenses in the fractionating column, which separates the two liquids.

Purpose
The purpose of this investigation is to use the concept and technique of fractional distillation to separate two hydrocarbon components from a liquid mixture.

Problem
What is the percent by volume of each of the two hydrocarbon components?

Design
A mixture of the two liquids is heated in a fractional distillation apparatus while the temperature is measured at regular intervals in a fume hood.

Materials
lab apron
eye protection
round-bottom distillation flask
electric heating mantle
2 small collecting flasks
stopwatch
fume hood
thermometer or temperature probe to fit column stopper
pentane, C_5H_{12}(l)
2-methylpropan-2-ol, $CH_3C(CH_3)OHCH_3$(l)
fractionating column
stand and clamp
50 mL graduated cylinder
large 400–600 mL beaker
condenser with tubing and fittings

Procedure

1. Obtain a 50 mL sample of the mixture of unknown composition.
2. Heat the flask slowly, taking the vapour temperature at regular 30 s intervals.
3. After most of the pentane has boiled off, when the column temperature rises noticeably, change collection flasks at the outflow of the condenser (**Figure 1**).
4. Measure the volume of pentane collected.
5. As soon as most of the alcohol has boiled off, remove the heating mantle.
6. Dispose of the organic chemicals into an organic waste container.

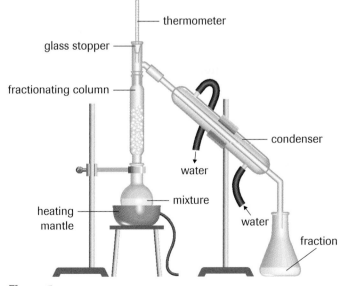

Figure 1
A fractional distillation apparatus

INVESTIGATION 9.4

Bitumen from Oil Sands

Report Checklist
- ○ Purpose
- ○ Problem
- ○ Hypothesis
- ○ Prediction
- ○ Design
- ○ Materials
- ○ Procedure
- ● Evidence
- ● Analysis
- ● Evaluation (1, 3)

In 1920, Canadian Karl Clark invented the hot water and caustic soda, NaOH(s), recovery process for extracting bitumen from oil sands. In 1923, Clark and associates at the Alberta Research Council set up a pilot project in the basement of the power plant at the University of Alberta. After decades of work, the first economically feasible world-scale plant for the extraction of bitumen from oil sands was opened in 1967 near Fort McMurray by what is now Suncor Energy.

Surface mining of oil sands allows for over 90% recovery of the bitumen. This percentage compares very favourably with in-situ oil sands at 25-75%, conventional heavy oil at less than 20%, and conventional light oil at about 30%. Fortunately for the environment, over 80% of the oil sands will have to be mined in-situ rather than from a surface mine. Unfortunately, the recovery rate of in-situ operations is significantly lower using current technologies.

Recent advances in bitumen extraction have called into question the need for caustic soda. This investigation involves testing the effects of using various amounts of caustic soda (sodium hydroxide).

Purpose
The purpose of this investigation is to test the need for caustic soda in the extraction of bitumen from oil sands.

Problem
What is the percentage, by mass, of bitumen in an oil sands sample?

Design
The volume of caustic soda solution added to a hot water and oil sands mixture is systematically varied. Each research group uses a different volume of caustic soda and reports the percentage mass of bitumen obtained to the other groups.

Materials
eye protection
lab apron
ring stand
wire gauze
250 mL beaker
400 mL beaker
ring clamp
laboratory burner or hot plate
beaker tongs
dropper bottle of 1.0 mol/L NaOH(aq)
small disposable plastic spoon
100 g of oil sands
laboratory balance
disposable clear plastic container
paper towel

 Sodium hydroxide is corrosive. Avoid skin and eye contact.

Procedure
1. Add about 250 mL of tap water to the 400 mL beaker.
2. Add 0, 4, 8, and 12 (as assigned to your research group) drops of 1.0 mol/L sodium hydroxide solution to the water.
3. Heat the water to boiling.
4. Place the disposable plastic container into a paper-towel-lined 250 mL beaker (or equivalent).
5. Add a measured mass of oil sands to the disposable container and set the container on a paper towel.
6. Use the beaker tongs to pour water onto the oil sands to make a slurry.
7. Use the plastic spoon to stir the slurry for about 5 min.
8. Use the beaker tongs again to add about 100 mL more hot water to nearly fill the container.
9. Allow the contents of the container to settle.
10. Measure the mass of a double layer of paper towel.
11. Use the plastic spoon to remove the bitumen from the container onto a double layer of paper towel.
12. Dry and measure the mass of the bitumen plus paper towel.

INVESTIGATION 9.5

Solvent Extraction (Demonstration)

Report Checklist
- ○ Purpose
- ○ Problem
- ○ Hypothesis
- ○ Prediction
- ● Design
- ● Materials
- ● Procedure
- ● Evidence
- ● Analysis
- ● Evaluation (1, 2, 3)

Solvent extraction of bitumen from oil sands has significant importance for the development of the oil sands chemical industry in Alberta. Some in-situ bitumen extraction processes use a solvent. Most, if not all, of the surface mining bitumen extractions use a solvent to assist with bitumen extraction during the hot-water extraction process. This investigation asks you to predict which of water, propan-2-ol (an alcohol), or naphtha (a distillation fraction) is the best solvent for bitumen.

Purpose
The purpose of this investigation is to test solubility theory to predict which solvents can be used to dissolve bitumen.

Problem 1
Which of water, propan-2-ol, or naphtha is the best solvent for extracting bitumen from oil sands?

Problem 2
Which of water, propan-2-ol, or naphtha is the best solvent for extracting bitumen from water?

INVESTIGATION 9.6

Complete and Incomplete Combustion

Report Checklist
- ○ Purpose
- ○ Problem
- ○ Hypothesis
- ○ Prediction
- ● Design
- ● Materials
- ● Procedure
- ● Evidence
- ● Analysis
- ● Evaluation (1, 3)

Complete combustion has several benefits over incomplete combustion. Even complete combustion can create problems for the environment, however. In this investigation, you have an opportunity to start a research program (a major experimental design) on improving combustion. This is an open entry investigation. You create it and then possibly carry it out.

Purpose
The purpose of this investigation is to create a research program to improve the combustion of, for example, candle wax.

Problem
How can the combustion of a hydrocarbon be improved?

Chapter 9 SUMMARY

Outcomes

Knowledge

- define organic compounds, recognizing inorganic exceptions (9.1)
- identify and describe organic compounds in everyday life, as well as their origins and applications (all sections)
- name and draw structures for saturated and unsaturated aliphatic and aromatic hydrocarbons (9.2 to 9.6)
- classify organic compounds from their structural formulas (9.2 to 9.5)
- define and use the concept of structural isomerism and relate to properties of isomers (9.2 to 9.4)
- compare boiling points and solubility of organic compounds (9.2 to 9.4)
- describe fractional distillation and solvent extraction (9.1, 9.2, 9.3, 9.5)
- define, provide examples of, predict products, and write and interpret balanced equations for combustion reactions (9.1, 9.2, 9.5, 9.6)
- describe major reactions for producing energy and economically important compounds from fossil fuels (all sections)

STS

- illustrate how science and technology are developed to meet societal needs and expand human capabilities (all sections)
- describe interactions of science, technology, and society (all sections)
- illustrate how science and technology have both intended and unintended consequences (9.1, 9.2, 9.5, 9.6)

Skills

- initiating and planning: describe procedures for safe handling, storing, and disposal of materials used in the laboratory, with reference to WHMIS and consumer product labelling information (9.5)
- performing and recording: separate a mixture of organic compounds based on boiling point differences (9.5); build molecular models depicting the structures of selected organic and inorganic compounds (9.3)
- analyzing and interpreting: follow IUPAC guidelines for naming and formulas and by compiling evidence to compare the properties of structural isomers (all sections); compile and organize data to compare the properties of structural isomers (9.3); investigate sources of greenhouse gases, that is, methane, carbon dioxide, water, and dinitrogen oxide (nitrous oxide) and the issue of climate change (all sections)
- communication and teamwork: work cooperatively in addressing problems and apply skills and conventions of science in communicating information and ideas and in assessing results by preparing reports on topics related to organic chemistry (all sections)

Key Terms

9.1
organic chemistry
hydrocarbon
refining

9.2
alkane
homologous series
saturated hydrocarbon
structural isomer
alkyl branch
cycloalkane

9.3
unsaturated compound
hydrogenation
alkene
alkyne
cycloalkene
cracking
aliphatic hydrocarbon

9.4
aromatic
phenyl group

9.5
fractionation
fraction
solvent extraction
hydrocracking
catalytic reforming

MAKE a summary

1. Create a concept map starting with "Fossil Fuels" in the centre of a page. Include all of the Key Terms, Key STS, and Key Skills, as well as any other information you think may be useful in studying for a test on this chapter.

2. Refer back to your answers to the Starting Points questions at the beginning of this chapter. How has your thinking changed?

Go To

The following components are available on the Nelson Web site. Follow the links for *Nelson Chemistry Alberta 20–30*.

- an interactive Self Quiz for Chapter 9
- additional Diploma Exam-style Review questions
- Illustrated Glossary
- additional IB-related material

There is more information on the Web site wherever you see the Go icon in this chapter.

Chapter 9 REVIEW

Many of these questions are in the style of the Diploma Exam. You will find guidance for writing Diploma Exams in Appendix H. Exam study tips and test-taking suggestions are on the Nelson Web site. Science Directing Words used in Diploma Exams are in bold type.

DO NOT WRITE IN THIS TEXTBOOK.

Part 1

1. Organic chemistry is defined as the study of
   A. all carbon compounds
   B. all carbon compounds except cyanide
   C. all carbon compounds except oxides and ionic compounds
   D. carbon-containing compounds produced by living organisms

2. Chemists have developed classification systems for the compounds they study. The empirical classification of hydrocarbons is based on
   A. physical properties only
   B. chemical properties only
   C. physical and chemical properties
   D. physical and chemical properties and empirically determined molecular formulas

3. Hydrocarbons that contain a benzene ring in their molecules are called
   A. aromatics
   B. alkanes
   C. alkenes
   D. alkynes

Use this information to answer questions 4 to 6.

1. hex-1-ene
2. hex-1-yne
3. 1,2-dimethylbenzene
4. hex-2-ene
5. 2-methylpent-2-ene
6. benzene
7. cyclohexane
8. cyclohexene
9. methylbenzene

4. In numerical order, the alkenes are
   NR ___ ___ ___ ___

5. In numerical order, the aromatics are
   NR ___ ___ ___ ___

6. In numerical order, the isomers of C_6H_{12} are
   NR ___ ___ ___ ___

7. The IUPAC name for the following hydrocarbon is

   $$CH_3 \qquad CH_3$$
   $$| \qquad\qquad |$$
   $$CH_2 - CH - CH - CH_3$$
   $$\qquad\qquad |$$
   $$\qquad\qquad CH_3$$

   A. 3,4-dimethylpentane
   B. 2,3-dimethylpentane
   C. 1,2,3-trimethylbutane
   D. 4,5,6-trimethylbutane

8. Differences in boiling points are the basis of
   A. catalytic cracking
   B. solvent extraction
   C. catalytic reforming
   D. fractional distillation

9. The natural gas used by consumers to heat their homes contains mostly
   A. methane
   B. ethane
   C. propane
   D. butane

10. For a related series of hydrocarbons, such as unbranched alkanes, a property that decreases as molar mass increases is
    A. boiling point
    B. melting point
    C. volatility
    D. viscosity

11. The change in boiling point of alkanes with increasing molar mass is primarily due to increasing
    A. London forces
    B. dipole–dipole forces
    C. covalent bonds
    D. hydrogen bonds

12. Separating the components of natural gas can involve all of the following processes *except*
    A. fractionator
    B. Claus converter
    C. condenser
    D. amine scrubber

13. The technology of cracking and reforming hydrocarbons was developed primarily to meet the societal demand for
    A. asphalt for paving highways
    B. lubricating oils for machines
    C. gasoline for automobiles
    D. propane for home heating

14. The chemical reaction that produces coke is
    A. combustion
    B. catalytic reforming
    C. catalytic cracking
    D. alkylation

15. Identify the missing product in the following catalytic cracking reaction: $C_{17}H_{36}(l) \rightarrow C_8H_{18}(l) + \underline{} + C(s)$
    A. $CH_3C(CH_3)_2(CH_2)_3CH_3(l)$
    B. $C_8H_{16}(l)$
    C. $C_8H_{18}(l)$
    D. $C_6H_5(CH_3)_2(l)$

Part 2

16. Write correct IUPAC names for the following structures:
    (a) $CH_3CH_2CH=CHCHCH_3$
          |
          CH_2CH_3
    (b)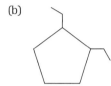
    (c)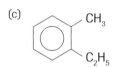

17. Draw a structural formula for each of the following compounds, and write the IUPAC name for each:
    (a) ethylene
    (b) propylene
    (c) acetylene
    (d) toluene, the toxic solvent used in many glues
    (e) the three isomers of xylene (dimethylbenzene), used in the synthesis of other organic compounds such as dyes

18. Using your knowledge of elements and compounds, **explain** why carbon forms more compounds than all other elements combined.

19. List some examples of carbon-containing compounds that are not classified as organic. What do these compounds have in common?

20. State the name and formula of the simplest member of each of the four hydrocarbon families and include one important use of each.

21. Can a molecular formula for an aliphatic hydrocarbon be used to predict the specific family to which it belongs? **Explain**, using examples and noting limitations.

22. **Compare** the physical processes of distillation and solvent extraction.

23. Draw a line structural formula for each hydrocarbon.
    (a) 2-methylnonane
    (b) methylcyclopentane
    (c) hex-3-yne
    (d) 1,2,4-trimethylbenzene
    (e) 4-propyloctane
    (f) 2-phenylpropane
    (g) 3-methylpent-2-ene
    (h) 1,4-diethylbenzene
    (i) 5-ethyl-2-methylhept-2-ene
    (j) 4-methylpent-2-yne

24. Classify each hydrocarbon in question 23 into its family of hydrocarbons and **identify** which ones are saturated and which ones are unsaturated.

25. Draw structural formulas and write the IUPAC names for all the isomers of C_4H_8.

26. In addition to alkanes, cracking reactions may also involve alkenes, alkynes, and aromatics. For each of the following reactions, draw a condensed structural formula equation. (You may draw a line structural formula for benzene.) Include all reactants and products.
    (a) but-1-ene $\rightarrow$ ethyne + ethane
    (b) 3-methylheptane $\rightarrow$ but-2-ene + butane
    (c) 3-methylheptane $\rightarrow$ propene + 2-methylbut-1-ene + hydrogen
    (d) propylbenzene $\rightarrow$ methylbenzene + ethene

27. Classify the reaction type and write condensed structural formula (except for benzene) equations for the following organic reactions.
    (a) ethane $\rightarrow$ ethene + hydrogen
    (b) but-2-ene + hydrogen $\rightarrow$ butane
    (c) 4,4-dimethylpent-2-yne + hydrogen $\rightarrow$ propene + methylpropane
    (d) methylbenzene + (excess) oxygen $\rightarrow$
    (e) decane $\rightarrow$ 2,2,4-trimethylheptane
    (f) octane $\rightarrow$ 1,3-dimethylbenzene
    (g) nonane $\rightarrow$ propane + pentane + carbon

28. Camping fuel is composed of alkanes with five or six carbon atoms per molecule. Name the following hydrocarbons found in this light naphtha fuel.
    (a) $CH_3-CH(CH_3)-CH_2-CH_3$
    (b) $CH_3-CH_2-CH_2-CH_2-CH_3$
    (c) $CH_3-CH(CH_3)-CH_2-CH_2-CH_3$
    (d) $(CH_3)_4C(l)$
    (e) $(CH_3)_2CHCH(CH_3)_2(l)$

29. Write IUPAC names for the following hydrocarbons.
    (a) $CH\equiv C-CH_2-CH_2-CH_3$
    (b) $CH_3-CH=CH-CH_2-CH_3$
    (c) $CH\equiv C-CH(CH_3)-CH_3$
    (d)

(e)

(f)

30. Coal is the most abundant fossil fuel in Alberta. The main use of coal in Alberta is to burn it to produce steam for driving turbines for turning electrical generators.
    (a) If bituminous coal has an empirical (simplest ratio) formula of $C_9H_6O(s)$, **predict** a balanced chemical equation for the complete combustion of this coal.
    (b) The sulfur and nitrogen content of coal from Alberta is relatively low, which puts Alberta coal in demand in Canada and beyond. If a coal molecule has a molecular formula of $C_{100}H_{40}O_{10}N_2S(s)$, **predict** a balanced chemical equation for the complete combustion of this molecule.

31. The safe storage and handling of hydrocarbon fuels requires knowledge of their boiling points. The trends in boiling points within a hydrocarbon family make it possible to predict boiling points that are unknown. The purpose of this activity is to **identify** the trend in the boiling points for each family, and then use the trend to **predict** the boiling point of the next member of the series. Complete the Prediction, Analysis, and Evaluation (2, 3) of the investigation report.

Problem
What are the boiling points of hexane, hex-1-ene, and hex-1-yne?

Design
For three families (classes) of aliphatic hydrocarbons, the boiling points are plotted versus the number of carbon atoms per molecule. The graphs are extrapolated to predict the boiling point of the next member of each family. The predictions are evaluated by looking up the measured values in the *CRC Handbook of Chemistry and Physics*.

Evidence
See **Table 1** below.

32. Copy and complete **Table 2**.

33. For each of the following compounds, write its structural formula, and state its origin and a technological application.
    (a) methane
    (b) ethene
    (c) ethyne
    (d) 2,2,4-trimethylpentane
    (e) 1,2-dimethylbenzene

34. Draw condensed structural formulas for and name the non-cyclic structural isomers of C_5H_{10}.

Table 1 Boiling Points of Aliphatic Hydrocarbons

Alkanes	Boiling point (°C)	Alkenes	Boiling point (°C)	Alkynes	Boiling point (°C)
ethane	−89	ethene	−104	ethyne	−84
propane	−42	propene	−47	propyne	−23
butane	−0.5	but-1-ene	−6.3	but-1-yne	8.1
pentane	36	pent-1-ene	30	pent-1-yne	40

Table 2 Some Important Hydrocarbons

IUPAC name	Condensed structural formula	Family/Class	Common use
	$CH_3-C \equiv C-H$	alkynes	cutting/welding torch
	$CH_2=CH_2$		petrochemical feedstock
propene			making plastics
methylpropane			butane lighters
2,3,3-trimethylpentane			gasoline
heptane		alkane	early anesthetic
methylbenzene			solvent

35. Write the IUPAC name of each of the following line structural formulas.
    (a)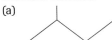
    (b)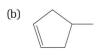
    (c)
    (d)
    (e)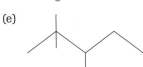
    (f)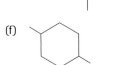

36. Complete the Design, Materials, and Procedure for the following investigation report.

    Purpose
    The purpose of this investigation is to use the properties of organic and inorganic chemicals to identify unknown chemicals.

    Problem
    Which of the chemicals provided is organic and which is inorganic?

37. Solvent extraction is used in chemical research and in the chemical industry to separate chemical components of a mixture. Researchers have found solvents that work best in specific situations. **Explain** the effectiveness of the solvent in each of the following processes.
    (a) Naphtha is used to extract bitumen from a water, sand, and bitumen mixture.
    (b) Toluene (methylbenzene) is used to separate asphaltenes (large aromatic molecules) from bitumen.
    (c) Water does not dissolve bitumen but is used to extract it from oil sands.

38. **Outline** the major processes used in the refining of crude oil. Your response should include
    - some details of the physical and chemical changes at each stage of the process
    - chemical reaction equations to illustrate some of the chemical changes
    - a consideration of the impact of crude oil refining from several perspectives

39. Mining and oil development, due to their extractive nature, inevitably leave an environmental footprint, even before drilling occurs. For example, in its search for oil in the Edmonton area, Anadarko Petroleum has tested seismic lines near the Kehewin Cree people's reserve. Despite the assertion by Anadarko officials that the band will be consulted before drilling, some of the Aboriginal residents say the search itself may ruin traditional hunting grounds. Research this issue or another similar dispute from the perspective of the energy company, a local Aboriginal band, and an environmentalist. On what specific issues do these groups agree and disagree?

    www.science.nelson.com

Extension

40. Some natural gas is sold without the hydrocarbon components being fractionated.
    (a) Write a balanced chemical equation for the burning of natural gas containing the smallest four alkanes (see Table 1, page 363). Include a realistic mole ratio of the hydrocarbons in the balanced chemical equation.
    (b) Why is it that more and more of the natural gas being burned is nearly pure methane, even though it is expensive to remove the other components?
    (c) Why do some importers, such as the U.S., insist that the natural gas *not* be fractionated?
    (d) At one time in the past, the second-largest hydrocarbon component of natural gas that was most difficult to remove was not removed. Identify the component, and speculate as to why it was not removed and why the situation has changed.

41. Do a literature/media/Internet search for information on methane trapped in the Arctic ice of Canada. Present your findings in an innovative manner.

42. Quite often in the media, you hear or read stories about "proven oil reserves." The most quoted source of this information is the U.S. Department of Energy. What is meant by the term "proven oil reserves"? Does the Middle East really have most of the oil in the world? What influence do politics and science have in determining oil reserves? Are the media giving a true picture of the world's oil reserves? Investigate these questions and present your findings, with illustrations and graphs, in a short article suitable for publication in a daily newspaper or on a Web page.

    www.science.nelson.com

43. Cracking and reforming of crude oil are very complex chemical processes. Read about catalytic reforming and then list some of the other kinds of processes involved, such as polymerization.

    www.science.nelson.com

44. In 1992, the International Institute for Sustainable Development (IISD) published the report *Our Responsibility to the Seventh Generation*. This report highlights the value of Indigenous knowledge and contributions to sustainable development. The report includes, for example, the quotation "We cannot simply think of our survival; each new generation is responsible to ensure the survival of the seventh generation."
    (a) What is the definition of sustainable development provided in this report?
    (b) In the Overview of the report starting on page 7, what are the one or two most prominent perspectives presented in the first three chapters? (See Appendix D.1.)

    www.science.nelson.com

chapter 10
Hydrocarbon Derivatives, Organic Reactions, and Petrochemicals

In this chapter

- Exploration: Burning Fossil Fuels
- Investigation 10.1: Substitution and Addition Reactions
- Lab Exercise 10.A: Synthesis of an Organic Halide
- Investigation 10.2: Isomers of Butanol
- Web Activity: Cellulosic Ethanol
- Lab Exercise 10.B: Explaining Physical Property Trends
- Biology Connection: Natural Esters
- Investigation 10.3: Synthesis of an Ester
- Investigation 10.4: Testing with Models
- Web Activity: Teflon—Healthy or Hazardous?
- Web Activity: Recycling Plastics
- Biology Connection: Correlation versus Cause and Effect
- Web Activity: Molecular Modelling
- Investigation 10.5: Preparing Nylon 6,10 (Demonstration)
- Explore an Issue: Natural or Artificial Polymers?
- Biology Connection: Glycogen
- Mini-Investigation: Starch and Cellulose
- Web Activity: Cellulose Acetate

Alberta's resources contribute to making this province one of the best places in the world to live and work. Alberta's resource wealth includes a diverse, well-educated population, an environment that brings tourists in all seasons, and petroleum reserves. You may not realize it, but all of these resources depend on organic chemicals and the reactions that this unique class of chemicals can undergo.

When the price of natural gas, crude oil, and gasoline increases, the price of almost everything else likewise increases—evidence for the strong connection between fossil fuels and many other products. Chemicals that are produced from petroleum are called *petrochemicals*. Petrochemicals include plastics, asphalt shingles, paints, dyes, cosmetics, antifreezes, lubricants, adhesives, carpets, pesticides, fertilizers, and pharmaceuticals.

Much of the petrochemical industry in Alberta is located near Sherwood Park, Fort Saskatchewan, and Red Deer. Both natural gas and crude oil are the raw materials that, when refined, supply the petrochemical feedstock. Ninety-five percent of refined crude oil and natural gas is burned. The other five percent is used as petrochemical *feedstock*. Feedstock is a variety of chemical substances used in the chemical processes to produce petrochemicals. The end products are in demand by industry, commerce, and consumers.

Some of the main petrochemical feedstocks derived from natural gas are methane, ethane, propane, and butane. The main feedstocks from crude oil are naphtha and gas oil. Organic reactions convert these hydrocarbon feedstocks to petrochemicals. In this chapter, you will learn how to describe, explain, and predict organic reactions. As with any undertaking, there are risks and benefits to the development of petrochemical industries in Alberta. Throughout this chapter, you are asked to adopt a multiperspective view and to gather information for and against this industrial development.

STARTING Points

Answer these questions as best you can with your current knowledge. Then, using the concepts and skills you have learned, you will revise your answers at the end of the chapter.

1. What reactions, other than combustion and cracking, do organic compounds undergo?
2. List some important consumer, commercial and/or industrial petrochemical products of organic reactions.
3. What are polymers? Give some specific examples.

Career Connection:
Chemical Engineer; Chemical Employee

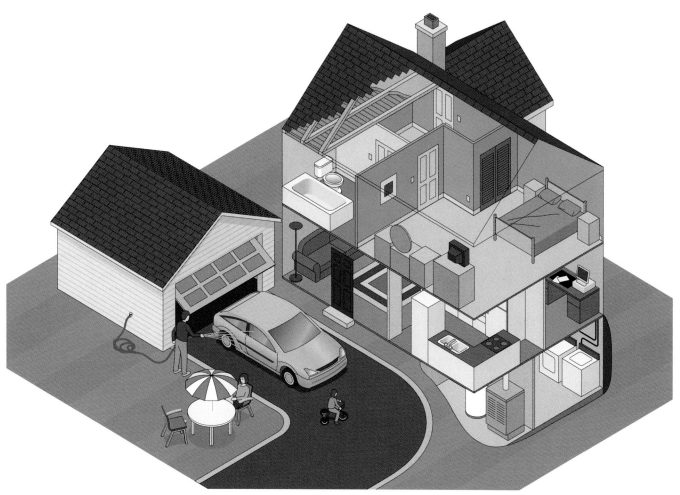

Figure 1
Chemicals from petroleum—petrochemicals—are everywhere around us. There are at least twenty different petrochemical products in this illustration, such as synthetic fabrics in bedding, countertops, rugs, vinyl flooring, garden hose, chairs, and siding. What are some other examples?

▶ Exploration *Burning Fossil Fuels*

Issue Checklist

○ Issue ● Design ● Analysis
○ Resolution ● Evidence ● Evaluation

This Exploration continues from Chapter 9. In this chapter (which looks at the production of petrochemicals from petroleum), the emphasis is on the second, boldfaced, statement below.

Scientists agree that fossil fuels are a nonrenewable, finite resource. Fossil fuels in Alberta include coal, natural gas, crude oil, oil sands, and heavy oil. Studies have indicated that about 95% of the fossil fuels used in Alberta are used for energy production, including natural gas and propane for home heating, and gasoline and diesel fuel for cars, trucks, and buses. The evidence indicates that about 5% of fossil fuels are converted into petrochemicals (such as plastics) through technologies based on chemical reactions. Two issues related to burning fossil fuels for energy production are

- whether we should be burning a very valuable resource at the rate that we are
- **whether we should be saving more of this resource for petrochemical use in the future**

Resolution: We should be saving more fossil fuels for future use as petrochemicals.
While progressing through Chapter 10, complete the Design, Evidence, Analysis, and Evaluation components of this second issue report. Specific to this chapter, the evidence includes a list of important organic reactions that produce valued petrochemicals.

Governments, industry, and citizens alike speak of sustainable development. The United Nations Commission on Environment and Development defines sustainable development as development that meets the needs of the present without compromising the ability of future generations to meet their own needs. Keep this concept in mind while completing this unit-long exploration.

10.1 Petrochemicals in Alberta

The petrochemical industry in Alberta is a secondary and tertiary industry that takes raw materials, such as crude oil and natural gas, and feedstocks, such as ethane, and converts them into value-added chemicals. This industry is an important economic and social addition to the fossil fuel, agriculture, and tourism industries of Alberta. The petrochemical industry in Alberta exports to other parts of Canada and the world. **Figure 1** depicts the raw materials, feedstock, and primary petrochemicals that start the chemical processes you will learn about in this chapter.

CAREER CONNECTION

Chemical Engineer
The chemical industry is Canada's most knowledge-based industry. About 30% of workers in the chemical industry have university degrees. The salaries in the chemical industry are, as expected, among the highest in Canada. What types of chemical industries employ chemical engineers?

www.science.nelson.com GO

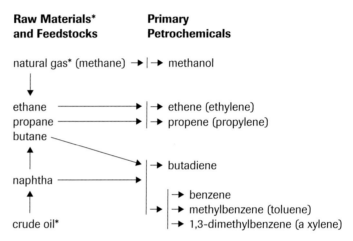

Figure 1
The sources of primary petrochemicals

Some of these primary petrochemicals have immediate uses, but most of them undergo further chemical reactions to produce chemical derivatives. For example, methanol has an immediate use as a fuel and as a windshield washer fluid and gas line antifreeze. Xylene is a chemical name that you find listed as an ingredient of solvents found in a hardware or lumber store. Ethene may be used directly for helping to ripen some varieties of fruits, but most of this primary petrochemical is used in chemical processes to produce polymers, such as polyethene.

DID YOU KNOW ?

A Keystone Industry
The chemical industry uses natural raw materials from the earth, water, and air to manufacture everyday products. The chemical industry produces more than 70 000 chemical products, including papers, pesticides, cleaning compounds, cosmetics, toiletries, paints, pharmaceuticals, and safe water.

> ### ▸ Practice
>
> 1. The following chemicals are feedstocks for a petrochemical industry. Draw full structural formulas for each.
>   (a) methane (c) propane
>   (b) ethane   (d) butane
> 2. Naphtha is a distillation fraction from crude, heavy, or synthetic oil.
>   (a) What range of hydrocarbons is expected to be found in naphtha?
>   (b) In what technological application(s) is naphtha burned?
> 3. The following chemicals are primary petrochemicals in a petrochemical industry. Draw line structural formulas for each:
>   (a) methanol   (e) benzene
>   (b) ethene   (f) methylbenzene
>   (c) propene   (g) 1,3-dimethylbenzene
>   (d) buta-1,3-diene ($CH_2CHCHCH_2(g)$)

Primary Petrochemicals

Chemical engineers use the primary petrochemicals in **Figure 1** to produce the intermediates and derivatives in **Figure 2**. Notice the extensive petrochemical industry that can be created from a few basic chemical building blocks—from primary petrochemicals obtained from crude oil and natural gas. Many new compounds can be made by adding other substituents (such as chlorine and oxygen), changing the bonding, or joining small molecules together. You will learn about polymers (listed mostly in the last column of Figure 2) in Section 10.5.

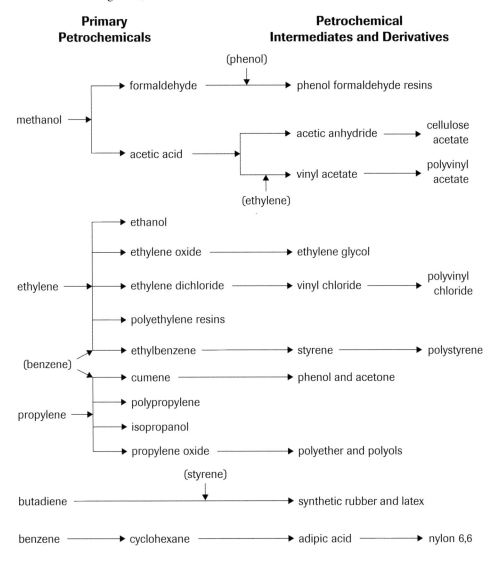

Figure 2
This figure, provided by industry, communicates primary petrochemicals, intermediates, and derivatives. You do not have to memorize this figure, or the common industrial names (except acetic acid, ethylene, and propylene). Most of these common names are translated into systematic IUPAC names in the exercise below.

DID YOU KNOW ?

Natural Gas Consumption
Chemical operators account for over half of the natural gas consumption in Alberta. This consumption includes the extraction and refining of fossil fuels, as well as the processing of refined fossil fuels into petrochemicals. As oil sands projects multiply, the consumption of natural gas by industry is likely to increase significantly. Most of the natural gas is used for heat and power. Only about 5% is converted directly into petrochemicals.

Practice

4. Draw structural formulas for the following list of intermediate, reagent, and derivative chemicals from **Figure 2**. Follow the bonding capacity of each element. The molecular formula is not given if you should know it. You might want to reconstruct Figure 2 in your notebook using full structural formulas. The chemicals marked with an asterisk (*) are ones that you will know by the end of this chapter.
   (a) formaldehyde is methanal, $HCHO(l)$
   (b) phenol is hydroxybenzene, $C_6H_5OH(s)$*
   (c) acetic acid is ethanoic acid, $CH_3COOH(l)$*
   (d) acetic anhydride is $(CH_3CO)_2O(l)$

(e) vinyl acetate is $CH_2CHOOCCH_3(l)$
(f) ethanol
(g) ethylene oxide is $(CH_2)_2O(g)$
(h) ethylene glycol is 1,2-dihydroxyethane, $CH_2OHCH_2OH(l)$*
(i) ethylene dichloride is 1,2-dichloroethane, $CH_2ClCH_2Cl(l)$
(j) vinyl chloride is chloroethene, $C_2H_3Cl(g)$*
(k) ethylbenzene (a full and line structural formula hybrid is okay for (k) and (l))
(l) styrene is vinylbenzene (phenylethene)*
(m) cumene is 2-phenylpropane
(n) acetone is propan-2-one, $CH_3COCH_3(l)$
(o) isopropanol is propan-2-ol, $CH_3CHOHCH_3(l)$*
(p) propylene oxide is $CH_3CHOCH_2(l)$
(q) cyclohexane
(r) adipic acid is hexane-1,6-dioic acid, $HOOC(CH_2)_4COOH(s)$

5. All of the chemicals in question 4 are involved in organic chemical reactions. Choose any two of the reactions from **Figure 1** or **Figure 2** and hypothesize a balanced chemical equation for the reactions; e.g., methanol to acetic acid, ethene to ethanol, ethene to styrene, and/or benzene to cyclohexane.

6. Polymers are long-chain molecules that can be made into plastics and synthetic fabrics. List six chemicals from Figure 2 that you think might be classified as polymers (e.g., polyethylene).

7. In 2002, Statistics Canada listed the sales of chemicals in Canada, in billions of dollars (**Table 1**). Create a pie chart to represent the percent sales for each category.

8. Use a reference to find some physical properties of two of the chemicals listed in **Figure 1** or **Figure 2**; e.g., styrene, ethylene dichloride, acetic acid, vinyl chloride, and/or formaldehyde.

Table 1 Chemical Sales in Canada

Chemical	Sales (G$)
basic*	20.2
soaps	1.8
fertilizers	2.5
paints	2.0
toiletries	1.3
pharmaceuticals	7.5
adhesives	0.6
inks	0.4
other	3.9

* basic chemicals and resins (for plastics)

DID YOU KNOW ?

Largest in Canada
After food production, the petrochemical industry is the second largest manufacturing sector in Alberta. Alberta is Canada's largest petrochemical producing area. The availability of energy, feedstock, and a highly educated population give Alberta an advantage over other regions in Canada.

Ethene and Its Derivatives

About 5% of our fossil fuels is used to produce petrochemicals. The economic importance of petrochemicals lies in the fact that basic raw materials are processed and reprocessed many times. For example, **Figure 3** shows the numbers of jobs in various industries that rely on petrochemicals. Ethene (ethylene) is one of the most important petrochemicals. For every 11 jobs involved in the manufacture of ethylene, 116 jobs are created in manufacturing the intermediate, vinyl chloride (chloroethene), 600 jobs in manufacturing the plastic polyvinyl chloride (PVC), and about 6000 jobs in manufacturing other commercial and consumer products, such as pipes and tiles.

As you learned in Chapter 9, ethene is produced in Alberta by cracking ethane that is extracted from natural gas (page 378). At one time, we burned the ethane in the natural gas without removing it. The natural gas that is burned now in consumer, commercial, and industrial settings, however, is likely to have had the ethane (and propane and butane) extracted. These gases are extracted as liquids and are called natural gas liquids (NGLs) or liquefied petroleum gases (LPGs). The remaining natural gas contains mostly methane.

In **Figure 2**, ethene (ethylene) is shown as undergoing reactions to produce ethanol, ethylene oxide, ethylene dichloride, polyethylene resins, and ethylbenzene. You will study some of these reactions, and others, in the pages that follow. **Table 2** gives the locations of some of the chemical plants that produce products related to ethane and ethene.

Section 10.1

Number of Jobs in Ethene (Ethylene) Industries
(based on 45 000 t/a of ethylene per plant)

```
                              ethylene
                                 11
        ┌─────────────┬─────────────┬─────────────┐
   ethylene        styrene      polyethylene    vinyl
    oxide         monomer           73        chloride
     31             100                          116
      │              │              │              │
      ▼              ▼              ▼              ▼
   glycols      polystyrene                   polyvinyl
     and           and                         chloride
   amines        resins                           600
     37            322
      │
 ┌────┼────┬────────┬────────┐    │              │              │
 ▼    ▼    ▼        ▼        ▼    ▼              ▼              ▼
others gas polyester antifreeze insulation,  moulding      pipe, tile
     treating fibre            paint, and     and             and
                                 tires      extrusion     baseboards
                                              film
  │    │    │        │        │              │              │
 801   7  1353      32      6568           1660           6000
```

Figure 3
Number of jobs in ethene industries, based on 45 000 t/a of ethene per plant. This is called a multiplier effect. For every 11 jobs in the ethene plant, there are thousands of jobs created elsewhere.

CAREER CONNECTION

Chemical Employees

According to Statistics Canada, the chemical industry in Alberta employs 8000 full-time people directly. Each job in the chemical industry creates 2.5 additional jobs in the local economy; e.g., engineers, scientists, technologists, and trades and support personnel.

Research a chemical career that you might be interested in at the Alberta Learning Information Service (ALIS) Web site. Report on the type and years of education required, and the expected salary.

www.science.nelson.com GO

Table 2 Alberta Chemical Operations (2006)

Chemical	Owner	Location in Alberta
ethane	Taylor NGL	Joffre (east of Red Deer)
	ATCO/Altagas	Edmonton
	Inter Pipeline (IPL)	Cochrane (west of Calgary)
	ATCO, EnCana, IPL, BP Canada, and Duke	Empress (6 plants straddling gas pipelines, northeast of Medicine Hat)
ethylene (ethene)	Dow Chemical	Fort Saskatchewan
	Nova Chemicals	Joffre (3 plants)
polyethylene	Dow Chemical	Prentiss (northeast of Red Deer), Fort Saskatchewan
	AT Plastics	Edmonton
	Nova Chemicals	Joffre (east of Red Deer)
styrene	Shell Chemicals	Scotford (east of Edmonton)
polystyrene	Dow Chemical	Fort Saskatchewan
vinyl chloride	Dow Chemical	Fort Saskatchewan
polyvinyl chloride	Oxy Vinyls	Fort Saskatchewan
ethylene glycol	Dow Chemical	Fort Saskatchewan, Prentiss
	Shell Chemical	Scotford (east of Edmonton)
ethylene dichloride	Dow Chemical	Fort Saskatchewan

Learning Tip

You are not expected to learn the names of the chemical companies nor their locations in Alberta. **Table 2** provides specific evidence that over 50% of Canada's petrochemical capacity is in Alberta. **Table 2** serves to reinforce the point that Alberta is a major petrochemical centre in Canada and could become a major centre in the world. All of the chemicals mentioned in **Table 2** are presented in greater detail later in this chapter.

The End of Oil?
Are we close to running out of easily available petroleum resources, or do we have enough to meet our needs for centuries? Which would be the "better" scenario? Two writers voice their perspectives on the topic. Either way, you may be worried.

www.science.nelson.com

SUMMARY Petrochemicals

- The petrochemical industry in Alberta and Canada contributes significantly to our economy.
- There are risks and benefits associated with every chemical—natural or manufactured.
- Ethane is obtained from natural gas, is cracked into ethene, and then the ethene is converted by chemical reactions into many other petrochemicals.
- Petrochemicals include plastics, paints, dyes, cosmetics, antifreezes, adhesives, carpets, fertilizers, and pharmaceuticals.
- There are many well-paying careers in the petrochemical industry.

Section 10.1 Questions

1. The chemicals listed in **Table 2**, Alberta Chemical Operations, are also listed in **Table 3**, along with the formulas of the chemicals involved. Many industry publications use common or alternative names for products. From the molecular formula provided, draw the structural formula. You will learn more about naming these compounds later in this chapter.

   Table 3 Alberta Chemical Operations—Names and Formulas

Chemical	Alternative name(s)	Molecular formula
(a) ethane*	ethyl hydride	CH_3CH_3
(b) ethylene	ethene*	CH_2CH_2
(c) polyethylene	polyethene*	$(CH_2CH_2)_n$
(d) styrene	vinylbenzene* or phenylethene*	$CH_2CHC_6H_5$
(e) polystyrene	polyphenylethene* or Styrofoam®	$(CH_2CHC_6H_5)_n$
(f) vinyl chloride	chloroethene*	CH_2CHCl
(g) polyvinyl chloride	PVC, polychloroethene*, or polychloroethylene	$(CH_2CHCl)_n$
(h) ethylene glycol	ethane-1,2-diol* or 1,2-dihydroxyethane	CH_2OHCH_2OH
(i) ethylene dichloride	1,2-dichloroethane*	CH_2ClCH_2Cl

   Note: The n is an indication that the structure repeats itself in a chain n units long. For this exercise, draw a full structural formula where $n = 3$.
   *preferred (IUPAC) name

2. **Figure 3**, on the previous page, illustrates the job multiplier effect that results from creating ethene from ethane.
   (a) For every 11 jobs in an ethene plant, how many potential jobs are created by using ethene?
   (b) Alberta citizens and petrochemical companies have objected to the export of natural gas from which ethane has not been removed. Why?
   (c) As the cost of natural gas increases, the value of ethane as a heating fuel increases. Consequently, the ethane may be burned as part of natural gas rather than being removed. How does burning ethane in natural gas affect the cost of ethane to petrochemical plants, the cost of products produced from ethene, and the number of jobs in the ethene industry?

3. Create a graph of your choice to communicate the relative quantities of petrochemicals produced in Canada (**Table 4**). Follow the link below to update the data from Statistics Canada.

   Table 4 Petrochemical Production in Canada, 2004

Petrochemical	Production (Mt)
benzene	10
ethene (ethylene)	57
buta-1,3-diene	3
butenes (butylenes)	3
propene (propylene)	11
methylbenzene (toluene)	3
dimethylbenzenes (xylenes)	4
phenylethene (styrene)	9

   www.science.nelson.com

4. Use the information above and from your own research to continue gathering perspective statements concerning the statement that we should be saving more fossil fuels for petrochemical use in the future.

Organic Halides and Addition and Substitution Reactions 10.2

Chemists divide organic compounds into two main classes: hydrocarbons and hydrocarbon derivatives. As you learned in Chapter 9, hydrocarbons contain only carbon and hydrogen atoms. **Hydrocarbon derivatives** are molecular compounds of carbon, usually hydrogen, and at least one other element. (See the diagram of organic compound families in Chapter 9, **Figure 3** on page 382.) For ease of classification, such compounds are named as if they had been produced by the modification of a hydrocarbon molecule. The hydrocarbon-derivative families studied in this chapter are organic halides, alcohols, carboxylic acids, esters, and polymers.

Organic Halides

Organic halides are organic compounds in which one or more hydrogen atoms have been replaced by halogen (group 17) atoms. The functional group for organic halides is the halogen atom. A **functional group** is a characteristic arrangement of atoms within a molecule that determines the most important chemical and physical properties of a class of compounds. Organic halides include many common products, such as freons (chlorofluorocarbons) used in refrigerators and air conditioners, and Teflon® (polytetrafluoroethylene), used in cookware and labware.

Many organic halides are toxic and many are also carcinogenic, so their benefits must be balanced against potential hazards. Two such compounds are the insecticide DDT (dichlorodiphenyltrichloroethane) and PCBs (polychlorinated biphenyls), used in electrical transformers. PCBs are no longer being produced but are still in circulation. DDT is banned in many countries but is still being produced in some countries, such as China.

According to solution theory, organic halides are solvents that can dissolve nonpolar hydrocarbons and/or polar hydrocarbons. Organic halide molecules may be polar or nonpolar, or they may have a relatively nonpolar hydrocarbon end and a polar halide end.

IUPAC nomenclature for halides follows the same format as that for branched-chain hydrocarbons. The branch is named by shortening the halogen name to *fluoro-*, *chloro-*, *bromo-*, or *iodo-* (**Table 1**). For example, $CHCl_3(l)$ is trichloromethane (chloroform), an important solvent, and $C_6H_5Br(l)$ is bromobenzene, an additive to motor oils.

When translating IUPAC names for organic halides into full structural formulas, draw the parent chain and add branches at locations specified in the name. For example, 1,2-dichloroethane indicates that this compound has a two-carbon (eth-), single bonded parent chain (-ane), with one chlorine atom on each carbon (1,2-dichloro-). 1,2-dichloroethane is a liquid organic halide produced by Dow Chemical in Fort Saskatchewan. 1,2-dichloroethane is used to produce chloroethene, which in turn is used to produce polyvinyl chloride (PVC).

$$\begin{array}{c} Cl\ \ Cl \\ |\ \ \ \ | \\ H-C-C-H \\ |\ \ \ \ | \\ H\ \ H \end{array} \rightarrow \begin{array}{c} H\ \ \ \ \ \ H \\ \diagdown\ \ \diagup \\ C=C \\ \diagup\ \ \ \diagdown \\ H\ \ \ \ \ \ Cl \end{array} + H-Cl$$

1,2-dichloroethane → chloroethene

DID YOU KNOW ?

Alkyl Halides
Organic halides are also known as alkyl halides and halogenated hydrocarbons. The term alkyl halides does not include aryl (aromatic) halides, such as bromobenzene. The more general term, organic halides, is used in this chapter.

Learning Tip

Each of the classes of hydrocarbon derivatives studied in this chapter is associated with a type of chemical reaction. For example,
- organic halides undergo addition and substitution reactions
- alcohols undergo elimination reactions
- carboxylic acids and esters are involved in esterification reactions

Try to use this organization to help you learn these concepts.

Table 1 Halide Functional Groups

Chemical symbol	Prefix
—F	fluoro
—Cl	chloro
—Br	bromo
—I	iodo

DID YOU KNOW ?

Ozone Depletion

In 1982, a 30% decrease in the ozone layer—referred to as an ozone "hole"—was noticed for the first time by a team of British researchers working in Halley Bay, Antarctica. The British team's results surprised American researchers who had been measuring ozone levels by weather satellite since 1978. American satellite data are transmitted to Earth and are automatically processed by computers before scientists examine them. The Americans had not noticed the decrease in ozone levels because their computers were programmed to reject low measurements as invalid anomalies. The British scientists had also been monitoring atmospheric concentrations of chlorofluorocarbons (CFCs) and they raised the possibility that the decreasing ozone levels and the increasing CFC concentrations in the atmosphere were related. Since 1982, American computers processing total ozone mapping spectrophotometer (TOMS) data no longer reject low values, and alarming depletions of 60% to 70% in ozone levels over Antarctica have been detected.

The ozone layer has now stopped shrinking because of the measures we have taken to reduce our use of CFCs, but it will take many years to regenerate.

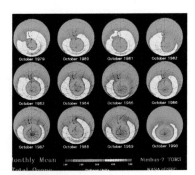

Figure 1
These NASA satellite photos show ozone levels over Antarctica as measured by the ozone-detecting device called TOMS. Look for recent photos on the Internet.

SAMPLE problem 10.1

Draw a structural formula for 2,2,5-tribromo-5-methylhexane.

First, draw and number the parent alkane chain, the hexane:

C—C—C—C—C—C
1 2 3 4 5 6

Next, add two Br atoms to carbon 2, one Br atom to carbon 5, and a methyl group to carbon 5.

Finally, complete the bonding by adding H atoms to the C atoms.

COMMUNICATION example 1

Write the IUPAC name for the organic halide

$CH_3-CH_2-CH_2-CH-CH_2-CH-CH_3$
                | |
                Cl   Br

Solution
This compound is 2-bromo-4-chloroheptane.

COMMUNICATION example 2

Draw line structural formulas for the three isomers of dichlorobenzene. Name the isomers.

Solution

1,2-dichlorobenzene   1,3-dichlorobenzene   1,4-dichlorobenzene

Practice

1. Draw a structural formula for each of the following organic halides:
   (a) 1,2-dichloroethane (solvent for rubber)
   (b) tetrafluoroethene (used in the manufacture of Teflon)
   (c) 1,2-dichloro-1,1,2,2-tetrafluoroethane (refrigerant)
   (d) 1,4-dichlorobenzene (moth repellent)

Section **10.2**

2. Write IUPAC names for each of the formulas given.
   (a) CHI$_3$ (antiseptic)
   (b) CH$_2$=C—CH$_2$Cl
              |
              CH$_3$
       (insecticide)
   (c) CH$_2$Cl$_2$ (paint remover)
   (d) CH$_2$Br—CHBr—CH$_2$Br (soil fumigant)
   (e) C$_6$H$_5$Cl(l) (insoluble in water)
   (f) C$_6$H$_5$—CH=CH$_2$ (monomer)

3. Use your knowledge of intermolecular forces to predict, with reasoning,
   (a) the relative boiling points of ethene and chloroethene
   (b) the relative solubility of ethene and chloroethene in water

4. Chlorofluorocarbons (CFCs) are organic halides that have chlorine or chlorine and fluorine atoms present. Until the 1990s, chlorofluorocarbons (CFCs) were used as refrigerants, foaming agents, and aerosol sprays. Besides depleting the ozone layer, CFCs are excellent absorbers of thermal infrared (IR) radiation; in other words, they are greenhouse gases (GHGs). Provide structural formulas for the following CFCs.
   (a) CCl$_4$(l)
   (b) CH$_3$CCl$_3$(l)
   (c) CFCl$_3$(l)
   (d) CF$_2$ClCFCl$_2$(l)

5. Hydrochlorofluorocarbons (HCFCs) have hydrogen and fluorine atoms present, in addition to chlorine atoms. HCFCs were used as a temporary replacement for CFCs in the 1990s. HFCs (hydrofluorocarbons) are replacing both CFCs and HCFCs. Draw structural formulas for these HCFCs.
   (a) CHF$_2$Cl(l)
   (b) CH$_3$CFCl$_2$(l)
   (c) CHF$_3$(l)
   (d) CF$_3$CH$_2$F(l)

+ **EXTENSION**

CFCs, HFCs, and the Ozone Layer

As a result of Nobel Prize research on stratospheric ozone depletion by a group of chemists, CFCs were phased out of production, worldwide. Find out what the CFCs were being used for, and what follow-up agreements are in the works.

www.science.nelson.com [GO]

Learning Tip

Recall that, for IUPAC nomenclature of substitution groups on a hydrocarbon chain, list the substituents in alphabetical order; e.g., 2-bromo-4-chloroheptane (as in Communication Example 1).

Addition Reactions

Unsaturated hydrocarbons react with small diatomic molecules, such as bromine and hydrogen. This type of reaction is an **addition reaction**. Addition reactions usually occur in the presence of a catalyst. Recall from Section 9.3 that addition of a sufficient quantity of hydrogen, called *hydrogenation*, converts unsaturated hydrocarbons to saturated ones. This reaction requires a catalyst.

but-1-ene + hydrogen → butane

propyne + excess hydrogen → propane

It seems logical that the addition of halogen or hydrogen halide molecules to the carbons of a double or triple bond would be an effective method of preparing halides. Experiment supports this expectation. Chemists explain the rapid rate of these reactions by the concept that a compound with a carbon–carbon double or triple bond can become more stable by achieving an octet of electrons in a tetrahedral structure of single bonds. For example, ethene reacts with chlorine, producing 1,2-dichloroethane. 1,2-dichloroethane is used as a solvent for fats, oils, and particularly for rubber. It is also

+ **EXTENSION**

Reactions of Hydrocarbons with Bromine

Watch a video in which bromine, dissolved in an inert solvent, is added to pentane and pent-1-ene in two dishes. Do these two hydrocarbons react in the same way?

www.science.nelson.com [GO]

DID YOU KNOW ?

Ethylene Dichloride
Why is ethylene dichloride the classical name for 1,2-dichoroethane? You can, perhaps, see the reason in the reaction equation for the production of 1,2-dichloroethane. Ethylene dichloride is produced from ethylene—it is a dichloride from (not of) ethylene. The name, ethylene dichloride, is misleading, which is why its IUPAC name is 1,2-dichloroethane.

used as a fumigant. More importantly for the Alberta chemical industry, 1,2-dichloroethane is used in huge quantities to produce chloroethene, a monomer used to produce the polymer PVC. (See the last equation on page 417 for the next step.)

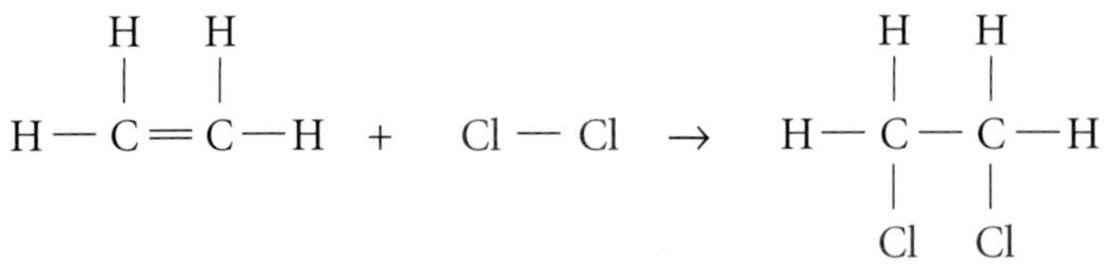

$$H_2C=CH_2 + Cl_2 \rightarrow CH_2Cl-CH_2Cl$$

ethene + chlorine → 1,2-dichloroethane

The addition of halogens to alkynes results in alkenes or alkanes. For example, the initial reaction of ethyne with bromine produces 1,2-dibromoethene.

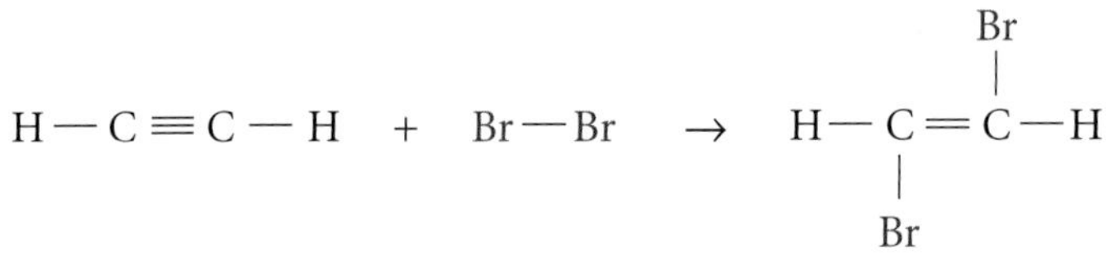

$$HC \equiv CH + Br_2 \rightarrow CHBr=CHBr$$

ethyne + bromine → 1,2-dibromoethene

Since addition reactions involving multiple bonds are very rapid, the alkene product, 1,2-dibromoethene, can easily undergo a second addition step to produce 1,1,2,2-tetrabromoethane. Excess bromine promotes this second step.

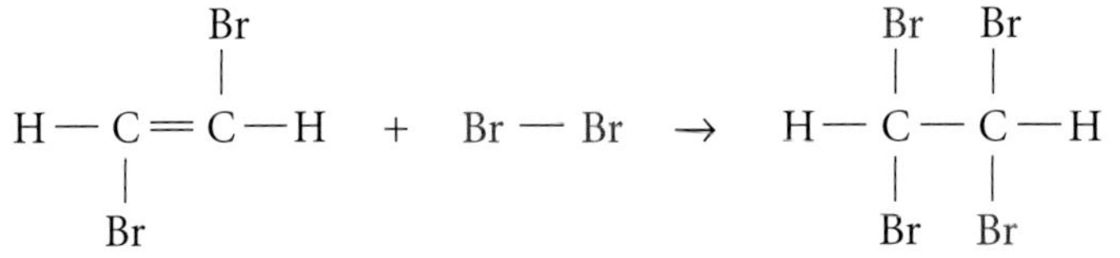

$$CHBr=CHBr + Br_2 \rightarrow CHBr_2-CHBr_2$$

1,2-dibromoethene + bromine → 1,1,2,2-tetrabromoethane

The addition of hydrogen halides (HF, HCl, HBr, or HI) to unsaturated compounds can produce structural isomers, since the hydrogen halide molecules can add in different orientations. If you were to create the hypothesis that the addition might occur equally with orientations of H—Cl and Cl—H, then you would predict the following reaction.

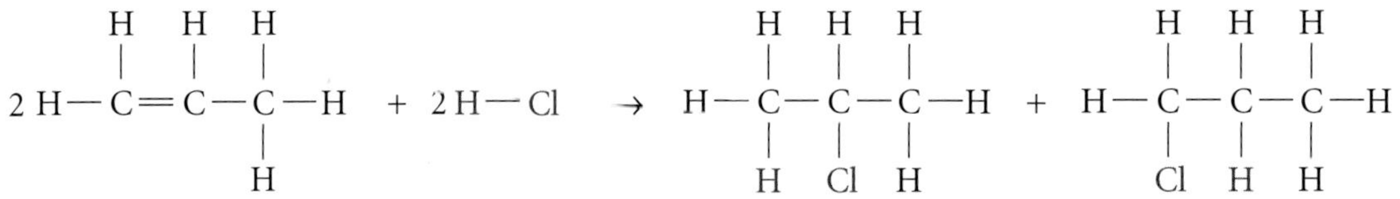

$$2\,CH_2=CH-CH_3 + 2\,HCl \rightarrow CH_3-CHCl-CH_3 + CH_2Cl-CH_2-CH_3$$

propene + hydrogen chloride → 2-chloropropane + 1-chloropropane

A laboratory test of this prediction would provide evidence to falsify this prediction. Only a tiny proportion of 1-chloropropane is produced compared to the 2-chloropropane.

Learning Tip

The general formula for an organic halide is R—X, where X is a halogen atom, the functional group.

COMMUNICATION example 3

Chloroethene (vinyl chloride) is made industrially in Alberta from a two-step reaction starting with ethene. An alternative method is to react ethyne (acetylene) with limited hydrogen chloride in the presence of a $HgCl_2$ catalyst. Write a structural formula equation to communicate this alternative reaction.

Solution

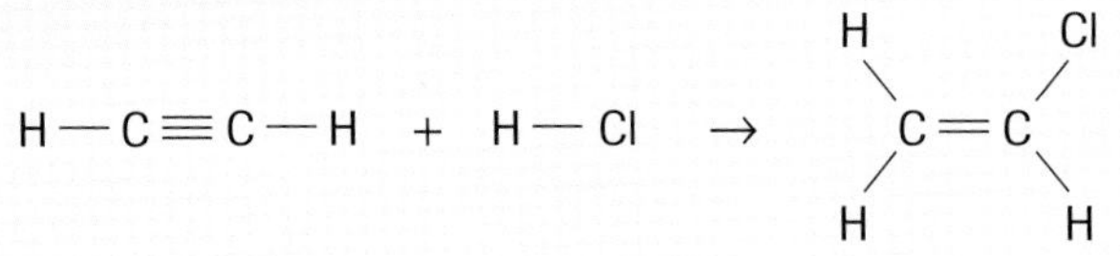

$$H-C\equiv C-H \;+\; H-Cl \;\rightarrow\; \begin{array}{c} H \\ \diagdown \\ C=C \\ \diagup \\ H \end{array} \begin{array}{c} Cl \\ \diagup \\ \\ \diagdown \\ H \end{array}$$

Substitution Reactions

According to laboratory work, another reaction that produces organic halides is a **substitution reaction**. The theoretical description of substitution reactions is that they involve breaking a carbon–hydrogen bond in an alkane or aromatic ring and replacing the hydrogen atom with another atom or group of atoms. These reactions often occur slowly at room temperature, indicating that very few of the molecular collisions at room temperature are energetic enough to break carbon–hydrogen bonds. Electromagnetic radiation (light) may be necessary for the substitution reaction to proceed at a noticeable rate. Consider the following example, the reaction of propane with bromine vapour:

$$C_3H_8(g) + Br_2(g) \xrightarrow{light} C_3H_7Br(l) + HBr(g)$$

In this reaction, a hydrogen atom on the propane molecule is substituted with a bromine atom. Propane contains hydrogen atoms bonded in two different locations—those on an end-carbon atom and those on the middle-carbon atom—so two different products are formed, in unequal proportions. (The following reaction equation does not accurately represent this.)

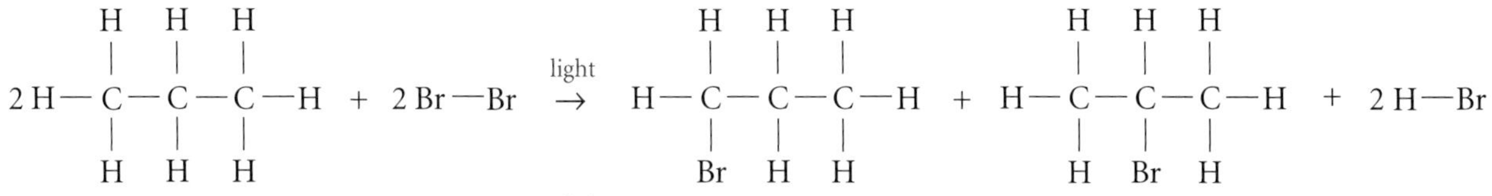

$$2\,C_3H_8 + 2\,Br_2 \xrightarrow{light} CH_3CH_2CH_2Br + CH_3CHBrCH_3 + 2\,HBr$$

propane + bromine $\xrightarrow{light}$ 1-bromopropane + 2-bromopropane + hydrogen bromide

Laboratory evidence indicates that benzene rings are stable structures and, like alkanes, react slowly with halogens, even in the presence of light. For example, the reaction of benzene with chlorine is shown to produce chlorobenzene and hydrogen chloride. As with alkanes, further substitution can occur in benzene rings until all hydrogen atoms are replaced by halogen atoms in the presence of excess halogens.

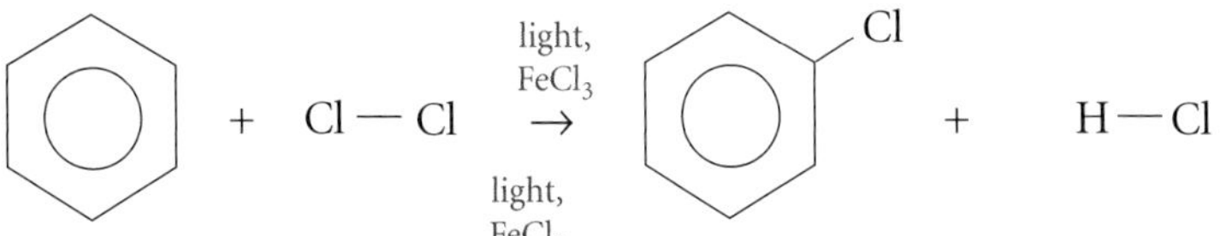

benzene + chlorine $\xrightarrow{light,\,FeCl_3}$ chlorobenzene + hydrogen chloride

The reaction of benzene and chlorine is so slow that it requires light and a catalyst such as $FeCl_3$. This reaction produces HCl, as evidenced by wet blue litmus turning red in or above the reaction mixture. This evidence dictates the reaction equation given above, and leads to the interpretation that this reaction is a substitution reaction.

> **COMMUNICATION** example 4

CFCs were used until the 1990s as refrigerants (for refrigerators and air conditioners) and to create bubbles in plastic foams (used as foam insulation, foam trays, and cartons). CFC-12, which is pure CF_2Cl_2, was made by reacting carbon tetrachloride with gaseous hydrogen fluoride. The carbon tetrachloride was produced by reacting methane with chlorine. Draw structural formula equations to represent these two reactions.

Solution

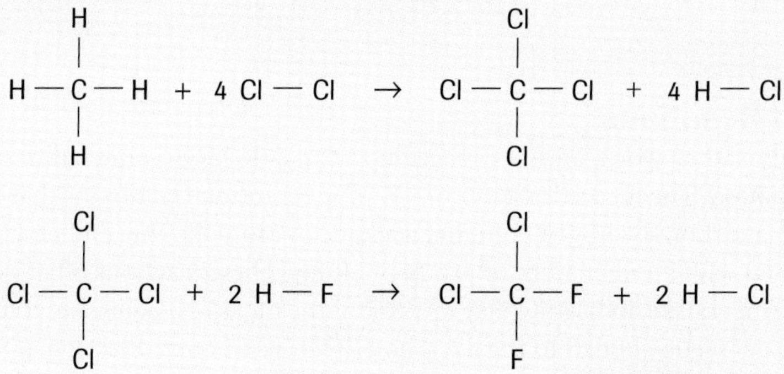

```
      H                              Cl
      |                              |
  H — C — H  + 4 Cl — Cl  →     Cl — C — Cl  + 4 H — Cl
      |                              |
      H                              Cl

      Cl                             Cl
      |                              |
  Cl — C — Cl + 2 H — F  →      Cl — C — F  + 2 H — Cl
      |                              |
      Cl                             F
```

INVESTIGATION 10.1 Introduction

Substitution and Addition Reactions

Substitution and addition reactions are two common classes of organic reactions. Substitution reactions are common to saturated hydrocarbons. Addition reactions are common to unsaturated (alkene and alkyne) hydrocarbons. In general, substitution reactions are slower and addition reactions are faster.

Purpose

The purpose of this investigation is to test the generalization that addition reactions are faster than substitution reactions.

Report Checklist
- ○ Purpose
- ○ Problem
- ○ Hypothesis
- ● Prediction
- ○ Design
- ○ Materials
- ○ Procedure
- ● Evidence
- ● Analysis
- ● Evaluation (1, 2, 3)

Problem

Which compound reacts faster with aqueous bromine: cyclohexane or cyclohexene?

Design

Aqueous bromine is added to samples of cyclohexane and cyclohexene in both the presence and absence of light.

To perform this investigation, turn to page 461.

> **Practice**

6. The halide derivatives of hydrocarbons form a wide variety of useful products. Draw structural formulas for the following examples:
   (a) 1,2-dimethylbenzene (solvent)
   (b) 1,1,1-trichloroethane (a banned CFC)

7. Classify each of the following reactions as substitution or addition reactions. Predict all possible products for only the initial reaction. Complete the word equation and the structural formula equation in each case. Assume that light is present in all cases and catalysts are used as required.
   (a) trichloromethane + chlorine →
   (b) propene + bromine →
   (c) ethylene + hydrogen iodide →
   (d) ethane + chlorine →
   (e) Cl — C ≡ C — Cl + F — F (excess) →

(f) 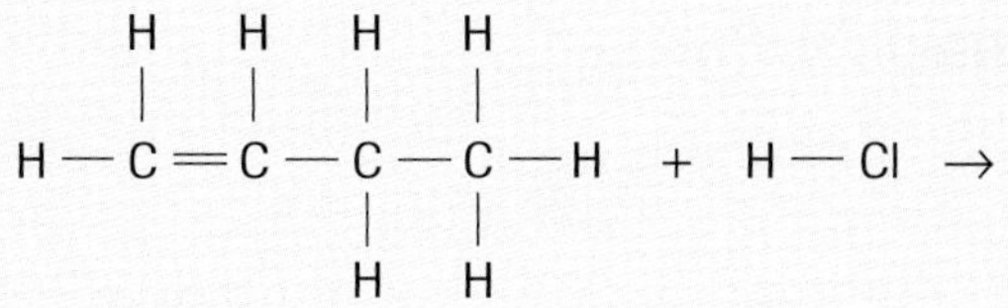
```
     H   H   H   H
     |   |   |   |
H — C = C — C — C — H  +  H — Cl  →
             |   |
             H   H
```

(g) 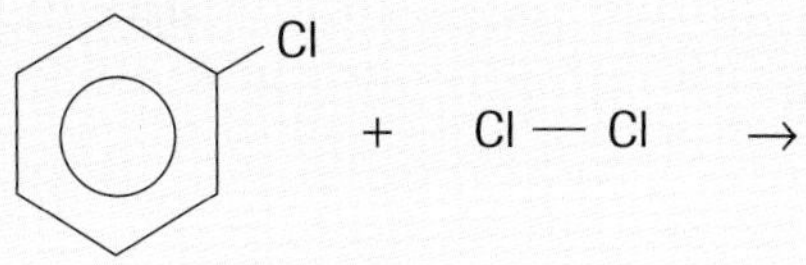 [benzene ring with Cl substituent] + Cl — Cl →

8. Classify and write balanced full structural formula equations for each of the following organic reactions. Assume that light is present in all cases and catalysts are used as required.
   (a) propane + chlorine →
                       1-chloropropane + 2-chloropropane + hydrogen chloride
   (b) propene + bromine → 1,2-dibromopropane
   (c) benzene + iodine → iodobenzene + hydrogen iodide
   (d) but-2-ene + hydrogen chloride → 2-chlorobutane
   (e) bromobenzene + chlorine →

9. The synthesis of an organic compound typically involves a series of reactions, for example, some substitutions and some additions.
   (a) Design an experiment beginning with a hydrocarbon to prepare 1,1,2-trichloroethane.
   (b) Why do chemists and chemical engineers invent chemical processes?
   (c) What other perspectives, besides scientific and technological, can be taken into account when inventing a new technology?

10. Test the explanatory power of intermolecular force theory by explaining why the two organic halide isomers, 1-bromopropane and 2-bromopropane, have different boiling points: 71 °C and 59 °C, respectively.

11. Petrochemicals are chemicals produced from petroleum (assume hydrocarbons) and from other petrochemicals, such as organic halides. Use your list of perspectives to classify the following statements about petrochemicals. (See STS Problem Solving in Appendix D.)
    (a) Many human lives have been saved by the use of organic halide pesticides.
    (b) Organic halides are used successfully as refrigerants and foaming agents.
    (c) Research is on-going to find a replacement for CFCs and HCFCs.
    (d) Farmers have saved considerable money by using organic halide pesticides.
    (e) Getting elected may depend on promising that controls will be placed on the use of organic halides.
    (f) There is a need for laws to control the release of organic halides.
    (g) There are claims that workers in the past were not informed about the toxic or carcinogenic character of organic halides with which they worked.
    (h) Communities have acted together to recycle petrochemical products.

DID YOU KNOW ?

Cis and Trans Isomerism

There is always a new concept to learn that extends what you have already learned. This applies to research chemists as much as it applies to you as a student. One example for you at this point is the introduction of evidence for the existence of *cis* and *trans* isomers, in addition to the structural isomers that you know about. Examples of *cis* and *trans* isomers are 1,2-dichloroethene, shown below as structural formulas.

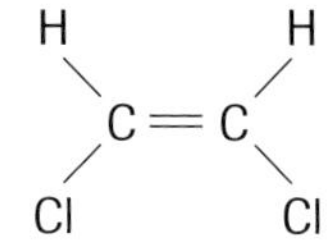

cis-1,2-dichloroethene

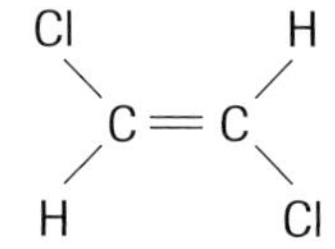

trans-1,2-dichloroethene

Cis isomers involve special bonding on the same side of the double bond. *Trans* isomers have the special bonding across the double bond. See more on *cis* and *trans* isomerism in relation to transfats in Sections 10.4 and 10.5.

LAB EXERCISE 10.A

Synthesis of an Organic Halide

1,2-dichloroethane is also commonly called ethylene dichloride. Chemical engineers have invented ways to convert ethane to ethene to 1,2-dichloroethane to chloroethene to the plastic PVC (polyvinylchloride). Finding economically and environmentally sound ways to manufacture chemicals is the task of chemists and chemical engineers. Your task is to propose a reaction series for at least two different ways to manufacture 1,2-dichloroethane (ethylene dichloride) from hydrocarbons.

Report Checklist
- ● Purpose
- ● Problem
- ● Hypothesis
- ○ Prediction
- ○ Design
- ○ Materials
- ○ Procedure
- ○ Evidence
- ○ Analysis
- ○ Evaluation

SUMMARY: Organic Halides

Properties

Organic halides
- may be polar or nonpolar molecules or may have a relatively nonpolar (hydrocarbon) end and a polar (halide) end
- have higher boiling points than similar hydrocarbons
- have very low solubility in water but higher solubility (especially for small molecules) than similar hydrocarbons
- are typically good solvents for organic materials such as fats, oils, waxes, gums, resins, and/or rubber

Preparation

- Addition reactions with halogens or hydrogen halides
  alkenes and alkynes → organic halides
  e.g., $CH_2=CH_2 + Cl_2 \rightarrow CH_2Cl-CH_2Cl$
- Substitution reactions with halogens
  alkanes and aromatics $\xrightarrow{light}$ organic halides
  e.g., $CH_3-CH_2Cl + Cl_2 \xrightarrow{light} CH_2Cl-CH_2Cl + H-Cl$

Section 10.2 Questions

1. Addition and substitution reactions typically require catalysts to produce a reasonable rate of reaction in a chemical plant. Substitution reactions occur (slowly) in the laboratory due to the presence of light. Design an experiment to test the last statement.

2. Organic halides include alkyl halides and halogenated hydrocarbons. Draw line structural formulas for the following organic halides.
   (a) 1,4-dichlorobenzene (moth repellant)
   (b) 1-bromo-4-methylbenzene (chemical reagent)
   (c) 1,2-dibromoethane (suspected carcinogen)
   (d) chloroethene (to produce PVC)

3. The relative boiling points of three halogen substituted ethane compounds are:
   chloroethane: 12.3 °C
   bromoethane: 38.2 °C
   iodoethane: 72.3 °C
   (a) Explain the relative boiling points of these organic halides.
   (b) Predict whether these compounds are more soluble in water or in ethanol.

4. Classify and write full structural formula equations for the following organic reactions:
   (a) propane + chlorine →
       1-chloropropane + 2-chloropropane + hydrogen chloride
   (b) propene + bromine → 1,2-dibromopropane
   (c) benzene + iodine → iodobenzene + hydrogen iodide

Extension

5. Shortly after the connection was made between the "hole" in the ozone layer and the release of chlorofluorocarbons, many manufacturers stopped using CFCs as propellants in aerosol cans.
   (a) Research what alternatives were developed, and the effectiveness of each in the marketplace. Are the alternatives still in use? Have any of them been found to cause problems?
   (b) Design a product (one that must be sprayed under pressure) and its packaging. Plan a marketing strategy that highlights the way in which your product is sprayed from the container.

   www.science.nelson.com

6. Use the information in this section and from your own research to continue gathering perspective statements concerning the statement that we should be saving more fossil fuels for petrochemical use in the future.

7. Long-term replacements for CFCs and HCFCs include hydrofluorocarbons (HFCs), which have no chlorine atoms. HFCs are now being used in refrigerators, automobile air conditioners, aerosol cans, and as foaming agents.
   Draw the structural formula for $CH_2FCF_3(l)$ (also called HFC-134a).

8. Why are some organic halides toxic whereas others are not? And why are some organisms affected more than others? Find out, using the following key words: bioaccumulation, fat soluble, food chain. Report on your findings in a short article for a popular science magazine or Web site.

   www.science.nelson.com

Alcohols and Elimination Reactions 10.3

Alcohols

Alcohols have many consumer, commercial, and industrial applications. They have very different properties compared to their hydrocarbon cousins with the same number of carbon atoms, such as relatively high solubility in water and clean burning properties. Chemists first classified alcohols by their empirical properties and their empirically determined formulas. Laboratory work showed that **alcohols** all contain one or more **hydroxyl groups**, –OH. The –OH group is the functional group for alcohols.

Alcohols have characteristic empirical properties that can be explained theoretically by the presence of a hydroxyl (–OH) functional group attached to a hydrocarbon chain. Alcohols boil at much higher temperatures than do hydrocarbons of comparable molar mass. Chemists explain that alcohol molecules, because of the –OH functional group, form hydrogen bonds (see Section 3.4, page 111) and, thus, liquid alcohols are less volatile than similar hydrocarbons. Shorter-chain alcohols are very soluble in water because of their size, polarity, and hydrogen bonding.

Because the hydrocarbon portion of the molecule of long-chain alcohols is non-polar, larger alcohols are less soluble in water and are good solvents for non-polar molecular compounds as well. (See the Solubility in Water Generalizations, Chapter 5, page 222.) Alcohols are frequently used as solvents in organic reactions because they are effective for both polar and non-polar compounds. Alcohols are also used as starting materials in the synthesis of other organic compounds.

Methanol and Ethanol

Two of the most common alcohols are methanol, $CH_3OH(l)$, and ethanol, $C_2H_5OH(l)$. Increasingly, one of the most important technological applications of alcohols is as a gasoline additive. Alcohol is added to gasoline in some jurisdictions, such as Manitoba, to increase the octane number, reduce harmful emissions (such as carbon monoxide), and conserve crude oil (see Chapter 9, page 399). The alcohols typically added are ethanol and methanol. These alcohols (as pure substances) have octane numbers of 113 and 108, respectively. They are often called oxygenators because they provide oxygen to the combustion reaction. The result is that less carbon monoxide is released in the exhaust. The combustion reaction of gasoline is, therefore, more complete.

An old method for manufacturing methanol involved decomposing and distilling the cellulose in wood, which is how it got its alternative name: wood alcohol. The modern method of preparing methanol involves two major processes. First, methane reacts catalytically with water (steam) to produce carbon monoxide and hydrogen:

$$CH_4(g) + H_2O(g) \xrightarrow{\text{catalyst}} CO(g) + 3\,H_2(g)$$

Next, carbon monoxide and hydrogen react at high temperature and pressure in the presence of a catalyst:

$$CO(g) + 2\,H_2(g) \xrightarrow{\text{catalyst}} CH_3OH(l)$$

A common consumer use of alcohols (including methanol) is as antifreezes (**Figure 2** on the next page).

Methanol is toxic to humans. Drinking even small amounts of it or inhaling the vapour for prolonged periods can lead to blindness or death. All alcohols are not the same. Methanol is even added to non-beverage ethanol to denature it. Denatured ethanol is toxic.

DID YOU KNOW ?

Ethanol in Alberta

At the time of writing, there are only three chemical plants in Alberta that produce ethanol from grain. The API Grain Processing Plant in Red Deer, Alberta produces industrial alcohol and Highwood Distillers in High River, Alberta produces beverage alcohol. There is also a large plant near the Alberta–Saskatchewan border at Lloydminster. In future, there will likely be more and more ethanol plants in western Canada to supply this gasoline additive. Grain-based ethanol plants are adding another significant market for grain farmers (**Figure 1**).

Figure 1
Grain is a source of ethanol, used as a gasoline additive to reduce GHG emissions.

Learning Tip

The general formula for an alcohol is R–OH. The alcohol functional group is the –OH (hydroxyl) group.

Figure 2
Alcohols available to consumers as antifreezes include, from left to right, gas line antifreeze (methanol), windshield de-icer (methanol), windshield washer antifreeze (methanol), and radiator and recreation vehicle antifreezes (ethane-1,2-diol and propane-1,2-diol).

Ethanol can be prepared by the fermentation of sugars from starch, such as corn and grains; hence its alternative name—grain alcohol. In the fermentation process, enzymes produced by yeast cells act as catalysts in the breakdown of sugar (glucose) molecules:

starch → glucose → ethanol

$$(C_6H_{10}O_5)_n(aq) \xrightarrow{} C_6H_{12}O_6(aq) \xrightarrow{yeast} 2\,CO_2(g) + 2\,C_2H_5OH(aq)$$

In terms of industrial applications, ethanol is an important synthetic organic chemical. It is a solvent in lacquers, varnishes, perfumes, and flavourings, and is a raw material in the synthesis of other organic compounds.

Some oil companies in Alberta put ethanol in gasoline, for example, as a winter additive, for increased vaporization, and as a gas line antifreeze. Currently there is no legislation requiring ethanol in gasoline in Alberta.

There is an increasing number of ethanol plants being built in western Canada to produce ethanol as a gasoline additive. Provincial legislation to include 10% ethanol in gasoline, and tax incentives, will accelerate the expansion of the ethanol industry.

DID YOU KNOW ?

Fill Up with Methanol
Alcohols have many uses, one of the more recent being a fuel for motor vehicles. The problem with methanol as a pure fuel for cars is that it is less volatile than the hydrocarbons that make up gasoline, and the low volatility makes it difficult to ignite. In our cold Canadian winters, there is little methanol vapour in the engine and an electrical spark is insufficient to start the car. Canadian scientists are investigating a variety of dual ignition systems, one of which is a plasma jet igniter that is 100 times more energetic than conventional ignition systems.

▶ Practice

1. Chemical engineers have introduced alcohol into gasoline to play multiple roles. Describe at least three of these roles.
2. Write pro and con statements from social, political, and economic perspectives relative to the resolution that legislation should require the addition of alcohol to gasoline (gasohol) in Alberta.
3. Write the chemical equations for
   (a) photosynthesis to produce glucose
   (b) the conversion of glucose to starch
   (c) the fermentation of starch to glucose
   (d) the fermentation of glucose to ethanol
4. Volatile compounds evaporate or vaporize relatively easily. Volatile organic compounds (VOCs) can cause atmospheric smog. For molecules of the same molar mass, list hydrocarbons, organic halides, and alcohols, in general, in order of increasing volatility.

Naming Alcohols

Simple alcohols are named from the alkane of the parent chain. The *-e* is dropped from the end of the alkane name and is replaced with *-ol*. For example, the simplest alcohol, with one carbon atom, has the IUPAC name "methanol." The number of carbon atoms in the alcohol is communicated by the standard prefixes: *meth-*, *eth-*, *prop-*, etc. Single bonds between the carbon atoms are communicated by the "an" in the middle of the name, for example, ethanol rather than ethenol.

> ### SAMPLE problem 10.2
>
> Write the IUPAC name for the following alcohol.
>
> ```
>      CH₃  H
>       |   |
> CH₃—CH—C—CH₃
>           |
>           OH
> ```
>
> 1. Identify the longest continuous carbon chain on which the alcohol functional group—the hydroxyl (–OH) group—is located; e.g., four carbon atoms.
> 2. Write the prefix for the name to indicate the number of carbons in the longest chain; e.g., "but" for a four-carbon chain.
> 3. Add an "an" to the prefix to indicate that all of the carbon–carbon bonds are single bonds; e.g., "butan-".
> 4. Number the carbon chain from the end nearest the hydroxyl group; e.g., 1–4 from the right (in this case).
> 5. If necessary, add the locant (number) for the hydroxyl group to the name; e.g., "butan-2".
> 6. Complete the parent name with "-ol" to indicate that the compound is an alcohol; e.g., "butan-2-ol".
> 7. Indicate the presence of any substituent as a prefix; e.g., "methylbutan-2-ol".
> 8. If necessary, add the locant (number) to indicate the location of the substituent; e.g., "3-methylbutan-2-ol".
> 9. Double-check to make sure that the numbers are necessary to differentiate this compound from a similar compound; e.g., 2-methyl-1-butanol.
>
> The name of this alcohol is 3-methylbutan-2-ol.

Learning Tip

The nomenclature of alcohols is very similar to that of alkanes and alkenes. Look for the longest continuous chain, including the functional group in the case of alkenes and alcohols. Identify the location of any functional group and any substituent(s). Communicate all of this information in the format specified by IUPAC. For example,

- 2-methylbutane
- 2-methylbut-2-ene
- 2-methylbutan-2-ol

Try drawing the structural formula for each of the above compounds.

> ### COMMUNICATION example 1
>
> Due to their physical and chemical properties, alcohols have many technological applications. Provide the molecular and structural formulas for alcohols (a) and (b), and the molecular and condensed formulas for alcohols (c) and (d).
> (a) methanol (race car fuel)   (c) propan-1-ol (solvent for organic compounds)
> (b) ethanol (gasoline additive)   (d) butan-1-ol (used for manufacturing rayon)
>
> **Solution**
>
> (a) methanol: CH₃OH(l)
>
>
>
> (b) ethanol: C₂H₅OH(l)
>
>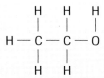
>
> (c) propan-1-ol: C₃H₇OH(l)
>   CH₃—CH₂—CH₂—OH
>
> (d) butan-1-ol: C₄H₉OH(l)
>   CH₃—(CH₂)₃—OH

Learning Tip

The carbon atoms can be numbered from either end of the chain, whichever gives a lower number for the location of the hydroxyl group.

DID YOU KNOW ?

Alcohol Poisoning
Like other short-chain alcohols, ethanol is poisonous, and a concentration of only 0.40% in the blood can cause death. According to Canadian law, individuals whose blood alcohol levels are greater than 0.08% are considered impaired. Driving while impaired is a serious threat to the driver, the passengers, and anyone else on or near the road.

Primary, Secondary, and Tertiary Alcohols

By convention, when we write the molecular formula or the condensed structural formula (rather than the structural formula) for alcohols, we write the –OH group at the end, for example, C_2H_5OH or $CH_3—CH_2—OH$ for ethanol.

The position of the –OH group can vary, however, and make alcohols quite different, in terms of their chemical and physical properties. For example, the 1 in propan-1-ol indicates that the hydroxyl group is bonded to a carbon atom at the end of the carbon chain, that is, $CH_3CH_2CH_2OH(l)$ rather than $CH_3CHOHCH_3(l)$.

Structural models of alcohols with four or more carbon atoms suggest that three structural types of alcohols exist.

- *primary* (1°) *alcohols*, in which the carbon atom carrying the –OH group is bonded to one other carbon atom, as in $CH_3CH_2CH_2CH_2OH(l)$, butan-1-ol
- *secondary* (2°) *alcohols*, in which the carbon atom carrying the –OH group is bonded to two other carbon atoms, as in $CH_3CHOHCH_2CH_3(l)$, butan-2-ol
- *tertiary* (3°) *alcohols*, in which the carbon atom carrying the –OH group is bonded to three other carbon atoms, as in $(CH_3)_3COH(l)$, 2-methylpropan-2-ol

$$CH_3-CH_2-CH_2-\underset{\underset{OH}{|}}{\overset{\overset{H}{|}}{C}}-H \qquad CH_3-CH_2-\underset{\underset{OH}{|}}{\overset{\overset{H}{|}}{C}}-CH_3 \qquad CH_3-\underset{\underset{OH}{|}}{\overset{\overset{CH_3}{|}}{C}}-CH_3$$

butan-1-ol, a 1° alcohol   butan-2-ol, a 2° alcohol   methylpropan-2-ol, a 3° alcohol

When naming alcohols with more than two carbon atoms, we indicate the position of the hydroxyl group. For example, there are two isomers of propanol, C_3H_7OH: propan-1-ol is used as a solvent for lacquers and waxes, as a brake fluid, and in the manufacture of propanoic acid; propan-2-ol, or isopropanol, is sold as rubbing alcohol and is used to manufacture oils, gums, and acetone. (The prefix *iso-* indicates that the hydroxyl group is bonded to the central carbon atom.) Both isomers of propanol are toxic to humans if taken internally.

▶ SAMPLE problem 10.3

Name the following alcohol from its condensed structural formula and indicate whether it is a primary, secondary, or tertiary alcohol.

$$CH_3-\underset{\underset{OH}{|}}{\overset{\overset{CH_3}{|}}{C}}-CH_2-CH_2-CH_3$$

1. Identify the longest C chain. Since it is five Cs long, the alcohol is a pentanol.
2. Look at where the hydroxyl groups are attached. An —OH group is located on the second C atom, so the alcohol is a pentan-2-ol.
3. Look to see where any other group(s) are attached. A methyl group is located on the second C atom, so the alcohol is 2-methylpentan-2-ol.
4. Since the second C atom, to which the OH is attached, is attached to three other carbon atoms, the alcohol is a tertiary alcohol.

▶ **COMMUNICATION** example 2

The following alcohol is used as a toxic denaturant for ethanol, as an octane booster in gasoline, as a paint remover, and for manufacturing perfumes. Name this alcohol from its condensed structural formula.

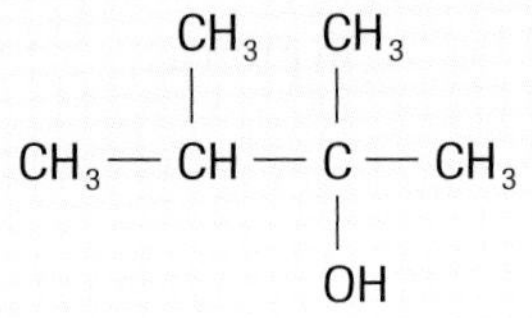

$$\text{CH}_3-\text{CH}-\overset{\overset{\text{CH}_3}{|}}{\underset{\underset{\text{OH}}{|}}{\text{C}}}-\text{CH}_3$$
$$\quad\quad\quad\; \overset{\text{CH}_3}{|}$$

Solution

This tertiary alcohol is 2,3-dimethylbutan-2-ol.

Polyalcohols

Alcohols that contain more than one hydroxyl group are called *polyalcohols*; their names indicate the number and positions of the hydroxyl groups. For example, ethane-1,2-diol (ethylene glycol) is used as antifreeze for car radiators. Propane-1,2,3-triol (glycerol) is safe to consume, in limited quantities. It is a base material in many cosmetics and functions as a moisturizer in foods such as chocolates. Glycerol, also called glycerin and glycerine, is sold in drugstores. Fats and oils are derived from glycerol.

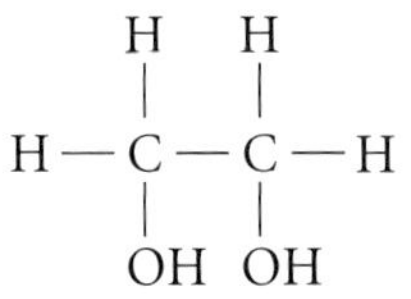

ethane-1,2-diol
(ethylene glycol)

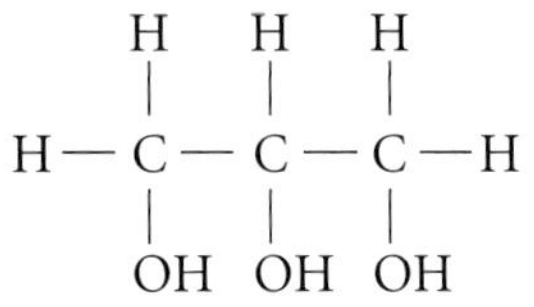

propane-1,2,3-triol
(glycerol)

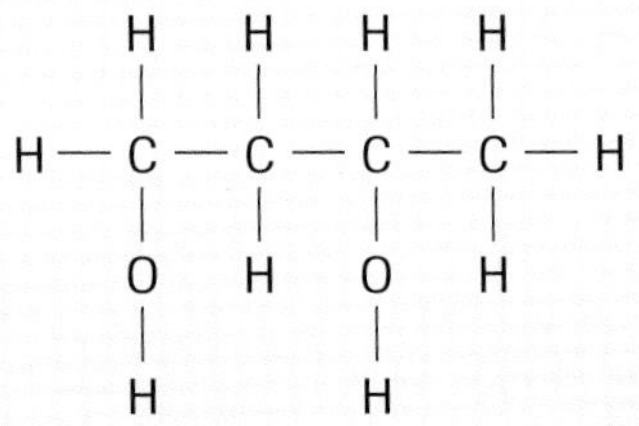

> **DID YOU KNOW ?**
>
> **Glycerol: An Everyday Polyalcohol**
> The moisturizing effect of glycerol (propane-1,2,3-triol) is related to its multiple hydroxyl groups, each capable of hydrogen bonding with water molecules. When glycerol, commonly sold as glycerine in drugstores, is added to a soap bubble solution, the soap film formed contains more fixed water molecules and thus does not readily disintegrate from drying out (evaporation of water molecules).

▶ **SAMPLE** problem **10.4**

Draw a structural formula for butane-1,3-diol.

First, write the C skeleton for the parent molecule, butane.

C — C — C — C

Next, attach an OH group to the first and third C atoms.

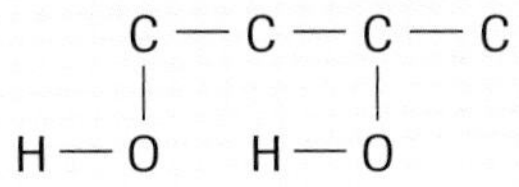

Finally, complete the remaining C bonds with H atoms.

Learning Tip
Ask your teacher whether you will need to learn about cyclo-alcohols or aromatic alcohols. If so, ask about how complex the ones you need to know will be.

Cyclic and Aromatic Alcohols

Chemists have discovered alcohols whose parent compounds are cycloalkanes, cycloalkenes, and benzene. These compounds can become very complex quickly, so you only need to know some of the simplest examples:

- $C_6H_{11}OH(l)$ is cyclohexanol
- $C_6H_5OH(s)$ is phenol (an aromatic alcohol)

You may have heard of phenols when chemists talk about the odour and taste of spring run-off water. The phenols come from the decaying plant matter over which the water runs. Phenols are slightly soluble in water, more soluble in benzene, and very soluble in alcohols.

DID YOU KNOW ?

Grape Expectations
The cosmetics industry has developed a line of "anti-aging" creams that purport to make the skin look younger. The active ingredient is listed as polyphenols (extracted from grape seeds and skins (**Figure 3**), and also from olives). Manufacturers claim that these compounds, described as antioxidants, reverse signs of aging in the skin.

▸ COMMUNICATION example 3

Draw a line structural formula for cyclopentane-1,2-diol.

Solution

[structure: cyclopentane ring with two adjacent OH groups]

Figure 3
Grapes contain polyphenols—antioxidants that can be made into cosmetics.

▸ Practice

5. Write IUPAC names for the following compounds:
    (a) $CH_3-CH-CH_2-CH_3$
              $|$
              OH
    (b) $CH_3-CH-CH_2-CH_2-CH_3$
              $|$
              OH
    (c) [cyclohexane with OH groups at 1,3 positions]

6. Draw a structural formula for
    (a) 3-methylbutan-1-ol
    (b) methylpropan-2-ol
    (c) cyclohexanol
    (d) phenol

7. Draw line structural formulas and name:
    (a) an isomer of butanol that is a secondary alcohol
    (b) the isomers of $C_5H_{11}OH$ that are pentanols

8. Explain briefly why methanol has a higher boiling point than methane.

9. Predict the order of increasing boiling points for the following compounds, and give reasons for your answer.
    (a) ethane, methanol, and fluoromethane
    (b) butan-1-ol, pentane, and 1-chlorobutane

10. Glycerol is more viscous than water, and can lower the freezing point of water. When added to biological samples, it helps to keep the tissues from freezing, thereby reducing damage. From your knowledge of the molecular structure of glycerol, suggest reasons to account for these properties of glycerol.

11. Use your knowledge of intermolecular forces to predict, with reasoning,
    (a) the relative boiling points of ethane, chloroethane, and ethanol.
    (b) the relative solubility in water of ethane, chloroethane, and ethanol.

430 Chapter 10

12. Methanol is produced from methane from natural gas in a two-step process. Write a molecular formula equation to communicate each of the following steps:
    (a) Methane reacts with water to produce carbon monoxide and hydrogen.
    (b) Carbon monoxide and hydrogen then react to produce methanol.
13. Ethanol can be produced by a two-step chemical process. Communicate the process using structural formula equations.
    (a) Ethane from natural gas is cracked to produce ethene.
    (b) Ethene reacts with water in an addition reaction to produce ethanol.
14. Only a few of the simpler alcohols are used in combustion reactions. Alcohol–gasoline mixtures, known as gasohol, are the most common examples. Write a balanced chemical equation, using structural formulas, for the complete combustion of each of the following alcohols:
    (a) ethanol (in gasohol)
    (b) methanol (in gas line antifreeze, **Figure 4**)
15. Ethanol is a beverage alcohol. Calculate the volume of ethanol in each of the drinks in (a), (b), and (c):
    (a) 355 mL of beer (5.0% ethanol by volume)
    (b) 150 mL of wine (12% ethanol by volume)
    (c) 45 mL of rum (40% ethanol by volume)
    (d) How does the quantity of ethanol compare for each of these standard-size drinks?
16. Why is industrial (and school) ethanol denatured (made toxic with, for example, methanol)? Include an economic and a societal perspective in your answer.
17. Alcohols have varying physical properties that may be attributed to intermolecular bonding. Test intermolecular bonding theories by predicting the relative boiling points of the alcohols listed below. Support your prediction with reasoning, based on intermolecular bonding theories. Collect the evidence from a named reference source such as the *CRC Handbook of Chemistry and Physics* or the Internet. Complete Step 2 of an Evaluation.

    Purpose
    The purpose of this investigation is to test intermolecular bonding theories.

    Problem
    What are the relative boiling points of butan-1-ol, methanol, propan-1-ol, and ethanol?

Figure 4
Methanol, sold as methyl hydrate, is used throughout Canada as gas line antifreeze and windshield washer fluid.

Elimination Reactions

The production of alkenes is a very important laboratory and industrial reaction. As you know from Section 10.1, ethene is the cornerstone of the Alberta petrochemical industry. As you will learn in Section 10.5, alkenes are necessary to produce one of the two main classes of polymers. Besides cracking reactions mentioned in Chapter 9 and reviewed below, elimination reactions are a primary source of alkenes—derived from either alcohols or alkyl halides.

Producing Ethene by Cracking Ethane

As indicated in Chapter 9, chemical engineers have devised several methods for producing ethene on an industrial scale. Over time, high-temperature cracking of ethane, as illustrated below, became the preferred technological process. As you can see, molecules of hydrogen are "eliminated" from the ethane.

$$\underset{\text{ethane}}{\text{H}_3\text{C}-\text{CH}_3} \rightarrow \underset{\text{ethene}}{\text{H}_2\text{C}=\text{CH}_2} + \underset{\text{hydrogen}}{\text{H}-\text{H}}$$

> **Learning Tip**
>
> Recall that the ethane to ethene reaction is called cracking, although it is quite different from cracking that involves breaking carbon–carbon bonds. Sometimes tradition determines the name of something against usual reasoning. This process is also sometimes called elimination or dehydrogenation.

Figure 5
A gas generator produces ethene, which speeds the ripening of fruits such as bananas. Ethanol reacts in the gas generator in the presence of an acid catalyst—an elimination reaction.

In review, ethane is cracked into ethene in a world-scale plant at, for example, Joffre, east of Red Deer, Alberta. The ethane used in the cracking process in Alberta comes from natural gas.

Producing Ethene by Elimination Reactions

Ethene was discovered serendipitously in the 17th century by heating ethanol in the presence of sulfuric acid (as a catalyst). Eventually, chemists experimented with many catalysts other than sulfuric acid and found that aluminium oxide and phosphoric acid were effective in this reaction (**Figure 5**).

$$\text{ethanol} \rightarrow \text{ethene} + \text{water}$$

This reaction, called an **elimination reaction**, involves eliminating atoms and/or groups of atoms from adjacent carbon atoms in an organic molecule. In the case of the synthesis of ethene from ethanol, a hydrogen atom and a hydroxyl group on adjacent carbon atoms are eliminated, forming water as a by-product. This particular kind of elimination reaction is also called *dehydration*, because of the apparent removal of water from the alcohol.

Another example of an elimination reaction is the dehydrohalogenation (removal of hydrogen and halogen atoms) of an organic halide to produce an alkene. Ethene can, for example, be produced in the laboratory by reacting chloroethane with potassium hydroxide:

$$\text{chloroethane} + \text{hydroxide ion} \rightarrow \text{ethene} + \text{chloride ion} + \text{water}$$

In this reaction, a hydrogen atom and a halogen atom are eliminated from the alkyl halide to produce the alkene plus a halide ion and a water molecule.

Chemists have identified other elimination reactions and other ways to synthesize ethene and other alkenes. Research has indicated, however, that the above two methods (starting with ethanol or chloroethane) are the best choices for the laboratory. From an industrial perspective, the cracking of abundant fossil fuel alkanes to produce alkenes is the preferred method.

> ### ▶ COMMUNICATION example 4
>
> The elimination of a halide ion from an organic halide is the most common method for preparing a specific alkene in the laboratory. Write a structural formula equation for the preparation of but-2-ene from 2-chlorobutane, in the presence of a strong base.
>
> **Solution**

Section 10.3

▶ COMMUNICATION example 5

The elimination of the elements of a water molecule from an alcohol is the second most common method for preparing a specific alkene in the laboratory. Write a structural formula equation for the preparation of but-2-ene from butan-2-ol, in the presence of a catalyst.

Solution

$$\begin{array}{c} \text{H H H H} \\ \text{H}-\text{C}-\text{C}-\text{C}-\text{C}-\text{H} \\ \text{H H OH H} \end{array} \rightarrow \begin{array}{c} \text{H H H H} \\ \text{H}-\text{C}-\text{C}=\text{C}-\text{C}-\text{H} \\ \text{H} \quad\quad \text{H} \end{array} + H_2O$$

▶ Practice

18. Write a structural formula equation to represent the synthesis of ethene by reacting bromoethane with a strong base.
19. Alkenes can be manufactured by three different chemical processes. Write structural formula equations for each of the following hypothetical chemical processes:
    (a) Butane is cracked into but-2-ene.
    (b) 1-chlorobutane undergoes an elimination reaction in a strongly basic solution to produce but-1-ene, chloride ions, and water.
    (c) butan-1-ol undergoes an elimination reaction to produce but-1-ene and water.
20. Write structural formula equations for the synthesis of propene by three different processes.
21. Chemical engineers face many difficulties during the technological design of ethane cracking plants (**Figure 6**). Suggest a different solution to each of the following problems encountered when ethane is thermally cracked:
    (a) 40% of the ethane remains uncracked after coming from the ethane cracker
    (b) ethyne (acetylene) is produced in addition to ethene
    (c) hydrogen and methane are produced

Figure 6
Alberta is Canada's top producer of petrochemicals and home to the world's two largest ethane-based processing plants—at Joffre, east of Red Deer, and Fort Saskatchewan, northeast of Edmonton. This photo shows the largest of the three ethylene plants located at the Joffre site. Significant infrastructure supports the Joffre site facilities: feedstock pipelines, an ethane extraction plant, and a cogeneration power plant.

INVESTIGATION 10.2 Introduction

Isomers of Butanol

Report Checklist
- ○ Purpose
- ○ Problem
- ● Hypothesis
- ● Prediction
- ○ Design
- ● Materials
- ○ Procedure
- ● Evidence
- ● Analysis
- ● Evaluation (1, 2, 3)

Organic halides have many uses, but they are not commonly found in nature. Organic halides must, therefore, be synthesized from other compounds. On an industrial scale, organic halides are produced by addition and substitution reactions of hydrocarbons. A common laboratory-scale process is the halogenation of alcohols in which the hydroxyl group of the alcohol is replaced by a halogen. This reaction requires the presence of a strong acid that contains a halogen, such as hydrochloric acid.

$$R-OH(l) + HX(aq) \rightarrow R-X(l) + H_2O(l)$$

Organic halides have much lower solubilities in water compared with their corresponding alcohols.

Purpose

The purpose of this investigation is to test a personal hypothesis about the relative reactivity of the alcohol isomers of butan-1-ol.

Problem

What is the difference in reactivity, if any, of the alcohol isomers of butan-1-ol with concentrated hydrochloric acid?

Design

Samples of butan-1-ol, butan-2-ol, and 2-methylpropan-2-ol are mixed with concentrated hydrochloric acid. Evidence of reaction is obtained by looking for a low-solubility organic halide product (cloudy mixture).

To perform this investigation, turn to page 462.

Alcohols and Elimination Reactions

Alcohols

Functional group:

–OH, hydroxyl group

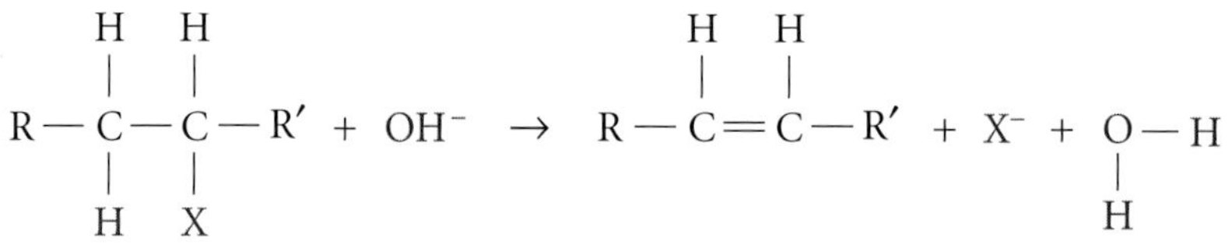

Learning Tip

The types of organic reactions that you are responsible for learning to this point in the unit are:

1. photosynthesis and respiration
2. hydrocracking
3. catalytic reforming
4. complete and incomplete combustion
5. substitution
6. addition
7. elimination

Naming alcohols:

- Drop the "e" from the alkane name and add "ol"; e.g., ethane becomes ethanol.
- If necessary, add a number (or numbers) to communicate where the hydroxyl group(s) is (are) located; e.g., propan-1-ol and propan-2-ol.
- If the alcohol has two or three hydroxyl groups, it is a diol or a triol, respectively; e.g., ethane-1,2-diol and propane-1,2,3-triol. For diols and triols, do not drop the "e" from the alkane name.

Preparation:

Addition reactions with water

- alkenes + water → alcohols

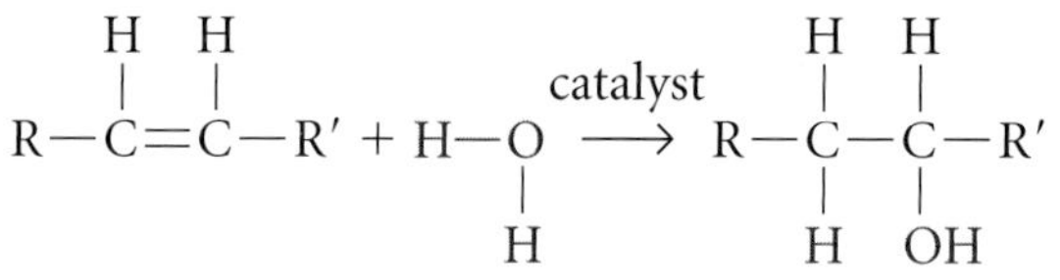

$$\underset{\substack{|\ \ |\\H\ \ H}}{R-C=C-R'} + H-\underset{H}{O} \xrightarrow{\text{catalyst}} \underset{\substack{|\ \ |\\H\ \ OH}}{R-\underset{\ }{C}-\underset{\ }{C}-R'}$$

(R and R' can be the same or different alkyl groups. Isomers often result.)

Elimination Reactions

- alcohols → alkenes + water

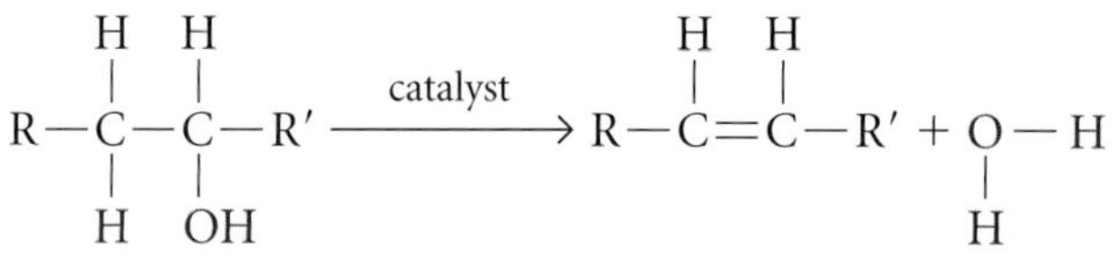

$$\underset{\substack{|\ \ |\\H\ \ OH}}{R-\underset{\substack{|\\H}}{C}-\underset{\ }{C}-R'} \xrightarrow{\text{catalyst}} \underset{\substack{|\ \ |\\H\ \ H}}{R-C=C-R'} + \underset{H}{O}-H$$

- organic halides + OH⁻ → alkenes + halide ion + water

$$\underset{\substack{|\ \ |\\H\ \ X}}{R-\underset{\substack{|\\H}}{C}-\underset{\ }{C}-R'} + OH^- \rightarrow \underset{\substack{|\ \ |\\H\ \ H}}{R-C=C-R'} + X^- + \underset{H}{O}-H$$

 WEB Activity

Web Quest—Cellulosic Ethanol

This Web Quest focuses on the technological, environmental, and ecological impacts of building a cellulosic ethanol plant. In the past, ethanol was widely used as a fuel, only to be replaced by abundant and inexpensive gasoline. What are the issues with creating and using cellulosic ethanol? Does it have the potential to provide a realistic alternative to gasoline? You will examine the cellulosic ethanol plant from the point of view of a consulting firm, and present your supported conclusions to the community.

www.science.nelson.com

Section 10.3 Questions

1. Write structural formulas and IUPAC names for all saturated alcohols with four carbon atoms and one hydroxyl group.
2. Explain why the propane that is used as fuel in a barbecue is a gas at room temperature, but propan-2-ol, used as rubbing alcohol, is a liquid at room temperature.
3. Draw the structural formulas and write the IUPAC names of the two alkenes that are formed when hexan-2-ol undergoes an elimination reaction in the presence of an acid catalyst.
4. Write an equation using structural formulas to show the production of each of the following alcohols from an appropriate alkene:
   (a) butan-2-ol
   (b) methylpropan-2-ol
5. A major use of organic halides is in the preparation of unsaturated compounds. Predict the products of the following elimination reaction. Write a word equation and a structural formula equation.

   ```
       H   H
       |   |
   H — C — C — H  +  OH⁻  →
       |   |
       H   Cl
   ```

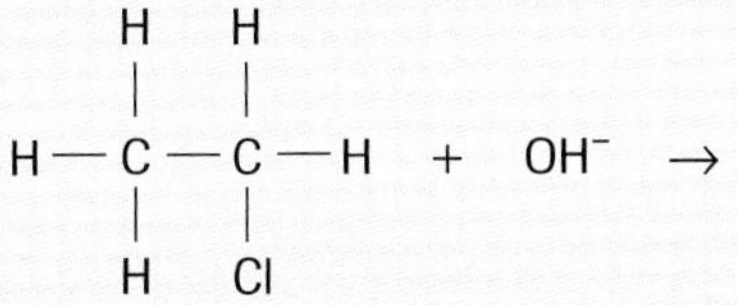

6. Classify and write structural formula equations for the following catalyzed organic reactions:
   (a) ethene + water → ethanol
   (b) butan-2-ol → but-1-ene + but-2-ene + water
   (c) ethene + hypochlorous acid (HOCl(aq)) → 2-chloroethanol
   (d) cyclohexanol + oxygen →
7. Draw structural formulas to represent the elimination reaction of 2-chloropentane to form an alkene. Include reactants, reaction conditions, and all possible products and their IUPAC names.
8. The nomenclature of organic compounds is similar across classes of compounds. Name the following alkanes, alkenes, and alcohols.
   (a) $CH_3—CH_2—CH_2—CH_3$
   (b) $CH_3—CH(CH_3)—CH_3$
   (c) $CH_3—CH=CH—CH_3$
   (d) $CH_3—C(CH_3)=CH_2$
   (e) $CH_3—CH_2—CH_2—CH_2—OH$
   (f) $CH_3—CH_2—CH(OH)—CH_3$
   (g) $CH_3—CH(CH_3)—CH_2—OH$
   (h) $CH_3—C(CH_3)(OH)—CH_3$
9. Draw line structural formulas for the following alcohols.
   (a) propan-2-ol
   (b) phenol (hydroxybenzene)
   (c) propane-1,2,3-triol (glycerol)
   (d) cyclohexanol
10. Complete the Hypothesis (including reasoning), Analysis, and Evaluation (2, 3) for the following investigation report:

    Purpose
    The purpose of this investigation is to test a hypothesis concerning the relative boiling points and solubility of alcohols. Assume that solubility means in water.

    Problems
    (a) What is the trend in boiling points for $C_1–C_6$ primary alcohols?
    (b) What is the trend in solubility for $C_1–C_6$ primary alcohols?

    Evidence

    Table 1 Boiling Points and Solubilities of Various Alcohols

Alcohol	Boiling point (°C)	Solubility (mL/100 mL)
methanol	65	miscible
ethanol	78	miscible
propan-1-ol	97	miscible
butan-1-ol	117	9.1
pentan-1-ol	138	3.0
hexan-1-ol	157	slight

11. Use the information in this section and from your own research to continue gathering perspective statements concerning the statement that we should be saving more fossil fuels for petrochemical use in the future.

Extensions

12. Alcohols have gained increased popularity as an additive to gasoline, as a fuel for automobiles. "Gasohols" may contain up to 10% methanol and ethanol, and are considered more environmentally friendly than gasoline alone.
    (a) Write balanced chemical reaction equations for the complete combustion of methanol and ethanol.
    (b) When small amounts of water are present in the gasoline in the gas lines of a car, the water may freeze and block gasoline flow. Explain how using a gasohol would affect this problem.

    www.science.nelson.com

13. Radiator antifreeze and coolant is ethane-1,2-diol (ethylene glycol). Ethylene glycol is usually mixed 50:50 with water for use in the radiator. Large quantities of this chemical product are produced near Fort Saskatchewan and Scotford, Alberta. Ethene reacts with oxygen in one reactor to produce ethylene oxide ($C_2H_4O(g)$). The ethylene oxide then reacts with water in a second reactor to produce ethylene glycol.
    (a) Write structural formula equations for these reactions.
    (b) Explain why ethylene glycol is a better choice as a radiator antifreeze and coolant than ethanol.
    (c) Explain why methanol, rather than ethylene glycol, is used as a windshield wiper antifreeze.
    (d) Which is more toxic, methanol or ethylene glycol?
    (e) Use a reference to find the freezing point of a 50:50 mixture of ethylene glycol and water.

    www.science.nelson.com

10.4 Carboxylic Acids, Esters, and Esterification Reactions

Learning Tip

The general formula for a carboxylic acid is

$$\text{R or (H)} - \overset{\overset{\displaystyle O}{\|}}{C} - OH$$

The functional group for carboxylic acids is the carboxyl group, –COOH.

Natural products and processes involve a wide variety of organic chemicals. In this section, you will learn what chemists and chemical engineers have discovered about some organic products and processes. Chemists, like you, started off learning the basics of organic chemistry—just as you have learned about inorganic chemistry. The study of most topics in science begins with classification based on the physical and chemical properties of the chemicals involved. Nomenclature—providing names for the classes of chemicals and for individual chemicals—usually follows quickly after the classification process. The parallel process is most often the creation of theories to explain the properties of each of the classes of compounds.

Carboxylic Acids

The family of organic compounds known as **carboxylic acids** contain the **carboxyl group**, –COOH, which includes both the carbonyl and hydroxyl functional groups.

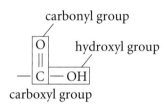

Note that, because the carboxyl group involves three of the carbon atom's four bonds, the carboxyl group is always at the end of a carbon chain or branch.

Chemists explain the characteristic properties of carboxylic acids by the presence of the carboxyl group. Carboxylic acids occur naturally in citrus fruits, crabapples, rhubarb, and other foods characterized by the sour, tangy taste of acids. Carboxylic acids also have distinctive odours (**Figure 1**).

As we might predict from the structure of carboxylic acids, the molecules of these compounds are polar and form hydrogen bonds both with each other and with water molecules. These acids exhibit the same solubility behaviour as alcohols. The smaller members (one to four carbon atoms) of the acid series are miscible with water, whereas larger ones are virtually insoluble. Aqueous solutions of carboxylic acids have the properties of acids; a litmus test can distinguish these compounds from other hydrocarbon derivatives. The smaller carboxylic acids are all liquids at room temperature. The dicarboxylic acids (even the small ones) are solids at room temperature, as are the larger-molecule carboxylic acids.

Carboxylic acids are named by replacing the -e ending of the corresponding alkane name with -oic, followed by the word "acid." The first member of the carboxylic acid family is methanoic acid, HCOOH, commonly called formic acid (**Figure 2**). Methanoic acid is used in removing hair from hides and in coagulating and recycling rubber.

Figure 1
Tracking dogs, with their acute sense of smell, are trained to identify odours in police work. As carboxylic acids have distinctive odours, the dogs may follow the characteristic blend of carboxylic acids in a person's sweat. Trained dogs are also used to seek out illegal drug laboratories by the odour of acetic acid. Acetic acid is formed as a byproduct when morphine, collected from opium poppies, is treated to produce heroin.

Figure 2
(a) Most ants and ant larvae are edible and are considered quite delicious. They have a vinegary taste because they contain methanoic acid, HCOOH, commonly called formic acid. In some countries, large ants are squeezed directly over a salad to add the tangy ant juice as a dressing.
(b) This ball-and-spring model represents the evidence chemists have collected on methanoic acid.

Ethanoic acid, CH₃COOH(l), commonly called acetic acid, is the compound that makes vinegar taste sour (**Figure 3(a)**). Wine vinegar and cider vinegar are produced naturally when sugar in fruit juices is fermented first to alcohol, and then to ethanoic acid. This acid is employed extensively in the textile dyeing process and as a solvent for other organic compounds.

glucose → ethanol → ethanoic acid (acetic acid)

fruit juice → wine → vinegar

Some acids contain two or three carboxyl groups. For example, oxalic acid, used in commercial rust removers and in copper and brass cleaners, consists of two carboxyl groups bonded together. Oxalic acid (**Figure 3(b)**) occurs naturally in rhubarb, tartaric acid occurs in grapes, and citric acid occurs in citrus fruits. Due to the extra hydrogen bonding, all three of these acids are solids as pure substances at room temperature.

```
                                          CH₂—COOH
                                           |
    COOH        HO—CH—COOH        HO—C—COOH
     |               |                    |
    COOH        HO—CH—COOH            CH₂—COOH
  oxalic acid    tartaric acid         citric acid
```

Figure 3
Oxalic acid (**b**), found in rhubarb, differs from pure acetic (ethanoic) acid, CH₃COOH(l) (**a**), in that its molecule contains an additional carboxyl group. This increased polarity and hydrogen bonding explain why oxalic acid is a solid while pure acetic acid is a liquid at the same temperature.

You do not have to memorize these names and condensed structural formulas, or those of any other polycarboxylic acids. However, you should recognize that they are carboxylic acids (from the carboxyl functional groups). You should also recognize that multiple hydrogen bonding is possible, which increases the melting and boiling points and solubility in water of these compounds.

LAB EXERCISE 10.B

Explaining Physical Property Trends

Report Checklist

- ○ Purpose   ○ Design   ● Analysis
- ○ Problem   ○ Materials   ● Evaluation (3)
- ○ Hypothesis   ○ Procedure
- ○ Prediction   ○ Evidence

Many physical properties of organic compounds can be understood using knowledge of the functional groups present and of intermolecular forces. Complete the Analysis and Evaluation of the investigation report.

Purpose

The purpose of this investigation is to use your knowledge of organic structures and intermolecular forces to explain some physical properties of alcohols and carboxylic acids.

Problem

What is the explanation for the trends in boiling points and solubilities in water within and between series of simple alcohols and carboxylic acids?

Evidence

Table 1 Boiling Points and Solubilities of Various Alcohols and Carboxylic Acids

Alcohol	Boiling point (°C)	Solubility in water (g/100 mL)	Carboxylic acid	Boiling point (°C)	Solubility in water (g/100 mL)
methanol	65	miscible	methanoic acid	101	miscible
ethanol	79	miscible	ethanoic acid	118	miscible
propan-1-ol	97	miscible	propanoic acid	141	miscible
butan-1-ol	117	8.0	butanoic acid	164	miscible
pentan-1-ol	137	2.7	pentanoic acid	186	2.4

DID YOU KNOW ?

Water-Soluble Vitamins
With its many polar hydroxyl groups, vitamin C (ascorbic acid) is highly water-soluble. The water-soluble vitamins are not stored in the body; rather, they are readily excreted in the urine. It is, therefore, important that we include these vitamins in our daily diet. Although taking too much vitamin C is not dangerous, taking excessive amounts is truly sending money down the drain.

ascorbic acid
(Vitamin C)

Learning Tip

The technological applications of the following chemicals should be memorized as common knowledge:

$CH_4(g)$	natural gas
$C_3H_8(g)$	barbecue fuel
$C_4H_{10}(g)$	lighter fluid
$C_8H_{18}(l)$	in gasoline
$C_2H_4(g)$	polyethylene monomer
$C_2H_2(g)$	cutting torch fuel
$C_6H_6(l)$	in gasoline
$CH_3OH(l)$	gasoline antifreeze
$C_2H_5OH(l)$	beverage alcohol
CH_2OHCH_2OH	radiator antifreeze
$CH_3COOH(l)$	in vinegar

▶ SAMPLE problem 10.5

Write the IUPAC name and the full structural formula for propanoic acid.

The prefix *propan-* indicates that the acid contains three C atoms; the end C atom is in the carboxyl group.
The structural formula for propanoic acid is

$$H-\underset{\underset{H}{|}}{\overset{\overset{H}{|}}{C}}-\underset{\underset{H}{|}}{\overset{\overset{H}{|}}{C}}-\overset{\overset{O}{\|}}{C}-OH$$

▶ COMMUNICATION example 1

What is the IUPAC name for this carboxylic acid, which has an unpleasant, rancid odour?

$$\underset{\underset{CH_3}{|}}{CH_2}-CH_2-COOH$$

Solution

This condensed structural formula represents butanoic acid.

▶ Practice

1. Draw a structural formula for each of the following compounds:
   (a) octanoic acid
   (b) benzoic acid
   (c) ethanoic (acetic) acid

2. Give IUPAC and, if applicable, common names for these molecules:
   (a) $$H-\overset{\overset{O}{\|}}{C}-OH$$
   (b) $HOOC-CH_2-CH_2-CH_2-CH_3$
   (c) $CH_3-CH_2-CH_2-CH_2-CH_2-COOH$

Esterification Reactions

Carboxylic acids undergo a variety of organic reactions. In a **condensation reaction**, a carboxylic acid combines with another reactant, forming two products—an organic compound and a compound such as water. For example, a carboxylic acid can react with an alcohol, forming an **ester** and water. This type of condensation reaction is known as an **esterification reaction**.

$$CH_3-\overset{\overset{O}{\|}}{C}-OH \;+\; HO-CH_3 \;\rightarrow\; CH_3-\overset{\overset{O}{\|}}{C}-O-CH_3 \;+\; H-\overset{\overset{}{}}{\underset{\underset{}{}}{O}}-H$$

carboxylic acid + alcohol → an ester + water

Esters

The **ester functional group**, –COO–, is similar to that of an acid, except that the hydrogen atom of the carboxyl group has been replaced by a hydrocarbon branch. Esters occur naturally in many plants and are responsible for some of the odours of fruits and flowers. Esters are often added to foods to enhance aroma and taste. Other commercial applications include cosmetics, perfumes, synthetic fibres, and solvents.

The odour of food strongly influences its flavour. Artificial flavourings are made by mixing synthetic esters to give the approximate odour (such as raspberry or banana) of the natural substance. For artificial fruit flavours, organic acids are usually added to give the sharp taste characteristic of fruit. Artificial flavours can only approximate the real thing, because it would be too costly to include all the components of the complex mixture of compounds present in the natural fruit or spice. **Table 2** shows the main esters used to create the odours of certain artificial flavours. The names are covered next.

> **Learning Tip**
> The general formula for an ester is
> $$R \text{ or } (H) - \overset{\overset{O}{\|}}{C} - O - R'$$
> where R′ is the branch from the alcohol and R (or H) is from the acid.

Table 2 Odours of Selected Esters

Odour	Name	Formula
apple	methyl butanoate	$CH_3-CH_2-CH_2-COO-CH_3$
apricot	pentyl butanoate	$CH_3-(CH_2)_2-COO-(CH_2)_4-CH_3$
banana	3-methylbutyl ethanoate	$CH_3-COO-CH_2-CH_2-\underset{\underset{CH_3}{\|}}{CH}-CH_3$
cherry	ethyl benzoate	$C_6H_5-COO-CH_2-CH_3$
orange	octyl ethanoate	$CH_3-COO-(CH_2)_7-CH_3$
pineapple	ethyl butanoate	$CH_3-CH_2-CH_2-COO-CH_2-CH_3$
red grape	ethyl heptanoate	$CH_3-(CH_2)_5-COO-CH_2-CH_3$
rum	ethyl methanoate	$H-COO-CH_2-CH_3$
wintergreen	methyl salicylate	(OH-substituted benzene ring)—C(=O)—O—CH₃

> **Learning Tip**
> Note that the names for esters have a space between the prefix and the principal group, for example, ethyl butanoate. These "organic salts" are named like inorganic salts; e.g., sodium chloride.

Naming and Preparing Esters

Esters are organic "salts" formed from the reaction of a carboxylic acid and an alcohol. Consequently, the name of an ester has two parts. The first part is the name of the alkyl group from the alcohol used in the esterification reaction. The second part comes from the acid. The ending of the acid name is changed from *-oic* acid to *-oate*. For example, in the reaction of ethanol and butanoic acid, the ester formed is ethyl butan*oate*, an ester with a banana odour. A strong acid catalyst, such as $H_2SO_4(aq)$, is used to increase the rate of this organic reaction, along with some careful heating.

> **Learning Tip**
> Note the colour coding of the reactant segments that show the movement of the O atoms and that also assist with naming the ester.

$$CH_3-CH_2-CH_2-\overset{\overset{O}{\|}}{C}-OH + CH_3-CH_2-OH \longrightarrow CH_3-CH_2-CH_2-\overset{\overset{O}{\|}}{C}-O-CH_2-CH_3 + H-O-H$$

| butanoic acid | + | ethanol | → | ethyl butanoate | + | water |
| carboxylic acid | + | alcohol | → | ester | + | water |

DID YOU KNOW?

Radioactive Tracing

Radioactive tracing can be used by medical personnel to trace the movement of chemicals in your body. Chemists use a similar approach to trace the movement of atoms in a chemical reaction. For example, if the O atom in propan-1-ol is manipulated to be radioactive (O-18), and the resulting ester is radioactive, then the radioactive O-18 atom has moved to the ester. The hydroxyl (–OH) group from the acid, therefore, must move to the water. Radioactive tracing is not important for predicting the products, but it provides chemists with deeper theoretical understanding.

The general formula of an ester is written as RCOOR′. When read from left to right, RCO– comes from the carboxylic acid, and –OR′ comes from the alcohol. Hence, $CH_3COOCH_2CH_2CH_3$ is propyl ethanoate. Note that, for an ester, the acid is the first part of its formula as drawn, but is the second part of its name.

$$C_3H_7COOH + CH_3OH \rightarrow C_3H_7COOCH_3 + H_2O$$
butanoic acid   methanol   methyl butanoate

▶ **SAMPLE** problem **10.6**

Draw a line structural formula equation and write the IUPAC name for the ester formed in the reaction between propan-1-ol and benzoic acid.

To name the ester:
- The first part of the name comes from the alcohol: propyl.
- The second part of the name comes from the acid: benzoate.
- The IUPAC name of the ester is propyl benzoate.

To draw the line structural formula:
- Draw structural formulas of the reactants and complete the condensation reaction.

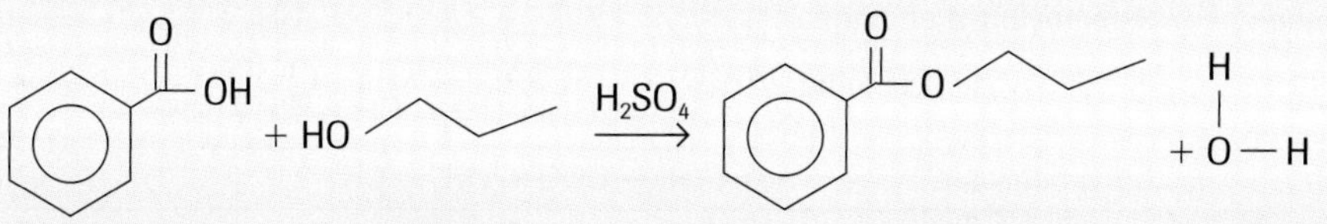

    benzoic acid   propan-1-ol   propyl benzoate   water

▶ **SAMPLE** problem **10.7**

Write a condensed structural formula equation for the esterification reaction to produce the ester $CH_3CH_2CH_2COOCH_2CH_3$. Write IUPAC names for each reactant and product.

First, identify the acid (four carbons—butanoic acid) and the alcohol (two carbons—ethanol) that may be used in the synthesis of the ester. Then write condensed structural formulas and include the conditions in the chemical equation.

$$CH_3-CH_2-CH_2-COOH + HO-CH_2-CH_3 \xrightarrow{H_2SO_4}$$
           butanoic acid   ethanol

$$CH_3-CH_2-CH_2-COO-CH_2-CH_3 + H_2O$$
                            ethyl butanoate   water

▶ **COMMUNICATION** example **2**

Name the ester CH_3COOCH_3 and the acid and alcohol from which it can be prepared.

What conditions are necessary?

Solution

The ester is methyl ethanoate, and it can be prepared from methanol and ethanoic acid.

A strong acid catalyst, such as $H_2SO_4(aq)$, is required, along with some heating.

Section 10.4

> **Practice**

3. Write a structural formula equation to illustrate the synthesis of each of the following esters from an alcohol and an acid. Refer to **Table 2** and identify the odour of each ester formed.
   (a) ethyl methanoate
   (b) ethyl benzoate
   (c) methyl butanoate
   (d) 3-methylbutyl ethanoate

4. Name each of the following esters, and the acids and alcohols from which they could be prepared:
   (a) $CH_3CH_2COOCH_2CH_3$
   (b) $CH_3CH_2CH_2COOCH_3$
   (c) $HCOOCH_2CH_2CH_2CH_3$
   (d) $CH_3COOCH_2CH_2CH_3$

5. Name the carboxylic acid and alcohol required to produce each of the following ester odours:
   (a) apricot
   (b) orange

6. The graph in **Figure 4** communicates evidence gathered for the boiling points of straight-chain carboxylic acids, methyl esters, and alkanes. (Methyl esters are prepared with methanol and their names begin with "methyl.") The boiling points are plotted against the molar masses of these organic compounds. The scientific purpose of this task is to test the explanatory power of the intermolecular force theories known to you.

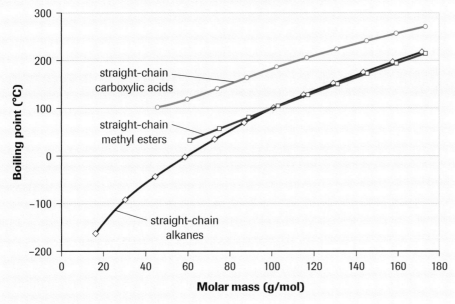

Figure 4
Boiling points of straight-chain carboxylic acids, methyl esters, and alkanes

(a) Explain the same upward trend in boiling points of the three classes of straight-chain organic compounds.
(b) Explain the relatively close boiling point values for the alkanes and the methyl esters.
(c) Explain the difference in boiling points for carboxylic acids versus alkanes and methyl esters with the same molar masses.

BIOLOGY CONNECTION

Natural Esters
Esters occur naturally. Beeswax, for example, is a natural ester containing chains 40–46 carbon atoms long. Plants' leaves are coated in waxy esters to minimize water loss; insects and animals may also have a surface coating (including skin) that includes esters. Chemical analysis of the coating on leaves shows that these natural esters (sometimes called wax esters) contain 34–62 carbon atoms. Of the natural waxes present on the surface of apples, for example, about 18% are esters.

www.science.nelson.com

Learning Tip

You will find it helpful to memorize the functional groups for the classes of organic compounds that you have studied. A short summary (with the functional group in parentheses) is:

alkenes; e.g., $CH_2{=}CH_2$
(double bond)

alkynes; e.g., $CH{\equiv}CH$
(triple bond)

aromatics; e.g., $C_6H_5{-}CH_3$
(benzene ring(s))

organic halides; e.g., $CH_3{-}Cl$; $R{-}X$
($X{=}F$, Cl, Br and/or I)

alcohols; e.g., $CH_3{-}OH$; $R{-}OH$ (hydroxyl group, $-OH$)

carboxylic acids; e.g., $CH_3{-}COOH$; $R(H){-}COOH$ (carboxyl group, $-COOH$)

esters; e.g., $CH_3{-}COO{-}C_2H_5$; $R(H)COOR'$ (ester group, $-COO-$)

R is a hydrocarbon chain/group. R(H) means that the R can be an H.

(d) Do your intermolecular force theories pass the test of being able to explain the relative boiling points of these straight-chain compounds? Give reasons for your answer.
(e) Why is the test restricted to comparing straight-chain compounds?
(f) Predict the position of the line representing the boiling point of alcohols, placed on this same set of axes.
(g) Referenced values of boiling points for some straight-chain alcohols include 64 °C for methanol, 97 °C for propan-1-ol, 138 °C for pentan-1-ol, and 176 °C for heptan-1-ol. Is your prediction in (f) verified or falsified? If falsified, critique your prediction and/or the hypothesis upon which you based your prediction.

7. What are some pros and cons, from at least two perspectives, for using artificial flavouring (manufactured esters) instead of natural flavours?

INVESTIGATION 10.3 Introduction

Synthesis of an Ester

Report Checklist
- ○ Purpose
- ○ Problem
- ○ Hypothesis
- ○ Prediction
- ○ Design
- ○ Materials
- ○ Procedure
- ● Evidence
- ● Analysis
- ○ Evaluation

Purpose
The purpose of this investigation is to use the esterification generalization and diagnostic tests to synthesize and observe the properties of two esters.

Problem
What are some physical properties of ethyl ethanoate (ethyl acetate) and methyl salicylate?

Design
The esters are produced by the reaction of appropriate alcohols and acids, using sulfuric acid as a catalyst. The solubility and the odours of the esters are observed.

To perform this investigation, turn to page 463.

INVESTIGATION 10.4 Introduction

Testing with Models

Report Checklist
- ○ Purpose
- ○ Problem
- ○ Hypothesis
- ○ Prediction
- ○ Design
- ● Materials
- ○ Procedure
- ● Evidence
- ○ Analysis
- ○ Evaluation

Molecular models, either physical or electronic, are based on empirically known molecular structures. These models can then be considered as representing evidence collected in a variety of experiments. Recall stereochemical formulas from Chapter 3.

Purpose
The purpose of this investigation is to test stereochemical formula equations by using molecular models.

Problem
What are the stereochemical formula equations for the following reactions?
(a) methane undergoes complete combustion
(b) ethane is cracked into ethene
(c) propane reacts with chlorine
(d) but-2-ene reacts with water
(e) ethanol eliminates water to produce ethene
(f) 1-chloropropane undergoes an elimination reaction with hydroxide ions to produce propene, a chloride ion, and water
(g) ethanol reacts with methanoic acid

Design
Physical or electronic models of the reactants and products for each reaction are constructed. The bonding and shapes of the molecules from the models are compared with the predictions.

To perform this investigation, turn to page 464.

SUMMARY: Carboxylic Acids, Esters, and Esterification

Functional groups
- carboxylic acid: –COOH carboxyl group

$$-\overset{\overset{O}{\|}}{C}-OH$$

- ester: –COO– alkylated carboxyl group

$$-\overset{\overset{O}{\|}}{C}-O-$$

Esterification reaction

carboxylic acid + alcohol → ester + water

$$R-\overset{\overset{O}{\|}}{C}-OH + R'-OH \xrightarrow[\text{heat}]{\text{conc. } H_2SO_4} R-\overset{\overset{O}{\|}}{C}-O-R' + H-\overset{H}{\underset{}{O}}$$

carboxylic acid   alcohol   ester   water

(For esterification, R can be replaced by an H, but R' cannot be replaced by an H.)

e.g., $CH_3-CH_2-CH_2-\overset{\overset{O}{\|}}{C}-OH + CH_3-OH \rightarrow CH_3-CH_2-CH_2-\overset{\overset{O}{\|}}{C}-O-CH_3 + H-O-H$

butanoic acid   + methanol →   methyl butanoate   + water

Learning Tip

Evidence of high boiling points and solubility indicates that, of the classes of organic compounds studied in this section, the following classes exhibit hydrogen bonding:

alcohols   R—OH
carboxylic acids
            R or H—COOH

Esters, R^1COOR^2, do not exhibit hydrogen bonding.

Section 10.4 Questions

1. Prepare a table with the headings: Family, General Formula, and Naming System. Complete the table for the following organic families:
   (a) alcohols
   (b) carboxylic acids
   (c) esters

2. In what way is the functional group of an ester different from that of a carboxylic acid? How does this difference account for any differences in properties?

3. Design an experimental procedure for the synthesis of an ester, given ethanol and acetic acid. Describe the steps in the procedure, the safety equipment required, and the precautions needed in the handling and disposal of the materials.

4. Esters are often referred to as organic salts, and the esterification reaction is sometimes considered a neutralization reaction. An ester is similar, in some ways, to an ionic compound.
   (a) Use chemical formulas and equations to identify similarities and differences between esters and inorganic salts.
   (b) Design an investigation to determine which of two unlabelled pure samples provided to you is an ester and which is an ionic compound. List several diagnostic tests.

5. Many organic compounds have more than one functional group in a molecule. Copy the following structural formulas. Circle and label the functional groups for an alcohol, a carboxylic acid, and/or an ester. Suggest either a source or a use for each of these substances.

(a) vanillin

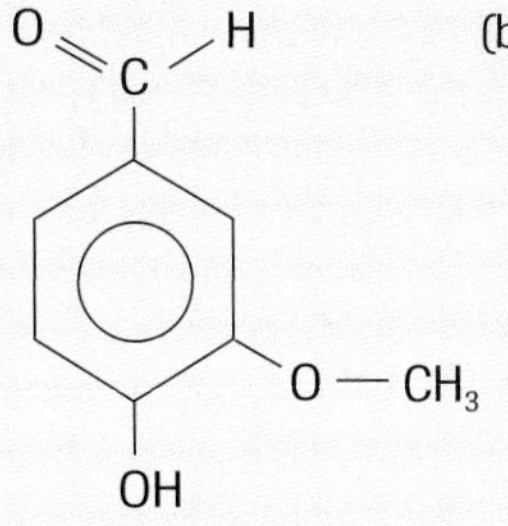

(b) acetylsalicylic acid (ASA)

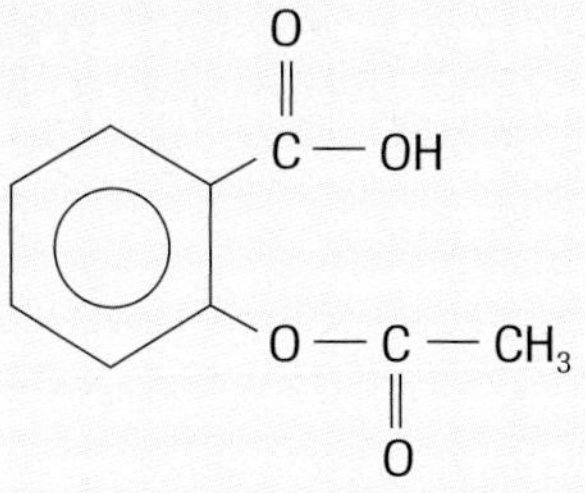

(c) $HOOC-CH_2-\underset{\underset{COOH}{|}}{\overset{\overset{OH}{|}}{C}}-CH_2-COOH$

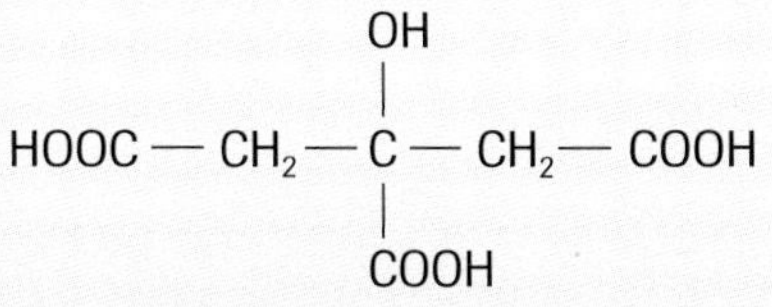

citric acid

6. Carboxylic acids, like inorganic acids, can be neutralized by bases. Carboxylic acids also undergo organic reactions. Classify the following reactions as neutralization or esterification. Write the complete condensed structural formulas and word equations for the reactions.

   (a) $CH_3-\underset{\underset{O}{\|}}{C}-OH \;+\; NaOH \rightarrow$

   (b) $CH_3-CH_2-\underset{\underset{O}{\|}}{C}-OH \;+\; CH_3-OH \rightarrow$

   (c) benzoic acid + potassium hydroxide →
   (d) ethanol + methanoic acid →

7. Fats and oils are naturally occurring esters that store chemical energy in plants and animals. Fatty acids, such as octadecanoic acid (also known as stearic acid), typically combine with propane-1,2,3-triol, known as glycerol, to form fat, a triester. Complete the following chemical equation by predicting the structural formula for the ester product.

   $$\begin{array}{l} CH_2OH \\ | \\ CHOH \\ | \\ CH_2OH \end{array} \;+\; CH_3(CH_2)_{16}COOH \rightarrow$$

   glycerol + stearic acid

8. Classify the chemicals and write a complete balanced structural formula equation for each of the following predicted organic reactions. Where possible, classify the reactions and name the compounds as well. Assume that catalysts are used where required (for most of the reactions).
   (a) $C_2H_6 + Br_2 \rightarrow C_2H_5Br + HBr$
   (b) $C_3H_6 + Cl_2 \rightarrow C_3H_6Cl_2$
   (c) $C_6H_6 + I_2 \rightarrow C_6H_5I + HI$
   (d) $CH_3CH_2CH_2CH_2Cl + OH^- \rightarrow CH_3CH_2CHCH_2 + H_2O + Cl^-$
   (e) $C_3H_7COOH + CH_3OH \rightarrow C_3H_7COOCH_3 + H_2O$
   (f) $C_2H_5OH \rightarrow C_2H_4 + H_2O$
   (g) $C_6H_5CH_3 + O_2 \rightarrow CO_2 + H_2O$

9. Complete the Prediction, Materials, Analysis, and Evaluation (2, 3) of the investigation report.

   Purpose
   The purpose of this investigation is to test the esterification reaction generalization.

   Problem
   What is the product of the reaction between benzoic acid and ethanol?

   Design
   Small quantities of benzoic acid and ethanol are mixed. A few drops of sulfuric acid catalyst are added. The mixture is heated and the product is tested for odour and solubility.

   Evidence
   The odour of cherries is observed.
   The product has low solubility in water.

10. Bees build honeycomb cells from beeswax. Qualitative chemical analysis indicates that beeswax is an ester formed from mainly C_{26} and C_{28} straight-chain carboxylic acids and C_{30} and C_{32} straight-chain primary alcohols.
    (a) Write a molecular formula equation for the formation of beeswax from the C_{26} carboxylic acid and the C_{32} alcohol.
    (b) Use a molecular formula equation to represent the synthesis of beeswax from the C_{28} carboxylic acid and the C_{30} alcohol.

11. Use the information in this section and from your own research to continue gathering perspective statements concerning the statement that we should be saving more fossil fuels for petrochemical use in the future.

    Extension

12. Tannic acid, originally obtained from the wood and bark of certain trees, has for centuries been used to "tan" leather.
    (a) Give the structural formula for tannic acid.
    (b) What effect does tannic acid have on animal hides? Explain your answer with reference to the chemical reactions that take place.

    www.science.nelson.com GO

13. Transfats are formed when glycerol reacts with a carboxylic acid that has at least one double bond with a *trans* orientation. In a *trans* orientation, the hydrogen atoms are across the double bond from one another (compared to a *cis* orientation, where the hydrogens are on the same side of the double bond).
    (a) Identify the number of *cis* and *trans* orientations found in the following two carboxylic acids. (See *cis* and *trans* isomers on page 423.)

    (b) Transfats have been removed from most foods. What is the issue?

    www.science.nelson.com GO

Polymerization Reactions— Monomers and Polymers 10.5

If you take a look around you, you will likely find that you are surrounded by plastic products of many shapes and sizes. They may include pens, buttons, buckles, and parts of your shoes, chair, and lamp. There are plastic components in your calculator, telephone, computer, sporting equipment, and even the building in which you live. What are plastics and what makes them such desirable and versatile materials?

Plastics belong to a group of substances called **polymers**: large molecules made by linking together many smaller molecules, called **monomers**, much like paper clips in a long chain. Different types of small molecules form links in different ways, by either addition or condensation reactions. The types of small units and linkages can be manipulated to produce materials with desired properties such as strength, flexibility, high or low density, transparency, and chemical stability. As consumer needs change, new polymers are designed and manufactured.

Plastics are synthetic polymers, but many natural polymers have similar properties recognized since early times. Amber from tree sap, and tortoise shell, for example, can be processed and fashioned into jewellery or ornaments. Rubber and cotton are plant polymers, and wool and silk are animal polymers that have been shaped and spun into useful forms. In fact, our own cells manufacture several types of polymers—molecules so large and varied that they make us the unique individuals we are. Proteins and carbohydrates are all very different natural polymers.

Polymerization is the formation of polymers from the reaction of monomer subunits. These compounds have long existed in nature but were only synthesized by technological processes in the 20th century. They have molar masses up to millions of grams per mole.

Addition Polymers

Many plastics are produced by the polymerization of alkenes. For example, polyethene (polyethylene) is made by polymerizing ethene molecules in a reaction known as **addition polymerization**. Polyethylene is used to make plastic insulation for wires, and containers such as plastic milk bottles (**Figure 1**), refrigerator dishes, and laboratory wash bottles. Addition polymers are formed when monomer units join each other in a process that involves the rearranging of electrons in double or triple bonds in the monomer. The polymer is the only product formed.

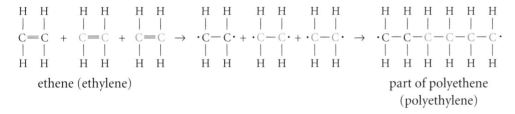

ethene (ethylene)   part of polyethene (polyethylene)

The monomers form dimers (from two monomers) and trimers (from three monomers), and continue reacting to form polymers (from many monomers, dimers, and trimers).

Polypropylene, polyvinyl chloride (PVC), Plexiglas, polystyrene (Styrofoam®), and natural rubber are also addition polymers. The production process for polypropylene is illustrated in **Figure 2**.

EXTENSION

CBC radioONE
QUIRKS & QUARKS

Plastics Under Pressure
Most plastics need to be heated and melted during recycling, but that same heat is damaging to the plastic, and degrades it into a less useful material. Researchers may have a solution to the problem: a new plastic that can be reformed under pressure instead of heat, and can be recycled many times. This technology could make recycling cheaper and much more environmentally friendly.

www.science.nelson.com

Figure 1
For recycling purposes, plastics (especially containers) are coded with a symbol for the type of plastic incorporated.
1. polyethylene terephthalate (PETE)
2. high-density polyethylene (HDPE)
3. polyvinyl chloride (PVC)
4. low-density polyethylene (LDPE)
5. polypropylene (PP)
6. polystyrene (PS)
7. other or mixtures

Look for these symbols on plastic products and, of course, recycle the product when you are finished with it.

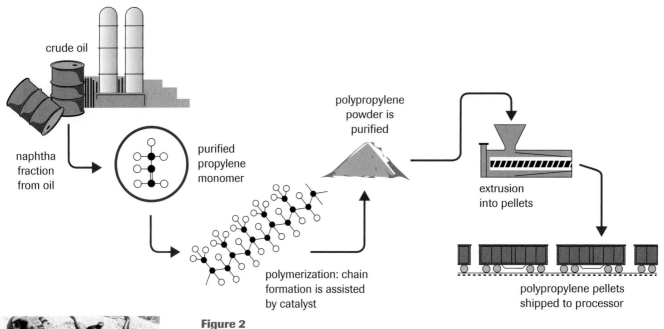

Figure 2
The production of polypropylene from crude oil

Figure 3
Polypropylene rope is one of many petrochemical products.

Other Addition Polymers: Carpets, Raincoats, and Insulated Cups

There are hundreds of different industrial polymers, all with different properties, and formed from different reactants.

Polypropene

Propene also undergoes addition polymerization, producing polypropene, commonly called polypropylene. You may have used polypropylene rope (**Figure 3**), or walked on polypropylene carpet.

$$\underset{\text{propene monomers}}{\overset{H}{\underset{H}{C}}=\overset{H}{\underset{CH_3}{C}} + \overset{H}{\underset{H}{C}}=\overset{H}{\underset{CH_3}{C}} + \overset{H}{\underset{H}{C}}=\overset{H}{\underset{CH_3}{C}} +} \longrightarrow \underset{\text{polypropene (polypropylene)}}{-\overset{H}{\underset{H}{C}}-\overset{H}{\underset{CH_3}{C}}-\overset{H}{\underset{H}{C}}-\overset{H}{\underset{CH_3}{C}}-\overset{H}{\underset{H}{C}}-\overset{H}{\underset{CH_3}{C}}-} \text{ or } \left[-\overset{H}{\underset{H}{C}}-\overset{H}{\underset{CH_3}{C}}-\right]_n$$

The polymerization reaction in the formation of polypropene is very similar to that of polyethene. The propene molecule can be considered as an ethene molecule with the substitution of a methyl group in place of a hydrogen atom. The polymer formed contains a long carbon chain, with methyl groups attached to every other carbon atom in the chain.

Polyvinyl Chloride

Ethene molecules with other substituted groups produce other polymers. For example, polyvinyl chloride, commonly known as PVC, is an addition polymer of chloroethene. A common name for chloroethene is vinyl chloride. PVC is used as insulation for electrical wires and as a coating on fabrics used for raincoats and upholstery materials (**Figure 4**).

Figure 4
Polychloroethene or polyvinyl chloride is used for upholstery.

Polystyrene

When a benzene ring is attached to an ethene molecule, the molecule is vinyl benzene, commonly called styrene. An addition polymer of styrene is polystyrene, often used to make cups and containers. Look for the recycle symbol with a 6 and PS.

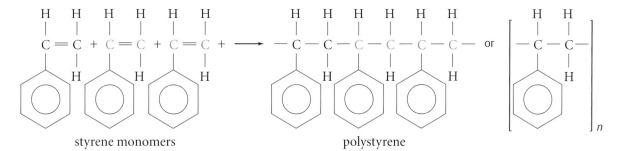

Teflon

Teflon® is the common name for an addition polymer with nonstick properties that are much desired in cookware (**Figure 5**).

The monomer used to synthesize Teflon is the simple molecule tetrafluoroethene, $CF_2=CF_2$, an ethene molecule in which all hydrogen atoms are replaced with fluorine atoms. The absence of carbon–hydrogen bonds and the presence of the very strong carbon–fluorine bonds make Teflon highly unreactive with almost all reagents. It is this unreactivity that allows it to be in contact with foods at relatively high temperatures without "sticking." The inert (unreactive) nature of Teflon makes it a very useful polymer in a wide variety of applications. There is, however, some controversy about its safety, as you will discover in the following Web Activity.

Figure 5
The invention of Teflon has made life easier in many ways.

DID YOU KNOW ?

Addition Monomers
Ethene and its double-bonded derivatives are monomers for polymers with a wide variety of properties. Some of these monomers include:

monomer of PVC

```
 H    Cl
 |    |
 C == C
 |    |
 H    Cl
```
monomer of Saran™ wrap

```
 H    H
 |    |
 C == C
 |    |
 H    CN
```
monomer of acrylic

```
 H    COOCH3
 |    |
 C == C
 |    |
 H    CN
```
monomer of instant glue

WEB Activity

Web Quest—Teflon: Healthy or Hazardous?

Cooks used to coat pots and pans with butter or oil to keep food from sticking, but the result was greasy high-fat food. The invention of Teflon®, a non-stick cooking surface, changed all this. Recently, concern has been growing regarding the possibility that Teflon® releases toxic chemicals if it gets too hot. Research the pros and cons of Teflon®, and decide for yourself if the risks are worth the benefits.

www.science.nelson.com

Practice

1. Draw a structural formula of three repeating units of
   (a) a polymer of but-1-ene
   (b) a polymer of vinyl fluoride
   (c) a polymer of 1-chloro-1,2,2-trifluoroethene
   (d) Predict the properties of the polymer in (c) in terms of solubility in organic solvents, rigidity, and resistance to heating.

2. Draw a structural formula of the monomer of the following polymer:

```
    H    F    H    F    H    F
    |    |    |    |    |    |
 —C —  C —  C —  C —  C —  C—
    |    |    |    |    |    |
    H   CH3   H   CH3   H   CH3
```

3. What monomer could be used to produce each of these polymers?

   (a)
   ```
      — CH — CH — CH — CH — CH — CH —
          |    |    |    |    |    |
         CH3  CH3  CH3  CH3  CH3  CH3
   ```

   (b)
   ```
          F    F    F    F    F    F
          |    |    |    |    |    |
      — C —  C —  C —  C —  C —  C —
          |    |    |    |    |    |
         CH3  Cl   CH3  Cl   CH3  Cl
   ```

4. What functional group(s), if any, must be present in a monomer that undergoes an addition polymerization reaction?

5. Addition polymers may be produced from two different monomers, called co-monomers. Saran™, the polymer used in a brand of food wrap, is made from the monomers vinyl chloride and 1,1-dichloroethene. Draw structural formulas for each monomer, and for three repeating units of the polymer, with alternating co-monomers.

6. (a) What are the typical properties of a plastic?
   (b) What types of bonding would you expect to find within and between the long polymer molecules?
   (c) Explain the properties of plastics by referring to their bonding.

7. Polymerization is a term used for a crude-oil refining process that converts small alkenes into gasoline-sized molecules. The polymerization process is controlled to produce dimers and trimers only (not large polymers), using phosphoric acid as a catalyst. Provide a structural formula equation for the reaction of
   (a) three molecules of ethene
   (b) propene and but-1-ene
   (c) but-1-ene and but-2-ene

8. Find out from your community recycling facility what types of plastic products are accepted for recycling in your area. If there are some plastics that are not accepted, find out the reason. Summarize your findings in a well-organized table.

Section **10.5**

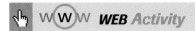 **WEB** *Activity*

Case Study—Recycling Plastics
Much of the plastic we use today is collected. We sometimes think of collecting as "recycling," but true recycling only happens if the plastic is used to make a product similar to the original one. How is recycling done? Investigate several Web sites and gather information to write a procedure for processing recycled plastic. Also, collect some pros and cons from multiple perspectives concerning the use of plastics.

 www.science.nelson.com

Condensation Polymers
Some polymers, including the synthetic fibres nylon and polyester, and natural polymers like proteins and DNA, are produced by **condensation polymerization** reactions. These reactions involve the formation of a small molecule (such as H_2O, NH_3, or HCl) from the functional groups of two different monomer molecules. The small molecule is said to be "condensed out" of the reaction. The monomer molecules bond at the site where atoms are removed from their functional groups. To form a condensation polymer, the monomer molecules must each have at least two functional groups, one on each end.

Comparing Natural with Synthetic Polymers
Research chemists have found that Nature has many well-designed molecular structures that are polymers. Many synthetic condensation polymers are structural analogs of the nutrient molecules found in foods: they have structures similar to those of lipids (fats and oils), proteins, and carbohydrates. Over the next few pages, there is a classification and description of some of these natural polymers: proteins and carbohydrate polymers. Most of us recognize these classes of compounds as categories of foodstuffs. Food labels usually list the mass of fat, protein, and carbohydrate per serving.

A synthetic compound that has a similar chemical structure to a naturally occurring substance is called a structural analog. For example, nylon is a structural analog of protein, but not a functional analog (**Figure 6**). Functional analogs are synthetic compounds that perform the same function as a naturally occurring substance but do not necessarily have similarities in chemical structure. For example, synthetic sweeteners are functional analogs of sweet carbohydrates: sugars. Chemists who study natural chemicals in order to prepare synthetic copies are called natural-product chemists.

Natural Product (food nutrient molecules)	Lipids	Proteins	Carbohydrates
	↓	↓	↓
Structural Analog (synthetic polymers)	Polyesters	Nylon	Cellulose Polymers

Lipids and Polyesters
Lipids (fats and oils) are formed by esterification reactions between glycerol (propane-1,2,3-triol) and fatty acids (long-chain carboxylic acids). Since glycerol has three hydroxyl (–OH) groups, three molecules of fatty acid can react with each glycerol molecule to form a tri-ester. This reaction is a condensation reaction that is not, strictly speaking, a polymerization reaction: The largest molecule formed is a tri-ester.

DID YOU KNOW
Photodegradable Polymers
James Guillet (1927–2005), a Canadian green-chemist, used his fascination with photosynthesis to inspire his study of the photochemistry of synthetic systems. While on holidays in the Bahamas, he was disturbed by the quantity of plastic litter and realized that, with his knowledge, he could "make it disappear." Guillet registered three patents for photodegradable polymers in 1970. Unfortunately, his invention was not widely embraced. Self-interests lobbied against his "human-made plastics."

DID YOU KNOW
Aboriginal Technologies
Aboriginal peoples used many natural materials to create useful technologies. For example, Aboriginal peoples invented the birch-bark canoe, seal-skin kayak, rawhide and willow snowshoes, and animal skin tipis. The natural polymers, such as cellulose and protein, in these Aboriginal technologies have, in some cases, been replaced by synthetic polymers.

Figure 6
Food nutrients and their synthetic structural analogs are an example of natural product chemistry, where chemists copy and modify the structure of chemicals found in nature.

Learning Tip

You are not required to learn the names of carboxylic acids beyond ten carbons. You also do not have to name the triesters representing fats and oils. You should, however, learn the general structures of
- polyalcohols (especially propane-1,2,3-triol),
- saturated and unsaturated fatty acids, and
- triester fats and oils.

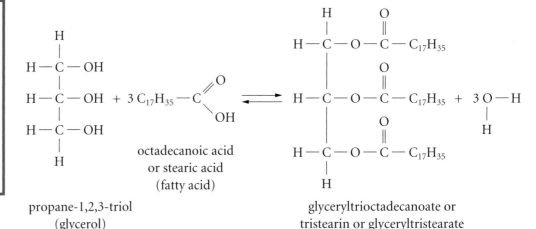

glycerol + saturated fatty acid ⇌ saturated fat + water

propane-1,2,3-triol (glycerol) + octadecanoic acid or stearic acid (fatty acid) ⇌ glyceryltrioctadecanoate or tristearin or glyceryltristearate (saturated fat present in butter and lard)

BIOLOGY CONNECTION

Correlation versus Cause and Effect

Many biochemistry studies related to food are correlational studies. These studies look for a correlation (relationship) between two variables (without being able to control all other variables—as a cause–effect study would). For example, scientists found a strong correlation between saturated fats (and transfats), and artery and heart diseases. Chemists responded with creative solutions (**Figure 7**).

Figure 7
Butter and margarine are examples of saturated and unsaturated fats (lipids). There are many criteria that can be used to evaluate which is the best product.

www.science.nelson.com **GO**

glycerol + unsaturated fatty acid ⇌ unsaturated oil + water

propane-1,2,3-triol (glycerol) + 3 CH_3—$(CH_2)_7$—C=C—$(CH_2)_7$—C(=O)OH

9-octadecenoic acid (oleic acid)

glyceryltri (9-octadecenoate) (triolein, an unsaturated *cis* fat present in olive oil)

At one time, unsaturated fats were considered to be healthier in the diet than saturated fats and oils. Unsaturated oils are generally liquids at room temperature. To change the oil into a solid (such as margarine), the unsaturated fat is partially hydrogenated to remove some of the double bonds and increase the melting point. Subsequent research showed that, although unsaturated fats were healthier, the hydrogenation process created unhealthy transformed fats.

Practice

9. Unsaturated lipids are generally more liquid (i.e., oils), and saturated lipids are generally more solid (i.e., fats). Create a hypothesis to explain the difference between the melting points of unsaturated versus saturated lipids.
10. The fatty acid $C_{19}H_{39}COOH$ reacts with glycerol, $C_3H_5(OH)_3$, to produce peanut oil. Write a structural formula equation to communicate this reaction.
11. The fatty acid $CH_3-(CH_2)_7-CH=CH-(CH_2)_7-COOH$ is hydrogenated. Write a structural formula equation for this chemical reaction.

DID YOU KNOW ?

Transfat
A transfat has a double bond where the hydrogen atoms are on the opposite side of the double bond. *Trans* means across, so the hydrogen atoms are across the double bond from one another.

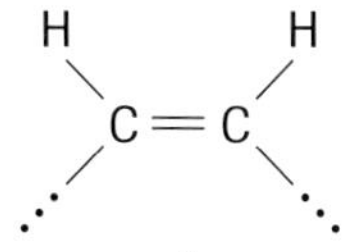

cis orientation

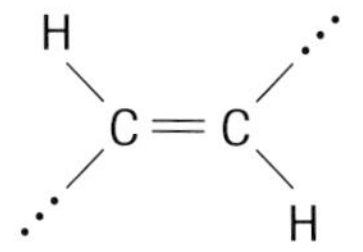

trans orientation

Polyesters

When a carboxylic acid reacts with an alcohol in an esterification reaction, a water molecule is eliminated and a single ester molecule is formed. The two reactant molecules are linked together into a single ester molecule. This esterification reaction can be repeated to form not just one ester molecule, but many esters joined in a long chain, a **polyester**. This is accomplished using a dicarboxylic acid (an acid with a carboxyl group at each end of the molecule), and a diol (an alcohol with a hydroxyl group at each end of the molecule). Ester linkages can then be formed end to end between alternating acid molecules and alcohol molecules.

$$\overline{HO}OC-CH_2-CH_2-CO\overline{OH}+\overline{H}O-CH_2-CH_2-O\overline{H}+\overline{HO}OC-CH_2-CH_2-CO\overline{OH}+\rightarrow$$
ethane-1,2-dicarboxylic acid   ethane-1,2-diol

$$-OC-CH_2-CH_2-CO-OCH_2-CH_2O-OC-CH_2-CH_2-CO-+H_2O(l)$$

If we were to depict the acid with the symbol $\Delta-\Delta$, the alcohol with $o-o$, and the ester linkage with $\boxed{o\Delta}$, we could represent the polymerization reaction like this:

$$\Delta-\Delta+o-o+\Delta-\Delta+o-o \rightarrow -\Delta-\boxed{\Delta o}-\boxed{o\Delta}-\boxed{\Delta o}-o-+\text{water}$$
polyester

Dacron®, another polyester, is made from *p*-phthalic acid (*p*-dibenzoic acid) and ethane-1,2-diol (ethylene glycol). Note the two carboxyl groups in the dicarboxylic acid and the two hydroxyl groups in the polyalcohol that start the chain reaction.

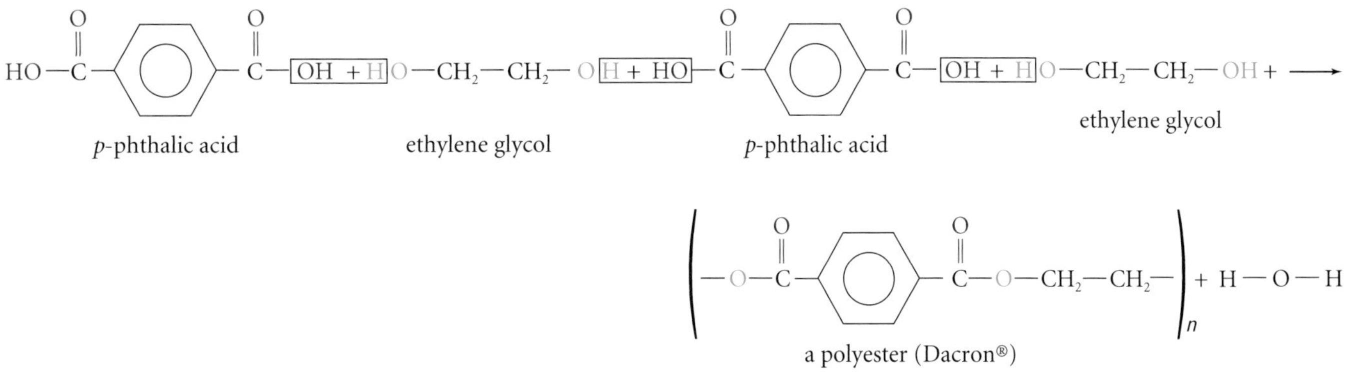

a polyester (Dacron®)

DID YOU KNOW ?

Amines and Amides
Two classes of organic compounds that are not studied in detail in this textbook are amines and amides. The simplest amines have the $-NH_2$ (amino) functional group, as in $R-NH_2$. The simplest amides have the $-CONH_2$ functional group in a molecule of $R-CONH_2$. A polyamide may be formed naturally or technologically by repeatedly reacting a carboxyl group with an amino group. The $-CONH-$ group is called an amide (or peptide) linkage. Proteins and nylon are polymers that are called polyamides. Protein polymers are specifically called polypeptides.

Practice

12. Draw a structural formula equation to show a repeating unit of a condensation polymer formed from the following compounds:

    $HOOC-CH_2-CH_2-COOH$ and $HO-CH_2-CH_2-CH_2-CH_2-OH$

13. What functional groups must be present to form a polyester?

14. How are the reactions to form fats and to form polyesters similar and different?

WEB Activity

Simulation—Molecular Modelling

Molecular modelling has become more of a computer activity than a physical activity. Chemists routinely use interactive models to better understand how the shapes of compounds affect their properties. Check out some of the molecular modelling Web sites.

www.science.nelson.com GO

Proteins and Nylon

Proteins are a fundamental structural material in plants and animals. Scientists estimate that there are more than ten billion different proteins in living organisms on Earth. Remarkably, all of these proteins are constructed from only about 20 amino acids. Through a reaction that involves the carboxyl group and the amine ($-NH_2$) group, amino acids polymerize into peptides (short chains of amino acids) or proteins (long chains of amino acids). The condensation reaction of the amino acids glycine and alanine illustrates the formation of a dipeptide. Condensation polymerization produces a protein, with a molar mass tens of thousands to millions of grams per mole—thousands of monomers long.

glycine + alanine → a dipeptide + water

The dipeptide reacts with itself or with more glycine and/or alanine. Of course, other amino acids may be present to produce an even more complex peptide. The polypeptide produced by polymerization is a protein with peptide ($-CONH-$) linkages. The following general equation illustrates the formation of a protein from amino acids.

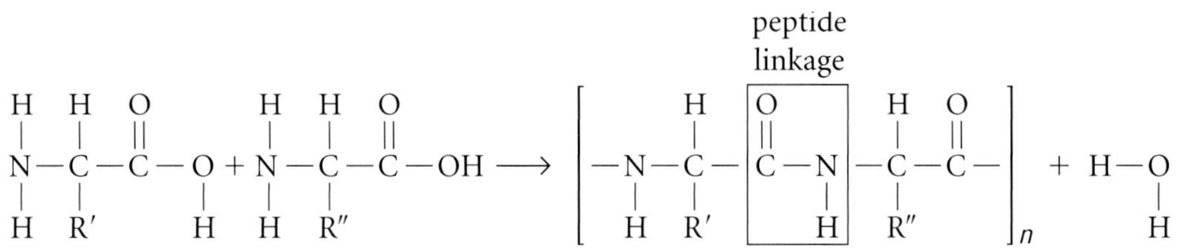

Section 10.5

[Diagram: amino acid 1 + amino acid 2 → polypeptide segment + H—O—H, showing peptide linkage formation]

amino acid 1   amino acid 2   polypeptide segment

Many synthetic polymers, such as nylon, form in similar ways to proteins. Nylon is a synthetic condensation polymer. For both the natural polymer (protein) and the synthetic polymer (nylon), the polymer forms by the reaction of a carboxyl group (–COOH) with a –NH$_2$ group with amide linkages (–CONH–). Polymers with amide linkages are called **polyamides**. Amide linkages in proteins are called peptide linkages and the polymers are called **polypeptides**.

DID YOU KNOW?
Nylon
Nylon was designed in 1935 by Wallace Carothers, a chemist who worked for DuPont; the name *nylon* is a contraction of New York and London.

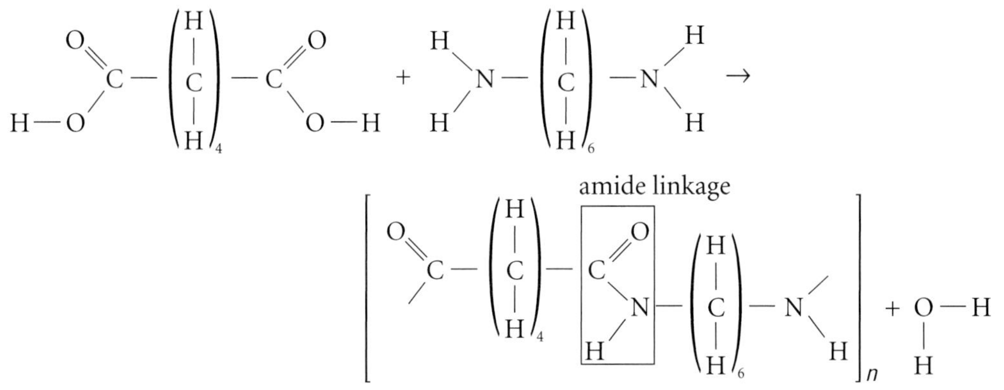

[Diagram showing the reaction of hexane-1,6-dioic acid with 1,6-diaminohexane to form part of the nylon 6,6 polymer chain with amide linkage, plus water]

hexane-1,6-dioic acid + 1,6-diaminohexane → part of the nylon 6,6 polymer chain + water

---

⚗ INVESTIGATION 10.5 Introduction

Preparing Nylon 6,10 (Demonstration)

Nylon was considered the "miracle" fibre when it was discovered in 1935. At the time, nylon was unique because it was the first synthetic fibre produced from petrochemicals. It initiated an entire new world of manufactured fibres. In this demonstration, sebacoyl chloride, COCl(CH$_2$)$_8$COCl(l), reacts with 1,6-diaminohexane to produce one type of nylon. For the prediction, write a condensed or line structural formula equation showing the formation of the polymer.

Purpose
The purpose of this demonstration is to use your knowledge of condensation polymerization to explain the formation of a nylon polymer.

To perform this investigation, turn to page 464.

Report Checklist
- ○ Purpose
- ○ Problem
- ○ Hypothesis
- ● Prediction
- ○ Design
- ○ Materials
- ○ Procedure
- ● Evidence
- ○ Analysis
- ○ Evaluation

Problem
How does the combination of sebacoyl chloride and 1,6-diaminohexane form the polymer known as nylon 6,10?

Design
A solution of 1,6-diaminohexane is carefully poured on top of an aqueous solution of sebacoyl chloride. The nylon that forms at the interface of the two layers is slowly pulled out of the mixture using forceps.

Nylon was synthesized as a substitute for silk, a natural polyamide whose structure nylon mimics. The onset of the Second World War speeded up nylon production to make parachutes, ropes, cords for aircraft tires, and even shoelaces for army boots. It is the –CONH groups that make nylon such a strong fibre. When spun, the long polymer

DID YOU KNOW ?

Pulling Fibres
When a polymer is to be made into a fibre, the polymer is first heated and melted. The molten polymer is then placed in a pressurized container and forced through a small hole, producing a long strand, which is then stretched (**Figure 8**). The process, called extrusion, causes the polymer chains to orient themselves lengthwise along the direction of the stretch. Covalent or hydrogen bonds form between the chains, giving the fibres added strength.

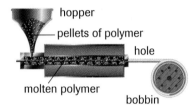

Figure 8
Making a polymer

DID YOU KNOW ?

Paintball: A Canadian Invention
The sport of paintball was invented in Windsor, Ontario. Paintballs were first used to mark cattle for slaughter and trees for harvesting, using oil-based paints in a gelatin shell. When recreational paintball use demanded a water-based paint, the water-soluble gelatin shell was modified by adjusting the ratio of the synthetic and natural polymers used.

---

chains line up parallel to each other, and the –CONH groups form hydrogen bonds with –CONH groups on adjacent chains.

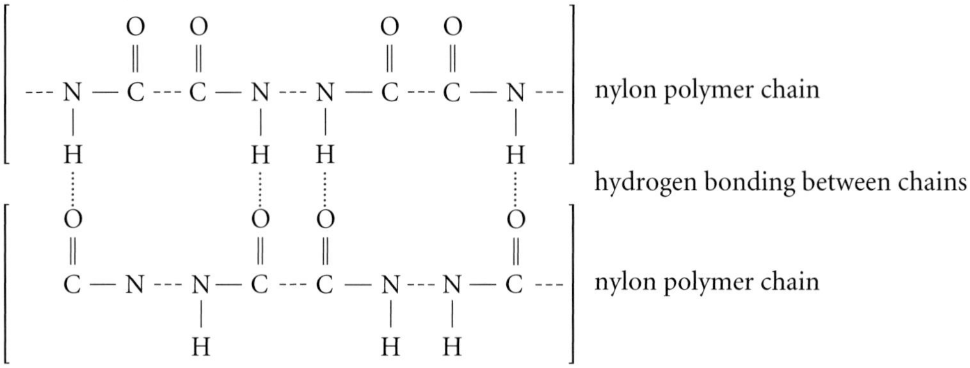

$$\begin{bmatrix} \cdots N - C \cdots C - N \cdots N - C \cdots C - N \cdots \\ | | | | \\ H H H H \end{bmatrix}$$ nylon polymer chain

hydrogen bonding between chains

$$\begin{bmatrix} C - N \cdots N - C \cdots C - N \cdots N - C \cdots \\ | | | \\ H H H \end{bmatrix}$$ nylon polymer chain

Kevlar

To illustrate the effect of hydrogen bonding in polyamides, consider a polymer with very special properties. It is stronger than steel and heat resistant, yet is lightweight enough to wear. This material is called Kevlar® (**Figure 9**) and is used to make products such as aircraft, sports equipment, protective clothing for firefighters, and bulletproof vests for police officers. What gives Kevlar these special properties? The polymer chains form a strong network of hydrogen bonds holding adjacent chains together in a sheet-like structure. The sheets are in turn stacked together to form extraordinarily strong fibres. When woven together, these fibres are resistant to damage, even that caused by a speeding bullet.

Figure 9
The chemical structure of Kevlar, with hydrogen bonding between the shaded polymer chains

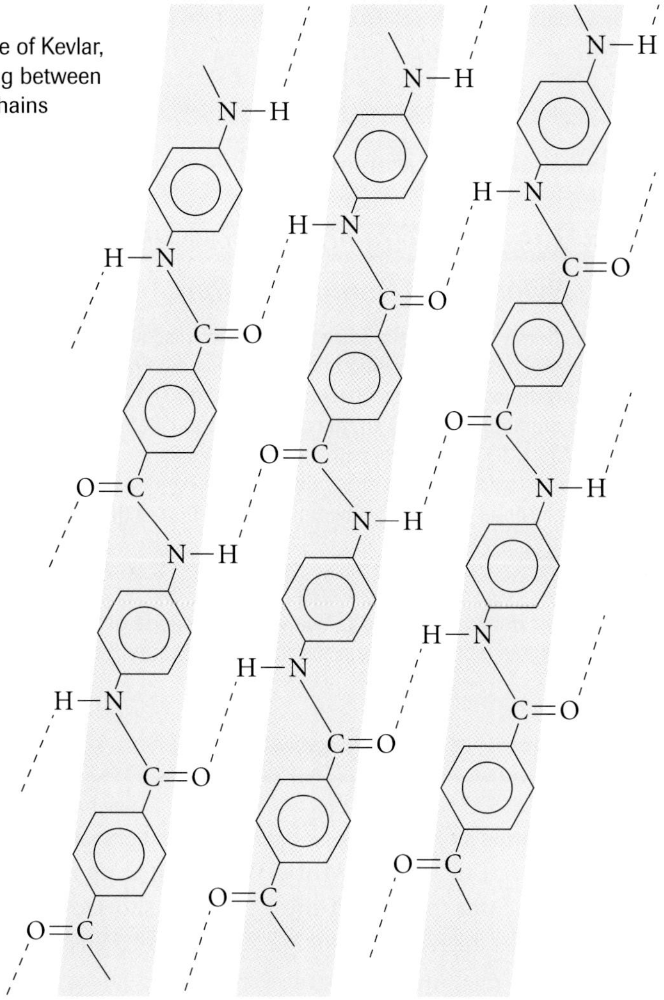

Section 10.5

Practice

15. What kind of polymerization does the formation of a polyamide (natural or synthetic) involve?
16. Provide one example each of a natural and of a similar synthetic condensation polymer.
17. Draw a structural formula representation of the nylon 6,8 repeating unit, formed by reacting $H_2N(CH_2)_6NH_2$ and $HOOC(CH_2)_6COOH$.

EXPLORE an issue

Natural or Artificial Polymers?

Issue Checklist
- ○ Issue   ○ Design   ● Analysis
- ○ Resolution   ● Evidence   ● Evaluation

From the time your grandparents were babies to the time you were born, diapers have been made entirely of polymers. Cotton cloth diapers were, and still are, made of cellulose, a natural polymer. Nowadays, disposable diapers, made mainly of synthetic polymers, are a popular choice. Which is better for the baby? How do they affect our environment? The typical disposable diaper has many components, mostly synthetic plastics, that are designed with properties particularly desirable for its function.

- *Polyethylene film*: The outer surface is impermeable to liquids, to prevent leakage. It is treated with heat and pressure to appear cloth-like for consumer appeal.
- *Hot melts*: Different types of glue are used to hold components together. Some glues are designed to bond elastic materials.
- *Polypropylene sheet*: The inner surface at the leg cuffs is designed to be impermeable to liquids and soft to the touch. The main inner surface is designed to be porous, to allow liquids to flow through and be absorbed by the bulk of the diaper.

- *Polyurethane, rubber, and Lycra*: Any or all of these stretchy substances may be used in the leg cuffs and the waistband.
- *Cellulose*: Basically processed wood pulp, this natural polymer is obtained from pine trees. It forms the fluffy filling of a diaper, absorbing liquids into the capillaries between the fibres.
- *Polymethylacrylate*: This crystalline polymer of sodium methylacrylate absorbs water through osmosis and hydrogen bonding. The presence of sodium ions in the polymer chains draws water that is held between the chains. The presence of water results in attractions between the chains and causes the formation of a gel (**Figure 10**).

Manufacturers claim that grains of sodium polymethylacrylate can absorb up to 400 times their own mass in water. If sodium ions are present in the liquid, they act as contaminants, reducing absorbency because the attraction of water to the polymer chains is diminished. Urine always contains sodium ions, so the absorbency of diapers for urine is actually less than the advertised maximum.

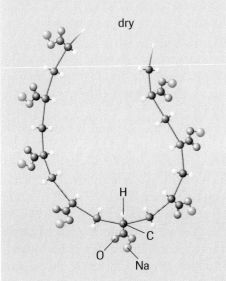
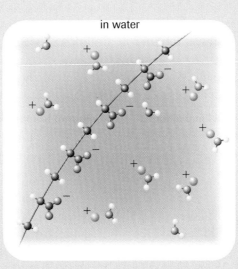

Figure 10
Polymethylacrylate

Hydrocarbon Derivatives, Organic Reactions, and Petrochemicals   455

Proponents of natural products argue that disposable diapers pose a long-term threat to our environment, filling waste disposal sites with non-biodegradable plastics for future centuries. The industry has developed some new "biodegradable" materials—a combination of cellulose and synthetic polymers, for example. Some of these materials have proven too unstable to be practical, while others appear to biodegrade in several years, which gives the diapers a reasonable shelf life. The energy and raw materials needed to manufacture the huge quantity of disposable diapers used is also of concern. Diaper manufacturers and their supporters argue, however, that using the natural alternative—cloth diapers—consumes comparable amounts of energy in the laundry. In addition, the detergents used in the cleaning process are themselves non-biodegradable synthetic compounds made from petroleum products.

Issue
Do cloth diapers pose less of a threat to the environment than disposable diapers?

Resolution
Consumers should return to using cloth diapers in order to protect the environment.

Design
Within small groups, research the pros and cons of using cloth diapers. Gather information from a wide variety of perspectives.

www.science.nelson.com

Carbohydrates and Cellulose Acetate

The monomers of carbohydrates—compounds with the general formula $C_x(H_2O)_y$—are simple sugar molecules. The sugar monomers undergo a condensation polymerization reaction in which a water molecule is formed, and the monomers join together to form a larger molecule. For example, the sugar monomers glucose and fructose can form sucrose (table sugar) and water. Both starch and cellulose consist of long chains of glucose molecules. Simple sugar monomers are sometimes called monosaccharides; dimers are called disaccharides; and polymers are called polysaccharides.

glucose + fructose → sucrose + water

DID YOU KNOW?

Maple Sugar and Corn
Canadian Aboriginal peoples introduced Europeans to maple sugar. Sugar (a disaccharide) can be obtained from sugar cane (from South America) and sugar beets (from southern Alberta). The greatest contribution of Aboriginal agriculture is corn—one of the most prolific and widespread crops in human history—from which we obtain corn syrup.

Starch for Energy; Cellulose for Support

Starchy foods such as rice, wheat, corn, and potatoes provide us with readily available energy. They are also the main method of energy storage for the plants that produce them, as seeds or tubers. Starches are polymers of glucose, in either branched or unbranched chains; they are, thus, polysaccharides.

We have, in our digestive tracts, very specific enzymes. One breaks down starch. However, the human digestive system does not have an enzyme to break down the other polymer of glucose: cellulose. Cellulose is a straight-chain, rigid polysaccharide with glucose–glucose linkages different from those in starch or glycogen. It provides structure and support for plants, some of which tower tens of metres in height. Wood is mainly cellulose; cotton fibres and hemp fibres are also cellulose. Indeed, it is because cellulose is indigestible that whole grains, fruits, and vegetables are good sources of dietary fibre.

It is the different glucose–glucose linkages that make cellulose different from starch. When glucose forms a ring structure, the functional groups attached to the ring are fixed in a certain orientation above or below the ring. Our enzymes are specific to the

Section **10.5**

orientation of the functional groups, and cannot break the glucose–glucose linkages found in cellulose. Herbivores such as cattle, rabbits, termites, and giraffes rely on some friendly help to do their digesting: They have specially developed stomachs and intestines that house enzyme-producing bacteria or protozoa to aid in the breakdown of cellulose.

In starch, glucose monomers are added at angles that lead to a helical structure, which is maintained by hydrogen bonds between –OH groups on the same polymer chain (**Figure 11(a)**). The single chains are sufficiently small to be soluble in water. Thus, starch molecules are both mobile and soluble—important properties in their role as readily available energy storage for the organism.

In cellulose, glucose monomers are added to produce linear polymer chains that can align side by side, favouring interchain hydrogen bonding (**Figure 11(b)**). These interchain links produce a rigid structure of layered sheets of cellulose. This bulky and inflexible structure not only imparts exceptional strength to cellulose; it also renders it insoluble in water. It is, of course, essential for plants that their main building material does not readily dissolve in water.

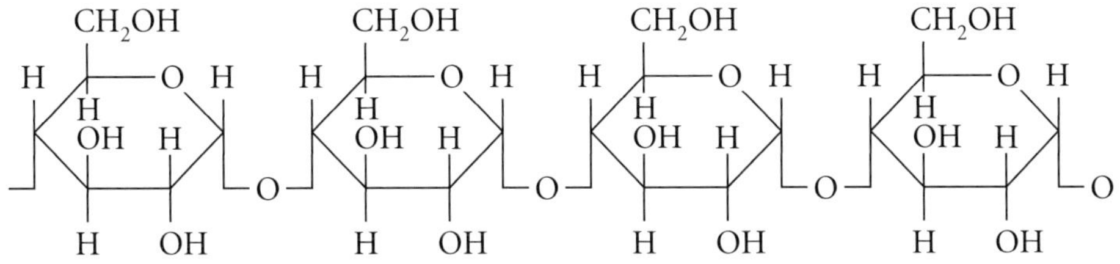

BIOLOGY CONNECTION

Glycogen
Animals also produce a starch-like substance, called glycogen, that performs an energy storage function. Glycogen is stored in the muscles as a ready source of energy, and also in the liver, where it helps to regulate blood glucose levels.

www.science.nelson.com GO

DID YOU KNOW

Cotton—A Natural Polymer
North American Aboriginal peoples made extensive use of cotton for clothing. This technology was exported to Europe, where it contributed to starting the industrial revolution in Britain. Cotton is a natural carbohydrate (cellulose) polymer. There are now regenerated (manufactured) fibres from cellulose, such as cellulose acetate and cellulose nitrate. In 1850, long-fibre cotton comprised over 50% of Britain's exports. The mechanization and infrastructure developed for processing cotton, and the resulting capital gained, helped to fuel and shape the Industrial Revolution.

(a) starch

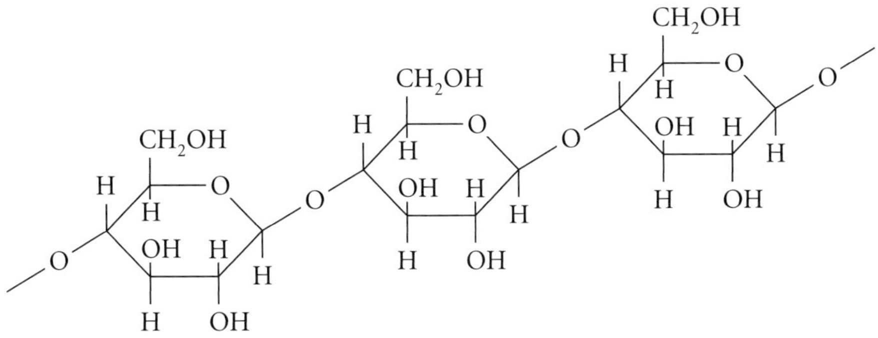

(b) cellulose

Figure 11
The difference in linkages between glucose monomers gives very different three-dimensional structures.
(a) In starch, the polymer takes on a tightly coiled helical structure.
(b) In cellulose, the linked monomers can rotate, allowing formation of straight fibres.

 **Starch and Cellulose**

Use a molecular model kit and/or a computer program to construct and/or investigate molecular models of glucose, sucrose, starch, and cellulose.

www.science.nelson.com

Hydrocarbon Derivatives, Organic Reactions, and Petrochemicals 457

DID YOU KNOW ?

The Centre of the Chocolate
Sucrose, a disaccharide, is slightly sweeter than glucose but only half as sweet as fructose. The enzyme sucrase, also called invertase, can break sucrose down into glucose and fructose—a mixture that is sweeter and more soluble than the original sucrose. The centres of some chocolates are made by shaping a solid centre of sucrose and invertase, and coating it with chocolate (**Figure 12**). Before long, the enzyme transforms the sucrose centre into the sweet syrupy mixture of glucose and fructose.

Figure 12
Invertase breaks sucrose down to glucose and fructose in some chocolates.

▶ Practice

18. Identify the functional groups present in a molecule of glucose and in a molecule of fructose.
19. Describe several functions of polysaccharides and how their molecular structures serve these functions in plants.
20. Compare the following pairs of compounds, referring to their structure and properties:
    (a) sugars and starch
    (b) starch and cellulose
21. What are the distinctive features of carbohydrate molecules that, given a structural formula, would allow you to classify them as carbohydrates?
22. Explain in terms of molecular structure why sugars have a relatively high melting point compared with hydrocarbons of comparable size.
23. Describe, and then explain, the relative solubility of glucose, starch, and cellulose in water.

Cellulose Acetates—Structural Analogs of Polysaccharides

Chemists have researched and developed polymers that are modifications of natural polymers. These natural–synthetic polymers are called **biopolymers**. For example, starch and cellulose have been made into synthetic polymers. An example is cellulose triacetate. Cellulose, a polysaccharide, is modified by reacting it with acetic acid, $CH_3COOH(l)$, and acetic anhydride, $(CH_3CO)_2O(l)$, with sulfuric acid as a catalyst.

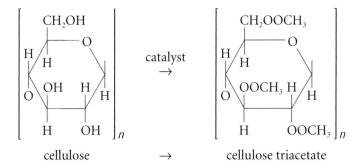

cellulose → cellulose triacetate

Cellulose triacetate is a polymer that is used in fabrics such as permanently pleated garments, textured knits, and sportswear. There are other, similar, cellulose acetates.

WEB Activity

Case Study—Cellulose Acetate

This activity leads you through an exploration of the historical and current work done by polymer chemists to use natural polymers to produce synthetic polymers with desired characteristics and applications.

SUMMARY Polymers—Plastics, Fibres, and Food

Polymers are a common part our everyday world and include natural polymers (lipids (triesters), proteins, and carbohydrates) and synthetic polymers (e.g., nylon, cellulose acetate, and polyesters). For the structural formula equations of these polymers, see previous pages.

Learning Tip
Try to equate the classification of condensation polymers with the three food groups when organizing your knowledge of polymers:
- polyesters include **lipids** and polyester fabrics
- polyamides include **proteins** and polyamide fabrics
- polysaccharides include **carbohydrates** and, for example, cellulose acetate.

Chemists create classification systems to help them organize their knowledge. Use the same approach to your advantage.

Addition Polymers
Synthetic Addition Polymer (e.g., polypropene)

$$\begin{array}{c} H \quad H \\ | \quad\quad | \\ C=C \\ | \quad\quad | \\ H \quad CH_3 \end{array} \rightarrow \left[\begin{array}{c} H \quad H \\ | \quad\quad | \\ -C-C- \\ | \quad\quad | \\ H \quad CH_3 \end{array} \right]_n$$

Condensation Polymers (Structural Analogs)

Polyesters
- Natural "Polyester" (e.g., butter):
  glycerol + fatty acid → fat or oil + water
  e.g., $C_3H_5(OH)_3 + 3\,C_{17}H_{35}COOH \rightarrow C_3H_5(OOCC_{17}H_{35})_3 + 3\,H_2O$

- Synthetic Polyester (e.g., Dacron)
  dicarboxylic acid + polyalcohol → polyester + water
  e.g., $HOOC-C_6H_4-COOH + HO-CH_2-CH_2-OH \rightarrow$
  $[-OOC-C_6H_4-COO-CH_2-CH_2-]_n + n\,H_2O$

Polyamides
- Natural (proteins; polypeptides):
  amino acid + amino acid + ... → protein + water
  e.g., $H_2N-CH_2-COOH + H_2N-CHCH_3-COOH \rightarrow$
  $[-NH-CH_2-CONH-CHCH_3-CO-]_n + n\,H_2O$

- Synthetic (e.g., nylon 6,6):
  dicarboxylic acid + diamine → nylon + water
  e.g., $HOOC-(CH_2)_4-COOH + H_2N-(CH_2)_6-NH_2 \rightarrow$
  $[-OC-(CH_2)_4-CONH-(CH_2)_6-NH-]_n + n\,H_2O$

Polysaccharides
- Natural (e.g., starch):
  glucose + glucose → starch + water
  glucose + glucose → cellulose + water
  (The glucose molecules that polymerize to produce starch or cellulose have slightly different stereochemical formulas.)
  $C_6H_{12}O_6 + C_6H_{12}O_6 \rightarrow [-C_6H_{10}O_5-]_n + n\,H_2O$

- Synthetic (e.g., cellulose triacetate):
  cellulose + acetic acid + acetic anhydride → cellulose acetate + ...
  $[-C_6H_{10}O_5-]_n + CH_3COOH + (CH_3CO)_2O \rightarrow [-C_6H_7O_5(OCH_3)_3-]_n + ...$

Section 10.5 Questions

1. Teflon®, made from tetrafluoroethene monomer units, is a polymer that provides a non-stick surface on cooking utensils. Write a structural formula equation to represent the formation of polytetrafluoroethene.

2. Polyvinyl chloride, or PVC plastic, has numerous applications. Write a structural formula equation to represent the polymerization of chloroethene (vinyl chloride).

3. Using a molecular model kit, construct models of the following monomers and polymers:
   (a) starch
   (b) polyvinyl chloride
   (c) nylon
   (d) polypropene
   (e) a polyester
   (f) cellulose acetate

4. As with most consumer products, the use of polyethylene has benefits and problems. What are some beneficial uses of polyethylene and what problems result from these uses? Suggest alternative substances for each application.

5. Use the information in this section and from your own research to continue gathering perspective statements concerning the statement that we should be saving more fossil fuels for petrochemical use in the future.

6. The first nylon product that was introduced to the public, in 1937, was a nylon toothbrush called Dr. West's Miracle-Tuft toothbrush. Earlier toothbrush bristles were made of hair from animals such as boar. From your knowledge of the properties of nylon, suggest some advantages and drawbacks of nylon toothbrushes compared with their natural counterparts.

7. Oxalic acid is a dicarboxylic acid found in rhubarb and spinach. Its structure is shown below.

$$\text{HOC}\overset{\overset{\displaystyle O}{\|}}{}-\overset{\overset{\displaystyle O}{\|}}{}\text{COH}$$

   Draw three repeating units of the condensation polymer made from oxalic acid and ethan-1,2-diol.

8. What is the synthetic polymer analog of each of the following foods?
   (a) pasta
   (b) meat
   (c) butter

9. Suppose that two new polymers have been designed and synthesized for use as potting material for plants.
   (a) List and describe the properties of an ideal polymer to be used to hold and supply water and nutrients for a plant over an extended period of time.
   (b) Design an experiment to test and compare the two polymers for the properties you listed. Write a brief description of the procedures followed, and possible interpretations of experimental results.

10. When we purchase a product, we may want to consider not only the source of its components, but also the requirements for its use and maintenance. In your opinion, how valid is the use of the terms "organic," "natural," and "chemical" in the promotion of consumer goods?

11. Starch and cellulose have the same caloric value when burned, but very different food values when eaten by humans. Explain.

12. Explain why the sugars in a maple tree dissolve in the sap but the wood in the tree trunk does not.

13. Many consumer products are available in natural or synthetic materials: paper or plastic shopping bags, wood or plastic lawn furniture, cotton or polyester clothing. Choose one consumer product and list the advantages and disadvantages of the natural and synthetic alternatives, with particular reference to structure and properties of the material used as it relates to the function of the product.

14. Classify the type of polymerization for each of the following polymers:
    (a) polyethenes
    (b) polyesters
    (c) polyamides
    (d) polysaccharides

15. Research or use what you know to write a combination of a word, molecular formula, and structural formula equation for each of the following chemical transformations:
    (a) starch (from grain) to glucose to ethanol
    (b) cellulose (from aspen trees) to glucose to methanol

Extension

16. Alkyd (oil-based) paint is a polyester formed by reacting glycerol ($CH_2OHCHOHCH_2OH$) with 1,2-benzenedicarboxylic acid ($C_6H_4(COOH)_2$). Communicate this reaction using a structural formula equation.

17. What functional group(s), if any, must be present in a monomer of a condensation polymer?

18. In an attempt to reduce body weight, many Canadians are considering low-carbohydrate or low-sugar diets. Based on your knowledge of the chemistry of carbohydrates and sugars, comment on which of these diets might be more effective.

19. Natural rubber is made from resin produced by the rubber tree, *Hevea brasiliensis*.
    (a) Research the history of the use of rubber by Aboriginal peoples long before Europeans came to the Americas.
    (b) Research the commercial invention, production, and use of natural rubber, and the circumstances that stimulated the development of synthetic rubber.
    (c) Write a brief report on your findings.

20. Raymond Lemieux, from Lac La Biche, Alberta, was the first chemist to artificially synthesize sucrose (in 1953). Consult the biographical information of Lemieux. Describe his interest in science and one of his other accomplishments (other than the synthesis of sucrose).

21. Check the latest Alberta Chemical Operations Directory to see if there are any new monomer and polymer chemical plants in Alberta.

Chapter 10 INVESTIGATIONS

INVESTIGATION 10.1

Substitution and Addition Reactions

Report Checklist
- ○ Purpose   ○ Design   ● Analysis
- ○ Problem   ○ Materials   ● Evaluation (1, 2, 3)
- ○ Hypothesis ○ Procedure
- ● Prediction ● Evidence

Substitution and addition reactions are two common classes of organic reactions. Substitution reactions are common to saturated hydrocarbons. Addition reactions are common to unsaturated (alkene and alkyne) hydrocarbons. In general, substitution reactions are slower and addition reactions are faster.

Purpose
The purpose of this investigation is to test the generalization that addition reactions are faster than substitution reaction.

Problem
Which compound reacts faster with aqueous bromine: cyclohexane or cyclohexene?

Design
Aqueous bromine is added to samples of cyclohexane and cyclohexene in both the presence and absence of light.

Materials
lab apron
eye protection
4 small test tubes with stoppers
test tube rack (or beaker)
aluminium foil (approx. 10 cm × 10 cm)
dropper bottles of
　cyclohexane
　cyclohexene
　aqueous bromine

 Cyclohexane and cyclohexene are flammable. Work in a well-ventilated area. Keep away from sparks or flames.

 Bromine is toxic and corrosive. Avoid skin and eye contact. Wash hands thoroughly when finished.

Procedure
1. Add 20 drops of cyclohexane to each of two test tubes and stopper immediately.
2. Cover one of the test tubes up to the rim with aluminium foil.
3. Repeat steps 1 and 2 using cyclohexene and two more test tubes.
4. For each of the four test tubes: remove the stopper briefly to add 10 drops of aqueous bromine.
5. Placing a thumb or finger on the stopper, briefly shake each test tube.
6. Record your observations immediately and again after several minutes. (You will need to momentarily slide the covered test tubes out of the foil to observe.)
7. Dispose of the contents of all test tubes into labelled waste containers.

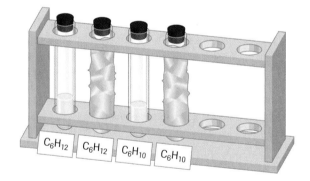

Figure 1
Set-up for reactions of cyclohexane and cyclohexene with and without light

Hydrocarbon Derivatives, Organic Reactions, and Petrochemicals

INVESTIGATION 10.2

Isomers of Butanol

Organic halides have many uses, but they are not commonly found in nature. Organic halides must, therefore, be synthesized from other compounds. On an industrial scale, organic halides are produced by addition and substitution reactions of hydrocarbons. A common laboratory-scale process is the halogenation of alcohols in which the hydroxyl group of the alcohol is replaced by a halogen. This reaction requires the presence of a strong acid that contains a halogen, such as hydrochloric acid.

$$R\text{-}OH(l) + HX(aq) \rightarrow R\text{-}X(l) + H_2O(l)$$

Organic halides have much lower solubilities in water compared with their corresponding alcohols.

Purpose
The purpose of this investigation is to test a personal hypothesis about the relative reactivity of the alcohol isomers of butan-1-ol.

Problem
What is the difference in reactivity, if any, of the alcohol isomers of butan-1-ol with concentrated hydrochloric acid?

Design
Samples of butan-1-ol, butan-2-ol, and 2-methylpropan-2-ol are mixed with concentrated hydrochloric acid. Evidence of reaction is obtained by looking for a low-solubility organic halide product (cloudy mixture).

 Concentrated hydrochloric acid is corrosive and the vapour is very irritating to the respiratory system. Avoid contact with skin, eyes, clothing, and the lab bench. Wear eye protection, gloves, and a laboratory apron.

 All three alcohols are highly flammable. Do not use near an open flame.

Report Checklist
- ○ Purpose
- ○ Problem
- ● Hypothesis
- ● Prediction
- ○ Design
- ● Materials
- ○ Procedure
- ● Evidence
- ● Analysis
- ● Evaluation (1, 2, 3)

Procedure

1. Add two drops each of butan-1-ol, butan-2-ol, and 2-methylpropan-2-ol into separate small labelled test tubes in a test-tube rack.

2. In the fume hood, add 10 drops of concentrated HCl(aq) to each of the three isomers in the test tubes.

3. Stopper and swirl each of the test tubes gently and cautiously.

4. Back at your laboratory bench, observe the test tubes over a period of a few minutes. Look for cloudiness in the water layer—an indication of the formation of an organic halide.

5. Dispose of the contents of all test tubes in the fume hood into a labelled waste container.

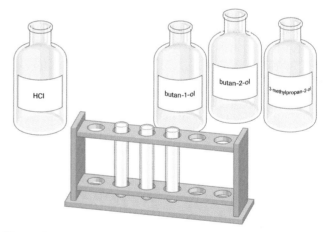

Figure 2
Set-up for testing the reactivity of three butanol isomers

INVESTIGATION 10.3

Synthesis of an Ester

Report Checklist
- ○ Purpose
- ○ Problem
- ○ Hypothesis
- ○ Prediction
- ○ Design
- ○ Materials
- ○ Procedure
- ● Evidence
- ● Analysis
- ○ Evaluation

Purpose
The purpose of this investigation is to use the esterification generalization and diagnostic tests to synthesize and observe the properties of two esters.

Problem
What are some physical properties of ethyl ethanoate (ethyl acetate) and methyl salicylate?

Design
The esters are produced by the reaction of appropriate alcohols and acids, using sulfuric acid as a catalyst. The solubility and the odours of the esters are observed.

Materials
lab apron
eye protection
dropper bottles of ethanol, methanol, glacial ethanoic (acetic) acid, and concentrated sulfuric acid
vial of salicylic acid (2-hydroxybenzoic acid)
two 25 × 250 mm test tubes
250 mL beaker or polystyrene cup
two 50 mL beakers
two 10 mL graduated cylinders
laboratory scoop
balance
hot plate or hot tap water
thermometer
ring stand with test tube clamp

 Concentrated ethanoic and sulfuric acids are dangerously corrosive. Protect your eyes, and do not allow the acids to come into contact with skin, clothes, or lab desks.

 Both methanol and ethanol are flammable; do not use near an open flame.

 Excessive inhalation of the products may cause headaches or dizziness. Use your hand to waft the odour from the beaker toward your nose. The laboratory should be well ventilated.

Procedure
1. Add about 5 mL of ethanol and 6 mL of ethanoic acid to one of the test tubes.
2. Ask your teacher to add 8 to 10 drops of concentrated sulfuric acid to the mixture.
3. Set up a hot water bath using the 250 mL beaker. (The temperature of the water should not exceed 70 °C.)
4. Clamp the test tube so that the reaction mixture is completely immersed in hot water (**Figure 3**).
5. As a safety precaution to block any eruption of the volatile mixture, invert a 50 mL beaker above the end of the test tube (**Figure 3**).
6. After the reaction mixture heats for about 10 min, rinse the second 50 mL beaker with cold tap water and add about 30 mL of cold water to this beaker.
7. Cool the test tube by running cold tap water on the outside of the tube.
8. Pour the contents of the test tube into the cold water in the 50 mL beaker. Observe and smell the mixture carefully, using the correct technique for smelling chemicals.
9. Repeat steps 1 to 7, using 3.0 g of salicylic acid, 10 mL of methanol, and 20 drops of sulfuric acid.
10. Dispose of all mixtures into the sink with lots of cold running water.

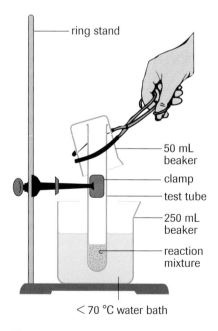

Figure 3
Set-up for synthesis of an ester

INVESTIGATION 10.4

Testing with Models

Report Checklist
- ○ Purpose
- ○ Problem
- ○ Hypothesis
- ○ Prediction
- ○ Design
- ● Materials
- ○ Procedure
- ● Evidence
- ○ Analysis
- ○ Evaluation

Molecular models, either physical or electronic, are based on empirically known molecular structures. These models can then be considered as representing evidence collected in a variety of experiments.

Purpose
The purpose of this investigation is to test stereochemical formula (see Chapter 3) equations by using molecular models.

Problem
What are the stereochemical formula equations for the following reactions?

(a) methane undergoes complete combustion
(b) ethane is cracked into ethene
(c) propane reacts with chlorine
(d) but-2-ene reacts with water
(e) ethanol eliminates water to produce ethene
(f) 1-chloropropane undergoes an elimination reaction with hydroxide ions to produce propene, a chloride ion, and water
(g) ethanol reacts with methanoic acid

Design
Physical or electronic models of the reactants and products for each reaction are constructed. The bonding and shapes of the molecules from the models are compared with the predictions.

INVESTIGATION 10.5

Preparing Nylon 6,10 (Demonstration)

Report Checklist
- ○ Purpose
- ○ Problem
- ○ Hypothesis
- ● Prediction
- ○ Design
- ○ Materials
- ○ Procedure
- ● Evidence
- ○ Analysis
- ○ Evaluation

Nylon was considered the "miracle" fibre when it was discovered in 1935. At the time, nylon was unique because it was the first synthetic fibre produced from petrochemicals. It initiated an entire new world of manufactured fibres. In this demonstration, sebacoyl chloride, $COCl(CH_2)_8COCl(l)$, reacts with 1,6-diaminohexane to produce one type of nylon. For the prediction, write a condensed or line structural formula equation showing the formation of the polymer.

Purpose
The purpose of this demonstration is to use your knowledge of condensation polymerization to explain the formation of a nylon polymer.

Problem
How does the combination of sebacoyl chloride and 1,6-diaminohexane form the polymer known as nylon 6,10?

Design
A solution of 1,6-diaminohexane is carefully poured on top of an aqueous solution of sebacoyl chloride. The nylon that forms at the interface of the two layers is slowly pulled out of the mixture using forceps.

 The reactants and solvent used are irritants. Use a fume hood for this demonstration.

Chapter 10 SUMMARY

Outcomes

Knowledge
- identify and describe significant organic compounds in daily life, demonstrating generalized knowledge of their origins and applications (all sections)
- name and draw structural, condensed structural and line diagrams and formulas for organic halides, alcohols, carboxylic acids, and esters (all sections)
- identify types of compounds from their functional groups, given the structural formula and name the functional groups (10.2, 10.3, 10.4)
- compare, both within a homologous series and between compounds with different functional groups, the boiling points and solubility of examples of alcohols and carboxylic acids (10.3, 10.4)
- define, illustrate, and provide examples of simple addition, substitution, elimination, and esterification (condensation) reactions (10.2, 10.3, 10.4, 10.5)
- predict products and write and interpret balanced equations for the above reactions (10.2, 10.3, 10.4, 10.5)
- define, illustrate, and give examples of monomers, polymers, and polymerization in living and non-living systems (10.5)
- relate the reactions described above to major reactions for producing economically important compounds from fossil fuels (all sections)

STS
- illustrate how science and technology are developed to meet societal needs and expand human capabilities (all sections)
- describe interactions of science, technology, and society (all sections)
- illustrate how science and technology have both intended and unintended consequences (10.5)

Skills
- initiating and planning: predict the ester formed from an alcohol and an organic acid (10.4); describe procedures for safe handling, storing and disposal of materials used in the laboratory, with reference to labelling information (10.2, 10.3, 10.4, 10.5); design an experiment to compare the properties of organic to inorganic compounds (10.4)
- performing and recording: build molecular models depicting the structures of selected organic and inorganic compounds (10.4); perform an experiment to investigate the reactions of organic compounds (10.2, 10.3, 10.4, 10.5)
- analyzing and interpreting: follow appropriate IUPAC guidelines in writing the names and formulas of organic compounds (all sections); compile and organize data to compare the properties of structural isomers (10.3); interpret results of a test to distinguish between a saturated and an unsaturated aliphatic using aqueous bromine solution (10.2); use appropriate chemical symbols and nomenclature in writing organic chemical reactions (all sections); use models to illustrate polymerization (10.5)
- work cooperatively in addressing problems and apply the skills and conventions of science in communicating information and ideas and in assessing results (10.3, 10.4, 10.5)

Key Terms

10.2
hydrocarbon derivative
organic halide
functional group
addition reaction
substitution reaction

10.3
alcohol
hydroxyl group
elimination reaction

10.4
carboxylic acid
carboxyl group

condensation reaction
ester
esterification reaction
ester functional group

10.5
polymer
monomer
polymerization
addition polymerization
condensation polymerization
polyester
polyamide
polypeptide

MAKE a summary

1. Draw a map of Alberta and indicate at least one location where each of the following chemicals is produced: ethane, ethene, ethylene glycol, methanol, polyethene, polyvinyl chloride, styrene. For each chemical, indicate the reaction type and illustrate the reaction with a structural formula equation.
2. Refer back to your answers to the Starting Points questions at the beginning of this chapter. How has your thinking changed?

Go To

The following components are available on the Nelson Web site. Follow the links for *Nelson Chemistry Alberta 20–30*.
- an interactive Self Quiz for Chapter 10
- additional Diploma Exam-style Review questions
- illustrated Glossary
- additional IB-related material

There is more information on the Web site wherever you see the Go icon in this chapter.

Chapter 10 REVIEW

Many of these questions are in the style of the Diploma Exam. You will find guidance for writing Diploma Exams in Appendix H. Exam study tips and test-taking suggestions are on the Nelson Web site. Science Directing Words used in Diploma Exams are in bold type.

DO NOT WRITE IN THIS TEXTBOOK.

Part 1

1. There are arguments pro and con from a variety of perspectives for saving our fossil fuels for petrochemical production for generations to come. Which of the following statements is *not* classified correctly?
   A. economic: We need the corporate profit and individual salaries now.
   B. legal: It isn't right for us to be burning these irreplaceable fossil fuels.
   C. ethical: This resource doesn't just belong to us; it belongs to future generations.
   D. social: Families and communities need the resource development now.

2. The organic compound ethane-1,2-diol is manufactured in Alberta from the primary petrochemical
   A. ethene
   B. propene
   C. methanol
   D. benzene

3. The type of chemical reaction that involves breaking a carbon–hydrogen bond and replacing the hydrogen atom with another atom or group of atoms is called
   A. addition
   B. substitution
   C. condensation
   D. esterification

Use the names of these organic compounds to answer questions 4 and 5.
   1. gasoline additive, ethanol
   2. glycerol, propane-1,2,3-triol
   3. PVC monomer, chloroethene
   4. organic solvent, ethyl ethanoate
   5. pineapple odour, ethyl butanoate
   6. radiator antifreeze, ethane-1,2-diol
   7. wintergreen odour, methyl salicylate
   8. banana odour, 3-methylbutyl ethanoate
   9. used to manufacture acetone, propan-2-ol

4. In numerical order, the alcohols are
   NR ____, ____, ____, and ____.

5. In numerical order, the esters are
   NR ____, ____, ____, and ____.

6. Chemists were initially surprised that the reaction of benzene and chlorine to form chlorobenzene and hydrogen chloride is an example of a(n)
   A. addition reaction
   B. elimination reaction
   C. esterification reaction
   D. substitution reaction

7. Chemists predict, based on a generalization, that the reaction between methanol and propanoic acid produces water and
   A. propanol
   B. propyl methanoate
   C. methyl propanoate
   D. methanoic acid

8. A generalization supported by laboratory evidence suggests that bromine and propene react to produce
   A. 1-bromopropane
   B. 2-bromopropane
   C. 1,1-dibromopropane
   D. 1,2-dibromopropane

There are more compounds of carbon than compounds of all other elements combined. Use this short list of carbon compounds to answer questions 9 to 12.
   1. butanoic acid   6. methylpropane
   2. butan-1-ol   7. octan-1-ol
   3. ethanoic acid   8. propanoic acid
   4. ethanol   9. pentanoic acid
   5. hexan-1-ol

9. From lowest to highest, the order of boiling points of the
   NR alcohols is ____, ____, ____, and ____.

10. From lowest to highest, the order of solubility of the
    NR alcohols is ____, ____, ____, and ____.

11. From lowest to highest, the order of solubility of the acids
    NR is ____, ____, ____, and ____.

12. The two compounds that must react to form ethyl
    NR butanoate are ___ and ___ ___ ___.

13. The functional group, —COOH, is characteristic of
    A. esters
    B. alcohols
    C. aromatics
    D. carboxylic acids

14. Which of the following substances is an addition polymer?
    A. protein
    B. polyester
    C. polypropylene
    D. polysaccharide

Part 2

15. **Identify** the category, and draw a line structural formula, for each of the following organic compounds.
    (a) triiodomethane
    (b) 1,2-difluoroethene
    (c) propane-1,2,3-triol
    (d) 2-methylpropan-2-ol
    (e) propanoic acid
    (f) methyl butanoate
    (g) phenol
    (h) 1,3-dibromobenzene
    (i) benzoic acid
    (j) ethyl benzoate

16. Complete the following word equations and draw complete structural formula equations for each of the following organic reactions.
    (a) butane + chlorine →
    (b) but-1-ene + hydrogen chloride →
    (c) chloroethane + hydroxide ion →
    (d) benzene + fluorine →
    (e) benzoic acid + ethanol →
    (f) propene + bromine →
    (g) propan-1-ol + ethanoic acid →

17. Alcohols, carboxylic acids, and esters are generally more soluble in water than the corresponding alkanes. Draw structural formulas for simple 3-carbon molecules from each of these families and **explain** why they are more soluble in water than propane.

18. **Explain** the difference between addition and condensation polymers. Your response should include
    - characteristics of the monomers
    - products of the polymerization reaction

19. **Identify** what special precautions must be taken when working with and disposing of organic compounds.

20. Because of side reactions in organic reactions, the yield of desired product is often low, and is usually expressed as a % yield. Complete the Analysis of the investigation report.

    Problem
    What is the percent yield in the esterification reaction between pentanol and butanoic acid to form pentyl butanoate?

    Design
    An excess of pentanol reacts with a measured mass of butanoic acid in the presence of a sulfuric acid catalyst. The product is separated from the reaction mixture by fractional distillation.

    Evidence
    mass of butanoic acid reacted = 2.00 g
    mass of pentyl butanoate obtained = 1.32 g

21. **Design** an experiment to distinguish alkanes from alkenes. Assume the hydrocarbons are in the liquid phase.

22. Write IUPAC names for the following organic compounds.
    (a) $CH_3—CH_2—CH_2—CHF—CH_2F$
    (b) $CH_3—CH_2—CH_2—CH_2OH$
    (c) $CH_3—CH_2—CH_2—CH_2—COOH$
    (d) $CH_3—CH_2—COO—CH_2—CH_3$
    (e) $CH_2OH—CHOH—CH_2OH$
    (f) $CH_3—CH_2—COOH$
    (g) $CH_3—CH_2—COO—CH_2—CH_2—CH_3$
    (h) $CH_3—CHBr—CH_2—CH_2Br$

23. Draw a line structural formula for each of the following aromatic compounds.
    (a) 1,4-dichlorobenzene
    (b) benzoic acid
    (c) phenol
    (d) methyl benzoate
    (e) 1,2,3-tribromobenzene
    (f) ethyl benzoate

Use the following information to answer questions 24 to 26.

The field of polymer chemistry has blossomed since Carothers' discovery of nylon in 1935. Thousands of new polymers have now been developed, including Teflon, polystyrene, polypropylene, and polyvinyl chloride.

Polymers have been very beneficial in many ways. They have also, however, introduced problems for many societies around the world.

24. (a) **Identify** the class of synthetic polymers into which all of the named compounds (above) fall.
    (b) Name and draw structural formulas for the monomers used to produce each of the four polymers.
    (c) What feature do these monomers have in common? **Explain** how this enables polymerization to occur. **Illustrate** your explanation with a structural formula reaction equation.

25. **Explain** why the monomers used to produce condensation polymers must be bifunctional (have two functional groups).

26. **Compare** the impacts of polymers on two different societies: one urban and relatively wealthy; one rural and relatively poor. Your response should include
    - a variety of perspectives
    - benefits and drawbacks of the development of synthetic polymers
    - suggestions for alternatives to synthetic polymers
    - an answer to the question: On balance, do synthetic polymers benefit humanity?

27. Consider the Resolution in the Exploration that began this chapter. Respond to the Resolution with the production and use of polymers in mind.

Unit 5 REVIEW

Many of these questions are in the style of the Diploma Exam. You will find guidance for writing Diploma Exams in Appendix H. Exam study tips and test-taking suggestions are on the Nelson Web site. Science Directing Words used in Diploma Exams are in bold type.

www.science.nelson.com

DO NOT WRITE IN THIS TEXTBOOK.

Part 1

1. Alberta refines and manufactures many organic compounds. The Alberta product that is not considered an organic compound is
   A. gas line antifreeze, $CH_3OH(l)$
   B. vinegar, $CH_3COOH(aq)$
   C. limestone, $CaCO_3(s)$
   D. natural gas, $CH_4(g)$

2. Which one of the following organic compounds is *least* likely obtained from a natural gas well?
   A. methane
   B. ethane
   C. propane
   D. benzene

3. There are many organic compounds that are important in daily life. Match the compounds listed with the applications.
   NR
   1 gasoline additive
   2 plastic wrap
   3 barbecue fuel
   4 natural sweetener

   When the common applications for propane, ethanol, sucrose, and polyethylene are listed in order, the order is __, __, __, and __.

4. The IUPAC name for the following possible component of gasoline is

   ```
            CH3    CH2 — CH3
             |      |
   CH3 — CH — C = CH — CH3
   ```
   A. 2-methyl-3-ethylpent-3-ene
   B. 3-ethyl-4-methylpent-2-ene
   C. 2-methyl-3-ethylpent-2-ene
   D. 2,3,3-trimethylpent-2-ene

5. The boiling points, from smallest to largest members of an aliphatic family
   A. increase
   B. decrease
   C. stay relatively constant
   D. increase initially, and then decrease

Use this information to answer questions 6 and 7.

Xylenes are important aromatics used in a variety of applications, such as solvents and as raw materials for the production of polyester fibres, dyes, and insecticides. The structure of one particular xylene is

6. The IUPAC name for this xylene is
   A. 1,3-dimethylcyclohexane
   B. ethylbenzene
   C. 2,6-dimethylbenzene
   D. 1,3-dimethylbenzene

7. A structural isomer of this xylene is
   A. octane
   B. ethylcyclohexane
   C. ethylbenzene
   D. 1,3-dimethylhex-1-ene

8. Which one of the following technological processes is *not* used to physically separate organic compounds from a mixture?
   A. solubility
   B. hydrocracking
   C. solvent extraction
   D. fractional distillation

9. The most likely reaction between an alkene and chlorine is
   A. addition
   B. combustion
   C. elimination
   D. substitution

10. Which of the following organic reactions is most likely to produce an alkene?
    A. $C_3H_6 + Cl_2 \rightarrow$
    B. $C_4H_8 + H_2 \rightarrow$
    C. $CH_3(CH_2)_2COOH + CH_3(CH_2)_2OH \rightarrow$
    D. $CH_2ClCH_3 + OH^- \rightarrow$

11. Which of the following chemical equations represents a catalytic reforming reaction?
    A. $C_6H_6 + I_2 \rightarrow C_6H_5I + HI$
    B. $CH_3(CH_2)_5CH_3 \rightarrow C_6H_5CH_3 + 4\,H_2$
    C. $C_2H_5COOH + C_2H_5OH \rightarrow C_2H_5COOC_2H_5 + H_2O$
    D. $2\,C_4H_{10} + 13\,O_2 \rightarrow 8\,CO_2 + 10\,H_2O$

468 Unit 5

Use this information to answer questions 12 to 15.

Scientists create classification systems to help them organize their knowledge and develop generalizations to make predictions. Classify the following organic reactions as requested.

1. $C_3H_6 + Cl_2 \rightarrow C_3H_6Cl_2$
2. $CH_3CH_2ClCH_3 + OH^- \rightarrow CH_3CHCH_2 + H_2O + Cl^-$
3. $C_3H_8 \rightarrow C_3H_6 + H_2$
4. $C_3H_8 + Br_2 \rightarrow CH_3CHBrCH_3 + HBr$
5. $C_{12}H_{26} + H_2 \rightarrow C_7H_{16} + C_5H_{12}$
6. $C_2H_5OH \xrightarrow{H^+} C_2H_4 + H_2O$
7. $C_8H_{18} + H_2 \rightarrow 2\,C_4H_{10}$
8. $C_4H_8 + H_2 \rightarrow C_4H_{10}$
9. $C_6H_6 + Cl_2 \rightarrow C_6H_5Cl + HCl$

12. In numerical order, the substitution reactions are
NR ___ ___ ___ ___.

13. In numerical order, the addition reactions are
NR ___ ___ ___ ___.

14. In numerical order, the hydrocracking reactions are
NR ___ ___ ___ ___.

15. In numerical order, the elimination reactions are
NR ___ ___ ___ ___.

Use this information to answer questions 16 to 19.

The most severe test of a generalization is its ability to predict. Predict the correct reaction product(s) for each organic reaction below. Some of the reactions have only one product. You may use the same number more than once.

1. 1-chloropropene
2. 2-chloropropane
3. prop-1-ene
4. propan-2-ol
5. carbon dioxide
6. hydrogen chloride
7. water vapour

16. propane + chlorine → ___ + ___
NR
17. propene + hydrogen chloride → ___ + ___
NR
18. propane + oxygen → ___ + ___
NR
19. propene + steam → ___ + ___
NR

Part 2

20. Describe, in words or using a chart, how you would classify an organic compound given its structural formula. Include all classes of organic compounds discussed in this unit.

21. Design an experiment to distinguish pure organic compounds from pure inorganic compounds. Include as many different methods as possible.

22. Fossil fuels are of major importance to Alberta. List the different kinds of fossil fuels found in Alberta. For each kind, state one or two major uses.

Use this information to answer questions 23 to 26.

Dow Chemical Canada Inc. (**Figure 1**) has extensive chemical operations in Prentiss and Fort Saskatchewan, Alberta. The major products produced are: hydrochloric acid, Styrofoam® (extruded polyphenylethene (polystyrene) insulation), vinyl chloride monomer (chloroethene), 1,2-dichloroethane, sodium hydroxide, chlorine, ethylene (ethene), and polyethylene. The sites are also home to MEGlobal Canada, which produces ethylene glycol (ethane-1,2-diol).

Figure 1
Dow Chemical Canada Inc.

23. Classify each of the products produced by Dow Chemical as organic or inorganic.

24. Draw full structural formulas for as many of the organic products produced at Dow Chemical's sites as possible. Circle functional groups and identify the organic family.

25. Suggest a possible chemical reaction for the production of 1,2-dichloroethane using only the substances produced by Dow Chemical.

26. Some of the substances produced by Dow Chemical are intermediates used to manufacture other substances, and some substances may show up in final products. From the list given, **identify** any substances that consumers are able to buy in stores. For each of these substances, state a typical use.

Use this information to answer questions 27 to 29.

Catalytic reforming is an important process in the modern oil refining industry. Catalytic reforming is used to produce aromatics to improve the quality of gasoline and as a starting material for the petrochemical industry. An overview of the catalytic reforming process is

C_6 to C_8 alkanes → cycloalkanes/enes → simple aromatics
(naphtha feedstock)   (intermediates)   (final products)

27. Write the IUPAC name for each of the following hydrocarbons in a naphtha feedstock:

    (a) $CH_3 — (CH_2)_6 — CH_3$

    (b) $CH_3 — CH — CH_2 — CH_2 — CH_2 — CH_2 — CH_3$
             |
           CH_3

    (c) $CH_3 — CH_2 — CH_2 — CH — CH_3$
                             |
                          CH_3

28. Draw a line structural formula for each of the following intermediates from catalytic reforming:
    (a) cyclohexane
    (b) 1-methylcyclopentene
    (c) 1,2-dimethylcyclohexane

29. Write the IUPAC name for each of the following aromatic products from catalytic reforming:

    (a)

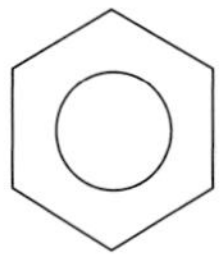

    (b)

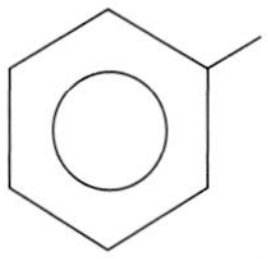

    (c)

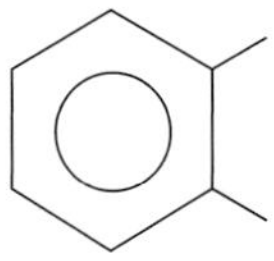

    (d)

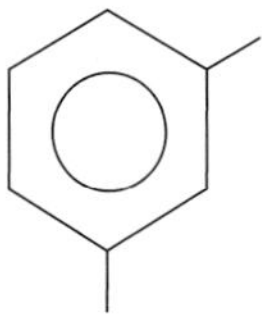

    (e)

    (f)

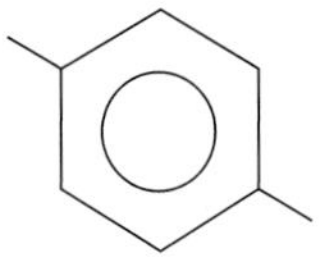

Use this information to answer questions 30 to 32.

Many new technologies are being developed for oil refining. One example is alkylation (isomerization). This process is used to produce more branching in alkanes and also to produce isomers of simple aromatics.

30. Using your knowledge of the term "isomers," **describe**, in your own words, the process of alkylation (isomerization).

31. For each of the following isomerization reaction equations, write IUPAC names or draw condensed structural formulas as required:

    (a) $CH_3 — (CH_2)_2 — CH_3 \rightarrow CH_3 — \underset{\underset{CH_3}{|}}{CH} — CH_3$

    (b) $CH_3 — CH_2 — \underset{\underset{CH_3}{|}}{CH} — CH_2 — CH_2 — CH_3 \rightarrow$
    $CH_3 — \underset{\underset{CH_3}{|}}{\overset{\overset{CH_3}{|}}{C}} — \underset{}{CH} — CH_3$
                               |
                             CH_3

    (c) heptane → 2,4-dimethylpentane
    (d) octane → 2-methyl-3-ethylpentane

32. What chemical process increases the octane rating by increasing the aromatic character of the gasoline?

33. The following compounds contain molecules that have approximately the same number of electrons: chloroethene, ethanoic acid, propan-1-ol, and butane.
    (a) Draw a full structural formula of each molecule.
    (b) Using the theoretical rules of intermolecular bonding, arrange these substances in order of increasing boiling points. Briefly explain the order.
    (c) **Predict** which substances would likely have a high solubility in water and which would have a low solubility. **Justify** your answer.

34. For each of the following reactions of one mole of an organic reactant, **predict** the chemical amount of hydrogen required for a complete reaction:
    (a) $CH_2{=}CH{—}CH{=}CH_2 + H_2 \rightarrow$
    (b) $CH{\equiv}C{—}CH_2{—}CH_2{—}C_6H_5 + H_2 \rightarrow$
    (c) $C_7H_{14} + H_2 \rightarrow$

35. Cycloalkanes have structural isomers that are alkenes. **Design** an experiment to distinguish between an alkene and a cycloalkane that are structural isomers of each other. Provide the Design, including diagnostic tests.

Problem
Which sample of $C_6H_{12}(l)$ is an alkene isomer, and which is a cycloalkane isomer?

Use this information to answer questions 36 and 37.

Petroleum products with double and triple bonds can polymerize, and form waxy deposits that plug pipe systems and pipelines. One purpose of hydrotreating is to remove multiple bonds to stabilize the liquids, so that polymerization in pipes does not occur.

36. Write a structural formula equation for the addition polymerization of ethene (ethylene) molecules, showing how two molecules can attach to each other and have a free bond at each end for further attachments.

37. Write a structural formula for a section of a polymer of propene (propylene) showing at least four monomer units in the section.

38. Suggest a reaction equation or a sequence of reaction equations to synthesize each of the following organic compounds:
    (a) ethanol (common solvent; component of alcoholic beverages and gasohol)
    (b) propene (monomer for polypropylene plastic)
    (c) dichlorodifluoromethane (CFC-12, a refrigerant)
    (d) Teflon® polymer
    (e) ethyl ethanoate (solvent)

39. For each of the following reactions, draw structural formulas for all reactants and products, name the products, and classify the reaction type.
    (a) cyclohexane + chlorine →
    (b) 2-phenylpropane + oxygen →
    (c) 2,3-dimethylbut-2-ene + hydrogen chloride →
    (d) octane → ethylbenzene
    (e) butan-1-ol + methanoic acid →
    (f) 2-bromohexane + hydroxide ion →
    (g) decane + hydrogen → butane + hexane

40. (a) Given that octane has a density of 0.70 kg/L and assuming that gasoline is octane, **predict** the volume of octane that will produce 1.0 t of carbon dioxide by complete combustion in an automobile engine.
    (b) How far could you drive a typical car on that much gasoline? (Assume 10 L/100 km.)

41. Copy and complete **Table 1**.

    Table 1 Some Important Organic Compounds

IUPAC name	Condensed structural formula	Organic family	Common use
	$CH_2 = CH_2$		
ethane-1,2-diol			
tetrachloroethene			degreaser for metal components
	$CH_3 - (CH_2)_6 - CH_3$		
ethyl ethanoate			fingernail polish solvent
methylbenzene			gasoline component; solvent
			vinegar
		alkyne	welding and cutting torch

42. Side reactions and byproducts are common for organic reactions. The yield of a product is often expressed as a % yield. Complete the Analysis of the investigation report.

    Problem
    What is the percent yield in the initial substitution reaction between methane and chlorine?

    Design
    A quantity of methane reacts with chlorine gas. The products of the reaction are separated by condensation of the gaseous products into separate fractions.

    Evidence
    mass of methane reacted = 1.00 kg
    mass of chloromethane produced = 2.46 kg

43. Fossil fuels come with various physical and chemical properties. As a result of these properties, they are refined by different processes and used in different ways.
    (a) List the fossil fuels of Alberta, from most dense to least dense.
    (b) **Describe** two differences between the refining of natural gas and the refining of crude oil.
    (c) **Evaluate**, with reasoning and evidence, the statement: "Gasoline must contain significant quantities of nitrogen because nitrogen oxides are found in the exhaust of cars and trucks."

44. Methane is produced naturally by a number of sources and becomes part of the hydrocarbon component in the atmosphere of Earth. Human technologies also contribute to atmospheric hydrocarbons.
    (a) **Identify** two natural sources and two technological sources of atmospheric methane.
    (b) Methane is a greenhouse gas. **Define** a greenhouse gas.
    (c) **Identify** some other greenhouse gases and their sources.
    (d) **Describe** actions you can take to reduce the quantity of hydrocarbon molecules that you may allow to escape into the atmosphere.

www.science.nelson.com

45. Copy and complete **Table 2**.

Table 2 Common Monomers

IUPAC name	Chemical formula	Polymer name	Type of polymer
tetrafluoroethene		Teflon®	
benzene-1,4-dicarboxylic acid		Dacron®	
ethane-1,2-diol			
chloroethene		PVC	
benzene-1,3-dicarboxylic acid		alkyd resin	
propane-1,2,3-triol			
phenylethene		styrene	
amino acids		proteins	
1,1-dichloroethene		Saran®	
glucose		starch	
propene		polypropylene	

Use this information to answer questions 46 to 49.

Greenhouse gas (GHG) emissions from internal combustion engines in transportation vehicles are a major concern for environmental chemists. Statistics Canada gathers data over time to provide citizens and scientists with information such as that presented in **Table 3**. Use this table, or a more recent one from StatsCan, to answer the following questions.

Table 3 Greenhouse Gas Emissions by Road Transportation Source

Transportation source	CO_2 (kt) 1990	CO_2 (kt) 2001	CH_4 (kt) 1990	CH_4 (kt) 2001	N_2O (kt) 1990	N_2O (kt) 2001	CO_2^1 (kt) 1990	CO_2^1 (kt) 2001	Change %
Gasoline automobiles	51 600	46 400	9.0	4.6	6.3	7.3	53 700	48 700	−9.3
Light duty gasoline trucks	20 400	36 400	4.0	4.8	4.2	9.0	21 800	39 400	80.7
Heavy duty gasoline vehicles	2 990	3 930	0.4	0.6	0.4	0.6	3 140	4 130	31.5
Motorcycles	225	236	0.2	0.2	0.0	0.0	230	242	5.2
Diesel automobiles	657	583	0.0	0.0	0.0	0.0	672	596	−11.3
Light duty diesel trucks	577	629	0.0	0.0	0.0	0.0	591	643	8.8
Heavy duty diesel vehicles	24 300	38 200	1.2	1.9	0.7	1.1	24 500	38 600	57.6
Propane and natural gas vehicles	2 160	1 100	1.7	1.7	0.0	0.0	2 210	1 140	−48.4
Transportation total	103 000	127 000	16.0	14.0	12.0	19.0	107 000	134 000	25.2

[1] CO_2-equivalent emissions are the weighted sum of all greenhouse gas emissions. The following global warming potentials are used as the weights: $CO_2 = 1$; $CH_4 = 21$; $N_2O = 310$.

46. **Identify** the top three road transportation sources for each of carbon dioxide, methane, and nitrous oxide.

47. **Identify** the road transportation source that had the greatest increase and greatest decrease in carbon-dioxide-equivalent emissions over the period.

48. **Determine** the percentage contribution of the chemical with the largest contribution to GHG emissions in each year presented.

49. List, in increasing weighted order, the least to the most individually problematic chemicals as greenhouse gases.

50. Who owns what? Traditional Aboriginal environmental views are generally that resources are gifts and that land cannot be owned, whereas Canadian law allows for ownership and exploitation of the land and its resources. Even ownership is subject to change as laws change, however. Who should change their views or practices? Is there a compromise that can or should be considered? Consider the issue from several perspectives. Briefly state your views relative to the fossil fuel industry.

51. Sulfur in gasoline is a pollution problem. Research indicates that refined gasoline may contain anywhere from 30 ppm to 1000 ppm sulfur content. Some jurisdictions have legislated a maximum sulfur content at 30 ppm, which adds one to three cents per litre to the cost of gasoline.
    (a) Nonanethiol, $CH_3(CH_2)_8SH(l)$, is one of many possible sulfur-containing components of gasoline. Write the balanced chemical equation for the complete combustion of this compound.
    (b) Briefly **describe** the pollution problems that sulfur in gasoline causes.
    (c) Are you willing to pay the extra price for low-sulfur gasoline? **Justify** your answer.

52. Oil sands extraction and upgrading requires chemical engineers to use many chemical concept. The concepts presented in this unit get you started on understanding the chemistry of the oil sands.
    Solubility is an important concept used in the oil sands industry. The following three processes are technological applications of solubility.
    - Water is used to separate the bitumen from the sand.
    - When oil sands dry, the water layer between the bitumen and the sand is removed and the hot-water separation of the bitumen from the sand is no longer possible.
    - Naphtha is used to dilute the bitumen for transportation.

    Another technological process used by the oil sands industry is the thermal coking of bitumen to crack large hydrocarbon molecules into smaller molecules and coke (carbon). Many of the products of coking are hydrotreated to remove sulfur, nitrogen, and multiple bonds from the molecules.
    (a) Use the concept of solubility to **explain** why the three processes, listed above, work.
    (b) Draw line structural formulas for the following products of coking.
        (i) ethylbenzene
        (ii) 4-methylhex-2-ene
        (iii) 6-ethyl-2,3,5-trimethyloctane
    (c) Draw structural formulas to complete the following catalyzed hydrotreating reactions.
        (i) 2,4,5-trimethyloct-2-ene + hydrogen →
        (ii) hept-3-yne + excess hydrogen →

53. Complete the Exploration of the two issues identified at the beginning of Chapters 9 and 10. Decide on the type of report and complete this report. (See Appendix D.2.)

54. Review the focusing questions on page 352. Using the knowledge you have gained from this unit, briefly **outline** a response to each of these questions.

Extension

55. Ethene has many uses, including the ripening of fruit and vegetables. List six examples of fruit and vegetables that you eat, along with ethylene production by the fruit or vegetable, and the ethylene sensitivity of the fruit or vegetable.

    www.science.nelson.com GO

56. Alberta has an extensive chemical industry that is the second-largest manufacturing industry in the province, producing about ten billion dollars' worth of products annually. List the four main segments of the Alberta chemical industry. Which one is the largest of the four? What is the main basis of this segment? **Describe** some employment opportunities in the Alberta petrochemical industry.

57. BP Amoco has a linear alpha olefin (LAO) plant in Joffre, Alberta that is producing 250 kt of LAO annually. What are LAOs? Write the IUPAC name and draw the structural formula for a few simple LAOs. **Describe** some major applications for small, medium, and large LAOs.

    www.science.nelson.com GO

58. Research matters. Research is at the heart of all advances and innovations in science and engineering. Search the Internet for information on E10 and E85 ethanol blends, oxydiesel, and/or P-series fuels to gain an understanding of the current research programs involving gasohols.

    www.science.nelson.com GO

59. In-situ oil sands processes include steam-assisted gravity drainage (SAGD) and hydrocarbon-gas injection (e.g., Vapex). In hydrocarbon-gas injection, propane and/or butane is injected (with or without steam) into the oil sands seam. The hydrocarbon acts as a solvent to reduce the viscosity of and extract the bitumen. The bitumen solution is collected in a recovery pipe below the injection pipe. View and/or find an animation to view this process. **Outline** several advantages of the hydrocarbon gas process over the SAGD process.

    www.science.nelson.com GO

60. Solvent extraction can be used to remove waxes from crude oil. Two solvents may be used: toluene (methylbenzene), which dissolves the hydrocarbon-oil, and methyl ethyl ketone ($CH_3COC_2H_5$), which dissolves the ester-wax but precipitates it when the temperature is dropped. **Explain**, using structural formulas and solubility theory, the solubility of hydrocarbon oil in toluene and the solubility of wax in methyl ethyl ketone.

    www.science.nelson.com GO

unit 6
Chemical Energy

In your life you encounter many different kinds of energy: radiant energy from our sun warms you on a spring day; electrical energy provides the energy to run your computer; chemical energy from food keeps your body functioning. In each of these cases energy is transformed from one form to another.

Energy transformations are the basis for all change, everywhere. Photosynthesis is a chemical process that takes radiant energy from the sun and stores it within molecules of glucose. This stored chemical energy provides the energy for life. Fossil fuels also contain stored chemical energy that can be transformed into other forms of energy. Energy links together all components of the universe as the energy is emitted, captured, stored, and used. All physical, chemical, and nuclear changes involve changes in energy, but this unit will focus on chemical changes involving energy: thermochemical changes, or *thermochemistry*.

Consider the Alberta photo on the right. Solar energy from the nuclear reactions in our sun is being used through photosynthesis by the barley plants. Carbohydrates are being chemically produced, which may be used to fuel us as human beings. The pumpjack is removing conventional crude oil from below the surface. The oil, we believe, was produced from the decay of plants. The oil is then processed by chemical technologies to, for example, run our cars and trucks. Solar energy, directly or indirectly, is the source of most energy available to us on Earth.

Everything humans do requires energy. It is a major factor in social change on our planet. Technologies, which inevitably consume energy, are developed for a social purpose. These technologies, however, often have drawbacks related to their use of energy. The control and use of our present sources of energy, as well as the development of new sources, will continue to have far-reaching environmental, economic, social, technological, and political implications for many years.

As you progress through the unit, think about these focusing questions:

- How does our society use the energy of chemical changes?
- How do chemists determine how much energy will be produced or absorbed for a given chemical reaction?

Unit **6**

GENERAL OUTCOMES

In this unit, you will
- determine and interpret energy changes in chemical reactions
- explain and communicate energy changes in chemical reactions

Unit 6
Chemical Energy

ARE YOU READY?

These questions will help you find out what you already know, and what you need to review, before you continue with this unit.

Knowledge

1. Most of the energy available on Earth comes from our sun. For each of the Alberta-based technological devices in **Figure 1**, briefly trace the energy produced back to its source, our sun, Sol.

(a)  (b)

(c)  (d)

Figure 1
(a) photovoltaic (solar) cells: Students and teachers have installed photovoltaic technology at Cochrane High School, Alberta.
(b) wind turbine: The Castle River Wind Farm is one of many in Alberta, which (along with Quebec) leads Canada in wind-generated electricity.
(c) hydroelectric power plant: Belly River Hydroelectric Plant is a low-impact generating station near Glenwood, Alberta.
(d) coal-fired power plant: Sundance Powerplant is on the south shore of Lake Wabamun, Alberta.

2. Write balanced chemical equations for the complete combustion of the following fuels:
   (a) coal (assume C(s))
   (b) propane (C_3H_8(g))
   (c) gasoline (assume C_8H_{18}(l))

3. List five energy sources that do not directly produce carbon dioxide.

4. The overall reaction for photosynthesis can be written as
   $6\ CO_2(g) + 6\ H_2O(l) + \text{sunlight} \rightarrow C_6H_{12}O_6(aq) + 6\ O_2(g)$
   (a) What additional substance must be present for this reaction to occur?
   (b) Is photosynthesis an endothermic or exothermic process?
   (c) What industries depend heavily on the chemistry of photosynthesis?

5. The overall reaction for cellular respiration can be written as
   $C_6H_{12}O_6(aq) + 6\ O_2(g) \rightarrow 6\ CO_2(g) + 6\ H_2O(l)$
   (a) Cellular respiration is similar to what common type of reaction?
   (b) If energy were to be included as a term in this reaction equation, should it be written on the right or left side of the arrow?

Prerequisites

Concepts
- sources of energy (e.g., solar and chemical)
- kinetic and potential energy
- thermal energy
- temperature and phase changes
- heat
- transfer of energy
- conservation of energy
- kinetic molecular theory
- photosynthesis
- cellular respiration
- carbon cycle
- endothermic and exothermic reactions

Skills
- measurement
- graphing

You can review prerequisite concepts and skills on the Nelson Web site, in the Chemistry Review unit, and in the Appendices.
   A Unit Pre-Test is also available online.

www.science.nelson.com

(c) Is cellular respiration an endothermic or exothermic process?
(d) Give two examples of cellular respiration.

6. Explain why fossil fuel energy can be considered as stored solar energy.

7. The carbon cycle involves the interrelationship of living things with the environment and with technologies that use fossil fuels. Draw a diagram of the carbon cycle using examples from the natural world, (such as plants and animals) and from the technological world (such as transportation and petrochemicals). Use arrows to illustrate the relationships (by the processes of cellular respiration, photosynthesis, or combustion) among these aspects of life in Alberta.

8. From a theoretical perspective, kinetic energy is related to the motion of an entity. The *temperature* of an object is a measure of the average kinetic energy of its entities, while the *thermal energy* of an object is the total kinetic energy of its entities. *Heat* is energy being transferred between two objects or materials (**Figure 2**). What happens to the average kinetic energy of water molecules when the water is
   (a) heated
   (b) cooled
   (c) maintained at a constant temperature
   (d) boiled

9. Potential energy is a stored form of energy. Chemical potential energy involves the energy stored in the bonds between atoms and molecules. State what happens to the potential energy of water molecules when the water is
   (a) heated   (b) boiled   (c) frozen

10. The motion of molecules can be translational (straight-line), rotational, and/or vibrational. Describe the dominant types of motion of water molecules in
    (a) ice   (b) liquid water   (c) water vapour

11. A pot of water at room temperature is placed on the burner of a gas stove and heated until 15% of the water has evaporated due to boiling.
    (a) Copy **Figure 3**, and complete it to show a temperature–time graph for the heating (including boiling) of the water.
    (b) Describe the changes that take place in the kinetic and potential energy of the water molecules as the water is heated, and then when it boils.
    (c) Describe the changes that take place in the kinetic and potential energy of the gas molecules as they undergo combustion in the stove.
    (d) Use the law of conservation of energy to explain the graph that you have drawn.

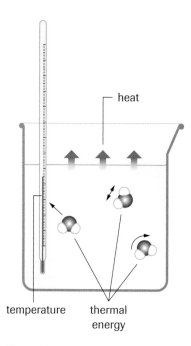

Figure 2
A thermometer (or a temperature probe) measures the average kinetic energy of the molecules in a beaker of water.

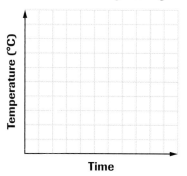

Figure 3
Temperature–time axes

STS Connections

12. List three energy sources that are renewable and three that are nonrenewable.

13. What is your reasoned response to someone trying to sell you shares in a corporation that is manufacturing a perpetual motion machine?

Skills

14. Laboratory thermometers are usually read to ±0.2 °C.
    (a) Interpret a thermometer reading of 22.4 °C on this thermometer.
    (b) Critique the communication when someone interprets a reading of 22.0 °C as "exactly twenty-two degrees Celsius" and records the temperature as 22 °C.

chapter 11

Enthalpy Change

In this chapter

- Exploration: Burning Oil
- Case Study: Personal Use of Chemical Energy
- Investigation 11.1: Designing and Evaluating a Calorimeter
- Web Activity: Thermal Insulation
- Lab Exercise 11.A: Molar Enthalpy of Neutralization
- Web Activity: Calorimetry
- Investigation 11.2: Molar Enthalpy of Reaction
- Web Activity: Hess' Law
- Lab Exercise 11.B: Testing Hess' Law
- Lab Exercise 11.C: Analysis Using Hess' Law
- Investigation 11.3: Applying Hess' Law
- Explore an Issue: Alternative Energy Sources and Technologies
- Lab Exercise 11.D: Testing $\Delta_r H°$ from Formation Data
- Web Activity: Rocket Fuel Thermochemistry

We have invented many technologies to use energy: campfires and wood stoves use energy stored in wood; stoves, heaters, and cars used energy stored in liquid hydrocarbons; industries use energy stored in natural gas; and power plants convert the chemical energy stored in natural gas, oil, and coal into electrical energy. Trucks, trains, and airplanes use energy stored in fuels refined from crude oil, heavy oil, and oil sands.

The future could look quite different from the present. Transportation technologies may derive their energy from the combustion of alternative fuels such as hydrogen, ethanol, and methanol. Hybrid cars, operating on both gasoline and batteries, are already becoming more common. Science and technology work in parallel with one another—although sometimes one is ahead of the other—often driven by social pressures. Sometimes a scientific concept is applied by technology; other times science is used to explain a technology after it is invented (independent of the science).

Consumers will have many choices to make about energy technologies. Knowing the science and technology of energy is advantageous. Energy-related issues are important from many perspectives. Will you support politicians who advocate further development of Alberta's oil sands? Will your next car be gasoline powered? This chapter will help you make these decisions in a more informed manner.

STARTING Points

Answer these questions as best you can with your current knowledge. Then, using the concepts and skills you have learned, you will revise your answers at the end of the chapter.

1. Consider the following changes: photosynthesis, cellular respiration, and combustion of gasoline. Classify each of these changes as exothermic or endothermic.
2. How is electrical energy produced in Alberta? What are the sources of energy that produce this electricity?
3. Select one energy source for producing electrical energy, and list three benefits and three risks.
4. Trace the transformations of energy from solar radiation through to using a fluorescent light bulb at home.
5. (a) How can we measure how much energy is released or absorbed by a chemical reaction?
   (b) How can we communicate how much energy is released or absorbed by a chemical reaction?

Career Connection:
Chemical Engineering Technologist

Figure 1
Our lives depend on energy transformations: from solar energy to wind energy to electrical energy; and from solar energy to chemical energy (food for cattle and ourselves).

▶ Exploration — Burning Oil

Engineers who design furnaces to heat homes and nutritionists who calculate the energy value of different foods need to analyze the energy supplied by different fuels. In experiments in which thermal energy is absorbed by water, they use the equation

$$Q = mc\Delta t$$

(thermal energy = mass × specific heat capacity × temperature change)
($c = 4.19$ J/(g·°C) for water)

The general concept of what these technology experts do is illustrated by the following low-technology experiment.

Materials: eye protection; lab apron; centigram balance; unshelled pecan or other nut; paper clip; small tin can, open at one end and punctured under the rim on opposite sides; pencil; thermometer; beaker; matches; laboratory stand with ring

- Measure the mass of the nut or assume that an average pecan has a mass of about 0.5 g.
- Bend a paper clip so that it forms a stand that will support a nut above the lab bench (**Figure 2**).
- Place 50 mL of tap water in the tin can. Measure the temperature of the water.
- Suspend the can of water inside the ring and above the nut by putting the pencil through the holes and under the rim of the can.
- Using a match, light the nut. When it has finished burning, measure the final temperature of the water.

(a) Calculate how much energy was absorbed by the water.
(b) Where did this energy come from?
(c) Calculate the quantity of energy produced per gram of fuel (nut) burned.
(d) Compare this combustion reaction to the reaction that would happen if you were to eat the nut instead of burning it.
(e) This experiment uses low-tech equipment. How can you improve on the technology and, therefore, improve on the certainty of the experimental results?

Figure 2
Experimental set-up

 Students with sensitivity to nuts or nut products should not perform this activity.

11.1 Energy Demands and Sources

Photosynthesis is a major natural contributor to stored chemical energy on Earth. Research has indicated that photosynthesis takes energy from our sun, Sol, and converts this radiant (electromagnetic) energy into chemical energy (**Figure 1**). Carbon dioxide and water react in plants, in the presence of sunlight, to produce glucose and oxygen. This chemical (potential) energy is consumed by animals, such as ourselves, and is stored or used to fuel life. The chemical energy stored by photosynthesis is accessed through cellular respiration, in which glucose and oxygen react to produce carbon dioxide and water and to supply energy. (This overall reaction is the reverse of photosynthesis.)

As presented in the carbon cycle in Chapter 9, the most accepted hypothesis for the origin of fossil fuels is the biogenic hypothesis. This hypothesis suggests that fossil fuels had their origins in deposits of plant and animal material. Solar energy was captured and stored by photosynthesis. The plant matter was either eaten by animals and stored as chemical potential energy (in carbohydrates, protein, and fat) in the body of the animal before the animal died, or the plant itself died. In certain geological areas, time, temperature, and pressure converted the dead animal and plant material, with its stored chemical potential energy, into hydrocarbons. This line of reasoning is used to support the claim that fossil-fuel energy has its origins in solar energy from our sun.

Fossil fuels are a major source of stored chemical energy. Humans have invented technologies to extract and process fossil fuels (**Figure 1**). In Alberta, huge technological advances have been made for recovering chemical energy from, in chronological order, coal, natural gas, crude oil, heavy oil, oil sands, and coal-bed methane. The demand for energy has challenged the technological inventiveness of engineers, scientists, and others in the energy workplace. In Alberta, for example, nearly 20% of all jobs are in the energy sector.

DID YOU KNOW ?

Energy Production and Consumption
In the 1960s Canada moved from being a net consumer of energy to a net producer of energy. By 2002 Canada was producing 45% more energy than it consumed. In the same year, according to Statistics Canada, Canadian energy consumption per capita reached a record high. More recent statistics may, however, rewrite that record.

CAREER CONNECTION

Chemical Engineering Technologist
This career requires problem-solving skills and creativity, and there are many different options for jobs. Chemical engineering technologists may work in production facilities, where they operate process control equipment and supervise construction activity, or they may work in engineering offices, where they redesign process equipment and perform computer-based simulations. Explore the variety of tasks performed in this career, and learn what training is needed.

www.science.nelson.com

Figure 1
Crude oil, pumped near Longview, Alberta, is an indirect source of solar energy.

Transportation and building designers have to use just as much ingenuity as the engineers and scientists working in the energy sector. On the production side, the processes have been made more efficient and economical. On the consumption side, the technologies (such as cars and homes) also need to be more efficient and economical. Of course, there are other perspectives, such as environmental, legal, ethical, and social, that also need to be considered when designing a technology for converting and using energy.

Chemists have created some concepts that help us explain chemical technologies and the transformation of energy to and from chemical energy. These concepts are conveyed using chemical energy terms, some of which are presented in the Summary on page 493. These definitions are simplified. If you take more in-depth chemistry courses you will learn more elaborate definitions, but these will suit us very nicely for now.

There are four major demands for energy from fossil fuels—heating, transportation, industry, and commercial and institutional. What are some alternatives to using fossil fuels? Options include both the use of different fuels and more economical management of fossil fuels (**Table 1**).

Table 1 Alternatives to Current Fossil Fuel Uses

Energy demands	Alternative energy sources and practices
heating residential (~15%)	• solar heating (**Figure 2**), heat pumps, geothermal energy, biomass gas, and electricity from hydroelectric and nuclear plants • improved building insulation and design
transportation (~30%)	• alcohol/gasohol and hydrogen fuels (**Figure 3**), and electric and hybrid (electric and gasoline) vehicles • mass transit, bicycles, and walking
industry (~40%)	• solar energy, nuclear energy, and hydroelectricity • improved efficiency and waste heat recovery (**Figure 4**)
commercial and institutional (~15%)	• solar water pre-heating and heat pumps, including geothermal • water and heat conservation

DID YOU KNOW ?
Electric Cars
Although electric cars are making a comeback in popularity, at one time they were more prevalent than gasoline-powered cars. For example, in 1900, in the cities of New York, New York, Chicago, Illinois, and Boston, Massachusetts, there was a total of 2370 automobiles: 400 gasoline-powered, 800 battery-powered, and 1170 steam-driven.

www.science.nelson.com

DID YOU KNOW ?
Living Lightly on the Earth
Many Aboriginal lifestyle philosophies embrace the idea of taking as little as possible from the environment and leaving more for future generations. In the 21st century, we are using energy at an accelerating rate, especially nonrenewable energy in the form of fossil fuels. To live more lightly on the Earth should we be looking for a technological fix or personal lifestyle changes?

Figure 3
Fuel cell vehicles, such as the DaimlerChrysler F-Cell pictured here, are powered by Ballard® fuel cell technology and use hydrogen as a fuel. Fuel cell vehicles emit only heat and water vapour, making them true zero-emission vehicles.

Figure 2
A well-insulated home with the majority of windows facing south is the main requirement for obtaining heat from direct sunlight. Solar heating and heat generated by people and appliances can reduce heating bills by as much as 90%. (A model home in the Drake Landing Solar Community, Okotoks, Alberta.)

Figure 4
A great deal of energy is wasted by motors, compressors, and exhaust emissions. Recovering this energy is one way in which industries improve their energy efficiency. This greenhouse uses heat from the compressors on the TransCanada Pipeline.

Case Study

Personal Use of Chemical Energy

We rely on chemical energy for our very lives, as well as to heat our homes, cook our food, and move us around. Chemical energy for personal use comes from two main sources: combustion reactions and electrochemical reactions. The most personal use of chemical energy is eating food. The human body converts the chemical energy in food into mechanical energy and body heat through the process of cellular respiration, a form of combustion. A typical high-calorie fast-food meal (burger, fries, and drink) provides about 6 MJ of energy. (Research indicates that daily intakes would be around 9 to 12 MJ for a normally active young adult, and up to 29 MJ for an extreme high-performance athlete.) When you buy food, you are buying energy to keep your body running.

First Nations peoples of Alberta traditionally consumed much less energy in their balanced, self-sustaining mode of life. It has been generally noted by anthropologists that earlier lifestyles, barring accidents and foreign disease, were much more healthful for individuals than current lifestyles, which too often include diets high in fat and sugar.

In Alberta, nearly all the energy for home heating and cooking comes from the combustion of fossil fuels, either directly or indirectly. For example, many homes are heated directly by burning methane or propane on-site, or by electricity, which may be produced by burning fossil fuels off-site. Propane and butane are convenient fuels for motor homes and campers because these hydrocarbons can easily be compressed, stored, and transported. A century ago in Alberta, firewood and coal were important fuels for heating and cooking, but now firewood is used mainly by wilderness campers and people living in remote areas.

Our transportation needs are also supplied by chemical energy (**Figures 5** and **6**). The family car and motorcycle use gasoline, which stores a large quantity of energy in a small volume. The combustion of a litre of gasoline produces about 34 MJ of energy.

Gasohol is a mixture of gasoline and methanol/ethanol, designed to reduce pollution and our reliance on fossil fuels. Renewable fuels derived from biomass (such as wood pulp, grain, and crop by-products) are used together with fossil fuels.

The technology for using hydrogen gas to run automobiles has been developed but its use is not widespread. Chemical energy stored in batteries is used to power golf carts, passenger cars, and delivery vans. A battery-powered vehicle can travel only a relatively short distance before the battery needs to be recharged. Hybrid automobiles combine a gasoline engine with a battery-powered electric motor that becomes a generator and recharges as you drive. Hybrid vehicles are also able to use gravitational potential energy when going downhill, and kinetic energy when braking, to help recharge the battery.

The portability of small batteries makes them valuable for radios, cameras, music players, phones, and personal computers.

Most of our personal use of chemical energy, however, comes from burning fossil fuels—even the energy used to recharge batteries likely comes from an electrical energy plant that burns coal or natural gas. Our dependence on fossil fuels raises concerns about depleting our energy reserves and increasing greenhouse gas emissions. On a personal level, a simple solution is to drive a car less and walk, bike, or ride a bus more often. The chemical energy for self propulsion comes from food, and the carbon dioxide produced by cellular respiration is significantly less than that produced by a car.

Case Study Questions

1. What are the two main sources of chemical energy for personal use?
2. When 1 mol of glucose is burned, 2802.5 kJ of energy is released. Calculate the quantity of energy released to a person by eating 5.00 g of glucose in a candy.
3. How could technology help us meet our need for energy? Suggest at least four ways.
4. Technological solutions (fixes) are often chosen over changing personal habits or behaviour. Some people argue that saving fossil-fuel energy for future generations is important. Provide three examples of technological fixes versus changes in personal behaviour for reducing chemical energy consumption. Briefly evaluate, for each example, which would be the better solution.
5. Identify and list the perspectives represented in this Case Study. See Appendix D.1.

Figure 5
Cycling uses food energy.

Figure 6
Motorcycling uses fossil fuel energy.

Section 11.1 Questions

Use **Table 2** to answer questions 1 and 2. Alternatively, go to the Statistics Canada Web site and obtain the latest data.

www.science.nelson.com

Table 2 Canadian Energy Production by Fuel Type (Source) in 2003 (in terajoules (TJ), or 10^{12} J)

Coal (TJ)	Crude oil (TJ)	Natural gas (TJ)	Natural gas liquids (NGLs) (TJ)	Primary electricity, hydro and nuclear (TJ)	Total production (TJ)
1 326 114	5 679 573	7 024 602	642 897	1 457 467	16 130 653

Source: Statistics Canada

1. (a) Convert the Canadian energy production into a percentage of the total for each fuel type for the latest year provided. For consistency of use in the pie chart in part (b), give each percentage to one decimal place.
   (b) Display these percentages as a pie chart. Use graphing software if available.

2. (a) Go to the Statistics Canada Web site and calculate the percentage of the production of each fuel type that is exported from Canada.
   (b) Display these percentages as a bar graph. Use graphing software if available.

www.science.nelson.com

Use **Table 3** to answer questions 3 to 6. Alternatively, go to the Statistics Canada Web site and obtain the latest data.

www.science.nelson.com

Table 3 Canadian Energy Use/Demand (in terajoules (TJ), or 10^{12} J)

Energy Use/Demand by Year	1999	2000	2001	2002	2003
Total industrial (TJ)	2 177 297	2 268 624	2 166 287	2 229 541	2 313 106
Total transportation (TJ)	2 307 283	2 279 845	2 240 367	2 250 130	2 242 042
Agriculture (TJ)	229 865	231 927	218 075	206 753	211 866
Residential (TJ)	1 232 263	1 287 825	1 239 970	1 286 677	1 348 041
Public administration (TJ)	124 522	131 288	126 813	125 164	127 678
Commercial and other institutional (TJ)	1 061 446	1 176 423	1 184 065	1 286 657	1 362 441
Total energy use (TJ)	**7 132 504**	**7 375 967**	**7 175 442**	**7 384 682**	**7 604 948**

Source: Statistics Canada

3. For the latest year listed, what are the percentages of use (to one decimal place) by the sectors listed?

4. Create a line graph for the total energy use (in petajoules (PJ; $\times 10^{15}$ J)) over the years for which data are provided. Use graphing software if available.

5. Create a line graph for residential use (in terajoules (TJ)) over the years for which data are provided. Use graphing software if available.

6. Go to the Statistics Canada Web site and obtain the breakdown of residential use by fuel type.
   (a) What percentage of residential energy use (to one decimal place) is from natural gas?
   (b) Quote three other statistics that you can extract from the data. Be sure to cite your sources.

7. There is an increasing number of energy choices for operating a car. List at least four of these choices.

8. Homes in Alberta are heated by a variety of energy sources. List at least four of these sources.

9. The Drake Landing Solar Community in Okotoks, Alberta, is an experimental housing project that uses solar energy to heat the homes. Run the animation through the Nelson Web site, and write a short description of the project from a technological perspective.

www.science.nelson.com

www.science.nelson.com

10. Energy consumption for keeping our homes warm (and supplied with hot water) varies fairly predictably during the year (**Table 4**).

Table 4 Average Monthly Energy Use in Alberta per Household

Month	Energy (GJ)
May	5
June	3
July	3
August	4
September	5
October	9
November	18
December	22
January	23
February	18
March	16
April	9
Annual total	**135**

(a) Prepare a bar graph to express the monthly data. Use graphing software if available.
(b) From energy bills, how does your household energy use compare with these values?

11. Several energy sources are available for heating: oil, gas, coal, wood, solar, geothermal, and nuclear. These sources produce heat directly. Indirect sources of heat include electric motors and lights.
    (a) Which of the direct energy sources originated in our sun, and which originated from the formation of our planet?
    (b) Which two energy sources are currently, in your area, most commonly used for heating?
    (c) Which two energy sources do you think will be the most widely used for heating by the year 2030?

12. Engineers and scientists have invented technologies for using locally available sources of energy to produce electricity. In **Figure 7**, the kinetic energy of falling water or of steam is used to turn a turbine attached to an electric generator.
    (a) What fuels can be burned to produce the steam?
    (b) What other sources of energy (other than hydro and fossil fuels) are available to produce electricity?

13. Some retailers have advertised titanium necklaces as an energy source. As evidence to support their claim, they ask customers to lift a bag containing a brick. They then drape the necklace over the customer's arm and have the person lift the bag again. Customers find that lifting the bag with the necklace present is easier, and they often buy the necklace as a result. Critique this experimental design.

Extension

14. Energy consumers were briefly excited when successful "cold fusion" was announced by two scientists in 1989. Successful cold fusion would represent an inexpensive, clean, readily available source of energy. Do some library research or search the Internet to learn more about the cold fusion experiments.
    (a) Were the two scientists empiricists or theoreticians?
    (b) Were they qualified to make theoretical interpretations of their experimental work?
    (c) Are there cold-fusion research programs still being funded? If so, where?
    (d) Evaluate your literature research strategies.

    www.science.nelson.com

15. The caption for Figure 3 (page 481) says that these cars are "true zero-emission vehicles." Consider this claim. Research the environmental effects of water vapour. Think about the research and production involved in developing such vehicles. Write or present a brief rebuttal or supporting statement for the "zero-emissions" claim.

    www.science.nelson.com

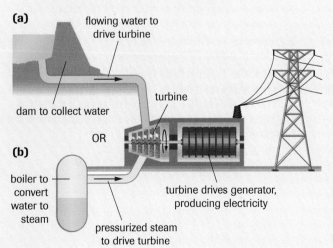

Figure 7
There are many energy sources that can be used to spin a turbine. If the turbine is connected to a generator, the energy of motion is converted to electrical energy.

Calorimetry 11.2

According to the law of conservation of energy, energy is neither created nor destroyed in any physical or chemical change. In other words, energy is only converted from one form to another. The study of energy changes (that is, energy produced or absorbed) by a chemical system during a chemical reaction is called **thermochemistry**. To study energy changes chemists require an **isolated system**, that is, one in which neither matter nor energy can move in or out. They also need carefully designed experiments and precise measurements. **Calorimetry** is the technological process of measuring energy changes of an isolated system called a **calorimeter** (**Figure 1**). Within this isolated system the chemical system (reactants and products) being studied is surrounded by a known quantity of liquid (generally water) inside the calorimeter. Energy, as **heat**, is transferred between the chemical system and the water. (Note that heat as a form of energy can only be transferred between substances; heat is never possessed by them.) The energy gained by the chemical system is equal to the energy lost by the calorimeter and its contents, as long as both the chemical system and the calorimeter (the surroundings) are part of an isolated system. In other words, for the measurement to be accurate, no energy may be transferred between the inside of the calorimeter and the environment outside the calorimeter.

The main assumption is that no heat is transferred between the calorimeter and the outside environment. Although chemists know this to be a false assumption, it provides a close approximation. The better the calorimeter technology, the closer the assumption is to being valid. A simplifying assumption is that any heat absorbed or released by the calorimeter materials, such as the container, is negligible. Another assumption is that a dilute aqueous solution in a calorimeter has the same physical properties as pure water (for example, the same volume and specific heat capacity).

Figure 1
A simple laboratory calorimeter consists of an insulated container made of three nested polystyrene cups, a measured quantity of water, and a thermometer. The chemical system is placed in or dissolved in the water of the calorimeter. Energy transfers between the chemical system and the surrounding water are monitored by measuring changes in the water temperature.

Analyzing Energy Changes

When methane reacts with oxygen in a laboratory burner, enough energy is transferred to the surroundings to increase the temperature (**Figure 2**). How is this energy quantified? Calorimetry requires careful measurements of masses and temperature changes. When a fuel such as methane burns, energy is released and heat is transferred from the chemical system to the surroundings (which include the water in the beaker). Given the same quantity of heat, a small quantity of water will undergo a greater increase in temperature than a large quantity of water. In addition, different substances vary in their ability to absorb quantities of thermal energy (**Figure 3**). Finally, if more heat is transferred, the observed temperature rise in the water is greater.

Calorimetry work in the laboratory has shown that, keeping the type of substance constant (for example, using only water in the calorimeter), the quantity of energy transferred is directly proportional to the mass of the substance and its change in temperature. The chemical energy lost by the chemical system is calculated from the thermal energy gained by the surroundings. **Thermal energy** is the total kinetic energy of the entities of a substance (for example, the molecules of water in a calorimeter).

These three factors—mass (m), type of substance, and temperature change (Δt)—are combined in an equation to represent the quantity of thermal energy (Q) transferred:

$$Q = mc\Delta t$$

where c is the **specific heat capacity**, the quantity of energy required to raise the temperature of a unit mass (for example, one gram) of a substance by one degree Celsius or one kelvin. For example, an average specific heat capacity of water around room temperature is 4.19 J/(g·°C). (Recall that the SI unit for energy is the **joule**, J.)

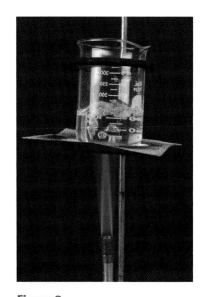

Figure 2
The fuel in the burner releases energy that is absorbed by the surroundings, which include the beaker, water, and air.

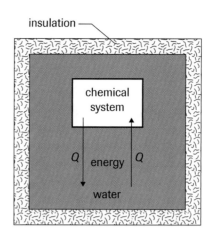

Figure 3
The chemical system inside the calorimeter undergoes either a phase change, such as fusion, or a chemical change, such as a double replacement reaction. Energy is either absorbed from the water or released to the water. An increase in the temperature of the water indicates an exothermic change of the system; a decrease in the temperature of the water indicates an endothermic change of the chemical system.

Table 1 Specific Heat Capacities of Substances

Substance	Specific heat capacity, c (J/(g·°C))
ice	2.00
water	4.19
steam	2.02
aluminium	0.897
iron	0.449

Specific heat capacities vary from substance to substance, and even for different states of the same substance (**Table 1**). The specific and volumetric heat capacities listed in Appendix I were determined by means of calorimetry.

COMMUNICATION example 1

Determine the change in thermal energy when 115 mL of water is heated from 19.6 °C to 98.8 °C.

Solution

$Q = mc\Delta t$

$= 115 \text{ g} \times \dfrac{4.19 \text{ J}}{\text{g·°C}} \times (98.8 - 19.6) \text{ °C}$

$= 38.2 \text{ kJ}$

Based upon the assumption that the temperature change is a complete measure of the change in thermal energy, the change in thermal energy of the water is 38.2 kJ.

DID YOU KNOW?

A Kilojoule of Energy
Completely burning a wooden match releases about 1 kJ of energy. This quantity of energy could heat $\frac{1}{4}$ cup of water by about 1 °C, and is equivalent to about a quarter of a calorie of food energy. There are two definitions of calorie but only one for kilojoule: 1000 J.

+ EXTENSION

Food Energy
Labels on food packaging report the number of kilojoules per serving of the contents. How do scientists determine this number? This audio clip will help you understand how thermochemistry is used to determine the quantity of energy available in our food.

Some of the calorimetry assumptions are that
- all the energy lost or gained by the chemical system is gained or lost (respectively) by the calorimeter; that is, the total system is isolated.
- all the material of the system is conserved; that is, the total system is isolated.
- the specific heat capacity of water over the temperature range is 4.19 J/(g·°C).
- the specific heat capacity of dilute aqueous solutions is 4.19 J/(g·°C).
- the density of a dilute aqueous solution is the same as that of water; that is, 1.00 g/mL. (see Communication Example 1)
- the thermal energy gained or lost by the rest of the calorimeter (other than water) is negligible; that is, the container, lid, thermometer, and stirrer do not gain or lose thermal energy.

INVESTIGATION 11.1 Introduction

Designing and Evaluating a Calorimeter

A calorimeter can be as simple as an uninsulated beaker or can. Literature research indicates that very sophisticated calorimeters, such as bomb and flow calorimeters, are also available. There are many laboratory calorimeters between these extremes. In this investigation you will create your own calorimeter. Use modern materials and/or consider what Aboriginal peoples might have used for insulating a calorimeter-like vessel.

Purpose

The purpose of this investigation is to test a hypothetical design of a calorimeter that is improved (based upon your own scientific and technological criteria) over the simplest calorimeter.

To perform this investigation, turn to page 516.

Report Checklist
- ○ Purpose
- ○ Problem
- ● Hypothesis
- ○ Prediction
- ● Design
- ○ Materials
- ● Procedure
- ● Evidence
- ● Analysis
- ● Evaluation (1, 2, 3)

Problem

What is the best design for a simple calorimeter?

Design

Each research group in class uses a different calorimeter design of their choosing. One group uses an uninsulated calorimeter as a control. An individual group and a class Analysis are employed.

Practice

1. In your own words, write the law of conservation of energy.
2. Why do we call a calorimeter an "isolated system"?
3. A calorimeter is used to measure (somewhat indirectly) what kind of energy?
4. Experiments have shown that a thermal energy change is affected by mass, specific heat capacity, and change in temperature. What happens to the thermal energy if
   (a) the mass is doubled?
   (b) the specific heat capacity is divided by two?
   (c) the change in temperature is tripled?
   (d) all three variables are doubled?
5. If the same quantity of energy were added to individual 100 g samples of water, aluminium, and iron, which substance would undergo the greatest temperature change? Explain. (See Appendix I for specific heat capacities.)
6. After coming in from outside, a student makes a cup of instant hot chocolate by heating water in a microwave. What is the gain in thermal energy if a cup (250 mL) of tap water is increased in temperature from 15 °C to 95 °C?
7. A backpacker uses an uncovered pot to heat 1.0 L of lake water on a single-burner stove. If the water temperature rises from 5.0 °C to 97.0 °C, what is the gain in thermal energy of the water?
8. To test the efficiency of an energy transfer, a chemist supplies 30.0 kJ of external energy to a simple calorimeter. If 150 mL of water in the calorimeter is heated from 20.6 °C to 52.8 °C, what is the percent efficiency of the calorimeter?

DID YOU KNOW?

Honeymoon Inspiration

During his honeymoon, while sitting beside a waterfall, James Joule speculated on the temperature increase of the water as it fell over the waterfall. Joule tenaciously pursued the relationship between work and heat, testing his hypothesis that the work needed to produce a certain quantity of heat is always the same. This realization led to the idea that heat is a form of energy rather than a fluid, as scientists believed at the time.

Joule is known for his patient and precise quantitative measurements that allowed him to calculate the specific heat capacity of water. He also studied the relationship between heat and temperature and helped establish the law of conservation of energy.

Heat Transfer and Enthalpy Change

Chemical systems have many different forms of energy, both kinetic and potential. These include the kinetic energies (energy of motions) of

- moving electrons within atoms
- the vibration of atoms connected by chemical bonds
- the rotation and translation of molecules that are made up of these atoms.

The chemical potential energy of a chemical system includes energy stored in

- covalent and/or ionic bonds between the entities
- intermolecular forces between entities

Figure 4
James Prescott Joule (1818–1889)

Learning Tip

In searching through references you may find the terms *enthalpy of reaction*, *heat of reaction*, *enthalpy change*, and ΔH. You can assume, at this point, that they all mean the same thing.

DID YOU KNOW ?

Setting Hard

The setting of concrete is quite exothermic, and the rate at which it sets or cures determines the hardness of the concrete. If the concrete sets too quickly (for example, if the enthalpy of reaction is not dissipated quickly enough into the air), the concrete may expand and crack. Doctors are cautious of similar heating when making a plaster cast for a broken limb.

The total of the kinetic and potential energy within a chemical system is called its enthalpy. The enthalpy of a chemical system is an expression of the chemical energy possessed by the chemical system. However, if the reactants and products are at the same initial and final temperature, then the kinetic energy of the system does not change.

Just as electrical and gravitational potential difference are most often communicated as a difference from a reference point, enthalpy is most often communicated as a difference in enthalpy between reactants and products for a particular chemical system. An **enthalpy change**, ΔH, communicates the difference between the enthalpy (assume the difference in chemical potential energy) of the products and the enthalpy of the reactants.

$$\Delta H = H_{\text{products}} - H_{\text{reactants}} \quad \text{or} \quad \Delta H = H_\text{P} - H_\text{R} \quad \text{(IUPAC definition)}$$

Analysis of calorimetric evidence is based on the law of conservation of energy and on several assumptions. The law of conservation of energy may be expressed in several ways, for example, "The total energy change of the chemical system is equal to the total energy change of the calorimeter surroundings." Using this method, both the enthalpy change, ΔH, and the quantity of thermal energy, Q, are calculated as absolute values, without a positive or negative sign.

$$\underset{\text{(system)}}{\Delta H} = \underset{\text{(calorimeter)}}{Q}$$

Energy loss by the chemical system is equal to energy gain by the surroundings. Consider the reaction that occurs when zinc metal is added to hydrochloric acid in an open flask:

$$\text{Zn(s)} + 2\,\text{HCl(aq)} \rightarrow \text{H}_2\text{(g)} + \text{ZnCl}_2\text{(aq)}$$

Research on this reaction indicates that the solution increases in temperature. In other words, there is an increase in thermal energy of the surroundings. Chemists infer from the temperature increase that the chemical system (in this case, Zn(s) and HCl(aq)) must have lost chemical energy to the surroundings (**Figure 5**). At this stage in your understanding, you can make the assumptions that $\Delta E_\text{p} = \Delta H$ and $\Delta E_\text{k} = Q$.

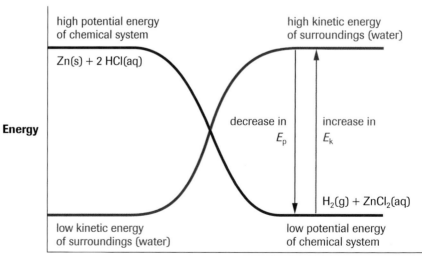

Figure 5

In this example of an exothermic change, the change in potential energy of the chemical system (ΔE_p or ΔH) equals the change in kinetic energy of the surroundings (ΔE_k or Q). This is consistent with the law of conservation of energy.

▶ **SAMPLE** problem **11.1**

In a simple calorimetry experiment involving a burning candle and a can of water, the temperature of 100 mL of water increases from 16.4 °C to 25.2 °C when the candle is burned for several minutes. What is the enthalpy change?

The first step in solving this problem is to use the law of conservation of energy. According to this law, the energy lost by the chemical system (the burning candle) is equal to the energy gained by the surroundings (assume the calorimeter water). The change in enthalpy ($\Delta_c H$) of the chemical system can be determined from the change in thermal energy (Q) of the surroundings, as expressed by the equation

$\Delta_c H = Q$

Since $Q = mc\Delta t$

$\Delta_c H = mc\Delta t$

Substituting the evidence gathered into the conservation of energy equation, and using the density of water as 1.00 g/mL

$\Delta_c H = mc\Delta t$
$= 100 \text{ g} \times \frac{4.19 \text{ J}}{\text{g} \cdot °\text{C}} \times (25.2 - 16.4) °\text{C}$
$= 3.69 \text{ kJ}$

Based upon the evidence gathered in this experiment, the approximate value for the enthalpy change of the chemical system is *recorded* as -3.69 kJ.

Learning Tip

The following symbols are used by IUPAC to denote a chemical process or reaction.

Table 2 Symbols for a Reaction

Subscript	Meaning
r	any reaction specified
c	complete combustion
f	formation
d	decomposition
sol	solution
dil	dilution

For example, $\Delta_c H$ denotes the change in enthalpy (ΔH) of combustion (c) for a specified combustion reaction.

The value for the enthalpy change in Sample Problem 11.1 is given as a negative value ($\Delta_c H = -3.69$ kJ) because the change is a decrease. If the surroundings gained energy (i.e., $Q > 0$), then, according to the law of conservation of energy, the chemical system must have decreased in energy (i.e., $\Delta_c H < 0$). Use this reasoning to add the sign after doing the calculation.

Sample Problem 11.1 illustrates how we can infer the size of an enthalpy change from the size of a thermal energy change. Even though we cannot measure the value of an enthalpy change directly, we can determine it indirectly.

Chemists call a chemical change that produces energy an **exothermic reaction**. The enthalpy change (ΔH) is negative. A reaction that absorbs energy is called an **endothermic reaction**. The enthalpy change (ΔH) is positive.

Learning Tip

Remember **En**dothermic and **ex**othermic by associating the first two letters of **en**ter and **ex**it. Just remember that the energy is entering or exiting the chemical system, not the surroundings.

▶ **COMMUNICATION** example **2**

When 50 mL of 1.0 mol/L hydrochloric acid is neutralized completely by 75 mL of 1.0 mol/L sodium hydroxide in a polystyrene cup calorimeter, the temperature of the total solution changes from 20.2 °C to 25.6 °C. Determine the enthalpy change that occurs in the chemical system.

Solution

$\Delta_r H = Q$
$= mc\Delta t$
$= (50 + 75) \text{ g} \times \frac{4.19 \text{ J}}{\text{g} \cdot °\text{C}} \times (25.6 - 20.2) °\text{C}$
$= 2.83 \text{ kJ}$

Based upon the evidence available, the enthalpy change for the neutralization of hydrochloric acid in this context is recorded as -2.83 kJ.

Learning Tip

There are several methods for solving these conservation of energy problems. The method shown here uses absolute values (with no positive or negative values in the equation). As a result, you have to provide the positive or negative sign to the final answer for ΔH.

WEB Activity

Case Study—Thermal Insulation

Living in Alberta in the winter involves dressing for the weather. Investigate what technological "advances" have been made by chemists creating synthetic materials. Compare these modern materials with traditional materials, such as animal skins, used for footwear and warm clothing.

www.science.nelson.com

Molar Enthalpies and Calorimetry

Enthalpy of reaction (change in enthalpy) refers to the energy change for a whole chemical system when reactants change to products. At this stage in your study of thermochemistry you are looking only at energy change as a result of a chemical change in the chemical system. (You are not, for example, investigating energy changes due to phase changes, nuclear changes, or the dissolving of ionic compounds.) For all chemical changes a stated enthalpy change is for the total quantity of chemical involved in the reaction as specified by the reaction equation coefficients. Chemists sometimes want to know the enthalpy change per unit chemical amount (per mole) because they want to relate the enthalpy change to a chemical reaction equation that is balanced in moles.

Theoretically, **molar enthalpy of reaction** is the enthalpy change in a chemical system per unit chemical amount (per mole) of a specified chemical undergoing change in the system at constant pressure. Although enthalpy changes can be for any amount or mass of chemical undergoing a chemical change, molar enthalpy is the energy lost or gained expressed in units of, for example, kilojoules per mole (kJ/mol). The IUPAC symbol for molar enthalpy for a specified reaction is $\Delta_r H_m$.

DID YOU KNOW ?

Measuring Enthalpy Change
Chemists and chemical engineers object to talking about *measuring* a change in enthalpy. They agree that we can measure a change in temperature (Δt) with a thermometer. Then we can *calculate* the change in thermal energy from $Q = mc\Delta t$. However, we must then use the law of conservation of energy to *infer* the molar enthalpy from the calculated thermal energy. In doing so, we *assume* that the change in enthalpy of the chemicals involved in a reaction is equal to the change in thermal energy of the surroundings.

Learning Tip

Note the logically consistent pattern for calculating quantities from their chemical amount (*n*) and molar quantity (Table 3). For example,

$m = n M$
$\quad = \text{mol} \times \text{g/mol}$

$V = n V_m$
$\quad = \text{mol} \times \text{L/mol}$

$\Delta_r H = n \Delta_r H_m$
$\quad = \text{mol} \times \text{kJ/mol}$

We can see that the equations are correct because the units work out.

Table 3 Quantities and Molar Quantities

Quantity (unit)	Molar quantity (unit)
mass, m (g)	molar mass, M (g/mol)
volume, V (L)	molar volume, V_m (L/mol)
enthalpy change, $\Delta_r H$ (kJ)	molar enthalpy, $\Delta_r H_m$ (kJ/mol)

In order to calculate enthalpy change from their respective molar quantities (**Table 3**), we have the following equation:

enthalpy change of reaction — chemical amount — molar enthalpy of reaction

$$\Delta_r H = n \Delta_r H_m$$

for example, $\Delta_c H = n \Delta_c H_m$ (c for combustion)

The latter equation is read as "the enthalpy of combustion is equal to the chemical amount of the fuel times the molar enthalpy of combustion of the fuel."

Molar quantities are often referenced (looked up in a table) or memorized: atomic molar masses are obtained from a periodic table; the molar volume of any gas at STP is memorized or referenced as 22.4 L/mol. We can use these molar quantities to predict, for example, the enthalpy of combustion ($\Delta_c H$) of an organic compound (as in Sample Problem 11.2).

Section 11.2

> ### SAMPLE problem 11.2

Predict the change in enthalpy due to the combustion of 10.0 g of propane used in a camp stove (**Figure 6**).

First, you have to determine the change in enthalpy for the combustion, $\Delta_c H$, using the equation

$$\Delta_c H = n \, \Delta_c H_m$$

To calculate $\Delta_c H$ you need to know the chemical amount (n) and the molar enthalpy of combustion ($\Delta_c H_m$). You can calculate the chemical amount from the mass and molar mass. The molar enthalpy of combustion of propane (producing water vapour) is $-2\,043.9$ kJ/mol.

$$\Delta_c H = n \, \Delta_c H_m$$
$$= 10.0 \text{ g} \times \frac{1 \text{ mol}}{44.11 \text{ g}} \times \frac{-2\,043.9 \text{ kJ}}{1 \text{ mol}}$$
$$= -463 \text{ kJ}$$

Based on the concept of molar enthalpy, the change in enthalpy due to the combustion of 10.0 g of propane is -463 kJ.

Figure 6
Propane is a camp stove fuel. It can also be used for barbecues, lanterns, and recreation vehicle refrigerators.

> ### COMMUNICATION example 3

Predict the enthalpy change due to the combustion of 10.0 g of butane in a camp heater. The molar enthalpy of combustion to produce water vapour is $-2\,657.3$ kJ/mol.

Solution

$$\Delta_c H = 10.0 \text{ g} \times \frac{1 \text{ mol}}{58.14 \text{ g}} \times \frac{-2\,657.3 \text{ kJ}}{1 \text{ mol}}$$
$$= -457 \text{ kJ}$$

Based upon the concept of molar enthalpy, the enthalpy change for the combustion of 10.0 g of butane is recorded as -457 kJ.

The molar enthalpy of reaction can be calculated from or used in calorimetry investigations. Again, the law of conservation of energy is used to produce an equality. The change in enthalpy (ΔH) of the chemical system is equal to the change in thermal energy (Q) of the calorimeter.

$$\Delta H = Q$$
$$n \, \Delta_c H_m = mc \Delta t$$

From this equation, any one of the five variables can be determined as an unknown. See Communication Example 4. The IUPAC symbols used in the calorimetry equation need to be interpreted carefully.

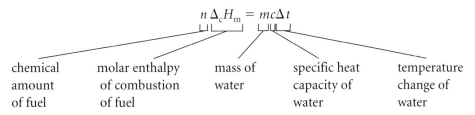

Learning Tip

The law of conservation of energy, as it relates to calorimetry, can be communicated as

$$\Delta H = Q$$
$$n \Delta_c H_m = mc \Delta t$$

However, the law of conservation of energy can be used without equating the two equations. Rather, we can first calculate the change in thermal energy, Q.

$$Q = mc \Delta t$$

then use the Q value as the $\Delta_r H$ value to calculate either the chemical amount or the molar enthalpy of reaction (as required); i.e.,

$$n = \frac{\Delta_c H}{\Delta_c H_m}$$

or

$$\Delta_c H_m = \frac{\Delta H}{n}$$

Your teacher may indicate whether you should write out these last two equations, or just substitute into them.

Learning Tip

There are several methods for solving these conservation of energy problems. The method shown in this textbook uses absolute values (with no positive or negative values in the equation). As a result, you have to provide the positive or negative sign to the final answer for ΔH. Negative values indicate exothermic changes and positive values indicate endothermic changes.

COMMUNICATION example 4

Ethanol is often added to gasoline as a renewable component that reduces harmful emissions. The mixture is known as gasohol. In a research laboratory, the combustion of 3.50 g of ethanol in a sophisticated calorimeter causes the temperature of 3.63 L of water to increase from 19.88 °C to 26.18 °C. Use this evidence to determine the molar enthalpy of combustion of ethanol.

Solution

$Q = mc\Delta t$

$= 3.63 \text{ kg} \times \dfrac{4.19 \text{ J}}{\text{g} \cdot °\text{C}} \times (26.18 - 19.88) \text{ °C}$

$= 95.8 \text{ kJ}$

$\Delta_c H_{m_{C_2H_5OH}} = \dfrac{95.8 \text{ kJ}}{3.50 \text{ g} \times \dfrac{1 \text{ mol}}{46.08 \text{ g}}}$

$= 1.26 \text{ MJ/mol}$

Based upon the evidence gathered, the molar enthalpy of combustion of ethanol is recorded as -1.26 MJ/mol.

EXTENSION

Stronger Hurricanes

The theory of global warming is now accepted by a majority of scientists. But global warming doesn't just make our winters milder and our summers hotter. Find out how global warming might be completely changing weather patterns around the world.

www.science.nelson.com

Practice

9. Write the symbols for the following terms.
   (a) enthalpy of formation
   (b) enthalpy of decomposition
   (c) molar enthalpy of formation
   (d) molar enthalpy of decomposition

10. In your own words, describe the similarities and differences between the enthalpy change and molar enthalpy symbols and meanings.

11. What is the significance of a positive or negative sign for an enthalpy change or molar enthalpy value?

12. Predict the enthalpy change for the combustion of every 100 g of methane in a natural gas water heater. ($\Delta_c H_m$ (CH_4) = -802.5 kJ/mol (to produce water vapour))

13. Methanol is one type of fuel that is used in fondue heaters. In an experiment using a simple tin can calorimeter, 2.98 g of methanol was burned to raise the temperature of 650 g of water by 20.9 °C. Using this evidence, calculate the molar enthalpy of combustion of methanol (to produce water as a vapour).

LAB EXERCISE 11.A

Molar Enthalpy of Neutralization

Report Checklist
- ○ Purpose
- ● Problem
- ○ Hypothesis
- ○ Prediction
- ○ Design
- ● Materials
- ○ Procedure
- ○ Evidence
- ● Analysis
- ○ Evaluation

The molar enthalpy of neutralization, $\Delta_n H_m$, for hydrochloric acid when it reacts with a base can be determined by calorimetry. When two aqueous solutions are mixed in a calorimeter, the final aqueous mixture is considered to be the water in the calorimeter when calculating Q.

Purpose

The purpose of this exercise is to use calorimetry to determine the molar enthalpy of neutralization for hydrochloric acid in its reaction with aqueous ammonia.

Design

1.00 L of 0.100 mol/L hydrochloric acid and 1.10 L of 0.100 mol/L aqueous ammonia are combined in a calorimeter.

Evidence

initial temperature of each solution = 23.20 °C
final temperature of mixture = 23.82 °C
volume of mixture = 2.10 L

Section **11.2**

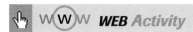 **WEB** *Activity*

Simulation—Calorimetry

The purpose of this activity is to use a calorimetry computer model to investigate the quantity of heat produced by combustion of various masses of different compounds in excess oxygen. Keep all variables constant, other than changing a reactant.

www.science.nelson.com GO ◀▶

 INVESTIGATION 11.2 *Introduction*

Molar Enthalpy of Reaction

The accuracy (percent difference) of the value for molar enthalpy obtained in this investigation is used to evaluate the calorimeter and the assumptions made in the analysis, rather than to evaluate the prediction and its authority. The ultimate authority in this experiment is considered to be the reference value used in the prediction.

Purpose

The purpose of this investigation is to test the calorimeter design and calorimetry procedure by comparing experimental evidence with a widely accepted value for the molar enthalpy of a neutralization reaction.

Report Checklist

- ○ Purpose
- ○ Problem
- ○ Hypothesis
- ○ Prediction
- ● Design
- ● Materials
- ● Procedure
- ● Evidence
- ● Analysis
- ● Evaluation (2, 1, 3)

Problem

What is the molar enthalpy of neutralization for sodium hydroxide when 50 mL of aqueous 1.0 mol/L sodium hydroxide reacts with an excess quantity of 1.0 mol/L sulfuric acid?

Prediction

The molar enthalpy of neutralization for sodium hydroxide is -57 kJ/mol, as per *The CRC Handbook of Chemistry and Physics*.

To perform this investigation, turn to page 516.

SUMMARY

- Thermal energy—the kinetic energy of entities (e.g., molecules of the surroundings, around a chemical system). Thermal energy increases with temperature.
- Change in thermal energy (Q in kJ)—measured by a temperature change and calculated using the relationship (equation) $Q = mc\Delta t$
- Heat—thermal energy transferred between systems. Heat is not possessed by a system. Heat is energy in transition (flowing) between systems.
- Kinetic energy (E_k in kJ)—a form of energy related to the motion of a chemical entity; measured as an average kinetic energy by the temperature of a chemical system
- Chemical potential energy (E_p in kJ)—the energy present in the chemical bonds of a substance
- Enthalpy change ($\Delta_r H$ in kJ)—a change in the enthalpy of a chemical system under constant pressure; also loosely called heat of reaction
- Molar enthalpy ($\Delta_r H_m$ in kJ/mol)—the change in enthalpy expressed per mole of a substance undergoing a specified reaction
- Endothermic change—a change in chemical energy where energy/heat enters (is absorbed by) the chemical system under consideration; results in an increase in chemical potential energy (and enthalpy); the $\Delta_r H$ is a positive value
- Exothermic change—a change in chemical energy where energy/heat exits (is released by) the chemical system under consideration; results in a decrease in chemical potential energy (and enthalpy); the $\Delta_r H$ is a negative value

Section 11.2 Questions

1. Record whether each of these statements is true or false. Provide your reasoning.
   (a) Heat is possessed by a chemical system.
   (b) For an exothermic change, energy exits from the surroundings and enters the chemical system.
   (c) The change in enthalpy of a chemical system is measured in kilojoules per mole (kJ/mol).
   (d) The change in thermal energy of the surroundings can be calculated using $Q = mc\Delta t$.
   (e) Temperature is the measure of the average kinetic energy of the entities in a chemical system.

2. It is commonly assumed in calorimetry labs with polystyrene calorimeters that a negligible quantity of heat is absorbed or released by the solid calorimeter materials such as the cup, stirring rod, and thermometer. Use the empirical data in **Table 4** to evaluate this assumption.
   (a) For a temperature change of 5.0 °C, calculate the energy change of the water only.
   (b) For a temperature change of 5.0 °C, calculate the total energy change of the water, polystyrene cups, stirring rod, and thermometer.
   (c) Calculate the percent error introduced by using only the energy change of the water.
   (d) Evaluate the assumption of negligible heat transfer to the solid calorimeter materials.

Table 4 Typical Quantities for Materials in a Simple Calorimeter

Material	Specific heat capacity (J/(g·°C))	Mass (g)
water	4.19	100.00
polystyrene cups	0.30	3.58
glass stirring rod	0.84	9.45
thermometer	0.87	7.67

3. Ethane is a component of natural gas. If the molar enthalpy of combustion of ethane is −1.56 MJ/mol, calculate the quantity of thermal energy gained by the burning of
   (a) 5.0 mol of ethane
   (b) 40.0 g of ethane

4. Decane is one of the hundreds of compounds in gasoline. The molar enthalpy of combustion of decane ($C_{10}H_{22}$) is −6.78 MJ/mol. What mass of decane would have to be burned in order to raise the temperature of 500.0 mL of water from 20.0 °C to 55.0 °C?

5. During sunny days, chemicals can store solar energy in homes for later release. Certain hydrated salts dissolve endothermically in their water of hydration when heated to decomposition and release heat when they solidify. For example, Glauber's salt, $Na_2SO_4 \cdot 10\ H_2O_{(s)}$, solidifies at 32 °C, releasing 78.0 kJ/mol of salt. What is the enthalpy change for the solidification of 1.00 kg of Glauber's salt used to supply energy to a home?

6. In a laboratory investigation into the neutralization reaction
   $$HNO_3(aq) + KOH(s) \rightarrow KNO_3(aq) + H_2O(l)$$
   a researcher adds solid potassium hydroxide to nitric acid solution in a polystyrene calorimeter. Complete the Analysis section of the lab report, including a calculation of the molar enthalpy of neutralization of potassium hydroxide.

   Evidence
   mass KOH = 5.2 g
   volume of nitric acid solution = 200 mL
   $t_1 = 21.0$ °C $t_2 = 28.1$ °C

7. In a laboratory investigation into the reaction
   $$Ba(NO_3)_2(s) + K_2SO_4(aq) \rightarrow BaSO_4(s) + 2\ KNO_3(aq)$$
   a researcher adds a 261 g sample of barium nitrate to 2.0 L of potassium sulfate solution in a polystyrene calorimeter. Complete the Analysis section of the lab report, including a calculation of the molar enthalpy of reaction of barium nitrate.

   Evidence
   As the barium nitrate dissolves, a precipitate is immediately formed.
   $t_1 = 26.0$ °C $t_2 = 29.1$ °C

8. A student noticed that chewing fast-energy dextrose tablets made her mouth feel cold. Write the Purpose, Problem, Hypothesis, and Design for an investigation to find out whether there really is a temperature change.

9. Plan an investigation to compare the molar enthalpies of combustion for butane and propane. Include a Problem statement, a Design, and a list of materials. In your Design, identify major variables and how you will control them, as well as any necessary precautions.

10. Describe the features of a calorimeter that make it appropriate for its purpose. What drawbacks does it have?

11. The molar enthalpy of neutralization, $\Delta_n H_m$, for a reactant in an aqueous acid–base reaction can be determined by calorimetry. Write the Problem, Procedure, and Analysis for the lab report.

    Purpose
    The purpose of this exercise it to use calorimetry to determine the molar enthalpy of neutralization for nitrous acid or for aqueous potassium hydroxide.

    Design
    1.00 L of 0.400 mol/L nitrous acid and 1.00 L of 0.400 mol/L aqueous potassium hydroxide are combined in a calorimeter.

    Evidence
    inital temperature of solutions = 22.08 °C
    final temperature of mixture = 24.04 °C

12. Research the technological applications of propane and butane. Identify the chemical properties of each that make these appropriate technologies for each of these hydrocarbons.

www.science.nelson.com

Communicating Enthalpy Changes 11.3

Scientists and engineers have a variety of methods for communicating the energy change for chemical reactions. It is not uncommon to find different communication systems in different contexts. Each of these systems has pros and cons, but understanding all of them provides a deeper understanding of the topic of reaction enthalpies.

Most information about energy changes comes from the experimental method of calorimetry (page 485). These studies yield molar enthalpies that can be communicated in at least four ways:

1. by stating the molar enthalpy of a specific reactant in a reaction
2. by stating the enthalpy change for a balanced reaction equation
3. by including an energy value as a term in a balanced reaction equation
4. by drawing a chemical potential energy diagram

All four of these methods of expressing energy changes are equivalent. The first three are closer to empirical descriptions, and the fourth method is a theoretical description. Each of these methods of communicating energy changes in chemical reactions is described in the following sub-sections.

Method 1: Molar Enthalpies of Reaction, $\Delta_r H_m$

The molar enthalpy of reaction for a substance is determined empirically from the measurement of the energy released or absorbed per unit chemical amount to or from the surroundings of the chemical system at constant pressure. The energy measurements usually involve calorimetry. To communicate a molar enthalpy, both the substance and the reaction must be specified. The substance is conveniently specified by its chemical formula. Some chemical reactions are well known and specific enough to be identified by name only. For instance, reference books often list standard molar enthalpies of formation ($\Delta_f H_m^\circ$) and combustion ($\Delta_c H_m^\circ$).

Standard enthalpy values are expressed with a ° superscript, as in $\Delta_f H_m^\circ$. The reactants and products are in their standard state: a pressure of 100 kPa, an aqueous concentration of 1 mol/L, and liquids and solids in their pure state. For liquid and solid elements the pure state must be at 25 °C. Liquid and solid compounds must only have the same initial and final temperature (most often 25 °C). For our purposes we often refer to SATP as the standard conditions for both elements and compounds, unless special conditions exist for the liquid and solid compounds.

For some well-known reactions, such as formation and combustion, no chemical equation is necessary, since formation and combustion refer to specific, unambiguous chemical reactions indicated by the subscript letter after the delta symbol. For example, the molar enthalpy of formation for methanol at standard conditions is communicated internationally as

$$\Delta_f H_m^\circ{}_{CH_3OH} = -239.2 \text{ kJ/mol}$$

This means that 239.2 kJ of energy is released to the surroundings when 1 mol of methanol is formed from its elements when they are in their standard states at SATP. The following chemical equation communicates the formation reaction assumed to occur:

$$C(s) + 2 H_2(g) + \tfrac{1}{2} O_2(g) \rightarrow CH_3OH(l)$$

DID YOU KNOW ?
Communication
Communication is a very important part of any discipline. Here are some words of wisdom from famous scientists to their students.

No amount of experiments can ever prove me right; a single experiment may at any time prove me wrong.
  Imagination is more important than knowledge.
  It is the theory which decides what we can observe.
— Albert Einstein

Truth in science can be defined as the working hypothesis best suited to open the way to the next better one.
— Konrad Lorentz

Discovery is seeing what everybody else has seen, and thinking what nobody else has thought.
— Albert Szent-Györgyi

The foolish reject what they see, not what they think; the wise reject what they think, not what they see.
— Huang Po

Science is a way to teach how something gets to be known, what is not known, to what extent things are known (for nothing is known absolutely), how to handle doubt and uncertainty, what the rules of evidence are, how to think about things so that judgments can be made, and how to distinguish truth from fraud, and from show.
— Richard Feynman

Where observation is concerned, chance favours only the prepared mind.
— Louis Pasteur

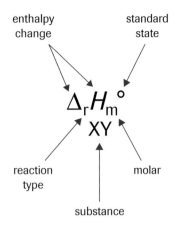

Figure 1
An explanation of the parts of an enthalpy symbol

Figure 2
Methanol burns more completely than gasoline and produces lower levels of some pollutants. The technology of methanol-burning vehicles was originally developed for racing cars because methanol burns faster than gasoline. However, its energy content is lower so that it takes twice as much methanol as gasoline to drive a given distance.

DID YOU KNOW ?

Renewable Energy Sources
Partly in reaction to the health hazards of some components of fossil fuels, scientists and engineers around the world are searching for alternative energy sources. Energy sources with low health, environmental, and resource impact include hydroelectric, wind, solar, biomass, and geothermal energy, along with small-scale hydro and tidal. Research by Statistics Canada indicates that, at the beginning of the 21st century, wind is the fastest-growing form of renewable energy in Canada.

As you know, we can find molar enthalpies from measurements involving systems at various conditions of temperature and pressure. If, however, chemists determine a molar enthalpy when the initial and final conditions of the chemical system are standard conditions, they call it a **standard molar enthalpy of reaction**. The symbol $\Delta_r H_m^\circ$ distinguishes standard molar enthalpies from molar enthalpies, $\Delta_r H_m$ (**Figure 1**). Standard molar enthalpies allow chemists to create tables to compare enthalpy values and to increase the precision of frequently used values by careful experimentation. For an exothermic reaction, the standard molar enthalpy is calculated by taking into account all the energy required to change the reaction system from standard conditions in order to initiate the reaction *and* all the energy released following the reaction, as the products are cooled to standard conditions. For example, the standard molar enthalpy of combustion of methanol is

$$\Delta_c H_m^\circ = -725.9 \text{ kJ/mol}$$
$$\text{CH}_3\text{OH}$$

This means that the complete combustion of 1 mol of methanol (**Figure 2**) releases 725.9 kJ of energy according to the following balanced equation:

$$\text{CH}_3\text{OH}(l) + \tfrac{3}{2}\text{O}_2(g) \rightarrow \text{CO}_2(g) + 2\,\text{H}_2\text{O}(l)$$

For a *standard* value, the initial and final conditions of the chemical system must be standard state. In this case, the carbon dioxide and liquid water are produced at a high temperature. They would be allowed to cool in a special calorimeter to standard conditions before experimenters take the final measurement in order to calculate the energy produced.

If a chemical reaction is not well known or if the equation for the reaction is not obvious, then the chemical equation must be stated along with the molar enthalpy. For example, methanol is produced industrially by the high-pressure reaction of carbon monoxide and hydrogen gases.

$$\text{CO}(g) + 2\,\text{H}_2(g) \rightarrow \text{CH}_3\text{OH}(l)$$

Chemists have determined that, in this rection, the standard molar enthalpy of reaction for methanol, $\Delta_r H_m^\circ$, is -128 kJ/mol. Note that this is not a formation reaction since the reactants are not all elements. Note also that you must always state which substance you are considering for this method of communication.

$$\Delta_r H_m^\circ = -128 \text{ kJ/mol}$$
$$\text{CH}_3\text{OH}$$

Method 2: Enthalpy Changes, $\Delta_r H$

A second method for communicating an energy change is to write an enthalpy change ($\Delta_r H$) beside the chemical equations. For example,

$$\text{CO}(g) + 2\,\text{H}_2(g) \rightarrow \text{CH}_3\text{OH}(l) \qquad \Delta_r H^\circ = -128 \text{ kJ}$$

Molar enthalpies of reaction can be used to calculate the enthalpy change during a chemical reaction; a molar enthalpy and a balanced chemical equation are required for the calculation. The enthalpy change is calculated using the empirical definition,

$$\Delta_r H^\circ = n\,\Delta_r H_m^\circ$$

where *n* is the chemical amount (in moles) of the substance whose molar enthalpy is known. *The enthalpy change for the equation as written equals the chemical amount (from the coefficient in the equation) times the molar enthalpy of reaction (for a specific chemical).* See Sample Problem 11.3.

▶ SAMPLE problem 11.3

Sulfur dioxide and oxygen react to form sulfur trioxide (**Figure 3**). The standard molar enthalpy of combustion of sulfur dioxide, in this reaction, is -98.9 kJ/mol. What is the enthalpy change for this reaction? First write the balanced chemical equation.

$$2\,SO_2(g) + O_2(g) \rightarrow 2\,SO_3(g)$$

Then obtain the chemical amount of sulfur dioxide from its coefficient in the balanced equation and use $\Delta_c H° = n\,\Delta_c H_m°$.

$$\Delta_c H° = n\,\Delta_c H_m°$$
$$= 2\text{ mol} \times \frac{-98.9\text{ kJ}}{1\text{ mol}} = -197.8\text{ kJ}$$

Report the enthalpy change for the reaction by writing it next to the balanced equation, as follows:

$$2\,SO_2(g) + O_2(g) \rightarrow 2\,SO_3(g) \qquad \Delta_c H° = -197.8\text{ kJ}$$

Note that the 2 in *2 mol* is an exact number; it does not affect the certainty of the answer.

Figure 3
Most sulfuric acid is produced in plants like this by the contact process, which includes two exothermic combustion reactions. Sulfur reacts with oxygen, forming sulfur dioxide; sulfur dioxide, in contact with a catalyst, reacts with oxygen, forming sulfur trioxide.

Note that the enthalpy change is not, in this method, a molar value, so does not require the "m" subscript and is not in kJ/mol. Also, it is not necessary to state which substance is under consideration because the quantity refers to the reaction equation *as written*.

The enthalpy change depends on the actual chemical amount of reactants and products in the chemical reaction. Therefore, if the balanced equation for the reaction is written differently, the enthalpy change should be reported differently. For example,

$$SO_2(g) + \tfrac{1}{2}O_2(g) \rightarrow SO_3(g) \qquad \Delta_c H° = -98.9\text{ kJ}$$
$$2\,SO_2(g) + O_2(g) \rightarrow 2\,SO_3(g) \qquad \Delta_c H° = -197.8\text{ kJ}$$

Both chemical equations agree with the empirically determined *molar* enthalpy of combustion for sulfur dioxide in this reaction.

$$\Delta_c H_m°_{SO_2} = \frac{-197.8\text{ kJ}}{2\text{ mol}} = -98.9\text{ kJ/mol}$$

Unlike molar enthalpies of formation or combustion, the enthalpy changes for most reactions must be accompanied by a balanced chemical equation so that we know what reaction is being described.

Learning Tip
If a chemical equation is altered by multiplying or dividing by some factor, then the $\Delta_r H$ is altered in exactly the same way.

Learning Tip
Fractions are convenient in many thermochemical equations but are not recommended by IUPAC. Note that use of fractions applies to fractions of moles of substances (e.g., $\tfrac{3}{2}$ mol represents 1.5 mol) rather than fractions of actual molecules.

Oxygen is often the reactant given a fractional coefficient in combustion equations. This is because it occurs as a diatomic molecule and the total number of oxygen atoms in the products is often an odd number.

▶ COMMUNICATION example 1

Wild natural gas wells are sometimes lit on fire to eliminate the very toxic hydrogen sulfide gas. The standard molar enthalpy of combustion of hydrogen sulfide is -518.0 kJ/mol. Express this value as a standard enthalpy change for the following reaction equation:

$$2\,H_2S(g) + 3\,O_2(g) \rightarrow 2\,H_2O(g) + 2\,SO_2(g) \qquad \Delta_c H° = ?$$

Solution

$$\Delta_c H = 2\text{ mol} \times -518.0\text{ kJ/mol}$$
$$= -1\,036.0\text{ kJ.}$$

$$2\,H_2S(g) + 3\,O_2(g) \rightarrow 2\,H_2O(g) + 2\,SO_2(g) \qquad \Delta_c H° = -1\,036.0\text{ kJ}$$

Figure 4
Combustion reactions are the most familiar exothermic reactions.

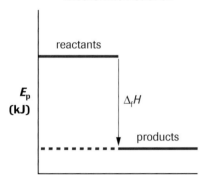

Figure 5
During an exothermic reaction, the enthalpy of the system decreases and heat flows into the surroundings. We observe a temperature increase in the surroundings.

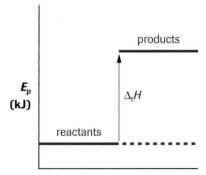

Figure 6
During an endothermic reaction, heat flows from the surroundings into the chemical system. We observe a temperature decrease in the surroundings.

Method 3: Energy Terms in Balanced Equations

Another way to report the enthalpy change in a chemical reaction is to include it as a term in a balanced equation. If a reaction is endothermic, it requires a certain quantity of additional energy for the reactants to continuously react. This energy (like the reactants) is transformed as the reaction progresses and is listed along with the reactants. For example,

$$H_2O(l) + 285.8 \text{ kJ} \rightarrow H_2(g) + \tfrac{1}{2} O_2(g)$$

If a reaction is exothermic, energy is released as the reaction proceeds (**Figure 4**) and is listed along with the products. For example,

$$Mg(s) + \tfrac{1}{2} O_2(g) \rightarrow MgO(s) + 601.6 \text{ kJ}$$

In order to specify the initial and final conditions for measuring the enthalpy change of the reaction, the temperature and pressure may be specified at the end of the equation.

$$Mg(s) + \tfrac{1}{2} O_2(g) \rightarrow MgO(s) + 601.6 \text{ kJ} \qquad \text{(under standard conditions)}$$

▶ COMMUNICATION example 2

Ethane is cracked into ethene in world-scale quantities in Alberta. Communicate the enthalpy of reaction as a term in the equation representing the cracking reaction.

$$C_2H_6(g) \rightarrow C_2H_4(g) + H_2(g) \qquad \Delta_{cr}H = +136.4 \text{ kJ}$$

Solution

$$C_2H_6(g) + 136.4 \text{ kJ} \rightarrow C_2H_4(g) + H_2(g)$$

Note that, in Communication Example 2, the enthalpy change has a positive value. This tells us that the reaction is endothermic, so the energy term is written with the reactants.

Method 4: Chemical Potential Energy Diagrams

Chemists use the law of conservation of energy to describe what happens during a chemical reaction. They explain their observations theoretically: observed energy changes are due to changes in chemical potential energy that occur during a reaction. This energy is a stored form of energy that is related to the relative positions of particles and the strengths of the bonds between them. As bonds break and re-form and the positions of atoms are altered, changes in potential energy occur. Evidence of a change in enthalpy of a chemical system is provided by a temperature change of its surroundings.

A **chemical potential energy diagram** shows the potential energy of both the reactants and the products of a chemical reaction (**Figures 5** and **6**). The difference between the initial and final energies in a chemical potential energy diagram is the enthalpy change, obtained from calorimetry by measuring the temperature change of the calorimeter. A temperature change is caused by a flow of heat into or out of the chemical system.

The vertical axis on the diagram represents the potential energy of the system, E_p. The reactants are written on the left and the products on the right, and the horizontal axis is called the reaction coordinate or reaction progress. In an exothermic change (**Figure 5**), the products have less potential energy than the reactants: energy is released to the surroundings as the products form. In an endothermic change (**Figure 6**), the products have more potential energy than the reactants: energy is absorbed from the surroundings. Neither of the axes is numbered; only the numerical change in potential energy (enthalpy change, $\Delta_r H$) of the system is shown in the diagrams.

▶ COMMUNICATION example 3

Communicate the following enthalpies of reaction as chemical potential energy diagrams.
(a) The burning of magnesium to produce a very bright emergency flare.

$$Mg(s) + \tfrac{1}{2}O_2(g) \rightarrow MgO(s) \qquad \Delta_c H = -601.6 \text{ kJ}$$

(b) The decomposition of water by electrical energy from a solar cell.

$$\Delta_d H_m = +285.8 \text{ kJ/mol}$$

Solution

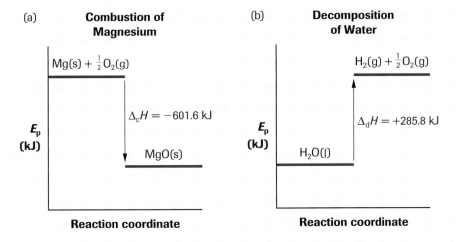

(a) **Exothermic Reaction**

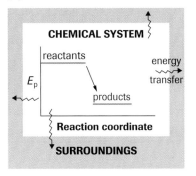

Surroundings are warmed as chemical system releases energy.

(b) **Endothermic Reaction**

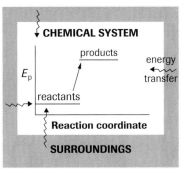

Surroundings are cooled as chemical system absorbs energy.

Figure 7
According to the law of conservation of energy, energy is neither created nor destroyed, only transformed. The diagrams above convey the notion that energy exits from an exothermic reaction system and enters into an endothermic reaction system.

Each of the four methods of communicating the molar enthalpy or change in enthalpy of a chemical reaction has advantages and disadvantages. To best understand energy changes in chemical reactions, you should learn all four methods. Communication Example 4, on the next page, illustrates these methods for an exothermic and an endothermic reaction (**Figure 7**).

SUMMARY   Four Ways of Communicating Energy Changes

	Exothermic Changes	Endothermic Changes
1. Molar Enthalpy	$\Delta_r H_m < 0$	$\Delta_r H_m > 0$
2. Enthalpy Change	reactants → products; $\Delta_r H < 0$	reactants → products; $\Delta_r H > 0$
3. Term in a Balanced Equation	reactants → products + energy	reactants + energy → products
4. Chemical Potential Energy Diagram	E_p (reactants) > E_p (products)	E_p (reactants) < E_p (products)

Learning Tip

Remember that the unit of $\Delta_r H$ is the kJ and the unit of $\Delta_r H_m$ is the kJ/mol, and that when multiplying by an exact number, you use the precision rule. For example, 2×40.65 kJ $= 81.30$ kJ, retaining the same precision as the measured value.

Figure 8
Energy can be captured by and from grain. The energy may be obtained from the grain in the form of, for example, foodstuffs, ethanol, and/or biodiesel. Grain grows well during the long summer days in Alberta, when the photosynthesis process exceeds the cellular respiration process. A grain field is an example of an open system, in which mass and energy flow freely into and out of the system.

COMMUNICATION example 4

Energy is transformed in cellular respiration and in photosynthesis (**Figure 8**). Cellular respiration, a series of exothermic reactions, is the breakdown of foodstuffs, such as glucose, that takes place within cells. Photosynthesis, a series of endothermic reactions, is the process by which green plants use light energy to make glucose from carbon dioxide and water.

Express the standard enthalpy changes for cellular respiration and for photosynthesis by using the four different methods of communication.

(a) One mole of glucose is consumed, during cellular respiration, to release 2802.5 kJ of energy.
(b) Glucose is produced during photosynthesis.

Solution

(a)
1. The standard molar enthalpy for cellular respiration of glucose:
$\Delta_r H_m^\circ = -2802.5$ kJ/mol
$C_6H_{12}O_6$

2. $C_6H_{12}O_6(g) + 6O_2(g) \rightarrow 6CO_2(g) + 6H_2O(l)$ $\Delta_r H^\circ = -2802.5$ kJ

3. $C_6H_{12}O_6(s) + 6O_2(g) \rightarrow 6CO_2(g) + 6H_2O(l) + 2802.5$ kJ

4. Potential energy diagram for cellular respiration:

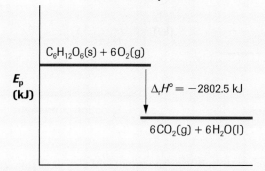

Cellular Respiration

(b)
1. The standard molar enthalpy for photosynthesis of glucose:
$\Delta_r H_m^\circ = +2802.5$ kJ/mol
$C_6H_{12}O_6$

2. $6CO_2(g) + 6H_2O(l) \rightarrow C_6H_{12}O_6(s) + 6O_2(g)$ $\Delta_r H^\circ = +2802.5$ kJ

3. $6CO_2(g) + 6H_2O(l) + 2802.5$ kJ $\rightarrow C_6H_{12}O_6(s) + 6O_2(g)$

4. Potential energy diagram for photosynthesis:

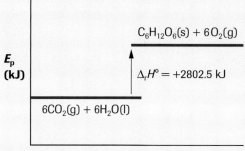

Photosynthesis

Section 11.3 Questions

1. Translate into words all parts of the following symbols:
   (a) $\Delta_c H_m°$
       CH_4
   (b) $n \Delta_f H_m°$
       C_3H_8
   (c) Q
       water

2. Communicate the standard enthalpy change by using the four methods described in this section for each of the following chemical reactions. Assume standard conditions for the measurements of initial and final states.
   (a) The formation of acetylene (ethyne, C_2H_2) fuel from solid carbon and gaseous hydrogen
       ($\Delta_f H_m° = +227.4$ kJ/mol acetylene)
   (b) The simple decomposition of aluminium oxide powder
       ($\Delta_{sd} H_m° = +1675.7$ kJ/mol aluminium oxide)
   (c) The complete combustion of pure carbon fuel
       ($\Delta_c H_m° = -393.5$ kJ/mol CO_2)

3. For each of the following balanced chemical equations and standard enthalpy changes, write the symbol and calculate the standard molar enthalpy of combustion for the substance that reacts with oxygen.
   (a) $2H_2(g) + O_2(g) \rightarrow 2H_2O(g)$ $\Delta_r H° = -483.6$ kJ
   (b) $4 NH_3(g) + 7 O_2(g) \rightarrow 4 NO_2(g) + 6 H_2O(g) + 1272.1$ kJ
   (c) $2 N_2(g) + O_2(g) + 163.2$ kJ $\rightarrow 2 N_2O(g)$
   (d) $3 Fe(s) + 2 O_2(g) \rightarrow Fe_3O_4(s)$ $\Delta_c H° = -1118.4$ kJ

4. The neutralization of a strong acid and a strong base is an exothermic process.

   $H_2SO_4(aq) + 2 NaOH(aq) \rightarrow Na_2SO_4(aq) + 2 H_2O(l) + 114$ kJ

   (a) What is the enthalpy change for this reaction?
   (b) Write this chemical equation, using the $\Delta_r H$ notation not under standard conditions.
   (c) Calculate the molar enthalpy of neutralization for sulfuric acid.
   (d) Calculate the molar enthalpy of neutralization for sodium hydroxide.

5. Translate the empirical molar enthalpies given below into a balanced chemical equation, including the standard enthalpy change; for example,

   $CH_4(g) + 2 O_2(g) \rightarrow CO_2(g) + 2 H_2O(g)$ $\Delta_c H° = -802.5$ kJ

   (a) The standard molar enthalpy of combustion for methanol to produce water vapour is -725.9 kJ/mol.
   (b) The standard molar enthalpy of formation for liquid carbon disulfide is 89.0 kJ/mol.
   (c) The standard molar enthalpy of roasting (combustion) for zinc sulfide is -441.3 kJ/mol.
   (d) The standard molar enthalpy of simple decomposition, $\Delta_{sd}H_m°$, for iron(III) oxide is 824.2 kJ/mol.

6. For each of the following reactions, translate the given standard molar enthalpy into a balanced chemical equation using the $\Delta_r H°$ notation, and then rewrite the equation including the energy as a term in the equation. Assume water is produced as a vapour.
   (a) Propane obtained from natural gas is used as a fuel in barbecues and vehicles (**Figure 9**).
       $\Delta_c H_m° = -2\,043.9$ kJ/mol C_3H_8
   (b) Nitrogen monoxide forms at the high temperatures inside an automobile engine. $\Delta_f H_m° = +91.3$ kJ/mol NO
   (c) Some advocates of alternative fuels have suggested that cars could run on ethanol.
       $\Delta_c H_m° = -1\,234.8$ kJ/mol C_2H_5OH

Figure 9
Propane-fuelled vehicles are not allowed to park in underground parking lots. Propane is denser than air, and a dangerous quantity of propane could accumulate in the event of a leak.

7. The standard molar enthalpy of combustion for hydrogen to produce liquid water is -285.8 kJ/mol. The standard molar enthalpy of decomposition for liquid water is $+285.8$ kJ/mol.
   (a) Write both chemical equations using the $\Delta_r H°$ notation.
   (b) How does the standard enthalpy change for the combustion of hydrogen compare with the standard enthalpy change for the simple decomposition of liquid water? Suggest a generalization to include all pairs of chemical equations that are the reverse of one another.

8. Combustion reactions are very useful to society, but also have some drawbacks. Create a list of possible drawbacks of our reliance on combustion reactions, and suggest some possible solutions.

11.4 Hess' Law

DID YOU KNOW?

Germain H. Hess and Enthalpy
An amazing part of the history of Hess' law is that it was created before the law of conservation of energy was well formulated. Hess' law also preceded the use of the term *enthalpy*, which did not appear until nearly fifty years later. Hess is known as the founder of thermochemistry for his extensive empirical work with heat produced by neutralization and combustion reactions.

Calorimetry is the basis for most information about chemical energy changes. However, not every reaction of interest to scientists and engineers can be studied by means of a calorimetric experiment. For example, the rusting of iron is extremely slow and, therefore, results in temperature changes too small to be measured using a conventional calorimeter. It is similarly impossible to measure, with a calorimeter, the temperature change involved in the energy of formation of carbon monoxide because the combustion of carbon produces carbon dioxide and carbon monoxide simultaneously. Chemists have devised a number of methods to predict an enthalpy change for reactions that are inconvenient to study experimentally. All the methods are based on the law of conservation of energy and the experimentally established principle that *net changes in all properties of a system are independent of the way the system changes from the initial state to the final state*. A temperature change is an example of a property that satisfies this principle. A net (overall) temperature change $(t_f - t_i)$ does not depend on whether the temperature changed slowly, quickly, or rose and fell several times between the initial temperature and the final temperature. This same principle applies to enthalpy changes. If several reactions occur in different ways but the initial reactants and final products are the same, the net enthalpy change is the same as long as the reactions have the same initial and final conditions.

Predicting $\Delta_r H$: Hess' Law

Based on experimental measurements and calculations of enthalpy changes, Swiss chemist G.H. Hess suggested in 1840 that the addition of chemical equations yields a net chemical equation whose enthalpy change is the sum of the individual enthalpy changes. This generalization has been tested in many experiments and is now accepted as the law of additivity of enthalpies of reaction, also known as **Hess' law**. Hess' law can be written as an equation using the uppercase Greek letter Σ (pronounced "sigma") to mean "the sum of."

$$\Delta_r H = \Delta_1 H + \Delta_2 H + \Delta_3 H + \ldots$$

or

$$\Delta_r H = \Sigma \Delta_r H$$

or, in standard conditions,

$$\Delta_r H° = \Sigma \Delta_r H°$$

Figure 1
In this illustration, bricks on a construction site are being moved from the ground up to the second floor, but there are two different paths that the bricks could follow. In one path, they could move from the ground up to the third floor, and then be carried down to the second floor. In the other path, they would be carried up to the second floor in a single step. In both cases, the overall change in position—one floor up—is the same.

An analogy for this concept is shown in **Figure 1**. The net vertical distance that the bricks rise is the same whether they go up in one stage or in two stages. The same principle applies to enthalpy changes: If a set of reactions occurs in different steps but the initial reactants and final products are the same, the overall enthalpy change is the same.

Hess' discovery allowed the determination of the enthalpy change of a reaction without direct calorimetry (see Sample Problem 11.4), using two rules for chemical equations and enthalpy changes that you already know:

- If a chemical equation is reversed, then the sign of $\Delta_r H$ changes.

- If the coefficients of a chemical equation are altered by multiplying or dividing by a constant factor, then the $\Delta_r H$ is altered by the same factor.

SAMPLE problem 11.4

Use Hess' law to determine the standard enthalpy change for the formation of carbon monoxide.

$$C(s) + \tfrac{1}{2} O_2(g) \rightarrow CO(g) \qquad \Delta_f H° = ?$$

This reaction cannot be studied calorimetrically since the combustion of carbon produces carbon dioxide as well as carbon monoxide. However, the standard enthalpy of complete combustion for carbon and for carbon monoxide can be calculated by calorimetric measurements, and the standard enthalpy of formation for carbon monoxide can be determined using Hess law as follows:

(1) $C(s) + O_2(g) \rightarrow CO_2(g) \qquad \Delta_c H° = -393.5$ kJ
(2) $2 CO(g) + O_2(g) \rightarrow 2 CO_2(g) \qquad \Delta_c H° = -566.0$ kJ

Rearrange these two equations, and then add them together to obtain the chemical equation for the formation of carbon monoxide. The first term in the formation equation for carbon monoxide is 1 mol of solid carbon. Therefore, leave equation (1) unaltered so that C(s) will appear on the reactant side when we add the equations. However, we want 1 mol of CO(g) to appear as a product, so reverse equation (2) and divide each of its terms (including the standard enthalpy change) by 2.

$$C(s) + O_2(g) \rightarrow CO_2(g) \qquad \Delta_1 H° = -393.5 \text{ kJ}$$
$$CO_2(g) \rightarrow CO(g) + \tfrac{1}{2} O_2(g) \qquad \Delta_2 H° = +283.0 \text{ kJ}$$

Note that the sign of the standard enthalpy change in equation (2) has changed, since the equation has been reversed. Now add the reactants, products, and standard enthalpy changes to get a net reaction equation. Note that $CO_2(g)$ can be cancelled because it appears on both sides of the net equation. Similarly, $\tfrac{1}{2} O_2(g)$ can be cancelled from each side of the equation, resulting in:

$$C(s) + O_2(g) \rightarrow CO_2(g) \qquad \Delta_1 H° = -393.5 \text{ kJ}$$
$$CO_2(g) \rightarrow CO(g) + \tfrac{1}{2} O_2(g) \qquad \Delta_2 H° = +283.0 \text{ kJ}$$
$$\overline{C(s) + \tfrac{1}{2} O_2(g) \rightarrow CO(g) \qquad \Delta_f H° = -110.5 \text{ kJ}}$$

While manipulating equations (1) and (2), you should check the desired equation and plan ahead to ensure that the substances end up on the correct sides and in the correct chemical amounts.

Sketching a chemical potential energy diagram (**Figure 2**) might help you ensure that you have made the appropriate additions and subtractions.

Learning Tip

The process for using Hess' law is a combination of being systematic and using trial and error. The given equations must be manipulated (like a system of equations in mathematics) to get what you want.

In Sample Problem 11.4 (as the accompanying audio clip explains):
- leave equation (1) as is
- divide equation (2) by 2 and reverse it.

Do what needs to be done to the given equations so they add to get the equation you want.

DID YOU KNOW ?

Carbon Monoxide Poisoning
Carbon monoxide is known as the silent killer. Incomplete combustion can occur with deadly effects in the following circumstances:
- using charcoal briquettes in a home fireplace, recreational vehicle, or tent
- operating a car with a leaky exhaust system
- operating a fireplace in a home with inadequate ventilation (causing furnace exhaust to reverse into the home).

Carbon monoxide is odourless. Know how to prevent exposure to it, and install a carbon monoxide detector in your home or recreation vehicle.

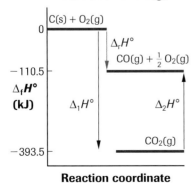

Figure 2
Carbon and oxygen react, forming carbon dioxide ($\Delta_1 H° = -393.5$ kJ), which theoretically could react to form carbon monoxide and oxygen ($\Delta_2 H° = +283.0$ kJ). The net standard enthalpy change of the two-step reaction is -393.5 kJ $+ 283.0$ kH $= -110.5$ kJ.
Note that the label on the y-axis must be "standard enthalpy of formation, $\Delta_f H°$."

Figure 3
The molar enthalpy of combustion for butane can be determined calorimetrically. The molar enthalpy of formation, however, is best determined with Hess' law.

Learning Tip

To obtain the formation equation and its enthalpy change in this Communication Example:
- reverse equation (1) and change the sign of the ΔH
- multiply equation (2) and its ΔH by 4
- multiply equation (3) and its ΔH by $\frac{5}{2}$

> **COMMUNICATION** *example*

In an experiment to find the standard molar enthalpy of formation of butane (from its elements), the following values were determined by calorimetry:

(1) $C_4H_{10}(g) + \frac{13}{2} O_2(g) \rightarrow 4\, CO_2(g) + 5\, H_2O(g)$ $\quad \Delta_c H° = -2657.4$ kJ

(2) $C(s) + O_2(g) \rightarrow CO_2(g)$ $\quad \Delta_f H° = -393.5$ kJ

(3) $2\, H_2(g) + O_2(g) \rightarrow 2\, H_2O(g)$ $\quad \Delta_f H° = -483.6$ kJ

What is the standard molar enthalpy of formation of butane (**Figure 3**)?

Solution

Net $\quad 4\, C(s) + 5\, H_2(g) \rightarrow C_4H_{10}(g) \quad\quad \Delta_f H° = ?$

(1) $\cancel{4\, CO_2(g)} + \cancel{5\, H_2O(g)} \rightarrow C_4H_{10}(g) + \cancel{\frac{13}{2} O_2(g)} \quad \Delta_1 H° = +2657.4$ kJ

(2) $\quad 4\, C(s) + \cancel{4\, O_2(g)} \rightarrow \cancel{4\, CO_2(g)} \quad\quad \Delta_2 H° = -1574.0$ kJ

(3) $\quad 5\, H_2(g) + \cancel{\frac{5}{2} O_2(g)} \rightarrow \cancel{5\, H_2O(g)} \quad\quad \Delta_3 H° = -1209.0$ kJ

Net $\quad 4\, C(s) + 5\, H_2(g) \rightarrow C_4H_{10}(g) \quad\quad \Delta_f H° = -125.6$ kJ

$$\Delta_f H_m°_{C_4H_{10}} = \frac{-125.6 \text{ kJ}}{1 \text{ mol}}$$

$$= -125.6 \text{ kJ/mol}$$

According to Hess' law and the data provided, the standard molar enthalpy of formation of butane is -125.6 kJ/mol.

> **SUMMARY**   *Enthalpy of Reaction and Hess' Law*

To determine an enthalpy change of a reaction by using Hess' law, follow these steps:
1. Write the net reaction equation, if it is not given.
2. Manipulate the given equations so they will add to yield the net equation.
3. Multiply, divide, and/or reverse the sign of the enthalpy of reaction.
4. Cancel and add the remaining reactants and products to yield the net equation.
5. Add the component enthalpy changes to obtain the net enthalpy change.
6. Determine the molar enthalpy for a reactant or product, if required.

Figure 4
The reaction of powdered aluminium and iron(III) oxide is known as the "thermite" reaction and is very exothermic. Molten, white-hot iron is produced. In this apparatus, widely used in the past, the thermite reaction occurs in the upper chamber, and molten iron flows down into a mold between the ends of two rails, welding them together.

> **Practice**

1. The standard enthalpy changes for the formation of aluminium oxide and iron(III) oxide (**Figure 4**) are

   $2\, Al(s) + \frac{3}{2} O_2(g) \rightarrow Al_2O_3(s) \quad\quad \Delta_f H° = -1675.7$ kJ

   $2\, Fe(s) + \frac{3}{2} O_2(g) \rightarrow Fe_2O_3(s) \quad\quad \Delta_f H° = -824.2$ kJ

   Use Hess' law to predict the standard enthalpy change for the following reaction.

   $Fe_2O_3(s) + 2\, Al(s) \rightarrow Al_2O_3(s) + 2\, Fe(s) \quad\quad \Delta_r H° = ?$

Section 11.4

Use this information about coal gasification to answer questions 2 to 4.

Question 2:
Coal gasification converts coal into a combustible mixture of carbon monoxide and hydrogen, called *coal gas*, in a gasifier (**Figure 5**).

(1) $H_2O(g) + C(s) \rightarrow CO(g) + H_2(g)$ $\Delta_r H° = ?$

Question 3:
The coal gas can be used as a fuel, for example, in a combustion turbine.

(2) $CO(g) + H_2(g) + O_2(g) \rightarrow CO_2(g) + H_2O(g)$ $\Delta_c H° = ?$

Question 4:
As an alternative to combustion, coal gas can undergo a process called *methanation*.

(3) $3 H_2(g) + CO(g) \rightarrow CH_4(g) + H_2O(g)$ $\Delta_r H° = ?$

2. Calculate the standard enthalpy change for reaction (1)—the gasification of coal—from the following chemical equations and standard enthalpy changes.

   $2 C(s) + O_2(g) \rightarrow 2 CO(g)$ $\Delta_f H° = -221.0$ kJ
   $2 H_2(g) + O_2(g) \rightarrow 2 H_2O(g)$ $\Delta_f H° = -483.6$ kJ

3. Calculate the standard enthalpy change for reaction (2)—the combustion of coal gas—from the following information.

   $2 C(s) + O_2(g) \rightarrow 2 CO(g)$ $\Delta_f H° = -221.0$ kJ
   $C(s) + O_2(g) \rightarrow CO_2(g)$ $\Delta_f H° = -393.5$ kJ
   $2 H_2(g) + O_2(g) \rightarrow 2 H_2O(g)$ $\Delta_f H° = -483.6$ kJ

4. Calculate the standard enthalpy change for reaction (3)—the methanation of coal gas—using the following information.

   $2 H_2(g) + O_2(g) \rightarrow 2 H_2O(g)$ $\Delta_f H° = -483.6$ kJ
   $2 C(s) + O_2(g) \rightarrow 2 CO(g)$ $\Delta_f H° = -221.0$ kJ
   $CH_4(g) + 2 O_2(g) \rightarrow CO_2(g) + 2 H_2O(g)$ $\Delta_c H° = -802.7$ kJ
   $C(s) + O_2(g) \rightarrow CO_2(g)$ $\Delta_f H° = -393.5$ kJ

DID YOU KNOW

Coal Gasification
The gasification of coal has three potential applications. Besides the electric power generation described below, there is the possibility of using steam to gasify deep coal to bring the energy to the surface. Another possibility is to use the excess coke produced by oil sands refining to produce electric power or methane.

DID YOU KNOW

Cogeneration
Cogeneration, also called CHP (combined heat and power), is a modern technology for the simultaneous production of electricity and heat using a single fuel. The heat is sometimes used to generate more electricity. Cogeneration provides greater efficiencies than traditional generating methods. The process illustrated in Figure 5 is one example of cogeneration.

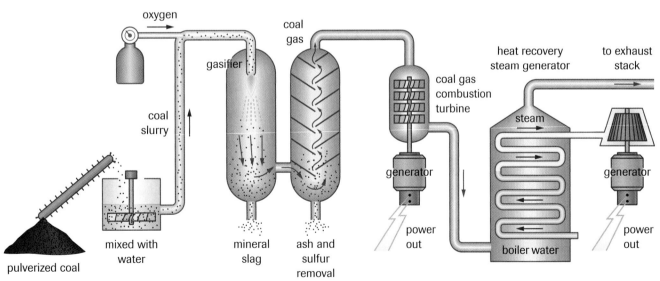

Figure 5
Electric power-generating stations that use coal as a fuel are 30% to 40% efficient. Coal gasification and combustion of the coal gas provide one alternative to burning coal. Efficiency is improved by about 10% by using both a combustion turbine and a steam turbine to produce electricity.

WEB Activity

Simulation—Hess' Law

Titanium(IV) chloride is an unusual compound: it is a liquid metal halide. Chemists explain this anomaly by theorizing that it is a molecular substance, rather than ionic. It is also very corrosive, forming hydrochloric acid when it comes into contact with water, so is only used with very strict safety precautions. We can, however, discover some of its properties by using simulations.

www.science.nelson.com

LAB EXERCISE 11.B

Testing Hess' Law

Report Checklist
- ○ Purpose
- ○ Problem
- ○ Hypothesis
- ● Prediction
- ○ Design
- ○ Materials
- ○ Procedure
- ○ Evidence
- ● Analysis
- ● Evaluation (2, 3)

The following data are from a test of Hess' law using a calorimeter. Use these data in your prediction, assuming that combustion produces carbon dioxide gas and liquid water.

$5\,C(s) + 6\,H_2(g) \rightarrow C_5H_{12}(l)$ $\Delta_f H° = -173.5$ kJ

$C(s) + O_2(g) \rightarrow CO_2(g)$ $\Delta_f H° = -393.5$ kJ

$H_2(g) + \tfrac{1}{2} O_2(g) \rightarrow H_2O(g)$ $\Delta_f H° = -241.8$ kJ

$H_2O(l) \rightarrow H_2O(g)$ $\Delta_{vap} H° = +40.65$ kJ

Then complete the Prediction, Analysis, and Evaluation of the investigation report.

Purpose

The purpose of this investigation is to test Hess' law.

Problem

What is the standard molar enthalpy of combustion of pentane?

Design

Hess' law is used to predict the standard molar enthalpy of combustion of pentane. To test the prediction and the acceptability of the law, the standard molar enthalpy of combustion of pentane is determined calorimetrically.

Evidence

mass of pentane reacted = 2.15 g
mass of water equivalent to calorimeter = 1.24 kg
initial temperature of calorimeter and contents = 18.4 °C
final temperature of calorimeter and contents = 37.6 °C

(Note that the calorimeter has the same thermal energy gain or loss as the water equivalent stated.)

LAB EXERCISE 11.C

Analysis Using Hess' Law

Report Checklist
- ○ Purpose
- ○ Problem
- ○ Hypothesis
- ○ Prediction
- ○ Design
- ○ Materials
- ○ Procedure
- ○ Evidence
- ● Analysis
- ○ Evaluation

Most natural gas is burned as fuel to provide heat. However, natural gas is also a source of hydrogen gas for producing ammonia-based fertilizers. Complete one possible Analysis for the investigation report. Not all the evidence need be used for a particular analysis.

Purpose

The purpose of this exercise is to use Hess' law to determine an enthalpy change.

Problem

What is the standard enthalpy change for the production of hydrogen from methane and steam?

$CH_4(g) + H_2O(g) \rightarrow CO(g) + 3\,H_2(g)$ $\Delta_r H° = ?$

Evidence

(1) $2\,C(s) + O_2(g) \rightarrow 2\,CO(g)$ $\Delta_c H° = -221.0$ kJ

(2) $CH_4(g) + 2\,O_2(g) \rightarrow CO_2(g) + 2\,H_2O(g)$ $\Delta_c H° = -802.5$ kJ

(3) $CO(g) + H_2O(g) \rightarrow CO_2(g) + H_2(g)$ $\Delta_r H° = -41.2$ kJ

(4) $2\,H_2(g) + O_2(g) \rightarrow 2\,H_2O(g)$ $\Delta_c H° = -483.6$ kJ

(5) $C(s) + 2\,H_2(g) \rightarrow CH_4(g)$ $\Delta_f H° = -74.6$ kJ

(6) $C(s) + H_2O(g) \rightarrow CO(g) + H_2(g)$ $\Delta_r H° = +131.3$ kJ

(7) $2\,CO(g) + O_2(g) \rightarrow 2\,CO_2(g)$ $\Delta_c H° = -566.0$ kJ

(8) $CO(g) + H_2(g) + O_2(g) \rightarrow CO_2(g) + H_2O(g)$ $\Delta_r H° = -524.8$ kJ

Section **11.4**

INVESTIGATION 11.3 *Introduction*

Applying Hess' Law

Magnesium burns rapidly, releasing heat and light. The enthalpy change of this reaction can be calculated from measurements involving a highly sophisticated calorimeter, but not a polystyrene cup calorimeter. The enthalpy change for the combustion of magnesium can, however, be determined by applying Hess' law to a series of chemical reaction equations (see page 517).

Report Checklist

- ○ Purpose
- ○ Problem
- ○ Hypothesis
- ○ Prediction
- ○ Design
- ○ Materials
- ● Procedure
- ● Evidence
- ● Analysis
- ● Evaluation (1, 2, 3)

Purpose

The purpose of this investigation is to use Hess' law to determine a molar enthalpy of combustion.

Problem

What is the standard molar enthalpy of combustion for magnesium?

Prediction

The standard molar enthalpy of combustion for magnesium is −601.6 kJ/mol, as per the table of standard molar enthalpies of formation (Appendix I). (The standard molar enthalpy of combustion of magnesium is the same as the standard molar enthalpy of formation of magnesium oxide because both processes have the same chemical reaction equation.)

Design

The enthalpy changes for the first two reactions in the series (on page 517) are determined empirically using a polystyrene calorimeter, and the third enthalpy change is obtained from Appendix I. The three $\Delta_r H$ values are used, along with Hess' law, to obtain the molar enthalpy of combustion for magnesium.

To perform this investigation, turn to page 517.

EXPLORE an issue

Alternative Energy Sources and Technologies

Issue Checklist

- ○ Issue
- ○ Resolution
- ○ Design
- ● Evidence
- ● Analysis
- ● Evaluation

Photosynthesis is the key to life on Earth. For millions of years green plants have used sunlight, water, and carbon dioxide to produce biomass. When prehistoric plants and the tiny animals that ate them died, the organic materials they contained settled on the bottom of swamps, lakes, and seas, where rocky sediments eventually buried them. Fossil fuels (coal, oil, and natural gas) are the remains of these ancient organisms, transformed by a combination of heat and pressure. While living organisms are composed mainly of carbon, hydrogen, and oxygen, the process of being converted into fossil fuels removed the oxygen from the organic matter so the remaining compounds are referred to as *hydrocarbons*. Although Alberta's hydrocarbon reserves are very large, they are also finite because of the time and conditions necessary to produce fossil fuels (**Figure 6**).

For several decades researchers have been looking for alternatives to fossil fuels. One promising alternative involves using "renewable" plant material from managed forests or farms, which takes a few months or years to grow, rather than using fossil fuels, which take about 300 million years to form. Another option takes waste agricultural or industrial material.

A bio-based economy uses renewable energy and renewable materials instead of fossil fuels to provide the energy, chemicals, and material for society. Proponents of a bio-based economy see a fuel source in the waste carbon compounds produced by agriculture and forestry (**Figure 7**). These waste plant and animal materials could supply the biomass to produce ethanol or biodiesel, or even hydrogen gas

Figure 6
A Suncor Energy Inc. shovel fills the bed of a heavy hauler truck in the oil sands region near Fort McMurray.

Figure 7
The new Grande Prairie EcoPower™ Centre uses wood waste to co-generate both electricity and steam for use in two nearby sawmills. The Government of Alberta and the City of Grande Prairie have committed to a long-term purchase of electricity from the EcoPower™ Centre.

for use in fuel cells. This is not a new concept. Wood has been and is being used for heating and cooking. When wood biomass was unavailable, people made use of other biomass sources. For example, prairie First Nations people used buffalo dung for heating, and the Inuit have made efficient use of seal oil for heating and light.

Biomass could also produce the feedstock chemicals used to make the wide array of synthetic materials, including plastics, that are currently synthesized from petrochemicals. Proponents insist that such an economy has the potential to be both renewable and sustainable.

Many of the processes needed to utilize the biomass from agriculture and forestry are already in place, but would need to be adapted and expanded. For example, ethanol is now made largely from starch, which could be used as a food source for people or animals. It may be more efficient to convert the waste parts of crop plants, or the waste from sawmills, into ethanol. Similarly, the technology exists to turn waste oil from the fast-food industry into diesel fuel. The source of biomass could be increased by planting crops specifically for that purpose, using farmland that is not currently being utilized. Some researchers estimate that we could obtain up to 25% of our current energy demands by converting waste biomass into useable fuels.

Many people in the energy industry are skeptical about the practicality of a bio-based economy. They argue that the technology for finding and using fossil fuels is already well established, while the technologies of a bio-based economy will take considerable time and money to develop. In their view, research would be better directed at finding cleaner and more efficient ways of extracting and using traditional fossil fuels. For example, three projects aimed at developing cleaner fossil fuels are
- enhanced coal bed methane
- heavy oil recovery using vapour extraction
- paste technology for oil sands tailings treatment

Energy issues have pros and cons from many perspectives (see Appendix D). Your minimum task is to write an energy Issue with a technology emphasis and complete the Resolution and Design. (An example is provided.) Your teacher may also give you instructions to complete the Evidence, Analysis, and Evaluation components of this issue resolution process.

Issue
What is the most effective use of government funding for research in energy technology?

Resolution
The provincial and federal governments should direct all their research funding to bio-based energy technology.

Design
Within small groups, research the pros and cons of using public money to fund bio-based versus fossil-fuel-based energy technology.

www.science.nelson.com GO

▶ Section 11.4 Questions

Use the following information to answer questions 1 to 5.

Energy may be produced or consumed by the production and synthesis of fuels such as alcohols. Use Hess' law to determine the quantity of energy required or released during the production of various alcohol fuels.

1. Ethanol (grain alcohol) is a renewable energy gasoline additive that can be made from glucose obtained from the fermentation of grain. Determine the standard enthalpy of reaction for

   $C_6H_{12}O_6(s) \rightarrow 2 C_2H_5OH(l) + 2 CO_2(g) \quad \Delta_rH = ?$

   from the following reaction equations.

   $C_6H_{12}O_6(s) + 6 O_2(g) \rightarrow 6 CO_2(g) + 6 H_2O(l) + 2\,803.1$ kJ
   $C_2H_5OH(l) + 3 O_2(g) \rightarrow 2 CO_2(g) + 3 H_2O(l) + 1\,366.8$ kJ

2. Ethanol can also be made from the nonrenewable resource ethene. First, ethane from natural gas is cracked to produce ethene, as is done in several ethene plants in Alberta.

   $C_2H_6(g) \rightarrow C_2H_4(g) + H_2(g) \quad \Delta_rH = ?$

   The second step is the production (synthesis) of ethanol from ethene by an addition reaction with water.

   $C_2H_4(g) + H_2O(l) \rightarrow C_2H_5OH(l) \quad \Delta_rH = ?$

   (a) Using the following enthalpies of reaction, predict the enthalpy of reaction for the ethane cracking reaction equation given.

   $2 C_2H_6(g) + 7 O_2(g) \rightarrow$
   $\qquad 4 CO_2(g) + 6 H_2O(l) + 3\,120.8$ kJ
   $C_2H_4(g) + 3 O_2(g) \rightarrow 2 CO_2(g) + 2 H_2O(l)$
   $\qquad \Delta_cH = -1\,411.0$ kJ
   $H_2(g) + \tfrac{1}{2} O_2(g) \rightarrow H_2O(l) \qquad \Delta_cH = -285.8$ kJ

   (b) Using the following enthalpies of reaction, predict the enthalpy of reaction for the addition reaction of ethene to water to produce ethanol.

   $C_2H_4(g) + 3 O_2(g) \rightarrow 2 CO_2(g) + 2 H_2O(l)$
   $\qquad \Delta_cH° = -1\,411.0$ kJ
   $C_2H_5OH(l) + 3 O_2(g) \rightarrow$
   $\qquad 2 CO_2(g) + 3 H_2O(l) + 1\,366.8$ kJ

   (c) If the reaction equations that you just used (to determine the enthalpies of reaction for the cracking of ethane and for the addition of water to ethene) are not the actual steps used in these chemical processes, then why does this Hess' law technique work?

3. Methanol (wood alcohol) is a fuel that can be made from biomass (such as cellulose obtained from aspen trees). Methanol can also be made from natural gas. The overall synthesis reaction for the latter case is

$CH_4(g) + H_2O(l) \rightarrow CH_3OH(l) + H_2(g)$ $\Delta_r H° = ?$

Use the reactions below and Hess' law to predict the standard molar enthalpy of synthesis of methanol by this process.

$CO(g) + 3 H_2(g) \rightarrow CH_4(g) + H_2O(l)$
$\Delta_r H° = -249.9$ kJ

$CO(g) + 2 H_2(g) \rightarrow CH_3OH(l)$ $\Delta_r H° = -128.7$ kJ

4. Consider each of the net reactions in questions 1, 2, and 3 for the synthesis of ethanol and methanol fuels: ethanol from glucose, ethanol from ethane, and methanol from methane. Create a summary, such as a table, in which to present your answers to the following questions.
   (a) Is each net synthesis reaction endothermic or exothermic?
   (b) Reference (look up) and record the molar enthalpies of combustion of each fuel.
   (c) Compare the molar enthalpy of synthesis of each fuel (ethanol and methanol) to the molar enthalpy of combustion of the fuel. Is the net chemical process, considering the synthesis and the combustion of the fuel, endothermic or exothermic?
   (d) Quantitatively, what is the net standard molar enthalpy of reaction for the combined chemical process of synthesis and combustion for each fuel in each of the synthesis processes?
   (e) What implications does this have for the production of these alcohols as fuels?

5. Of course, the net energy is less than that calculated in the previous question. What other kinds of energy costs might be associated with the manufacture of fuels such as ethanol and methanol?

6. Describe some other technological solutions to the problem of our growing per-capita and per-country demand for energy.

7. Ethyne gas may react with hydrogen gas to form ethane gas in the following reaction:

$C_2H_2(g) + 2 H_2(g) \rightarrow C_2H_6(g)$

Predict the standard enthalpy change for the reaction of 200 g of ethyne, using the following information. After using Hess' law, refer back to Sample Problem 11.2 on page 491.

(1) $C_2H_2(g) + \frac{5}{2} O_2(g) \rightarrow 2 CO_2(g) + H_2O(l)$
$\Delta_1 H° = -1299$ kJ

(2) $H_2(g) + \frac{1}{2} O_2(g) \rightarrow H_2O(l)$ $\Delta_2 H° = -286$ kJ

(3) $C_2H_6(g) + \frac{7}{2} O_2(g) \rightarrow 2 CO_2(g) + 3 H_2O(l)$
$\Delta_3 H° = -1560$ kJ

8. As an alternative to combustion, coal gas can undergo a process called *methanation*.

$3 H_2(g) + CO(g) \rightarrow CH_4(g) + H_2O(g)$ $\Delta_r H° = ?$

Determine the standard enthalpy change involved in the reaction of 300 g of carbon monoxide in this methanation reaction.

9. $C(s) + O_2(g) \rightarrow CO_2(g)$ $\Delta_c H° = -393.5$ kJ
$2 H_2(g) + O_2(g) \rightarrow 2 H_2O(g) + 483.6$ kJ
$\Delta_c H°_m = -5074.1$ kJ/mol
C_8H_{18}

Use Hess' law and combustion reaction information above to predict the standard enthalpy change for the formation of octane from its elements.

$8 C(s) + 9 H_2(g) \rightarrow C_8H_{18}(l)$ $\Delta_f H° = ?$

10. Bacteria sour wines and beers by converting ethanol, C_2H_5OH, into acetic acid, CH_3COOH. Assume that the reaction is

$C_2H_5OH(l) + O_2(g) \rightarrow CH_3COOH(l) + H_2O(l)$

The standard molar enthalpies of combustion of ethanol and acetic acid to produce liquid water are, respectively, -1367 kJ/mol and -875 kJ/mol. Write chemical equations for the combustions, and use Hess' law to determine the standard enthalpy change for the conversion of ethanol to acetic acid.

11. A series of calorimetric experiments is performed to test Hess' law. Complete the Purpose, Prediction, Analysis, and Evaluation (2, 3) sections of the investigation report. Evaluation should include a calculation of a percent difference in the experiment, given that the $\Delta_r H$ for equation (1) should equal the sum of of the $\Delta_r H$'s for equations (2) and (3).

Problem
What is the enthalpy change for the reaction of aqueous potassium hydroxide with hydrobromic acid?

Design
Three calorimetry experiments are performed: Experiment 1 for the Prediction and Experiments 2 and 3 for the Analysis. The three thermochemical reactions are

(1) $HBr(aq) + KOH(aq) \rightarrow H_2O(l) + KBr(aq)$ $\Delta_1 H = ?$ kJ
(2) $KOH(s) \rightarrow KOH(aq)$ $\Delta_2 H = ?$ kJ
(3) $KOH(s) + HBr(aq) \rightarrow H_2O(l) + KBr(aq)$ $\Delta_3 H = ?$ kJ

Evidence

Table 1 Reaction of KOH(aq) and HBr(aq)

Observation	Experiment 1	Experiment 2	Experiment 3
quantity of reactant 1	100.0 mL of 1.00 mol/L HBr(aq)	5.61 g KOH(s) in 200.0 mL of solution	5.61 g KOH(s)
quantity of reactant 2	100.0 mL of 1.00 mol/L KOH(aq)	N/A	200.0 mL of 0.50 mol/L HBr(aq)
initial temperature	20.0 °C	20.0 °C	20.0 °C
final temperature	22.5 °C	24.1 °C	26.7 °C

11.5 Molar Enthalpies of Formation

Chemists rely on conventions to simplify explanations and communication. For example, SATP is a set of internationally accepted conditions that define a *standard state* for formation reactions. It is convenient to set at zero the value for the enthalpies of elements in their most stable form at SATP (the standard conditions for elements). This convention, defining elements as the reference point at which the potential energy is zero, is the **reference energy state**. This does not mean that the enthalpy of an element is *always* considered to be zero; in another situation, a different convention might be more convenient. (Similarly, the Celsius temperature scale sets 0 °C at the freezing point of pure water. This is a convenient reference point but it does not mean that water molecules have zero thermal or kinetic energy at that temperature.)

A formation reaction is a reaction in which a compound is formed from its constituent elements. The enthalpy change calculated from measurements of a formation reaction under standard conditions is called the **standard enthalpy of formation**, $\Delta_f H°$. We can now describe $\Delta_f H°$ theoretically: as a change in enthalpy from zero (the reference enthalpy of formation of the elements) to some final value determined by the enthalpy change. For example,

$$H_2(g) + \tfrac{1}{2} O_2(g) \rightarrow H_2O(l) \qquad \Delta_f H° = -285.8 \text{ kJ}$$

$\Delta_f H°$ (kJ): 0 0 -285.8

The potential energy decreases from 0 kJ for the reactants to –285.8 kJ for the product. In other words, the reactants are at a higher chemical potential energy than the product. This potential energy is transferred to the surroundings and appears as thermal energy or other forms of energy. Suppose you were seated on a bicycle at the top of a hill and coasted downhill. Your potential energy at the top of the hill is converted into kinetic energy as you move from a point of higher potential energy (top) to one of lower potential energy (bottom) (**Figure 1**). Of course, to return to the top of the hill you must supply the energy to move to a higher potential energy. Similarly, to convert the water back into hydrogen and oxygen requires that energy be added, specifically 285.8 kJ/mol of water.

We can use tables of standard molar enthalpies of formation (Appendix I) to compare the stabilities of compounds. Most compounds are formed exothermically from their elements, and the molar enthalpies of formation are negative. This means that the compounds are more stable than their elements (defined at 0 kJ/mol). **Thermal stability** is the tendency of a compound to resist decomposition when heated. The standard molar enthalpies of formation give an indication of thermal stability. The lower the value, the greater the thermal stability. By analogy, you are physically more stable when you are at a lower gravitational potential energy—closer to the ground. Another example is a comparison of the standard molar enthalpies of formation for tin(II) oxide and tin(IV) oxide, which can be obtained in Appendix I.

$\Delta_f H_m° = -280.7$ kJ/mol
SnO

$\Delta_f H_m° = -577.6$ kJ/mol
SnO_2

Chemists explain that tin(IV) oxide has more thermal stability than tin(II) oxide because tin(IV) oxide has a more negative molar enthalpy of formation. The lower the energy, the greater the stability.

Enthalpies of reaction and molar enthalpies of reaction for many chemical reactions can be predicted from the molar enthalpies of formation of elements and compounds. Sample Problem 11.5 illustrates the derivation of the molar enthalpies of formation method from the Hess' law method.

Figure 1
As a cyclist coasts downhill, his potential energy decreases—it is converted to kinetic energy.

DID YOU KNOW ?

Zero Energy
Setting a reference point for the energy (especially potential energy) being zero is common practice. Examples of "zero points" include
• zero (0 °C) on the Celsius scale
• zero gravitational potential energy at the surface of Earth
• zero electrical potential energy for the ground
• zero electrode potential for the hydrogen electrode (see Chapter 13)
• zero enthalpy of formation for elements under standard conditions

Ideally, these reference points are set (and sometimes changed) by international agreement (for example, by IUPAC) to facilitate international communication.

SAMPLE problem 11.5

How can molar enthalpies of formation be used to calculate enthalpies of reaction? Consider the slaking of lime, calcium oxide, represented by the following chemical reaction equation (**Figure 2**).

$$CaO(s) + H_2O(l) \rightarrow Ca(OH)_2(s) \qquad \Delta_r H° = ?$$

What is the standard enthalpy of reaction for this reaction?

First, write the formation equation and corresponding standard enthalpy of formation (Appendix I) for each compound in the given equation. To find the enthalpy of formation, $\Delta_f H°$, multiply the chemical amount by the molar enthalpy of formation, $\Delta_f H° = n \Delta_f H_m°$.

$$Ca(s) + \tfrac{1}{2} O_2(g) \rightarrow CaO(s) \qquad \Delta_f H° = 1 \text{ mol} \times -634.9 \text{ kJ/mol}$$

$$H_2(g) + \tfrac{1}{2} O_2(g) \rightarrow H_2O(l) \qquad \Delta_f H° = 1 \text{ mol} \times -285.8 \text{ kJ/mol}$$

$$Ca(s) + O_2(g) + H_2(g) \rightarrow Ca(OH)_2(s) \qquad \Delta_f H° = 1 \text{ mol} \times -986.1 \text{ kJ/mol}$$

By adding the third equation to the reverse of the first two equations, the chemical equation required for the slaking of lime is obtained.

$$CaO(s) \rightarrow \cancel{Ca(s)} + \cancel{\tfrac{1}{2} O_2(g)} \qquad -\Delta_f H°$$

$$H_2O(l) \rightarrow \cancel{H_2(g)} + \cancel{\tfrac{1}{2} O_2(g)} \qquad -\Delta_f H°$$

$$\cancel{Ca(s)} + \cancel{O_2(g)} + \cancel{H_2(g)} \rightarrow Ca(OH)_2(s) \qquad \Delta_f H°$$

Net: $CaO(s) + H_2O(l) \rightarrow Ca(OH)_2(s) \qquad \Delta_r H° = ?$

Applying Hess' law gives the following equation:

$$\Delta_r H° = \underset{Ca(OH)_2}{\Delta_f H°} + \underset{CaO}{(-\Delta_f H°)} + \underset{H_2O}{(-\Delta_f H°)}$$

Notice that the net enthalpy change is equal to the enthalpy of formation for the product minus the enthalpies of formation of the reactants.

$$\Delta_r H° = \underset{Ca(OH)_2}{\Delta_f H°} - (\underset{CaO}{\Delta_f H°} + \underset{H_2O}{\Delta_f H°})$$

Substituting the definition $\Delta_f H° = n \Delta_f H_m°$ (Section 11.1) and combining terms results in the following formula, where $\Sigma n \Delta_{fP} H_m°$ is the standard enthalpy change of all the products (P), and $\Sigma n \Delta_{fR} H_m°$ is the standard enthalpy change of all the reactants (R).

$$\Delta_r H° = \underset{Ca(OH)_2}{\Sigma n \Delta_{fP} H_m°} - \underset{CaO + H_2O}{\Sigma n \Delta_{fR} H_m°}$$

$$= \underset{Ca(OH)_2}{n \Delta_{fP} H_m°} - (\underset{CaO}{n \Delta_{fR} H_m°} + \underset{H_2O}{n \Delta_{fR} H_m°})$$

$$= \left(1 \text{ mol Ca(OH)}_2 \times \frac{-986.1 \text{ kJ}}{1 \text{ mol Ca(OH)}_2}\right) - \left(1 \text{ mol CaO} \times \frac{-634.9 \text{ kJ}}{1 \text{ mol CaO}} + 1 \text{ mol H}_2\text{O} \times \frac{-285.8 \text{ kJ}}{1 \text{ mol H}_2\text{O}}\right)$$

$$= -986.1 \text{ kJ} - (-634.9 + (-285.8)) \text{ kJ}$$

$$= -65.4 \text{ kJ}$$

According to Hess' law and empirically determined standard molar enthalpies of formation, the standard enthalpy change for the slaking of lime is reported as follows:

$$CaO(s) + H_2O(l) \rightarrow Ca(OH)_2(s) \qquad \Delta_r H° = -65.4 \text{ kJ}$$

Figure 2
Limestone, in this case from the mountains near Exshaw, Alberta, can be decomposed to lime. The lime is then converted to slaked lime. The limestone, lime, and slaked lime all have technological applications.

Learning Tip

The symbols for this method of calculating the enthalpy of reaction are meant to help but, initially, can be confusing.

sum   chemical   change
of   amount   in   enthalpy
(Σ) (n)   (Δ)   (H)

$$\Sigma n \Delta_{fP} H_m$$

(f)   (P)   (m)
formation products molar

This set of symbols translates into: "the sum of the chemical amounts times the molar enthalpies of formation of the products." This set of symbols changes from P for the products to R for the reactants. Learn the concept (that is, the change in enthalpy is the energy of the products minus the energy of the reactants), and the equation then makes sense. See the Summary on the next page.

Section 11.5 Questions

1. Which of the following fuels has the greater thermal stability: methanol or ethanol? Provide your reasoning.

2. Methane, the major component of natural gas, is used as a source of hydrogen gas to produce ammonia. Ammonia is used as a fertilizer and a refrigerant, and is used to manufacture fertilizers, plastics, cleaning agents, and prescription drugs. The following questions refer to some of the chemicals involved.
   (a) The first step in the production of ammonia is the reaction of methane with steam using a nickel catalyst. Use the molar enthalpies of formation method to predict the $\Delta_r H°$ for the following reaction:
   $$CH_4(g) + H_2O(g) \rightarrow CO(g) + 3 H_2(g)$$
   (b) Another step of this process is the further reaction of carbon monoxide to produce more hydrogen. Iron and zinc–copper catalysts are used. Predict the $\Delta_r H°$.
   $$CO(g) + H_2O(g) \rightarrow CO_2(g) + H_2(g)$$
   (c) After the carbon dioxide gas is removed, the hydrogen reacts with nitrogen obtained from the air. Predict the $\Delta_r H°$ to form 2 mol of ammonia.
   $$N_2(g) + 3 H_2(g) \rightarrow 2 NH_3(g) \qquad \Delta_f H° = ?$$

3. Nitric acid, required in the production of nitrate fertilizers, is produced from ammonia by the Ostwald process (**Figure 7**). Predict the standard enthalpy change for each reaction in the process, as written, and then predict the standard molar enthalpy of reaction for the first reactant listed in each equation.
   (a) $4 NH_3(g) + 5 O_2(g) \rightarrow 4 NO(g) + 6 H_2O(g)$
   (b) $2 NO(g) + O_2(g) \rightarrow 2 NO_2(g)$
   (c) $3 NO_2(g) + H_2O(l) \rightarrow 2 HNO_3(l) + NO(g)$

Figure 7
An Ostwald process plant converts ammonia to nitric acid cleanly and efficiently. Unreacted gases and energy from the exothermic reactions are recycled. Catalytic combustors burn noxious fumes to minimize environmental effects and to supply additional energy to operate the plant.

4. Ammonium nitrate fertilizer is produced by the reaction of ammonia with the nitric acid resulting from the series of reactions given in question 3. Ammonium nitrate is one of the most important fertilizers for increasing crop yields.
   (a) Predict the standard enthalpy change of the reaction used to produce ammonium nitrate.
   $$NH_3(g) + HNO_3(l) \rightarrow NH_4NO_3(s) \qquad \Delta_r H° = ?$$
   (b) Sketch an enthalpy change diagram for the reaction of ammonia and nitric acid.

5. During World War II an oil embargo on Germany left that country short of fuels. Germany responded by producing and burning ammonia as a fuel for cars, trucks, and tanks.
   (a) What is the standard molar enthalpy of combustion of ammonia gas? Assume that nitrogen dioxide and liquid water are the products of combustion.
   (b) Draw an enthalpy change diagram for the burning of 4 mol of ammonia, including labelling the y-axis with the energy involved.
   (c) Evaluate this technological solution from two perspectives.

6. The refining of sour natural gas removes toxic hydrogen sulfide from the natural gas at many of the gas plants in Alberta. One of the reactions during the conversion of the hydrogen sulfide to sulfur is the combustion of some of this gas.
   (a) What is the standard molar enthalpy of combustion of hydrogen sulfide to produce sulfur dioxide gas and liquid water?
   (b) Communicate the standard enthalpy of combustion of hydrogen sulfide in the four ways described earlier in this chapter.
   (c) Should the products of the combustion of hydrogen sulfide be released into the atmosphere? Provide your reasoning.

7. You now have several ways of knowing the enthalpy of a chemical reaction. The simplest are to be given, to reference, or to memorize the enthalpy change for selected chemical reactions. The law of conservation of energy has so far provided you with three other methods for calculating an enthalpy change. How would you recognize when you are supposed to use each of these methods, listed below, in a question? That is, what would the question look like? What information would you be given in the question?
   (a) calorimetry
   (b) Hess' law
   (c) molar enthalpies of formation

Use this information to answer questions 8 and 9.

There is a whitish-grey mountain near Exshaw, Alberta from which limestone is mined (Figure 2, page 511). This limestone is then thermally decomposed into lime and carbon dioxide. Some of the lime is then slaked: the lime reacts with water to produce slaked lime. The lime is used in a complex cement mixture to create concrete. The lime can also be used to make mortar for bricklaying.

8. (a) Use standard molar enthalpies of formation to predict the standard reaction enthalpy for the first step of the chemical process to produce slaked lime.

    $CaCO_3(s) \rightarrow CaO(s) + CO_2(g)$ $\Delta_r H° = ?$

    (b) Now predict the $\Delta_r H°$ for the second step.

    $CaO(s) + H_2O(l) \rightarrow Ca(OH)_2(s)$ $\Delta_r H° = ?$

    (c) Write the overall (net) reaction equation for the two steps, and use Hess' law to predict the net standard enthalpy of reaction.
    (d) Use the net reaction equation and molar enthalpies of formation to predict the net standard reaction enthalpy.
    (e) Compare your answers in (c) and (d).
    (f) List some further technological applications of lime and slaked lime.

    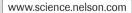 www.science.nelson.com

9. Slaked lime is used in mortar, for building brick walls (**Figure 8**). As the wall "cures," the mortar is strengthened by the reaction of slaked lime with carbon dioxide from the air to produce calcium carbonate and water. Look back at your answers for the previous question to determine the standard enthalpy of reaction without having to use Hess' law or standard molar enthalpies of reaction.

Figure 8
The rate at which mortar cures can affect its strength and longevity.

Extension

10. The prediction of enthalpy changes using standard molar enthalpies of formation depends entirely on the availability of tables of standard molar enthalpies of formation. Many of these $\Delta_f H_m°$ values can be initially determined from $\Delta_c H_m°$ values by using the formation method equation. In this problem you will use a known $\Delta_c H_m°$ value to calculate a corresponding $\Delta_f H_m°$ value. Complete the Analysis of the investigation report. Work out your own problem-solving approach here.

Problem
What is the standard molar enthalpy of formation for hexane, $C_6H_{14}(l)$?

Design
The standard molar enthalpy of formation for hexane is determined using the concept
$$\Delta_c H° = \Sigma n \Delta_{fp} H_m° - \Sigma n \Delta_{fr} H_m°$$
to describe the hexane combustion reaction. The standard molar enthalpy of combustion of hexane is obtained from a reference source.

Evidence
The CRC Handbook of Chemistry and Physics lists the calorimetric value for the standard molar enthalpy of combustion for hexane as −4 162.9 kJ/mol. Carbon dioxide gas and liquid water are the only products of the combustion.

11. Besides specific heat capacity (e.g., 4.19 J/(g•°C) for water), there are other energy descriptions for chemical substances. Based upon the units used for these quantities, describe as best you can what the quantity is about and where it might be of importance or of use.
    (a) volumetric heat capacity (e.g., 1.2 kJ/(m³•°C) for air at SATP)
    (b) molar heat capacity (e.g., 75.5 kJ/(mol•°C) for water)
    (c) heat capacity (e.g., 9.12 kJ/°C for a particular bomb calorimeter)
    (d) specific heat (e.g., 13.4 kJ/g of water for the formation of water vapour from hydrogen and oxygen)

12. The bomb calorimeter and flow calorimeter are high-tech calorimeters used to determine, for example, the energy value of foodstuffs. In a test of the molar enthalpies of formation method, a Prediction is made of the standard molar enthalpy of respiration of glucose. Use calorimetry in the Analysis to obtain an empirical test value. Complete the Evaluation (2 and 3). (Assume liquid water is produced.)

Problem
What is the standard molar enthalpy of respiration of glucose?

Evidence
mass of glucose = 3.00 g
heat capacity of calorimeter = 8.52 kJ/°C
initial temperature = 25.02 °C
final temperature = 30.49 °C

13. There are many commercial products sold in automotive parts and service centres that claim to increase gas mileage, for example, oil and gasoline additives. Your company has been chosen to test one of these claims. Design an investigation to determine the validity of the claim.

14. Research an R-2000 (energy-efficient) home on the Internet, and write a brief description.

www.science.nelson.com

Chapter 11 INVESTIGATIONS

INVESTIGATION 11.1

Designing and Evaluating a Calorimeter

Report Checklist
- ○ Purpose
- ○ Problem
- ● Hypothesis
- ○ Prediction
- ● Design
- ○ Materials
- ● Procedure
- ● Evidence
- ● Analysis
- ● Evaluation (1, 2, 3)

A calorimeter can be as simple as an uninsulated beaker or can. Literature research indicates that very sophisticated calorimeters, such as bomb and flow calorimeters, are also available. There are many laboratory calorimeters between these extremes. In this investigation you will create your own calorimeter. Use modern materials and/or consider what Aboriginal peoples might have used for insulating a calorimeter-like vessel.

Purpose
The purpose of this investigation is to test a hypothetical design of a calorimeter that is improved (based upon your own scientific and technological criteria) over the simplest calorimeter.

Problem
What is the best design for a simple calorimeter?

Design
Each research group in class uses a different calorimeter design of their choosing. One group uses an uninsulated calorimeter as a control. An individual group and a class Analysis are employed.

Materials
lab apron	centigram balance
eye protection	stirring rod
small soup can	ring and stand
insulation	paraffin candle
aluminium foil	matches
duct tape	watch glass
thermometer or temperature probe	100 mL graduated cylinder
	timer

Take care around lit matches and candles. Tie back loose hair and clothing. Never leave a flame unattended.

INVESTIGATION 11.2

Molar Enthalpy of Reaction

Report Checklist
- ○ Purpose
- ○ Problem
- ○ Hypothesis
- ○ Prediction
- ● Design
- ● Materials
- ● Procedure
- ● Evidence
- ● Analysis
- ● Evaluation (2, 1, 3)

Evaluating evidence (dependent upon an experimental design, materials, procedure, and skill) and estimating the certainty of empirically determined values are important skills in interpreting scientific statements. The accuracy (percent difference) of the value for molar enthalpy obtained in this investigation is used to evaluate the calorimeter and the assumptions made in the analysis, rather than to evaluate the prediction and its authority. The ultimate authority in this experiment is considered to be the reference value used in the prediction.

Purpose
The purpose of this investigation is to test the calorimeter design and calorimetry procedure by comparing experimental evidence with a widely accepted value for the molar enthalpy of a neutralization reaction.

Problem
What is the standard molar enthalpy of neutralization for sodium hydroxide when 50 mL of aqueous 1.0 mol/L sodium hydro-xide reacts with an excess quantity of 1.0 mol/L sulfuric acid?

Prediction
The standard molar enthalpy of neutralization for sodium hydroxide is −57 kJ/mol, as per *The CRC Handbook of Chemistry and Physics*.

 Wear eye protection. Both sodium hydroxide and sulfuric acid are corrosive chemicals. Rinse with lots of cold water if these chemicals contact your skin.

INVESTIGATION 11.3

Applying Hess' Law

Report Checklist
- ○ Purpose   ○ Design   ● Analysis
- ○ Problem   ○ Materials   ● Evaluation (1, 2, 3)
- ○ Hypothesis ● Procedure
- ○ Prediction ● Evidence

Magnesium burns rapidly, releasing heat and light.

$$Mg(s) + \tfrac{1}{2} O_2(g) \rightarrow MgO(s) \qquad \Delta_c H = ?$$

The enthalpy change of this reaction can be calculated from measurements involving a highly sophisticated calorimeter, but not a polystyrene cup calorimeter. The enthalpy change for the combustion of magnesium can, however, be determined by applying Hess' law to the following three chemical equations.

$$MgO(s) + 2\,HCl(aq) \rightarrow MgCl_2(aq) + H_2O(l)$$
$$\Delta_r H = ?$$
$$Mg(s) + 2\,HCl(aq) \rightarrow MgCl_2(aq) + H_2(g) \quad \Delta_r H = ?$$
$$H_2(g) + \tfrac{1}{2} O_2(g) \rightarrow H_2O(l) \qquad \Delta_f H° = -285.8\text{ kJ}$$

Purpose
The purpose of this investigation is to use Hess' law to determine a standard molar enthalpy of combustion.

Problem
What is the standard molar enthalpy of combustion for magnesium?

Prediction
The standard molar enthalpy of combustion for magnesium is −601.6 kJ/mol, as per the table of standard molar enthalpies of formation (Appendix I). (The standard molar enthalpy of combustion of magnesium is the same as the standard molar enthalpy of formation of magnesium oxide because both processes have the same chemical reaction equation.)

$$\Delta_c H_m°_{\,Mg} = \Delta_f H_m°_{\,MgO} = -601.6 \text{ kJ/mol}$$

$$Mg(s) + \tfrac{1}{2} O_2(g) \rightarrow MgO(s)$$
$$\Delta_c H° = \Delta_f H° = -601.6 \text{ kJ}$$

Design
The enthalpy changes for the first two reactions in the series are determined empirically using a polystyrene calorimeter (**Figure 1**), and the third enthalpy change is obtained from Appendix I. The three $\Delta_r H$ values are used, along with Hess' law, to obtain the molar enthalpy of combustion for magnesium.

Materials
- lab apron
- eye protection
- magnesium ribbon (maximum 15 cm strip)
- magnesium oxide powder (maximum 1.00 g sample)
- 1.00 mol/L hydrochloric acid (use 50 mL each time)
- polystyrene calorimeter with lid and thermometer or temperature probe
- 50 mL or 100 mL graduated cylinder
- laboratory scoop or plastic spoon
- steel wool
- weighing boat or paper
- centigram balance
- ruler

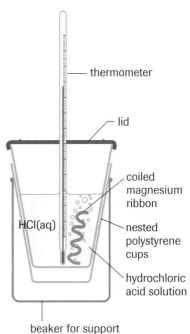

Figure 1
Magnesium ribbon reacts rapidly in dilute hydrochloric acid. With nested polystyrene cups, the enthalpy change can be determined by measuring the temperature change of the hydrochloric acid solution.

- Hydrochloric acid is corrosive. Avoid contact with skin, eyes, and clothing. If you spill this acid on your skin, wash immediately with lots of cool water.
- Hydrogen gas, produced in the reaction of hydrochloric acid and magnesium, is flammable. Ensure that there is adequate ventilation and that there are no open flames in the classroom.
- Magnesium oxide is an extremely fine dust. Do not inhale magnesium oxide powder because it is an irritant.

Chapter 11 SUMMARY

Outcomes

Knowledge
- explain how the sun is a major source of stored chemical energy on Earth (11.1)
- apply the equation $Q = mc\Delta t$ to the analysis of energy transfer (11.2)
- define enthalpy and molar enthalpy for chemical reactions (11.2, 11.3)
- use calorimetry evidence to determine enthalpy changes in chemical reactions (11.2, 11.3, 11.4, 11.5)
- describe photosynthesis, cellular respiration, and hydrocarbon combustion reactions, and understand that combustion and cellular respiration are similar and the reverse of photosynthesis (11.3)
- write and interpret balanced chemical equations incorporating $\Delta_r H$ notation (11.3)
- classify chemical reactions as endothermic or exothermic (11.3, 11.4, 11.5)
- explain and use Hess' law to calculate energy changes for a net reaction from a series of reactions (11.4)
- predict the standard enthalpy change for chemical equations using standard molar enthalpies of formation (11.5)

STS
- state that a goal of technology is to solve practical problems (11.1, 11.2)
- recognize that solving technological problems may require various solutions and have both intended and unintended consequences (11.1, 11.3, 11.4)

Skills
- initiating and planning: design a method to compare molar enthalpy changes when burning fuels (11.2); describe procedures for safe handling, storage, and disposal of materials used in the laboratory, with reference to WHMIS and consumer product labelling information (11.2, 11.4)
- performing and recording: perform calorimetry experiments to determine molar enthalpy changes and use thermometers appropriately (11.2, 11.4)
- analyzing and interpreting: compare energy changes by analyzing data and energy diagrams (11.3, 11.4, 11.5)
- communication and teamwork: work collaboratively using appropriate notation and units for enthalpy changes and molar enthalpies (11.2, 11.3, 11.4, 11.5)

Key Terms

11.2
thermochemistry
isolated system
calorimetry
calorimeter
heat

thermal energy, Q
specific heat capacity, c
joule (J)
enthalpy change, ΔH or $\Delta_r H$
exothermic reaction
endothermic reaction

enthalpy of reaction
molar enthalpy of reaction, $\Delta_r H_m$

11.3
standard molar enthalpy of reaction, $\Delta_r H_m°$
chemical potential energy diagram

11.4
Hess' law

11.5
reference energy state
standard enthalpy of formation $\Delta_f H°$
thermal stability

Key Equations

$Q = mc\Delta t$

$\Delta_r H = n\Delta_r H_m$

$n\Delta H = mc\Delta t$ (calorimetry)

$\Delta_r H = \Sigma \Delta_r H$ (Hess' law)

$\Delta_r H° = \Sigma n \Delta_{fP} H_m° - \Sigma n \Delta_{fR} H_m°$ (enthalpies of formation)

1. Prepare a table to summarize the four methods for communicating enthalpies of reaction, the quantities and symbols used, the units used. Prepare a second table showing the three methods for determining enthalpy changes. Use the adjectives "molar" and "standard" (and symbols) where appropriate.
2. Refer back to your answers to the Starting Points questions at the beginning of this chapter. How has your thinking changed?

▶ **Go** To   www.science.nelson.com

The following components are available on the Nelson Web site. Follow the links for *Nelson Chemistry Alberta 20–30*.
- an interactive Self Quiz for Chapter 11
- additional Diploma Exam-style Review questions
- Illustrated Glossary
- additional IB-related material

There is more information on the Web site wherever you see the Go icon in this chapter.

A New Kind of Rocket Science
Scientists are looking beyond chemical reactions for rocket propulsion systems. They are investigating nuclear thermal rockets, nuclear electricity, and even solar sailing.

www.science.nelson.com

Chapter 11 REVIEW

Many of these questions are in the style of the Diploma Exam. You will find guidance for writing Diploma Exams in Appendix H. Exam study tips and test-taking suggestions are on the Nelson Web site. Science Directing Words used in Diploma Exams are in bold type.

www.science.nelson.com

DO NOT WRITE IN THIS TEXTBOOK.

Part 1

1. A source of energy that is an alternative to fossil fuels is
   A. coal
   B. solar
   C. kerosene
   D. natural gas

2. Over 95% of the electricity in Alberta is generated by burning fossil fuels. In thermal electric generating stations, energy moves through several forms. List these forms in order.
   | 1 | chemical | 3 | mechanical |
   | 2 | electrical | 4 | thermal |
   ___ ___ ___ ___

3. The quantity of thermal energy required to raise the temperature of a unit mass of a substance by one degree Celsius is referred to as
   A. specific heat
   B. heat capacity
   C. specific heat capacity
   D. volumetric heat capacity

4. A propane camp stove is used to heat 1.50 L of water for making tea. If the water is heated from 5.5 °C to 98.6 °C, the energy absorbed by the water is _____ kJ.

5. Enthalpy change (ΔH) is defined as the change in the
   A. thermal energy in a system
   B. property of a substance or system that relates to its ability to do work
   C. state of a closed system in which all measurable properties are constant
   D. chemical energy of a system when the pressure of the system is held constant

6. Some coal deposits contain significant quantities of sulfur, which forms sulfur dioxide when the coal is burned. Sulfur dioxide reacts with oxygen in the atmosphere to produce sulfur trioxide, which in turn reacts with rainwater to produce sulfuric acid. Use the following equation to determine the molar enthalpy of reaction for sulfur dioxide.
   $2\ SO_2(g) + O_2(g) \rightarrow 2\ SO_3(g) + 197.8\ kJ$
   -_____ kJ/mol

7. The correct reaction equation for the formation of water vapour is
   A. $2\ H_2(g) + O_2(g) \rightarrow 2\ H_2O(g) + 120.9\ kJ$
   B. $2\ H_2(g) + O_2(g) \rightarrow 2\ H_2O(g) + 241.8\ kJ$
   C. $2\ H_2(g) + O_2(g) \rightarrow 2\ H_2O(g) + 483.6\ kJ$
   D. $2\ H_2(g) + O_2(g) \rightarrow 2\ H_2O(g) + 571.6\ kJ$

8. The equation that correctly represents the enthalpy change involved in the simple decomposition of dinitrogen tetroxide, $N_2O_4(g)$, is
   A. $N_2(g) + 2\ O_2(g) \rightarrow N_2O_4(g)$ $\Delta_r H = +11.1\ kJ$
   B. $N_2(g) + 2\ O_2(g) \rightarrow N_2O_4(g)$ $\Delta_r H = -11.1\ kJ$
   C. $N_2O_4(g) \rightarrow N_2(g) + 2\ O_2(g)$ $\Delta_r H = +11.1\ kJ$
   D. $N_2O_4(g) \rightarrow N_2(g) + 2\ O_2(g)$ $\Delta_r H = -11.1\ kJ$

Use the following statements to answer questions 9 and 10.
1. Carbon dioxide is a reactant.
2. Carbon dioxide is a product.
3. Water is a reactant.
4. Water is a product.
5. The reaction is exothermic.
6. The reaction is endothermic.

9. Which of the above statements are true for photosynthesis?
   ___ ___ ___ ___

10. Which of the above statements are true for cellular respiration?
    ___ ___ ___ ___

Consider these fuels when answering questions 11 to 13.
1. methane
2. propane
3. octane
4. methanol

11. List the fuels in order of increasing standard molar enthalpy of formation.
    ___ ___ ___ ___

12. List the fuels in order of increasing standard molar enthalpy of combustion.
    ___ ___ ___ ___

13. List the fuels in order of increasing thermal stability.
    ___ ___ ___ ___

14. The standard enthalpy change for the world-scale catalytic cracking of ethane into ethene, for example, near Red Deer and near Fort Saskatchewan, Alberta, is
    +_____ kJ

Part 2

15. **Explain**, in general terms, how stored energy in the bonds of hydrocarbons originated from the sun.

16. **Distinguish** between an enthalpy change and a molar enthalpy including general symbols and word descriptions.

Enthalpy Change 519

17. The energy content of foods can be determined by combustion analysis using a calorimeter. The combustion of 1.25 g of peanut oil caused the temperature of 2.0 kg of water to increase by 5.3 °C.
    (a) **Determine** the increase in thermal energy of the water.
    (b) **Determine** the energy content of peanut oil in kilojoules per gram.

18. Methylpropane, $C_4H_{10}(g)$, is used as a lighter fluid. When 1.52 g of methylpropane is burned in a calorimeter, the temperature of 2.23 kg of water changed by 8.04 °C.
    (a) Calculate the change in thermal energy of the water.
    (b) **Determine** the enthalpy change for the combustion of 1.52 g of methylpropane under these conditions.
    (c) If you were using this evidence to report the molar enthalpy of combustion for methylpropane, what symbol, value, and sign would you use? **Justify** your answer.

19. The enthalpy of neutralization of an acid–base reaction can be determined using a graduated cylinder, a thermometer, and a simple polystyrene calorimeter. Complete the Analysis of the investigation report.

    Problem
    What is the molar enthalpy of neutralization for ethanoic acid?

    Design
    Measured volumes of ethanoic acid and sodium hydroxide are mixed in a polystyrene calorimeter. The temperatures of the solutions are measured before and after mixing.

    Evidence

volume of 1.00 mol/L ethanoic acid	= 50.0 mL
volume of 1.00 mol/L sodium hydroxide	= 50.0 mL
initial temperature of ethanoic acid	= 22.5 °C
initial temperature of sodium hydroxide	= 22.5 °C
final temperature of mixture	= 27.8 °C

20. As an alternative to chemical energy, heating water is one of the most cost-effective uses of solar energy, providing hot water for showers, dishwashers, and washing machines.
    (a) **Determine** the thermal energy acquired (and chemical energy saved) when the temperature of 100 L of water is raised from 6.5 °C to 58.5 °C.
    (b) What mass of natural gas (methane) would have to be burned to heat the same volume of water by the same temperature change?
    (c) A domestic solar water heater saves 32.5 MJ of chemical (combustion) energy while increasing the thermal energy of some water from 12.3 °C to 45.5 °C. What mass and volume of water did the water heater contain?

21. Prepare a general comparison of photosynthesis, cellular respiration, and hydrocarbon combustion. Your response, which could be a paper or electronic graphic or table, should include
    - reactants
    - products
    - endothermic/exothermic description
    - sign of the enthalpy change

22. The flame of an oxyacetylene torch is hot enough to melt most metals. Calculate and then communicate the energy changes for the combustion of acetylene (ethyne) in four different ways.

23. Although calorimetry is the basic method of obtaining enthalpy changes, this method cannot always be directly used for many reactions. **Describe** two other methods commonly used to obtain an enthalpy change for a reaction. Include what information is required, where this information is obtained, and any mathematical equations.

24. Vinyl chloride monomer is produced at Fort Saskatchewan, Alberta, to produce, in turn, polyvinyl chloride (PVC). Use Hess' law to **determine** the enthalpy of reaction for the conversion of ethene into chloroethene (vinyl chloride).

    $C_2H_4(g) + Cl_2(g) \rightarrow C_2H_3Cl(g) + HCl(g)$ $\Delta_rH = ?$
    $C_2H_4(g) + Cl_2(g) \rightarrow CH_2ClCH_2Cl(l)$ $\Delta_rH = -179.3$ kJ
    $C_2H_3Cl(g) + HCl(g) \rightarrow CH_2ClCH_2Cl(l)$ $\Delta_rH = -71.9$ kJ

25. To meet the demand for gasoline, hydrocarbons of large molar mass are broken into smaller fragments in a process called *cracking*. An overall reaction equation for producing octane from hexadecane is

    $C_{16}H_{34}(l) + H_2(g) \rightarrow 2 C_8H_{18}(l)$ $\Delta_rH° = ?$

    Use the following reaction equations to **determine** the standard enthalpy change for the above reaction.

    $2 C_{16}H_{34}(l) + 49 O_2(g) \rightarrow 32 CO_2(g) + 34 H_2O(l)$
    $\Delta_rH° = -21\,446$ kJ

    $2 C_8H_{18}(l) + 25 O_2(g) \rightarrow 16 CO_2(g) + 18 H_2O(l)$
    $\Delta_rH° = -10\,940$ kJ

    $2 H_2(g) + O_2(g) \rightarrow 2 H_2O(l)$ $\Delta_rH° = -572$ kJ

26. In the metal refining industry, sulfide ores are usually roasted in air to produce metal oxides and sulfur dioxide. For example, copper(II) sulfide is converted to copper(II) oxide as part of the refining of copper.
    (a) Use referenced data to **determine** the standard molar enthalpy of roasting for copper(II) sulfide.
    (b) Draw a chemical potential energy diagram for this roasting reaction.

27. Calculate the standard enthalpy of reaction for the following organic reactions in the Alberta chemical industry studied in Unit 5.
    (a) octane reacts with hydrogen (i.e., undergoes hydrocracking) to produce pentane and propane
    (b) ethene reacts with benzene to produce phenylethene (polystyrene monomer) and hydrogen

28. The following investigation was carried out to find the enthalpy of formation, $\Delta_f H$, for calcium oxide. Complete the Prediction and the Analysis of the investigation report.

Problem
What is the enthalpy change for the reaction?
$$Ca(s) + \tfrac{1}{2} O_2(g) \rightarrow CaO(s)$$

Design
Calcium metal reacts with hydrochloric acid in a calorimeter and the enthalpy change is determined. Similarly, the enthalpy change for the reaction of calcium oxide with hydrochloric acid is determined. These two chemical equations are combined with the formation equation for water to determine the required enthalpy change.

Evidence
concentration of HCl(aq) = 1.0 mol/L
volume of HCl(aq) = 100 mL

Table 1

Reactant	Mass (g)	Initial temperature (°C)	Final temperature (°C)
Ca(s)	0.52	21.3	34.5
CaO(s)	1.47	21.1	28.0

29. Cars and furnaces often do not run at their highest energy efficiency.
    (a) Calculate the standard molar enthalpy of complete combustion of propane (to produce liquid water).
    (b) Calculate the standard molar enthalpy of complete combustion of propane (to produce water vapour).
    (c) Calculate the standard molar enthalpy of incomplete combustion for propane. Use the following chemical equation for the incomplete combustion of propane.
    $$2\,C_3H_8(g) + 8\,O_2(g) \rightarrow$$
    $$C(s) + 2\,CO(g) + 3\,CO_2(g) + 8\,H_2O(g)$$
    (d) List the above standard molar enthalpies of combustion of propane from lowest to highest.
    (e) Research the features of a high-efficiency (condensing) furnace, and relate these features to your calculations above.

    www.science.nelson.com

30. According to an Inuit Elder, the refrigerator has damaged the Inuit community and lifestyle because families do not have to share in the same way that they used to. Suggest possible community interactions before and after the arrival of refrigerators. What does this imply about our notion of technological progress?

Extension

31. Choose an "alternative" fuel to research. Select and integrate information (including graphs, tables, and graphics) from several sources. Your information should illustrate at least three perspectives on the use of this fuel. Decide on the target audience, and present your findings attractively and persuasively.

    www.science.nelson.com

32. Thermal-electric power stations burn coal (various types), fuel oil (heavy and light), natural gas, or wood to produce steam to drive a turbine and generator. The efficiency of converting the available chemical energy into electrical energy can be expressed as a percentage. The Statistics Canada publication *Human Activity and the Environment* provides this kind of data shown in **Table 2**.
    (a) **Analyze** the data in Table 2 (or obtain more recent data) to find any trends or other insights.
    (b) **Identify** other non-combustible energy sources that can be used to produce steam in a power station.

    www.science.nelson.com

Table 2 Percent Efficiency of Thermal-Electric Power Stations by Fuel Type (%)

| Year | Coal | | | | | Fuel oil | | Gas | Wood |
	CB	IB	CSB	ISB	Lig	HFO	LFO	NG	W
1998	34.01	37.21	32.51	34.91	29.71	33.01	32.21	34.71	40.01
1999	33.91	35.61	32.71	35.11	29.11	33.21	30.41	35.41	35.41
2000	31.31	34.81	32.81	33.21	30.11	33.01	28.31	35.11	31.31
2001	32.41	32.81	32.41	28.51	30.11	33.01	29.31	34.41	26.81
2002	30.21	33.91	32.81	33.01	30.21	33.51	29.41	35.91	27.11

CB: Canadian bituminous; IB: imported bituminous; CSB: Canadian sub-bituminous; ISB: imported sub-bituminous; Lig: lignite; HFO: heavy fuel oil; LFO: light fuel oil and diesel; NG: natural gas; W: wood

chapter 12

Explaining Chemical Changes

▶ In this chapter

- Exploration: Starting, Comparing, and Altering Reactions
- Biology Connection: "Cold-Blooded" Animals?
- Web Activity: Collisions and Reactions
- Web Activity: Collision–Reaction Theory
- Web Activity: Neurotransmitters and Nerve Agents
- Web Activity: John Polanyi
- Investigation 12.1: Iodine Clock Reaction
- Web Activity: A Catalyzed Reaction
- Investigation 12.2: Evidence for an Activated Complex

It is part of human nature to try to make sense of the world around us and to develop technologies to solve practical problems, especially problems involving survival in a harsh environment. Fire is one of the earliest technologies, and a good example of the many energy changes that are an important part of the natural world. What is fire, and how can it be used? Almost every culture has had its stories or explanations about the origin and use of fire. From the Greek culture we have the story of Prometheus, and from North American Aboriginal peoples there is the story of the coyote stealing fire from the fire beings to give to the humans. Western science also has its "fire story" in the form of theories of chemical reactions, such as combustion. Scientists have considerable empirical knowledge of fire, or combustion. They can take measurements and calculate the energy changes, and represent them quantitatively in various forms as shown in the previous chapter. How does fire, or any other reaction, get started? How does it progress? And how can it be manipulated to alter its rate? The answers to these questions form part of Western science's story or theory of fire and other reactions. This chapter will start to answer these questions.

Not surprisingly, the empirical and theoretical descriptions about the nature of chemical energy changes are interwoven with the technologies of energy changes. Technologies that are researched and developed lead to advances in science, and scientific research leads to improved technologies. An important example is the study and application of catalysts—substances used extensively to increase the rate of chemical reactions. Industries that use catalysts include oil refining (catalytic cracking), petrochemical production (for polymerization), and chemical technologies (such as automobile catalytic converters, **Figure 1**). The story or theory of chemical energy changes in this chapter will also include numerous important technological applications.

💡 STARTING Points

Answer these questions as best you can with your current knowledge. Then, using the concepts and skills you have learned, you will revise your answers at the end of the chapter.

1. What is necessary for a successful collision to occur between reacting entities?
2. How do bond energies affect the rate and enthalpy of reaction?
3. (a) What is the purpose of using a catalyst in a chemical process?
   (b) Why does a catalyst have this effect?
4. List some technological products and processes that employ catalysts.

Career Connection:
Field Production Operator

Figure 1
A catalytic converter in a car exhaust reduces the toxic emissions by converting the harmful exhaust by-products into relatively harmless by-products. Oxidation catalysts convert hydrocarbons into carbon dioxide and water. Three-way catalysts also convert oxides of nitrogen back into nitrogen.

▶ Exploration Starting, Comparing, and Altering Reactions

We have seen many chemical reactions and know that they can be classified into different categories (reaction types). We can use this knowledge for predicting reaction equations and analyzing the stoichiometry of reactions. Although this is an important starting point, it leaves many questions unanswered. How do reactions start, and why do they keep going? Why do apparently similar reactions proceed at very different rates? How can the rate of a reaction be altered? The three parts of this Exploration will help you think about answers to these three questions.

Materials: eye protection; lab apron; 5 small test tubes; test tube rack; tweezers; distilled water wash bottle; 1 mol/L HCl(aq); 3% H_2O_2(aq); pieces of Mg(s), Zn(s), Fe(s), and fruits such as bananas and apples; some flakes of rust; cut open D cell

Hydrochloric acid is corrosive; hydrogen peroxide is both corrosive and a strong oxidizer. The chemicals in an open dry cell are likely to be corrosive and possibly toxic. Handle all substances with care. If they contact the skin, wash thoroughly with cool water. If they contact the eyes, flush with cool water for at least 15 min and inform your teacher.

Part A: Starting Reactions
Baking generally starts with measuring and mixing the ingredients, which include some potential reactants.
(a) Is mixing the ingredients all that you need to do? What is the usual next step in the baking procedure?
(b) List some observations that would indicate that a chemical reaction is occurring or has occurred.

When using a laboratory burner, natural gas and air are mixed together at the bottom of the tube and emerge out the top.
(c) Does the combustion start when the two reactants mix? What needs to be done next?
(d) List some observations that would indicate that a chemical reaction is occurring.

Part B: Comparing Reactions
One of the empirical properties of acids is that they react with active metals to produce hydrogen gas.
(e) When comparing the reactions of different metals with an acid, what variables need to be kept constant?
- Set up three test tubes containing about 1 cm depth of hydrochloric acid. Add a piece of Mg(s) to the first test tube, Zn(s) to the second, and Fe(s) to the third. Observe for a few minutes, and remove the metal with tweezers.
(f) List the similarities and differences in the evidence for the reactions of the three metals.

Part C: Altering Reactions
Hydrogen peroxide is produced in both natural and technological systems. When hydrogen peroxide decomposes it produces oxygen gas and water.
- Set up several clean test tubes with equal volumes (about 1 cm depth) of hydrogen peroxide.
- To all but one of the test tubes containing the hydrogen peroxide, add a small quantity of MnO_2(s) from the inside of a dry cell, some rust particles, or some small pieces of fruit. Leave one test tube containing only hydrogen peroxide.
(g) Record the evidence of reaction including comparisons of different test tubes.
(h) Why was one test tube set up with only hydrogen peroxide?

12.1 Activation Energy

Learning Tip

When using the stoichiometric law to make predictions, we normally assume that the chemical reaction in question is
- quantitative (reacts 100%)
- spontaneous (happens without the input of energy)
- stoichiometric (has a fixed ratio)
- fast (reacts rapidly)

DID YOU KNOW ?

Thought Experiments

Chemists can communicate evidence in a number of ways, one being by a thought experiment. Thought experiments are done in the mind. With prompting, we visualize the design, materials, procedure, and evidence collection, although no actual experiment is carried out.

DID YOU KNOW ?

Rate of Reaction

As you may have learned in past courses, there is evidence to show that the rate of a chemical reaction is generally increased by
- increasing the temperature
- increasing the pressure of a gas
- increasing the surface area of a liquid or solid
- changing one of the reactants
- adding a catalyst

Chemical technologies, like most technologies, are evaluated on the basis of whether they are simple, economical, reliable, and efficient. Chemists, chemical engineers, and chemical technologists work on chemical processes for years—sometimes even decades—to perfect a technology that meets these criteria. Each chemical process, past and present, has its own story of people and technologies. For example, the refining of crude oil has changed significantly over the years, as has the design of automobile engines that use the gasoline refined from the crude oil. This chapter tells some of the inside stories of how chemists explain chemical reactions and how chemical engineers use and control chemical reactions.

Reaction Progress

From your own experience, including the Exploration on page 523, you can see that some reactions do not proceed spontaneously at room temperature unless additional energy is added to start them off. A lit match, for example, is needed to start the combustion of candle wax in a candle, wood in a campfire, and propane in a barbeque. Consider a thought experiment for a dangerous situation in which methane gas is escaping into a closed room and is mixing with the air. Based on what you know about methane and oxygen, the predicted reaction is highly exothermic. Why is this highly exothermic reaction not spontaneous—why does an immediate reaction not occur to form the more thermally stable products carbon dioxide and water? A spark from a light switch being turned on, however, can cause an explosive reaction of the methane–air mixture. Why is this initiating energy source necessary to cause the reaction?

From your experience and from the Exploration, you may also have noticed that different reactants appear to react at different rates, even when all other variables are controlled. Why does changing one of the reactants in a similar chemical reaction make such a difference in the rate of the reaction? For example, different metals in contact with the same acid react at varying rates, and the same metal in contact with different acids reacts at varying rates. Consider another thought experiment where the flame of a propane torch is applied for the same period of time to different types of wood cut into identical shavings. In this case, the initiating energy is the same; however, the resulting rates of reaction are different. Some of the wood shavings burn very quickly while others burn very slowly. What is there about the different wood shavings that produces a varying rate of reaction? How can we explain this phenomenon?

Some reactants have different temperatures at which they start reacting. Also, some reactants at the same temperature react at different rates. What theoretical explanation is there for the different ignition temperatures and for the varying rates of reaction for different reactants?

Collision–Reaction Theory

Although empirical knowledge is the foundation of science, an ultimate aim of science is to understand processes by creating simple, consistent, and logical theories. Scientists develop theoretical knowledge, and communicate it by way of models and analogies. They test the resultant knowledge by judging its ability to describe, explain, and predict. Theories are often better at explaining than at predicting. The collision–reaction theory is an excellent example of this type of theory. It is easier to explain reactions using the theory than it is to make predictions of specific reactions.

Chemists created the collision–reaction theory to describe, explain, and predict characteristics of chemical reactions. Some of the main ideas of the collision–reaction theory are the following:

- A chemical sample consists of entities (atoms, ions, or molecules) that are in constant random motion at various speeds, rebounding elastically from collisions with each other. (Kinetic energy is conserved during elastic collisions.)
- A chemical reaction must involve collisions of reactant entities.
- An effective collision requires sufficient energy. Collisions with the required minimum energy have the potential to react.
- An effective collision also requires the correct orientation (positioning) of the colliding entities so that bonds can be broken and new bonds formed.
- Ineffective collisions involve entities that rebound elastically from the collision.

When looking at different reactions, chemists have noticed that some reactions appear to occur more readily than others. According to collision–reaction theory, reactions can only take place when entities collide, but not all collisions result in a reaction. If the orientation is correct and the energy is sufficient, then a reaction can occur (**Figure 1**). In other reactions, however, the collisions of reactant entities may involve insufficient energy or the collision may not have the correct orientation (**Figure 2**).

To use the collision–reaction theory more effectively, we need to consider other factors, such as the nature of the bonds present and the type of reaction occurring.

DID YOU KNOW?
Reaction Kinetics
In May 1884, Svante Arrhenius presented his doctoral thesis on a theory of ionic dissociation. His study jump-started branches of chemistry called *physical chemistry* and *reaction kinetics*. However, it wasn't until the 1920s and 1930s that these topics became part of organic chemistry, until the 1940s and 1950s for inorganic chemistry, and, finally, completing the reform, into high school chemistry textbooks in the 1960s.

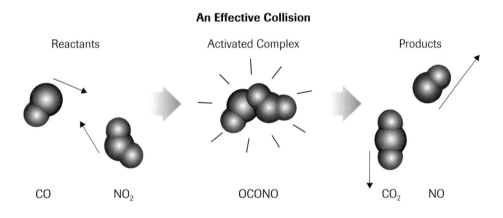

Figure 1
These molecules have both the correct orientation and sufficient energy, so the collision is effective: the atoms are rearranged and products are formed.

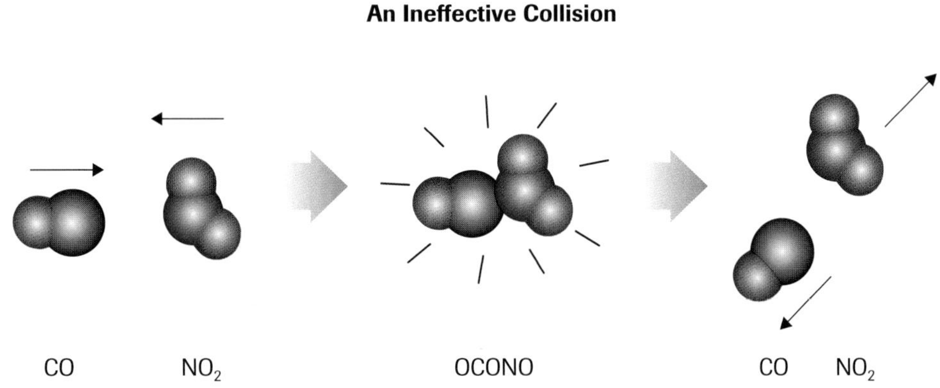

An Ineffective Collision

CO   NO₂   OCONO   CO   NO₂

Figure 2
These molecules collide with a "wrong" orientation, so the bonds do not rearrange and no new substances form. The collision is ineffective.

BIOLOGY CONNECTION

"Cold-Blooded" Animals?
Many biological reactions take place at around 37 °C. How, then, do cold-blooded animals manage to live and function, despite being cooler than this? In your Biology course, you may discover that cold-blooded animals aren't always quite as cold as we think.

www.science.nelson.com

In summary, there are two sources of evidence that need explaining. First, why do some chemicals react faster than others, when all other variables except the type of chemical are controlled (for example, why does magnesium react faster than zinc with hydrochloric acid)? Second, why do some reactions require an initial input of external energy to react (for example, why is a match needed to start the combustion of a hydrocarbon)? Empirical chemists gather the evidence and establish empirical concepts, such as generalizations about reaction rates. Theoretical chemists create explanations for the evidence gathered by the empirical chemists. The empirical and theoretical chemists may work as a team in the same university or may be around the world from each other. Likewise, the theory might be developed ahead of the evidence. The empirical chemist may have the role of testing the theory created by the theoretician. So what concepts did the theoretical chemists come up with?

Simulation—Collisions and Reactions

View the animations to see models of effective and ineffective collisions. Draw models to represent the orientation of the atoms for both types of collision.

www.science.nelson.com

Practice

1. What two conditions must be met for two molecules to react?
2. Provide one scientific reason and one technological reason for understanding how chemical reactions occur.
3. Consider a thought experiment in which nitrogen and oxygen gases are mixed, as in air. No spontaneous reaction occurs under normal conditions or even by supplying high temperature and pressure. Then consider a lightning strike, which provides evidence for the production of nitrogen oxides. The evidence for this reaction is so strong that scientists list this reaction as part of the nitrogen cycle.
   (a) Write a balanced chemical equation for the reaction of nitrogen and oxygen to produce nitrogen dioxide, including the enthalpy change.
   (b) Draw and label a chemical potential energy diagram to express the enthalpy change for this reaction.
   (c) Create a hypothesis to explain why lightning is necessary to initiate this reaction.

DID YOU KNOW?

Creating a Concept
Svante Arrhenius, in 1889, created the theoretical concept of activation energy to explain his laboratory work involving the effect of temperature on rates of reaction. This followed his theory of ionic dissociation (1884) and preceded his hypothesis that carbon dioxide variations in the atmosphere may have caused the extinction of dinosaurs and the advent of ice ages. Of course, you know about his contributions to acid–base theory.

Activation Energy of a Reaction

Since empirically measured reaction rates are often relatively slow, and in many cases too slow to be detectable (as in the nitrogen–oxygen reaction of air), chemists are forced to interpret the evidence as showing that normally only an extremely tiny fraction of the collisions actually produce products. The theoretical explanation for this evidence involves the concept of **activation energy**—an energy barrier that must be overcome for a chemical reaction to occur. Entities must reach this minimum energy before they can react. The input energy (which supplies the activation energy) may be in the form of heat, light, or electricity.

The concept of chemical activation energy, E_a, can be illustrated by an analogy with gravitational potential energy. Consider the analogy of a billiard ball rolling on a smooth track shaped as shown in **Figure 3**. The ball leaves point A moving to the right. As it travels on the uphill portion of the track it slows down—as kinetic energy is converted to gravitational potential energy. The ball can only successfully overcome the rise of the track to reach point B if it has a large enough initial speed (kinetic energy). We could call this situation an *effective* trip. The minimum kinetic energy required is analogous to the *activation energy* for a reaction. If the ball does not have enough kinetic energy, it will not reach the top of the track and will just roll back to point A. This is like two molecules colliding without enough energy to rearrange their bonds—they just rebound elastically.

Note that a ball that returns to point A will have the same kinetic energy it began with, but a ball that makes it to point B will have more kinetic energy (but less potential energy) because it will be moving faster. The example above is also analogous to the enthalpy change for an exothermic reaction. The enthalpy change (net chemical energy change) results in energy being immediately released to neighbouring entities. These entities then move faster, collide with more energy, and are more likely to react. For example, the energy released when the first few molecules of hydrogen and oxygen react (initiated by a spark or flame) is quickly transferred to other molecules, allowing the reaction to proceed unaided by external sources of energy (**Figure 4**). The reaction, once begun, is self-sustaining as long as enough reactants remain to make collisions likely. Exothermic reactions, once begun, often drive themselves.

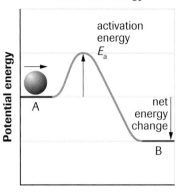

Figure 3
On a trip from A to B, there is a net decrease in potential energy, but there must be an initial increase in potential energy (activation energy) for the trip to be possible.

Figure 4
The self-sustaining exothermic reaction of hydrogen and oxygen requires only a single spark to provide the initial energy of activation and start the reaction. This is useful when desired, as in the operation of the space shuttle main engines **(a)**, and disastrous in other cases, as in the explosion of the German airship *Hindenburg* in Lakehurst, New Jersey, in 1937 **(b)**. The exothermic reactions generate a continuous supply of energy to sustain the reaction.

Consider the reaction of carbon monoxide with nitrogen dioxide, plotted as potential energy of the molecules versus progress of reaction; that is, the progress over time of the molecular activity that constitutes the reaction (**Figure 5**, page 528).

$$CO(g) + NO_2(g) \rightarrow CO_2(g) + NO(g) \qquad \Delta_r H° = -224.9 \text{ kJ}$$

In Figure 5, the molecular collision follows an energy (or reaction) pathway along the plot from left to right. The *energy pathway* is the relative potential energy of the chemical system as it moves from reactants through activated complex to products. The *activated complex* is the chemical entity containing the collided reactants. Along the flat

Figure 5
The energy pathway and the models represent a theoretical explanation of the reaction. The activation energy explains both the initiating energy required and the rate for this particular reaction.

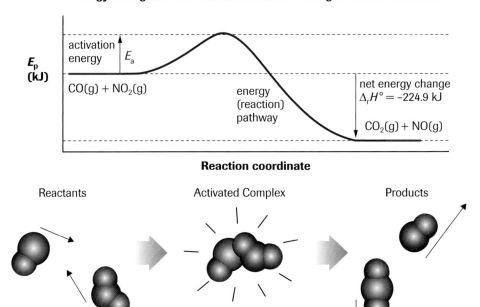

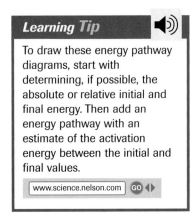

Learning Tip

To draw these energy pathway diagrams, start with determining, if possible, the absolute or relative initial and final energy. Then add an energy pathway with an estimate of the activation energy between the initial and final values.

www.science.nelson.com

region on the left, the molecules are moving toward each other, but are still distant from each other. As the molecules approach more closely, they are affected by repulsion forces and begin to slow down. Some of their kinetic energy is changed to chemical potential energy, and is stored as a repelling electric field between them. If the molecules have enough kinetic energy—which means more energy than is required to get to the energy level of the activated complex—they can approach closely enough for their bond structure to rearrange. The activated complex occurs at the maximum potential energy point in the change along the energy pathway.

In the carbon monoxide–nitrogen dioxide example, the energy initially absorbed to break the nitrogen–oxygen bond is less than the energy released when a new carbon–oxygen bond forms. Repulsion forces push the product molecules apart, converting potential energy to kinetic energy. If the overall energy change is obtained by comparing reactants and products at the same temperature, the difference is observed as a net release of energy to the surroundings. If the energy change is measured at constant pressure (the usual situation for a reaction open to the atmosphere), it is called the enthalpy change, $\Delta_r H$, which is -224.9 kJ at standard conditions for this particular (exothermic) example (**Figure 6**).

+ EXTENSION

Reaction Coordinate Diagrams
This simulation will take you a little further in your understanding of communicating energy changes during chemical reactions.

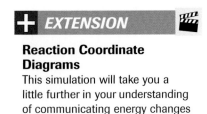

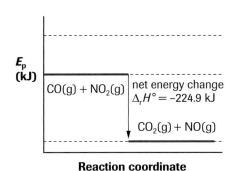

Figure 6
Chemical potential energy diagram for the carbon monoxide–nitrogen dioxide reaction

Chemists infer that, if a large quantity of energy is needed to start a reaction and if the reaction progresses relatively slowly, then the activation energy is large. A spontaneous reaction at room temperature and a higher rate of reaction is interpreted as a relatively small activation energy.

Recall the chemical potential energy diagrams showing enthalpy changes in Chapter 11 and Figure 6. The new energy pathway diagrams (such as Figure 5) show the energy pathway between the initial and the final energy states. You can think of the new energy pathway diagrams as being an expanded form of chemical potential energy diagram, with the approximate energy of the activated complex also represented.

The carbon monoxide–nitrogen dioxide reaction is exothermic: the potential energy of the products is less than that of the reactants. If the reverse were true—if the potential energy of the products were greater than that of the reactants—the reaction would be endothermic. A continuous input of energy, usually heat, would be needed to keep the reaction going, and the enthalpy change would be positive (**Figure 7**).

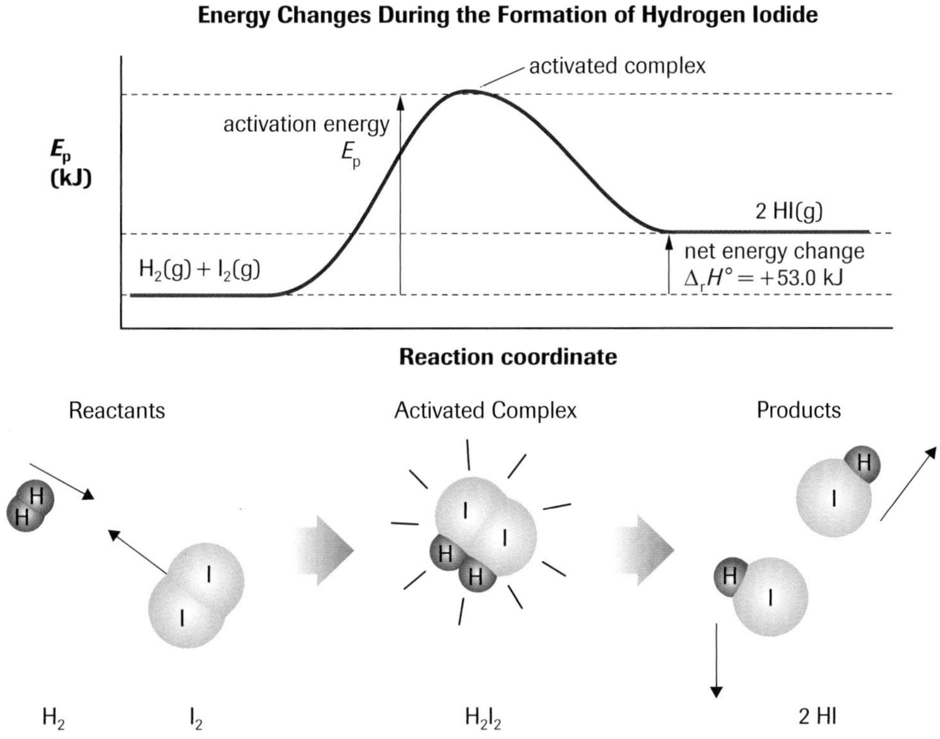

Learning Tip

In Chapter 11 you drew mostly chemical potential energy diagrams with E_p on the y-axis (with no values). You also saw some enthalpy change diagrams with $\Delta_f H$ and numerical values on the y-axis. For energy pathway diagrams, E_p (with no values) is a more appropriate axis label because it is not possible to measure an enthalpy change, under standard conditions, for reactants to activated complex, and from activated complex to products.

Figure 7
In this example the graph and the models represent the theory that chemists use to explain the "kick-start" that a reaction needs and the continuing rate of reaction afterwards.

 **WEB** *Activity*

Simulation—Collision–Reaction Theory

View a simulation to review some key points of the collision–reaction theory including activation energy and enthalpy changes.

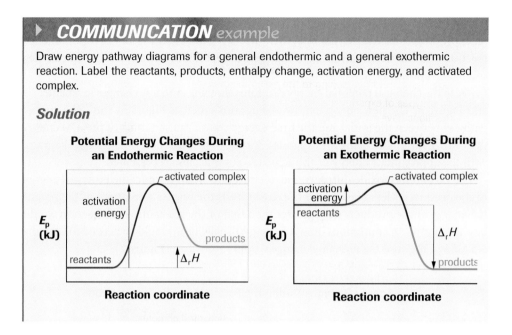

> ▶ **COMMUNICATION** *example*
>
> Draw energy pathway diagrams for a general endothermic and a general exothermic reaction. Label the reactants, products, enthalpy change, activation energy, and activated complex.
>
> **Solution**

Activation Energy and Ozone

Research indicates that the energy of a photon of electromagnetic radiation is inversely proportional to the wavelength. The longer the wavelength, the lower the energy of the radiation. Ozone (O_3) is formed and destroyed naturally in the stratosphere by ultraviolet (UV) radiation from our sun providing the activation energy. Ozone is formed by converting molecular oxygen (O_2) into atomic oxygen (O). The atomic oxygen then reacts with molecular oxygen to produce ozone:

$$O_2(g) + \sim 240 \text{ nm UVC} \rightarrow O(g) + O(g)$$

$$O(g) + O_2(g) \rightarrow O_3(g)$$

The ozone molecule is then split when it absorbs skin-cancer causing UVB radiation.

$$O_3(g) + 240\text{--}320 \text{ nm UVB} \rightarrow O_2(g) + O(g) \quad \text{(which then reacts to replenish the ozone)}$$

The energy supplied by the sun's UV radiation is a measure of the activation energy for these reactions; that is, the activation energy (from UVB) for destroying ozone is lower than the activation energy (from UVC) for converting molecular oxygen to atomic oxygen. In the laboratory chemists can perform controlled experiments with electromagnetic radiation to determine the activation energy for reactions.

DID YOU KNOW ?

Ozone Depletion Hypothesis

Over thirty years of research supports the ozone depletion hypothesis. Parallel chemical research efforts show that the natural concentration of ozone in the stratosphere is being reduced by technology-produced chemicals. For example, air conditioner and fire-fighting CFCs (chlorofluorocarbons) are decomposed by radiation to produce atomic chlorine. This atomic chlorine reacts with ozone to reduce the ozone concentration. Evidence for this reaction is the existence of ClO(g) in the stratosphere.

Web Quest—Neurotransmitters and Nerve Agents

This Web Quest requires you to actively research how chemistry could be used in a dangerous and deadly way. Work as a team to develop informational posters for emergency response personnel, outlining how nerve agents work and how to treat individuals who have been exposed to nerve agents. You will explain the normal functions of nerves, referring to the action of biological catalysts. You can also research how nerve agents affect nerve impulses to cause symptoms, and outline appropriate treatment.

Section 12.1 Questions

1. Describe, in your own words, what happens in a successful reaction of two molecules. Include as many key points as possible from the beginning to the end of the reaction.
2. What types of experiments led to the hypothesis that reactions have an activation energy?
3. (a) Is the reaction in **Figure 8** endothermic or exothermic?
   (b) What do (i) and (ii) represent?
   (c) Explain the energy changes that occur at different points during the reaction progress, referring to potential and kinetic energy.

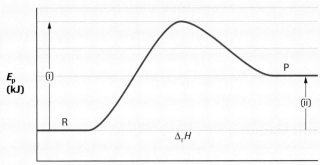

Figure 8
An energy pathway diagram

4. Enthalpy change and activation energy are two important concepts used when describing a chemical reaction. How are these two terms similar, and how are they different?
5. The combustion of hydrogen to produce water is a well-studied chemical reaction. Hydrogen may become one of the most important fuels in the future—for heat, transportation, and electricity. If possible, present your responses to the following questions electronically.
   (a) What theoretical conditions are necessary for this reaction to occur?
   (b) Draw an enthalpy change diagram for the combustion of hydrogen to form liquid water. Include actual enthalpy of formation values. Use whole-number coefficients in the chemical reaction equation.
   (c) Assuming a one-step reaction, draw an energy pathway diagram.
   (d) Label the reactants, products, enthalpy change, activation energy, and activated complex.

Use this information to answer questions 6 and 7.

You may have seen a video or a demonstration of the relative rates of reaction of the alkali metals—sodium, potassium, and cesium—with water (**Figure 9**). This evidence suggests that there is an increasing rate of reaction from sodium through cesium.

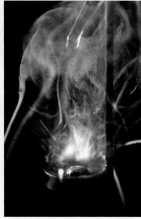

Figure 9
Different reactants, like the alkali metals placed in water, have a different rate of reaction. Sodium fizzes while potassium bursts into flames. Cesium is even more reactive.

6. Complete the Purpose, Problem, Hypothesis, Prediction, Design, Materials (including cautions), and Procedure for an investigation to test the hypothesis that the activation energy decreases from sodium to cesium.
7. Draw three energy pathway diagrams with relative activation energies, for the reactions.

Extension

8. Aboriginal peoples used fire for many purposes. What technologies did they use to start a fire?

   www.science.nelson.com

9. List the ignition temperature of a few combustible materials that interest you. Write a brief theoretical explanation for the different ignition temperatures.

   www.science.nelson.com

10. For thousands of years, peoples living in cold climates have taken advantage of the low temperatures to help them preserve food.
    (a) Apply the concepts learned in this section to create a hypothesis to explain why food lasts longer at low temperatures.
    (b) Research a technology that was developed to capitalize on the benefits of cold storage. Consider the risks and benefits of this technology.
    (c) Create an illustrated article entitled "Cold Storage: Theory and Application" suitable for publication in a popular science magazine or e-zine.

    www.science.nelson.com

12.2 Bond Energy and Reactions

Energy transfer is an important factor in all chemical changes. Exothermic reactions, such as the combustion of gasoline in a car engine and the metabolism of fats and carbohydrates in the human body (**Figure 1**), release energy into the surroundings. Endothermic reactions, such as photosynthesis (**Figure 2**) and the decomposition of water into hydrogen and oxygen, remove energy from the surroundings. Knowledge of energy and energy changes is important to society and to industry; the study of energy changes provides chemists with important information about chemical bonds.

Just as glue holds objects together, electrical forces hold atoms together. In order to pull apart objects that are glued together, you have to supply some energy. Similarly, if atoms or ions are bonded together, energy (in the form of heat, light, or electricity) is required to separate them. In other words, bond breaking—when two entities (such as atoms or ions) move apart—requires energy.

bonded particles + energy → separated particles

In contrast, bond making—when two entities move together—releases energy.

separated particles → bonded particles + energy

Figure 1
In the human body, the metabolism of fats and carbohydrates is an exothermic reaction. Athletes like Jordin Tootoo and Todd Simpson need to "fuel up" before competing.

By analogy it requires energy to lift an object away from Earth, while energy is released when an object is dropped to Earth.

The stronger the bond holding the particles together, the greater the quantity of energy required to separate them. **Bond energy** is the energy required to break a chemical bond. It is also the energy released when a bond is formed. Even the simplest of chemical reactions involves the breaking and forming of several individual bonds. The terms *endothermic* and *exothermic* are empirical descriptions of overall changes that scientists explain by their knowledge of bond changes.

Some reactions might involve several steps, with several unstable intermediate products along the way. For simplicity, however, we will just consider the energy changes between the initial reactants and the final products.

Endothermic Reactions

Consider the decomposition of water:

$$2\,H_2O(l) \rightarrow 2\,H_2(g) + O_2(g) \qquad \Delta_r H° = +571.6 \text{ kJ}$$

In this reaction, hydrogen–oxygen bonds in the water molecules must be broken, and the hydrogen–hydrogen and oxygen–oxygen bonds must be formed.

Since the overall change is endothermic, the energy required to break the O—H bonds must be greater than the energy released when the H—H and O=O bonds form. In any endothermic reaction, more energy is needed to break bonds in the reactants than is released by bonds formed in the products.

Exothermic Reactions

For exothermic reactions, the opposite is true. More energy is released by bonds formed in the products than is needed to break bonds in the reactants. The reaction between hydrogen and chlorine (**Figure 3**) illustrates the energy of bond breaking and bond making. Although an activated complex may be involved in the transition from bond breaking to bond making, for simplicity, we will only look at the individual bonds that are changing. Energy is required to break the bonds in hydrogen molecules (H_2) to create hydrogen

Figure 2
In some Aboriginal cultures, the visual effects of electricity are seen as energy transfers. In Western science, we describe similar energy transfers. Plants, such as this fiddlehead fern, grow by increasing the complexity and number of electron bonds between atoms in the process of photosynthesis. When we eat food, we break down the chemical bonds and reassemble them, resulting in energy being released, the growth of cells, and the production of heat. Our nerves and brain function all occur because of the transfer of energy by electrons.

Section 12.2

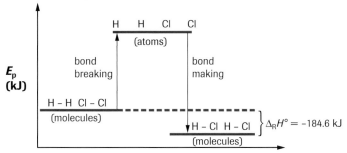

Figure 3
Energy is absorbed in order to break the H–H and Cl–Cl bonds, but more energy is released when the H–Cl bonds form. The overall result is an exothermic reaction.

atoms (H). The two hydrogen atoms have higher chemical potential energy than a hydrogen molecule. Likewise, the chlorine atoms have higher potential energy than the chlorine molecules. When the hydrogen and chlorine atoms make bonds to create hydrogen chloride molecules, energy is released. Since this reaction is exothermic, the logically consistent explanation is that more energy is released by bond making than is required for bond breaking. We can communicate this explanation using a potential energy diagram.

The explanation above assumes that sufficient energy is initially added to equal or exceed the required activation energy. This energy may be provided in a variety of forms: heat, light, or electrical. In all cases, the exothermic nature of this reaction means that heat is transferred to the surrounding molecules, thus increasing their kinetic energy.

Bond energies are the fourth method that you have encountered for predicting or explaining a change in enthalpy for a chemical reaction. Although there are quantitative values for bond energies, and scientists can predict a change in enthalpy from bond energies, this is not part of our study in this course. The methods that you have studied for predicting and/or explaining a change in enthalpy are the following:

1. calorimetry: the change in enthalpy equals the change in thermal energy
2. Hess' law: the change in enthalpy equals the sum of component enthalpy changes
3. molar enthalpies of formation: the change in enthalpy equals the enthalpies of formation of the products minus the enthalpies of formation of the reactants
4. bond energies: the change in enthalpy equals the energy released from bond making minus the energy required for bond breaking

All of these methods are logically consistent with each other: they all support each other, moving from more empirical to more theoretical. Every method also yields the same answer to the question "What is the change in enthalpy for a particular reaction?"

DID YOU KNOW?

Unanswerable Questions
In chemistry we use the concept that energy can neither be created nor destroyed. Biology has a similar concept: that all cells come from preexisting cells. These concepts are not intended to address the ultimate question of where the original energy (or cells) came from. They do, however, give us a starting point for our studies. Western science asks specific questions, often about isolated parts of nature. Other cultures look at these big questions in other ways. Aboriginal cultures, for example, often take a more holistic approach to their investigation of nature and natural processes.

+ EXTENSION

Bond Length and Bond Energy
Using a specific example of bond length versus bond energy, this audio clip will enrich your understanding of bond energy as it relates to bond length and multiple bonds.

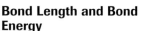

SUMMARY   Bond Energy and Enthalpy Changes

- Bond energy is the energy required to break a chemical bond; it is also the energy released when a bond is formed.
- The change in enthalpy represents the net effect from breaking and making bonds.
  $\Delta_r H =$ (energy released from bond making) $-$ (energy required for bond breaking)
  Exothermic reaction: making $>$ breaking ($\Delta_r H$ is negative.)
  Endothermic reaction: breaking $>$ making ($\Delta_r H$ is positive.)

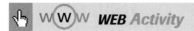 **WEB** Activity

Canadian Achievers—John Polanyi

Dr. John C. Polanyi (**Figure 4**), Canadian Nobel laureate, has refined the empirical and theoretical descriptions of molecular motions during chemical reactions and has published many scientific papers on this subject, known as *reaction dynamics*. He predicted the conditions required for the operation of a chemical laser, and saw his prediction verified by subsequent experiments.

"Those of us who are scientists, and those of us who are not, should rid ourselves of the absurd notion that science stands apart from culture. Today, science permeates our lives—our doing and our thinking."

John Polanyi

1. A laser is a technology that allowed new science to be done. Find some examples of chemistry that were allowed due to the invention of the laser.
2. If a laser is chosen by a chemist to initiate and/or maintain a chemical reaction, what does this say about the activation energy of the reaction?
3. What else is Polanyi well known for, besides his chemistry research?

www.science.nelson.com

Figure 4
John Polanyi (1929–) of the University of Toronto is a Nobel laureate in chemistry.

Section 12.2 Questions

1. Write two definitions of bond energy.
2. Explain how bond energies are related to the activation energy for a reaction.
3. Why must chemical reactions include both the breaking of bonds and the forming of bonds?
4. Draw a chemical potential energy diagram (similar to Figure 3 on page 533) for the decomposition of liquid water into hydrogen and oxygen gases. Illustrate the energy change for the breaking of water molecules into atoms and then for the subsequent formation of hydrogen and oxygen molecules.
5. (a) Explain why the decomposition of water is an endothermic process.
   (b) Explain why the burning of hydrogen is an exothermic process.
6. Predict whether the reaction of hydrogen with bromine is endothermic or exothermic. Making reference to bond energies, explain your prediction.
7. In the Exploration activity (page 523) you looked at the reactions of metals with hydrochloric acid, and in Investigation 11.3 (pages 507 and 517) you determined the enthalpy change for the reaction of magnesium with hydrochloric acid.
   (a) Draw a chemical potential energy diagram, including the enthalpy change, for the reaction of magnesium metal with hydrochloric acid.
   (b) Using the net ionic equation for this reaction, list the bonds that must be broken in the reactants and the bonds that must be formed in the products.
   (c) Suggest a possible theoretical hypothesis to explain the empirical differences in the rates of reaction of zinc, magnesium, and iron with hydrochloric acid. Identify any assumptions you are making.

8. The reaction of hydrogen with chlorine at room temperature, in the absence of light, is undetectable owing to the very slow rate of reaction. The same reaction is explosively fast, however, if exposed to sunlight. The following mechanism has been suggested for this reaction.

   $Cl_2(g)$ + light energy → $Cl(g) + Cl(g)$
   $Cl(g) + H_2(g) → HCl(g) + H(g)$
   $H(g) + Cl_2(g) → HCl(g) + Cl(g)$
   $Cl(g) + Cl(g) → Cl_2(g)$

   (a) Write the net overall reaction equation.
   (b) Identify the intermediate entities that are formed during the step-by-step reaction process.
   (c) Discuss the activation energy for the collision of molecular chlorine with molecular hydrogen, and for the collision of atomic chlorine with molecular hydrogen. Which reaction has the greater activation energy, and what evidence can you use to support your argument?

9. The reaction of hydrogen and oxygen (see Figure 4, page 527) is exothermic and self-sustaining. Write the equation for this reaction, and provide a reason why it is not likely that the reaction occurs as a single step.

Extension

10. Ozone formation and destruction are natural bond-making and bond-breaking processes in the stratosphere. (See page 530.)
    (a) What is the source of the energy for breaking bonds?
    (b) Why is shorter wavelength radiation required to break the oxygen bond(s) than to break the ozone bond(s)?
    (c) If energy is produced by making ozone from molecular and atomic oxygen, why can't this energy be used to break more molecular oxygen into atomic oxygen in the stratosphere?

Catalysis and Reaction Rates 12.3

Empirical Effect of Catalysis

Catalysis deals with the properties and development of catalysts, and the effects of catalysts on the rates of reaction. A **catalyst** is a substance that increases the rate of a chemical reaction without being consumed itself in the overall process. The chemical composition and amount of a catalyst are identical at the start and at the end of a reaction. A catalyst reduces the quantity of energy required to start the reaction, and results in a catalyzed reaction producing a greater yield in the same period of time (even at a lower temperature) than an uncatalyzed reaction. Research indicates that the use of a catalyst does not alter the net enthalpy change for a chemical reaction. In green plants, for example, the process of photosynthesis can take place only in the presence of the catalyst chlorophyll (**Figure 1**). Most catalysts significantly accelerate reactions, even when present in very tiny amounts compared with the amount of reactants present.

The action of catalysts seriously perplexed early chemists, who had problems with the concept of something obviously being involved in a chemical reaction but not being changed by that reaction. Effective catalysts for reactions have almost all been discovered by purely empirical methods—trying everything to see what worked.

Chemists learned early that metals prepared with a large surface area (powder or shavings) catalyze many reactions, including the decomposition of hydrogen peroxide in the presence of a platinum catalyst. Many people who wear contact lenses know about this reaction. They use it daily to clean their lenses.

A common consumer example of catalysis today is the use of platinum, palladium, and rhodium in catalytic converters in car exhaust systems (**Figure 2**). These catalysts speed the combustion of the exhaust gases so that a higher proportion of the exhaust will be the relatively harmless, completely oxidized products. Catalysts are extremely important in chemical technology and industry because they allow the use of lower temperatures. This not only reduces energy consumption but also prevents the decomposition of reactants and products and decreases unwanted side reactions. The result is an increase in the efficiency and economic benefits of many industrial reactions. For many industrial processes the difference between success and failure depends on the use of catalysts because they make the reactions fast enough to be profitable. (For an example of this, see The Haber Process case study, in Chapter 8.)

Compounds that act as catalysts in living systems are called enzymes. **Enzymes** are usually extremely complex molecules (proteins). A great many physiological reactions, such as metabolism, are actually controlled by the amount of enzyme present. Enzymes are also of great importance for catalyzing reactions in the food, beverage, cleaner, and pharmaceutical industries.

Figure 1
Green plants contain chlorophyll, which acts as a catalyst in a reaction that splits water molecules into oxygen and hydrogen ions. This reaction is the first step in the process of photosynthesis, which converts carbon dioxide and water into glucose and oxygen.

Figure 2
Many commercial and industrial catalysts are heterogeneous, which means that they provide a solid surface upon which the reactants can be absorbed and reacted. The catalytic converter in a car is a good example of a heterogeneous catalyst. The inner surface of the catalytic converter is coated with an alloy containing platinum, rhodium, and palladium.

▶ Practice

1. Do technologists normally use empirical or theoretical knowledge when doing work that involves catalysts?
2. List some technological inventions that involved the use of catalysts.
3. List some natural (biological) processes that involve the use of catalysts.
4. Enzymes in your body are generally present in extremely small quantities, but any substances that affect your enzymes are almost always very toxic and dangerous. Explain why this should be so, referring to reaction rates in your explanation.

INVESTIGATION 12.1 Introduction

Iodine Clock Reaction

Report Checklist
- ○ Purpose
- ○ Problem
- ○ Hypothesis
- ○ Prediction
- ● Design
- ● Materials
- ● Procedure
- ● Evidence
- ● Analysis
- ● Evaluation (1, 3)

Technological problem solving often involves a systematic trial-and-error approach that is guided by knowledge and experience. Thomas Edison, the greatest inventor of all time, used this method, for example, to invent the incandescent light bulb and when making batteries. Usually, one variable at a time is manipulated. In this investigation, a variety of potential catalysts are added one at a time to a reaction mixture. The rate of the reaction, as determined by the time required to get to a blue endpoint, is the responding variable.

Use MSDS to find cautions. Include safety and disposal information at the end of the Materials section.

Purpose
The purpose of this investigation is to create a generalization about the relative effect that potential catalysts have on the rate of a chemical reaction.

Problem
What is the relative effect of the chemicals tested as potential catalysts for increasing the rate of the iodine clock reaction?

To perform this investigation, turn to page 543.

DID YOU KNOW ?

Bread Making
Bread making is an example of chemical kinetics. Yeast (one-celled fungi) is added to produce enzymes, which catalyze several reactions to produce, for example, carbon dioxide. Gases cause the bread to rise and be less dense.

Figure 3
Bread machines are designed to allow the reactions to take place at the optimum temperatures, which change throughout the process.

DID YOU KNOW ?

Inhibitors
Some substances are known to slow or stop reactions. Chemists and biochemists believe that these inhibitor substances act differently from catalysts, controlling the reaction rate in a completely different way. Preservatives are inhibitors added to foodstuffs to delay deterioration.

Theoretical Explanation of Catalysis

Chemists believe that catalysts accelerate a reaction by providing an alternative lower-energy pathway from reactants to products. That is, a catalyst allows the reaction to occur by a different activated complex, but resulting in the same products overall. If the new pathway has a lower activation energy, a greater fraction of molecules possess the minimum required energy and the reaction rate increases. Since the activation energy is lowered by exactly the same amount for the reverse reaction, the rate of any reverse reaction increases as well (**Figure 4**).

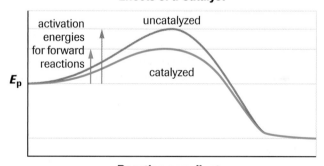

Figure 4
The catalyzed pathway has a lower activation energy, so more collisions lead to a successful reaction.

Scientists do not really understand the actual mechanism by which catalysis occurs for most reactions, and discovering effective catalysts has traditionally been an empirical process involving trial-and-error. Chemists have studied a few catalyzed reactions in detail, so they believe they understand the changes involved. Most of the catalysts (enzymes) for biological reactions work by shape and orientation. They fit substrate proteins into locations on the enzyme as a key fits into a lock, enabling only specific molecules to link or detach on the enzyme, as shown in **Figure 5**.

Section **12.3**

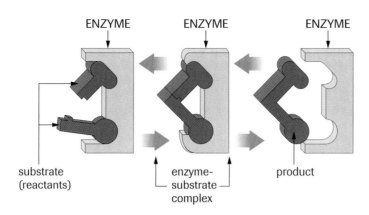

Figure 5
Note that the enzyme shape is specific for the reactant (substrate) molecules. Almost all enzymes catalyze only one specific reaction.

One well-understood non-biological reaction is the decomposition of methanoic (formic) acid in aqueous solution. At room temperature this reaction is very slow, with no noticeable activity (**Figure 6**). When strong acid is added, the solution begins to bubble. Testing indicates that carbon monoxide gas is being produced. If this reaction proceeds until all the formic acid has been consumed, the solution will still contain the same quantity of acid as was initially added. It does not seem to matter which strong acid is added, so we assume that the acting catalyst is the aqueous hydrogen ion, which is common to all aqueous acidic solutions. **Figure 8** shows the energy pathway diagram of the reaction with a catalyst. Note that, like all catalysts, the H^+ is regenerated and, therefore, is not consumed during the reaction. In a catalyzed reaction, catalysts react with one or more of the reactants, but then are regenerated by the end of the reaction.

DID YOU KNOW ?

Hot Beetle Juice
The bombardier beetle of South America has a unique way of protecting itself. When it is threatened, it pumps stored hydroquinone and hydrogen peroxide from one chamber in its body into another chamber. In the second chamber the enzymes catalase and peroxidase mix with the stored compounds. The resulting catalyzed exothermic reaction produces a hot, corrosive liquid that is sprayed at the attacker.

Figure 7
The beetle's spray can cause a skin burn that lasts for several days.

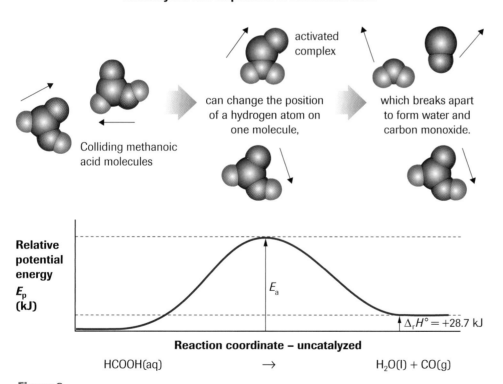

Figure 6
The uncatalyzed reaction proceeds too slowly to notice at room temperature. Note that this reaction is endothermic, requiring an overall increase in (input of) energy.

EXTENSION

CBC radioONE
QUIRKS & QUARKS

Bombardier Beetles
A British professor of thermodynamics is researching the bombardier beetle's squirting ability, with the aim of developing a device that will relight extinguished jet engines.

www.science.nelson.com

EXTENSION

Catalysts and Reaction Mechanisms

With more details on the steps during catalysis, this audio clip will enrich your understanding of how catalysts can alter a reaction mechanism and hence speed up a reaction.

www.science.nelson.com GO

DID YOU KNOW

Communication Through Journals

The German chemist Justus von Liebig wrote in 1834: "Chemical literature is not to be found in books, it is contained in journals." Refereed journals are regularly published reports of research that have passed a panel of judges. (See Appendix B.4.) For example, Johann Döbereiner discovered in 1823 that platinum catalyzes the burning of hydrogen. Within three months this discovery was reported in about a dozen European science journals.

DID YOU KNOW

Chemical Marriage

The Chinese symbol for *catalyst* is the same as that used for *marriage broker*.

觸媒

Catalyzed Decomposition of Methanoic Acid

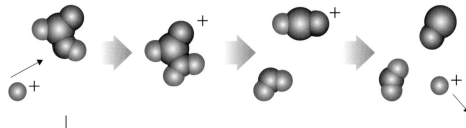

Formic acid, HCOOH, can also react with an aqueous H^+ ion

to form the positive intermediate ion product $HCOOH_2^+$,

which reacts to form water, H_2O, and the intermediate ion product HCO^+.

Finally, the HCO^+ reacts to form CO and to *regenerate* the H^+ ion.

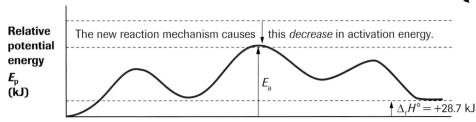

(overall) $HCOOH(aq) \xrightarrow{H^+(aq)} H_2O(l) + CO(g)$

Figure 8

The catalyzed reaction proceeds rapidly enough to produce bubbles of carbon monoxide gas at room temperature. A higher fraction of the molecular collisions become effective because the activation energy is lower. Catalysts are normally written above the reaction arrow in a reaction equation.

Catalysis and the Nature of Science

The practice of science uses two important kinds of reasoning—inductive and deductive. *Inductive reasoning* involves extending specific examples to obtain a general statement; for example, using the evidence from an experiment to form a hypothesis in the analysis section of an investigation report. *Deductive reasoning* involves applying a general concept such as a theory, law, or generalization to obtain (deduce) a specific instance.

Chemists use indirect and direct evidence to hypothesize the reaction pathway for uncatalyzed and catalyzed reactions. They create hypotheses for the structures of the activated complexes, and hypotheses for the individual reaction steps in an overall reaction. A *reaction mechanism* describes the individual reaction steps and the intermediates formed during the reaction, starting with reactants and finishing with products. **Intermediates** are chemical entities that form with varying stability at the end of a step in a reaction mechanism. They are more stable (have lower potential energy) than the activated complexes, but less stable than the reactants and the products. The intermediate then reacts in a subsequent step and does not appear in the final reaction mixture. For example, a three-step reaction mechanism (Figure 8) can be generalized as follows:

Step 1: reactant(s) → activated complex 1 → intermediate 1
Step 2: intermediate 1 → activated complex 2 → intermediate 2
Step 3: intermediate 2 → activated complex 3 → product(s)

The inductive reasoning of using the evidence from an experiment to hypothesize a reaction mechanism is usually accompanied by deductive reasoning to test the logic of the mechanism. Chemists constantly ask themselves if a hypothesis makes sense based upon all the evidence they have collected and all of their chemical experience. If the mechanism has internal consistency, all the intermediates and reaction steps seem to be correct/logical. If the mechanism has external consistency, none of the thinking behind the proposed mechanism contradicts accepted theories, laws, or generalizations.

For example, evidence from sophisticated technologies (such as spectrometers) suggests that there are two intermediates in the catalyzed decomposition of methanoic acid. The spectrographs indicate that the chemical formulas of the intermediates are $HCOOH_2^+(aq)$ and $HCO^+(aq)$. The same equipment provides evidence that the products are water and carbon monoxide. Other diagnostic tests on the final products provide supporting evidence.

A consistency check of the mechanism indicates that all chemical equations must be balanced and that all reactants, intermediates, and products can be accounted for by writing balanced chemical equations. Furthermore, the chemical formulas of all chemicals described are consistent with chemical formulas of known chemicals and/or with the bonding capacity of each atom in each entity. From an energy perspective, the law of conservation of energy seems to be obeyed, the chemical potential energies of the entities involved are consistent with known values, and the enthalpy change remains the same as for the uncatalyzed reaction. It all fits together in a nice logical package. There are no contradictions, no inconsistencies.

Uses of Catalysts
The Oil Industry

The oil industry uses catalysts in the cracking and reforming of crude oil and bitumen to produce more marketable fractions (such as gasoline). The science and engineering of crude oil refining has advanced tremendously, due mostly to the use of catalysts that increase the rate of the reaction while decreasing the energy (which often means decreasing the temperature) required for the chemical process. **Table 1** lists the typical catalysts for cracking and reforming of crude oil.

Table 1 Catalysts Used in the Oil Industry

	Process	Description	Typical catalysts
cracking	fluid catalytic cracking (high temperature)	heavy gas oil to diesel oils and gasoline	zeolite (silicates)
	hydrocracking (lower temperature, higher pressure, presence of H_2)	heavy oil to gasoline and kerosene	Pt(s), Pt(s)/Re(s)
reforming	alkylation	smaller to larger molecules plus increased branching	H_2SO_4(aq), HF(aq)
	catalytic reforming	naphtha to high octane hydrocarbons	Pt(s), Pd(s)

 **WEB** *Activity*

Simulation—A Catalyzed Reaction

Hydrogenation is an important industrial process. Scientific studies of hydrogenation typically start with simple reactions in order to understand the molecular process (mechanism). In this animation, you will investigate the hydrogenation of ethene using a platinum catalyst.

DID YOU KNOW?
Syrup Digesters
High-fructose corn syrup is a key ingredient in many processed foods: pop, fruit-flavoured drinks, cookies, jam, candy ... It is made by treating starch (extracted from corn) with enzymes to break it down into glucose monomers. Glucose, however, is not as sweet as the sucrose (table sugar) that has traditionally been used to sweeten foods. Fortunately, there are enzymes that convert some of the syrup's glucose into super-sweet fructose. It is this combination of glucose and fructose that is so widely used as corn syrup. The mixture can even be tailored to the needs of the food and beverage industries. While 42% fructose syrup is perfectly satisfactory for many foods, a sweeter syrup is required by the soft drink industry: up to 55% fructose.

Learning Tip
You are not expected to memorize which catalyst goes with which chemical process. However, you should know that major reactions, for example, cracking, reforming, and hydrodesulfurization, use catalysts. In fact, the message is that most major chemical processes in industry involve catalysis.

DID YOU KNOW?
Gasoline Production
Gasoline is the most marketable of all products from crude oil. In the 1920s about one-quarter of a barrel of oil could be converted to gasoline. Due to the use of modern catalysts, about half of a barrel of oil can now be converted to gasoline.

DID YOU KNOW ?

Greenhouse Gases and Global Warming Potentials

The major greenhouse gases, with their contribution to global warming, are as follows:

$H_2O(g)$ ... 36%–70%
$CO_2(g)$ ... 9%–26%
$CH_4(g)$ ... 4%–9%
$O_3(g)$ 3%–7%

Note that the mixing of these gases provides a range of contribution. The higher end of the range is for the gas alone.

Although accumulated carbon dioxide contributes most to global warming, its global warming potential (GWP) per unit mass is low. The relative GWPs for some chemicals (over a period of one hundred years) are as follows:

GWP	Chemical
1	$CO_2(g)$
21	$CH_4(g)$
310	$N_2O(g)$
12 000	$CHF_3(g)$
22 200	$SF_6(g)$

CAREER CONNECTION

Field Production Operator

The primary responsibility for field production operators is to remove impurities from the oil, gas, or both before transporting to refineries and markets. They may also ensure that the wells and flowlines are maintained and efficient, and that safety regulations are enforced.

www.science.nelson.com

Upgrading of Bitumen from Oil Sands

Oil sand is about 84% bitumen, over 90% of which is recovered from the sand. Over 93% of the Alberta oil sands are located too deep for surface mining and must be recovered *in situ*, that is, the bitumen is recovered from the oil sand without removing the sand. Regardless of whether the oil sands are surface mined or mined *in situ*, the composition of the bitumen is about the same. Bitumen is composed of about 80% aromatic hydrocarbons, which have an average H/C ratio of less than 1. Upgraded bitumen (synthetic crude) contains mostly alkanes and small aromatics with an average H/C ratio of about 2. The ratio is increased during bitumen upgrading by coking (removing carbon) and/or hydrocracking (adding hydrogen). Hydrotreating is used to remove sulfur and nitrogen contaminants and to convert alkynes and alkenes into alkanes. Catalysts play a significant role in two of these chemical processes (**Table 2**).

Table 2 Catalysts Used during Bitumen Upgrading

Stage	Process	Description	Typical catalyst
1a. hydrocracking	cracking and hydrogenation	creates smaller molecules; increases the H/C ratio	(Ni-Mo) sulfide(s) on alumina ($Al_2O_3(s)$)
1b. and/or coking	remove carbon	creates smaller molecules; increases the H/C ratio	no catalyst
2. hydrotreating	hydrogenation to remove S and N	...S + $H_2(g) \rightarrow H_2S(g)$ + ...N + $H_2(g) \rightarrow NH_3(g)$ + ...	(Ni-Mo) sulfide(s) on alumina ($Al_2O_3(s)$)

Emissions Control

Emissions control is another use of catalysts. These emissions may be nitrogen oxides (from power plants), sulfur (from gas plants), and chemicals that contribute to smog (from internal combustion engines). **Table 3** shows some of the emission control reactions and their catalysts. (You do not need to memorize these reactions and catalysts.)

Table 3 Catalysts Used for Emissions Control

Process	Description	Typical catalysts
NO_x removal from exhaust streams of industrial heat and power generators	NO_x is selectively reduced by reacting with a reagent such as ammonia or urea; e.g., NO(g) and $NO_2(g)$ to $N_2(g)$	V(s), Ti(s), silica, and/or zeolite
sulfur removal using hydrodesulfurization (HDS)	sulfur compounds + $H_2(g) \rightarrow H_2S(g)$ $H_2S(g) \rightarrow S_8(s)$ using Claus process	$MoS_2(s)/Co(s)$ alumina
automobile catalytic converter (3-way)[1]	reduction catalyst: $2 NO(g) \rightarrow N_2(g) + O_2(g)$ $2 NO_2(g) \rightarrow N_2(g) + 2 O_2(g)$ oxidation catalyst: $2 CO(g) + O_2(g) \rightarrow 2 CO_2(g)$ $C_xH_y(g) + n O_2(g) \rightarrow x CO_2(g) + m H_2O(g)$	Pt(s)/Rh(s) Pt(s)/Pd(s)

[1] Modern catalytic converters are called 3-way converters because they use catalysts to convert three pollutants—NO_x, CO, and hydrocarbons—into less harmful substances.

As you can see from the above reactions, catalysis is an extremely important branch of chemistry and chemical engineering. Catalysts play a very important role in chemical industries—for producing chemicals that we use in our work and in our homes and for controlling environmentally harmful emissions.

Section **12.3**

Enzymes

Natural product chemists have discovered many naturally occurring catalysts. Most of these catalysts are enzymes that increase the rate of specific reactions (**Table 4**).

Chemists are now using enzymes as catalysts for the production of chemicals not found in nature, such as pharmaceuticals and agricultural chemicals. These enzymes are designed to be highly selective in the reaction each catalyzes, effective under ambient conditions, and convenient and safe to dispose.

Table 4 Natural Enzymes as Catalysts

Technological process	Description	Catalyst(s)
detergents with enzymes (widest application of enzymes today) (**Figure 9**)	hydrolyzes (breaks down) starch	amylases
	attacks cellulose fibres to remove tiny fibres and prevent pilling	cellulases
	breaks down oily and fatty stains	lipases
	degrades proteins	proteases
brewing (fermentation)	$C_6H_{12}O_6(s) \rightarrow 2\ C_2H_5OH(l) + 2\ CO_2(g)$	zymase (yeast)
cleaning contact lenses	decomposes $H_2O_2(aq)$ to $O_2(g)$, to disinfect contact lenses	catalase
high-fructose corn syrup	three enzymatic steps: liquefies corn syrup, hydrolyzes sugar, isomerizes glucose	amylase glucoamylase glucose isomerase
Natural process	**Description**	**Catalyst**
nitrogen fixation	converts nitrogen into nitrogen compounds	nitrogenase
photosynthesis	$6\ CO_2(g) + 6\ H_2O(l) \rightarrow C_6H_{12}O_6(s) + 6\ O_2(g)$	chlorophyll

Figure 9
Many laundry detergents contain enzymes to enhance their effectiveness.

INVESTIGATION 12.2 Introduction

Evidence for an Activated Complex

Evidence for an activated complex is difficult to obtain. According to reaction theory, catalyzed and uncatalyzed reactions have different activated complexes. Most activated complexes are colourless and fairly unstable. A few, however, are coloured and exist long enough to be observed.

Purpose
The purpose of this investigation is to test the theoretical concept that a catalyzed reaction involves an activated complex different from that in an uncatalyzed reaction.

Report Checklist

- ○ Purpose
- ○ Problem
- ● Hypothesis
- ○ Prediction
- ○ Design
- ○ Materials
- ○ Procedure
- ● Evidence
- ● Analysis
- ● Evaluation (1, 2, 3)

Problem
What evidence is there for the existence of a different activated complex for a catalyzed reaction when compared to an uncatalyzed reaction?

Design
An uncatalyzed reaction is observed, and then a catalyst is added to the chemical system. Evidence for the formation of a different activated complex is sought.

To perform this investigation, turn to page 543.

SUMMARY   Catalysis

- A catalyst is a substance that increases the rate of a reaction without being consumed in the overall process.
- According to theory, catalysts accelerate a reaction by providing an alternative pathway with a lower activation energy.
- A catalyst does not alter the net enthalpy change of a reaction. Both catalyzed and uncatalyzed versions of the same reaction have the same $\Delta_r H$.
- Catalysts are widely used in industry, consumer technologies, and biological processes.

Section 12.3 Questions

1. Compare two reactions with the same reactants. One reaction involves a catalyst; the other does not.
   (a) What is *the same* for both reactions?
   (b) What is *different* for both reactions?

2. Consider the following reaction mechanism, in which A, B, and E may be elements or compounds, and C, D, and F are compounds:
   (1) 2 A + B → C
   (2) C → D
   (3) D + E → F
   (a) Which entities are intermediates?
   (b) Which entities are reactants?
   (c) Which entities are products?
   (d) What is the net reaction equation?

Use this information to answer questions 3 to 5.

Consider the energy pathway diagram for the hypothetical reaction in **Figure 10**. This reaction is reversible because, under certain conditions, products will re-form the reactants.

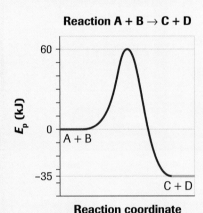

Figure 10 Energy pathway diagram for questions 3 to 5.

3. (a) What is the activation energy for the following net forward reaction?
      A + B → C + D
   (b) What is the activation energy for the following net reverse reaction?
      C + D → A + B
   (c) What is the change in enthalpy (net energy change) for the net forward reaction?
   (d) What is the change in enthalpy (net energy change) for the net reverse reaction?
   (e) Which reaction (forward or reverse) is exothermic?

4. Explain what you would expect to occur if the original collision of particles in the forward reaction has a total available kinetic energy equivalent to 55 kJ.

5. Explain what you would expect to occur if the original collision of particles in the reverse reaction has a total available kinetic energy equivalent to 55 kJ.

6. How do catalysts provide solutions to technological problems? Give at least four examples.

7. Complete the Purpose, Design, Materials, Analysis, and Evaluation (2, 3) sections of the following lab report.

   Problem
   What is the molar enthalpy of decomposition of 20 mL of 2 mol/L aqueous hydrogen peroxide without and with an iron(III) catalyst being added?

   Hypothesis
   The molar enthalpy of reaction does not change when a catalyst is added.

   Evidence
   Trial 1:
   volume of aqueous hydrogen peroxide = 20.0 mL
   volume of 0.1 mol/L sodium hydroxide = 5.0 mL
   volume of aqueous iron(III) nitrate = 0 mL
   initial temperature of hydrogen peroxide = 21.4 °C
   final temperature of hydrogen peroxide = 23.4 °C
   time = 10 min
   Trial 2:
   volume of aqueous hydrogen peroxide = 20.0 mL
   volume of 0.1 mol/L sodium hydroxide = 5.0 mL
   volume of aqueous iron(III) nitrate = 0.3 mL
   initial temperature of hydrogen peroxide = 21.4 °C
   final temperature of hydrogen peroxide = 23.4 °C
   time = 2 min

   Extension

8. Find at least two examples of enzymes and industrial catalysts. Summarize your research to these questions.
   (a) For each of the enzymes or catalysts, record
      • the reaction that is catalyzed
      • how the catalyst was discovered
      • where and how the catalyst or enzyme acts
      • for an enzyme, physiological implications of its presence or deficiency, whether such a condition exists, and, if so, how it is currently treated
      • economic implications of the industrial catalysts' use
   (b) Which of the industrial catalysts has the greatest effect on your own life? Explain.

   www.science.nelson.com

9. Catalysts make it possible for us to modify the components of crude oil to meet our many needs and wants. Despite growing research and development into alternative sources of energy, we rely on crude oil for energy more and more each year. What are the risks and benefits of relying on fossil fuels as energy sources?

10. An ad claims wonderful health benefits from a new form of water created with a catalyst. What kinds of authorities would you accept for this claim?

    www.science.nelson.com

Chapter 12 INVESTIGATIONS

INVESTIGATION 12.1

Iodine Clock Reaction

Report Checklist
- ○ Purpose
- ○ Problem
- ○ Hypothesis
- ○ Prediction
- ● Design
- ● Materials
- ● Procedure
- ● Evidence
- ● Analysis
- ● Evaluation (1, 3)

Technological problem solving often involves a systematic trial-and-error approach that is guided by knowledge and experience. Thomas Edison, the greatest inventor of all time, used this method, for example, to invent the incandescent light bulb and for inventing new batteries. Usually, one variable at a time is manipulated. In this investigation, a variety of potential catalysts are added one at a time to a reaction mixture. The rate of the reaction, as determined by the time required to get to a blue endpoint, is the responding variable.

Use MSDS to find cautions. Include safety and disposal information at the end of the Materials section.

Purpose
The purpose of this investigation is to create a generalization about the relative effect that potential catalysts have on the rate of a chemical reaction.

Problem
What is the relative effect of the chemicals tested as potential catalysts for increasing the rate of the iodine clock reaction?

INVESTIGATION 12.2

Evidence for an Activated Complex

Report Checklist
- ○ Purpose
- ○ Problem
- ● Hypothesis
- ○ Prediction
- ○ Design
- ○ Materials
- ○ Procedure
- ● Evidence
- ● Analysis
- ● Evaluation (1, 2, 3)

Evidence for an activated complex is difficult to obtain. According to reaction theory, catalyzed and uncatalyzed reactions have different activated complexes. Most activated complexes are colourless and fairly unstable. A few, however, are coloured and exist long enough to be observed.

Purpose
The purpose of this investigation is to test the theoretical concept that a catalyzed reaction involves an activated complex that is different from that in an uncatalyzed reaction.

Problem
What evidence is there for the existence of a different activated complex for a catalyzed reaction when compared to an uncatalyzed reaction?

Prediction
According to the theory of catalysis, a catalyst increases the rate of a reaction by creating an alternative activated complex with a lower activation energy. The evidence should show that the catalyzed reaction had a different activated complex than did the uncatalyzed reaction.

Design
An uncatalyzed reaction is observed, and then a catalyst is added to the chemical system. Evidence for the formation of a different activated complex is sought.

Materials
eye protection	40 mL of 0.30 mol/L sodium potassium tartrate
lab apron	
250 mL beaker	40 mL of 3% H_2O_2(aq)
watch glass for beaker	5 mL of 0.30 mol/L $CoCl_2$(aq)
10 mL graduated cylinder	hot plate
two 100 mL graduated cylinders	thermometers or temperature probes

 Sodium potassium tartrate, hydrogen peroxide, and cobalt(II) chloride may irritate skin and eyes.

 Cobalt(II) chloride is also toxic by ingestion and a possible carcinogen. Handle with care.

Procedure
1. Add 40 mL of 0.30 mol/L sodium potassium tartrate solution to the beaker followed by 40 mL of hydrogen peroxide solution.
2. Heat the solution in the beaker (covered with a watch glass) to 45 °C to 50 °C. Record any evidence of a reaction and of an activated complex.
3. Remove the beaker from the hot plate and add 5 mL of 0.30 mol/L cobalt(II) chloride as a catalyst.
4. Again, cover the beaker with the watch glass and observe the reaction.

Explaining Chemical Changes **543**

Chapter 12 SUMMARY

Outcomes

Knowledge
- analyze and label energy diagrams for a chemical reaction, including reactants, products, enthalpy change, and activation energy (all sections)
- define activation energy as the energy barrier that must be overcome for a chemical reaction to occur (12.1)
- explain the energy changes that occur during chemical reactions referring to bonds breaking and forming and changes in potential and kinetic energy (12.2)
- explain that catalysts increase reaction rates by providing alternative pathways for changes without affecting the net energy involved (12.3)

STS
- recognize the values and limitations of technological products and processes (12.1, 12.3)
- state that a goal of technology is to solve practical problems (12.2)
- evaluate technologies from a variety of perspectives (12.3)

Skills
- initiating and planning: describe procedures for safe handling, storage, and disposal of materials used in the laboratory, with reference to WHMIS and consumer product labelling information (12.3)
- performing and recording: plot chemical potential energy diagrams, enthalpy diagrams, and energy pathway diagrams indicating changes in energy for chemical reactions (all sections);
- analyzing and interpreting: interpret energy diagrams for chemical reactions (all sections)
- communication and teamwork: work collaboratively in addressing problems and apply the skills and conventions of science in communicating information and ideas and in assessing results by using appropriate SI notation, and fundamental and derived units for calculating and communicating enthalpy changes (all sections)

Key Terms

12.1
activation energy, E_a

12.2
bond energy

12.3
catalysis
catalyst
enzyme
intermediate

▶ MAKE a summary

1. Draw a combined potential energy diagram and energy pathway diagram to illustrate the uncatalyzed and catalyzed reaction of hydrogen and fluorine. Add labels and explanations to communicate what you have learned in this chapter, including the concept of catalysis.
2. Refer back to your answers to the Starting Points questions at the beginning of this chapter. How has your thinking changed?

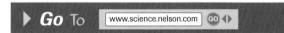

The following components are available on the Nelson Web site. Follow the links for *Nelson Chemistry Alberta 20–30*.
- an interactive Self Quiz for Chapter 12
- additional Diploma Exam-style Review questions
- Illustrated Glossary
- additional IB-related material

There is more information on the Web site wherever you see the Go icon in this chapter.

Chapter 12 REVIEW

Many of these questions are in the style of the Diploma Exam. You will find guidance for writing Diploma Exams in Appendix H. Exam study tips and test-taking suggestions are on the Nelson Web site. Science Directing Words used in Diploma Exams are in **bold** type.

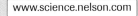

DO NOT WRITE IN THIS TEXTBOOK.

Part 1

1. Which of the following statements concerning collision-reaction theory is *false*?
   A. A chemical reaction must involve collisions of reactant entities.
   B. An effective collision requires sufficient energy.
   C. All collisions with the required minimum energy form new bonds.
   D. All collisions without the required minimum energy rebound elastically.

2. Catalysts increase the rate of a chemical reaction by providing a reaction pathway that has
   A. lower activation energy
   B. higher activation energy
   C. lower net energy change
   D. higher net energy change

3. Which of the following statements regarding bond energy is *false*?
   A. Bond energy is the energy released when a bond is formed.
   B. Bond energy is the energy required to break a chemical bond.
   C. The change in enthalpy represents the net effect from breaking and making bonds.
   D. In an endothermic reaction the energy released from bond making is greater than the energy required for bond breaking.

4. The following statements are *not* all true.
   NR
   1. Catalysts are consumed in the net reaction.
   2. Catalysts are not consumed in the net reaction.
   3. Catalysts decrease the enthalpy of the products.
   4. Catalysts increase the enthalpy of the products.
   5. Catalysts decrease the overall rate of a reaction.
   6. Catalysts increase the overall rate of a reaction.
   7. Catalysts provide an alternate pathway for the reaction.
   8. Catalysts are an essential part of most industrial chemical processes.
   List, in order, the *true* statements about catalysts.
   ____ ____ ____ ____.

5. The ignition system in an automobile is designed to provide the engine with
   A. activation energy
   B. enthalpy
   C. oxygen
   D. fuel

6. The catalytic converter in an automobile is designed to
   A. provide more energy
   B. provide more oxygen
   C. improve fuel efficiency
   D. remove nitrogen oxides from the exhaust

7. Which of the following is an example of a catalyst?
   A. enzymes in the production of sugar from starch
   B. ethene in the production of ethylene glycol
   C. heat in the decomposition of water
   D. light in the formation of hydrogen chloride

8. The structural arrangement of entities representing the highest potential energy point in a chemical reaction step is called the
   A. product
   B. catalyst
   C. reactant
   D. activated complex

9. The change in enthalpy of a chemical reaction can be predicted by all the following methods *except*
   A. Hess' law
   B. stoichiometry
   C. bond energies
   D. molar enthalpies of formation

10. All the alkali metals react with water at room temperature to produce hydrogen gas and hydroxide ions. The tendency to react increases as the atomic number of the alkali metal increases. Based on this evidence, which of the alkali metals has the highest activation energy in this reaction?
    A. cesium
    B. lithium
    C. potassium
    D. sodium

Part 2

11. Carbon tetrachloride, $CCl_4(l)$, is a toxic solvent for oils, fats, lacquers, and varnishes. Carbon tetrachloride can be produced by the reaction of methane and chlorine gas.

    $$CH_4(g) + 4\,Cl_2(g) \rightarrow CCl_4(l) + 4\,HCl(g)$$

    (a) **Describe** the conditions required for a successful collision between methane and chlorine molecules.
    (b) **How** would a catalyst change the rate of the reaction?
    (c) **Why** does a catalyst have this effect?
    (d) **Explain** how bond energies affect the rate of this reaction.
    (e) **Explain** how bond energies affect the net energy change of this reaction.

12. Using the concept of bond energies,
    (a) **explain** why the oxidation of aluminium is exothermic.
    (b) **explain** why the decomposition of aluminium oxide is endothermic.

13. Draw energy pathway diagrams for the following reactions. Label the reactants, products, enthalpy change, activation energy, and activated complex.
    (a) $C(s) + O_2(g) \rightarrow CO_2(g) + 393.5$ kJ
    (b) $N_2(g) + 2 O_2(g) + 66.4$ kJ $\rightarrow 2 NO_2(g)$

14. **Explain** why it is important to avoid producing static electricity while refuelling an automobile.

15. Catalytic reforming technology is used to improve the quality of gasoline by converting straight-chain hydrocarbons to aromatic molecules, for example, heptane + energy → methylbenzene + hydrogen (**Figure 1**). **Analyze** this reaction. Your response should include
    - a balanced equation including structures
    - your justification of which is greater: the energy released from bond making or the energy required for bond breaking
    - an energy pathway diagram for the reaction, with labels for the reactants, products, enthalpy change, and activation energy for the catalyzed and uncatalyzed reactions

Figure 1
A catalytic reforming unit

16. Use **Figure 2** to answer the following questions.

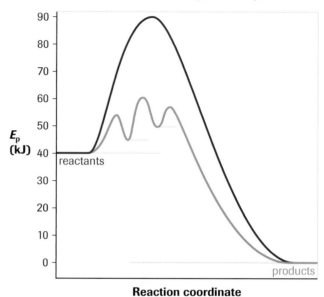

Figure 2
Energy pathway diagram for the synthesis of an organic compound

(a) What is the activation energy for the uncatalyzed forward reaction?
(b) What is the activation energy for the uncatalyzed reverse reaction?
(c) What is the activation energy for the catalyzed forward reaction?
(d) What is the activation energy for the catalyzed reverse reaction?
(e) What is the enthalpy change for the uncatalyzed forward reaction?
(f) What is the enthalpy change for the uncatalyzed reverse reaction?
(g) What is the enthalpy change for the catalyzed forward reaction?
(h) What is the enthalpy change for the catalyzed reverse reaction?

17. **Illustrate** each of the following technological processes with examples. For each example, provide a word equation and suggest a suitable catalyst.
    (a) hydrocracking heavy oil to gasoline
    (b) pollution control in an automobile catalytic converter
    (c) production of high-fructose corn syrup

Unit 6 REVIEW

Many of these questions are in the style of the Diploma Exam. You will find guidance for writing Diploma Exams in Appendix H. Exam study tips and test-taking suggestions are on the Nelson Web site. Science Directing Words used in Diploma Exams are in bold type.

DO NOT WRITE IN THIS TEXTBOOK.

Part 1

1. The correct units for specific heat capacity are
   A. J/(g·°C)
   B. kJ/g·°C
   C. kg/(J·°C)
   D. °C/(g·J)

2. Water can be decomposed by electrical energy as shown in the following equation:
   $2 H_2O(l) \rightarrow 2 H_2(g) + O_2(g) \quad \Delta_r H° = ?$
   The value of $\Delta_r H°$ for the equation as written is _____ kJ (to three digits).

3. There are several electrical generating stations in Alberta that use the energy from burning coal to generate electricity. Assuming coal is pure carbon, the temperature increase in 1.00 t of water when it absorbs all the energy produced by the complete combustion of 1.00 kg of coal is calculated to be _____ °C.

Use the following statements to answer questions 4 and 5.
1. Chemical bonds are broken.
2. Chemical bonds are formed.
3. Energy is lost to the surroundings.
4. Energy is gained from the surroundings.
5. The chemical potential energy of the products is less than the enthalpy of the reactants.
6. The chemical potential energy of the products is greater than the enthalpy of the reactants.

4. Which statements are true for exothermic reactions?
   ___, ___, ___, and ___

5. Which statements are true for endothermic reactions?
   ___, ___, ___, and ___

6. The energy barrier that must be overcome in order for a reaction to occur is referred to as the
   A. activation energy
   B. bond energy
   C. enthalpy change
   D. molar enthalpy

7. Catalysts are an essential part of most industrial chemical processes. Which of the following statements about catalysts is *false*?
   A. Catalysts are not consumed in the net reaction.
   B. Catalysts increase the enthalpy of the products.
   C. Catalysts increase the overall rate of a reaction.
   D. Catalysts provide an alternative pathway for the reaction.

8. Using a catalyst in a chemical reaction
   A. increases the enthalpy change of the reaction
   B. increases the activation energy of the reaction
   C. decreases the enthalpy change of the reaction
   D. decreases the activation energy of the reaction

9. A chemistry student determines in the laboratory that the molar enthalpy of combustion of methanol is −721 kJ/mol.
   $2 CH_3OH(l) + 3 O_2(g) \rightarrow 2 CO_2(g) + 4 H_2O(g) + energy$
   When the energy term is written as a term in the reaction equation, for the same conditions, the value is
   _____ kJ.

10. An automobile catalytic converter is a technological invention for the conversion of noxious exhaust gases. For example,
    $2 NO(g) + 2 CO(g) \rightarrow N_2(g) + 2 CO_2(g) \quad \Delta_r H° = ?$
    The standard enthalpy change for this reaction is (to three digits).
    −_____ kJ.

11. A much studied chemical reaction is communicated by the reaction equation
    $H_2(g) + I_2(g) \rightarrow 2 HI(g) \quad \Delta_r H° = +53.0 \text{ kJ}$
    This equation is representative of an
    A. endothermic formation reaction
    B. endothermic decomposition reaction
    C. exothermic formation reaction
    D. exothermic decomposition reaction

12. Glucose is an important natural product. Which of the following reactions involving glucose is endothermic?
    A. Formation of glucose from its elements
    B. Combustion of glucose
    C. Cellular respiration of glucose
    D. Photosynthesis involving glucose

NEL

Chemical Energy 547

Use this graph to answer questions 13 and 14.

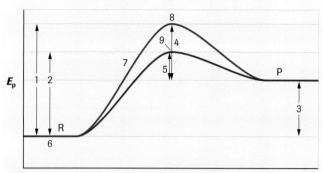

Potential Energy Diagram for System R→P

13. Match the appropriate energy change, labelled 1 to 5 in the graph, with the following reactions. Give your answers in the same order as the energy changes listed.
    - enthalpy change for the forward reaction
    - enthalpy change for the reverse reaction
    - activation energy for the uncatalyzed reaction
    - activation energy for the catalyzed reaction

    ___, ___, ___, and ___

14. From the graph above, match the appropriate sections, numbered 6 to 9, with the following statements. Give your answers in the same order as the statements.
    - the activated complex for the uncatalyzed reaction
    - the activated complex for the catalyzed reaction
    - the conversion of kinetic energy into potential energy
    - the same average chemical potential energy of entities with varying kinetic energy

    ___, ___, ___, and ___

15. Combustion of a hydrocarbon, cellular respiration, and photosynthesis have much in common. Which of the following is *not* common to all three reactions?
    A. energy
    B. $CO_2(g)$
    C. $H_2O(l)$
    D. $H_2O(g)$

16. Part of the character of chemicals is their thermal stability. List the following organic compounds in order of increasing stability.
    1. sucrose, $C_{12}H_{22}O_{11}(s)$
    2. ethanoic acid, $CH_3COOH(l)$
    3. ethanol, $C_2H_5OH(l)$
    4. glucose, $C_6H_{12}O_6(s)$

    ___ ___ ___ ___

Part 2

17. **Enumerate** four factors that influence the rate of a chemical reaction.

18. The reaction equation
    $$A + B \rightarrow C + D \qquad \Delta_r H = -250 \text{ kJ}$$
    represents a hypothetical reaction. **Illustrate** the reaction with an energy pathway diagram showing the catalyzed and uncatalyzed reactions, given that the activation energy is twice as large for the uncatalyzed reaction as for the catalyzed reaction.

19. The reaction between ammonia and hydrogen chloride gas produces a white smoke that was once widely used for "special effects" in dramatic productions.
    $$NH_3(g) + HCl(g) \rightarrow NH_4Cl(s) \qquad \Delta_r H° = ?$$
    (a) Use a table of standard molar enthalpies of formation to **predict** the standard enthalpy change for the reaction.
    (b) **Predict** which is greater: the energy required for breaking the bonds in the reactant molecules or the energy released by forming bonds in the product. Explain your reasoning.

20. The modern method of preparing methanol combines carbon monoxide and hydrogen at high temperature and pressure, in the presence of a catalyst.
    $$CO(g) + 2 H_2(g) \rightarrow CH_3OH(l) \qquad \Delta_r H° = ?$$
    (a) **Explain** the purpose of the catalyst.
    (b) **Predict** the standard enthalpy change for the reaction.
    (c) **Determine** the quantity of energy released by the production of 1.00 kg of methanol.

21. Some of the natural gas fields in Alberta contain hydrogen sulfide, a very poisonous gas. The sulfur is removed by burning some of the hydrogen sulfide to form sulfur dioxide, which is then reacted with the remaining hydrogen sulfide to form elemental sulfur. The overall reaction equation for the process is
    $$16 H_2S(g) + 8 SO_2(g) \rightarrow 16 H_2O(g) + 3 S_8(s) \qquad \Delta_r H° = ?$$
    (a) Use a table of standard molar enthalpies of formation to **predict** the enthalpy change of the above reaction.
    (b) **Determine** the standard molar enthalpy of reaction for sulfur, $S_8(s)$.
    (c) **Determine** the quantity of energy released by the production of 1.00 t of sulfur, $S_8(s)$.

22. In addition to being a fuel, ethane is used to produce ethene and ethyne, the starting points for the synthesis of a host of useful compounds. Use a table of standard molar enthalpies of formation to **predict** the enthalpy change of the following reactions:
    (a) $C_2H_6(g) \rightarrow C_2H_4(g) + H_2(g) \qquad \Delta_r H° = ?$
    (b) $C_2H_6(g) \rightarrow C_2H_2(g) + 2 H_2(g) \qquad \Delta_r H° = ?$

23. Cracking (especially to produce gasoline and ethene) is the most common industrial chemical reaction in Alberta. For catalytic cracking of crude oil fractions into gasoline hydrocarbons, synthetic zeolites (aluminosilicates, $Al_xSi_yO_z$) are the preferred catalysts.
    (a) Use the following information (only) to **determine** the enthalpy of cracking for the following reaction:

    $C_{10}H_{22}(l) + H_2(g) \rightarrow C_4H_{10}(g) + C_6H_{14}(l)$ $\Delta_{cr}H = ?$

    $2 C_{10}H_{22}(l) + 31 O_2(g) \rightarrow 20 CO_2(g) + 22 H_2O(l)$
    $\Delta_cH = -13\ 555.8$ kJ

    $2 C_4H_{10}(g) + 13 O_2(g) \rightarrow 8 CO_2(g) + 10 H_2O(l)$
    $\Delta_cH = -5\ 754.6$ kJ

    $2 C_6H_{14}(l) + 19 O_2(g) \rightarrow 12 CO_2(g) + 14 H_2O(l)$
    $\Delta_cH = -8\ 325.8$ kJ

    $C(s) + O_2(g) \rightarrow CO_2(g)$ $\Delta_cH = -393.5$ kJ

    $H_2(g) + \frac{1}{2}O_2(g) \rightarrow H_2O(l)$ $\Delta_cH = -285.8$ kJ

    (b) **Explain** how the addition of the zeolite changes the overall
        (i) enthalpy of reaction
        (ii) rate of reaction
        (iii) activation energy
    (c) View an animation of natural and synthetic zeolite.
        (i) What cation is primarily used to enrich the synthetic zeolite to make an even more effective catalyst than the natural crystal?
        (ii) What other cations may be present in the synthetic zeolite catalyst?

    www.science.nelson.com GO

24. Catalysts are one of many technological solutions used to reduce greenhouse gas (GHG) emissions.
    (a) List the four GHGs that research shows have the most effect on global warming.
    (b) From research, which of carbon dioxide or methane is found to have the larger global warming potential (GWP)?
    (c) The catalysts used in catalytic converters in cars are from what class of elements in the periodic table?

    www.science.nelson.com GO

Use this information to answer questions 25 to 27.

The Kyoto Protocol was officially ratified in Canada on February 16, 2005. Canada initially committed to reduce greenhouse gas emissions to 6% below the 1990 level. Part of Canada's initial strategy for reducing greenhouse gases is to have Canadians make a personal commitment to reduce their greenhouse gas (GHG) emissions by 1 t/a. Industry, commerce, and individuals are being challenged to reduce their GHG emissions to help us reach the Canadian target. There is more information on Canada's effort to meet our commitments on the Statistics Canada Web site.

www.science.nelson.com GO

Table 1 Per Capita Carbon Dioxide Emissions from Fossil Fuel Combustion and Production, 1990 to 2001

Year	Mass of CO_2 emissions (t/a)
1990	15.60
1991	15.04
1992	15.36
1993	15.13
1994	15.45
1995	15.73
1996	15.97
1997	16.22
1998	16.38
1999	16.87
2000	17.50
2001	17.02

Sources: Environment Canada, 2003, Canada's Greenhouse Gas Inventory, 1990-2001, Ottawa Statistics Canada, CANSIM, tables 051-0001 and 380-0017 Human Activity and the Environment, Annual Statistics 2004, p. 25, Table 2

25. (a) **Table 1** shows the increase in per capita GHG emissions from fossil fuel combustion and production since 1990. What is the percentage increase in this time?
    (b) Research recent GHG emissions in Environment Canada's Greenhouse Gas Inventory. Graph the per capita emissions over the last fifteen years or so. Comment on the trends you observe.
    (c) What is the Canadian GHG emission target in tonnes per capita? If every one of us meets this target, what will be the percentage reduction in GHG emissions? Relate this to the graph you plotted in (b).

    www.science.nelson.com GO

26. List five ways by which you can personally reduce GHG emissions.

27. **Why** do most Canadians (as indicated by polls) want to reduce the emission of GHGs?

28. At 300 K the activation energy, E_a, for the uncatalyzed decomposition of hydrogen peroxide into water and oxygen is empirically determined to be 75.3 kJ/mol. For a particular concentration of the catalyst potassium iodide, also at 300 K, the activation energy is 56.5 kJ/mol.
    (a) **Predict** the standard molar enthalpy of decomposition of hydrogen peroxide.
    (b) Draw a potential energy diagram, including reaction pathways, to communicate all the information above.

29. **Compare** three different organic fuels, such as methane, propane, and butane.
    Your response should include
    - a calculation of the standard molar enthalpy of combustion for each fuel to produce liquid water
    - a calculation of the quantities of energy released per kilogram of fuel
    - a calculation of the chemical amounts of carbon dioxide released per kilogram of fuel
    - an overall evaluation of the three fuels

30. Ethanoic acid (acetic acid) is an important industrial chemical that is manufactured from the reaction of methanol and carbon monoxide using rhodium or iridium catalysts.
    (a) Write the balanced chemical equation for the industrial production of ethanoic acid.
    (b) **Sketch** a chemical potential energy diagram, and label the enthalpy change.
    (c) **Sketch** an energy pathway diagram, and label the activation energies for the uncatalyzed and catalyzed reactions.
    (d) Provide the scientific explanation and the technological purpose for using a catalyst.

31. Ethanol can be oxidized (to carbon dioxide and water) in a laboratory fume hood through a series of reactions by using a platinum or palladium wire catalyst. Use the reaction mechanism below to answer the questions.

    $CH_3CH_2OH(l) \rightarrow CH_3CHO(l) + H_2(g)$
    $\Delta_r H = +85.3$ kJ

    $H_2(g) + \frac{1}{2} O_2(g) \rightarrow H_2O(l)$
    $\Delta H = -285.8$ kJ

    $CH_3CHO(l) + \frac{5}{2} O_2(g) \rightarrow 2 CO_2(g) + 2 H_2O(l)$
    $\Delta_r H = -1\,166.3$ kJ

    (a) **Determine** the enthalpy change for the overall reaction resulting from the above catalyzed reaction mechanism.
    (b) **Explain** how the enthalpy change for the overall reaction is affected by catalyzing the reaction and creating an alternative pathway.

32. Statistics Canada lists the percent hydroelectric power generation by province and territory (**Table 2**).

    Table 2 Hydroelectric Power Generation by Province and Territory, 1994 and 2002

Province	1994	2002
NL	97.7	93.9
PE	0.0	0.0
NS	10.4	8.5
NB	17.4	12.6
QC	96.5	96.3
ON	25.6	24.9
MB	99.0	97.9
SK	21.9	15.8
AB	3.4	2.8
BC	87.5	90.1
YT	88.9	87.5
NT	32.6	37.9

    Note: NT includes NU.
    Source: Table B35 on page 80 of HAatE by StatCan

    (a) **Analyze** this data (or more recent data from the Statistics Canada Web site) to find any trends and insights concerning the percentage of hydroelectric power generation in Canada.
    (b) **Illustrate** the percentage generation of hydroelectric power by province on a bar graph.
    (c) **Explain** the difference between hydro use in Alberta and hydro use in one other province or territory.
    (d) From an environmental perspective **enumerate** pros and cons for hydroelectric power stations.
    (e) If a goal of technology is to provide solutions to practical problems, **evaluate** the solution of replacing fossil fuel power plants with hydroelectric power. Provide at least one pro and one con from at least two perspectives other than environmental.

    www.science.nelson.com

33. People living in different regions of Canada use (or in the past have used) different fuels for cooking and for heating their homes. **Enumerate** five different fuels used by peoples in different parts of Canada—north to south and west to east.

34. **Compare** methane and propane as fuels. Establish at least four criteria for evaluating these fuels. Provide quantitative and/or qualitative information for each criterion. Finally, provide a personal choice based upon the information compiled and upon your values. Share your findings as an audio-visual presentation.

35. **Table 3** lists ignition temperatures for various fuels. The piloted ignition temperature is the temperature at which the fuel will ignite when provided with a spark or flame (such as a pilot light in a furnace). The spontaneous ignition temperature is the temperature at which the fuel ignites without a spark or flame.

    Table 3 Ignition Temperatures

Fuel	Piloted ignition temperature (°C)	Spontaneous ignition temperature (°C)
ammonia[1]	—	651
asphalt	204	485
corn oil	254	393
ethanol	13	390
gasoline	−43	280
methane[1]	—	537
olive oil	225	343
paraffin wax	199	245
propane[1]	—	450
wood	350	600

    1 Gas that ignites with a spark or flame at any temperature.
    Source:
    http://encarta.msn.com/media_461551262/Ignition_Temperatures.html

    (a) Use concepts from Chapter 12 to **explain** (theoretically) why gasoline ignites at or above −43 °C when a spark is provided (for example, by a spark plug).
    (b) **Explain** where the energy comes from for spontaneous ignition of propane at 450 °C.
    (c) Methane ignites with a spark or flame at any temperature. Gas companies advise people to "Call before you dig." What might produce a spark to ignite a broken natural gas line?
    (d) Methane, propane, gasoline, and asphalt are hydrocarbons. Consult Table 3 to develop a hypothesis about their relative activation energies. What certainties and uncertainties do you have about the hypothesis you have written?

36. The molar enthalpy of neutralization, $\Delta_n H_m$, for aqueous acid–base reactions can be determined by calorimetry. If one of the reactants is in excess, then the molar enthalpy of neutralization is determined for the other reactant. Complete the Analysis portion of the lab report.

    Purpose
    The purpose of this exercise it to use calorimetry to determine the molar enthalpy of neutralization for hypoiodous acid when reacted with excess aqueous sodium hydroxide.

    Evidence
    1.00 L of 0.400 mol/L HIO(aq)
    1.00 L of 0.500 mol/L NaOH(aq)
    initial temperature of solutions = 21.22 °C
    final temperature of mixture = 22.44 °C

37. Over the years there have been many "free energy" devices or "energy machines" advertised on the Internet. The energy output of these devices is claimed to be greater than the energy input. As a skeptic, what kind of authorities would you consult and/or what kinds of tests would you advocate to test these claims?

38. Review the focusing questions on page 474. Using the knowledge you have gained from this unit, briefly **outline** a response to each of these questions.

Extension

39. When investigating reactions to produce ammonia, Haber and Bosch tested some 20 000 candidate catalysts in a trial-and-error process before discovering an iron ore from Sweden that worked well for fertilizer production. (The ore happened to contain traces of alumina and alkali metal compounds.) In the catalyzed reaction of nitrogen and hydrogen, the gases are adsorbed onto the solid iron ore catalyst.
    (a) **Compare** the activation energy for the uncatalyzed reaction (red pathway) with the largest activation energy for the catalyzed reaction (green pathway).
    (b) **Explain** why the catalyzed reaction is much faster.
    (c) **Compare** the enthalpy changes for the catalyzed and uncatalyzed reactions. Comment on the magnitude and sign.

    www.science.nelson.com

40. Use a greenhouse gas (GHG) calculator or table of values to see how your energy emissions compare with the national average, and what you can do to reduce your emissions. Determine how you could reduce your GHGs by one tonne per year.

    www.science.nelson.com

41. Green chemistry is a term that describes chemical industrial work involving environmental stewardship. Research two examples of green chemistry associated with energy production or use.

    www.science.nelson.com

unit 7
Electrochemistry

You have previously classified chemical reactions using empirical generalizations such as single replacement. This classification of reactions is useful because it helps you to make predictions that generally can be verified in a laboratory. However, these predictions do not require any theoretical knowledge about how the reaction actually occurs. Chemists classify most common reactions based on a theoretical understanding of the reaction process. One very important example is the electron–transfer reaction, known as an electrochemical reaction. The study of chemical reactions associated with the transfer of electrons is known as electrochemistry.

Figure 1
Dr. Viola Birss

Electrochemical reactions may be the most important reactions on Earth. Living things use electrochemical reactions for photosynthesis and metabolism. Nonliving things undergo electrochemical reactions such as corrosion, (as shown by the metal door in the photograph), metallurgy, and combustion.

Electrochemistry can be applied to solve various technological problems. For example, the research conducted by Dr. Birss (**Figure 1**) and her research team at the University of Calgary is rooted in electrochemistry. According to Dr. Birss, "One of the main goals of our research over the last 15 years has been to produce useful thin film materials that can serve as electrocatalysts in fuel cells, to protect metals from corrosion, as a matrix for enzymes for biosensor applications, and to create useful nanoarchitectures."

Research into fuel cells involves developing new catalysts of nanometre-sized metal or metal oxide particles, researching new low-cost cathode materials for the reduction of oxygen in fuel cells, and investigating factors that affect the performance of these cells. Dr. Birss' team is also researching the development of novel electrochemical methods of coating metals with protective oxide films, which is of particular interest to the aerospace and oil and gas industries, and the development of glucose biosensors for people with diabetes.

As you progress through the unit, think about these focusing questions:

- What is an electrochemical change?
- How have scientific knowledge and technological innovation been integrated in the field of electrochemistry?

Unit 7

GENERAL OUTCOMES

In this unit, you will
- explain the nature of oxidation–reduction reactions
- apply the principles of oxidation–reduction to electrochemical cells

Electrochemistry

Unit 7 Electrochemistry

ARE YOU READY?

These questions will help you find out what you already know, and what you need to review, before you continue with this unit.

Prerequisites

Concepts
- atoms and ions
- ionic and covalent bonding
- balanced chemical reaction equations
- dissociation and ionization
- mole concept and stoichiometry

Skills
- safe handling and disposal based on WHMIS
- diagnostic tests
- solution preparation
- reaction equipment and procedures
- chemical amount and stoichiometry calculations
- designing and evaluating experiments
- SI and IUPAC communication conventions

You can review prerequisite concepts and skills in the Chemistry Review unit, on the Nelson Web site, and in the Appendices.
A Unit Pre-Test is also available online.

www.science.nelson.com

Knowledge

1. When a metal atom forms an ion, the atom _____ electrons to form a _____ charged ion.

2. When a nonmetal atom forms an ion, the atom _____ electrons to form a _____ charged ion.

3. According to trends in the periodic table and your general chemistry knowledge, copy and complete **Table 1**.

Table 1 Reactivity of Elements

Category	Groups or examples
most reactive metals	
least reactive metals	
most reactive nonmetals	
least reactive nonmetals	

4. Complete the following chemical equations using Lewis symbols or formulas for the products.

   (a) K· + ·C̈l̈: →
   potassium + chlorine →

   (b) ·P̈· + ·C̈l̈: →
   phosphorus + chlorine →

   (c) Compare electron rearrangement in (a) to electron rearrangement in (b).

   (d) According to the bonding theory you have studied, what is believed to determine whether two atoms transfer or share electrons?

5. (a) Write the generalized chemical equation for a single replacement reaction.
   (b) How do you know what class of element—metal or nonmetal—forms in a single replacement reaction?

6. Complete and balance the chemical equation for each of the following reactions:
   (a) __ Zn(s) + __ AgNO$_3$(aq) →
   (b) __ Cl$_2$(aq) + __ KBr(aq) →
   (c) __ Al(s) + __ HCl(aq) →
   (d) __ C$_3$H$_8$(g) + __ O$_2$(g) → __ CO$_2$(g) + __

STS Connections

7. When lead ion solutions are used in a laboratory, disposal of reactant solutions and products requires special care. The recommended method for lead disposal is to precipitate all lead(II) ions in the form of the insoluble lead(II) silicate. This solid can then be disposed of as regular solid waste as long as it is buried and not incinerated.
   (a) Why should lead compounds not be incinerated or disposed in a soluble form down the drain?

(b) Briefly describe the issue related to lead in the environment. What was the most common source of lead pollution until legislation forced a change in the substance sold?
(c) Suggest a reason why it is acceptable to bury a lead compound that is not soluble in water.
(d) When excess sodium silicate solution was added to 150 mL of lead(II) nitrate solution, 2.41 g of dried precipitate was obtained. Calculate the amount concentration of the lead(II) nitrate solution.

8. Compare the nature of science and the nature of technology, noting the key differences in the characteristics of these two forms of human endeavours.

Skills

9. What should you check before plugging in electrical equipment?

10. List some safety precautions for operating electrical equipment.

11. When studying chemical reactions, diagnostic tests are important to identify products (see Appendix C.4). Interpret the evidence in the photos below to identify the product or type of product formed in a chemical reaction.

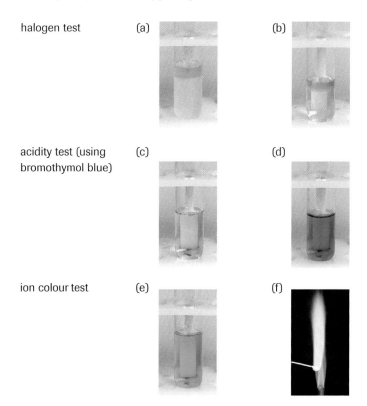

halogen test (a) (b)

acidity test (using bromothymol blue) (c) (d)

ion colour test (e) (f)

12. List the general steps of a procedure to prepare a standard solution starting with a pure solid.

13. Titration is a common technological process used in different types of chemical analysis. List the main laboratory equipment needed to conduct a titration.

chapter 13
Redox Reactions

In this chapter

- Exploration: Cleaning Silver
- Case Study: Early Metallurgy
- Investigation 13.1: Single Replacement Reactions
- Web Activity: Henry Taube and Rudolph Marcus
- Web Activity: Redox Reaction
- Biology Connection: Metabolic Redox
- Investigation 13.2: Spontaneity of Redox Reactions
- Web Activity: Piercings: A Rash Decision
- Lab Exercise 13.A: Building a Redox Table
- Investigation 13.3: Predicting the Reaction of Sodium Metal (Demonstration)
- Lab Exercise 13.B: Oxidation States of Vanadium
- Web Activity: Catalytic Converters
- Case Study: Bleaching Wood Pulp
- Lab Exercise 13.C: Analyzing for Tin
- Web Activity: Imants Lauks
- Lab Exercise 13.D: Analyzing for Chromium in Steel
- Investigation 13.4: Analyzing a Hydrogen Peroxide Solution

Of all chemical changes, electrochemical (electron transfer) reactions are the most common in both living and nonliving systems. Photosynthesis, cellular respiration, and metabolism are all electrochemical processes. Technologies involving electrochemistry, such as combustion and the production of metals from their ores, have been used for thousands of years. In the development of these applications, the technology was successfully developed long before there was any scientific understanding of the processes. In other words, technology led science. More recently, however, a sophisticated understanding of modern electrochemistry has led to the invention of modern technologies such as aureate (gold-like) plating of coins (**Figure 1**), and the development of fuel cells and biosensors. Today, science and technology nurture each other in a symbiotic relationship.

Knowledge of electrochemistry will help you connect and clarify many seemingly unrelated reactions, and understand the interactions of science and technology. For example, paper for this book is produced from trees that used photosynthesis reactions to grow. Harvesting trees requires machinery made from steel, which is produced by the electrochemical reduction of iron ore. The energy used to run the machines in a pulp mill comes from the combustion of fossil fuels. Electrochemical reactions play a role in the production of paper from wood pulp. Photographs used in books may involve the reduction of silver ions to silver metal to form a negative image, which is printed using metal plates made by electrochemical reactions. All of the people involved, from those who harvest the trees to those who read the book, metabolize food to live and work. An understanding of electrochemistry will give you a broader comprehension of many chemical reactions and their importance in both living and nonliving systems.

STARTING Points

Answer these questions as best you can with your current knowledge. Then, using the concepts and skills you have learned, you will revise your answers at the end of the chapter.

1. What are the key theoretical concepts that distinguish electrochemical reactions from other kinds of chemical reactions?
2. How can you predict whether or not a mixture of chemicals will react?
3. What are the similarities and differences between redox stoichiometry and other stoichiometry that you have learned?

Career Connections:
Materials/Metallurgical Engineer; Conservator

Figure 1
The aureate plating on the $1 and $2 coins gives them their golden colour. This plating process was invented in Fort Saskatchewan, Alberta, at the Sherritt Gordon plant.

▶ Exploration Cleaning Silver

Have you ever noticed how silver becomes darkened and cloudy over time? This dark tarnish is mainly silver sulfide. The sulfur may come from several sources: air, which contains small amounts of hydrogen sulfide from industrial operations or from the decomposition of organic matter; natural gas, which contains a sulfur compound to give the odourless methane a noticeable odour; and some food items, including egg whites, mustard, and mayonnaise.

There are several technological solutions to restore the silvery appearance of tarnished silverware. Many silver-cleaning mixtures contain abrasives that are used to scour off the tarnish. However, these cleaners can remove the silver as well. A better technological process is to convert the silver sulfide back to silver metal. A common household remedy based on this approach is to react the tarnish with aluminium in a hot solution of baking soda and table salt.

Materials: tarnished silverware such as a small spoon, 400 or 600 mL beaker, piece of aluminium foil about 5 cm by 5 cm, hot plate, stirring rod, measuring spoon, tongs, baking soda, table salt, tap water

 Do not touch the surface of the hot plate. Switch off the hot plate immediately after use.

- Fill the beaker about three-quarters full with tap water.
- Add about one teaspoon each of baking soda and table salt to the water. Stir to dissolve.
- Heat the solution on the hot plate to near boiling and then turn off the hot plate.

(a) Describe the appearance of the tarnished silver object and the aluminium foil.

- Use the tarnished silver object to push the aluminium foil to the bottom of the beaker. Make sure the silver object remains in contact with the foil.
- Observe any changes immediately and after several minutes.
- Use the tongs to remove the silver object and the aluminium foil.

(b) Describe the final appearance of the silver object and the aluminium foil.

(c) Write a balanced chemical equation for the single replacement reaction of solid silver sulfide and aluminium. How well does this equation explain your observations?

(d) Evaluate this technological process for cleaning tarnished silver.

- When cool, dispose of the solution down the sink and put the aluminium foil into the regular garbage.

13.1 Oxidation and Reduction

In prehistoric times, people learned to extract metals from rocks and minerals (**Figure 1**). This discovery initiated both the technology of metallurgy and humanity's progression from the Stone Age, through the Bronze Age and the Iron Age, to our increasingly technological modern age. Only a few metals, such as gold and silver, exist naturally in the form of a pure element. Most metals exist in a variety of compounds mixed with other substances in rocks called ores. *Metallurgy* is the science and technology of extracting metals from their naturally occurring compounds and adapting these metals for useful purposes. For some metals, the basic procedures are quite simple and were developed early in human history; for others, more complex procedures have been developed more recently. Making steel requires higher temperatures than those provided by a simple wood fire; therefore, this technology developed much later than the making of copper and bronze. A relatively recent example is the production of aluminium. This requires the technology to produce current electricity, which was not discovered until the 1800s. In all of these examples, from copper to aluminium, the technology was developed to solve practical problems before there was any scientific understanding of the processes involved.

Figure 1
The technology of metallurgy has a long history, preceding by thousands of years the scientific understanding of the processes.

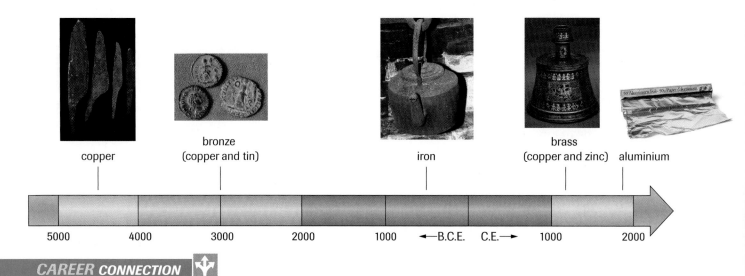

CAREER CONNECTION

Materials/Metallurgical Engineer
Being a Materials/Metallurgical Engineer is a challenging and rewarding career. The job involves learning the properties of materials, as well as developing new materials to meet specific requirements. Materials/Metallurgical Engineers also study and develop solutions for material fractures and breakages.

Read about the experiences of Nalaine Morin of the Tahltan First Nation, a metallurgical engineer.

www.science.nelson.com

From metallurgy, the term *reduction* came to be associated with producing metals from their compounds. For example, iron(III) oxide is "reduced" by carbon monoxide gas to iron metal. Tin and copper metals are other examples where a metal compound is reduced to the metal.

$$Fe_2O_3(s) + 3\ CO(g) \rightarrow 2\ Fe(s) + 3\ CO_2(g)$$
$$SnO_2(s) + C(s) \rightarrow Sn(s) + CO_2(g)$$
$$CuS(s) + H_2(g) \rightarrow Cu(s) + H_2S(g)$$

As you can see from these chemical equations, a substance called a *reducing agent* causes or promotes the reduction of a metal compound to an elemental metal. In the preceding examples, carbon monoxide is the reducing agent for the production of iron from iron(III) oxide, carbon (charcoal) is the reducing agent for the production of tin from tin(IV) oxide, and hydrogen is the reducing agent for the production of copper from copper(II) sulfide. These are three of the most common reducing agents used in metallurgical processes.

Although the discovery of fire occurred much earlier than that of metal refining, both discoveries advanced the development of civilization significantly. There are also important similarities in the chemistry behind these technological developments. Of course, it does not require a detailed scientific understanding of the processes to use fire to refine metals (**Figure 2**). Only in the 18th century did we realize the role of oxygen in burning. Understanding the connection between corrosion and burning is an even more recent development. Scientists now understand that corrosion, including the rusting of metals, is similar to combustion, although corrosion reactions occur more slowly. Historically, *corrosion* was considered to be the spontaneous reaction of metals with air (oxygen) to form metal compounds (such as rust). In effect, corrosion was returning the metal to its natural state as a compound and, therefore, can be considered to be the opposite of metallurgy. Chemists eventually called the reactions of substances with oxygen, whether they were the explosive combustion of gunpowder, the burning of wood, or the slow corrosion of iron, *oxidation*. As the study of chemistry developed, chemists realized that oxygen was not the only substance that could cause reactions similar to oxidation reactions. For example, metals can be converted to compounds by most nonmetals and by some other substances as well. The rapid reaction process we call burning may even take place with gases other than oxygen, such as chlorine or bromine (**Figure 3**). The term "oxidation" has been extended beyond reactions with oxygen to include a wide range of combustion and corrosion reactions, such as the following:

Figure 2
Making samurai swords requires a special type of fire (hardwood coals and bellows) to produce the higher temperatures required. This was a technological development long before any scientific understanding.

$$2\,Mg(s) + O_2(g) \rightarrow 2\,MgO(s)$$
$$2\,Al(s) + 3\,Cl_2(g) \rightarrow 2\,AlCl_3(s)$$
$$Cu(s) + Br_2(g) \rightarrow CuBr_2(s)$$

A substance that causes or promotes the oxidation of a metal to produce a metal compound is called an *oxidizing agent*. In the reactions shown above, the oxidizing agents are oxygen, chlorine, and bromine. As you will see, an understanding of reduction and oxidation is necessary to explain many seemingly unrelated chemical reactions.

▶ Practice

1. Using the historical context of metallurgy and corrosion, write an empirical definition for each of the following terms:
   (a) reduction
   (b) oxidation
   (c) oxidizing agent
   (d) reducing agent
   (e) metallurgy
   (f) corrosion

2. For each of the following, classify the reaction of the metal or metal compound in the historical context of reduction or oxidation, and identify the oxidizing agent or the reducing agent.
   (a) $4\,Fe(s) + 3\,O_2(g) \rightarrow 2\,Fe_2O_3(s)$
   (b) $2\,PbO(s) + C(s) \rightarrow 2\,Pb(s) + CO_2(g)$
   (c) $NiO(s) + H_2(g) \rightarrow Ni(s) + H_2O(l)$
   (d) $Sn(s) + Br_2(l) \rightarrow SnBr_2(s)$
   (e) $Fe_2O_3(s) + 3\,CO(g) \rightarrow 2\,Fe(s) + 3\,CO_2(g)$
   (f) $Cu(s) + 4\,HNO_3(aq) \rightarrow Cu(NO_3)_2(aq) + 2\,H_2O(l) + 2\,NO_2(g)$

3. List three reducing agents used in metallurgy.
4. What class of elements behaves as oxidizing agents for metals?
5. In the history of metallurgy, which came first, technological applications or scientific understanding? Elaborate on your answer.
6. Technologies that are intended to provide useful products and processes usually have unintended consequences. Using previous knowledge, list some unintended consequences that result from the production and refining of metals.

Figure 3
Copper metal is oxidized by reactive nonmetals such as bromine.

DID YOU KNOW?

Putting Out Class D Fires
A Class D fire includes combustible metals such as magnesium and the alkali metals. These metals burn at very high temperatures and water can make the fire much worse because of violent reactions. Carbon dioxide extinguishers don't help either since magnesium burns very well in carbon dioxide. So how can you put out such a fire? Sand is a simple option, but a special fire extinguisher such as MET-L-X, which contains sodium chloride, is the preferred method.

Case Study

Early Metallurgy

Metallurgy, which is the process of extracting metals from ores and forming them into useful objects, is an example of an early technology that was developed to solve practical problems. There is considerable archeological evidence that early humans discovered that gold, silver, and copper occurred in their natural state as nuggets or veins in rocks thousands of years ago. Stone tools were used to extract these metals from rocks and to hammer them into various objects. In particular, North American Aboriginal peoples developed the technology to extract copper from large copper pits in the Lake Superior region. The copper was used to make various ornaments, tools, and weapons. The problem was that the use of naturally occurring metals was limited to areas in which these metals were easily unearthed.

Over six thousand years ago, a new technology emerged when humans discovered that an extremely hot fire built on certain greenish rocks produced nuggets of solid copper in the cooling ashes. The extraction of metals from ores by heating is called *smelting*. Smelting copper ore to separate the copper from copper-bearing ore minerals, such as green malachite, meant that copper became much more widely used. There is evidence that smelting was discovered around 4500 B.C.E. in the Middle East because copper ornaments became more common after that time.

Since the ores of different metals often occur together, humans discovered that the metal smelted from a mixture of ores was much harder than pure copper. Smelting a mixture of copper and tin ores formed a hard copper–tin alloy known as *bronze*. Blades made from bronze could hold an edge very well and, if necessary, could be pounded back into shape. Bronze was one of the first metallic alloys widely used for containers, tools, weapons (**Figure 4**), and armour. The technology of smelting copper and formulating bronze diffused across Europe during the period of history known as the Bronze Age.

Figure 4
Bronze spearheads from the third millennium B.C.E.

The Bronze Age continued until humans developed the technology for smelting iron, which ushered in the Iron Age.

Case Study Questions

1. Is the discovery of metallurgy a scientific or technological achievement? Justify your answer.
2. What kinds of practical problems were solved or improved with the use of metals versus stone or wood? What new problems were created with the development of metal implements?
3. The following reaction equation represents the ancient process for smelting copper from malachite, $Cu_2CO_3(OH)_2(s)$, in a wood fire. Identify the reducing agent and the substance reduced.

$$Cu_2CO_3(OH)_2(s) + 2\, CO(g) \rightarrow 2\, Cu(s) + H_2O(g) + 3\, CO_2(g)$$

Extension

4. How was the process of trial and error used by early peoples to extract metals from their ores?

www.science.nelson.com

INVESTIGATION 13.1 Introduction

Single Replacement Reactions

This investigation is a review of single replacement reactions in preparation for the development of a theory of oxidation and reduction. As part of the Design, include diagnostic tests (as in Appendix C.4) for the predicted products.

Purpose

The purpose of this investigation is to use the single replacement reaction generalization to predict and analyze the reactants and products.

Report Checklist

- ○ Purpose
- ○ Problem
- ○ Hypothesis
- ● Prediction
- ● Design
- ○ Materials
- ○ Procedure
- ● Evidence
- ● Analysis
- ● Evaluation (1)

Problem

What are the products of the single replacement reactions for the following pairs of reactants?

(a) copper and aqueous silver nitrate
(b) aqueous chlorine and aqueous sodium bromide
(c) magnesium and hydrochloric acid
(d) zinc and aqueous copper(II) sulfate
(e) aqueous chlorine and aqueous potassium iodide

To perform this investigation, turn to page 601.

Section **13.1**

 WEB Activity

Canadian Achievers—Henry Taube and Rudolph Marcus

Henry Taube (**Figure 5(a)**) was born and educated in Saskatchewan. Rudolph Marcus (**Figure 5(b)**) was born and educated in Quebec. Taube and Marcus both contributed to the theory of electron transfer reactions. Each received a Nobel Prize in Chemistry, in 1983 and 1992, respectively, for their work.

1. In one sentence, describe the work done by each chemist as cited by the Nobel Committee.
2. Describe one difference between Taube's and Marcus' work.

www.science.nelson.com GO

Figure 5
(a) Henry Taube (1915–2005)
(b) Rudolph Marcus (1923–)

Electron Transfer Theory

Single replacement reactions are reactions in which one element replaces another element in a compound. These reactions are useful to investigate first because they provide a relatively simple introduction to the modern theoretical definitions of oxidation and reduction. Imagine that a reaction is a combination of two parts called *half-reactions*. A half-reaction represents what is happening to one reactant in an overall reaction. It tells only part of the story. Another half-reaction is required to complete the description of the reaction. Splitting a chemical reaction equation into two parts not only makes the explanations simpler, but also leads to some important applications, which are discussed later in this unit.

For example, when zinc metal is placed into a hydrochloric acid solution, gas bubbles form as the zinc slowly dissolves (**Figure 6**). Diagnostic tests show that the gas is hydrogen and that zinc ions are present in the resulting solution. Notice that zinc metal is oxidized to zinc chloride. This is a corrosion of zinc caused by the hydrochloric acid.

$$Zn(s) + 2\,HCl(aq) \rightarrow ZnCl_2(aq) + H_2(g)$$

What happens to the zinc and what happens to the hydrochloric acid? We can look at half-reactions to answer these questions. The zinc atoms in the solid, $Zn(s)$, are converted to zinc ions in solution, $Zn^{2+}(aq)$. Atomic theory requires that the zinc atoms lose two electrons, as shown by the following half-reaction equation:

$$Zn(s) \rightarrow Zn^{2+}(aq) + 2\,e^-$$

Simultaneously, hydrogen ions in the solution gain electrons and are converted into hydrogen gas, as shown below:

$$2\,H^+(aq) + 2\,e^- \rightarrow H_2(g)$$

Notice that both of these half-reaction equations, or half-reactions, are balanced by mass (same number of atoms/ions of each element on both sides) and by charge (same total charge on both sides). A **half-reaction** is a balanced chemical equation that represents either a loss or gain of electrons by a substance.

In a laboratory, single replacement reactions in aqueous solution are easier to study than the metallurgy or corrosion reactions discussed earlier in this chapter. However, all of these reactions share a common feature: ions are converted to atoms and atoms are converted to ions. For example, consider the reduction of aqueous silver nitrate to silver metal in the presence of solid copper (**Figure 7** on page 562). According to atomic theory, silver atoms are electrically neutral particles ($47p^+$, $47e^-$) and silver ions are

Figure 6
A common single replacement reaction is the reaction of active metals with an acid, such as zinc with hydrochloric acid.

Figure 7

A piece of copper before it is placed into a beaker of silver nitrate solution (left). Note the changes after the reaction has occurred (right). The blue colour of the solution indicates Cu^{2+}(aq) ions are present.

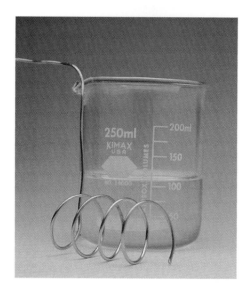
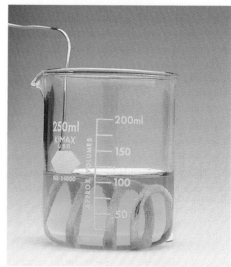

Learning Tip

A common difficulty in writing half-reaction equations is deciding on which side of the equation the electrons should appear. There are two ways to approach this. You can use your theoretical knowledge of atoms and ions to determine whether the species is losing or gaining electrons. The alternative is to add up the electric charges on both sides of the equation arrow. The total must be the same on both sides. Try this for the half-reaction equations on this page.

DID YOU KNOW?

Redox Mnemonics

A mnemonic is a word or group of words used to help remember information. "LEO the lion says GER" is a mnemonic to help remember that "Loss of Electrons is Oxidation" and "Gain of Electrons is Reduction." Another mnemonic is "OIL RIG," which translates as "Oxidation Is Loss, Reduction Is Gain."

charged particles ($47p^+$, $46e^-$). Therefore, in this reaction, a single electron is required to convert a silver ion into a silver atom. The following half-reaction equation explains the reduction of silver ions using the theoretical rules for atoms and ions:

$$Ag^+(aq) + e^- \rightarrow Ag^0(s) \qquad \text{(reduction)}$$

*According to modern theory, the gain of electrons is called **reduction**.*

Although this theoretical definition of reduction agrees with current atomic theory, it does not explain where the electrons come from. As crystals of silver metal are produced, the solution becomes blue, indicating that copper atoms are being converted to copper(II) ions. According to atomic theory, copper atoms ($29p^+$, $29e^-$) must each be losing two electrons as they form copper(II) ions ($29p^+$, $27e^-$):

$$Cu^0(s) \rightarrow Cu^{2+}(aq) + 2\,e^- \qquad \text{(oxidation)}$$

*According to modern theory, the loss of electrons is called **oxidation**.*

Evidence shows that the silver-coloured solid and the blue colour of the solution are simultaneously formed near the surface of the copper metal. Scientists believe, therefore, that the electrons required by the silver ions are supplied when silver ions collide with copper atoms on the metal surface.

The theory of electron transfer requires that the *total number of electrons gained in a reaction must equal the total number of electrons lost*, and that oxidation and reduction are separate processes. This theoretical description requires oxidation and reduction to occur simultaneously rather than sequentially. Oxidation–reduction reactions are often simply called "redox" reactions. A **redox** (**red**uction–**ox**idation) **reaction** is a chemical reaction in which electrons are transferred between entities.

The equations for reduction and oxidation half-reactions, and the overall (net) ionic equation summarize the electron transfer that is believed to take place during a redox reaction. As in other chemical reactions, the net equation must be balanced. We will learn how to write balanced half-reactions and net equations in Sample Problem 13.1.

Section **13.1**

Simulation—Redox Reaction

Use the computer simulation of the reaction between lead atoms and silver ions to determine which entity loses electrons and which gains electrons. Note the number of electrons lost or gained by each kind of atom/ion. Write the oxidation and reduction half-reaction equations.

> ### ▶ SAMPLE problem 13.1
>
> Write the balanced half-reaction equations and a balanced net equation for the reaction of copper metal with aqueous silver nitrate.
>
> To show that the number of electrons gained equals the number of electrons lost in two half-reaction equations, it may be necessary to multiply one or both half-reaction equations by a coefficient to balance the electrons. In this example, the silver half-reaction equation must be multiplied by 2.
>
> $$Cu(s) \rightarrow Cu^{2+}(aq) + 2\,e^- \quad \text{(two electrons lost by one atom)}$$
> $$2\,[Ag^+(aq) + e^- \rightarrow Ag(s)] \quad \text{(two electrons gained by two ions)}$$
>
> Now, add the half-reaction equations and cancel the terms that appear on both sides of the equation to obtain the net ionic equation.
>
> $$2\,Ag^+(aq) + \cancel{2\,e^-} + Cu(s) \rightarrow 2\,Ag(s) + Cu^{2+}(aq) + \cancel{2\,e^-}$$
> $$2\,Ag^+(aq) + Cu(s) \rightarrow 2\,Ag(s) + Cu^{2+}(aq)$$
>
> Silver ions are reduced to silver metal by reaction with copper metal. Simultaneously, copper metal is oxidized to copper(II) ions by reaction with silver ions (**Figure 8**).
>
> ```
>              oxidized to metal ion
>           ┌──────────────────────┐
>           ↓                      ↓
>    2 Ag⁺(aq) + Cu(s) → 2 Ag(s) + Cu²⁺(aq)
>           ↑                      ↑
>           └──────────────────────┘
>              reduced to metal
> ```

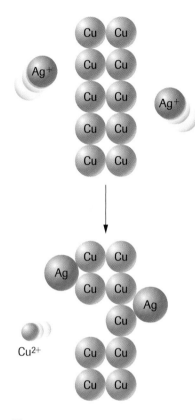

Figure 8
A model of the reaction of copper metal and silver nitrate solution illustrates aqueous silver ions reacting at the surface of a copper strip.

> ### ▶ COMMUNICATION example 1
>
> Write and label two balanced half-reaction equations to describe the reaction of zinc metal with aqueous lead(II) nitrate, as given by the following chemical equation:
>
> $$Zn(s) + Pb(NO_3)_2(aq) \rightarrow Pb(s) + Zn(NO_3)_2(aq)$$
>
> #### Solution
>
> $$Zn(s) \rightarrow Zn^{2+}(aq) + 2\,e^- \quad \text{(oxidation)}$$
> $$Pb^{2+}(aq) + 2\,e^- \rightarrow Pb(s) \quad \text{(reduction)}$$

To evaluate the theory of oxidation and reduction, you should look at its logical consistency with other accepted theories and definitions. The theoretical definitions of oxidation and reduction are consistent with the historical, empirical definitions presented earlier in this chapter; for example, a compound is reduced to a metal and a metal is oxidized to form a compound. Redox theory is also consistent with accepted atomic theory and the collision–reaction theory. Most importantly, redox theory explains the observations made by scientists.

> ### Learning *Tip*
>
> When you are cancelling terms for a net ionic equation, the terms must be identical, including their states of matter. Electrons must always cancel completely. This is because the electrons that appear in each half-reaction equation are the same electrons. They are the electrons that transfer from one entity to another.

> **DID YOU KNOW?**
>
> **Compartmentalization**
> The tradition in Western science is generally to break large systems down into smaller and smaller pieces to describe and explain them. Science is broken down into various areas such as biology, chemistry, and physics, which can then be further subdivided. One compartment of chemistry is redox, which includes several subcategories.
>
> In contrast, Aboriginal science is more holistic: the whole of nature, from the smallest parts to the cosmos, is connected. Viewing nature this way gives us a different perspective on current problems such as pollution and climate change. Recently, there is a trend in Western science toward a more holistic view after decades of compartmentalization.

Figure 9
The reduction of iron(III) oxide by aluminium is called the "thermite" reaction. Because this reaction is rapid and very exothermic, molten white-hot iron is produced. Here a falling aluminium wrench momentarily sparks a thermite reaction when it strikes a rusted iron block.

SUMMARY Electron Transfer Theory

- A *redox reaction* is a chemical reaction in which electrons are transferred between entities.
- The total number of electrons gained in the reduction equals the total number of electrons lost in the oxidation.
- *Reduction* is a process in which electrons are gained by an entity.
- *Oxidation* is a process in which electrons are lost by an entity.
- Both reduction and oxidation are represented by balanced half-reaction equations.

▶ Practice

7. Write a theoretical definition for each of the following terms:
   (a) redox reaction   (b) reduction   (c) oxidation

8. Write a pair of balanced half-reaction equations—one showing a gain of electrons and one showing a loss of electrons—for each of the following reactions:
   (a) $Zn(s) + Cu^{2+}(aq) \rightarrow Zn^{2+}(aq) + Cu(s)$
   (b) $Mg(s) + 2\,H^+(aq) \rightarrow Mg^{2+}(aq) + H_2(g)$

9. For each of the following, write and label the oxidation and reduction half-reaction equations. Ignore spectator ions.
   (a) $Ni(s) + Cu(NO_3)_2(aq) \rightarrow Cu(s) + Ni(NO_3)_2(aq)$
   (b) $Pb(s) + Cu(NO_3)_2(aq) \rightarrow Cu(s) + Pb(NO_3)_2(aq)$
   (c) $Ca(s) + 2\,HNO_3(aq) \rightarrow H_2(g) + Ca(NO_3)_2(aq)$
   (d) $2\,Al(s) + Fe_2O_3(s) \rightarrow 2\,Fe(l) + Al_2O_3(s)$ **(Figure 9)**

10. We have only looked at one type of single replacement reaction—a metal displacing another metal from an ionic compound. A nonmetal can also displace another nonmetal from an ionic compound. For example,

    $Cl_2(aq) + 2\,NaI(aq) \rightarrow I_2(s) + 2\,NaCl(aq)$

    Using your knowledge of atoms and ions and the ideas presented in this chapter, write a pair of balanced half-reaction equations for this reaction—one showing a gain of electrons and one showing a loss of electrons.

11. Ionic compounds can react in double replacement reactions. For example,

    $FeCl_3(aq) + 3\,NaOH(aq) \rightarrow Fe(OH)_3(s) + 3\,NaCl(aq)$

    According to ideas discussed in this chapter, has a redox reaction taken place in the reaction above? Explain your answer.

Writing Complex Half-Reaction Equations

Although most metals and nonmetals have relatively simple half-reaction equations, polyatomic ions and molecular compounds undergo more complicated oxidation and reduction processes. In most of these processes, the reaction takes place in an aqueous solution that is very often acidic or basic. Experimental evidence shows that water molecules, hydrogen ions, and hydroxide ions play an important role in these half-reactions. A method of writing half-reactions for polyatomic ions and molecular compounds requires that water molecules and hydrogen or hydroxide ions be included. This method, illustrated in the following sample problem, is sometimes called the "half-reaction" or "ion–electron" method.

SAMPLE problem 13.2

Nitrous acid can be reduced in an acidic solution to form nitrogen monoxide gas. What is the reduction half-reaction for nitrous acid?

The first step is to write the reactants and products.

$HNO_2(aq) \rightarrow NO(g)$

If necessary, you should balance all atoms other than oxygen and hydrogen in this partial equation. In this example, there is only one nitrogen atom on each side.

Next, add water molecules, present in an aqueous solution, to balance the oxygen atoms.

$HNO_2(aq) \rightarrow NO(g) + H_2O(l)$

Because the reaction takes place in an acidic solution, hydrogen ions are available. These can be used to balance the hydrogen on both sides of the equation.

$H^+(aq) + HNO_2(aq) \rightarrow NO(g) + H_2O(l)$

At this stage, all of the atoms should be balanced, but the charge on both sides will not be balanced. Add an appropriate number of electrons to balance the charge. Because electrons carry a negative charge, they are always added to the less negative, or more positive, side of the half-reaction.

$e^- + H^+(aq) + HNO_2(aq) \rightarrow NO(g) + H_2O(l)$

This balanced half-reaction equation represents a gain of electrons, or a reduction of the nitrous acid. Check to make sure that both the atom symbols and the charge are balanced.

BIOLOGY CONNECTION
Metabolic Redox

There are many biological processes, including metabolism and cellular respiration, that are redox reactions. You will learn about them in a biology course.

www.science.nelson.com

In a basic solution, the concentration of hydroxide ions greatly exceeds that of hydrogen ions. For basic solutions, we will develop the half-reaction as if it occurred in an acidic solution, and then convert the hydrogen ions into water molecules using hydroxide ions. This trick works because a hydrogen ion and a hydroxide ion react in a 1:1 ratio to form a water molecule. The following sample problem illustrates the procedure for writing half-reaction equations that occur in basic solutions.

SAMPLE problem 13.3

Copper metal can be oxidized in a basic solution to form copper(I) oxide. What is the half-reaction for this process?

Following the same steps as before, we write the formula and balance the atoms, other than oxygen and hydrogen. Here, the copper atoms must be balanced.

$2\,Cu(s) \rightarrow Cu_2O(s)$

Next, balance the oxygen using water molecules and balance the hydrogen using hydrogen ions, assuming, for the moment, an acidic solution. Balance the charge using electrons.

$H_2O(l) + 2\,Cu(s) \rightarrow Cu_2O(s) + 2\,H^+(aq) + 2\,e^-$

Learning Tip

You can also balance half-reaction equations in a basic solution by following these steps.
1. Balance atoms other than oxygen and hydrogen.
2. Add hydroxide ions to balance the oxygen.
3. Add water to balance the hydrogen.
4. Add electrons to balance the charge.

The difficulty with this method is that adding hydroxide ions and water simultaneously affects both the oxygen and hydrogen count.

Because the half-reaction occurs in a basic solution, add the same number of hydroxide ions as there are hydrogen ions to both sides of the equation. This is done to maintain the balance of mass and charge.

$$2\,OH^-(aq) + H_2O(l) + 2\,Cu(s) \rightarrow Cu_2O(s) + 2\,H^+(aq) + 2\,e^- + 2\,OH^-(aq)$$

Combine equal numbers of hydrogen ions and hydroxide ions to form water molecules.

$$2\,OH^-(aq) + H_2O(l) + 2\,Cu(s) \rightarrow Cu_2O(s) + 2\,H_2O(l) + 2\,e^-$$

Finally, cancel equal amounts of $H_2O(l)$ and anything else that is the same from both sides of the equation. Check that the atom symbols and charge are balanced.

$$2\,OH^-(aq) + \cancel{H_2O(l)} + 2\,Cu(s) \rightarrow Cu_2O(s) + \cancel{2}H_2O(l) + 2\,e^-$$
$$2\,OH^-(aq) + 2\,Cu(s) \rightarrow Cu_2O(s) + H_2O(l) + 2\,e^-$$

Learning Tip
Recall that loss of electrons is oxidation (LEO) and gain of electrons is reduction (GER).

► COMMUNICATION *example* 2

Chlorine is converted to perchlorate ions in an acidic solution. Write the half-reaction equation. Is this half-reaction an oxidation or a reduction?

Solution

$$8\,H_2O(l) + Cl_2(aq) \rightarrow 2\,ClO_4^-(aq) + 16\,H^+(aq) + 14\,e^-$$

According to redox theory, this half-reaction is an oxidation.

► COMMUNICATION *example* 3

Aqueous permanganate ions are reduced to solid manganese(IV) oxide in a basic solution. Write the half-reaction equation. Is the half-reaction an oxidation or a reduction?

Solution

$$4\,OH^-(aq) + 4\,H^+(aq) + MnO_4^-(aq) + 3\,e^- \rightarrow MnO_2(s) + 2\,H_2O(l) + 4\,OH^-(aq)$$
$$4\,H_2O(l) + MnO_4^-(aq) + 3\,e^- \rightarrow MnO_2(s) + 2\,H_2O(l) + 4\,OH^-(aq)$$
$$MnO_4^-(aq) + 2\,H_2O(l) + 3\,e^- \rightarrow MnO_2(s) + 4\,OH^-(aq)$$

According to redox theory, this half-reaction is a reduction.

+ EXTENSION

Redox Reactions
This short movie and the accompanying exercise will help you check your understanding of redox reactions.

www.science.nelson.com GO ◀▶

► Practice

12. For each of the following, complete the half-reaction equation and classify it as an oxidation or a reduction.
    (a) dinitrogen oxide to nitrogen gas in an acidic solution
    (b) nitrite ions to nitrate ions in a basic solution
    (c) silver(I) oxide to silver metal in a basic solution
    (d) nitrate ions to nitrous acid in an acidic solution
    (e) hydrogen gas to water in a basic solution

SUMMARY: Writing Half-Reaction Equations

Step 1: Write the chemical formulas for the reactants and products.
Step 2: Balance all atoms, other than O and H.
Step 3: Balance O by adding $H_2O(l)$.
Step 4: Balance H by adding $H^+(aq)$.
Step 5: Balance the charge on each side by adding e^- and cancel anything that is the same on both sides.

For basic solutions only:
Step 6: Add $OH^-(aq)$ to both sides to equal the number of $H^+(aq)$ present.
Step 7: Combine $H^+(aq)$ and $OH^-(aq)$ on the same side to form $H_2O(l)$. Cancel equal amounts of $H_2O(l)$ from both sides.

Section 13.1 Questions

1. Compare the historical empirical definitions of oxidation and reduction with the modern theoretical definitions.
2. According to modern theory, explain what is meant by a redox reaction.
3. What is a half-reaction and how does it relate to the overall chemical reaction equation?
4. Describe what happens to the electrons that are lost by one entity in a redox reaction.
5. What common type of chemical reaction is generally not a redox reaction? Include a brief theoretical explanation.
6. Write and label a pair of balanced half-reaction equations for each of the following reactions.
   (a) $Pb(s) + Cu^{2+}(aq) \rightarrow Pb^{2+}(aq) + Cu(s)$
   (b) $Cl_2(aq) + 2\,Br^-(aq) \rightarrow 2\,Cl^-(aq) + Br_2(l)$
   (c) the removal of silver tarnish (**Figure 10**) in the single replacement reaction of silver sulfide solid with aluminium

Figure 10
If tarnished silver is placed on aluminium foil in a hot electrolyte solution, a redox reaction converts the tarnish back into silver metal.

7. For each of the following applications, complete the half-reaction equation and classify it as an oxidation or a reduction.
   (a) bacterial action in soil: ammonia to nitrite ions in an acidic environment
   (b) pulp and paper bleaching: hydrogen peroxide to water in an acidic solution
   (c) alkaline battery: manganese(IV) oxide to manganese(III) oxide in a basic environment

Extension

8. Photosynthesis is a complex electron-transfer process that is essential for almost all life forms on Earth. Write and label the half-reaction equations for the "light reaction" and the "dark reaction."

    www.science.nelson.com

9. One of the earliest technologies is the use of fire. There are various stories in many cultures about how people learned about fire, such as the North American Aboriginal story of how the coyote steals fire. These stories serve as metaphors for the technology that we use today. Summarize the legend of the coyote and fire. How does this metaphor relate to the goals and effects of technology today?

    www.science.nelson.com

10. Archaeometallurgy is the study of ancient metallurgy using modern analytical techniques (**Figure 11**). Give some examples of research in the field. What metals and time periods have been studied? Can a metal from one mine be distinguished from the same metal from another mine? How is this information used in archaeological studies?

     www.science.nelson.com

Figure 11
Aslihan Yener (left) pioneered the use of modern X-ray techniques to identify metals from as early as 8000 B.C.E.

13.2 Predicting Redox Reactions

A redox reaction may be explained as a transfer of valence electrons from one substance to another. Evidence indicates that the majority of atoms, molecules, and ions are stable and do not readily release electrons. Since two entities must be involved in an electron transfer, chemists explain this transfer as a competition for electrons. Using a tug-of-war analogy, each entity pulls on the same electrons. If one entity is able to pull electrons away from the other, a spontaneous reaction occurs (**Figure 1**). Otherwise, no reaction occurs (**Figure 2**). In the spontaneous reaction of copper(II) ions and zinc metal, the Cu^{2+} ion is electron deficient and pulls electrons from a Zn atom. We can explain the reaction evidence using the idea that Cu^{2+} "pulls harder" on Zn's electrons than Zn does. Cu^{2+} "wins" the two valence electrons from a Zn atom. A successful electron transfer has occurred.

Without mixing all possible reactants and observing any evidence of reaction, how can we predict if a reaction will occur? If a reaction occurs, what will be produced? The answers to these questions cannot be obtained easily using redox theory. By observing many successful and unsuccessful reactions, patterns emerge and empirical generalizations can be made.

Figure 1
Copper(II) ions react spontaneously with zinc metal. A copper(II) ion has a stronger attraction for the valence electrons of a zinc atom than zinc does.

Oxidizing and Reducing Agents

Before we look at these patterns, we need to define some terms commonly used by chemists. When discussing possible reactants and comparing their reactivities, it is customary and convenient to classify the reactants in a redox reaction. This classification originated historically, but is now defined theoretically in terms of an ability to lose or gain electrons. In any redox reaction, an electron transfer occurs. This means that one reactant is oxidized and one reactant is reduced.

$$\underset{\text{loses}\atop\text{(is oxidized)}}{X} \xrightarrow{e^-} \underset{\text{gains}\atop\text{(is reduced)}}{Y} \rightarrow \text{products}$$

electron change

Examples:
$Zn(s) + Cu^{2+}(aq) \rightarrow Zn^{2+}(aq) + Cu(s)$
$2\ Br^-(aq) + Cl_2(g) \rightarrow Br_2(l) + 2\ Cl^-(aq)$

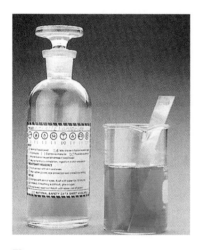

Figure 2
When a strip of copper is placed in a solution of nickel(II) ions, the green nickel(II) ion colour remains and the copper metal does not react. Collisions between copper atoms and nickel(II) ions apparently do not result in the transfer of electrons.

Rather than saying "the reactant that is oxidized" and "the reactant that is reduced," chemists use the terms reducing agent (RA) and oxidizing agent (OA). These terms originated in the early history of metallurgy and corrosion. For example, to "reduce" a larger volume of iron(III) oxide to a smaller volume of pure iron, a substance called a reducing agent was required, such as $CO(g)$.

$$\text{reducing agent} + \underbrace{Fe_2O_3(s) \rightarrow Fe(s)}_{\text{reduction}} + \text{other products}$$

Similarly, oxidation was originally associated with an oxidizing agent. For example, a metal could be oxidized by certain substances called oxidizing agents. At first, oxygen was the only known oxidizing agent, but others (e.g., halogens) can also oxidize, or corrode, metals.

$$\text{oxidizing agent} + \overbrace{Mg(s) \rightarrow MgO(s)}^{\text{oxidation}} + \text{other products}$$

The terms oxidizing and reducing agents developed separately, long before any redox theory of electron transfer emerged. Today, chemists routinely think in terms of electron transfer to explain redox reactions. A redox reaction is recognized as an electron transfer between an oxidizing agent and a reducing agent (**Figure 3**). Based on this idea, chemists say that a **reducing agent** causes reduction by donating (losing) electrons to another substance in a redox reaction. During this process, the reducing agent is oxidized. An **oxidizing agent** causes oxidation by removing (gaining) electrons from another substance in a redox reaction. During this process, the oxidizing agent is reduced. *It is important to note that oxidation and reduction are processes, and oxidizing agents and reducing agents are substances.* For example,

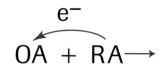

Figure 3
In all redox reactions, electrons are transferred from a reducing agent (RA) to an oxidizing agent (OA).

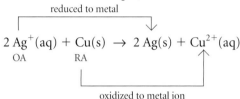

INVESTIGATION 13.2 Introduction

Spontaneity of Redox Reactions

In previous units in this textbook, we assumed that all chemical reactions are spontaneous; that is, they occur once the reactants are placed in contact, without a continuous addition of energy to the system. Spontaneous redox reactions in solution generally provide visible evidence of a reaction within a few minutes.

Purpose
The purpose of this investigation is to test the assumption that all single replacement reactions are spontaneous.

Report Checklist
- ○ Purpose
- ○ Problem
- ○ Hypothesis
- ○ Prediction
- ○ Design
- ○ Materials
- ● Procedure
- ● Evidence
- ● Analysis
- ● Evaluation (1, 2, 3)

Problem
Which combinations of copper, lead, silver, and zinc metals and their aqueous metal ion solutions produce spontaneous reactions?

Design
A drop of each solution is placed in separate locations on a clean area of each of the four metal strips.

To perform this investigation, turn to page 602.

Development of a Redox Table

Some redox reactions, such as single replacement reactions, are easy to study experimentally. The evidence of a reaction is immediately obvious and the interpretation of an electron transfer is relatively simple. In the past, you have generally assumed that all single replacement reactions are spontaneous. However, the evidence you obtained in Investigation 13.2 clearly shows that this assumption is unacceptable as only six of the combinations led to a reaction. The question that arises is, "How do you know when a chemical reaction will occur spontaneously without actually doing the reaction?"

Let's look at the combinations of metals and metal ions used in Investigation 13.2. Copper, lead, silver, and zinc metals were combined one at a time with each of copper(II), lead(II), silver, and zinc ion solutions. Based on the evidence collected, we can rank the ability of the metal ions to react with the metals (**Table 1**).

Table 1 Reactivities of Metal Ions with Metals

Ions	$Ag^+(aq)$	$Cu^{2+}(aq)$	$Pb^{2+}(aq)$	$Zn^{2+}(aq)$
reacted with	Cu(s), Pb(s), Zn(s)	Pb(s), Zn(s)	Zn(s)	none
number of reactions	3	2	1	0
reactivity order	most			least

> **Learning Tip**
>
> If a positively charged metal ion reacts, then it is usually converted to a metal atom. According to redox theory, this requires a gain of electrons and hence the metal ion is behaving as an oxidizing agent. Similarly, if a metal atom reacts, then it is always converted to a positively charged ion by losing electrons. Metals always behave as reducing agents.

The most reactive metal ion, $Ag^+(aq)$, has the greatest tendency to gain electrons. On the other hand, $Zn^{2+}(aq)$ shows no tendency to gain electrons in the combinations tested. Therefore, the order of reactivity is also the order of strength as oxidizing agents.

strongest oxidizing agent $\quad Ag^+(aq) + e^- \rightarrow Ag(s)$

$\quad\quad\quad\quad\quad\quad\quad\quad\quad Cu^{2+}(aq) + 2\,e^- \rightarrow Cu(s)$

$\quad\quad\quad\quad\quad\quad\quad\quad\quad Pb^{2+}(aq) + 2\,e^- \rightarrow Pb(s)$

weakest oxidizing agent $\quad Zn^{2+}(aq) + 2\,e^- \rightarrow Zn(s)$

The order of reactivity of the four metals can be obtained in a similar way (**Table 2**).

Table 2 Reactivities of Metals with Metal Ions

Metals	Zn(s)	Pb(s)	Cu(s)	Ag(s)
reacted with	$Ag^+(aq)$, $Cu^{2+}(aq)$, $Pb^{2+}(aq)$	$Ag^+(aq)$, $Cu^{2+}(aq)$	$Ag^+(aq)$	none
number of reactions	3	2	1	0
reactivity order	most			least

The most reactive metal, $Zn(s)$, has the greatest tendency to lose electrons and $Ag(s)$ shows no tendency to lose electrons in the combinations tested. Metals behave as reducing agents and so $Zn(s)$ is the strongest reducing agent among those tested.

strongest reducing agent $\quad Zn(s) \rightarrow Zn^{2+}(aq) + 2\,e^-$

$\quad\quad\quad\quad\quad\quad\quad\quad\quad Pb(s) \rightarrow Pb^{2+}(aq) + 2\,e^-$

$\quad\quad\quad\quad\quad\quad\quad\quad\quad Cu(s) \rightarrow Cu^{2+}(aq) + 2\,e^-$

weakest reducing agent $\quad Ag(s) \rightarrow Ag^+(aq) + e^-$

In these four reactions, the metal ions are the oxidizing agents and the silver ion is the strongest oxidizing agent (SOA) because it is the most reactive in our group. The two lists can be summarized using a single set of half-reactions (**Table 3**).

Table 3 Relative Strengths of Oxidizing and Reducing Agents

	OA	$+\ n\,e^-$	⇌ RA	
decreasing reactivity of oxidizing agents (SOA)	$Ag^+(aq)$ $Cu^{2+}(aq)$ $Pb^{2+}(aq)$ $Zn^{2+}(aq)$	$+\ e^-$ $+\ 2\,e^-$ $+\ 2\,e^-$ $+\ 2\,e^-$	⇌ $Ag(s)$ ⇌ $Cu(s)$ ⇌ $Pb(s)$ ⇌ $Zn(s)$	decreasing reactivity of reducing agents (SRA)

> **Learning Tip**
>
> The double arrows indicate that half-reactions may be read from left to right (top arrow) or from right to left (bottom arrow).

In **Table 3**, the metal ions are on the left side of the equations and the metal atoms are on the right side. For metal ions (the oxidizing agents), the half-reaction equations are read from left to right in the table. For metal atoms (the reducing agents), the half-reaction equations are read from right to left.

 **WEB Activity**

Web Quest—Piercings: A Rash Decision

Most Canadians wear some type of jewellery. However, some individuals react to their jewellery. This Web Quest explores the chemistry behind these reactions. What can you do to protect yourself from your jewellery?

Section **13.2**

SUMMARY Oxidizing and Reducing Agents

- An *oxidizing agent* causes oxidation by removing (gaining) electrons from another substance in a redox reaction. In this process, the oxidizing agent is reduced.
- A *reducing agent* promotes reduction by donating (losing) electrons to another substance in a redox reaction. In this process, the reducing agent is oxidized.
- A table of relative strengths of oxidizing and reducing agents—more simply known as a redox table—is, by convention, listed as reductions (from left to right) in the form: **OA** + $n\,e^-$ ⇌ **RA**, with the strongest oxidizing agent at the top left and strongest reducing agent at the bottom right of the table.

Learning Tip

If a nonmetal reacts, then it is usually converted to a negatively charged ion by gaining electrons. Nonmetals always behave as oxidizing agents. Negatively charged nonmetal ions lose electrons when they react; therefore, nonmetal ions behave as reducing agents.

Practice

1. Oxidation and reduction are processes, and oxidizing agents and reducing agents are substances. Explain this statement.
2. If a substance is a very strong oxidizing agent, what does this mean in terms of electrons?
3. If a substance is a very strong reducing agent, what does this mean in terms of electrons?
4. The terms, oxidizing agent and reducing agent, may be confusing especially when used in conjunction with the terms, oxidation and reduction. The key to avoiding confusion is to focus on the word "agent." Use a dictionary or write a definition in your own words of the term "agent." How can a banker (loans agent) be used as an analogy for a reducing agent?

Refer to **Table 1** (on page 569), and **Tables 2** and **3** (on page 570) to answer questions 5 to 9.

5. List the metal(s) that react spontaneously with a copper(II) ion solution.
6. Which metal(s) did not appear to react with a copper(II) ion solution?
7. Start with the position of Cu^{2+}(aq) in **Table 3** and note the position of the metal(s) that reacted and the metal(s) that did not react. For a metal that reacts spontaneously with Cu^{2+}(aq), where does the metal appear on a table of reduction half-reactions (**Table 3**)?
8. Repeat questions 5 to 7 for the Pb^{2+}(aq) ion.
9. Your answers to questions 7 and 8 form an empirical hypothesis that can be tested by making predictions for the other metal ions. Use **Table 3** to predict which of the reactions should be spontaneous. According to the evidence from Investigation 13.2, are your predictions correct? Is your hypothesis verified?

10. An experiment similar to the example of metals and metal ions was conducted using halogens and halide ions. Prepare a redox table of half-reaction equations similar to **Table 3** for the halogens.

Evidence
Only three combinations produced evidence of a reaction (**Figure 4**, **Table 4**).

Table 4 Reactions of Halogens with Solutions of Halides

	Br_2(aq)	Cl_2(aq)	I_2(aq)
Br^-(aq)	no reaction	yellow-brown	no reaction
Cl^-(aq)	no reaction	no reaction	no reaction
I^-(aq)	pink/purple	pink/purple	no reaction

(a)

(b)

(c)

Figure 4
None of the combinations of aqueous solutions of chlorine, bromine, and iodine with their corresponding halides show any evidence of reaction except for the reaction between **(a)** bromine and iodide ions, **(b)** chlorine and bromide ions, and **(c)** chlorine and iodide ions.

LAB EXERCISE 13.A

Building a Redox Table

Report Checklist
- ○ Purpose
- ○ Problem
- ○ Hypothesis
- ○ Prediction
- ○ Design
- ○ Materials
- ○ Procedure
- ○ Evidence
- ● Analysis
- ○ Evaluation

Suppose that a research team is developing a table of relative strengths of oxidizing and reducing agents. One team member had completed an investigation summarized in **Table 3**, page 570, and another had completed the investigation reported in Practice question 10, page 571. A third member used the combination of metals, nonmetals, and solutions shown below. By completing this exercise, you will see how scientists have developed extensive tables of relative strengths of oxidizing and reducing agents.

Purpose
The purpose of this lab exercise is to create an extended table of relative strengths of oxidizing and reducing agents.

Problem
What is the table of relative strengths of oxidizing and reducing agents for the combined results from three experiments?

Evidence

Table 5 Reactions of Metals and Nonmetals with Solutions of Ions

	$I_2(aq)$	$Cu^{2+}(aq)$	$Ag^+(aq)$	$Br_2(aq)$
$I^-(aq)$	X	X	✓	✓
Cu(s)	✓	X	✓	✓
Ag(s)	X	X	X	✓
$Br^-(aq)$	X	X	X	X

X no evidence of a redox reaction
✓ evidence redox reaction occurred

Analysis
(a) Use **Table 5** to construct a mini-redox table of just those substances.
(b) Compare **Table 3**, your analysis table from Practice question 10, and your table. Note that there are several substances that appear in two of these tables. Combine all three tables in one larger table showing the order of oxidizing and reducing agents.

DID YOU KNOW ?

Redox Table
The format of the redox table used in this textbook is very common. However, some books and references show a redox table as reversed (top to bottom), or list the oxidation and reduction half-reactions in alphabetical order or in order of strength of the agents.

Figure 5
The redox spontaneity rule

The Spontaneity Rule

Evidence obtained from the study of many redox reactions has been used to establish a generalization, called the **redox spontaneity rule**. The redox spontaneity rule states that a spontaneous redox reaction occurs only if the oxidizing agent (OA) is above the reducing agent (RA) in a table of relative strengths of oxidizing and reducing agents. **Figure 5** illustrates how you can use the rule, along with a table of oxidizing and reducing agents, to predict whether a reaction is spontaneous or nonspontaneous (no reaction).

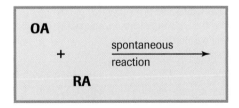
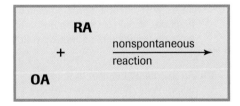

Another Method for Building Redox Tables

Once a spontaneity rule is developed from experimental evidence, the rule may be used to generate redox tables. The evidence to be analyzed in this case is a net ionic equation, accompanied by observations of spontaneity. In the following method, the spontaneity rule, rather than the number of reactions observed, is used to order the oxidizing and reducing agents to produce a redox table. The procedure for this type of analysis and synthesis is illustrated by Sample Problem 13.4.

SAMPLE problem 13.4

Three reactions among indium, cobalt, palladium, and copper were investigated. The reaction equations below indicate that two spontaneous reactions occurred and only one combination did not react. Using these equations, construct a redox table of half-reaction equations showing the relative strengths of the oxidizing and reducing agents.

$3 Co^{2+}(aq) + 2 In(s) \rightarrow 2 In^{3+}(aq) + 3 Co(s)$ (spontaneous)

$Cu^{2+}(aq) + Co(s) \rightarrow Co^{2+}(aq) + Cu(s)$ (spontaneous)

$Cu^{2+}(aq) + Pd(s) \rightarrow$ no evidence of reaction   (nonspontaneous)

To construct a redox table from this information, work with one equation at a time. Identify the oxidizing and reducing agents for the first reaction, and arrange them in two columns using the spontaneity rule. For the first reaction, this step is shown in **Figure 6(a)**. $Co^{2+}(aq)$ is the oxidizing agent and $In(s)$ is the reducing agent. Since the reaction is spontaneous, the oxidizing agent is above the reducing agent in the list.

In the second reaction, $Cu^{2+}(aq)$ is the oxidizing agent and $Co(s)$ is the reducing agent. This reaction is also spontaneous; therefore, $Cu^{2+}(aq)$ is above $Co(s)$ in the list. Since a metal appears on the same line as its ion in a redox table, add $Co(s)$ and extend the list as shown in **Figure 6(b)**.

No reaction occurs for the third pair of reactants. If a reaction had occurred, $Cu^{2+}(aq)$ would be the oxidizing agent and $Pd(s)$ would be the reducing agent. As this reaction is not spontaneous, the oxidizing agent appears below the reducing agent. **Figure 6(c)** shows the list extended to include $Pd(s)$. To complete the table, write balanced half-reaction equations for each oxidizing/reducing agent pair.

SOA $\quad Pd^{2+}(aq) + 2e^- \rightleftarrows Pd(s)$

$\quad\quad Cu^{2+}(aq) + 2e^- \rightleftarrows Cu(s)$

$\quad\quad Co^{2+}(aq) + 2e^- \rightleftarrows Co(s)$

$\quad\quad In^{3+}(aq) + 3e^- \rightleftarrows In(s)$ SRA

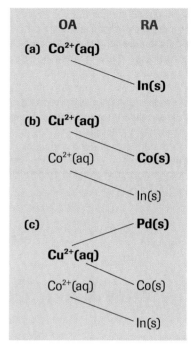

Figure 6
The relative position of a pair of oxidizing and reducing agents indicates whether a reaction will be spontaneous.

Practice

11. A student performed the following reactions. Construct a table of relative strengths of oxidizing and reducing agents.

    $Co^{2+}(aq) + Zn(s) \rightarrow Co(s) + Zn^{2+}(aq)$

    $Mg^{2+}(aq) + Zn(s) \rightarrow$ no evidence of reaction

12. In a school laboratory, four metals were combined with each of four solutions and the following reactions occurred:

    $Be(s) + Cd^{2+}(aq) \rightarrow Be^{2+}(aq) + Cd(s)$

    $Cd(s) + 2 H^+(aq) \rightarrow Cd^{2+}(aq) + H_2(g)$

    $Ca^{2+}(aq) + Be(s) \rightarrow$ no evidence of reaction

    $Cu(s) + H^+(aq) \rightarrow$ no evidence of reaction

    Construct a table of relative strengths of oxidizing and reducing agents.

13. Is the redox spontaneity rule empirical or theoretical? Justify your answer.

14. Use the relative strengths of nonmetals and metals as oxidizing and reducing agents, as indicated in the following unbalanced equations, to construct a table of half-reactions.

    $Ag(s) + Br_2(l) \rightarrow AgBr(s)$

    $Ag(s) + I_2(s) \rightarrow$ no evidence of reaction

    $Cu^{2+}(aq) + I^-(aq) \rightarrow$ no redox reaction

    $Br_2(l) + Cl^-(aq) \rightarrow$ no evidence of reaction

Learning Tip

The nonspontaneity of a reaction is communicated in several ways: "no evidence of reaction," "nonspontaneous," "no reaction," or "nonspont," written over the equation arrow.

+ EXTENSION

Explaining Relative Strengths of Agents
Why are some metals more effective as reducing agents than others? The strength of a reducing agent depends on several factors. This audio clip provides a bit more background on what makes some metals stronger reducing agents than others.

www.science.nelson.com

Learning Tip

Although you can consult the redox table in Appendix I, it is much more efficient to memorize which classes of substances tend to be oxidizing agents or reducing agents (question 15). This general pattern helps to speed up the process of recognizing oxidizing and reducing agents and is necessary to classify substances that don't appear in Appendix I.

Fe
↑   GER/OA
Fe^{2+}
↓   LEO/RA
Fe^{3+}

Figure 7
Iron(II) ions can either lose or gain electrons and, therefore, can act as either reducing agents or oxidizing agents.

DID YOU KNOW?

Science and Belief Systems
Traditional knowledge of Aboriginal peoples is based on observation, experience, testing, teaching, and recording. Although the methodology may differ, this empirical basis is not that much different from Western science. Traditional knowledge, such as traditional ecological knowledge, is sometimes criticized within Western culture because of its connection to Aboriginal beliefs. Yet Western science also operates under a belief system that sometimes makes it difficult for novel ideas to become accepted. For example, Arrhenius' ideas were disputed and overruled by the prominent scientists of the day in spite of Arrhenius' solid evidence.

An Extended Redox Table

Chemists have analyzed evidence collected in many experiments, to produce an extended redox table of oxidizing and reducing agents, such as the one found in Appendix I. This table represents the combined efforts of many people over many years. A redox table is an important reference for chemists. You can use this table to compare oxidizing and reducing agents, and to predict spontaneous redox reactions.

▶ Practice

Use the redox table in Appendix I to answer the following questions.

15. Arrange the following metal ions in order of decreasing strength as oxidizing agents: lead(II) ions, silver ions, zinc ions, and copper(II) ions. How does this order compare with **Table 3** on page 570?

16. What classes of substances (e.g., metals, nonmetals, acidic, basic) usually behave as
    (a) oxidizing agents?   (b) reducing agents?

17. Use atomic theory to explain why nonmetals behave as oxidizing agents and metals behave as reducing agents. Is there logical consistency between atomic theory and the empirically determined table of oxidizing and reducing agents?

18. Trends in the reactivity of elements show that fluorine is the most reactive nonmetal. How does this relate to the position of fluorine in the redox table of oxidizing and reducing agents? State one reason why this element is the most reactive nonmetal. Why is your reason an explanation? (Keep asking a series of "why" questions until your theoretical knowledge is expended. Does your theory pass the test of being able to explain the empirically determined table?)

19. (a) Repeat your answer to question 16 using a Venn diagram (two large intersecting circles).
    (b) Identify three oxidizing agents (other than $Fe^{2+}(aq)$, shown in **Figure 7**) from the redox table that can also act as reducing agents and record their symbols on your Venn diagram.
    (c) Try to explain this surprising behaviour.

20. Use the redox spontaneity rule to predict whether the following mixtures will show evidence of a reaction; that is, predict whether the reactions are spontaneous. (Do not write the equations for the reaction.)
    (a) nickel metal in a solution of silver ions
    (b) zinc metal in a solution of aluminium ions
    (c) an aqueous mixture of copper(II) ions and iodide ions
    (d) chlorine gas bubbled into a bromide ion solution
    (e) an aqueous mixture of copper(II) ions and tin(II) ions
    (f) copper metal in nitric acid

21. Complete the Prediction, Design, and Materials (including safety precautions) for the following investigation.

    Purpose
    The purpose of this investigation is to test the order of strengths of oxidizing and reducing agents given on the redox table (Appendix I).

    Problem
    What is the relative order of strengths of oxidizing and reducing agents for aluminium, nickel, lead, cobalt, and their respective aqueous ions?

22. Describe two designs or methods that can be used to build a redox table.

23. From your knowledge, list two metals that are found as elements and two that are never found as elements in nature. Test your answer by referring to the position of these metals in the table of oxidizing and reducing agents.

24. Has empirical or theoretical knowledge been the most useful to you in predicting the spontaneity of redox reactions? Explain.

Predicting Redox Reactions in Solution

Arrhenius' ideas about solutions provide an important starting point for predicting redox reactions. In solutions, molecules and ions behave approximately independently of each other. A first step in predicting redox reactions is to list all entities that are present. (Some helpful reminders are listed in **Table 6**.) For example, when copper metal is placed into an acidic potassium permanganate solution, copper atoms, potassium ions, permanganate ions, hydrogen ions, and water molecules are all present. Next, using your knowledge of oxidizing and reducing agents, and the redox table in Appendix I, label all possible oxidizing and reducing agents in the starting mixture. The permanganate ion is listed as an oxidizing agent only in an acidic solution. To indicate this combination, draw an arc between the permanganate and hydrogen ions as shown below, and label the pair as an oxidizing agent. This procedure of listing and identifying entities present is a crucial step in predicting redox reactions.

$$\text{Cu(s)} \quad \text{K}^+\text{(aq)} \quad \overset{\overset{\text{OA}}{\frown}}{\text{MnO}_4^-\text{(aq)} \quad \text{H}^+\text{(aq)}} \quad \overset{\text{OA}}{\text{H}_2\text{O(l)}}$$
$$\text{RA} \qquad\qquad\qquad\qquad\qquad\qquad\qquad\qquad \text{RA}$$

(OA labels above MnO$_4^-$, H$^+$, and H$_2$O; RA labels below Cu(s) and H$_2$O)

Table 6 Hints for Listing and Labelling Entities

- Aqueous solutions contain H$_2$O(l) molecules.
- Acidic solutions contain H$^+$(aq) ions.
- Basic solutions contain OH$^-$(aq) ions.
- Some oxidizing and reducing agents are combinations, for example, MnO$_4^-$(aq) and H$^+$(aq).
- H$_2$O(l), Fe^{2+}(aq), Cu$^+$(aq), Sn^{2+}(aq), and Cr^{2+}(aq) may act as either oxidizing or reducing agents. Label both possibilities in your list.

▶ Practice

25. List all entities initially present in the following mixtures, and identify all possible oxidizing and reducing agents.
    (a) A lead strip is placed in a copper(II) sulfate solution.
    (b) A gold coin is placed in a nitric acid solution.
    (c) A potassium dichromate solution is added to an acidic iron(II) nitrate solution.
    (d) An aqueous chlorine solution is added to a phosphorous acid solution.
    (e) A potassium permanganate solution is mixed with an acidified tin(II) chloride solution.
    (f) Iodine solution is added to a basic mixture containing manganese(IV) oxide.

We can use a redox table to identify the strongest oxidizing and reducing agents in a mixture, and then predict which reactions will occur. If we assume that collisions are completely random, the strongest oxidizing agent and the strongest reducing agent will react. (In some cases, further reactions may occur as well, but we will consider only the initial reaction, unless otherwise specified.) The following instructions allow you to make the correct and most efficient use of a redox table, such as the one in Appendix I.

- Choose the strongest oxidizing agent present in your mixture by starting at the top left corner of a redox table and going down the list until you find the oxidizing agent that is in your mixture.
- Choose the strongest reducing agent in your mixture by starting at the bottom right corner of the table and going up the list until you find the reducing agent that is in your mixture.
- Read reduction half-reaction equations from left to right (following the forward arrow).
- Read oxidation half-reaction equations from right to left (following the reverse arrow).
- Assume that any substances not present in the table are spectator ions. You do not need to label or consider these substances.

DID YOU KNOW ?

Getting Rid of Skunk Odour
The smell of a skunk (**Figure 8**) is caused by a thiol compound (R–SH). To deodorize a pet sprayed by a skunk, you need to convert the smelly thiol to an odourless compound. Hydrogen peroxide in a basic solution (usually from sodium bicarbonate) acts as an oxidizing agent to change the thiol to a disulfide compound (RS–SR), which is odourless.

Figure 8
A skunk's only defence is its ability to spray a smelly liquid a distance of 3 m.

Learning Tip

- If the half-reaction equation shows two or more entities present, then both must be in your list. If there is only one entity, then leave it unlabelled as a spectator ion.
- Be careful with a few entities that can act either as an OA or an RA, for example, the iron(II) ion in Sample Problem 13.5.

SAMPLE problem 13.5

Suppose a solution of potassium permanganate is slowly poured into an acidified iron(II) sulfate solution. Does a redox reaction occur and, if it does, what is the reaction equation? Describe two diagnostic tests of your prediction.

To make a prediction, the entities initially present are identified as oxidizing agents, reducing agents, or both, as shown below.

$$\text{OA} \quad \overbrace{\text{OA} \quad \text{OA}} \quad \text{OA} \quad \overbrace{\text{OA} \quad \text{OA}}$$
$$K^+(aq) \quad MnO_4^-(aq) \quad H^+(aq) \quad Fe^{2+}(aq) \quad SO_4^{2-}(aq) \quad H_2O(l)$$
$$\qquad\qquad\qquad\qquad\qquad\qquad\quad RA \qquad\qquad\qquad\quad RA$$

Use the redox table in Appendix I to choose the strongest oxidizing agent and the strongest reducing agent from your list and indicate them as SOA and SRA.

$$\text{OA} \quad \overbrace{\text{SOA} \quad \text{OA}} \quad \text{OA} \quad \overbrace{\text{OA} \quad \text{OA}}$$
$$K^+(aq) \quad MnO_4^-(aq) \quad H^+(aq) \quad Fe^{2+}(aq) \quad SO_4^{2-}(aq) \quad H_2O(l)$$
$$\qquad\qquad\qquad\qquad\qquad\qquad\quad SRA \qquad\qquad\qquad\quad RA$$

Now, write the half-reaction equation for the reduction of the SOA from the redox table.

$$MnO_4^-(aq) + 8\,H^+(aq) + 5\,e^- \rightarrow Mn^{2+}(aq) + 4\,H_2O(l)$$

Write the half-reaction equation for the oxidation of the SRA. Remember to read from right to left in the table.

$$Fe^{2+}(aq) \rightarrow Fe^{3+}(aq) + e^-$$

Before combining the half-reaction equations, balance the number of electrons transferred by multiplying one or both half-reaction equations by an integer so that the number of electrons gained by the oxidizing agent equals the number of electrons lost by the reducing agent.

In this case, the iron ion half-reaction must be multiplied by 5. Add the two equations, but remember to cancel any common terms. You can cancel terms as you add (e.g., 5e⁻) or after you add the two half-reactions.

$$MnO_4^-(aq) + 8\,H^+(aq) + 5\,e^- \rightarrow Mn^{2+}(aq) + 4\,H_2O(l)$$
$$5\,[Fe^{2+}(aq) \rightarrow Fe^{3+}(aq) + e^-]$$
$$\overline{MnO_4^-(aq) + 8\,H^+(aq) + 5\,Fe^{2+}(aq) \rightarrow 5\,Fe^{3+}(aq) + Mn^{2+}(aq) + 4\,H_2O(l)}$$

Finally, use the spontaneity rule to predict whether the net ionic equation represents a spontaneous redox reaction. Indicate this by writing "spont." or "nonspont." over the equation arrow.

$$MnO_4^-(aq) + 8\,H^+(aq) + 5\,Fe^{2+}(aq) \xrightarrow{\text{spont.}} 5\,Fe^{3+}(aq) + Mn^{2+}(aq) + 4\,H_2O(l)$$

In this case, we predict that the reaction is spontaneous. We can test this prediction by mixing the solutions (**Figure 9**) and performing some diagnostic tests. If the solutions are mixed and the purple colour of the permanganate ion disappears, then it is likely that the permanganate ion reacted. If the pH of the solution is tested before and after reaction, and the pH has increased, then the hydrogen ions likely reacted.

Figure 9
A solution of potassium permanganate is being added to an acidic solution of iron(II) ions. The dark purple colour of $MnO_4^-(aq)$ ions instantly disappears. The interpretation is that $MnO_4^-(aq)$ ions react with $Fe^{2+}(aq)$ ions to produce the yellow-brown $Fe^{3+}(aq)$ and $Mn^{2+}(aq)$ ions.

> **COMMUNICATION** *example 1*

In a chemical industry, could copper pipe be used to transport a hydrochloric acid solution? To answer this question,
(a) predict the redox reaction and its spontaneity, and
(b) describe two diagnostic tests that could be done to test your prediction.

Solution

(a)
```
           SOA              OA
  Cu(s)   H⁺(aq)   Cl⁻(aq)  H₂O(l)
  SRA              RA   RA  RA
```

$$2\,H^+(aq) + 2\,e^- \rightarrow H_2(g)$$
$$Cu(s) \rightarrow Cu^{2+}(aq) + 2\,e^-$$

$$2\,H^+(aq) + Cu(s) \xrightarrow{\text{nonspont.}} H_2(g) + Cu^{2+}(aq)$$

Since the reaction is nonspontaneous, it should be possible to use a copper pipe to carry hydrochloric acid.

(b) If no gas is produced when the mixture is observed, then it is likely that no hydrogen gas was produced (**Figure 10**). If the colour of the solution did not change to blue, then copper probably did not react to produce copper(II) ions. (If the solution is tested for pH before and after adding the copper, and the pH did not increase, then the hydrogen ions probably did not react.)

Figure 10
Copper in hydrochloric acid does not appear to react.

Disproportionation

Chemists believe that redox reactions are electron-transfer reactions. One reactant (OA) removes electrons from a second reactant (RA) if a spontaneous reaction takes place. Although the oxidizing and reducing agents that react are usually different entities, this is not a requirement. A reaction in which a species is both oxidized and reduced is called **disproportionation**. This type of reaction is often a redox reaction and occurs when a substance can act either as an oxidizing agent or as a reducing agent.

For example, we know that an iron(II) ion can behave either as an oxidizing agent or a reducing agent (**Figure 7**, page 574). What happens if two iron(II) ions in a solution collide? Will a spontaneous reaction occur as a result of an electron transfer from one iron(II) ion to another iron(II) ion?

$$Fe^{2+}(aq) + 2\,e^- \rightarrow Fe(s)$$
$$2[Fe^{2+}(aq) \rightarrow Fe^{3+}(aq) + e^-]$$
$$3\,Fe^{2+}(aq) \rightarrow Fe(s) + 2\,Fe^{3+}(aq)$$

Using a redox table and the spontaneity rule with iron(II) as the strongest oxidizing agent and iron(II) as the strongest reducing agent, we see that this reaction is nonspontaneous.

Not all disproportionation reactions are nonspontaneous. For example, check the disproportionation of the copper(I) ion on the redox table in Appendix I. Copper(I) as an oxidizing agent appears above copper(I) as a reducing agent. Therefore, an aqueous solution of copper(I) ions will spontaneously, but slowly, disproportionate into copper(II) ions and copper metal.

> **Learning Tip**
>
> Disproportionation reactions are also referred to by the more descriptive terms of "self oxidation–reduction" or "autoxidation."

COMMUNICATION example 2

Will a solution of chromium(II) chloride be stable? Predict the redox reaction and its spontaneity.

Solution

$$\underset{SRA}{\underset{SOA}{Cr^{2+}(aq)}} \quad \underset{RA}{Cl^{-}(aq)} \quad \underset{RA}{\underset{OA}{H_2O(l)}}$$

$$Cr^{2+}(aq) + 2e^- \rightarrow Cr(s)$$

$$\underline{2\,[Cr^{2+}(aq) \rightarrow Cr^{3+}(aq) + e^-]}$$

$$3\,Cr^{2+}(aq) \xrightarrow{nonspont.} Cr(s) + 2\,Cr^{3+}(aq)$$

According to the redox table and the spontaneity rule, a chromium(II) chloride solution should not react and, therefore, should be stable.

SUMMARY: Five-Step Method for Predicting Redox Reactions

Step 1: List all entities present and classify each as a possible oxidizing agent, reducing agent, or both. Do not label spectator ions.

Step 2: Choose the strongest oxidizing agent as indicated in a redox table, and write the equation for its reduction.

Step 3: Choose the strongest reducing agent as indicated in the table, and write the equation for its oxidation.

Step 4: Balance the number of electrons lost and gained in the half-reaction equations by multiplying one or both equations by a number. Then add the two balanced half-reaction equations to obtain a net ionic equation.

Step 5: Using the spontaneity rule, predict whether the net ionic equation represents a spontaneous or nonspontaneous redox reaction.

INVESTIGATION 13.3 Introduction

Predicting the Reaction of Sodium Metal (Demonstration)

Report Checklist

- ○ Purpose
- ○ Problem
- ○ Hypothesis
- ● Prediction
- ● Design
- ○ Materials
- ○ Procedure
- ● Evidence
- ● Analysis
- ● Evaluation (1, 2, 3)

The process of developing theories, laws, and generalizations requires that they be tested numerous times in as many different situations as possible. This process is necessary not only to determine their validity, but also to identify exceptions that may lead to new knowledge.

As part of the Design, include a list of diagnostic tests using the "If [procedure] and [evidence], then [analysis]" format for every product predicted. (This format is described in Appendix C.4.)

Purpose
The purpose of this demonstration is to test the five-step method for predicting redox reactions.

Problem
What are the products of the reaction of sodium metal with water?

To perform this investigation, turn to page 602.

Practice

26. Use the five-step method to predict the most likely redox reaction in each of the following situations. For any spontaneous reaction, describe one diagnostic test to identify a primary product.
    (a) During a demonstration, zinc metal is placed in a hydrochloric acid solution.
    (b) A gold ring accidentally falls into a hydrochloric acid solution.
    (c) Nitric acid is painted onto a copper sheet to etch a design.

27. In your previous chemistry course, you made predictions of reactions according to the single replacement generalization assuming the formation of the most common ion.
    (a) Use the generalization about single replacement reactions to predict the reaction of iron metal with a copper(II) sulfate solution.
    (b) Use the redox theory and table to predict the most likely redox reaction of iron metal with a copper(II) sulfate solution.
    (c) Can both predictions be correct? Which do you think is most likely correct and why?
    (d) Write one qualitative and one quantitative experimental design to test the two different predictions made for the reaction between iron metal and the copper(II) sulfate solution.

28. Write a Prediction, with your reasoning, and a Design (including safety precautions, diagnostic tests, and disposal instructions) for the following experiment.

    Problem
    What are the products of the reaction of tin(II) chloride with an ammonium dichromate solution acidified with hydrochloric acid?

29. When aluminium pots are used for cooking, small pits often develop in the metal. Use your knowledge of redox reactions to explain the formation of these pits. Suggest why this might be a slow process.

30. Oxygen gas is bubbled into an aqueous solution of iron(II) iodide containing excess hydrochloric acid. Predict all spontaneous reactions in the order in which they will occur.

DID YOU KNOW ?

Aluminium Oxide Clouds
The solid rocket boosters of the space shuttle contain the main reactants ammonium perchlorate and aluminium powder. Ammonium perchlorate is a powerful oxidizing agent and aluminium is a relatively strong reducing agent. Their very exothermic reaction produces finely divided aluminium oxide, which forms the billows of white smoke you see in **Figure 11**.

Figure 11
The space shuttle

Predicting Redox Reactions by Constructing Half-Reactions

A redox reaction includes both an oxidation and a reduction. In other words, one substance has to lose electrons as another substance gains electrons. Choosing and writing half-reaction equations is commonly done using a table of relative strengths of oxidizing and reducing agents. But what if this table does not provide the half-reaction equations needed? In this case, use your knowledge about constructing your own half-reaction equations (refer to the Summary on page 567), and then balance electrons to obtain the overall redox reaction equation.

For a particular reaction, chemists know the main starting materials and the reaction conditions (such as acidic or basic). A chemical analysis of the products determines the oxidized and reduced species produced in the reaction. This provides a *skeleton equation* showing only the main reactants and products. Chemists can then determine the details of the final redox equation by looking at the individual balanced half-reaction equations.

Figure 12
A breathalyzer is a technological device that measures the alcohol content in exhaled air based on the colour change from orange (dichromate ion) to green (chromium(III) ion) when an acidic dichromate ion solution reacts with ethanol in a breath sample.

CAREER CONNECTION

Conservator
The career of conservator is highly varied, allowing these professionals to tailor their work to their own special interests. Conservators work with a wide range of objects, including antique metals, furniture, and ceramics. The conservator makes recommendations on how the objects should be preserved, as well as performs chemical analyses to treat existing damage. Conservators often use chemistry to restore historical artifacts. Find out more about the requirements for this interesting and challenging career.

www.science.nelson.com

SAMPLE problem 13.6

A common example of the application of redox reactions is the technology of a breathalyzer (**Figure 12**). A person suspected of being intoxicated blows into this device and the alcohol in the person's breath reacts with an acidic dichromate ion solution to produce acetic acid (ethanoic acid) and aqueous chromium(III) ions. Predict the balanced redox reaction equation.

The information provided tells you the skeleton equation for the major reactants and products.

$$C_2H_5OH(aq) + Cr_2O_7^{2-}(aq) \rightarrow CH_3COOH(aq) + Cr^{3+}(aq)$$

The first step is to separate the entities into the start of two half-reaction equations, keeping related entities together.

$$C_2H_5OH(aq) \rightarrow CH_3COOH(aq)$$
$$Cr_2O_7^{2-}(aq) \rightarrow Cr^{3+}(aq)$$

Now you can complete each half-reaction equation using the same procedure you used in Section 13.1: Balance atoms other than O and H; balance O by adding $H_2O(l)$; balance H by adding $H^+(aq)$; and finally balance the charge by adding electrons. For the ethanol half-reaction,

$$H_2O(l) + C_2H_5OH(aq) \rightarrow CH_3COOH(aq) + 4\,H^+(aq) + 4\,e^-$$

Follow the same procedure to construct the second half-reaction equation.

$$6\,e^- + 14\,H^+(aq) + Cr_2O_7^{2-}(aq) \rightarrow 2\,Cr^{3+}(aq) + 7\,H_2O(l)$$

Recall that the total number of electrons lost must equal the total number of electrons gained. Using the simplest whole numbers, multiply one or both half-reaction equations so that the electrons will be balanced. In this example, the simplest solution is a total of twelve electrons transferred.

$$3\,[H_2O(l) + C_2H_5OH(aq) \rightarrow CH_3COOH(aq) + 4\,H^+(aq) + 4\,e^-]$$
$$2\,[6\,e^- + 14\,H^+(aq) + Cr_2O_7^{2-}(aq) \rightarrow 2\,Cr^{3+}(aq) + 7\,H_2O(l)]$$

Add the two half-reaction equations. Cancel the electrons and anything else that is the same on both sides of the equation. Note that it is not unusual to have unequal chemical amounts of some entities, such as hydrogen ions and water, on the two sides of the equations. The lower amount will always cancel completely. For example, 3 $H_2O(l)$ from the first half-reaction equation cancels 3 mol out of the 14 $H_2O(l)$ in the second equation leaving 11 $H_2O(l)$ in the net equation.

$$3\,C_2H_5OH(aq) + 2\,Cr_2O_7^{2-}(aq) + 16\,H^+(aq) \rightarrow 3\,CH_3COOH(aq) + 4\,Cr^{3+}(aq) + 11\,H_2O(l)$$

Check the final redox equation to make sure that both the symbols and the charges are balanced.

For reactions that occur in basic solutions, you can follow the procedure outlined in Sample Problem 13.6, and then convert to a basic solution. In other words, create the balanced redox equation for an acidic solution, and then add $OH^-(aq)$ to convert the $H^+(aq)$ to water molecules. An example for a basic solution is shown in the following Communication Example.

▶ COMMUNICATION example 3

Permanganate ions and oxalate ions react in a basic solution to produce carbon dioxide and manganese(IV) oxide.

$$MnO_4^-(aq) + C_2O_4^{2-}(aq) \rightarrow CO_2(g) + MnO_2(s)$$

Write the balanced redox equation for this reaction.

Solution

$$2\,[3\,e^- + 4\,H^+(aq) + MnO_4^-(aq) \rightarrow MnO_2(s) + 2\,H_2O(l)]$$
$$3\,[C_2O_4^{2-}(aq) \rightarrow 2\,CO_2(g) + 2\,e^-]$$
$$\overline{8\,H^+(aq) + 2\,MnO_4^-(aq) + 3\,C_2O_4^{2-}(aq) \rightarrow 2\,MnO_2(s) + 4\,H_2O(l) + 6\,CO_2(g)}$$

$$8\,OH^-(aq) + 8\,H^+(aq) + 2\,MnO_4^-(aq) + 3\,C_2O_4^{2-}(aq) \rightarrow$$
$$2\,MnO_2(s) + 4\,H_2O(l) + 6\,CO_2(g) + 8\,OH^-(aq)$$

$$4\,H_2O(l) + 2\,MnO_4^-(aq) + 3\,C_2O_4^{2-}(aq) \rightarrow 2\,MnO_2(s) + 6\,CO_2(g) + 8\,OH^-(aq)$$

> **Learning Tip**
>
> Always check your final answer by counting the number of each kind of atom on both sides of the equation, and by checking the net charge on each side.

▶ Practice

31. Balance the following redox equations using the half-reaction method. All reactions occur in an acidic solution.
  (a) $Zn(s) + NO_3^-(aq) \rightarrow NH_4^+(aq) + Zn^{2+}(aq)$
  (b) $Cl_2(aq) + SO_2(g) \rightarrow Cl^-(aq) + SO_4^{2-}(aq)$

32. Balance the following skeleton redox equations using the half-reaction method. All reactions occur in a basic solution.
  (a) $MnO_4^-(aq) + I^-(aq) \rightarrow MnO_2(s) + I_2(s)$
  (b) $CN^-(aq) + IO_3^-(aq) \rightarrow CNO^-(aq) + I^-(aq)$
  (c) $OCl^-(aq) \rightarrow Cl^-(aq) + ClO_3^-(aq)$

33. Balance the following redox equation.
  $KMnO_4(aq) + H_2S(aq) + H_2SO_4(aq) \rightarrow K_2SO_4(aq) + MnSO_4(aq) + S(s)$

SUMMARY — Predicting Balanced Redox Equations by Constructing Half-Reactions

Step 1: Use the information provided to start two half-reaction equations.
Step 2: Balance each half-reaction equation.
Step 3: Multiply each half-reaction equation by simple whole numbers to balance the electrons lost and gained.
Step 4: Add the two half-reaction equations, cancelling the electrons and anything else that is exactly the same on both sides of the equation.

For basic solutions only
Step 5: Add $OH^-(aq)$ to both sides equal in number to the number of $H^+(aq)$ present.
Step 6: Combine $H^+(aq)$ and $OH^-(aq)$ on the same side to form $H_2O(l)$, and cancel the same number of $H_2O(l)$ on both sides.

Section 13.2 Questions

1. What is the key idea used to explain a redox reaction?
2. Distinguish between oxidation and oxidizing agent in terms of modern theory.
3. Distinguish between reduction and reducing agent in terms of modern theory.
4. Write and label two half-reaction equations to describe each of the following reactions:
   (a) $Co(s) + Cu(NO_3)_2(aq) \rightarrow Cu(s) + Co(NO_3)_2(aq)$
   (b) $Cd(s) + Zn(NO_3)_2(aq) \rightarrow Zn(s) + Cd(NO_3)_2(aq)$
   (c) $Br_2(l) + 2 KI(aq) \rightarrow I_2(s) + 2 KBr(aq)$
5. Using the redox table in Appendix I, predict the spontaneity of each of the reactions shown in 4(a) to (c).
6. Prepare a redox table showing the relative strengths of oxidizing and reducing agents in **Table 7**.

Table 7 Reactions of Group 13 Elements and Ions

	Al^{3+}(aq)	Tl^+(aq)	Ga^{3+}(aq)	In^{3+}(aq)
Al	X	✓	✓	✓
Tl	X	X	X	X
Ga	X	✓	X	✓
In	X	✓	X	X

X no evidence of a redox reaction   ✓ a spontaneous reaction

7. What is the relative strength of oxidizing and reducing agents for strontium, cerium, nickel, hydrogen, platinum, and their aqueous ions? Use the following information to construct a table of relative strengths of oxidizing and reducing agents.

   $3 Sr(s) + 2 Ce^{3+}(aq) \rightarrow 3 Sr^{2+}(aq) + 2 Ce(s)$
   $Ni(s) + 2 H^+(aq) \rightarrow Ni^{2+}(aq) + H_2(g)$
   $2 Ce^{3+}(aq) + 3 Ni(s) \rightarrow$ no evidence of reaction
   $Pt(s) + 2 H^+(aq) \rightarrow$ no evidence of reaction

8. Write an experimental design to determine a mini-redox table for the first three metals and metal ions of Group 12. Include safety and disposal information.
9. For each of the following mixtures, use the complete five-step method to predict the most likely redox reaction. Include one diagnostic test to test your predicted reaction.
   (a) Solutions of nickel(II) nitrate and iron(II) chloride are mixed.
   (b) Oxygen gas is bubbled over a solid silver mesh immersed in a solution of sodium iodide.
   (c) An acidic solution of potassium dichromate is added to a sodium iodide solution.
10. A chemical technician prepares several solutions for use in a chemical analysis. Will each of the solutions listed below be stable if stored for a long time? Justify your answer.
    (a) acidic tin(II) chloride in an inert glass container
    (b) copper(II) nitrate in a tin can
11. In the industrial production of iodine, chlorine gas is bubbled into seawater. Using only water and iodide ions in seawater as the possible reactants, predict the most likely redox reaction, including equations for the half-reactions.
12. The steel of an automobile fender is exposed to acid rain. (Assume that steel is made mainly of iron.) Predict the most likely redox reactions, including the equations for the relevant half-reactions.
13. Define disproportionation and illustrate it using half-reaction equations of tin(II) ions.
14. An excess of cobalt metal was left in an aqueous mixture containing silver ions, iron(III) ions, and copper(II) ions for an extended time. Write a balanced redox equation for every reaction that occurs.
15. Balance the following equation representing a reaction that occurs in an acidic solution:
    $Mn^{2+}(aq) + HBiO_3(aq) \rightarrow Bi^{3+}(aq) + MnO_4^-(aq)$
16. Balance the following equations representing reactions that occur in a basic solution:
    (a) $Cr(OH)_3(s) + IO_3^-(aq) \rightarrow CrO_4^{2-}(aq) + I^-(aq)$
    (b) $Ag_2O(s) + CH_2O(aq) \rightarrow Ag(s) + CHO_2^-(aq)$
17. Many commercially available drain cleaners contain a basic solution of sodium hydroxide, which helps to remove grease in the drains. Some solid drain cleaners contain solid sodium hydroxide and finely divided aluminium metal. When mixed with water, this produces a very vigorous, exothermic reaction shown by the following skeleton equation:

    $Al(s) + H_2O(l) \rightarrow Al(OH)_4^-(aq) + H_2(g)$

    (a) Complete the balanced redox equation for this reaction.
    (b) How does this chemical technology provide a solution to a practical problem?
    (c) Describe and discuss some possible health and safety issues associated with the use of solid drain cleaners.
18. What is the WHMIS symbol for oxidizing materials? What Household Hazardous Product Symbols would be used for oxidizing materials present in a consumer product?

Extension

19. The subject of antioxidants may be controversial, but there are many groups and companies that want you to use antioxidant products. What does the term *antioxidant* suggest? Define antioxidant and list three important examples. Briefly summarize some of the controversy that is associated with antioxidants.

    www.science.nelson.com

20. Many biologically important molecules contain a metal ion surrounded by and bonded to a large organic molecule (such as porphyrin). One very important example is chlorophyll. Prepare a report (paper, poster, or electronic) that contains information about the structure of the main form of chlorophyll, the general role that chlorophyll plays in electron transfer reactions, the close relatives of chlorophyll that contain iron and cobalt, and a technological application inspired by the chlorophyll molecule.

    www.science.nelson.com

Oxidation States 13.3

Historically, oxidation and reduction were considered to be separate processes, more of interest for technology than for science. With modern atomic theory came the idea of an electron transfer involving both a gain of electrons by one entity and a loss of electrons by another entity. This theory of redox reactions is most easily understood for atoms or monatomic ions. Metals and monatomic anions tend to lose electrons (become oxidized), whereas nonmetals and monatomic cations tend to gain electrons (become reduced).

More complex redox reactions, such as the reduction of iron(III) oxide by carbon monoxide in iron production, the oxidation of glucose in cellular respiration, and the use of dichromate ions in chemical analysis, are not adequately described or explained with simple redox theory.

To describe the oxidation and reduction of molecules and polyatomic ions, chemists developed a method of "electron bookkeeping" to keep track of the loss and gain of electrons. The idea is similar to determining the electric charge of a simple ion by counting electrons and protons; for example, a sodium ion ($11p^+$, $10e^-$) has an ion charge of $1+$. For atoms in molecules and polyatomic ions, chemists count a shared pair of electrons in a covalent bond as if it belongs entirely to the more electronegative atom in the bond. For example, in a water molecule (**Figure 1**), the shared pairs of electrons are assigned to the oxygen atom because oxygen has a higher electronegativity (3.5) compared with hydrogen (2.1). Now we can count the electrons around the oxygen and compare this with the number of protons in the nucleus (just like we do for simple ions). In this system, the *oxidation state* of an atom in an entity is defined as the apparent net electric charge that an atom would have if electron pairs in covalent bonds belonged entirely to the more electronegative atom. An oxidation state is a useful idea for keeping track of electrons, but it does not represent an actual charge on an atom—oxidation states are arbitrary charges and should not be confused with actual electric charges.

An **oxidation number** is a positive or negative number corresponding to the oxidation state assigned to an atom in a covalently bonded entity. For example, in a water molecule, the oxidation number of the oxygen atom is -2 and the oxidation number of each hydrogen atom is $+1$.

To distinguish oxidation numbers from actual electrical charges, oxidation numbers are written in this textbook as positive or negative numbers; that is, with the sign preceding the number. Chemists use this method to assign oxidation numbers to many common atoms and ions (**Table 1**), which can then be used to determine the oxidation numbers of other atoms.

Learning Tip

Although the meaning of the terms *oxidation state* and *oxidation number* are slightly different, some people use these terms interchangeably.

Figure 1
An oxygen atom has $8p^+$ and $8e^-$. If the oxygen atom gets to count the two hydrogen electrons in the two shared pairs of electrons, then $8p^+$ and $10e^-$ results in an apparent net charge of $2-$. Each hydrogen atom with $1p^+$ has no additional electrons. Its one electron has already been counted by the oxygen atom. Therefore, the hydrogen has an apparent net charge of $1+$.

Learning Tip

Oxidation numbers are simply positive or negative numbers assigned on the basis of a set of arbitrary rules. It is important for you to realize that these are not electric charges. For this reason, chemists use the term *oxidation number*. For example, we assign oxidation numbers of -2 and $+1$ to the oxygen and hydrogen atoms in a water molecule.

Table 1 Common Oxidation Numbers

Atom or ion	Oxidation number	Examples
all atoms in elements	0	Na is 0
		Cl in Cl_2 is 0
hydrogen in all compounds,	$+1$	H in HCl is $+1$
except hydrogen in hydrides	-1	H in LiH is -1
oxygen in all compounds,	-2	O in H_2O is -2
except oxygen in peroxides	-1	O in H_2O_2 is -1
all monatomic ions	charge on ion	Na^+ is $+1$
		S^{2-} is -2

Oxidation numbers are simply a systematic way of counting electrons. Therefore, the sum of the oxidation numbers in a compound or ion must equal the total charge—zero for neutral compounds and the ion charge for ions.

Learning Tip

1. The sum of the oxidation numbers of any entity must equal the net charge on that entity: zero for neutral compounds, the ion charge for polyatomic ions.
2. The method only works if there is just one unknown after referring to **Table 1**. If there are two or more unknowns, a Lewis formula and electronegativities are required.

> **SAMPLE problem 13.7**

What is the oxidation number of carbon in methane, CH_4?

This is determined by assigning an oxidation number of $+1$ to hydrogen (**Table 1**).

$$\overset{x\ +1}{CH_4}$$

Now solve for x. Since a methane molecule is electrically neutral, the oxidation numbers of the one carbon atom and the four hydrogen atoms (4 times $+1$) must equal zero.

$$x + 4(+1) = 0$$
$$x = -4$$

$$\overset{x\ +1}{CH_4} \quad \text{becomes} \quad \overset{-4\ +1}{CH_4}$$

Carbon in methane has an oxidation number of -4.

> **SAMPLE problem 13.8**

What is the oxidation number of manganese in a permanganate ion, MnO_4^-.

The oxidation number of manganese in the permanganate ion, MnO_4^-, is determined using the oxidation number of oxygen as -2 (**Table 1**) and the knowledge that the charge on the ion is $1-$. The total of the oxidation numbers of the one manganese atom (x) and the four oxygen atoms (4 times -2) must equal the charge on the ion ($1-$).

$$x + 4(-2) = -1$$
$$x = +7$$

$$\overset{x\ -2}{MnO_4^-} \quad \text{becomes} \quad \overset{+7\ -2}{MnO_4^-}$$

The oxidation number of manganese in MnO_4^- is $+7$.

Learning Tip

Alternatively, you can always split an ionic formula into the cation and anion before solving for an unknown oxidation number.

For example, if you want to know the oxidation number for sulfur in sodium sulfate, start with the sulfate ion:

$$\overset{x\ -2}{SO_4^{2-}}$$

$$x + 4(-2) = -2 \quad x = +6$$

> **COMMUNICATION example 1**

What is the oxidation number of sulfur in sodium sulfate?

Solution

$$\overset{+1\ \ \ x\ -2}{Na_2SO_4} \quad\quad 2(+1) + x + 4(-2) = 0$$
$$x = +6$$

According to the concept of oxidation states, the oxidation number of sulfur in sodium sulfate is $+6$.

Section 13.3

SUMMARY Determining Oxidation Numbers

Step 1: Assign common oxidation numbers (**Table 1** on page 583).
Step 2: The total of the oxidation numbers of atoms in a molecule or ion equals the value of the net electric charge on the molecule or ion.
  (a) The sum of the oxidation numbers for a compound is zero.
  (b) The sum of the oxidation numbers for a polyatomic ion equals the charge on the ion.
Step 3: Any unknown oxidation number is determined algebraically from the sum of the known oxidation numbers and the net charge on the entity.

▶ Practice

1. Determine the oxidation number of
   (a) S in SO_2
   (b) Cl in $HClO_4$
   (c) S in SO_4^{2-}
   (d) Cr in $Cr_2O_7^{2-}$
   (e) I in MgI_2
   (f) H in CaH_2

2. Determine the oxidation number of nitrogen in
   (a) $N_2O(g)$
   (b) $NO(g)$
   (c) $NO_2(g)$
   (d) $NH_3(g)$
   (e) $N_2H_4(g)$
   (f) $NaNO_3(s)$
   (g) $N_2(g)$
   (h) $NH_4Cl(s)$

3. Determine the oxidation number of carbon in
   (a) graphite (elemental carbon)
   (b) glucose
   (c) sodium carbonate
   (d) carbon monoxide

4. Bruderheim, Alberta, is the site of several companies that produce sodium chlorate. Almost all of the sodium chlorate produced is sold to pulp and paper mills to produce chlorine dioxide as a bleaching agent (**Figure 2**). Determine the oxidation number of every atom or ion in the following chemical equation for the industrial production of chlorine dioxide.

   $2 ClO_3^-(aq) + 2 Cl^-(aq) + 4 H^+(aq) \rightarrow 2 ClO_2(g) + Cl_2(g) + 2 H_2O(l)$

5. Carbon can be progressively oxidized in a series of organic reactions. Determine the oxidation number of carbon in each of the compounds in the following series of oxidations:

   methane→ methanol→ methanal→ methanoic acid→ carbon dioxide
   CH_4 CH_3OH CH_2O $HCOOH$ CO_2

Figure 2
The bleaching of wood pulp to produce white paper is now mostly done using chlorine dioxide in place of the more environmentally damaging elemental chlorine.

Oxidation Numbers and Redox Reactions

Although the concept of oxidation states is somewhat arbitrary, because it is based on assigned charges, it is self-consistent and allows predictions of electron transfer. Chemists believe that if the oxidation number of an atom or ion changes during a chemical reaction, then an electron transfer (that is, an oxidation–reduction reaction) occurs. Based on oxidation numbers, an increase in the oxidation number is defined as an **oxidation** and a decrease in the oxidation number is a **reduction**. If oxidation numbers are listed as positive and negative numbers on a number line (**Figure 3**), then the process of oxidation involves a change to a more positive value ("up" on the number line) and reduction is a change to a more negative value ("down" on the number line). If the oxidation numbers do not change, this is interpreted as no transfer of electrons. A reaction in which all oxidation numbers remain the same is not a redox reaction.

Coal is a fossil fuel that is burned in huge quantities in some electrical power generating stations. If we assume pure carbon and complete combustion, carbon is converted to carbon dioxide. In this reaction, the oxidation number of carbon changes from 0 in

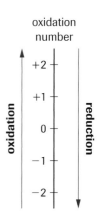

Figure 3
In a redox reaction, both oxidation and reduction occur.

C(s) to +4 in $CO_2(g)$. Simultaneously, oxygen is reduced from 0 in $O_2(g)$ to −2 in $CO_2(g)$.

$$\underset{\text{reduction}}{\overset{\text{oxidation}}{\overset{0}{C}(s) + \overset{0}{O_2}(g) \rightarrow \overset{+4\;-2}{CO_2}(g)}}$$

The main purpose of assigning oxidation numbers is to see how these numbers change as a result of a chemical reaction. In any redox reaction, like the combustion of carbon, there will always be both an oxidation and a reduction. We will use these changes to balance redox equations, but first we will look at some additional examples.

LAB EXERCISE 13.B

Oxidation States of Vanadium

Vanadium is a transition metal that forms many different ions (**Table 2**). Vanadium and its compounds have many different uses, including colouring for glass, ceramics, and plastics.

Report Checklist
- ○ Purpose
- ○ Problem
- ○ Hypothesis
- ○ Prediction
- ○ Design
- ○ Materials
- ○ Procedure
- ○ Evidence
- ● Analysis
- ○ Evaluation

Purpose
The purpose of this lab exercise is to use the concept of oxidation states to investigate some redox chemistry of vanadium compounds.

Problem
What are the oxidation numbers and changes in oxidation number for vanadium ions?

Table 2 Colours of Vanadium Ions

Ion name	Ion formula	Colour
vanadate(V)	VO_3^-(aq)	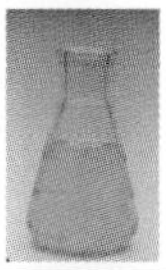
vanadate(IV)	VO^{2+}(aq)	
vanadium(III)	V^{3+}(aq)	
vanadium(II)	V^{2+}(aq)	

Analysis
(a) Using **Table 2**, identify the vanadium ions in the sequence of reactions in **Table 3**.
(b) In each case, is the vanadium being oxidized or reduced? Justify your answer, using oxidation numbers.
(c) Explain the observations made in (3) to (6) in **Table 3**. Suggest what is causing these changes.

Evidence

Table 3 Reactions of Vanadium Ions

Procedure	Final solution colours
(1) ammonium vanadate(V) dissolved in sulfuric acid	yellow
(2) yellow solution with three subsequent additions of small quantities of zinc dust	yellow turned blue, then green, then violet
(3) violet solution left sitting in an open container	slowly turned green
(4) yellow solution mixed with potassium iodide solution	very dark blue, almost black
(5) blue solution mixed with potassium iodide solution	stayed blue; no change
(6) violet solution slowly mixed with acidic potassium permanganate	violet to green to blue to yellow

SAMPLE problem 13.9

You have seen the reaction of active metals such as zinc with an acid. Identify the oxidation and reduction in the reaction of zinc metal with hydrochloric acid.

First, you need to write the chemical equation, as it is not provided. Net ionic equations are best, but the procedure will still work if you write a nonionic equation.

$$Zn(s) + 2\,H^+(aq) \rightarrow Zn^{2+}(aq) + H_2(g)$$

After writing the equation, determine all oxidation numbers.

$$\overset{0}{Zn}(s) + 2\,\overset{+1}{H^+}(aq) \rightarrow \overset{+2}{Zn^{2+}}(aq) + \overset{0}{H_2}(g)$$

Now look for the oxidation number of an atom/ion that increases as a result of the reaction and label the change as oxidation. There must also be an atom/ion whose oxidation number decreases. Label this change as reduction.

$$\underbrace{\overset{0}{Zn}(s) + 2\,\overset{+1}{H^+}(aq) \rightarrow \overset{+2}{Zn^{2+}}(aq) + \overset{0}{H_2}(g)}_{\text{reduction}}^{\text{oxidation}}$$

SAMPLE problem 13.10

When natural gas burns in a furnace, carbon dioxide and water form. Identify oxidation and reduction in this reaction.

First, write the equation.

$$CH_4(g) + 2\,O_2(g) \rightarrow CO_2(g) + 2\,H_2O(g)$$

Now we can insert the oxidation numbers and arrows.

$$\overset{-4\ +1}{CH_4}(g) + 2\,\overset{0}{O_2}(g) \rightarrow \overset{+4\ -2}{CO_2}(g) + 2\,\overset{+1\ -2}{H_2O}(g)$$

Carbon is oxidized from −4 in methane to +4 in carbon dioxide as it reacts with oxygen. Simultaneously, oxygen is reduced from 0 in oxygen gas to −2 in both products.

Notice that the oxygen atoms in the reactant are distributed between the two products. This does not change our procedure because we are only looking for the change from reactant to product. We say that "oxygen is reduced" in this reaction and it does not matter where the reduced oxygen appears in the products.

DID YOU KNOW ?

Redox in Biological Systems
Biologists often classify oxidation and reduction in terms of the addition or removal of oxygen or hydrogen.

- Removal of oxygen decreases the oxidation number of carbon (i.e., a reduction), e.g.,

  HCOOH → HCHO

  C is +2   C is 0

  The reverse also applies. Addition of oxygen corresponds to an oxidation.

- Removal of hydrogen increases the oxidation number of carbon (i.e., an oxidation), e.g., in the nicotinamide coenzyme

  NADH → NAD$^+$

  C is +2   C is −1

  The reverse also applies. Addition of hydrogen corresponds to a reduction.

Figure 4
Since sour natural gas often contains hydrogen sulfide, burning it can be a significant source of pollutants. It is also a waste of energy when many of these flares operate in a large oil field.

Learning Tip

You can adjust the number of electrons per atom to the number per molecule by multiplying the number per atom by the subscript of the atom in the chemical formula.

SAMPLE problem 13.11

Hydrogen sulfide is an unpleasant constituent of "sour" natural gas. Hydrogen sulfide is not only very toxic, but it also smells terrible, similar to rotting eggs. It is common practice to "flare," or burn, relatively small quantities of sour natural gas that occur with oil deposits (**Figure 4**). The gas is burned because it is not worth recovering and treating a small quantity of gas. When this gas is burned, hydrogen sulfide is converted to sulfur dioxide. Use oxidation numbers to balance this equation.

$$H_2S(g) + O_2(g) \rightarrow SO_2(g) + H_2O(g)$$

The first step is to assign oxidation numbers to all atoms/ions and look for the numbers that change. Circle or highlight the oxidation numbers that change.

$$\overset{+1\ -2}{H_2S(g)} + \overset{0}{O_2(g)} \longrightarrow \overset{+4\ -2}{SO_2(g)} + \overset{+1\ -2}{H_2O(g)}$$

(oxidation: S from −2 to +4; reduction: O from 0 to −2)

Notice that a sulfur atom is oxidized from −2 to +4. This is a change of 6 and means 6 e⁻ have been transferred. An oxygen atom is reduced from 0 to −2, a change of 2 or 2 e⁻ transferred. Because the substances in the equation are molecules, not atoms, we need to specify the change in the number of electrons per molecule.

$$\overset{+1\ -2}{H_2S(g)} + \overset{0}{O_2(g)} \rightarrow \overset{+4\ -2}{SO_2(g)} + \overset{+1\ -2}{H_2O(g)}$$

6 e⁻/S atom   2 e⁻/O atom
6 e⁻/H₂S   4 e⁻/O₂

One H₂S molecule contains one sulfur atom. Therefore, the number of electrons transferred per sulfur atom is the same number per H₂S molecule. An O₂ molecule contains two O atoms. Therefore, when one O₂ molecule reacts, two oxygen atoms transfer 2 e⁻ each for a total of 4 e⁻.

The next step is to determine the simplest whole numbers that will balance the number of electrons transferred for each reactant. The numbers become the coefficients for the reactants.

$$\overset{+1\ -2}{H_2S(g)} + \overset{0}{O_2(g)} \rightarrow \overset{+4\ -2}{SO_2(g)} + \overset{+1\ -2}{H_2O(g)}$$

6 e⁻/S atom   2 e⁻/O atom
6 e⁻/H₂S   4 e⁻/O₂
× 2   × 3
12   12

Now you have the coefficients for the reactants.

2 H₂S(g) + **3** O₂(g) → SO₂(g) + H₂O(g)

The coefficients of the products can easily be obtained by balancing the atoms whose oxidation numbers have changed, and then any other atoms. The final balanced equation is shown below:

2 H₂S(g) + **3** O₂(g) → **2** SO₂(g) + **2** H₂O(g)

Sometimes you may not know all of the reactants and products of a redox reaction. The main reactants and oxidized/reduced products will always be given and you will know if the reaction took place in an acidic or basic solution. Experimental evidence shows that water molecules, hydrogen ions, and hydroxide ions play important roles in reactions in such solutions. The procedure for balancing such equations is initially the same as the one used in Sample Problem 13.11, but you will need to add water molecules, hydrogen ions, and/or hydroxide ions to finish the balancing of the overall equation. The following two sample problems illustrate this procedure.

▶ SAMPLE problem 13.12

Chlorate ions and iodine react in an acidic solution to produce chloride ions and iodate ions. Balance the equation for this reaction.

$$ClO_3^-(aq) + I_2(aq) \rightarrow Cl^-(aq) + IO_3^-(aq)$$

Assign oxidation numbers to each atom/ion and note which numbers change.

$$\overset{+5\ -2}{ClO_3^-}(aq) + \overset{0}{I_2}(aq) \rightarrow \overset{-1}{Cl^-}(aq) + \overset{+5\ -2}{IO_3^-}(aq)$$

A chlorine atom is reduced from +5 to −1, a change of 6. Simultaneously, an iodine atom is oxidized from 0 to +5, a change of 5. Record the change in the number of electrons per atom, and per molecule or polyatomic ion.

$$\overset{+5\ -2}{ClO_3^-}(aq) + \overset{0}{I_2}(aq) \rightarrow \overset{-1}{Cl^-}(aq) + \overset{+5\ -2}{IO_3^-}(aq)$$

6 e⁻/Cl   5 e⁻/I

6 e⁻/ClO₃⁻   10 e⁻/I₂

The total number of electrons transferred by each reactant must be the same. Multiply the numbers of electrons by the simplest whole numbers to make the totals equal, in this case, 30 e⁻. You can now write the coefficients for the reactants and the products.

$$5\ ClO_3^-(aq) + 3\ I_2(aq) \rightarrow 5\ Cl^-(aq) + 6\ IO_3^-(aq)$$

6 e⁻/Cl   5 e⁻/I
6 e⁻/ClO₃⁻   10 e⁻/I₂
× 5   × 3

Although the chlorine and iodine atoms are now balanced, notice that the oxygen atoms are not; 15 on the left versus 18 on the right. Because this reaction occurs in an aqueous solution, we can add H₂O molecules to balance the O atoms. The reactant side requires three oxygen atoms (from three water molecules) to equal the total of 18 oxygen atoms on the product side.

$$3\ H_2O(l) + 5\ ClO_3^-(aq) + 3\ I_2(aq) \rightarrow 5\ Cl^-(aq) + 6\ IO_3^-(aq)$$

In adding water molecules, we are also adding H atoms. Because this reaction occurs in an acidic solution, we will add H⁺(aq) to balance the hydrogen.

$$3\ H_2O(l) + 5\ ClO_3^-(aq) + 3\ I_2(aq) \rightarrow 5\ Cl^-(aq) + 6\ IO_3^-(aq) + \mathbf{6\ H^+(aq)}$$

The redox equation should now be completely balanced. Check your work by checking the total numbers of each atom/ion on each side and checking the total electric charge, which should also be balanced.

Learning Tip

A balanced chemical reaction equation includes both a mass and charge balance. Mass is balanced using the atomic symbols. *If the symbols balance, but not the charge, the equation is not balanced.* Be sure to check both the symbols and charges.

DID YOU KNOW ?

Aboriginal Food Preservation Technology
Pemmican, from the Cree word "pimikan," is a mixture of dried, ground red meat, dried berries, and fat or grease from bone marrow (**Figure 5**). When stored in skin or intestine bags (essentially becoming vacuum-sealed), the pemmican can be kept unspoiled for months or even years. This Aboriginal technology solved a problem that Western scientists recognize today as the oxidation of lipids and proteins by aerobic bacteria, the main culprit in the spoilage of meat.

Figure 5
Pemmican

Case Study

Bleaching Wood Pulp

The production of pulp and paper is one of Canada's major industries (**Figure 7**), employing several thousand people across the country and contributing billions of dollars to Canada's export market. The processes used in the pulp and paper industry illustrate how a successful technology can have both intended and unintended consequences.

Modern chemistry and chemical technology make it possible for cellulose fibres from wood pulp to be bleached, dyed, coated, and treated to manufacture paper, as well as cellophane and explosives. The production of white paper involves a bleaching process in which a strong oxidizing agent oxidizes coloured organic compounds. For many years, the oxidizing agent was elemental chlorine, which breaks down and removes lignin, an organic polymer that binds the wood fibres together. Over 300 reaction by-products, including chloroform, carbon tetrachloride, chlorophenols, and furans, are produced during the process.

Many of these by-products are potentially harmful. Research indicates that, although only one of 75 dioxin isomers is extremely toxic, some dioxins can cause immune system suppression and severe reproductive disorders, including birth defects and sterility. Also, certain dioxins are potent carcinogens. Once in the ecosystem, dioxins resist breakdown and bioaccumulate in animal tissues. Traces of dioxins have been found in bleached paper products such as diapers, sanitary products, paper plates, toilet paper, coffee filters, food packaging, and writing paper.

In the 1980s, growing concern over toxic chemicals entering the ecosystem from pulp mill effluent prompted the Canadian government to implement stricter environmental regulations. As a result, the pulp and paper industry underwent the largest environmental upgrade in its history. The use of elemental chlorine for bleaching was reduced or replaced, which produced a corresponding reduction in the emission of dioxins and furans. In addition, the recycling of paper increased and the water consumption per tonne of paper decreased. Overall, the effluent quality has vastly improved, although its adverse effects on fish and other aquatic organisms are still being seen.

Industry researchers proposed multiple solutions, many of which were implemented to reduce the quantities of toxic chemicals produced by pulp mills. Each solution involved different designs, materials, and processes. Some of the measures introduced to reduce the emission of organochlorines include: more efficient washing of pulp; oxygen pre-bleaching; using chlorine dioxide, hydrogen peroxide, or ozone instead of elemental chlorine; and bleaching to a lesser degree. While these new technologies are reducing the negative impact of pulp mills on the environment, there is concern that they, too, will have some unintended effects on the environment.

Figure 7
There are over 150 pulp and paper mills in Canada. Canada is the fourth largest producer in the world of pulp and paper products, and the world's largest supplier of newsprint.

Case Study Questions

1. What were the unintended consequences of using chlorine to bleach wood pulp?

2. List some new processes that were developed to reduce the problems created by bleaching wood pulp with chlorine.

3. Describe the perspectives that need to be considered when selecting the chemical to replace elemental chlorine in the bleaching process.

4. The most common alternative to the use of elemental chlorine as a bleaching agent is chlorine dioxide gas. In addition to having a less negative environmental impact, chlorine dioxide has a greater oxidizing ability. Assuming both chlorine and chlorine dioxide are converted to chloride ions, use oxidation numbers to show the greater oxidizing ability of chlorine dioxide.

Extension

5. Investigate at least two pulp and paper companies (one in Canada and one outside of North America) to see what actions they have taken to reduce their impact on the environment. Present your findings in a medium of your choice.

www.science.nelson.com

6. Research several reactions involved in the bleaching of paper, and balance or verify the balancing of the reaction equations using oxidation numbers.

www.science.nelson.com

Section 13.3 Questions

1. Copy and complete the following table to distinguish between oxidation and reduction:

	Electron transfer	Oxidation states
oxidation		
reduction		

2. Define an oxidation number.

3. State two ways in which you can recognize a redox reaction, using a chemical reaction equation.

4. Write the oxidation number of each atom/ion in the following substances:
   (a) carbon monoxide, $CO(g)$, a toxic gas
   (b) ozone, $O_3(g)$, ozone layer
   (c) ammonium chloride, $NH_4Cl(s)$, used in dry cells (batteries)
   (d) phosphoric acid, $H_3PO_4(aq)$, in cola soft drinks
   (e) sodium thiosulfate, $Na_2S_2O_3(s)$, antidote for cyanide poisoning
   (f) sodium tripolyphosphate, $Na_5P_3O_{10}(s)$, in laundry detergents

5. Assigning oxidation numbers using the rules we have established may occasionally produce some unusual results. For example, consider Fe_3O_4.
   (a) Determine the oxidation number of iron in Fe_3O_4.
   (b) What is unusual about your answer? Suggest a reason for your answer.

6. Redox reactions are common in organic chemistry. For example, carboxyl groups can be oxidized to form carbon dioxide. In the following chemical equation, permanganate ions convert oxalate ions to carbon dioxide in an acidic solution.

   $$2\,MnO_4^-(aq) + 5\,C_2O_4^{2-}(aq) + 16\,H^+(aq) \rightarrow 2\,Mn^{2+}(aq) + 8\,H_2O(l) + 10\,CO_2(g)$$

   (a) Assign oxidation numbers to all atoms/ions.
   (b) Which atom is oxidized? State the change.
   (c) Which atom is reduced? State the change.
   (d) Identify the oxidizing and reducing agents.

7. When carbon dioxide is released into the atmosphere from natural or human activities, some of it reacts with water to form carbonic acid. This accounts for the natural acidity of rainwater and may also contribute to acid rain.
   (a) Write the balanced chemical equation for the reaction of carbon dioxide with water to form carbonic acid.
   (b) Is this a redox reaction? Justify your answer.

8. Balance the following equations representing reactions that occur in an acidic solution:
   (a) $Cu(s) + NO_3^-(aq) \rightarrow Cu^{2+}(aq) + NO_2(g)$
   (b) $H_2O_2(aq) + Cr_2O_7^{2-}(aq) \rightarrow Cr^{3+}(aq) + O_2(g) + H_2O(l)$
   (c) $Mn^{2+}(aq) + HBiO_3(aq) \rightarrow Bi^{3+}(aq) + MnO_4^-(aq)$

9. Balance the following equations representing a reaction that occurs in a basic solution:
   (a) $Cr(OH)_3(s) + IO_3^-(aq) \rightarrow CrO_4^{2-}(aq) + I^-(aq)$
   (b) $Ag_2O(s) + HCHO(aq) \rightarrow Ag(s) + HCO_2^-(aq)$
   (c) $S_2O_4^{2-}(aq) + O_2(g) \rightarrow SO_4^{2-}(aq)$

10. Hydrogen peroxide decomposes in the presence of a catalyst to form water and oxygen gas. Use oxidation numbers to show that this reaction is a disproportionation.

11. State two general experimental designs that could help determine the balancing of the main species in a redox reaction.

12. Evidence shows that iron(III) sulfide reacts according to the following balanced redox equation:

    $$2\,Fe_2S_3(s) + 6\,H_2O(l) + 11\,O_2(g) \rightarrow 4\,FeO(s) + 6\,H_2SO_4(aq)$$

    (a) List the oxidation numbers for all entities in this reaction equation.
    (b) Identify the entities that are oxidized and reduced.
    (c) How does this reaction equation illustrate the limitations of the method presented to balance redox reaction equations?

Extension

13. How is the making of pemmican by First Nations peoples an example of technology providing solutions to practical problems? List some advantages of pemmican and state its importance to early European explorers and settlers.

    www.science.nelson.com

14. Most organisms derive their metabolic energy from cellular respiration, making this one of the most important biological redox processes. Outline the chemistry of cellular respiration. Your response should include:
    - the overall chemical reaction equation for cellular respiration, with the oxidation, reduction, oxidizing agent, reducing agent, and overall direction of electron transfer indicated
    - brief descriptions of the three main stages of aerobic respiration

    www.science.nelson.com

15. Science terms and concepts are often used to help promote a variety of new technologies marketed to consumers. One recent example is the titanium necklace or bracelet. Refer to Appendix B.4 and use the Internet to answer the following questions.
    (a) List some science terms and concepts that are mentioned in the promotion of this product.
    (b) Briefly summarize the claims implied by the manufacturer.
    (c) What kind of evidence is presented to justify the claims?
    (d) Write a brief experimental design to conduct a scientific test of the claims and collect more reliable evidence.

    www.science.nelson.com

13.4 Redox Stoichiometry

The stoichiometric method can be used to predict or analyze the quantity of a chemical involved in a chemical reaction. You encountered many applications of stoichiometry in Chapter 7 involving masses, volumes, and concentrations of reactants and products. For the stoichiometry calculations in Chapter 7, you assumed that all the reactions were spontaneous, fast, stoichiometric, and quantitative. These same assumptions apply to redox stoichiometry.

There are many industrial and laboratory applications of redox stoichiometry. For example, a mining engineer must know the concentration of iron in a sample of iron ore in order to decide whether or not a mine would be profitable. Chemical technicians in industry, monitoring the quality of their companies' products, must determine the concentration of substances such as sodium hypochlorite (NaClO) in bleach, or hydrogen peroxide (H_2O_2) in disinfectants. Hospital laboratory technicians and environmental chemists detect tiny traces of chemicals by a variety of methods. Although much analytical chemistry involves sophisticated equipment, the basic technological process of titration still has an important role (Appendix C.4).

Figure 1
Titration is a common experimental design for quantitative chemical analysis.

In a titration, one reagent (the *titrant*) is slowly added to another (the *sample*) until an abrupt change in a solution property (the *endpoint*) occurs (**Figure 1**). In acid–base titrations, the titrant is generally a strong acid or base. In redox titrations, the titrant is always a strong oxidizing or reducing agent. Two oxidizing agents commonly used in redox titrations are acidic solutions of permanganate ions or dichromate ions. They are both strong oxidizing agents and undergo a colour change when they oxidize a reducing agent in a sample being titrated. The permanganate ion, which has an intense purple-pink colour in solution, changes to the essentially colourless manganese(II) ion in a reaction with a reducing agent that is usually colourless (**Figure 2**).

$$\underset{\text{purple-pink}}{MnO_4^-(aq)} + 8\,H^+(aq) + 5\,e^- \rightarrow \underset{\text{colourless}}{Mn^{2+}(aq)} + 4\,H_2O(l)$$

Once the reducing agent in the sample has completely reacted, the next drop of permanganate added remains unreacted and causes a pink colour in the mixture. The colour change of the reaction mixture (colourless to pink) is the endpoint and corresponds to a slight excess of unreacted permanganate ion. The volume of permanganate solution added when the endpoint is reached is a measurement of the point at which stoichiometric quantities of reactants have been combined.

The dichromate ion is also commonly used in redox titrations; however, its colour change is not as easy to see: the orange dichromate solution gradually changes to a green chromium(III) solution. A redox indicator is usually added to produce a sharper endpoint.

$$\underset{\text{orange}}{Cr_2O_7^{2-}(aq)} + 14\,H^+(aq) + 6\,e^- \rightarrow \underset{\text{green}}{2\,Cr^{3+}(aq)} + 7\,H_2O(l)$$

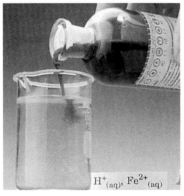

Figure 2
A solution of potassium permanganate is being added to an acidic solution of iron(II) ions. The dark purple-pink colour of $MnO_4^-(aq)$ ions instantly disappears as they react with iron(II) ions to produce the almost colourless $Mn^{2+}(aq)$.

When a titration is used to analyze the concentration of a sample, the concentration of the titrant used must be accurately known. If the titrant is not a standard solution, the titrant is standardized by calculating its concentration using evidence from an analysis with a primary standard. A *primary standard* is a chemical that can be used directly to prepare a standard solution—a solution of precisely known concentration (refer to Unit 4).

SAMPLE problem 13.14

A solution of potassium permanganate cannot be directly prepared with a precisely known concentration because the permanganate ion reacts with organic and inorganic impurities in the water and with the water itself. Thus, potassium permanganate is not used as a primary standard. Complete the Analysis of the investigation report.

Problem
What is the concentration of the potassium permanganate solution?

Design
A freshly prepared solution of potassium permanganate is titrated against samples of acidic tin(II) chloride solution, which has a known concentration. The tin(II) chloride solution is the primary standard.

Evidence
Table 1 Titration of 10.00 mL of Acidic 0.0500 mol/L $SnCl_2(aq)$ with $KMnO_4(aq)$

Trial	1	2	3	4
final burette reading (mL)	18.4	35.3	17.3	34.1
initial burette reading (mL)	1.0	18.4	0.6	17.3
volume of $KMnO_4(aq)$ (mL)	17.4	16.9	16.7	16.8
endpoint colour	dark pink	light pink	light pink	light pink

Analysis
The Analysis for a titration experiment follows the same general stoichiometry steps that you practiced in Unit 4. The main difference is that you have a much more sophisticated way of writing the chemical reaction equation.

The first endpoint was overshot and was not used in the average for the analysis. At the endpoint, an average of 16.8 mL of permanganate solution was used.

```
   OA         SOA    OA         OA           OA
  K⁺(aq)   MnO₄⁻(aq) H⁺(aq)   Sn²⁺(aq)   Cl⁻(aq)  H₂O(l)
                SRA            RA     RA     RA
```

$2 \,[MnO_4^-(aq) + 8\, H^+(aq) + 5\, e^- \rightarrow Mn^{2+}(aq) + 4\, H_2O(l)]$

$5 \,[Sn^{2+}(aq) \rightarrow Sn^{4+}(aq) + 2\, e^-]$

$2\, MnO_4^-(aq) + 16\, H^+(aq) + 5\, Sn^{2+}(aq) \rightarrow 2\, Mn^{2+}(aq) + 8\, H_2O(l) + 5\, Sn^{4+}(aq)$

16.8 mL   10.00 mL
c 0.0500 mol/L

$n_{Sn^{2+}} = 10.00 \text{ mL} \times \dfrac{0.0500 \text{ mol}}{1 \text{ L}} = 0.500 \text{ mmol}$

$n_{MnO_4^-} = 0.500 \text{ mmol} \times \dfrac{2}{5} = 0.200 \text{ mmol}$

$[MnO_4^-] = \dfrac{0.200 \text{ mmol}}{16.8 \text{ mL}} = 0.0119 \text{ mol/L or } 11.9 \text{ mmol/L}$

or $[MnO_4^-] = 10.00 \text{ mL Sn}^{2+} \times \dfrac{0.0500 \text{ mol Sn}^{2+}}{1 \text{ L Sn}^{2+}} \times \dfrac{2 \text{ mol MnO}_4^-}{5 \text{ mol Sn}^{2+}} \times \dfrac{1}{16.8 \text{ mL MnO}_4^-}$

$= 0.0119 \text{ mol/L}$

According to the evidence gathered and the stoichiometric analysis, the amount concentration of the potassium permanganate solution is 0.0119 mol/L or 11.9 mmol/L.

DID YOU KNOW ?
Scientific Credibility
For credibility, scientific claims must be testable empirically. The results of any tests must be replicated by further experimentation. The same person might repeat the measurements, in a titration, for example, or members of the same research team might repeat the experiment. The scientific community accepts new scientific discoveries only if different scientists in different laboratories are able to reproduce the results.

Learning Tip
The general stoichiometry procedure is as follows:
1. Write a balanced chemical equation with measurements and conversion factors.
2. Convert given measurements into a chemical amount.
3. Calculate the amount of the required substance using the mole ratio.
4. Convert this calculated amount to the final requested quantity.

DID YOU KNOW ?
Standardized Solution
The *standardized* potassium permanganate solution can be used as a strong oxidizing agent in further titrations. A laboratory technician might standardize the solution in the morning, and then re-standardize at noon and at the end of the day to increase the certainty of the results.

LAB EXERCISE 13.C

Analyzing for Tin

Report Checklist
- ○ Purpose
- ○ Problem
- ○ Hypothesis
- ○ Prediction
- ○ Design
- ○ Materials
- ○ Procedure
- ○ Evidence
- ● Analysis
- ○ Evaluation

Extensive long-term research has found that treating children's teeth with fluoride significantly reduces tooth decay. When this was first discovered, toothpastes were produced containing tin(II) fluoride. Complete the Analysis of the investigation report.

Purpose
The purpose of this lab exercise is to use the stoichiometric method in a redox chemical analysis.

Problem
What is the amount concentration of tin(II) ions in a solution prepared for research on toothpaste?

Evidence

Table 2 Titration of 10.00 mL of acidic Sn^{2+}(aq) with 0.0832 mol/L $KMnO_4$(aq)

Trial	1	2	3
final burette reading (mL)	15.8	28.1	40.6
initial burette reading (mL)	3.4	15.8	28.1
volume of $KMnO_4$(aq) (mL)	12.4	12.3	12.5

Figure 3
Imants Lauks (1952–)

WEB Activity

Canadian Achievers—Imants Lauks

Dr. Imants Lauks (**Figure 3**) is a world leader in developing biochips for clinical diagnostic products. He invented the silicon chip blood analyzer in 1986. Biochips combine silicon chip technology with chemical reactions, many of which are electrochemical (redox) reactions.

1. Describe the FlexCard™ technology that Dr. Lauks' current company, Epocal, is developing.
2. What practical problems does this technology solve?

www.science.nelson.com

Practice

1. Titration is a common experimental procedure for the quantitative analysis of chemical substances. What are the four requirements for titration experiments?
2. Titration is one of several experimental procedures that can be used to determine the quantity of a chemical in a sample. What are some alternative designs available for this purpose?
3. Silver metal can be recycled by reacting nickel metal with waste silver ion solutions. What volume of 0.10 mol/L silver ion solution will react completely with 25.0 g of nickel metal?
4. In a chemical analysis of a chromium alloy, all of the chromium is first converted to chromate ions. A 50.0 mL sample of the chromate ion solution is then reduced in a basic solution to chromium(III) hydroxide by reaction with 22.6 mL of 1.08 mol/L sodium sulfite. In this reaction, the sulfite ions are oxidized to sulfate ions. What is the amount concentration of the chromate ion solution?
5. Pure iron metal may be used as a primary standard for permanganate solutions. A 1.08 g sample of pure iron wire was dissolved in acid, converted to iron(II) ions, and diluted to 250.0 mL. In the titration, an average volume of 13.6 mL of permanganate solution was required to react with 10.0 mL of the acidic iron(II) solution. Calculate the amount concentration of the permanganate solution.

Section **13.4**

LAB EXERCISE 13.D

Analyzing for Chromium in Steel

Report Checklist
- ○ Purpose   ○ Design   ● Analysis
- ○ Problem   ○ Materials   ○ Evaluation
- ○ Hypothesis   ○ Procedure
- ○ Prediction   ○ Evidence

Stainless steel is a corrosion-resistant, esthetically pleasing alloy, normally composed of nickel, chromium, and iron. Complete the Analysis of the investigation report.

Purpose
The purpose of this lab exercise is to use the stoichiometric method in a redox chemical analysis.

Problem
What is the amount concentration of chromium(II) ions in a solution obtained in the analysis of a stainless steel alloy?

Design
A standard potassium dichromate solution is used as an oxidizing agent to oxidize chromium(II) ions to chromium(III) ions in an acidic solution (**Figure 4**).

Evidence
Table 3 Titration of 10.00 mL of acidic Cr^{2+}(aq) with 0.125 mol/L $K_2Cr_2O_7$(aq)

Trial	1	2	3
final burette reading (mL)	17.5	34.9	18.9
initial burette reading (mL)	0.1	17.5	1.5
volume of $K_2Cr_2O_7$(aq) (mL)	17.4	17.4	17.4

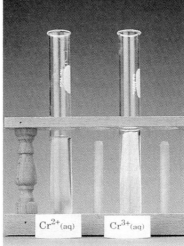

Figure 4
The blue Cr^{2+}(aq) solution is oxidized to a green Cr^{3+}(aq) solution.

INVESTIGATION 13.4 *Introduction*

Analyzing a Hydrogen Peroxide Solution

Report Checklist
- ○ Purpose   ○ Design   ● Analysis
- ○ Problem   ● Materials   ● Evaluation (1, 2, 3)
- ○ Hypothesis   ○ Procedure
- ● Prediction   ● Evidence

In this investigation, you assume the role of a laboratory technician working in a consumer advocacy laboratory, testing the concentration of a hydrogen peroxide solution. The labelled percent concentration of hydrogen peroxide on the commercial product is used for the Prediction.

Purpose
The technological purpose of this investigation is to test and evaluate the percent concentration of the consumer solution of hydrogen peroxide.

Problem
What is the percent concentration of hydrogen peroxide in a consumer product?

Design
An acidic solution of the primary standard, iron(II) ammonium sulfate–water (1/6), is prepared and the potassium permanganate solution is standardized by a titration with this primary standard. A 25.0 mL sample of a consumer solution of hydrogen peroxide is diluted to 1.00 L with water (that is, it is diluted by a factor of 40). The standardized potassium permanganate solution is used to titrate the diluted and acidified hydrogen peroxide. The amount concentration of the original hydrogen peroxide is obtained by analysis of the titration evidence, and by using a graph, prepared with a graphing calculator or computer spreadsheet program, of the amount and percent concentration of aqueous hydrogen peroxide.

To perform this investigation, turn to page 603.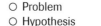

Section 13.4 Questions

1. State the similarities and differences between the method for redox stoichiometry and other examples of stoichiometry.
2. In acid–base titrations, one reactant (usually the titrant) is a strong acid or strong base. Similarly, in a redox stoichiometry, the reactant used to analyze a sample of unknown concentration is either a strong oxidizing or strong reducing agent.
   (a) State two common strong oxidizing agents commonly used in a redox titration.
   (b) Using the redox table, suggest a strong reducing agent that might be suitable for a redox titration analysis.
   (c) What are some examples of other strong reducing agents that might be used in an analysis? What type of experimental design would be appropriate?
3. Why is it necessary to standardize a potassium permanganate solution to be used in a chemical analysis?
4. In a chemical analysis, 10.00 mL samples of aqueous hydrogen peroxide are acidified and then titrated with 0.200 mol/L sodium perchlorate solution. From the evidence, an average volume of 24.0 mL of aqueous sodium perchlorate is required to reach the endpoint. Calculate the amount concentration of the hydrogen peroxide solution.
5. Complete the Analysis and Evaluation (of the prediction and, thus, of the metallurgical process) of the investigation report.

   Purpose
   The purpose of this lab exercise is to use redox stoichiometry to evaluate a technological process.

   Problem
   What is the amount concentration of iron(II) ions in a solution obtained in an iron ore analysis?

   Prediction
   According to the required standards for the metallurgical process, the concentration of the iron(II) ions should be 80.0 mmol/L.

   Design
   The iron(II) solution is titrated to iron(III) with a standard cerium(IV) ion solution, which is reduced to cerium(III). The indicator shows, as the endpoint, a sharp colour change from red to pale blue.

   Evidence

   Table 4 Titration of 25.0 mL of Fe^{2+}(aq) with 0.125 mol/L Ce^{4+}(aq)

Trial	1	2	3	4
final burette reading (mL)	15.7	30.7	45.6	40.2
initial burette reading (mL)	0.6	15.7	30.7	25.3

6. A scientist used the titration method to analyze for tin(II) chloride. Complete the two steps of the Analysis.

   Problem
   What is the amount concentration of a tin(II) chloride solution prepared from a sample of tin ore?

   Design
   The potassium dichromate solution is first standardized by titration with 10.00 mL of an acidified 0.0500 mol/L solution of the primary standard, $FeSO_4 \cdot (NH_4)_2SO_4 \cdot 6 H_2O$(s). The standardized dichromate solution is then titrated against 10.00 mL of the acidified tin(II) chloride solution.

   Evidence

   Table 5 Titration of 10.00 mL of 0.0500 mol/L Fe^{2+}(aq) with $K_2Cr_2O_7$(aq)

Trial	1	2	3	4
final burette reading (mL)	13.8	24.4	35.2	45.9
initial burette reading (mL)	2.3	13.8	24.4	35.2

   Table 6 Titration of 10.00 mL of Sn^{2+}(aq) with $K_2Cr_2O_7$(aq)

Trial	1	2	3	4
final burette reading (mL)	11.8	22.9	33.9	45.0
initial burette reading (mL)	0.3	11.8	22.9	33.9

   Extension
7. Potassium dichromate is a common reagent used in the analysis of the iron content of iron ore samples. If each analysis begins with the same mass of the ore, a redox titration can be designed such that the volume of dichromate required corresponds to the percent iron in the ore. This design eliminates the need for any calculations, so rapid, efficient analyses can be carried out by technicians. Starting with a 1.00 g sample of iron ore, the sample is treated to convert all the iron into iron(II) ions, and then acidified. Predict the concentration of potassium dichromate required so that the volume (in millilitres) equals the percentage of iron in the original sample.

Chapter 13 INVESTIGATIONS

INVESTIGATION 13.1

Single Replacement Reactions

Report Checklist

○ Purpose ● Design ● Analysis
○ Problem ○ Materials ● Evaluation (1)
○ Hypothesis ○ Procedure
● Prediction ● Evidence

This investigation is a review of single replacement reactions in preparation for the development of a theory of oxidation and reduction. As part of the Design, include diagnostic tests (as in Appendix C.4) for the predicted products.

Purpose
The purpose of this investigation is to use the single replacement reaction generalization to predict and analyze the reactants and products.

Problem
What are the products of the single replacement reactions for the following pairs of reactants?

(a) copper and aqueous silver nitrate
(b) aqueous chlorine and aqueous sodium bromide
(c) magnesium and hydrochloric acid
(d) zinc and aqueous copper(II) sulfate
(e) aqueous chlorine and aqueous potassium iodide

Materials
lab apron	chlorine water
eye protection	magnesium ribbon
five small test tubes	zinc strip
two test tube stoppers	aqueous silver nitrate
test tube rack	aqueous sodium bromide
steel wool	hydrochloric acid
wash bottle	aqueous copper(II) sulfate
matches	aqueous potassium iodide
copper strip	hexane

Toxic, corrosive, and irritant chemicals are used in this investigation. Avoid skin contact. Wash any splashes on the skin or clothing with plenty of water. If any chemical is splashed in the eye, rinse for at least 15 min and inform your teacher.

Keep the hexane sealed to avoid evaporation. Dispose of the hexane as directed by your teacher. Hexane is highly flammable. Keep away from open flame. Make sure matches are extinguished by dipping in water. Do not inhale the vapours.

Procedure
1. Set up five test tubes, each filled to a depth of 2–3 cm with one of the five aqueous solutions.
2. Add the element indicated to each test tube.
3. Perform diagnostic tests on each of the five mixtures. Record your evidence.
4. Dispose of the solutions as directed by your teacher.

INVESTIGATION 13.2

Spontaneity of Redox Reactions

Report Checklist
- ○ Purpose
- ○ Problem
- ○ Hypothesis
- ○ Prediction
- ○ Design
- ○ Materials
- ● Procedure
- ● Evidence
- ● Analysis
- ● Evaluation (1, 2, 3)

In previous units in this textbook, we assumed that all chemical reactions are spontaneous; that is, they occur once the reactants are placed in contact, without a continuous addition of energy to the system. Spontaneous redox reactions in solution generally provide visible evidence of a reaction within a few minutes.

Purpose
The purpose of this investigation is to test the assumption that all single replacement reactions are spontaneous.

Problem
Which combinations of copper, lead, silver, and zinc metals and their aqueous metal ion solutions produce spontaneous reactions?

Design
A drop of each solution is placed in separate locations on a clean area of each of the four metal strips.

Materials
lab apron
eye protection
reusable strips of copper, lead, silver, and zinc metals
 (*Note that the lead strips bend much more easily than the zinc strips, which look similar.*)
0.10 mol/L solutions of copper(II) nitrate, lead(II) nitrate, silver nitrate, and zinc nitrate in dropper bottles
steel wool or sandpaper

These chemicals are toxic—especially the lead solution—and irritants. Avoid skin contact. Remember to wash your hands before leaving the laboratory. Rinse all of the metal strips thoroughly and return them so they can be used again.

INVESTIGATION 13.3

Predicting the Reaction of Sodium Metal (Demonstration)

Report Checklist
- ○ Purpose
- ○ Problem
- ○ Hypothesis
- ● Prediction
- ● Design
- ○ Materials
- ○ Procedure
- ● Evidence
- ● Analysis
- ● Evaluation (1, 2, 3)

The process of developing theories, laws, and generalizations requires that they must be tested numerous times in as many different situations as possible. This process is necessary not only to determine their validity, but also to identify exceptions that may lead to new knowledge.

As part of the Design, include a list of diagnostic tests using the "If [procedure] and [evidence], then [analysis]" format for every product predicted. (This format is described in Appendix C.4.)

Purpose
The purpose of this demonstration is to test the five-step method for predicting redox reactions.

Problem
What are the products of the reaction of sodium metal with water?

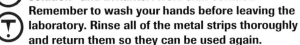
This reaction of sodium metal must be demonstrated with great care, because a great deal of heat is produced. Use only a piece the size of a small pea, use a safety screen, wear a lab apron, eye protection, and face shield, and keep observers at least two metres away.

INVESTIGATION 13.4

Analyzing a Hydrogen Peroxide Solution

Report Checklist
- ○ Purpose
- ○ Problem
- ○ Hypothesis
- ● Prediction
- ○ Design
- ● Materials
- ○ Procedure
- ● Evidence
- ● Analysis
- ● Evaluation (1, 2, 3)

In this investigation, you assume the role of a laboratory technician working in a consumer advocacy laboratory, testing the concentration of a hydrogen peroxide solution. The labelled percent concentration of hydrogen peroxide on the commercial product is used for the Prediction (**Figure 1**).

Purpose
The technological purpose of this investigation is to test and evaluate the percent concentration of the consumer solution of hydrogen peroxide.

Problem
What is the percent concentration of hydrogen peroxide in a consumer product?

Design
An acidic solution of the primary standard, iron(II) ammonium sulfate–water (1/6), is prepared and the potassium permanganate solution is standardized by a titration with this primary standard. A 25.0 mL sample of a consumer solution of hydrogen peroxide is diluted to 1.00 L with water (that is, it is diluted by a factor of 40). The standardized potassium permanganate solution is used to titrate the diluted and acidified hydrogen peroxide. The amount concentration of the original hydrogen peroxide is obtained by analysis of the titration evidence, and by using a graph, prepared with a graphing calculator or computer spreadsheet program, of the information in **Table 1**.

Materials
lab apron
eye protection
FeSO$_4$·(NH$_4$)$_2$SO$_4$·6H$_2$O(s)
2 mol/L H$_2$SO$_4$(aq)
diluted H$_2$O$_2$(aq)
KMnO$_4$(aq)
(list to be completed by student)

Sulfuric acid is corrosive. Iron(II) ammonium sulfate, hydrogen peroxide, and potassium permanganate are irritants. Avoid inhaling any solid, and avoid skin or eye contact.

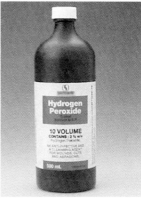

Figure 1
In drugstores, hydrogen peroxide is usually sold as a 3% solution. Hairdressers use a 6% solution.

Table 1 H$_2$O$_2$(aq) Concentration

Amount concentration (mol/L)	Percent concentration (%)
0.73	2.5
0.76	2.6
0.79	2.7
0.82	2.8
0.85	2.9
0.88	3.0
0.91	3.1
0.94	3.2
0.97	3.3
1.0	3.4

Procedure
1. (Pre-lab) Calculate the required mass of FeSO$_4$·(NH$_4$)$_2$SO$_4$·6H$_2$O(s) to prepare 100.0 mL of a 0.0500 mol/L solution.
2. Measure the required mass of the iron(II) compound in a clean, dry 100 mL beaker.
3. Dissolve the solid in about 40 mL of H$_2$SO$_4$(aq).
4. Transfer this solution into a clean 100 mL volumetric flask, rinsing and adding pure water to complete the preparation of the standard solution.
5. Transfer 10.00 mL of the standard iron(II) solution by pipette into a clean 250 mL Erlenmeyer flask.
6. Titrate the acidic iron(II) sample with KMnO$_4$(aq).
7. Repeat steps 5 and 6 until three consistent volumes (within 0.1 mL) are obtained.
8. Transfer 10.00 mL of the diluted hydrogen peroxide solution by pipette into a clean 250 mL Erlenmeyer flask.
9. Using a 10 mL graduated cylinder, add 5 mL of H$_2$SO$_4$(aq) to the hydrogen peroxide solution.
10. Titrate the acidic hydrogen peroxide solution with KMnO$_4$(aq).
11. Repeat steps 8 to 10 until three consistent volumes (within 0.1 mL) are obtained.
12. Dispose of all solutions into a labelled waste container.

Chapter 13 SUMMARY

Outcomes

Knowledge
- define oxidation and reduction operationally (historically) and theoretically (13.1, 13.2, 13.3)
- define the following terms: oxidizing agent, reducing agent, oxidation number, half-reaction, disproportionation (13.1, 13.2, 13.3)
- differentiate between redox reactions and other reactions by identifying half-reactions and changes in oxidation number (13.1, 13.2, 13.3)
- identify electron transfer, oxidizing agents, and reducing agents in redox reactions that occur in everyday life in both living and nonliving systems (all sections)
- compare the relative strengths of oxidizing and reducing agents from empirical data (13.2)
- predict the spontaneity of a redox reaction based on a redox table, and compare predictions to experimental results (13.2)
- write and balance equations for redox reactions in acidic, basic, and neutral solutions, including disproportionation reactions, by using half-reaction equations, developing simple half-reaction equations, and assigning oxidation numbers (13.2, 13.3, 13.4)
- perform calculations to determine quantities of substances involved in redox titrations (13.4)

STS
- state that a goal of technology is to solve practical problems (all sections)
- recognize that technological problems may require various solutions and have both intended and unintended consequences (13.1, 13.2, 13.3)

Skills
- initiating and planning: design an experiment to determine the reactivity of various metals (13.1, 13.2); and describe procedures for safe handling, storing, and disposal of materials used in the laboratory, with reference to WHMIS and consumer product labelling information (13.1, 13.2, 13.4)
- performing and recording: select and use appropriate equipment to perform a redox titration (13.4); use a standard redox table to predict the spontaneity of redox reactions (13.2, 13.4); and create charts, tables, or spreadsheets related to redox reactions (13.1, 13.2, 13.4)
- analyzing and interpreting: analyze evidence from an experiment to derive a simple redox table (13.2); interpret patterns and trends in redox reactions (all sections); and evaluate redox experiments, including identifying the limitations of the evidence (13.1, 13.2, 13.4)
- communication and teamwork: work collaboratively in addressing problems, and select and use appropriate modes of representation for redox reactions and answers to redox problems (all sections)

Key Terms

13.1
half-reaction
reduction
oxidation
redox reaction

13.2
reducing agent
oxidizing agent
redox spontaneity rule
disproportionation

13.3
oxidation number
oxidation
reduction

▶ MAKE a summary

1. Start with "redox" and make a flow chart or concept map that includes all of the Key Terms listed, plus any important generalizations and procedures that will help you learn the material in this chapter.
2. Refer back to your answers to the Starting Points questions at the beginning of this chapter. How has your thinking changed?

▶ Go To

The following components are available on the Nelson Web site. Follow the links for *Nelson Chemistry Alberta 20–30*.
- an interactive Self Quiz for Chapter 13
- additional Diploma Exam-style Review questions
- Illustrated Glossary
- additional IB-related material

There is more information on the Web site wherever you see the Go icon in this chapter.

➕ EXTENSION

Electric Universe
Electricity is everywhere. Only two centuries ago, people developed technologies to release and control electricity. Since then we have come to rely on it for almost every aspect of our lives. A science writer talks about some of the characters who helped to bring electricity into our lives.

www.science.nelson.com

Chapter 13 REVIEW

Many of these questions are in the style of the Diploma Exam. You will find guidance for writing Diploma Exams in Appendix H. Exam study tips and test-taking suggestions are on the Nelson Web site. Science Directing Words used in Diploma Exams are in **bold** type.

www.science.nelson.com

DO NOT WRITE IN THIS TEXTBOOK.

Part 1

1. Historically, __i__ meant producing __ii__ from their naturally occurring compounds. According to modern theory, this process involves a __iii__ of electrons.

   The above statement is completed by the information in which row?

Row	i	ii	iii
A.	oxidation	metals	gain
B.	reduction	metals	gain
C.	oxidation	nonmetals	loss
D.	reduction	nonmetals	loss

2. A reducing agent can be described as a substance that
   A. loses electrons and causes reduction
   B. loses electrons and becomes reduced
   C. gains electrons and causes oxidation
   D. gains electrons and becomes reduced

3. Which one of the following general reaction types will *not* be a redox reaction?
   A. combustion
   B. simple decomposition
   C. disproportionation
   D. double replacement

4. Some natural and technological processes that involve redox reactions are
   1. corrosion
   2. metallurgy
   3. cellular respiration
   4. photosynthesis
   5. rusting of iron
   6. magnesium metal flares

   The processes in which oxygen behaves as an oxidizing agent are, **in numerical order**,

   ___, ___, ___, and ___.

5. Which of the following combinations would produce a spontaneous redox reaction?
   A. nitric acid and iron(III) chloride solution
   B. chromium metal and aqueous cobalt(II) chloride
   C. oxygen gas bubbled into a sodium bromide solution
   D. aqueous tin(II) nitrate and potassium iodide solutions

Use this information to answer questions 6 and 7.

Four different metals and their corresponding metal ion solutions are mixed to determine if a spontaneous reaction occurs.

Table 1 Metal-Ion Reactions

	X^{2+}(aq)	Y^{2+}(aq)	Z^{2+}(aq)	W^{2+}(aq)
X(s)	X	X	X	✓
Y(s)	✓	X	✓	✓
Z(s)	✓	X	X	✓
W(s)	X	X	X	X

✓ spontaneous reaction
X no evidence of reaction

6. The strongest oxidizing agent is
   A. X^{2+}(aq)
   B. Y^{2+}(aq)
   C. Z^{2+}(aq)
   D. W^{2+}(aq)

7. The metal that has the weakest attraction for its electrons is
   A. W(s)
   B. Z(s)
   C. Y(s)
   D. X(s)

8. Which of the following solutions should *not* be stored in a tin-plated container?
   I $NaNO_3$(aq)   III $SnBr_2$(aq)
   II $AgNO_3$(aq)   IV Cl_2(aq)

   A. I only
   B. II and III
   C. II and IV
   D. III and IV

Use this information to answer questions 9 and 10.

Ozone is a strong oxidizing agent that will oxidize aqueous iodide ions in an acidic solution to iodate ions. The unbalanced redox reaction equation is

___O_3(g) + ___I^-(aq) → ___IO_3^-(aq) + ___O_2(g)

9. When the half-reaction equations are constructed using coefficients to balance electrons, the number of electrons transferred is
   A. 6
   B. 3
   C. 2
   D. 1

10. The coefficients of the net redox equation, in the order of substances in the equation, are

    ___, ___, ___, and ___.

Redox Reactions **605**

11. The oxidation number of the carbon atom in the carbonate ion is
    A. +6
    B. +4
    C. −2
    D. 0

Use this information to answer questions 12 to 14.

Many natural gas wells, called "sour" gas wells, contain considerable quantities of hydrogen sulfide gas as well as methane. When this mixture burns (**Figure 1**), hydrogen sulfide is converted to sulfur dioxide. Once in the atmosphere, sulfur dioxide may be converted to sulfur trioxide.

Figure 1
A flaring gas well

12. The oxidation numbers of the sulfur atoms in hydrogen sulfide, sulfur dioxide, and sulfur trioxide are, in order of these compounds,
    A. 0, 0, 0
    B. −2, +4, +6
    C. −2, +4, +4
    D. +2, −4, −6

13. In the two-step conversion from hydrogen sulfide to sulfur trioxide, the sulfur atoms are
    A. oxidized in both steps
    B. reduced in both steps
    C. oxidized first, then reduced
    D. reduced first, then oxidized

14. An unintended consequence of this process may be
    A. depletion of the ozone layer
    B. natural gas shortages
    C. altering local climate
    D. acid rain (deposition)

15. In a standardization experiment, 25.0 mL of an acidic 0.100 mol/L tin(II) chloride solution required an average volume of 12.7 mL of potassium dichromate solution for complete reaction. The amount concentration of the potassium dichromate solution is _____ mmol/L.

Part 2

16. Write a theoretical description of a redox reaction. Include the following terms in your answer: electrons, oxidation, reduction, oxidizing agent, and reducing agent.

17. Use a table of relative strengths of oxidizing and reducing agents, such as the one in Appendix I, to answer the following questions.
    (a) How can you predict whether or not a combination of substances will react spontaneously?
    (b) If a spontaneous redox reaction occurs, what kinds of evidence might be observed?

18. **Define** each of the following in terms of both electrons and oxidation numbers.
    (a) oxidation
    (b) reduction
    (c) redox reaction

19. For each of the following, complete the half-reaction equation and classify as an oxidation or reduction.
    (a) $HClO_2(aq) \rightarrow HClO(aq)$ (acidic)
    (b) $Al(OH)_4^-(aq) \rightarrow Al(s)$ (basic)
    (c) $Br^-(aq) \rightarrow BrO_4^-(aq)$ (acidic)
    (d) $ClO^-(aq) \rightarrow Cl_2(g)$ (basic)

20. Various pairs of metals and metal ions were combined and the evidence interpreted, as shown below:
    $2\,Ga(s) + 3\,Cd^{2+}(aq) \rightarrow 2\,Ga^{3+}(aq) + 3\,Cd(s)$
    $Ga(s) + Mn^{2+}(aq) \rightarrow$ no evidence of reaction
    $3\,Mn^{2+}(aq) + 2\,Ce(s) \rightarrow 3\,Mn(s) + 2\,Ce^{3+}(aq)$
    (a) Use this information and the redox spontaneity rule to develop a table of oxidizing and reducing agents for these metals and their ions.
    (b) **Identify** the strongest oxidizing and the strongest reducing agent in your table.

21. For the following solutions, list the entities believed to be present, and classify them as possible oxidizing or reducing agents.
    (a) aqueous chlorine solution
    (b) tin(II) nitrate solution
    (c) acidic potassium iodate solution

22. For each of the following mixtures, list and classify the entities present, **predict** the half-reaction and net ionic reaction equations, and **predict** whether or not a spontaneous reaction will be observed.
    (a) Chlorine gas is bubbled into an iron(II) sulfate solution.
    (b) Nickel(II) nitrate solution is mixed with a tin(II) sulfate solution.
    (c) A zinc coating on a drain pipe is exposed to air and water.
    (d) An acidic solution of sodium sulfate is spilled on a steel laboratory stand. (Consider only the iron in the steel.)
    (e) For use in a titration, a sodium hydroxide solution is added to a potassium sulfite solution to make it basic.

23. The reactivity of metals varies considerably from very low reactivity (noble metals) to explosively reactive. **Design** an experiment to study the reactivity of calcium metal in water. Your response should include:
    - Purpose
    - Problem
    - Prediction, including half-reaction and net ionic equations
    - Design, including diagnostic tests
    - Materials, including WHMIS safety cautions
    - Procedure, including disposal instructions

24. **Predict** the balanced redox reaction equation by constructing and labelling oxidation and reduction half-reaction equations.
    (a) $Pt(s) + NO_3^-(aq) + Cl^-(aq) \rightarrow PtCl_6^{2-}(aq) + NO_2(g)$ (acidic)
    (b) $CN^-(aq) + ClO_2^-(aq) \rightarrow CNO^-(aq) + Cl^-(aq)$ (basic)
    (c) $PH_3(g) + CrO_4^{2-}(aq) \rightarrow Cr(OH)_4^-(aq) + P_4(s)$ (basic)

25. Assign oxidation numbers to all atoms/ions and indicate which atom/ion is oxidized and which is reduced.
    (a) $2 Al(s) + Fe_2O_3(s) \rightarrow 2 Fe(s) + Al_2O_3(s)$
    (b) $In(s) + 3 Tl^+(aq) \rightarrow In^{3+}(aq) + 3 Tl(s)$
    (c) $2 Cr^{3+}(aq) + Sn^{2+}(aq) \rightarrow 2 Cr^{2+}(aq) + Sn^{4+}(aq)$
    (d) $Cl_2(aq) + 2 I^-(aq) \rightarrow 2 Cl^-(aq) + I_2(aq)$
    (e) $UCl_4(s) + 2 Ca(s) \rightarrow 2 CaCl_2(s) + U(s)$

26. Balance the following chemical equations using the oxidation number method.
    (a) $C_6H_{12}O_6(s) + O_2(g) \rightarrow CO_2(g) + H_2O(l)$
    (b) $Au^{3+}(aq) + SO_2(aq) \rightarrow SO_4^{2-}(aq) + Au(s)$ (acidic)
    (c) $BrO_3^-(aq) + C_2H_6O(aq) \rightarrow CO_2(g) + Br^-(aq)$
    (d) $Ag(s) + NO_3^-(aq) \rightarrow Ag^+(aq) + NO(g)$ (acidic)
    (e) $HNO_3(aq) + SO_2(g) \rightarrow H_2SO_4(aq) + NO(g)$
    (f) $Zn(s) + BrO_4^-(aq) \rightarrow Zn(OH)_4^{2-}(aq) + Br^-(aq)$ (basic)

27. Balanced redox equations can be obtained using three different methods, other than trial-and-error for simple equations.
    (a) Briefly **describe** each method.
    (b) Which method is also used to predict the products? What other information can also be predicted?
    (c) For which methods do you need to know the primary products?
    (d) Which method do you prefer to use? Why?

28. A commercial kit is available to clean silver by removing the tarnish using a redox reaction. (Assume that silver tarnish is silver sulfide.) A zinc strip is placed in a water softener solution and the tarnished silver is placed so that it is in contact with the zinc strip.
    (a) Write the overall chemical equation and balance it using the simplest possible method.
    (b) **Verify**, using oxidation numbers, that the chemical equation is balanced.
    (c) Write oxidation and reduction half-reaction equations.

29. Magnesium metal reacts rapidly in hot water. **Predict** the mass of precipitate that will form if a 2.0 g strip of magnesium reacts completely with water.

30. A student uses a redox titration to determine the concentration of iron(II) ions in an acidic solution. The evidence in **Table 2** shows the volume of 7.50 mmol/L $MnO_4^-(aq)$ that reacted with 10.0 mL of $Fe^{2+}(aq)$. Calculate the amount concentration of the iron(II) ions.

    Table 2 Titration of 10.0 mL of $Fe^{2+}(aq)$ with 7.50 mmol/L $MnO_4^-(aq)$

Trial	1	2	3
final burette reading (mL)	16.4	31.4	46.3
initial burette reading (mL)	1.3	16.4	31.4

31. Three chemistry teachers developed a problem to test students' understanding of redox concepts. The challenge is to identify three unknown solutions (labelled A, B, and C) using any of the materials listed below in your procedure. Assuming all possible spontaneous reactions are rapid and that the nitrate ion is a spectator ion, write a procedure to identify which solution is sodium nitrate, which is lead(II) nitrate, and which is calcium nitrate. **Describe** the expected results.
    0.25 mol/L solutions of A, B, and C
    silver, zinc, and magnesium strips
    dropper bottles of 0.25 mol/L aqueous solutions of: sodium sulfate; sodium carbonate; and sodium hydroxide
    steel wool
    test tubes and test tube rack
    50 mL beakers
    400 mL waste beaker

32. Photofinishing laboratories (**Figure 2**) often produce a waste solution containing silver ions. The CEO of a company wants to know if it is economical to recover this silver. The first step in the study is to determine the concentration of silver ions in the waste solution.
    (a) Write an experimental design based on redox concepts to determine the concentration of silver ions.
    (b) Using appropriate terms and chemical equations, **explain** the redox chemistry in your proposed design.

Figure 2
In film processing, the film is placed in a developing agent.

33. Methanol is used as a windshield-washer antifreeze; containers are usually labelled with the freezing point of the solution (**Figure 3**). A chemical technician can test the validity of the claim using various experimental designs. The experimental design chosen below is the titration of a basic solution of methanol with a standardized solution of potassium permanganate based on the following (unbalanced) chemical equation.
$CH_3OH(aq) + MnO_4^-(aq) \rightarrow CO_3^{2-}(aq) + MnO_4^{2-}(aq)$
Use the information in **Tables 3**, **4**, and **5** and complete the Analysis of the investigation report.

Purpose
The purpose of this investigation is to use redox stoichiometry for a chemical analysis.

Problem
What is the freezing point of a sample of windshield-washer fluid?

Design
A potassium permanganate solution is prepared and standardized against an acidic 0.331 mol/L solution of iron(II) ammonium sulfate (**Table 3**). The standardized permanganate solution is then titrated against a basic methanol solution, which has been diluted by a factor of 1000 (**Table 4**).

Evidence

Table 3 Titration of Potassium Permanganate Solution with 10.00 mL Acidic FeSO$_4$·(NH$_4$)SO$_4$(aq)

Trial	1	2	3	4
final burette reading (mL)	13.3	25.8	38.1	12.9
initial burette reading (mL)	0.2	13.3	25.8	0.5

Table 4 Titration of Standardized Potassium Permanganate Solution with 10.00 mL CH$_3$OH(aq)

Trial	1	2	3
final burette reading (mL)	12.4	24.1	35.8
initial burette reading (mL)	0.7	12.4	24.1

Table 5 Concentrations and Freezing Points of Aqueous Solutions of Methanol

Amount concentration (mol/L)	Percent by mass (%)	Freezing point (°C)
0	0	0
6.035	20.00	−15.0
11.672	40.00	−38.6
16.754	60.00	−74.5

Figure 3
This methanol windshield-washer antifreeze can be used at temperatures above −45 °C.

Extension

34. For the production of pulp from wood, a variety of methods are used, including mechanical and chemical processes. These have advantages and disadvantages that have been widely debated. Prepare an argument for or against the following statement: "The immediate economic value of using technology to produce a product far outweighs any possible future adverse effects."
Your response should also include
    - researched information about a variety of mechanical and chemical processes
    - an evaluation of these processes from technological, economic, and ecological perspectives
    - reference to redox chemistry

www.science.nelson.com

35. Vanadium (**Figure 4**) is a very versatile element in terms of its reactivity. Vanadium metal reacts with fluorine to form VF$_5$, with chlorine to form VCl$_4$, with bromine to form VBr$_3$, with iodine to form VI$_2$, with oxygen to form V$_2$O$_5$, and with hydrochloric acid to form VCl$_2$.
    (a) **Identify** the oxidation states of vanadium in each of these compounds.
    (b) What interpretation can be made about the oxidizing power of chemicals that react with vanadium metal?
    (c) **Describe** how the oxidation state of vanadium relates to the colours of the compounds formed.
    (d) Briefly **describe** some technological applications of vanadium and its compounds.

Figure 4
Vanadinite is the most common vanadium ore found in large reserves in many countries, including Canada.

Chapter 13

36. The earliest metallurgy would be classified as pyrometallurgy. Other processes such as hydrometallurgy and electrometallurgy are more recent inventions. **Define** each of these types of metallurgy and provide one common example of each. Why are all three of these processes examples of redox reactions? What do these processes illustrate about the goal of technology and the interaction between science and technology?

   www.science.nelson.com

37. The nitrogen cycle (**Figure 5**) is a very important and complex biological system that includes nitrogen fixation, nitrification, and denitrification. **Describe** the chemical reactions in the nitrogen cycle.
   Your response should include:
   - definitions of the terms *nitrogen fixation, nitrification*, and *denitrification*.
   - a summary of the main changes of nitrogen-containing entities and their oxidation numbers in the nitrogen cycle
   - some examples of the half-reaction equations, labelled as oxidation or reduction
   - descriptions of some positive and negative environmental impacts of the processes in the nitrogen cycle.

   www.science.nelson.com

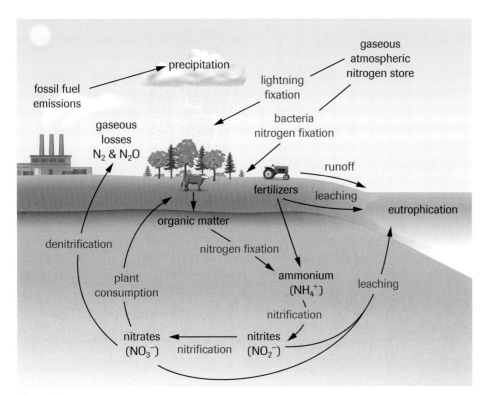

Figure 5
The nitrogen cycle

Redox Reactions

chapter 14
Electrochemical Cells

In this chapter

- Exploration: A Simple Electric Cell
- Investigation 14.1: Designing an Electric Cell
- Web Activity: Hydrogen: Wonderfuel or Hype?
- Web Activity: Lewis Urry
- Case Study: The Ballard Fuel Cell
- Investigation 14.2: A Voltaic Cell (Demonstration)
- Web Activity: Voltaic Cells Under Standard Conditions
- Investigation 14.3: Testing Voltaic Cells
- Biology Connection: Reduction Potentials
- Lab Exercise 14.A: Developing a Redox Table
- Mini Investigation: Home Corrosion Experiment
- Web Activity: Galvanizing Steel
- Investigation 14.4: A Potassium Iodide Electrolytic Cell
- Investigation 14.5: Electrolysis (Demonstration)
- Web Activity: Electrolytic Cell Stoichiometry

Since their invention in 1888, vehicles powered entirely by electricity have drifted in and out of fashion (**Figure 1**). Many experts predict that in the next decade, electric vehicles will finally make a breakthrough. Electric power is slowly becoming a viable alternative to gasoline power, thanks to a combination of political, economic, and environmental factors. One advantage of electric cars over gasoline-fuelled cars is that they produce less pollution. Also, while cars powered by gasoline engines are about 15% efficient, many electric cars are 90% efficient. (Of course, overall efficiency and environmental impacts depend on how the electricity and gasoline are produced.) Other attractive features of electric vehicles are that they are nearly silent and require minimal maintenance.

The biggest obstacle to the widespread use of electric cars is the lack of a powerful, lightweight, inexpensive battery. Scientists and engineers are researching alternatives to the common lead–acid battery. Perhaps the most promising alternative is a battery that runs continuously as fuel is supplied. One such alternative is the aluminium–air fuel cell, which uses aluminium metal as the fuel and oxygen from the air to produce electricity. Another possibility is a fuel cell in which a hydrogen-rich fuel and oxygen from the air produce electricity.

Redox reactions can produce electricity and, conversely, electricity can cause redox reactions. Many materials that we take for granted were virtually unknown until the process of electrolysis made their production possible. Aluminium, chlorine, hydrogen, sodium hydroxide, magnesium, and copper are produced in large quantities by electrolytic processes. In this chapter, you will learn how batteries are made, how electricity can be used to produce chemicals, and how science and technology work together in the development of electrochemical processes.

STARTING Points

Answer these questions as best you can with your current knowledge. Then, using the concepts and skills you have learned, you will revise your answers at the end of the chapter.

1. What concepts can we use to explain how electrochemical cells work?
2. Describe the relationship between science and technology in the development of electrochemical cells.
3. List the types and uses of a variety of common electric cells. Include an assessment of the impact of each one on our lives.

Career Connections:
Materials Engineering Technologist; Chemical Technologist

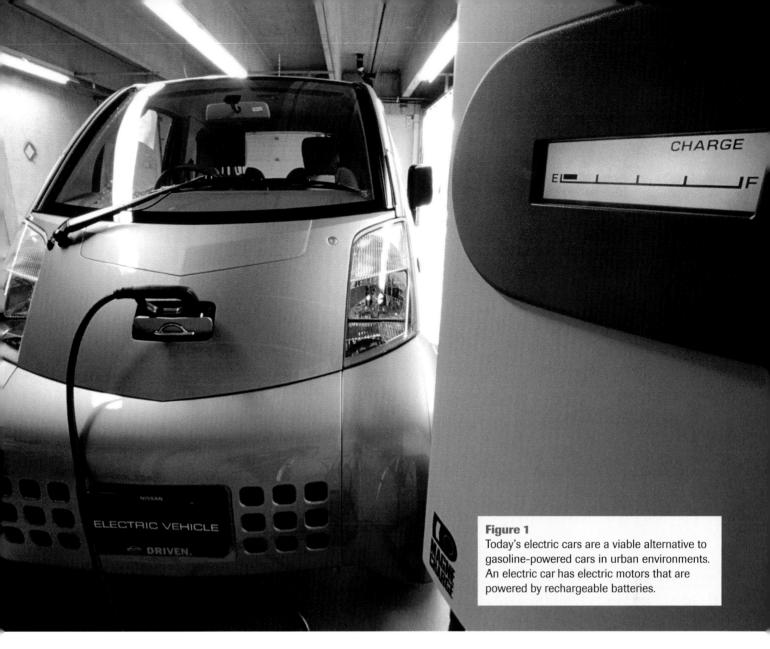

Figure 1
Today's electric cars are a viable alternative to gasoline-powered cars in urban environments. An electric car has electric motors that are powered by rechargeable batteries.

▶ Exploration   *A Simple Electric Cell*

A cell that produces electricity can be amazingly simple because it uses very common materials and requires no technical expertise to construct. Anyone can make one and then improve its efficiency without much understanding of the scientific principles involved. That is why the electric cell was used for more than 100 years before scientists understood how it worked.

Materials: copper and zinc metal strips (or any two different metals); steel wool; orange, apple, and potato (and other fruits or vegetables); LCD clock; voltmeter (or multimeter) with leads

- Clean the metal strips with steel wool to remove any coating or oxides.
- Stick both metal strips into the orange. Make sure that the metal strips are not in contact inside or outside the orange.
- Momentarily touch the leads (red—positive; black—negative) from the voltmeter, one to each metal strip. Now reverse the leads and test again.

(a) Record and describe what happened in each case.

- Connect the leads to the LCD clock, paying attention to positive and negative connections.

(b) Does the clock work? If it does not, suggest a solution to make it work. Try it.
(c) Explain, in your own words, what you think happened in (b).

- Repeat the process using other fruits and vegetables.

(d) Which fruit or vegetable seemed to be the best at producing electricity?
(e) What do all fruits and vegetables have in common?
(f) How could you improve upon your electric cell?

14.1 Technology of Cells and Batteries

Before 1800, scientists knew that static electricity was produced by the friction created by two moving objects in contact. They discovered ways of storing the charges temporarily, but when the energy was released in the form of an electrical spark, it could not be put to practical use. Practical applications of electricity were developed after 1800, the year in which Alessandro Volta announced his invention of the electric cell.

Volta invented the first electric cell but he got his inspiration from the work, almost 30 years earlier, of the Italian physician Luigi Galvani. Galvani noticed that the muscles in a frog's leg would twitch when a spark hit the leg. Galvani's crucial observation was that two different metals could make the muscle twitch. Unfortunately, Galvani thought his discovery was due to some mysterious "animal electricity." It was Volta who recognized that this effect had nothing to do with animals or muscle tissue, and everything to do with conductors and electrolytes, as you observed in the Exploration at the beginning of this chapter.

Cells and Batteries

Although an *electric cell* is a device that continuously converts chemical energy into electrical energy, the electric cells that Volta invented produced very little electricity. Eventually, he came up with a better design by joining several cells together. A *battery* is a group of two or more electric cells connected to each other in series, like railway cars in a train. Volta's first battery consisted of several bowls of brine (aqueous sodium chloride) connected by metals that dipped from one bowl into the next (**Figure 1**). This arrangement of metal strips and electrolytes produced a steady flow of electric current.

> **DID YOU KNOW?**
> **Shocking Personal Experiments**
> "I introduced into my ears two metal rods with rounded ends and joined them to the terminals of the apparatus. At the moment the circuit was completed, I received a shock in the head—and began to hear a noise—a crackling and boiling. This disagreeable sensation, which I feared might be dangerous, has deterred me so that I have not repeated the experiment."
> — Alessandro Volta (1745–1827)

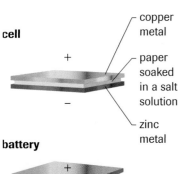

Figure 1
A version of Volta's first battery. Each beaker contains two different metals, copper and zinc, in an electrolyte, salt water. A series of beakers forms a series of cells (a battery) whose total voltage is the sum of the individual voltages of all cells.

Figure 2
Volta's revised cell design, simpler than the first, consisted of a sandwich of two metals separated by paper soaked in salt water (the electrolyte). A cell consisted of a layer of zinc metal separated from a layer of copper metal by the brine-soaked paper. A large pile of cells could be constructed to give more electrical energy per unit charge.

Volta improved the design of this battery by replacing the strips of metal with flat sheets, and replacing the bowls with paper or leather soaked in brine. This produced more electric current for a longer time. As shown in **Figure 2**, Volta stacked cells on top of each other to form a battery, known as a voltaic pile. When a loop of wire was attached to the top and bottom of this voltaic pile, a steady electric current flowed. Volta assembled voltaic piles containing more than 100 cells.

Volta's invention was an immediate technological success because it produced an electric current more simply and more reliably than methods that depended on static charges. It also produced a steady electric current—something no other device could do. The development of this technology led to many advances in physics (for example, the theory and description of current electricity), in chemistry (for example, the discovery of Groups 1 and 2 metals), and in electrical and chemical engineering.

Basic Cell Design and Properties

Each electric cell is composed of two **electrodes** (solid electrical conductors) and one **electrolyte** (aqueous electrical conductor) (**Figure 3**). In the cells we buy for home use, the electrolyte is usually a moist paste, containing only enough conducting solution to make the cell function. The electrodes are usually two metals, or graphite and a metal. In some designs, one of the electrodes is the container of the cell. One of the electrodes is marked positive (+) and the other is marked negative (−). *In an electric cell or battery, the cathode is the positive electrode and the anode is the negative electrode.*

According to theory, electricity is the flow of electrons. Electrons move from the anode of a battery through the external circuit to the cathode. A battery produces electricity only when there is an external conducting path, such as a wire, through which electrons can move. Disconnecting the wire from the battery immediately stops the flow of electrons.

A voltmeter is a device that can be used to measure the energy difference, per unit electric charge, between any two points in an electric circuit (**Figure 4(a)**). The energy difference per unit charge is called the **electric potential difference** or the voltage, and is measured in volts (V). For example, the electrons transferred via a 1.5 V cell release only one-sixth as much energy compared with the electrons from a 9 V battery.

Since the voltage is a ratio of energy to charge, it is not dependent on the size of the cell. You may have noticed that you can buy the same type and brand of 1.5 V cells in a variety of sizes, such as AA, B, C, and D. All are rated at 1.5 V. The larger cells can store more energy at the same time as transferring more charge, but the ratio of energy to charge is the same as the smaller cells. *The voltage of a cell depends mainly on the chemical composition of the reactants in the cell.*

Electric current, measured by an ammeter in amperes (A), is a measure of the rate of flow of charge past a point in an electrical circuit (**Figure 4(b)**). The larger the electric cell of a particular type, the greater the current that can be produced by the cell. The charge transferred by a cell or battery is measured in coulombs (C) and expresses the total charge transferred by the movement of charged particles.

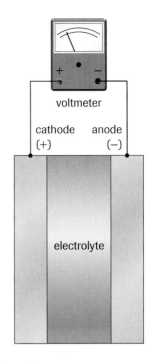

Figure 3
An electric cell always contains two electrodes—an anode and a cathode—and an electrolyte. When testing the voltage of a cell or battery, the red (+) lead of the voltmeter is connected to the positive electrode (cathode), and the black (−) lead is connected to the negative electrode (anode).

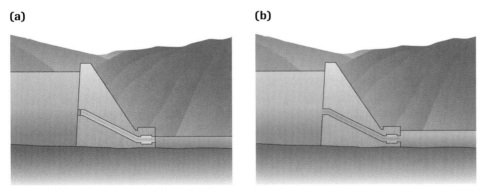

Figure 4
(a) A dam built across a stream or river stops the flow of water. There is a potential energy difference between a kilogram of water at the top of the dam and a kilogram of water at the bottom of the dam. A disconnected cell or battery is like a dam holding back electrons. There is a potential energy difference between the electrons at the anode and the electrons at the cathode. A voltmeter measures this potential energy difference.
(b) If water is released from behind the dam, it naturally flows from the region of higher potential energy (behind the dam) to a lower potential energy below the dam. Similarly, when an electric circuit is connected with a cell or battery, the electrons naturally flow because there is a difference in potential energy between the anode and the cathode.

DID YOU KNOW ?

Electric Charge and Current Analogy
We can extend the analogy in **Figure 4**. The mass (in kg) of water stored behind the dam is similar to the available charge (in C) stored in the chemicals of a cell. When released, the current, or flow, of water (in kg/s) is similar to the current, or flow, of electrons (in C/s).

The *power* of a cell or battery is the rate at which it produces electrical energy. Power is measured in watts (W), and is calculated as the product of the current and the voltage of the battery. The *energy density*, or *specific energy* of a battery, is a measure of the quantity of energy stored or supplied per unit mass. Energy density may be measured in joules per kilogram (J/kg). **Table 1** summarizes some important electrical quantities and their units of measurement.

Table 1 Electrical Quantities and SI Units

Quantity	Symbol	Meter	Unit	Unit symbol
charge	Q	–	coulomb	C
current	I	ammeter	ampere	A (1 A = 1 C/s)
potential difference	V	voltmeter	volt	V (1 V = 1 J/C)
power	P	–	watt	W (1 W = 1 J/s)
energy density	–	–	joules per kilogram	J/kg

SUMMARY: Components of an Electric Cell

- An electric cell must have two electrodes and an electrolyte.
- An electrode is a solid conductor.
- An electrolyte is an aqueous conductor.
- The cathode is the electrode labelled positive.
- The anode is the electrode labelled negative.
- The electrons flow through the external circuit from the anode to the cathode.

Practice

1. What are the components of a simple electric cell?
2. Write an empirical definition of electrode and electrolyte, and a conventional definition of anode and cathode.
3. If a DVD player requires 9 V to operate, how many 1.5 V "dry" cells connected in series would it need?
4. Differentiate between electric current and voltage.
5. Why do manufacturers of battery-operated devices print a diagram showing the correct orientation of the batteries? (Supply two answers to this question: one from a scientific perspective and one from a technological perspective.)
6. List several examples illustrating how a new technology (electric cells) led to new scientific discoveries.

DID YOU KNOW?

Aboriginal Science and Technology

Aboriginal peoples have a long history of technological development in many areas such as agriculture, food preservation, medicine, and transportation. Just as in Western societies, Aboriginal technology had a practical purpose, was developed through a trial-and-error process, and was interdependent on Aboriginal science using traditional or indigenous knowledge.

Technological Problem Solving

The initial development of cells and batteries preceded much of the current scientific understanding of these devices. Cells and batteries existed almost 100 years before the electron was discovered. The study of electric cells is a good illustration of tremendous advances in technology based on very limited scientific knowledge. Technological development or problem solving is similar in some ways to scientific problem solving, but its purpose differs. The purpose of technological problem solving is to find a realistic way around a practical difficulty—to make something work—while the purpose of scientific problem solving is to describe, explain, and ultimately understand natural and technological phenomena. Technology and science are dependent on each other. Although scientific knowledge can be used to guide the creation of a technology, the technology may create new scientific understanding.

A systematic trial-and-error process is often used in technological problem solving (Appendix C.1). This is not a new process. Aboriginal peoples used a systematic trial-and-error process to ensure the survival of the tribe. An example of a trial-and-error process is as follows:

- Develop a general design for problem-solving trials; for example, select which variables to manipulate and which to control.
- Follow several prediction–procedure–evidence–analysis cycles, manipulating and systematically studying one variable at a time.
- Complete an evaluation based on criteria such as efficiency, reliability, cost, and simplicity.

This technological problem-solving model was important in the early development of practical electric cells.

> **DID YOU KNOW?**
>
> **A "Not Quite Dry" Cell**
> The electrolyte in the "dry cell" is actually a moist paste (**Figure 5**). If the cell were completely dry, it would not work because the ions in the electrolyte must be able to carry the electric current to complete the circuit. Just enough water is added so that the ions can move, but not enough to make the mixture liquid.

Figure 5

INVESTIGATION 14.1 Introduction

Designing an Electric Cell

In this cell, an aluminium soft-drink can is one of the electrodes. The other electrode is a solid conductor, such as a piece of copper wire or pipe, an iron nail, or graphite from a pencil. The electrolyte may be a salt solution, or an acidic or basic solution. Although many characteristics of a cell are important, only one characteristic, voltage, is investigated here. Check with your teacher if you want to evaluate other designs and materials.

When evaluating the Purpose (Part 3), include your opinion about the reliability, cost, and simplicity of your final electric cell.

Purpose
The purpose of this investigation is to make an electric cell.

Report Checklist
- ○ Purpose
- ○ Problem
- ○ Hypothesis
- ● Prediction
- ○ Design
- ● Materials
- ● Procedure
- ● Evidence
- ● Analysis
- ● Evaluation (1, 3)

Problem
What combination of electrodes and electrolyte gives the largest voltage for an aluminium-can cell?

Design
In a trial-and-error procedure, different variables (second electrode, electrolyte) are modified one at a time while all other variables are held constant. The voltage of each cell is measured as the responding variable.

To perform this investigation, turn to page 658.

Consumer, Commercial, and Industrial Cells

Since Volta's invention of the electric cell and battery, there have been many advances in electrochemistry and technology. Invented in 1865, the zinc chloride cell is commonly referred to as a dry cell because this design was the first to use a sealed container. The first 9 V battery was made up of a series of 1.5 V dry cells (**Figure 6**). Both the 1.5 V dry cell and the 9 V battery are simple, reliable, and relatively inexpensive. Other cells, such as the alkaline dry cell and the mercury cell (**Table 2** on page 616), were developed to improve the performance of the original dry cell. One problem with all of these cells is that the chemicals eventually become depleted and irreversible reactions prevent these cells from being recharged. Cells that cannot be recharged are called *primary cells*.

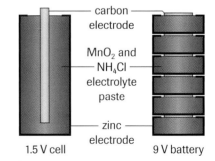

Figure 6
Like a flashlight D cell, the zinc chloride dry cell on the left has a voltage of 1.5 V. The 9 V battery on the right is made up of six 1.5 V dry cells in series.

Table 2 Primary, Secondary, and Fuel Cells

Type	Name of cell	Half-reactions	Characteristics and uses
primary cells	dry cell (1.5 V)	$2 MnO_2(s) + 2 NH_4^+(aq) + 2 e^- \rightarrow Mn_2O_3(s) + 2 NH_3(aq) + H_2O(l)$ $Zn(s) \rightarrow Zn^{2+}(aq) + 2 e^-$	• inexpensive, portable, many sizes • flashlights, radios, many other consumer items
	alkaline dry cell (1.5 V)	$2 MnO_2(s) + H_2O(l) + 2 e^- \rightarrow Mn_2O_3(s) + 2 OH^-(aq)$ $Zn(s) + 2 OH^-(aq) \rightarrow ZnO(s) + H_2O(l) + 2 e^-$	• longer shelf life; higher currents for longer periods compared with dry cell • same uses as dry cell
	mercury cell (1.35 V)	$HgO(s) + H_2O(l) + 2 e^- \rightarrow Hg(l) + 2 OH^-(aq)$ $Zn(s) + 2 OH^-(aq) \rightarrow ZnO(s) + H_2O(l) + 2 e^-$	• small cell; constant voltage during its active life • hearing aids, watches
secondary cells	Ni–Cd cell (1.25 V)	$2 NiO(OH)(s) + 2 H_2O(l) + 2 e^- \rightarrow 2 Ni(OH)_2(s) + 2 OH^-(aq)$ $Cd(s) + 2 OH^-(aq) \rightarrow Cd(OH)_2(s) + 2 e^-$	• can be completely sealed; lightweight but expensive • all normal dry cell uses, as well as power tools, shavers, portable computers
	lead–acid cell (2.0 V)	$PbO_2(s) + 4 H^+(aq) + SO_4^{2-}(aq) + 2 e^- \rightarrow PbSO_4(s) + 2 H_2O(l)$ $Pb(s) + SO_4^{2-}(aq) \rightarrow PbSO_4(s) + 2 e^-$	• very large currents; reliable for many recharges • all vehicles
fuel cells	hydrogen–oxygen cell (1.2 V)	$O_2(g) + 2 H_2O(l) + 4 e^- \rightarrow 4 OH^-(aq)$ $2 H_2(g) + 4 OH^-(aq) \rightarrow 4 H_2O(l) + 4 e^-$	• lightweight; high efficiency; can be adapted to use hydrogen-rich fuels • vehicles and space shuttle
	aluminium–air cell (2 V)	$3 O_2(g) + 6 H_2O(l) + 12 e^- \rightarrow 12 OH^-(aq)$ $4 Al(s) \rightarrow 4 Al^{3+}(aq) + 12 e^-$	• very high energy density; made from readily available aluminium alloys • designed for electric cars

EXTENSION

The Molicel
Find out about a made-in-Canada technological development that has had a big impact on portable power.

Secondary Cells

Secondary cells can be recharged by using electricity to reverse the chemical reaction that occurs when electricity is produced by the cell. Secondary cells and batteries include the nickel–cadmium (Ni–Cd) cell and the lead–acid battery (**Table 2** and **Figure 7**). A relatively recently developed secondary cell with a unique design is a lithium-ion cell, called the Molicel.

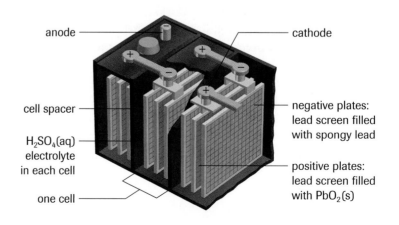

Figure 7
The anodes of a lead–acid car battery are composed of spongy lead and the cathodes are composed of lead(IV) oxide on a metal screen. The large electrode surface area is designed to deliver sufficient current to start a car engine.

One of the most common and reliable secondary cells is the lead–acid cell in a typical car battery. The discharging of this cell (see lead–acid cell, **Table 2**) produces approximately 2.0 V based on the following net equation.

$$\text{Pb(s)} + \text{PbO}_2\text{(s)} + 2\,\text{H}_2\text{SO}_4\text{(aq)} \xrightarrow{\text{discharging}} 2\,\text{PbSO}_4\text{(s)} + 2\,\text{H}_2\text{O(l)}$$

This cell requires the input (from the car's alternator) of at least 2.0 V to force the products to change back to the reactants to recharge it. The half-reactions for the lead–acid cell listed in **Table 2** must both be reversed to obtain the following net equation.

$$2\,\text{PbSO}_4\text{(s)} + 2\,\text{H}_2\text{O(l)} \xrightarrow{\text{charging}} \text{Pb(s)} + \text{PbO}_2\text{(s)} + 2\,\text{H}_2\text{SO}_4\text{(aq)}$$

A battery can be recharged if the products are stable with no further reactions occurring and if the products are able to travel through the electrolyte toward the appropriate electrode.

> **Learning Tip**
>
> Discharging a cell or battery is like letting the water spontaneously run out from the higher level behind a dam. Charging (or recharging) is like pumping the water up behind the dam. This is not a spontaneous process and requires energy.

Fuel Cells

A fuel cell is a different solution to the problem of the limited life of a primary cell. **Fuel cells** produce electricity by the reaction of a fuel that is continuously supplied to keep the cell operating. In principle, the fuel cell could be used forever, provided the fuel is continuously supplied. The fuel cell offers several advantages over methods that produce electricity by the combustion of fossil fuels. For example, fuel cells generate electricity more efficiently (**Table 3**), without producing greenhouse gases or substances that contribute to acid rain. The development of a cost-effective fuel cell is currently the focus of much scientific study and technological research and development.

William Grove accidentally invented the first fuel cell in 1839 using platinum electrodes, hydrogen, and oxygen as fuels, and sulfuric acid as the electrolyte. Grove was actually studying the reverse process—using electricity to convert water into hydrogen and oxygen. After one experiment, he reconnected the two electrodes without a power supply attached, and found that a small current was produced spontaneously as hydrogen and oxygen combined to form water. Grove continued to work on this cell, but eventually decided it was not a practical device because the characteristics, such as voltage, current, and energy capacity, were quite poor. Although many scientists, including Nobel Prize winners Fritz Haber and Walther Nernst, worked on improving the cell, they were largely unsuccessful. They manipulated many variables, such as different electrodes and electrolytes, but the reaction rates were too low and the electrodes became corroded.

Finally, in 1955, Francis Bacon succeeded where many others had failed. He produced a practical hydrogen–oxygen fuel cell using an alkaline electrolyte and electrodes constructed of porous nickel (**Figure 8**). Although the idea had been around for a long time, Bacon's cell was really the first practical fuel cell. NASA quickly adopted the hydrogen–oxygen fuel cell as an electrical power source for space flights because hydrogen and oxygen are already available for propulsion systems and the product, water, can be purified for drinking. NASA's fuel cell, a modification of the original Bacon cell, is an alkaline cell using potassium hydroxide as the electrolyte (**Table 2** on page 616). It produces 12 kW of electricity and operates at 70% efficiency. Unfortunately, NASA's fuel cell is expensive and has a relatively short working life, primarily due to the corrosive electrolyte. As a result, NASA's cell is not economically viable for general or commercial applications.

Table 3 Efficiencies of Different Technologies*

Technology	Efficiency*
fuel cells	40–70%
electric power plants	30–40%
automobile engines	17–23%
gasoline lawn mower	about 12%

*Efficiency is the fraction of the maximum available energy that is actually usable.

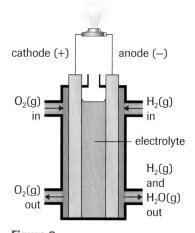

Figure 8
Hydrogen and oxygen gases are continuously pumped into this hydrogen–oxygen fuel cell. Each gas reacts at a different electrode. Unused gases are recycled.

EXTENSION

Fuel Cells

Car designers are working on futuristic cars, powered by fuel cells, that would cut down on pollution. This video shows practical applications of science and technology that may, some day, bring us cheap, clean, efficient transportation.

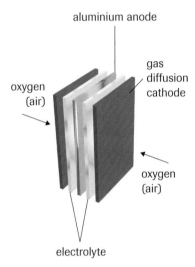

Figure 9
Several Canadian companies have done extensive research in the development of the aluminium–air solid fuel cell.

Figure 10
The world's first commercial phosphoric acid fuel cell was produced by ONSI/International Fuel Cells. It has been available since 1992 and uses natural gas, waste methane, propane, or hydrogen as fuels. This unit produces 200 kW electricity and 200 kW heat at a total system efficiency of 80%.

WEB Activity

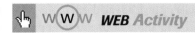

Web Quest—Hydrogen: Wonderfuel or Hype?

Taking the role of a research company, you and your group will investigate and prepare an illustrated presentation on the subject of hydrogen as a fuel. Is it clean? Is it inexpensive? Is it widely available? Will hydrogen become the wonderfuel of the 21st century or is hydrogen just power hype?

Aluminium–Air Cell

Another type of fuel cell is the metal–air fuel cell, the most common of which is the aluminium–air cell (**Table 2**). This is actually an aluminium–oxygen cell and has been developed for possible use in electric cars. Air is pumped into the cell and oxygen reacts at the cathode while a replaceable mass of aluminium reacts at the anode (**Figure 9**). The fuel is solid aluminium metal and the product, aluminium hydroxide, can be recycled back to aluminium metal. The simple design means that this cell can be assembled in almost any size. The high energy density of these cells results from the fact that three moles of electrons are released from every mole of aluminium, and aluminium is a very lightweight metal. Unlike hydrogen, storage and transportation of a solid fuel do not pose a problem. Estimates from prototypes suggest that the aluminium anode will need replacement every 2500 km in an electric car.

Large-Scale Commercial and Industrial Fuel Cells

The requirements for electrical power fuel cells for large-scale use in businesses and industry are similar with regard to the fuel, but there is less concern about volume, weight, or energy density. However, businesses and industries need cells with much longer lifetimes. Fuel cells for large-scale commercial and industrial use are almost always co-generation units. This means that they produce electricity as well as heat for space heating. Co-generation means that the overall efficiencies can be as high as 90%. Commercially viable fuel cells today are usually acid electrolyte cells such as the phosphoric acid fuel cell, which can produce 400 MW of power, sufficient for meeting the electrical energy needs of a small city (**Figure 10**). These cells usually use natural gas as a source of hydrogen for the fuel cell and operate at temperatures of 200 °C.

Section 14.1

▸ Practice

7. Compare scientific and technological problem solving.
8. What steps are involved in technological problem solving?
9. Suppose you decided to develop and market an aluminium-can cell (see Investigation 14.1.) How and why would you alter the electrolyte?
10. Distinguish between primary and secondary cells, including a common example of each.
11. What are some advantages and disadvantages of the zinc chloride dry cell?
12. What do the designs of a dry-cell container and an ice-cream cone have in common?
13. One of the most successful batteries has been the lead–acid car battery.
    (a) Identify the anode, cathode, and electrolyte.
    (b) How are the large currents produced that are necessary to start a car?
    (c) What has been the social impact of this battery?
    (d) What are some possible environmental impacts of this battery?
14. For both the hydrogen–oxygen fuel cell and the aluminium–air fuel cell,
    (a) write the two half-reaction equations (refer to **Table 2**, page 616)
    (b) label each equation from (a) as an oxidation or a reduction
    (c) write the net ionic equation for each cell
15. Using several perspectives, state some advantages and disadvantages of a fuel cell.
16. Assess the possible future importance of fuel cells in society.
17. Experimental cell phones that run on miniature hydrogen–oxygen fuel cells exist today. What must be in place before the average consumer can buy one?

DID YOU KNOW ?

Success or Failure?
Thomas Edison, the American inventor, set out to invent a secondary battery. His plan was to use an alkaline electrolyte and iron as the anode, but he needed to find a suitable cathode material. After thousands of experiments, his friend and associate Walter Mallory commented on Edison's lack of results. Edison replied that he had lots of results and he knew fifty thousand things that did not work! Eventually he discovered that a thin metallic nickel film produced a battery that was lighter and more powerful than the existing lead–acid batteries.

 **WEB Activity**

Canadian Achievers—Lewis Urry

After graduating in chemical engineering from the University of Toronto in 1950, Lewis Urry (**Figure 11**) went to work for Eveready Battery Company where he developed the first practical, long-life electric cell.

1. Identify the unique feature in Urry's cell ("battery") and explain how it works.
2. List three familiar devices that depend on the alkaline cell.
3. Name one other cell developed by Urry.

www.science.nelson.com

Figure 11
Lewis Urry (1927–2004), inventor of the alkaline battery

Case Study

The Ballard Fuel Cell

A variation of a hydrogen–oxygen fuel cell, also known as the hydrogen fuel cell, was developed for commercial applications by Ballard Power Systems in Burnaby, BC. The Ballard fuel cell uses a proton exchange membrane (PEM) in place of a liquid electrolyte. Normal electric cells use the ions in the liquid electrolyte to transfer electric charge within the cell. In a hydrogen fuel cell, the PEM is made from a solid proton-conducting polymer that transfers charge within the cell (**Figure 12**). The PEM is simple, robust, eliminates corrosive liquids, and permits a high energy density.

The Ballard fuel cell consists of an anode and a cathode separated by a PEM. Hydrogen fuel admitted through a porous anode is converted into hydrogen ions (protons) and free electrons in the presence of a catalyst at the anode. An external circuit conducts the free electrons and produces the desired electrical current. Water and heat are produced when the protons, after migrating through the polymer membrane to the cathode, react both with oxygen molecules from the air and with the free electrons from the external circuit.

Fuel cells can be connected in series (stacked) to increase the voltage and power output. For example, an experimental transit bus uses an electric motor powered by a Ballard fuel cell that is capable of 205 kW (or 250 hp).

Ballard has development agreements with most major car manufacturers to use its cells in future electric cars. The zero-emission engines convert hydrogen, or hydrogen-rich fuels such as natural gas and even methanol, into electricity, producing water and heat as the main byproducts.

Although the Ballard hydrogen fuel cell looks very promising, there are several problems yet to be solved. Cost is a major factor, which may be partially solved by mass production. The fuel is also under debate. If hydrogen gas is used, where does it come from? Electrolysis of water uses a lot of electrical energy and, if this energy comes from a fossil-fuel generating plant, a lot of pollution is produced along with the hydrogen gas. How would the hydrogen gas be distributed and stored on board the vehicle? There are important safety concerns associated with the handling and storage of hydrogen, which is flammable. Many scientists and engineers believe that the solution is not to use hydrogen gas directly, but to use hydrogen-rich fuels. We have a lot of knowledge of reforming hydrocarbons to produce hydrogen. If natural gas or even gasoline were reformed as needed on board the vehicle, then we would have a familiar fuel source and an infrastructure in place to supply this fuel. Not everyone agrees that this is a good solution.

Hydrogen fuel cells may change the way that we use energy, especially for transportation. Fuel cells may be used not only to power cars, buses, and boats, but also many portable devices.

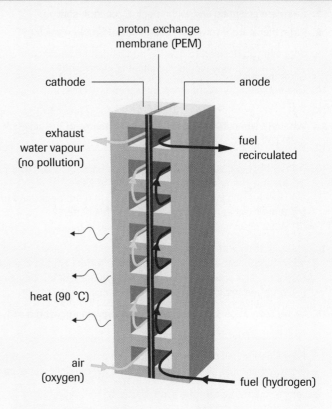

Figure 12
The hydrogen fuel cell has the same design as Volta's original cell, but the electrolyte is a conducting solid.

Case Study Questions

1. Is hydrogen a "clean" fuel? Explain your answer.
2. Make a list, in order of importance, of the advantages of a Ballard hydrogen fuel cell for urban buses compared with those of a diesel engine for buses. Make a similar list of possible disadvantages.
3. How has the development of fuel cells been influenced by society and how might they influence society in the future? Include a variety of perspectives. Use information from various print and electronic sources.

Extension

4. One solution to the hydrogen storage problem is to make hydrogen as it is needed using a reformer. Research how reformers can be used to produce hydrogen.

5. Iceland has ambitious plans to be the first nation to replace all its fossil fuels with "clean" hydrogen. Research how Iceland plans to produce the hydrogen it will need.

Section 14.1 Questions

1. Draw a simple diagram of an electric cell and label: electrodes, electrolyte, cathode, anode, signs for cathode and anode, and direction of electron flow through an external wire.

2. State the evidence that an electric cell involves a redox reaction.

3. List three types of electric cells used in consumer and commercial operations. Briefly describe the main feature of each cell.

4. State two common examples of consumer cells and where they may be used.

5. A silver oxide cell is often used when a miniature cell or battery is required, as in watches, calculators, and cameras. The following half-reaction equations occur in the cell:
   $Ag_2O(s) + H_2O(l) + 2 e^- \rightarrow 2 Ag(s) + 2 OH^-(aq)$
   $Zn(s) + 2 OH^-(aq) \rightarrow Zn(OH)_2(s) + 2 e^-$
   (a) In which direction does the electric current flow: silver to zinc or zinc to silver?
   (b) Which is the anode and which is the cathode?
   (c) Write the net redox equation for the discharging of the silver oxide cell.

6. Describe an example where a new technology led to scientific discoveries and an example in which scientific knowledge led to a better technology.

7. Possible solutions to technological problems often involve different designs, materials, and processes. List several examples that solve the problem of cells becoming depleted and unusable.

8. What are some unintended consequences of the widespread use of many different types of cells and batteries?

9. Describe how societal needs and expectations have affected the development of the electric cells available today.

10. Suppose cells and batteries did not exist. What impact would that have on your life?

11. State several reasons why it is important to recycle the components of cells and batteries when they have reached the end of their useful life.

12. (a) Why is there a great deal of interest in electric cars?
    (b) Suggest some reasons why we don't use lead–acid batteries as the only power source for electric cars.
    (c) How have advances in hydrogen fuel cells facilitated the development of electric cars?

Extension

13. Most people associate technological development with "progress" in Western societies and think that technology transfer to Aboriginal peoples only occurs in one direction. In fact, although not often recognized and valued, there has been significant technology transfer from Aboriginal peoples to Western society. Summarize several examples of significant Aboriginal technological knowledge that was transferred to early European settlers. What area is still a significant source of technological transfer from Aboriginal to Western societies? Why is this an area of dispute?

 www.science.nelson.com

14. Portable electronic devices can be found everywhere. Laptop computers, cellular telephones, mobile radios, cordless phones, portable disc and MP3 players, and digital cameras all require an electric cell.
    (a) What are some of the requirements for cells used in these applications?
    (b) Why are some rechargeable batteries used in various portable devices supposed to be totally "drained" (discharged) before recharging?

 www.science.nelson.com

15. People whose heart occasionally beats too slowly or too quickly often have a pacemaker fitted to keep the heart beating regularly (**Figure 13**). Pacemakers use a battery for electric power. What kind of battery is commonly used today? How long does it last? How does the doctor know when the battery is nearing the end of its life and needs to be replaced? Why are rechargeable batteries generally not used?

 www.science.nelson.com

Figure 13
A pacemaker contains electronics and a built-in battery. The whole unit is only a few centimetres in size and is implanted under the skin near the collarbone.

16. Plastic batteries were the dream of the 1980s, the disappointment of the 1990s, and the subject of the 2000 Nobel Prize for Chemistry. Now it appears that some commercial products will eventually result from the research and development invested in plastic batteries. Briefly describe the electrodes and electrolyte for a plastic battery. How is this battery similar to and different from an ordinary battery? What are some advantages and disadvantages?

 www.science.nelson.com

14.2 Voltaic Cells

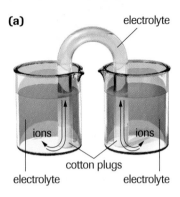

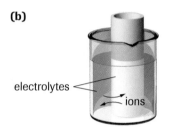

Figure 1
(a) A salt bridge is a U-shaped tube containing an inert (unreactive) aqueous electrolyte such as sodium sulfate.
(b) An unglazed porcelain (porous) cup containing one electrolyte sits in a container of a second electrolyte.

Electric cells were invented for practical purposes in about 1800, but they were not explained scientifically until about 100 years after their invention. Their use, however, contributed to the scientific understanding of redox reactions. Later this knowledge helped explain reactions inside the cell itself. Electric cells adapted for scientific study are often called *galvanic cells* (in recognition of Luigi Galvani), or *voltaic cells* (in recognition of Alessandro Volta).

You learned in Chapter 13 that a redox reaction involves a transfer of electrons from the reducing agent to the oxidizing agent. In a voltaic cell, electrons pass from the reducing agent to the oxidizing agent through an external circuit rather than passing directly from one substance to another. You have seen that the individual components of a cell—electrodes and electrolytes—determine characteristics such as voltage and current. Why is this so? What happens in different parts of a cell? To answer these questions, chemists use a cell that has two electrodes and their electrolytes separated. This is not a very practical arrangement, but it greatly facilitates the study of cells. Each electrode is in contact with an electrolyte. A **porous boundary** separates the two electrolytes, at least for a short time, while still permitting ions to move between the two solutions through tiny openings in the cotton plugs of a salt bridge (**Figure 1(a)**) or in the walls of a porcelain cup (**Figure 1(b)**).

Using this design modification, a cell can be split into two parts connected by a porous boundary. Each part, called a **half-cell**, consists of one electrode and one electrolyte. For example, the copper–zinc cell shown in **Figure 2** has two half-cells, copper metal in a solution of copper ions and zinc metal in a solution of zinc ions.

This cell can be represented using the following abbreviated ("shorthand") notation, called a cell notation:

$$Zn(s) \mid Zn(NO_3)_2(aq) \parallel Cu(NO_3)_2(aq) \mid Cu(s)$$

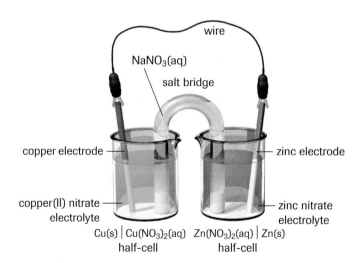

Figure 2
Each electrode is in its own electrolyte, forming a half-cell. The two half-cells are connected by a salt bridge (containing $NaNO_3(aq)$) and by an external wire to make a complete circuit.

In this notation, a single line (|) indicates a phase boundary, such as the interface of an electrode and an electrolyte in a half-cell. A double line (||) represents a physical boundary, such as a porous boundary between half-cells. A **voltaic cell** is an arrangement of two half-cells separated by a porous boundary. Voltaic cells, such as the one in **Figure 2**, are especially suitable for scientific study.

> **Learning Tip**
>
> It is a common practice to write the cell notation with the anode on the left-hand side and the cathode on the right-hand side. Diagrams or photos of cells can have any arrangement.

INVESTIGATION 14.2 Introduction

A Voltaic Cell (Demonstration)

An important characteristic of consumer, commercial, and industrial cells is a simple, efficient design that works for the intended application. For scientific research, this is not as important as a design that can be easily manipulated and studied.

Purpose

The purpose of this investigation is to test the design and operation of a voltaic cell used in scientific research.

Report Checklist
- ○ Purpose ○ Design ● Analysis
- ○ Problem ○ Materials ○ Evaluation
- ○ Hypothesis ○ Procedure
- ○ Prediction ● Evidence

Problem
What is the design and operation of a voltaic cell?

Design
An electric cell with only one electrolyte is compared with voltaic cells containing the same electrodes, but two electrolytes.

To perform this investigation, turn to page 659.

A Theoretical Description of a Voltaic Cell

Observation of a voltaic cell as it operates provides evidence to explain what is happening inside the cell. For example, during your study of the silver–copper cell in Investigation 14.2, you observed the evidence listed in **Table 1**. Note that the table also includes a theoretical interpretation of each point, which is also shown in **Figure 3**.

According to the electron transfer theory and the concept of relative strengths of oxidizing and reducing agents, silver ions are the strongest oxidizing agents in the cell; they undergo a reduction half-reaction at the cathode. The strongest oxidizing agent in the cell always undergoes a reduction at the cathode. Copper atoms, which are the strongest reducing agents in the cell, give up electrons in an oxidation half-reaction and enter the

> **Learning Tip**
>
> People often use acronyms or similar devices to help them remember important information. "OIL RIG (oxidation is loss, reduction is gain)" is an example. One way to help you remember important details of a cell is the expression SOAC/GERC, loosely read as "soak a jerk." Translated, this means the Strongest Oxidizing Agent at the Cathode Gains Electrons and is Reduced at the Cathode. Another example is "An ox ate a red cat," which helps to recall **a**node **o**xidation; **r**eduction **c**athode.

Table 1 Evidence and Interpretations of the Silver–Copper Cell

Evidence	Interpretation
The copper electrode decreases in size and the intensity of the blue colour of the electrolyte increases.	Oxidation of copper metal is occurring: $Cu(s) \rightarrow Cu^{2+}(aq) + 2\,e^-$ blue
The silver electrode increases in size as long, silver-coloured crystals grow.	Reduction of silver ions is occurring: $Ag^+(aq) + e^- \rightarrow Ag(s)$
A blue colour slowly moves up the U-tube from the copper half-cell to the silver half-cell.	Copper(II) ions (cations) move toward the cathode.
An ammeter shows that the electric current flows along a wire between the copper electrode and the silver electrode.	Electrons move through the wire from the copper electrode to the silver electrode.
A voltmeter indicates that the silver electrode cathode (positive) and the copper electrode is the anode (negative).	Electrons have a tendency to leave the copper half-cell and enter the silver half-cell.

Learning Tip

A variety of names can be used for cells based upon spontaneous redox reactions: electric, voltaic, galvanic, and electrochemical. In this book, *electric cell* is used for consumer cells and *voltaic cell* is used for scientific research cells. *Electrochemical cell* is a general term that includes both voltaic (spontaneous) and electrolytic (nonspontaneous) cells.

DID YOU KNOW ?

Electron Sources and Sinks
Chemists sometimes refer to the anode as the electron source and the cathode as the electron sink. Look at some half-reaction equations to see why these terms apply.

SUMMARY: Voltaic Cells

- A voltaic cell consists of two half-cells separated by a porous boundary with solid electrodes connected by an external circuit.
- The cathode is the positive electrode. The strongest oxidizing agent is reduced at the cathode.
- The anode is the negative electrode. The strongest reducing agent is oxidized at the anode.
- Electrons travel in the external circuit from the anode to the cathode.
- Internally, anions move toward the anode and cations move toward the cathode as the cell operates. The solutions remain electrically neutral.
- Cell notation: anode | electrolyte || electrolyte | cathode
  where a single vertical line represents a phase boundary and a double vertical line represents a porous boundary.

▶ Practice

1. Write an empirical description of each of the following terms: voltaic cell, half-cell, porous boundary, salt bridge, electrolyte, external circuit, and inert electrode.
2. Write a theoretical definition of a cathode and an anode.
3. Indicate whether the following processes occur at the cathode or at the anode of a voltaic cell.
   (a) reduction half-reaction
   (b) oxidation half-reaction
   (c) reaction of the strongest reducing agent
   (d) reaction of the strongest oxidizing agent
4. When is an inert electrode used? Give two common examples.
5. What are the characteristics of the solution in a salt bridge? Provide an example.
6. For each of the following cells, use the given cell notation to identify the strongest oxidizing and reducing agents. Write chemical equations to represent the cathode, anode, and net cell reactions. Draw a diagram of each cell, labelling the electrodes, electrolytes, direction of electron flow, and direction of ion movement.
   (a) $Zn(s) | Zn^{2+}(aq) || Ag^+(aq) | Ag(s)$
   (b) $Al(s) | Al^{3+}(aq) || NO_3^-(aq), H^+(aq) | Pt(s)$
7. Ions move through a porous boundary between the two half-cells of a voltaic cell.
   (a) In what direction do the cations and anions move?
   (b) Why do the ions move? Take your answer and convert it into another "why" question. Now answer this question.
   (c) For the copper–silver cell shown in **Figure 3** (page 624), we know that $Na^+(aq)$ and $NO_3^-(aq)$ in the salt bridge move into the cathode and anode compartments, respectively. Explain the evidence that the blue colour moves up the salt bridge towards the cathode (refer to **Table 1**, page 623)?
8. Draw and label a diagram for a voltaic cell constructed from some (not all) of the following materials:

strip of cadmium metal	voltmeter
strip of nickel metal	connecting wires
solid cadmium sulfate	glass U-tube
solid nickel(II) sulfate	cotton
solid potassium sulfate	various beakers
distilled water	porous porcelain cup

9. Redesign the voltaic cell in question 8 by changing at least one electrode and one electrolyte. The net reaction should remain the same for the redesigned cell.

Standard Cells and Cell Potentials

The investigations and activities you have completed show that the design and composition of a cell affect its operation. To make comparisons and scientific study easier, chemists specify the composition of a cell and the conditions under which the cell operates. A **standard cell** is a voltaic cell in which each half-cell contains all entities shown in the half-reaction equation at SATP conditions, with a concentration of 1.0 mol/L for the aqueous solutions. If a metal is not part of a half-cell, then an inert electrode is used to construct the standard cell. For example, for a standard zinc–dichromate cell, the cell description is

$$\text{Zn(s)} \mid \underset{1.0 \text{ mol/L}}{\text{Zn}^{2+}\text{(aq)}} \parallel \underset{1.0 \text{ mol/L}}{\text{Cr}_2\text{O}_7^{2-}\text{(aq)}}, \underset{1.0 \text{ mol/L}}{\text{H}^+\text{(aq)}}, \underset{1.0 \text{ mol/L}}{\text{Cr}^{3+}\text{(aq)}} \mid \text{C(s)} \quad \text{at SATP}$$

The **standard cell potential**, $E°_{cell}$, is the maximum electric potential difference (voltage) of the cell operating under standard conditions; $E°_{cell}$ represents the energy difference (per unit of charge) between the cathode and the anode. The degree sign (°) indicates that standard 1.0 mol/L and SATP conditions apply. Based on the idea of competition for electrons, a **standard reduction potential**, $E°_r$, represents the ability of a standard half-cell to attract electrons, thus undergoing a reduction. The half-cell with the greater attraction for electrons—that is, the one with the more positive reduction potential—gains electrons from the half-cell with the lower reduction potential. The standard cell potential is the difference between the reduction potentials of the two standard half-cells.

$$E°_{cell} = E°_{r\text{ cathode}} - E°_{r\text{ anode}}$$

It is impossible to determine experimentally the reduction potential of a single half-cell because electron transfer requires both an oxidizing agent and a reducing agent. Note that a voltmeter can only measure a potential difference, $E°_{cell}$. To assign values for standard reduction potentials, we measure the "reducing" strength of all possible half-cells relative to an accepted, standard half-cell. The half-cell used for this purpose is the standard hydrogen half-cell. A half-cell such as this, that is chosen as a reference and arbitrarily assigned an electrode potential of exactly zero volts, is called a **reference half-cell**.

Standard Hydrogen Half-Cell

The standard hydrogen half-cell (**Figure 6**) consists of an inert platinum electrode immersed in a 1.00 mol/L solution of hydrogen ions, with hydrogen gas at a pressure of 100 kPa bubbling over the electrode. The pressure and temperature of the cell are kept at SATP conditions. Standard reduction potentials for all other half-cells are measured relative to that of the standard hydrogen half-cell. The reduction potential of the hydrogen ion reduction half-reaction is defined to be exactly zero volts.

$$2 \text{ H}^+\text{(aq)} + 2 \text{ e}^- \rightleftarrows \text{H}_2\text{(g)} \qquad E°_r = 0.00 \text{ V}$$

As a result, we can assign a numerical value to the reduction potential associated with every other reaction. A reduction potential that has a positive value means that the oxidizing agent is a stronger oxidizing agent than hydrogen ions. A negative reduction potential means that the oxidizing agent is a weaker oxidizing agent than hydrogen ions. The choice of the standard hydrogen half-cell as a reference is the accepted convention. If a different half-cell had been chosen as the reference, individual reduction potentials would be different, but their relative values would remain the same.

Learning Tip

Think of the standard cell potential $E°_{cell}$ as representing the *difference* in ability of two half-cells to gain electrons. This potential difference can only be measured accurately if no current is allowed to flow. A good-quality voltmeter has a large internal resistance to prevent current flow.

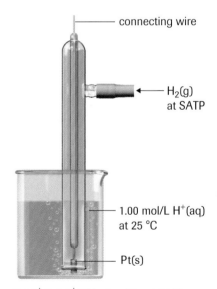

$\text{Pt(s)} \mid \text{H}_2\text{(g)} \mid \text{H}^+\text{(aq)} \quad E°_r = 0.00 \text{ V}$

Figure 6
The standard hydrogen half-cell is used internationally as the reference half-cell in electrochemical research. Notice that the second vertical line in the cell notation designates the phase boundary between the gas and the liquid.

Learning Tip

A voltmeter has two terminals, positive (red) and negative (black). Connect these to the electrodes of any cell so that the voltmeter gives a positive reading. Whatever electrode is connected to the positive terminal will be the cathode, and the other electrode will be the anode.

Measuring Standard Reduction Potentials

We can measure the standard reduction potential of a half-cell by constructing a standard cell using a hydrogen reference half-cell and the half-cell whose reduction potential you want to measure. There are two things you need to know: the voltage and the direction of the current. The magnitude of the voltage determines the numerical value of the half-cell potential and the direction of the current determines the sign of the half-cell potential. The cell potential is measured with a voltmeter, which also shows the direction the electrons tend to flow. If $E°_{cell}$ is positive, then the positive terminal on the voltmeter is connected to the cathode and the oxidizing agent at the cathode is stronger than hydrogen ions.

The cell shown in **Figure 7** can be represented as follows:

$$\underbrace{Pt(s) \mid H_2(g) \mid H^+(aq)}_{\text{anode}} \parallel \underbrace{Cu^{2+}(aq) \mid Cu(s)}_{\text{cathode}} \quad E°_{cell} = +0.34 \text{ V}$$

cathode: $Cu^{2+}(aq) + 2e^- \rightarrow Cu(s)$

anode: $H_2(g) \rightarrow 2H^+(aq) + 2e^-$

net: $Cu^{2+}(aq) + H_2(g) \rightarrow Cu(s) + 2H^+(aq)$

$$E°_{cell} = E°_r{}_{\text{cathode}} - E°_r{}_{\text{anode}}$$

$$= 0.34 \text{ V} - 0.00 \text{ V}$$

$$= +0.34 \text{ V}$$

The voltmeter shows that the copper electrode is the cathode and is 0.34 V higher in potential than the platinum anode. If you were to replace the voltmeter with a connecting wire so that the current is allowed to flow, the blue colour of the copper(II) ion disappears and the pH of the hydrogen half-cell decreases as more hydrogen ions are produced and the solution becomes more acidic. Based on this evidence, copper(II) ions are being reduced to copper metal and hydrogen molecules are being oxidized to hydrogen ions. Since this redox reaction is spontaneous, copper(II) ions are stronger oxidizing agents than are hydrogen ions. The standard cell potential, $E°_{cell} = 0.34$ V, is the difference between the reduction potentials of these two half-cells (**Figure 8**).

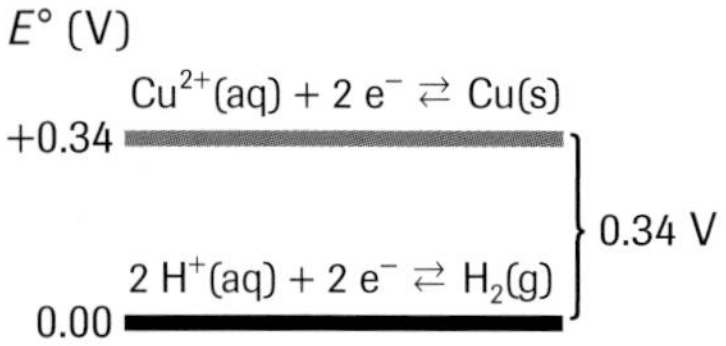

Suppose a standard aluminium half-cell was set up with a standard hydrogen half-cell (**Figure 9**). We can represent the cell as

$$\underbrace{Al(s) \mid Al^{3+}(aq)}_{\text{anode}} \parallel \underbrace{H^+(aq) \mid H_2(g) \mid Pt(s)}_{\text{cathode}} \quad E°_{cell} = +1.66 \text{ V}$$

According to the voltmeter, the platinum electrode is the cathode and the aluminium electrode is the anode. This indicates that hydrogen ions are stronger oxidizing agents than aluminium ions by 1.66 V. Since the reduction potential of hydrogen ions is defined as 0.00 V, the reduction potential of the aluminium ions must be 1.66 V below that of hydrogen, or -1.66 V (**Figure 10**).

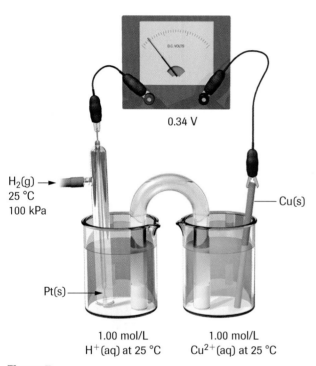

Figure 7
A copper–hydrogen standard cell

Figure 8
Copper(II) ions are stronger oxidizing agents than hydrogen ions. The cell potential provides a quantitative measurement of how much stronger.

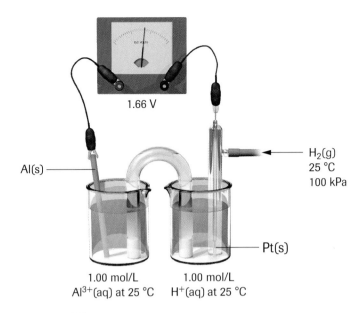

Figure 9
An aluminium–hydrogen standard cell

$$2\,H^+(aq) + 2\,e^- \rightleftarrows H_2(g) \qquad E°_r = 0.00\,V$$
$$Al^{3+}(aq) + 3\,e^- \rightleftarrows Al(s) \qquad E°_r = -1.66\,V$$

The standard cell potential, $E°_{cell} = 1.66\,V$, is the difference between the reduction potentials of these two half-cells. To obtain the net or overall cell reaction, add the reduction and oxidation half-reactions, but remember to balance and cancel the electrons.

cathode: $\quad 3\,[2\,H^+(aq) + 2\,e^- \rightarrow H_2(g)]$
anode: $\quad\quad\quad\quad 2\,[Al(s) \rightarrow Al^{3+}(aq) + 3\,e^-]$
net: $\quad\quad 6\,H^+(aq) + 2\,Al(s) \rightarrow 3\,H_2(g) + 2\,Al^{3+}(aq)$

$$E°_{cell} = E°_{r\,cathode} - E°_{r\,anode}$$
$$= 0.00\,V - (-1.66\,V)$$
$$= +1.66\,V$$

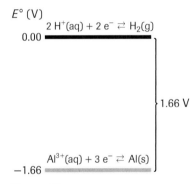

Figure 10
On a redox table, hydrogen ions are stronger oxidizing agents than aluminium ions. The cell potential tells us the hydrogen ions are 1.66 V above aluminium ions.

Notice that the half-reaction equations were multiplied by appropriate factors to balance the electrons, but the *reduction potentials were not altered by the factors used to balance the electrons*. Electric potential represents energy per coulomb of charge (1 V = 1 J/C). Multiplying the aluminium half-reaction by a factor of 2 doubles both the energy and the charge transferred, so that the ratio of energy (J) to charge (C), that is the voltage, is unaffected.

In both of these examples, the strongest oxidizing agent is reduced at the cathode and the strongest reducing agent is oxidized at the anode. The measured cell potential is the difference between the reduction potentials at the cathode and at the anode.

A positive cell potential ($E°_{cell} > 0$) indicates that the net reaction is spontaneous—a requirement for all voltaic cells.

> **Learning Tip**
>
> Altering the coefficients in a half-reaction equation does **not** affect the reduction potential.

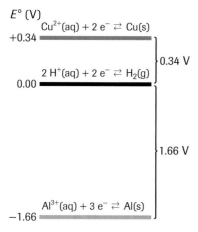

Figure 11
Measurements of standard cell potentials show that the reduction potential of $Cu^{2+}(aq)$ is $+0.34$ V greater than that of $H^+(aq)$, which is $+1.66$ V greater than that of $Al^{3+}(aq)$. If you set up a standard cell using copper and aluminium, what would be the cell potential, $E°_{cell}$? (Answer: $+2.00$ V)

Figure 11 shows the combined results from the copper–hydrogen and aluminium–hydrogen standard cells. This process of measuring standard cell potentials can quickly be extended to more and more oxidizing agents. Notice that this process, although started with the hydrogen reference cell, does not require that it be used for all cell measurements. For example, knowing that the reduction potential of copper(II) ions is 0.34 V, we can now set up many cells that include a standard copper half-cell. A more extensive list of reduction potentials is found in the redox table in Appendix I. You can predict the reaction that occurs spontaneously in any voltaic cell operating under standard conditions using the redox table in Appendix I. The standard cell potential is predicted as follows:

$$E°_{cell} = E°_{r\ cathode} - E°_{r\ anode}$$

This order of subtraction is necessary to find out whether the reaction is spontaneous under standard conditions. If $E°_{cell}$ is positive, the reaction is spontaneous.

SUMMARY Rules for Analyzing Standard Cells

You can analyze a standard cell knowing the contents of both half-cells using one or more of the following steps:

- Determine which electrode is the cathode. The cathode is the electrode where the strongest oxidizing agent present in the cell reacts, i.e., the oxidizing agent that is closest to the top on the left side of the redox table. If required, copy the reduction half-reaction for the strongest oxidizing agent and its reduction potential.
- Determine which electrode is the anode. The anode is the electrode where the strongest reducing agent present in the cell reacts, i.e., the reducing agent that is closest to the bottom on the right side of the redox table. If required, copy the oxidation half-reaction (reverse the half-reaction by reading from right to left) for the strongest reducing agent and its reduction potential.
- Determine the overall cell reaction. Balance the electrons for the two half-reaction equations (but do not change the $E°_r$) and add the half-reaction equations.
- Determine the standard cell potential, $E°_{cell}$, using the equation:

$$E°_{cell} = E°_{r\ cathode} - E°_{r\ anode}$$

WEB Activity

Simulation—Voltaic Cells Under Standard Conditions

In this computer simulation, you will construct voltaic cells using different half-cells, and then analyze their components and processes. A series of questions will prompt you as you form generalizations about the spontaneity of various reactions.

www.science.nelson.com

COMMUNICATION example 2

A standard lead–dichromate cell is constructed. Write the cell notation, label the electrodes, and calculate the standard cell potential.

Solution

$$\text{Pb(s)} \mid \text{Pb}^{2+}\text{(aq)} \parallel \text{Cr}_2\text{O}_7^{2-}\text{(aq)}, \text{H}^+\text{(aq)}, \text{Cr}^{3+}\text{(aq)} \mid \text{C(s)}$$
anode   cathode

$E°_{cell} = 1.23 \text{ V} - (-0.13 \text{ V})$
       $= +1.36 \text{ V}$

COMMUNICATION example 3

A standard scandium–copper cell is constructed and the cell potential measured. The voltmeter indicates that copper electrode is positive.

$$\text{Sc(s)} \mid \text{Sc}^{3+}\text{(aq)} \parallel \text{Cu}^{2+}\text{(aq)} \mid \text{Cu(s)} \quad\quad E°_{cell} = +2.36 \text{ V}$$

Write and label the half-reaction and net equations, and calculate the standard reduction potential of the scandium ion.

Solution

cathode:	$3 [\text{Cu}^{2+}\text{(aq)} + 2 e^- \rightarrow \text{Cu(s)}]$	$E°_r = +0.34 \text{ V}$
anode:	$2 [\text{Sc(s)} \rightarrow \text{Sc}^{3+}\text{(aq)} + 3 e^-]$	$E°_r = ?$
net:	$3 \text{Cu}^{2+}\text{(aq)} + 2 \text{Sc(s)} \rightarrow 3 \text{Cu(s)} + 2 \text{Sc}^{3+}\text{(aq)}$	$E°_{cell} = +2.36 \text{ V}$

$2.36 \text{ V} = 0.34 \text{ V} - E°_r$
$E°_r = -2.02 \text{ V}$

Learning Tip

To ensure a correct interpretation, always write the cathode half-reaction of the SOA first. This will help you to remember to subtract the reduction potentials in the correct order.

INVESTIGATION 14.3 Introduction

Testing Voltaic Cells

Report Checklist
- ○ Purpose
- ○ Problem
- ○ Hypothesis
- ● Prediction
- ● Design
- ○ Materials
- ● Procedure
- ● Evidence
- ● Analysis
- ● Evaluation (1, 2, 3)

Testing is a procedure that is common to both technology and science. In technology, testing is necessary to determine how a product or process works using criteria such as efficiency, reliability, and cost. In science, testing is a key part in the advancement of knowledge. Scientific concepts are developed and then tested to determine their validity and limitations. New ideas that fail the test then need to be restricted, revised, or replaced.

In your Evaluation, pay particular attention to sources of error or uncertainty, and to limitations of the evidence collected.

Purpose

The purpose of this investigation is to test the predictions of cell potentials and the identity of the electrodes of various cells.

Problem

In cells constructed from various combinations of copper, aluminium, silver, and zinc half-cells, what are the standard cell potentials, and which is the anode and cathode in each case?

To perform this investigation, turn to page 660.

Practice

10. Standard cells are very important in the scientific study of voltaic cells.
    (a) Describe the contents and conditions of a standard cell.
    (b) Define the standard cell potential in words and in symbols.

11. For each of the following cells, write the equations for the reactions occurring at the cathode and at the anode, and an equation for the overall or net cell reaction. Calculate the standard cell potential. (Use the redox table in Appendix I.)
    (a) $\text{Cr(s)} \mid \text{Cr}^{2+}\text{(aq)} \parallel \text{Sn}^{2+}\text{(aq)} \mid \text{Sn(s)}$
    (b) $\text{Co(s)} \mid \text{Co}^{2+}\text{(aq)} \parallel \text{SO}_4^{2-}\text{(aq)}, \text{H}^+\text{(aq)}, \text{H}_2\text{SO}_3\text{(aq)} \mid \text{C(s)}$
    (c) $\text{Pt(s)} \mid \text{H}_2\text{(g)} \mid \text{OH}^-\text{(aq)} \parallel \text{OH}^-\text{(aq)} \mid \text{O}_2\text{(g)} \mid \text{Pt(s)}$

BIOLOGY CONNECTION

Reduction Potentials

Electron transfer in biological systems is governed by the same concepts as electron transfer in simpler chemical systems. In both systems, reduction potentials are used to rank important half-reactions. Cytochromes are proteins that are part of the electron transfer process in cellular respiration; ferredoxins are proteins that are an important part of the electron transfer in photosynthesis. These proteins attach to a metal ion such as Fe^{3+}, greatly influencing its ability to transfer electrons.

(free)
$Fe^{3+}(aq) + e^- \rightleftarrows Fe^{2+}(aq)$
$E°_r = +0.77$ V

(in cytochrome c)
$Fe^{3+}(aq) + e^- \rightleftarrows Fe^{2+}(aq)$
$E°_r = +0.25$ V

(in ferredoxin)
$Fe^{3+}(aq) + e^- \rightleftarrows Fe^{2+}(aq)$
$E°_r = -0.43$ V

www.science.nelson.com

Cell Potentials Under Nonstandard Conditions

The electric potential difference or voltage of a cell decreases slowly as the cell operates. Simultaneously, observable changes such as colour changes and precipitate formation occur. If the cell is left for a very long time, the voltage would eventually become zero and no further changes would be observed in the cell. When people refer to a "dead" cell or battery, this is often what they mean.

The electric potential difference of a cell is a measure of the tendency for electrons to flow. Ideally, during a measurement of the cell potential, a voltmeter should not allow any electrons to flow. If electrons flow, oxidation and reduction reactions occur which, in turn, change the concentrations from the standard 1.0 mol/L value. The value that is measured by a voltmeter represents an electric potential or stored energy, just as the water behind a lock in a canal has gravitational potential energy (**Figure 12(a)**). Connecting the electrodes of a cell in a circuit allows the electrons to flow from the anode to the cathode. This is analogous to opening the outlet and allowing the water to flow from behind the gates to a lower point in front of the gates (**Figure 12(b)**). In both cases, stored potential energy is converted to kinetic energy of electrons or water. If the water available behind a lock is allowed to flow out, then eventually no more water will flow. The level (potential energy) of the water on the two sides of the gate is equalized. An equilibrium is reached with no potential energy difference (**Figure 12(c)**). A similar situation occurs with an operating cell. If electrons are allowed to flow, eventually an equilibrium will be reached when the flow ceases. The rate of the forward reaction, which predominates initially, decreases as the rate of the reverse reaction increases, until the two rates become equal. This is the equilibrium condition and no net flow of electrons will occur. At this time, the electric potential difference as measured by a voltmeter becomes zero.

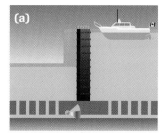 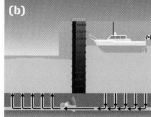 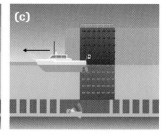

Figure 12
(a) The water behind the gates in a lock has a certain potential energy, ΔE, relative to the bottom of the closed outlet.
(b) When the outlet is opened, water spontaneously flows to the lower level on the other side of the gates. Potential energy, ΔE, is converted to kinetic energy of the flowing water. The water flowing through the outlet is analogous to electron flow.
(c) The flow of water ceases when the levels on both sides of the gates become equal. The gates open, and the ship can then exit to the next lock.

Discrepancies between Measured and Predicted Cell Potentials

Nonstandard conditions for concentrations, temperature, and pressure will cause differences between the cell potentials predicted from a standard redox table and ones measured in a laboratory. However, these differences are generally small if the conditions are relatively close to standard values. Other more important reasons for discrepancies include the purity of the substances, the presence of oxide coatings on metals, and the type and size of the porous boundary.

▸ Practice

12. For each of the following standard cells, refer to the redox table in Appendix I, to represent the cell using the standard cell notation (listing the anode first). Identify the cathode and anode and calculate the standard cell potential without writing half-reaction equations.
    (a) copper–lead standard cell
    (b) zinc–nickel standard cell
    (c) iron(III)–hydrogen standard cell

13. One experimental design for determining the position of a half-cell reaction that is not included in a redox table is shown below. Use the following standard cell, refer to the standard reduction potential of gold in Appendix I, and calculate the reduction potential for the indium(III) ion.

    $$\text{In(s)} \mid \text{In}^{3+}(aq) \parallel \text{Au}^{3+}(aq) \mid \text{Au(s)} \qquad E°_{cell} = +1.84 \text{ V}$$
    anode ⟵⟶ cathode

14. Any standard half-cell could have been chosen as the reference half-cell, the zero point of the reduction potential scale. What would be the standard reduction potentials for copper and zinc half-cells, assuming that the standard lithium cell were chosen as the reference half-cell with its reduction potential defined as 0.00 V? Justify your answer.

15. List some reasons for differences that might be observed between cell potentials predicted from a table of reduction potentials and cell potentials measured in a laboratory.

16. A zinc–iron cell is constructed and allowed to operate until the measured potential difference becomes zero. What interpretation can be made about the chemical system at this point?

LAB EXERCISE 14.A

Developing a Redox Table

Report Checklist

- ○ Purpose
- ○ Problem
- ○ Hypothesis
- ○ Prediction
- ○ Design
- ○ Materials
- ○ Procedure
- ○ Evidence
- ● Analysis
- ○ Evaluation

We can use standard cells and their measured cell potentials to develop a redox table.

Purpose
The purpose of this exercise is to use the concepts and rules of standard cells to develop a redox table.

Problem
What is the table of relative strengths of oxidizing and reducing agents based on measured cell potentials?

Design
Several cells are investigated; each cell has at least one half-cell in common with one of the other cells. The cell potentials are measured, and the positive and negative electrodes of each cell are identified.

Evidence

Negative electrode ⟶ Positive electrode

$$\text{Pd(s)} \mid \text{Pd}^{2+}(aq) \parallel \text{Cr}_2\text{O}_7^{2-}(aq), \text{H}^+(aq) \mid \text{C(s)} \qquad E°_{cell} = +0.28 \text{ V}$$

$$\text{Ti(s)} \mid \text{Ti}^{2+}(aq) \parallel \text{Tl}^+(aq) \mid \text{Tl(s)} \qquad E°_{cell} = +1.29 \text{ V}$$

$$\text{Tl(s)} \mid \text{Tl}^+(aq) \parallel \text{Pd}^{2+}(aq) \mid \text{Pd(s)} \qquad E°_{cell} = +1.29 \text{ V}$$

Corrosion

Human history is often divided into different "ages" such as the Copper, Bronze, and Steel ages. These descriptions are based on when these metals were first widely refined and used for tools and weapons. The process of refining a metal is electrochemical in nature and requires energy to recover the pure metal from its naturally occurring compounds (ores). **Corrosion** is also an electrochemical process in which a metal reacts with substances in the environment, returning the metal to an ore-like state. Because we live in an oxidizing (oxygen) environment, oxidation (corrosion) of some metals occur spontaneously. In fact, we need to produce metals such as iron continually to replace the metals lost to corrosion. Preventing corrosion and dealing with the effects of corrosion are major economic and technological problems for our society (**Figure 13**).

As a metal is oxidized, metal atoms lose electrons to form positive ions. A redox table of relative strengths of oxidizing and reducing agents provides the evidence that metals vary greatly in their ability to be oxidized. Some metals, such as gold and silver, are "noble" because they are relatively weak reducing agents. On the other hand, Group 1 and 2 metals are very strong reducing agents and are, therefore, easily oxidized. In general, any metal appearing below the various oxygen half-reactions in a redox table will be oxidized in our environment. Iron (including steel) and aluminium are such metals, and are extensively used as structural materials. Why is the corrosion or rusting of iron such a major problem, but the corrosion of aluminium, which is a much stronger reducing agent, is not? The answer lies primarily in the nature of the oxide that forms on the surface of the metal. A freshly cleaned surface of aluminium rapidly oxidizes in air to form aluminium oxide.

$$4\,Al(s) + 3\,O_2(g) \rightarrow 2\,Al_2O_3(s)$$

The aluminium oxide adheres tightly to the surface of the metal. This prevents further corrosion by effectively sealing any exposed surfaces.

Unfortunately, the iron compounds that form on the surface of exposed iron do not adhere very well. They flake off, exposing new iron to be corroded. In addition, the corrosion of iron is a complex process that is significantly affected by the presence of substances other than oxygen.

Rusting of Iron

Studies of the corrosion of iron have shown that the presence of both oxygen and water is required and the iron is converted into iron hydroxides and oxides. The first step of the mechanism is thought to be the oxidation of iron at a wet exposed surface (**Figure 14**).

$$Fe(s) \rightarrow Fe^{2+}(aq) + 2\,e^-$$

Iron(II) ions diffuse through the water on the iron surface while the electrons easily travel through the iron metal, which is an electrical conductor. The electrons are picked up by oxygen molecules dissolved in water on the surface at a point away from the original oxidation site (**Figure 14**).

$$\tfrac{1}{2}O_2(g) + H_2O(l) + 2\,e^- \rightarrow 2\,OH^-(aq)$$

The combination of iron(II) ions and hydroxide ions forms a low-solubility precipitate of iron(II) hydroxide, which is further oxidized by oxygen and water to form iron(III) hydroxide, a yellow-brown solid. The familiar red-brown rust is formed by the dehydration of iron(III) hydroxide to form a mixture of iron(III) hydroxide and hydrated iron(III) oxide. The amount of the hydroxide and the oxide varies, so rust is referred to as a hydrated oxide of indeterminate formula, $Fe_2O_3\cdot x\,H_2O(s)$.

Figure 13
Large ships have steel hulls. The rusting of steel involves the oxidation of iron in the steel and is a constant headache for shipping companies.

DID YOU KNOW?
Rates of Corrosion
A tin can (tin on steel) will corrode completely in about 100 a; an aluminium can in about 400 a; and a glass bottle in about 100 ka. (a, or annum, is the SI unit for year.)

DID YOU KNOW?
Hydrated Oxide
Iron(III) hydroxide can be converted to iron(III) oxide trihydrate as shown below:

$$2\,Fe(OH)_3(s) \rightarrow Fe_2O_3\cdot 3\,H_2O(s)$$

It is difficult to determine how much of the iron(III) exists in rust as the hydroxide or hydrated oxide.

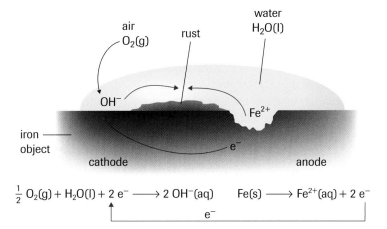

Figure 14
The corrosion of iron is a small electrochemical cell with iron oxidation at one location (the anode) and oxygen reduction at another location (the cathode).

This simplified mechanism for the rusting of iron can be used to explain why certain conditions promote rusting. If the iron is kept in a dry environment (low humidity) or if air has been removed from the water, little or no corrosion occurs (**Figure 15**). Eliminating either water or the oxygen in the water makes the reduction of aqueous oxygen impossible. Iron cannot be oxidized unless a suitable oxidizing agent is present. If oxidizing agents other than oxygen are present, such as certain metal ions, nonmetals, or hydrogen ions, the iron can still be corroded through spontaneous redox reactions. This helps to explain the corrosion of iron in acidic environments; for example, why acid rain corrodes iron faster than natural rain does.

In general, electrolytes accelerate rusting. Ships rust more rapidly in seawater than in fresh water and cars rust more rapidly in places where salt is used on roads. Chloride ions from salt are known to inhibit the adherence of protective oxide coatings on many metals, thus exposing more metal to be corroded. Electrolytes like sodium chloride conduct electricity and improve charge transfer, accelerating the rusting process. Plumbers know that you cannot use steel (iron) straps or nails to hold copper pipes in place because corrosion of the iron will be accelerated. Any moisture that is present sets up an electric cell similar in principle to Volta's original discovery of electricity from dissimilar metals (Section 14.1). As the cell operates, the iron corrodes to form rust. In general, *the rusting of iron requires the presence of oxygen and water and is accelerated by the presence of acidic solutions, electrolytes, mechanical stresses, and contact with less active metals.*

Figure 15
Rusting of exposed iron is almost negligible when the relative humidity is less than 50%. This iron pillar in Delhi, India, has existed for about 1500 years because of the very dry and unpolluted environment.

▶ mini Investigation — Home Corrosion Experiment

Soft drinks are acidic and contain electrolytes. Would different types of soft drink corrode iron at different rates?

Materials: soft drinks (cola, lemon-lime flavoured soda), 2 identical steel nails, 2 plastic glasses

(a) Predict which drink will cause the nails to corrode faster.
- Test your prediction. Place a clean steel nail in a plastic cup filled with the cola (**Figure 16**). Put the other nail in the other cup filled with the other soft drink.
- After at least one day, examine the nails for evidence of any changes.

(b) Record and explain your results.
- Dispose of all materials as directed by your teacher.

Figure 16
What happens to a steel nail in clear soda?

CAREER CONNECTION

Materials Engineering Technologist
How does materials engineering relate to corrosion? Research the entrance requirements, job prospects, and typical salaries for technologists in this field.

+ EXTENSION

Corrosion Prevention
You might have seen shiny, reddish copper roofs turn green-grey, after they have been installed for a while. This is a natural patina, that actually helps prevent further corrosion. Scientists are working on developing an artificial patina, that protects the metal much more quickly. Listen to this audio clip to find out more about this process.

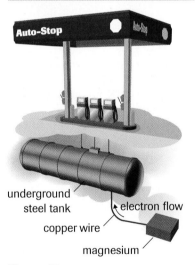

Figure 17
Corrosion of iron involves the oxidation of iron at the anode of a cell. If the iron is attached directly or connected electrically to a metal that is more easily oxidized (a sacrificial anode), then a spontaneous cell develops in which iron is the cathode. The electrolyte of the cell is the moisture in the ground.

Corrosion Prevention

Corrosion is such an important problem in our society that many technologies have been developed, and continue to be developed, to minimize this problem. There are also many career opportunities as engineers or technologists. Methods used for preventing or minimizing the corrosion of iron can be divided into two categories: barrier methods that employ *protective coatings*, and the method of *cathodic protection*. In some critical situations, such as a large fuel tank, both methods may be used.

Protective Coatings

Paint and other similar coatings are a simple method of corrosion prevention. This method works well as long as the surface is completely covered and the coating remains intact. Unfortunately, a scratch or chip in the surface can easily expose a small surface of iron and corrosion begins.

Both tin and zinc are used as metallic coatings. Tin, as in the familiar tin can, adheres well to the iron and provides a strong, shiny coating. The outer surface of the tin coating has a thin, strongly adhering layer of tin oxide that protects the tin as long as the food stored in the can is not too acidic. If a crack or break occurs in the tin layer, moisture can collect in the crack and an electric cell with tin and iron electrodes is established. Since iron is more easily oxidized than tin, iron becomes the anode in this cell. The electrons released by the oxidation of iron flow to the tin and corrosion is accelerated. Evidence of this is the typical iron rust on tin cans that have been crushed and left outside.

A spontaneous electric cell also arises when a zinc coating on an iron object is broken. However, in this case, the zinc is more easily oxidized than the iron. The zinc is preferentially oxidized, preventing corrosion of the iron. Zinc plating (galvanizing) of steel or iron provides double protection: a protective layer and preferential corrosion of the zinc.

Cathodic Protection

According to the redox theory of a cell, oxidation is the loss of electrons and occurs at the anode of a cell. Therefore, an effective method of preventing corrosion of iron is **cathodic protection** in which the iron is forced to become the cathode by supplying the iron with electrons, using either an impressed current or a sacrificial anode.

For a battery or DC generator connected in a circuit, electrons flow out of the negative terminal and into the positive terminal. If the negative terminal is connected to the iron object and the positive terminal is connected to an inert carbon electrode, an electric current is forced to flow to the iron through an electrolyte, such as ground water, from the carbon electrode. The iron is forced to become the cathode and is prevented from corroding. An *impressed current* is an electric current forced to flow toward an iron object by an external potential difference. This method of corrosion prevention requires a constant electric power supply (typically 8 mV) and is used as cathodic protection for pipelines and culverts.

A simpler method of cathodic protection is the use of a sacrificial anode. A *sacrificial anode* is a metal more easily oxidized than iron and connected to the iron object to be protected. The practice of zinc plating (galvanizing) iron objects is a common example of this method. Sacrificial zinc anodes are also connected to the exposed underwater metal surfaces of ships and boats to prevent the corrosion of the iron in the steel. Blocks of magnesium can also be used as sacrificial anodes (**Figure 17**). In all cases, the more active metal (appearing below iron in a half-reaction table) is slowly consumed or sacrificed at the anode, forcing the iron object to be the cathode of the cell.

Practice

17. What are the minimum requirements for the corrosion of iron?
18. List some factors that accelerate or promote the corrosion of iron.
19. Write the balanced net ionic equation for the corrosion of iron to iron(II) ions in the presence of oxygen and water.
20. Although the corrosion of iron is a serious problem, other metals are also corroded in air or other environments. For each of the following situations, use your knowledge of writing and balancing redox equations to write and label the half-reaction and net ionic equations:
    (a) Zinc is an active metal that oxidizes when exposed to air and water.
    (b) A lead pipe corrodes if it is used to transport acidic solutions that also contain dissolved oxygen.
    (c) In dry air, minute quantities of hydrogen sulfide gas can slowly react with silver objects to produce hydrogen gas and silver sulfide, recognized by the dark tarnish on the surface of the silver.
21. You may have noticed that when a car body rusts, the rust appears around the break or chip in the paint, but the damage may extend under the painted surface for some distance.
    (a) What is the evidence for damage extending well beyond the break in the paint?
    (b) Suggest an explanation why the damage may extend far from the break in the paint.
22. Would a basic solution prevent or slow down the corrosion of iron? Provide your reasoning.
23. Why is a zinc coating on iron better than a tin coating?
24. What are the two methods of cathodic protection and how are they similar?

 **WEB Activity**

Case Study—Galvanizing Steel

A galvanized (zinc-coated) chain-link fence can last a long time without deteriorating (**Figure 18**). In this computer simulation, you will learn about the steps involved in the galvanizing process. After you observe the computer simulations, describe the method and purpose of each of the cleaning and galvanizing steps in the production of galvanized steel. Evaluate the corrosion resistance of a galvanized pipe compared with a painted pipe.

www.science.nelson.com

Figure 18
The metal in this fence is galvanized steel: steel coated with zinc. As the zinc surface layer corrodes, it forms a protective coating that prevents the steel from corroding.

Section 14.2 Questions

1. In the context of a voltaic cell, write a definition of each of the following terms: half-cell, cathode, anode, cation, anion, and inert electrode.
2. State the function of a porous boundary in a voltaic cell and describe two common examples.
3. Distinguish between the external circuit and the internal circuit of a cell.
4. Define a standard cell.
5. (a) For a given cell, how is the cell potential predicted?
   (b) What are the restrictions on this prediction?
6. How does the cell potential indicate spontaneity of the reaction?
7. Why are the reactions in voltaic cells always spontaneous? What does this imply about the cell potential?
8. Define the hydrogen reference cell, including contents and conditions. Include the half-cell notation.
9. Why is a reference half-cell necessary?
10. (a) What is the cell potential of a standard cobalt–zinc cell?
    (b) What is the theoretical interpretation of this cell potential?
    (c) List some factors that may account for the differences you would see between the experimental and predicted values of the cell potential for the cobalt–zinc cell.

11. For each of the following cells
    - identify the strongest oxidizing and reducing agents
    - write chemical equations to represent the cathode, anode, and overall (net) cell reactions (include the half-cell and cell potentials)
    - draw a diagram of each cell, labelling the electrodes, polarity (signs) of electrodes, electrolytes, direction of electron flow, and direction of ion movement
    (a) Zn(s) | Zn^{2+}(aq) || Cu^{2+}(aq) | Cu(s)
    (b) Sn(s) | Sn^{2+}(aq) || Cr$_2$O$_7^{2-}$(aq), H$^+$(aq) | C(s)

12. You can determine a possible identity of an unknown half-cell from the cell potential involving a known half-cell. Use the following evidence and the redox table in Appendix I to determine the reduction potential and possible identity of the unknown X^{2+}(aq) | X(s).

    2 Ag$^+$(aq) + X(s) → 2 Ag(s) + X^{2+}(aq) $E°_{cell}$ = +1.08 V

13. An important goal of technology is to provide solutions to practical problems in society. Illustrate this goal by identifying a common problem and the technological solution.

Extension

14. A zinc wire is connected to and buried with a pipeline when it is built (**Figure 19**).

Figure 19
When this pipeline was being constructed, a zinc wire was attached to and buried with the pipe.

(a) Why is this done? Include a brief description of the principles involved.
(b) Is this the only type of corrosion protection used with major pipelines?
(c) Discuss the environmental and safety issues associated with protecting and also not protecting pipelines.

15. Complete the Prediction for the following investigation. Include your reasoning.

    Problem
    What is the total electric potential difference of two cells connected in series?

    Design
    Copper–silver and copper–zinc standard cells are connected as shown in **Figure 20**. The total electric potential difference of the two cells is measured with a voltmeter connected to the silver and zinc electrodes.

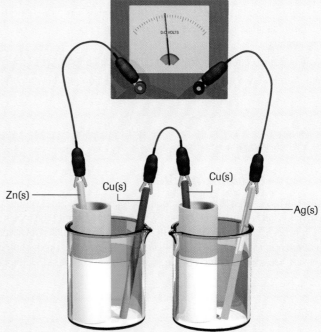

Zn(s) | Zn^{2+}(aq) || Cu^{2+}(aq) | Cu(s) — Cu(s) | Cu^{2+}(aq) || Ag$^+$(aq) | Ag(s)

Figure 20
Two standard cells in series

Electrolytic Cells 14.3

Electric cells and batteries used by consumers contain reactants chosen to react spontaneously to convert their chemical energy into electrical energy. These cells or batteries can be used to power a portable music player, start a car, or plate silver metal on jewellery.

A scientific research cell, or voltaic cell, produces electricity spontaneously because each half-cell contains both oxidized and reduced entities. For example,

$$Zn(s) \mid Zn^{2+}(aq) \parallel Pb^{2+}(aq) \mid Pb(s)$$

The cell potential, $E°_{cell}$, is always greater than zero. In a redox table, the strongest oxidizing agent present in the cell will always be above the strongest reducing agent present.

If a cell does not contain all oxidized and reduced species shown in the half-reaction equation, it is possible that the reactants (electrodes and electrolyte) present will not react spontaneously. For example, if lead electrodes are placed in a solution of zinc sulfate and the electrodes are connected with a wire, there is no evidence of any reaction.

$$Pb(s) \mid ZnSO_4(aq) \mid Pb(s)$$

The strongest oxidizing agent present in this cell is $Zn^{2+}(aq)$ and the strongest reducing agent present is Pb(s). A quick check in the redox table shows that the oxidizing agent, $Zn^{2+}(aq)$, is well below the reducing agent, Pb(s), and the $E°_{cell}$ is negative (**Figure 1**).

$Pb^{2+}(aq)$	+	$2e^-$	⇌	Pb(s)
$Ni^{2+}(aq)$	+	$2e^-$	⇌	Ni(s)
$Fe^{2+}(aq)$	+	$2e^-$	⇌	Fe(s)
$Zn^{2+}(aq)$	+	$2e^-$	⇌	Zn(s)

We can calculate the $E°_{cell}$ for the only reaction that could occur.

$$Zn^{2+}(aq) + 2e^- \rightarrow Zn(s) \qquad E°_r = -0.76 \text{ V}$$
$$Pb(s) \rightarrow Pb^{2+}(aq) + 2e^- \qquad E°_r = -0.13 \text{ V}$$
$$\overline{Zn^{2+}(aq) + Pb(s) \rightarrow Zn(s) + Pb^{2+}(aq)}$$

$$E°_{cell} = E°_{r\,cathode} - E°_{r\,anode}$$
$$= -0.76 \text{ V} - (-0.13 \text{ V})$$
$$= -0.63 \text{ V}$$

Since the $E°_{cell}$ for the reaction is negative, we conclude that the lead will not be oxidized spontaneously in the zinc sulfate solution. Note that the reverse reaction would be spontaneous, but could not occur because neither $Pb^{2+}(aq)$ nor Zn(s) is present initially.

Strictly speaking, the zinc sulfate cell is not a standard cell even if the concentration of the zinc sulfate were 1.0 mol/L because the cell does not contain all entities listed in the half-reaction equations. Therefore, the cell potential that is calculated is not accurate, but will be close enough for our purposes. This cell would not produce electricity because the reaction is nonspontaneous. Why would anyone be interested in a cell like this? Certainly not to use in a battery. However, by supplying electrical energy to a nonspontaneous cell, we can force the reaction to occur. This is especially useful for producing substances, particularly elements. For example, the zinc sulfate cell discussed above is similar to the cell used in the industrial production of zinc metal (**Figure 2**).

Learning Tip

Standard Cells
Recall that a standard cell contains all entities listed in the equation for the half-reaction. In addition, the concentration of aqueous entities is 1.0 mol/L and the conditions are SATP for all substances, including gases at 100 kPa.

Figure 1
According to the redox spontaneity rule, if the strongest oxidizing agent present is below the strongest reducing agent present, no spontaneous reaction will occur ($E°_{cell} < 0$).

Figure 2
Cominco in Trail, British Columbia, operates the world's largest zinc and lead smelter, producing almost 300 kt of zinc annually.

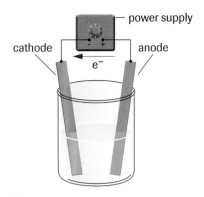

Figure 3
Electrons are pulled from the anode and pushed to the cathode by the battery or power supply.

The term *electrochemical cell* is often used in chemistry to refer to either a cell with a spontaneous reaction, such as electric and voltaic cells, or a cell with a nonspontaneous reaction, which we call an **electrolytic cell** (**Figure 3**). An electrolytic cell consists of a combination of two electrodes, an electrolyte, and an external battery or power source. It uses a process called **electrolysis**, which is the process of supplying electrical energy to force a nonspontaneous redox reaction to occur. The external power supply acts as an "electron pump;" the electric energy is used to do work on the electrons to cause an electron transfer inside the electrolytic cell. In an electrolytic cell, the chemical reaction is the reverse of that of a spontaneous cell. However, most of the scientific principles you have already studied also apply to electrolytic cells (**Table 1**).

reactants ⇌ products + electrical energy
(electric/voltaic cell → ; electrolytic cell ←)

Table 1 Comparing Electrochemical Cells: Voltaic and Electrolytic

	Voltaic cell	**Electrolytic cell**
spontaneity	spontaneous reaction	nonspontaneous reaction
standard cell potential, $E°_{cell}$	positive	negative
cathode	• strongest oxidizing agent present undergoes a *reduction* • positive electrode	• strongest oxidizing agent present undergoes a *reduction* • negative electrode
anode	• strongest reducing agent present undergoes an *oxidation* • negative electrode	• strongest reducing agent present undergoes an *oxidation* • positive electrode
direction of electron movement	anode → cathode	anode → cathode
direction of ion movement	anions → anode cations → cathode	anions → anode cations → cathode

Learning Tip

"Positive" and "negative" are labels that are used in many situations, mostly decided by general agreement (convention). For example, "positive" and "negative" can be used to label or describe attitudes, directions, axes on a graph, charges, and electrodes. By convention, a voltaic cell has a cathode labelled positive and an anode labelled negative. In an electrolytic cell, it is customary to reverse these labels. It is best to think of positive and negative for electrodes as labels, not charges.

Secondary Cells: Electric and Electrolytic

A secondary cell is a rechargeable cell such as a nickel–cadmium (Ni–Cd) cell. A secondary cell can be used to illustrate the difference between an electric cell and an electrolytic cell. As the cell discharges, electrical energy is spontaneously produced and the cell functions as an electric cell. When the cell is recharged, the electrical energy forces the products to react and re-form the original reactants. During recharging, the secondary cell is functioning as an electrolytic cell.

> ### ▶ Practice
>
> 1. Describe the type of agent reacting and the process occurring at the cathode and anode of an electrolytic cell.
> 2. Describe the differences between the cathode and anode of an electrolytic cell and a voltaic cell.
> 3. Describe the direction of movement of electrons and ions within an electrolytic cell.
> 4. In a town's water supply (**Figure 4**), water is pumped from a lake into a water tower, which supplies water to the town's houses. Describe how this is an analogy for a secondary cell.

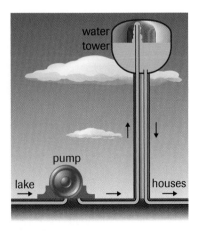

Figure 4
A town's water supply

Section 14.3

INVESTIGATION 14.4 Introduction

A Potassium Iodide Electrolytic Cell

Electrolytic cells were discovered before the science was understood. However, as with all successful technological inventions, the important criteria was that it worked, not why it worked. Eventually, chemists understood the science and were able to explain why electrolytic cells work.

In the Evaluation, suggest changes to the Design, Materials, and Procedure that would improve the Evidence.

Purpose

The purpose of this investigation is to use diagnostic tests to determine the reaction products of an electrolytic cell.

Report Checklist
- ○ Purpose
- ○ Problem
- ○ Hypothesis
- ○ Prediction
- ○ Design
- ○ Materials
- ○ Procedure
- ● Evidence
- ● Analysis
- ● Evaluation (1, 3)

Problem

What are the products of the reaction during the operation of an aqueous potassium iodide electrolytic cell?

Design

Inert electrodes are placed in a 0.50 mol/L solution of potassium iodide, and a battery or power supply provides a direct current of electricity to the cell. The litmus and halogen diagnostic tests are conducted to test the solution near each electrode before and after the reaction.

To perform this investigation, turn to page 661.

The Potassium Iodide Electrolytic Cell: A Synthesis

In the potassium iodide electrolytic cell (**Figure 5**), litmus paper does not change colour in the initial solution and turns blue only near the electrode from which gas bubbles. At the other electrode, a yellow-brown colour and a dark precipitate forms. The yellow-brown substance produces a purplish-red colour in a halogen test. This chemical evidence agrees with the interpretation supplied by the following half-reaction equations. According to the redox table of relative strengths of oxidizing and reducing agents, water is the stronger oxidizing agent present and iodide ions are the stronger reducing agents present in a potassium iodide solution.

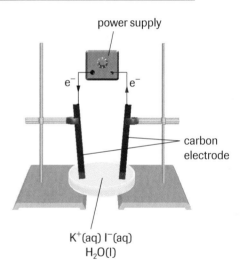

Figure 5
A power supply provides the energy for the chemical reactions at the two electrodes.

$$\underset{\text{OA}}{K^+(aq)} \quad \underset{\text{SRA}}{I^-(aq)} \quad \underset{\text{RA}}{\overset{\text{SOA}}{H_2O(l)}}$$

cathode: $2\ H_2O(l) + 2\ e^- \rightarrow \underset{\text{gas bubbles}}{H_2(g)} + \underset{\text{blue litmus}}{2\ OH^-(aq)}$

anode: $2\ I^-(aq) \rightarrow \underset{\text{purplish-red in hexane}}{I_2(s)} + 2\ e^-$

Evidence from the study of this and many other aqueous electrolytic cells suggests that the generalizations for voltaic cells also apply to electrolytic cells. From a theoretical perspective, the strongest oxidizing agent present in a particular mixture has the greatest attraction for electrons and gains electrons at the cathode. Notice that it does not matter where the electrons originate, from a power supply or directly from another electrode. The strongest reducing agent present in the mixture has the least attraction for electrons and loses electrons at the anode. In other words, the theoretical definitions of cathode and anode are the same for both voltaic and electrolytic cells (**Table 1**).

Observation of a potassium iodide cell indicates that the transfer of electrons is not spontaneous. When a voltage is supplied to the cell, electrons flow from the negative terminal of the battery toward the cathode of the electrolytic cell and are consumed (gained) by water molecules, which have the more positive reduction potential. Simultaneously, electrons flow from iodide ions on the surface of the anode to the positive terminal of the battery. This explanation agrees with previous redox concepts and with the observations, so we can accept the explanation. Predictions of cathode, anode, and overall cell reactions for electrolytic cells follow the same steps outlined for voltaic cells in Section 14.2.

Learning Tip

At first glance, the cell notation, C(s) | KI(aq) | C(s), does not show that water is involved in the reaction. Of course, the subscript (aq) means "dissolved in water." Don't forget to consider the presence of water because it is often a reactant in aqueous electrolytic cells.

▶ SAMPLE problem 14.1

What are the cell reactions and the cell potential of the aqueous potassium iodide electrolytic cell?

First, we identify the major entities in the solution and use the redox table in Appendix I to identify the strongest oxidizing and reducing agents, as we did in Chapter 13.

```
    OA              SOA
K⁺(aq),  I⁻(aq),  H₂O(l)
         SRA       RA
```

Now we can write the half-reaction equations and calculate the cell potential. The potassium iodide cell is not a standard cell because the products of the reactions are not present initially. Therefore, the reduction potentials given in the table of half-reactions are not strictly applicable, but we will use them to approximate the cell potential.

cathode: $2 H_2O(l) + 2 e^- \rightarrow H_2(g) + 2 OH^-(aq)$ $E°_r = -0.83 \text{ V}$

anode: $\underline{2 I^-(aq) \rightarrow I_2(s) + 2 e^-}$ $E°_r = +0.54 \text{ V}$

net: $2 H_2O(l) + 2 I^-(aq) \rightarrow H_2(g) + 2 OH^-(aq) + I_2(s)$

$$E°_{cell} = E°_{r\ cathode} - E°_{r\ anode}$$
$$= -0.83 \text{ V} - (+0.54 \text{ V})$$
$$= -1.37 \text{ V}$$

A negative sign for a cell potential indicates that the chemical process is nonspontaneous. The more negative the cell potential, the more energy is required. In Sample Problem 14.1, electrons must be supplied with a minimum of +1.37 V from an external battery or other power supply to force the cell reactions. In practice, however, a greater voltage is required, for example, to make the reaction occur at a reasonable rate.

SUMMARY: Procedure for Analyzing Electrolytic Cells

The procedure for analyzing electrolytic cells is essentially the same as for voltaic cells.

- Use the redox table (Appendix I) to identify the strongest oxidizing and reducing agents present. (Do not forget to consider water for aqueous electrolytes.)
- Write equations for the reduction (cathode) and oxidation (anode) half-reactions. Include the reduction potentials if required.
- Balance the electrons and write the net cell reaction including the cell potential.

$$E°_{cell} = E°_{r\ cathode} - E°_{r\ anode}$$

- If required, state the minimum electric potential (voltage) to force the reaction to occur. (The minimum voltage is the absolute value of $E°_{cell}$.)
- If a diagram is requested, use the general outline shown in **Figure 6** and add specific labels for chemical entities.

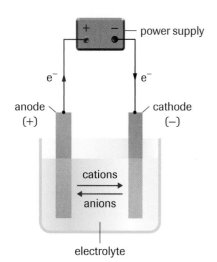

Figure 6
A generic electrolytic cell

COMMUNICATION example 1

An electrolytic cell containing cobalt(II) chloride solution and lead electrodes is assembled. The notation for the cell is as follows:

Pb(s) | Co^{2+}(aq), Cl^-(aq) | Pb(s)

(a) Predict the reactions at the cathode and anode, and in the overall cell.
(b) Draw and label a cell diagram for this electrolytic cell, including the power supply.
(c) What minimum voltage must be applied to make this cell work?

Solution

(a)   SRA   SOA
    Pb(s) | Co^{2+}(aq), Cl^-(aq) | Pb(s)

cathode: $Co^{2+}(aq) + 2\,e^- \rightarrow Co(s)$
anode: $Pb(s) \rightarrow Pb^{2+}(aq) + 2\,e^-$
net: $Co^{2+}(aq) + Pb(s) \rightarrow Co(s) + Pb^{2+}(aq)$

(b)

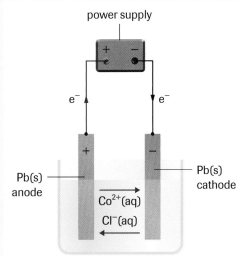

(c) $E°_{cell} = E°_{r\,cathode} - E°_{r\,anode}$

$= -0.28\text{ V} - (-0.13\text{ V})$

$= -0.15\text{ V}$

According to the redox table, a minimum voltage of 0.15 V is required.

DID YOU KNOW ?

Hydrometallurgy

Hydrometallurgy is the process of extracting metals or their compounds from ores using aqueous solutions. Sherritt Metals of Fort Saskatchewan, Alberta is a world leader in high pressure hydrometallurgy of nickel and cobalt ores. Sherritt uses an ammonia leaching of nickel and cobalt sulfides followed by a chemical reduction using hydrogen as a reducing agent. Other hydrometallurgical operations use electrolysis instead of a chemical reduction to obtain or purify metals.

COMMUNICATION example 2

An electrolytic cell is set up with a power supply connected to two nickel electrodes immersed in an aqueous solution containing cadmium nitrate and zinc nitrate. Predict the equations for the initial reaction at each electrode and the net cell reaction. Calculate the minimum voltage that must be applied to make the reaction occur.

Solution

SRA — Ni(s), H_2O(l), Cd^{2+}(aq), NO_3^-(aq), Zn^{2+}(aq) — SOA

cathode: $Cd^{2+}(aq) + 2\,e^- \rightarrow Cd(s)$ $E°_r = -0.40$ V

anode: $Ni(s) \rightarrow Ni^{2+}(aq) + 2\,e^-$ $E°_r = -0.26$ V

net: $Cd^{2+}(aq) + Ni(s) \rightarrow Cd(s) + Ni^{2+}(aq)$

$$E°_{cell} = E°_{r\,cathode} - E°_{r\,anode}$$
$$= -0.40\text{ V} - (-0.26\text{ V})$$
$$= -0.14\text{ V}$$

According to the redox table, a minimum voltage of 0.14 V is required.

Practice

5. Predict the cathode, anode, and net cell reactions for each of the following electrolytic cells. Calculate the minimum potential difference that must be applied to force the cell reaction to occur.
   (a) C(s) | Ni^{2+}(aq), I^-(aq) | C(s)   (b) Pt(s) | Na^+(aq), OH^-(aq) | Pt(s)

6. What is the minimum electric potential difference of an external power supply that produces chemical changes in the following electrolytic cells?
   (a) C(s) | Cr^{3+}(aq), Br^-(aq) | C(s)   (b) Cu(s) | Cu^{2+}(aq), SO_4^{2-}(aq) | Cu(s)

INVESTIGATION 14.5 Introduction

Electrolysis (Demonstration)

Scientific knowledge progresses by the experimental testing of ideas. The more rigorous the test, the more certain the knowledge or the better the chance of making new discoveries.

Report Checklist

○ Purpose   ○ Design   ● Analysis
○ Problem   ○ Materials   ● Evaluation (1, 2, 3)
○ Hypothesis   ○ Procedure
● Prediction   ● Evidence

Purpose

The purpose of this demonstration is to test the method of predicting the products of electrolytic cells.

Problem

What are the products of electrolytic cells containing one of the following aqueous solutions: copper(II) sulfate, sodium sulfate, and sodium chloride?

Design

The electrolysis of the aqueous copper(II) sulfate is carried out in a U-tube, and the electrolysis of aqueous sodium sulfate and sodium chloride is carried out in a Hoffman apparatus so that any gases produced can be collected. Diagnostic tests with necessary control tests (before electrolysis) are conducted to determine the presence of the predicted products.

To perform this investigation, turn to page 662.

Evaluation of Predicted Reactions—The Chloride Anomaly

As you know from Investigation 14.5, some redox reactions predicted using the strongest oxidizing and reducing agents from a redox table do not always occur in an electrolytic cell. Like other chemical processes, a half-cell reaction at an electrode has an activation energy (Unit 6) that varies for different half-reactions and conditions. Therefore, the *actual* reduction potential required for a particular half-reaction and the *reported* half-reaction reduction potential may be quite different. This difference is known as the half-cell overvoltage. It is generally much greater for the production of oxygen than for the production of chlorine, for example, in the electrolysis of aqueous chloride compounds. The explanation for overvoltage is well beyond the level of this book. *As an empirical rule, you should recognize that chlorine gas is produced instead of oxygen gas in situations where chloride and water are the only reducing agents present.*

Learning Tip

There are exceptions to all rules and generalizations. You only need to remember the chloride anomaly. This occurs during the electrolysis of solutions containing the chloride ion. Since water is the strongest reducing agent present, water should react at the anode. However, the chloride ions react preferentially to water molecules.

SUMMARY Electrolytic Cells

- An electrolytic cell is based upon a reaction that is nonspontaneous; the $E°_{cell}$ for the reaction is negative. An applied voltage of at least the absolute value of $E°_{cell}$ is required to force the reactions to occur.
- The strongest oxidizing agent undergoes reduction at the cathode (negative electrode).
- The strongest reducing agent undergoes oxidation at the anode (positive electrode).
- Electrons are forced by a power supply to travel from the anode to the cathode through the external circuit.
- Internally, anions move toward the anode and cations move toward the cathode.

Practice

7. List the main similarities between a voltaic cell and an electrolytic cell.
8. What is the key difference between voltaic and electrolytic cells?
9. Why is the procedure for analyzing voltaic and electrolytic cells so similar?
10. Explain why a power supply is necessary for an electrolytic cell.
11. Which of the following cells would produce a spontaneous reaction? Justify each answer, using the cell potential.
    (a) C(s) | Cr(NO$_3$)$_2$(aq) | C(s)   (b) Cu(s) | FeCl$_3$(aq) | Cu(s)
12. For each of the following electrolytic cells, write equations for the cathode and anode half-reactions and the net reaction. Determine the minimum potential difference that must be applied to make the cell operate.
    (a) C(s) | K$_2$SO$_4$(aq) | Cd(s)   (b) Pt(s) | SnBr$_2$(aq) | Pt(s)
13. Draw a diagram of an electrolytic cell containing a zinc iodide solution and inert carbon electrodes.
    - Label the power supply and electrodes, including signs, the electrolyte, and the directions of electron and ion movements.
    - Write half-reaction and net equations.
    - Calculate the cell potential, using standard values.
14. State the chloride anomaly, and include how it is recognized.
15. Describe a specific consumer product that you use sometimes as an electric cell and sometimes as an electrolytic cell. What practical problem does this technology solve?

Science and Technology of Electrolysis

Volta's invention of the electric cell in 1800 immediately resulted in many discoveries in chemistry. One discovery was that electric cells could be used as an electric power source for electrolytic cells. Many natural substances, such as soda (sodium carbonate) and potash (potassium carbonate) that were thought to be elements, were shown to be composed of the previously unknown elements sodium and potassium. Industrial applications of electrolytic cells include the production of elements, the refining of metals, and the plating of metals onto the surface of an object. The study of electrolysis in industry reveals the strong relationship between science and technology.

Production of Elements

Most elements occur naturally combined with other elements in compounds. For example, ionic compounds of sodium, potassium, lithium, magnesium, calcium, and aluminium are abundant, but these reactive metals are not found uncombined in nature. The explanation is that the reduction potentials for these metals are very negative. Consequently, the metals are easily oxidized by practically all other substances. Even water has a more positive reduction potential than any of these metal ions. If the metals did exist naturally, a spontaneous reaction would convert them into their ions.

SOA
$$Zn^{2+}(aq) + 2\,e^- \rightleftarrows Zn(s)$$
$$2\,H_2O(l) + 2\,e^- \rightleftarrows H_2(g) + 2\,OH^-(aq)$$
$$Mg^{2+}(aq) + 2\,e^- \rightleftarrows Mg(s)$$
$$Na^+(aq) + e^- \rightleftarrows Na(s)$$

Figure 7
In his youth, Sir Humphry Davy (1778–1829) worked as an assistant to a physician who was interested in the therapeutic properties of gases. Davy studied nitrous oxide (laughing gas) by conducting experiments on himself. He was eventually fired from his job, supposedly because of his liking for explosive chemical reactions. Davy's main fame came from his experiments with electricity. He constructed a voltaic pile with over 250 metal plates. He used this powerful cell to decompose stable compounds and discovered the elements sodium, potassium, barium, strontium, calcium, and magnesium. Given his habit of tasting, inhaling, and exploding new chemicals, it is not surprising that he was an invalid in his early thirties and died in middle age, probably of chemical poisoning.

Many metals can be produced by electrolysis of solutions of their ionic compounds, but two difficulties arise. First, many naturally occurring ionic compounds have a low solubility in water and second, water is a stronger oxidizing agent than many active metal cations. To overcome these difficulties, we can use a technological design in which water is not present. Fortunately, ionic compounds can be melted. These molten ionic compounds are good electrical conductors and can function as the electrolyte in a cell.

The production of active metals (strong reducing agents) from their minerals typically involves the electrolysis of molten compounds of the metal, a technology first used in the scientific work of Humphry Davy (**Figure 7**). Strontium metal was one of many active metals discovered by Davy using the electrolysis of molten salts. Strontium chloride was first melted in an electrolytic cell with inert electrodes. In this cell, there are only two kinds of ions present, $Sr^{2+}(l)$ and $Cl^-(l)$. You may recall from the previous chapter that metal cations generally tend to undergo a reduction and nonmetal anions tend to undergo an oxidation. In this cell, there are no other competing substances. Therefore, the strontium ions will consume (gain) electrons at the cathode to form strontium metal:

$$Sr^{2+}(l) + 2\,e^- \rightarrow Sr(s) \qquad \text{(reduction at the cathode)}$$

At the anode, chloride ions will give up (lose) electrons to form chlorine gas:

$$2\,Cl^-(l) \rightarrow Cl_2(g) + 2\,e^- \qquad \text{(oxidation at the anode)}$$

The electrons are balanced. Adding the two equations gives the overall reaction in the cell.

$$Sr^{2+}(l) + 2\,Cl^-(l) \rightarrow Sr(s) + Cl_2(g)$$

This reaction would not be possible in an aqueous solution because water is a stronger oxidizing agent (has a more positive reduction potential) than aqueous strontium ions.

In molten-salt electrolysis, metal cations move to the cathode and are reduced to metals, and nonmetal anions move to the anode and are oxidized to nonmetals.

> ▶ **COMMUNICATION** *example 3*
>
> Lithium is the least dense of all metals and is a very strong reducing agent; both qualities make it an excellent anode for batteries. Lithium can be produced by electrolysis of molten lithium chloride at a temperature greater than 605 °C, the melting point of lithium chloride. Write the equations for the cathode and anode half-reactions, and the net cell reaction.
>
> **Solution**
>
> cathode: $\quad 2\,[Li^+(l) + e^- \rightarrow Li(s)]$
> anode: $\quad\quad\quad 2\,Cl^-(l) \rightarrow Cl_2(g) + 2\,e^-$
> net: $\quad\quad 2\,Li^+(l) + 2\,Cl^-(l) \rightarrow 2\,Li(s) + Cl_2(g)$

> **Learning Tip**
>
> No reduction potentials can be listed for the electrolysis of a molten salt. The redox table in Appendix I lists only electric potentials for half-reactions in 1.0 mol/L aqueous solutions at SATP.

Production of Aluminium

Aluminium is the third most abundant element on Earth. It was discovered in France in the early 1800s. At the time, aluminium was more expensive than gold. The wonderful properties of aluminium—shiny, light, strong, and corrosion resistant—made it ideal for jewellery and cutlery, so there was a high demand for the metal, especially among the aristocracy. However, the supply of aluminium was limited because the technology for producing aluminium was not yet practical or economically viable for mass production.

Initial efforts to produce aluminium by electrolysis were unproductive because its common ore, $Al_2O_3(s)$, has a high melting point, 2072 °C. No material could be found to hold the molten compound. In 1886, two scientists working independently and knowing nothing of each other's work made the same discovery. Charles Hall in the United States and Paul Héroult in France discovered that $Al_2O_3(s)$ dissolves in a molten mineral called cryolite, Na_3AlF_6. In this design, the cryolite acts as an inert solvent for the electrolysis of aluminium oxide and forms a molten conducting mixture with a melting point around 1000 °C. Aluminium (m.p. 660 °C) can be produced electrolytically from this molten mixture (**Figure 8**). This discovery had an immediate effect on the supply and cost of aluminium. Around 1855, aluminium was sold for $45 000 per kilogram; a few years after the Hall–Héroult invention, the cost was about 90 cents per kilogram.

> **DID YOU KNOW ?**
>
> **Aluminium Production in Canada**
>
> The production of aluminium is important to Canada's economy, although Canada does not have large deposits of aluminium ore. Hydroelectric power is used to produce aluminium metal from concentrated, imported bauxite in an electrolytic cell. Recycling aluminium from soft drink and beer cans requires only 5% of the energy required to produce aluminium by electrolysis.

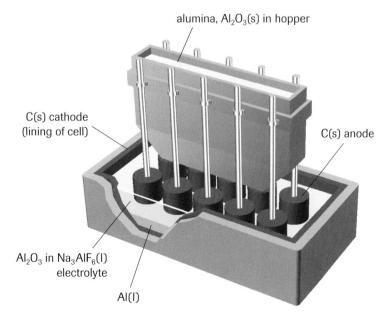

Figure 8
The Hall–Héroult cell for the production of aluminium. The cathode is the carbon lining of the steel cell. At the cathode, aluminium ions are reduced to produce liquid aluminium, which collects at the bottom of the cell and is periodically drained away. At the carbon anode, oxide ions are oxidized to produce oxygen gas. The oxygen produced at the anode reacts with the carbon electrodes, producing carbon dioxide, so these electrodes must be replaced frequently.

Aluminium Energy Source

Laptops and even some cars of the future may use aluminium as an energy source. It takes a great deal of electrical energy to produce aluminium and some of this energy can be recovered. Two University of British Columbia researchers have devised a way to produce hydrogen from aluminium using only salt and water. The hydrogen that is produced can be used to power a fuel cell for portable electronics or even a car.

www.science.nelson.com

Figure 9
Large salt deposits sit hundreds of metres below Fort Saskatchewan, Alberta. This means that Fort Sakatchewan is a good location for Dow's chlor–alkali plant because the main raw material is available on site.

Aluminium oxide is obtained from bauxite, an aluminium ore. Once the ore is purified, the aluminium oxide is dissolved in molten cryolite and it dissociates into individual ions. The reactions occurring at the electrodes in a Hall–Héroult cell are summarized below:

cathode: $\quad 4[Al^{3+}(cryolite) + 3\ e^- \rightarrow Al(l)]$

anode: $\quad 3[2\ O^{2-}(cryolite) \rightarrow O_2(g) + 4\ e^-]$

$$4\ Al^{3+}(cryolite) + 6\ O^{2-}(cryolite) \rightarrow 4\ Al(l) + 3\ O_2(g)$$

The overall cell reaction is a decomposition of aluminium oxide.

$$2\ Al_2O_3(s) \rightarrow 4\ Al(s) + 3\ O_2(g)$$

The Chlor–Alkali Process

The most important nonmetal produced by electrolysis is chlorine. More than 95% of the world production of chlorine and almost 100% of the world production of sodium hydroxide is done using the chlor–alkali process, which is a reaction that you studied in Investigation 14.5. The chlor–alkali process is the electrolysis of aqueous sodium chloride (brine) to produce chlorine, hydrogen, and sodium hydroxide:

$$2\ NaCl(aq) + 2\ H_2O(l) \rightarrow \underset{\text{(anode product)}}{Cl_2(g)} + \underset{\text{(cathode products)}}{H_2(g) + 2\ NaOH(aq)}$$

You may recall that the anode reaction in this electrolysis was unexpected in Investigation 14.5 because water is a slightly stronger reducing agent than chloride ions. This is the chloride anomaly and is an important exception to the rules for predicting half-reactions.

Dow Chemical in Fort Saskatchewan, Alberta, has an ideal location for its chlor–alkali plant (**Figure 9**). The sodium chloride is extracted from large underground deposits using hot water and pumped to electrolytic cells at the surface. Dow uses diaphragm electrolytic cells, but newer membrane electrolytic cells are becoming more common in the chlor–alkali industry. These two technologies are similar, but the membrane cell has the advantage of producing sodium hydroxide with very little sodium chloride contamination. This is accomplished by using an ion-exchange membrane that allows sodium ions, but not chloride ions, to move from the anode to the cathode compartments (**Figure 10**).

Hydrogen gas is used to make ammonia, hydrogen peroxide, and margarine, and to crack petroleum. It may also be used on site as a fuel to produce electricity. Chlorine is used as a disinfectant for drinking water and to manufacture bleach (sodium hypochlorite), plastics, pesticides, and solvents. Sodium hydroxide is used on a large scale in industry to make cellophane, pulp and paper, aluminium, and detergents.

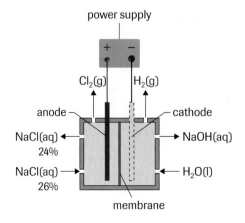

Figure 10
Membrane cells are a relatively recent technology. Most companies throughout the world are switching to these cells from the older diaphragm and mercury cells.

Section 14.3

▶ **Practice**

16. Describe two difficulties associated with the electrolysis of aqueous ionic compounds in the production of active metals. What is a technological solution that overcomes both difficulties?

17. Scandium is a metal with a low density and a melting point that is higher than that of aluminium. These properties are of interest to engineers who design space vehicles. Scandium metal is produced by electrolysis of molten scandium chloride. List all ions present in the electrolysis cell, and write the equations for the reactions that occur at the cathode and anode, and the net cell reaction.

18. Write the cathode, anode, and cell reaction equations for the chlor–alkali process.

19. The following statements summarize the steps in the chemical technology of obtaining magnesium from seawater. Write a balanced equation to represent each reaction.
    (a) Slaked lime (solid calcium hydroxide) is added to seawater (ignore all solutes except $MgCl_2(aq)$) in a double displacement reaction to precipitate magnesium hydroxide.
    (b) Hydrochloric acid is added to the magnesium hydroxide precipitate.
    (c) After the magnesium chloride product is separated and dried, it is melted in preparation for electrolysis. List all ions present in the electrolysis, and write the equations for the reactions that occur at the cathode and anode, and the net cell reaction.
    (d) An alternative process produces magnesium from dolomite, a mineral containing $CaCO_3$ and $MgCO_3$. Suggest some technological advantages and disadvantages of the dolomite process compared with the seawater process.

20. What products in your home may have originated from substances produced in the chlor–alkali process?

21. Why should we recycle metals such as aluminium? State several arguments that you might use in a debate.

DID YOU KNOW ?

The Pidgeon Process
Lloyd Pidgeon (1903–1999) was born in Markham, Ontario, studied undergraduate chemistry at the University of Manitoba, and obtained his Ph.D. from McGill University in 1929. Later, while working at the National Research Council in Ottawa, Pidgeon developed the first process for producing high-quality magnesium metal from dolomite (calcium magnesium carbonate). This led to the formation of Dominion Magnesium Ltd. using dolomite mined in the Ottawa Valley. The Pidgeon Process is still used in many countries to produce magnesium.

Refining of Metals

In the production of metals, the initial product is usually an impure metal. Impurities are often other metals that come from various compounds in the original ore. To purify or refine a metal, a variety of methods are used. However, a common method, known as *electrorefining*, uses an electrolytic cell to obtain high-grade metals at the cathode from an impure metal at the anode.

A good example is the electrorefining of copper. The presence of impurities in copper lowers its electrical conductivity, not a desirable property considering that one of the most common uses of copper is in electrical wiring. The initial smelting process produces copper that is about 99% pure, containing some silver, gold, platinum, iron, and zinc. These valuable impurities can be recovered and sold to help pay for the process. As shown in **Figure 11**, a slab of impure copper is the anode of an electrolytic cell that contains copper(II) sulfate dissolved in sulfuric acid. The cathode is a thin sheet of very pure copper. As the cell operates, copper and some of the other metals in the anode are oxidized, but only copper is reduced at the cathode. An understanding of oxidation, reduction, and reduction potentials allows precise control over what is oxidized and what is reduced; after electrorefining, the copper is about 99.98% pure. The half-reactions are:

cathode: reduction of copper $\quad Cu^{2+}(aq) + 2\,e^- \rightarrow Cu(s)$
anode: oxidation of copper $\quad\quad\quad Cu(s) \rightarrow Cu^{2+}(aq) + 2\,e^-$
oxidation of zinc $\quad\quad\quad\quad\; Zn(s) \rightarrow Zn^{2+}(aq) + 2\,e^-$
oxidation of iron $\quad\quad\quad\quad\;\; Fe(s) \rightarrow Fe^{2+}(aq) + 2\,e^-$

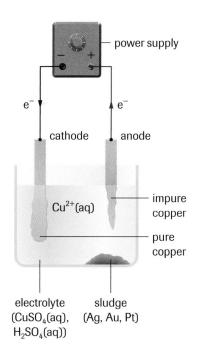

Figure 11
Only copper and metals more easily oxidized than copper, such as iron and zinc, are oxidized to ions and dissolve at the anode. Only copper is reduced at the cathode. Other impurities in the anode, such as silver, gold, and platinum, do not react; these fall to the bottom of the cell as a sludge called anode mud.

Another method of purifying metals is to reduce metal cations from a molten or aqueous electrolyte at the cathode of an electrolytic cell, much like the production of elements discussed previously. This method is known as *electrowinning*. Using a molten salt is the only way to obtain some active metals, such as those in Group 1. Many other metals, such as zinc, can be produced by electrowinning an aqueous solution. For example, Cominco's operation at Trail, British Columbia, uses the electrolysis of an acidic zinc sulfate solution with a specially treated lead anode to deposit very pure zinc metal at the cathode.

cathode: $\quad 2[Zn^{2+}(aq) + 2\,e^- \rightarrow Zn(s)]$

anode: $\quad\quad\quad\quad 2\,H_2O(l) \rightarrow O_2(g) + 4\,H^+(aq) + 4\,e^-$

net: $\quad 2\,Zn^{2+}(aq) + 2\,H_2O(l) \rightarrow 2\,Zn(s) + O_2(g) + 4\,H^+(aq)$

Electroplating

Several metals, such as silver, gold, zinc, and chromium, are valuable because of their resistance to corrosion. However, products made from these metals in their pure form are either too expensive or they lack suitable mechanical properties, such as strength and hardness. To achieve the best compromise among price, mechanical properties, appearance, and corrosion resistance, utensils and jewellery may be made of a relatively inexpensive, yet strong alloy such as steel, and then coated (plated) with another metal or alloy to improve appearance or corrosion resistance. Plating of a metal at the cathode of an electrolytic cell is called *electroplating* and is a common technology that is used to cover the surface of an object with a thin layer of metal.

The process for plating metals is obtained by systematic trial and error, involving the careful manipulation of one possible variable at a time. In this situation, a scientific perspective helps identify variables, but cannot usually provide successful predictions.

The development and use of electric cells preceded scientific understanding of the processes involved. Today, we still have examples of technological processes that are not fully understood, such as chromium plating (**Figure 12**) and silver plating. For example, there is no satisfactory explanation for why silver deposited during electrolysis of a silver nitrate solution does not adhere well to any surface, whereas silver plated from silver cyanide solution does.

Electroplating is only one method used to cover the surface of an object with a metal. One other method is vapour deposition, in which metal vapour is condensed on the surface. Another method is dipping, in which an object is dipped into a molten metal that solidifies on the surface. Zinc-plated nails for exterior use are made by dipping.

CAREER CONNECTION

Chemical Technologist
Chemical technologists may work closely with scientists and engineers studying electroplating processes. What education and training is required for this job? Outline some current job opportunities in this area, including typical salaries.

www.science.nelson.com

Figure 12
Chromium is best plated from a solution of chromic acid. A thin layer of chromium metal is very shiny and, like aluminium, protects itself from corrosion by forming a tough oxide layer.

SUMMARY Applications of Electrolytic Cells

- In molten-salt electrolysis, metal cations are reduced to metal atoms at the cathode and nonmetal anions are oxidized at the anode.
- Electrorefining is a process used to obtain high-grade metals at the cathode from an impure metal at the anode.
- Electroplating is a process in which a metal is deposited on the surface of an object placed at the cathode of an electrolytic cell.

Section 14.3 Questions

1. Define an electrolytic cell.
2. List the key similarities and differences in the typical laboratory construction of a voltaic cell and an electrolytic cell.
3. The rules for predicting the chemical reactions, and ion and electron movements are essentially the same for both voltaic and electrolytic cells. What is different about the cell reaction equations and characteristics of each type of cell?
4. Two tin electrodes are placed in an aqueous solution containing potassium nitrate and magnesium iodide.
   (a) If a power supply is connected to force any reactions to occur, what would be the reactions at the cathode, anode, and in the overall cell?
   (b) Draw and label a cell diagram, including electrodes, electrolyte, power supply, and the direction of movement of electrons and ions.
   (c) Would a 1.5 V cell be suitable as a power supply? Justify your answer.
5. For the electrolysis of aqueous solutions, describe the common exception to the rules for predicting products of an electrolytic cell. Clearly identify the circumstances when this exception is used.
6. List three uses of electrolytic cells in industry.
7. Why were many metals discovered only after the invention of the electric cell?
8. How does the occurrence of metals in nature relate to the redox table?
9. Draw and label a simple cell for the electrolysis of molten potassium iodide (m.p. 682 °C). Label electrodes and power supply, directions of electron and ion flow, and write half-reaction and net equations.
10. When refining metals in an electrolytic cell, why must the metal product form at the cathode?
11. High-purity copper metal is produced using electrorefining.
    (a) At which electrode is the impure copper placed? Why?
    (b) What is the minimum electric potential difference required for this cell?
    (c) Why is it unlikely that your answer to (b) would be used? Discuss briefly.
12. How can you predict which metals might be refined from an aqueous solution?
13. Describe the relationship of science and technology in the area of electrolysis. Include several examples in your description.
14. "German silver" is an alloy containing copper, zinc, and nickel. A piece of German silver is used as the anode in an electrolytic cell containing aqueous sodium sulfate. The other electrode is platinum metal.
    (a) As the applied voltage is slowly increased, in what order will the half-reactions occur at the anode? Write an equation for each half-reaction.
    (b) Describe what happens at the cathode.
    (c) German silver does not contain any silver metal. Why is it called German silver? Why is it a very useful alloy?

    www.science.nelson.com

15. Design a cell to electroplate zinc onto an iron spoon. In your cell diagram, include:
    - ions in the solution
    - substances used for the electrodes
    - anode and cathode labels
    - power supply, showing signs and connections
    - directions of ion and electron flow

16. Suppose you work for a mining company and you are given a job to design a process that will recover nickel metal from a waste solution containing nickel(II) ions.
    (a) Propose an experimental design involving electrolysis that could be tested in the laboratory on a small scale.
    (b) What are some possible complications or factors that need to be considered? List these as questions.

Extension

17. The one-dollar coin, or the loonie (**Figure 13**) replaced the one-dollar bill, which typically wore out in a few months. Sherritt Gordon of Fort Saskatchewan, Alberta developed a unique process for plating the loonie coin.
    (a) Research the production and composition of the loonie.
    (b) What is the golden "aureate" finish on the loonie? Describe the materials and process for producing this finish.
    (c) Why did the coin end up with a loon stamped on it?

    Figure 13
    The Royal Canadian Mint in Winnipeg produces the Canadian loonie for general circulation.

    www.science.nelson.com

18. Aluminium cans are widely used to contain beverages. Write a short report about the production of aluminium cans, including how the can is made, how the top is attached to the can, how the construction of the can has changed since the first model, and the advantages of using recycled aluminium instead of new aluminium.

19. "Cold fusion" made front-page news when it was announced in 1989 as a new practical source of energy.
    (a) Briefly describe cold fusion.
    (b) How is cold fusion related to electrochemistry?
    (c) Outline some theoretical and empirical arguments in the controversy about the existence of cold fusion.

    www.science.nelson.com

14.4 Cell Stoichiometry

In the production of elements, the refining of metals, and electroplating, the quantity of electricity that passes through a cell determines the masses of substances that react or are produced at the electrodes. As you know from oxidation and reduction half-reactions, a specific number of electrons are lost or gained. For example, when zinc is plated onto a steel pipe to galvanize it, two moles of electrons must be gained by one mole of zinc ions to deposit one mole of zinc atoms as metal.

$$Zn^{2+}(aq) + 2\,e^- \rightarrow Zn(s)$$

As in all stoichiometry, this relationship establishes a mole ratio of electrons to some other substance in the half-reaction equation. Unfortunately, there is no meter or instrument for measuring directly (or counting directly) the number of electrons. The number of electrons (as moles of electrons) is determined indirectly. In the past, you have measured mass and then converted to a chemical amount of a substance; a similar procedure is necessary for determining the amount of electrons.

Before we can look at the amount of electrons, we need to see how the charge is determined. Charge, Q, in coulombs, is determined from the electric current, I, in amperes (coulombs per second), and the time, t, in seconds, according to the following definition:

$$Q = It$$

One coulomb (C) is the quantity of charge transferred by a current of one ampere (A) during a time of one second

DID YOU KNOW?

Science versus Technology
Jack Kilby (1923–2005) was a junior engineer when he invented the microchip that is the foundation of all modern electronics. He was awarded a Nobel Prize for Physics in 2000. On the topic of his prize, Kilby commented:

"Those big prizes are for the advancement of understanding. They are for scientists who are motivated by pure knowledge. But I'm an engineer. I'm motivated by a need to solve problems, to make something work. For guys like me, the prize is seeing a successful solution."

▶ SAMPLE problem 14.2

The technology of the Hall–Héroult cell for producing aluminium has improved considerably since the first industrial factory. Modern electrolytic cells may use up to 300 kA of current. What is the charge that passes through one of these cells in a 24 h period?

By definition, a current in amperes (A) is the number of coulombs per second, 1 A = 1 C/s. You always need to convert the time into seconds before time can be used in the calculation of charge.

$$t = 24\,\cancel{h} \times \frac{3600\,s}{1\,\cancel{h}} = 8.6 \times 10^4\,s$$

Now the charge in coulombs can be calculated as follows:

$$Q = It$$
$$= 300 \times 10^3 \,\frac{C}{\cancel{s}} \times 8.6 \times 10^4\,\cancel{s}$$
$$= 2.6 \times 10^{10}\,C$$

Therefore, a current of 300 kA for 24 h transfers 2.6×10^{10} C of charge. This is a huge quantity of charge. For comparison, the charge passing through a 100 W light bulb in 24 h is about 7.2×10^4 C.

> **Practice**
>
> 1. Calculate the charge transferred by a current of 1.5 A flowing for 30 s.
> 2. In an electrolytic cell, 87.6 C of charge is transferred in 22.5 s. Determine the electric current.
> 3. Calculate the charge transferred by a current of 250 mA in a time of 28.5 s.
> 4. How long, in minutes, does it take a current of 1.60 A to transfer a charge of 375 C?

Faraday's Law

The relationship between electricity and electrochemical changes was first investigated by Michael Faraday (**Figure 1**) in the 1830s. Faraday continued Humphry Davy's work in electrochemistry, coining the terms electrolysis, electrolyte, electrode, anode, cathode, cation, and anion. His quantitative study of electrolysis identified the factors that determine the mass of an element produced or consumed at an electrode. He discovered that this mass was directly proportional to the time the cell operated, as long as the current was constant (**Faraday's law**). Furthermore, he found that *9.65 × 10⁴ C of charge is transferred for every mole of electrons that flows in the cell.* In modern terms, this value is the molar charge of electrons, also called the **Faraday constant**, F.

$$F = 9.65 \times 10^4 \frac{C}{\text{mol e}^-}$$

This constant can be used as a conversion factor in converting electric charge to an amount in moles of electrons—in the same way that molar mass is used to convert mass to a chemical amount.

$$n_{e^-} = \frac{Q}{F}$$

Since $Q = It$, the amount of electrons can now be written as

$$n_{e^-} = \frac{It}{F}$$

> **SAMPLE problem 14.3**
>
> What amount of electrons is transferred in a cell that operates for 1.25 h at a current of 0.150 A?
>
> Recall from the calculation of electric charge that the time must always be in seconds because the ampere is defined as coulombs per second (1 A = 1 C/s).
>
> $$t = 1.25 \text{ h} \times \frac{3600 \text{ s}}{1 \text{ h}} = 4.50 \times 10^3 \text{ s}$$
>
> Now you can calculate the amount, in moles, of electrons using the Faraday constant as a conversion factor:
>
> $$n_{e^-} = \left(0.150 \frac{C}{s} \times 4.50 \times 10^3 \text{ s}\right) \times \frac{1 \text{ mol}}{9.65 \times 10^4 \text{ C}}$$
> $$= 6.99 \times 10^{-3} \text{ mol}$$
>
> The amount of electrons transferred is 6.99 mmol.

Figure 1
As a young man, Michael Faraday (1791–1867) taught himself chemistry and convinced the famous English chemist Humphry Davy to hire him as his assistant. Faraday proved himself more than worthy of this trust. He eventually made important contributions in the study of gases, low temperatures, the discovery of benzene, quantitative aspects of electrolysis, electric motors, generators, and transformers. A deeply religious man, Faraday had strong convictions about the appropriate uses of science and technology. He refused to help Britain produce a poison gas for use against the Russians in the Crimean War (1854–56).

Learning Tip

Note the similarity between various definitions of amounts using molar quantities:

$$n_{substance} = \frac{m}{M}$$

$$n_{gas} = \frac{v}{V}$$

$$n_{e^-} = \frac{Q}{F}$$

▶ COMMUNICATION example 1

Convert a current of 1.74 A for 10.0 min into an amount of electrons.

Solution

$$t = 10.0 \text{ min} \times \frac{60 \text{ s}}{1 \text{ min}} = 600 \text{ s}$$

$$n_{e^-} = (1.74 \frac{C}{s} \times 600 \text{ s}) \times \frac{1 \text{ mol}}{9.65 \times 10^4 \text{ C}}$$

$$= 0.0108 \text{ mol}$$

According to Faraday's law, the amount of electrons transferred is 0.0108 mol or 10.8 mmol.

The same method can be used to calculate electric current or time if the other variables are known, as shown in Communication Example 2.

Learning Tip

An alternative solution is to use the mathematical formula and solve for time, t.

$$n_{e^-} = \frac{It}{F}$$

$$t = \frac{n_{e^-}F}{I}$$

You can either substitute values first and then solve for time, or rearrange the mathematical formula and then substitute the values. Either way is acceptable.

▶ COMMUNICATION example 2

How long, in minutes, will it take a current of 3.50 A to transfer 0.100 mol of electrons?

Solution

$$t = (0.100 \text{ mol} \times 9.65 \times 10^4 \frac{C}{mol}) \times \frac{1 \text{ s}}{3.50 \text{ C}}$$

$$= 2.76 \times 10^3 \text{ s} \times \frac{1 \text{ min}}{60 \text{ s}}$$

$$= 46.0 \text{ min}$$

According to Faraday's law, it would require 46.0 min to transfer 0.100 mol e$^-$ using a current of 3.50 A.

▶ Practice

5. An electroplating cell operates for 35 min with a current of 1.9 A. Calculate the amount, in moles, of electrons transferred.
6. A cell transferred 0.146 mol of electrons with a constant current of 1.24 A. How long, in hours, did this take?
7. Calculate the current required to transfer 0.015 mol of electrons in 20 min.

Section **14.4**

Half-Cell Calculations

Since the mass of an element produced at an electrode depends on the amount of transferred electrons, a half-reaction equation showing the number of electrons involved is necessary to do stoichiometric calculations. This applies to all electrochemical cells, whether voltaic or electrolytic. Separate calculations are carried out for each electrode, although the *same charge and, therefore, the same amount of electrons passes through each electrode in a cell* or a group of cells in series. As the following examples show, concepts of stoichiometry used in other calculations also apply to half-cell calculations. The only new part of the stoichiometry is the calculation of the amount of electrons based on the Faraday constant.

> ### ▶ SAMPLE problem 14.4
>
> What is the mass of copper deposited at the cathode of a copper electrorefining cell (**Figure 2**) operated at 12.0 A for 40.0 min?
>
> First identify and write the appropriate half-cell equation. Because copper is being deposited at the cathode, copper(II) ions must be gaining electrons to form copper metal. Write the equation for this reduction and list all information given, including constants such as molar mass and Faraday.
>
> $$Cu^{2+}(aq) + 2\,e^- \rightarrow Cu(s)$$
>
> 40.0 min m
> 12.0 A   63.55 g/mol
> 9.65×10^4 C/mol
>
> Notice that we have all of the information necessary to calculate the amount of electrons. Don't forget to make sure the time is converted to units of seconds, if necessary.
>
> $$n_{e^-} = (12.0\,\tfrac{C}{s} \times 40.0\,\text{min} \times \tfrac{60\,s}{1\,\text{min}}) \times \tfrac{1\,\text{mol}}{9.65 \times 10^4\,C}$$
>
> $= 0.298$ mol
>
> The procedure that is common to all stoichiometry is the use of the mole ratio from a balanced equation. The mole ratio is what allows us to convert from a chemical amount in moles of one substance to another. In the reduction half-reaction given, notice that 1 mol of copper metal is formed when 2 mol of electrons are transferred.
>
> $$n_{Cu} = 0.298\,\text{mol} \times \tfrac{1}{2}$$
>
> $= 0.149$ mol
>
> The final step is to convert to the quantity requested in the question, in this case, the mass of copper metal.
>
> $$m_{Cu} = 0.149\,\text{mol} \times \tfrac{63.55\,g}{\text{mol}}$$
>
> $= 9.48$ g
>
> or
>
> $$m_{Cu} = 40.0\,\text{min} \times \tfrac{60\,s}{1\,\text{min}} \times 12.0\,\tfrac{C}{1\,s} \times \tfrac{1\,\text{mol}\,e^-}{9.65 \times 10^4\,C} \times \tfrac{1\,\text{mol}\,Cu}{2\,\text{mol}\,e^-} \times \tfrac{63.55\,g\,Cu}{1\,\text{mol}\,Cu}$$
>
> $= 9.48$ g
>
> According to Faraday's law and the stoichiometric method, the mass of copper metal deposited is 9.48 g.

Figure 2
These pure copper cathodes from the electrolytic refining cells are ready for melting and processing into copper wire and other copper products.

Learning Tip

Note the similarity of the procedure for stoichiometry calculations of half-cells to all other stoichiometry calculations you have done in the past. Essentially, the only difference is a new relationship (formula based on Faraday's law) to convert to and from the amount of electrons.

SUMMARY: Procedure for Half-Cell Stoichiometry

Step 1: Write the balanced equation for the half-cell reaction of the substance produced or consumed. List the measurements and conversion factors for the given and required entities.

Step 2: Convert the given measurements to an amount in moles by using the appropriate conversion factor (M, c, F).

Step 3: Calculate the amount of the required substance by using the mole ratio from the half-reaction equation.

Step 4: Convert the calculated amount to the final quantity by using the appropriate conversion factor (M, c, F).

▶ COMMUNICATION example 3

Silver is deposited on objects (**Figure 3**) in a silver electroplating cell. If 0.175 g of silver is to be deposited from a silver cyanide solution in a time of 10.0 min, predict the current required.

Solution

$$Ag^+(aq) + e^- \rightarrow Ag(s)$$

10.0 min	0.175 g
I	107.87 g/mol
9.65×10^4 C/mol	

$$n_{Ag} = 0.175 \text{ g} \times \frac{1 \text{ mol}}{107.87 \text{ g}}$$

$$= 1.62 \times 10^{-3} \text{ mol}$$

$$n_{e^-} = 1.62 \times 10^{-3} \text{ mol} \times \frac{1}{1}$$

$$= 1.62 \times 10^{-3} \text{ mol}$$

$$I = \frac{1.62 \times 10^{-3} \text{ mol} \times 9.65 \times 10^4 \frac{C}{mol}}{10.0 \text{ min} \times \frac{60 \text{ s}}{\text{min}}}$$

$$= 0.261 \text{ C/s}$$

or

$$I = 0.175 \text{ g Ag} \times \frac{1 \text{ mol Ag}}{107.87 \text{ g Ag}} \times \frac{1 \text{ mol } e^-}{1 \text{ mol Ag}} \times \frac{9.65 \times 10^4 \text{ C } e^-}{1 \text{ mol } e^-} \times \frac{1}{10 \text{ min}} \times \frac{1 \text{ min}}{60 \text{ s}}$$

$$= 0.261 \text{ C/s}$$

According to the stoichiometry and Faraday's law, the current required to plate 0.175 g of silver in 10.0 min is 0.261 A.

Figure 3
A silver electroplating cell uses silver cyanide to silver plate objects, such as this tray.

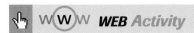

Simulation—Electrolytic Cell Stoichiometry

In this activity, you will simulate the operation of an electrolytic cell. You will follow a series of steps to calculate the approximate cost of producing a given mass of aluminium. These calculations are similar to ones that scientists and engineers do when designing and monitoring an industrial cell.

www.science.nelson.com

Section 14.4 Questions

1. A battery delivers 0.300 A for 15.0 min. What amount of electrons, in moles, is transferred?

2. A current of 55 kA passes through a chlor–alkali cell. What mass of chlorine is formed during 8.0 h?

3. A family wishes to plate an antique teapot with 10.00 g of silver. If the current to be used is 1.80 A, what length of time, in minutes, is required?

4. A typical Hall–Héroult cell produces 425 kg of molten aluminium in 24.0 h. Calculate the current used.

5. Magnesium metal is produced in an electrolytic cell containing molten magnesium chloride. A current of 2.0×10^5 A is passed through the cell for 18.0 h.
   (a) Determine the mass of magnesium produced.
   (b) What mass of chlorine is produced at the same time?

6. Cobalt metal is plated from 250.0 mL of cobalt(II) sulfate solution. What is the minimum concentration of cobalt(II) sulfate required for this cell to operate for 2.05 h with a current of 1.14 A?

7. A 25.72 g piece of copper metal is the anode in a cell in which a current of 0.876 A flows for 75.0 min. Determine the final mass of the copper electrode.

8. A student reconstructs Volta's electric battery using sheets of copper and zinc, and a current of 0.500 A is produced for 10.0 min. Calculate the mass of zinc oxidized to aqueous zinc ions.

9. Electroplating is a common technological process for coating objects with a metal to enhance the appearance of the object or its resistance to corrosion.
   (a) A car bumper is plated with chromium using chromium(III) ions in solution. If a current of 54 A flows in the cell for 45 min 30 s, determine the mass of chromium deposited on the bumper.
   (b) For corrosion resistance, a steel bolt is plated with nickel from a solution of nickel(II) sulfate. If 0.250 g of nickel produces a plating of the required thickness and a current of 0.540 A is used, predict how long in minutes the process will take.

10. During the electrolysis of molten aluminium chloride in an electrolytic cell, 5.40 g of aluminium is produced at the cathode. Predict the mass of chlorine produced at the anode.

11. Chromium metal can be plated onto an object from an acidic solution of dichromate ions. What average current is required to plate 17.8 g of chromium metal in a time of 2.20 h? (You will need to construct your own equation for the half-reaction.)

12. The purpose of this experiment is to test the method of stoichiometry in cells. Complete the Prediction, Analysis, and Evaluation (Part 2 only) of the investigation report.

    Problem
    What is the mass of tin electroplated at the cathode of a tin-plating cell by a current of 3.46 A for 6.00 min?

    Design
    A steel can is placed in an electroplating cell as the cathode. An electric current of 3.46 A flows through the cell, which contains a 3.25 mol/L solution of tin(II) chloride, for 6.00 min.

    Evidence
    initial mass of can = 117.34 g
    final mass of can = 118.05 g

13. Using a specific example of an electrolytic cell, describe how Faraday's law is useful in designing and controlling the process.

Extension

14. A rapidly developing technology is the production of less expensive, more durable, and more energy-dense electrochemical cells, that is, cells with a high energy-to-mass ratio.
    (a) A car battery has a rating of 125 A•h (ampere-hours). What does this tell you about the electrical capacity of this battery?
    (b) Why is this a useful way to rate batteries?
    (c) What mass of lead is oxidized as this battery discharges?
    (d) If an aluminium–oxygen fuel cell has the same rating as the car battery in (a), what mass of aluminium metal would be oxidized?
    (e) Comment on the implications of your answers to (c) and (d).

Chapter 14 INVESTIGATIONS

INVESTIGATION 14.1

Designing an Electric Cell

Report Checklist
- ○ Purpose
- ○ Problem
- ○ Hypothesis
- ● Prediction
- ○ Design
- ● Materials
- ● Procedure
- ● Evidence
- ● Analysis
- ● Evaluation (1, 3)

In this cell, an aluminium soft-drink can is one of the electrodes (**Figure 1**). The other electrode is a solid conductor, such as a piece of copper wire or pipe, an iron nail, or graphite from a pencil. The electrolyte may be a salt solution, or an acidic or basic solution. Although many characteristics of a cell are important, only one characteristic, voltage, is investigated here. Check with your teacher if you want to evaluate other designs and materials.

When evaluating the Purpose (Part 3), include your opinion about the reliability, cost, and simplicity of your final electric cell.

Be careful when handling acidic and basic solutions used for electrolytes, as they are corrosive. Wear eye protection and work near a source of water. Some electrolytes may be toxic or irritant; follow all safety precautions. Avoid eye and skin contact.

Purpose
The purpose of this investigation is to make an electric cell.

Problem
What combination of electrodes and electrolyte gives the largest voltage for an aluminium-can cell?

Design

Part 1
Using the same electrolyte and aluminium can as the controlled variables, two or three different materials are used as the second electrode. The voltage of each cell is measured as the responding variable.

Part 2
Using the same two electrodes as the controlled variables, two or three possible electrolytes are tested. The voltage of each cell is measured as the responding variable.

Part 3
Additional combinations are tested, based on the analysis of the initial trials.

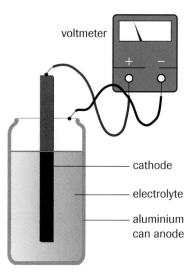

Figure 1
An aluminium-can cell is an efficient design since one of the electrodes also serves as the container.

INVESTIGATION 14.2

A Voltaic Cell (Demonstration)

Report Checklist
- ○ Purpose
- ○ Problem
- ○ Hypothesis
- ○ Prediction
- ○ Design
- ○ Materials
- ○ Procedure
- ● Evidence
- ● Analysis
- ○ Evaluation

An important characteristic of consumer, commercial, and industrial cells is a simple, efficient design that works for the intended application. For scientific research, this is not as important as a design that can be easily manipulated and studied.

Purpose
The purpose of this investigation is to test the design and operation of a voltaic cell used in scientific research.

Problem
What is the design and operation of a voltaic cell?

Design
An electric cell with only one electrolyte is compared with voltaic cells containing the same electrodes, but two electrolytes.

   Solutions used are toxic and irritant. Avoid contact with skin and eyes.

Procedure
1. Construct the three cells shown in **Figure 2**.
2. For each design, use a voltmeter to determine which electrode is positive and which is negative (see Appendix C.3), and measure the electric potential difference of each cell.
3. With the voltmeter connected, remove and then replace the various parts of the cell.
4. For each cell, connect the two electrodes with a wire. Record any evidence of a reaction after several minutes, and after one or two days. Measure the electric potential difference after several days.

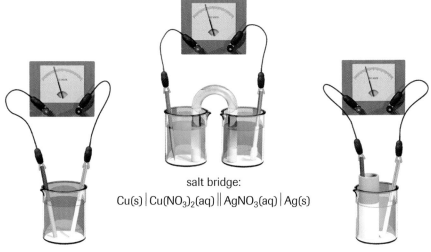

no porous boundary:
$Cu(s) | NaNO_3(aq) | Ag(s)$

salt bridge:
$Cu(s) | Cu(NO_3)_2(aq) \| AgNO_3(aq) | Ag(s)$

porous cup:
$Cu(s) | Cu(NO_3)_2(aq) \| AgNO_3(aq) | Ag(s)$

Figure 2
In Investigation 14.2, you compare three different cell designs.

INVESTIGATION 14.3

Testing Voltaic Cells

Report Checklist
- ○ Purpose
- ○ Problem
- ○ Hypothesis
- ● Prediction
- ● Design
- ○ Materials
- ● Procedure
- ● Evidence
- ● Analysis
- ● Evaluation (1, 2, 3)

Testing is a procedure that is common to both technology and science. In technology, testing is necessary to determine how a product or process works using criteria such as efficiency, reliability, and cost. In science, testing is a key part in the advancement of knowledge. Scientific concepts are developed and then tested to determine their validity and limitations. New ideas that fail the test then need to be restricted, revised, or replaced.

In your Evaluation, pay particular attention to sources of error or uncertainty, and to limitations of the evidence collected.

Purpose
The purpose of this investigation is to test the predictions of cell potentials and the identity of the electrodes of various cells.

Problem
In cells constructed from various combinations of copper, aluminium, silver, and zinc half-cells, what are the standard cell potentials, and which is the anode and cathode in each case?

Materials
lab apron
eye protection
voltmeter and connecting wires
U-tube with cotton plugs, porous cups, or filter paper
four 100 mL beakers (or well plate)
distilled water
steel wool
Cu(s), Al(s), Ag(s), and Zn(s) strips
1.0 mol/L each of:
$CuSO_4(aq)$
$Al(NO_3)_3(aq)$
$AgNO_3(aq)$
$NaNO_3(aq)$
$ZnSO_4(aq)$

 The materials used are toxic and irritant. Avoid skin and eye contact.

INVESTIGATION 14.4

A Potassium Iodide Electrolytic Cell

Report Checklist
- ○ Purpose
- ○ Problem
- ○ Hypothesis
- ○ Prediction
- ○ Design
- ○ Materials
- ○ Procedure
- ● Evidence
- ● Analysis
- ● Evaluation (1, 3)

Electrolytic cells were discovered before the science was understood. However, as with all successful technological inventions, the important criteria was that it worked, not why it worked. Eventually, chemists understood the science and were able to explain why electrolytic cells work.

In the Evaluation, suggest changes to the Design, Materials, and Procedure that would improve the Evidence.

Purpose
The purpose of this investigation is to use diagnostic tests to determine the reaction products of an electrolytic cell.

Problem
What are the products of the reaction during the operation of an aqueous potassium iodide electrolytic cell?

Design
Inert electrodes are placed in a 0.50 mol/L solution of potassium iodide, and a battery or power supply provides a direct current of electricity to the cell. The litmus and halogen diagnostic tests (Appendix C.4) are conducted to test the solution near each electrode before and after the reaction.

Materials
lab apron
eye protection
petri dish
two carbon electrodes
two connecting wires
3 V to 9 V battery or power supply
red and blue litmus paper
ring stand and two utility clamps
small test tube with stopper
dropper bottle of hexane
0.50 mol/L KI(aq)

 Hexane is highly flammable. Do not use near an open flame and avoid inhaling the fumes. Use only in a well-ventilated area.

Procedure
1. Set up the KI(aq) cell as shown in **Figure 3**, (or with one ring stand), but with a single wire connecting the electrodes (i.e., no power supply).
2. Observe the cell and test the solution with litmus paper and hexane.
3. Use two connecting wires to hook up the power supply to the electrodes.
4. Turn on the power supply.
5. Record all observations at each electrode.
6. Perform both diagnostic tests at each electrode.
7. Dispose of the solutions by putting the hexane mixture in a labelled waste container and washing the potassium iodide solution down the sink drain.

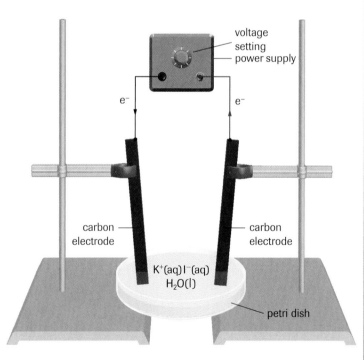

Figure 3
Apparatus for the electrolysis of potassium iodide

INVESTIGATION 14.5

Electrolysis (Demonstration)

Report Checklist
- ○ Purpose
- ○ Problem
- ○ Hypothesis
- ● Prediction
- ○ Design
- ○ Materials
- ○ Procedure
- ● Evidence
- ● Analysis
- ● Evaluation (1, 2, 3)

Scientific knowledge progresses by the experimental testing of ideas. The more rigorous the test, the more certain the knowledge or the better the chance of making new discoveries.

Purpose
The purpose of this demonstration is to test the method of predicting the products of electrolytic cells.

Problems
What are the products of electrolytic cells containing one of the following solutions:
- aqueous copper(II) sulfate?
- aqueous sodium sulfate?
- aqueous sodium chloride?

Design
The electrolysis of the aqueous copper(II) sulfate is carried out in a U-tube, and the electrolysis of aqueous sodium sulfate and sodium chloride is carried out in a Hoffman apparatus (**Figure 4**) so that any gases produced can be collected. Diagnostic tests with necessary control tests (before electrolysis) are conducted to determine the presence of the predicted products.

Copper(II) sulfate is toxic and irritant. Avoid skin and eye contact. If you spill copper(II) sulfate solution on your skin, wash the affected area with lots of cool water. During electrolysis, corrosive substances are produced; avoid skin and eye contact.

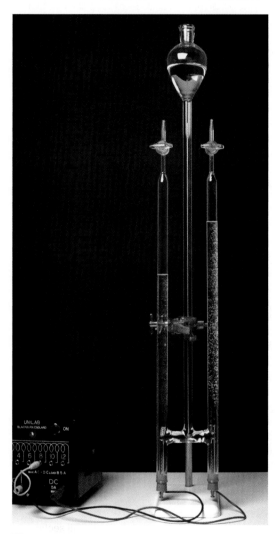

Figure 4
Hoffman apparatus

Chapter 14 SUMMARY

Outcomes

Knowledge

- define anode, cathode, anion, cation, salt bridge/porous cup, electrolyte, voltaic cell, and electrolytic cell (14.2, 14.3)
- predict and write the half-reaction equation that occurs at each electrode in an electrochemical cell (14.2, 14.3, 14.4)
- explain that the values of standard reduction potential are all relative to $E°_r = 0.00$ V set for the hydrogen electrode at standard conditions (14.2)
- calculate the standard cell potential for electrochemical cells (14.2, 14.3)
- predict the spontaneity of redox reactions based on standard cell potentials (14.2, 14.3)
- identify the similarities and differences between a voltaic cell and an electrolytic cell (14.3)
- recognize that predicted reactions do not always occur (14.3)
- calculate mass, amounts, current, and time in single voltaic and electrolytic cells by applying Faraday's law and stoichiometry (14.4)

STS

- state examples of science leading technology and technology leading science (14.1, 14.3)
- recognize the values and limitations of technological products and processes (14.1, 14.3)
- describe the interactions of science and technology (14.1, 14.2, 14.3)

Skills

- initiating and planning: design an experiment, including a labelled diagram, to test predictions for reactions occurring in electrochemical cells (14.2, 14.3); describe procedures for safe handling, storage, and disposal of materials used in the laboratory (14.1, 14.2); and develop a plan to build a cell (battery) using a trial-and-error procedure (14.1)
- performing and recording: construct and observe electrochemical cells (14.1, 14.2, 14.3); investigate the issue of disposal of batteries (14.1); and compile and display information about electrochemical cells (all sections)
- analyzing and interpreting: identify the products of electrochemical cells (all sections); compare predictions with observations (all sections); identify the limitations of evidence collected (14.1, 14.2, 14.3); explain discrepancies between predicted and measured cell potentials (14.2, 14.3); assess the practicalities of different cell designs (14.1); and evaluate experimental designs for cells (14.1, 14.2, 14.3)
- communication and teamwork: work collaboratively in addressing problems and use appropriate SI notation and significant digits (all sections); and select and integrate information from various sources about technological applications of cells (all sections)

Key Terms

14.1
electrode
electrolyte
electric potential difference
electric current
fuel cell

14.2
porous boundary
half-cell
voltaic cell
cathode
anode
inert electrode

standard cell
standard cell potential
standard reduction potential
reference half-cell
corrosion
cathodic protection

14.3
electrolytic cell
electrolysis

14.4
Faraday's law
Faraday constant

Key Equations

- $E°_{cell} = E°_{r\,cathode} - E°_{r\,anode}$ (14.2, 14.3)
- $n_{e^-} = \dfrac{It}{F}$ (14.4)

▶ MAKE a summary

1. Draw a diagram of a simple, general voltaic cell. Show and label all components, including names and signs where appropriate. Show the directions of electron and ion movement. State the process occurring at each electrode and how the cell potential is determined. List several specific technological examples.
2. Repeat question 1 for a simple, general electrolytic cell.
3. Refer back to your answers to the Starting Points questions at the beginning of this chapter. How has your thinking changed?

▶ Go To   www.science.nelson.com

The following components are available on the Nelson Web site. Follow the links for *Nelson Chemistry Alberta 20-30*.

- an interactive Self Quiz for Chapter 14
- additional Diploma Exam-style Review questions
- Illustrated Glossary
- additional IB-related material

There is more information on the Web site wherever you see the Go icon in this chapter.

Chapter 14 REVIEW

Many of these questions are in the style of the Diploma Exam. You will find guidance for writing Diploma Exams in Appendix H. Exam study tips and test-taking suggestions are on the Nelson Web site. Science Directing Words used in Diploma Exams are in bold type.

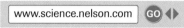

DO NOT WRITE IN THIS TEXTBOOK.

Part 1

1. Which one of the following is the best example of a new technology that led to scientific discoveries?
   A. electric cell
   B. Faraday constant
   C. reference half-cell
   D. hydrogen fuel cell

2. Which of the following statement applies more to science than to technology?
   A. It involves the development of devices/process that have a practical purpose.
   B. It seeks explanations about the natural world.
   C. It applies mainly to the human-designed processes.
   D. It applies trial-and-error strategies.

3. The main advantage of fuel cells is that they
   A. are reliable
   B. are portable
   C. are inexpensive to build
   D. run continuously as reactants are added

4. In a voltaic cell, the cathode is labelled as the __i__ electrode where the __ii__ half reaction of the strongest __iii__ agent occurs.
   The above statement is completed by the information in which row?

Row	i	ii	iii
A.	negative	oxidation	oxidizing
B.	negative	oxidation	reducing
C.	positive	reduction	oxidizing
D.	positive	reduction	reducing

Use the following information to answer questions 5 and 6.
1. net reaction is spontaneous
2. net reaction is nonspontaneous
3. anions migrate toward the anode
4. cell potential is positive
5. cell potential is negative
6. electrons flow from the cathode to the anode
7. ions complete the internal circuit in the cell

5. In numerical order, the statements that are true for voltaic cells are ___ ___ ___ ___.

6. In numerical order, the statements that are true for electrolytic cells are ___ ___ ___ ___.

7. The cell potential of a standard lead–chlorine cell is predicted to be
   +_____ V

8. The main products formed in the electrolysis of an aqueous nickel(II) chloride solution are

	at the anode	at the cathode
A.	$O_2(g)$, $H^+(aq)$	$Ni(s)$
B.	$Cl_2(g)$	$Ni(s)$
C.	$Cl_2(g)$	$H_2(g)$, $OH^-(aq)$
D.	$Ni^{2+}(aq)$	$Cl^-(aq)$

9. If chemists had decided to use the standard zinc half-cell as the reference half-cell, the new standard reduction potential for $Hg^{2+}(aq)$ would be
   A. -1.61 V
   B. -0.09 V
   C. $+0.09$ V
   D. $+1.61$ V

10. What is the standard reduction potential for $Ce^{4+}(aq)$ using the following net reaction and standard cell potential?

    $Ce^{4+}(aq) + Fe^{2+}(aq) \rightarrow Ce^{3+}(aq) + Fe^{3+}(aq)$
    $E°_{cell} = +0.67$ V

    A. $+1.44$ V
    B. $+0.10$ V
    C. -0.10 V
    D. -1.44 V

11. The industrial production of alkali metals involves
    A. electrolysis of molten salts
    B. electrolysis of aqueous salts
    C. reduction with strong reducing agents
    D. oxidation with strong oxidizing agents

12. Which of the following is **not** an effective method of corrosion prevention?
    A. galvanizing
    B. sacrificial anode
    C. sacrificial cathode
    D. impressed current

13. In an electroplating operation, the mass of copper metal produced from a copper(II) ion solution by a 0.325 A current flowing for 3.00 h is predicted to be
    _____ g.

14. The following electrolytic cell is set up:
    $C(s) | Co(NO_3)_2(aq) | C(s)$
    The minimum voltage that must be applied to operate this cell is
    +_____ V

Part 2

15. **Explain** the function of the following components of voltaic cells.
    (a) electrolyte
    (b) salt bridge
    (c) electrode
    (d) connecting wire

16. For each of the following cells, write the equations for the reactions occurring at the cathode and at the anode, and the equation for the net cell reaction.
    (a) $Ni(s) \mid Ni^{2+}(aq) \parallel Cu^{2+}(aq) \mid Cu(s)$
    (b) $Zn(s) \mid Zn^{2+}(aq) \parallel Cr_2O_7^{2-}(aq), H^+(aq) \mid C(s)$
    (c) $Pt(s) \mid H_2(g) \mid OH^-(aq) \parallel Ag^+(aq) \mid Ag(s)$

17. Draw and label a diagram of a voltaic cell constructed from some (not all) of the following materials:

strip of silver metal	voltmeter
strip of lead metal	connecting wires
aqueous silver nitrate	glass U-tube
aqueous lead(II) nitrate	cotton
aqueous sodium nitrate	various beakers
distilled water	porous porcelain cup

18. Provide the following information for the cell constructed in question 17.
    (a) anode half-reaction
    (b) cathode half-reaction
    (c) net cell reaction
    (d) standard cell potential

19. Cathodic protection is widely used in industry for preventing the corrosion of steel tanks and pipes.
    (a) **Describe** two forms of cathodic protection including the chemical principles involved.
    (b) **Assess** the importance of corrosion protection using at least two perspectives.

20. A student constructs an electrolytic cell to nickel plate a carbon rod. Draw and label a diagram of an electrolytic cell constructed using the following materials:

strip of nickel metal	power source
carbon rod	connecting wires
aqueous nickel(II) sulfate	large beaker
voltmeter	

21. Provide the following information for the cell constructed in question 20.
    (a) anode half-reaction
    (b) cathode half-reaction
    (c) net cell reaction
    (d) minimum voltage required

22. As part of a class project, a student needs to construct a cell that produces a voltage sufficient to operate a 1.25 V clock.
    (a) **Outline** the design of a suitable cell including a fully labelled cell diagram.
    (b) **Justify** your design using appropriate chemical terms, equations, and calculations.
    (c) Suggest some possible discrepancies or limitations of this cell for the chosen task.

23. In the Hall–Héroult process, an electric current is passed through the electrolyte at low voltage, but very high current. Calculate the mass of aluminium metal that would be produced from aluminium ions by a 150 kA current flowing in a Hall–Héroult cell for 1.00 h (**Figure 1**).

Figure 1
Molten aluminium from an electrolytic cell

24. A teacher constructs a demonstration voltaic cell, $Zn(s) \mid Zn^{2+}(aq) \parallel Cu^{2+}(aq) \mid Cu(s)$, and allows it to operate for 24.0 h. The mass of the cathode was measured before and after the demonstration. Use the evidence collected to calculate the average current produced by this cell.

    final mass of the cathode: 24.68 g
    initial mass of the cathode: 21.12 g

25. Briefly **explain** the process of corrosion. Your response should include
    - several examples of metal corrosion of manufactured materials
    - any environmental, health, or safety issues associated with your examples
    - brief descriptions of some technological solutions to the problem of corrosion
    - at least one example of corrosion that is desirable

Unit 7 REVIEW

Many of these questions are in the style of the Diploma Exam. You will find guidance for writing Diploma Exams in Appendix H. Exam study tips and test-taking suggestions are on the Nelson Web site. Science Directing Words used in Diploma Exams are in bold type.

www.science.nelson.com

DO NOT WRITE IN THIS TEXTBOOK.

Part 1

1. A redox reaction involves a transfer of electrons
   A. from the oxidizing agent to the reducing agent
   B. from the reducing agent to the oxidizing agent
   C. through a porous barrier
   D. between metals only

2. A general reaction type that is **not** a redox reaction is
   A. neutralization
   B. disproportionation
   C. combustion
   D. formation

3. When solutions
   NR
   1. sulfuric acid
   2. lithium hydroxide
   3. gold(III) fluoride
   4. chromium(II) nitrate
   are ranked in order of strength of oxidizing agents, the order, from **strongest to weakest** oxidizing agent, is __, __, __, and __.

4. During the process of photosynthesis,
   $6\ CO_2(g) + 6\ H_2O(g) \rightarrow C_6H_{12}O_6(aq) + 6\ O_2(g)$
   A. carbon in carbon dioxide is oxidized
   B. hydrogen in water is reduced
   C. oxygen in carbon dioxide and/or water is oxidized
   D. oxygen in glucose is oxidized

5. Which of the following reaction equations describes a redox reaction?
   A. $C_2H_4(g) + 3\ O_2(g) \rightarrow 2\ CO_2(g) + 2\ H_2O(g)$
   B. $H^+(aq) + OH^-(aq) \rightarrow H_2O(l)$
   C. $Ag^+(aq) + Cl^-(aq) \rightarrow AgCl(s)$
   D. $HMnO_4(aq) \rightarrow H^+(aq) + MnO_4^-(aq)$

6. The metal molybdenum, Mo(s), reacts to form $MoO_2(s)$. The half-reaction equation that explains the change in oxidation state of molybdenum can be written as
   A. $Mo(s) + 2\ e^- \rightarrow Mo^{2+}(s)$
   B. $Mo(s) \rightarrow Mo^{2+}(s) + 2\ e^-$
   C. $Mo^{4+}(s) + 4\ e^- \rightarrow Mo(s)$
   D. $Mo(s) \rightarrow Mo^{4+}(s) + 4\ e^-$

7. A high school laboratory's waste container is used to dispose of aqueous solutions of sodium nitrate, potassium sulfate, hydrochloric acid, and tin(II) chloride. The most likely net redox reaction predicted to occur inside the waste container is represented by the equation:

   A. $2\ H^+(aq) + 2\ K^+(aq) \rightarrow H_2(g) + K(s)$
   B. $Sn^{2+}(aq) + 2\ NO_3^-(aq) + 4\ H^+(aq) \rightarrow N_2O_4(g) + 2\ H_2O(l) + Sn^{4+}(aq)$
   C. $SO_4^{2-}(aq) + 4\ H^+(aq) + 2\ Cl^-(aq) \rightarrow H_2SO_3(aq) + H_2O(l) + Cl_2(g)$
   D. $2\ Cl^-(aq) + Sn^{2+}(aq) \rightarrow 2\ Cl_2(g) + Sn(s)$

Use this information to answer questions 8 to 11.

Steel is the most widely used alloy in the world. However, an estimated 20% of the iron and steel produced annually is used to replace that lost by corrosion through exposure to air and water. Therefore, corrosion prevention is of considerable importance. The empirical and theoretical chemistry of corrosion, and technological research and development together provide solutions to this important practical problem.

8. In the corrosion of steel objects in the natural environment, the most likely reducing agent is
   A. $Fe^{2+}(aq)$
   B. $Fe^{3+}(aq)$
   C. $Fe(s)$
   D. $O_2(g)$

9. The metals
   NR
   1. Fe(s)   3. Zn(s)
   2. Sn(s)   4. Mg(s)
   listed in order from *least to most* likely to corrode under similar atmospheric conditions are __, __, __, and __.

10. The reduction half-reaction that is generally involved in the corrosion of iron is
    A. $Fe^{2+}(aq) + 2\ e^- \rightarrow Fe(s)$
    B. $Fe(s) \rightarrow Fe^{3+}(aq) + 3\ e^-$
    C. $2\ H_2O(l) \rightarrow O_2(g) + 4\ H^+(aq) + 4\ e^-$
    D. $O_2(g) + 2\ H_2O(l) + 4\ e^- \rightarrow 4\ OH^-(aq)$

11. The net cell potential, under standard conditions, for the
    NR iron–oxygen cell in an aqueous environment is
    +____ V.

12. In a titration experiment, 10.0 mL samples of 0.650 mol/L
    NR chromium(II) ion solution reacted with an average volume of 12.4 mL of acidic potassium dichromate solution. The amount concentration of the potassium dichromate solution is

    _____ mmol/L.

13. All voltaic and electrolytic cells require
    A. one electrode and two electrolytes
    B. two electrodes and one or two electrolytes
    C. an external power supply
    D. a voltmeter

14. Standard reduction potentials for half-cells are based on the strengths of
    A. oxidizing agents relative to hydrogen ions
    B. oxidizing agents relative to hydrogen gas
    C. reducing agents relative to hydrogen ions
    D. reducing agents relative to a standard acidic solution

666 Unit 7

15. In a voltaic cell, the reduction potentials of two standard half-cells are +0.35 V and −1.13 V. The predicted cell potential of the cell constructed from these two half-cells is
    A. +0.35 V
    B. +0.78 V
    C. +1.13 V
    D. +1.48 V

16. If the electrodes of a standard copper–silver cell are connected with a wire, then
    A. silver is plated at the anode
    B. a voltmeter would show a reading of 1.14 V
    C. the solution at the anode becomes darker blue
    D. electrons flow from the silver to the copper electrodes

17. The electrolysis of brine, NaCl(aq), is an important industrial process. The major products formed at each electrode are

	Cathode	Anode
A.	$H_2(g)$, $OH^-(aq)$	$O_2(g)$, $H^+(aq)$
B.	$H_2(g)$, $OH^-(aq)$	$Cl_2(g)$
C.	$Na(s)$	$O_2(g)$, $H^+(aq)$
D.	$Cl_2(g)$	$OH^-(aq)$

Use this information to answer questions 18 to 20.

An aqueous solution of potassium hydroxide undergoes electrolysis using 5.9 A of current for a total time of 22 min.

18. Electrons are transferred through the
    A. solution from the anode to the cathode
    B. solution from the cathode to the anode
    C. external wire from the anode to the cathode
    D. external wire from the cathode to the anode

19. The product(s) at the anode will be
    A. $O_2(g)$, $H^+(aq)$
    B. $K(s)$
    C. $H_2(g)$, $OH^-(aq)$
    D. $O_2(g)$, $H_2O(l)$

20. The mass of the gas produced at the anode is
    NR g.

Part 2

21. **Define** oxidation and reduction in three different contexts: empirical (historical), theoretical (in terms of electrons), and theoretical (in terms of oxidation numbers).

22. Using a general reaction equation (A + B → C + D), label the agents and processes for any redox reaction.

23. **Define** disproportionation and provide one simple example.

24. From the information in this unit, list two or three examples of situations in which technology preceded scientific explanations.

25. Name two common reactions that occur in living and nonliving systems. For each, identify the oxidizing agent, reducing agent, and the direction of electron transfer.

26. Distinguish, in as many ways as possible, between anode and cathode. Does your answer apply equally to voltaic and electrolytic cells? Explain briefly.

27. Briefly **describe** two technological solutions to the problem of batteries "going dead."

28. **Explain** why corrosion often occurs in places where two different metals (such as copper and iron) are joined together.

29. Electrochemical cells are very important technological devices in our society. **Compare** the main differences between voltaic and electrolytic cells in terms of their purpose and the chemical reactions that occur in them.

30. **Predict** whether a spontaneous redox reaction will occur in the following situations:
    (a) A copper penny is dropped into hydrochloric acid.
    (b) A nickel is dropped into nitric acid.
    (c) A silver earring is dropped into sulfuric acid.

31. While working on the development of a new electrochemical cell, a research chemist places selected Period 4 transition metal strips into aqueous solutions of their ionic compounds. She observes that the following combinations of metal and cations react spontaneously:
    $$V(s) + Mn^{2+}(aq) \rightarrow V^{2+}(aq) + Mn(s)$$
    $$V^{2+}(aq) + Ti(s) \rightarrow V(s) + Ti^{2+}(aq)$$
    $$Co^{2+}(aq) + Mn(s) \rightarrow Co(s) + Mn^{2+}(aq)$$
    (a) Use this information to develop a table of oxidizing and reducing agents for these metals and their ions.
    (b) **Identify** the strongest oxidizing and the strongest reducing agent in your table.

32. Make a list of everything that must be balanced in a net ionic equation representing a redox reaction.

33. Write and label balanced half-reaction equations for each of the following redox reactions.
    (a) $2\ Fe^{3+}(aq) + Ni(s) \rightarrow 2\ Fe^{2+}(aq) + Ni^{2+}(aq)$
    (b) $Br_2(aq) + 2\ I^-(aq) \rightarrow 2\ Br^-(aq) + I_2(s)$
    (c) $Pd^{2+}(aq) + Sn^{2+}(aq) \rightarrow Pd(s) + Sn^{4+}(aq)$
    (d) Label each reactant in (a), (b), and (c) as an oxidizing or a reducing agent.

34. Use your knowledge of electrochemistry and some brainstorming to **describe** at least three methods for determining or approximating the position of the beryllium half-reaction in a table of half-reactions.

35. Potassium metal spontaneously reacts with water (**Figure 1**).

Figure 1
Potassium reacts vigorously with water.

(a) Write the half-reaction and net ionic reaction equations for this reaction.
(b) **Describe** diagnostic tests (procedure, evidence, analysis) that could be done to test for the predicted products.

36. An acidic solution of potassium dichromate is added to a sodium iodide solution. **Predict** the net ionic equation and show all of your work. State two diagnostic tests that could be used to test your reaction prediction.

37. What is the oxidation number of
    (a) I in $I_2(s)$?
    (b) I in $CaI_2(s)$?
    (c) I in $HIO(aq)$?
    (d) H in NH_3?
    (e) H in AlH_3?
    (f) O in CH_3OH?

38. **Predict** the balanced redox reaction equation by constructing and labelling oxidation and reduction half-reaction equations.
    (a) $MnO_4^{2-}(aq) \rightarrow Mn^{2+}(aq) + MnO_4^{-}(aq)$ (acidic)
    (b) $ClO^{-}(aq) \rightarrow ClO_2^{-}(aq) + Cl_2(g)$ (basic)

39. Use the oxidation number method to balance the reaction equations for the following redox reactions in acidic solutions:
    (a) $MnO_4^{-}(aq) + H_2C_2O_4(aq) \rightarrow Mn^{2+}(aq) + CO_2(g) + H_2O(l)$
    (b) $KIO_3(aq) + KI(aq) + HCl(aq) \rightarrow KCl(aq) + I_2(s) + H_2O(l)$

40. Chromium steel alloys are analyzed using a series of redox reactions.
    Step 1: The alloy is initially reacted with perchloric acid, which converts the chromium metal into dichromate ions, while the perchloric acid is reduced to chlorine gas.
    Step 2: The dichromate ions are then reduced to chromium(III) ions by adding an excess of iron(II) solution.
    Step 3: The unreacted iron(II) is then titrated with a solution of cerium(IV) ions, which reduces them to cerium(III) ions.
    Write a balanced redox equation for each step of this procedure.

41. **Predict** the mass of gold formed when a gold(III) nitrate solution reacts completely with 125 mL of 0.352 mol/L sulfurous acid (aqueous hydrogen sulfite).

42. Complete the Materials (including precautions) and Analysis of the following investigation report.

Purpose
The purpose of this exercise is to use redox stoichiometry to standardize a potassium dichromate solution.

Problem
What is the amount concentration of a potassium dichromate solution?

Evidence
Volume of iron(II) ammonium sulfate solution = 10.0 mL
Concentration of acidic $Fe(NH_4)_2(SO_4)_2(aq)$ = 0.0625 mol/L

Trial	1	2	3
final burette reading (mL)	13.4	25.5	37.6
initial burette reading (mL)	1.1	13.4	25.5

43. Electrolysis is used in the industrial production of several important elements and compounds.
    (a) **Define** electrolysis.
    (b) In an electrolytic cell, what type of half-reaction occurs at the anode? at the cathode?
    (c) **Compare** the electrolysis of molten compounds with the electrolysis of aqueous solutions. State some similarities and differences.
    (d) Briefly **describe** three important industrial applications of electrolysis.

44. In which of the following mixtures must an external voltage be applied to inert electrodes to observe evidence of a redox reaction?
    (a) a solution of cadmium nitrate
    (b) a solution of iron(III) iodide
    (c) solutions of iron(III) chloride and tin(II) sulfate in separate half-cells connected by a salt bridge
    (d) solutions of potassium iodide and zinc nitrate in separate half-cells

45. **Determine** the minimum potential difference that must be applied to the following electrolytic cells to cause a chemical reaction. (You do not need to write the half-cell reaction equations.)
    (a) iron(II) sulfate electrolyte with inert electrodes
    (b) hydrochloric acid electrolyte with silver electrodes
    (c) tin(II) chloride electrolyte with tin electrodes

46. Write the equations for reactions at the cathode and anode, and the net cell reaction. **Determine** the minimum potential difference that must be applied to each of the following electrolytic cells to cause a reaction.
    (a) $C(s) \mid NaBr(aq) \mid C(s)$
    (b) $Pt(s) \mid KOH(aq) \mid Pt(s)$
    (c) $C(s) \mid CoCl_2(aq) \mid C(s)$

47. Volta's invention of the electric battery in 1800 led to a flurry of scientific research using this new technology. A few weeks after he heard about Volta's battery, William Nicholson, an English chemist, built his own battery and passed a current through slightly acidified water. With the current flowing, bubbles of colourless gases formed at each electrode. This was the first demonstration that an electric current could bring about a chemical reaction.
    (a) Write equations for the cathode, anode, and net reactions that occurred in Nicholson's demonstration.
    (b) Determine the minimum potential difference needed for the reaction.

48. Potassium hydroxide is obtained commercially by the electrolysis of aqueous potassium chloride.
    (a) **Sketch** a diagram of a cell that could be used to electrolyze an aqueous solution of potassium chloride. Label electrodes, electrolyte, power supply, and the directions of the electron and ion flow.
    (b) **Predict** the cathode, anode, and net reactions, and calculate the minimum potential difference for the electrolysis of aqueous potassium chloride.

49. One technological process for refining zinc metal involves the electrolysis of a zinc sulfate solution.
    (a) Write equations for the cathode, anode, and net reactions, and calculate the minimum potential difference for the electrolysis of a zinc sulfate solution.
    (b) Calculate the time required to produce 1.00 kg of zinc using a 5.00 kA current.

50. **Determine** the current required to produce 1.00 kg of aluminium per hour in a single Hall–Héroult cell for the production of aluminium.

51. The electrolysis of copper(II) sulfate using carbon electrodes is demonstrated to a chemistry class. In the demonstration, a 1.50 A current passes through 75.0 mL of 0.125 mol/L copper(II) sulfate solution. How long, in minutes, would it take to plate all of the copper from the solution?

52. Given that the typical current used in the chlor–alkali plant is 55 kA, **predict** the rate at which chlorine gas is produced (in moles per hour).

53. Battery technology is a very active area of research. One proposal that shows some promise is a vanadium redox flow cell, also known as the All Vanadium Redox Battery. **Describe** the general construction of this battery, including electrodes, electrolytes, porous boundary, and external tanks. What redox reactions occur at the electrodes within this cell? List some unique aspects of this technology and some advantages and proposed uses.

www.science.nelson.com

54. Rechargeable nickel-metal hydride (NiMH) batteries have twice the energy density of Ni–Cd batteries and a similar operating voltage. The cells in NiMH batteries use NiO(OH)(s) as one electrode, a hydrogen-absorbing alloy as the other, and an alkaline electrolyte. In the following reduction half-equations, M indicates a hydrogen-absorbing alloy and H_{ab} indicates absorbed hydrogen.

$NiO(OH)(s) + H_2O(l) + e^- \rightarrow Ni(OH)_2(s) + OH^-(aq)$ $E°_r = +0.49$ V

$M(s) + H_2O(l) + e^- \rightarrow MH_{ab} + OH^-(aq)$ $E°_r = -0.71$ V

(a) Write balanced equations for the anode, cathode, and net reactions occurring during the cell's operation.
(b) **Determine** the cell potential.
(c) Research and list some of the technological, economic, and environmental considerations involved in evaluating the NiMH battery.

www.science.nelson.com

55. Electroplating finishes are often layered. For example, chromium plating does not work well on a zinc base, so a layer of copper is applied to the zinc, then a layer of nickel is added, and then the top chromium layer is plated on.
    (a) Propose a general design of an experiment to place a final chromium layer onto a galvanized metal. Include a labelled diagram and general plan.
    (b) In any electroplating process a particular thickness of metal is desired. **Outline** the experimental variables and the type of calculations necessary to plan a particular thickness of metal plating.

56. Review the focusing questions on page 552. Using the knowledge you have gained from this unit, briefly **outline** a response to each of these questions.

Extension

57. The Alberta government is considering making carbon monoxide detectors mandatory for new residential homes as is done in Ontario. What are the sources and the health effects of carbon monoxide? Briefly **describe** how a modern CO detector works including the reaction equations and catalyst involved. Should CO detectors be required in all homes? Justify your opinion.

www.science.nelson.com

58. Moli Energy of Maple Ridge, British Columbia, was the first company in the world to develop a commercial, rechargeable lithium ion cell, called a Molicel (**Figure 2**). Research the characteristics and advantages of Molicels compared with other secondary cells.

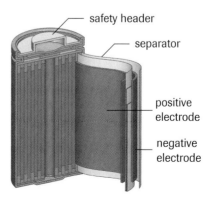

Figure 2
The Molicel is a high-energy, rechargeable lithium-ion cell in a unique, jellyroll design.

unit 8
Chemical Equilibrium Focusing on Acid–Base Systems

Equilibrium describes any condition or situation of balance. We recognize equilibrium in a chemical reaction system, oddly enough, by noticing nothing—we see no change in any property of the system. The easiest conclusion to draw would be that nothing is happening, but closer study reveals that, at the molecular level, a lot of change is going on.

Chemical reaction equilibrium is always a *dynamic balance* between two *opposing* changes, which are balanced because they are occurring at *equal* rates, within a *closed* system. What we observe directly is the net effect—neither an increase nor a decrease in any measurable property.

Chemistry involves the study of change in chemical substances. To predict and control chemical change, we must better understand the nature of the system at the molecular level. For instance, to understand why and how bubbles of gas form or dissolve in a liquid, we must take into account the nature of the gas, the nature of the liquid, and the actions of their invisible entities—all at the same time.

This unit explores the nature of dynamic equilibrium in chemical systems. It explains much more thoroughly and completely many of the chemical change concepts you have already learned. You will examine some very important reactions—those involving acids and bases in solution—at a higher conceptual level. This knowledge will allow you to describe, explain, and predict many new chemical systems and situations. The continual exploration and improvement of concepts such as these is a critical part of the nature of science.

As you progress through the unit, think about these focusing questions:
- What is happening in a system at equilibrium?
- How do scientists predict shifts in the equilibrium position of a system?
- How do Brønsted–Lowry acids and bases illustrate equilibrium?

Unit 8

GENERAL OUTCOMES

In this unit, you will

- explain that there is a dynamic balance of opposing reactions in chemical systems at equilibrium
- determine quantitative relationships in acid–base ionization and other simple systems at equilibrium

Chemical Equilibrium Focusing on Acid–Base Systems

Unit 8
Chemical Equilibrium Focusing on Acid–Base Systems

ARE YOU READY?

These questions will help you find out what you already know, and what you need to review, before you continue with this unit.

Knowledge

1. A large number of chemical reactions (most notably redox reactions) will *only* occur in aqueous solution, at least at a rate great enough to be observable in a laboratory. In terms of collision–reaction theory (**Figure 1**), state
   (a) why a given reaction might only occur in solution
   (b) why collisions between entities do not necessarily result in reaction
   (c) the two effects that an increase in temperature has on collisions between entities involved in a reaction
   (d) the effect that an increase in concentration of one kind of entity has on collisions between all entities involved in a reaction

An Ineffective Collision

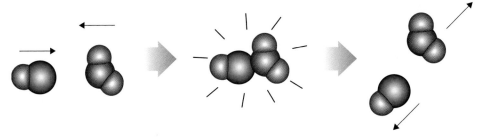

Figure 1
Collision–reaction theory considers the numbers, speeds, and orientation of colliding entities to explain the progress of chemical reactions.

2. Assume that each of the following substances is placed in water. Rewrite the formula and the physical state to indicate whether it is very soluble or only slightly soluble in water at SATP. For those substances predicted to produce ions in solution, write symbols for all aqueous ions present after the substance dissociates upon dissolving.
   (a) $MgSO_4 \cdot 7H_2O(s)$
   (b) $CH_3COCH_3(l)$
   (c) $CH_2CH_2(g)$
   (d) $CaCO_3(s)$
   (e) $PbCl_2(s)$
   (f) $FeCl_3(s)$
   (g) $C_3H_5(OH)_3(l)$

3. Each of the following substances is mixed with water to form an aqueous solution. Write an equation (using the modified Arrhenius theory) to explain the acidity or basicity of the solution in terms of hydronium or hydroxide ions. Your equations should, when necessary, represent the "ionizing" of the substance as a reaction with water.
   (a) $HCl(g)$
   (b) $CH_3COOH(l)$
   (c) $H_2SO_4(l)$
   (d) $HNO_3(l)$
   (e) $NaCl(s)$
   (f) $NH_3(g)$
   (g) $NaOH(s)$

Prerequisites

Concepts
- collision–reaction theory
- dissociation and ionization
- amount concentration
- ion concentration
- percent reaction
- stoichiometric calculation
- net ionic equations
- acids and bases
- indicators

Skills
- laboratory safety
- scientific problem solving

You can review prerequisite concepts and skills in the appendices and on the Nelson Web site.
   A Unit Pre-Test is also available online.

www.science.nelson.com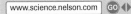

4. The concentration of chemical substances in solution can vary widely (**Figure 2**). Concentration affects solution properties such as colour, conductivity, freezing point, and viscosity. Concentration also affects the frequency of particle (entity) collisions; and, thus, will usually affect the observed rate of a reaction. Complete **Table 1**.

Table 1 Concentration of Entities and Quantities of Reagents in Solution

Reagent	Mass dissolved (g)	Solution volume (L)	Entity	Amount Concentration (mol/L)
$NaOH(s)$	1.74	0.500	$OH^-(aq)$	
$Al_2(SO_4)_3(s)$		2.00	$SO_4^{2-}(aq)$	0.100
$Al_2(SO_4)_3(s)$		2.00	$Al^{3+}(aq)$	0.100
$CaCl_2(s)$	1.00		$Cl^-(aq)$	0.00440

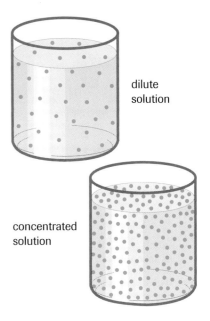

Figure 2
Concentration affects the rate of chemical reaction as well as physical properties.

5. Chemical substances may also have widely varying concentrations in the gaseous state. Calculate the amount concentration of
   (a) 24.0 g of hydrogen in a 2.00 L container
   (b) 500 kPa of oxygen in a 10.0 L container at 0 °C
   (c) 4.40 mol of carbon dioxide at SATP
   (d) 0.227 mol of methane at STP

6. Understanding chemical equilibrium theory often involves using a net ionic equation. For each of the following combinations of reagents, predict the product(s), and write a net ionic reaction equation. Balance the equation with simplest integer coefficients, and include physical states for all substances.
   (a) Copper(II) chloride and potassium carbonate solutions are mixed.
   (b) Ethene (ethylene) reacts with hydrogen chloride to form chloroethane.
   (c) Aluminium foil reacts with hydrochloric acid.
   (d) Ammonia undergoes simple decomposition.
   (e) Magnesium metal is placed in an aqueous solution of gold(III) chloride.
   (f) Mixing sulfurous acid solution with sodium hypochlorite solution results in a spontaneous redox reaction.

7. The acidity of solutions often has a considerable effect on the type, rate, and extent of the reactions they will undergo. Acids may be classed as strong or weak, depending on the extent of their reaction with water. Complete **Table 2**, identifying the acids as strong or weak.

Table 2 Solution Acidity and Basicity

Acid	Strength S/W	[Acid] (mol/L)	% Reaction in/with water	pH	$[H_3O^+(aq)]$ (mol/L)
$HCl(aq)$		0.016	>99		
$HBr(aq)$			>99		0.024
$HNO_3(aq)$			>99	4.0	
$CH_3COOH(aq)$		0.100	1.3		
$HCN(aq)$		0.200		5.0	
$HNO_2(aq)$		0.010	7.0		

chapter 15
Equilibrium Systems

In this chapter

- Exploration: Shakin' the Blues
- Investigation 15.1: The Extent of a Chemical Reaction
- Web Activity: Equilibrium State
- Mini Investigation: Modelling Dynamic Equilibrium
- Lab Exercise 15.A: The Synthesis of an Equilibrium Law
- Web Activity: Paul Kebarle
- Lab Exercise 15.B: Determining an Equilibrium Constant
- Web Activity: Writing Equilibrium Expressions
- Investigation 15.2: Equilibrium Shifts (Demonstration)
- Biology Connection: CO_2 Transport
- Investigation 15.3: Testing Le Châtelier's Principle
- Case Study: Urea Production in Alberta
- Lab Exercise 15.C: The Nitrogen Dioxide–Dinitrogen Tetraoxide Equilibrium
- Investigation 15.4: Studying a Chemical Equilibrium System
- Web Activity: Poison Afloat

Equilibrium in Chemical Systems

The simplest equilibrium systems are *static*: Nothing is moving or changing to create the balance. A textbook sitting on a level desktop is an example of *static equilibrium*. It stays motionless because two equal and opposite forces act on it simultaneously. The downward pull of Earth (gravity) on the book is exactly balanced by the upward push by the desktop. A laboratory balance is a common technology that uses this kind of equilibrium.

Chemical equilibrium is also a balance between two opposing agents of change, but always in a *dynamic* system. An expert juggler in performance (**Figure 1**) is similar to a chemical system at equilibrium. The juggler's act is a *dynamic equilibrium*, with some balls moving upward and some moving downward at any given moment. There is no net change because the rates of upward movement and downward movement are equal at any given moment.

Chemical systems at equilibrium have constant observable properties. Nothing appears to be happening because the internal movement involves entities that are too small to see. A critical task of chemical engineers is to disturb (unbalance) chemical equilibria in industrial reactions. Production of specific desired products is controlled by manipulating the conditions under which reactions occur. Some general concepts apply to all chemical equilibrium systems; these concepts are the focus of this chapter.

STARTING Points

Answer these questions as best you can with your current knowledge. Then, using the concepts and skills you have learned, you will revise your answers at the end of the chapter.

1. What, precisely, is happening to the chemical entities involved in a reaction while observation shows that products are being formed?
2. Is anything happening to the chemical entities involved in a reaction when observation shows the reaction appears to have stopped, with no more products being formed?
3. Why do some reactions seem to occur partially, and apparently stop while some of all of the reactants are still present, while in other reactions all of the limiting reagent appears to be consumed?
4. Can the chemical amount of product be predicted successfully for reactions that are not quantitative?

Career Connection:
Food Science Technologist; Chemical Process Engineer

Figure 1 How is a juggler similar to a chemical system in equilibrium?

▶ Exploration *Shakin' the Blues*

When small pieces of zinc are added to a dilute solution of excess hydrochloric acid in an open beaker (**Figure 2**), a vigorous reaction occurs with lots of gas and heat given off. The zinc continues to react and, when it is completely consumed, the visible signs of reaction come to an end. After seeing many reactions such as this one in your science studies, you may have come to think that all chemical reactions only go one way: from reactants to products. But do they always?

Materials: lab apron; eye protection; 400 mL flask and stopper; 250 mL water; 5.0 g potassium hydroxide (KOH(s)); 3.0 g glucose or dextrose; 2% methylene blue; stirring rod

- Pour 250 mL of water into the flask.
- Add 6 drops of methylene blue and all of the potassium hydroxide and glucose to the flask.
- Stir the mixture with the stirring rod until the solids have dissolved.
- Stopper the flask and set it on the bench. Observe the colour of the solution.
- Shake the solution vigorously and note any changes (**Figure 3**).
- Set the flask on the table and leave it standing until another change is noticed.
- Repeat the previous two steps many times. Make observations each time.

(a) Describe the reaction in the flask in relation to the discussion at the beginning of this activity.
(b) What evidence do you have to substantiate your answer to question (a)?
(c) Predict whether the colour changes will continue forever.
- Test your prediction over a reasonable period of time.
(d) Evaluate your prediction.

 Potassium hydroxide is poisonous and corrosive.

 Keep potassium hydroxide away from skin and eyes. Wear eye protection.

Figure 2 Zinc reacts rapidly and quantitatively with hydrochloric acid.

Figure 3 How many times does this reaction happen?

15.1 Explaining Equilibrium Systems

Scientists describe chemical systems in terms of empirical properties such as temperature, pressure, volume, and amounts of substances present. Chemical systems are simpler to study when separated from their surroundings by a definite boundary. This separation gives an experimenter control over the system so that no matter can enter or leave. Such a physical arrangement is called a **closed system**. A reaction in solution in a test tube or a beaker can be considered a closed system, as long as no gas is used or produced in the reaction. Systems involving gases must be closed on all sides by a solid container. Separating a chemical reaction system from its surroundings makes studying its properties, conditions, and changes much simpler. The use of controlled systems is an integral part of scientific study.

Figure 1
When the pressure on this equilibrium system changes, the equilibrium is disturbed.

Closed Systems at Equilibrium

One example of a chemical system at equilibrium is a soft drink in a closed bottle—a closed system in equilibrium. Nothing appears to change, until the bottle is opened. Removing the bottle cap and reducing the pressure alters the equilibrium state, as the carbon dioxide is allowed to leave the system (**Figure 1**). Carbonated drinks that have gone "flat" because of the decomposition of carbonic acid can be carbonated again by the addition of pressurized carbon dioxide to the solution to reverse the reaction, and then capping the container to restore the original equilibrium.

Collision–reaction theory is fundamental to the study of chemical systems. As originally introduced in this textbook to provide a basis for stoichiometric calculations, this theory required us to initially assume, for simplicity, that reactions are always spontaneous, rapid, quantitative, and stoichiometric. Common experience, however, shows that this assumption is not always true. Not all reactions are rapid; for example, corrosion of a car body may take years. A study of oxidation–reduction reactions soon provides evidence that many reactions are not spontaneous. This chapter will examine and test the assumption of quantitative reaction in detail to significantly increase your understanding of chemical systems.

INVESTIGATION 15.1 Introduction

The Extent of a Chemical Reaction

In Chapter 8, you performed experiments that produced evidence that reactions are quantitative. In a quantitative reaction, the limiting reagent is completely consumed. To identify the limiting reagent, you can test the final reaction mixture for the presence of the original reactants. For example, in a diagnostic test, you might try to precipitate ions from the final reaction mixture that were present in the original reactants.

Purpose
The purpose of this investigation is to test the validity of the assumption that chemical reactions are quantitative.

Report Checklist
- ○ Purpose
- ○ Problem
- ○ Hypothesis
- ● Prediction
- ○ Design
- ○ Materials
- ● Procedure
- ● Evidence
- ● Analysis
- ● Evaluation (1, 2, 3)

Problem
What are the limiting and excess reagents in the chemical reaction of selected quantities of aqueous sodium sulfate and aqueous calcium chloride?

Design
Samples of sodium sulfate solution and calcium chloride solution are mixed in different proportions and the final mixture is filtered. Samples of the filtrate are tested for the presence of excess reagents, using diagnostic tests.

To perform this investigation, turn to page 700.

Practice

1. The evidence gathered in Investigation 15.1 may be classified as an anomaly—an unexpected result that contradicts previous rules or experience.
   (a) Write the balanced equation for the double replacement reaction of sodium sulfate and calcium chloride solutions.
   (b) Write a statement describing the anomaly that occurred, using chemical names from the equation.
   (c) Write the net ionic equation for the reaction.
   (d) Use chemical names from the net ionic equation to write a statement about the anomaly.
   (e) Which of the previous statements more accurately describes the chemical system, according to collision–reaction theory?

2. When scientists first encounter an apparent anomaly, they carefully evaluate the design, procedure, and technological skills involved in an investigation. One important consideration is the reproducibility of the evidence. Compare your evidence in Investigation 15.1 with the evidence collected by other groups. Is there support for the reproducibility of this evidence?

DID YOU KNOW ?

Anomalies—Signals for Change

Anomalies, or discrepant events, are important as scientists acquire and develop scientific knowledge. Sometimes these events have been ignored, discredited, or elaborately explained away by scientists who do not wish to question or reconsider accepted laws and theories. Investigating anomalies sometimes leads to the restriction, revision, or replacement of scientific laws and theories.

Evidence obtained from many reactions contradicts the assumption that reactions are always quantitative. In Investigation 15.1, there is direct evidence for the presence of both reactants after the reaction appears to have stopped. This apparent anomaly can be explained, in terms of collision–reaction theory, by the idea that a reverse reaction can occur: the products, calcium sulfate and sodium chloride, can react to re-form the original reactants. The final state of this chemical system can be explained as a competition between collisions of reactants to form products and collisions of products to re-form reactants.

$$Na_2SO_4(aq) + CaCl_2(aq) \underset{\text{reverse}}{\overset{\text{forward}}{\rightleftarrows}} CaSO_4(s) + 2\,NaCl(aq)$$

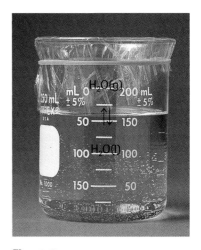

Figure 2
According to the theory of dynamic equilibrium, as long as the container remains closed at constant temperature, the rate at which molecules in the liquid state evaporate is equal to the rate at which molecules in the gas state condense.

$$H_2O(l) \rightleftarrows H_2O(g)$$

This competition requires that the system be closed so that reactants and products cannot escape from the reaction container. The chemical system in Investigation 15.1 can be considered a closed system, bounded by the volume of the liquid phase.

We assume that any closed chemical system with constant macroscopic properties (no observable change occurring) is in a state of **equilibrium**, usually classified, for convenience, as one of three types. **Phase equilibrium** involves a single chemical substance existing in more than one phase in a closed system. Water placed in a sealed container evaporates until the water vapour pressure (concentration of water in the gas phase) rises to a maximum value, and then remains constant (**Figure 2**). **Solubility equilibrium** involves a single chemical solute interacting with a solvent substance, where excess solute is in contact with the saturated solution (**Figure 3**). A **chemical reaction equilibrium** involves several substances: the reactants and products of a chemical reaction. All three types of equilibrium are explained by a theory of **dynamic equilibrium**—a balance between two opposite processes occurring at the same rate.

The terms *forward* and *reverse* are used to identify which process is being referred to, and are specific to a written equilibrium equation. When any equation is written with arrows to show that the change occurs both ways, the left-to-right change is called the **forward reaction**, and the right-to-left change is called the **reverse reaction**.

Figure 3
For excess solid copper(II) sulfate in equilibrium with its saturated aqueous solution, the rates of dissolving and crystallization are equal.

$$CuSO_4(s) \rightleftarrows Cu^{2+}(aq) + SO_4^{2-}(aq)$$

Simulation—Equilibrium State

This simulation illustrates the establishment of a simple dynamic equilibrium by showing how the concentrations of entities change over time, beginning from an initial condition. A textbook can only represent this process as a sequence of static (still) diagrams (such as those in **Figure 4**) or graphs (such as that in **Figure 5**).

www.science.nelson.com

mini Investigation — Modelling Dynamic Equilibrium

This activity models the progress of a chemical reaction to equilibrium, representing concentrations of reactants and products as volumes of water. The establishment of equilibrium is graphed as volume versus number of transfers. This graph then represents a typical concentration–time graph of a chemical reaction to equilibrium.

Materials: two drinking straws of different diameters; two 25 mL graduated cylinders; graph paper; water; meniscus finder

- Label one 25 mL graduated cylinder **R** (for reactants), and fill with water to the 25.0 mL mark.
- Label the other cylinder **P** (for products) and leave it empty.
- Holding a straw in each hand, place one straw in each graduated cylinder so that each straw rests on the bottom of its cylinder. Use the larger straw in the **R** cylinder.
- Transfer water simultaneously from each cylinder to the other by placing an index finger over the open end of each straw (to seal it). Then lift the straws out of their original cylinders, move the bottom of each straw over the *other* cylinder, and lift your index fingers to allow the water in each straw to drain into the cylinder below. Replace the straws in their original cylinders.
- After each transfer, measure and record the volume of water in each cylinder to 0.1 mL. Also record the number of the transfer.
- Repeat the transfer step until no significant change in water volumes has occurred for at least three transfers.
- Graph the volume of water in each cylinder as a dependent (responding) variable against the number of transfers as the independent (manipulated) variable.
- Repeat the activity, switching the straws used in each cylinder. Plot the values on the same axes as for the first trial.

(a) How does the graph change, and how is it the same, when the smaller straw is initially in the **R** cylinder?

(b) How would doing the transfers more slowly affect the final volumes in each cylinder?

(c) How would replacing the larger straw with an even larger one affect the final volumes in each cylinder?

(d) Express the equilibria of the water transfer trials as percent yields; that is, the percentage of "reactant" water that is converted to "product".

(e) Predict the graph's shape if both of the straws chosen had the same diameter.

Chemical Reaction Equilibrium

Chemical reaction equilibria are more complex than phase or solubility equilibria, due to the variety of possible chemical reactions and the greater number of substances involved. To explain chemical equilibrium systems, we need to combine ideas from atomic theory, kinetic molecular theory, collision–reaction theory, and the concepts of reversibility and dynamic equilibrium. Although this synthesis is successful as a description, and also as an explanation, it has only limited application in predicting quantitative properties of an equilibrium system.

The Hydrogen–Iodine Reaction System

Chemists have studied the reaction of hydrogen gas and iodine gas extensively, because the molecules are simple in structure and the reaction takes place in the gas phase. Once hydrogen and iodine are mixed, the reaction proceeds rapidly at first. The initial dark purple colour of the iodine vapour gradually fades, and then remains constant (**Figure 4**).

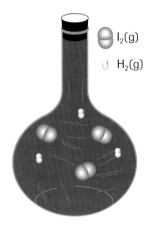 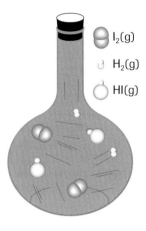 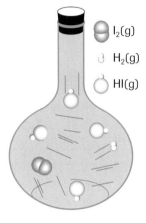

Figure 4
When hydrogen and iodine are added to the flask, the colour of the iodine vapour is the only easily observable (empirical) property. An equilibrium equation describes this evidence theoretically.
$H_2(g) + I_2(g) \rightleftarrows 2\,HI(g)$, $t = 448\ °C$

Initially, hydrogen (in excess) and iodine are added to the flask. The colour of the iodine vapour is the only easily observable property.

Early in the reaction, hydrogen and iodine form hydrogen iodide faster than hydrogen iodide forms hydrogen and iodine. Overall, the amount of iodine decreases, so the colour of the flask contents appears to lighten. Both hydrogen and hydrogen iodide are colourless.

At equilibrium, analysis shows that the flask contains all three substances. The purple colour shows that some iodine remains. The constancy of the colour is evidence that equilibrium exists. Forward and reverse reactions are occurring at equal rates.

Table 1 contains data from three experiments with the hydrogen–iodine system: one in which hydrogen and iodine are mixed; one in which hydrogen, iodine, and hydrogen iodide are mixed; and one in which only hydrogen iodide is present initially. At a temperature of 448 °C, the system quickly reaches an observable equilibrium each time. Chemists use evidence such as that in Table 1 to describe a state of equilibrium in two ways: in terms of percent reaction and in terms of an equilibrium constant. Percent reaction describes the equilibrium for one specific system example only, whereas an equilibrium constant describes all systems of the same reaction at a given temperature. Alternatively, you can draw a graph of the reaction progress, plotting quantity (or concentration) of the reagents versus time (**Figure 5**).

Table 1 The Hydrogen–Iodine System at 448 °C

System	Initial system concentrations (mmol/L)			Equilibrium system concentrations (mmol/L)		
	$H_2(g)$	$I_2(g)$	$HI(g)$	$H_2(g)$	$I_2(g)$	$HI(g)$
1	5.00	5.00	0	1.10	1.10	7.80
2	0.50	0.50	1.70	0.30	0.30	2.10
3	0	0	3.20	0.35	0.35	2.5

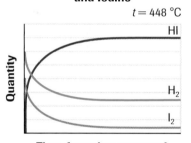

Figure 5
The graph of the quantity of each substance against time shows that the rate of reaction of the reactants decreases as the number of reactant molecules decreases, and the rate at which the product changes back to reactants increases as the number of product molecules increases. These two rates must become equal at some point, after which the quantity of each substance present will not change.

DID YOU KNOW?

Lavoisier and Closed Systems
Antoine Laurent Lavoisier (1743–1794) is recognized as the father of modern chemistry for many reasons. He created the basic nomenclature system we still use for compounds, demonstrated the law of conservation of mass, and explained and clarified the theory of combustion. Most importantly, perhaps, his successes convinced the chemical community of the critical importance of his methods of careful measurement, and of carrying out experiments in closed systems—carefully accounting for, and preventing, invisible reactants and products (notably, gases) from escaping. When chemists began to understand the importance of maintaining closed systems in order to draw correct conclusions about reactions, chemistry could finally move forward as a true science.

Learning Tip

When a reaction is shown to be quantitative (as written in the equation), it means that the reverse reaction happens so little that it can be ignored for all normal purposes. Another way to think of a quantitative reaction is that if the products shown (as written) were mixed together as reactants, there would be no apparent reaction. A reaction that is quantitative in the forward direction is necessarily nonspontaneous in the reverse direction. In this unit, we will (arbitrarily) assume that "quantitative" specifies a reaction that, at equilibrium, is more than 99.9% complete. Another way to think of it is that less than one part per thousand (0.1%) of an original reactant remains unreacted, at equilibrium, in a quantitative reaction.

A percent yield is defined as the yield of product measured at equilibrium compared with the maximum possible yield of product. In other words, percent yield can be useful for communicating the *position of an equilibrium*. The maximum possible yield of product is calculated using the method of stoichiometry, assuming a quantitative forward reaction with no reverse reaction. Percent yield provides an easily understood way to refer to quantities of chemicals present in equilibrium systems. For example, analysis of the evidence in System 1, Table 1, shows that, at 448 °C, this particular hydrogen–iodine system reaches an equilibrium with a percent yield of 78.0% (**Table 2**).

Table 2 Percent Yield of the Hydrogen–Iodine System at 448 °C

System	Equilibrium [HI]* (mmol/L)	Maximum possible [HI]* (mmol/L)	Percent yield (%)
1	7.80	10.0	78.0
2	2.10	2.70	77.8
3	2.50	3.20	78.1

*Square brackets [ ] indicate amount concentration.

Equilibrium arrows ($\rightleftarrows$) communicate that an equilibrium exists. To communicate the extent of a reaction, a percent yield may be written above the equilibrium arrows in a chemical equation. The following equation describes the position of a hydrogen–iodine equilibrium in System 1, Table 2, at 448 °C.

$$H_2(g) + I_2(g) \; \overset{78\%}{\rightleftarrows} \; 2\,HI(g) \qquad t = 448 \, °C$$

Scientists now think of all chemical reactions as occurring in both forward and reverse directions. Any reaction falls loosely into one of four categories. Reactions that favour reactants very strongly, that is, reactions that normally have a percent yield of much less than 1%, are simply observed as being nonspontaneous. In these reactions, mixing reactants has no observable result. Reactions producing observable equilibrium conditions may react less or more than 50%, favouring reactants or products respectively. Significant amounts of both reactants and products are always present. Finally, reactions that favour products very strongly, much more than 99%, are observed to be complete (*quantitative*). The chemical equations for quantitative reactions are generally written with a single arrow to indicate that the effect of the reverse reaction is negligible. **Table 3** shows how percent yield may be used to classify equilibrium systems and how the classification may be communicated in reaction equations.

Table 3 Classes of Chemical Reaction Equilibria

Percent yield	Description of equilibrium	Position of equilibrium
negligible	nonspontaneous (no apparent reaction)	
< 50%	reactants favoured	< 50% $\rightleftarrows$
> 50%	products favoured	> 50% $\rightleftarrows$
> 99.9%	quantitative	$\rightarrow$

When considering equilibrium systems, we cannot use the simple assumption of quantitative reaction. When there is no limiting reagent for a reaction, and when we cannot assume complete reaction, stoichiometric calculations require a little more thought. Such calculations may conveniently be set up as an **ICE table**, meaning that the *initial*, *change*, and *equilibrium* values are arranged in tabular form.

▶ SAMPLE problem 15.1

Consider the reaction equation for the formation of hydrogen iodide at 448 °C. Assume the reaction is begun with 1.00 mmol/L concentrations of both $H_2(g)$ and $I_2(g)$. Construct an ICE table to determine equilibrium concentrations of the reagents. The equilibrium concentration of $I_2(g)$ (determined by colour intensity) is 0.22 mmol/L.

Set up the ICE table as follows:

Table 4 The $H_2(g) + I_2(g) \rightleftarrows 2HI(g)$ Equilibrium

Concentration	[H_2(g)] (mmol/L)	[I_2(g)] (mmol/L)	[HI(g)] (mmol/L)
Initial	1.00	1.00	0.00
Change			
Equilibrium		0.22	

Begin by calculating the change (*decrease*) in concentration of iodine.

(0.22 − 1.00) mmol/L = −0.78 mmol/L

The changes of the other concentrations may be calculated directly from this value, using stoichiometric ratios from the balanced equation.

For hydrogen, the change is also a *decrease* (negative value) of

0.78 mmol/L × $\frac{1}{1}$ = −0.78 mmol/L

For hydrogen iodide, the change is an *increase* (positive value) of

0.78 mmol/L × $\frac{2}{1}$ = +1.6 mmol/L

Complete the ICE table, using these values to enter the concentrations at equilibrium.

Table 5 The $H_2(g) + I_2(g) \rightleftarrows 2HI(g)$ Equilibrium

Concentration	[H_2(g)] (mmol/L)	[I_2(g)] (mmol/L)	[HI(g)] (mmol/L)
Initial	1.00	1.00	0.00
Change	−0.78	−0.78	+1.6
Equilibrium	0.22	0.22	1.6

In Sample Problem 15.1, notice that *every* substance in the reaction is a gas. Therefore, all of the stoichiometric calculations can use concentrations directly, rather than chemical amounts, because the volume must be the same for every gaseous substance in a closed container. The volume is a common factor in the calculation step that uses the stoichiometric ratio. This same reasoning means that concentrations can also be used directly for stoichiometric calculation whenever every substance in a reaction is an aqueous entity dissolved in the same volume of solvent.

DID YOU KNOW ?

Diabetes: Blood Sugar Equilibrium
For people with diabetes, reaction equilibrium established by sugar in the human body is critically important. Recent advances in the technology allow testing of blood sugar concentration with personal devices such as the One Touch® SureStep® blood-glucose meter (**Figure 6**). The meter displays the concentration in mmol/L after analyzing a single drop of blood extracted from a fingertip. Multiple readings, including date and time, are stored electronically and can be displayed at any time. The data can be downloaded to a computer.

Figure 6
This device helps diabetics monitor their blood glucose levels.

CAREER CONNECTION

Food Science Technologist
The development and analysis of food products for individuals with diabetes is crucial. Food science technologists carefully measure and conduct tests on carbohydrates so that patients can control their blood sugar equilibrium.

Learn more about the many food industries that employ food science specialists.

www.science.nelson.com

Practice

3. For a chemical system at equilibrium:
   (a) What are the observable characteristics?
   (b) Why is the equilibrium considered "dynamic"?
   (c) What is considered "equal" about the system?

4. In a gaseous reaction system, 2.00 mol of methane, $CH_4(g)$, is initially added to 10.00 mol of chlorine, $Cl_2(g)$. At equilibrium the system contains 1.40 mol of chloromethane, $CH_3Cl(g)$, and some hydrogen chloride, $HCl(g)$.
   (a) Write a balanced reaction equation for this equilibrium and calculate the maximum possible yield of chloromethane product.
   (b) Calculate the percent yield at this equilibrium and state whether products or reactants are favoured.

5. Combustion reactions, such as the burning of methane, often favour products so strongly that they are written with a single arrow. Assuming the forward reaction has a very low activation energy (Chapter 12) and the reverse reaction has a very high activation energy (to account for the difference in the tendency to occur), sketch a possible potential energy diagram representing the progress of such a reaction.

6. After 4.0 mol of $C_2H_4(g)$ and 2.50 mol of $Br_2(g)$ are placed in a sealed container, the reaction

   $$C_2H_4(g) + Br_2(g) \rightleftarrows C_2H_4Br_2(g)$$

   establishes an equilibrium. **Figure 7** shows the concentration of $C_2H_4(g)$ as it changes over time at a fixed high temperature until equilibrium is reached.

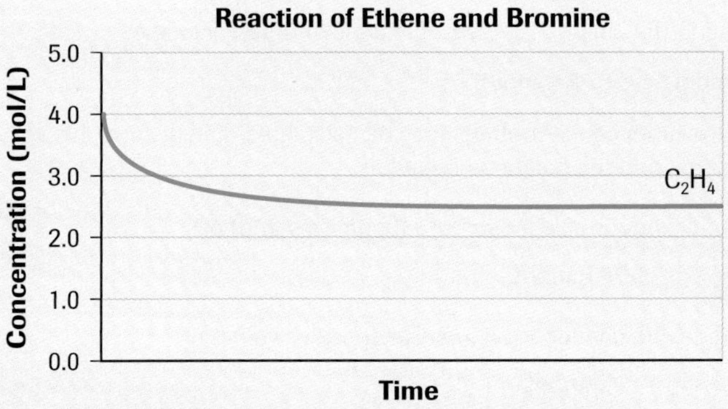

Figure 7
A graph of the reaction of ethene with bromine

   (a) Sketch this graph. Draw lines on your copy to show how the concentration of each of the other two substances changes.
   (b) Create an ICE table, using reagent amount concentrations.
   (c) What is the volume of the container?
   (d) Calculate the percent yield of dibromoethane.

7. Write a balanced net ionic equation for each of the following described reactions, showing appropriate use of equilibrium "arrow" symbols where appropriate, and indicating (with symbols) whether products or reactants are favoured.
   (a) An excess of solid copper reacts with virtually all of the silver ions in a sample solution.
   (b) When a solution containing calcium ions is mixed with a solution containing a large excess of sulfate ions, a precipitate forms, but tests indicate that a small quantity of calcium ions remains in solution.
   (c) When acetic acid is dissolved in water, the acetic acid molecules react with water molecules to form hydronium and acetate ions. Careful pH testing shows that about 980 of every 1000 acetic acid molecules remain in their molecular form, at equilibrium.

Learning Tip

Chemists often refer to "homogeneous" reaction systems. This term means that every entity involved in the reaction exists in the same physical state. The reaction from Sample Problem 15.1, where all reactants and products are gases, is a typical case. The other common case involves reactions where all entities involved are in aqueous solution. A system that has more than one phase, such as solid copper reacting in aqueous silver nitrate solution, is a "heterogeneous" reaction system.

Section 15.1

LAB EXERCISE 15.A

The Synthesis of an Equilibrium Law

Report Checklist
- ○ Purpose
- ○ Problem
- ○ Hypothesis
- ○ Prediction
- ○ Design
- ○ Materials
- ○ Procedure
- ○ Evidence
- ● Analysis
- ○ Evaluation

The following chemical equation represents a chemical equilibrium:

$$Fe^{3+}(aq) + SCN^-(aq) \rightleftharpoons FeSCN^{2+}(aq)$$

This equilibrium is convenient to study because the colour of the system characterizes the equilibrium position of the system (**Figure 8**).

Purpose
The purpose of this investigation is the synthesis of an equilibrium law.

Problem
What mathematical formula, using equilibrium concentrations of reactants and products, gives a constant for the iron(III)–thiocyanate reaction system?

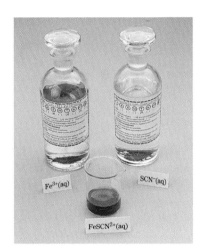

Figure 8
The two reactants combine to form a dark red equilibrium mixture. The red colour of the solution is due to the aqueous thiocyanate–iron(III) product, $FeSCN^{2+}(aq)$.

Design
Reactions are performed using various initial concentrations of iron(III) nitrate and potassium thiocyanate solutions. The equilibrium concentrations of the reactants and the product are determined from the measurement and analysis of the colour intensity using a spectrophotometer. Possible mathematical relationships among the concentrations are tried and analyzed to determine if the mathematical formula gives a constant value.

Evidence

Table 6 Iron(III)–Thiocyanate Equilibrium at SATP

Trial	$[Fe^{3+}(aq)]$ (mol/L)	$[SCN^-(aq)]$ (mol/L)	$[FeSCN^{2+}(aq)]$ (mol/L)
1	3.91×10^{-2}	8.02×10^{-5}	9.22×10^{-4}
2	1.48×10^{-2}	1.91×10^{-4}	8.28×10^{-4}
3	6.27×10^{-3}	3.65×10^{-4}	6.58×10^{-4}
4	2.14×10^{-3}	5.41×10^{-4}	3.55×10^{-4}
5	1.78×10^{-3}	6.13×10^{-4}	3.23×10^{-4}

Analysis
Test the following mathematical relationships for constancy:

1. $[Fe^{3+}(aq)][SCN^-(aq)][FeSCN^{2+}(aq)]$
2. $[Fe^{3+}(aq)] + [SCN^-(aq)] + [FeSCN^{2+}(aq)]$
3. $\dfrac{[FeSCN^{2+}(aq)]}{[Fe^{3+}(aq)][SCN^-(aq)]}$
4. $\dfrac{[Fe^{3+}(aq)]}{[FeSCN^{2+}(aq)]}$
5. $\dfrac{[SCN^-(aq)]}{[FeSCN^{2+}(aq)]}$

 WEB Activity

Canadian Achievers—Paul Kebarle

Paul Kebarle (**Figure 9**) pioneered the measurements of gas-phase ion-molecule equilibria. Kebarle's findings, now significantly expanded by other workers, constitute a central database that is of fundamental importance in many diverse fields of scientific research.

1. What fundamental data did Kebarle and his co-workers obtain from their research?
2. List three fields of research that Kebarle's work has aided.

 www.science.nelson.com

Figure 9
Paul Kebarle

DID YOU KNOW ?

Computers
Scientists often use computers to analyze numerical evidence in order to establish mathematical relationships among experimental variables. The mathematical formulas derived are useful in understanding chemical processes and in applying these processes to technology.

DID YOU KNOW ?

Related Interests
Two Norwegian chemists, Cato Maximilian Guldberg and Peter Waage, conducted detailed empirical studies of many equilibrium systems in the mid-1800s (**Figure 10**). By 1864, they had proposed a mathematical description of the equilibrium condition that they called the "law of mass action." Analyzing the results of their experiments, Guldberg and Waage noticed that, when they arranged the equilibrium concentrations into a specific form of ratio, the resulting value was the same no matter what combinations of initial concentrations were mixed.

Figure 10
Cato Maximilian Guldberg (1836–1902) and Peter Waage (1833–1900) were related by more than their interest in chemistry: They were brothers-in-law!

Learning Tip

It is common practice (convention) to ignore units and list only the numerical value for an (amount concentration) equilibrium constant. The expression of units is often very complex for K_c relationships. But, because each entity concentration is always entered with mol/L units, any entity concentration we calculate from a K_c value will always give an answer having mol/L units, so you need only memorize this (simplifying) rule.

The Equilibrium Constant, K_c

Analysis of the evidence from many experiments such as those in Lab Exercise 15.A (page 683) reveals a mathematical relationship that provides a constant value for a chemical system over a range of amount concentrations. This constant value is called the **equilibrium constant**, K_c, for the reaction system. Evidence and analysis of many equilibrium systems have resulted in the following **equilibrium law**.

For the reaction $a\,A + b\,B \rightleftarrows c\,C + d\,D$,

the equilibrium law expression is $K_c = \dfrac{[C]^c[D]^d}{[A]^a[B]^b}$

In this mathematical expression, A, B, C, and D represent chemical entity formulas and a, b, c, and d represent their coefficients in the balanced chemical equation. The relationship holds *only* when amount concentrations are observed to remain constant, in a closed system, at a given temperature.

▸ COMMUNICATION example 1

Write the equilibrium law expression for the reaction of nitrogen monoxide gas with oxygen gas to form nitrogen dioxide gas.

Solution

$$2\,NO(g) + O_2(g) \rightleftarrows 2\,NO_2(g)$$

$$K_c = \frac{[NO_2(g)]^2}{[NO(g)]^2[O_2(g)]}$$

We use a balanced chemical equation with whole-number coefficients to write the mathematical expression of the equilibrium law. The coefficients of the balanced equation become the exponents of the amount concentrations. If the equation were to be written in reverse, the equilibrium law expression would simply be the reciprocal of the expression above, and the equilibrium constant would be the reciprocal of the one for the reaction as written here. Using the *products over reactants* convention results in a *relationship* between the numerical value of K_c and the forward extent of the equilibrium that is easier to visualize for the equation as written. The higher the numerical value of the equilibrium constant, the greater the tendency of the system to favour the forward direction; that is, the greater the equilibrium constant, the more the products are favoured at equilibrium.

▸ COMMUNICATION example 2

The value of K_c for the formation of HI(g) from H_2(g) and I_2(g) is 40, at temperature t. Determine the value of K_c for the decomposition of HI(g) at the same temperature.

Solution

$$2\,HI(g) \rightleftarrows H_2(g) + I_2(g)$$

$$K_c = \frac{[H_2(g)][I_2(g)]}{[HI(g)]^2} = \frac{1}{40} = 0.025$$

Note that the decomposition reaction equation is the reverse of the formation reaction equation, and the value of K_c for decomposition is the reciprocal of the K_c for formation.

Experiments have shown that the value of the equilibrium constant depends on temperature. The value is also affected by very large changes in the equilibrium concentration of a reactant or a product. A moderate change in the concentration of any one of the reactants or products results in a change in the other concentrations, so that the equilibrium constant remains the same. The equilibrium constant provides only a measure of the equilibrium position of the reaction; it does not provide any information on the rate of the reaction. Because they hold for a significant range of different concentrations, equilibrium constant expressions have been found to be very useful, and K_c values for reactions are in common use throughout the scientific community.

Equilibrium constants are adjusted to reflect the fact that pure substances in solid or liquid (condensed) states have concentrations that are essentially fixed—the chemical amount (number of moles) per unit volume is a *constant* value. For example, a litre of liquid water at SATP has a mass of 1.00 kg (a chemical amount of 55.5 mol) and, thus, a fixed amount concentration of 55.5 mol/L. The concentration of condensed states is not included in a K_c expression—we assume that these constant values become part of the expressed equilibrium constant. Substances in a gaseous or dissolved state have *variable* concentrations, and must *always* be shown in an equilibrium law expression.

▶ COMMUNICATION example 3

Write the equilibrium law expression for the decomposition of solid ammonium chloride to gaseous ammonia and gaseous hydrogen chloride.

Solution

$$NH_4Cl(s) \rightleftarrows NH_3(g) + HCl(g)$$

$$K_c = [NH_3(g)][HCl(g)]$$

The concentration of solid $NH_4Cl(s)$ is omitted from the equilibrium law expression.

The role of temperature in equilibrium constant expressions is critical, although the temperature is not written in the expression directly. The value of the equilibrium constant, K_c, always depends on the temperature. Any stated numerical value for an equilibrium constant, or any calculation using an equilibrium constant expression, *must* specify the reaction temperature at equilibrium.

Since equilibrium depends on the concentrations of reacting substances, these substances must be represented in the expression as they actually exist—meaning that ions in solution must be represented as individual entities. Equilibrium constant expressions are always written from the *net ionic* form of reaction equations, balanced with simplest whole-number (integral) coefficient values unless otherwise specified.

▶ COMMUNICATION example 4

Write the equilibrium law expression for the reaction of zinc in copper(II) chloride solution.

Solution

$$Zn(s) + Cu^{2+}(aq) \rightleftarrows Cu(s) + Zn^{2+}(aq)$$

$$K_c = \frac{[Zn^{2+}(aq)]}{[Cu^{2+}(aq)]}$$

Again, note that the solids, as well as the spectator ions (the chloride ions in this example), are omitted from the equilibrium law expression.

DID YOU KNOW ?

Using Constant Relationships

You are already familiar with the usefulness of some other constant mathematical relationships about real phenomena. Finding a relationship that is constant for equilibrium concentrations is just another example.

If you examine circles of different sizes carefully, you discover that the distance around any circle divided by the distance across it (at the widest part) *always* gives the same (constant) answer, no matter how big or small the circle. This constant value, 3.14159..., is so useful that it has been given its own symbol, the Greek letter pi, π. This relationship is most usefully expressed as $C = \pi d$ because measuring a diameter is much easier than measuring a circumference.

Many other relationship expressions produce this same constant, including those usually used to calculate the area of a circle and the time period of a pendulum's swing.

➕ EXTENSION

The Meaning of the Equilibrium Constant

Try this simulation to deepen your understanding of the equilibrium constant.

www.science.nelson.com

LAB EXERCISE 15.B

Determining an Equilibrium Constant

Report Checklist
- ○ Purpose
- ○ Problem
- ○ Hypothesis
- ○ Prediction
- ○ Design
- ○ Materials
- ○ Procedure
- ○ Evidence
- ● Analysis
- ○ Evaluation

Determining the equilibrium constant for a reaction at specific conditions is often essential for an industrial chemist. This knowledge is necessary before adjusting the reaction conditions in order to optimize production of the desired substances. Complete the Analysis of the investigation report.

Purpose
The purpose of this investigation is to use the equilibrium law to determine the equilibrium constant at 200 °C for the decomposition reaction of the molecular compound phosphorus pentachloride.

Problem
What is the value of the equilibrium constant for the decomposition of phosphorus pentachloride gas to phosphorus trichloride gas and chlorine gas, at a temperature of 200 °C?

Evidence
equilibrium temperature = 200 °C
equilibrium concentrations:
$[PCl_3(g)] = [Cl_2(g)] = 0.014$ mol/L
$[PCl_5(g)] = 4.3 \times 10^{-4}$ mol/L

SUMMARY: Writing Equilibrium Law Expressions

Write an equilibrium law expression based on a balanced equation for the reaction system. Use single whole-number coefficients, written in net ionic form, and ignore concentrations of pure solid or liquid phases:

If: $aA + bB \rightleftarrows cC + dD$

then: $K_c = \dfrac{[C]^c[D]^d}{[A]^a[B]^b}$

An equilibrium constant value
- always depends on the system temperature
- is independent of the reagent concentrations
- is independent of any catalyst present
- is independent of the time taken to reach equilibrium
- is normally stated as a numerical value, ignoring any units
- is greater, the more the system favours the formation of products

Predicting Final Equilibrium Concentrations

For simple homogeneous systems, it is possible to algebraically predict reagent concentrations at equilibrium using a known value for K_c and initial reactant concentration values. For more complex systems, the calculation becomes more difficult—such systems are left for more advanced chemistry courses.

▶ SAMPLE problem 15.2

In a 500 mL stainless steel reaction vessel at 900 °C, carbon monoxide and water vapour react to produce carbon dioxide and hydrogen. Evidence indicates that this reaction establishes an equilibrium with only partial conversion of reactants to products. Initially, 2.00 mol of each reactant is placed in the vessel. K_c for this reaction is 4.20 at 900 °C. What amount concentration of each substance will be present at equilibrium?

Write a balanced equation for the reaction equilibrium.

$CO(g) + H_2O(g) \rightleftarrows CO_2(g) + H_2(g)$ $K_c = 4.20$ at 900 °C

Use the balanced equation to write the equilibrium law expression.

$$K_c = 4.20 = \frac{[CO_2(g)][H_2(g)]}{[CO(g)][H_2O(g)]}$$

The *initial* amount concentrations of the CO(g) and the H_2O(g) are the same:

$$c = \frac{2.00 \text{ mol}}{0.500 \text{ L}} = 4.00 \text{ mol/L} = [CO(g)] = [H_2O(g)]$$

An ICE table makes it easier to keep track of amount concentration changes that occur during a reaction, and to find the amount concentrations at equilibrium.

At equilibrium, let the final amount concentration of the product H_2(g) be x mol/L (any convenient symbol could be used). Then $[CO_2(g)]$ must also be x mol/L, since the stoichiometric ratio of the reaction is 1:1:1:1. By this same reasoning, at equilibrium, the initial concentrations of CO(g) and H_2O(g) must have *decreased* by x mol/L; so $[CO(g)] = [H_2O(g)] = (4.00 - x)$ mol/L. When you enter values for "Change" into the ICE table, you must show increases as "+", and decreases as "−".

Table 7 The CO(g) + H_2O(g) ⇌ CO_2(g) + H_2(g) Equilibrium

Concentration	[CO(g)] (mol/L)	[H_2O(g)] (mol/L)	[CO_2(g)] (mol/L)	[H_2(g)] (mol/L)
Initial	4.00	4.00	0	0
Change	−x	−x	+x	+x
Equilibrium	(4.00 − x)	(4.00 − x)	x	x

Substitute equilibrium concentrations in the equilibrium law expression.

$$4.20 = \frac{[CO_2(g)][H_2(g)]}{[CO(g)][H_2O(g)]} = \frac{x^2}{(4.00 - x)^2}$$

Since the right side of the equation is a perfect square, solving for x is quite straightforward.

$$\sqrt{4.20} = \sqrt{\frac{x^2}{(4.00 - x)^2}}$$

$$2.05 = \frac{x}{4.00 - x}$$

$$x = (2.05)(4.00 - x)$$

$$= 8.20 - 2.05x$$

$$3.05x = 8.20$$

$$x = 2.69$$

Assign positive and negative signs and complete the ICE table.

Table 8 The CO(g) + H_2O(g) ⇌ CO_2(g) + H_2(g) Equilibrium

Concentration	[CO(g)] (mol/L)	[H_2O(g)] (mol/L)	[CO_2(g)] (mol/L)	[H_2(g)] (mol/L)
Initial	4.00	4.00	0	0
Change	−2.69	−2.69	+2.69	+2.69
Equilibrium	1.31	1.31	2.69	2.69

At equilibrium, at 900 °C, $[CO(g)] = [H_2O(g)] = 1.31$ mol/L and
$[CO_2(g)] = [H_2(g)] = 2.69$ mol/L

+ **EXTENSION**

Equilibrium Constant and Reaction Quotient
Is there any way of knowing whether or not a reaction is at equilibrium? If we know the concentrations of reactants and products, and the equilibrium constant at the appropriate temperature, we can use a concept called the "reaction quotient" to predict which way the reaction will proceed, or if it is already at equilibrium.

www.science.nelson.com

Note that, if both initial reactant concentrations are not the same, solving the equation for x is more complicated, requiring use of the quadratic formula. Questions in this text are restricted to examples that do not require the quadratic formula for solution.

 **WEB Activity**

Simulation—Writing Equilibrium Expressions

This simulation allows you to select a reaction type and the initial reactant concentrations, which the program uses to plot the resulting equilibrium graph. You will then be guided through a series of questions.

www.science.nelson.com

Section 15.1 Questions

1. Write a balanced equation with integer coefficients and the expression of the equilibrium law for each of the following reaction systems at fixed temperature.
   (a) Hydrogen gas reacts with chlorine gas to produce hydrogen chloride gas in the industrial process that eventually produces hydrochloric acid.
   (b) In the Haber process (Chapter 8), nitrogen reacts with hydrogen to produce ammonia gas.
   (c) At some time in the future, industry and consumers may make more extensive use of the combustion of hydrogen as an energy source.
   (d) When aqueous ammonia is added to an aqueous nickel(II) ion solution, the $Ni(NH_3)_6^{2+}(aq)$ complex ion is formed (**Figure 11**).

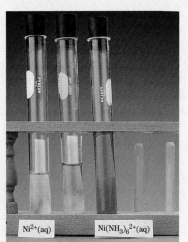

Figure 11
A $Ni^{2+}(aq)$ solution is green. Ammonia reacts with the nickel(II) ion to form the intensely blue hexaamminenickel(II) ion, $Ni(NH_3)_6^{2+}(aq)$.

   (e) In the Solvay process for making washing soda (Chapter 7), one reaction involves heating solid calcium carbonate (limestone) to produce solid calcium oxide (quicklime) and carbon dioxide.
   (f) In Investigation 15.1, aqueous solutions of sodium sulfate and calcium chloride are mixed. (Remember to use a net ionic equation.)
   (g) In a sealed can of soda, carbonic acid, $H_2CO_3(aq)$, decomposes to liquid water and carbon dioxide gas.

2. You can apply the empirical and theoretical concepts of equilibrium to many different chemical reaction systems. Use the generalizations from your study of organic chemistry to predict the position of equilibrium for bromine placed in a reaction container with ethylene at a high temperature.

3. Interpret the graph in **Figure 12** to answer the questions about the reaction. Hydrogen and iodine were placed in a reaction vessel, which was then sealed, and heated to 450 °C.

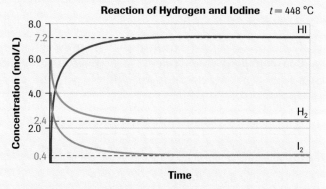

Figure 12
The progress of a hydrogen–iodine reaction

   (a) All three substances are gases. If the container has a volume of 2.00 L, what chemical amount of each substance was present initially?
   (b) What chemical amount of hydrogen iodide had formed at equilibrium? (Create an ICE table.)
   (c) Describe the rate at which hydrogen is reacting from the moment the reactants are mixed to the time when equilibrium has been established, in terms of collision–reaction theory.

4. For each of the following, write the chemical reaction equation with appropriate equilibrium arrow, as shown in **Table 3** (page 680).
   (a) The Haber process is used to manufacture ammonia fertilizer from hydrogen and nitrogen gases. Under less-than-desirable conditions, only an 11% yield of ammonia is obtained at equilibrium.
   (b) A mixture of carbon monoxide and hydrogen, known as water gas, is used as a supplementary fuel in many large industries. At high temperatures, the reaction of coke and steam forms an equilibrium mixture in which the products (carbon monoxide and hydrogen gases) are favoured. (Assume that coke is pure carbon.)
   (c) Because of the cost of silver, many high school science departments recover silver metal from waste solutions containing silver compounds or silver ions. A quantitative reaction of waste silver ion solutions with

copper metal results in the production of silver metal and copper(II) ions.

(d) One step in the industrial process used to manufacture sulfuric acid is the production of sulfur trioxide from sulfur dioxide and oxygen gases. Under certain conditions, the reaction produces a 65% yield of products.

5. Write the expression of the equilibrium law for the hydrogen–iodine–hydrogen iodide system at 448 °C. Using the evidence for System 1 as reported in **Table 1** on page 679, calculate the value of the equilibrium constant.

6. In the Haber process for synthesizing ammonia gas from nitrogen and hydrogen, the value of K_c is 6.0×10^{-2} for the reaction at 500 °C. In a sealed container at equilibrium at 500 °C, the concentrations of $H_2(g)$ and of $N_2(g)$ are measured to be 0.50 mol/L and 1.50 mol/L, respectively. Write the equilibrium law expression and calculate the equilibrium concentration of $NH_3(g)$.

7. At a certain constant (very high) temperature, 1.00 mol of HBr(g) is introduced into a 2.00 L container. Decomposition of this gas to hydrogen and bromine gases quickly establishes an equilibrium, at which point the amount concentration of HBr(g) is measured to be 0.100 mol/L.
   (a) Write a balanced equation for the reaction.
   (b) Write the equilibrium law expression.
   (c) Calculate the chemical amount of HBr(g) present at equilibrium.
   (d) Calculate the chemical amount of HBr(g) that has reacted to form $H_2(g)$ and $Br_2(g)$ products when equilibrium is established.
   (e) Calculate the chemical amounts of $H_2(g)$ and $Br_2(g)$ that have been produced, and, thus, are present, when equilibrium is established.
   (f) Calculate the amount concentration of all substances present at equilibrium.
   (g) Calculate K_c for this reaction at this temperature.

8. To a heated reaction vessel with a volume of 1.00 L, a lab technician adds 6.23 mmol $H_2(g)$, 4.14 mmol of $I_2(g)$, and 22.40 mmol of HI(g). At equilibrium, a spectrophotometer is used to determine that the concentration of iodine vapour is 2.58 mmol/L. Construct an ICE table and find K_c for the reaction system
$$H_2(g) + I_2(g) \rightleftarrows 2\,HI(g).$$

9. Consider the system
$$CO_2(g) + H_2(g) \rightleftarrows CO(g) + H_2O(g)$$
Initially, 0.25 mol of water and 0.20 mol of carbon monoxide are placed in a 1.0 L reaction vessel. At equilibrium, spectroscopic evidence shows that 0.10 mol of carbon dioxide is present. Construct an ICE table and find K_c for this system.

10. Consider the system
$$2\,HBr(g) \rightleftarrows H_2(g) + Br_2(g)$$
Initially, 0.25 mol of hydrogen and 0.25 mol of bromine are placed into a 500 mL electrically heated reaction vessel. K_c for the reaction at the temperature used is 0.020.
    (a) Find the concentrations of the substances at equilibrium.
    (b) Calculate the chemical amount of each substance present at equilibrium.

11. Explain briefly how atomic theory, kinetic molecular theory, collision–reaction theory, and the concepts of reaction rate and reversible reactions are all necessary to explain chemical reaction equilibrium observations.

Extension

12. In a *very* long-term sense, Earth may be considered a closed system. One equilibrium of concern to scientists is the same one involved in carbonation of soft drinks, on a vastly larger scale. Scientists believe that over time, the carbon dioxide gas in the atmosphere should be in equilibrium with carbon dioxide dissolved in the oceans. They also know that the concentration of $CO_2(g)$ in the atmosphere has been increased significantly (by about 20%) in the last century, which, they believe, is mostly due to the burning of fossil fuels. Concerns about the consequences of global warming make it imperative that scientists improve their theories about the various cycles, processes, and equilibria involving this greenhouse gas. Research and summarize currently accepted theory about carbon dioxide dissolved in the oceans, and list some other cycles and systems involving reaction or production of $CO_2(g)$.

www.science.nelson.com

15.2 Qualitative Change in Equilibrium Systems

Observing the effects of varying system properties on the equilibrium of systems contributes greatly to our understanding of the equilibrium state. From a technological perspective, controlling the extent of equilibrium by manipulating properties is very desirable because control leads to more efficient and economic processes. From a scientific perspective, observing systems at equilibrium leads to improved theories that describe, explain, and predict the nature of equilibrium, thus increasing our understanding.

Equilibrium is an area of study where, historically, technology has led science. Reactions were first manipulated in response to some human need, although the reactions' responses were not explained until much later by successive theories of increasing validity. This section of the chapter will examine equilibrium manipulation in the same way, with empirical descriptions of equilibrium manipulation given first, followed by theoretical explanations of the observed results.

According to **Le Châtelier's principle**, when a chemical system at equilibrium is disturbed by a change in a property of the system, the system always appears to react in the direction that opposes the change, until a new equilibrium is reached (**Figures 1** to **3**). The application of Le Châtelier's principle involves a three-stage process: an initial equilibrium state, a shifting non-equilibrium state, and a new equilibrium state.

Le Châtelier's principle provides a method of predicting the response of a chemical system to an imposed change. Using this simple and completely empirical approach, chemical engineers could produce more of the desired products, making technological processes more efficient and more economical. For example, Fritz Haber used Le Châtelier's principle to devise a process for the economical production of ammonia from atmospheric nitrogen. (See the Haber process, Chapter 8, page 325.)

Figure 1
Henri Louis Le Châtelier (1850–1936), French chemist and engineer, worked in chemical industries. To maximize the yield of products, Le Châtelier used systematic trial and error. After measuring properties of equilibrium states in chemical systems, he discovered a pattern and stated it as a generalization. This generalization has been supported extensively by evidence and is now considered a scientific law. By convention, it is known as Le Châtelier's principle.

Figure 2
$Fe^{3+}(aq) + SCN^-(aq) \rightleftharpoons FeSCN^{2+}(aq)$
The test tube on the left is at equilibrium, as shown by the constant colour of the $FeSCN^{2+}(aq)$ ion. The equilibrium is disturbed by the addition of $Fe^{3+}(aq)$ ions. The system shifts and some of the additional $Fe^{3+}(aq)$ reacts to produce more $FeSCN^{2+}(aq)$, thus establishing a new equilibrium state. When the shift is complete, the concentration of $Fe^{3+}(aq)$ is higher than before (only some of the added ions react), the concentration of $SCN^-(aq)$ is lower, and the concentration of $FeSCN^{2+}(aq)$ is higher. The higher concentration of $FeSCN^{2+}(aq)$ is evident from the more intense colour of the solution observed in the test tube on the right.

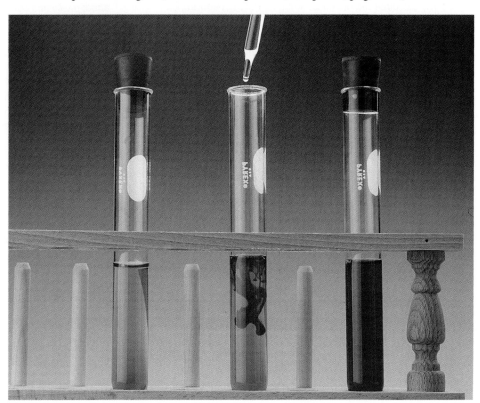

$Fe^{3+}(aq) + SCN^-(aq) \rightleftharpoons FeSCN^{2+}(aq)$

Section **15.2**

INVESTIGATION 15.2 Introduction

Equilibrium Shifts (Demonstration)

In this investigation, you will be looking at two equilibrium systems:

$$N_2O_4(g) + \text{energy} \rightleftharpoons 2\,NO_2(g)$$
colourless — reddish brown

$$CO_2(g) + H_2O(l) \rightleftharpoons H^+(aq) + HCO_3^-(aq)$$

The second equilibrium system, produced by the reaction of carbon dioxide gas and water, is commonly found in the human body and in carbonated drinks. A diagnostic test is necessary to detect shifts in this equilibrium. Bromothymol blue, an acid–base indicator, can detect an increase or decrease in the hydrogen ion concentration in this system. Bromothymol blue turns blue when the hydrogen ion concentration decreases, and yellow when the hydrogen ion concentration increases.

To perform this investigation, turn to page 700.

Report Checklist
- ○ Purpose
- ○ Problem
- ○ Hypothesis
- ● Prediction
- ○ Design
- ○ Materials
- ○ Procedure
- ● Evidence
- ● Analysis
- ● Evaluation (2, 3)

Purpose

The purpose of this demonstration is to test Le Châtelier's principle by studying two chemical equilibrium systems: the equilibrium between two oxides of nitrogen, and the equilibrium of carbon dioxide gas and carbonic acid.

Problem

How does a change in temperature affect the nitrogen dioxide–dinitrogen tetraoxide equilibrium system? How does a change in pressure affect the carbon dioxide–carbonic acid equilibrium system?

Le Châtelier's Principle and Concentration Changes

Le Châtelier's principle predicts that if the addition of a reactant to a system at equilibrium increases the concentration of that substance, then that system will undergo an **equilibrium shift** forward (to the right). The effect of the shift is that, temporarily, we observe the reactant concentration decreasing, as some of the added reactant changes to products. This period of change ends with the establishment of a *new* equilibrium state where, once again, there are no observable changes. The system has changed in such a way as to *oppose* the change introduced. For example, the production of freon-12, a CFC refrigerant, involves the following equilibrium reaction taking place at a fixed temperature:

$$CCl_4(l) + 2\,HF(g) \rightleftharpoons \underset{\text{freon-12}}{CCl_2F_2(g)} + 2\,HCl(g)$$

To improve the yield of the primary product, freon-12, more hydrogen fluoride is added to the initial equilibrium system. The additional concentration of reactant disturbs the equilibrium state and the system shifts to the right, consuming some of the added hydrogen fluoride by reaction with carbon tetrachloride. As a result, more freon-12 is produced and a new equilibrium state is reached. In chemical reaction equilibrium shifts, an imposed concentration change is normally only *partially* counteracted, and the final equilibrium state concentrations of the reactants and products are usually different from the values at the original equilibrium state. See **Figure 3** for a graphic interpretation of the freon-12 equilibrium shift.

Note that adding more carbon tetrachloride, $CCl_4(l)$, would have no effect on the equilibrium state in the container. This reactant is (and stays) in liquid form, so its concentration is constant and would not be increased by increasing the amount of $CCl_4(l)$ present.

Adjusting an equilibrium state by adding and/or removing a substance is by far the most common application of Le Châtelier's principle. For industrial chemical reactions, engineers strive to design processes where reactants are added continuously and products are continuously removed, so that an equilibrium is never allowed to establish. If the reaction is always shifting forward, the process is always making product (and, presumably, the industry is always making money).

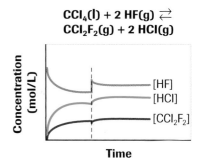

Figure 3
The reaction establishes an equilibrium that is disturbed (at the time indicated by the vertical dotted line) by the addition of HF(g). Some of the added HF reacts, decreasing in concentration, while the concentration of both products increases until a new equilibrium is established and concentrations become constant again. Note that the concentration of HF(g) at the new equilibrium is greater than at the original equilibrium, so the imposed change is only partly counteracted. The initial K_c value and final K_c value are the same.

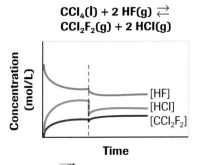

Figure 4
The reaction establishes an equilibrium that is disturbed (at the time indicated by the vertical dotted line) by the removal of HCl(g). The equilibrium shifts forward, increasing the concentration of both products while decreasing HF(g) concentration, until a new equilibrium is established. The initial K_c value and the final K_c value are the same.

The removal of a product (if the removal decreases concentration as well as chemical amount) will also shift an equilibrium forward, producing more product to counteract the change imposed. The freon-12 reaction can be shifted forward by removing either gaseous product, since decreasing the amount of a gas lowers its concentration in any reaction container of fixed size (**Figure 4**).

The following equation represents the final step in the production of nitric acid:

$$3\,NO_2(g) + H_2O(l) \rightleftarrows 2\,HNO_3(aq) + NO(g)$$

In this industrial process, nitrogen monoxide gas is removed from the chemical system by a reaction with oxygen gas. The removal of the nitrogen monoxide causes the system to shift to the right—some nitrogen dioxide and water react, replacing some of the removed nitrogen monoxide. As the system shifts, more of the desired product, nitric acid, is produced.

Although equilibrium systems are important in industrial chemical production, they are even more vital in biological systems. A particularly important biological equilibrium is that of hemoglobin (a protein in red blood cells), oxygen, and oxygenated hemoglobin.

$$Hb + O_2 \rightleftarrows HbO_2$$

As blood circulates to the lungs, the high concentration of oxygen shifts the equilibrium to the right and the blood becomes oxygenated (**Figure 5**). As the blood circulates throughout the body, cell reactions consume oxygen. This removal of oxygen shifts the equilibrium to the left and more oxygen is released.

Collision–Reaction Theory and Concentration Changes

Collision–reaction theory provides a simple explanation of the equilibrium shift that occurs when a reactant concentration is increased. We assume that the number of reactant entities per unit volume suddenly increases, so that collisions are suddenly much more frequent for the forward reaction. The forward reaction rate, therefore, increases significantly. Since the reverse reaction rate is not changed, the opposing rates are no longer equal, and, for a time, the difference in rates results in an observed increase of products.

Of course, as the concentration of products increases, so does the reverse reaction rate. At the same time, the new forward rate decreases as reactant is consumed, until eventually the two rates become equal to each other again. The rates at the new equilibrium are faster than those at the original equilibrium, because the system now contains a larger number of particles (and, therefore, a higher concentration) in dynamic equilibrium. If a substance is removed, causing an equilibrium shift, the explanation is similar except that the initial effect is to suddenly decrease either the forward or the reverse rate by *decreasing* the concentration.

Addition or removal of a reagent present in pure solid or pure liquid state does not change the *concentration* of that substance. The reaction of condensed phases (solids and liquids) takes place only at an exposed surface—and if the surface area exposed is changed, it is always exactly the same change in available area for both forward and reverse reaction collisions. The forward and reverse rates change by exactly the same amount if they change at all, so equilibrium is not disturbed and no shift occurs.

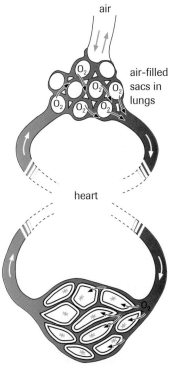

Figure 5
Oxygenated blood from the lungs is pumped by the heart to body tissues. The deoxygenated blood returns to the heart and is pumped to the lungs. Shifts in equilibrium occur over and over again as oxygen is picked up in the lungs and released throughout the body.

Le Châtelier's Principle and Temperature Changes

The energy in a chemical equilibrium equation is treated as though it were a reactant or a product.

reactants + energy ⇌ products   (endothermic in the forward direction)
reactants ⇌ products + energy   (exothermic in the forward direction)

Heating or cooling a system adds or removes thermal energy from the system. In either situation, the equilibrium shifts to minimize the change. If the system is cooled, the equilibrium shifts so that more energy is produced. If the system is heated, the equilibrium shifts in the direction in which energy is absorbed.

For example, in the salt–sulfuric acid process used to produce hydrochloric acid, the system is heated in order to increase the percent yield of hydrogen chloride gas.

$$2 \text{ NaCl(s)} + \text{H}_2\text{SO}_4\text{(l)} + \text{energy} \rightleftharpoons 2 \text{ HCl(g)} + \text{Na}_2\text{SO}_4\text{(s)}$$

Adding energy shifts the system to the right, absorbing some of the added energy.

In the production of sulfuric acid, the key reaction step is the equilibrium represented by the following equation. Percent yield of the product is increased at low temperature.

$$2 \text{ SO}_2\text{(g)} + \text{O}_2\text{(g)} \rightleftharpoons 2 \text{ SO}_3\text{(g)} + \text{energy}$$

Removing energy causes the system to shift to the right. This shift yields more sulfur trioxide while partially replacing the energy that was removed.

Collision–Reaction Theory and Energy Changes

Collision–reaction theory explains the equilibrium shift (that occurs when the thermal energy of a system at equilibrium is changed) as the result of an imbalance of reaction rates. Consider the previously mentioned reaction equation—a typical exothermic reaction. The reaction energy is shown this time in standard $\Delta_r H$ notation.

$$2 \text{ SO}_2\text{(g)} + \text{O}_2\text{(g)} \rightleftharpoons 2 \text{ SO}_3\text{(g)} \qquad \Delta_r H = -198 \text{ kJ}$$

We explain the result of cooling the system by assuming that *both* forward and reverse reaction rates are slower at lower temperatures, because the particles move more slowly and collide less frequently. The reverse rate decreases *more* than the forward rate, however. While the rates remain unequal, the observed result is the production of more product and the release of more heat energy. The shift causes concentration changes that will increase the reverse rate and decrease the forward rate until they become equal again, at a new, lower temperature (**Figure 6**).

Note that industrial exothermic equilibrium reactions are often carried out at high temperatures, even though adding heat energy shifts the equilibrium toward reactants (lowers the percent yield). The Haber process (Case Study, Section 8.3) is a good example. Energy is added because the forward and reverse reaction rates are too slow at lower temperatures to allow the reaction to reach equilibrium in a reasonable time. Making large quantities of the marketable product in a short time is much more important to a manufacturer than creating a small increase in the yield of each batch. Whenever possible, chemical engineers try to design a continuous process for an industrial reaction—one that shifts the reaction forward by constantly adding reactants, and constantly removing products. This system is no longer a closed system, so the reaction never establishes equilibrium. Such an industrial reaction may run continuously for months or even years.

BIOLOGY CONNECTION

CO₂ Transport

Many biological processes depend on equilibria. For example, the transportation of carbon dioxide depends on the $\text{CO}_2 \rightleftharpoons \text{H}_2\text{CO}_3$ equilibrium system. In the tissues of the body where carbon dioxide is produced, the equilibrium shifts so that more carbonic acid is formed. In the lungs, the shift is in the reverse direction, as carbon dioxide is released. You will discover other examples of equilibria in a biology textbook.

www.science.nelson.com

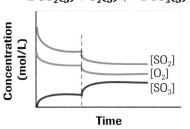

Figure 6
The reaction establishes an equilibrium that is disturbed (at the time indicated by the vertical dotted line) by a decrease in temperature. The equilibrium shifts forward, increasing the concentration of SO₃ product while decreasing the concentration of both reactants, until a new equilibrium is established. Because the temperature is changed, the final K_c value for this example is greater than the initial K_c value because the shift favours the forward reaction.

CAREER CONNECTION

Chemical Process Engineer

What role do chemical engineers play in designing systems and sequences of reactions for industrial chemical production? Research the education requirements, job prospects, and work assignments of people in this profession.

www.science.nelson.com

Le Châtelier's Principle and Gas Volume Changes

According to Boyle's law, the pressure of a gas in a container is inversely proportional to the volume of the container. Since the amount concentration of a gas is directly proportional to its pressure, we can predict the possible effect of container volume change on the equilibrium position of homogeneous gaseous systems. Decreasing the volume by half doubles the concentration of every gas in the container. To predict whether a change in pressure will affect a system's equilibrium, you must consider the total chemical amount of gas reactants and the total chemical amount of gas products. For example, in the equilibrium reaction of sulfur dioxide and oxygen, three moles of gaseous reactants produce two moles of gaseous products.

$$2\,SO_2(g) + O_2(g) \rightleftarrows 2\,SO_3(g)$$

If the volume is decreased, the overall pressure is increased. Increased pressure causes a shift to the right, which decreases the total number of gas molecules (three moles to two moles) and, thus, reduces the pressure. If the volume is increased, the pressure is decreased, and the shift is in the opposite direction. A system with equal numbers of gas molecules on each side of the equation, such as the equilibrium reaction between hydrogen and iodine (page 679), is not affected by a change in volume. Similarly, systems involving only liquids or solids are not affected by changes in pressure. Note that adding a gas that is not involved in the equilibrium (such as an inert gas) to the container will increase the overall pressure in the container, but will not cause a shift in equilibrium. Adding or removing gaseous substances not involved in the reaction does not change the concentrations of the reactant and product gases.

Collision–Reaction Theory and Gas Volume Changes

When a system involving gaseous reactants and products is changed in volume, the resulting equilibrium shift is again explained as an imbalance of reaction rates.

$$2\,SO_2(g) + O_2(g) \rightleftarrows 2\,SO_3(g) + 198\,kJ$$

Collision–reaction theory explains the result of decreasing the volume of this system by assuming that *both* forward and reverse reaction rates become faster because the concentrations of reactants and products both increase. For this example, however, the forward rate increases *more* than the reverse rate because there are more particles involved in the forward reaction. Consequently, the increase in the total number of collisions is greater for the forward reaction process. Again, while the rates remain unequal, the observed result is the production of more product. The shift causes concentration changes that gradually increase the reverse rate and decrease the forward rate until they become equal again (**Figure 7**).

Catalysts and Equilibrium Systems

Catalysts are used in most industrial chemical systems. A catalyst decreases the time required to reach an equilibrium position, but does not affect the final position of equilibrium. The presence of a catalyst in a chemical reaction system lowers the activation energy for both forward and reverse reactions by an equal amount (Chapter 12), so the equilibrium establishes much more rapidly but at the same position as it would without the catalyst present. Forward and reverse rates increase equally. The final equilibrium concentrations are reached in a shorter time compared with the same, but uncatalyzed, reaction. The value of catalysts in industrial processes is to decrease the time required for equilibrium shifts created by manipulating other variables, allowing a more rapid overall production of the desired product.

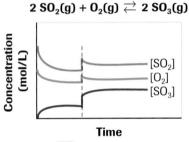

Figure 7
The reaction equilibrium is disturbed by a decrease in container volume (at the time indicated by the vertical dotted line). The equilibrium shifts forward, increasing the concentration of SO_3 while decreasing the concentration of reactants, until a new equilibrium is established. The initial K_c value and the final K_c value are the same.

> **Learning Tip**
>
> Since increasing the temperature of any exothermic reaction at equilibrium always shifts the equilibrium left (toward reactants), an increase in temperature must decrease the value of K_c for such a reaction. Similarly, increasing the temperature will increase the value of K_c for any endothermic reaction. You have already learned that the value of K_c is temperature dependent. No other change imposed on a system at equilibrium changes the numerical value of the equilibrium constant.

SUMMARY: Variables Affecting Chemical Equilibria

Variables	Imposed Change	Response of System
concentration	increase	shifts to consume some of the added reactant or product
	decrease	shifts to replace some of the removed reactant or product
temperature	increase	shifts to absorb some of the added energy
	decrease	shifts to replace some of the removed energy
volume (gaseous systems only)	increase (decrease in pressure)	shifts toward the side with the larger total chemical amount of gaseous entities
	decrease (increase in pressure)	shifts toward the side with the smaller total chemical amount of gaseous entities

Practice

1. What three types of changes shift the position of a chemical equilibrium?
2. For each of the following chemical systems at equilibrium, use Le Châtelier's principle to predict the effect of the change imposed on the chemical system. Indicate the direction in which the equilibrium is expected to shift. For each example, sketch the graph of concentrations versus time, plotted from just before the change to the established new equilibrium.
   (a) $H_2O(l) + energy \rightleftharpoons H_2O(g)$
   The container is heated.
   (b) $H_2O(l) \rightleftharpoons H^+(aq) + OH^-(aq)$
   A few crystals of NaOH(s) are added to the container.
   (c) $CaCO_3(s) + energy \rightleftharpoons CaO(s) + CO_2(g)$
   $CO_2(g)$ is removed from the container.
   (d) $CH_3COOH(aq) \rightleftharpoons H^+(aq) + CH_3COO^-(aq)$
   A few drops of pure $CH_3COOH(l)$ are added to the system.
3. Much methanol is produced industrially by the exothermic reaction $CO(g) + 2H_2(g) \rightleftharpoons CH_3OH(l)$, carried out at high pressure (5–10 MPa) and temperature (250 °C) in the presence of several catalyst substances. Methanol is less flammable than gasoline, and so it is a safer fuel. It is the fuel used in open-wheel Champ Car racing, and also in the Indianapolis 500.
   (a) State, in terms of forward and reverse reaction rates, why using a very high pressure of the reactant gases is economically desirable for the manufacturer.
   (b) State in which direction a high temperature will shift this reaction equilibrium.
   (c) Explain why using a high temperature is desirable, in terms of the time required for the reaction to reach equilibrium.
   (d) Explain, in terms of equilibrium position and equilibrium shift, why this reaction is done in an open system, where reactants are continually added to the pressure vessel and liquid product is continually removed.

INVESTIGATION 15.3 Introduction

Testing Le Châtelier's Principle

Report Checklist		
○ Purpose	○ Design	● Analysis
○ Problem	○ Materials	● Evaluation (2, 3)
○ Hypothesis	○ Procedure	
● Prediction	● Evidence	

The equilibria chosen for this investigation involve chemicals that provide coloured solutions. The investigation tests predictions about equilibrium shifts (made using Le Châtelier's principle) by observing colour changes.

In order to complete the Prediction section of the report, you must read the Design, Materials, and Procedure carefully. Then make a Prediction about the result of each change made in the Procedure.

Purpose

The purpose of this investigation is to test Le Châtelier's principle by applying stress to four different chemical equilibria.

To perform this investigation, turn to page 701.

Problem

How does applying changes to conditions of particular chemical equilibria affect the systems?

Design

Stresses are applied to four chemical equilibrium systems and evidence is gathered to test predictions made using Le Châtelier's principle. Control samples are used in all cases.

Case Study

Urea Production in Alberta

Canada has a vast wealth of natural resources that we can use to produce many chemicals with an incredible variety of uses. Because plants require nitrogen for growth, the primary purpose of some of these chemicals is for agricultural use, such as nitrogen fertilizers. You have already learned that Alberta produces large amounts of ammonia, which can be used directly as a fertilizer. Ammonia is stored as a liquid at high pressure, and injected directly into the soil (see Section 8.3, Figure 7), but it is toxic and corrosive to human tissue, which makes it dangerous to use. Special equipment and care are required. Many food producers prefer to use a high-nitrogen, nontoxic, solid compound.

Urea (**Figure 8**) is another simple molecular chemical—also a nitrogen fertilizer—that is inexpensive, simple to produce, easy to transport, and extraordinarily useful. This chemical is used in the millions of tonnes, for applications as diverse as

- wastewater plants (for treating effluent)
- air transportation (for de-icing runways)
- forestry and agriculture (for fertilizer)
- livestock feeding (for a protein supplement)
- woodworking (for making glues and resins)
- construction supplies (for making insulation)
- furniture (for making particle board and chipboard)
- clothing (for making certain dyes)

Until the 1900s, the common source of this chemical was stale animal urine. The body forms this compound as its principal means of removing excess nitrogen. In fact, until the early 1800s, it was thought that this, or any other "organic" chemical extracted from living things, *always* had to be produced by a living organism. In 1828, however, Friedrich Wöhler, in a famous classic experiment, synthesized urea in his laboratory. His work forever changed the "organic" concept of chemistry. Early in the last century, an efficient industrial method was

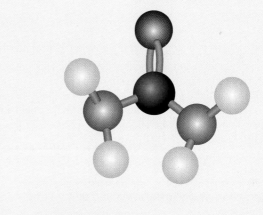
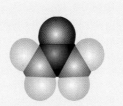
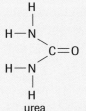

Figure 8
Urea is a small and simple molecule, but a critically important nitrogen-containing compound. It has a very high nitrogen percentage by mass, and is very soluble in water—both very important points for a compound to be useful as a plant fertilizer. Three different representations of the urea molecule are shown here.

developed, and has since been used for mass production of this versatile and valuable substance.

Urea is produced through the reaction of ammonia, $NH_3(g)$, with carbon dioxide, $CO_2(g)$, at high temperature and pressure. Most of the reactants exist in liquid form at the pressures used. A hot concentrated solution, containing about 80% urea by mass, results from this reaction. This hot solution is further concentrated and cooled through evaporation of the water content, to form either granules (angular crystals) or prills (small round pellets) of white solid urea (**Figure 9**). The overall reaction may be written as

$$CO_2(l) + 2\,NH_3(l) \rightleftarrows NH_2CONH_2(aq) + H_2O(l)$$
$$t = 150\text{--}200\ °C$$
$$P = 12\text{--}20\ MPa$$

The science and technology for urea production developed in response to a strong demand for this compound. The process depends on an understanding of reaction equilibrium, and of the effect (on equilibrium) of high temperature and pressure conditions. Also key was the design and construction of reaction containers (vessels) able to withstand high pressures and temperatures.

Sometimes, as well as building on existing scientific knowledge, new technology results in scientific advances. A technology, such as equipment to create very high pressures, may allow great leaps forward in science, such as the synthesis of completely new forms of crystalline solids. New observations then lead to new theories and sometimes even new laws.

Figure 9
Prills form when sprayed droplets of very hot urea solution are made to fall through air in a huge tower, cooling and evaporating to dryness on the way down.

Case Study Questions

1. There are two phase equilibria that shift during the reaction of carbon dioxide and ammonia. Write equilibrium equations to represent these phase changes, and use Le Châtelier's principle to explain how they are continuously being shifted, both by pressure, and also by the effect of the chemical reaction that is occurring in the vessel.

2. Crystallization of urea from aqueous solution can be expressed as a solubility equilibrium, according to the equation

$$NH_2CONH_2(aq) \rightleftarrows NH_2CONH_2(s)$$

   (a) On which side of this equation would heat energy be written? Explain your reasoning.
   (b) Based on molecular structure and bonding theory, explain why you would expect urea to be a very highly soluble compound.
   (c) Urea granules that are bagged and sold for fertilizer use are quite uniform in size (**Figure 10**). Explain what physical process would likely be used to separate these granules from any larger or smaller granules that form during crystallization.

Figure 10
Urea fertilizer is a nitrogen source for crops.

Extension

3. Research Friedrich Wöhler's classic experiment of 1828. Write the balanced equation for the reaction he performed.

4. Find out how much urea is produced annually in Alberta, where it is produced, and its current price per tonne. Assemble your findings into an attractive presentation.

www.science.nelson.com GO

LAB EXERCISE 15.C

The Nitrogen Dioxide–Dinitrogen Tetraoxide Equilibrium

Report Checklist
- ○ Purpose
- ○ Problem
- ○ Hypothesis
- ● Prediction
- ○ Design
- ○ Materials
- ○ Procedure
- ○ Evidence
- ● Analysis
- ● Evaluation (1, 2, 3)

Complete the Prediction, Analysis, and Evaluation sections of the report.

Purpose
The purpose of this problem is to use Le Châtelier's principle to predict the response of an equilibrium to an introduced change in conditions.

Problem
How does increasing the pressure affect the nitrogen dioxide–dinitrogen tetraoxide equilibrium?

Design
A sample of nitrogen dioxide gas is compressed in a syringe and the intensity of the colour is used as evidence to test the Prediction.

Evidence
The orange-brown nitrogen dioxide gas colour increases in intensity when the plunger on the syringe is depressed, and then decreases in intensity (**Figure 11**). The final colour is slightly more intense than the original colour (before moving the plunger).

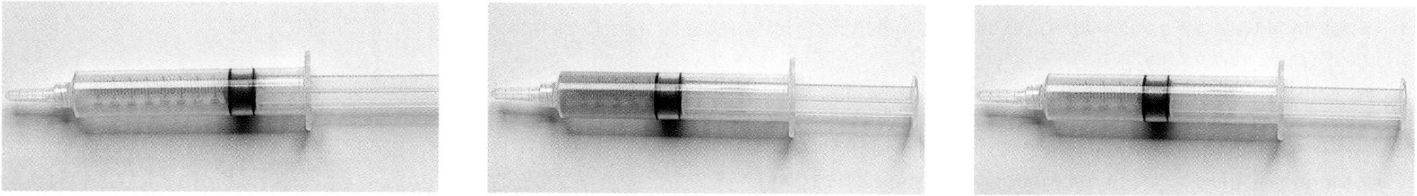

Figure 11
$2\,NO_2(g) \rightleftharpoons N_2O_4(g)$
An increase in pressure on the nitrogen dioxide–dinitrogen tetraoxide equilibrium in the closed system results initially in a more intense colour followed by a decrease in colour intensity.

INVESTIGATION 15.4 Introduction

Studying a Chemical Equilibrium System

Report Checklist
- ○ Purpose
- ● Problem
- ○ Hypothesis
- ● Prediction
- ● Design
- ● Materials
- ● Procedure
- ● Evidence
- ● Analysis
- ● Evaluation (1, 2, 3)

Figure 2, page 690, shows the colours of aqueous solutions of iron(III), thiocyanate, and iron(III) thiocyanate ions. Use your knowledge of Le Châtelier's principle to write a Problem statement, and then design and carry out a simple investigation to determine whether the reaction as written is exothermic or endothermic.

Purpose
The purpose of this investigation is to use Le Châtelier's principle to solve a problem concerning the effect of an energy change on the following equilibrium system.

$$Fe^{3+}(aq) \;+\; SCN^-(aq) \rightleftharpoons FeSCN^{2+}(aq)$$
almost colourless — colourless — red

To perform this investigation, turn to page 703.

WEB Activity

Web Quest—Poison Afloat

Have you ever considered becoming a crime scene investigator? In this Web Quest, you are promoted to Chief Chem Crime Investigator. You are presented with a body and a series of clues... The detecting is up to you. You will have a chance to use your knowledge of chemistry to solve this puzzle and gather the evidence to unravel what happened.

www.science.nelson.com

Section 15.2 Questions

1. The following equation represents part of the industrial production of nitric acid. Predict the direction of the equilibrium shift for each of the following changes. Explain any shift in terms of the changes in forward and reverse reaction rates.

   $4 NH_3(g) + 5 O_2(g) \rightleftarrows 4 NO(g) + 6 H_2O(g) + energy$

   (a) $O_2(g)$ is added to the system.
   (b) The temperature of the system is increased.
   (c) $NO(g)$ is removed from the system.
   (d) The pressure of the system is increased by decreasing the volume.

2. The following chemical equilibrium system is part of the Haber process for the production of ammonia.

   $N_2(g) + 3 H_2(g) \rightleftarrows 2 NH_3(g) + energy$

   Suppose you are a chemical process engineer. Use Le Châtelier's principle to predict five specific changes that you might impose on the equilibrium system to increase the yield of ammonia.

3. In a solution of copper(II) chloride, the following equilibrium exists:

   $CuCl_4^{2-}(aq) + 4 H_2O(l) \rightleftarrows Cu(H_2O)_4^{2+}(aq) + 4 Cl^-(aq)$
   dark green   blue

   For the following stresses put on the equilibrium, predict the shift in the equilibrium and draw a graph of concentration versus time to communicate the shift.
   (a) Concentrated hydrochloric acid is added.
   (b) Saturated aqueous silver nitrate is added, causing a precipitation reaction.

4. Identify the nature of the changes imposed on the following equilibrium system at the four times indicated by coordinates A, B, C, and D (**Figure 12**).

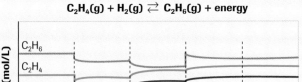

Figure 12
Graph showing four disturbances to an equilibrium system

5. In which of the following cases would an increase in temperature increase the percent yield at equilibrium?
   (a) $H_2O(l) \rightleftarrows H_2O(g)$
   (b) $N_2(g) + 3 H_2(g) \rightleftarrows 2 NH_3(g)$ $\Delta_r H = -91$ kJ
   (c) $KOH(s) \rightleftarrows K^+(aq) + OH^-(aq) + heat$
   (d) $2 C(s) + 2 H_2(g) \rightleftarrows C_2H_4(g)$ $\Delta_r H = +53$ kJ

6. Chloromethane (methyl chloride) is manufactured by "chlorinating" methane. For this reaction system at equilibrium, explain the effect of each of the imposed changes on the position of reaction equilibrium.

   $CH_4(g) + Cl_2(g) \rightleftarrows CH_3Cl(g) + HCl(g)$ $\Delta_r H$ is negative

   (a) More methane is injected into the reaction vessel.
   (b) The container volume is increased.
   (c) The temperature is lowered.
   (d) A catalyst is introduced into the system.

7. Ethyne (acetylene) is manufactured by a high-temperature combustion of methane, using a large excess of methane. For this endothermic reaction system at equilibrium, explain the effect of each of the imposed changes on the value of the equilibrium constant.

   $6 CH_4(g) + O_2(g) \rightleftarrows 2 C_2H_2(g) + 10 H_2(g) + 2 CO(g)$

   (a) More methane is injected into the reaction vessel.
   (b) The container volume is decreased.
   (c) The temperature is lowered.
   (d) A catalyst is introduced into the system.

Extension

8. In a deep lake or in the ocean, the pressure that a human considers "normal" has doubled by the time a diver reaches a depth of 10 m, and increases by about one atmosphere for every extra 10 m, to a maximum of about 100 MPa at the deepest points in Earth's oceans. This fact is of major concern to scuba divers for several reasons. Pressure inside a scuba diver's lungs must constantly be adjusted to equal outside water pressure, otherwise the lungs could collapse upon diving, and could explode upon rising. In fact, the development of the pressure regulator (**Figure 13**) was the technology that made scuba diving possible.

   The gases in a diver's lungs are dissolved to some extent in the blood that passes through. Pressure changes can change several solubility equilibria, with some serious effects. Research the gases dissolved in human blood, and what equilibrium shifts cause the diving conditions called nitrogen narcosis, and the "bends."

   www.science.nelson.com

Figure 13
The mouthpiece pressure from a diver's tank automatically adjusts for changes in depth.

Chapter 15 INVESTIGATIONS

INVESTIGATION 15.1

The Extent of a Chemical Reaction

Report Checklist

- ○ Purpose
- ○ Problem
- ○ Hypothesis
- ● Prediction
- ○ Design
- ○ Materials
- ● Procedure
- ● Evidence
- ● Analysis
- ● Evaluation (1, 2, 3)

In Chapter 8, you performed experiments that produced evidence that reactions are quantitative. In a quantitative reaction, the limiting reagent is completely consumed. To identify the limiting reagent, you can test the final reaction mixture for the presence of the original reactants. For example, in a diagnostic test, you might try to precipitate ions from the final reaction mixture that were present in the original reactants.

Purpose

The purpose of this investigation is to test the validity of the assumption that chemical reactions are quantitative.

Problem

What are the limiting and excess reagents in the chemical reaction of selected quantities of aqueous sodium sulfate and aqueous calcium chloride?

Design

Samples of sodium sulfate solution and calcium chloride solution are mixed in different proportions and the final mixture is filtered. Samples of the filtrate are tested for the presence of excess reagents, using the following diagnostic tests:

- If a few drops of $Ba(NO_3)_2(aq)$ are added to the filtrate and a precipitate forms, then sulfate ions are present.
  $Ba^{2+}(aq) + SO_4^{2-}(aq) \rightarrow BaSO_4(s)$
- If a few drops of $Na_2CO_3(aq)$ are added to the filtrate and a precipitate forms, then calcium ions are present.
  $Ca^{2+}(aq) + CO_3^{2-}(aq) \rightarrow CaCO_3(s)$

Materials

lab apron	two 50 mL or 100 mL beakers
eye protection	two small test tubes
25 mL of 0.50 mol/L $CaCl_2(aq)$	10 mL or 25 mL graduated cylinder
25 mL of 0.50 mol/L $Na_2SO_4(aq)$	filtration apparatus
1.0 mol/L $Na_2CO_3(aq)$ in dropper bottle	filter paper
saturated $Ba(NO_3)_2(aq)$ in dropper bottle	wash bottle
	stirring rod

 Soluble barium compounds are toxic. Remember to wash your hands before leaving the laboratory.

 Barium nitrate in solid form is a strong oxidizing substance.

INVESTIGATION 15.2

Equilibrium Shifts (Demonstration)

Report Checklist

- ○ Purpose
- ○ Problem
- ○ Hypothesis
- ● Prediction
- ○ Design
- ○ Materials
- ○ Procedure
- ● Evidence
- ● Analysis
- ● Evaluation (2, 3)

In this investigation, you will be looking at two equilibrium systems:

$$N_2O_4(g) + \text{energy} \rightleftharpoons 2\,NO_2(g)$$
$$\text{colourless} \qquad\qquad\qquad \text{reddish brown}$$

$$CO_2(g) + H_2O(l) \rightleftharpoons H^+(aq) + HCO_3^-(aq)$$

The second equilibrium system, produced by the reaction of carbon dioxide gas and water, is commonly found in the human body and in carbonated drinks. A diagnostic test is necessary to detect some shifts in this equilibrium. Bromothymol blue, an acid–base indicator, can detect an increase or decrease in the hydrogen ion concentration in this system. Bromothymol blue turns blue when the hydrogen ion concentration decreases, and yellow when the hydrogen ion concentration increases.

Purpose

The purpose of this demonstration is to test Le Châtelier's principle by studying two chemical equilibrium systems: the equilibrium between two oxides of nitrogen, and the equilibrium of carbon dioxide gas and carbonic acid.

INVESTIGATION 15.2 continued

Problem
How does a change in temperature affect the nitrogen dioxide–dinitrogen tetraoxide equilibrium system? How does a change in pressure affect the carbon dioxide–carbonic acid equilibrium system?

Materials
lab apron
eye protection
two $NO_2(g)/N_2O_4(g)$ sealed flasks
carbon dioxide–hydrogen carbonate ion equilibrium mixture (pH = 7)
bromothymol blue indicator in dropper bottle
small syringe with needle removed (5 to 50 mL)
solid rubber stopper to seal end of syringe
beaker of ice–water mixture
beaker of hot water

Be careful with the flasks containing nitrogen dioxide: this gas is highly toxic. Use in a fume hood in case of breakage.

Procedure
1. Place the sealed $NO_2(g)/N_2O_4(g)$ flasks in hot and cold water baths (**Figure 1**) and record your observations.
2. Place two or three drops of bromothymol blue indicator in the carbon dioxide–hydrogen carbonate ion equilibrium mixture.
3. Draw some of the carbon dioxide–hydrogen carbonate ion equilibrium mixture into the syringe, and then block the end with a rubber stopper.
4. Slowly move the syringe plunger and record your observations.

Figure 1
Each of these flasks contains an equilibrium mixture of dinitrogen tetraoxide and nitrogen dioxide. Shifts in equilibrium can be seen when one of the flasks is heated or cooled.

INVESTIGATION 15.3

Testing Le Châtelier's Principle

Report Checklist
- ○ Purpose
- ○ Problem
- ○ Hypothesis
- ● Prediction
- ○ Design
- ○ Materials
- ○ Procedure
- ● Evidence
- ● Analysis
- ● Evaluation (2, 3)

The equilibria chosen for this investigation involve chemicals that provide coloured solutions. This investigation tests predictions about equilibrium shifts (made using Le Châtelier's principle) by observing colour changes. For example, if the colour in Reaction 3 becomes more intensely red, the equilibrium has shifted right to produce more $FeSCN^{2+}(aq)$ ions, increasing their concentration. See **Table 1**.

In order to complete the Prediction section of the report, you must read the Design, Materials, and Procedure carefully. Then, make a Prediction about the result of each change made in the Procedure.

Purpose
The purpose of this investigation is to test Le Châtelier's principle by applying stress to four different chemical equilibria.

Table 1 Solution Colours

Ion	Colour
$CoCl_4^{2-}$(alc)	blue
$Co(H_2O)_6^{2+}$(alc)	pink
H_2Tb(aq)	red
HTb^-(aq)	yellow
Tb^{2-}(aq)	blue
Fe^{3+}(aq)	pale yellow
SCN^-(aq)	colourless
$FeSCN^{2+}$(aq)	red
$Cu(H_2O)_4^{2+}$(aq)	pale blue
$Cu(NH_3)_4^{2+}$(aq)	deep blue

INVESTIGATION 15.3 continued

Problem
How does applying stresses to particular chemical equilibria affect the systems?

Part I
$$CoCl_4^{2-}(alc) + 6 H_2O(alc) \rightleftharpoons Co(H_2O)_6^{2+}(alc) + 4 Cl^-(alc) + energy$$

Part II
$$H_2Tb(aq) \rightleftharpoons H^+(aq) + HTb^-(aq)$$
$$HTb^-(aq) \rightleftharpoons H^+(aq) + Tb^{2-}(aq)$$

Part III
$$Fe^{3+}(aq) + SCN^-(aq) \rightleftharpoons FeSCN^{2+}(aq)$$

Part IV
$$Cu(H_2O)_4^{2+}(aq) + 4 NH_3(aq) \rightleftharpoons Cu(NH_3)_4^{2+}(aq) + 4 H_2O(l)$$

Design
Stresses are applied to four chemical equilibrium systems and evidence is gathered to test predictions made using Le Châtelier's principle. Control samples are used in all cases. For example, before adding sodium hydroxide to a new equilibrium solution, split the solution into two samples in order to have a control sample for colour comparison.

Part I Cobalt(II) Complexes
Water, saturated silver nitrate, and heat are added to, and heat is removed from, samples of the provided equilibrium mixture. Note: This reaction equilibrium is in solution using an alcohol solvent, shown as (alc), so the concentration of water is a variable in this system.

Part II Thymol Blue Indicator
Hydrochloric acid and sodium hydroxide are added to samples of the provided equilibrium mixture.

Part III Iron(III)–Thiocyanate Equilibrium
Iron(III) nitrate, potassium thiocyanate, and sodium hydroxide are added to samples of the provided equilibrium system.

Part IV Copper(II) Complexes
Aqueous ammonia and hydrochloric acid are added to samples of the provided equilibrium mixture.

Materials
lab apron
100 mL beaker
6 to 12 small test tubes
distilled water
eye protection
large waste beaker
test-tube rack
crushed ice
hot water bath
cobalt(II) chloride equilibrium mixture in ethanol
dropper bottles containing

0.2 mol/L $AgNO_3$(aq)	thymol blue indicator
0.1 mol/L HCl(aq)	0.1 mol/L NaOH(aq)

iron(III) thiocyanate equilibrium mixture

0.2 mol/L $Fe(NO_3)_3$(aq)	0.2 mol/L KSCN(aq)
6.0 mol/L NaOH(aq)	0.1 mol/L $CuSO_4$(aq)
1.0 mol/L NH_3(aq)	1.0 mol/L HCl(aq)

The chemicals used may be corrosive or poisonous, and may cause other toxic effects. Exercise great care when using the chemicals and avoid skin and eye contact. Immediately rinse the skin if there is any contact. If any chemicals get in the eyes, flush eyes for a minimum of 15 min and inform the teacher. Ethanol is flammable. Make sure there are no open flames in the laboratory when using the ethanol solution of cobalt(II) chloride.

Procedure

Part I Cobalt(II) Complexes

1. Obtain 25 mL of the equilibrium mixture with the cobalt(II) chloride complex ions.

2. Place a small amount of the mixture into each of five small test tubes. Use the fifth test tube as a control for comparison purposes.

3. Add drops of water to one test tube until a change is evident. Record the evidence.

4. Add drops of 0.2 mol/L silver nitrate to another test tube and record the evidence.

5. Heat another equilibrium mixture in a hot water bath and record the evidence.

6. Cool an equilibrium mixture in an ice bath and record the evidence.

Part II Thymol Blue Indicator

7. Add about 5 mL of distilled water to each of two small test tubes.

8. Add 1 to 3 drops of thymol blue indicator to the water in each test tube to obtain a noticeable colour. Use one test tube of solution as a control.

9. Add drops of 0.1 mol/L HCl(aq) to the experimental test tube to test for the predicted colour changes.

10. Add drops of 0.1 mol/L NaOH(aq) to the same tube to test for the predicted colour changes.

INVESTIGATION 15.3 continued

Part III Iron(III)–Thiocyanate Equilibrium

11. Obtain about 20 mL of the iron(III)–thiocyanate equilibrium mixture.
12. Place about 5 mL of the equilibrium mixture in each of three test tubes. Use one test tube as a control.
13. Add drops of $Fe(NO_3)_3(aq)$ to one test tube until a change is evident.
14. Add drops of 6.0 mol/L NaOH(aq) to this new equilibrium mixture until a change occurs. (Iron(III) hydroxide has very low solubility.)
15. Add drops of KSCN(aq) to another equilibrium mixture until a change is evident.

Part IV Copper(II) Complexes

16. Obtain 2 mL of 0.1 mol/L $CuSO_4(aq)$ in a small test tube.
17. Add three drops of 1.0 mol/L $NH_3(aq)$ to establish the equilibrium mixture.
18. Add more 1.0 mol/L $NH_3(aq)$ to the above equilibrium mixture and record the results.
19. Add 1.0 mol/L HCl(aq) to the equilibrium mixture from step 18 and record the results.

Dispose of the chemicals as directed by your teacher. Identify each as toxic (to be collected) or nontoxic (disposable in the sink).

Ensure that all equipment and surfaces are clean and wash your hands thoroughly before leaving the laboratory.

INVESTIGATION 15.4

Studying a Chemical Equilibrium System

Report Checklist

- ○ Purpose
- ● Problem
- ○ Hypothesis
- ● Prediction
- ● Design
- ● Materials
- ● Procedure
- ● Evidence
- ● Analysis
- ● Evaluation (1, 2, 3)

Figure 1 shows the colours of aqueous solutions of iron(III), thiocyanate, and iron(III) thiocyanate ions. Use your knowledge of Le Châtelier's principle to write a Problem statement, and then design and carry out a simple investigation to determine whether the reaction as written is exothermic or endothermic.

Purpose

The purpose of this investigation is to use Le Châtelier's principle to solve a problem concerning the effect of an energy change on the following equilibrium system.

$$Fe^{3+}(aq) + SCN^-(aq) \rightleftarrows FeSCN^{2+}(aq)$$

almost colourless — colourless — red

> **Iron(III) compounds are irritants. Thiocyanate ion solutions are toxic. Avoid skin and eye contact. If there is any skin or eye contact, immediately rinse with plenty of water. Flush the eyes for at least 15 min and inform the teacher.**

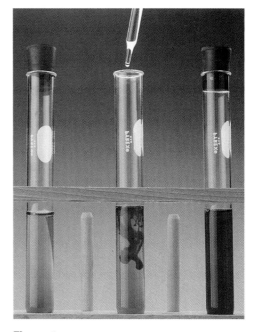

Figure 1
The iron–thiocyanate reaction

Chapter 15 SUMMARY

Outcomes

Knowledge
- define equilibrium and state the criteria that apply to a chemical system in equilibrium (15.1)
- identify, write, and interpret chemical equations for systems at equilibrium (15.1, 15.2)
- predict, qualitatively, using Le Châtelier's principle, shifts in equilibrium caused by changes in temperature, pressure, volume, concentration, or the addition of a catalyst, and describe how these changes affect the equilibrium constant (15.2)
- define K_c and write equilibrium law expressions for given chemical equations, using lowest whole-number coefficients (15.1)
- calculate equilibrium constants and concentrations for homogeneous systems when concentrations at equilibrium are known, when initial concentrations and one equilibrium concentration are known, and when the equilibrium constant and one equilibrium concentration are known (15.1)

STS
- state that the goal of science is knowledge about the natural world (15.1, 15.2)
- list the characteristics of empirical and theoretical knowledge (15.2)
- state that a goal of technology is to solve practical problems (15.2)

Skills
- initiating and planning: predict variables that can cause a shift in equilibrium (15.2); design an experiment to show equilibrium shifts (15.2); describe procedures for safe handling, storage, and disposal of materials used in the laboratory (15.1, 15.2)
- performing and recording: perform an experiment to test, qualitatively, predictions of equilibrium shifts (15.2)
- analyzing and interpreting: write the equilibrium law expression for a given equation (15.1); analyze, qualitatively, the changes in concentrations of reactants and products after an equilibrium shift (15.2); interpret data from a graph to determine when equilibrium is established, and determine the cause of a stress on the system (15.2)
- communication and teamwork: work collaboratively in addressing problems and communicate effectively (15.1, 15.2)

➕ EXTENSION

Plague and the Little Ice Age
A Dutch researcher discusses his theory that the atmospheric carbon dioxide equilibrium shifted when 40% of Europe's human population was killed by a plague in the 14th century. Could this have caused "the little ice age"?

www.science.nelson.com

Key Terms

15.1
closed system
equilibrium
phase equilibrium
solubility equilibrium
chemical reaction equilibrium
dynamic equilibrium theory
forward reaction
reverse reaction

ICE table
equilibrium constant, K_c
equilibrium law

15.2
Le Châtelier's principle
equilibrium shift

Key Equations

For the reaction $aA + bB \rightleftharpoons cC + dD$,

the equilibrium law is $K_c = \dfrac{[C]^c[D]^d}{[A]^a[B]^b}$

▶ MAKE a summary

1. Make a concept map, beginning with the word "Equilibrium" in the centre of a page. Link all of the Key Terms from this chapter, together with points of your own, to explain and illustrate how connections among these terms include the equilibrium law expression, ICE table format, Le Châtelier's principle, and points from the section Summaries.

2. Refer back to your answers to the Starting Points questions at the beginning of this chapter. How has your thinking changed?

▶ Go To www.science.nelson.com

The following components are available on the Nelson Web site. Follow the links for *Nelson Chemistry Alberta 20-30*.
- an interactive Self Quiz for Chapter 15
- additional Diploma Exam-style Review questions
- Illustrated Glossary
- additional IB-related material

There is more information on the Web site wherever you see the Go icon in this chapter.

Chapter 15 REVIEW

Many of these questions are in the style of the Diploma Exam. You will find guidance for writing Diploma Exams in Appendix H. Exam study tips and test-taking suggestions are on the Nelson Web site. Science Directing Words used in Diploma Exams are in bold type.

DO NOT WRITE IN THIS TEXTBOOK.

Part 1

1. A "closed" system for a chemical equilibrium means that
   A. the reaction container must be solid and sealed, with a fixed volume
   B. no substance involved in the equilibrium must be able to enter or leave
   C. pressure and temperature in the reaction vessel must be constant
   D. no chemical of any kind may be added to the reaction vessel

2. The possible equilibrium that is *not* dynamic is the one between
   A. the liquid and gas phases of octane
   B. chromate and dichromate ions in aqueous solution
   C. the downward force of gravity exerted by Earth on an object, and the upward force of a balance exerted on the object
   D. the oxygen dissolved in water in a lake and the oxygen dissolved in nitrogen in the atmosphere

3. In a reaction, 4.22 mol of product has formed but there are no more visible signs of change. Calculations show that as much as 6.00 mol of product could have formed from the chemical amounts of reactants used. The percent yield is _____ %.

Use this information to answer questions 4 to 6.

Sulfuric acid is the most common commercial acid, with millions of tonnes produced each year (**Figure 1**). The second step in the "contact" process for industrial production of sulfuric acid involves the oxidation of sulfur dioxide gas catalyzed by contact with $V_2O_5(s)$ powder. The reaction equation is

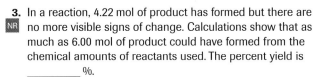

Figure 1
Most sulfuric acid is produced in plants such as this one by the contact process, which includes two exothermic combustion reactions. Sulfur reacts with oxygen, forming sulfur dioxide; then, sulfur dioxide, in contact with a catalyst, reacts with oxygen, forming sulfur trioxide. Sulfur trioxide and water form sulfuric acid.

4. Which is the correct form of the equilibrium law for this reaction, as the equation is written?

   A. $K_c = \dfrac{[SO_2(g)]^2[O_2(g)]}{[SO_3(g)]}$

   B. $K_c = [SO_2(g)]^2[O_2(g)][SO_3(g)]^2$

   C. $K_c = \dfrac{[SO_2(g)]^2[O_2(g)]}{[SO_3(g)]^2}$

   D. $K_c = \dfrac{[SO_3(g)]^2}{[SO_2(g)]^2[O_2(g)]}$

5. The imposed condition that does *not* shift equilibrium toward the product is
   A. adding the vanadium pentoxide catalyst
   B. decreasing the temperature
   C. decreasing the container volume
   D. adding oxygen to the system container

6. If the equilibrium constant, K_c, for the above oxidation reaction at a given temperature is 2.4×10^3, then the constant for the reverse reaction (written as the decomposition of sulfur trioxide) is
   A. 2.4×10^3
   B. 4.2×10^{-4}
   C. 2.4×10^{-3}
   D. 4.2×10^4

7. If excess copper reacts in a solution of silver nitrate, it is correct to state that
   A. the reaction is not considered quantitative
   B. the equilibrium constant at SATP will have a numerical value between 1 and 100
   C. water is not written in the equilibrium law expression because it is a spectator species
   D. silver nitrate is not written in the equilibrium law expression because it is a species with a constant concentration

Use this information to answer questions 8 to 10.

Consider the following reaction to produce hydrogen, done as a first step in the industrial process to make ammonia. Methane (from natural gas) reacts with steam over a nickel powder catalyst. The reaction equation is

$$CH_4(g) + 2\,H_2O(g) + \text{heat energy} \rightleftarrows CO_2(g) + 4\,H_2(g)$$

8. This reaction is done in a laboratory autoclave (a stainless steel pressure vessel) and allowed to reach equilibrium. More methane is then injected into the autoclave. We can predict that, when a new equilibrium is reached (at the same temperature), the concentration of every reagent in the equation will have increased, *except* that of
   A. methane
   B. water
   C. carbon dioxide
   D. hydrogen

9. When this gaseous reaction system at equilibrium is disturbed by heating the autoclave, we theorize that
   A. both forward and reverse reaction rates increase, but the forward rate increases more
   B. the forward reaction rate does not change, but the reverse reaction rate increases
   C. the reverse reaction rate does not change, but the forward reaction rate increases
   D. both reaction rates increase equally, so the equilibrium position is unchanged

10. A test reaction is done starting with only methane and excess water in the autoclave. If the initial concentration of methane is 0.110 mol/L, and the methane concentration at equilibrium is 0.010 mol/L, then the equilibrium concentration of hydrogen is _____ mol/L.
    NR

Part 2

11. **Define** chemical equilibrium empirically.

12. What main idea explains chemical equilibrium?

13. What phrase is used to **describe** a reaction equilibrium in which the proportion of reactants to products is quite high?

14. **Describe** and **explain** a situation in which a carbonated soft drink is in
    (a) a non-equilibrium state
    (b) an equilibrium state

15. **Predict** whether adding a catalyst affects a state of equilibrium. What does the catalyst do?

16. For each of the following descriptions, write a chemical equation for the system at equilibrium. Communicate the position of the equilibrium with equilibrium arrows. Then write a mathematical expression of the equilibrium law for each chemical system.
    (a) A combination of low pressure and high temperature provides a percent yield of less than 10% for the formation of ammonia in the Haber process.
    (b) At high temperatures, the formation of water vapour from hydrogen and oxygen is quantitative.
    (c) The reaction of carbon monoxide with water vapour to produce carbon dioxide and hydrogen has a percent yield of 67% at 500 °C.

17. Scientists and technologists are particularly interested in the use of hydrogen as a fuel. **Interpret** this reaction equation by predicting the relative proportions of reactants and products in this system at equilibrium.

$$2\,H_2(g) + O_2(g) \rightleftarrows 2\,H_2O(g) \quad K_c = 1 \times 10^{80} \text{ at SATP}$$

Use this information to answer questions 18 to 24.

The solubility of pure oxygen in contact with liquid water at SATP is very low: only about 42 ppm, or 42 mg/L. The solubility equilibrium when water is in contact with air (21% oxygen) at SATP, is even lower: about 8.7 mg/L. The equilibrium equation is

$$O_2(g) \rightleftarrows O_2(aq)$$

18. Write the equilibrium law expression for a saturated solution of oxygen in water.

19. If the gas in the closed system is pure oxygen, the solubility is higher than it is if the gas is air. Express the solubility of pure oxygen in water at SATP as an amount concentration.

20. Express the concentration of pure oxygen gas at SATP as an amount concentration. (Recall that the molar volume of gases at SATP is 24.8 L/mol.)

21. Use the answers to the previous two questions to **determine** the value of K_c for an equilibrium of pure oxygen gas in contact with its saturated solution at SATP.

22. **Predict** whether the value of K_c for this equilibrium will be different for an equilibrium of air in contact with water at 25 °C. **Predict** which system condition changes will, and which will not, change the value of an equilibrium constant.

23. In terms of the equilibrium law, **explain** why more oxygen dissolves in water when the gas above it is pure oxygen than when the gas above it is air.

24. The quality of surface water in lakes and streams (**Figure 2**) is of critical importance to society. Dissolved oxygen content of the surface fresh water in Canada averages 10 ppm. **Explain** what stress is placed upon fish and other organisms in streams, lakes, and wetlands if climate change increases the average temperature of the water.

Figure 2
The Bow River begins as a cold, glacier-fed mountain stream with a higher-than-average oxygen content, and is world famous for its trout fly fishery. It also supplies the city of Calgary with water for a million people daily, as well as providing water for agriculture in southern Alberta.

25. In many processes in industry, engineers try to maximize the yield of a product. **Outline** how concentration can be manipulated in order to increase the yield of a product.

26. In a container at high temperature, ethyne (acetylene) and hydrogen react to produce ethene (ethylene). No ethene is initially present.

$$C_2H_2(g) + H_2(g) \rightleftarrows C_2H_4(g)$$

The initial concentrations of both acetylene and hydrogen are 1.00 mol/L. Later, at equilibrium, the concentration of ethene is 0.060 mol/L.
(a) Use an ICE table to **determine** the equilibrium constant.
(b) **Sketch** a reaction progress graph to show the change in concentration values over time, from the beginning of the reaction to equilibrium.

27. $H_2(g) + Br_2(g) \rightleftarrows 2\,HBr(g)$ $K_c = 12.0$ at t °C
(a) 8.00 mol of hydrogen and 8.00 mol of bromine are added to a 2.00 L reaction container. Construct an ICE table and use it to **predict** the concentrations at equilibrium.
(b) 12.0 mol of hydrogen and 12.0 mol of bromine are added to a 2.00 L reaction container. Construct an ICE table and use it to **predict** the concentrations at equilibrium.

28. $CO(g) + H_2O(g) \rightleftarrows CO_2(g) + H_2(g)$ $K_c = 4.00$ at 900 °C

In a container, carbon monoxide and water vapour react to produce carbon dioxide and hydrogen. The equilibrium concentrations are
$[H_2O(g)] = 2.00$ mol/L, $[CO_2(g)] = 4.00$ mol/L, and $[H_2(g)] = 2.00$ mol/L.
Determine the equilibrium concentration of carbon monoxide.

29. Write a statement of Le Châtelier's principle.

30. What variables are commonly manipulated to shift a chemical equilibrium system?

31. **Describe** how a change in volume of a closed system containing a gaseous reaction at equilibrium affects the pressure of the system.

32. In a sealed container, nitrogen dioxide is in equilibrium with dinitrogen tetraoxide.

$$2\,NO_2(g) \rightleftarrows N_2O_4(g) \quad K_c = 1.15,\ t = 55\ °C$$

(a) Write the mathematical expression for the equilibrium law applied to this chemical system.
(b) If the equilibrium concentration of nitrogen dioxide is 0.050 mol/L, **predict** the concentration of dinitrogen tetraoxide.
(c) **Predict** the shift in equilibrium that will occur when the concentration of nitrogen dioxide is increased.

33. **Predict** the shift in the following equilibrium system resulting from each of the following changes:

$$4\,HCl(g) + O_2(g) \rightleftarrows 2\,H_2O(g) + 2\,Cl_2(g) + 113\ kJ$$

(a) an increase in the temperature of the system
(b) a decrease in the system's total pressure due to an increase in the volume of the container
(c) an increase in the concentration of oxygen
(d) the addition of a catalyst

34. Chemical engineers use Le Châtelier's principle to predict shifts in chemical systems at equilibrium resulting from changes in the reaction conditions. **Predict** the changes necessary to maximize the yield of product in each of the following industrial chemical systems:
(a) the production of ethene (ethylene)
$$C_2H_6(g) + \text{energy} \rightleftarrows C_2H_4(g) + H_2(g)$$
(b) the production of methanol
$$CO(g) + 2\,H_2(g) \rightleftarrows CH_3OH(g) + \text{energy}$$

35. Apply Le Châtelier's principle to **predict** whether, and in which direction, the following established equilibrium would be shifted by the change imposed:

    $2 CO(g) + O_2(g) \rightleftarrows 2 CO_2(g) + \text{heat energy}$

    (a) temperature is increased
    (b) vessel volume is increased
    (c) oxygen is added
    (d) platinum catalyst is added
    (e) carbon dioxide is removed

36. For each example, **predict** whether, and in which direction, an established equilibrium would be shifted by the change imposed. **Explain** any shift in terms of changes in forward and reverse reaction rates.

    (a) $Cu^{2+}(aq) + 4 NH_3(g) \rightleftarrows Cu(NH_3)_4^{2+}(aq)$
    $CuSO_4(s)$ is added
    (b) $CaCO_3(s) + \text{energy} \rightleftarrows CaO(s) + CO_2(g)$
    temperature is decreased
    (c) $Na_2CO_3(s) + \text{energy} \rightleftarrows Na_2O(s) + CO_2(g)$
    sodium carbonate is added
    (d) $H_2CO_3(aq) + \text{energy} \rightleftarrows CO_2(g) + H_2O(l)$
    vessel volume is decreased
    (e) $KCl(s) \rightleftarrows K^+(aq) + Cl^-(aq)$
    $AgNO_3(s)$ is added
    (f) $CO_2(g) + NO(g) \rightleftarrows CO(g) + NO_2(g)$
    vessel volume is increased
    (g) $Fe^{3+}(aq) + SCN^-(aq) \rightleftarrows FeSCN^{2+}(aq)$
    $Fe(NO_3)_3(s)$ is added

37. **Predict** in which of the following equilibria a decrease in temperature favours the forward reaction.

    (a) $Br_2(l) \rightleftarrows Br_2(g)$
    (b) $N_2(g) + 3 H_2(g) \rightleftarrows 2 NH_3(g)$ $\Delta_r H$ is negative
    (c) $LiCl(s) \rightleftarrows Li^+(aq) + Cl^-(aq) + \text{heat}$
    (d) $6 C(s) + 3 H_2(g) \rightleftarrows C_6H_6(l)$ $\Delta_r H = +49$ kJ
    (e) $CaCO_3(s) + \text{energy} \rightleftarrows CaO(s) + CO_2(g)$

Use this information to answer questions 38 and 42.

Alberta's petroleum industry has a chronic problem—all fossil fuels found in the province contain some sulfur. If not removed, this sulfur will react upon burning to release $SO_2(g)$. Sulfur dioxide is very irritating to lung tissue and is highly corrosive; thus, it is a major contributor to air pollution, acid rain, and respiratory disease. Furthermore, sulfur impurities may damage the fuel injection and anti-pollution systems of modern internal combustion engines if not removed from gasoline and diesel fuels. For these reasons, recent Canadian legislation requires that sulfur content in diesel fuels sold for on-road use must be less than 15 ppm (15 mg/kg) by July, 2006.

In a refinery or bitumen upgrader, the sulfur is first removed from fossil fuel feedstock by cracking and/or hydrogenation, resulting in reaction of the sulfur to produce (*extremely* toxic) $H_2S(g)$. Standard industry technology to remove hydrogen sulfide gas from petrochemical gas stream mixtures involves the use of an amine scrubber unit, in a two-step process that depends on two kinds of equilibrium. For simplicity, assume that an amine scrubber reaction vessel contains a 25% aqueous solution of diethanolamine $(C_2H_4OH)_2NH(aq)$, which approximates the actual solution used in the various oil sands plants in Alberta.

38. The scrubbing operation involves a gas mixture of low-molar-mass hydrocarbons, carbon dioxide, and hydrogen sulfide. This mixture of gases is injected into the scrubber unit at high pressure, with the scrubber solution at about 40 °C. For hydrogen sulfide, first the toxic gas dissolves

    $H_2S(g) \rightleftarrows H_2S(aq)$ (negative $\Delta_r H$)

    which is followed immediately by the chemical reaction.

    $(C_2H_4OH)_2NH(aq) + H_2S(aq) \rightleftarrows$
    $(C_2H_4OH)_2NH_2^+(aq) + HS^-(aq)$ (negative $\Delta_r H$)

    (a) Draw a structural formula for diethanolamine.
    (b) **Explain**, using Le Châtelier's principle, how the relatively low temperature and high pressure act to help the scrubber solution "absorb" toxic hydrogen sulfide.

39. When the unabsorbed hydrocarbon gases are removed from the scrubber unit, the sulfur atoms remain behind, trapped in solution as hydrogen sulfide ions. In the next process step, the scrubber solution is "regenerated": the absorbed toxic compound is now emitted from solution and removed from the reaction vessel. Use Le Châtelier's principle to **describe** how conditions should be altered in the scrubber vessel, to shift equilibrium positions to make this process as efficient as possible.

40. When carbon dioxide gas "dissolves" in water, the process is more correctly thought of as being an exothermic chemical reaction *with* water, to form aqueous carbonic acid. Soft drink beverages are "carbonated" in this way, at high pressure. Write and balance an equilibrium equation for this reaction.

41. Carbonic acid, H_2CO_3(aq), reacts exothermically with basic aqueous diethanolamine in essentially the same way that hydrosulfuric acid, H_2S(aq), does in question 38. Write and balance the equilibrium equation for this reaction, and use it to **explain** whether the process conditions for scrubbing H_2S(g) should work to remove CO_2(g) as well.

Extension

42. Work cooperatively to research, assemble, and present a more complete and accurate summary of the operation of a typical Alberta industrial amine "scrubber" system.
    Your presentation should include
    - a description of the role of chemical equilibrium in the system
    - information on applications of this process throughout Alberta
    - the use of the best features of any available word processing or slideshow software

    www.science.nelson.com

43. A halogen light bulb contains a tungsten (wolfram) filament, W(s), in a mixed atmosphere of a noble gas and a halogen; for example, Ar(g) and I_2(g) (**Figure 3**). The operation of a halogen lamp depends, in part, on the equilibrium system

    $$W(s) + I_2(g) \rightleftharpoons WI_2(g)$$

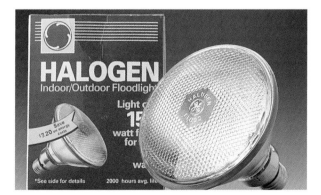

Figure 3
Halogen light bulbs are more efficient than ordinary incandescent bulbs, but they burn so hot they may require special fixtures, and must be treated with extra care.

Research and **describe** the role of temperature in the operation of a halogen lamp. For example, **how** is it possible for a halogen lamp to operate with the filament at 2700 °C when the tungsten normally would not last very long at this high temperature? **Why** is such a high temperature desirable?

www.science.nelson.com

44. When the Olympic Games were held in Mexico in 1968, many athletes arrived early to train in the higher altitude (2.3 km) and lower atmospheric pressure of Mexico City. Exertion at high altitudes, for people who are not acclimatized, may make them dizzy or "lightheaded" from lack of oxygen. **Explain** this observation. Your explanation should include
    - the theory of dynamic equilibrium
    - Le Châtelier's principle
    - a description of how people who normally live at high altitudes are physiologically adapted to their reduced-pressure environment

    www.science.nelson.com

chapter 16

Equilibrium in Acid–Base Systems

In this chapter

- Exploration: Salty Acid or Acidic Salt?
- Lab Exercise 16.A: The Chromate-Dichromate Equilibrium
- Web Activity: Edgar Steacie
- Investigation 16.1: Creating an Acid–Base Strength Table
- Lab Exercise 16.B: Predicting Acid–Base Equilibria
- Web Activity: Pool Chemistry
- Lab Exercise 16.C: Aqueous Bicarbonate Ion Acid–Base Reactions
- Investigation 16.2: Testing Brønsted-Lowry Reaction Predictions
- Lab Exercise 16.D: Creating an Acid–Base Table
- Case Study: Changing Ideas on Acids and Bases—The Evolution of a Scientific Theory
- Web Activity: Titration of Polyprotic Acids and Bases
- Biology Connection: Homeostasis
- Web Activity: Preparation of Buffer Solutions
- Web Activity: Maud Menten
- Investigation 16.3: Testing a Buffer Effect

The nature of science involves constant questioning and testing of theories—and so it is with theories about acids and bases. A great many chemical reaction systems involve acids in some way, including the one that begins the decomposition (digestion) of the food you eat, and, thus provides you with energy. Many other types of aqueous reaction systems have rates that are easily controlled by adjusting the level of acidity. Because such systems are commonly found both in nature and in industry, scientists seek to work with the most complete and successful acid–base concepts—which in turn means seeking new ways to test previously accepted theories. Theories that do not describe, explain, and predict the chemistry of acids and bases well enough will initially be restricted to only those situations where they work. As a result of further testing, scientists will either revise the theory, or replace it altogether.

This chapter presents new hypotheses, evidence, and analyses to help you develop a more comprehensive understanding of both the aqueous reaction environment, and the activity of acids and bases within that environment. Your knowledge of chemical equilibrium (Chapter 15) allows you to explore these questions from a new perspective, and, in turn, will allow you to form a more complete and less restrictive theory of acids and bases. In fact, few topics in chemistry illustrate this scientific principle of ongoing theory testing and development so well.

These underlying principles—that theories must be supported by evidence, and that understanding is increased by always questioning and testing existing knowledge—are the basis of the uniquely productive "way of knowing" about the natural world that we call science. To these principles, and to the enormous accumulation of knowledge they have made possible, we owe most aspects of our present technological civilization.

STARTING Points

Answer these questions as best you can with your current knowledge. Then, using the concepts and skills you have learned, you will revise your answers at the end of the chapter.

1. How can some substances neutralize *both* acids and bases?
2. Can acid–base reactions and their products be predicted? Explain.
3. Can pH curves for titrations of weak acids and weak bases predict equivalence points as strong acid–strong base pH curves do? Explain.
4. How is the buffering of some medications related to stomach fluid acidity?

Career Connection:
Environmental Engineer; Chemistry Researcher; Microbiologist

Figure 1
Testing your concepts is a continual part of "doing" science.

▶ Exploration — Salty Acid or Acidic Salt?

In Chapter 5, you learned about saturated solutions as examples of dynamic equilibrium, and prepared a saturated sodium chloride solution. The strong acid, HCl(aq), shares half of its chemical formula with NaCl(aq). How might the two solutions interact?

Materials: two 10 mL graduated cylinders; saturated sodium chloride solution, NaCl(aq); concentrated hydrochloric acid, HCl(aq)

(a) Write and balance a chemical reaction equation for any reaction you can predict when sodium chloride solution and concentrated hydrochloric acid are mixed.

(b) Write an equation to express the nature of a saturated sodium chloride aqueous solution as a dynamic equilibrium.

(c) Write the equilibrium law expression for a saturated aqueous sodium chloride solution. Identify the substance with a constant concentration.

 Hydrochloric acid is corrosive. Wear appropriate eye protection, lab gloves, and a lab apron.

- Measure approximately 8 mL of saturated sodium chloride solution into the larger graduated cylinder.

- Measure approximately 2 mL of concentrated hydrochloric acid into the smaller graduated cylinder.

- Pour the concentrated hydrochloric acid into the cylinder containing the saturated sodium chloride solution. Record your observations.

- Dispose of all substances down the sink, using lots of water.

(d) Explain what has apparently happened to the saturated NaCl(aq) equilibrium.

(e) Explain which ion concentration in the large cylinder *must* have changed, whether it was increased or decreased, and what principle you use to "know" this answer.

(f) Which initial solution was more concentrated? Use Le Châtelier's principle to explain how you "know" this answer.

(g) Was the initial HCl(aq) a saturated solution? Was it at equilibrium? Was it at equilibrium before being removed from its closed storage container? Explain how you can use the basic principles of equilibrium to "know" these answers.

16.1 Water Ionization and Acid–Base Strength

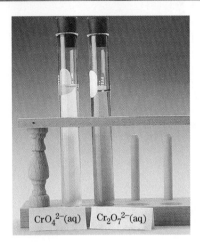

Figure 1
The chromate–dichromate aqueous ion system, in equilibria at high and low pH

In many of the preceding units, you have studied examples of chemical reactions and systems that in some way depend on the nature of acids and bases and/or the pH of solution. For example, aqueous permanganate ions are powerful oxidizing agents, but can act to oxidize other reagents only in the presence of hydrogen (hydronium) ions. Almost any soluble R–COOH organic compound will make an aqueous solution with a pH below 7, because the hydrogen atom of such a group is relatively easily removed. The electrolysis of potassium iodide solution produces a solution that is strongly basic. Understanding these connections involves considering equilibrium effects, as shown by the chromate–dichromate aqueous ion system (**Figure 1**).

You have already learned several concepts about the strengths and properties of acids and bases (see Chapter 6 Summary, page 262). If we now combine equilibrium concepts with these acid–base concepts, we can develop a much more comprehensive understanding of acids and bases. This understanding, in turn, will allow you to better explain and predict how acids and bases behave. For instance, the reaction examined in Lab Exercise 16.A is an easily explained example of how the acidity of a solution can directly affect other ion equilibria.

LAB EXERCISE 16.A

The Chromate–Dichromate Equilibrium

Report Checklist
- ○ Purpose
- ● Design
- ○ Analysis
- ○ Problem
- ○ Materials
- ○ Evaluation
- ○ Hypothesis
- ○ Procedure
- ● Prediction
- ○ Evidence

In an aqueous solution, chromate ions are in equilibrium with dichromate ions (Figure 1).

$$2\,CrO_4^{2-}(aq) + 2\,H^+(aq) \rightleftarrows Cr_2O_7^{2-}(aq) + H_2O(l)$$

Complete the Prediction and Design (including diagnostic tests) of the investigation report.

Purpose
The scientific purpose of this investigation is to test a Design for varying the acidity of an equilibrium.

Problem
How does changing the hydrogen ion concentration affect the chromate–dichromate equilibrium?

Hypothesis
The position of this equilibrium depends on the acidity of the solution.

Learning Tip

For discussing acid–base reactions involving weak acids or bases, or when comparing acid or base strengths in aqueous solution, it is normally necessary to represent the "hydrogen" ion as a "hydronium" ion, $H_3O^+(aq)$. Chapter 16 will use this ion representation exclusively, from this point on.

For simplicity, a great many chemical reaction equations use $H^+(aq)$ to represent an aqueous hydrogen ion. This representation often works very well, as in the redox equations used in Unit 7, or the titration analysis equations used for stoichiometric calculations in Chapter 8. For other purposes, however, it is necessary to represent this ion more accurately in order to understand and explain the theoretical nature of the reaction. As you learned in Chapter 6, it is more consistent with evidence to think of this entity as a hydronium ion, and to represent it as $H_3O^+(aq)$. Acid–base equilibrium theory necessarily involves collision–reaction theory for a variety of entities. This chapter will, therefore, use the hydronium ion convention almost exclusively. Before aqueous acidic or basic solution equilibrium can be investigated further, the equilibrium nature of the ions of the solvent (water) must first be examined, understood, and taken into consideration.

The Water Ionization Constant, K_w

Even highly purified water has a very slight conductivity that is only observable if measurements are made with very sensitive instruments (**Figure 2**). According to Arrhenius' theory, conductivity is due to the presence of ions (**Figure 3**). Therefore, the conductivity observed in pure water must be the result of ions produced by the ionization of some water molecules into hydronium ions and hydroxide ions. Because the conductivity is so slight, the equilibrium at SATP must greatly favour the water molecules.

Figure 2
A sensitive multimeter is required to detect the electrical conductivity of the highly purified water that is typically used in a chemistry laboratory. (See Appendix C.3 for guidance on using a multimeter.)

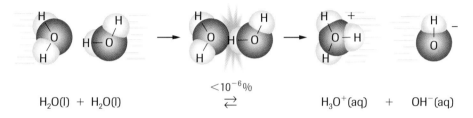

$$H_2O(l) + H_2O(l) \underset{<10^{-6}\%}{\rightleftarrows} H_3O^+(aq) + OH^-(aq)$$

Figure 3
Collisions of water molecules very occasionally produce this cation–anion pair.

$$K_c = \frac{[H_3O^+(aq)][OH^-(aq)]}{[H_2O(l)][H_2O(l)]} \quad \text{or} \quad K_c = [H_3O^+(aq)][OH^-(aq)]$$

Evidence indicates that, at 25 °C, at any given moment, fewer than two of every billion molecules in pure liquid water exist in ionized form! When we write an equilibrium constant expression for this ionization, the value of K_c is an extremely small number. Recall that, because pure liquid water (or the water in any dilute aqueous solution) has an essentially constant concentration, it does not appear in the expression; it is simply incorporated into the equilibrium constant value.

The water ionization equilibrium relationship is so important in chemistry that this particular K_c constant is given its own special symbol and name. This new constant is called the ion product or **ionization constant for water, K_w**.

$$K_w = [H_3O^+(aq)][OH^-(aq)] = 1.00 \times 10^{-14} \text{ at SATP}$$

The equilibrium equation for the ionization of water shows that hydronium ions and hydroxide ions form in a 1:1 ratio. Therefore, the concentration of hydronium ions and hydroxide ions in pure water must be equal. This equality must also be true for any neutral aqueous solution. Using the mathematical expression for K_w, and the value of K_w at SATP, the concentrations of $H_3O^+(aq)$ and $OH^-(aq)$ can be calculated by taking the square root of the K_w value. Recall (from Chapter 15) that in all entity concentration calculations from any K_c value, the entity concentration is simply assumed to have units of mol/L.

$$[H_3O^+(aq)] = [OH^-(aq)] = \sqrt{1.00 \times 10^{-14}} = 1.00 \times 10^{-7} \text{ mol/L}$$

The ionization of water is especially important in the empirical and theoretical study of acidic and basic solutions. Recall from Chapter 6 that, according to the modified Arrhenius theory, an acid is a substance that reacts with water to produce hydronium ions. The additional hydronium ions provided by the acid increase the hydronium ion concentration. Since the hydronium ion concentration is greater than 10^{-7} mol/L, the solution is *acidic*. A *basic* solution is one in which the hydroxide ion concentration is greater

DID YOU KNOW ?

Successful Collisions
A collision that successfully forms hydronium and hydroxide ions is very rare. This is not because *collisions* are rare—each water molecule collides with others tens of trillions of times every second! But, the chance that any given collision will both have sufficient energy, and also be at exactly the right orientation, is very, very small indeed. Ordinarily, an equilibrium so strongly favouring the reverse reaction would just be ignored—thought of as not happening at all—but, in this case, the ions are uniquely important because of the effects they have on all reactions in aqueous solution.

Conversely, you already know that the reverse reaction (hydronium ions with hydroxide ions) is quantitative. Almost every such ion collision will be effective because the required energy is very low, and the attraction of opposite ion charges acts to orient the entities correctly.

Modelling water molecules this way—as space-filling models with superimposed atomic symbols and lines to represent bonds—gives an overall representation of the process that is logically consistent with experimental evidence.

Learning Tip

Keep in mind that the value for K_w is subject to the same restrictions as any other equilibrium constant: one being that it will change if the temperature changes. For example, in pure water at 20 °C, K_w has a numerical value of 6.76×10^{-15}, whereas at 30 °C its value is 1.47×10^{-14}.

K_w will also change enough to be invalid in any aqueous solution with a very high solute concentration—because then the assumption that the concentration of the water solvent is at or near a constant value (55.5 mol/L) no longer holds true.

CAREER CONNECTION

Environmental Engineer

Monitoring and protecting water quality is an essential part of the work of environmental engineers. They design systems and treatment processes that control pH levels, temperature, and amounts of dissolved oxygen for industries that use water.

Would you like to become an environmental engineer? Look into salaries, employment opportunities, and educational programs at at least two different educational institutions.

www.science.nelson.com

Learning Tip

These communication examples show the application of the usual units convention for simplifying calculations from equilibrium constants. Units for the constant are ignored, and other concentration units are always entered in mol/L. Then, since the units for the calculated value are always mol/L, they are just written in with the answer.

---

than 10^{-7} mol/L. Basic solutions are produced in two ways: either by the complete dissociation (upon dissolving in water) of an ionic hydroxide, or by partial reaction of some weak base entity (ion or molecule) with water to produce hydroxide ions.

The most important point about K_w is that it applies to pure water, and also to any solution that is mostly water. This means that this ionization equilibrium will be involved in any other reaction going on in aqueous solution, if that reaction involves hydronium ions or hydroxide ions in any way. As an example, consider the equilibrium reaction just studied in Lab Exercise 16.A. Since that reaction involves hydronium (hydrogen) ions, the chromate–dichromate equilibrium can be controlled (shifted) easily, simply by adjusting either the hydronium ion or the hydroxide ion concentration; which really just means deliberately shifting the water ionization equilibrium. A great many chemical reactions are dependent in this way on the water ionization equilibrium. Many of them (those containing $H_3O^+(aq)$ or $OH^-(aq)$ ions) are evident in the Relative Strengths of Oxidizing and Reducing Agents table (Appendix I).

Note that, since the mathematical relationship is simple, we can easily use K_w to calculate either the hydronium ion amount concentration or the hydroxide ion amount concentration in an aqueous solution, if the other concentration is known.

Since $[H_3O^+(aq)][OH^-(aq)] = K_w$

then $[H_3O^+(aq)] = \dfrac{K_w}{[OH^-(aq)]}$

and $[OH^-(aq)] = \dfrac{K_w}{[H_3O^+(aq)]}$

In ordinary dilute aqueous acidic or basic solutions, the presence of substances other than water decreases the certainty of the K_w value at 25 °C to two significant digits.

All questions and examples in this text assume temperatures of 25 °C, and aqueous solutions that are not highly concentrated, with a K_w value of 1.0×10^{-14}, unless specifically stated otherwise.

▶ COMMUNICATION *example 1*

A 0.15 mol/L solution of hydrochloric acid at 25 °C is found to have a hydronium ion concentration of 0.15 mol/L. Calculate the amount concentration of the hydroxide ions.

Solution

$HCl(aq) + H_2O(l) \rightleftharpoons H_3O^+(aq) + Cl^-(aq)$

$[OH^-(aq)] = \dfrac{K_w}{[H_3O^+(aq)]}$

$= \dfrac{1.0 \times 10^{-14}}{0.15 \text{ mol/L}}$ (entity concentration units assumed to be mol/L)

$= 6.7 \times 10^{-14}$ mol/L

Using the K_w relationship, the hydronium ion concentration is 6.7×10^{-14} mol/L.

► COMMUNICATION example 2

Calculate the amount concentration of the hydronium ion in a 0.25 mol/L solution of barium hydroxide.

Solution

$$Ba(OH)_2(s) \longrightarrow Ba^{2+}(aq) + 2\,OH^-(aq)$$

$$[OH^-(aq)] = 2 \times [Ba(OH)_2(aq)]$$
$$= 2 \times 0.25 \text{ mol/L}$$
$$= 0.50 \text{ mol/L}$$

$$[H_3O^+(aq)] = \frac{K_w}{[OH^-(aq)]}$$
$$= \frac{1.0 \times 10^{-14}}{0.50 \text{ mol/L}}$$
$$= 2.0 \times 10^{-14} \text{ mol/L}$$

Using the K_w relationship, the hydronium ion concentration is 2.0×10^{-14} mol/L.

► COMMUNICATION example 3

Determine the hydronium ion and hydroxide ion amount concentrations in 500 mL of an aqueous solution for home soap-making containing 2.6 g of dissolved sodium hydroxide.

Solution

$$n_{NaOH} = 2.6 \text{ g} \times \frac{1 \text{ mol}}{40.00 \text{ g}} = 0.065 \text{ mol}$$

$$[NaOH(aq)] = \frac{0.065 \text{ mol}}{0.500 \text{ L}} = 0.13 \text{ mol/L}$$

$$NaOH(s) \rightarrow Na^+(aq) + OH^-(aq)$$

$$[OH^-(aq)] = [NaOH(aq)] = 0.13 \text{ mol/L}$$

$$[H_3O^+(aq)] = \frac{K_w}{[OH^-(aq)]}$$
$$= \frac{1.0 \times 10^{-14}}{0.13 \text{ mol/L}}$$
$$= 7.7 \times 10^{-14} \text{ mol/L}$$

Using the K_w relationship, the hydronium ion concentration is 7.7×10^{-14} mol/L, and the hydroxide ion concentration is 0.13 mol/L.

DID YOU KNOW ?

Warnings on Packaging
Many households, in previous centuries, kept supplies of lye (sodium hydroxide) on hand for making soap. Because it looked like sugar, it was occasionally swallowed by curious children, causing terrible injuries to their throats. A prominent American physician, Dr. Chevalier Jackson (1865–1958) realized that warnings on the packaging would encourage parents to keep this dangerous substance out of the reach of their children. This is one of the earliest instances of warning labelling on packaging. Partially because of Dr. Jackson's efforts, the United States Congress passed the Federal Caustic Labelling Act in 1927. In Canada, we have the Consumer Chemicals and Containers Regulations.

Figure 4
Many solutions sold for cleaning windows contain ammonia.

COMMUNICATION example 4

Calculate the amount concentration of hydronium ions in a 0.100 mol/L aqueous solution of ammonia that is used in a spray bottle for window cleaning solution (**Figure 4**). A reference states that there is 2.1% reaction of dissolved ammonia with water (ionization) at this concentration, at SATP.

Solution

$$NH_3(aq) + H_2O(l) \rightleftarrows NH_4^+(aq) + OH^-(aq)$$

$$[OH^-(aq)] = \frac{2.1}{100} \times 0.100 \text{ mol/L}$$

$$= 0.0021 \text{ mol/L}$$

$$[H_3O^+(aq)] = \frac{K_w}{[OH^-(aq)]}$$

$$= \frac{1.0 \times 10^{-14}}{0.0021 \text{ mol/L}}$$

$$= 4.8 \times 10^{-12} \text{ mol/L}$$

Using the K_w relationship, in 0.100 mol/L aqueous ammonia, the hydronium ion concentration is 4.8×10^{-12} mol/L.

Practice

1. The hydronium ion concentration in an industrial effluent is 4.40 mmol/L. Determine the concentration of hydroxide ions in the effluent.
2. The hydroxide ion concentration in a household cleaning solution is 0.299 mmol/L. Calculate the hydronium ion concentration in the cleaning solution.
3. Calculate the hydroxide ion amount concentration in a solution prepared by dissolving 0.37 g of hydrogen chloride in 250 mL of water.
4. Calculate the hydronium ion amount concentration in a saturated solution of calcium hydroxide (limewater) that has a solubility of 6.9 mmol/L.
5. What is the hydronium ion amount concentration in a solution made by dissolving 20.0 g of potassium hydroxide in water to form 500 mL of solution?
6. Calculate the percent ionization of water at SATP. Recall that 1.000 L of water has a mass of 1000 g.

Communicating Concentrations: pH and pOH

Recall (Chapter 6) that the enormous range of aqueous solution hydronium ion concentrations is more easily expressed using the logarithmic pH scale (**Figure 5**). Mathematically,

$$\text{pH} = -\log[H_3O^+(aq)] \text{ and, inversely, } [H_3O^+(aq)] = 10^{-\text{pH}}$$

For basic solutions, it is sometimes more useful to use a scale based on the amount concentration of hydroxide ions. Recall that the definition of pOH follows the same format and the same certainty rule as pH.

$$\text{pOH} = -\log[OH^-(aq)] \text{ and, inversely, } [OH^-(aq)] = 10^{-\text{pOH}}$$

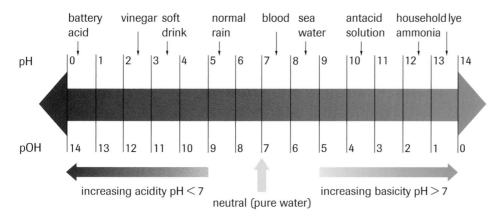

Figure 5
The pH scale

The mathematics of logarithms allows us to express a simple relationship between pH and pOH. According to the rules of logarithms,

$$\log(ab) = \log(a) + \log(b)$$

Using the equilibrium law for the ionization of water,

$$[H_3O^+(aq)][OH^-(aq)] = K_w$$

$$\log[H_3O^+(aq)] + \log[OH^-(aq)] = \log(K_w)$$

$$(-pH) + (-pOH) = -14.00$$

$$pH + pOH = 14.00 \quad \text{(at SATP)}$$

This relationship allows for a quick conversion between pH and pOH values.

SUMMARY: Water Ionization Conversions and Values (at SATP)

$$K_w = [H_3O^+(aq)][OH^-(aq)] = 1.0 \times 10^{-14}$$

$$pH = -\log[H_3O^+(aq)] \quad pOH = -\log[OH^-(aq)]$$

$$[H_3O^+(aq)] = 10^{-pH} \quad [OH^-(aq)] = 10^{-pOH}$$

$$pH + pOH = 14.00$$

Learning Tip

Recall this convenient "rule of thumb" for keeping track of significant digits in (logarithmic) calculations of pH or pOH.

The number of digits following the decimal point in the pH or pOH (logarithm) value should be equal to the number of significant digits shown in the amount concentration of the ion.

For example, a hydronium ion concentration of 2.7×10^{-3} mol/L is expressed as a pH of 2.57, and a pOH of 4.3 is expressed as a hydroxide ion concentration of 5×10^{-5} mol/L.

▶ Practice

7. Food scientists and dieticians measure the pH of foods when they devise recipes and special diets.

   (a) Copy and complete **Table 1**.

   Table 1 Acidity of Foods

Food	[H_3O^+(aq)] (mol/L)	[OH^-(aq)] (mol/L)	pH	pOH
oranges	5.5×10^{-3}			
asparagus				5.6
olives		2.0×10^{-11}		
blackberries				10.6

   (b) Based on pH only, predict which of the foods would taste most sour.

8. To clean a clogged drain, 26 g of sodium hydroxide is added to water to make 150 mL of solution. What are the pH and pOH values for the solution?

9. What mass of potassium hydroxide is contained in 500 mL of solution that has a pH of 11.5? Comment on the degree of certainty of your answer.

Acid Strength as an Equilibrium Position

In Chapter 6, you learned that acidic solutions of different substances at the same concentration do not possess acid properties to the same degree. The pH of a 1.00 mol/L solution of an acid can vary anywhere from a value of nearly 7 to a value of nearly 0, depending on the specific acid in the solution. Other properties can also vary. For example, acetic acid does not conduct an electric current nearly as well as hydrochloric acid of equal concentration (**Figure 6**). When we observe chemical reactions of these acids, it is apparent that acetic acid, although it reacts in the same manner and amount as hydrochloric acid, does not react as quickly. The concepts of strong and weak acids were developed to describe and explain these differences in properties of acids.

An acid is described as *weak* if its characteristic properties are less than those of a common strong acid, such as hydrochloric acid. Weak acids are weaker electrolytes and react at a slower rate than strong acids do; the pH of solutions of weak acids is closer to 7 than the pH of strong acids of equal concentration.

In Chapter 6, strong acids were explained as ionizing quantitatively by reacting with water to form hydronium ions, whereas weak acids were explained as ionizing only partially (usually <50%). The empirical distinction between strong and weak acids can be explained much more completely now by combining the modified Arrhenius theory with equilibrium theory. A strong acid is explained as an acid that reacts quantitatively with water to form hydronium ions. For example, the reaction of dissolved hydrogen chloride (hydrochloric acid) with water is virtually complete. Even though the equation *could* be written with double equilibrium arrows and the extent shown with a >99.9% note, it is simpler, and much more common, to just use a single arrow to show that the reaction is quantitative.

$$HCl(aq) + H_2O(l) \rightarrow H_3O^+(aq) + Cl^-(aq)$$

A weak acid is an acid that reacts partially with water to form hydronium ions. Measurements of pH indicate that most weak acids react less than 50%. For example, acetic acid reacts only 1.3% in solution at 25 °C and 0.10 mol/L concentration.

Learning Tip

For any aqueous solution of an acid, percent reaction of that acid with water is also commonly called its percent ionization in chemistry references, because the net effect is the same as if the acid molecules simply ionize, as was once (simplistically) assumed in Arrhenius' original theory.

You should consider "reaction with water" and "ionization in water" to be equivalent terms for acid–base solution theory, and may expect to see either term used routinely in text and in questions.

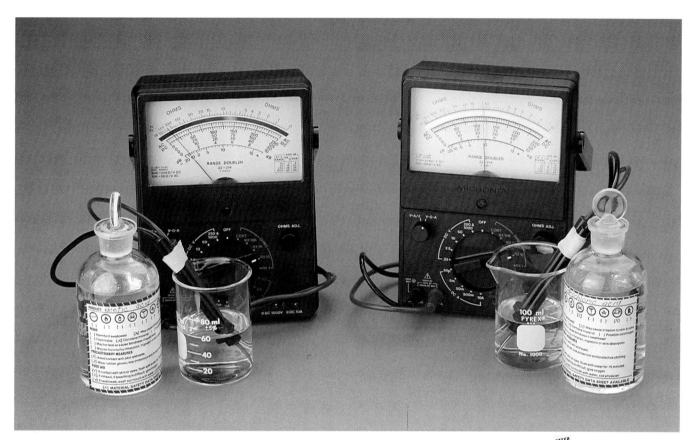

Figure 6
In solutions of equal concentration, a weak acid such as acetic acid conducts electricity to a lesser extent than does a strong acid such as hydrochloric acid.

Recall, from Chapter 15, that any equilibrium position depends on concentration(s) as well as on temperature; so, this 1.3% ionization value for acetic acid is *only* valid for a 0.10 mol/L solution at 25 °C. Laboratory pH experiments show that the higher the concentration of a weak acid solution, the lower its percent ionization becomes.

$$CH_3COOH(aq) + H_2O(l) \underset{}{\overset{1.3\%}{\rightleftarrows}} H_3O^+(aq) + CH_3COO^-(aq)$$

The hydronium ion concentration of any acid solution can be calculated by multiplying the percent reaction by the initial amount concentration of the acid solute. For example, in HCl(aq) solution, virtually 100% of the HCl molecules react with water molecules at equilibrium.

$$HCl(aq) + H_2O(l) \rightarrow H_3O^+(aq) + Cl^-(aq)$$

$$[H_3O^+(aq)] = \frac{100}{100} \times 0.10 \text{ mol/L}$$

$$= 0.10 \text{ mol/L}$$

There are six acids ordinarily classed as "strong": hydrochloric, nitric, sulfuric, hydrobromic, hydroiodic, and perchloric acid solutions. Only the first three are common.

For any aqueous strong acid, we can simply assume that the concentration of hydronium ions in solution is equal to the initial concentration of the acid dissolved. For weak acids (the majority of examples), we always need to calculate the hydronium ion concentration in solution, because only a small proportion of the initial acid concentration will be converted to ions at equilibrium.

> ## COMMUNICATION example 5

In a 0.10 mol/L solution of acetic acid, only 1.3% of the CH_3COOH molecules have reacted at equilibrium to form hydronium ions. Calculate the hydronium ion amount concentration.

Solution

$$CH_3COOH(aq) + H_2O(l) \xrightleftharpoons{1.3\%} H_3O^+(aq) + CH_3COO^-(aq)$$

$$[H_3O^+(aq)] = \frac{1.3}{100} \times 0.10 \text{ mol/L}$$

$$= 1.3 \times 10^{-3} \text{ mol/L}$$

The hydronium ion concentration in 0.10 mol/L acetic acid is 1.3×10^{-3} mol/L.

As explained in Chapter 6, we can easily compare the strengths of different acids by comparing the measured pH values for aqueous solutions of equal concentration. The lower the pH, the higher the hydronium ion concentration, the greater the percent reaction, and thus the stronger the acid. Furthermore, we can find the percent reaction for ionization of any weak acid solution from the measured pH of a solution of known initial concentration.

Learning Tip

Percent ionization:

$$p = \frac{[H_3O^+(aq)]}{[HA(aq)]} \times 100$$

Rearrange the equation to solve for the equilibrium hydronium ion concentration:

$$[H_3O^+(aq)] = \frac{p}{100} \times [HA(aq)]$$

where p = percent ionization
and
$[HA(aq)]$ = initial concentration of weak acid

> ## COMMUNICATION example 6

The pH of a 0.10 mol/L methanoic acid solution is 2.38. Calculate the percent reaction for ionization of methanoic acid.

Solution

$$[H_3O^+(aq)] = 10^{-pH}$$
$$= 10^{-2.38} \text{ mol/L}$$
$$= 4.2 \times 10^{-3} \text{ mol/L}$$

$$[H_3O^+(aq)] = \frac{p}{100 \times [HCOOH(aq)]}$$

$$p = \frac{[H_3O^+(aq)]}{[HCOOH(aq)]} \times 100$$

$$= \frac{4.2 \times 10^{-3} \text{ mol/L}}{0.10 \text{ mol/L}} \times 100$$

$$= 4.2\%$$

The percent ionization of 0.10 mol/L aqueous methanoic acid is 4.2%.

Section **16.1**

WEB *Activity*

Canadian Achievers—Edgar Steacie

Edgar Steacie (**Figure 7**) was an internationally acclaimed research scientist and a senior administrator of the National Research Council.

1. What was Steacie's main area of research?
2. Why was Steacie known as a statesman of science for Canada?
3. What is a Steacie Fellowship?

www.science.nelson.com

Figure 7
Edgar Steacie (1900–1962)

Section 16.1 Questions

1. How does the hydronium ion concentration compare with the hydroxide ion concentration if a solution is
   (a) neutral?
   (b) acidic?
   (c) basic?
2. What two diagnostic tests can distinguish a weak acid from a strong acid?
3. According to Arrhenius' original theory, what do all bases have in common?
4. Hydrocyanic acid is a very weak acid.
   (a) Write an equilibrium reaction equation for the ionization of 0.10 mol/L HCN(aq). The percent ionization at SATP is 7.8×10^{-3}%.
   (b) Calculate the hydronium ion concentration and the pH of a 0.10 mol/L solution of HCN(aq).
5. At 25 °C, the hydronium ion concentration in vinegar is 1.3 mmol/L. Calculate the hydroxide ion concentration.
6. At 25 °C, the hydroxide ion concentration in normal human blood is 2.5×10^{-7} mol/L. Calculate the hydronium ion concentration and the pH of blood.
7. Acid rain has a pH less than that of normal rain. The presence of dissolved carbon dioxide, which forms carbonic acid, gives normal rain a pH of 5.6. What is the hydronium ion concentration in normal rain?
8. If the pH of a solution changes by 3 pH units as a result of adding a weak acid, by how much does the hydronium ion concentration change?

9. If 8.50 g of sodium hydroxide is dissolved to make 500 mL of cleaning solution, determine the pOH of the solution.
10. What mass of hydrogen chloride gas is required to produce 250 mL of a hydrochloric acid solution with a pH of 1.57?
11. Determine the pH of a 0.10 mol/L hypochlorous acid solution, which has 0.054% ionization at 25 °C.
12. Calculate the pH and pOH of a hydrochloric acid solution prepared by dissolving 30.5 kg of hydrogen chloride gas to make 806 L of solution. What assumption is made when doing this calculation?
13. Acetic (ethanoic) acid is the most common weak acid used in industry. Determine the pH and pOH of an acetic acid solution prepared by dissolving 60.0 kg of pure, liquid acetic acid to make 1.25 kL of solution. The percent reaction with water at this concentration is 0.48%.
14. Determine the mass of sodium hydroxide that must be dissolved to make 2.00 L of a solution with a pH of 10.35.
15. Write an experimental design for the identification of four colourless solutions: a strong acid solution, a weak acid solution, a neutral molecular solution, and a neutral ionic solution. Write sentences, create a flow chart, or design a table to describe the required diagnostic tests.
16. Sketch a flow chart or concept map that summarizes the conversion of $[H_3O^+(aq)]$ to and from $[OH^-(aq)]$, pH, and percent reaction (ionization) of acid solute. Make your flow chart large enough that you can write the procedure between the quantity symbols in the diagram.

16.2 The Brønsted–Lowry Acid–Base Concept

By now, our much-revised acid and base theory includes concepts of hydronium ions, reaction with water, reaction equilibrium, and the ionization equilibrium of water. Our modified theory is, thus, much more comprehensive, and much better at describing, explaining, and predicting acid–base reactions, than the original theory proposed by Arrhenius. We find, however, that there are still problems, and our theory must still be considered too restrictive. There is no provision in it for reactions that do not occur in aqueous solution. In addition, we find that there are some substances that seem to have both acid and base properties.

As a common example, sodium hydrogen carbonate (sodium bicarbonate, baking soda) forms a basic solution (raises the pH) in water, but the same compound will partly neutralize (lower the pH of) a sodium hydroxide (lye) solution. When we observe that $NaHCO_3(s)$ forms a basic aqueous solution, we conclude that this happens because some hydrogen carbonate ions react with water molecules to produce hydroxide ions. We know from many other observations that sodium ions have no acidic or basic properties. The following reaction equation seems to explain our observation easily:

$$HCO_3^-(aq) + H_2O(l) \rightleftarrows H_2CO_3(aq) + OH^-(aq)$$

But, if adding sodium hydrogen carbonate makes a (strongly basic) sodium hydroxide solution less basic, the concepts we are using lead us to conclude that something must be reacting to decrease the concentration of the hydroxide ions. It must be the hydrogen carbonate ions because there are no other entities in this system except sodium ions and water molecules. We can easily write an equation to explain this observation, as well:

$$HCO_3^-(aq) + OH^-(aq) \rightleftarrows CO_3^{2-}(aq) + H_2O(l)$$

This equation seems to explain how the hydroxide ions can be partially consumed. But, if it is correct, it begs the question, "Is sodium bicarbonate a base, or is it an acid?" Clearly, our theory still needs some work! We need a broader, more comprehensive concept to successfully explain these kinds of observations.

The Proton Transfer Concept

New theories in science usually result from looking at the evidence in a way that has not occurred to other observers. In 1923, two European scientists independently developed a new approach to acids and bases (**Figure 1**). These scientists focused on the role of an acid and a base *in a reaction* rather than on the acidic or basic properties of their aqueous solutions. An acid, such as hydrogen chloride, functions in a way opposite to a base, such as ammonia. According to the Brønsted–Lowry concept, hydrogen chloride, upon dissolving, *donates* a proton to a water molecule:

$$\underset{\text{acid}}{HCl(aq)} + H_2O(l) \rightarrow H_3O^+(aq) + Cl^-(aq) \quad (H^+)$$

and ammonia, upon dissolving, *accepts* a proton from a water molecule.

$$\underset{\text{base}}{NH_3(aq)} + H_2O(l) \rightleftarrows OH^-(aq) + NH_4^+(aq) \quad (H^+)$$

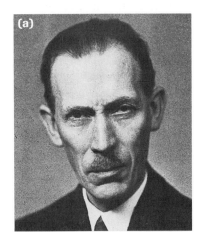

Figure 1
Johannes Brønsted (1879–1947) **(a)** and Thomas Lowry (1874–1936) **(b)** independently created new theoretical definitions for acids and bases, based upon proton transfer during a reaction.

Water does not have to be one of the reactants. For example, hydronium ions present in a hydrochloric acid solution can react directly with dissolved ammonia molecules.

$$\underset{\text{acid}}{H_3O^+(aq)} + \underset{\text{base}}{\overset{\overset{H^+}{\longleftarrow}}{NH_3(aq)}} \rightarrow H_2O(l) + NH_4^+(aq)$$

We can describe this reaction as NH_3 molecules removing protons from H_3O^+ ions. Hydronium ions act as the acid, and ammonia molecules act as the base. Water is present as the solvent but not as a primary reactant. In fact, water does not even have to be present, as evidenced by the reaction of hydrogen chloride and ammonia gases (**Figure 2**).

$$\underset{\text{acid}}{HCl(g)} + \underset{\text{base}}{\overset{\overset{H^+}{\longleftarrow}}{NH_3(g)}} \rightarrow NH_4Cl(s)$$

Figure 2
Invisible fumes of ammonia gas and hydrogen chloride gas mix and react above these beakers, producing tiny visible particles of solid ammonium chloride that are suspended in the air.

According to the Brønsted–Lowry concept, a **Brønsted–Lowry acid** is a proton donor and a **Brønsted–Lowry base** is a proton acceptor. A Brønsted–Lowry neutralization is a competition for protons that results in a proton transfer from the strongest acid present to the strongest base present. A **Brønsted–Lowry reaction equation** is an equation written to show an acid–base reaction involving the transfer of a proton from one entity (an acid) to another (a base).

The Brønsted–Lowry concept does away with defining a *substance* as being an acid or base. Only an *entity* that is involved in a proton transfer in a reaction can be defined as an acid or base—and only for that particular reaction. This last point is extremely important: protons may be gained in a reaction with one entity, but lost in a reaction with another entity. For example, in the reaction of HCl with water shown on the previous page, water acts as the base; whereas, in its reaction with NH_3, water acts as the acid. A substance that appears to act as a Brønsted–Lowry acid in some reactions and as a Brønsted–Lowry base in other reactions is called amphoteric, or sometimes (incorrectly) amphiprotic. The empirical term, **amphoteric**, properly refers to a chemical *substance* with the ability to react as either an acid or a base. The theoretical term, **amphiprotic**, describes an *entity* (ion or molecule) having the ability to either accept or donate a proton. The hydrogen carbonate ion in baking soda (**Figure 3**), like every other hydrogen polyatomic ion, is amphiprotic, as shown by the following reactions:

$$\underset{\text{base}}{HCO_3^-(aq)} + \underset{\text{acid}}{H_2O(l)} \rightleftarrows OH^-(aq) + H_2CO_3(aq) \qquad K_c = 2.2 \times 10^{-8} \text{ mol/L}$$

$$\underset{\text{acid}}{HCO_3^-(aq)} + \underset{\text{base}}{H_2O(l)} \rightleftarrows H_3O^+(aq) + CO_3^{2-}(aq) \qquad K_c = 4.7 \times 10^{-11} \text{ mol/L}$$

Figure 3
Baking soda is a common household chemical, but it requires an uncommonly sophisticated theory to describe, explain, or predict all of its properties.

When bicarbonate ions are in aqueous solution, some react with the water molecules by acting as an acid, and some react by acting as a base. K_c values given for these two equilibriums show that one of them predominates—so much so that the other reaction is simply inconsequential. The number of ions reacting as a base is over 2000 times more than the number reacting as an acid. As you might expect, the resulting solution is basic.

We can get a clearer idea of the amphoteric nature of baking soda (caused by the amphiprotic nature of the hydrogen carbonate ion) by noting the evidence expressed in the next two equations. Note that bicarbonate ions partly neutralize a strong acid, but they can also partly neutralize a strong base.

$$\underset{\text{base}}{HCO_3^-(aq)} + \underset{\text{acid}}{H_3O^+(aq)} \rightleftarrows H_2CO_3(aq) + H_2O(l) \qquad \text{(raises the pH of a strong acid solution)}$$

$$\underset{\text{acid}}{HCO_3^-(aq)} + \underset{\text{base}}{OH^-(aq)} \rightleftarrows CO_3^{2-}(aq) + H_2O(l) \qquad \text{(lowers the pH of a strong base solution)}$$

DID YOU KNOW ?

Superacids

In aqueous solution, all strong acids are equal in strength, since they all react instantly and completely with water to form the strongest possible acid entity that can exist in water—the hydronium ion. Scientists have long suspected, however, that the well-known strong acids, HCl and H_2SO_4, may not be equal in strength when no water is present and that much stronger acids could exist. Dr. Ronald Gillespie of McMaster University has done extensive research on non-aqueous strong acids. His definition of a superacid—one that is stronger than pure sulfuric acid—is now generally accepted by scientists. Perchloric acid, the only common superacid, easily loses protons to $H_2SO_4(l)$ molecules. Fluorosulfonic acid, $HSO_3F(l)$, is more than one thousand times stronger than $H_2SO_4(l)$. It is the strongest Brønsted-Lowry acid known.

Figure 4
White vinegar is a 5% acetic (ethanoic) acid solution. The amount concentration is 0.83 mol/L. Such a solution is only about 0.43% ionized at 25 °C, making acetic acid a typical weak acid, by definition.

Scientists consider the Brønsted–Lowry concept to be a theoretical definition. It falls short of being a comprehensive theory because it does not explain *why* a proton is donated or accepted, and cannot predict theoretically *which* reaction will occur for a given entity in any given new situation. *The advantage of the Brønsted–Lowry definitions is that they enable us to define acids and bases in terms of chemical reactions rather than simply as substances that form acidic and basic aqueous solutions.* A definition of acids and bases in terms of chemical reactions allows us to describe, explain, and predict a great many more reactions, whether they take place in aqueous solution, in solution in some other solvent, or between two undissolved entities in pure chemical states.

▶ Practice

1. Theories in science develop over a period of time. Illustrate this development by writing a theoretical definition of an acid, using the following concepts. Begin your answer with, "According to [authority], acids are substances that...."
   (a) Arrhenius' original theory
   (b) the modified Arrhenius theory
   (c) the Brønsted–Lowry concept
2. How does the definition of a base according to the modified Arrhenius theory compare with the Brønsted-Lowry definition?
3. Classify each reactant in the following equations as a Brønsted-Lowry acid or base.
   (a) $HF(aq) + SO_3^{2-}(aq) \rightleftarrows F^-(aq) + HSO_3^-(aq)$
   (b) $CO_3^{2-}(aq) + CH_3COOH(aq) \rightleftarrows CH_3COO^-(aq) + HCO_3^-(aq)$
   (c) $H_3PO_4(aq) + OCl^-(aq) \rightleftarrows H_2PO^{4-}(aq) + HOCl(aq)$
   (d) $HCO_3^-(aq) + HSO_4^-(aq) \rightleftarrows SO_4^{2-}(aq) + H_2CO_3(aq)$
4. Evidence indicates that the hydrogen sulfite ion is amphiprotic. A sodium hydrogen sulfite solution can partly neutralize either a sodium hydroxide spill or a hydrochloric acid spill.
   (a) Write a net ionic equation for the reaction of aqueous hydrogen sulfite ions with the hydroxide ions in solution. Label the reactants as acids or bases.
   (b) Write a net ionic equation for the reaction of hydrogen sulfite ions with the hydronium ions from a hydrochloric acid solution. Label the reactants as acids or bases.
5. What restrictions to acid–base reactions do the Brønsted-Lowry definitions remove?
6. Why is the Brønsted-Lowry concept labelled a theoretical definition rather than a theory?

Conjugate Acids and Bases

According to the Brønsted–Lowry concept, acid–base reactions involve the transfer of a proton. These reactions are universally reversible and always result in an acid–base equilibrium.

In a proton transfer reaction at equilibrium, both forward and reverse reactions involve Brønsted-Lowry acids and bases. For example, in an acetic acid solution (**Figure 4**), the forward reaction is explained as a proton transfer from acetic acid to water molecules, and the reverse reaction is a proton transfer from hydronium to acetate ions.

$$\underset{\text{acid}}{CH_3COOH(aq)} + \underset{\text{base}}{H_2O(l)} \rightleftarrows \underset{\text{base}}{CH_3COO^-(aq)} + \underset{\text{acid}}{H_3O^+(aq)}$$

This equilibrium is typical of all acid–base reactions. There will always be two acids (in the example, CH_3COOH and H_3O^+) and two bases (in the example, H_2O and CH_3COO^-) in any acid–base reaction equilibrium. Furthermore, the base on the right (CH_3COO^-) is formed by removal of a proton from the acid on the left (CH_3COOH). The acid on the right (H_3O^+) is formed by the addition of a proton to the base on the left (H_2O). A pair of substances with formulas that differ only by a proton is called a **conjugate acid–base pair**. An acetic acid molecule and an acetate ion are a conjugate acid–base pair. Acetic acid is the conjugate acid of the acetate ion, and the acetate ion is the conjugate base of acetic acid. The equation shows that there must always be two conjugate acid–base pairs in any proton transfer reaction. The hydronium ion and water are the second conjugate acid–base pair in this equilibrium. Conjugate acid–base pairs appear opposite each other in a table of acids and bases, such as that in Appendix I.

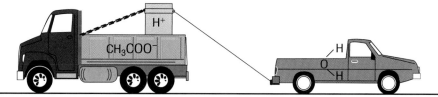

$$CH_3COOH(aq) + H_2O(l) \rightleftharpoons CH_3COO^-(aq) + H_3O^+(aq) \quad (0.10 \text{ mol/L at } 25\,°C)$$

(conjugate pair, 1.3%)

Figure 5
This pictorial analogy is used to explain why acetic acid is a weak acid in aqueous solution. We believe that the proton (H^+) of the carboxyl group is attracted more strongly by the rest of the acetic acid molecule than it is by the water molecule. We infer this conclusion because, at equilibrium, pH evidence indicates that very few of the acetic acid molecules have lost protons in their collisions with water molecules.

At equilibrium, only 1.3% of the CH_3COOH molecules have reacted with water in a 0.10 mol/L solution at SATP. It appears that the ability of the CH_3COO^- part of the acetic acid molecule to keep its proton (H^+) is much greater than the ability of H_2O to attract the proton away (**Figure 5**). This means that CH_3COO^- is a stronger base (it has a greater attraction for protons) than H_2O.

When HCl molecules react with water (**Figure 6**), the Cl^- of each HCl molecule has a much weaker attraction for its proton (H^+) than any colliding water molecule has. The water molecules "win" this "competition" for protons overwhelmingly. At equilibrium, essentially all of the HCl molecules have lost protons to water molecules. In this case, the transfer of protons is quantitative, and, because its molecules have an extremely weak attraction for their protons, HCl(aq) is called (somewhat confusingly) a strong acid.

Remember: The stronger the base, the more it attracts another proton. The stronger an acid, the less it attracts its own proton.

Figure 6
When gaseous hydrogen chloride dissolves in water, the HCl molecules are thought to collide and react quantitatively with water molecules to form hydronium and chloride ions.

Learning Tip

Acids were studied much earlier than bases were, and sometimes the old terminology used to describe acid-base situations can be misleading—because it always focuses on the nature of the acid. Strong acids are very reactive, but we believe it is because they are very weak proton attractors. It is common to speak of acids "donating" protons, and of bases "accepting" them, but this terminology gives an unrealistic view of what we believe is really happening. It is like saying that a bank robber "accepts" money "donated" by a bank teller. The tug-of-war analogy in Figure 5 is a more realistic way to think of proton transfer reactions—as a competition between two bases, both attracting the same proton.

The terms *strong* acid and *weak* acid can be explained in terms of the Brønsted–Lowry concept, and also by comparing the reactions of different acids with the same base—for example, water. Using HA as the general symbol for any acid and A^- as its conjugate base, the empirically derived Relative Strengths of Aqueous Acids and Bases table (Appendix I) lists the position of equilibrium of aqueous solutions of many different acids. They are ordered according to how much they ionize in (react with) the water solvent.

$$HA(aq) + H_2O(l) \rightleftarrows H_3O^+(aq) + A^-(aq)$$

The extent of the proton transfer between HA and H_2O determines the relative strength of HA(aq). In Brønsted–Lowry terms, when a strong acid reacts with water, an almost complete transfer of protons results for the forward reaction and almost no transfer of protons occurs for the reverse reaction; a nearly 100% reaction with water. The equilibrium constant value is found to be very large. Theoretically, a strong acid holds its proton very weakly, and easily loses the proton to any base, even very weak bases such as water. This result leads to the interpretation that the conjugate base, A^-, of a strong acid must have a very weak (negligible) attraction for protons. A useful generalization regarding the relative strengths of an acid–base conjugate pair is: *the stronger an acid, the weaker its conjugate base; and conversely, the weaker an acid, the stronger its conjugate base.* (See the Relative Strengths of Aqueous Acids and Bases table in Appendix I.)

Chemists have no simple explanation, in terms of forces or bonds, for the differing abilities of acids to donate protons or of bases to accept them. The inability to predict acid and base strengths for an entity not already included in an empirically determined table of acids and bases is a major deficiency of all acid–base theories.

SUMMARY *Brønsted–Lowry Definitions*

- An acid is a proton donor and a base is a proton acceptor, in a specific reaction.
- An acid–base reaction involves a single proton transfer from one entity (the acid) to another (the base).
- An amphiprotic entity (amphoteric substance) is one that acts as a Brønsted–Lowry acid in some reactions and as a Brønsted–Lowry base in other reactions.
- A conjugate acid–base pair consists of two entities with formulas that differ only by a proton.
- A strong acid has a very weak attraction for protons. A strong base has a very strong attraction for protons.
- The stronger an acid, the more weakly it holds its proton. The stronger a base, the more it attracts another proton.
- The stronger an acid, the weaker is its conjugate base. The stronger a base, the weaker is its conjugate acid.

▶ Practice

7. Use the Brønsted-Lowry definitions to identify the two conjugate acid-base pairs in each of the following acid-base reactions.
   (a) $HCO_3^-(aq) + S^{2-}(aq) \rightleftarrows HS^-(aq) + CO_3^{2-}(aq)$
   (b) $H_2CO_3(aq) + OH^-(aq) \rightleftarrows HCO_3^-(aq) + H_2O(l)$
   (c) $HSO_4^-(aq) + HPO_4^{2-}(aq) \rightleftarrows H_2PO_4^-(aq) + SO_4^{2-}(aq)$
   (d) $H_2O(l) + H_2O(l) \rightleftarrows H_3O^+(aq) + OH^-(aq)$

8. Some ions can form more than one conjugate acid-base pair. List the two conjugate acid-base pairs involving a hydrogen carbonate ion in the reactions in question 7.

Section **16.2**

INVESTIGATION 16.1 *Introduction*
Creating an Acid–Base Strength Table

An acid–base table organizes common acids (and their conjugate bases) in order of decreasing acid strength. Acid strength can be tested several ways, including by a carefully designed use of indicators. Predict the order of strengths using the Relative Strengths of Aqueous Acids and Bases table (Appendix I). Use the indicators provided to create a valid and efficient Design, in which you clearly identify the relevant variables. Evaluate the Design (only), and suggest improvements if any problems are identified.

Report Checklist

- ○ Purpose
- ○ Problem
- ○ Hypothesis
- ● Prediction
- ● Design
- ○ Materials
- ● Procedure
- ● Evidence
- ● Analysis
- ● Evaluation (1)

Purpose
The purpose of this investigation is to test an experimental design for using indicators to create a table of relative strengths of acids and bases.

Problem
Can the indicators available be used to rank the acids and bases provided in order of strength?

To perform this investigation, turn to page 768.

Predicting Acid–Base Reaction Equilibria

The Brønsted–Lowry concept unfortunately does not include any theoretical explanation about why any given entity attracts a proton more or less strongly. To predict the outcome of any acid–base combination, we must rely on empirical evidence, gained by measuring and recording the relative strengths of acid and base entities. Predictions must be restricted to only those acid–base combinations for which we already have data. To help us, we can now look for a simple generalization that might allow us to predict the approximate position of equilibrium in an acid–base proton transfer.

LAB EXERCISE 16.B
Predicting Acid–Base Equilibria

Is it possible to predict how far a reaction will proceed? Use the table of Relative Strengths of Aqueous Acids and Bases in Appendix I and the evidence of position of equilibrium to complete the Analysis of the investigation report.

Purpose
The purpose of this investigation is to develop a generalization for predicting the position of acid–base equilibria.

Problem
How do the positions of the reactant acid and base in the acid–base table relate to the position of equilibrium?

Report Checklist

- ○ Purpose
- ○ Problem
- ○ Hypothesis
- ○ Prediction
- ○ Design
- ○ Materials
- ○ Procedure
- ○ Evidence
- ● Analysis
- ○ Evaluation

1. $CH_3COOH(aq) + H_2O(l) \rightleftharpoons H_3O^+(aq) + CH_3COO^-(aq)$ (<50%)

2. $HCl(aq) + H_2O(l) \rightleftharpoons H_3O^+(aq) + Cl^-(aq)$ (>99%)

3. $CH_3COO^-(aq) + H_2O(l) \rightleftharpoons CH_3COOH(aq) + OH^-(aq)$ (<50%)

4. $H_3PO_4(aq) + NH_3(aq) \rightleftharpoons H_2PO_4^-(aq) + NH_4^+(aq)$ (>50%)

5. $HCO_3^-(aq) + SO_3^{2-}(aq) \rightleftharpoons HSO_3^-(aq) + CO_3^{2-}(aq)$ (<50%)

6. $H_3O^+(aq) + OH^-(aq) \rightarrow H_2O(l) + H_2O(l)$

 **WEB** *Activity*

Web Quest—Pool Chemistry

A swimming pool can be an enjoyable place to spend a hot summer afternoon. Most of us are familiar with the dangers of pools. This Web Quest introduces an unfamiliar risk: pool gas. What is it and how does it form? What are the dangers of pool gas, and how can they be reduced?

www.science.nelson.com

CAREER CONNECTION

Chemistry Researcher

Farideh Jalilehvand (**Figure 7**) is an associate professor of chemistry at the University of Calgary. Her research involves a lot of X-ray absorption spectroscopy. Find out how acidity due to sulfur accumulation in water-logged wood is connected to ancient shipwrecks by visiting her information site, and check out her curriculum vitae (CV). Science is international in scope!

Figure 7
Farideh Jalilehvand

www.science.nelson.com

Predicting Acid–Base Reactions

When making complex predictions, scientists often combine a variety of empirical and theoretical concepts. We need to use a combination of concepts to predict *both* the products and the extent of acid–base reactions. According to the collision–reaction theory, a proton transfer may result from a collision between an acid and a base. In a system that is a mixture of several different acid and/or base entities, there are countless random collisions of all the entities present. So, in theory, in any such system, there are many different possible acid–base reactions, and all of these reactions occur (to some extent) all the time. Evidence indicates that one reaction predominates: it occurs to an extent so much greater than the other reactions that it is the only observable reaction. We will be able to explain which reaction will predominate if we use a combination of the collision–reaction theory and the Brønsted–Lowry concept.

According to collision–reaction theory, in an acid–base system, collisions of all entities present are constantly occurring. According to the Brønsted–Lowry concept, a proton will only transfer if an acid entity collides with a base entity that is a better proton attractor than itself. A proton could theoretically transfer several times (if there were several different acids and bases in the system), transferring each time to a stronger proton attractor. Once gained by an entity of the strongest base present, a proton will remain there, because there is no other base present in the system that can attract that proton strongly enough to remove it. By the same logic, once any entity of the strongest acid present has lost its proton, its (remaining) conjugate base cannot gain one back from any other entity present, because it is a weaker proton attractor than anything else in the system. Overall, theory suggests that the predominant acid–base reaction should be the one that involves proton transfer from the strongest acid to the strongest base present in the system. Experimental evidence indicates that this explanation is correct. Other possible proton transfer reactions occur to such a small extent that they have a negligible effect on the reactants and products, so they are normally ignored.

We theorize that the only significant reaction in any acid–base system involves proton transfer from the strongest acid present to the strongest base present. For an aqueous solution system, we first must list all the entities present *as they exist in aqueous solution*. **Table 1** summarizes how to represent the entities that are present.

Table 1 Predominant Entities Present in Aqueous Solution (other than $H_2O(l)$)

Substance dissolved (example)	Kinds of entities in solution	Predominant entities present (example)
Ionic compounds $Ca(HCO_3)_2(aq)$	cations, anions	$Ca^{2+}(aq)$ and $HCO_3^-(aq)$
Ionic oxides $Na_2O(aq)$, $CaO(aq)$	cations, hydroxide ions	$Na^+(aq)$, $OH^-(aq)$ or $Ca^{2+}(aq)$, $OH^-(aq)$
Strong acids $HCl(aq)$ or $HNO_3(aq)$	$H_3O^+(aq)$, conjugate base	$H_3O^+(aq)$, $Cl^-(aq)$ or $H_3O^+(aq)$, $NO_3^-(aq)$
Weak acids $HF(aq)$ or $HSO_3^-(aq)$	molecules or ions	$HF(aq)$ or $HSO_3^-(aq)$
Weak bases $NH_3(aq)$ or $HCO_3^-(aq)$	molecules or ions	$NH_3(aq)$ or $HCO_3^-(aq)$

The process for determining the nature and extent of the predominant proton transfer reaction in an aqueous acid–base system can be thought of as five distinct steps, for convenience, as shown in the following Sample Problem.

SAMPLE problem 16.1

What will be the predominant reaction if spilled drain cleaner (sodium hydroxide) solution is neutralized with vinegar (**Figure 8**)? Are reactants or products favoured in this reaction?

Step 1: List all entities present as they exist in aqueous solution. Refer to **Table 1** if necessary. The entity list for this situation is

Na$^+$(aq)   OH$^-$(aq)   CH$_3$COOH(aq)   H$_2$O(l)

Step 2: Use the entity lists of the Relative Strengths of Aqueous Acids and Bases table to identify and label each entity present as a Brønsted–Lowry acid or base. Amphiprotic entities are labelled for *both* possibilities. Conjugate bases of strong acids are *not* included or labelled as bases because they cannot *act* as bases in aqueous solution. Metal ions are treated as spectators.

```
                                A                A
Na⁺(aq)      OH⁻(aq)      CH₃COOH(aq)     H₂O(l)
                B                                 B
```

Step 3: Use the *order* of the entities in the Relative Strengths of Aqueous Acids and Bases table to identify and label the *strongest* Brønsted–Lowry acid (the highest one on the table) and the *strongest* Brønsted–Lowry base (the lowest one on the table) that are present in the solution.

```
                                SA               A
Na⁺(aq)      OH⁻(aq)      CH₃COOH(aq)     H₂O(l)
                SB                                B
```

Step 4: Write a balanced equation to show a proton transfer from the strongest acid to the strongest base, assuming that their respective conjugates are the reaction products.

$$\text{CH}_3\text{COOH(aq)} + \overset{\overset{H^+}{\frown}}{\text{OH}^-}\text{(aq)} \rightleftarrows \text{CH}_3\text{COO}^-\text{(aq)} + \text{H}_2\text{O(l)}$$

Step 5: Predict the position of equilibrium using the generalization developed in Lab Exercise 16.B and illustrated in the margin Learning Tip. For this reaction, the strongest acid is positioned higher on the table than the strongest base, so products are favoured. *Under certain assumed conditions* (see the following discussion), the equilibrium percent reaction may be labelled as greater than 50%.

$$\text{CH}_3\text{COOH(aq)} + \text{OH}^-\text{(aq)} \overset{>50\%}{\rightleftarrows} \text{CH}_3\text{COO}^-\text{(aq)} + \text{H}_2\text{O(l)}$$

Figure 8
The drain cleaner shown is a concentrated solution of a very strong base. A spill would be highly corrosive and quite hazardous. Excess weak acid, such as the vinegar (5% acetic acid), is the best choice to "neutralize" a spilled strong base. The final solution will not actually be neutral, but it will be only mildly acidic, and much safer to handle.

The nature of water complicates the prediction of outcomes of acid–base reactions in aqueous solution. We need to consider the specific restrictions that apply when predicting proton transfer reactions *in aqueous solution*, because they are much more common than acid–base reactions in any other environment.

- Hydronium ion is the strongest acid entity that can exist in aqueous solution. If a stronger acid than hydronium ion is dissolved in water, it reacts instantly and completely with water molecules to form hydronium ions. For this reason, the six strong acids are all written as H$_3$O$^+$(aq) when in aqueous solution.

- Hydroxide ion is the strongest base entity that can exist in aqueous solution. If a stronger base than hydroxide ion is dissolved in water, it reacts instantly and completely to form hydroxide ions. The only common example is the dissolving of soluble ionic oxide compounds such as Na$_2$O(s). In such cases, the oxide ion is written as OH$^-$(aq) when in aqueous solution.

Learning Tip

The relative positions of the strongest acid and the strongest base on an acid–base table can be used to approximately determine the position of an acid–base equilibrium.

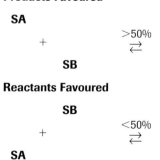

> **Learning Tip**
>
> Since hydroxide ion is the strongest possible base in aqueous solution, reaction of any acid with this ion in aqueous solution is automatically one that favours products. Similarly, the reaction of any base with the strongest possible acid, hydronium ion, must favour products. Of course, the reaction of hydroxide ion with hydronium ion is always quantitative.

- No entity in aqueous solution can react as a base if it is a weaker base than water. For this reason, the conjugate bases of the six strong acids are not considered as bases, in aqueous solution.

Even though you find that products are favoured (as in Sample Problem 16.1), for the predominant reaction, you cannot accurately predict the actual equilibrium position of all cases of any specific acid–base reaction by this method. It simply means you know that the forward reaction happens more readily than the reverse reaction, all things being equal. But, recall (from Chapter 15) that the initial concentration of entities plays a very large part in determining the percent reaction at equilibrium. To be able to predict something meaningful about the position of an acid–base reaction equilibrium, we must restrict the reaction conditions. Unless you are specifically informed otherwise in a question, assume in this text that, for the predominant reaction in an acid–base system,

- the equation represents a single proton transfer between two entities, neither of which is water, and has a stoichiometric ratio of 1:1:1:1
- the strongest acid and the strongest base present in the reaction system are both present in significant chemical amounts, in approximately equal concentrations

Under these circumstances (and only then), you may assume that a Brønsted–Lowry acid–base reaction where products are favoured has a percent reaction greater than 50%, or that a reaction where reactants are favoured may be labelled as "< 50%." It is also correct (given these restrictions) to state that the K_c value for any such reaction is > 1 if products are favoured, and < 1 if reactants are favoured. It is not possible to be more precise about the equilibrium position, however, at our current level of theory.

Note that the above restrictions exclude any reaction equation written to describe the ionization (reaction with water) of any single acid or base entity dissolved to make an aqueous solution. In these cases, water is acting as one of the reactants, and the amount concentration of the water is very much greater than that of the other entity.

Finally, recall that the reaction of hydronium ions with hydroxide ions is always quantitative. This equation should always be written with a single arrow. There are many other acid–base reactions that are quantitative, particularly those where either hydronium or hydroxide ions are involved, but it is not possible to easily predict whether any acid–base reaction will be quantitative from the table of Relative Strengths of Aqueous Acids and Bases. In this text, *other quantitative reactions will always be identified for you*, as in the following Communication Example.

> ### ▶ COMMUNICATION example
>
> Ammonium nitrate fertilizer is produced by the quantitative reaction of aqueous ammonia with nitric acid. Write a balanced acid–base equilibrium equation.
>
> **Solution**
>
>   A SA
> $NH_3(aq), H_2O(l), H_3O^+(aq), NO_3^-(aq)$
> SB B
>
> $NH_3(aq) + H_3O^+(aq) \rightarrow NH_4^+(aq) + H_2O(l)$

SUMMARY: A Five-Step Method for Predicting the Predominant Acid–Base Reaction

1. List all entities (ions, atoms, or molecules including $H_2O(l)$) initially present as they exist in aqueous solution. (Refer to Table 1, page 728.)
2. Identify and label all possible aqueous acids and bases, using the Brønsted–Lowry definitions.
3. Identify the strongest acid and the strongest base present, using the table of Relative Strengths of Aqueous Acids and Bases (Appendix I).
4. Write an equation showing a transfer of one proton from the strongest acid to the strongest base, and predict the conjugate base and the conjugate acid to be the products.
5. Predict the approximate position of the equilibrium, using the generalization developed in Lab Exercise 16.B on page 727, the Learning Tip on page 729, and the table of Relative Strengths of Aqueous Acids and Bases (Appendix I).

Practice

Use the five-step method to predict the predominant reactions in the following chemical systems:

9. Hydrofluoric acid and an aqueous solution of sodium sulfate are mixed to test the five-step method of predicting acid–base reactions.
10. Strong acids, such as perchloric acid, have been shown to react quantitatively with strong bases, such as sodium hydroxide.
11. Methanoic acid is added to an aqueous solution of sodium hydrogen sulfide.
12. A student mixes solutions of ammonium chloride and sodium nitrite in a chemistry laboratory.
13. Empirical work has shown that nitric acid reacts quantitatively with a sodium acetate solution.
14. A consumer attempts to neutralize an aqueous sodium hydrogen sulfate cleaner with a solution of lye. (See Appendix J if you do not remember what lye is.)
15. Can ammonium nitrate fertilizer, added to water, be used to neutralize a muriatic acid (hydrochloric acid) spill?
16. Predict the acid–base reaction of bleach with vinegar (**Figure 9**).
17. Commercial laundry bleach (Figure 9) is made by reacting chlorine gas with a sodium hydroxide solution, according to the equation

    $Cl_2(g) + 2\,OH^-(aq) \rightleftarrows OCl^-(aq) + Cl^-(aq) + H_2O(l)$

    As sold, the pH of laundry bleach solutions is always well above 8. You know that the element chlorine has very low solubility in pure water. Explain, using Le Châtelier's principle, why bleach bottle labels warn so strongly against mixing bleach with other cleaning agents, such as acidic toilet bowl cleaners.

    Note: Household bleach also produces toxic chloramines if mixed with basic ammonia cleaning solutions. The best rule is, NEVER mix bleach directly with any other cleaning powder or solution. It is arguably the most dangerous common household chemical.

Figure 9
Bottles of household bleach display a warning against mixing the bleach (aqueous sodium hypochlorite) with acids. Does your prediction of the reaction between vinegar and hypochlorite ions provide any clues about the reason for the warning?

LAB EXERCISE 16.C

Aqueous Bicarbonate Ion Acid–Base Reactions

Report Checklist
- ○ Purpose
- ○ Problem
- ○ Hypothesis
- ● Prediction
- ○ Design
- ○ Materials
- ○ Procedure
- ○ Evidence
- ● Analysis
- ● Evaluation (2, 3)

An acid–base table organizes common acids (and their conjugate bases) in a way that enables us to predict predominant acid–base reactions. Evaluate the reliability of the five-step method (only) using information from this investigation, by comparing your analysis of the evidence with your predictions.

Purpose
The purpose of this investigation is to test the five-step method for predicting reactions in acid–base systems.

Problem
What are the products and position of the equilibrium for sodium hydrogen carbonate (**Figure 10**) with stomach acid, vinegar, household ammonia, and lye, respectively?

Evidence

Table 2 The Addition of Baking Soda to Various Solutions

Reactant	Bubbles	Odour	pH
HCl(aq)	yes	none	increases
$CH_3COOH(aq)$	yes	disappears	increases
$NH_3(aq)$	no	remains	decreases
NaOH(aq)	no	none	decreases

Figure 10
The versatility of baking soda is demonstrated by its use in extinguishing fires, in baking biscuits, and in neutralizing excess stomach acid. It is also used as a medium for local anesthetics—baking soda reduces stinging sensations by neutralizing the acidity of the anesthetic, with the result that the speed and efficiency of the anesthetic are improved. The broad range of uses for baking soda results, in part, from its amphiprotic character.

INVESTIGATION 16.2 Introduction

Testing Brønsted–Lowry Reaction Predictions

Report Checklist
- ○ Purpose
- ○ Problem
- ○ Hypothesis
- ● Prediction
- ○ Design
- ● Materials
- ● Procedure
- ● Evidence
- ● Analysis
- ● Evaluation (1, 2, 3)

When predicting products for this investigation, list all entities present as they normally exist in an aqueous environment. For those reactants that are added in solid state, assume that they will dissolve. Use the resulting entities for prediction. Evaluate the predictions, the Brønsted-Lowry concept, and the five-step method for acid–base reaction prediction.

Purpose
The purpose of this investigation is to test the Brønsted–Lowry concept and the five-step method for reaction prediction from a table of relative acid–base strength.

Problem
What reactions occur when various pairs of substances are mixed?

Design
A prediction is made for each of eleven pairs of substances. The prediction is then tested using one or more diagnostic tests, complete with controls. Additional diagnostic tests increase the certainty of the evaluation.

To perform this investigation, turn to page 768.

LAB EXERCISE 16.D

Creating an Acid–Base Table

Report Checklist
- ○ Purpose
- ○ Problem
- ○ Hypothesis
- ○ Prediction
- ○ Design
- ○ Materials
- ○ Procedure
- ○ Evidence
- ● Analysis
- ○ Evaluation

Complete the Analysis of the investigation report, including a short table of the four acids and bases involved. Use entity position generalizations in the acid–base table for reactions that favour products or reactants. The evidence for reaction 1 is interpreted as:

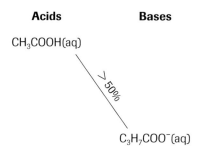

Purpose
The purpose of this investigation is to test an experimental design for using equilibrium position to create a table of relative strengths of acids and bases.

Problem
What is the order of acid strength for the first four members of the carboxylic acid family?

Evidence

$$CH_3COOH(aq) + C_3H_7COO^-(aq) \underset{}{\overset{>50\%}{\rightleftarrows}} CH_3COO^-(aq) + C_3H_7COOH(aq)$$

$$HCOOH(aq) + CH_3COO^-(aq) \underset{}{\overset{>50\%}{\rightleftarrows}} HCOO^-(aq) + CH_3COOH(aq)$$

$$C_2H_5COOH(aq) + C_3H_7COO^-(aq) \underset{}{\overset{<50\%}{\rightleftarrows}} C_2H_5COO^-(aq) + C_3H_7COOH(aq)$$

$$C_2H_5COOH(aq) + HCOO^-(aq) \underset{}{\overset{<50\%}{\rightleftarrows}} C_2H_5COO^-(aq) + HCOOH(aq)$$

Case Study

Changing Ideas on Acids and Bases—The Evolution of a Scientific Theory

Historically, chemists have known the empirical properties of substances long before any theory was developed to explain and predict those properties. For example, chemists were familiar with the distinguishing properties of several acids and bases, and used them routinely, by the middle of the 17th century. Early attempts at an acid–base theory tended to focus on acids and to ignore bases. Over time, several theories passed through cycles of formulation, testing, acceptance, further testing, and eventual rejection. Following is a brief historical summary of acid–base theories and the evidence that led to their revision.

- **Antoine Lavoisier** (1743–1794) (**Figure 11**) believed that the properties of acids could be traced back to a single substance. Lavoisier studied the combustion of phosphorus and sulfur and determined that these elements combine with something in the atmosphere to produce compounds that form acidic solutions when dissolved in water. When Joseph Priestley identified the component of the atmosphere that actively supports combustion, Lavoisier (1777) named the gas *oxygen*, meaning "acid maker." Lavoisier assumed that oxygen was the substance responsible for the generic properties of acids. There were soon problems with this theory when it was found that several acids, such as muriatic acid (HCl), do not contain oxygen. Furthermore, many substances that form basic solutions (such as lime, CaO) *were* found to contain oxygen. This evidence led to the rejection of the oxygen theory, but it is historically important because it is the first systematic attempt to chemically characterize acids and bases. The generalization that nonmetallic oxides form acidic solutions is still useful in chemistry.

- **Sir Humphry Davy** (1778–1829) (**Figure 12**) conducted the experiments that demonstrated the absence of oxygen in muriatic acid (HCl), which led to the rejection of Lavoisier's theory. Davy (1810) advanced his own theory that the

Figure 11
Antoine Lavoisier

Figure 12
Humphry Davy

presence of hydrogen gives compounds acidic properties. This theory, however, did not explain why many compounds containing hydrogen have neutral properties (for example, CH_4) or basic properties (for example, NH_3). Justus von Liebig (1803–1873) revised Davy's idea to define acids as substances in which the hydrogen could be replaced by a metal. This revision meant that acids could be thought of as ionic compounds in which hydrogen had replaced the metal ion. Liebig had no corresponding theoretical definition for bases, which were still identified empirically as substances that neutralized acids.

- **Svante Arrhenius** (1859–1927) (**Figure 13**) developed a theory in 1887 that provided the first useful theoretical definitions of acids and bases. He defined acids as *substances that ionize in aqueous solution to form hydrogen ions*, $H^+(aq)$. Similarly, he defined bases as *substances that dissociate to form hydroxide ions*, $OH^-(aq)$, *in solution*. This theory explained the process of neutralization as the combination of hydrogen ions and hydroxide ions to form water

Figure 13 Svante Arrhenius

$$H^+(aq) + OH^-(aq) \rightarrow H_2O(l)$$

Arrhenius explained the strengths of acids in terms of degrees of ionization. While these ideas were a major development in chemists' understanding of the properties of acids and bases, there were some problems with Arrhenius' theory.

- In Arrhenius' theory, an acid is expected to be an acid in any solvent, which was found not to be the case. For example, when dissolved in water, HCl supposedly breaks up into hydrogen ions and chloride ions, but when it is dissolved in benzene, the HCl remains as intact molecules. The nature of the solvent, therefore, had to play a critical role in acid–base properties of substances.
- The need for hydroxide to always be the base led Arrhenius to propose formulas such as $NH_4OH(aq)$ for the formula for ammonia in water, which led to the misconception that $NH_4OH(aq)$ was the base, not $NH_3(aq)$.
- According to Arrhenius' theory, all salts (ionic solids) should produce neutral solutions, but many do not. For example, solutions of ammonium chloride are acidic and solutions of sodium acetate are basic.
- Bonding theory suggested that it was very unlikely that a single proton could exist in aqueous solution without being bonded to at least one water molecule.

- **Johannes Brønsted** (1879–1947) (**Figure 14**) and **Thomas Lowry** (1874–1936) (**Figure 15**) independently published (1923) essentially the same concept about how acids and bases behave. These scientists focused on the

Figure 14 Johannes Brønsted

Figure 15 Thomas Lowry

role of an acid and a base in a reaction rather than on the properties of their aqueous solutions. According to the Brønsted–Lowry concept, *an acid is a proton (H^+) donor* and *a base is a proton acceptor*. The solvent has a central role in the Brønsted–Lowry concept. Water can be considered an acid or a base since it can lose a proton to form a hydroxide ion (OH^-) or accept a proton to form a hydronium ion (H_3O^+). In their view, neutralization is a competition for protons that results in a proton transfer from the strongest acid present to the strongest base present. For example, in the reaction

$$HSO_4^-(aq) + NH_3(aq) \rightarrow SO_4^{2-}(aq) + NH_4^+(aq)$$

the hydrogen sulfate ion acts as the acid (because it donates a proton) and ammonia acts as the base (because it accepts a proton). A weakness of the Brønsted–Lowry concept is its limitation to solutions (gaseous or liquid).

- **Gilbert Lewis** (1875–1946) (**Figure 16**) developed an acid–base theory in 1923 that includes all previous theories and definitions of acids and bases. He viewed acids and bases in terms of the covalent bond, a theory he had developed in 1916. Lewis defined acids as *electron-pair acceptors* and bases as *electron-pair donors*. It is important to note that in Lewis acid–base theory, no hydrogen ion and no solvent need be involved. The Lewis definition is broader than all previous definitions and explains more inorganic and organic reactions (**Figure 17**). For example, in the reaction

Figure 16 Gilbert Lewis

$$BF_3(g) + NH_3(g) \rightarrow H_3NBF_3(g)$$

boron trifluoride acts as a Lewis acid because it accepts (forms a bond with) a pair of electrons from the ammonia, which acts as the Lewis base.

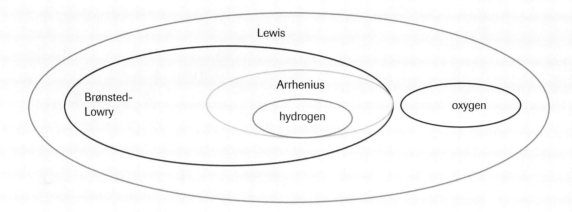

Figure 17
This Venn diagram shows how comprehensive some different acid–base theories are considered to be.

In science, it is unwise to assume that any scientific concept is complete. Whenever scientists assume that they understand a concept, two things usually happen. First, conceptual knowledge tends to remain static for a while because little conflicting evidence exists, or because conflicting evidence (being somewhat discomfiting) is ignored. Second, when enough conflicting evidence accumulates, a change in thinking occurs within the scientific community in which the current theory is drastically revised or entirely replaced. Revolutionary concepts most often tend to be formed in a moment of insight, usually following a long period of incredibly hard work done by a great many people.

Case Study Questions

1. Identify a word or phrase that describes the central idea in each of the theories described above.
2. For each theory, describe the weakness that led to its being replaced.
3. What was the first theoretical definition of a base, and who developed it?
4. Write Lewis formulas for each substance in the reaction of boron trifluoride with ammonia, to illustrate the "electron pair transfer" concept.

Section 16.2 Questions

1. Aqueous solutions of nitric acid and nitrous acid of the same concentration are prepared.
   (a) Predict how their pH values compare.
   (b) Explain your answer using the Brønsted–Lowry concept.
2. Briefly state the five steps involved in predicting the predominant reaction and the approximate position of equilibrium in an acid–base reaction system.
3. According to the Brønsted–Lowry concept, what determines the position of equilibrium in an acid–base reaction?
4. What generalization from the table of Relative Strengths of Aqueous Acids and Bases (Appendix I) can be used to predict the position of an acid–base equilibrium?
5. State two examples of conjugate acid–base pairs, each involving the hydrogen sulfite ion.
6. Predict, with reasoning, including Brønsted–Lowry equations, whether each of the following chemical systems will be acidic, basic, or neutral.
   (a) aqueous hydrogen bromide
   (b) aqueous potassium nitrite
   (c) aqueous ammonia
   (d) aqueous sodium hydrogen sulfate
   (e) carbonated beverages
   (f) limewater
   (g) vinegar
7. Write an experimental design to test each of the predictions in question 6.
8. Write two experimental designs to rank a group of bases in order of strength.

9. Ammonia molecules and hydronium ions have remarkably similar shapes and structures (**Figure 18**). Both have three hydrogen atoms bonded in a pyramidal structure to a small central atom, which has one lone pair of electrons. Oxygen and nitrogen atoms both have very high electronegativities, and they are adjacent on the periodic table. Explain why an ammonia molecule is a good proton attractor, whereas a hydronium ion is an extremely poor proton attractor.

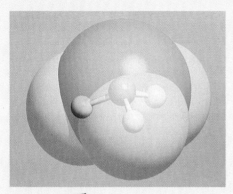

Figure 18
Evidence indicates that the ammonia molecule, NH_3, modelled here, has the same pyramidal shape as the hydronium ion, H_3O^+.

10. Many household cleaning solutions claim to "dissolve away" rust stains and hard water "lime" deposits (**Figure 19**). These cleaners are simply acidic solutions (as is vinegar) that react with slightly soluble substances such as $Fe_2O_3(s)$ and $CaCO_3(s)$. The word "react," however, sounds dangerous to consumers, and is hardly ever used in advertising. These cleaning solutions *must* be used carefully, and you should *always* read *all* of the label instructions, or you may find the cleaner also reacting with the item you are trying to clean, or even with you!

Figure 19
Commercial rust removing solutions must be used with care, but can be a very effective illustration of the old TV slogan, "better living, through chemistry."

These cleaners have a variety of formulations, but most contain hydroxyacetic acid (glycolic acid), $CH_2OHCOOH(aq)$. A 0.10 mol/L glycolic acid solution is 3.9% ionized at 25 °C, compared to 0.10 mol/L acetic acid, which is 1.3% ionized at the same temperature.
(a) Write a Brønsted-Lowry equation for the reaction of aqueous glycolic acid with carbonate ions (from hard water "scum"), and label both conjugate acid–base pairs.
(b) Explain whether the equilibrium would favour products more, or less, if acetic acid (vinegar) were used to do the same cleaning job. Is it likely that any difference would be significant? State your reasoning.
(c) Explain whether glycolic acid would react with a given chemical amount of rust more quickly or more slowly than an equal volume of equally concentrated acetic acid solution.
(d) Explain whether glycolic acid would react with a greater, or lesser, chemical amount of rust than an equal volume of equally concentrated acetic acid solution.

11. Since ancient times, people have known that strong heating of natural limestone (calcium carbonate) will produce quicklime (calcium oxide). This very useful substance is both reactive and corrosive with many organic substances because it has a high affinity for water. Adding water to quicklime causes the formation of slaked lime (calcium hydroxide)—the common name was derived from the idea that the lime was "slaking" its thirst. This reaction also produces a very considerable quantity of heat!
(a) Write a balanced chemical equation for the decomposition of calcium carbonate to carbon dioxide and calcium hydroxide.
(b) Write a balanced chemical equation for the addition of water to solid calcium oxide to produce solid calcium hydroxide.
(c) Write a Brønsted-Lowry equation for the instantaneous and quantitative (and highly exothermic) reaction of aqueous oxide ions with water.

Acid–Base Strength and the Equilibrium Law 16.3

Modifying Arrhenius' original concept of acids and bases allowed us to explain the behaviour of many substances in solution that could not be explained by his original theory. Then, problems with varying degrees of properties led to considering reactions of acids and bases as examples of equilibrium, which allowed a much more comprehensive understanding of relative acid or base strengths. Apparent conflicts in the definition of acids and bases then led us to define them by *what they do in a reaction* (in terms of proton transfer), rather than by *what they are* (in terms of some typical structure). This process of continually testing (identifying problems) and then expanding and improving concepts is typical of science—with the goal always being the formation of a more complete and comprehensive theory. Science assumes that no theory ever formed by the human mind can be "perfect," and that any theory, no matter how well supported by evidence, must always be questioned and tested. We also realize that every step moves us closer to a more complete understanding. Testing, and then restricting, revising, and (perhaps) replacing concepts, and then testing again is the constant cycle of the scientific method. Next, we explore the possible value of using the equilibrium law to further increase our understanding of acid–base behaviours.

The Acid Ionization Constant, K_a

One important way that chemists communicate the strength of any weak acid is by using the equilibrium constant expression for a Brønsted–Lowry reaction equation showing the acid ionizing in (reacting with) water. This expression is another very important *special case* of equilibrium. The constant is given its own unique name and symbol: the **acid ionization constant, K_a**. See the K_a values listed in the Relative Strengths of Aqueous Acids and Bases data table (Appendix I).

Note that in this table, the first six acids have K_a values given only as "very large." These acids are collectively called the *strong* acids because they all react quantitatively (>99.9%) with water to form hydronium ions. Because no acid stronger than H_3O^+(aq) can exist in aqueous solution, all of these acids are considered equivalent. For each of them, the actual acid species present is H_3O^+(aq).

All of the other acids listed on the table are considered to be *weak* acids, and they vary greatly in extent of reaction with water at equilibrium. To empirically rank any weak acid as stronger or weaker than any other, we normally compare the ionization of solutions *of the same concentration*, to see which is more strongly acidic (has a lower pH). This is how Appendix I was created.

By considering the equilibrium established in solutions of weak acid to be a Brønsted–Lowry proton transfer, we gain a more complete understanding about what is actually going on among the entities reacting. By using K_a equilibrium constants, we can derive and predict quantities that are very useful when working with acid–base reactions. Consider a solution of hydrofluoric acid, HF(aq). The equation for the reaction in water (ionization) of this weak acid is

$$HF(aq) + H_2O(l) \rightleftharpoons H_3O^+(aq) + F^-(aq)$$

The equilibrium law expression is

$$K_a = \frac{[H_3O^+(aq)][F^-(aq)]}{[HF(aq)]}$$

Note that the concentration of liquid water is omitted from the K_a equilibrium law expression. We make an assumption that this value will remain *essentially* constant for

Acid–Base Reactions
This quick simulation lets you explore the percent reaction when various acids and bases react together. The acid ionization constants are provided for two strong and three weak acids.

www.science.nelson.com

Learning *Tip*

Use of ionization terminology (*strong* and *weak*) can create a problem, caused by the common use of the word "strong." For an acid, *strength* refers only to the ease with which a base can remove its proton, and has nothing to do with the quantity present in the solution. *Concentration* refers to the chemical quantity per unit volume of solution, and has nothing to do with strength.

Both the strength and the concentration of an acid affect how rapidly, and to what extent, that acid will react.

> **Learning Tip**
>
> The great advantage of K_a values over percent ionization values is that, once determined, K_a values are valid over a wide range of acid solution concentrations. A percent ionization value is useful *only* for one specific concentration of one specific weak acid.

aqueous solutions *that are not highly concentrated*. This assumption is not *exactly* true, because the water is not a separate liquid phase in an aqueous solution. However, as long as the amount of water solvent is *much greater* than the amount of acid solute, making this assumption will cause negligible change in (and will not add uncertainty to) the answers for any calculations.

All K_a values are calculated by making this assumption. For this reason (and for other measurement reasons too involved to explain here), K_a values are necessarily somewhat inaccurate, and become less accurate as the acid concentration increases. It is important to know that K_a values given in tables are normally given to only two significant digits *because* they are usually only known to a certainty of about ± 5%.

Two calculations involving the K_a constant are common for weak acid solutions:

- calculating a K_a value from measured (empirical) amount concentration data
- using a K_a value to predict a concentration of hydronium ions for an aqueous solution where the initial weak acid amount concentration is known

Calculating K_a from Amount Concentrations

> **SAMPLE problem 16.2**
>
> The pH of a 1.00 mol/L solution of acetic acid is carefully measured to be 2.38 at SATP. What is the value of K_a for acetic acid?
>
> Use the balanced equation to write the equilibrium law expression.
>
> $$CH_3COOH(aq) + H_2O(l) \rightleftharpoons H_3O^+(aq) + CH_3COO^-(aq)$$
>
> $$K_a = \frac{[H_3O^+(aq)][CH_3COO^-(aq)]}{[CH_3COOH(aq)]}$$
>
> First, find the equilibrium concentration of aqueous hydronium ion from the pH.
>
> $$[H_3O^+(aq)] = 10^{-pH}$$
> $$= 10^{-2.38}$$
> $$= 0.0042 \text{ mol/L}$$
>
> Next, use the hydronium ion concentration and the balanced chemical equation to calculate the concentration of the acetate ion. Since one acetate ion forms for each hydronium ion, the equilibrium concentration of acetate ion must be identical to that of the hydronium ion. The stoichiometric ratio is 1:1.
>
> $$[CH_3COO^-(aq)] = [H_3O^+(aq)] = 0.0042 \text{ mol/L}$$
>
> An ICE table is useful to find the equilibrium concentration of acetic acid molecules.
>
> **Table 1** ICE Table for $CH_3COOH(aq) + H_2O(l) \rightleftharpoons H_3O^+(aq) + CH_3COO^-(aq)$
>
> | Concentration | [$CH_3COOH(aq)$] (mol/L) | [$H_3O^+(aq)$] (mol/L) | [$CH_3COO^-(aq)$] (mol/L) |
> |---|---|---|---|
> | Initial | 1.00 | 0 | 0 |
> | Change | − 0.0042 | + 0.0042 | + 0.0042 |
> | Equilibrium | 1.00 | 0.0042 | 0.0042 |
>
> $$K_a = \frac{(0.0042 \text{ mol/L})(0.0042 \text{ mol/L})}{(1.00 \text{ mol/L})}$$
>
> $$= 0.000\ 017$$
>
> Regardless of numerical size, K_a values are usually expressed in scientific notation. The calculated K_a for acetic acid is 1.7×10^{-5}.

> **Learning Tip**
>
> Because the value for the initial acid concentration is precise only to two decimal places, the calculation of the *equilibrium* concentration of the acid (1.00 mol/L − 0.0042 mol/L) *rounds* to 1.00 mol/L. (See the Precision Rule for Calculations, Appendix F.3.) In other words, the decrease in the initial acid concentration is negligible, *in this particular case*. A negligible change in concentration is *quite often* the case for K_a calculations for weak acids, *but not always*.
>
> Recall that amount concentrations must be used to calculate any K_c value and, by convention, units for the equilibrium constant value are simply ignored.

COMMUNICATION example 1

A student measures the pH of a 0.25 mol/L solution of carbonic acid to be 3.48. Calculate the K_a for carbonic acid from this evidence.

Solution

$$H_2CO_3(aq) + H_2O(l) \rightleftharpoons H_3O^+(aq) + HCO_3^-(aq)$$

$$K_a = \frac{[H_3O^+(aq)][HCO_3^-(aq)]}{[H_2CO_3(aq)]}$$

At equilibrium,

$$[H_3O^+(aq)] = 10^{-pH}$$
$$= 10^{-3.48}$$
$$= 3.3 \times 10^{-4} \text{ mol/L}$$

$[HCO_3^-(aq)] = [H_3O^+(aq)] = 3.3 \times 10^{-4}$ mol/L

$[H_2CO_3(aq)] = (0.25 - 0.00033)$ mol/L
$= 0.25$ mol/L

Table 2 ICE Table for $H_2CO_3(aq) + H_2O(l) \rightleftharpoons H_3O^+(aq) + HCO_3^-(aq)$

Amount concentration	$[H_2CO_3(aq)]$ (mol/L)	$[H_3O^+(aq)]$ (mol/L)	$[HCO_3^-(aq)]$ (mol/L)
Initial	0.25	0	0
Change	− 0.000 33	+ 0.000 33	+ 0.000 33
Equilibrium	0.25	0.000 33	0.000 33

$$K_a = \frac{(0.000\ 33 \text{ mol/L})(0.000\ 33 \text{ mol/L})}{(0.25 \text{ mol/L})} = 0.000\ 000\ 44$$

According to the equilibrium law, the K_a for carbonic acid is 4.4×10^{-7}.

Learning Tip

Because a Brønsted-Lowry equation is written to show a single proton transfer, the stoichiometric ratio will always be 1:1:1:1. Thus, the units for K_a values will always be mol/L. Because we can make this assumption, units are often not included with K_a values in reference tables or in final answer statements.

COMMUNICATION example 2

The pH of a 0.400 mol/L solution of sulfurous acid is measured to be 1.17. Calculate the K_a for sulfurous acid from this evidence.

Solution

$$H_2SO_3(aq) + H_2O(l) \rightleftharpoons H_3O^+(aq) + HSO_3^-(aq)$$

$$K_a = \frac{[H_3O^+(aq)][HSO_3^-(aq)]}{[H_2SO_3(aq)]}$$

At equilibrium,

$$[H_3O^+(aq)] = 10^{-pH}$$
$$= 10^{-1.17}$$
$$= 0.068 \text{ mol/L}$$

$[HSO_3^-(aq)] = [H_3O^+(aq)] = 0.068$ mol/L

$[H_2SO_3(aq)] = (0.400 - 0.068)$ mol/L
$= 0.332$ mol/L

Table 3 ICE Table for $H_2SO_3(aq) + H_2O(l) \rightleftharpoons H_3O^+(aq) + HSO_3^-(aq)$

Amount concentration	$[H_2SO_3(aq)]$ (mol/L)	$[H_3O^+(aq)]$ (mol/L)	$[HSO_3^-(aq)]$ (mol/L)
Initial	0.400	0	0
Change	− 0.068	+ 0.068	+ 0.068
Equilibrium	0.332	0.068	0.068

$$K_a = \frac{(0.068 \text{ mol/L})(0.068 \text{ mol/L})}{(0.332 \text{ mol/L})} = 0.014$$

According to the equilibrium law, the K_a for sulfurous acid is 1.4×10^{-2}.

Calculating $[H_3O^+(aq)]$ from K_a

The second type of calculation involving a K_a value allows us to *predict* the acidity of any weak acid solution. We can calculate the concentration of hydronium ions from the initial acid concentration and the K_a value as follows.

▶ SAMPLE problem 16.3

Predict the $[H_3O^+(aq)]$ and pH for a 0.200 mol/L aqueous solution of methanoic acid.

Look up the value of K_a for methanoic (formic) acid in the Relative Strengths of Aqueous Acids and Bases data table (Appendix I). The value is given as 1.8×10^{-4}.

Use the balanced equation to write the equilibrium law expression.

$$HCOOH(aq) + H_2O(l) \rightleftharpoons H_3O^+(aq) + HCOO^-(aq)$$

$$K_a = 1.8 \times 10^{-4} = \frac{[H_3O^+(aq)][HCOO^-(aq)]}{[HCOOH(aq)]}$$

Use x to represent *the numerical value* for the equilibrium amount concentration of aqueous hydronium ions.

$[H_3O^+(aq)] = x$ mol/L

An ICE table is useful to keep track of the concentration values.

Using the 1:1:1:1 stoichiometric ratio, calculate concentration increases and decreases for the reaction to equilibrium, and note them in the ICE table.

Table 4 ICE Table for $HCOOH(aq) + H_2O(l) \rightleftharpoons H_3O^+(aq) + HCOO^-(aq)$

Concentration	$[HCOOH(aq)]$ (mol/L)	$[H_3O^+(aq)]$ (mol/L)	$[HCOO^-(aq)]$ (mol/L)
Initial	0.200	0	0
Change	− x	+ x	+ x
Equilibrium	(0.200 − x)*	x	x

Substituting into the equilibrium law expression, ignoring units,

$$1.8 \times 10^{-4} = \frac{[H_3O^+(aq)][HCOO^-(aq)]}{[HCOOH(aq)]}$$

$$= \frac{(x)(x)}{(0.200 - x)}$$

> *** Note**:
> Technically, solving for x (the amount concentration of hydronium ion) from this expression requires the use of the quadratic formula, $x = \dfrac{-b \pm \sqrt{b^2 - 4ac}}{2a}$, which is quite tedious, unless you have a calculator that is preprogrammed to solve such formulas.
>
> Conveniently, for many (but not all) weak acid solutions, we can use an approximation that greatly simplifies solving for x. If the percent reaction of the weak acid is quite small, we can assume that the initial concentration of the acid decreases by so little that the numerical value of the acid amount concentration will remain effectively unchanged at equilibrium.
>
> The problem, of course, is deciding when a percent reaction is small enough to allow this assumption to be used. In this text, we will apply the mathematically simplest (and most convenient) "rule of thumb" that is commonly used to make this decision.
>
> *If the initial amount concentration of the acid is numerically at least 1000 times its K_a value, then you may assume that the initial and equilibrium acid amount concentrations are numerically equal.*
>
> This limiting assumption is consistent with the ±5% certainty of most K_a values, meaning it ignores those amount concentration changes that would change the K_a value by less than the uncertainty that already exists.

DID YOU KNOW ?
Rule of Thumb
The very old English phrase "rule of thumb," meaning a rule for any quick way to approximate a quantity without measuring, probably comes directly from use of human thumbs. Since the width of an adult's thumb is approximately one inch, and the length of a thumb is approximately four inches, you can make a fairly good estimate of the length of something (in Imperial units) by simply placing your thumbs along the object, one after the other. What is the length of this textbook in (your) thumb widths? What do equestrians mean when they say a horse stands 16 "hands" high?

A practical rule of thumb for Americans driving their cars into Mexico or Canada is to multiply the posted speed by 6 and drop the last digit, in order to quickly convert the km/h speeds on road signs into a fairly close miles-per-hour equivalent.

Note that the equation to be solved contains a squared value.

$$0.000\,18 = \dfrac{x^2}{(0.200 - x)}$$

Multiplying both sides by $(0.200 - x)$, collecting terms, and equating to zero gives $x^2 + 0.000\,18x - 0.000\,036 = 0$ (expressed in the form of a quadratic equation).

Notice what happens when we test our present example for the simplifying assumption.

$$\dfrac{[HCOOH(aq)]_{initial}}{K_a} = \dfrac{0.200}{0.000\,18} = 1111, \text{ which is greater than } 1000$$

The test allows us to use the assumption (initial acid concentration = equilibrium acid concentration, or $[HA(aq)]_{initial} = [HA(aq)]_{equilibrium}$). In this specific case, so little acid has reacted with water at equilibrium that the initial acid concentration is decreased by a negligible amount—less than the uncertainty of the initial value.

Numerically, for this example, we may assume that $(0.200 - x) = 0.200$, and we can then simplify the equation to

$$0.000\,18 = \dfrac{x^2}{0.200}, \text{ which makes solving for } x \text{ quick and easy.}$$

Isolate x and take the square root of both sides of the equation:

$$x = \sqrt{0.000\,18 \times 0.200}$$
$$= 0.0060$$

so, $[H_3O^+(aq)] = 0.0060$ mol/L or 6.0×10^{-3} mol/L

and $pH = -\log[H_3O^+(aq)]$
$$= -\log(0.0060)$$
$$= 2.22$$

DID YOU KNOW ?

pK_a

For purposes of easy comparison, the K_a values of different acids are sometimes expressed as pK_a values. A pK_a value is the negative logarithm of the K_a. A reported pK_a value of 12 would represent an acid with a K_a of 1×10^{-12}, a very weak acid. In comparing any two weak acids, the one with the lower pK_a value is the stronger of the two acids.

As shown in the next Communication Example, when the negligible ionization (percent reaction) assumption holds true, predictions of [H_3O^+(aq)] using K_a values become simple and straightforward. Note that you can perform the 1000:1 ratio test for the simplifying assumption *before starting the problem*, which is very convenient.

▶ COMMUNICATION *example 3*

Predict the [H_3O^+(aq)] and pH for a 0.500 mol/L aqueous solution of hydrocyanic acid.

Solution

Test the simplifying assumption:

$$\frac{[HCN(aq)]_{initial}}{K_a} = \frac{0.500}{6.2 \times 10^{-10}} = 8.1 \times 10^7 \text{ (greater than 1000)}$$

The assumption holds, and may be used.

$$K_a = 6.2 \times 10^{-10} = \frac{[H_3O^+(aq)][CN^-(aq)]}{[HCN(aq)]}$$

At equilibrium:

Let $x = [H_3O^+(aq)] = [CN^-(aq)]$

Then $[HCN(aq)] = (0.500 - x) = 0.500$ (using the assumption)

Table 5 ICE Table for $HCN(aq) + H_2O(l) \rightleftarrows H_3O^+(aq) + CN^-(aq)$

Amount concentration	[HCN(aq)] (mol/L)	[H_3O^+(aq)] (mol/L)	[CN^-(aq)] (mol/L)
Initial	0.500	0	0
Change	$-x$	$+x$	$+x$
Equilibrium	$(0.500 - x) = 0.500$	x	x

$$6.2 \times 10^{-10} = \frac{x^2}{0.500}$$

$$x = \sqrt{6.2 \times 10^{-10} \times 0.500} = 1.8 \times 10^{-5}$$

so, $[H_3O^+(aq)] = 1.8 \times 10^{-5}$ mol/L

and $pH = -\log[H_3O^+(aq)] = -\log(1.8 \times 10^{-5}) = 4.75$

According to the equilibrium law, the hydronium ion amount concentration of 0.500 mol/L hydrocyanic acid is 1.8×10^{-5} mol/L, and the pH of the solution is 4.75.

Note: No Practice, Section, or Review question in this text will require you to use the quadratic formula to solve acid ionization constant problems. However, you may be asked to state whether a given problem would require use of this formula for solution (in other words, whether the ionization assumption test fails). You should develop the habit of always initially testing the assumption and making a qualifying statement before completing the solution to any K_a question where this assumption holds.

Practice

1. (a) Write a *theoretical* definition for the strength of an acid.
   (b) State the *empirical* properties that provide evidence for differing acid strengths.
   (c) Explain the difference in meaning between *strength* and *concentration*, as these terms are used to refer to aqueous acids.
   (d) Does the stronger of two acids, when dissolved, necessarily make a more acidic aqueous solution? Explain.

2. Refer to Appendix I for required information to make the following predictions. For each case, first decide and state whether a solution would require use of the quadratic formula. For all cases where use of the quadratic formula is *not* required, communicate the full solution.
   (a) Predict $[H_3O^+(aq)]$ for 0.20 mol/L hydrobromic acid.
   (b) Predict $[H_3O^+(aq)]$ for 0.20 mol/L hydrofluoric acid.
   (c) Predict $[H_3O^+(aq)]$ for 0.20 mol/L ethanoic acid.
   (d) Predict the pH of 2.3 mmol/L nitric acid.
   (e) Predict the pH of 2.3 mmol/L nitrous acid.
   (f) Predict the pH of 2.3 mmol/L hydrosulfuric acid.

3. For all of the cases (a–f) in question 2 *where a prediction could be calculated*, rank the acid solutions in order of decreasing acidity.

4. For all of the cases (a–f) in question 2, rank the acids in order of decreasing acid strength. Which two acids have essentially the same strength?

5. The hydronium ion concentration in 0.100 mol/L propanoic acid is determined (from a pH measurement) to be 1.16×10^{-3} mol/L.
   (a) Calculate the percent reaction (ionization) of this particular weak acid solution.
   (b) Calculate K_a for aqueous propanoic acid.
   (c) Is this K_a value constant for propanoic acid? Explain.

6. A 0.10 mol/L solution of lactic acid, a weak acid found in milk, has a measured pH of 2.43. The chemical name for this very common organic compound is 2-hydroxypropanoic acid.
   (a) Find the percent reaction of this lactic acid solution.
   (b) Calculate the K_a value for aqueous lactic acid.
   (c) Compare the K_a values for 2-hydroxypropanoic acid and for propanoic acid. What effect does adding an OH group to the central carbon atom of this molecule have on the ability of the COOH group to attract a proton?

7. Unlike all other aqueous hydrogen halides, hydrofluoric acid is not a strong acid. It does, however, have a special chemical property: it reacts with glass. This property is used to etch frosted patterns on glassware (**Figure 1**).
   (a) Write the K_a expression for hydrofluoric acid.
   (b) Calculate the hydronium and fluoride ion amount concentrations, the pH, and the percent reaction in a commercial 2.0 mol/L HF(aq) solution at 25 °C.

8. Phosphoric acid is used in rust-remover solutions. The aqueous acid is available for purchase by high schools in concentrations of about 15 mol/L.
   (a) Predict the hydronium ion amount concentration, the pH, and the percent reaction of a 10 mol/L solution of phosphoric acid.
   (b) Suggest a reason (having to do with K_a values and solution concentrations) why you might well suspect that these values could be inaccurate.

9. Ascorbic acid is the chemical name for Vitamin C (**Figure 2**). A student prepares a 0.200 mol/L aqueous solution of ascorbic acid, tests its pH, and reads a value of 2.40 from the pH meter.
   (a) From the student's evidence, calculate the K_a for ascorbic acid.
   (b) Compare your result to the value listed in the table of Relative Strengths of Aqueous Acids and Bases (Appendix I). The K_a value for ascorbic acid increases with increasing temperature. State whether the student's aqueous solution was likely warmer or colder than standard temperature (25 °C) when the pH was tested.

Figure 1
Hydrofluoric acid is a weak acid but it etches glass. The etching is not due to the presence of hydronium ions, because, as you know, even the strongest acids are routinely stored in glass containers.

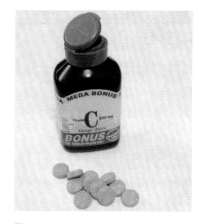

Figure 2
Vitamin C (ascorbic acid) is present in many fresh foods, particularly citrus fruits. It is essential to the body to promote healing of injuries and to fight infection.

In the winter of 1535, the men of Jacques Cartier's expedition were suffering from scurvy (vitamin C deficiency) in their fort at Stadacona (now Montreal). Cartier lost 25 men before learning of a simple cure from the Iroquois. The Aboriginal people prepared a tea made from white cedar needles—rich in ascorbic acid—that cured sufferers within days.

DID YOU KNOW ?

Sailors and Scurvy

Vitamin C deficiency (from a dietary lack of fresh fruits and vegetables) causes scurvy, a disease where the body cannot fight infections, and cuts, sores, and bruises do not heal. Until the last half of the 18th century, this disease was a terrible problem for sailors, and a major cause of death on long voyages. For example, on his famous voyage of 1498, Vasco da Gama lost 100 of 160 crewmen to scurvy. The diet of European sailors was primarily salted meat and hard biscuits—items selected for their resistance to spoilage.

In 1747, James Lind, a Scottish physician, wrote that feeding citrus fruit to scurvy victims effected an amazingly rapid cure. Captain Cook tried this remedy on his crew during his famous voyages of exploration in the 1770s, and reported that he was astounded to have lost only one man on a three-year voyage!

By 1795, the British Royal Navy (after decades of deliberation) formally adopted the practice of providing lemon juice (incorrectly referred to as lime juice) for all hands. Scurvy in the fleet was wiped out, and British sailors have been called "limeys" ever since.

Note that this story illustrates another case of technology leading science: The cure for scurvy was known long before the cause of the disease was explained.

Interestingly, sailors on the long sea voyages undertaken by the Chinese during the Ming Dynasty (1368–1644) had no problems with scurvy because their traditional diet included fresh germinated soya beans, the shoots of which are naturally rich in Vitamin C.

Base Strength and the Ionization Constant, K_b

The Brønsted–Lowry concept specifies that the strongest base possible in aqueous solution must be hydroxide ions. So, hydroxide ion is considered to be "the" strong base—somewhat parallel to the strong acids. A difference is that many ionic substances contain hydroxide ions initially, before being dissolved; whereas hydronium ions are always formed by the reaction of some entity with water, after a substance dissolves. Recall that for solutions of all ionic hydroxides, such as NaOH(aq) or Ca(OH)$_2$(aq), we assume that the compound dissociates completely upon dissolving. Therefore, finding the hydroxide ion concentration does not involve any reaction with water or any reaction equilibrium. Rather, in these cases, we find the hydroxide ion concentration directly from a dissociation equation, as shown in the next Communication Example.

▶ COMMUNICATION example 4

Find the hydroxide ion concentration of a 0.064 mol/L solution of barium hydroxide.

Solution

Ba(OH)$_2$(aq) → Ba^{2+}(aq) + 2 OH$^-$(aq)   (complete dissociation)

$$[OH^-(aq)] = [Ba(OH)_2(aq)] \times \frac{2}{1}$$

$$= 0.064 \text{ mol/L} \times \frac{2}{1}$$

$$= 0.13 \text{ mol/L}$$

According to the stoichiometric ratio, the hydroxide ion concentration is 0.13 mol/L.

Other entities that act as bases are collectively called weak bases, because empirical evidence indicates that they attract protons less than hydroxide ions do. Turn to the table of Relative Strengths of Aqueous Acids and Bases (Appendix I), and observe that, to the right of every acid entity formula listed, there is a formula for an entity that is identical to the acid's formula, *except* that one proton (represented as H$^+$) is missing. As you have learned, such an entity is called the conjugate base of the acid. While the list of these entities (reading upward) can be taken to be a list of bases and their (decreasing) relative strengths, note that there is no corresponding value given to show the extent of reaction of these entities with water. This convention is common in chemistry—information about bases in solution must be derived from a table of acid values.

Bases vary in strength much as acids do, and the system used to identify, classify, and rank the strength of bases is quite similar to the one you have just learned for acids. Note that the table in Appendix I shows that the weaker an acid is, the stronger its conjugate base (and vice versa). This observation is common sense. If it is very easy to remove the proton from an acid, then the entity remaining (the conjugate base) must not attract protons very well.

As you know, weak bases react only partially (usually much less than 50%) with water in aqueous solution to produce hydroxide ions. We can communicate the strength of any weak base using the equilibrium constant expression for its reaction with water in dilute aqueous solution. This equilibrium constant for weak bases is another special case, where the K_c constant is called the **base ionization constant, K_b**.

Consider a solution of sodium citrate, Na$_3$C$_6$H$_5$O$_7$(aq). We assume that in aqueous solution, any ionic compound dissociates completely into its component ions. In this case, the ions are aqueous sodium and citrate ions: $Na^+(aq)$ and $C_6H_5O_7^{3-}(aq)$. Sodium ions are "spectators," with no apparent acidic or basic properties. The citrate ions, however, are found on the Relative Strengths of Aqueous Acids and Bases data table (Appendix I) in the conjugate bases column. So the citrate ion is a weak Brønsted–Lowry base and reacts with water to form a basic solution at equilibrium.

The reaction equilibrium equation for aqueous sodium citrate is

$$C_6H_5O_7^{3-}(aq) + H_2O(l) \rightleftharpoons HC_6H_5O_7^{2-}(aq) + OH^-(aq)$$

The equilibrium law expression is

$$K_b = \frac{[HC_6H_5O_7^{2-}(aq)][OH^-(aq)]}{[C_6H_5O_7^{3-}(aq)]}$$

Just as with K_a values, there are two kinds of weak base solution calculations that involve the K_b constant:

- calculating a K_b value from empirical (measured) amount concentration data
- using a K_b value to predict an amount concentration of hydroxide ions for an aqueous solution where the initial weak base concentration is known

Calculating K_b from Amount Concentrations

Calculating K_b for a weak base uses essentially the same method as calculating K_a for a weak acid. Often, an extra (simple) calculation step must be included because measured pH values may need to be converted to find the equilibrium hydroxide ion concentration, as shown in the following Communication Example.

> **Learning Tip**
>
> The concentration of liquid water is omitted from K_b equilibrium law expressions for the same reasons that it is omitted from K_a expressions. As with K_a values, K_b values are also only certain to about ±5%. For weak base K_b values, we also assume that aqueous solutions are not highly concentrated, just as we do for weak acids. Just as with K_a values, K_b value units are ignored, by convention.

▶ COMMUNICATION example 5

A student measures the pH of a 0.250 mol/L solution of aqueous ammonia and finds it to be 11.32.

Calculate the K_b for ammonia from this evidence. Show the establishment of the reaction equilibrium using an ICE table.

Solution

$$NH_3(aq) + H_2O(l) \rightleftharpoons NH_4^+(aq) + OH^-(aq)$$

$$K_b = \frac{[NH_4^+(aq)][OH^-(aq)]}{[NH_3(aq)]}$$

At equilibrium:

$$[H_3O^+(aq)] = 10^{-pH}$$
$$= 10^{-11.32}$$
$$= 4.8 \times 10^{-12} \text{ mol/L}$$

$$K_w = [H_3O^+(aq)][OH^-(aq)]$$

$$[OH^-(aq)] = \frac{K_w}{[H_3O^+(aq)]}$$

$$= \frac{1.0 \times 10^{-14}}{4.8 \times 10^{-12} \text{ mol/L}}$$

$$= 0.0021 \text{ mol/L}$$

$$[NH_4^+(aq)] = [OH^-(aq)]$$
$$= 0.0021 \text{ mol/L}$$
$$[NH_3(aq)] = (0.250 - 0.0021) \text{ mol/L}$$
$$= 0.248 \text{ mol/L}$$

Table 6 ICE Table for $NH_3(aq) + H_2O(l) \rightleftharpoons NH_4^+(aq) + OH^-(aq)$

Concentration	[NH₃ (aq)] (mol/L)	[NH₄⁺(aq)] (mol/L)	[OH⁻(aq)] (mol/L)
Initial	0.250	0	0
Change	− 0.0021	+ 0.0021	+ 0.0021
Equilibrium	0.248	0.0021	0.0021

$$K_b = \frac{[0.0021 \text{ mol/L}][0.0021 \text{ mol/L}]}{[0.248 \text{ mol/L}]} = 0.000\,018$$

From this evidence, and according to the equilibrium law, the K_b for aqueous ammonia is 1.8×10^{-5}.

Figure 3
Dyes with molecular structures based on aniline make possible a variety of bright, stable colours for fabrics and leathers.

Learning Tip

The use of chemical (IUPAC) and common names can be confusing if you have not memorized a few examples. In this textbook, the normal practice is to give the chemical name, followed by a common name in parentheses. A few substances are so widely used, however, that the common name is ubiquitous (used everywhere, even by chemists). Baking soda (sodium bicarbonate, sodium hydrogen carbonate) and vinegar (acetic acid, ethanoic acid) are two examples of such chemicals; the common name is often given, rather than the IUPAC name. You are expected to be familiar with such examples, and others listed in Appendix J.

▶ Practice

10. List some empirical properties that would be useful when distinguishing strong bases from weak bases.
11. For each of the following weak bases, write the chemical equilibrium equation and the equilibrium law expression for K_b.
    (a) $CN^-(aq)$
    (b) $SO_4^{2-}(aq)$
12. The hydroxide ion concentration in a 0.157 mol/L solution of sodium propanoate, $NaC_2H_5COO(aq)$, is found to be 1.1×10^{-5} mol/L. Calculate the base ionization constant for the propanoate ion.
13. Aniline, $C_6H_5NH_2$, is used to make a wide variety of drugs and dyes (**Figure 3**). It has the structure of an ammonia molecule, with a phenyl group substituted for one hydrogen, and, like ammonia, acts as a weak base. If the pH of a 0.10 mol/L aniline solution was found to be 8.81, what is its K_b?

Calculating [OH⁻(aq)] from K_b

The second type of calculation involving a K_b value is the *prediction* of the basicity of any weak base solution. We can calculate the concentration of hydroxide ions from the initial weak base concentration and the K_b value. There is an automatic problem, however, because tables of relative acid–base strengths do not normally list K_b values. We use the automatic relationship between conjugate acid–base pairs to deal with this problem.

The K_a–K_b Relationship for Conjugate Acid–Base Pairs

To develop the next useful concept, we use the common weak acid, acetic acid, $CH_3COOH(aq)$, and its conjugate base, the acetate ion, $CH_3COO^-(aq)$. Of course, it would be just as correct to state that we will use the weak base, acetate ion, and its conjugate acid, acetic acid.

The equilibrium reaction and law expression for aqueous acetic acid are

$$CH_3COOH(aq) + H_2O(l) \rightleftharpoons H_3O^+(aq) + CH_3COO^-(aq)$$

$$K_a = \frac{[H_3O^+(aq)][CH_3COO^-(aq)]}{[CH_3COOH(aq)]}$$

For the weak base, aqueous acetate ion,

$$CH_3COO^-(aq) + H_2O(l) \rightleftharpoons CH_3COOH(aq) + OH^-(aq)$$

$$K_b = \frac{[CH_3COOH(aq)][OH^-(aq)]}{[CH_3COO^-(aq)]}$$

Notice what happens when we multiply these equilibrium constant expressions:

$$K_a \times K_b = \frac{[H_3O^+(aq)][\cancel{CH_3COO^-(aq)}]}{[\cancel{CH_3COOH(aq)}]} \times \frac{[\cancel{CH_3COOH(aq)}][OH^-(aq)]}{[\cancel{CH_3COO^-(aq)}]}$$

$$= [H_3O^+(aq)][OH^-(aq)] \quad \text{(Does this expression look familiar?)}$$

$$= K_w$$

So, for any conjugate acid–base pair, $K_w = K_a K_b$ and $K_b = \dfrac{K_w}{K_a}$.

This last equation is particularly useful because now, to find a K_b value for any base listed on the Relative Strengths of Aqueous Acids and Bases table (Appendix I), you need only identify its conjugate acid, and then divide K_w, 1.0×10^{-14}, by the K_a value given for that conjugate acid.

▶ SAMPLE problem 16.4

Solid sodium benzoate forms a basic solution. Determine K_b for the weak base present.

$$NaC_6H_5COO(s) \rightarrow Na^+(aq) + C_6H_5COO^-(aq) \text{ (complete dissociation)}$$

First identify, from the Relative Strengths of Aqueous Acids and Bases table, and the entities present in solution, which entity is reacting as a base.

The benzoate ion must be the weak base entity. Its equilibrium reaction with water is

$$C_6H_5COO^-(aq) + H_2O(l) \rightleftharpoons C_6H_5COOH(aq) + OH^-(aq)$$

From the equilibrium equation, the conjugate acid of the benzoate ion is identified as benzoic acid. From the acid–base table,

$$K_a \text{ for } C_6H_5COOH(aq) = 6.3 \times 10^{-5}$$

Using the K_w relationship for conjugate acid–base pairs, find K_b for $C_6H_5COO^-(aq)$.

$$K_b = \frac{K_w}{K_a}$$

$$= \frac{1.0 \times 10^{-14}}{6.3 \times 10^{-5}}$$

$$= 1.6 \times 10^{-10}$$

According to the K_w relationship, the benzoate ion has a K_b value of 1.6×10^{-10}.

DID YOU KNOW ?

Why Acidity?
It should be obvious by now that pH values are used much more than pOH values, and K_a constants are listed in tables rather than K_b constants. You might wonder why we choose to emphasize the hydronium ion properties of aqueous solutions—after all, hydroxide ions play exactly the opposite (and an equally important) role. But historically, chemists have always been much more concerned with acidic properties than with basic properties. Acids dissolve (react with) many metals to form common ionic compounds and hydrogen. Hydrogen gas was a fascinating product to early chemists, both because it is lighter than air and because it is violently explosive. The stronger acids tend to have painfully sharp odours and flavours, and react destructively with human tissue—all properties that make acids quite noticeable. Even when it became clearly understood that acids and bases were opposite aspects of the same concept, it seemed easier to choose acid properties as the initial basis for studying acid–base relationships. We talk about the "acidity" of something routinely; it is a common term, found in any pocket dictionary. What is the equivalent term for how *basic* something is? There is an accepted word for it, but you'll need a *really* good (very big) dictionary to find it. Or a good Chemistry text, of course...

> **COMMUNICATION** example 6

Calculate K_b for the weak base aqueous ammonia, commonly used in commercial window cleaning solutions. Write the equation for the equilibrium.

Solution

$NH_3(aq) + H_2O(l) \rightleftharpoons NH_4^+(aq) + OH^-(aq)$

For $NH_3(aq)$, $NH_4^+(aq)$ is the conjugate acid:

$K_a = 5.6 \times 10^{-10}$

$K_b = \dfrac{K_w}{K_a}$

$= \dfrac{1.0 \times 10^{-14}}{5.6 \times 10^{-10}}$

$= 1.8 \times 10^{-5}$

According to the table of Relative Strengths of Aqueous Acids and Bases, and the K_a–K_b relationship, the K_b value for aqueous ammonia is 1.8×10^{-5}.

Now we can easily predict how basic a solution of known concentration will be from its K_b value. Predicting basicity always involves calculating the hydroxide ion concentration, and may also involve calculating a value for pOH, pH, and/or percent reaction. The first step may be determining the K_b value by using a table of K_a values. The form of the calculation then follows the same format as the equivalent type of question for weak acids, with a similar assumption made, and the same restrictions.

- Assume that the initial weak base aqueous concentration decreases so little that it is numerically unchanged at equilibrium, but *only* if the initial amount concentration of the weak base is at least 1000 times greater than its K_b value.
- For questions in this text, you are not required to do calculations using the quadratic formula.

The Sample Problem that follows shows the format for calculations of this type, and also an example of each conversion that may be required.

Recall, from Section 16.2, that, in aqueous solution, the amphiprotic ion $HCO_3^-(aq)$ reacts with water as a base to a *very* much greater extent than it does as an acid. You may, therefore, assume that the solution is basic because the ion reacts as a base, and ignore its negligible reaction extent as an acid.

> **SAMPLE** problem 16.5

Find the hydroxide ion amount concentration, pOH, pH, and percent reaction (ionization) of a 1.20 mol/L solution of baking soda.

The compound is sodium hydrogen carbonate, $NaHCO_3(s)$. The weak base is the $HCO_3^-(aq)$ ion, produced by dissolving $NaHCO_3(s)$ in water.

For $HCO_3^-(aq)$, $H_2CO_3(aq)$ is the conjugate acid:

$K_a = 4.5 \times 10^{-7}$

Find the K_b value using the K_a–K_b–K_w relationship.

$K_b = \dfrac{K_w}{K_a}$

$= \dfrac{1.0 \times 10^{-14}}{4.5 \times 10^{-7}}$

$= 2.2 \times 10^{-8}$

Write the equation for the reaction and the equilibrium expression for K_b for hydrogen carbonate ion.

$$HCO_3^-(aq) + H_2O(l) \rightleftarrows H_2CO_3(aq) + OH^-(aq)$$

$$K_b = \frac{[H_2CO_3(aq)][OH^-(aq)]}{[HCO_3^-(aq)]}$$

At equilibrium, let x be the numerical value of $[OH^-(aq)]$ and also $[H_2CO_3(aq)]$.

Table 7 ICE Table for $HCO_3^-(aq) + H_2O(l) \rightleftarrows H_2CO_3(aq) + OH^-(aq)$

Amount concentration	$[HCO_3^-(aq)]$ (mol/L)	$[H_2CO_3(aq)]$ (mol/L)	$[OH^-(aq)]$ (mol/L)
Initial	1.20	0	0
Change	$-x$	$+x$	$+x$
Equilibrium	$(1.20 - x)$	x	x

Then $[HCO_3^-(aq)] = (1.20 - x)$

Test the assumption that $(1.20 - x)$ is numerically equal to 1.20.

$$\frac{[HCO_3^-(aq)]_{initial}}{K_b} = \frac{1.20}{2.2 \times 10^{-8}} = 5.5 \times 10^7 \text{ (much greater than 1000)}$$

The assumption holds, so substitute into the K_b expression and solve for x.

$$2.2 \times 10^{-8} = \frac{x^2}{1.20}$$

$$x = \sqrt{2.2 \times 10^{-8} \times 1.20}$$

$$= 1.6 \times 10^{-4}$$

$[OH^-(aq)] = 1.6 \times 10^{-4}$ mol/L

Use $[OH^-(aq)]$ to find the value of pOH.

$$pOH = -\log [OH^-(aq)]$$
$$= -\log (1.6 \times 10^{-4})$$
$$= 3.80$$

Use the pOH value to find pH.

$$pH = 14.00 - pOH$$
$$= 14.00 - 3.80$$
$$= 10.20$$

Percent reaction is the percent ratio of the equilibrium concentration of produced hydroxide ions to the initial concentration of hydrogen carbonate ions.

$$\text{percent reaction} = \frac{[OH^-(aq)]_{equilibrium}}{[HCO_3^-(aq)]_{initial}} \times 100\%$$

$$= \frac{1.6 \times 10^{-4} \text{ mol/L}}{1.20 \text{ mol/L}} \times 100\%$$

$$= 0.013\%$$

A 1.20 mol/L sodium hydrogen carbonate solution has a hydroxide ion concentration of 1.6×10^{-4} mol/L, a pOH of 3.80, a pH of 10.20, and a percent reaction of 0.013%.

The Effect of Amphoteric Entities

Finally, we deal with the problem of amphoteric entities. If an entity can react as either a Brønsted–Lowry acid or base, how do you know which will be the predominant reaction in its aqueous solution? You can determine the answer with one simple calculation.

DID YOU KNOW ?

Visualizing Reaction Extents
From any percent reaction value, you can get a better idea of the reaction extent at equilibrium by converting mentally to a whole-number ratio. If bicarbonate ions react with water 0.013% at equilibrium, the ratio is 0.013:100, or 13:100 000. This means that at any given instant, about 13 of every 100 000 initial bicarbonate ions have changed to carbonic acid molecules, but 99 987 of them have not. The few ions that have changed are responsible for all the basicity of the solution!

If you *really* want to test your powers of imagination, try to visualize that for every 100 000 bicarbonate ions in this aqueous solution, there are approximately 5 500 000 water molecules. Now think about all of these entities constantly moving at speeds averaging about 1500 km/h, and then imagine every one of them colliding with another molecule or ion at a rate of approximately 7 000 000 000 times *every second*...

Read the K_a value for the entity from the table of Relative Strengths of Aqueous Acids and Bases. Calculate the K_b value for the entity from K_w and the K_a value (from the table) of its conjugate acid, as shown in Sample Problem 16.5. The higher of the "K" values tells you which reaction predominates, and, therefore, whether the entity reacts as an acid or as a base in aqueous solution.

COMMUNICATION example 7

Which reaction predominates when $NaHSO_3(s)$ is dissolved in water to produce $HSO_3^-(aq)$ solution? Will the solution be acidic or basic?

Solution

$K_a = 6.3 \times 10^{-8}$

The conjugate acid is $H_2SO_3(aq)$, with $K_a = 1.4 \times 10^{-2}$.

So, for $HSO_3^-(aq)$,

$$K_b = \frac{K_w}{K_a}$$

$$= \frac{1.0 \times 10^{-14}}{1.4 \times 10^{-2}}$$

$$= 7.1 \times 10^{-13}$$

The K_a value for $HSO_3^-(aq)$ far exceeds its calculated K_b value.

According to the K_w relation and the equilibrium law, an aqueous solution of this substance will be acidic because a hydrogen sulfite ion will react predominately as a Brønsted–Lowry acid.

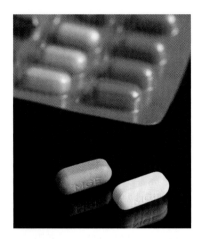

Figure 4
Codeine (methylmorphine, $C_{18}H_{21}NO_3$) is available by prescription as an ingredient in some analgesic (pain relief) medicines. This narcotic alkaloid was initially extracted from opium. Other alkaloids include nicotine, quinine, cocaine, strychnine, and caffeine. All are common plant-derived organic compounds based on nitrogen-containing ring structures.

Section 16.3 Questions

1. Codeine, an ingredient in these migraine pills (**Figure 4**), has a K_b of 1.73×10^{-6}. Calculate the pH of a 0.020 mol/L codeine solution (use Cod as a chemical shorthand symbol for this complex weak base entity.).

2. What is the pH of a 0.18 mol/L cyanide ion solution?

3. Acetylsalicylic acid (ASA) is a painkiller used in many headache tablets. This drug forms an acidic solution that attacks the digestive system lining. *The Merck Index* lists its K_a at 25 °C to be 3.27×10^{-4}. Explain the difficulty in calculating the pH of a saturated 0.018 mol/L solution of acetylsalicylic acid, $C_6H_4COOCH_3COOH(aq)$.

4. Boric acid is used for weatherproofing wood and fireproofing fabrics. Very dilute aqueous boric acid is used in eyewash solution as a preservative. Predict the pH of a 0.50 mol/L solution of boric acid.

5. Salicylic acid (2-hydroxybenzoic acid, $C_6H_5OHCOOH(s)$) is an active ingredient of medications, such as Clearasil®, that are used to treat acne. Since the K_a for this acid was not listed in any convenient references, a student tried to determine the value experimentally. She kept adding water slowly to a 1.00 g sample, with constant stirring, until the crystals all dissolved. The volume of solution was 460 mL. She found that the pH of this solution of salicylic acid was 2.40 at 25 °C. Calculate the ionization constant for this acid.

6. Sodium hydrogen ascorbate is often used as an antioxidant in packaged non-fat foods. The pH of a 0.15 mol/L laboratory solution of sodium hydrogen ascorbate, $NaHC_6H_6O_6(aq)$ is measured to be 8.65. From this evidence, calculate the K_b value for the hydrogen ascorbate ion.

7. Sodium hypochlorite is a strong oxidizing agent that is a fire hazard when in contact with organic materials. Solutions of sodium hypochlorite are used as bleach and disinfectant. Determine the hydroxide ion amount concentration of a sodium hypochlorite solution sold as household bleach. The bleach bottle label reads "5.25% (by mass) when packed."

8. Write the equilibrium law expression for acetic acid reacting (ionizing) in water. Use Le Châtelier's principle to explain, in terms of equilibrium shift, why dilute acetic acid solutions have a higher percent reaction than more concentrated solutions. Consider that diluting a solution containing reacting aqueous entities is very similar to increasing the volume of a container of reacting gaseous entities.

9. Predict and write the predominant Brønsted–Lowry reaction in aqueous solutions of the following substances:
    (a) $K_2HPO_4(s)$ (b) $NaH_2PO_4(s)$ (c) $Na_2HC_6H_5O_7(s)$

10. Calculate the pH of 5.0×10^{-2} mol/L solutions of each of the reagents in question 9.

16.4 Interpreting pH Curves

For many acid–base reactions, the appearance of the products resembles that of the reactants, so we cannot directly observe the progress of a reaction. Also, acids cannot easily be distinguished from bases except by measuring pH. As you first learned in Chapter 8, a graph showing the continuous change of pH during an acid–base titration, continued until the titrant is in great excess, is called a **pH curve** (titration curve) for the reaction. The pH values and changes provide important information about the nature of acids and bases, the properties of conjugate acid–base pairs and indicators, and the stoichiometric relationships in acid–base reactions.

Buffering Regions, Endpoints, and Indicators

The pH curves for acid–base reactions have characteristic shapes. You have learned that a common reason for plotting a pH curve is to determine the value of the solution's pH when an acid–base reaction reaches its equivalence point. We can then use this information to select an appropriate indicator—one that will produce an easily seen endpoint when a solution reaches this pH value. Then we can use the same acid–base reaction, done as a titration to that indicator's endpoint, for titration analysis.

In all titration analyses, the critical measured quantity is the titrant volume required to reach the titration endpoint. For accurate analysis, this value needs to be as close as possible to the theoretically exact value needed to reach the reaction equivalence point. Analysis titrations, like all empirical work, necessarily involve some uncertainty. We call any variation between endpoint values and equivalence point values the titration error of an experiment. Good technique and careful application of knowledge can ensure that this variation is as small as possible—preferably negligible.
Remember:

- *Endpoint* refers to that point in a titration analysis where the addition of titrant is stopped. The endpoint is defined (empirically) by the observed colour change of an indicator.

- *Equivalence point* refers to that point in any chemical reaction where chemically equivalent amounts of the reactants have been combined. The equivalence point is defined (theoretically) by the stoichiometric ratio from the reaction equation.

Endpoints are easily detectable because pH changes a great deal, and very abruptly, as the reaction solution changes (at the equivalence point) from a tiny excess of acid to a tiny excess of base (or vice versa). But this effect begs the question: Why, for most of a titration, does the pH hardly change at all?

To better understand the information that a pH curve provides, we will use a familiar reaction about which we already know a great deal. The pH curve for the strong acid–strong base reaction (Figure 2) has three regions of interest. In the course of this excess titration, the pH first changes very slowly, then very rapidly, and finally very slowly again, as titrant is steadily added. Notice that the addition of acid titrant to the base sample initially has very little effect on the pH of the solution in the flask. In fact, the high pH has not changed much even after enough acid has been added to react with 90% of the original amount of the base sample. This nearly level region on a pH curve identifies a buffering region. **Buffering** is the property of some solutions (often called buffer solutions) of resisting (counteracting) any significant change in pH when an acid or a base is added. In this particular case, buffering occurs because any strong acid added immediately reacts with excess hydroxide ions. Consider that before *any* acid is added, the sample solution is primarily water molecules and hydroxide ions.

Learning Tip

When you began to study acids and bases, the term "neutralize" was understood to refer to an acid–base reaction that had the same final pH as neutral pure water: a pH of 7. Later, you learned that the pH at the equivalence point for an acid–base reaction is almost never 7, except for the strong acid–strong base reaction of hydronium ions with hydroxide ions.

As commonly used, the word "neutralize" is taken to have a more general meaning: that one substance can completely or partially lessen the acidic or basic properties of another substance. When we say sodium carbonate can be used to neutralize spilled nitric acid (**Figure 1**), we mean that adding the base brings the pH value up much closer to neutral.

Figure 1
Soda ash (sodium carbonate) is blown onto nitric acid that has spilled from some railway tank cars, to neutralize the highly reactive and dangerous acid.

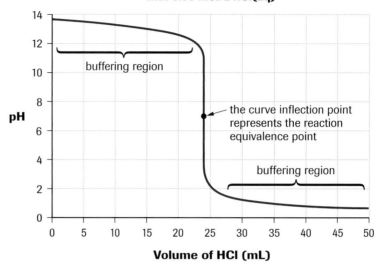

Figure 2
This pH curve for the continuous and excess addition of 0.50 mol/L HCl(aq) to a 25 mL sample of 0.48 mol/L NaOH(aq) helps chemists to understand the nature of acid–base reactions.

Learning Tip

Recall from Unit 4 that the inflection point of any plotted curve is the point where the curvature of the line changes direction. For instance, on the strong acid–strong base pH curve in Figure 2, the line curves downward until about 24 mL of acid has been added. At some observable point, the line curvature changes to upward.

For this specific titration, the change in pH is so large and so rapid that the pH plot becomes a nearly vertical line. In such a case the equivalence point can also be identified as the "midpoint" of the steep drop in the plotted line.

Also recall that, for this particular reaction (only), we already know (from memory) that the equivalence point pH must be 7.0 (at SATP), without actually needing to plot a titration curve: The reaction of hydroxide and hydronium ions is highly quantitative, and the only product is water.

As long as a large amount of hydroxide ions is present, any added acid is immediately converted to water, producing a solution that *still* consists primarily of water molecules and hydroxide ions. The hydroxide ion concentration decreases a little, but that affects the solution pH value only very slightly. This pH "levelling effect" finally fails near the equivalence point—when the hydroxide ions in the solution become almost completely consumed. But, until then, the solution maintains a (nearly) constant pH. This buffering property turns out to be critically important for a great number of reactions in applied chemistry and biology. Note that once *excess* acid has been added, the solution consists primarily of water molecules and hydronium ions. Again, the pH remains stable because adding more hydronium ions now does not change the nature of the solution; it only increases the hydronium ion concentration slightly. Another buffering region is established—this time at a low pH. You will learn more about the causes of buffering regions and the nature of buffer solutions later in this section.

Following the first buffering region there is a very rapid change in pH for a very small additional volume of the titrant. The inflection point on the pH curve represents the equivalence point of this reaction: At this point, we know from theory that the pH must be 7. We also know from experience (Chapter 8) that this reaction can be used for titration analysis, if an indicator is selected that changes colour in solution at a pH very close to 7. As shown in Figure 3, bromothymol blue indicator has a colour change pH range with a middle value *very* close to (just slightly below) pH 7, which makes it an appropriate indicator for this particular reaction. The values below show that the (theoretical) titrant volume required for complete reaction, as predicted by calculation, *should* be 24 mL. Figure 3 shows that if the colour change of bromothymol blue to a green intermediate colour is the endpoint, then the titrant volume *actually* measured at the endpoint (24 mL) is negligibly different from the theoretical predicted value. So, this titration analysis gives very accurate results.

$$OH^-(aq) \quad + \quad H_3O^+(aq) \quad \rightarrow \quad 2\,H_2O(l)$$

$25\ \text{mL} \times 0.48\ \text{mol/L} \qquad 24\ \text{mL} \times 0.50\ \text{mol/L}$

$12\ \text{mmol} \qquad\qquad\qquad\quad 12\ \text{mmol}$

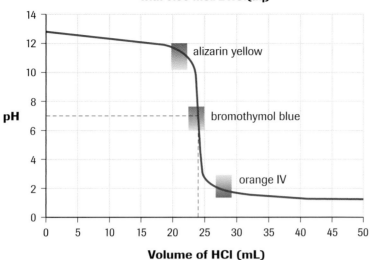

Figure 3
Alizarin yellow is not a suitable indicator because it will change colour long before the equivalence point of this strong acid–strong base reaction, which theoretically has a pH of exactly 7. Orange IV is also unsuitable; its colour change would occur too late. The pH at the middle of the colour change range for bromothymol blue is 6.8, which very closely matches the equivalence point pH; so, a titration analysis endpoint for this reaction, as indicated by bromothymol blue, should give accurate results.

Acid–Base Indicator Equilibrium

Acid–base indicators may be better understood and explained by using Brønsted–Lowry definitions. Any acid–base indicator is really two entities for which we use the same name: a Brønsted–Lowry conjugate acid–base pair. At least one (most often both) of the entities is visibly coloured, so you can tell simply by looking when it forms or is consumed in a reaction. Phenolphthalein is a common example (**Figure 4**), where the conjugate base form is bright red and the conjugate acid form is colourless. Typically, common indicators such as methyl red (**Figure 5**) have quite complex molecular formulas, so we use a (very simplified) shorthand to identify them, for convenience. For example, we symbolize the (invisible) acid form of phenolphthalein as HPh, and the (bright red) base form as Ph$^-$. When an indicator equilibrium is shifted to the point where equal concentrations of both forms exist, an intermediate colour may be observed. For example, equal concentrations of the blue and yellow forms of bromothymol blue (at a pH of 6.8) appear green to the human eye.

The explanation of the behaviour of acid–base indicators depends, in part, on both the Brønsted–Lowry concept and the equilibrium concept. *An indicator is a conjugate weak acid–weak base pair formed when an indicator dye dissolves in water.* Using HIn(aq) to

Figure 4
Sodium hydroxide solution has been added to hydrochloric acid containing phenolphthalein indicator, which is colourless in acids and red in bases. The red colour, indicating the temporary presence of some unreacted sodium hydroxide, is rapidly disappearing as the flask contents swirl and mix.

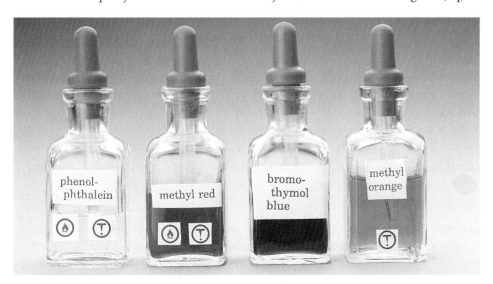

Figure 5
A few common acid–base indicators are shown here. Each indicator has its own pH range over which it changes colour from the acid form (HIn) at the lower pH value to the base form (In$^-$) at the higher pH value. Material Safety Data Sheets are available for these chemicals.

represent the acid form and In⁻(aq) to represent the base form of any indicator, we can write the following equilibrium equation. (Litmus colours are given below the equation as an example.)

$$\underbrace{\text{HIn(aq)} + \text{H}_2\text{O(l)} \rightleftharpoons \text{In}^-\text{(aq)}}_{\text{conjugate pair}} + \text{H}_3\text{O}^+\text{(aq)}$$

acid   base   base   acid
red   (litmus colour)   blue

According to Le Châtelier's principle, an increase in the hydronium ion concentration will shift the equilibrium to the left. Then more indicator will change to the colour of the acid form (HIn(aq)). This change happens, for example, when litmus is added to an acidic solution. Similarly, in basic solutions the hydroxide ions remove hydronium ions by reacting with them to form water, with the result that the equilibrium shifts to the right. Then the base colour of the indicator (In⁻) predominates. Since different indicators have different acid strengths, the acidity or pH of the solution at which an indicator changes colour varies (**Figures 5** and **6**). These pH values have been measured and are reported in the table of acid–base indicators on the inside back cover of this book.

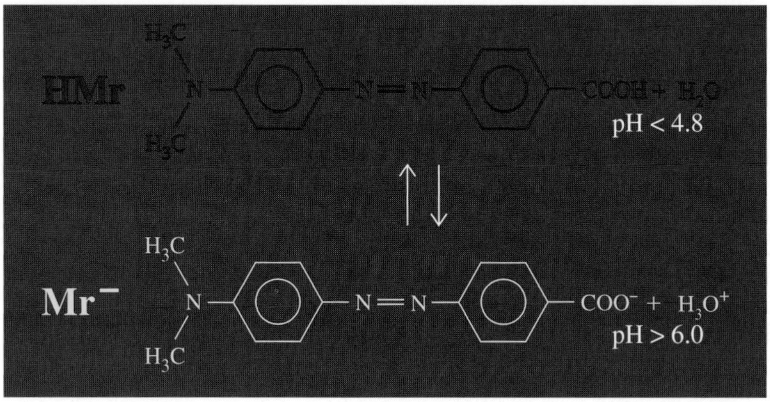

> **Learning Tip**
>
> Typically, acid–base indicators such as methyl red (**Figure 6**) are large molecules with quite complex formulas. Smaller molecules and ions are less likely to interact with light waves in the visible spectrum.
>
> The shorthand symbolism we use to represent indicators is partly for convenience, but also partly to reinforce the concept that indicators are conjugate acid–base pairs. Consider the actual formulas for the methyl red entities, which we represent as HMr for the acid form, and Mr⁻ for the base form. This simplified symbolism makes the proton transfer, which causes the colour change, more obvious.

Figure 6
The visible colour of methyl red indicator depends on the equilibrium proportions of its two coloured forms at the pH of the solution in which it is placed. Methyl red exists predominantly in its red (acid) form at pH values less than 4.8, and in its yellow (base) form at pH values greater than 6.0. Between pH values of 4.8 and 6.0, intermediate orange colours occur, as both forms of the indicator are present in detectable quantities.

> ▶ **Practice**
>
> 1. The shape of a pH curve is interpreted to describe the change of properties throughout the course of an acid–base reaction.
>   (a) In terms of curve shape, describe the characteristics of a buffering region.
>   (b) In terms of pH change and titrant volume, explain what a buffering region represents.
>   (c) In terms of curve shape, describe the characteristics near an equivalence point.
>   (d) In terms of pH change and titrant volume, explain what an equivalence point represents.
>
> Use this information to answer questions 2 to 4.
>
> According to the Brønsted–Lowry concept, acid–base indicators are simply coloured conjugate acid–base pairs. As for all other weak acids, the conjugate acid forms of these indicators must differ in strength for different entities.
>
> 2. For each indicator following, write a Brønsted–Lowry equilibrium equation for the reaction of the acid form of the indicator with water, to form a hydronium ion and the

base form of the indicator. Use the same indicator symbolism as used in the table of Acid–Base Indicators on the inside back cover of this book. Identify both conjugate acid–base pairs for each equation, and label all entities "acid" or "base."
(a) methyl orange
(b) indigo carmine
(c) thymol blue (the entity with two "H's" in the symbolized formula)
(d) thymol blue (the entity with one "H" in the symbolized formula)
(e) bromothymol blue

3. Which indicator entity, in your answers to question 2, can behave as both an acid and a base? What chemical term is used to describe this property of an entity?

4. (a) Refer to the table of Acid–Base Indicators to determine which of the five conjugate acid forms for the indicators in question 2 is the strongest acid. (Note the colour change pH ranges.) Then rank the other four indicators beneath it in order of decreasing strength, as in a relative strengths of acids table.

   (b) Suppose that five identical samples of an aqueous hydroxide ion solution each have four drops of a different indicator (from question 2) added. If each sample is now titrated with the same HCl(aq) titrant, in which indicator/sample combination will the least amount of titrant cause a colour change?

Polyprotic Entities and Sequential Reactions

Expanding on what you learned in Chapter 8, and adding the Brønsted–Lowry concept, we now examine some acids and bases that can lose (or gain) more than one proton. Polyprotic acids can lose more than one proton, and polyprotic bases can gain more than one proton, in Brønsted–Lowry transfers. If more than one proton transfer actually occurs in the course of a titration, chemists believe the process occurs as a series of single-proton transfer reactions. Typical pH curves for reactions of diprotic or triprotic acids and bases differ from those of monoprotic acids and bases—and can provide useful information about the reactions going on.

We observe (**Figure 7**) that the pH curve for the addition of HCl(aq) to Na_2CO_3(aq) shows two equivalence points—two significant changes in pH. Chemists interpret such curves as indicating how many quantitative reactions have occurred, sequentially, for that particular acid–base titration. Here, for example, two successive quantitative reactions have occurred. The two equivalence points evident in Figure 7 can be explained by two different proton transfer equations.

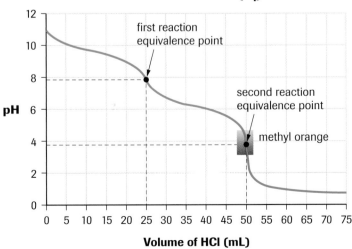

Figure 7
A pH curve for the addition of 0.50 mol/L HCl(aq) to a 25.0 mL sample of 0.50 mol/L Na_2CO_3(aq) can be used to select an appropriate indicator for this reaction done as a titration analysis. The colour change of methyl orange (from pH 4.4 to pH 3.2) means it will show an endpoint that corresponds closely to the second reaction equivalence point.

First, protons transfer from hydronium ions to carbonate ions, the strongest base present in the initial sample. The first significant drop in pH is evidence that the first reaction is quantitative. We write the Brønsted–Lowry equation in the usual way, starting with the acid and base entities present in the sample solution, and also listing the $H_3O^+(aq)$ that is being added.

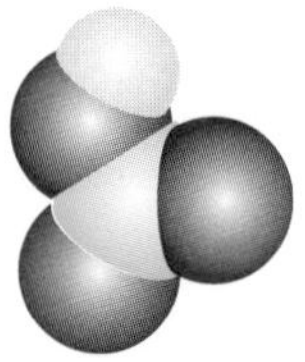

$$\underset{SA}{H_3O^+(aq)}, Cl^-(aq), Na^+(aq), \underset{SB}{CO_3^{2-}(aq)}, \underset{B}{\overset{A}{H_2O(l)}}$$

$$H_3O^+(aq) + CO_3^{2-}(aq) \rightarrow H_2O(l) + HCO_3^-(aq)$$

Because the first reaction is quantitative, essentially all the carbonate ions will have been consumed at the equivalence point, and an equal quantity of (new) hydrogen carbonate ions will have been formed. Then, in a second reaction, protons transfer from additional (continually added) hydronium ions to the hydrogen carbonate ions that were formed in the first reaction.

$$\underset{SA}{H_3O^+(aq)}, Cl^-(aq), Na^+(aq), \underset{B}{\overset{A}{H_2O(l)}}, \underset{SB}{\overset{A}{HCO_3^-(aq)}}$$

$$H_3O^+(aq) + HCO_3^-(aq) \rightarrow H_2O(l) + H_2CO_3(aq)$$

Because the second reaction is also quantitative, all of the hydrogen carbonate ions have reacted at the equivalence point, and only newly formed carbonic acid molecules remain in the solution. Continuing to add more hydronium ion after this point will cause no significant further change (no third reaction), because the strongest base in the system now is water.

The carbonate ion is a *diprotic base* because it can accept a total of two protons. Other polyprotic bases include sulfide ions and phosphate ions. The conjugate entities formed when these bases gain successive protons are:

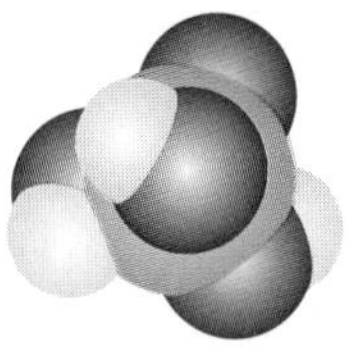

$$S^{2-}(aq) — HS^-(aq) — H_2S(aq)$$

$$PO_4^{3-}(aq) — HPO_4^{2-}(aq) — H_2PO_4^-(aq) — H_3PO_4(aq)$$

Other polyprotic acids include oxalic acid and phosphoric acid (**Figure 8**).

$$HOOCCOOH(aq) — HOOCCOO^-(aq) — OOCCOO^{2-}(aq)$$

$$H_3PO_4(aq) — H_2PO_4^-(aq) — HPO_4^{2-}(aq) — PO_4^{3-}(aq)$$

Evidence from pH measurements clearly shows that, for every proton transferred by a polyprotic entity, the strength of the new acid or base entity formed *greatly* decreases. The new entity may be anywhere from 100 to 100 000 times weaker! Chemists believe that electric charge and electrostatic attraction explain this effect. For example, it should logically be much easier to pull a *positively charged* proton away from a *neutral* oxalic acid molecule, than it is to pull one away from a *negatively charged* hydrogen oxalate ion.

Figure 9 shows the pH curve for phosphoric acid titrated with sodium hydroxide. Phosphoric acid is triprotic, so three reactions should be possible. However, note that only two equivalence points are evident. We describe the process as before. At the first equivalence point (pH 4), equal chemical amounts of $H_3PO_4(aq)$ and $OH^-(aq)$ have been combined in solution. As the last of the $H_3PO_4(aq)$ reacts, the pH rises abruptly, because when the reaction moves past the equivalence point, only newly formed $H_2PO_4^-(aq)$ is now present, and it is a much weaker acid than $H_3PO_4(aq)$.

nitric acid, $HNO_3(aq)$

sulfuric acid, $H_2SO_4(aq)$

phosphoric acid, $H_3PO_4(aq)$

Figure 8
For oxyacids, the protons that transfer in Brønsted–Lowry reactions are covalently bonded to oxygen atoms (identified by a red colour in the models above). Of these three examples, the monoprotic and diprotic acids are strong acids, and the triprotic acid is a weak acid.

Acid formulas were originally derived from stoichiometric evidence. Arguably, it would be more informative to write the molecular formulas differently, as NO_2OH, $SO_2(OH)_2$, and $PO(OH)_3$, to clarify the bonding concept. In both science and society, however, once conventions are established, they are difficult to change.

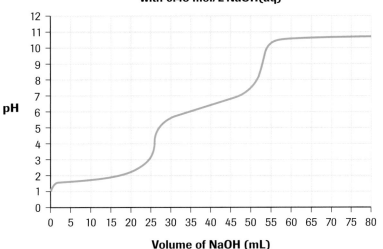

Figure 9
A pH curve for the addition of NaOH(aq) to a sample of H₃PO₄(aq) displays only two rapid changes in pH. This result is interpreted as indicating that there are only two quantitative proton transfer reactions of phosphoric acid with hydroxide ions. The third proton transfer reaction never goes to completion, but instead establishes an equilibrium.

To write the reaction for the transfer of the first proton, start by listing the entities.

Na⁺(aq), OH⁻(aq), H₃PO₄(aq), H₂O(l)
         SA   A
  SB   B

OH⁻(aq) + H₃PO₄(aq) → H₂O(l) + H₂PO₄⁻(aq)

Since all the H₃PO₄(aq) has reacted when the reaction has passed the first equivalence point, the second plateau must represent the reaction of OH⁻(aq) with H₂PO₄⁻(aq). The second equivalence point (pH 9) corresponds to the completion of the reaction of H₂PO₄⁻(aq) with additional OH⁻(aq) solution. As the last of the H₂PO₄⁻(aq) reacts, the pH begins to rise abruptly again because, when the reaction has passed this (second) equivalence point, only newly formed HPO₄²⁻(aq) is left, and it is an even weaker acid.

Na⁺(aq), OH⁻(aq), H₂O(l), H₂PO₄⁻(aq)
          A   SA
  SB   B   B

OH⁻(aq) + H₂PO₄⁻(aq) → H₂O(l) + HPO₄²⁻(aq)

No pH endpoint is apparent for the possible reaction of HPO₄²⁻(aq) with additional OH⁻(aq). A clue to this lack of a third endpoint can be obtained from the table of acids and bases. The hydrogen phosphate ion is an extremely weak acid and apparently does not quantitatively lose protons to OH⁻(aq). For the third reaction, an equilibrium is established that gradually shifts right as more base is added. There is no pH increase at a third equivalence point because the reaction never goes to completion.

Na⁺(aq), OH⁻(aq), H₂O(l), HPO₄²⁻(aq)
          A   SA
  SB   B   B

$$HPO_4^{2-}(aq) + OH^-(aq) \underset{}{\overset{>50\%}{\rightleftarrows}} PO_4^{3-}(aq) + H_2O(l)$$

As a general rule, *only quantitative reactions produce detectable equivalence points in an acid–base titration.*

Learning Tip

When any weak polyprotic acid or base is initially dissolved in water, we always assume that the ionization equilibrium only involves the first proton transfer—to or from water. This is essentially true because the tendency of any entity that is formed by a first ionization to lose or gain a (second) proton is very much less than that of the original polyprotic entity dissolved; so much so that it can safely be considered negligible.

DID YOU KNOW ?

Sulfuric Acid

Sulfuric acid (**Figure 10**) is probably the world's most important industrial chemical. It is used for so many things that the industrial development and standard of living of a country may be thought of as being proportional to its sulfuric acid production. Canada produces approximately four million tonnes every year. U.S. production (the world's largest) is about 10 times as much. The main direct consumer use of sulfuric acid is the 4.5 mol/L solution inside every standard car and truck battery.

Sulfuric acid plants (**Figure 11**) can be quite compact, and are assembled from "off-the-shelf" technology. They involve only a few simple reactions, starting with sulfur (in Alberta, mostly extracted from fossil fuels) and oxygen.

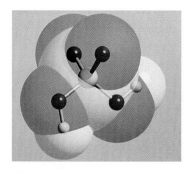

Figure 10
A model of sulfuric acid, showing relative atom size, bonding, and molecular shape

Figure 11
More sulfuric acid is manufactured in North America than any other chemical.

Sulfuric acid, H_2SO_4(aq), is a unique polyprotic acid because it is the only common one for which the first proton loss is already quantitative in aqueous solution; that is, it is the only strong acid that is polyprotic. Sulfuric acid acts like any of the other five common strong acids, except that its 100% ionization produces hydrogen sulfate ions, HSO_4^-(aq), along with hydronium ions, H_3O^+(aq). Recall that of all the anions of this type (like HCO_3^-(aq), or HPO_4^{2-}(aq)), the hydrogen sulfate ion is the only one that is a weaker base than water, so it cannot react as a base in aqueous solution. When reacting as an acid, however, hydrogen sulfate ion is one of the stronger weak acids, and usually reacts quantitatively with bases (except for very weak bases, of course). So sulfuric acid will usually, but not always, react with a base (assuming the base is in excess) in two complete proton transfer reactions, one after the other. When you first began a study of reactions of acids and bases, you would have written a complete reaction of sulfuric acid with sodium hydroxide as

$$H_2SO_4(aq) + 2\,NaOH(aq) \rightarrow Na_2SO_4(aq) + 2\,H_2O(l)$$

This reaction equation, if you think about it now, implies that you assumed that both of this acid's "hydrogens" (protons) were "replaced" (transferred). You had no information at that time, however, about how this transfer occurs. For simple stoichiometric calculations based on this reaction, this limited understanding (and simplified equation) worked perfectly well. Explaining and predicting other reactions of sulfuric acid, however, is a different story. To do so, understanding how the neutralization process occurs (as two distinct and sequential reactions) is necessary, because both reactions may not be quantitative when a different (weaker) base is involved, and stoichiometric calculation can only be done for a reaction that is quantitative.

There are almost no overall reactions of common polyprotic acids or bases that have more than two definite endpoints. For any such reactions that are actually done as analysis titrations, an indicator can be selected so that the titration can be stopped at either chosen equivalence point. For the titration of sodium carbonate solution with hydrochloric acid, this would mean titrating through two sequential quantitative reactions to the second equivalence point, as shown in Figure 7. For this titration, the second equivalence point involves a greater pH change, and thus is easier to detect (more accurate). As we found earlier, methyl orange indictor is suitable for detecting this second equivalence point.

For purposes of stoichiometric calculation, we can combine sequential quantitative Brønsted–Lowry reaction equations into a single "overall" reaction equation. This example is for the titration of sodium carbonate solution with hydrochloric acid.

1. $H_3O^+(aq) + CO_3^{2-}(aq) \rightarrow \cancel{HCO_3^-(aq)} + H_2O(l)$ followed by
2. $H_3O^+(aq) + \cancel{HCO_3^-(aq)} \rightarrow H_2CO_3(aq) + H_2O(l)$ "totals" to

$$2\,H_3O^+(aq) + CO_3^{2-}(aq) \rightarrow H_2CO_3 + 2\,H_2O(l) \quad \text{(overall reaction)}$$

For purposes of stoichiometric calculation, this equation is equivalent to writing

$$2\,HCl(aq) + Na_2CO_3(aq) \rightarrow H_2CO_3(aq) + 2\,NaCl(aq)$$

because, either way, the reactant acid–base stoichiometric ratio is found to be 2:1. Chemists normally use the simplest representation that will be useful, so Brønsted–Lowry equations are more often used for correctly predicting products and equilibria, whereas standard chemical formula equations are more often used for "doing" stoichiometry as, for example, in titration analyses.

Section **16.4**

SUMMARY: pH Curve Reaction Information

Empirical pH curves provide a wealth of information:
- initial pH of sample solutions
- pH when excess titrant is added
- number of quantitative reactions
- non-quantitative (equilibrium) reactions
- equivalence point(s) for indicator selection
- buffering regions

▶ Practice

5. How is buffering action displayed on a pH curve?
6. How are quantitative reactions displayed on a pH curve?
7. How is a pH curve used to choose an indicator for a titration?
8. An acetic acid sample is titrated with sodium hydroxide (**Figure 12**.)

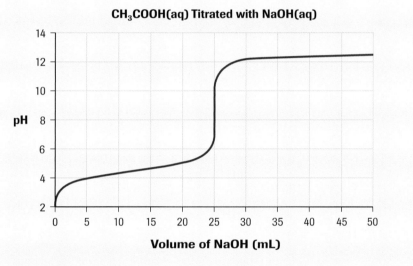

Figure 12
The pH curve for the addition of 0.48 mol/L NaOH(aq) to 25.0 mL of 0.49 mol/L CH_3COOH(aq) illustrates pH changes during the reaction of a weak acid with a strong base.

   (a) Based on Figure 12, estimate the pH at the equivalence point.
   (b) Choose an appropriate indicator for this titration.
   (c) Write a Brønsted–Lowry equation for this reaction.
   (d) At the very beginning of the titration, before the curve levels off, it rises. Explain this rise in terms of entities present in the mixture before and after beginning the titration.

9. A sodium phosphate solution is titrated with hydrochloric acid (**Figure 13** on next page).
   (a) Why are only two equivalence points evident?
   (b) Write three Brønsted–Lowry equations for the sequential reactions shown on the pH curve in Figure 13. Communicate the position of equilibrium for each of the three reactions.

10. Oxalic acid reacts quantitatively in a two-step overall reaction with a sodium hydroxide solution. Assuming that an excess of sodium hydroxide is added, sketch a pH curve (without any numbers) for all possible reactions.

Equilibrium in Acid–Base Systems

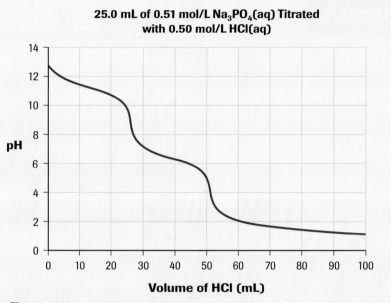

Figure 13
The pH curve for the addition of HCl(aq) to Na_3PO_4(aq) can be interpreted using the Brønsted–Lowry acid–base concept.

pH Curve Shape versus Acid and Base Strength

You have learned that the reaction of the strongest possible acid in aqueous solution, H_3O^+(aq), and the strongest possible base in aqueous solution, OH^-(aq), is quantitative, and always results in an equivalence point pH of 7.00 at 25 °C. Not all titrations involve the hydronium and the hydroxide ions, however.

pH curves show that some weak acids (such as acetic acid) react quantitatively with OH^-(aq) (**Figure 12**). The reactions of some weak bases (such as PO_4^{3-}(aq) and HPO_4^{2-}(aq)) with H_3O^+(aq), can also produce a quantitative reaction (**Figure 13**). The strongest of the weak acids also react quantitatively with the strongest of the weak bases. For example, the reaction of HSO_4^-(aq) with CO_3^{2-}(aq) is a quantitative reaction. The farther apart the reacting acid and base are on the Relative Strengths of Aqueous Acids and Bases table (i.e., the larger the difference in K_a), the more likely it is that the reaction will be quantitative. After plotting many pH curves, chemists have developed the general pH curve reference diagram shown in **Figure 14**.

Learning Tip

Recall that, for the examples you have studied, bromothymol blue has been an appropriate indicator for the SA–SB reaction, because it detects an endpoint with a pH of 7 very well. Methyl orange detects endpoint pH values around 4 well, and is, therefore, useful for many SA–WB reactions. Finally, phenolphthalein detects endpoints with pH values around 10 well, and is, therefore, useful for many WA–SB titrations.

Weak and Strong Acids and Bases

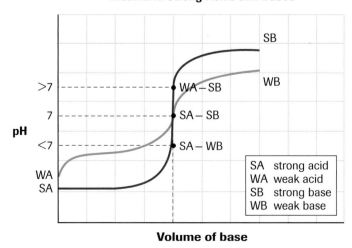

Figure 14
Laboratory evidence is generalized here to show approximations of the pH curves for quantitative reactions of weak and strong acids and bases. Note the initial change in pH for curves involving "weak" entities, due to the immediate change of the kinds of entities present in solution when the titration begins.

The (stoichiometric) equivalence point pH for a strong acid–strong base (SA–SB) reaction is seven (7). The equivalence point pH of any quantitative strong acid–weak base (SA–WB) reaction is less than seven (<7), whereas that of any quantitative weak acid–strong base reaction (WA–SB) is greater than seven (>7).

The composite curves in Figure 14 are not acceptable (do not predict well) for any reactions between a weak acid and a weak base. Weak acid–weak base reactions often do not *have* a detectable equivalence point because they are usually not quantitative. For all intents and purposes, analysis reactions and stoichiometric calculations are not done for weak acid–weak base reactions, so a pH curve for such a reaction is simply irrelevant.

The pH of a solution at the equivalence point of an acid–base reaction may be explained by considering the nature of those entities that are present in significant quantities at that point. The entities present at an equivalence point can be predicted either by following the five-step Brønsted–Lowry method, or simply by converting a chemical substance formula equation into a net ionic equation.

Now let's examine a case of each of the common acid–base reaction types, and explain the equivalence point pH in each case. Each of the examples used is a quantitative reaction.

Strong Acid–Strong Base Reaction

For example, nitric acid reacts with potassium hydroxide: SA–SB pH = 7.

$$\overset{SA}{H_3O^+(aq)}, NO_3^-(aq), K^+(aq), \underset{SB}{OH^-(aq)}, \overset{A}{\underset{B}{H_2O(l)}}$$

$$H_3O^+(aq) + OH^-(aq) \rightarrow 2\,H_2O(l)$$

or

$$HNO_3(aq) + KOH(aq) \rightarrow H_2O(l) + KNO_3(aq)$$

$$H^+(aq) + \cancel{NO_3^-(aq)} + \cancel{K^+(aq)} + OH^-(aq) \rightarrow H_2O(l) + \cancel{K^+(aq)} + \cancel{NO_3^-(aq)}$$

$$H^+(aq) + OH^-(aq) \rightarrow H_2O(l)$$

At the equivalence point, water is the only acid or base entity present, which produces a neutral solution (pH = 7).

Strong Acid–Weak Base Reaction

For example, hydrochloric acid reacts with aqueous ammonia: SA–WB pH < 7.

$$\overset{SA}{H_3O^+(aq)}, Cl^-(aq), \underset{SB}{NH_3(aq)}, \overset{A}{\underset{B}{H_2O(l)}}$$

$$H_3O^+(aq) + NH_3(aq) \rightarrow H_2O(l) + NH_4^+(aq)$$

or

$$HCl(aq) + NH_3(aq) \rightarrow NH_4Cl(aq)$$

$$H^+(aq) + \cancel{Cl^-(aq)} + NH_3(aq) \rightarrow NH_4^+(aq) + \cancel{Cl^-(aq)}$$

$$H^+(aq) + NH_3(aq) \rightarrow NH_4^+(aq)$$

At the equivalence point, the only acid or base entity present (other than water) is the ammonium ion, which is a weak acid, resulting in an acidic solution (pH <7).

> **Learning Tip**
>
> Recall that entities common to both sides of the equation are cancelled out and the coefficients of the remaining entities are reduced to the simplest ratio, in a net ionic equation.

Weak Acid–Strong Base Reaction

For example, acetic acid reacts with barium hydroxide: WA–SB pH > 7.

$$\underset{SB}{\underset{SA}{CH_3COOH(aq),}\ Ba^{2+}(aq),\ OH^-(aq),}\ \underset{B}{\overset{A}{H_2O(l)}}$$

$$CH_3COOH(aq) + OH^-(aq) \rightarrow H_2O(l) + CH_3COO^-(aq)$$

or

$$2\ CH_3COOH(aq) + Ba(OH)_2(aq) \rightarrow 2\ H_2O(l) + Ba(CH_3COO)_2(aq)$$

$$2\ CH_3COOH(aq) + \cancel{Ba^{2+}(aq)} + 2\ OH^-(aq) \rightarrow$$
$$2\ H_2O(l) + \cancel{Ba^{2+}(aq)} + 2\ CH_3COO^-(aq)$$

$$CH_3COOH(aq) + OH^-(aq) \rightarrow H_2O(l) + CH_3COO^-(aq)$$

At the equivalence point, the only acid or base entity present (other than water) is the acetate ion, which is a weak base, resulting in a basic solution (pH >7).

> **Learning Tip**
>
> In an equation for a titration, a single arrow, ⟶, is used. The reaction represented by the equation must be quantitative in order for any stoichiometric calculations based on the equation to be valid. In such cases, chemical formula equations are often the most convenient form.

SUMMARY Titration Generalizations

- Strong acid–strong base reactions are quantitative (100%) and have an equivalence point pH = 7.
- Strong acid–weak base quantitative reaction equivalence points have a pH <7.
- Weak acid–strong base quantitative reaction equivalence points have a pH >7.
- Polyprotic entity samples produce sequential reactions in titrations, each of which may or may not be quantitative.

 **WEB Activity**

Simulation—Titration of Polyprotic Acids and Bases

These two simulations enable you to explore the pH curves that result from various titrations. Of course, polyprotic acids and bases each have more than one K value, so the shapes of their titration curves will reflect this. Predict the shapes of the curves before working through the exercises.

www.science.nelson.com

▶ Practice

11. For the first quantitative reaction in each of the following acid–base titrations, predict (where possible) whether the equivalence point pH will be greater than, less than, or equal to 7.
    (a) hydroiodic acid + aqueous sodium hydrogen phosphate ⟶
    (b) boric acid + aqueous sodium hydroxide ⟶
    (c) aqueous sodium hydrogen sulfate + aqueous potassium hydroxide ⟶
    (d) hydrochloric acid + solid magnesium hydroxide ⟶
    (e) hydrosulfuric acid + aqueous sodium hydrogen carbonate ⟶
    (f) sulfuric acid + aqueous ammonia ⟶

12. Assume only three indicator solutions are available in your lab: methyl orange, bromothymol blue, and phenolphthalein. Choose the most suitable indicator for detecting the equivalence point of each of the following acid–base titrations (for the first quantitative reaction only).
    (a) HBr(aq) + Ca(OH)$_2$(s) ⟶
    (b) HNO$_3$(aq) + Na$_2$CO$_3$(aq) ⟶
    (c) KOH(aq) + HNO$_2$(aq) ⟶

13. The following formulas each represent a solution at an acid–base reaction equivalence point. From the entities present in each solution, predict the observed pH as being greater than, less than, or equal to 7.
    (a) NH$_4$Cl(aq)
    (b) Na$_2$S(aq)
    (c) KNO$_3$(aq)
    (d) NaHSO$_4$(aq)

14. Use the five-step Brønsted–Lowry method to predict the overall reaction net ionic equation when the following chemicals are mixed.
    (a) solutions of perchloric acid and sodium carbonate (A pH curve shows two protons are transferred quantitatively in successive reactions.)
    (b) solutions of nitrous acid and potassium hydroxide
    (c) solutions of phosphoric acid and sodium hydroxide (A pH curve shows two protons are transferred quantitatively in successive reactions.)

15. Write a standard chemical formula equation, and also the net ionic equation, for the predominant reaction that occurs when each of the following pairs of reagents mix.
    (a) Stomach acid is neutralized by solid magnesium hydroxide.
    (b) Aqueous ammonia is added to sulfurous acid.
    (c) A solution of sulfuric acid is neutralized by sodium hydroxide. (A pH curve indicates two quantitative reactions.)

pH Curve Buffering Regions and Buffer Solutions

A titration pH curve for any quantitative acid–base reaction shows at least one region where buffering action occurs between the beginning of the titration and the equivalence point. If the titration is continued much past the equivalence point, it enters another region of buffering. Such a region represents a solution in which partial reaction has occurred, and, therefore, contains significant amounts of both entities of a conjugate acid–base pair. The term **buffer** refers specifically to the combination of any weak acid with its conjugate base, in the same solution. Buffer solutions have a specific pH due to the nature and concentration of the entities present, and change pH very little when other acids or bases are added—provided that the quantity added is much less than the quantities of the conjugate pair entities present in the buffer.

For example, consider the titration of acetic acid with sodium hydroxide (**Figure 15**). The solution pH is approximately 4.8 when 10 mL of sodium hydroxide solution has been added. This volume represents one-half of the (calculated) equivalence point volume of titrant. At this point in the reaction, one-half of the acetic acid originally present will have reacted (changed to acetate ions), according to the following equation:

$$\text{OH}^-(aq) + \text{CH}_3\text{COOH}(aq) \rightarrow \text{H}_2\text{O}(l) + \text{CH}_3\text{COO}^-(aq)$$

The mixture, therefore, now contains approximately equal amounts of the remaining unreacted weak acid, CH$_3$COOH, and of the conjugate base, CH$_3$COO$^-$, produced in the reaction. Chemists would describe this mixture as an acetic acid–acetate ion buffer.

Learning Tip

You have observed (Figure 2, page 752) that the strong acid–strong base reaction pH curve shows that the conjugate pair H$_3$O$^+$(aq)–H$_2$O(l) will maintain a stable (very low) pH; and that the H$_2$O(l)–OH$^-$(aq) pair will stabilize pH at a very high value.

Even though these solutions certainly do produce a buffering effect, they are not technically considered to be buffers, because in each case one of the two conjugate pair entities (water) is also the solution solvent, and is present in great excess.

A buffer is normally thought of as an aqueous solution mixture of a weak acid and its conjugate base, with both entities of the conjugate pair present in significant quantity, but with both quantities much less than the quantity of the solution solvent (water) present.

Figure 15
The highlighted plateau shows an effective buffering region during this titration of aqueous acetic acid with sodium hydroxide. At the point halfway to the equivalence point, the solution is a 1:1 ratio mix of an acetic acid–acetate ion buffer.

> **Learning Tip**
>
> Buffers are normally prepared as mixtures of a weak acid with a solution of some salt of that weak acid. Thus, dissolving solid sodium acetate in a solution of acetic acid would produce a buffer solution with properties the same as those of the highlighted buffering region of Figure 15.

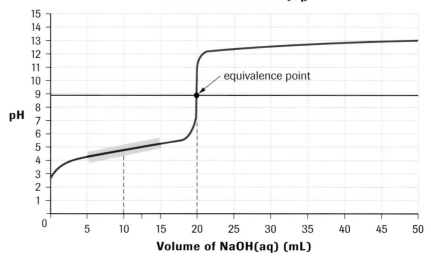

(a) Acetic Acid Buffer

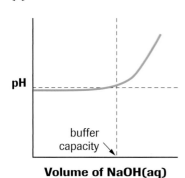

(b) Acetic Acid Buffer

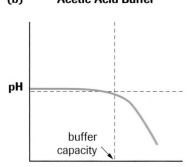

Figure 16
When an initial chemical amount of a buffer is eventually depleted, the pH then changes very quickly.

Buffering can be readily explained using Brønsted–Lowry equations. Suppose a small amount of NaOH(aq) is added to the acetic acid–acetate ion buffer. Using the five-step method for predicting the predominant acid–base reaction (page 731), the following equation is obtained:

$$\underset{SB}{\text{Na}^+(aq)}, \underset{}{\overset{SA}{\text{OH}^-(aq)}}, \underset{B}{\text{CH}_3\text{COOH}(aq)}, \underset{B}{\overset{A}{\text{CH}_3\text{COO}^-(aq)}}, \text{H}_2\text{O}(l)$$

$$\text{OH}^-(aq) + \text{CH}_3\text{COOH}(aq) \rightarrow \text{H}_2\text{O}(l) + \text{CH}_3\text{COO}^-(aq)$$

Figure 15 shows that this reaction is quantitative. A small amount of OH^- would convert a small amount of acetic acid to acetate ions. The overall effect will be just a small decrease in the ratio of acetic acid to acetate ions in the buffer and a slight increase in the pH. The very small change in concentration of each entity of the acid–base conjugate pair present, and the complete consumption of the added hydroxide ions in the process, explains why the pH change is small. This buffer would work equally well if a small amount of a strong acid, such as HCl(aq), were to be added. Evidence indicates that the reaction that occurs in that case is also quantitative.

$$\overset{SA}{\text{H}_3\text{O}^+(aq)}, \overset{A}{\text{Cl}^-(aq)}, \text{CH}_3\text{COOH}(aq), \underset{SB}{\overset{A}{\text{CH}_3\text{COO}^-(aq)}}, \underset{B}{\text{H}_2\text{O}(l)}$$

$$\text{H}_3\text{O}^+(aq) + \text{CH}_3\text{COO}^-(aq) \rightarrow \text{H}_2\text{O}(l) + \text{CH}_3\text{COOH}(aq)$$

Added hydronium ion is consumed and the mixture then has a slightly higher ratio of acetic acid to acetate ions and a slightly lower pH. **Figure 16** illustrates the concept of **buffer capacity**—the limit of the ability of a buffer to maintain a pH level. When the entity of the conjugate acid–base pair that reacts with an added reagent is completely consumed, the buffering fails and the pH changes dramatically.

The ability of buffers to maintain a relatively constant pH is important in many biological processes where certain chemical reactions can only occur at a specific pH value. Many aspects of cell functions and metabolism in living organisms are very sensitive to pH changes. For example, each enzyme carries out its function optimally over a small pH range. One essential buffer operates to maintain a stabilized pH in the internal fluid of all living cells. This critically important buffer is a mixture of dihydrogen phosphate ions and hydrogen phosphate ions. In mammals, cellular fluid has a pH in the range of 6.9 to 7.4. The $H_2PO_4^-$(aq)–HPO_4^{2-}(aq) buffer system has a pH of 7.2 when the concentrations of these two conjugate entities are equal. With small variations in concentration, this buffer is effective in maintaining optimum pH levels for the innumerable reactions going on within any given cell. The major buffer system in the blood and other body fluids (except the cytoplasm within cells) is the conjugate acid–base pair H_2CO_3(aq)–HCO_3^-(aq). Blood plasma has a remarkable buffering ability, as shown by the empirical results in **Table 1**. Since production of the entities of this buffer in the body is a *continuous* process, the buffer is not a limiting reagent, and you do not ordinarily have to worry about using up your buffering capacity.

Table 1 Buffering Action of Neutral Saline Solution and of Blood Plasma

Solution (1.0 L)	Initial pH of mixture	Final pH after adding 1 mL of 10 mol/L HCl
neutral saline	7.0	2.0
blood plasma	7.4	7.2

Human blood plasma normally has a pH of about 7.4. Any change of more than 0.4 pH units, induced by poisoning or disease, can be lethal. If the blood were not buffered, the acid absorbed from a glass of orange juice would probably be fatal.

Simulation—Preparation of Buffer Solutions

This simulation shows you how to select an appropriate acid–base conjugate pair to make a buffer of the desired pH.

www.science.nelson.com

Buffers are also important in many consumer, commercial, and industrial applications. Fermentation and the manufacture of antibiotics require buffering to optimize yields and to avoid undesirable side reactions. The production of various cheeses, yogurt, and sour cream are very dependent on controlling pH levels, since an optimum pH is needed to control the growth of micro-organisms and to allow enzymes to catalyze fermentation processes. Sodium nitrite and vinegar are widely used to preserve food; part of their function is to prevent the fermentation that takes place only at certain pH values.

The *CRC Handbook of Chemistry and Physics* provides recipes for preparing buffer solutions. For example, a buffer with a pH of 10.00 can be prepared by mixing 50 mL of 0.050 mol/L $NaHCO_3$(aq) with 10.7 mL of 0.10 mol/L NaOH(aq). The reaction to establish the buffer is described in the same way as all other acid–base reactions.

$$\underset{B}{Na^+(aq)},\ \underset{SB}{\overset{SA}{HCO_3^-(aq)}},\ \underset{B}{\overset{A}{OH^-(aq)}},\ H_2O(l)$$

$$HCO_3^-(aq) + OH^-(aq) \rightarrow H_2O(l) + CO_3^{2-}(aq)$$

The buffer preparation recipe reaction converts about $\frac{3}{5}$ of the initial hydrogen carbonate ions into carbonate ions. An aqueous mixture of a 3:2 ratio of these two acid–base conjugates at the concentrations specified has been found empirically to have an equilibrium pH of precisely 10.00 at 25 °C. Buffer mixtures like this one—with highly precise pH values—are routinely used to calibrate pH meters for accuracy.

Section 16.4

BIOLOGY CONNECTION
Homeostasis

Many reactions in the human body produce or consume carbonic acid or hydrogen carbonate ions. When one of these entities is depleted in the blood, Le Châtelier's principle comes into effect, and more of that entity is produced by an equilibrium shift to keep everything in balance.

The carbonic acid in solution in blood is really in equilibrium with dissolved carbon dioxide:

$$CO_2(aq) + H_2O(l) \rightleftarrows H_2CO_3(aq)$$

The enzyme carbonic anhydrase acts as a catalyst to allow this equilibrium to establish rapidly, or reestablish quickly, once shifted. It is possible to make your blood temporarily too basic by deliberately breathing heavily for a long time, thereby causing your body to lose too much carbon dioxide. Doctors call this condition respiratory alkalosis.

There are countless equilibrium reactions interconnected in this way in and around living cells. Biologists refer to this interdependent network of reaction equilibria as an example of homeostasis—the condition of automatic continual readjustment of cells, systems, and whole organisms to very specific conditions. Biology courses will tell you a lot more about homeostasis and the factors that affect it.

www.science.nelson.com

 EXTENSION

The Hydrogen Carbonate Buffer System

The pH of your blood is maintained at a constant level, courtesy of a series of equilibrium reactions, including a buffer system. This extension outlines the reactions, and gives an example of how climbing at high altitude can upset this delicate balance.

www.science.nelson.com

Figure 17
Maud Menten (1879–1960)

DID YOU KNOW?
Commercial Buffers
Buffers are fairly common, commercially. You can buy buffer mixtures to adjust the pH of aquarium water to suit the type of fish you have. Buffer mixes are sold to add to hot tubs to control the pH of the water. Many food recipes mix ingredients that form natural buffers within the dough, batter, or sauce.

WEB Activity

Canadian Achievers—Maud Menten

Maud Menten (**Figure 17**) was a world-renowned pioneer in explaining the chemical action of enzymes. What key concept was her chief contribution to understanding enzyme activity?

www.science.nelson.com

Practice

16. Give both an empirical and a theoretical definition of a buffer.
17. List two buffers that help maintain a normal pH level in your body.
18. Use the five-step method to predict the quantitative reaction of a carbonic acid–hydrogen carbonate ion buffer
    (a) when a small amount of HCl(aq) is added
    (b) when a small amount of NaOH(aq) is added
19. What happens if a large (excess) amount of a strong acid or base is added to a buffer?
20. Use Le Châtelier's principle to predict what will happen to a benzoic acid–benzoate ion buffer when a small amount of each of the following substances is added:
    (a) HCl(aq) (b) NaOH(aq)
21. Which of the following solution pairs, when mixed in equal quantities, will not form an effective buffer?
    (a) $HNO_3(aq)$ and $NaNO_3(aq)$
    (b) $C_6H_5COOH(aq)$ and $NaC_6H_5COO(aq)$
    (c) $NH_3(aq)$ and $NH_4Cl(aq)$
    (d) HCl(aq) and NaOH(aq)

INVESTIGATION 16.3 Introduction

Testing a Buffer Effect

References provide "recipes" for preparing standard buffer solutions of any desired pH from 1.0 to 13.0. The one used in this investigation has a pH of precisely 7.0, and might be used to calibrate a pH meter, for example. If our theory of buffers is correct, this solution should resist significant change in pH upon gradual addition of outside acid or base entities. Write the Design, Materials, and table of evidence to match the Procedure that is provided. The buffer is prepared by a reaction communicated by the following chemical equation:

$H_2PO_4^-(aq)$ + $OH^-(aq)$ → $HPO_4^{2-}(aq)$ + $H_2O(l)$
excess   limiting   (base part
(acid part   reagent   of buffer)
of buffer)

To perform this investigation, turn to page 769.

Report Checklist

○ Purpose   ● Design   ● Analysis
○ Problem   ● Materials   ● Evaluation (1, 2, 3)
○ Hypothesis   ○ Procedure
● Prediction   ● Evidence

Purpose
The purpose of this investigation is to test our concept of buffers.

Problem
How does the pH change when a strong acid and a strong base are slowly added separately to an $H_2PO_4^-(aq)$–$HPO_4^{2-}(aq)$ buffer?

CAREER CONNECTION

Microbiologist
Microbiologists study, test for, and isolate microorganisms such as bacteria, fungi, and viruses. As part of a medical or research team, their work is essential for public health and safety.

www.science.nelson.com

SUMMARY Brønsted–Lowry Is a Unifying Concept

The five-step Brønsted–Lowry method to explain and predict acid–base reactions is a preferred, acceptable method because it works for all quantitative and non-quantitative reactions studied so far:

- neutralization reactions
- indicator reactions
- buffer reactions
- polyprotic reactions
- excess reactions
- titration reactions

Section 16.4 Questions

1. Draw a pH curve to illustrate the following information about acid–base reaction systems.
   (a) What is buffering?
   (b) Where does buffering appear on a pH curve?
   (c) How are quantitative reactions represented on a pH curve?
   (d) Define endpoint and equivalence point.
   (e) How is a suitable indicator chosen for a titration?
   (f) Do non-quantitative reactions have an endpoint? Explain your answer briefly.

2. Sketch and label generalized pH curves on a single set of axes to illustrate the addition of a strong and a weak base to a strong and a weak acid.

3. Sketch a generalized pH curve for the addition of a strong base to a weak acid. Label the approximate equivalence point pH, and suggest an indicator that could be suitable for titration analysis using this reaction.

4. If the pH of a solution is 6.8, what is the colour of each of the following indicators in this solution?
   (a) methyl red
   (b) chlorophenol red
   (c) bromothymol blue
   (d) phenolphthalein
   (e) methyl orange

5. Use **Figure 18** to answer the following questions.
   (a) How many quantitative reactions have occurred?
   (b) Write the chemical equation for each quantitative reaction.
   (c) State the equivalence point pH for each quantitative reaction.
   (d) Choose a suitable indicator to correspond to the equivalence point pH value(s).
   (e) Identify the buffering region(s) and state the chemical formulas for the entities present in each region.

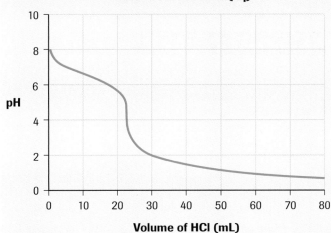

Figure 18
pH curve for the titration of sodium sulfite solution with hydrochloric acid

6. Write a net ionic equation to represent the following acid–base reactions. (Use any method that you find convenient to derive the net ionic equation.)
   (a) Oxalic acid is titrated with aqueous sodium hydroxide. (The indicator detects the second equivalence point.)
   (b) Sodium phosphate is titrated with hydrochloric acid. (The titration is stopped at the first equivalence point.)
   (c) Sodium hydrogen phosphate is titrated with hydrochloric acid. (An indicator is chosen to detect the first equivalence point.)
   (d) Nitric acid is titrated with aqueous barium hydroxide.
   (e) A sulfuric acid spill is neutralized by adding excess lye.

7. Write a formula (non-ionic) equation, and also the corresponding Brønsted–Lowry equation, to represent the following quantitative acid–base reactions.
   (a) Acetic acid is titrated with aqueous sodium hydroxide.
   (b) Nitrous acid is titrated with aqueous barium hydroxide.
   (c) Sodium carbonate is titrated with hydrobromic acid. (The titration is stopped after the first quantitative reaction.)
   (d) Carbonic acid is titrated with aqueous sodium hydroxide. (An indicator endpoint is used to stop the reaction with a stoichiometric ratio of 1:1.)
   (e) A sulfuric acid spill is neutralized by adding excess lye (caustic soda).

8. State two different applications of buffers.

9. Suggest a compound that could be dissolved in a sulfurous acid solution to make an effective buffer.

10. Assume that you have a hydrogen carbonate ion solution. Suggest substances to add to make
    (a) a buffer with a lower pH than the original solution
    (b) a buffer with a higher pH than the original solution

11. Complete the Design of the following investigation report.
    Purpose
    The purpose of this investigation is to test buffer concepts by quantitatively measuring equilibrium shifts for a buffer system.
    Problem
    Does the Brønsted–Lowry concept, as applied to buffers, predict changes in equilibrium concentration of $H_3O^+(aq)$ in a standard buffer solution, when strong acids or bases are added?
    Materials
    pH 7.00 $H_2PO_4^-/HPO_4^{2-}$ buffer solution
    pH meter
    1.00 mol/L HCl(aq)
    1.00 mol/L NaOH(aq)

Extension

12. From a CRC Handbook or any other reference, locate and copy at least one "recipe" for preparing a buffer solution that will have a pH of precisely 5.00 at 25 °C.

Chapter 16 INVESTIGATIONS

INVESTIGATION 16.1

Creating an Acid–Base Strength Table

An acid–base table organizes common acids (and their conjugate bases) in order of decreasing acid strength (**Figure 1**). Acid strength can be tested several ways, including by a carefully designed use of indicators. Predict the order of strengths using the Relative Strengths of Aqueous Acids and Bases Table (Appendix I). Use the indicators provided to create a valid and efficient Design, in which you clearly identify the relevant variables. Evaluate the Design (only), and suggest improvements if any problems are identified.

Purpose
The purpose of this investigation is to test an experimental design for using indicators to create a table of relative strengths of acids and bases.

Problem
Can the indicators available be used to rank the acids and bases provided in order of strength?

Materials
lab apron
eye protection
0.10 mol/L HCl(aq)
0.10 mol/L NaHSO$_4$(aq)
0.10 mol/L CH$_3$COOH(aq)
0.10 mol/L NaHSO$_3$(aq)
0.10 mol/L Na$_2$CO$_3$(aq)
0.10 mol/L NaOH(aq)
indicators, for example:
methyl orange, methyl violet, bromothymol blue, phenolphthalein
(6) 13 × 100 mm test tubes for each indicator or spot plate (microchem)
test-tube rack

Report Checklist
- ○ Purpose
- ○ Problem
- ○ Hypothesis
- ● Prediction
- ● Design
- ○ Materials
- ● Procedure
- ● Evidence
- ● Analysis
- ● Evaluation (1)

Chemicals used include toxic, corrosive, and irritant materials. Avoid eye and skin contact. If you spill any of the chemical solutions on your skin, immediately rinse the area with lots of cool water. In the unlikely situation of getting some of the chemicals in your eye, immediately rinse your eye for at least 15 min and inform your teacher.

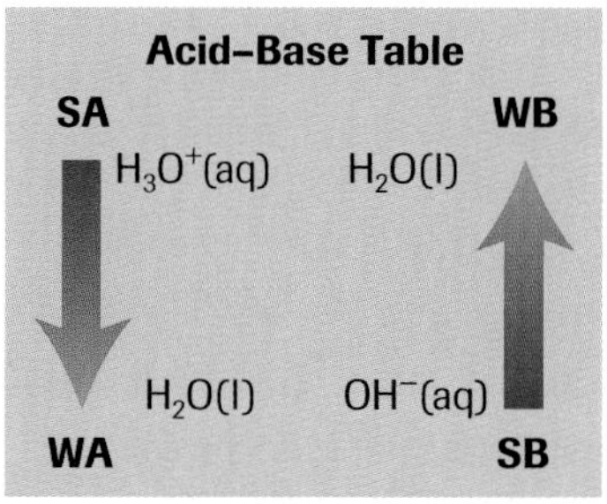

Figure 1 Simplified acid–base table

INVESTIGATION 16.2

Testing Brønsted–Lowry Reaction Predictions

When predicting products for this investigation, since at least one reactant is always in solution, list all entities present as they normally exist in an aqueous environment. For those reactants that are added in solid state, assume that they will dissolve. Use the resulting entities for prediction. Evaluate the predictions, the Brønsted–Lowry concept, and the five-step method for acid–base reaction prediction.

Purpose
The purpose of this investigation is to test the Brønsted–Lowry concept and the five-step method for reaction prediction from a table of relative acid–base strength.

Report Checklist
- ○ Purpose
- ○ Problem
- ○ Hypothesis
- ● Prediction
- ○ Design
- ● Materials
- ● Procedure
- ● Evidence
- ● Analysis
- ● Evaluation (1, 2, 3)

Problem
What reactions occur when the following substances are mixed? (Hints for diagnostic tests are in parentheses.)

1. ammonium chloride and sodium hydroxide solutions (odour)
2. hydrochloric acid and sodium acetate solutions (odour)
3. sodium benzoate and sodium hydrogen sulfate solutions (benzoic acid has low solubility)
4. hydrochloric acid and aqueous ammonium chloride (odour)

INVESTIGATION 16.2 *continued*

5. solid sodium chloride added to water (litmus)
6. solid aluminium sulfate added to water (litmus)
7. solid sodium phosphate added to water (litmus)
8. solid sodium hydrogen sulfate added to water (litmus)
9. solid sodium hydrogen carbonate added to hydrochloric acid (pH)
10. solid sodium hydrogen carbonate added to sodium hydroxide solution (pH)
11. solid sodium hydrogen carbonate added to sodium hydrogen sulfate solution (pH)

Design

A prediction is made for each of eleven pairs of substances. The prediction is then tested using one or more diagnostic tests, complete with controls. Additional diagnostic tests increase the certainty of the evaluation.

Chemicals used include toxic, corrosive, and irritant materials. Avoid eye and skin contact. If you spill any of the chemical solutions on your skin, immediately rinse the area with lots of cool water. In the unlikely situation of getting some of the chemicals in your eye, immediately rinse your eye for at least 15 min and inform your teacher. Remember to detect odours cautiously by wafting air toward your nose from the container.

INVESTIGATION 16.3

Testing a Buffer Effect

Report Checklist

○ Purpose   ● Design   ● Analysis
○ Problem   ● Materials   ● Evaluation (1, 2, 3)
○ Hypothesis ○ Procedure
● Prediction ● Evidence

References provide "recipes" for preparing standard buffer solutions of any desired pH from 1.0 to 13.0. The one used in this investigation has a pH of precisely 7.0, and might be used to calibrate a pH meter, for example. If our theory of buffers is correct, this solution should resist significant change in pH upon gradual addition of outside acid or base entities. Write the Design, Materials, and table of evidence to match the Procedure that is provided. The buffer is prepared by a reaction communicated by the following chemical equation:

$$\underset{\substack{\text{excess} \\ \text{(acid part} \\ \text{of buffer)}}}{H_2PO_4^-(aq)} + \underset{\substack{\text{limiting} \\ \text{reagent}}}{OH^-(aq)} \longrightarrow \underset{\substack{\text{(base part} \\ \text{of buffer)}}}{HPO_4^{2-}(aq)} + H_2O(l)$$

Purpose

The purpose of this investigation is to test our concept of buffers.

Problem

How does the pH change when a strong acid and a strong base are slowly added separately to an $H_2PO_4^-(aq)$–$HPO_4^{2-}(aq)$ buffer?

Acids and bases are corrosive and toxic. Avoid skin and eye contact. If you spill any of the chemical solutions on your skin, immediately rinse the area with lots of cool water. In the unlikely situation of getting some of the chemicals in your eye, immediately rinse your eye for at least 15 min and inform your teacher.

Procedure

1. Obtain 50 mL of 0.10 mol/L KH_2PO_4(aq) and 29 mL of 0.10 mol/L NaOH(aq) in separate graduated cylinders.
2. Pour the KH_2PO_4(aq) and then the NaOH(aq) into a beaker to prepare a buffer with a pH of 7.
3. Pour an equal volume of the buffer into two test tubes.
4. Add 0.10 mol/L NaCl(aq) as a control into a third and a fourth test tube.
5. Add two drops of bromocresol green to one buffer test tube and one control test tube.
6. Add and count drops of 0.10 mol/L HCl(aq) until the colour changes.
7. Repeat steps 5 and 6 with phenolphthalein and 0.10 mol/L NaOH(aq) using the other two test tubes.
8. Dispose of all solutions down the drain with running water.

Chapter 16 SUMMARY

Outcomes

Knowledge
- describe Brønsted–Lowry acids as proton donors and bases as proton acceptors (16.2)
- write Brønsted–Lowry equations and predict whether reactants or products are favoured for acid–base equilibrium reactions (including indicators and polyprotic acids and bases) (16.2, 16.4)
- identify polyprotic acids, polyprotic bases, conjugate pairs, and amphiprotic entities (16.2, 16.4)
- define a buffer as relatively large amounts of a conjugate acid–base pair in equilibrium that maintain a relatively constant pH when small amounts of acid or base are added (16.4)
- sketch and qualitatively interpret titration curves of monoprotic and polyprotic acids and bases, identifying equivalence points and regions of buffering for weak acid–strong base, strong acid–weak base, and strong acid–strong base reactions (16.4)
- define K_w, K_a, and K_b and use them to determine pH, pOH, $[H_3O^+]$, and $[OH^-]$ of acidic and basic solutions (16.1, 16.3)
- calculate equilibrium constants and concentrations for homogeneous systems and Brønsted–Lowry acids and bases (excluding buffers) when concentrations at equilibrium are known, when initial concentrations and one equilibrium concentration are known, and when the equilibrium constant and one equilibrium concentration are known (16.1, 16.3)

STS
- state that the goal of science is knowledge about the natural world (16.1, 16.2, 16.3, 16.4)
- state that a goal of technology is to solve practical problems (16.2, 16.3, 16.4)

Skills
- initiating and planning: design an experiment to show quantitative equilibrium shifts in concentration under a given set of conditions (16.4); describe procedures for safe handling, storage, and disposal of materials used in the laboratory, with reference to WHMIS and consumer product labelling information (16.2, 16.4)
- performing and recording: prepare a buffer to investigate the relative abilities of a buffer and a control to resist a pH change when a small amount of strong acid or strong base is added (16.4)
- analyzing and interpreting: use experimental data to calculate equilibrium constants (16.3)
- communication and teamwork: work cooperatively in addressing problems and apply the skills and conventions of science in communicating information and ideas and in assessing results (16.2, 16.4)

Key Terms

16.1
ionization constant for water, K_w

16.2
Brønsted–Lowry acid
Brønsted–Lowry base
Brønsted–Lowry reaction equation
amphiprotic
amphoteric
conjugate acid–base pair

16.3
acid ionization constant, K_a
base ionization constant, K_b

16.4
pH curve
buffering
buffer
buffer capacity

Key Equations

(All at SATP)

$$K_w = [H_3O^+(aq)][OH^-(aq)] = 1.0 \times 10^{-14} \quad (16.1)$$

$$pH + pOH = 14.00 \quad (16.1)$$

For any aqueous acid, HA(aq), $K_a = \dfrac{[H_3O^+(aq)][A^-(aq)]}{[HA(aq)]}$ (16.3)

For any aqueous base, B(aq), $K_b = \dfrac{[HB^+(aq)][OH^-(aq)]}{[B(aq)]}$ (16.3)

For any conjugate acid–base pair, $K_a K_b = K_w$ (16.3)

▶ MAKE a summary

1. Use a blank page to create a "star" (radial) chart for quantities and calculations related to Brønsted–Lowry acid–base reaction systems. Begin by writing $H_3O^+(aq)$ and $OH^-(aq)$ symbols about 10 cm apart horizontally, centred on the page and connected by a line. Draw lines from each of these symbols to other quantity symbols that are calculated from them (and from each other, in some cases). Along each connecting line, indicate what information the calculation requires and how it is performed. Include K_a, K_b, pH, and pOH quantities, but not percent ionization (because it is not specific to acid–base reactions).

2. Revisit your answers to the Starting Points questions at the beginning of this chapter. How would you answer the questions differently now? Why?

▶ Go To

The following components are available on the Nelson Web site. Follow the links for *Nelson Chemistry Alberta 20-30*.

- an interactive Self Quiz for Chapter 16
- additional Diploma Exam-style Review questions
- Illustrated Glossary
- additional IB-related material

There is more information on the Web site wherever you see the Go icon in this chapter.

➕ EXTENSION

Lost Treasures of Tibet

What does acid–base chemistry have to do with ancient works of art? Watch this video to discover how archeologists and conservators made use of chemical reactions to restore the brilliance to Tibetan temple paintings.

Chapter 16 REVIEW

Many of these questions are in the style of the Diploma Exam. You will find guidance for writing Diploma Exams in Appendix H. Exam study tips and test-taking suggestions are on the Nelson Web site. Science Directing Words used in Diploma Exams are in bold type.

www.science.nelson.com

DO NOT WRITE IN THIS TEXTBOOK.

Part 1

1. A hydrated proton is referred to as a
   A. hydronium ion
   B. hydroxide ion
   C. hydroxyl group
   D. Brønsted–Lowry base

2. The hydroxide ion concentration in a solution of window cleaner is 2.1 mmol/L. The hydronium ion concentration in this solution is calculated to be
   A. 2.1×10^{-15} mol/L
   B. 4.8×10^{-15} mol/L
   C. 2.1×10^{-12} mol/L
   D. 4.8×10^{-12} mol/L

3. A pH meter indicates that the pH of a soft drink is 3.46. The pOH of the soft drink is calculated to be
   A. 10.46
   B. 10.54
   C. 14.46
   D. 14.54

4. The recipe for preparing a cleaning solution calls for dissolving 5.0 g of sodium hydroxide in 4.0 L of water. The pH of the resulting solution is calculated to be
   A. 1.51
   B. 12.49
   C. 13.10
   D. 13.90

5. Acid–base theories have developed over the last two centuries. Which of the following chemists did *not* make a significant contribution to acid–base theory?
   A. Gilbert Lewis
   B. Svante Arrhenius
   C. Ernest Rutherford
   D. Johannes Brønsted

Use the names given and Appendix G to answer questions 6 and 7. Note that some of the ions listed are amphiprotic.
1. phenol
2. sulfide ion
3. cyanide ion
4. ammonium ion
5. hydrogen sulfate ion
6. hydrogen carbonate ion

6. Listed from strongest to weakest, the acids are
 ___ ___ ___ ___ .

7. Listed from strongest to weakest, the bases are
 ___ ___ ___ ___ .

8. The nearly level region on a pH curve represents
   A. the endpoint
   B. a region of buffering
   C. an indicator point
   D. the equivalence point

9. The titration of a weak acid with a strong base has an equivalence point pH of 9.0. Which of the following indicators would be most suitable for this titration, done as an analysis?
   A. litmus
   B. methyl red
   C. phenolphthalein
   D. alizarin yellow R

10. Which of the following combinations of ions does *not* represent a conjugate acid–base pair?
    A. $H_3PO_4(aq)$ and $H_2PO_4^-(aq)$
    B. $H_2PO_4^-(aq)$ and $HPO_4^{2-}(aq)$
    C. $HPO_4^{2-}(aq)$ and $PO_4^{3-}(aq)$
    D. $H_3PO_4(aq)$ and $HPO_4^{2-}(aq)$

11. Which of the following statements about acid–base titrations is *not* true?
    A. Strong acid–strong base reactions have an equivalence point pH of 7.
    B. Strong acid–weak base reaction equivalence points have a pH > 7.
    C. Weak acid–strong base reaction equivalence points have a pH > 7.
    D. Strong acid–strong base reactions are quantitative.

12. Which of the following statements about buffers is *not* true?
    A. Buffers maintain a solution pH at approximately 7.
    B. Buffers can be formed by partially neutralizing a weak acid with a strong base.
    C. Buffers maintain a relatively constant pH when small amounts of acid or base are added.
    D. Buffers contain relatively large amounts of a conjugate acid–base pair in equilibrium.

Part 2

13. **Define** each of the following substances, according to the Brønsted–Lowry theory.
    (a) an acid
    (b) a base
    (c) an acid–base reaction
    (d) an amphiprotic entity
    (e) a strong acid
    (f) a strong base

14. Write net ionic equations for each of the following reactions in aqueous solution. Label the reactants as Brønsted–Lowry acids or bases. **Identify** any amphiprotic ions.
    (a) Solutions of sodium hydrogen sulfate and sodium carbonate are mixed.
    (b) Aqueous ammonia is added to a solution of potassium hydrogen sulfite.
    (c) Solutions of sodium hydrogen phosphate and acetic acid are mixed.
    (d) Aqueous sodium hydroxide is added to a solution of sodium hydrogen phosphate.

15. **Identify** all acids, bases, and conjugate pairs, and **predict** the position of equilibrium (reactants or products favoured) in each of the following reactions.
    (a) $CH_3COOH(aq) + CN^-(aq) \rightleftarrows CH_3COO^-(aq) + HCN(aq)$
    (b) $HSO_3^-(aq) + HPO_4^{2-}(aq) \rightleftarrows SO_3^{2-}(aq) + H_2PO_4^-(aq)$
    (c) $NH_4^+(aq) + CO_3^{2-}(aq) \rightleftarrows NH_3(aq) + HCO_3^-(aq)$

16. Write equilibrium law expressions for each of the following equilibrium situations.
    (a) $HF(aq) + H_2O(l) \rightleftarrows H_3O^+(aq) + F^-(aq)$
    (b) $NH_3(aq) + H_2O(l) \rightleftarrows NH_4^+(aq) + OH^-(aq)$
    (c) $H_2SO_3(aq) + H_2O(l) \rightleftarrows H_3O^+(aq) + HSO_3^-(aq)$

17. Refer to Appendix I for required information to **predict** $[H_3O^+(aq)]$ and pH for each of these 0.10 mol/L solutions.
    (a) hydroiodic acid
    (b) methanoic acid (optional: enrichment)
    (c) hydrosulfuric acid

18. A student measures the pH of a 0.25 mol/L solution of potassium hydrogen citrate to be 3.42.
    (a) **Determine** the K_a of the hydrogen citrate ion from this evidence.
    (b) How could a student find the pH of a solution without access to a pH meter? **Design** a procedure, involving three indicators, that would establish the approximate pH of this solution. **Criticize** your design.

19. If a sample of acid rain has a pH of 5.0, **predict** the colour of each of the following indicators in this solution.
    (a) litmus
    (b) methyl red
    (c) methyl orange
    (d) phenolphthalein

20. Aqueous 0.10 mol/L solutions of potassium sulfate and potassium benzoate are prepared. **Predict** how the pH values of the solutions would compare, using the Brønsted–Lowry proton transfer concept to **justify** your answer.

21. Separate samples of a household cleaning solution were tested with two indicators. Indigo carmine was blue and phenolphthalein was red in the solution. Estimate the approximate pH and approximate hydroxide ion concentration in the solution.

22. **Determine** the percent reaction, pH, pOH, and hydroxide concentration in a 0.015 mol/L solution of sodium acetate.

23. **Sketch** a pH (titration) curve for each of the following titrations, assuming all the acids and bases have a concentration of 0.10 mol/L, and all reaction proton transfers are quantitative.
    (a) A diprotic acid is titrated with sodium hydroxide.
    (b) A diprotic base is titrated with hydrochloric acid.

24. **Design** an experiment to determine the relative strengths of four weak acids.

25. Use the five-step Brønsted–Lowry method to **predict** the net ionic equation for the overall reaction when the following chemicals are mixed.
    (a) solutions of sodium sulfate and potassium benzoate
    (b) aqueous ammonium nitrate fertilizer and aqueous sodium phosphate
    (c) solutions of citric acid and sodium hydrogen carbonate

Unit 8 REVIEW

Many of these questions are in the style of the Diploma Exam. You will find guidance for writing Diploma Exams in Appendix H. Exam study tips and test-taking suggestions are on the Nelson Web site. Science Directing Words used in Diploma Exams are in bold type.

www.science.nelson.com

DO NOT WRITE IN THIS TEXTBOOK.

Part 1

1. The following aqueous solutions are each at equilibrium in sealed containers that are half-full of liquid. For which one of these may the volume of the liquid solution be reasonably considered to be the boundary of the closed system for the equilibrium?
   A. HCl(aq)
   B. NaOH(aq)
   C. H_2CO_3(aq)
   D. NH_3(aq)

2. Basicity of solutions has been explained by a variety of theories, each one more comprehensive than its predecessors. The Brønsted–Lowry concept defines a base as an entity that is a
   A. proton attractor in a specific acid–base reaction
   B. proton donor in a specific acid–base reaction
   C. hydronium attractor in any acid–base reaction
   D. hydroxide producer in any acid–base reaction

3. For which of the following aqueous solution titrations would you expect to measure a pH of 7 at the equivalence point, at SATP?
   A. acetic acid titrated with sodium hydroxide
   B. nitric acid titrated with potassium carbonate
   C. hydrochloric acid titrated with lithium hydroxide
   D. ammonia titrated with hydrochloric acid

4. An industrial reaction process designer has a choice of several different equilibrium reactions, all of which produce the same desired chemical substance. These reactions each have different reaction characteristics. The designer wishes to produce product as fast as is safely possible. The best choice from an economic perspective is the reaction that
   A. has a high rate (speed) of reaction and a low percent yield
   B. has a low rate (speed) of reaction and a high percent yield
   C. requires extreme high pressure to improve the percent yield
   D. requires an expensive catalyst to react noticeably

5. What energy condition must be met to maintain a dynamic equilibrium system?
   A. The temperature must be constant.
   B. The forward reaction must be exothermic.
   C. The container must not conduct heat.
   D. The activation energy must be high.

Use this information to answer questions 6 to 8.

Pure solid arsenic acid, also called orthoarsenic acid, is a hydrated substance at room temperature with a formula that may be written as $H_3AsO_4 \cdot \tfrac{1}{2}H_2O$(s), or alternatively as $(H_3AsO_4)_2 \cdot H_2O$(s). This substance is used in commercial glassmaking and in making wood-preservative solutions. Arsenic acid is soluble in water and ionizes according to the following equation:

$$H_3AsO_4(aq) + H_2O(l) \rightleftharpoons H_3O^+(aq) + H_2AsO_4^-(aq)$$
$$K_a = 5.6 \times 10^{-3} \ (18\ °C)$$

6. The conjugate base of arsenic acid in the ionization reaction shown is
   A. H_3AsO_4(aq)
   B. $H_2AsO_4^-$(aq)
   C. $HAsO_4^{2-}$(aq)
   D. AsO_4^{3-}(aq)

7. Which is the correct form of the equilibrium law for this acid ionization reaction?

   A. $K_a = \dfrac{[H_3O^+(aq)][H_2AsO_4^-(aq)]}{[H_3AsO_4(aq)]}$

   B. $K_a = \dfrac{[H_3O^+(aq)][H_2AsO_4^-(aq)]}{[H_3AsO_4(aq)][H_2O(aq)]}$

   C. $K_a = \dfrac{[H_3O^+(aq)]}{[H_3AsO_4(aq)][H_2AsO_4^-(aq)]}$

   D. $K_a = \dfrac{[H_3O^+(aq)][H_2AsO_4^-(aq)]}{[H_3AsO_4(aq)][HAsO_4^{2-}(aq)]}$

8. An empirical pH curve for a titration of a sample of aqueous arsenic acid with a standardized sodium hydroxide solution shows two distinct rapid changes in pH. From this evidence, it can be inferred that
   A. hydrogen arsenate ion is a very weak acid, with a relatively strong conjugate base
   B. arsenate ion is a very weak base, with a strong conjugate acid
   C. an arsenic acid molecule can lose two protons simultaneously to one hydroxide ion
   D. the third sequential proton transfer reaction is quantitative

9. **NR** A student discovers that a pure liquid substance that is sold as a common agricultural insecticide forms an acidic solution in water. The measured pH of a 1.00 mol/L aqueous solution of this weak acid is 5.22. The calculated K_a value for this acid at this temperature may be expressed numerically as $a.b \times 10^{-cd}$. The values (in order) of a, b, c, and d are ____, ____, ____, and ____.

774 Unit 8

10. In writing an equilibrium law expression, condensed (pure solid and liquid) phases are ignored because
    A. forward and reverse reaction rates must be equal if they occur at the surface of a solid or liquid
    B. solids and liquids are completely separate physically from the rest of the reaction components, which are dissolved in each other
    C. solid or liquid pure substances have essentially constant amount concentration values
    D. other entities cannot move between molecules of solids or liquids to collide with their entities

11. Consider the reaction $N_2(g) + 3H_2(g) \rightleftarrows 2NH_3(g)$, at equilibrium. If the system container volume were to be suddenly decreased, the concentration of nitrogen gas in the system would immediately
    A. increase, then decrease to a new constant value
    B. remain unchanged, because the reaction is at equilibrium
    C. increase, then remain steady at a new constant value
    D. decrease, then remain steady at a new constant value

Use this information to answer questions 12 to 14.

Titration of acid samples with hydroxide ion solution produces very different pH curves depending on the strength of the acid sample (**Figure 1**). These curves represent data from two separate titrations plotted on the same axes. The two acid samples were standardized hydrochloric acid and acetic acid solutions, both with the same initial concentration and volume. The same sodium hydroxide titrant was used in both titrations.

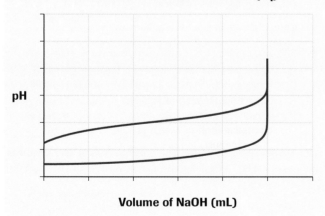

Figure 1
Titration curves for $HCl(aq)$ and $CH_3COOH(aq)$ with $NaOH(aq)$

12. The strong acid titration curve does not increase noticeably when the very *first* addition of NaOH(aq) is made, because
    A. $OH^-(aq)$ was already present in the original sample solution
    B. $OH^-(aq)$ does not initially react with entities present in the sample solution
    C. reaction of the added $OH^-(aq)$ creates no new entities in the sample solution
    D. hydronium ions resist any change to their structure

13. During most of these titrations, the pH curves remain nearly level. The conjugate acid–base pair that is responsible for creating this buffering region during the strong acid–strong base titration is
    A. $H_3O^+(aq)-H_2O(l)$
    B. $H_3O^+(aq)-CH_3COOH(aq)$
    C. $CH_3COOH(aq)-CH_3COO^-(aq)$
    D. $CH_3COO^-(aq)-OH^-(aq)$

14. The *incorrect* statement about the weak acid–strong base titration is:
    A. Initial addition of $OH^-(aq)$ produces a new entity in the sample solution.
    B. The titrant volume at the reaction equivalence point is the same as for the strong acid–strong base titration.
    C. The solution pH at the reaction equivalence point will be higher than for the strong acid–strong base titration.
    D. At the equivalence point, the solution pH will be controlled by the concentration of hydroxide ion, which is the strongest base present.

Part 2

15. Formal concepts of acids have existed since the 18th century. **Explain** the main idea and the limitations of each of the following: the oxygen concept; the hydrogen concept; Arrhenius' concept; and the Brønsted–Lowry concept of acids.

16. What happens when scientists find a theory, such as Arrhenius' original theory of acids, to be unacceptable?

17. **Describe** two main ways in which a theory or a theoretical definition may be tested, in terms of what an *acceptable* theory is required to do.

18. According to modern evidence, what is the nature of a "hydrogen ion" in aqueous solution?

19. Briefly **outline** how the theoretical definition of a base has changed from Arrhenius' original concept to the modified Arrhenius concept, and subsequently to the Brønsted–Lowry concept.

20. Aqueous solutions of sodium sulfite and of sodium carbonate of equal concentrations are prepared. **Explain** how the pH values would compare, using the Brønsted–Lowry proton transfer concept to **justify** your answer.

21. Use the *If* [procedure], *and* [evidence], *then* [analysis] format (Appendix C.4) to **describe** a diagnostic test that could be used to determine which of two solutions (of equal concentration) is sodium benzoate, and which is sodium hydroxide.

22. **Identify** two different examples of conjugate acid–base pairs, each involving the dihydrogen phosphate ion.

23. Which of the following statements are necessarily true when products are strongly favoured in an acid–base equilibrium, assuming equal amount concentrations and chemical amounts of the initial reactants?
    (a) The stronger of the two Brønsted–Lowry bases is a product.
    (b) The equilibrium constant is greater than one.
    (c) The forward reaction is exothermic.
    (d) The stronger Brønsted–Lowry acid is a reactant.
    (e) The percent reaction is greater than 50%.
    (f) The pH of the final solution is greater than 7.
    (g) The reactant acid is above the reactant base in an acid–base table.

24. From the equation information provided, **predict** all of the system changes you might introduce that, according to Le Châtelier's principle, will act to shift the equilibria to maximize the percent yield of the specified product.
    (a) production of propene (propylene)
    $C_3H_8(g) + energy \rightleftharpoons C_3H_6(g) + H_2(g)$
    (b) production of iodine
    $5\,Sn^{2+}(aq) + 2\,IO_3^-(aq) + 12\,H_3O^+(aq) \rightleftharpoons$
    $5\,Sn^{4+}(aq) + I_2(s) + 18\,H_2O(l)$

25. Consider this system at equilibrium.
    $PCl_5(g) \rightleftharpoons PCl_3(g) + Cl_2(g)$; $K_c = 0.40$ at 170 °C
    (a) One mole of phosphorus pentachloride was initially placed into a 1.0 L container. Once equilibrium had been reached, it was found that the equilibrium concentration of PCl_5 was 0.54 mol/L. **Figure 2** describes the change in $[PCl_5]$ over time. Copy the graph. **Sketch** lines to indicate the changing concentrations of PCl_3 and Cl_2 over the same time period.

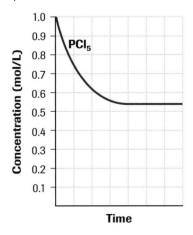

Figure 2
Reaction progress: $PCl_5(g) \rightleftharpoons PCl_3(g) + Cl_2(g)$

(b) 0.50 mol of PCl_3 and 0.30 mol of Cl_2 were placed into a 1.0 L container. Once equilibrium had been reached, it was found that the equilibrium concentration of PCl_3 was 0.36 mol/L. **Figure 3** describes the change in $[PCl_3]$ over time. Copy the graph. **Sketch** how the concentrations of PCl_5 and Cl_2 change over the same time period.

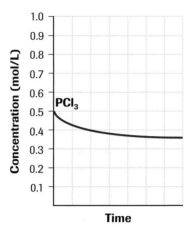

Figure 3
Reaction progress: $PCl_3(g) + Cl_2(g) \rightleftharpoons PCl_5(g)$

26. Consider the equilibrium reaction,
    $CO(g) + H_2O(g) \rightleftharpoons CO_2(g) + H_2(g)$ $K_c = 5.0$ at 650 °C
    In a rigid 1.00 L laboratory reaction vessel, a technician places 1.00 mol of each of the four substances involved in this equilibrium. The vessel is heated to 650 °C.
    Determine the equilibrium amount concentrations of each substance, organizing your values in an ICE table.

27. Write chemical formulas and net ionic equations for the following overall reactions. (If necessary, refer to Appendix J.)
    (a) Vinegar is used to neutralize a drain cleaner spill.
    (b) Baking soda solution is used to neutralize spilled rust remover containing oxalic acid.
    (c) An antacid tablet (flavoured calcium carbonate) is used to neutralize excess stomach acid.

28. Identify all acids, bases, and conjugate pairs, and **predict** the position of equilibrium (reactants or products favoured) in each of the following reactions:
    (a) $HCOOH(aq) + CN^-(aq) \rightleftharpoons HCOO^-(aq) + HCN(aq)$
    (b) $HPO_4^{2-}(aq) + HCO_3^-(aq) \rightleftharpoons$
    $H_2PO_4^-(aq) + CO_3^{2-}(aq)$
    (c) $Al(H_2O)_6^{3+}(aq) + H_2O(l) \rightleftharpoons$
    $H_3O^+(aq) + Al(H_2O)_5OH^{2+}(aq)$
    $K_a = 1 \times 10^{-5}$
    (d) $C_6H_5NH_2(aq) + H_2O(l) \rightleftharpoons$
    $C_6H_5NH_3^+(aq) + OH^-(aq)$
    $K_b = 4 \times 10^{-10}$

29. Separate samples of an unknown solution were tested with two indicators. Congo red was red and chlorophenol red was yellow in the solution. **Predict** the approximate pH and approximate hydronium ion concentration of the solution.

30. The test for acceptability of any theory is its ability to explain and predict a wide range of phenomena: to be a unifying theory. Test the Brønsted-Lowry concept by using the five-step procedure to write chemical equations describing and predicting each of the following acid–base reactions. **Design**, where possible, one diagnostic test that could be used to verify each prediction.
    (a) the addition of hydrofluoric acid to a solution of potassium sulfate
    (b) the addition of a solution of sodium hydrogen sulfate to a solution of sodium hydrogen sulfide
    (c) the titration of methanoic acid with sodium hydroxide solution
    (d) the addition of a small amount of a strong acid to a hydrogen phosphate ion–phosphate ion buffer solution
    (e) the addition of colourless phenolphthalein indicator, HPh(aq), to a strong base
    (f) sodium sulfate is dissolved in water
    (g) the addition of blue bromothymol blue, Bb^-(aq), to vinegar
    (h) the addition of washing soda to water
    (i) the addition of baking soda to water

31. Many acid–base phenomena can be described and/or predicted by using Le Châtelier's principle. **Evaluate** this generalization's ability to describe or predict a variety of reactions.
    Your response should include
    - reference to the following three examples:
        NaOH(aq) is added to the following buffer equilibrium:
        $$NH_3(aq) + H_2O(l) \rightleftharpoons NH_4^+(aq) + OH^-(aq)$$
        NaOH(aq) is added to a bromothymol blue indicator solution:
        $$HBb(aq) + H_2O(l) \rightleftharpoons H_3O^+(aq) + Bb^-(aq)$$
        NaOH(aq) titrant is added to a vinegar sample:
        $$CH_3COOH(aq) + H_2O(l) \rightleftharpoons H_3O^+(aq) + CH_3COO^-(aq)$$
    - descriptions of diagnostic tests that would provide evidence for your predictions

32. One way to evaluate a theory is to test predictions with new substances. Sodium methoxide, $NaCH_3O$(s), is dissolved in water. **Predict** whether the final solution will be acidic, basic, or neutral. **Explain** your answer using a net ionic equation. (Hint: Think of the methoxide ion as a hydroxide ion, with a methyl group substituted for the hydrogen.)

33. In an experimental investigation of amphoteric substances, samples of baking soda were added to a solution of sodium hydroxide and to a solution of hydrochloric acid. The pH of the sodium hydroxide changed from 13.0 to 9.5 after the addition of the baking soda. The pH of the hydrochloric acid changed from 1.0 to 4.5 after the addition of baking soda. **Explain** these results.
    Your response should include
    - chemical reaction equations to describe the reactions
    - identification of the amphiprotic entity

34. Each of seven unlabelled beakers was known to contain one of the following 0.10 mol/L solutions: CH_3COOH(aq), $Ba(OH)_2$(aq), NH_3(aq), $C_2H_4(OH)_2$(aq), H_2SO_4(aq), HCl(aq), and NaOH(aq). **Describe** diagnostic test(s) required to distinguish the solutions and label the beakers. Use the "If ____, and ____, then ____" format (Appendix C.4), a flow chart, or a table to communicate your answer.

35. Use **Figure 4** to answer the following questions.
    (a) **Infer** how many quantitative reactions have occurred.
    (b) Write the Brønsted-Lowry equation for each successive proton transfer reaction, with appropriate "arrows" to indicate the extent of each reaction.
    (c) **Determine** the titrant volume at each quantitative reaction equivalence point.
    (d) **Identify** a suitable indicator to correspond to the equivalence point pH values.
    (e) **Identify** the buffering region(s), and state the chemical formulas for the entities present in solution in each buffering region.

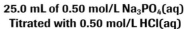
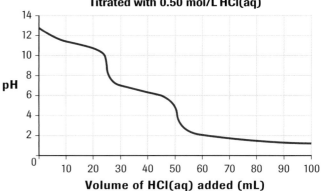

Figure 4
Graph for question 35

36. Because chlorine–oxygen compounds are toxic to microorganisms and react readily with organic materials in food stains, they find wide application in disinfectants and bleaches. The reaction of hypochlorous acid with molecules of coloured substances in stains often produces colourless products, making the stain "disappear." Hypochlorous acid may be produced by the following reaction:
$$H_2O(g) + Cl_2O(g) \rightleftharpoons 2\,HOCl(g)$$
$K_c = 0.090$ (25 °C)
Determine the concentrations of each reagent at equilibrium at 25 °C if the initial concentrations of both water vapour and chlorine monoxide were 4.00 mol/L.

37. Ethyl acetate (ethyl ethanoate) is an ester with a great many different uses as an organic solvent—for everything from paint to perfume. It is a product of the following equilibrium reaction equation:
$CH_3COOH(os) + C_2H_5OH(os) \rightleftharpoons CH_3COOC_2H_5(os) + H_2O(os)$
The equation shows the reaction taking place at 25 °C, with all components dissolved in a complex non-reacting liquid organic solvent, which we have symbolized as "os" for convenience. **Determine** K_c for this reaction if, at equilibrium, the reagent concentrations are
   [$CH_3COOH(os)$] = 2.5 mol/L
   [$C_2H_5OH(os)$] = 1.7 mol/L
   [$CH_3COOC_2H_5(os)$] = 3.1 mol/L
   [$H_2O(os)$] = 3.1 mol/L

38. Phenolphthalein indicator was used for a titration of several 10.00 mL samples of hypochlorous acid with 0.350 mol/L barium hydroxide solution. An average titrant volume of 12.6 mL was required to reach the observed endpoint of these trials. According to this evidence, **predict** the amount concentration of the hypochlorous acid solution.

39. In a titration, 0.20 mol/L HCl(aq) titrant is to be gradually added to a 20.0 mL sample of 0.10 mol/L NaOH(aq).
    (a) **Sketch** a theoretical (calculated) pH curve.
    (b) Include the following information as labels on your sketch. Show all relevant calculations.
       (i) the equivalence point pH and titrant volume for the reaction
       (ii) the initial pH of the sodium hydroxide sample solution
       (iii) the pH after adding 5.0 mL of HCl(aq) (treat this part as a limiting reagent calculation)
       (iv) the entities present at the equivalence point
       (v) the pH after adding 9.0 mL of HCl(aq)
       (vi) the pH after adding 11.0 mL of HCl(aq)
    (c) Suggest a suitable indicator for an endpoint determination for this titration. Indicate the pH values for the indicator colour change range on your pH curve.

Use this information to answer questions 40 to 43.

Lime is a very simple substance, and very easily prepared by strongly heating natural chalk. It has been used throughout recorded history, and by now the word "lime" appears in common names for many substances. Lime is calcium oxide, CaO(s), which is often sold as quicklime, or unslaked lime. This oxide is hazardous to handle because it reacts very readily and rapidly with water, releasing a lot of heat. When "slaked" with water, it forms calcium hydroxide, $Ca(OH)_2(s)$, which is much less dangerous to handle. This compound is widely sold in garden stores as agricultural, or horticultural, lime—meaning a rather impure form. It is commonly added to soils to raise the soil pH value, because the absorption by plants of some nutrients and trace elements depends heavily on soil pH levels.

Hydrochloric acid has also been in common use for a very long time. It is still sold under its historical name, muriatic acid, usually as a solution of about 30–35% HCl. It is very useful as a powerful rust remover, or to adjust pH levels in swimming pools. It will also "etch" the surface of concrete by reacting with carbonate ions in the solid mixture, with the result that paint can adhere to the clean, rough concrete surface. A spill of this very corrosive acid is always a serious problem (**Figure 5**).

Figure 5
Horticultural lime (calcium hydroxide) is quickly sprinkled on a spill of concentrated muriatic (hydrochloric) acid.

40. The solubility of calcium hydroxide is expressed in the following equation, representing a saturated solution equilibrium at 25 °C:
    $Ca(OH)_2(s) \rightleftharpoons Ca^{2+}(aq) + 2\,OH^-(aq)$
    $K_c = 1.3 \times 10^{-6}$
    (a) **Predict** whether calcium hydroxide is highly soluble or only slightly soluble in water.
    (b) What is the common name for a saturated calcium hydroxide solution, and what diagnostic test can be performed with it?
    (c) **Explain** how the solubility makes this compound safe to handle, even though hydroxide ion is the strongest base possible in aqueous solution.
    (d) **Explain** why "liming" of a (moist) garden soil will only raise the pH by a small amount, but then will continue to keep it elevated for months.

41. In a concentrated commercial hydrochloric acid solution, about one of every four molecules is HCl, as calculated from the mass percent printed on the label. **Predict** what fraction of these HCl molecules is actually in the "ready to react" form of $H_3O^+(aq)$.

42. When neutralizing this acid spill with calcium hydroxide, **identify** which compound you would want to be in excess, and **explain** why.

43. Write a chemical equation and a Brønsted–Lowry equation for this neutralization.

44. A series of experiments with a non-aqueous solvent determined that the products are highly favoured in each of the following acid–base equilibria, as written.
$(C_6H_5)_3C^- + C_4H_4NH \rightleftarrows (C_6H_5)_3CH + C_4H_4N^-$
$CH_3COOH + HS^- \rightleftarrows CH_3COO^- + H_2S$
$O^{2-} + (C_6H_5)_3CH \rightleftarrows OH^- + (C_6H_5)_3C^-$
$C_4H_4N^- + H_2S \rightleftarrows C_4H_4NH + HS^-$

(a) Identify the Brønsted–Lowry acids, bases, and conjugate acid–base pairs in each of these chemical reactions.
(b) Arrange the acids in these four chemical reactions in order of decreasing acid strength (in the solvent used), as a standard table of acids and conjugate bases.

45. The hydronium ion concentration of a 0.100 mol/L *n*-butanoic (butyric) acid solution was measured to be 1.24×10^{-3} mol/L.

Determine the percent reaction (ionization) of this particular weak acid solution, and a K_a value for aqueous *n*-butanoic acid at this ambient temperature.

46. A 0.10 mol/L solution of the essential amino acid tryptophan (1-α-amino-3-indolepropanoic acid) has a measured pH of 5.19 at 25 °C. **Predict** the percent reaction of this tryptophan solution, and the K_a value for aqueous tryptophan. The molecular formula for tryptophan may be written as $C_{10}H_{11}N_2COOH$.

47. Glycine, $H_2NCH_2COOH(aq)$, is a nonessential amino acid, with the simplest structure of all the amino acids, and has a $K_a = 4.5 \times 10^{-7}$ at 25 °C. Calculate the hydronium ion concentration, the pH, and the percent reaction in a 0.050 mol/L aqueous solution of glycine at 25 °C.

48. Thioacetic acid, $CH_3COSH(aq)$, has a $K_a = 4.7 \times 10^{-4}$ at 25 °C.
(a) Write the K_a expression for thioacetic acid.
(b) Calculate the hydronium and thioacetate ion concentrations, the pH, and the percent reaction in a 2.00 mol/L $CH_3COSH(aq)$ solution at 25 °C.
(c) Calculate K_b for the conjugate base, the thioacetate ion.

49. Predict the percent reaction, pH, pOH, and hydroxide ion concentration of a 0.012 mol/L solution of sodium benzoate.

50. A 0.100 mol/L laboratory solution of sodium propanoate, $NaC_2H_5COO(aq)$, has a measured pH of 8.95 at 25 °C. Calculate K_b for the propanoate ion.

51. Chloro-substituted acetic acids are used in organic synthesis, cleaners, and herbicides. These acids are prepared by the chlorination of acetic acid in the presence of small amounts of phosphorus. This process is known as the Hell–Volhard–Zelinsky reaction. As is typical of organic synthesis reactions, at equilibrium, the reaction vessel contains a mixture of all the different possible reaction products as well as some unreacted acetic acid and chlorine. When these acids were studied separately, the data in **Table 1** were obtained.

Table 1 Comparison of 0.100 mol/L Solutions of Acetic Acid and the Chloroacetic Acids

Substance	Formula	pH
acetic acid	$CH_3COOH(aq)$	2.89
chloroacetic acid	$CH_2ClCOOH(aq)$	1.94
dichloroacetic acid	$CHCl_2COOH(aq)$	1.30
trichloroacetic acid	$CCl_3COOH(aq)$	1.14

(a) Calculate acid ionization constants for each chloroacetic acid. Write chemical equations to describe the reactions of these acids with water.
(b) Suggest a theoretical explanation for the relative strengths of this series of acids. (Hint: Would adding chlorines to the *other* end of the molecule make the –COOH end of the molecule more, or less, negative?)
(c) A chemical technician is assigned the design of an acid–base titration to determine the amounts of each acid present in the mixture. She knows from experience that the reaction of acetic acid with a strong base is quantitative. **Sketch** a simplified pH curve for titration of a mixture of the four acids produced by chlorinating acetic acid. How could the technician determine relative chemical amounts of the acids present in the mixture?

52. Liquid ammonia can be used as a solvent for acid–base reactions.
(a) **Predict** the strongest acid entity that could be present in this solvent. (Consider the parallel with liquid water, $H_2O(l)$. Also consider the likely reaction of a very strong proton donor, such as hydrogen chloride, when it dissolves and reacts quantitatively in liquid ammonia.)
(b) The ionization equilibrium of pure liquid ammonia is similar to that of pure liquid water. Write the equilibrium equation for the ionization of liquid ammonia.
(c) **Predict** the strongest base that could be present in liquid ammonia solution.
(d) **Sketch** a titration curve for the addition of the strongest acid in ammonia solution to the strongest base. Suggest what value, instead of pH, might be used on the vertical axis of your graph.

53. Review the focusing questions on page 670. Using the knowledge you have gained from this unit, briefly **outline** a response to each of these questions.

Chemistry Appendices

contents

A.	**Numerical Answers to Questions**	**783**
B.	**Scientific Problem Solving**	**790**
B.1	Scientific Problem-Solving Model	790
B.2	Investigation Report Outline	790
B.3	Sample Investigation Report	793
B.4	The Nature of Scientific Research	794
C.	**Technological Problem Solving**	**796**
C.1	Technological Problem-Solving Model	796
C.2	Investigation Reports	796
C.3	Laboratory Equipment	797
C.4	Laboratory Processes	802
D.	**STS Problem Solving**	**806**
D.1	STS Decision-Making Model	806
D.2	Types of Reports	806
E.	**Safety Knowledge and Skills**	**807**
E.1	Laboratory Safety	807
E.2	Safety Symbols and Information	809
E.3	Waste Disposal	810
F.	**Communication Skills**	**811**
F.1	Scientific Language	811
F.2	SI Symbols and Conventions	811
F.3	Quantitative Descriptions and Rules	813
F.4	Tables and Graphs	815
F.5	Problem-Solving Methods	816
G.	**Review of Chemistry 20**	**817**
H.	**Diploma Exam Preparation**	**823**

I. Data Tables — 826

- Thermodynamic Properties of Selected Elements — 826
- Thermodynamic Properties of Selected Compounds — 826
- Miscellaneous Specific and Volumetric Heat Capacities — 826
- Standard Molar Enthalpies of Formation — 827
- Relative Strengths of Oxidizing and Reducing Agents — 828
- Relative Strengths of Aqueous Acids and Bases — 829

J. Common Chemicals — 830

Appendix A NUMERICAL ANSWERS TO QUESTIONS

Chemistry Review Unit
Are You Ready? (pp. 4–5)
2. (b) 4.2 g

Chapter 1
Section 1.4
Section 1.4 Questions (p. 26)
11. 1+, 2+, 3+, 3−, 2−, 1−

Chapter 2
Section 2.4
Section 2.4 Questions (p. 57)
1. (a) 18.02 g/mol
   (b) 44.01 g/mol
   (c) 58.44 g/mol
   (d) 342.34 g/mol
   (e) 252.10 g/mol
2. (a) 4 sig.dig.
   (b) 2 sig.dig.
   (c) 2 sig.dig.
   (d) 3 sig.dig.
   (e) 1 sig.dig.
   (f) 4 sig.dig.
3. (a) 0.117 mol
   (b) 24 g
   (c) 50.0 mmol
   (d) 12.49 g
4. (a) 1.72×10^3 g or 1.72 kg
   (b) 50.0 L
   (c) 1.55 mol/L
   (d) 13.6 mL
   (e) 2%
   (f) 3.94 kJ
5. (a) 0.907 mol
   (b) 8.56 mol
   (c) 29.21 mol
   (d) 1.80 mmol
   (e) 2.45 mol
6. (a) 71.9 g
   (b) 8.96 g
   (c) 1.03 g
   (d) 1.49 Mg
   (e) 0.10 kg
7. (a) 2.02 g, 70.90 g, 92.92 g
   (b) 64.10 g, 96.00 g, 88.02 g, 72.08 g

Unit 2
Chapter 4
Section 4.1
Lab Exercise 4.1 (p. 148)
1, 5, 3, 2, 4
Practice (p. 150)
2. (a) 0.952 atm, 724 mm Hg
   (b) 110 kPa, 1.09 atm
   (c) 253 kPa, 1.90×10^3 mm Hg
Practice (p. 152)
6. 263 kPa
7. 137 L
8. (b) 0.17 L.
9. 0.15 MPa
Practice (p. 154)
11. −273 °C, 0 K
12. (a) $T = (0 + 273)$ K = 273 K
    (b) $T = (100 + 272)$ K = 373 K
    (c) $T = (-30 + 273)$ K = 243 K
    (d) $T = (25 + 273)$ K = 298 K
13. (a) $t = (0 - 273)$ °C = −273 °C
    (b) $t = (100 - 273)$ °C = −173 °C
    (c) $t = (300 - 273)$ °C = 27 °C
    (d) $t = (373 - 273)$ °C = 100 °C
Practice (p. 156)
14. (a) 16 mL
15. 0.12 L
16. 26%
17. 79 °C
Practice (p. 159)
20. 87.6 kPa
21. 7.9 L
23. 25 °C
Section 4.1 Questions (pp. 161–162)
1. (a) 8.87 kPa, 66.5 mm Hg
   (b) 0.247 atm, 188 mm Hg
   (c) 112 kPa, 1.11 atm
2. (a) 298 K
   (b) 238 K
   (c) 39 °C
   (d) −65 °C
3. 384 mm Hg
4. 0.16 L
5. (a) 62 L
   (b) 2.3 times larger
6. 231 °C

11. (a) 3.82:1
15. (a) 2.8 L
Section 4.2
Practice (p. 166)
5. 25.0 L
6. 0.60 L
7. (a) 1.5 ML
Section 4.2 Questions (p. 168)
4. (a) 186 kL
   (b) 124 kL
5. (a) oxygen, 125 L; nitrogen monoxide, 100 L; water vapour, 150 L
   (b) 375 L
   (c) 33.3 L
Section 4.3
Section 4.3 Questions (p. 171)
5. 186 L
6. 2.0 mmol
7. 50.4 L
8. 0.18 mol
9. 73 mL
10. 9.80 L
11. 0.539 ML (or 539 kL)
12. 0.727 g
13. 2.58 g
14. (a) 1.8 g/L
Section 4.4
Practice (pp. 174–175)
3. 1.6 MPa
4. 41.0 mmol
5. 34 kL or 34 m^3
Lab Exercise 4.B (p. 175)
R, 8.48 kPa·L/(mol·K)
Section 4.4 Questions (p. 176)
5. 5.6 mol
6. 225 °C
8. $0.0821 \frac{\text{atm·L}}{\text{mol·K}}$
9. (a) 34.0 g/mol
10. 22.4 L, 24.8 L
11. (a) 1.1 g/L
    (d) 1.74 g/L
Unit 2 Review (pp. 180–183)
18. (a) 273 K
    (b) 294 K
    (c) 0 K

19. (a) 0.405 MPa
    (b) 102 kPa
    (c) 45.6 MPa
20. (a) 0.21 mol
    (b) 0.924 mol
21. (a) 12.4 kL
    (b) 1.4 ML
26. 8.23 L
27. -14 °C
28. 0.33 mol
29. 4.73 L
31. (a) 1.78 atm
    (b) 196 kPa
32. 317 °C
33. (a) 302 kPa
    (b) 30.7 kg
35. (a) ammonia, 1.00 L; oxygen, 1.25 L
36. (a) 50 mL
37. methane, 0.647 g/L; nitrogen, 1.13 g/L
38. (a) 150 mL
39. (b) 7.5 mmol
    (c) 167 mL

Unit 3
Chapter 5
Section 5.3
Practice (pp. 205–206)
2. 7.5% V/V
3. 32% W/V
4. 4.9% W/W
5. 5.4 ppm
6. 1.8 mol/L

Practice (p. 208)
7. 350 mL
8. 7.5 kg
9. 4.1 mol
10. 0.25 mol
11. 403 mL
12. 54 mL

Practice (p. 210)
13. 15.0 g
14. 0.16 kg
    (a) 355 mg
    (b) 8.07 mmol/L
    (a) 7.83% W/V
    (b) 1.34 mol/L

Practice (p. 212)
17. (a) $[Na^+(aq)] = 0.82$ mol/L;
    $[S^{2-}(aq)] = 0.41$ mol/L

(b) $[Sr^{2+}(aq)] = 1.2$ mol/L;
    $[NO_3^-(aq)] = 2.4$ mol/L
(c) $[NH_4^+(aq)] = 0.39$ mol/L;
    $[PO_4^{3-}(aq)] = 0.13$ mol/L
18. $[Fe^{3+}(aq)] = 49.6$ mol/L;
    $[Cl^-(aq)] = 149$ mmol/L
19. (a) 11.1 g
    (b) 18.5 g

Section 5.3 Questions (p. 214)
3. (a) 5% W/V
4. 20 ppm
5. (a) 0.32 mol/L
   (b) $[NH_4^+(aq)] = 0.64$ mol/L;
       $[CO_3^{2-}(aq)] = 0.32$ mol/L
6. 79.0 g
7. 11 mg
8. 4.3 mol/L
9. (a) 0.58 g
   (b) 0.75 g
   (c) 1.1 g
10. 0.20 L
11. (a) 0.143 mol/L
    (b) 0.429 mol/L
12. 2.3 g
17. (a) $1:10^9$
    (b) 0.001 ppm
    (c) $\dfrac{1 \times \text{g solute}}{\text{kg solution}}$
    (d) 30 µg/kg

Section 5.4
Practice (p. 216)
1. 3.10 g
2. 200 g
4. (a) 33.2 g
5. (a) 5.93 g

Practice (pp. 218–219)
6. 42%
7. 22.5 mL
8. (a) 0.250 mmol/L
   (b) 0.399 mg

Section 5.4 Questions (p. 219)
3. 3.27 g
4. 43.5 L
5. 6.85 g/L
6. 1.51 g
7. 25.0 mL

8. (a) 1.30 g
   (b) initial volume = 10.0 mL

Chapter 5 Review (pp. 231–233)
22. 51 g, 150 g, 250 g
23. 0.3 L
24. (a) 0.70 mol/L
    (b) 0.125 mol/L
    (c) 2.0 mol/L
    (d) 0.66 mmol/L
25. 12.6 g
26. 42.8 mL
27. 6.6 g
28. (a) 56 mg
32. 28.1 mL
34. (a) $[K^+(aq)] = 0.14$ mol/L;
       $[NO_3^-(aq)] = 0.14$ mol/L
    (b) $[Ca^{2+}(aq)] = 0.14$ mol/L;
       $[Cl^-(aq)] = 0.28$ mol/L
    (c) $[NH_4^+(aq)] = 0.42$ mol/L;
       $[PO_4^{3-}(aq)] = 0.14$ mol/L

Chapter 6
Section 6.2
Practice (p. 239)
1. (a) 10^{-7} mol/L
   (b) 10^{-11} mol/L
   (c) 10^{-2} mol/L
   (d) 10^{-4} mol/L
   (e) 10^{-14} mol/L
2. (a) 3
   (b) 5
   (c) 7
   (d) 10
3. 100

Practice (p. 242)
5. (a) 2.68
   (b) 5.0
   (c) 6.602
   (d) 8.14
6. (a) 5×10^{-12} mol/L
   (b) 2.2×10^{-3} mol/L
   (c) 6×10^{-5} mol/L
   (d) 1.76×10^{-14} mol/L

Practice (p. 243)
9. -0.65
10. 1.6×10^{-7} mol/L

Section 6.2 Questions (p. 244)
6. 5.00

8. increase of 3 pH units
9. (a) −1.00
   (b) 14.80

Section 6.3
Lab Exercise 6.B (p. 247)
4.4 - 4.8; 6.0 - 6.6; 3.2 - 3.8

Section 6.3 Questions (p. 247)
2. (a) < 4.8
   (b) > 12.0
   (c) > 5.4
   (d) between 6.0 and 7.6
3. (a) between 5.4 and 6.0
   (b) 1×10^{-6} mol/L

Section 6.5
Practice (p. 255)
4. (a) 0.15 mol/L
   (b) 0.82

Chapter 6 Review (pp. 263–264)
14. (a) 2.74×10^{-12} mol/L
    (b) 3×10^{-4} mol/L
15. 10×
17. (a) between 5.4 and 6.0
    (b) 1×10^{-6} mol/L to 4×10^{-6} mol/L

Unit 3 Review (pp. 265–269)
22. (a) 75.0 ppm
    (b) 1.87 mmol/L
23. 1.2 L
24. (a) 0.23 mol/L
    (b) 20 mmol
    (c) 0.103 L
    (d) 0.155 mol/L
    (e) 1.21 g
25. 16.9 mg/L
26. (a) 14 L
27. (a) $[Na^+(aq)] = 4.48$ mol/L;
        $[S^{2-}(aq)] = 2.24$ mol/L
    (b) $[Fe^{2+}(aq)] = 0.44$ mol/L;
        $[NO_3^-(aq)] = 0.88$ mol/L
    (c) $[K^+(aq)] = 0.525$ mol/L;
        $[PO_4^{3-}(aq)] = 0.175$ mol/L
    (d) $[Co^{3+}(aq)] = 0.0862$ mol/L;
        $[SO_4^{2-}(aq)] = 0.129$ mol/L
30. (a) 2.12
    (b) 2.60
31. (a) 2.74×10^{-12} mol/L
    (b) 3×10^{-4} mol/L
35. (a) 10^{-3} mol/L (fruit juice); 10^{-12} mol/L (cleaning solution)
    (b) 10^9:1
37. (a) 6.3

(b) 5.10
39. 2.82 g
49. 0.012 mol/L
51. 0
56. (b) 4×10^{-3} mol/L

Unit 4
Are You Ready? (pp. 272–273)
3. (a) 2×10^{-5} mol/L
4. $(NH_4)_3PO_4(s)$, 0.2951 mol; $CH_3COOH(l)$, 3.5 g
5. NaOH(aq), 1.10 mol; HCl(aq), 0.00345 L; Na_2SO_4(aq), 1.13 mol/L
6. CH_4(g), 0.611; UF_6(g), 0.120; CO_2(g), 2.48×10^3 L; Ar(g) 0.161

Chapter 7
Section 7.2
Practice (p. 290)
9. 12 g
10. 66.2 g
11. 36.3 g
12. 6.11 g
13. 0.307 g
14. 2.32 g

Lab Exercise 7.A (p. 291)
6.68 g

Lab Exercise 7.B (p. 293)
9.12 g

Section 7.2 Questions (p. 293)
6. 3.88 g
8. (a) 2.93 g
   (b) 95.9%
10. (a) 11 g
    (b) 91%

Section 7.3
Practice (p. 296)
1. 16 L
2. 60.1 kL
3. 150 L

Section 7.3 Questions (p. 298–299)
2. 6.0 L
3. 2.62 ML
4. 232 kg
5. 0.21 kL, or 0.21 m^3
6. 0.77 L
7. 1.94 kg
8. 0.57 L

Section 7.4
Practice (p. 302)
1. 0.537 mol/L
2. 375 mL
3. 90.0 mL

Lab Exercise 7.C (p. 302)
210 mg; 0.20 g

Lab Exercise 7.D (p. 303)
0.351 mol/L

Section 7.4 Questions (p. 303)
1. 0.35 L
2. 17.8 mol/L
3. 23.9 mmol/L
4. (b) 624 mg or 0.624 g
   (c) 98.0%
5. 639 g

Chapter 7 Review (p. 309–311)
20. (a) $NaHCO_3$(s), 2.4 mol; Na_2CO_3(s), 1.2 mol; CO_2(g), 1.2 mol; H_2O(l), 1.2 mol
    (b) 0.631 kg
21. 46 mL
    2.93 g
23. (a) 95.0%
26. (a) 14.7 g
    (b) 10.7 g
28. 2.08 kg
29. 1.91 kg
30. 1.17 kg
33. 0.668 mol/L

Chapter 8
Section 8.2
Lab Exercise 8.A (p. 317)
5.40 g

Section 8.2 Questions (p. 319)
1. 97 g sodium sulfate
2. 0.135 mol/L
3. 0.210 g predicted; 0.20 g obtained; 5%
4. 0.351 mol/L

Section 8.3
Practice (p. 321)
1. (a) 1.1 g to 1.2 g
2. 33.6 mL

Practice (p. 324)
3. (a) 5.0 mol
   (b) 0.55 mol
   (c) 0.26 mol
   (d) 5.46 mmol

4. (a) 6.25 g
(b) 2.86 g
(c) 36.mg
(d) 4.2 kg

Section 8.3 Questions (p. 327)
4. 3 g
5. (a) 2.77 g
(b) 3.89 g
6. 97.9%
7. (b) 2.3 g $BaCl_2$
(c) 8.4 g
8. (b) 3.5 g Zn
(c) 83%
9. (b) 125 mL
(c) 138 mL

Section 8.4
Section 8.4 Questions (p. 332)
1. $[NH_3] = 20.7$ mol/L
4. 0.660 mmol

Section 8.5
Web Activity: Web Quest—Blood Alcohol Content (p. 333)
Analysis
(a) 3.00×10^{-5} mol
(b) 2.22×10^{-6} mol
(c) 2.778×10^{-5} mol
(d) 0.0139 mol/L; BAC = 0.064 g/100 mL

Practice (p. 336)
2. (b) 7

Section 8.5 Questions (p. 339)
8. (a) 1.27 g
9. 0.140 mol/L

Chapter 8 Review (pp. 346–348)
13. (a) 0.075 mol lead (II) nitrate
(b) 1.00 mol propane
(c) 0.50 mol zinc
(d) 50 mmol sulfuric acid
14. 79.7%
15. 52.4 mL
16. 2.70 g
17. (a) 20 mL, 20 mL, 20 mL, 20 mL
(b) 7, 9, 5, 7
19. 21.3 mmol/L

Unit 4 Review (pp. 349–51)
24. (a) 106%
25. 699 kg
26. (a) 0.20 mol Zn(s)
(b) 5.0 mmol Cl_2(aq)
(c) 0.05 mol NaOH(aq)
27. 0.12 kg
31. 13.0 mL
33. 2.93 g
34. (b) 1.07 g
36. 1.22 g predicted; 1.27 g obtained; 4%

Unit 5
Chapter 10
Section 10.3
Practice (p. 431)
15. (a) 18 mL
(b) 18 mL
(c) 18 mL

Chapter 10 Review (pp. 466-467)
20. 36.8%

Unit 5 Review (pp. 119-120)
40. (a) 0.46 kL
(b) 4.6×10^3 km
42. 78.2%
48. 96.3%; 94.8%

Unit 6
Chapter 11
Section 11.1
Case Study, (p. 482)
2. 77.8 kJ

Section 11.1 Questions (pp. 483–484)
1. (a) respectively, for 2003: 8.22%, 35.21%, 43.55%, 3.99%, 9.04%
2. (a) respectively, for 2003: 58%, 63%, 56%, 42%, 8%
3. respectively, for 2003: 30.4%, 29.5%, 2.8%, 17.7%, 1.7%, 17.9%

Section 11.2
Practice (p. 487)
6. 84 kJ
7. 0.39 MJ
8. 67.7%

Practice (p. 492)
12. −5.00 MJ
13. −612 kJ

Lab Exercise 11.A (p. 492)
−54.6 kJ/mol

Section 11.2 Questions (p. 494)
2. (a) 2.1 kJ
(b) 2.2 kJ
(c) 4%
3. (a) −7.8 MJ
(b) −2.07 MJ
4. 1.54 g
5. −0.242 MJ
6. −64 kJ/mol
7. −26 kJ/mol
11. −41.1 kJ/mol

Section 11.3
Section 11.3 Questions (p. 501)
3. (a) 241.8 kJ/mol
(b) −318.0 kJ/mol
(c) +81.6 kJ/mol
(d) −372.8 kJ/mol
4. (a) −114 kJ
(c) −114 kJ/mol
(d) −57 kJ/mol

Section 11.4
Practice in 11.4 (pp. 504–505)
1. −851.5 kJ
2. +131.3 kJ
3. −524.8 kJ
4. −205.7 kJ

Lab Exercise 11.B (p. 506)
−3488.7 kJ/mol; -3.35 MJ/mol; 4.0%

Lab Exercise 11.C (p. 506)
+205.9 kJ

Section 11.4 Questions (pp. 508–509)
1. −69.5 kJ
2. (a) +136.4 kJ
(b) −44.2 kJ
3. +121.2 kJ/mol
7. −2.39 MJ
8. −2.20 MJ
9. −250.1 kJ/mol
10. −492 kJ
11. −21.8 kJ

Section 11.5
Lab Exercise 11.D (p. 513)
−725.9 kJ/mol; −598 kJ/mol; 17.6%

Section 11.5 Questions (pp. 514–515)
2. (a) +205.9 kJ
(b) −41.2 kJ
(c) −91.8 kJ

3. (a) −225.5 kJ/mol
   (b) −58.1 kJ/mol
   (c) −23.6 kJ/mol
4. (a) −145.6 kJ/mol
5. (a) −349.6 kJ/mol
6. (a) −562.0 kJ
8. (a) +179.2 kJ
   (b) −64.5 kJ
   (c) +114.7 kJ
   (d) +114.7 kJ
9. −114.7 kJ
10. −198.7 kJ/mol
12. −2803.1 kJ/mol; −2.80 MJ/mol; 0.143%

Chapter 11 Review (pp. 519–521)
17. (a) 44 kJ
    (b) 36 kJ/g
18. (a) 75.1 kJ
    (b) −75.1 kJ
    (c) −2.87 MJ/mol
19. −44.4 kJ/mol
20. (a) 21.8 MJ
    (b) 0.436 kg
    (c) 234 L
22. −1 256.2 kJ
24. −107.4 kJ
25. −69 kJ
26. (a) −401.0 kJ/mol
27. (a) −27.2 kJ
    (b) +2.3 kJ
28. −426 kJ; −110 kJ; −602 kJ
29. (a) −2 219.9 kJ/mol
    (b) −2 043.9 kJ/mol
    (c) −1 564.2 kJ/mol

Chapter 12
Section 12.3 Questions (p. 70)
3. (a) +60 kJ
   (b) +95 kJ
   (c) −35 kJ
   (d) +35 kJ
7. 5 kJ/mol; 6 kJ/mol; 11%

Chapter 12 Review (pp. 545–546)
16. (a) 52 kJ
    (b) 90 kJ
    (c) 22 kJ
    (d) 60 kJ
    (e) −38 kJ
    (f) 38 kJ
    (g) −38 kJ
    (h) 38 kJ

Unit 6 Review (pp. 547–551)
19. (a) −176.2 kJ
20. (b) −128.7 kJ
    (c) −4.02 MJ
21. (a) −1164.8 kJ
    (b) −388.3 kJ/mol
    (c) −1.51 GJ
22. (a) +136.4 kJ
    (b) +311.4 kJ
23. (a) −23.5 kJ
28. (a) −98.0 kJ/mol
29. −890.5 kJ/mol; −1560.4 kJ/mol; −2219.9 kJ/mol
    −55.48 MJ/kg; 51.88 MJ/kg; 50.33 MJ/kg
    62.31 mol/kg; 66.49 mol/kg; 68.01 mol/kg
30. (b) −134.6 kJ
31. (a) −1366.8 kJ
36. −25.6 kJ/mol

Unit 7
Chapter 13
Section 13.3
Practice (p. 585)
1. (a) +4
   (b) +7
   (c) +6
   (d) +6
   (e) −1
   (f) −1
2. (a) +1
   (b) +2
   (c) +4
   (d) −3
   (e) −2
   (f) +5
   (g) 0
   (h) −3
3. (a) 0
   (b) 0
   (c) +4
   (d) +2

5. −4; −2; 0; +2; +4

Section 13.3 Questions (p. 595)
4. (a) C +2; O −2
   (b) O 0
   (c) N −3; H +1; Cl −1
   (d) H +1; P +5; O −2
   (e) Na +1; S +2; O −2
   (f) Na +1; P +5; O −2
5. (a) $+\dfrac{8}{3}$

Section 13.4
Lab Exercise 13.C (p. 598)
0.258 mol/L

Practice (p. 598)
3. 8.5 L
4. 0.325 mol/L
5. 11.4 mmol/L

Lab Exercise 13.D (p. 599)
1.31 mol/L

Investigation 13.4 (p. 603)
Analysis: 2.94%

Section 13.4 Questions (p. 600)
4. 1.92 mol/L
5. Analysis: 74.9 mmol/L
   Evaluation: 6.4%
6. 25.9 mmol/L
7. 29.8 mmol/L

Chapter 13 Review (pp. 606–609)
29. 4.8 g
30. 56.3 mmol/L
33. Analysis: −33 °C

Chapter 14
Section 14.1
Practice (p. 614)
3. 6

Section 14.2
Practice (p. 631)
11. (a) +0.77 V
    (b) +0.45 V
    (c) +1.23 V

Practice (p. 633)
12. (a) +0.47 V
    (b) +0.50 V
    (c) +0.77 V
13. −0.34 V
14. Cu: +3.38 V; Zn: +2.28 V

Lab Exercise 14.A (p. 633)

$Cr_2O_7^{2-}$: +1.23 V; Pd^{2+}: +0.95 V; Tl^+: −0.34 V; Ti^{2+}: −1.63 V

Section 14.2 Questions (pp. 637–638)

10. (a) +0.48 V
12. −0.28 V
15. +1.56 V

Section 14.3
Practice (p. 644)

5. (a) +0.80 V
   (b) +1.23 V
6. (a) +1.48 V
   (b) 0.00 V

Practice (p. 645)

12. (a) +0.43 V
    (b) +0.29 V
13. −1.30 V

Section 14.4
Practice (p. 653)

1. 45 C
2. 3.89 A
3. 7.13 C
4. 3.91 min

Practice (p. 654)

5. 41 mmol
6. 3.16 h
7. 1.2 A

Section 14.4 Questions (p. 657)

1. 2.80 mmol
2. 0.58 t
3. 82.8 min
4. 52.8 kA
5. (a) 1.63 t
   (b) 4.76 t
6. 0.174 mol/L
7. 24.42 g
8. 0.102 g
9. (a) 26 g
   (b) 25.4 min
10. 21. 3 g
11. 25.0 A
12. 0.766 g; 0.71 g; 7%
14. (c) 483 g
    (d) 41.9 g

Chapter 14 Review (p. 665)

18. (d) +0.93 V
21. (d) 0.00 V
23. 50.3 kg
24. 0.125 A

Unit 7 Review (pp. 667–669)

41. 5.78 g
42. 8.56 mmol/L
45. (a) 1.22 V
    (b) 0.80
    (c) 0.00 V
46. (a) 1.90 V
    (b) 1.23 V
    (c) 1.64 V
47. (b) 1.23 V
48. (b) 2.19 V
49. (a) 1.99 V
    (b) 590 s (or 9.84 min)
50. 2.98 kA
51. 20.1 min
52. 1.0 kmol/h
54. (b) +1.20 V

Unit 8
Are You Ready? (pp. 672–673)

5. (a) 5.94 mol/L
   (b) 0.220 mol/L
   (c) 0.0403 mol/L
   (d) 0.0446 mol/L

Chapter 15
Section 15.1
Practice (p. 682)

4. (a) 2.00 mol
   (b) 70.0%
6. (c) 1.00 L
   (d) 60%

Lab Exercise 15.B (p. 686)

0.46

Section 15.1 Questions (pp. 688–689)

3. (b) 14 mol
5. 51
6. 0.11 mol/L
7. (c) 0.200 mol
   (d) 0.80 mol
   (e) 0.40 mol
   (f) [HBr(g)] = 0.100 mol/L;
       [H_2(g)] = 0.20 mol/L;
       [Br_2(g)] = 0.20 mol/L
   (g) 4.0
8. 54.1
9. 1.5
10. (a) 0.78 mol/L
    (b) 0.39 mol

Chapter 15 Review (p. 705–706)

19. 0.0013 mol/L
20. 0.0403 mol/L
21. 0.032
26. (a) 0.068
27. (a) 5.07 mol/L
    (b) 7.60 mol/L
28. 1.00 mol/L
32. (b) 0.0029 mol/L

Chapter 16
Section 16.1
Practice (p. 716)

1. 2.3×10^{-12} mol/L
2. 3.3×10^{-11} mol/L
3. 2.5×10^{-13} mol/L
4. 7.2×10^{-13} mol/L
5. 1.4×10^{-14} mol/L
6. 1.8×10^{-7} %

Practice (p. 718)

7. (a) 1.8×10^{-12}; 2.26; 11.74
       4×10^{-9}; 3×10^{-6}; 8.4
       5.0×10^{-4}; 3.30; 10.70
       4.0×10^{-4}; 2.5×10^{-11}; 3.40
8. 14.64; −0.64
9. 0.09 g

Section 16.1 Questions (p. 721)

4. (b) 7.8×10^{-6} mol/L; 5.11
5. 7.7×10^{-12} mol/L
6. 7.40
7. 3×10^{-6} mol/L
8. 1000 (10^3)
9. 0.372
10. 0.25 g
11. 4.27
12. −0.016; 14.02
13. 2.42; 11.58
14. 18 mg

Section 16.3
Practice (p. 743)

2. (a) 0.20 mol/L
   (c) 1.9×10^{-3} mol/L
   (d) 2.64
   (f) 4.84
5. (a) 1.16%
   (b) 1.36×10^{-5}
6. (a) 3.7%
   (b) 1.35×10^{-5}; 1.4×10^{-4}

7. (b) 3.5×10^{-2} mol/L; 3.5×10^{-2} mol/L; 1.45; 1.8%
8. (a) 0.26 mol/L; 0.58; 2.6%
9. (a) 8.2×10^{-5}

Practice (p. 746)
12. 7.7×10^{-10}
13. 4.2×10^{-10}

Section 16.3 Questions (p. 750)
1. 10.27
2. 11.23
4. 4.77
5. 1.4×10^{-3}
6. 1.3×10^{-10}
7. 4.2×10^{-4} mol/L
10. (a) 9.95
    (b) 4.25
    (c) 3.85

Section 16.4
Practice (pp. 762–763)
13. (a) < 7
    (b) > 7
    (c) ~ 7
    (d) < 7

Chapter 16 Review (p. 773)
17. (a) 0.10 mol/L; 1.00
    (c) 9.4×10^{-5} mol/L; 4.03
18. (a) 5.8×10^{-7}
20. 7; > 7
21. 10.0 to 11.4; 2.5×10^{-3} to 2.5×10^{-3} mol/L
22. 0.019%; 8.46; 5.54; 2.9×10^{-6} mol/L

Unit 8 Review (pp. 775–779)
26. 0.62 mol/L; 0.62 mol/L; 1.38 mol/L; 1.38 mol/L
29. 5.0 to 5.2; 1×10^{-5} mol/L to 6×10^{-6} mol/L
35. (c) 25 mL; 50 mL
36. 3.48 mol/L; 1.04 mol/L
37. 2.3
38. 0.882 mol/L
39. (a) (i) 7.00; 10 mL
        (ii) 13.00
        (iii) 12.60
        (v) 11.8
        (vi) 1.9
45. 1.24%; 1.6×10^{-5}
46. 6.5×10^{-3} %; 4.2×10^{-10}
47. 1.5×10^{-4} mol/L; 0.58; 0.30%
48. (b) 3.1×10^{-2} mol/L; 1.51; 1.5%
    (c) 2.1×10^{-11} mol/L
49. 0.012%; 8.14; 5.86; 1.4×10^{-6} mol/L
50. 7.9×10^{-10}
51. (a) 1.5×10^{-3}; 5.0×10^{-2}; 1.9×10^{-1}

Appendix B SCIENTIFIC PROBLEM SOLVING

B.1 Scientific Problem-Solving Model

Scientists ask questions and seek concepts to answer these questions by applying consistent, logical reasoning to describe, explain, and predict observations, and by doing experiments to test hypotheses or predictions from these hypotheses. In this way science progresses using a general model for solving problems and employing specific processes as part of a problem-solving strategy.

Every investigation in science has a *purpose*; for example:

- to *create* a scientific concept (a theory, law, generalization, or definition)
- to *test* a scientific concept
- to *use* a scientific concept, e.g., in chemical analysis

Once you know the purpose, you need a problem and a general design. For example, if the purpose is to perform a chemical analysis to determine the quantity of a substance, then possible designs include distillation and precipitation. Once you choose a design, there are many specific questions that you might ask, many possible reactants you might choose, and many other variables you might need to consider.

B.2 Investigation Report Outline

An investigation report is the final result of your problem solving. Your report should follow the model outlined in **Figure 1**. As a further guide, use the information and instructions for the specific processes listed below. The parts of the investigation report that you are to provide are indicated in the text in a checklist (**Figure 2**).

Report Checklist
- ○ Purpose   ○ Design   ● Analysis
- ○ Problem   ○ Materials   ● Evaluation (1, 2
- ○ Hypothesis ○ Procedure   and/or 3)
- ○ Prediction ● Evidence

Figure 2
Shaded circles indicate the parts you are expected to complete in a particular investigation report. One or more parts of an Evaluation may be required, as indicated by the numbers.

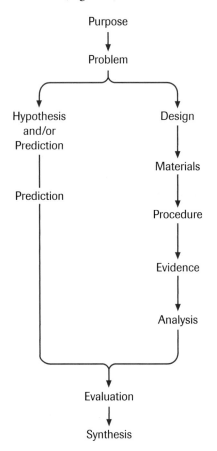

Figure 1
A scientific problem-solving model helps to guide your laboratory work, but does not illustrate the complexity of the work.

Purpose
Although this is usually provided, you will be expected to identify the purpose of an investigation before, during, and after your laboratory work. Most often, the purpose is to create, test, or use a chemistry concept.

Problem
The Problem is a specific question to be answered in the investigation. If appropriate, you should state the question in terms of manipulated and responding variables. In most cases, the problem is chosen for you. Only when creating a concept will the Purpose and the Problem be the same.

Hypothesis
The hypothesis is an (often untested) empirical or theoretical concept that provides a possible explanation for a natural or technological phenomenon. Only some kinds of investigations require a hypothesis, such as investigations that test a hypothesis using a general question as the Problem.

Prediction
The Prediction is the expected answer to the Problem according to a scientific concept (for example, a hypothesis, theory, law, or generalization) or another authority (for

example, a reference source or a label on a bottle). Write your Prediction using the format, "According to [an authority], [answer to the Problem]." Include your qualitative and quantitative reasoning (based upon the authority) with the Prediction.

Design
The Design provides a brief overview of the Procedure to obtain an answer to the Problem. Included in the Design are reacting chemicals and, if applicable, brief descriptions of diagnostic tests, variables, and controls. Write your Design as a paragraph of one to three sentences.

Materials
This section consists of a complete list of all equipment and chemicals, in columns, including sizes and quantities. Appendix C.3, page 797, shows laboratory equipment, including common sizes.

Procedure
The Procedure is a detailed set of instructions designed to obtain the evidence needed to answer the Problem. Write a list of numbered steps in the correct sequence, including any safety and waste disposal instructions (Appendix E, page 807). Whenever possible, repeat measurements several times.

Evidence
The Evidence includes all qualitative and quantitative observations relevant to answering the Problem. Organize your evidence in tables whenever possible (Appendix F.4, page 815). Be as precise as possible in your measurements and include any unexpected observations that may affect your answer and its certainty (in significant digits). Scientific honesty demands that you report all evidence collected and not just the evidence you think is correct or "normal."

Analysis
The Analysis includes manipulations, interpretations, and calculations based on the evidence. Tables and graphs that include or facilitate interpretations and calculations are included in the Analysis. You may need to differentiate between relevant and irrelevant observations. Communicate your work clearly and logically. Conclude the Analysis with a statement of your experimental answer to the Problem, including a phrase such as, "According to the evidence gathered in this experiment, [answer to the Problem statement]."

Evaluation
Part 1 of the Evaluation of an investigation that you actually perform usually involves judging the validity of the experiment (Design, Materials, Procedure, and Skills). Evaluation of Lab Exercises (simulated investigations) will not usually involve Part 1. Only if you are confident that no major flaws are present can you proceed to the second part. In Part 2, you use the results of the experiment to evaluate the Prediction (if one was made) and the Hypothesis (if there is one). This assumes that a prediction is being tested in an experiment. If it is the experimental design that is being tested, then the two parts of the evaluation would be reversed and the percent difference is used to judge the success of the design. The last part of the Evaluation, Part 3, comes back, full circle, to the Purpose of the investigation. Was the Purpose fulfilled?

The parts of the Evaluation you will be expected to complete are shown in parentheses after Evaluation in the Report Checklist.

Write your Evaluation in paragraph form, using the topic sentences suggested below or an adaptation of them. Some of the more important criteria for a judgment are listed as questions; use selected questions to guide your judgments. Show as much independent, critical, and creative thought as possible in support of your judgments.

Part 1. Evaluation of the Investigation

- "The design of the investigation [name or describe in a few words] is judged to be adequate/inadequate because ..."

  Were you able to answer the Problem using the chosen experimental design? Are there any obvious flaws in the design? What alternative designs (better or worse) are available? As far as you know, is this design the best available in terms of controls, variables, efficiency, and safety?

- "The materials are judged to be adequate/inadequate because ..."

  Did you have all of the necessary materials? Was the equipment of reasonable quality? What materials could be improved to obtain better results?

- "The procedure is judged to be adequate/inadequate because ..."

  Were the steps that you used in the laboratory correctly sequenced, and adequate to gather sufficient evidence? What improvements could be made to the procedure, such as more trials?

- "The technological skills are judged to be adequate/inadequate because ..."

  Which specialized skills, if any, might have the greatest effect on the evidence gathered? Was the evidence from repeated trials reasonably similar?

How can the evidence gathered be improved?

- "Based upon my evaluation of the experiment, I am not/moderately/very certain of my experimental evidence. The sources of uncertainty or error are …"

  State the sources and your confidence in the experimental evidence. What would be an acceptable total of the experimental error (1%, 5%, or 10%)?

Part 2. Evaluation of the Prediction and Authority Being Tested

- If applicable, "the percent difference between the experimental result and the predicted value is…"

  $$\% \text{ difference} = \frac{|\text{experimental value} - \text{predicted value}|}{\text{predicted value}} \times 100\%$$

  How does this difference compare with your estimated total uncertainty or experimental error?

- "The prediction is judged to be verified/inconclusive/falsified because …"

  Does the predicted answer clearly agree with the experimental answer reported in your analysis? Can the percent difference be accounted for by the sources of error (percent error) listed above?

- "The authority being tested [name the authority or hypothesis (reasoning)] is judged to be acceptable/unacceptable in this experiment because …"

  Was the prediction verified, inconclusive, or falsified? How confident do you feel about your judgment?

Part 3. Revisiting the Purpose

Did you accomplish the Purpose of this investigation? Is there a need for additional investigations to better achieve the Purpose?

Notes on Data and Evidence

Data

Data may be found on data sheets, in data tables, and in databases. Data from one of these sources can become evidence with the purpose to create, test, or use a scientific concept. Evidence is data with a scientific purpose.

Common Sources of Experimental Error

- conditions (e.g., SATP) not controlled
- impure reactants or products
- any measurement process
- incomplete reaction
- judgment of colour (e.g., indicator)
- loss of solid in a filtration (stuck to glass or passed through filter)
- incomplete drying of a product
- manipulative skill

Do not use "human error" as a source of uncertainly or experimental error.

Percent Yield versus Percent Difference

In most experiments, a percent difference is a measure of accuracy. In some experiments in which a product is collected and measured, a percent yield is used instead of a percent difference.

$$\% \text{ yield} = \frac{\text{actual quantity obtained}}{\text{predicted (maximum) quantity}} \times 100\%$$

Experimental Error and Percent Difference

Some people and books use the term "percent error" in place of "percent difference." In this textbook, we use two percentages: One is an estimate of the total expected *experimental error* (Evaluation, Part 1), and the other is the actual difference (the *percent difference*) you determined based on your prediction and analysis (Evaluation, Part 2). The crucial point is how these two percentages compare. No experiment can ever be expected to be perfect. For example, if the equipment you used is only precise to $+/-$ 5%, then any percent difference you obtain that is less than or equal to 5% is as good a result as can be expected. If the percent difference is larger than the reasonably acceptable experimental error, then the prediction is falsified.

If the percent difference is equal to or less than the experimental error, then the prediction is verified.

Replication

An authority may be judged unacceptable in one experiment. This does not mean the authority is immediately discarded. Replication by independent workers is always required to refute any accepted theory.

B.3 Sample Investigation Report

The Reaction of Hydrochloric Acid with Zinc

Purpose
The purpose of this investigation is to test one of the ideas of the collision–reaction theory.

Report Checklist
- ○ Purpose
- ○ Problem
- ○ Hypothesis
- ● Prediction
- ● Design
- ● Materials
- ● Procedure
- ● Evidence
- ● Analysis
- ● Evaluation (1, 2, 3)

Problem
How does changing the concentration of hydrochloric acid affect the time required for the reaction of hydrochloric acid with zinc?

Prediction
According to the collision–reaction theory, if the concentration of hydrochloric acid is increased, then the time required for the reaction with zinc will decrease. The reasoning that supports the prediction is that a higher concentration produces more collisions per second between the hydrochloric acid entities and the zinc atoms. More collisions per second would produce more reactions per second and, therefore, a shorter time required to consume the zinc.

Design
Different known concentrations of excess hydrochloric acid react with the same quantity of zinc metal. The time for the zinc to completely react is measured for each concentration of acid solution. The concentration of hydrochloric acid is the manipulated variable and time is the responding variable. The temperature, mass, and surface area of zinc, and volume of acid are the controlled variables.

Materials
lab apron
eye protection
(4) 10 mL graduated cylinders
(4) 18 × 150 mm test tubes and test-tube rack
clock or watch (precise to the nearest second)
four pieces of a zinc metal strip (5 mm × 5 mm)
HCl(aq): 2.0 mol/L, 1.5 mol/L, 1.0 mol/L, 0.5 mol/L
a weak base (e.g., baking soda)

Procedure
1. Transfer 10 mL of 2.0 mol/L HCl(aq) into an 18 × 150 mm test tube. Avoid contact with skin, eyes, clothing, or the desk. If you spill this acid on your skin, wash immediately with lots of cool water.

2. Carefully place a piece of Zn(s) into the hydrochloric acid solution and note the starting time of the reaction.

3. Measure and record the time required for all of the zinc to react.

4. Repeat steps 1 to 3 using 1.5 mol/L, 1.0 mol/L, and 0.5 mol/L HCl(aq).

5. Neutralize the acid with a weak base and then pour it down the sink with the water running.

Evidence
Gas bubbles formed immediately on the surface of the zinc strip when it was placed into the hydrochloric acid solution. The bubbles appeared to form more rapidly when the concentration of the acid was higher.

The Reaction of HCl(aq) with Zn(s)

Concentration of HCl(aq) (mol/L)	Time for reaction (s)
2.0	70
1.5	80
1.0	144
0.5	258

Analysis

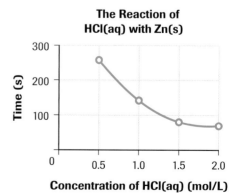

The Reaction of HCl(aq) with Zn(s)

According to the evidence obtained, increasing the concentration of hydrochloric acid decreases the time required for the complete reaction of a fixed quantity of zinc.

Evaluation

The design, reacting zinc with excess hydrochloric acid, is judged to be adequate because this experiment produced the type of evidence needed to answer the problem with a high degree of certainty. There are no obvious flaws in this design. In my judgment, the design is efficient and safe, and all necessary variables are controlled.

The materials appear to be adequate because the quality of the evidence was sufficient to give a clear answer to the problem. Using a pipette would provide better precision than a graduated cylinder but this would not change the overall result.

The procedure is also judged adequate because it produced sufficient evidence. Possible improvements include extending the range of concentrations and performing more trials for each concentration. The reaction could be done in a container that allowed better mixing.

The technological skills are judged adequate because no specialized skills were involved. Timing the start of the reaction may have some uncertainty but the lack of multiple trials makes this difficult to judge.

Based upon my evaluation of the experiment, I am very certain about the experimental results. Sources of uncertainty in this investigation include: measurement errors for volume and time, the purity and surface area of the zinc metal strip, the concentration of the acid, and a little uncertainty in estimating when the last bit of zinc had reacted. An estimate of the total effect of all experimental uncertainties is about 5%.

The prediction is judged to be verified because the qualitative observations and the graph clearly indicate that the reaction time decreases as the concentration increases. There is little deviation from a smooth curve in the graphed results.

The collision–reaction theory is judged to be acceptable in this experiment because the prediction was clearly verified. I am quite confident in this judgment because other groups in the class obtained similar results.

The purpose of this investigation—to test one idea of the collision–reaction theory—was accomplished but only for one reaction. Replication with many other reactions would need to be investigated to have a more valid test.

B.4 The Nature of Scientific Research

Citizens in a democratic society are often required to read and interpret media reports of scientific research. Health and environment research reports are, for example, commonly portrayed in the media. Sometimes the research reports appear to contradict each other and sometimes the reports promote more uncertainty than certainty. Understanding the terminology and concepts for describing a research study is increasingly important for responsible citizenry. Listed below are some of the terms and concepts that will help you both answer questions in this textbook and understand and critique media reports of research.

Types of Studies

correlational study—the connection or degree of agreement (e.g., –0.3, +1.0) is sought between two variables, often without controlling for other variables; correlational studies often lead to cause-and-effect studies

cause-and-effect study—one variable is manipulated and all other variables, other than the responding variable, are controlled

control experiment—see cause-and-effect study

clinical trial—a controlled study involving people; usually a final-stage, double-blind study

Design Factors

term of study—the duration of the experiment e.g., observations over 5 s, 30 min, 3 mon or 15 a; long-term studies are most often preferred

sample size—the number of entities or people in a study; generally large sample sizes are preferred

random sample—one chosen randomly from the population of entities (to reduce bias)

replication—repetition of a study, generally, by an independent research group

placebo—in medicinal experiments, an inactive item (e.g., sugar pill) or treatment given to the control group

placebo effect—a beneficial effect arising from a patient's expectations; present in both the control group and the experimental group

single blind—the subject (e.g., patient) does not know whether she/he has received the treatment or a placebo, but the experimenter knows

double blind—neither the subject (e.g., patient) nor the directly involved experimenter knows whether the subject has received the treatment or a placebo

control—a standard or comparison value, or procedure (e.g., leaving one of several identical samples unaltered for comparison), or a placebo *control group*—a comparison group that does not receive the experimental treatment

experimental group—a group that receives the experimental treatment

Nature of Evidence and Results

anecdotal—based upon personal experience or hearsay

reliable—reproducible or consistent time after time

valid—judged to be supported by adequate designs, materials, procedures, and skills

accurate—judged to be true or agreeing with the accepted value

precise—closely related or very similar; related to reproducibility of results

statistical bias—a sampling or testing error caused by systematically favouring some outcomes over others

random result—a result that could be expected on the basis of probability (e.g., 50% heads and tails when flipping a large number of coins)

coincidence—a result that is accidental and irrelevant to the study

significant difference—a difference that is greater than could be randomly expected when an experimental group and a control group are compared

certainty—the degree to which something is accepted by an individual or community (e.g., the evidence may have a high or low degree of certainty); measured by, for example, counting significant digits

Reporting Research

refereed journal—an academic journal for which research papers are sent to subject experts to determine whether the report is of sufficient quality to publish; also called *peer-reviewed journal*

abstract—a short summary describing the research processes and results

Science–Technology–Society (STS) Issues

risk–benefit analysis—a process of gathering and analyzing evidence that leads to decision making (and to an evaluation of the process itself)

stewardship—actively supervising and managing an entity or event (e.g, the environment)

perspective—a point of view or way of analyzing an object or event

multiperspective—based upon positive and negative evidence and arguments from many perspectives (e.g., scientific, technological, economic, environmental, political, legal, ethical, social, and emotional)

ways of knowing—methods used to obtain knowledge or information; examples include traditional (Aboriginal), empirical, theoretical, referenced, and memorized

Scientific Attitudes

scientific attitude—a disposition or demonstration of feelings or thoughts (e.g., honesty, objectivity, willingness to change, respect for evidence, critical mindedness, suspended judgment, open-mindedness, and questioning predisposition)

tolerance of uncertainty—the degree to which people and institutions tolerate uncertainty (without claiming absolute certainty), although they strive for greater and greater certainty

▶ COMMUNICATION example 1

Create an experimental design to test a new drug to promote weight loss.

Solution

Randomly selected control and experimental groups of 1000 volunteers are studied over two years. Both receive pills: The experimental group receives the drug; the control group unknowingly receives a placebo. Technicians, who do not know which group each person is in, record the weight of the subjects every month. Experimenters, who never meet or see the volunteers, analyze the evidence gathered.

▶ COMMUNICATION example 2

Act as a referee (peer-reviewer) to critique the following experimental design including, if necessary, suggestions for improvement.

Ten volunteers are provided with their horoscope for the test day. The volunteers orally respond in a group to the question: Does this horoscope describe your personality and life situation?

Solution

This experimental design is very inadequate because:

- ten volunteers is a very small sample size; the number needs to be at least 100 or more, randomly selected from thousands
- to control the horoscope variable, all subjects should be provided with the same horoscope
- to control subject interaction and influence, subjects should be isolated from one another
- to provide for accurate reporting of the subjects' responses, investigators should make audio or audiovisual recordings

Appendix C TECHNOLOGICAL PROBLEM SOLVING

The goal of technological problem solving is to solve practical problems by developing or revising a product or a process. The product or the process must fulfill its function, but it is not essential to understand why or how it works. Products are evaluated based on criteria such as simplicity, reliability, and cost. Technological processes are also evaluated by their efficiency. Technological products and processes may have both intended and unintended consequences. Therefore, it is important that various perspectives, such as ecological, economic, and political, are used in any assessment. For example, chlorofluorocarbons may be simple and inexpensive to make, and they may be useful for a particular function, but their effect on the ozone layer in the upper atmosphere must also be considered. Processes such as the chlorine bleaching of wood pulp may be efficient, but they may adversely affect an ecosystem. We often look to technological fixes as solutions for problems. Ecological, economic, political, legal, ethical, and/or social efforts by individuals or groups can often lead to more sustaining solutions than quick technological fixes.

Chemistry has always been closely associated with technology. Part of technology is the laboratory equipment, processes, and procedures used in both chemical and technological research and development. In modern chemistry, simple equipment and processes, such as beakers and filtration, are still used but chemistry also depends on sophisticated technology, such as computers, to store and manipulate the evidence collected.

C.1 Technological Problem-Solving Model

Technological problem solving is similar in some ways to scientific problem solving but its purpose differs. A characteristic of technological problem solving is a systematic, trial-and-error manipulation of variables (**Figure 3**). Variables are predicted and tested and the results are evaluated. When the cycle is repeated many times, the most effective set of conditions can be determined. Compare this model with the scientific problem-solving model in **Figure 1**, on page 790.

Technological problem-solving contexts include industrial, commercial, and consumer. The general process for technological problem solving is similar in all of these contexts. Technological problem solving is very common to us in our everyday lives. Learning more about this systematic trial-and-error approach can help us on an everyday basis.

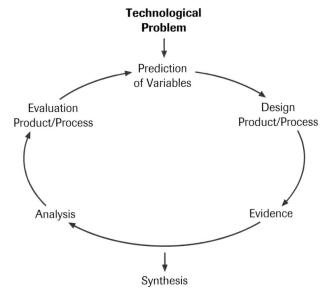

Figure 3
A technological problem-solving model

C.2 Investigation Reports

Investigation reports for technological problem-solving investigations can use the same headings as those for scientific problem-solving investigations (Appendix B.2, page 790). Some key differences between these two types of reports are

Purpose
The purpose will be to solve a specific, practical problem by developing or revising a product or process.

Evaluation
Evaluation criteria can be many and varied. In some cases, it will be sufficient to demonstrate that the product or process works for the materials used. You may also be asked to judge the simplicity, reliability, and efficiency of the product or process, recognize its value and limitations, and evaluate it from a variety of perspectives (Appendix D.2, page 806).

C.3 Laboratory Equipment

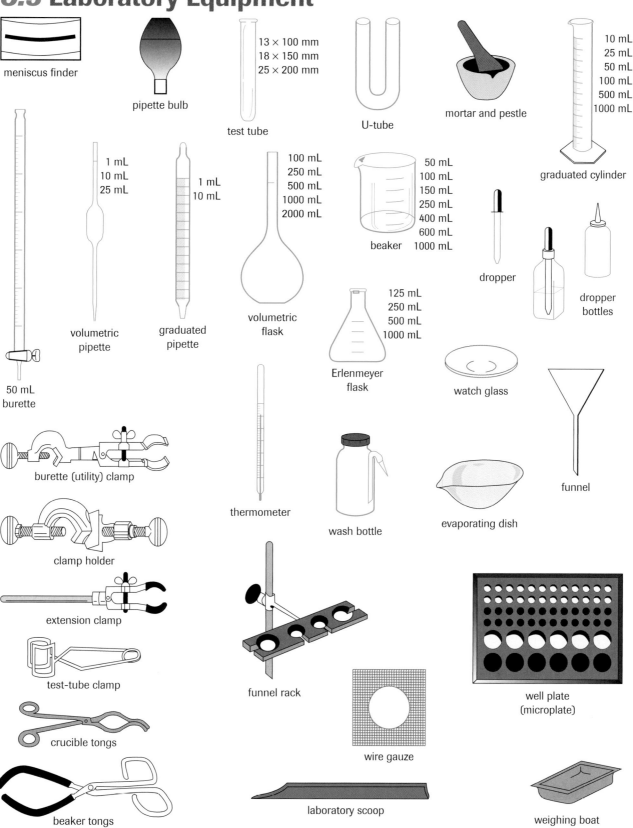

Figure 4
Common lab equipment

 Glassware is breakable and should always be handled with care.

Using a Laboratory Burner

The procedure outlined below should be practised and memorized. Note the safety caution. You are responsible for your safety and the safety of others near you.

1. Turn the air and gas adjustments to the off position (**Figure 5**).
2. Connect the burner hose to the gas outlet on the bench.
3. Turn the bench gas valve to the fully on position.
4. If you suspect that there may be any gas leaks, replace the burner. (Give the leaky burner to your teacher.)
5. While holding a lit match above and to one side of the barrel, open the burner gas valve until a small yellow flame results (**Figure 6**). If a striker is used instead of matches, generate sparks over the top of the barrel (**Figure 7**).
6. Adjust the air flow and obtain a pale blue flame with a dual cone (**Figure 8**). In most common types of laboratory burners, rotating the barrel adjusts the air intake. Rotate the barrel slowly. If too much air is added, the flame may go out. If this happens, immediately turn the gas flow off and relight the burner following the procedure outlined above. If your burners have a different kind of air adjustment, revise the procedure accordingly.
7. Adjust the gas valve on the burner to increase or decrease the height of the blue flame. The hottest part of the flame is the tip of the inner blue cone. Usually a 5 to 10 cm flame, which just about touches the object heated, is used.
8. Laboratory burners, when lit, should not be left unattended. If the burner is on but not being used, adjust the air and gas intakes to obtain a small yellow flame. This flame is more visible and, therefore, less likely to cause problems.

 When lighting or using a laboratory burner, never position your head or fingers directly above the barrel. Tie back long hair and sleeves.

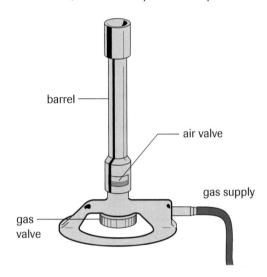

Figure 5
The parts of a common laboratory burner

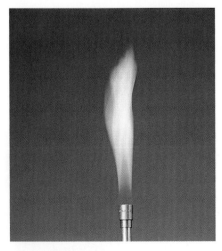

Figure 6
A yellow flame is a relatively cool flame and is easier to obtain than a blue flame when lighting a burner. A yellow flame is not used for heating objects because it contains a lot of black soot.

Figure 7
To generate a spark with a striker, pull up and across on the side of the handle containing the flint.

Figure 8
A pale, almost invisible flame is much hotter than a yellow flame. The hottest point is at the tip of the inner blue cone.

Using a Laboratory Balance

There are two types of balances: electronic (**Figure 9**) and mechanical (**Figure 10**). All balances must be handled carefully and kept clean. Always place chemicals into a container such as a beaker or plastic boat to avoid contamination and corrosion of the balance pan. To avoid error due to convection currents in the air, allow hot or cold samples to return to room temperature before placing them on the balance. Always record masses showing the correct precision. On a centigram balance, mass is measured to the nearest hundredth of a gram (0.01 g). When it is necessary to move a balance, hold the instrument by the base and steady the beam. Never lift a balance by the beams or pans.

Figure 9
An electronic balance

To avoid contaminating a whole bottle of reagent, do not scoop diectly from the original container of a chemical. Pour a quantity of the chemical into a clean, dry beaker or bottle, from which samples can be taken. Another acceptable technique for dispensing a small quantity of chemical is to rotate or tap the chemical bottle.

Using an Electronic Balance

Electronic balances are sensitive to small movements and changes in level; do not lean on the counter when using the balance.

1. Place a container or weighing paper on the balance.
2. Reset (tare) the balance so the mass of the container registers as zero.
3. Add chemical until the desired mass of chemical is displayed. Air currents or the high sensitivity of the balance may cause the last digit to vary.
4. Remove the container and sample.

There is a video demonstration of this technique on the Nelson Web site.

www.science.nelson.com GO

Using a Mechanical Balance

Different kinds of mechanical balances are shown in **Figures 10(a)** and **(b)**. Some general procedures apply to most of them.

1. Clean and zero the balance. (Turn the zero adjustment screw so that the beam is balanced when the instrument reads 0 g and no load is on the pan.)
2. Place the container on the pan.
3. Move the largest beam mass one notch at a time until the beam drops, and then move the mass back one notch.
4. Repeat this process with the next smaller mass and continue until all masses have been moved and the beam is balanced. If you are using a dial type balance, the final step will be to turn the dial until the beam balances, as shown in **Figure 10(c)**.
5. Record the mass of the container.
6. Set the masses on the beams to correspond to the total mass of the container plus the desired sample.
7. Add the chemical until the beam is once again balanced.
8. Remove the sample from the pan and return balance to the zero position.

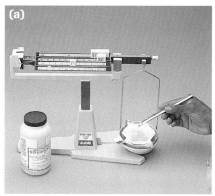
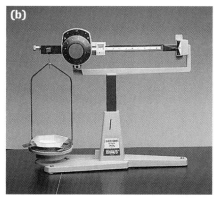
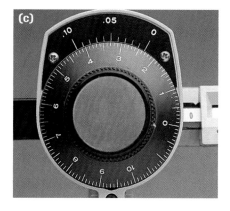

Figure 10
(a) On this type of mechanical balance, the sample is balanced by moving masses on several beams.
(b) Another type of mechanical balance has beams for the larger masses and a dial for the final adjustment.
(c) The dial reading on this balance with a vernier scale is 2.34 g. To read the hundredth of a gram, look below the zero on the vernier, and then look for the line on the vernier that lines up best with a line on the dial.

Using a Multimeter

A multimeter (**Figure 11**) is a device that measures a variety of electrical quantities, such as resistance, voltage, and current.

Figure 11
(a) An analog meter has a needle that moves in front of a labelled scale.
(b) A digital meter gives a direct reading with appropriate units.

Conductivity Measurements of Solutions

1. Set the dial on the meter to one of the higher values on the ohm (Ω) scale; for example, R $\times$ 100 or R $\times$ 1 K.
2. Touch the two metal probes together to check the battery. If the needle does not deflect significantly (more than one-half scale), have your teacher adjust the meter or replace the battery.
3. Test a sample of pure water as a control and note the movement of the needle.
4. Test your aqueous sample and record the deflection of the needle according to your teacher's instructions.
5. Rinse the probes with pure water before testing another sample.
6. Shut off the meter by using either the on/off switch or by turning the dial to any setting other than "Resistance."

Voltage Measurements of Cells and Batteries

1. Set the dial to the appropriate value on the direct current volts (DCV) scale; for example, 3 V.
2. The black lead (labelled negative or COM) is normally connected to the anode, and the red lead (positive) is connected to the cathode of a voltaic cell.
3. Make a firm contact between each metal probe and an electrode of the cell. (Press firmly with the pointed probe or use leads with an alligator clip.)
4. On analog meters (those with a needle), read the scale corresponding to the meter value you set in step 1.
5. If the needle attempts to move to the left off the scale or a digital meter registers a negative number, then switch the connections to the cell.

Using a Pipette

A pipette is a specially designed glass tube used to measure precise volumes of liquids. There are two types of pipettes and a variety of sizes for each type. A *volumetric pipette* (**Figure 12**) transfers a fixed volume, such as 10.00 mL or 25.00 mL, accurate to within 0.04 mL. A *graduated pipette* (**Figure 13**) measures a range of volumes, just as a graduated cylinder does. A 10 mL graduated pipette delivers volumes accurate to within 0.1 mL. There is a video demonstration of this technique on the Nelson Web site.

www.science.nelson.com

Figure 12
A volumetric pipette delivers the volume printed on the label if the temperature is near room temperature.

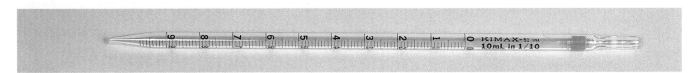

Figure 13
To use a graduated pipette, you must be able to start and stop the flow of the liquid.

1. Rinse the pipette with small volumes of distilled water using a wash bottle, and then with the sample solution.
   A clean pipette has no visible residue or liquid drops clinging to the inside wall. Rinsing with aqueous ammonia and scrubbing with a pipe cleaner might be necessary to clean the pipette.
2. Hold the pipette with your thumb and fingers near the top. Leave your index finger free.
3. Place the pipette in the sample solution, resting the tip on the bottom of the container if possible. Be careful that the tip does not hit the sides of the container.
4. Squeeze the bulb into the palm of your hand and place the bulb firmly and squarely on the end of the pipette (**Figure 14**) with your thumb across the top of the bulb.
5. Release your grip on the bulb until the liquid has risen above the calibration line.
   This may require bringing the level up in stages: remove the bulb, put your finger on the pipette, squeeze the air out of the bulb, re-place the bulb, and continue the procedure.
6. Remove the bulb, placing your index finger over the top.
   If you are using a dispensing bulb (**Figure 15**), it remains attached to the pipette.
7. Wipe all solution from the outside of the pipette using a paper towel.
8. While touching the tip of the pipette to the inside of a waste beaker, gently roll your index finger (or squeeze the valve of the dispensing bulb) to allow the liquid level to drop until the bottom of the meniscus reaches the calibration line (**Figure 16**).
   To avoid parallax errors, set the meniscus at eye level. Stop the flow when the bottom of the meniscus is on the calibration line. Use the bulb to raise the level of the liquid again if necessary.
9. While holding the pipette vertically, touch the pipette tip to the inside wall of a clean receiving container. Remove your finger or adjust the valve and allow the liquid to drain freely until the solution stops flowing.
10. Finish by touching the pipette tip to the inside of the container held at about a 45° angle (**Figure 17**). Do not shake the pipette. The delivery pipette is calibrated to leave a small volume in the tip.

 Never use your mouth to draw a liquid up a pipette. Always use a pipette bulb.

Figure 14
Release the bulb slowly. Pressing down with your thumb placed across the top of the bulb maintains a good seal. Setting the pipette tip on the bottom slows the rise or fall of the liquid.

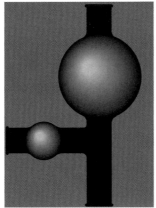

Figure 15
A dispensing pipette bulb uses a small valve in the side stem to control the flow of liquid in a pipette.

Figure 16
To allow the liquid to drop slowly to the calibration line, it is necessary for your finger and the pipette top to be dry. Also keep the tip on the bottom to slow down the flow.

Figure 17
A vertical volumetric pipette is drained by gravity and then the tip is placed against the inside wall of the container. A small volume is expected to remain in the tip.

C.4 Laboratory Processes

The processes or experimental procedures listed below are part of common designs used in scientific or technological laboratories.

Crystallization

Crystallization is used to separate a solid from a solution by evaporating the solvent or lowering the temperature. Evaporating the solvent is useful for quantitative analysis of a binary solution; lowering the temperature is commonly used to purify and separate a solid whose solubility is temperature-sensitive. Chemicals that have a low boiling point or decompose on heating cannot be separated by crystallization using a heat source.

1. Measure the mass of a clean beaker or evaporating dish.
2. Place an accurate volume of the solution in the container.
3. Set the container aside to evaporate the solution slowly, or warm the container gently on a hot plate or with a laboratory burner.
4. When the contents appear dry, measure the mass of the container and solid (**Figure 18**).

Figure 18
When the substance has crystallized, it may appear dry but small quantities of water may still be present. To be certain the solid is dry, it must be heated until the mass becomes constant.

5. Heat the solid with a hot plate or burner, cool it, and measure the mass again.
6. Repeat step 5 until the final mass remains constant. (Constant mass indicates that all of the solvent has evaporated.)

Filtration

In filtration, solid is separated from a mixture using a porous filter paper. The more porous papers are called qualitative filter papers. Quantitative filter papers allow only invisibly small particles through the pores of the paper. There is a video demonstration of this technique on the Nelson Web site.

www.science.nelson.com

1. Set up a filtration apparatus (**Figure 19**): stand, funnel holder, filter funnel, waste beaker, wash bottle, and a stirring rod with a flat end for scraping.

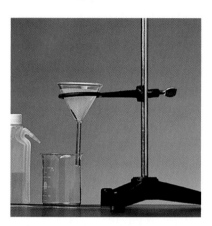

Figure 19
The tip of the funnel should touch the inside wall of the collecting beaker.

2. Fold the filter paper along its diameter, and then fold it again to form a cone. A better seal of the filter paper on the funnel is obtained if a small piece of the outside corner of the filter paper is torn off (**Figure 20**).

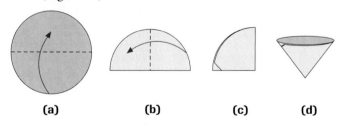

(a) (b) (c) (d)

Figure 20
To prepare a filter paper, fold it in half twice, and then remove the outside corner as shown.

3. Measure and record the mass of the filter paper after removing the corner.
4. While holding the open filter paper in the funnel, wet the entire paper and seal the top edge firmly against the funnel.

5. With the stirring rod touching the spout of the beaker, decant most of the solution into the funnel (**Figure 21**). Transferring the solid too soon clogs the pores of the filter paper. Keep the level of liquid about two-thirds up the height of the filter paper. The stirring rod should be rinsed each time it is removed.

Figure 21
Pouring along the stirring rod prevents drops of liquid from going down the outside of the beaker when you stop pouring.

6. When most of the solution has been filtered, pour the remaining solid and solution into the funnel. Use the wash bottle and the flat end of the stirring rod to clean any remaining solid from the beaker.
7. Rinse the stirring rod and the beaker.
8. Wash the solid two or three times to ensure that no solution is left in the filter paper. Direct a gentle stream of water around the top of the filter paper.
9. When the filtrate has stopped dripping from the funnel, remove the filter paper. Press your thumb against the thick (three-fold) side of the filter paper and slide the paper up the inside of the funnel.
10. Transfer the filter paper from the funnel onto a labelled watch glass and unfold the paper to let the precipitate dry.
11. Determine the mass of the filter paper and dry precipitate.

Preparation of Standard Solutions

Laboratory procedures often call for the use of a solution of specific, accurate concentration. The apparatus used to prepare such a solution is a volumetric flask. A meniscus finder is useful in setting the bottom of the meniscus on the calibration line (**Figure 22**).

Preparing a Standard Solution from a Solid Reagent

There is a video demonstration of this technique on the Nelson Web site.

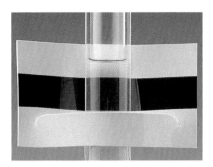

Figure 22
Raise the meniscus finder along the back of the neck of the volumetric flask until the meniscus is outlined as a sharp, black line against a white background.

1. (Prelab) Calculate the required mass of solute from the volume and concentration of the solution.
2. Measure the required mass of solute in a clean, dry beaker or weighing boat. (Refer to "Using a Laboratory Balance" on page 799.)
3. Pour less than one-half of the final volume of pure water into a beaker. Transfer the solute to the water. Stir to dissolve.
4. Transfer the solution and all water used to rinse the equipment into a clean volumetric flask. (The beaker and any other equipment should be rinsed two or three times with pure water.)
5. Add pure water, using a medicine dropper for the final few millilitres while using a meniscus finder to set the bottom of the meniscus on the calibration line.
6. Stopper the flask and mix the solution by slowly inverting the flask several times.

Preparing a Standard Solution by Dilution

There is a video demonstration of this technique on the Nelson Web site.

www.science.nelson.com GO

1. (Prelab) Calculate the volume of concentrated reagent required.
2. Add approximately one-half of the final volume of pure water to the volumetric flask.
3. Measure the required volume of stock solution using a pipette. (Refer to "Using a Pipette" on page 800).
4. Transfer the stock solution slowly into the volumetric flask while mixing.
5. Add pure water, and then use a medicine dropper and a meniscus finder to set the bottom of the meniscus on the calibration line (**Figure 22**).

6. Stopper and mix the solution by slowly inverting the flask several times.

 If water is added directly to some solids or concentrated liquids, there may be boiling or splattering. Always add a solid solute or concentrated liquids to water.

Titration

Titration is used in the volumetric analysis of an unknown concentration of a solution. Titration involves adding a solution (the titrant) from a burette to another solution (the sample) in an Erlenmeyer flask until a recognizable endpoint, such as a colour change, occurs. (See the video on the Nelson Web site.)

www.science.nelson.com GO

1. Rinse the burette with small volumes of pure water using a wash bottle. Using a burette funnel, rinse with small volumes of the titrant (**Figure 23**). (If liquid droplets remain on the sides of the burette after rinsing, scrub the burette with a burette brush. If the tip of the burette is chipped or broken, replace the tip or the whole burette.)

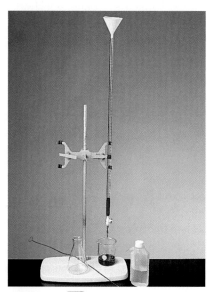

Figure 23
A burette should be rinsed with water and then the titrant before use.

2. Using a small burette funnel, pour the titrant solution into the burette until the level is near the top. Open the stopcock for maximum flow to clear any air bubbles from the tip and to bring the liquid level down to the scale.

3. Record the initial burette reading to the nearest 0.1 mL. Avoid parallax errors by reading volumes at eye level with the aid of a meniscus finder.

4. Pipette a known volume of the solution of unknown concentration into a clean Erlenmeyer flask. Place a white piece of paper beneath the Erlenmeyer flask to make it easier to detect colour changes.

5. Add an indicator if one is required. Add the smallest quantity necessary (usually 1 to 2 drops) to produce a noticeable colour change in your sample.

6. Add the solution from the burette quickly at first, and then slowly, drop-by-drop, near the endpoint (**Figure 24**). Stop as soon as a drop of the titrant produces a permanent colour change in the sample solution. A permanent colour change is considered to be a noticeable change that lasts for 10 s after swirling.

7. Record the final burette reading to the nearest 0.1 mL.

8. The final burette reading for one trial becomes the initial burette reading for the next trial. Three trials with results within 0.2 mL are normally required for a reliable analysis of an unknown solution.

9. Drain and rinse the burette with pure water. Store the burette upside down with the stopcock open.

Figure 24
Near the endpoint, continuous gentle swirling of the solution is particularly important.

Diagnostic Tests

The tests described in **Table 1** are commonly used to detect the presence of a specific substance. All diagnostic tests include a brief procedure, some expected evidence, and an interpretation of the evidence obtained. This is conveniently communicated using the format, "If [procedure] and [evidence], then [analysis]." Diagnostic tests can be constructed using any characteristic empirical property of a substance. For example, diagnostic tests for acids, bases, and neutral substances can be specified in terms of the pH of the solutions. For specific chemical reactions, properties of the products that the reactants do not have, such as the insolubility of a precipitate, the production of a gas, or the colour of ions in aqueous solutions, can be used to construct diagnostic tests.

If possible, you should use a control to illustrate that the test does not give the same results with other substances. For example, in the test for oxygen, inserting a glowing splint into a test tube that contains only air is used to compare the effect of air on the splint with a test tube in which you expect oxygen has been collected.

Communication of Diagnostic Tests

The procedure, evidence, and analysis information for a diagnostic test can be communicated in three different formats:
- "If ... and ... then ..." statement
- table
- flowchart

Table 1 Some Standard Diagnostic Tests

Substance tested	Diagnostic test
water	If cobalt(II) chloride paper is exposed to a liquid or vapour, and the paper turns from blue to pink, then water is likely present.
oxygen	If a glowing splint is inserted into the test tube, and the splint glows brighter or re-lights, then oxygen gas is likely present.
hydrogen	If a flame is inserted into the test tube, and a squeal or pop is heard, then hydrogen is likely present.
carbon dioxide	If the unknown gas is bubbled into a limewater solution, and the limewater turns cloudy, then carbon dioxide is likely present.
halogens	If a few millilitres of a hydrocarbon solvent is added, with shaking, to a solution in a test tube, and the colour of the solvent appears to be • light yellow-green, then chlorine is likely present • orange, then bromine is likely present • purple, then iodine is likely present
acid	If strips of blue and red litmus paper are dipped into the solution, and the blue litmus turns red, then an acid is present.
base	If strips of blue and red litmus paper are dipped into the solution, and the red litmus turns blue, then a base is present.
neutral solution	If strips of blue and red litmus paper are dipped into the solution, and neither litmus changes colour, then only neutral substances are likely present.
neutral ionic solution	If a neutral solution is tested for conductivity with a multimeter, and the solution conducts a current, then a neutral ionic substance is likely present.
neutral molecular solution	If a neutral solution is tested for conductivity with a multimeter, and the solution does not conduct a current, then a neutral molecular substance is likely present.
	There are thousands of diagnostic tests. You can create some of these using data from the periodic table (on the inside front cover of this book), and from the data tables in Appendix I and on the inside back cover.

Appendix D STS PROBLEM SOLVING

Science is a human endeavour, technology has a social purpose, and both have always been part of society. Science and technology together affect society in a myriad of ways. Society also affects science and technology by placing controls on them and expecting solutions to societal problems.

When controversial issues related to science and technology arise in our society, there is often heated debate among various special-interest groups. Often, little progress is made because different parties in the debate generally recognize only a single perspective on the issue. Many people now realize that an informed multi-perspective view is more defensible. The following model represents one possible procedure for making an informed decision on a social issue related to science and technology.

D.1 STS Decision-Making Model

1. *Identify an STS (science–technology–society) issue.* Newspapers, magazines, and news broadcasts are sources of current STS issues. However, some issues like acid rain have been current for some time and rarely appear in the news. When identifying an issue for debate, it is convenient to state the issue as a resolution (e.g., "Be it resolved that the use of fossil fuels for heating homes should be eliminated.").

2. *Design a plan to address the STS issue.* Possible designs include individual research, a debate, a town-hall meeting (or role-playing), or participation in an actual hearing or on a committee.

3. *Identify and obtain relevant information on as many perspectives as possible.* An STS issue will always have scientific and technological perspectives. Common perspectives are shown in **Table 2**.

Table 2 Perspectives on STS Issues

scientific	ethical
technological	social
ecological	militaristic
economic	esthetic
political	mystical
legal	emotional

Another perspective is the world view or perspective of Aboriginal peoples. In general, Aboriginal peoples believe that we are an integral part of our environment. Their holistic view includes not only a physical interdependence but also a spiritual one.

Information from different perspectives can be obtained from references and through group discussions. There are many sides to every issue. There can be positive and negative viewpoints about the resolution from every perspective.

4. *Generate a number of alternative solutions to the STS problem.* Some obvious solutions will arise from the resolution. Other creative solutions often arise from a brainstorming session within a group.

5. *Evaluate each solution and decide which is best.* One method is to rank the value of a particular solution from each perspective. For example, a solution might have little economic advantage and be ranked as 1 on a scale of 1 to 5; the solution might have a significant ecological benefit and be ranked as 5, for a total of 6. A different solution might be judged as 3 from the economic perspective and 1 from the ecological perspective, for a total of 4. The solution with the highest total is likely to be chosen. Although simplistic, this method facilitates evaluation and illustrates the tradeoffs that occur in any real issue.

D.2 Types of Reports

There are many ways to communicate the results of an investigation of an STS issue (**Table 3**). All methods will require some research about the issue and perspectives on the issue (including positive and negative viewpoints). Some methods can also include alternative solutions and the evaluation of these solutions. Working within a group and brainstorming is a useful process. No matter how the issue will be presented and reported, you need to be well prepared.

Table 3 STS Investigations and Reports

Plan	Reporting suggestions
individual or group research	• written report or poster • multimedia presentation
debate	• research notes • videotape of the debate
role-playing (e.g., town hall meeting)	• research notes • videotape of the meeting
survey	• survey form with tables and graphs
newspaper article	• published article

Appendix E SAFETY KNOWLEDGE AND SKILLS

E.1 Laboratory Safety

Safety is always important in a laboratory or in other settings that feature chemicals or technological devices. It is your responsibility to be aware of possible hazards, to know the rules—including ones specific to your classroom—and to behave appropriately. Always alert the teacher in case of any accident.

Alberta Education has an extensive document, "Safety in the Science Classroom," that deals with hazards and safety.

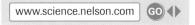

General Safety Rules

Safety in the laboratory is an attitude and a habit more than it is a set of rules. It is easier to prevent accidents than to deal with the consequences of an accident. Most of the following rules are common sense:

- Read all directions before doing any laboratory work, and follow all verbal instructions.
- Know the potential hazards, including the contents and location of MSDS, and the location of all safety equipment.
- Wear eye protection and lab aprons/coats.
- Behave responsibly. Avoid sudden or rapid motion that may interfere with someone carrying or working with chemicals.
- Wear closed shoes (not sandals or bare feet) when working in the laboratory.
- Place your books, bags, and purses away from the work area.
- Do not chew gum, eat, drink, or taste anything in the laboratory.
- Ask for assistance when you are not sure how to do a procedural step.
- Inform your teacher immediately if any problem or accident occurs.
- Never attempt any unauthorized or unsupervised experiments.
- Never handle any chemical with your hands. Use a laboratory scoop or spoon for solids.
- Never use the contents of a bottle that has no label or an illegible label. Always double check the label to ensure that you are using the chemical needed. Always pour from the side opposite the label.
- When leaving chemicals in containers, ensure that they are labelled.
- Do not take any more chemical than needed and never return excess chemicals to their original container.
- Hold larger bottles with both hands; one hand on the base.
- Do not inhale any vapours directly from any container. If smell is to be tested, fan the vapours toward your nose, keeping the container away from underneath your nose.
- Always use a pipette bulb, and never pipette by mouth.
- When heating a test tube over a burner, use a test-tube holder with the test tube at an angle, facing away from you and others. Gently move the test tube backwards and forwards through the flame.
- Clean up all spills, even spills of water, immediately. Clean up your work area at the end of an experiment.
- Dispose of chemicals appropriately as directed by your teacher.
- Always wash your hands with soap and water before you leave the laboratory.
- Do not forget safety procedures when you leave the laboratory. These same rules also apply at home or at work.

Glass Safety and Cuts

- Never use glassware that is cracked or chipped. Give such glassware to your teacher or dispose of it as directed. Do not put the item back into circulation.
- Never pick up broken glassware with your fingers. Use a broom and dustpan.
- Do not put broken glassware into garbage containers. Dispose of glass fragments in special containers marked "broken glass."
- If you cut yourself, inform your teacher immediately. Imbedded glass or continued bleeding requires medical attention.

Burns

- In a laboratory where burners or hot plates are being used, never pick up a glass object without first checking the temperature by lightly and quickly touching the item. Glass items that have been heated stay hot for a long time but do not appear to be hot. Metal items such as ring stands and hot plates can also cause burns; take care when touching them.
- Before using a laboratory burner, make sure that long hair is always tied back. Do not wear loose clothing. (Wide long sleeves should be tied back or rolled up.)

- Do not use a laboratory burner near wooden shelves, flammable liquids, or any other item that is combustible. Know how to use the type of burner in your laboratory. (See Using a Laboratory Burner, page 798)
- Never look down the barrel of a laboratory burner.
- Always pick up a burner by the base, never by the barrel.
- Never leave a lighted bunsen burner unattended.
- If you burn yourself, *immediately* run cold water over the burned area and inform your teacher.

Eye Safety

- Always wear approved eye protection in a laboratory, no matter how simple or safe the task appears to be. Keep the safety glasses over your eyes, not on top of your head. For certain experiments, full face protection may be necessary.
- Never look directly into the opening of flasks or test tubes.
- If, in spite of all precautions, you get a solution in your eye, quickly use the eyewash station or nearest running water. Continue to rinse the eye with water for at least 15 minutes. This is a very long time—have someone time you. Unless you have a plumbed eyewash system, you will also need assistance in refilling the eyewash container. Have another student inform your teacher of the accident. The injured eye should be examined by a doctor.
- It is recommended that you do not wear contact lenses in the laboratory. If you wear contact lenses in the laboratory, there is a danger that a chemical might get behind the lens where it cannot be rinsed out with water. If you must wear contact lenses in the chemistry laboratory, be extra careful. Tell your teacher if you are wearing contact lenses in the laboratory. Whether or not you wear contact lenses, do not touch your eyes without first washing your hands.
- If a piece of glass or other foreign object enters an eye, immediate medical attention is required.

Fire Safety

Immediately inform your teacher of any fires. Very small fires in a container may be extinguished by covering the container with a wet paper towel or a ceramic square, which would cut off the supply of air. If anyone's clothes or hair catch fire, the fire can be extinguished by smothering the flames with a blanket or a piece of clothing. Larger fires require a fire extinguisher. (Know how to use the fire extinguisher that is in your laboratory.) If the fire is too large to approach safely with an extinguisher, vacate the location and sound the fire alarm. (School staff will inform the fire department.)

If you use a fire extinguisher, direct the extinguisher at the base of the fire and use a sweeping motion, moving the extinguisher nozzle back and forth across the front of the fire's base. You must use the correct extinguisher for the kind of fire you are trying to control. Each extinguisher is marked with the class of fire for which it is effective. The fire classes are outlined below. Most fire extinguishers in schools are of the ABC type.

- Class A fires involve ordinary combustible materials that leave coals or ashes, such as wood, paper, or cloth. Use water or dry chemical extinguishers on Class A fires. (Carbon dioxide extinguishers are not satisfactory as carbon dioxide dissipates quickly and the hot coals can reignite.)
- Class B fires involve flammable liquids such as gasoline or solvents. Carbon dioxide or dry chemical extinguishers are effective on Class B fires. (Water is not effective on a Class B fire since the water splashes the burning liquid and spreads the fire.)
- Class C fires involve live electrical equipment, such as appliances, photocopiers, computers, or laboratory electrical apparatus. Carbon dioxide or dry chemical extinguishers are recommended for Class C fires. Carbon dioxide extinguishers are much cleaner than the dry chemical variety. (Using water on live electrical devices can result in severe electrical shock.)
- Class D fires involve burning metals, such as sodium, potassium, magnesium, or aluminium. Sand or salt are usually used to put out Class D fires. (Using water on a metal fire can cause a violent reaction.)
- Class E fires involve a radioactive substance. These involve special considerations at each site.

Electrical Safety

Water or wet hands should never be used near electrical equipment. When unplugging equipment, remove the plug gently from the socket (do not pull on the cord). Do not use any devices with electric motors when flammable liquids are present unless the area is well ventilated.

E.2 Safety Symbols and Information

Educational, Commercial, and Industrial Information

 Class A: Compressed gas

 Class B: Flammable and combustible material

 Class C: Oxidizing material

Class D: Poisonous and Infectious Materials

 Division 1 Materials causing immediate and serious toxic effect

 Division 2 Materials causing other toxic effects

 Division 3 Biohazardous infectious material

 Class E: Corrosive material

 Class F: Dangerously reactive material

Figure 25
WHMIS symbols

The Workplace Hazardous Materials Information System (WHMIS) provides workers and students with complete and accurate information regarding hazardous products. All chemical products supplied to schools, businesses, and industry must contain standardized labels and be accompanied by Material Safety Data Sheets (MSDS) providing detailed information about the product. Clear and standardized labelling is an important component of WHMIS (**Figure 25**). These labels must be present on the product's original container or be added to other containers if the product is transferred.

Although MSDS must be supplied with every product sold, current MSDS can also be obtained at several Internet sites, which are useful for researching information about chemicals.

www.science.nelson.com

Consumer Information

The Canadian Hazardous Products Act requires manufacturers of consumer products containing chemicals to include a symbol specifying both the nature and degree of the primary hazard, and to note any secondary hazards, first aid treatment, storage, and disposal. The symbols show the hazard by an illustration and the degree of the hazard by the type of border surrounding the illustration (**Figure 26**).

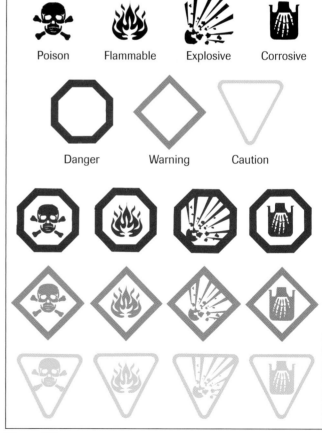

Figure 26
Household Hazardous Product Symbols

E.3 Waste Disposal

Disposal of chemical wastes at home, at school, or at work is a societal issue. We all need to be stewards of our planet; in other words, to behave as custodians or keepers. Some governments, institutions, and industries have begun to implement product stewardship programs. This is an environmental management plan based on the principle that whoever designs, produces, sells, or uses a product should take responsibility for minimizing the product's environmental impact over its complete life cycle. Governments have regulations for the handling, transportation, and disposal of chemicals, but each of us needs to take responsibility for the wastes we produce at home and at school.

Most laboratory waste can be washed down the drain, or, if it is in solid form, placed in ordinary garbage containers. However, some waste must be treated more carefully. Throughout this textbook, special waste disposal problems are noted, but it is your responsibility to dispose of waste in the safest possible manner.

Flammable Substances

Flammable liquids should not generally be washed down the drain. (The exceptions to this rule are aqueous solutions of non-toxic flammables such as alcohol–water solutions: they can safely be flushed.) Special fire-resistant containers are used to store flammable liquid waste. Waste solids that pose a fire hazard should be stored in fireproof containers. Care must be taken not to allow flammable waste to come into contact with any sparks, flames, other ignition sources, or oxidizing materials. The particular method of disposal depends on the nature of the substance.

Corrosive Solutions

Solutions that are corrosive but not toxic, such as acids, bases, or oxidizing agents, can usually be washed down the drain, but care should be taken to ensure that they are properly neutralized and diluted.

To neutralize diluted waste acids, use diluted waste bases, and vice versa. Or, use sodium bicarbonate for neutralizing the acid and use dilute hydrochloric acid for neutralizing the base. Oxidizing agents, such as potassium permanganate, should also be diluted with a 10% aqueous solution of sodium thiosulfate (reducing agent) before washing them into the drain.

Use large quantities of water and continue to pour water down the drain for a few minutes after all the substance has been washed away.

Heavy Metal Solutions

Heavy metal compounds (for example, lead, mercury, or cadmium compounds) should not be flushed down the drain. These substances are cumulative poisons and should be kept out of the environment. A special container is kept in the laboratory for heavy metal solutions. Pour any heavy metal waste into this container. Remember that paper towels used to wipe up solutions of heavy metals, as well as filter papers with heavy metal compounds imbedded in them, should be treated as solid toxic waste.

Disposal of heavy metal solutions is usually accomplished by precipitating the metal ion (for example, as lead(II) silicate) and disposing of the solid. Disposal may be by elaborate means such as deep well burial, or by simpler but accepted means such as delivering the substance to a landfill. Heavy metal compounds should not be placed in school garbage containers. Usually, waste disposal companies collect materials that require special disposal and dispose of them as required by law.

Toxic Substances

Solutions of toxic substances, such as oxalic acid, should not be poured down the drain, but should be disposed of in the same manner as heavy metal solutions. Solid toxic substances are handled similarly to precipitates of heavy metal. Chemicals should be stored in their original containers, with their labels clearly visible.

Appendix F COMMUNICATION SKILLS

Communication is essential in science. The international scope of science requires that quantities, chemical symbols, and mathematical tools such as numbers, operations, tables, and graphs, be understood by scientists in different countries with different languages. The way in which scientific knowledge is expressed also reflects the nature of scientific knowledge, and in particular, the certainty of the knowledge.

F.1 Scientific Language

Science deals with two types of knowledge: empirical (observable "facts") and theoretical (non-observable ideas). Directly observable knowledge is generally considered to be more certain than interpretations or theoretical concepts. Theories are subject to change and, therefore, are less certain than the observations upon which they are based.

When observations are interpreted or explained, the language used should reflect some uncertainty or tentativeness. Use phrases such as

- The evidence suggests that…
- According to the theory of…
- It appears likely that…
- Scientists generally believe that…
- One could hypothesize that…

Avoid the use of the word "prove." Scientific concepts cannot be proven. The evidence may be extensive and reliable, but a concept to explain the evidence will never be 100% certain.

In general, the language that you use should reflect the certainty of the information (observations are more certain than scientific concepts), and it should refer to the evidence available to you.

DID YOU KNOW?

Confidence in Empirical versus Theoretical Knowledge

A candle does not burn unless air is present. In a closed container, a candle flame is extinguished after a short period of time. These are simple and relatively certain facts that can be directly stated. At one time, scientists believed that burning releases a substance called phlogiston, which was absorbed by the air until it could hold no more phlogiston; this is what stopped the burning. This theory, which was firmly believed by many chemists until the 1800s, was eventually replaced by the oxygen theory of combustion. The facts (evidence) remained the same but the idea (theory) completely changed.

F.2 SI Symbols and Conventions

The International System of Units, known as SI from the French name, *Système international d'unités*, is the measurement and communication system used internationally by scientists; it is also the legal measurement system in Canada and most countries in the world. Physical quantities are ultimately expressed in terms of seven fundamental SI units, called base units, which cannot be expressed as combinations of simpler units (**Table 4**). Although the base unit for mass is the kilogram (kg), it is more common in a chemistry laboratory to use the gram (g). Similarly, although the base unit for temperature (T) is kelvin (K), the common temperature (t) unit is degree Celsius (°C).

All other quantities can be expressed in terms of these seven fundamental quantities. For convenience, a unit derived from a combination of base units may be assigned a symbol of its own. **Table 5** lists a few of the physical quantities and derived units most commonly encountered in chemistry.

Table 4 Quantities and Fundamental Base Units

Quantity	Symbol	Unit	Symbol
length	l	metre	m
time	t	second	s
mass	m	kilogram	kg
chemical amount	n	mole	mol
temperature	T	kelvin	K
electric current	I	ampere	A
luminous intensity	I_v	candela	cd

Quantities and their SI base units are listed in Table 5, on the next page. These are the units most widely used by scientists. For convenience, however, units such as tonne (T) for *mass* and annum (a) for *year* are sometimes used to represent quantities that would be inconveniently large when expressed as base units.

Table 5 Quantities and Base Units

Quantity	Symbol	Unit	Symbol
molar mass	M	grams per mole	g/mol
volume	V	litre	L
amount concentration	c	moles per litre	mol/L
pressure	P	pascal	Pa
energy	E	joule	J
heat capacity	C	joules per degree Celsius	J/°C
specific heat capacity	c	joules per gram per degree Celsius	J/(g·°C)
volumetric heat capacity	c	megajoules per cubic metre per degree Celsius	MJ/(m³·°C)
molar enthalpy	$\Delta_r H_m$	kilojoules per mole	kJ/mol
enthalpy change	$\Delta_r H$	kilojoules	kJ
electric charge	Q	coulomb	C
electric potential difference (voltage)	E	volt (joules per coulomb)	V

SI Prefixes

Next to universality, the most important feature of any system of units is convenience. SI has been designed to maximize convenience in a number of ways. A given quantity is always measured in the same base unit regardless of the context in which it is measured. For example, all forms of energy, including energy in food, are measured in joules. When a unit is too large or too small for convenient measurement, the unit is adjusted in size with a prefix. (See **Table 6**.) Prefixes allow units to be changed in size by multiples of ten. However, except for the use of "centi" in centimetre, we commonly use only prefixes that change the unit in multiples of a thousand.

Table 6 Some SI Prefixes

Prefix	Symbol	Fact
tera	T	10^{12}
giga	G	10^9
mega	M	10^6
kilo	k	10^3
milli	m	10^{-3}
micro	μ	10^{-6}
nano	n	10^{-9}
pico	p	10^{-12}

Scientific Notation

Scientific notation is a convenient method for expressing either a very large value or a very small value as a number between 1 and 10 multiplied by a power of 10. For example, the following numbers are expressed in regular notation and scientific notation:

Regular notation	Scientific notation
1200 L	1.200×10^3 L
0.000 000 998 mol/L	9.98×10^{-7} mol/L

On some calculators, the F ⇌ E key or the FSE key changes the number in the display into or from scientific notation. To enter a value in scientific notation in your calculator, the EXP or EE key is used to enter the power of ten. Note that the base 10 is not keyed into the calculator. For example, to enter

1.200×10^3 press [1] [.] [2] [EXP] [3]

9.98×10^{-7} press [9] [.] [9] [8] [EXP] [7] [+/−]

All mathematical operations and functions (such as +, −, ×, ÷, log) can be carried out with numbers in scientific notation.

Scientific notation is useful in calculations because it simplifies the cancellation of units and the totalling of powers of ten. However, scientific notation is sometimes overused. SI recommends that, wherever possible, prefixes be used to report measured values. Scientific notation should be reserved for situations where no prefix exists, or where it is essential to use the same unit (for example, comparing a wide range of energy values in kilojoules per gram). A reported value should use a prefix or scientific notation, but not both, unless you are comparing values. Scientific notation should usually use the base unit.

F.3 Quantitative Descriptions and Rules

Quantities that have *exact values* are either *defined* quantities (for example, 1 t is defined as exactly 1000 kg, and the SI prefix *kilo*, k, is exactly 1000), or quantities obtained by *counting* (for example, 32 people in a class or any coefficient in a balanced chemical equation). You can be certain about such quantities; there will be a small degree of uncertainty when counting very large numbers.

On the other hand, most quantities are measured by a person using some measuring instrument (for example, measuring the mass of a chemical using a balance). Since every instrument has its limitations and no one can perfectly measure a quantity, there is always some uncertainty about the number obtained. This uncertainty depends on the size of the sample measured, the particular instrument used, and the technological skill of the person doing the measurement.

Accuracy

Accuracy is an expression of how close an experimental value is to the accepted value. The comparison of the two values is often expressed as a percent difference. For example, the accuracy of a prediction based on some authority can be expressed as the absolute value of the difference divided by a predicted value and converted to a percent.

$$\% \text{ difference} = \frac{|\text{experimental value} - \text{predicted value}|}{|\text{predicted value}|} \times 100$$

This expression of accuracy is often used in the Evaluation section of investigation reports.

Precision

Accuracy is an expression of how close a value is to the accepted, expected, or predicted value, whereas **precision is a measure of the reproducibility or consistency of a result** (Figure 27). Accuracy is generally attributed to an error in the system (a *systematic error*); precision is associated with a *random error* of measurement. For example, if you used a balance without zeroing it, you might obtain measurements that have high precision (reproducibility) but low accuracy. The systematic error might be high (low accuracy), but the random error of the measurement is low (high precision).

Scientists define precision as the closeness of the agreement between independent measurements. We make the assumption that, if an instrument produces a certain decimal fraction (like a tenth of a unit), then all repeated measurements would be the same except for that last digit.

As long as an instrument is read correctly, for simplicity, we will assume that **precision is the place value of the last measurable digit and is determined by the instrument.** A mass of 17.13 g is more precise than 17.1 g. The precision is determined by the particular system or instrument used; for example, a centigram balance versus a decigram balance.

You may not know how uncertain the last measured digit is. On a centigram balance, the error of measurement in the last digit is usually ±0.01 g. Measurements such as 12.39 g, 12.40 g, and 12.41 g all have the same precision (hundredths), and may all be equally correct masses for the same object. The precision with which you read a thermometer might be ±0.2 °C (for example, 21.0 °C, 21.2 °C or 21.4 °C) and a ruler might be read to ±0.5 mm; you must decide, for example, whether to record 11.0 mm, 11.5 mm, or 12.0 mm.

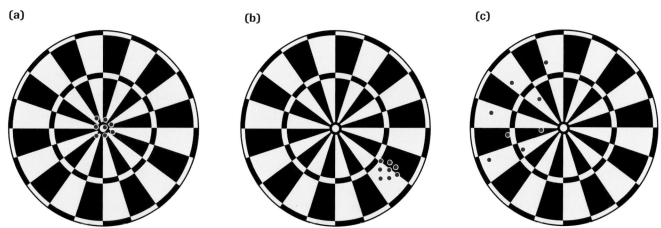

Figure 27
The positions of the darts in each of these figures are analogous to measured or calculated results in a laboratory setting.
The results in **(a)** are precise and accurate, in **(b)** they are precise but not accurate, and in **(c)** they are neither precise nor accurate.

Precision Rule for Calculations

A result obtained by adding or subtracting measured values is rounded to the same precision (number of decimal places) as the least precise value used in the calculation. For example, 12.6 g + 2.07 g + 0.142 g totals to 14.812 g on your calculator. This value is rounded to one-tenth of a gram and reported as 14.8 g because the first measurement limits the precision of the final result to tenths of a gram and the rounding rule suggests leaving the 8 as is. The final result is reported to the least number of decimal places in the values added or subtracted.

Precision Rule for pH

The precision rule for pH and pOH is a special case, although the logic is consistent with that for values expressed in scientific notation. A hydrogen or hydronium ion concentration of 1.0×10^{-7} mol/L converts to a pH of 7.00. Just as the ^{7} is not counted as a significant digit when communicating the scientific notation value, the 7 is also not counted when communicating the pH value. Therefore, the rule is
- The number of digits following the decimal point in a pH or pOH value is equal to the number of significant digits in the corresponding hydronium or hydroxide ion concentration.

and
- The number of significant digits in a hydronium or hydroxide ion concentration is equal to the number of digits following the decimal place in the corresponding pH or pOH.

Certainty

How certain you are about a measurement depends on two factors—the precision of the instrument and the value of the measured quantity. More precise instruments give more certain values; for example, 15.215 °C as opposed to 15 °C. Consider two measurements with the same precision, 0.4 g and 12.8 g. If the balance used is precise to ± 0.2 g, the value 0.4 g could vary by as much as 50%. However, 0.2 g of 12.8 g is a variation of less than 2%. For both factors—precision of instrument and value of the measured quantity—the more digits in a measurement, the more certain you are about the measurement. We communicate the **certainty** of any measurement by the number of significant digits. In a measured or calculated value, *significant digits are all those digits that are certain plus one estimated (uncertain) digit.* Significant digits include all digits correctly reported from a measurement, except leading zeros. Leading zeros are the zeros at the beginning of a decimal fraction and are written only to locate the decimal point. For example, 6.20 mL (3 significant digits) has the same number of significant digits as 0.00620 L.

For each of the following measurements, the certainty (number of significant digits) is stated beside the measured or calculated value:

0.41 mL	a certainty of 2 significant digits
700 mol	a certainty of 3 significant digits
0.020 50 km	a certainty of 4 significant digits
2×10^{40} m	a certainty of 1 significant digit

Certainty Rule for Calculations

Significant digits are primarily used to determine the certainty of a result obtained from calculations using several measured values. *A result obtained by multiplying or dividing measured values is rounded to the same certainty (number of significant digits) as the least certain value used in the calculation.* For example, 0.024 89 mol × 6.94 g/mol is displayed as 0.1727366 g on a calculator. This is correctly reported as 0.173 g or 173 mg because the second value used (6.94) limits the final result to a certainty of three significant digits.

Research in the News

News media often quote the results of surveys, such as the percentage of people who would vote for a certain political party. What does it mean when we hear, "The results were Yes 52%, No 42%, Undecided 5%, with a margin of error of 3% 19 times out of 20"? The *margin of error* (3%) is usually calculated as the reciprocal of the square root of the sample size. A larger sample size would therefore produce a smaller margin of error. The *confidence level* (19 times out of 20) is like the precision. If the survey were repeated 20 times, the result would be within the percent error 19 times and 1 time it would be very different. The pollster, in effect, claims to be (accurate) within 3% of the "real" answer 19 out of 20 times.

Rounding

When completing calculations that involve more than one step, there are two rules that are used for answers in this textbook:

- *Never* round off partial answers in your calculator.
- *Always* round off when communicating partial answers on paper.

When chained calculations involve both multiplication/division and addition/subtraction, you may be required to store the partial answers in your calculator memory or to use the bracket function on your calculator.

Rounding Rule

Calculations are usually based on measurements (for example, in the Analysis section of a report). To report a calculated result correctly, follow this procedure. *Check the first digit following the digit that will be rounded. If this digit is less than 5, it and all following digits are discarded. If this digit is 5 or greater, it and all following digits are discarded, and the preceding digit is increased by one.*

F.4 Tables and Graphs

Both tables and graphs are used to summarize information and to illustrate patterns or relationships. Preparing tables and graphs requires some knowledge of accepted practice and some skill in designing the table or graph to best describe the information.

Tables

1. Write a descriptive title (**Table 7**).
2. The row or column with the manipulated variable usually precedes the row or column with the responding variable.
3. Label all rows and/or columns with a heading, including units in parentheses where necessary. Units are not usually written in the main body of the table.

Table 7 The Reaction of HCl(aq) with Zn(s)

Concentration of HCl(aq) (mol/L)	Time for reaction (s)
2.0	70
1.5	80
1.0	144
0.5	258

Graphs

1. Write a descriptive title on the graph and label the axes (**Figure 28**).
   - Label the horizontal (*x*) axis with the name of the manipulated variable and the vertical (*y*) axis with the name of the responding variable.
   - Include the unit in parentheses on each axis label, for example, "Time (s)."
2. Assign numbers to the scale on each axis.
   - As a general rule, the data points should be spread out so that at least one-half of the graph paper is used.

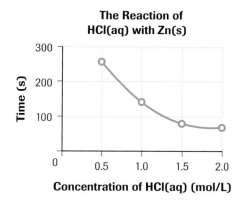

Figure 28
A sample graph

   - Choose a scale that is easy to read and has equal divisions. Each division (or square) must represent a small simple number of units of the variable; for example, 0.1, 0.2, 0.5, or 1.0.
   - It is not necessary to have the same scale on each axis or to start a scale at zero.
   - Do not label every division line on the axis. Scales on graphs are labelled in a way similar to the way scales on rulers are labelled.
3. Plot the data points.
   - Locate each data point by making a small dot in pencil. When all points are drawn and checked, draw an X over each point, or circle each point in ink.
   - Be suspicious of a data point that is obviously not part of the pattern. Double-check the location of such points, but do not eliminate the point from the graph if it does not align with the rest.
4. Draw the best-fitting line.
   - Using a sharp pencil, draw a line that best represents the trend shown by the collection of points. Do not force the line to go through each point. Uncertainty of experimental measurements may cause some of the points to be misaligned.

- If the collection of points appears to fall in a straight line, use a ruler to draw the line. Otherwise, draw a smooth curve that best represents the pattern of the points.
- Since the data points are in ink and the line is in pencil, it is easy to change the position of the line if your first curve does not fit the points to your satisfaction.

Using Your Graph
Although a graph is constructed using a limited number of measured values, the pattern may be used to extend the empirical information.

- *Interpolation* is used to find values between measured points on the graph.
- *Extrapolation* is used to find values beyond the measured points on a graph. A dotted line on a graph indicates an extrapolation.
- The scattering of points gives a visual indication of the uncertainty in the experiment. A point that is obviously not part of the pattern may require a remeasurement to check for an error or may indicate the influence of an unexpected variable.

F.5 Problem-Solving Methods

Definition of Terms
- **ratio**: comparison of two numbers; e.g., $\frac{2}{3}$, $\frac{1.5 \text{ mol}}{1 \text{ L}}$
- **proportion**: an equality of two ratios; e.g.
  $$\frac{n_{NH_3}}{2.0 \text{ L}} = \frac{1.5 \text{ mol}}{1 \text{ L}}$$
- **conversion factor**: a specific type of ratio that is used to convert a quantity from one unit to another unit; e.g.,
  $\frac{1000 \text{ m}}{1 \text{ km}}$ or $\frac{1 \text{ km}}{1000 \text{ m}}$, $\frac{40.00 \text{ g}}{1 \text{ mol}}$ or $\frac{1 \text{ mol}}{40.00 \text{ g}}$
- **formula**: a mathematical statement of a relationship using SI and IUPAC symbols and format; e.g., $m = nM$ (Note: m = n × M is not acceptable.)

Basic Problem-Solving Methods
A stoichiometry example is used to compare the three basic methods.

Copper metal is used to recover silver from a silver nitrate solution. Predict the mass of silver obtained from the complete reaction of 50.0 g of copper.

$$Cu(s) + 2\, AgNO_3(aq) \rightarrow 2\, Ag(s) + Cu(NO_3)_2(aq)$$
50.0 g   m
63.55 g/mol   107.87 g/mol

Proportion Method
$$\frac{n_{Cu}}{50.0 \text{ g}} = \frac{1 \text{ mol}}{63.55 \text{ g}} \qquad n_{Cu} = 0.787 \text{ mol}$$

$$\frac{n_{Ag}}{0.787 \text{ mol}} = \frac{2}{1} \qquad n_{Ag} = 1.57 \text{ mol}$$

$$\frac{m_{Ag}}{1.57 \text{ mol}} = \frac{107.87 \text{ g}}{1 \text{ mol}} \qquad m_{Ag} = 170 \text{ g}$$

Formula Method
$$n_{Cu} = \frac{m}{M} = \frac{50.0 \text{ g}}{63.55 \text{ g/mol}} = 0.787 \text{ mol}$$

$$n_{Ag} = 0.787 \text{ mol} \times \frac{n_{required}}{n_{given}} = 0.787 \text{ mol} \times \frac{2}{1} = 1.57 \text{ mol}$$

$$m_{Ag} = nM = 1.57 \text{ mol} \times 107.87 \text{ g/mol} = 170 \text{ g}$$

Conversion Factor Methods
(a) Step method

$$n_{Cu} = 50.0 \text{ g} \times \frac{1 \text{ mol}}{63.55 \text{ g}} = 0.787 \text{ mol}$$

$$n_{Ag} = 0.787 \text{ mol} \times \frac{2}{1} = 1.57 \text{ mol}$$

$$m_{Ag} = 1.57 \text{ mol} \times \frac{107.87 \text{ g}}{\text{mol}} = 170 \text{ g}$$

(b) Full method ("Factor Label")

$$m_{Ag} = 50.0 \text{ g Cu} \times \frac{1 \text{ mol Cu}}{63.55 \text{ g Cu}} \times \frac{2 \text{ mol Ag}}{1 \text{ mol Cu}} \times \frac{107.87 \text{ g Ag}}{1 \text{ mol Ag}} = 170 \text{ g}$$

Appendix G REVIEW OF CHEMISTRY 20

Your review of Chemistry 20 for Chemistry 30 will be more successful if you study the highlighted Summaries, Sample Problems, and Communication Examples in each chapter. By answering the following questions you will find out where you need to check your understanding before starting Chemistry 30.

Unit 1 Chemical Bonding
(Chapter 3)

1. Distinguish between the two important types of scientific knowledge.
2. Identify the characteristics of acceptable scientific theories.
3. Explain the octet rule and how it relates to chemical reactivity.
4. Copy and complete the following table.

Table 1 Theoretical Descriptions of Selected Elements

Element name	Lewis symbol	Group number	Number of valence electrons	Number of lone pairs	Number of bonding electrons
calcium					
aluminium					
arsenic					
oxygen					
bromine					
neon					

5. (a) State the types of elements expected to react to form compounds containing covalent bonds.
   (b) State the types of elements expected to react to form compounds containing ionic bonds.
   (c) Explain your answers to (a) and (b) using the concept of electronegativity.
   (d) Why is it difficult to predict the type of bonding in some compounds using only electronegativities?
6. The two major types of compounds are ionic and molecular.
   (a) Compare the naming of these compounds.
   (b) Given the name of an example of each compound, outline how the chemical formula is obtained. Use specific examples in your answer.
7. Theories are created to explain observations. For each of the following properties of ionic compounds, write a brief theoretical explanation.
   (a) Ionic compounds are hard solids with high melting and boiling points.
   (b) Ionic compounds are electrical conductors in molten and aqueous states.
8. Using Lewis symbols and formulas, write the formation equation for each of the following compounds.
   (a) potassium bromide
   (b) sodium oxide
   (c) calcium fluoride
9. Why are chemical formulas for ionic compounds always based the simplest whole number ratio of ions? Is the simplest whole number ratio also used for molecular formulas? Why or why not?
10. Compare ionic and covalent bonds, including how they are formed, according to theory and the nature of the bond.
11. For each of the following molecular formulas, draw the Lewis, structural, and stereochemical formulas, and state the shape around the central atom.
    (a) OCl_2
    (b) SiH_4
    (c) NCl_3
    (d) HCN
    (e) CH_2O
12. Classify each of the molecules represented in the previous question as polar or nonpolar. Justify your answer using the molecular shape and bond dipoles (charge distributions).
13. Methylisocyanate is a toxic pesticide that is manufactured using the following chemical reaction.
    $$CS_2 + CH_3NH_2 \rightarrow CH_3NCS + H_2S$$
    Rewrite this chemical equation using structural formulas for all reactants and products.
14. Define the three types of intermolecular forces. For each type of force, state how you would know if this type of force is likely present among molecules of a substance.
15. Each of the following four substances is either a liquid at SATP or converted to a liquid by changing the conditions: C_3H_7F, $C_3H_5(OH)_3$, $C_3H_7NH_2$, C_3H_8.
    (a) Construct a table to summarize the types of intermolecular forces believed to be present among molecules of each of these substances.
    (b) Predict the order of boiling points from lowest to highest. Justify your answer.

16. Why are boiling points often used as an indirect measure of the strength of intermolecular forces among molecules of a substance?

17. Explain each of the following observations in terms of the characteristics of molecules and intermolecular forces.
    (a) The boiling point of fluorine is significantly less than that of chlorine.
    (b) Drops of ethanol are attracted to an electrically charged strip.
    (c) Ice has a regular hexagonal structure.

18. A simple, but useful, distinction that is often made is to classify the water on Earth as either fresh water (as in most lakes and streams) or salt water (as in the oceans).
    (a) Contrast these two terms from a scientific perspective.
    (b) How is this distinction useful from a technological perspective?

19. Describe an example in which scientific research led to the development of a new technology.

Unit 2 Gases (Chapter 4)

1. List seven ways by which empirical knowledge is communicated.

2. List the three characteristics of acceptable scientific laws and generalizations.

3. Describe one natural phenomenon and one technological product that each depend on the properties of gases.

4. Complete the following statements.
    (a) At a constant temperature and chemical amount of gas, as the pressure increases, the volume _____.
    (b) At a constant pressure and chemical amount of gas, as the temperature decreases, the volume _____.
    (c) At a constant volume and temperature, if the chemical amount of gas inside a container is increased, the pressure _____.

5. Choose one of the statements in question 4 and write a general design for an experiment to test the statement. Include the identification of all important variables.

6. For each statement in question 4, sketch a graph of the relationship between the manipulated and responding variables.

7. Convert 95.8 kPa into units of millimetres of mercury and atmospheres.

8. A 1.5 L volume of gas is compressed at a constant temperature from 1.0 atm to 5.0 atm. Calculate the final volume.

9. A balloon can hold 800 mL of air before breaking. A balloon at 4.0 °C containing 750 mL of air is allowed to warm up. Assuming a constant pressure inside the balloon, determine the minimum Celsius temperature when the balloon breaks.

10. A sample of argon gas at 101 kPa and 22.0 °C occupies a volume of 150 mL. If the volume doubles at a temperature of 150 °C, determine the new pressure.

11. Using the kinetic molecular theory, explain Boyle's and Charles' laws.

12. Illustrate the law of combining volumes using a simple example. Describe the theory used to explain this law.

13. Many people use propane barbeques for outdoor cooking. Predict the volume of carbon dioxide produced when 15 L of propane completely burns at SATP.

14. Describe and compare the behaviour of real and ideal gases using the kinetic molecular theory.

15. Predict the volume that 25.0 g of oxygen gas would occupy at 22.0 °C and 98.1 kPa.

16. Compare the volume that 0.278 mol of hydrogen would occupy at STP and SATP.

17. An average bungalow requires about 400 kmol of methane per year for space heating.
    (a) Determine the volume of methane at SATP used in one year.
    (b) Predict the volume of methane used if the pressure is 98.5 kPa and the temperature is 12.7 °C.

Unit 3 Solutions, Acids and Bases (Chapters 5 & 6)

1. For each of the following perspectives write a brief statement describing the focus or concern of that point of view.
   - scientific
   - technological
   - economic
   - ecological
   - political

2. List three topics that are current STS issues.

3. Classify each of the following statements using one of the issue perspectives listed in question 1. All of the statements concern sulfur dioxide emissions.
   (a) An industry spokesman reported that emissions of sulfur dioxide were within the limits set by environmental legislation.
   (b) Laboratory research has provided evidence that sulfur dioxide from the combustion of fossil fuels is converted to sulfur trioxide in the presence of oxygen.
   (c) The cost of ending sulfur dioxide pollution of the atmosphere will be high. The longer we delay facing the problem, the greater will be the cost.
   (d) Sulfur oxides and their related dissolved acids are particularly damaging to soil microbes, water life forms, plants, building materials, and people.
   (e) One of the most promising scrubbers to remove sulfur dioxide gas from a smoke stack is the limestone–dolomite process.

4. Compare the goals of science and technology.

5. Describe a homogeneous mixture and provide several examples.

6. Define the two main parts of a solution. State an example using a chemical formula and identity the two parts in words.

7. In the exploration of outer space, scientists usually look for the presence of water as a strong indication of the existence of living things. Briefly explain this statement in terms of solutions and reactions.

8. List at least six examples of manufactured solutions found in the home and six examples of natural solutions found in the environment.

9. Distinguish between electrolytes and non-electrolytes, including examples of types of substances in each category.

10. Explain, in terms of breaking and forming bonds, why the dissolving of substances in water can be either exothermic or endothermic.

11. Compounds may be ionic or molecular and may also be acids, bases, or neutral compounds.
    (a) Design an experiment to classify the solute in each of a number of different solutions.
    (b) Outline the expected results.

12. Write dissociation or ionization equations for the following pure substances dissolving in water.
    (a) lithium phosphate solid
    (b) hydrogen chloride gas
    (c) aluminium sulfate solid

13. For each of the following pure substances, write the formulas for the entities present when each substance is placed in water.
    (a) $Sr(OH)_2(s)$
    (b) $HNO_3(l)$
    (c) $C_3H_8(g)$
    (d) $CH_3COOH(l)$
    (e) $AgCl(s)$
    (f) $CH_3OH(l)$

14. List the three advantages of solutions for technological applications.

15. Suppose you are given four unlabelled beakers, each containing a colourless aqueous solution of one solute. The possible solutions are NaCl(aq), HCl(aq), $BaCl_2$(aq), and CH_3Cl(aq). Write a series of diagnostic tests to distinguish each solution from the others.

16. Compare the ways in which solution concentrations are expressed in chemistry labs, consumer products, and environmental studies.

17. A household cleaner contains 12.5 g of sodium hypochlorite in 500 mL of solution. Determine the percentage mass by volume concentration of this solution.

18. A drain cleaner contains 2.75 mol/L sodium hydroxide. Calculate the volume of solution that contains 0.375 mol of sodium hydroxide.

19. A windshield washer solution was prepared by dissolving 100 g of methanol in water to form 2.00 L of solution. Calculate the amount concentration of the solution.

20. A 0.251 mol/L calcium chloride solution is required for an experiment.
    (a) Calculate the mass of calcium chloride that needs to be measured.
    (b) Write a specific procedure for an untrained laboratory technician to prepare this solution.
21. (a) Predict the volume of concentrated, 14.6 mol/L phosphoric acid required to prepare 250 mL of a 0.375 mol/L solution.
    (b) Write a specific procedure to prepare this solution.
22. Calculate the amount concentration of each ion in a 2.1 mol/L solution of iron(III) chloride?
23. How does the solubility of solids and gases change as the temperature increases?
24. Excess copper(II) sulfate is added to water in a closed system until no more solute dissolves at a constant temperature.
    (a) Describe some empirical properties of this mixture.
    (b) Provide a brief theoretical explanation of these properties.
25. Write the acid formula for each of the following substances.
    (a) aqueous hydrogen bromide
    (b) aqueous hydrogen sulfite
    (c) hydrofluoric acid
    (d) sulfuric acid
26. Copy and complete the following table.

Table 2 Hydroxide Concentrations and pHs

[H_3O^+(aq)] (mol/L)	pH	Acidic/basic/neutral
1.0×10^{-7}		
	8	
	3.7	
6.23×10^{-9}		

27. The pH of pure water is 7, of carbonated water about 5, and of a cola drink about 3. How many times more acidic is a cola drink than carbonated water and pure water?

28. Use the modified Arrhenius theory to write chemical equations explaining the following evidence.
    (a) A vinegar solution is acidic.
    (b) A baking soda (sodium hydrogen carbonate) solution has a pH of 8.
    (c) Some muriatic (hydrochloric) acid is neutralized with a lye (sodium hydroxide) solution.
29. A simple window cleaning solution containing 0.25 mol/L ammonia has a pOH of 2.5.
    (a) Convert the pOH into an amount concentration of hydroxide ions.
    (b) Write a balanced chemical equation to explain this basic solution.
    (c) Is ammonia a strong or weak base? Justify your answer.
30. Write a design for an experiment to identify strong and weak acids. Include three different diagnostic tests and identify important controlled variables.
31. Polyprotic acids and bases occur naturally and are manufactured for a variety of purposes.
    (a) Distinguish between monoprotic and polyprotic acids and bases.
    (b) Using boric acid (aqueous hydrogen borate) as an example, write a series of chemical equations showing successive reactions with water.
32. Most scientists agree that the increasing emission of carbon dioxide into the atmosphere from the burning of fossil fuels is the prime cause of global warming. This problem might be even worse if it were not for the fact that approximately half of the carbon dioxide produced is absorbed by the world's oceans. However, recent research has shown that this is making the oceans more acidic—about 30% more acidic over the past two hundred years.
    (a) Use the modified Arrhenius theory to write a chemical equation explaining the increased acidity of the world's oceans.
    (b) Scientists are not certain what effect the increased acidity will have. If we assume there will be a problem in the oceans, describe some solutions to reduce the addition of carbon dioxide to the oceans.
33. Using pesticides as an example, summarize the intended and unintended consequences of this chemical technology.

Unit 4 Quantitative Relationships
(Chapters 7 & 8)

1. Compare the fields of chemistry and chemical technology.
2. Describe two examples of chemical technologies, used by consumers, that are based on the stoichiometry of chemical reactions.
3. Distinguish between qualitative and quantitative chemical analysis and provide an example of each type of analysis.
4. For each of the following mixtures, write a balanced net ionic equation and identify all spectator ions. All reactant solutions are assumed to be at least 0.10 mol/L in concentration.
   (a) sodium hydroxide and cobalt(II) chloride solutions
   (b) silver nitrate and calcium iodide solutions
   (c) silver nitrate solution and zinc metal
   (d) hydrochloric acid and solid calcium hydroxide
   (e) the precipitation of aluminium hydroxide in qualitative analysis
5. In your own words, describe the meaning of stoichiometry.
6. List the three types of stoichiometry and describe how each type is recognized.
7. In general, how do chemical industries use the principles of stoichiometry to maximize yields and minimize waste?
8. In the steel industry, carbon reacts with iron(III) oxide (from iron ore) to produce molten iron and carbon dioxide.
   (a) Write a complete balanced chemical equation for this reaction.
   (b) Translate this chemical equation into an English sentence including all chemical amounts and states of matter.
   (c) Using the coefficients, calculate the mass of each reactant and product in this balanced chemical equation.
   (d) How does the total mass of reactants compare with the total mass of products? What principle does this illustrate?
9. Predict the mass of lead(II) iodide precipitate that forms when 2.93 g of potassium iodide in solution reacts with excess lead(II) nitrate.
10. In a hard water analysis, a calcium chloride solution is reacted with excess aqueous sodium oxalate to produce 0.452 g of calcium oxalate precipitate. Determine the mass of calcium chloride present in the original solution.
11. Analysis for sulfate ions is usually done by first precipitating barium sulfate from a sample. The filter paper containing the barium sulfate precipitate is then ignited. Carbon from the burnt filter paper then reacts with the barium sulfate as shown in the balanced chemical equation below.
    $BaSO_4(s) + 2\ C(s) \rightarrow BaS(s) + 2\ CO_2(g)$
    (a) Predict the mass of carbon required to react with 1.50 g of barium sulfate precipitate.
    (b) List the assumptions you have made in this calculation.
12. In a test of the stoichiometric method, an excess of sodium hydroxide solution is reacted with a solution containing 1.50 g of aluminium sulfate.
    (a) Predict the mass of precipitate expected in this reaction.
    (b) If the actual yield in this experiment was 0.96 g of precipitate, calculate the percent yield.
    (c) Outline at least three possible reasons for the discrepancy between the theoretical (predicted) yield and the actual yield.
13. Powdered aluminium metal is one of the fuels used in the solid rocket boosters for the NASA Space Shuttle. What volume of oxygen at SATP is required to react completely with 100 kg of aluminium?
14. A portable hydrogen generator uses the reaction of solid calcium hydride and water to form calcium hydroxide and hydrogen. Determine the volume of hydrogen at 96.5 kPa and 22 °C that can be produced from a 50 g cartridge of $CaH_2(s)$.
15. A volumetric analysis shows that it takes 32.0 mL of 2.12 mol/L NaOH(aq) to completely react with 10.0 mL of sulfuric acid from a car battery. Calculate the amount concentration of sulfuric acid in the battery solution.

16. In a laboratory, silver metal can be recycled to produce silver nitrate by the following reaction.

    3 Ag(s) + 4 HNO$_3$(aq) →

    3 AgNO$_3$(aq) + NO(g) + 2 H$_2$O(l)

    Predict the volume of 15.4 mol/L nitric acid required to react with 1.68 kg of silver metal.

17. Distinguish between limiting and excess reagents.

18. Describe the purpose of using an excess reagent in a quantitative analysis?

19. Calcium carbonate is commonly used in simple antacid products to counteract acidity in the stomach. Suppose you add a 750 mg tablet of calcium carbonate to 200 mL of 0.10 mol/L hydrochloric acid (representing the stomach acid).

    (a) Which reactant is in excess and by how much? Give your answer in moles.

    (b) Predict the mass of calcium chloride formed in this reaction.

20. Complete the Materials and Analysis of the following lab report.

    Problem

    What is the amount concentration of an unknown sodium carbonate solution?

    Design

    Samples of sodium carbonate solution were titrated with a standardized hydrochloric acid solution using methyl orange as the indicator.

    Evidence

    Table 3 Titration of 25.0 mL Samples of Na$_2$CO$_3$(aq) with 0.352 mol/L HCl(aq)

Trial	1	2	3	4
Final burette reading (mL)	16.5	31.8	47.0	16.4
Initial burette reading (mL)	0.6	16.5	31.8	1.2

22. Titration curves are useful in studying the progress of a reaction, such as an acid–base reaction.

    (a) Sketch a general curve for the titration of a strong base with a strong acid. Label the axes and provide a title for the graph. No numbers are required.

    (b) Place an "X" on the curve where the reaction is complete. At what pH should this occur?

    (c) Identify a suitable indicator for any strong base–strong acid titration and justify your answer.

    (d) Would your answers to (a), (b), and (c) change if a strong acid were titrated with a strong base? Note any differences.

Appendix H DIPLOMA EXAM PREPARATION

You have been preparing for the Diploma Exam throughout your high school career. In your final year, as you work through the Chemistry 30 course, here are some tips that will help you perform as well as you possibly can in the Diploma Exam.

- **Involve Yourself in Class**: Attend class regularly and be active in your learning by asking questions and completing assignments. If you work steadily, there will be no need to try to learn everything just before the exam.
- **Keep Up-to-Date with Chemistry 30 Material**: Schedule a regular review time every week and use this time to organize your notes, review the material, and ask yourself questions about what you have learned. Use the Self Quizzes, Chapter Summaries, and other study aids.
- **Read and Understand the Scoring Criteria for Diploma Exams**: The full scoring criteria for the different types of questions are available in the Chemistry 30 Information Bulletin found online. Read these criteria carefully and make sure you understand what they mean.

- **Practice Writing Old Exams**: Simulate the conditions of the exam to get used to sitting through an entire exam and the time constraints of writing the exam. You will also get used to the types of questions on the exam and, afterward, be able to compare your answers to the scoring criteria.
- **Read the Instructions**: Make sure you read the instructions, directions, and questions very carefully.
- **Become Familiar with the Types of Questions**: Read the information below and practice answering each type of question.

There are three types of questions on the Diploma Exam: multiple choice, numerical response, and written response.

Multiple Choice Questions

Multiple choice questions are a large part of the diploma exam. Most of the multiple choice questions on the diploma exam are context-dependent. The others are called "discrete."

Context-dependent multiple choice questions use information provided in addition to the actual question. Examples of this type of question include questions 10 and 11 in the Unit 2 Review.

Use this information to answer questions 9 to 11.

The empirical study of gases provided a number of laws that formed the basis for important developments in chemistry such as atomic theory and the mole concept.

Statements

1. The volume of a gas varies inversely with the pressure on the gas.
2. Volumes of reacting gases are always in simple, whole number ratios.
3. The volume of a gas varies directly with the absolute temperature of the gas.
4. The volume of a gas varies directly with the absolute temperature and inversely with the pressure.

10. Which statements require that the temperature be a controlled variable?
    A. 1, 2, 3, and 4
    B. 1, 3, and 4 only
    C. 1 and 2 only
    D. 3 and 4 only

11. Identify the statement that is best explained by Avogadro's theory.
    A. 1
    B. 2
    C. 3
    D. 4

Discrete multiple choice questions have no additional information or directions, such as questions 1 and 2 in the Chapter 7 Review.

1. A main goal of technology is to
    A. advance science
    B. identify problems
    C. explain natural processes
    D. solve practical problems

2. In the reaction of aqueous solutions of sodium sulfide and zinc nitrate in a chemical analysis, the spectator ions are
    A. sodium and nitrate ions
    B. sulfide and zinc ions
    C. sodium and zinc ions
    D. sulfide and nitrate ions

When answering multiple choice questions:
- Try to answer the question before looking at the choices.
- Eliminate any choices that are incorrect by crossing them out.
- Stay alert for key words: *most, least, not one of the following*, etc. Negative terms ("Which of the following is *not* a physical property of a gas?") will be italicized. Look out for these.
- Choose the correct answer on the question sheet and then fill in the corresponding circle on the answer sheet. It is important that you stay aware of time, so that you don't run out of time to transcribe your answers from the question sheet to the answer sheet.

Numerical Response Questions

Numerical response questions on the Diploma Exam are clearly indicated with the heading "NUMERICAL RESPONSE." Examples of these types of questions are clearly marked with the icon "NR" in this textbook. There are four types of numerical response questions on the Diploma Exam. They are

- calculation of numerical values,
- calculation of numerical values expressed in scientific notation,
- selecting numerical responses from diagrams or lists, and
- determining the sequence of listed events.

Specific instructions for recording the answer to each type of numerical response are given in the instructions of the Diploma Exam, as well as with each question. Read the instructions *carefully*.

- **Numerical calculations**: This category is fairly straightforward. You have to use the provided data to calculate an answer. The answer is a numerical response with a maximum of four digits (including the decimal point). The first digit of your answer goes in the left-hand box on the answer sheet. Depending on the number of digits in your answer, there may be unfilled boxes to the right. The decimal point, if there is one, occupies one of the boxes. If an answer has a value between 0 and 1, for example 0.25, make sure you record the '0' before the decimal point.

- **Calculations requiring an answer in scientific notation**: This category is similar to regular numerical calculations, except that the four digits of the response come from an answer in the form $a.b \times 10^{cd}$. When you have completed your calculation, just write the four digits represented by a, b, c, and d in order.

- **Numerical responses from diagrams or lists**: This category involves selecting numbers (usually representing a term or item from several provided) and writing them in the correct order.

- **Sequence of numbered events or data**: These questions ask you to rearrange variables, events, or data into a specified order. Pay particular attention to the instructions, which might specify, for example, "in order of increasing melting temperature."

Written Response Questions

There are two written response questions on the Chemistry 30 Diploma Exam. One written response question is a closed-response question (which has only one correct response) and the other is an open-response question (which has more than one correct response).

Learn to determine which type of question is being asked. **Closed response questions** have specific questions that must be directly answered. These questions are presented as sections and subsections (question 1. a, b, c, etc.). **Open response questions**, or holistic questions, typically begin with an instruction followed by a series of bullets. You are expected to write your answer in full sentences. Each bullet must be addressed and your responses combined into the answer.

When answering written response questions

- Carefully read the information box and make sure you fully understand the material *and* all of the question parts before beginning to answer.
- Identify each key piece of information and make notes about the meaning and implications of that information. If it helps, mark key words and phrases. Identify which unit of Chemistry 30 is being addressed. This will help you focus your attention to the correct material.
- Identify any irrelevant information.

- Identify the **directing words** in the question. These are usually highlighted in bold in the question. The directing words have specific meanings and are indicators of what the graders expect for an answer. Examples of directing words include **illustrate**, **analyze**, **explain**, and **predict**. A complete list of directing words and their meanings can be found online. These words are also included and defined in the Glossary. In your preparation, refer to the list of meanings for directing words. Make sure that you know what is expected for each directing word.

  www.science.nelson.com

- Read the question carefully and ask yourself what you are being asked to do. Write the question out in your own words if there are any doubts. Remember, if you don't understand the question, you will probably not be able to answer it correctly!

- Summarize your answers on scrap paper before writing them on the test answer page.

- Once you have answered the question, review your answer and make sure you have addressed all parts of the question. This is especially important for the open response question.

Answering Closed Response Questions

Closed response questions are often based on a summary of current research or a scenario, and data may be given in graph or table format. There are several parts to the question, but the number of parts depends on the context of the question. An example of a closed response question follows. This one is taken from the Chapter 13 Review.

> 35. Vanadium is a very versatile element in terms of its reactivity. Vanadium metal reacts with fluorine to form VF_5, with chlorine to form VCl_4, with bromine to form VBr_3, with iodine to form VI_2, with oxygen to form V_2O_5, and with hydrochloric acid to form VCl_2.
>
> (a) Identify the oxidation states of vanadium in each of these compounds.
>
> (b) What interpretation can be made about the oxidizing power of the chemicals that react with vanadium metal?
>
> (c) Describe how the oxidation state of vanadium relates to the colours of the compounds formed.
>
> (d) Report on some technological applications of vanadium and its compounds.

Complete examples of closed-response questions are online.

www.science.nelson.com

Each section of the question must be answered completely for full marks. The number of marks for the question is given in parentheses. Use this as a guide for how detailed your answer should be.

Answering Open Response Questions

Open response questions are often based on a situation or scenario. You are generally asked to write an essay-type response, guided by the directing word in the question. Following the initial question are several points. Your responses to these points should be integrated into your answer.

> 34. For the production of pulp from wood, a variety of methods are used, including mechanical and chemical processes. These have advantages and disadvantages that have been widely debated. Prepare an argument for or against the following statement: "The immediate economic value of using technology to produce a product far outweighs any possible future adverse effects."
>
> Your response should also include
>
> • researched information about a variety of mechanical and chemical processes
>
> • an evaluation of these processes from technological, economic, and ecological perspectives
>
> • reference to redox chemistry

You will do best answering these questions if you refer to current scientific advances in your answer, as you address the technological and societal aspects of the question. Try to stay up-to-date on current events by reading the newspaper, science magazines, or reliable science Web sites on a regular basis. Make sure that your answer includes both scientific and technology and society aspects.

Write your answer out in full. When you have completed your answer, **recheck** it against the bullets, making sure all parts of the question have been addressed.

The open response questions are scored against **two** separate scales: a **science** scale, and a **technology and society** scale. The scores are: 0 (insufficient), 1 (poor), 2 (limited), 3 (satisfactory), 4 (good) and 5 (excellent). The highest score (5) is given for clear, complete answers that address all of the directing words and give more than one example or piece of information for each bullet in the question. The lowest response (0) is given if the answer does not address the questions presented, or is too brief to assess.

Appendix I DATA TABLES

THERMODYNAMIC PROPERTIES OF SELECTED ELEMENTS*

Name	Formula	$\Delta_{fus}H_m^0$ (kJ/mol)	$\Delta_{vap}H_m^0$ (kJ/mol)	c (J/(g·°C))
aluminium	Al	10.79	294	0.897
argon	Ar	1.18	6.43	0.520
beryllium	Be	7.90	297	1.825
boron	B	50.2	480	1.026
bromine	Br_2	10.57	29.96	0.474
carbon (graphite)	C	117	–	0.709
chlorine	Cl_2	6.40	20.41	0.479
chromium	Cr	21.0	339.5	0.449
cobalt	Co	16.06	377	0.421
copper	Cu	12.93	300.4	0.385
fluorine	F_2	0.51	6.62	0.824
gallium	Ga	5.58	254	0.371
germanium	Ge	36.94	334	0.320
gold	Au	12.72	324	0.129
helium	He	0.014	0.08	5.193
hydrogen	H_2	0.12	0.90	14.304
iodine	I_2	15.52	41.57	0.214
iron	Fe	13.81	340	0.449
krypton	Kr	1.64	9.08	0.248
lead	Pb	4.78	179.5	0.129
magnesium	Mg	8.48	128	1.023
manganese	Mn	12.91	221	0.479
mercury	Hg	2.29	59.1	0.140
neon	Ne	0.33	1.71	1.030
nickel	Ni	17.04	377.5	0.444
nitrogen	N_2	0.71	5.57	1.040
oxygen	O_2	0.44	6.82	0.918
phosphorus	P_4	0.66	12.4	0.769
platinum	Pt	22.17	469	0.133
radon	Rn	3.25	18.10	0.094
scandium	Sc	14.1	332.7	0.568
selenium	Se	6.69	95.48	0.321
silicon	Si	50.21	359	0.705
silver	Ag	11.28	258	0.235
sulfur	S_8	1.72	45	0.710
tin	Sn	7.17	296.1	0.228
titanium	Ti	14.15	425	0.523
tungsten	W	52.31	806.7	0.132
uranium	U	9.14	417.1	0.116
vanadium	V	21.5	459	0.489
xenon	Xe	2.27	12.57	0.158
zinc	Zn	7.07	123.6	0.388

* molar enthalpies at 101.325 kPa (1 atm) and specific heat capacities for standard state at SATP

THERMODYNAMIC PROPERTIES OF SELECTED COMPOUNDS*

Name	Formula	$\Delta_{fus}H_m^0$ (kJ/mol)	$\Delta_{vap}H_m^0$ (kJ/mol)	c (J/(g·°C))
ice	$H_2O(s)$	6.01	–	2.00
water	$H_2O(l)$	–	40.65	4.19
steam	$H_2O(g)$	–	–	2.02
ammonia	$NH_3(g)$	5.66	23.33	2.06
methanol	$CH_3OH(l)$	3.22	35.21	2.53
ethanol	$C_2H_5OH(l)$	4.93	38.56	2.44
Freon-12	$CCl_2F_2(g)$	4.14	20.1	0.60

*at 101.325 kPa (1 atm)

MISCELLANEOUS SPECIFIC AND VOLUMETRIC HEAT CAPACITIES

Substance	Specific heat capacity, c (J/(g·°C))	Volumetric heat capacity, c (MJ/(m³·°C))
air	1.01	0.0012
water	4.19	4.19
wood	1.26	–
glass	0.84	–
polystyrene	0.30	–
brick/rock	–	1.9
concrete	–	2.1
ethylene glycol (50%)	–	3.7
aluminium	0.897	–
copper	0.385	–
tin	0.228	–

STANDARD MOLAR ENTHALPIES OF FORMATION

Chemical name	Formula	$\Delta_f H_m^\circ$ (kJ/mol)	Chemical name	Formula	$\Delta_f H_m^\circ$ (kJ/mol)
acetone	$(CH_3)_2CO(l)$	−248.1	manganese(II) oxide	$MnO(s)$	−385.2
aluminium oxide	$Al_2O_3(s)$	−1675.7	manganese(IV) oxide	$MnO_2(s)$	−520.0
ammonia	$NH_3(g)$	−45.9	mercury(II) oxide (red)	$HgO(s)$	−90.8
ammonium chloride	$NH_4Cl(s)$	−314.4	mercury(II) sulfide (red)	$HgS(s)$	−58.2
ammonium nitrate	$NH_4NO_3(s)$	−365.6	methanal (formaldehyde)	$CH_2O(g)$	−108.6
barium carbonate	$BaCO_3(s)$	−1213.0	methane	$CH_4(g)$	−74.6
barium chloride	$BaCl_2(s)$	−855.0	methanoic (formic) acid	$HCOOH(l)$	−425.0
barium hydroxide	$Ba(OH)_2(s)$	−944.7	methanol	$CH_3OH(l)$	−239.2
barium oxide	$BaO(s)$	−548.0	methylpropane	$C_4H_{10}(g)$	−134.2
barium sulfate	$BaSO_4(s)$	−1473.2	nickel(II) oxide	$NiO(s)$	−240.6
benzene	$C_6H_6(l)$	+49.1	nitric acid	$HNO_3(l)$	−174.1
bromine (vapour)	$Br_2(g)$	+30.9	nitrogen dioxide	$NO_2(g)$	+33.2
butane	$C_4H_{10}(g)$	−125.7	nitrogen monoxide	$NO(g)$	+91.3
calcium carbonate	$CaCO_3(s)$	−1207.6	nitromethane	$CH_3NO_2(l)$	−113.1
calcium chloride	$CaCl_2(s)$	−795.4	octane	$C_8H_{18}(l)$	−250.1
calcium hydroxide	$Ca(OH)_2(s)$	−985.2	ozone	$O_3(g)$	+142.7
calcium oxide	$CaO(s)$	−634.9	pentane	$C_5H_{12}(l)$	−173.5
calcium sulfate	$CaSO_4(s)$	−1434.5	phenylethene (styrene)	$C_6H_5CHCH_2(l)$	+103.8
carbon dioxide	$CO_2(g)$	−393.5	phosphorus pentachloride	$PCl_5(s)$	−443.5
carbon disulfide	$CS_2(l)$	+89.0	phosphorus trichloride (liquid)	$PCl_3(l)$	−319.7
carbon monoxide	$CO(g)$	−110.5	phosphorus trichloride (vapour)	$PCl_3(g)$	−287.0
chloroethene	$C_2H_3Cl(g)$	+37.3	potassium bromide	$KBr(s)$	−393.8
chromium(III) oxide	$Cr_2O_3(s)$	−1139.7	potassium chlorate	$KClO_3(s)$	−397.7
copper(I) oxide	$Cu_2O(s)$	−168.6	potassium chloride	$KCl(s)$	−436.5
copper(II) oxide	$CuO(s)$	−157.3	potassium hydroxide	$KOH(s)$	−424.6
copper(I) sulfide	$Cu_2S(s)$	−79.5	propane	$C_3H_8(g)$	−103.8
copper(II) sulfide	$CuS(s)$	−53.1	silicon dioxide (α–quartz)	$SiO_2(s)$	−910.7
1,2-dichloroethane	$C_2H_4Cl_2(l)$	−126.9	silver bromide	$AgBr(s)$	−100.4
dinitrogen tetraoxide	$N_2O_4(g)$	+11.1	silver chloride	$AgCl(s)$	−127.0
ethane	$C_2H_6(g)$	−84.0	silver iodide	$AgI(s)$	−61.8
ethane-1,2-diol	$C_2H_4(OH)_2(l)$	−454.8	sodium bromide	$NaBr(s)$	−361.1
ethanoic (acetic) acid	$CH_3COOH(l)$	−484.3	sodium chloride	$NaCl(s)$	−411.2
ethanol	$C_2H_5OH(l)$	−277.6	sodium hydroxide	$NaOH(s)$	−425.6
ethene (ethylene)	$C_2H_4(g)$	+52.4	sodium iodide	$NaI(s)$	−287.8
ethyne (acetylene)	$C_2H_2(g)$	+227.4	sucrose	$C_{12}H_{22}O_{11}(s)$	−2226.1
glucose	$C_6H_{12}O_6(s)$	−1273.3	sulfur dioxide	$SO_2(g)$	−296.8
hexane	$C_6H_{14}(l)$	−198.7	sulfur trioxide (liquid)	$SO_3(l)$	−441.0
hydrogen bromide	$HBr(g)$	−36.3	sulfur trioxide (vapour)	$SO_3(g)$	−395.7
hydrogen chloride	$HCl(g)$	−92.3	sulfuric acid	$H_2SO_4(l)$	−814.0
hydrogen fluoride	$HF(g)$	−273.3	tin(II) chloride	$SnCl_2(s)$	−325.1
hydrogen iodide	$HI(g)$	+26.5	tin(IV) chloride	$SnCl_4(l)$	−511.3
hydrogen perchlorate	$HClO_4(l)$	−40.6	tin(II) oxide	$SnO(s)$	−280.7
hydrogen peroxide	$H_2O_2(l)$	−187.8	tin(IV) oxide	$SnO_2(s)$	−577.6
hydrogen sulfide	$H_2S(g)$	−20.6	2,2,4-trimethylpentane	$C_8H_{18}(l)$	−259.2
iodine (vapour)	$I_2(g)$	+62.4	urea	$CO(NH_2)_2(s)$	−333.5
iron(II) oxide	$FeO(s)$	−272.0	water (liquid)	$H_2O(l)$	−285.8
iron(III) oxide	$Fe_2O_3(s)$	−824.2	water (vapour)	$H_2O(g)$	−241.8
iron(II, III) oxide	$Fe_3O_4(s)$	−1118.4	zinc oxide	$ZnO(s)$	−350.5
lead(II) bromide	$PbBr_2(s)$	−278.7	zinc sulfide	$ZnS(s)$	−206.0
lead(II) chloride	$PbCl_2(s)$	−359.4			
lead(II) oxide	$PbO(s)$	−219.0			
lead(IV) oxide	$PbO_2(s)$	−277.4			
magnesium carbonate	$MgCO_3(s)$	−1095.8			
magnesium chloride	$MgCl_2(s)$	−641.3			
magnesium hydroxide	$Mg(OH)_2(s)$	−924.5			
magnesium oxide	$MgO(s)$	−601.6			
magnesium sulfate	$MgSO_4(s)$	−1284.9			

- Standard molar enthalpies (heats) of formation are measured at SATP (25 °C and 100 kPa). The values were obtained from *The CRC Handbook of Chemistry and Physics*.
- The standard molar enthalpies of elements in their standard states are defined as zero.

RELATIVE STRENGTHS OF OXIDIZING AND REDUCING AGENTS

	Oxidizing agents		Reducing agents	$E°_r$ (V)
SOA Strongest Oxidizing Agents	$F_2(g) + 2\,e^-$	$\rightleftharpoons$	$2\,F^-(aq)$	+2.87
	$PbO_2(s) + SO_4^{2-}(aq) + 4\,H^+(aq) + 2\,e^-$	$\rightleftharpoons$	$PbSO_4(s) + 2\,H_2O(l)$	+1.69
	$MnO_4^-(aq) + 8\,H^+(aq) + 5\,e^-$	$\rightleftharpoons$	$Mn^{2+}(aq) + 4\,H_2O(l)$	+1.51
	$Au^{3+}(aq) + 3\,e^-$	$\rightleftharpoons$	$Au(s)$	+1.50
	$ClO_4^-(aq) + 8\,H^+(aq) + 8\,e^-$	$\rightleftharpoons$	$Cl^-(aq) + 4\,H_2O(l)$	+1.39
	$Cl_2(g) + 2\,e^-$	$\rightleftharpoons$	$2\,Cl^-(aq)$	+1.36
	$2\,HNO_2(aq) + 4\,H^+(aq) + 4\,e^-$	$\rightleftharpoons$	$N_2O(g) + 3\,H_2O(l)$	+1.30
	$Cr_2O_7^{2-}(aq) + 14\,H^+(aq) + 6\,e^-$	$\rightleftharpoons$	$2\,Cr^{3+}(aq) + 7\,H_2O(l)$	+1.23
	$O_2(g) + 4\,H^+(aq) + 4\,e^-$	$\rightleftharpoons$	$2\,H_2O(l)$	+1.23
	$MnO_2(s) + 4\,H^+(aq) + 2\,e^-$	$\rightleftharpoons$	$Mn^{2+}(aq) + 2\,H_2O(l)$	+1.22
	$2\,IO_3^-(aq) + 12\,H^+(aq) + 10\,e^-$	$\rightleftharpoons$	$I_2(s) + 6\,H_2O(l)$	+1.20
	$Br_2(l) + 2\,e^-$	$\rightleftharpoons$	$2\,Br^-(aq)$	+1.07
	$Hg^{2+}(aq) + 2\,e^-$	$\rightleftharpoons$	$Hg(l)$	+0.85
	$ClO^-(aq) + H_2O(l) + 2\,e^-$	$\rightleftharpoons$	$Cl^-(aq) + 2\,OH^-(aq)$	+0.84
	$Ag^+(aq) + e^-$	$\rightleftharpoons$	$Ag(s)$	+0.80
	$2\,NO_3^-(aq) + 4\,H^+(aq) + 2\,e^-$	$\rightleftharpoons$	$N_2O_4(g) + 2\,H_2O(l)$	+0.80
	$Fe^{3+}(aq) + e^-$	$\rightleftharpoons$	$Fe^{2+}(aq)$	+0.77
	$O_2(g) + 2\,H^+(aq) + 2\,e^-$	$\rightleftharpoons$	$H_2O_2(l)$	+0.70
	$MnO_4^-(aq) + 2\,H_2O(l) + 3\,e^-$	$\rightleftharpoons$	$MnO_2(s) + 4\,OH^-(aq)$	+0.60
	$I_2(s) + 2\,e^-$	$\rightleftharpoons$	$2\,I^-(aq)$	+0.54
	$Cu^+(aq) + e^-$	$\rightleftharpoons$	$Cu(s)$	+0.52
	$O_2(g) + 2\,H_2O(l) + 4\,e^-$	$\rightleftharpoons$	$4\,OH^-(aq)$	+0.40
	$Cu^{2+}(aq) + 2\,e^-$	$\rightleftharpoons$	$Cu(s)$	+0.34
	$SO_4^{2-}(aq) + 4\,H^+(aq) + 2\,e^-$	$\rightleftharpoons$	$H_2SO_3(aq) + H_2O(l)$	+0.17
	$Sn^{4+}(aq) + 2\,e^-$	$\rightleftharpoons$	$Sn^{2+}(aq)$	+0.15
	$Cu^{2+}(aq) + e^-$	$\rightleftharpoons$	$Cu^+(aq)$	+0.15
	$S(s) + 2\,H^+(aq) + 2\,e^-$	$\rightleftharpoons$	$H_2S(aq)$	+0.14
	$AgBr(s) + e^-$	$\rightleftharpoons$	$Ag(s) + Br^-(aq)$	+0.07
	$2\,H^+(aq) + 2\,e^-$	$\rightleftharpoons$	$H_2(g)$	0.00
	$Pb^{2+}(aq) + 2\,e^-$	$\rightleftharpoons$	$Pb(s)$	−0.13
	$Sn^{2+}(aq) + 2\,e^-$	$\rightleftharpoons$	$Sn(s)$	−0.14
	$AgI(s) + e^-$	$\rightleftharpoons$	$Ag(s) + I^-(aq)$	−0.15
	$Ni^{2+}(aq) + 2\,e^-$	$\rightleftharpoons$	$Ni(s)$	−0.26
	$Co^{2+}(aq) + 2\,e^-$	$\rightleftharpoons$	$Co(s)$	−0.28
	$H_3PO_4(aq) + 2\,H^+(l) + 2\,e^-$	$\rightleftharpoons$	$H_3PO_3(aq) + H_2O(l)$	−0.28
	$PbSO_4(s) + 2\,e^-$	$\rightleftharpoons$	$Pb(s) + SO_4^{2-}(aq)$	−0.36
	$Se(s) + 2\,H^+(aq) + 2\,e^-$	$\rightleftharpoons$	$H_2Se(aq)$	−0.40
	$Cd^{2+}(aq) + 2\,e^-$	$\rightleftharpoons$	$Cd(s)$	−0.40
	$Cr^{3+}(aq) + e^-$	$\rightleftharpoons$	$Cr^{2+}(aq)$	−0.41
	$Fe^{2+}(aq) + 2\,e^-$	$\rightleftharpoons$	$Fe(s)$	−0.45
	$NO_2^-(aq) + H_2O(l) + e^-$	$\rightleftharpoons$	$NO(g) + 2\,OH^-(aq)$	−0.46
	$Ag_2S(s) + 2\,e^-$	$\rightleftharpoons$	$2\,Ag(s) + S^{2-}(aq)$	−0.69
	$Zn^{2+}(aq) + 2\,e^-$	$\rightleftharpoons$	$Zn(s)$	−0.76
	$Te(s) + 2\,H^+(aq) + 2\,e^-$	$\rightleftharpoons$	$H_2Te(aq)$	−0.79
	$2\,H_2O(l) + 2\,e^-$	$\rightleftharpoons$	$H_2(g) + 2\,OH^-(aq)$	−0.83
	$Cr^{2+}(aq) + 2\,e^-$	$\rightleftharpoons$	$Cr(s)$	−0.91
	$Se(s) + 2\,e^-$	$\rightleftharpoons$	$Se^{2-}(aq)$	−0.92
	$SO_4^{2-}(aq) + H_2O(l) + 2\,e^-$	$\rightleftharpoons$	$SO_3^{2-}(aq) + 2\,OH^-(aq)$	−0.93
	$Al^{3+}(aq) + 3\,e^-$	$\rightleftharpoons$	$Al(s)$	−1.66
	$Mg^{2+}(aq) + 2\,e^-$	$\rightleftharpoons$	$Mg(s)$	−2.37
	$Na^+(aq) + e^-$	$\rightleftharpoons$	$Na(s)$	−2.71
	$Ca^{2+}(aq) + 2\,e^-$	$\rightleftharpoons$	$Ca(s)$	−2.87
	$Ba^{2+}(aq) + 2\,e^-$	$\rightleftharpoons$	$Ba(s)$	−2.91
	$K^+(aq) + e^-$	$\rightleftharpoons$	$K(s)$	−2.93
	$Li^+(aq) + e^-$	$\rightleftharpoons$	$Li(s)$	−3.04	**SRA** Strongest Reducing Agents

DECREASING STRENGTH OF OXIDIZING AGENTS ↓

DECREASING STRENGTH OF REDUCING AGENTS ↑

- 1.0 mol/L solutions at 25 °C and 1 atm
- Values in this table are taken from *The CRC Handbook of Chemistry and Physics*.

RELATIVE STRENGTHS OF AQUEOUS ACIDS AND BASES

	Equilibrium constant, K_a	Acid		Conjugate base	
		Name	Formula	Formula	Name
SA Strongest Acid	very large	perchloric acid	$HClO_4(aq)$	$ClO_4^-(aq)$	perchlorate ion
	very large	hydroiodic acid	$HI(aq)$	$I^-(aq)$	iodide ion
	very large	hydrobromic acid	$HBr(aq)$	$Br^-(aq)$	bromide ion
	very large	hydrochloric acid	$HCl(aq)$	$Cl^-(aq)$	chloride ion
	very large	sulfuric acid	$H_2SO_4(aq)$	$HSO_4^-(aq)$	hydrogen sulfate ion
	very large	nitric acid	$HNO_3(aq)$	$NO_3^-(aq)$	nitrate ion
	1.0	hydronium ion	$H_3O^+(aq)$	$H_2O(l)$	water
	5.6×10^{-2}	oxalic acid	$HOOCCOOH(aq)$	$HOOCCOO^-(aq)$	hydrogen oxalate ion
	1.4×10^{-2}	sulfurous acid ($SO_2 + H_2O$)	$H_2SO_3(aq)$	$HSO_3^-(aq)$	hydrogen sulfite ion
	1.0×10^{-2}	hydrogen sulfate ion	$HSO_4^-(aq)$	$SO_4^{2-}(aq)$	sulfate ion
	6.9×10^{-3}	phosphoric acid	$H_3PO_4(aq)$	$H_2PO_4^-(aq)$	dihydrogen phosphate ion
	5.6×10^{-4}	nitrous acid	$HNO_2(aq)$	$NO_2^-(aq)$	nitrite ion
	7.4×10^{-4}	citric acid*	$H_3C_6H_5O_7(aq)$	$H_2C_6H_5O_7^-(aq)$	dihydrogen citrate ion*
	6.3×10^{-4}	hydrofluoric acid	$HF(aq)$	$F^-(aq)$	fluoride ion
	1.8×10^{-4}	methanoic acid	$HCOOH(aq)$	$HCOO^-(aq)$	methanoate ion
	1.5×10^{-4}	hydrogen oxalate ion	$HOOCCOO^-(aq)$	$OOCCOO^{2-}(aq)$	oxalate ion
	1.4×10^{-4}	lactic acid	$CH_3CHOHCOOH(aq)$	$CH_3CHOHCOO^-(aq)$	lactate ion
	9.1×10^{-5}	ascorbic acid	$H_2C_6H_6O_6(aq)$	$HC_6H_6O_6^-(aq)$	hydrogen ascorbate ion
	6.3×10^{-5}	benzoic acid	$C_6H_5COOH(aq)$	$C_6H_5COO^-(aq)$	benzoate ion
	1.8×10^{-5}	ethanoic (acetic) acid	$CH_3COOH(aq)$	$CH_3COO^-(aq)$	ethanoate (acetate) ion
	1.7×10^{-5}	dihydrogen citrate ion*	$H_2C_6H_5O_7^-$	$HC_6H_5O_7^{2-}$	hydrogen citrate ion*
	1.5×10^{-5}	butanoic acid	$C_3H_7COOH(aq)$	$C_3H_7COO^-(aq)$	butanoate ion
	1.3×10^{-5}	propanoic acid	$C_2H_5COOH(aq)$	$C_2H_5COO^-(aq)$	propanoate ion
	4.5×10^{-7}	carbonic acid ($CO_2 + H_2O$)	$H_2CO_3(aq)$	$HCO_3^-(aq)$	hydrogen carbonate ion
	4.0×10^{-7}	hydrogen citrate ion*	$HC_6H_5O_7^{2-}$	$C_6H_5O_7^{3-}$	citrate ion*
	8.9×10^{-8}	hydrosulfuric acid	$H_2S(aq)$	$HS^-(aq)$	hydrogen sulfide ion
	6.3×10^{-8}	hydrogen sulfite ion	$HSO_3^-(aq)$	$SO_3^{2-}(aq)$	sulfite ion
	6.2×10^{-8}	dihydrogen phosphate ion	$H_2PO_4^-(aq)$	$HPO_4^{2-}(aq)$	hydrogen phosphate ion
	4.0×10^{-8}	hypochlorous acid	$HClO(aq)$	$ClO^-(aq)$	hypochlorite ion
	6.2×10^{-10}	hydrocyanic acid	$HCN(aq)$	$CN^-(aq)$	cyanide ion
	5.8×10^{-10}	boric acid	$H_3BO_3(aq)$	$H_2BO_3^-(aq)$	dihydrogen borate ion
	5.6×10^{-10}	ammonium ion	$NH_4^+(aq)$	$NH_3(aq)$	ammonia
	1.0×10^{-10}	phenol	$C_6H_5OH(aq)$	$C_6H_5O^-(aq)$	phenoxide ion
	4.7×10^{-11}	hydrogen carbonate ion	$HCO_3^-(aq)$	$CO_3^{2-}(aq)$	carbonate ion
	2.2×10^{-12}	hydrogen peroxide	$H_2O_2(aq)$	$HO_2^-(aq)$	hydrogen peroxide ion
	2.0×10^{-12}	hydrogen ascorbate ion	$HC_6H_6O_6^-(aq)$	$C_6H_6O_6^{2-}(aq)$	ascorbate ion
	4.8×10^{-13}	hydrogen phosphate ion	$HPO_4^{2-}(aq)$	$PO_4^{3-}(aq)$	phosphate ion
	1.3×10^{-13}	hydrogen sulfide ion	$HS^-(aq)$	$S^{2-}(aq)$	sulfide ion
	1.0×10^{-14}	water (55.5 mol/L)	$H_2O(l)$	$OH^-(aq)$	hydroxide ion
	very small	hydroxide ion	$OH^-(aq)$	$O^{2-}(aq)$	oxide ion

DECREASING STRENGTH OF ACIDS →
DECREASING STRENGTH OF BASES ←
SB Strongest Base

* The molecular formula representing (triprotic) citric acid has been compressed here to its simplest form for ease of use when writing proton transfer equations.

Values in this table are taken from the *CRC Handbook of Chemistry and Physics* for 25 °C.

Appendix J COMMON CHEMICALS

You live in a chemical world. As one bumper sticker asks, "What in the world isn't chemistry?" Every natural and technologically produced substance around you is composed of chemicals. Many of these chemicals are used to make your life easier or safer, and some of them have life-saving properties. Following is a list of selected common chemicals. The chemicals marked with an asterisk are to be memorized.

Common name	Recommended name	Formula	Common use/source
acetic acid*	ethanoic acid	$CH_3COOH(aq)$	vinegar
acetone*	propanone	$(CH_3)_2CO(l)$	nail polish remover
acetylene*	ethyne	$C_2H_2(g)$	cutting/welding torch
ASA (Aspirin®)	acetylsalicylic acid	$C_6H_4COOCH_3COOH(s)$	for pain relief medication
baking soda*	sodium hydrogen carbonate	$NaHCO_3(s)$	leavening agent
battery acid*	sulfuric acid	$H_2SO_4(aq)$	car batteries
bleach	sodium hypochlorite	$NaClO(s)$	bleach for clothing
bluestone	copper(II) sulfate–(1/5)-water	$CuSO_4 \cdot 5H_2O(s)$	algicide/fungicide
brine*	aqueous sodium chloride	$NaCl(aq)$	water-softening agent
citric acid	2-hydroxy-1,2,3-propanetricarboxylic acid	$C_3H_4OH(COOH)_3$	in fruit and beverages
CFC	chlorofluorocarbon	$C_xCl_yF_z(l)$; e.g., $C_2Cl_2F_4(l)$	refrigerant
charcoal/graphite*	carbon	$C(s)$	fuel/lead pencils
dry ice*	carbon dioxide	$CO_2(g)$	"fizz" in carbonated beverages
ethylene*	ethene	$C_2H_4(g)$	for polymerization
ethylene glycol*	ethane-1,2-diol	$C_2H_4(OH)_2(l)$	radiator antifreeze
freon-12	dichlorodifluoromethane	$CCl_2F_2(l)$	refrigerant
Glauber's salt	sodium sulfate–(1/10)-water	$Na_2SO_4 \cdot 10H_2O(s)$	solar heat storage
glucose*	D-glucose; dextrose	$C_6H_{12}O_6(s)$	in plants and blood
grain alcohol*	ethanol (ethyl alcohol)	$C_2H_5OH(l)$	beverage alcohol
gypsum	calcium sulfate–water	$CaSO_4 \cdot 2H_2O(s)$	wallboard
lime (quicklime)*	calcium oxide	$CaO(s)$	masonry
limestone*	calcium carbonate	$CaCO_3(s)$	chalk and building materials
lye (caustic soda)*	sodium hydroxide	$NaOH(s)$	oven/drain cleaner
malachite	copper(II) hydroxide carbonate	$Cu(OH)_2 \cdot CuCO_3(s)$	copper mineral
methyl hydrate*	methanol (methyl alcohol)	$CH_3OH(l)$	gas-line antifreeze
milk of magnesia	magnesium hydroxide	$Mg(OH)_2(s)$	antacid (for indigestion)
MSG	monosodium glutamate	$NaC_5H_8NO_4(s)$	flavour enhancer
muriatic acid*	hydrochloric acid	$HCl(aq)$	in concrete etching
natural gas*	methane	$CH_4(g)$	fuel
PCBs	polychlorinated biphenyls	$(C_6H_xCl_y)_2$; e.g., $(C_6H_4Cl_2)_2(l)$	in transformers
potash*	potassium chloride	$KCl(s)$	fertilizer
road salt*	calcium chloride or sodium chloride	$CaCl_2(s)$ or $NaCl_2(s)$	melts ice
rotten-egg gas*	hydrogen sulfide	$H_2S(g)$	in natural gas
rubbing alcohol	propan-2-ol	$CH_3CHOHCH_3(l)$	for massage
sand (silica)	silicon dioxide	$SiO_2(s)$	in glass making
slaked lime*	calcium hydroxide	$Ca(OH)_2(s)$	limewater
soda ash*	sodium carbonate	$Na_2CO_3(s)$	in laundry detergents
sugar*	sucrose	$C_{12}H_{22}O_{11}(s)$	sweetener
table salt*	sodium chloride	$NaCl(s)$	seasoning
washing soda*	sodium carbonate–(1/10)-water	$Na_2CO_3 \cdot 10H_2O(s)$	water softener
vitamin C	ascorbic acid	$H_2C_6H_6O_6(s)$	vitamin

Glossary

A

absolute temperature scale, T [e] a temperature scale based on absolute zero, whereby absolute zero ($-273.15\ °C$) is zero kelvin (0 K); **kelvin temperature scale**

absolute zero [e] the lowest possible temperature; $-273.15\ °C$ or 0 K

accuracy an expression of how close an experimental value is to the accepted value; often reported as a percent difference

acid [e] a compound that forms electrically conducting aqueous solutions that turn blue litmus red, neutralize bases, and react with active metals to produce hydrogen gas

acid (Arrhenius) [t] a substance that ionizes in water to produce hydrogen ions

acid (modified Arrhenius) [t] a substance that reacts with water to produce hydronium ions

acid (Brønsted–Lowry) [t] a proton donor

acid ionization constant, K_a [e] the equilibrium constant for the ionization of weak acids; also known as the acid dissociation constant

acid–base indicator [e, t] a substance that changes colour in solution when the acidity of the solution changes; a conjugate acid–base pair formed by an indicator dye (large complex molecule)

acidic [e] a solution that turns blue litmus to red and has a pH less than 7; $[H_3O^+(aq)] > [OH^-(aq)]$

activated complex [t] the structural arrangement of atoms/ions believed to represent the highest potential energy point in a chemical potential energy diagram

activation energy, E_a [t] an energy barrier that must be overcome for a chemical reaction to occur; the minimum energy that entities must reach before they can react

active metal [e] a metal that spontaneously reacts with water or an acid to produce hydrogen gas; a metal that is a stronger reducing agent than hydrogen

addition polymerization [t] a reaction in which many unsaturated monomer units with double or triple bonds join together, involving addition reactions

addition reaction [t] a type of organic reaction of unsaturated hydrocarbons in which a small molecule is added to the double or triple bond

alcohol [t] an organic compound containing one or more hydroxyl ($-OH$) groups as the functional group

aliphatic [t] a broad class of hydrocarbons including straight or branched chains or rings of alkanes, alkenes, and alkynes, but not aromatics

alkali metal [e] a soft, silver-coloured metal that reacts violently with water; a Group 1 element

alkaline see **basic**

alkaline-earth metal [e] a light, reactive metal that forms an oxide coating when exposed to air; in Group 2

alkane [t] one of a family of hydrocarbons whose molecules contain only carbon–carbon single bonds; general formula C_nH_{2n+2}

alkene [t] one of a family of hydrocarbons with one or more carbon–carbon double bonds; general formula of C_nH_{2n}

alkyl branch [t] a group of atoms, consisting of only singly bonded carbon and hydrogen, that is not part of the main structure of the molecule

alkyne [t] one of a family of hydrocarbons with one or more carbon–carbon triple bonds; general formula of C_nH_{2n-2}

ambient conditions surrounding or room conditions

amount concentration, c or [...] [e] the chemical amount of solute present in one litre of solution, in units of moles per litre (mol/L); formerly called molar concentration

amount of substance, n [e] see **chemical amount**

ampere SI base unit for electric current; 1 A = 1 C/s

amphiprotic [t] an entity having the ability to either accept or donate a proton

amphoteric [e] a chemical substance with the ability to react as either an acid or a base

analogy a comparison that communicates an idea in more familiar or recognizable terms, often related to everyday experiences

analysis the part of an investigation report that includes manipulations, interpretations, and calculations of the Evidence in order to answer the question stated in the Problem

analyze to make a mathematical, chemical, or methodical examination of parts to determine the nature, proportion, function, or interrelationship of the whole; e.g., evidence is analyzed

anecdotal evidence evidence that is based upon personal experience or hearsay

anhydrous the form of a substance without any water of hydration

anion [t] a negatively charged ion

anode [t] the electrode of a voltaic or an electrolytic cell where oxidation (loss of electrons) occurs

aqueous solution [e] a homogenous mixture of one or more solutes dissolved in water; designated as (aq)

aromatic [t] a description of an organic compound with a structure based on benzene or a benzene-like ring

assumption an untested statement presumed to be correct in order to develop or apply a theory, law, or generalization

atmosphere non-SI unit of pressure; 1 atm = 101.325 kPa

atmospheric pressure [e] the force per unit area exerted by air on all objects

atom [t] the smallest entity of an element that is still characteristic of that element

atomic number [t] the characteristic number of protons in the nucleus of an atom of a particular element

Avogadro's constant, N_A [e] the number of entities of a chemical substance in one mole of that substance; estimated at 6.02×10^{23}/mol

Avogadro's theory [t] the statement: *equal volumes of gases at the same temperature and pressure contain equal numbers of molecules*

B

balanced chemical equation [t] a chemical equation in which the total number of each kind of atom or ion in the reactants is equal to the total number of the same kind of atom or ion in the products

barometer a technological device for measuring atmospheric pressure

base [e] a compound that forms electrically conducting aqueous solutions that turn red litmus blue and neutralize acids

base (Arrhenius) [t] an ionic compound that dissociates in water to produce hydroxide ions

base (modified Arrhenius) [t] a substance that reacts with water to produce hydroxide ions

base (Brønsted–Lowry) [t] a proton acceptor

base ionization constant, K_b [e] the equilibrium constant for the ionization of weak bases; also known as the base dissociation constant

basic [e] a solution that turns red litmus to blue and has a pH greater than 7; [OH⁻(aq)] > [H₃O⁺(aq)]

battery a set of two or more electric (or voltaic) cells joined in series to produce a larger voltage

binary ionic compound [t] a compound that contains only two kinds of monatomic ions

binary molecular compound [t] a compound composed of only two kinds of atoms, both of which are nonmetals

bitumen [e] a tarry residue or heavy oil, often found mixed with sand

bond dipole [t] the charge separation that occurs when the electronegativity difference of two bonded atoms shifts the shared electrons, making one end of the bond partially positive and the other partially negative

bond energy [t] the energy required to break a chemical bond; also the energy released when a bond is formed

bonding capacity [t] the maximum number of single covalent bonds that an atom can form; determined by the number of bonding electrons in the atom

bonding electron [t] a single unpaired electron, in a valence orbital, that can be shared or exchanged with another atom

Boyle's law [e] the statement: *as the pressure on a gas increases, the volume of the gas decreases proportionally, provided that the temperature and amount of gas remain constant*; $P_1V_1 = P_2V_2$

branch any group of atoms that are not part of the main (parent) chain of an organic molecule

Brønsted–Lowry concept [t] the idea that the reaction of an acid with a base is a transfer of a proton, H⁺, from the acid to the base

Brønsted–Lowry reaction equation [t] a chemical reaction equation written to show an acid–base reaction involving the transfer of a proton from one entity (an acid) to another (a base)

buffer [e] a mixture of a conjugate acid–base pair that maintains a nearly constant pH when diluted or when a strong acid or base is added

buffer capacity [e] the limit of the ability of a buffer to maintain a nearly constant pH

buffer zone [e] a relatively flat region on the pH curve of an acid–base reaction where buffering action occurs; also called the buffering region

burette a long, graduated tube equipped with a stopcock; used to measure solution volumes in a titration

byproduct an additional product (other than the primary product) in a chemical process

C

calorimeter [e] an isolated system in which the chemical system (reactants and products) being studied is surrounded by a known quantity of liquid (generally water)

calorimetry [e] the technological process of measuring energy changes of an isolated system

carboxyl group [t] –COOH; the functional group of carboxylic acids

carboxylic acid [t] an organic compound that contains the carboxyl group (–COOH) as the functional group

catalysis the study of the properties and development of catalysts and their effects on rates of reaction

catalyst [e] a substance that increases the rate of a chemical reaction without being consumed itself in the overall process

catalytic reforming [e] a technological process for converting molecules in a naphtha (gasoline) fraction into aromatic gasoline molecules

cathode [t] the electrode of a voltaic or electrolytic cell where reduction (gain of electrons) occurs

cathodic protection [e] an effective method of preventing corrosion of iron in which the iron is forced to become the cathode by using either an impressed current or a sacrificial anode

cation [t] a positively charged ion

central atom [t] the atom in a molecule that has the most bonding electrons and, therefore, is likely to form the most bonds

certainty an expression of the level of confidence; communicated by a number of significant digits for quantitative values

Charles' law [e] the statement: *as the absolute temperature of a gas increases, the volume increases proportionally, provided that the pressure and amount of gas remain constant*; $\frac{V_1}{T_1} = \frac{V_2}{T_2}$

chemical amount, *n* [e] SI quantity for the number of entities in a substance, measured in units of moles (mol); **amount of substance**

chemical bond [t] the electrical attraction that holds atoms or ions together in an element or compound

chemical change [e] see **chemical reaction**

chemical decomposition [e] the breakdown of a compound into smaller components such as elements (as in **simple decomposition**)

chemical formula [t] a series of symbols representing the atoms/ions, and their proportions, present in a pure substance

chemical potential energy [t] the energy present in the chemical bonds of a substance

chemical potential energy diagram [t] a graphical representation of a chemical reaction in which the difference between the chemical potential energy of the reactants and that of the products is the enthalpy change

chemical reaction [e, t] a change in which one or more new substances with different properties are formed as evidenced by changes in colour, energy, odour, or state; a change in the chemical bonds resulting from the rearrangement of entities (electrons, ions, protons)

chemical reaction equilibrium [t] an apparently static state of a reaction system where the reactants and products may be favoured but the rates of the forward and reverse reactions are equal

chemical technology the study and application of skills, processes, and equipment for the production and use of chemicals

chemistry the physical science that deals with the composition, properties, and changes in matter; the study of chemicals and their reactions, associated technologies, and environmental effects

closed system [e] a chemical system that is separated from its surroundings by a definite boundary so that no matter can enter or leave, but energy can enter and leave (e.g., a capped pop bottle)

coefficient [t] the number, placed in front of a chemical formula in a balanced chemical equation, that represents the number of molecules or formula units of that substance involved in the reaction

collision–reaction theory [t] the idea that chemical reactions involve collisions and rearrangements of entities

colorimetry [e] chemical analysis by colour, using light emitted, absorbed, or transmitted by a solution

combined gas law [e] the statement: *the product of the pressure and volume of a gas sample is proportional to its absolute temperature in Kelvin*; $\frac{P_1 V_1}{T_1} = \frac{P_2 V_2}{T_2}$

combustion [e] the rapid reaction of a chemical with oxygen to produce oxides and heat; also known commonly as burning

commercial pertaining to companies that use manufactured products to produce other products and processes

compare examine the character or qualities of two things by providing characteristics of both that point out their similarities and differences

complete combustion reaction [e] the burning of a substance with sufficient oxygen available to produce the most common oxides of the elements making up the substance that is burned

compound [e, t] a pure substance that can be separated into its elements by heat or electricity; a substance containing atoms/ions of more than one element in a definite fixed proportion

concentration, *c* [e] a ratio of the quantity of solute to the quantity of solution or solvent; expressed in a variety of units

conclude state a logical end based on reasoning and/or evidence

condensation polymerization [t] a reaction that involves the "condensing out" or removal of a small molecule (such as H_2O, NH_3, or HCl) from the functional groups of two different monomer molecules to form a condensation polymer

condensation reaction [t] an organic reaction in which two smaller molecules combine to form a larger molecule, along with a small molecule such as water; e.g., **esterification reaction**

conductivity [e] a measure of the ability of a substance to conduct an electric current

conjugate acid–base pair [t] a pair of substances with chemical formulas that differ only by a proton

consumer pertaining to an individual in society

control a substance, condition, or procedure that does not change and is used as a comparison in an experiment

controlled variable a property or condition that could vary but is held constant so as not to affect the outcome of an experiment; also known as fixed or restrained variable

coordinate covalent bond [t] a type of covalent bond in which one of the atoms donates both electrons

corrosion [e] an electrochemical process in which a metal reacts with substances in the environment, returning the metal to an ore-like state

coulomb SI unit for the quantity of electric charge transferred in one second by one ampere of current; unit symbol, C

contrast point out the differences between two things that have similar or comparable natures; **distinguish**

covalent bond [t] the simultaneous attraction of two nuclei for a shared pair of bonding electrons

covalent network [t] a three-dimensional arrangement of covalent bonds between atoms throughout a crystal (e.g., quartz, diamond)

cracking [t] a chemical process in which larger (e.g., hydrocarbon) molecules are broken down at high temperature, with or without catalysts, to produce smaller molecules

create–test–use a three-stage process of laboratory work that results in scientific knowledge coming to be accepted with increased certainty

crystal lattice [t] a continuous, three-dimensional pattern of atoms, ions, or molecules in a crystalline solid

crystallization [e] the technological process of separating a solid from a solution by evaporating the solvent or cooling the solution

cycloalkane [t] a cyclic hydrocarbon (molecules have a closed ring structure) in which all the carbon–carbon bonds are single bonds; general formula C_nH_{2n}

cycloalkene [t] a cyclic hydrocarbon with at least one carbon–carbon double bond

D

define provide the essential qualities or meaning of a word or concept; make distinct and clear by marking out the limits

describe give a written account or represent the characteristics of something by a figure, model, or picture

design construct a plan for a specific purpose; for an investigation, an overview of the Procedure including reacting chemicals, diagnostic tests, variables, and controls

determine find a solution, to the required degree of accuracy, to a problem by showing appropriate formulas, procedures, and calculations

diagnostic test a short laboratory procedure with expected evidence and analysis; used to identify or classify substances

diatomic molecule [t] a molecule containing two atoms

dilution [e] the process of adding solvent to a solution to reduce its concentration

dipole [t] a partial separation of positive and negative charges within a molecule, due to electronegativity differences

dipole–dipole force [t] the simultaneous attraction between oppositely charged ends of polar molecules

discharging the spontaneous conversion of chemical energy into electrical energy by a cell or battery

disproportionation [e] the reaction of a single substance with itself to produce two different substances; a type of redox reaction in which an entity is both oxidized and reduced

dissociation [t] a separation and dispersal of previously bonded entities; the separation of ions that occurs when an ionic compound dissolves in water

distinguish point out the differences between two things that have similar or comparable natures; **contrast**

double blind an experimental design where neither the subject nor the experimenter knows whether the subject is in a control group or an experimental group

double bond [t] an attraction between atoms in a molecule due to the sharing of two pairs of electrons in a covalent bond

double replacement reaction [e] a reaction of two ionic compounds in which the cations and anions rearrange, producing two different compounds

dry cell [e] any sealed electric cell with semi-solid contents; originally used to describe the zinc–chloride cell

ductile [e] able to be drawn (stretched) into a wire or a tube

dynamic equilibrium theory [t] the idea that there is a balance between two opposing processes (forward and reverse) occurring at the same rate

E

ecological perspective pro or con statement relating to relationships between living organisms (including human beings) and the environment

economic perspective pro or con statement relating to the production, distribution, and consumption of wealth

electric cell a technological device for converting chemical energy into electrical energy

electric current, *I* a measure of the rate of flow of charge past a point in an electrical circuit; measured by an ammeter in ampères (A)

electric potential difference, *V* the energy difference, per unit charge, between any two points in an electrical circuit; measured in volts (V); also called **voltage**

electrochemical cell a cell that either converts chemical energy to electrical energy (electric or voltaic cell), or electrical energy into chemical energy (electrolytic cell)

electrochemistry the branch of chemistry that studies electron transfers in chemical reactions

electrode [e] a solid electrical conductor

electrolysis [e] the process of supplying electrical energy to force a nonspontaneous redox reaction to occur

electrolyte [e] a substance that conducts electricity in aqueous solution; also a liquid such as a molten solid that conducts electricity

electrolytic cell [e] a cell in which a nonspontaneous redox reaction is forced to occur; a combination of two electrodes, an electrolyte, and an external power source

electron [t] a small, negatively charged subatomic particle; has a specific energy within an atom

electron dot diagram see **Lewis symbol** and **Lewis formula**

electronegativity [t] a number that describes the relative ability of an atom to attract a pair of bonding electrons in its valence level

electroplating [e] a process for depositing a metal onto another object that is the cathode of a cell

electrorefining [e] a process for producing pure metals using an electrolytic cell to deposit the pure metal at the cathode and using the impure metal as the anode

element [e, t] a pure substance that cannot be broken down into simpler chemical substances by any physical or chemical means; consists of only one kind of atom

elimination reaction [t] an organic reaction that involves the removal of atoms and/or groups of atoms from adjacent carbon atoms in an organic molecule (e.g., the removal of H_2O from an alcohol)

emotional perspective a pro or con statement referring to feelings as opposed to logic or reasoning

empirical definition a statement that defines an object or a process in terms of observable properties

empirical formula the experimentally determined simplest whole-number ratio of atoms or ions in a compound

empirical hypothesis a preliminary generalization, regarding observable properties, that requires further testing

empirical knowledge knowledge gained through observation

endothermic reaction [e, t] a chemical reaction that absorbs energy from the surroundings; a change resulting in a higher enthalpy of the chemical system

endpoint [e] the point during a titration when there is a sudden change in some observable property of the solution, such as colour, pH, or conductivity

energy [e] a property of a substance or system that relates to its ability to do work

energy diagram see **chemical potential energy diagram**

energy level [t] a specific energy an electron can have in an atom or ion

enthalpy [t] the total kinetic and potential energy within a system at constant pressure

enthalpy change, ΔH or $\Delta_r H$ [t] the difference between the enthalpy of the products and the enthalpy of the reactants of a chemical reaction

enthalpy of reaction [e] the energy change for a whole chemical system when reactants change to products; also known as change in enthalpy

entity [t] a general term that includes atoms, ions, and molecules

enumerate specify one by one or list in concise form and according to some order

enzyme [e] a compound, usually a protein, that acts as a catalyst in a living system; often responsible for controlling a certain physiological reaction

equilibrium [e] the state of a closed system with no observable changes occurring; having constant macroscopic properties

equilibrium constant, K_c (e) a constant value for a reaction system over a range of concentrations

equilibrium law [e] a mathematical expression relating concentrations of reactants and products to the equilibrium constant, $K_c = [C]^c[D]^d/([A]^a[B]^b)$ where A, B, C, and D represent chemical entity formulas and a, b, c, and d represent their coefficients in the balanced chemical equation

equilibrium shift [e] the change in concentrations that occurs when a system at equilibrium is disrupted (e.g., by the addition of more reactant)

equivalence point [t] the point during a titration at which the exact theoretical chemical amount of titrant has been added to the sample

ester [e, t] a family of organic compounds characterized by the **ester functional group** (–COO–); formed from the reaction of a carboxylic acid and an alcohol

ester functional group [t] –COO–; the functional group of esters

esterification reaction [e] a condensation reaction in which a carboxylic acid and an alcohol combine to produce an ester and water

esthetic perspective a pro or con statement relating to beauty

ethical perspective a pro or con statement relating to whether an action is right or wrong

evaluate give the significance or worth of something by identifying the good and bad points or the advantages and disadvantages

evaluation the part of an investigation report that includes judging the validity of the experiment (Design, Materials, Procedure, and Skills), judging the Prediction and authority, and assessing whether the Purpose was achieved

evidence the part of an investigation report that includes all qualitative and quantitative observations related to answering the Problem; data or observations with a scientific purpose

excess reagent [e] the reactant that is still present after a reaction has gone to completion

exothermic reaction [e, t] a chemical reaction that releases energy to the surroundings; a change resulting in lower enthalpy of the chemical system

explain make clear what is not immediately obvious or entirely known; give the cause of or reason for; make known in detail

F

family [e] a set of elements with similar chemical properties; the elements in a vertical column in the main part of the periodic table; also called a **group**

Faraday constant, F [e] the quantity of charge transferred for every mole of electrons that flows in the cell; 9.65×10^4 C/mol

Faraday's law [e] the statement: *the mass of an element produced or consumed at an electrode is directly proportional to the time the cell operates, as long as the current is constant or vice versa*

filtrate [e] the solution that flows through a filter paper

filtration the technological process of separating a slightly soluble solid from a mixture using a porous filter paper or medium

flame test [e] a diagnostic test based on the characteristic colours of ions in a flame

formation reaction [e] the reaction of two or more elements to form either an ionic compound (from a metal and a nonmetal) or a molecular compound (from two or more nonmetals)

formula subscript [t] the number of ions or atoms present in one molecule or formula unit of a substance

formula unit [t] the simplest whole number ratio of ions in an ionic compound

forward reaction the left-to-right change in a written chemical equilibrium equation

fossil fuels an energy resource believed to be the accumulated remains of plants and animals from past geologic periods; includes petroleum and coal

fraction [e, t] a product in the fractionation process, identified by its boiling-point range or by the approximate number of carbon atoms in the component molecules

fractionation [e] a technological process in which the components of crude oil are physically separated by means of their differences in boiling points

fuel cell [e] an electric cell that produces electricity by the reaction of a fuel that is continuously supplied to keep the cell operating

functional group [t] a characteristic arrangement of atoms or bonds within a molecule that determines the most important chemical and physical properties of a class of compounds

G

gas stoichiometry [e] the method of predicting or analyzing the quantity of a gas involved in a chemical reaction

gasohol [e] a mixture of gasoline and alcohol used as an automobile fuel

generalization a statement that summarizes a limited number of empirical results

graduated pipette a precise device consisting of a narrow glass tube with regular markings

graphically using a drawing that is produced electronically or by hand and that shows a relationship between certain sets of numbers

gravimetric analysis [e] a chemical analysis that uses stoichiometric calculations based on a measured mass of a reagent

gravimetric stoichiometry [e] the method of predicting or analyzing the masses of reactants and/or products involved in a chemical reaction

green chemistry the design of chemical products and processes that reduce or eliminate the use and generation of hazardous substances

greenhouse gas a gas (e.g., carbon dioxide, chlorofluorocarbons, methane), that is released or present in the atmosphere, that serves to trap heat and reduce its transfer into outer space

group [e] a set of elements with similar chemical properties; the elements in a vertical column in the main part of the periodic table; also called a **family**

H

Haber process an industrial process, named after Fritz Haber, that produces ammonia from hydrogen and atmospheric nitrogen

half-cell [e] an electrode–electrolyte combination forming one-half of a complete cell

half-reaction [t] a balanced chemical equation that represents either a loss or gain of electrons by a substance

halogen [e] a reactive nonmetal element from Group 17

heat [e] a form of energy that can only be transferred between substances, never possessed by them

heat of reaction see **enthalpy change**

Hess' law [e] the statement: *the addition of chemical equations yields a net chemical equation whose enthalpy change is the sum of the individual enthalpy changes*; $\Delta_r H° = \Sigma \Delta_r H°$ (in standard conditions)

heterogeneous mixture [e] a mixture that is non-uniform, and may consist of more than one phase

homogeneous mixture [e] a mixture that is uniform and consists of only one phase; e.g., a solution

homologous series [e] a series of compounds, with similar structures, in which each member differs from the next by a constant unit (e.g., CH_2)

hydrate [e, t] a pure substance that decomposes at a relatively low temperature to produce water and another substance; a substance containing loosely bonded water molecules

hydrocarbon [t] a compound containing only carbon and hydrogen atoms

hydrocarbon derivative [t] a molecular compound of carbon and, usually, hydrogen plus at least one other element (e.g., organic halides, alcohols)

hydrocracking [e] a technological process involving a combination of catalytic cracking and hydrogenation that is used to reduce heavier feedstocks to, for example, gasoline-type molecules

hydrogen bond [t] the simultaneous attraction of a hydrogen nucleus (a proton), that is covalently bonded to a very electronegative atom, for a lone pair of electrons on an adjacent molecule; generally involves molecules with F–H, O–H, and/or N–H bonds

hydrogenation [t] an addition reaction in which hydrogen is added to an unsaturated hydrocarbon to reduce the number of multiple bonds

hydronium ion [t] a hydrated hydrogen ion (proton), $H_3O^+(aq)$; the entity responsible for acidic properties in aqueous solutions

hydroxyl group [t] –OH; the functional group for alcohols

hypothesis a preliminary empirical or theoretical concept that provides a possible explanation for a natural or technological phenomenon, and that requires further testing

I

ICE table a table used to record the *initial*, *change*, and *equilibrium* concentrations of reactants and products in solution

ideal gas [e, t] a gas that obeys all the gas laws perfectly under all conditions; a hypothetical gas with molecules of negligible size and no intermolecular forces between the molecules

ideal gas law [e] the statement: *the product of pressure and volume of a gas is directly proportional to the chemical amount and absolute temperature*; $PV = nRT$

identify recognize and select as having the characteristics of something

illustrate make clear by giving a specified example

indicator see **acid–base indicator**

industrial pertaining to large-scale companies that usually deal with natural raw materials

inert electrode [e] an unreactive solid conductor in a cell that provides a location to connect a wire and a surface on which a half-reaction can occur

infer form a generalization from sample data; arrive at a conclusion by reasoning from evidence

intermediate [e] a chemical entity that forms with varying stability at the end of a step in a reaction mechanism

intermolecular force [t] the relatively weak forces of attraction and repulsion between molecules

interpret tell the meaning of something; present information in a new form that adds meaning to the original data

interpretation an indirect form of knowledge that builds upon a concept or an experience to further describe or explain an observation

intramolecular force [t] the relatively strong bonds or forces of attraction and repulsion within a molecule; typically covalent bonds

ion [t] an entity with a net positive or negative electrical charge due to the loss or gain of one or more electrons

ionic bond [t] the simultaneous attraction among positive and negative ions

ionic compound [e, t] a pure substance formed from a metal and a nonmetal; a crystalline solid at SATP with a relatively high melting point and a conductor of electricity in aqueous or molten states; a compound containing positive and negative ions

ionic formula [t] a group of chemical symbols representing the simplest whole number ratio of ions in the compound

ionization [t] a process by which a neutral atom or molecule is converted to an ion; the reaction of substances in water to create ions

ionization constant for water, K_w [e] the equilibrium constant in the water ionization equilibrium; $K_w = [H_3O^+(aq)][OH^-(aq)]$, calculated with amount concentration units

isoelectronic molecules [t] molecules having the same total number of electrons

isolated system [e] a system in which neither matter nor energy can move in or out; used to study energy changes (e.g., a calorimeter)

isomers compounds with the same molecular formula but with different structures

IUPAC International Union of Pure and Applied Chemistry; the organization that establishes the conventions used by chemists

J

joule SI unit for energy of all kinds (J)

justify show reasons for, or give facts that support, a decision

K

Kelvin temperature scale see **absolute temperature scale**

kinetic energy [t] a form of energy related to the motion of a particle

kinetic molecular theory [t] the idea that the smallest entities of a substance—atoms, ions, or molecules—are in continuous motion, colliding with each other and objects in their path

L

law [e] see **scientific law**

law of combining volumes [e] the statement: *when measured at the same temperature and pressure, volumes of gaseous reactants and products of chemical reactions are always in simple ratios of whole numbers*

law of conservation of mass [e] the statement: *in any physical or chemical change, the total initial mass of reactant(s) is equal to the total final mass of product(s)*

Le Châtelier's principle [e] the statement: *when a chemical system at equilibrium is disturbed by a change in a property of the system, the system always appears to react in the direction that opposes the change, until a new equilibrium is reached*

legal perspective a pro or con statement relating to the laws of the land

Lewis formula [t] a chemical formula that shows all valence electrons in shared or lone pairs around each atom in a molecule or polyatomic ion

Lewis symbol [t] a simple model of the arrangement of valence electrons in atoms or monatomic ions in which the symbol for the element represents the nucleus plus all the inner electrons, and single or paired dots represent the valence electrons

limiting reagent [e] the reactant that is completely consumed in a reaction

line structural formula [t] a representation of an organic molecule using line segments to show the hydrocarbon portion where the intersections and ends of the line segments represent carbon atoms with the appropriate number of hydrogen atoms

London force [t] the simultaneous attraction between a momentary dipole in a molecule and the momentary dipoles in the surrounding molecules; strength depends on the number of electrons (and protons) in a molecule

lone pair [t] two valence electrons occupying the same orbital

M

main group element [e] an element in Groups 1, 2, or 12 to 18; best follows the periodic law

malleable [e] able to be hammered into a thin sheet; e.g., a metal

manipulated variable the property that is deliberately and systematically changed in an experiment to see what effect the change will have on another property; also known as the independent variable

mass number [t] the sum of protons and neutrons in the nucleus of an atom or monatomic ion

materials section of an investigation report that provides a complete list of equipment and chemicals, including sizes and quantities

matter [e] anything that has mass and occupies space; may be a pure substance or a mixture

metal [e] an element that is shiny, bendable, and a good conductor of electricity

metallic bond [t] the simultaneous attraction between positive nuclei and mobile valence electrons in a metal

metallurgy [e] the technological process of extracting metals from their naturally occurring compounds and adapting these metals for useful purposes

miscible [e] a term used to describe liquids that dissolve completely in each other in any proportion

mixture [e] matter whose composition includes two or more substances and may or may not be uniform throughout the sample; i.e., homogeneous or heterogeneous

model physical, graphic, or mental representation used to communicate an abstract idea

molar enthalpy of reaction, $\Delta_r H_m$ [e] the enthalpy change in a chemical system per mole of a specified chemical undergoing change in the system

molar mass, M [e] the mass of one mole of a substance in units of grams per mole (g/mol)

molar volume, V_m [e] the volume occupied by one mole of a gas at a specified temperature and pressure in units of litres per mole (L/mol)

mole [t] the SI base unit for the chemical amount or amount of a substance, where one mole is the number of entities corresponding to Avogadro's constant; unit symbol, mol

molecular compound [e, t] a pure substance that may be a solid, liquid, or gas at SATP, has a relatively low melting point, and is non-conducting in any state; a compound that consists of covalently bonded nonmetal atoms

molecular formula [t] a group of chemical symbols indicating the type and number of nonmetal atoms in a single molecule

molecular prefixes a list of prefixes recommended by IUPAC to specify the number of atoms in molecule or water molecules in a hydrate

molecule [t] an entity consisting of a group of nonmetal atoms held together by covalent bonds

momentary dipole [t] an uneven distribution of electrons around a molecule, resulting in a temporary charge difference between its ends

monatomic ion [t] a positively or negatively charged particle formed from a single atom by the loss or gain of electrons; also known as a simple ion

monomer [t] a small molecule that links with many other similar molecules, in an addition or a condensation reaction, to form a polymer

monoprotic acid [t] an acid that possesses only one ionizable (acidic) proton, HA, and can react only once with water to produce hydronium ions

monoprotic base [t] a base that can react with water only once to produce hydroxide ions

MSDS Material Safety Data Sheets; publications detailing the properties, safe handling, and disposal techniques for chemicals

multiperspective view a view that is based upon many perspectives on an issue, including both pro and con statements from each perspective

multi-valent [t] the ability of an atom to form a variety of ions

N

net ionic equation [t] a chemical reaction equation that includes only reacting entities (molecules, atoms, and/or ions) and omits any that do not change

neutral [t] having neither acidic nor basic properties; or having a net charge of zero (electrically neutral)

neutralization [e, t] a type of double replacement reaction in which an acid reacts with a base to produce water and an ionic compound; a reaction between hydronium and hydroxide ions to produce water

neutron [t] an uncharged subatomic particle present in the nuclei of most atoms

noble gas [e, t] a very unreactive gaseous element from Group 18; an element with a full shell of valence electrons

nomenclature a systematic method for naming something; chemical nomenclature is governed by the IUPAC

nonelectrolyte compounds that do not conduct electricity in aqueous solution nor in the molten/liquid state

nonmetal [e] an element that is not shiny, not bendable, and generally not a good conductor of electricity

nonpolar covalent bond [t] a bond in which the electrons are shared equally because the bonded atoms have the same electronegativity

nonpolar molecule [t] a molecule in which the negative (electron) charge is distributed symmetrically among the atoms making up the molecule

nuclear change [t] a change within the nucleus of an atom/ion that creates one or more new atoms/ions; represented by a change in atomic symbols

O

observation a direct form of knowledge obtained by means of one of the five human senses—sight, smell, taste, hearing, or touch—or with the aid of an instrument, such as a balance

octet rule [t] the idea that a maximum of eight electrons (four pairs) can occupy orbitals in the valence level of an atom

open system [e] a system in which both matter and energy can move in or out (e.g., an open bottle of pop)

orbital [t] a region of space around the nucleus of an atom where an electron of particular energy is likely to be found; may contain 0, 1, or 2 electrons

organic chemistry a major branch of chemistry that deals with compounds of carbon, excluding oxides and ionic compounds of carbon-based ions such as carbonate, cyanide, and carbide ions

organic halide [t] an organic compound in which one or more hydrogen atoms have been replaced by halogen atoms; the functional group is the halogen atom

outline give, in the requested format, the essential parts of something

oxidation [t] a chemical process involving the loss of electrons by an entity; an increase in the **oxidation number**

oxidation number a positive or negative number corresponding to the oxidation state assigned to an atom or ion

oxidizing agent [t] an entity that causes oxidation by removing (gaining) electrons from another substance in a redox reaction

P

Pascal SI unit of pressure; $1 \text{ Pa} = 1 \text{ N/m}^2$

percent difference the ratio of the difference between experimental and predicted values to the predicted value; a measure of the accuracy of the experimental value

percent yield the ratio of the difference between the actual or experimental quantity of product obtained (actual yield) and the maximum possible quantity of product (predicted or theoretical yield obtained from a stoichiometry calculation) to the maximum quantity

period [e] a horizontal row of elements in the periodic table whose properties gradually change from metallic to nonmetallic from left to right along the row

periodic law [e] the observation that chemical and physical properties of elements repeat themselves at regular intervals when the elements are arranged in order of increasing atomic number

peripheral atom [t] atoms surrounding the central atom in a molecule or polyatomic ion

perspective a point of view; a way of looking at an issue, pro or con

petrochemical a chemical made from petroleum

petroleum a complex mixture of hydrocarbons; natural gas and various types of oil

pH a measure of the acidity of a solution as the negative exponent to the base ten of the hydronium ion concentration; $\text{pH} = -\log[\text{H}_3\text{O}^+(\text{aq})]$

pH curve [e] a graph showing the continuous change of pH versus the volume of titrant during an acid–base reaction

phase equilibrium [e] a type of equilibrium that involves a single chemical substance existing in more than one phase (state) in a closed system

phenyl group [t] C_6H_5-; a benzene ring attached as a branch of a parent chain

photosynthesis [e] the formation of carbohydrates and oxygen from carbon dioxide, water, and sunlight, catalyzed by chlorophyll in the green parts of a plant

physical change [e] any change in the form of a substance in which the chemical composition does not change

pipette a glass tube used to measure precise volumes of solution

placebo in drug studies, an inactive item (e.g., sugar pill) or treatment given to the control group for comparison to the experimental group

pOH a measure of the basicity of a solution as the negative exponent to the base ten of the hydroxide ion concentration; $\text{pOH} = -\log[\text{OH}^-(\text{aq})]$

polar covalent bond [t] a bond in which the bonding electrons are not shared equally because the bonded atoms have different electronegativities

polar molecule [t] a molecule in which the negative charge is not distributed symmetrically among the atoms, resulting in partial positive and negative charges on opposite ends of the molecule

political perspective a pro or con statement relating to vote-getting actions or campaigning

polyamide [t] a polymer in which the monomers—one with two carboxyl groups (–COOH) and one with two amine groups (–NH$_2$), react to form amide linkages (–CONH–)

polyatomic ion [t] a group of atoms with a net positive or negative charge on the whole group

polyester [t] a polymer in which the monomers—one with two carboxyl groups (–COOH) and one with two alcohol groups (–OH)—react to form ester linkages (–COO–)

polymer [t] a large molecule made by linking together many smaller molecules called **monomers**

polymerization [e] the process of forming polymers from the reaction of monomers; e.g., addition and condensation polymerizations

polypeptide [t] a polymer made up of amino acid monomers joined with peptide (–CONH–) linkages; a protein is a polypeptide that is more than about fifty amino acids long

polyprotic acid [t] an acid with more than one acidic hydrogen available to react with water to form hydronium ions; an acid that can donate more than one proton

polyprotic base [t] a base that can react more than once with water to form hydroxide ions; a base that can accept more than one proton

potential energy [t] a stored form of energy dependent on the relative positions of particles in a system

porous boundary [e] a physical barrier separating two electrolytes while still permitting ions to move between the two solutions through tiny openings

ppm parts per million; the symbol for a unit of concentration corresponding to $1\ g/10^6\ g$ or $1\ mg/kg$

precipitate [e] a solid substance(s) formed during a reaction in solution

precipitation [e] a type of double replacement reaction in which a precipitate forms

precision a measure of the reproducibility or consistency of a result; in a simplified approach, precision can be measured as the place value of the last measurable digit from an instrument

predict tell in advance on the basis of empirical evidence and/or logic

prediction the expected answer (usually to the Problem in an investigation) according to some scientific concept or other authority, and including the reasoning

pressure, P the force per unit area measured in SI units of pascals (Pa); $1\ Pa = 1\ N/m^2$

primary cell a cell that cannot be recharged, usually due to irreversible side reactions

primary standard [e] a chemical that can be obtained at high purity, with a mass that can be measured to high accuracy and precision

problem the part of an investigation report that states the specific question to be answered, in terms of manipulated and responding variables where possible

procedure the part of an investigation report that is a numbered, sequential list of directions necessary to obtain the evidence needed to answer the problem

products the substances produced by a chemical reaction; substances whose chemical formulas appear to the right of the arrow in a chemical reaction equation

proteins natural polymers of amino acids forming the basic material of living things

proton a positively charged subatomic particle found in the nucleus

pseudo-scientific falsely represented as scientific knowledge or process

pure substance [e] matter whose composition is constant and uniform; composed of only one kind of chemical

purpose the stated reason for doing a scientific investigation, usually to gather evidence to create, test, or use a scientific concept

Q

qualitative analysis [e] the identification of a specific chemical substance

qualitative observation [e] an observation that describes qualities of matter or changes in matter (e.g., odour, colour, and physical state); observation with no numerical value

quantitative analysis [e] the determination of the quantity of a substance present

quantitative observation [e] an observation that involves one or more measurements of some property; observation with numerical value

quantitative reaction a reaction in which more than 99% of the limiting reagent is consumed

quantum mechanics [t] a current theory of atomic structure in which electrons are described in terms of their energy and the probability patterns

R

random error an uncertainty that is non-systematic and related to measuring or sampling errors

reactants the substances being consumed in a chemical reaction; substances whose chemical formulas appear to the left of the arrow in a chemical reaction equation

reaction mechanism [t] a sequential list of chemical reaction equations that describes the individual reaction steps and intermediates formed from the initial reactants to the final products

reagent a chemical, usually relatively pure, used in a reaction

real gas [e] an actual gas that condenses when cooled and deviates from the gas laws under certain conditions

recharging the nonspontaneous conversion of electrical energy to chemical energy in a cell

redox reaction [t] a re**d**uction–**ox**idation reaction; a chemical reaction in which electrons are transferred between entities

redox spontaneity rule [e] the generalization: *a spontaneous redox reaction occurs only if the oxidizing agent (OA) is above the reducing agent (RA) in a table of relative strengths of oxidizing and reducing agents with reduction potentials*

reducing agent [t] an entity that causes reduction by donating (losing) electrons to another substance in a redox reaction

reduction [t] a chemical process involving the gain of electrons by an entity; a decrease in the oxidation number

reference energy state a reference point at which the potential energy of elements in their most stable form at SATP is defined as zero

reference half-cell a half-cell that is chosen as a reference and arbitrarily assigned an electrode potential of exactly zero volts; a standard hydrogen electrode

refereed journal an academic publication where research papers are sent to peers (subject experts) for review and evaluation before publication

refining [e] the technology that includes physical and chemical processes for separating complex mixtures into simpler mixtures or near-pure components

reforming [e] the technological process of converting molecules into larger or more branched molecules

relate show logical or causal connection between things

responding variable the measured property that changes in response to the change in the manipulated variable; also known as dependent variable

reverse reaction usually in reference to the right–to–left change in an equilibrium equation

roasting [e] the technological process of reacting metal sulfides with oxygen to form metal oxides and sulfur dioxide

S

sacrificial anode [e] a metal that is more easily oxidized than iron and is connected to the iron object to be protected

salt bridge [e] a tube or connection containing an inert electrolyte that connects two half-cells in a voltaic cell

sample [e] a selected or measured part of the whole; e.g., the fixed volume of solution in a titration analysis

SATP standard ambient temperature and pressure; 25 °C and 100 kPa

saturated hydrocarbon [t] a compound of carbon and hydrogen whose molecules contain only carbon–carbon single bonds

saturated solution [e] a solution in which no more solute will dissolve at a specified temperature; at maximum solute concentration

science the study of the natural and technological world with the goal of describing, explaining, and predicting substances and changes

scientific perspective a pro or con statement relating to researching (describing, explaining, and predicting) natural and technological phenomena

scientific law a major empirical concept that is based on a large body of empirical knowledge

secondary cell a rechargeable cell

semi-metal a class of elements that are distributed along the "staircase line" in the periodic table; also called metalloids

significant digits all digits in a measured or calculated value that are certain plus one estimated (uncertain) digit

simple decomposition reaction [e] the breakdown of a compound into its component elements

single bond [t] an attraction between atoms in a molecule due to the sharing of a single pair of electrons in a covalent bond

single replacement reaction [e] the reaction of an element with a compound to produce a new element and a new compound

sketch provide a drawing that represents the key features of an object or graph

social perspective a pro or con statement referring to the effects on human communities

solubility [e] the concentration of a saturated solution at a specified temperature

solubility equilibrium [e, t] a constant state of a system in which excess solute is in contact with the saturated solution; the rate of dissolving equals the rate of crystallizing

solute [e] a substance that is dissolved in a solvent

solution [e] a homogenous mixture of substances composed of at least one solute and one solvent

solution stoichiometry [e] the method of predicting or analyzing the volume and concentration of solutions involved in a chemical reaction

solve give a solution for a problem

solvent a substance (usually a liquid) in which a solute is dissolved to form at solution

solvent extraction [e] a process in which a solvent is added to selectively dissolve and remove part of a mixture

specific heat capacity, c [e] the quantity of energy required to raise the temperature of a unit mass of a substance by one degree Celsius or one Kelvin

spectator ion [t] an ion that is present but does not take part in a chemical reaction

spontaneous occurring naturally without any external energy source

standard cell a voltaic cell in which each half-cell contains all entities shown in the half-reaction equation at SATP conditions, with a concentration of 1.0 mol/L for the aqueous solutions

standard cell potential, $E°_{cell}$ [e, t] the maximum electrical potential difference (voltage) of the cell operating under standard conditions; the energy difference (per unit of charge) between the cathode and the anode

standard enthalpy of formation $\Delta_f H°$ [e] the enthalpy change calculated from measurements of a formation reaction under standard conditions

standard molar enthalpy of reaction, $\Delta_r H_m°$ [e] the molar enthalpy of reaction determined when the initial and final conditions of the chemical system are standard conditions

standard reduction potential, $E°_r$ [t] the ability of a standard half-cell to attract/gain electrons under standard conditions

standard solution [e] a solution with a concentration that is known with considerable certainty; a solution of accurate concentration

standardizing [e] the process of determining the concentration of a solution by reacting it with another solution that has been prepared from a primary standard

state of matter [e] the physical form of a substance such as solid (s), liquid (l), gas (g), or aqueous solution (aq)

stereochemical formula a chemical formula that represents the three-dimensional shape of the molecule

stereochemistry the study of the 3-D spatial configuration of molecules and how this affects their reactions

stock solution [e] an initial, usually concentrated, solution from which samples are taken for dilution

stoichiometric a condition in which a reaction can be represented by a fixed and reliable mole ratio from a balanced chemical equation

stoichiometry [e] a method for predicting or analyzing the quantities of reactants and products involved in a chemical reaction

STP standard temperature and pressure; 0 °C and 1 atm (101.325 kPa)

strong acid [e, t] a substance that forms a solution with strong acidic properties, such as a low pH; a substance that reacts completely (>99%) with water to form hydronium ions

strong base [e, t] a substance that forms a solution with strong basic properties, such as high electrical conductivity and a very high pH; an ionic hydroxide that dissociates completely (>99%) to release hydroxide ions

structural formula [t] a representation of a molecule in which each covalent bond is shown as a single, double, or triple line (representing single or multiple bonds) between symbols representing bonding atoms

structural isomer [t] a compound with the same molecular formula but with a different structure as another compound

STS science, technology, and society

substitution reaction [t] an organic reaction that involves breaking a carbon–hydrogen bond in an alkane or aromatic ring and replacing the hydrogen atom with another atom or group of atoms

summarize give a brief account of the main points

surroundings [e] the environment around a chemical system

sustainable development development that meets the needs of the present generation without negatively affecting the ability of future generations to meet their own needs

systematic error an uncertainty that is inherently part of a measuring system, instrument, or design

T

technological perspective a pro or con statement referring to the development and use of machines, instruments, and processes that have a social purpose

technology the skills, processes, and equipment required to manufacture useful products or to perform useful tasks

temperature [t] an indirect measure of the average kinetic energy of molecules in a sample

theoretical definition a general statement that characterizes the nature of a substance or a process in terms of non-observables

theoretical description a specific descriptive statement based on theories or models

theoretical hypothesis a theoretical concept that is untested or extremely tentative

theoretical knowledge knowledge that explains and describes scientific observations in terms of non-observables

theoretical yield the maximum quantity of product that can be produced, predicted from the stoichiometry of the chemical reaction

theory a concept or set of ideas that explains a large number of observations in terms of non-observables

thermal energy, Q [t] the total kinetic energy of the entities of a substance; $Q = mc\Delta t$

thermal stability [e] the tendency of a compound to resist decomposition when heated

thermochemistry the study of the energy changes that take place during chemical reactions

titrant [e] the solution that is added from a burette during a titration analysis

titration [e] a technological process of carefully adding a solution (**titrant**) from a burette into a measured, fixed volume of another solution (**sample**) until the reaction is judged to be complete; often uses an indicator to indicate the endpoint

titration analysis [e] a chemical analysis that uses stoichiometric calculations based on a known concentration and measured solution volumes

transition element [e] an element in Groups 3 to 11

triple bond [t] an attraction between atoms in a molecule due to the sharing of three pairs of electrons in a covalent bond

U

universal gas constant, R [e] the constant used in the ideal gas law; usually 8.314 (kPa·L)/(mol·K)

unsaturated compound [t] an organic compound containing double or triple carbon–carbon bonds

V

valence electron [t] an electron in the highest energy level of the atom; an electron available for a covalent bond or electron exchange

valence orbital [t] the volume of space that can be occupied by valence electrons in the highest energy level of an atom

van der Waals forces [t] weak attractive forces that molecules exert on each other, including **London** and **dipole–dipole forces**

verify establish, by substitution for a particular case or by geometric comparison, the truth of a statement

volt SI unit of electric potential difference; 1 V = 1 J/C

voltage see **electric potential difference**

voltaic cell [e] an arrangement of two half-cells, separated by a porous boundary, that spontaneously produces electricity

volumetric pipette a glass tube used to measure precise volumes of a liquid; also known as a delivery pipette

VSEPR theory [t] the Valence-Shell-Electron-Pair-Repulsion theory; a theory for predicting and explaining the shape of a molecule based on the electrical repulsion of bonded and unbonded (lone) electron pairs in a molecule

W

weak acid [e, t] a substance that forms a solution with weak acidic properties, such as a pH slightly below 7; a substance that reacts incompletely (<50%) with water to form relatively few hydronium ions

weak base [e, t] a substance that forms a solution with weak basic properties, such as low electrical conductivity and pH slightly above 7; an ionic or molecular substance that reacts partially (<50%) with water to produce relatively few hydroxide ions

WHMIS Workplace Hazardous Materials Information System, used to communicate chemical hazards

---

[e] empirical definition

[t] theoretical definition

Index

A

Abegg, Richard, 78
Absolute temperature scale, 153–54
Absolute zero, 153–54
Acetate ions, 746–47
Acetic acid, 198, 436, 437
Acetylsalicylic acid (ASA), 339, 382
Acid–base equilibrium, 725
    collision–reaction theory and, 712
Acid–base indicators, 245–47, 336, 753–54.
    See also names of individual indicators
Acid–base reactions
    pH curves and, 761
    predicting of equilibria, 727–30
    and proton transfer, 724–26
    quantitative, 730
    titration of, 333
Acid–base theories, history of, 733–34
Acid–base titrations, 596
Acid deposition, 252, 393
Acidic solutions
    aqueous hydrogen ions and, 238
    hydrogen ions and, 237
    nature of, 236–35
    water ionization and, 713–14
Acid ionization constant, K_a, 737–43
Acid rain, 145, 190, 635, 721
Acids, 28
    in chemical reactions, 710
    conductivity of, 194, 200–201
    conjugate, 724–26
    defined, 249
    empirical definitions of, 28
    and hydronium ions, 249
    litmus test of, 194
    naming and writing formulas for, 34–36
    nomenclature, 248
    oxygen and, 733
    properties of, 198–99, 236–37
    reaction with water, 249
    strong. *See* Strong acids
    weak. *See* Weak acids
Actinoids, 16
Activated complex, 527–29
Activation energy, 526–30, 694
Addition polymerization, 445
Addition reactions, 419–21
Adhesion, 113
Aerospace engineers, 278
Air, 142, 148
Air bags, 146, 176, 351
Alberta Chemical Operations, 416
Alcohols, 399, 425, 427, 428, 430
Aliphatic hydrocarbons, 379
Alizarin yellow indicator, 336, 753
Alkali metals, 16
Alkaline-earth metals, 16
Alkanes, 362, 366–73, 367–69, 382, 540
Alkenes, 374–77, 375–78, 376
Alkylation, 390
Alkylbenzene, 382
Alkyl branches, 367
Alkyl halides, 417
Alkynes, 374–77, 375–78, 376

Alloys, 27, 83, 192
Aluminium, 290, 558, 634, 647–48, 652
Aluminium oxide, 432
Amides, 452
Amines, 363, 452
Amino acids, 112, 115, 452–53
    chirality of, 114–15
Ammeter, 613
Ammonia
    geometry of, 93
    Haber process for production of, 325–26
    uses of, 332
Amount concentration, 205–206, 694
    calculating K_b from, 745–46
Amount of substance, 51
Amperes, 613
Amphiprotic, defined, 723
Amphoteric, defined, 723
Amphoteric substances, 749–50
Analogies, 18
Analytic measurement technology, 337–38
Anions, 24, 120, 248, 624
Anodes, 613, 616, 624, 626, 628–29, 636
Anomalies, 677
Antifreezes, 425, 426
Antioxidants, 430
Aqueous acid solutions
    percent reactions, 718
    strengths of, 829
Aqueous bases, strengths of, 829
Aqueous hydrogen ions, 238
Aqueous hydroxide ions, 35, 238
Aqueous solutions, 29
    conductivity of, 193–94, 197
    properties of, 193–94
    proton transfer in, 729–30
Aristotle, 149
Aromatics, 381–85
Arrhenius, Svante, 19, 197, 198, 237, 248, 525, 526, 574, 575, 734
Arrhenius theory, 713, 722, 734
    modified, 248–51, 255, 256, 713, 718, 737
Ascorbic acid, 438
Astronomy, 315
Athabasca oil/tar sands, 361, 395–96
Atmospheric pressure, 149
Atomic molar masses, 490
Atomic numbers, 21
Atomic theory, 6, 18–19, 583
Atoms, 12, 18–19
    counting of individual, 337–38
Aureate plating, 556, 651
Avogadro, Amedeo, 165
Avogadro's number, 51, 85
Avogadro's theory, 165, 169

B

Bacon, Francis, 617
Balanced chemical equations, 48–50, 52–54
    energy terms in, 498
    and enthalpy changes for reactions, 497
    mole ratio from, 321–22
    and reaction stoichiometry, 287

    translating, 51–52
Balmer, Johann, 23
Base ionization constant, K_b, 744–50
Bases, 28, 35–36
    conjugate, 724–26, 728
    definition of, 250
    dissociation of, 198, 248
    empirical definitions of, 28
    litmus test on, 194
    properties of, 198–99, 236–37
    reaction with water, 249, 250
    strength of, 744
    strong. *See* Strong bases
    weak. *See* Weak bases
Basicity, prediction of, 746–49, 748–49
Basic solutions
    aqueous hydroxide ions and, 238
    hydroxide ions and, 237
    nature of, 236–37
    water ionization and, 713–14
Batteries
    See also Electric cells
    car, 616, 617
    dead, 632
    defined, 612
    for electric vehicles, 610
    energy density of, 614
    lead–acid, 610, 616, 617
    9 V, 615
    plastic, 621
    power of, 614
Bauxite, 290, 648
Becquerel, Henri, 46
Benzene, 381, 382, 430
Beryllium, 94
    dihydride, 92
Binary compounds, 29
    ionic, 120, 121
    molecular, 33
Bio-based economies, 507–508
Biochemists, 91
Biodegradability, 456
Biopolymers, 458
Birss, Viola, 552
Bitumen, 394, 395–96, 539–40
Bleaching, of wood pulp, 594
Bleach (sodium hypochlorite), 212–13
Blood
    alcohol content, 312, 588
    equilibrium of circulation, 692
Blood plasma, 190
    ions in, 200
Bohr's atomic theory, 21–23
Boiling points, 106, 107–108, 109, 121, 155
Bondar, Roberta Lynn, 278
Bond breaking, 532–33
Bond dipoles, 101
Bond energies, 99, 105, 532, 533
Bonding/bonds, 82–84
    carbon, 126–27
    tetrahedral, 92, 93, 94
    theories, 72, 76
    triple, 85
Bonding capacity, 87
Bonding electrons, 80

Bond polarity
    electronegativity and, 99–101
    molecular polarity and, 101–103
Boron, 94
Bosch, Carl, 325
Boyle, Robert, 150, 151, 163
Boyle's law, 151, 156, 160, 163, 169, 694
Breathalyzers, 312, 580
Bromothymol blue indicator, 245, 246, 336, 752, 753, 760
Brønsted, Johannes, 722, 734
Brønsted–Lowry acid, 723
Brønsted–Lowry base, 723
Brønsted–Lowry concept, 722–36, 723, 727, 728, 744, 755
Brønsted–Lowry equations, 756, 764
Brønsted–Lowry reaction equations, 723, 758
Bronze, 560
Brooks, Harriet, 25
Buffer capacity, 764
Buffering, 751, 752, 763–65
Buffers, 766
Bunsen, Robert Wilhelm, 23, 315
Burettes, 328
Buriak, Jillian, 123
Burning
    corrosion and, 559
    of fossil fuels, 190, 357
    oxygen and, 559

C

Calcium hydroxide. *See* Slaked lime
Calorimeters, 485, 486, 487, 498, 502, 507, 515
Calorimetry, 485–87, 495, 498, 502, 533
Capillary action, 105, 113, 114
Carbohydrates, 449
Carbon
    covalent networks of, 126–27
    oxidation number of, 585–88
Carbon cycle, 356–57
Carbon dioxide, 356, 585
    in combustion, 399
    crystal structure of, 124
    oil recovery and, 388
Carbon monoxide, 320, 399, 502, 503
Carboxyl group, 436
Carboxylic acids, 436–37
Carcinogens, 381, 417
Careers, 10. *See also* names of particular careers
Carothers, Wallace, 453
Catalysis, 535
    empirical effect of, 535–36
    theoretical explanation of, 536–39
Catalysts, 522, 523, 535, 539–41, 694
Catalytic converters, 522, 523, 535
Catalytic cracking, 389
Catalytic reforming, 390
Cathodes, 613, 616, 624, 626, 628–29
Cathodic protection, 636
Cations, 24, 120, 248, 624
Cavendish, Henry, 152
Cell notation, 622
Cellular respiration, 500, 556, 565

Cellulose, 455
    acetates, 458
    starches vs., 456–57
Celsius temperature scale, 153, 295
CFCs. *See* Chlorofluorocarbons (CFCs)
Charles, Jacques, 152, 153, 155, 163, 164
Charles' law, 154–55, 156, 163–64, 169
Chemical amount, 51, 55–57
Chemical analysis, 314–16
Chemical changes, 9, 46. *See also* Chemical reactions
    energy transformations and, 474
    at molecular level, 670
Chemical decomposition, 12
Chemical energy
    personal use of, 482
    transportation and, 482
Chemical engineering technologists, 480
Chemical engineers, 284, 412
Chemical formulas, 13
    states of matter in, 29
    subscripts, 49
Chemical industry employees, 415
Chemical potential energy diagrams, 498–99, 529
Chemical process engineers, 693
Chemical reaction equilibrium, 670, 677, 678–83
Chemical reactions, 9, 42, 47–50, 76. *See also* Chemical changes; Hydrogen–iodine reaction system
    acids in, 710
    assumptions regarding, 280–81
    calculating masses in, 288
    classification of, 58–60
    communicating, 48–49
    equations. *See* Reaction equations
    forward. *See* Forward reactions
    generalizations about, 58
    homogeneous vs. heterogeneous, 682
    quantitativeness of, 281, 677, 680
    rate of, 524
    reverse. *See* Reverse reactions
    in solution, 61–63
    speed of, 280–81
    spontaneity of, 280, 676
    as stoichiometric, 281
    stoichiometry of, 287
Chemicals, common, 830
Chemical systems, 676. *See also* Closed systems
Chemical technologies, 270, 274, 286. *See also* Technology/technologies
    evaluation of, 524
Chemical technologists, 286, 650
Chemistry, 6, 9
    defined, 9
    and technology, 270, 274
Chemistry teachers, 9
Chlor-alkali process, 648
Chloride anomaly, 645, 648
Chlorine
    gas, 645, 646
    uses of, 648
Chlorobenzene, 421–22

Chloroethene, 420, 447
Chloroethene (vinyl chloride), 421
Chlorofluorocarbons (CFCs), 184, 417, 418, 419, 422, 530
Chlorophyll, 535
Cholesterol, 375
CHP. *See* Cogeneration
Chromium, 128, 327, 350, 650
Chuang, Karl, 390
Citric acid, 255, 437
Clark, Karl, 394, 395
Claus converters, 363
Cleaning solutions/products, 119–20, 220, 238, 268
Closed systems, 676–78, 689
Coal, 358, 361, 505, 585
Coalbed methane, 358, 365
Coefficients, 49
Cogeneration, 505, 618
Cohesion, 113
Coking, 396, 540
Collision–reaction theory, 47, 120, 249, 281, 524–26, 563, 676, 728
    acid–base equilibrium theory and, 712
    concentration changes and, 692
    and energy changes, 693
    and gas volume changes, 694
    and reverse reactions, 677
Colorimetry, 314–16
Combined gas law, 156–57, 169
Combustion, 522, 556
    in automobiles, 146, 399
    complete, 398–400
    complete reactions, 58–59
    corrosion and, 559
    incomplete, 398–400
Communication, 10, 87, 811–16
Compounds, 12–13, 13, 76
    binary, 29
    classification of, 27–31
    empirical definitions of, 28
    molecular, 33–34
    thermodynamic properties, 826
Computer chip industry, 159, 338
Concentration changes
    collision–reaction theory and, 692
    and Le Châtelier's principle, 691–92
Concentration(s). *See also* Amount concentration
    acid ionization constant, K_a and, 738
    calculating, 206–210, 216–17
    dilution of solutions and, 216–18
    equilibrium and, 719
    and equilibrium constant, 685
    hydronium ion, 238–39
    of ions, 210–13
    maximum solute, 221
    parts per billion (ppb), 204–205, 214
    parts per million (ppm), 204–205, 207
    percentage, 203–204
    pH, 716–18
    pOH, 716–18
    predicting of, at equilibrium, 686–87
    solutions, 203
    strength vs., 737

Condensation polymerization, 449
Condensation reactions, 438
Conductivity, 28
    and acid strength, 718–20
    of aqueous solutions, 193–94, 197
        Arrhenius' theory and, 713
    of pure water, 713
Conjugate acid–base pairs, 725
    K_a–K_b relationship for, 746–49
Conservators, 580
Constant relationships, 685
Consumer products
    information on, 809
Cooley, Jean, 352
Coordinate covalent bonds, 87
Copper
    bonding in, 124
    early uses of, 560
    electrorefining of, 649
    in silver–copper cells, 623–24
Correlational studies, 450
Corrosion, 120, 559, 634–35, 636–37
Corrosiveness, 810
Cotton, 457
Coulombs, 613
Covalent bonds, 78, 82, 86, 111
    double, 86–87, 96
    triple, 87, 96
Covalent networks, 125
Cracking, 378, 389–90
    ethane, 414, 431–32
Crude oil
    cracking and reforming of, 539
    discovery of, 360
    refining of, 358, 386–87
Cryolite, 647
Crystal lattice, 122
Crystal lattices, 83, 121, 122
Crystallization, 76, 77, 802
Crystallography, 82, 91, 96
Crystals
    covalent network, 125–27
    diamond, 125
    ionic, 119–23
    metallic, 123–24
    molecular, 124
    quartz, 126
    three-dimensional models of, 121
    X-ray diffraction of, 121
Cycloalkanes, 371–73, 374, 375, 430
Cycloalkenes, 376–77, 430
Cycloalkynes, 376–77
Cyclohexanol, 430

D

Dacron, 451
Dalton, John, 13, 18–19, 46
Dalton's atomic theory, 18–19, 163
Davy, Sir Humphry, 197, 237, 646, 653, 733–34
DDT (dichlorodiphenyltrichloroethane), 277, 417
Deductive reasoning, 538–39
Dehydration, 432

Dehydrogenation, 379
Dehydrohalogenation, 432
Describing, 72, 78, 276
Detergents, 103, 210, 541
Dewaxing, 387
Diabetes, 681
Diagnostic tests, 28, 48, 805
    for ions, 284
Diamond(s), 125, 128
Diatomic molecules, 33, 85
Dicarboxylic acids, 436, 451
Dichromate ions, 596
Diethylbenzene, 382
Dihydrogen phosphate ions, 764
Dilutes, 203
Dilution, 216–18
Dimethylpropane, 368
Dioxins, 594
Diploma Exam, 823–25
Dipole–dipole forces, 106, 107
Diprotic bases, 335, 756
Disaccharides, 456
Disproportionation reactions, 577, 592
Dissociation, 197, 198, 210–211, 249, 250
DNA molecule, 113
Döbereiner, Johann, 538
Doping, 127
Double helix, 112
Double replacement reactions, 62–63, 251
Dry cleaning, 103
Ductile, defined, 15
Dynamic equilibrium, 224–26, 670, 674, 677

E

Ecological perspectives, 45
Ecologists, 240
Economic perspectives, 45
Edison, Thomas, 619
Einstein, Albert, 18, 21, 78
Electric cars. See Electric vehicles
Electric cells. See also Batteries
    alkaline, 615, 617
    dead, 632
    defined, 612
    design and properties, 613–14
    dry, 615
    galvanic, 622
    lithium-ion, 616
    membrane, 648
    mercury, 615
    nickel–cadmium, 616, 640
    power of, 614
    as power source for electrolytic cells, 646
    primary, 615
    rechargeable, 615, 640
    secondary, 616–17, 640
    silver–copper, 623–24
    stoichiometry of, 652–57
    voltaic cells vs., 626
    zinc chloride, 615
    zinc sulfate, 639
Electric current, 612–13, 613
    direction of, 628
    impressed, 636

Electricity
    current, 558
    as flow of electrons, 613
    fuel cells and, 617
    and redox reactions, 610
    static, 612
Electric potential difference, 613
Electric vehicles, 481, 610, 618
Electrochemical cells, 626, 640, 655
Electrochemical reactions, 552
Electrochemistry, 556
Electrodes, 613
Electrolysis, 297, 298, 610, 640, 646–51
Electrolytes, 193–94, 197, 613
    acids as, 198
    conductivity of, 197
    and rusting, 635
Electrolytic cells, 639–51
    electric cells as power source for, 646
    potassium iodide, 641–42
Electronegativities, 83, 86, 99–101
Electronegativity, 81
Electrons
    "bookkeeping," 583
    loss and gain of, 583, 652
Electron transfer(s), 78, 83, 85, 552, 583, 585, 589
    in biological systems, 632
    in electrolytic cells, 641
    theory, 561–62
Electroplating, 327, 650, 669
Electrorefining, 649
Electrowinning, 650
Elements, 12, 13
    classification of, 14–17
    empirical knowledge of, 18
    families of, 14
    IUPAC rules, 15–16
    molecular, 33, 85–86
    production of, 646–48
    rare earth, 16
    solubility of, 222
    thermodynamic properties, 826
Elimination reactions, 432
Emissions
    control, 540
Empirical definitions, 10
    of acids, 28
    of bases, 28
    of compounds, 28
    of neutral compounds, 28
Empirical formulas, 30, 88
Empirical hypotheses, 10
Empirical knowledge, 10. See also Evidence; Scientific knowledge
    observations and, 18
    theories vs., 524–26
Endothermic changes, 200, 486, 498–99
Endothermic reactions, 111, 489, 498, 532
Endpoint, 751
Endpoints, 328, 596, 751, 758
Energy
    alternative sources and technologies, 507–508
    changes, 199–98, 485–86, 488

consumption, 480, 481
density, of batteries, 614
industry/sector, 480, 508
kinds of, 474
pathway, 527–29
production, 481
quantum theory of, 21
renewable sources, 496, 507–508
saving for future use, 482
and technologies, 474
transfers, 532
transformations, 474, 499
Engine knock, 392, 393
Enthalpies of reaction, 490, 510
Enthalpy changes, 487–89, 488, 490, 496–97, 502, 527, 533
Entities, 12
Environment
    Aboriginal peoples and, 186, 252, 481
    damage to, 6, 45
    disposable diapers and, 455–56
    electric vehicles and, 610
    fossil fuels and, 360
    technology and, 277
Environmental engineers, 714
Enzymes, 91, 535, 541
Equilibrium, 670, 674, 677
    acid strength and, 718–20
    adjusting by adding/removing a substance, 691
    in biological systems, 692
    blood sugar, 681
    catalysts and, 694
    closed systems at, 676–78
    concentration and, 719
    and modified Arrhenius theory, 718
    position of, 680
    prediction of concentrations at, 686–87
    solutions in, 224
    temperature and, 719
    and volume vs. pressure, 694
Equilibrium constant, K_c, 684–87, 713
Equilibrium law, 684
Equilibrium shifts, 691
Equivalence points, 328, 333, 334–35, 751, 755, 756, 761
Erlenmeyer flask, 215
Ester functional group, 439
Esterification reactions, 438
Esters, 439–44
    as artificial flavourings, 439
    naming of, 439–40
    natural, 441
    wax, 441
Ethane, 378–79, 379, 414, 431–32
Ethanoic acid. *See* Acetic acid
Ethanol, 312, 399, 425–26
Ethene (ethylene), 374, 378, 414–15, 431–32
Ethylene dichloride, 420, 423
Ethyne, 97–98
Ethyne (acetylene), 375
Evaluations, of investigations, 791–92, 796
Evaporation, 113
Evidence, 9, 25, 163, 248, 710, 791, 795. *See also* Empirical knowledge

tables of, 10
Excess reagents, 284, 318
    calculating mass of, 320
    identifying limiting, 321–22
Exothermic changes, 200, 486, 498–99
Exothermic reactions, 111, 489, 496, 498, 527, 532–33
Experimental error, 792
Experimental uncertainties, 292
Experiments, 8, 18
    controlled, 148
    prediction of results, 249
Explaining, 72, 78, 276

F

Families, of elements, 14, 16
Faraday, Michael, 19, 197, 381, 625, 653
Faraday constant, 653, 655
Faraday's law, 653
Fatty acids, 449
Feedstocks, 389, 410, 508
Fermentation, 312
Fertilizers, 325–26, 349, 696
Field production operators, 540
Filtration, 291, 802–803
Fire(s), 522
    as early technology, 567
    extinguishers, 559
    metal refining and, 559
Flammability, 113, 810
Foods
    digestion of, 710
    energy content of, 486, 520
    labelling of product ingredients, 331, 349
    preparation, 274
Food science technologists, 681
Force, 149
Forensic laboratory analysts, 48
Formation reactions, 58, 510
Formic acid. *See* methanoic acid
Formulas
    molecular, 85–90
    simplest ratio, 86
    stereochemical, 88
    types of, 87–88
Formula units, 30
Forward reactions, 677, 680
    concentration changes and, 691–92
    proton transfer and, 724–25, 726
    temperature and, 693
    volume and, 694
Fossil fuels
    alternatives to, 481, 507–508
    burning of, 190, 357, 482
    chemical energy and, 480
    classifying, 359
    environment and, 360
    fuel cells vs., 617
    industries, 359–60
    origin of, 356, 480
    refining of, 358
Fractional distillation, 386, 388
Fractionation, 386–87

Fractions, 386–87, 497
Fractures, 119
Frankland, Edward, 78
Franklin, Ursula, 321
Freezing, 113, 155
Freons. *See* Chlorofluorocarbons (CFCs)
Fuel cells, 481, 508, 556, 617–20
    aluminium–air, 610, 618
    aluminium–oxygen, 618
    Ballard, 620
    hydrogen, 293, 296, 620
    hydrogen–oxygen, 617
    large-scale uses, 618
    phosphoric acid, 618
    research into, 552
Fuels
    alternative, 478
    hydrogen as, 481
    renewable, 482
Functional analogs, 449
Functional groups, 417

G

Galileo, 149
Galvani, Luigi, 612, 622
Galvanizing, 636, 637
Gases, 142
    air and, 142
    chemical safety hazards of, 159–60
    compressed, 159–60, 163, 170
    density of, 176
    empirical properties of, 148–62
    ideal. *See* Ideal gases
    liquefied, 170, 414
    molar volume of, 169–71
    motion of molecules in, 163
    noble. *See* Noble gases
    pressures of, 149, 159–60, 163
    properties of, 163–68
    quantity calculations, 295
    real vs. ideal, 172
    solubility of, 222
    technologies, 146
    uses of, 148
Gasohol, 311, 482
Gasoline, 113, 192
    ethanol in, 399, 426
    octane and, 392
    production, 539
    sulfur in, 393–94
Gas stoichiometry, 288, 294–97
Gay-Lussac, Joseph, 164–65
Gemmologists, 125
Generalizations, 11, 154, 222–23, 645
Geologists, 125
Giguère, Paul, 237
Gillespie, Ronald, 91, 93, 724
Glass, 119, 126
Global warming, 492, 540
Glucose–glucose linkages, 456–57
Glycerol, 429, 430, 449
Glycogen, 457
Gold, 15, 20, 560
Grain alcohol. *See* Ethanol

Graphite, 126–27
Graphs, 10, 317, 815–16
Gravimetric analysis, 314, 317–19
Gravimetric stoichiometry, 288
Gravitational potential energy, 482, 526–27
Grease, dissolving of, 103, 196
Greenhouse gases (GHGs), 400, 540
Groups of elements, 14, 16
Grove, William, 617
Guillet, James, 449
Guldberg, Cato Maximilian, 684

H

Haber, Fritz, 325, 617, 690
Haber process, 325–26, 693
Half-cells, 622
    overvoltage, 645
    reference, 627
    standard hydrogen, 627
    standard reduction potential of, 628–30
    stoichiometry calculations, 655–56
Half-reaction equations, 562, 564–67, 570, 589
Half-reactions, 561
    and predicting redox reactions, 579
Halides, 417, 423, 424, 433, 571
Hall, Charles, 647
Halogens, 16, 417
Harrison, Jed, 338
Hazards
    of gases, 159–60
    of household chemical solutions, 212–13
Heat
    energy as, 485
    transfer, 487–89
Heat capacities, 826
Heavy metal compounds, 810
Helium, 160, 170, 364
Hell–Volhard–Zelinsky reaction, 779
Hemoglobin, 130, 692
Heptane, 390
Héroult, Paul, 647
Herzburg, Gerhard, 109
Hess, Germain H., 502
Hess' law, 502–507, 510–13, 533
Heterogeneous mixtures, 12
Homeostasis, 765
Homogeneous mixtures, 12, 61, 192
Homologous series, 366
Hot-air balloons, 152, 156, 176
Household chemical solutions, 212–13
Huggins, Maurice, 111
Hybrid cars, 478
Hydrocarbon derivatives, 417
    hydrocarbons vs., 417
Hydrocarbons, 358
    aromatic, 540
    cyclic, 371
    hydrocarbon derivatives vs., 417
    saturated, 374
    solutions, 192
    unsaturated, 374
Hydrochloric acid, 200–201, 234, 255, 421–22, 778

Hydrochlorofluorocarbons (HCFCs), 184, 419
Hydrocracking, 390, 396, 540
Hydrogen
    and acidic properties, 237
    fuel cells, 293, 296
    production by steam reforming, 297
Hydrogenation, 374, 393, 419, 450
Hydrogen bonding, 111–13
Hydrogen chloride, 421–22
Hydrogen fuel cells, 297–98
Hydrogen gas, 747
    as fuel, 481
    in fuel cells, 507–508
    uses of, 648
Hydrogen–iodine reaction system, 679–82
Hydrogen ions, 237
Hydrogen phosphate ions, 764
Hydrogen sulfide gas, 362, 363
Hydrologists, 330
Hydrometallurgy, 644
Hydronium ions, 237, 238–39, 248, 255, 712, 713, 718–20, 729, 730, 744
    acids and, 249
    emphasis on, 747
    pH and, 240, 241
Hydrotreating, 393, 396, 540
Hydroxide ions, 198, 237, 238, 713, 729, 730, 744, 747
    pOH and, 243–44
Hydroxyl groups, 425, 428
Hypotheses, 8, 9, 98, 154, 790

I

ICE table, 681
Ideal concepts, 172
Ideal gases, 172
Ideal gas law, 172–76, 296
Immiscibility, 222
Indigenous knowledge (IK), 162, 163, 222. *See also* Traditional ecological knowledge
Inductive reasoning, 538–39
Inert electrodes, 625
Inflection point, 335
Inhibitors, 536
Intermediates, 538
Intermolecular bonding, 201
Intermolecular bonds, 111
Intermolecular forces, 105–18
    and liquids, 113
    research in, 114–15
International System of Units. *See* SI (Système International d'Unités)
International Union of Pure and Applied Chemistry (IUPAC), 14, 15–16, 29
Interpretations, 9–10
Intramolecular bonds, 111. *See also* Covalent bonds
Intramolecular forces, 105
Investigations, 8
    reports, 790–92, 796
Iodine. *See also* Hydrogen–iodine reaction system
    model of crystal of, 124

Ionic bonds, 78, 83
Ionic compounds, 27–28, 29–31, 58, 86
    binary, 120, 121
    empirical formulas, 88
    formation of, 120
    soluble, 119
Ionic crystals, 119–23
Ionic hydrates, 31
Ionic hydroxides, 194
Ionization, 198
Ionization constant for water, K_w, 713–16
Ions, 23–24
    concentration of, 210–11
    diagnostic tests for, 284
    electric particles as, 197
    origin of term, 625
Iron, 120, 634–35, 636
Isoelectronic molecules, 108
Isolated systems, 485
Isomers, 370
    cis, 423
    trans, 423
Isotopes, 21, 24
    radioactive. *See* Radioisotopes

J

Jackson, Chevalier, 715
Jalilehvand, Farideh, 728
Joule, James Prescott, 487
Joules, 485

K

Kebarle, Paul, 683
Kekulé, August, 381
Kekulé, Friedrich, 78, 82
Kelvin, Lord, 153, 154
Kelvin temperature scale, 153, 295
Kevlar, 454
Kilby, Jack, 652
Kilojoules, 486
Kilopascals, 149
Kinetic energy, 482, 487–88, 510, 527, 632
Kinetic molecular theory (KMT), 47, 163–64, 172, 224
Kirchhoff, Gustav Robert, 23, 315

L

Laboratory equipment, 797–801
Laboratory reports, 8
Laboratory safety, 807–808
Lactic acid, 255
Language, scientific, 49, 811
Lanthanoids, 16
Lauks, Imants, 598
Lavoisier, Antoine Laurent, 680, 733
Law of additivity of enthalpies of reaction. *See* Hess' law
Law of combining volumes, 164–66
Law of conservation of energy, 488, 49?, 499, 502
Law of conservation of mass, 11?

Law of definite composition, 18, 19
Law of mass action, 684
Law of multiple proportions, 18, 19
Laws, 9
Le Bel, Joseph, 78
Le Châtelier, Henri Louis, 690
Le Châtelier's principle, 690, 691–92, 693, 694
LEDs (light-emitting diodes), 126
Lemieux, Raymond, 460
Le Roy, Robert J., 114
Le Roy radius, 114
Lewis, Gilbert, 78, 81, 91, 734
Lewis formulas, 81–82, 85, 88, 89, 91, 92–96
Lewis symbols, 81
Liebig, Justus von, 237, 538, 734
Lime, 120, 511. See also Quicklime; Slaked lime
Limestone, 511
Limiting reagents, 284
Lind, James, 744
Lipids, 449–51
Liquefied petroleum gases (LPGs), 414
Liquids
    motion of molecules in, 163
    physical properties of, 113–14
    solubility of, 222
Litmus indicator, 194, 198, 236, 245, 246, 333, 641
Logarithmic scales, 240, 243, 716–18
London, Fritz, 106, 107
London forces, 106–10
Lone pairs, 80
Loschmidt, Josef, 381
Lowry, Thomas, 722, 734
LPGs (liquid petroleum gases), 378
Lye. See Sodium hydroxide

M

MacGill, Elizabeth ("Elsie"), 164
Main group elements, 16
Malleability, 15, 20
Mallory, Walter, 619
Manuel, Rob, 387
Marcus, Rudolph, 561
Margarine, 374
Marker, Russell, 392
Mass
    calculation of, 209
    numbers, 21
Materials engineering technologists, 636
Materials/metallurgical engineers, 558
Materials Safety Data Sheets (MSDS), 385, 809
Matter
    changes in, 46–50
    classification of, 12–13
    defined, 12
    laboratory technologists, 246
    states, 110, 121, 126
Mendeleev, Dmitri, 14, 23

Metal ions
    flame tests for, 315
Metallic bonding, 83, 124
Metallurgy, 558, 567
    early, 560
Metals, 14–15
    crystals of, 123–24
    electronegativities of, 81
    extraction of, 120, 558
    multi-valent, 29–30
    refining of, 559, 649–50
    toxicity of, 337
Meteorologists, 166, 167
Methane, 378
    coalbed, 358
    geometry of, 93
    mining of, 364
Methanoic acid, 436, 537, 539
Methanol, 425–26
Methylbenzene, 382, 390
Methylbutane, 367
Methyl orange indicator, 758, 760
Methyl red indicator, 753, 754
Microbiologists, 766
Microchips, 652
Micromachining techniques, 338
Microscopes, 25, 115
Miscibility, 222
Mixtures, 12, 192
Models, 18
    molecular, 49, 90, 377, 378
Mohs Hardness Scale, 128
Molar enthalpies of combustion, 495
Molar enthalpies of formation, 495, 510–15, 533
Molar enthalpies of reaction, 490, 495–96, 510
Molar mass, 55–57
Molar quantities, 490
Molar volume, 169–71, 490
Molecular compounds, 27–28, 33–34, 58, 86–87
    formulas, 88
    solubility of, 201
Molecular crystals, 124
Molecular elements, 85–86
Molecular formulas, 33, 85–90
Molecular polarity
    bond polarity and, 101–103
    dipole theory, 98–103
Molecular shapes, 91, 92–98
Molecules, 13, 33
Moles, 51, 52, 169, 209
Molicel, 616, 669
Momentary dipole, 106
Monatomic ions, 23–24
Monobasic bases, 258
Monomers, 445
Monoprotic acids, 258
Monoprotic bases, 258
Monosaccharides, 456
Multiple choice questions, 823–24
Muriatic acid. See Hydrochloric acid

N

Naphthas, 387
Natural gas
    consumption, 398, 413
    discovery of, 359–60
    emissions, 385
    extraction of, 362
    helium from, 364
    pipelines. See Pipelines
    processing plants, 364
    refining of, 358, 363–64
    sour, 363
    sweet, 363
Natural gas liquids (NGLs), 414
Natural resources, 356, 360, 410
    nonrenewable, 357
    renewable, 354
Nernst, Walther, 617
Net ionic equations, 281–84, 334, 562, 563
Neutral compounds, 28
    empirical definitions of, 28
    litmus test on, 194
Neutralization, 63, 251, 734, 751
Neutrons, 21
Newtons, 149
Nitric acid, 200–201, 255
Nitrogen
    cycle, 609
    liquefied, 170
Noble gases, 16, 120, 159, 174
Noddack, Ida, 15
Nomenclature, 29, 680, 746
Nonelectrolytes, 193–94, 197
Nonmetals, 15
    electronegativities of, 81
    oxides, 251, 252
    reactions of, 571
Nonpolar covalent bonds, 99
Nonpolar molecular compounds, solubility of, 201
Nonpolar molecules, 98, 101, 106
Nuclear changes, 46
Numerical response questions, 824
Nyholm, Ronald, 91, 93
Nylon, 449, 453–54

O

Observations, 9, 18, 163
    predictions about, 249
    redox theory and, 563
Octadecanoic acid. See Stearic acid
Octane numbers, 392–93, 425
Octet rule, 80, 85
Ohmmeter, 194
Oil industry
    catalysts in, 539–40
    pipelines. See Pipelines
Oil refining, 360, 522
    chemical processes in, 389–90
    physical processes in, 386–88
Oil sands, 358, 359, 392, 394
Orange IV indicator, 753
Orbitals, 80, 85

Organic chemistry, definition of, 358
Organic compounds
    as hydrocarbons vs. hydrocarbon
      derivatives, 417
    molecular formulas for, 370
Organic halides, 417
Oxalic acid, 255, 437
Oxidation, 281, 559, 562, 563, 583, 585
Oxidation numbers, 583–85, 589–93
Oxidation states, 583–84
Oxidizing agents (OAs), 559, 568–69, 569, 828
Oxyacetylene torches, 375
Oxygen, 142
    acids and, 733
    atomic, 530
    liquefied, 170
    molecular, 530
    naming of, 733
    role in burning, 559
Oxygenation, 399
Oxygenators, 425
Ozone, 418, 530

P

Particles, 12, 120
Pascals, 149
Pasteur, Louis, 82
Pauli, Wolfgang, 80
Pauli exclusion principle, 80
Pauling, Linus, 78, 81, 82, 91, 99–100, 107
PCBs (polychlorinated biphenyls), 417
Pentane, 367, 368
Peptides, 452
Percent difference, 792
Percent ionization, 718
Percent reaction, 718, 720
Percent yield, 292, 680, 792
Periodic law, 21
Periodic tables, 6, 14, 15, 21, 30
Periods, 14
Permafrost, 118
Perspectives, 45
Pesticides, 224
Petrochemicals, 356, 357, 410, 413–16, 423, 522
Petroleum, 381
Petroleum engineers, 387
Petroleum Human Resources Council of Canada, 389
pH, 236, 239
    of acid vs. normal rain, 721
    buffers and, 764–65
    calculations, 240–42
    changes during titration, 333
    dependence on, 241
    emphasis on, vs. pOH, 747
    hydronium ions and, 240, 241
    measurement of, 241, 242
    precision rule for, 814
    scale, 236, 237, 239
    of soils, 251
    test strips, 246
Phase equilibrium, 677

pH concentrations
    logarithmic scale, 716–18
pH curves, 751–67
    acid and base strength vs., 760–63
    for strong acid–strong base reaction, 751
    titration, 334–35
Phenolphthalein indicator, 245, 753, 760
Phenols, 430
Phenyl group, 382
pH meters, 333
Phosphoric acid, 432
Photodegradability, 449
Photosynthesis, 142, 480, 535, 556, 567, 588
Physical changes, 9, 46
Pidgeon, Lloyd, 649
Pidgeon process, 649
Pipelines, 360, 364
Pipettes, 218, 800–801
pK_a, 742
Planck, Max, 21
Plastics, 445, 449
pOH
    emphasis on pH vs., 747
    hydroxide ions and, 243–44
    precision rule for, 814
pOH concentrations
    logarithmic scale, 716–18
Polanyi, John, 87, 534
Polar compounds, solubility of, 201
Polar covalent bonds, 99–100
Polar molecules, 98, 103
Political perspectives, 45
Pollutants, 146, 207, 224, 251, 252, 332, 393, 400
Pollution, 160
    dilution and, 219
    electric vehicles and, 610
    technology and, 356
Polyalcohols, 429
Polyamides, 453
Polyatomic ions, 30, 121
Polyatomic molecules, 85
Polybasic bases, 258
Polycarboxylic acids, 437
Polyesters, 451
Polymerization, 116, 445, 448, 452, 456
Polymers, 286, 414, 445–48, 449
Polypeptides, 453
Polyphenols, 430
Polypropylene, 445, 446, 455
Polyprotic acids, 258, 755
Polyprotic bases, 258, 755
Polyprotic entities, 756, 757, 758
Polysaccharides, 456
Polystyrene (Styrofoam), 445, 447
Polyvinylchloride (PVC), 414, 420, 445, 447
Popper, Karl, 25
Porous boundaries, 622
Potassium hydrogen phthalate, 330
Potassium hydroxide, 432, 617, 669
Potential energy, 487–90, 510, 632
Precipitates, 61, 225, 291
Precipitation, 63, 252
    completeness of, 318–19

Predicting, 72, 78, 276
    acid–base reaction equilibrium, 727–30
    acid–base reactions, 728–30
Prediction(s), 8, 25, 96, 97, 163, 790–91
    about observations, 249
    of basicity, 746–49, 748–49
    of concentrations at equilibrium, 686–87
    of empirical formulas, 29
    quantitative, 286–85
    of results in new situations, 251
    of results of experiments, 249
    testing of concepts and, 290
    testing of stoichiometric, 291
Pressure, 148–50
    amount concentration and, 694
    of gases, 159–60, 163
    reaction equations and, 279
    volume and, 150–52, 694
Priestley, Joseph, 733
Primary standards, 330, 596
Problem solving, 8
    methods, 816
    scientific. *See* Scientific problem solving
    STS. *See* STS (science, technology, and society)
    technological. *See* Technological problem solving
Propane, 378, 482
Propene (propylene), 374, 379
Proteins, 112, 115, 449, 452
    formation from amino acids, 452–53
Proton exchange membrane (PEM), 620
Protons
    hydrated, 237
    loss or gain of, 755
Proton transfer, 722–26, 728–30, 737, 755
Pure substances, 12

Q

Qualitative analysis, 314
Qualitative observations, 9
Quantitative analysis, 314, 317
Quantitative observations, 9
Quantities
    descriptions and rules for, 813–15
    reaction equations and, 279
Quantum mechanics, 79, 86
Quarks, 23
Quartz, 125, 126
Quicklime, 736

R

Radioisotopes, 24
R&D (research and development). *See* scientific research and development (R&D)
Reaction coordinates, 498
Reaction dynamics, 534
Reaction equations, 278–79, 279
Reaction kinetics, 525
Reaction mechanisms, 538
Reaction progress, 498, 524
Reaction quotient, 687

Redox equations, balancing of, 589–93
Redox reactions, 562, 568
    in biological systems, 587
    electricity and, 610
    evaluation of predicted, 645
    in living organisms, 588
    oxidation numbers and, 585–86, 589–93
    predicting, 575–81, 579
    stoichiometry, 596–600, 676
    as theory, 563
    theory of, 583
Redox spontaneity rule, 569, 572, 628, 676
Redox tables, 569–70, 572–74, 575, 624
Redox titrations, 596
Reducing agents (RAs), 558, 568–69, 569, 828
Reduction, 558, 562, 583, 585
    half-reactions, 562
    theoretical vs. empirical definitions of, 563
Reduction–oxidation reaction. *See* Redox reactions
Reference energy state, 510
Reference half-cells, 627
Refining, 358
Replacement, of theories, 25, 72
Replication, 792
Restriction, of theories, 25, 72
Reverse reactions, 677, 680
    concentration changes and, 692
    proton transfer and, 724–25, 726
    temperature and, 693
    volume and, 694
Revision, of theories, 25, 72
Rounding, 814–15
Rubber, 445, 460
Rule of thumb, 741
Rusting. *See* Corrosion; Oxidation
Rutherford's atomic theory, 20–21, 78

S

Safety symbols, 809
Samples, 328, 596
SATP. *See* Standard ambient temperature and pressure (SATP)
Saturated fats and oils, 450
Saturated hydrocarbons, 366, 374
Saturated solutions, 221, 224–26
Schindler, David, 206
Science, 8
    as compartmentalized vs. holistic, 564
    reductionism of, 172
    and technology, 8–9
    technology and, 276, 478, 556, 558, 646, 690
Science, technology, and society (STS). *See* STS (science, technology, and society)
    attitudes, 9, 795
    knowledge, 9, 44. *See also* Empirical
    [illegible], 248
    from empirical to theoretical,
    knowledge (IK) and, 163
    technological problem solving and, 614–15
Scientific laws, 11, 154
Scientific notation, 812
Scientific perspectives, 45
Scientific problem solving, 614–15, 790–92
Scientific research, 8, 9, 794–95
Scientific research and development (R&D), 79, 359
Scurvy, 743, 744
Semiconductors, 127
Semi-metals, 14
    toxicity of, 337
Series of elements, 16
Silicon, 588
Silver, 560, 650
Silver sulfide (tarnish), 120, 557
Simple decomposition reactions, 58
Single replacement reactions, 62, 561–62
SI (Système International d'Unités), 55, 149, 811–12
Skeleton equations, 579
Slaked lime, 511
Smelting, 560, 649
Soap, 103
Sodium hydrogen carbonate, 257, 268, 722, 723
Sodium hydroxide, 216, 234, 256, 330, 648, 715
Soil scientists, 301
Solar energy, 480, 483
Solids
    motion of molecules in, 163
    physical properties of, 119–30
    solubility of, 221, 222
Solubility, 61, 221
    charts, 61, 200, 201
    generalizations about, 222–23
    and oil and gas extraction, 362
    of solids, 221
    tables, 222–23
    temperature and, 222
Solubility equilibrium, 677
Solutes, 61, 192
    calculating quantities of, 206
    conductivity for, 236
    constant quantity of, 216–17
    excess, 226
Solutions, 12, 13, 186, 192–93
    calculating quantities of, 206
    chemical reactions in, 61–63
    concentration of, 203
    energy changes and formation of, 199–100
    in equilibrium, 224
    and handling of chemicals, 220
    pH of, 239, 240
    preparation dilution, 216–18
    preparation of, 803–804
    reactions in, 281
    saturated. *See* saturated solutions
Solution stoichiometry, 288, 300–302
Solvent extractions, 387
Solvents, 61, 192
    water as, 190, 197
Sørensen, Søren, 239
Specific heat capacity, 485
Spectator ions, 283
Spectrophotometers, 315
Spectroscopy, 114, 315
Standard ambient temperature and pressure (SATP), 14–15, 149, 169, 295, 495, 510
Standard cell potential, 627
Standard cells, 627, 639
Standard enthalpy of formation, 510
Standardized solutions, 330–31, 597
Standard molar enthalpies of formation, 827
Standard molar enthalpies of reaction, 496
Standard reduction potentials, 627, 628–30
Standard solutions, 215–16, 330
Standard temperature and pressure (STP), 149, 169, 295
Starches, 456
Steacie, Edgar, 721
Stearic acid, 255
Steel, 120, 634, 636, 637
Stereochemical formulas, 88
Stereochemistry, 91
Steroids, 375
Stock solutions, 216
Stoichiometry, 287, 320–29
    applications of, 292–93
    of electric cells, 652–57
    of redox reactions, 596–600, 676
STP. *See* Standard temperature and pressure (STP)
Strong acids, 200–201, 254–55, 718–20, 725, 726, 737
Strong bases, 256–55
Strontium chloride, 646
Strontium metal, 646
Structural analogs, 449
Structural formulas, 78, 85, 88, 89, 368, 372, 376
Structural isomers, 366–67
STS (science, technology, and society), 44–45, 79, 795, 806
Styrene, 447
Substitution reactions, 421–22
Sugar, 106
Sulfur
    compounds, 394
    in gasoline, 393–94
    impurities in fuels, 332
Sulfur dioxide, 145, 252, 332, 363, 393
Sulfur dioxide gas, 171
Sulfuric acid, 145, 200–201, 255
Sulfur trioxide, 145
Superacids, 724
Sustainable development, 357
Syntheses, 107, 173

T

Tables, 815
Tables of evidence, 10
Tarnish. *See* Silver sulfide (tarnish)
Tartaric acid, 437
Taube, Henry, 561
Taylor, Richard, 23

Technological applications, 45
Technological perspectives, 45
Technological problem solving, 236, 796
    quality control tests, 293
    scientific knowledge and, 614–15
    trial-and-error approach to, 219, 276, 615
Technology/technologies. See also Chemical technologies
    of Aboriginal peoples, 276, 449, 614, 615, 621
    alternative energy, 507–508
    context of, 276
    contexts of applications, 236
    energy and, 474
    and environment, 277
    goal of, 276, 286
    implementation of, 277
    and nonrenewable resources, 277
    obsolescence of, 277
    pollution-reducing, 252
    science and, 8–9, 276, 478, 556, 558, 646, 690
    and social change, 277
    as solving practical problems, 276, 286
    sustainability of, 277
    testing in, 631
    time and, 277
    transfer, 621
Teflon (polytetrafluoroethylene), 447–48
Temperature
    changes, 498, 502
    equilibrium and, 719
    and equilibrium constant, 685
    and forward/reverse reactions, 693
    Le Châtelier's principle and, 693
    measurement technologies, 155–56
    reaction equations and, 279
    and solubility, 222
    volume and, 152–56
Testing, 248, 290, 291, 578, 597, 631, 710
Tetrahedral bonds, 92, 93, 94
Theoretical definitions, 18, 724
Theoretical descriptions, 18
Theoretical hypotheses, 18
Theoretical knowledge, 10
Theories, 9, 18, 98, 710
    empirical knowledge vs., 524–26
    evaluation of, 24–25
    evolution of, 733–35
    falsification of, 25
    and prediction of results in new situations, 42, 251
    proving of, 25
    replacement of, 25, 72
    restriction of, 25, 72
    revision of, 25, 72
    use of, 250
    validity of, 42
    verification of, 25
Theory of ionic dissociation, 526
Thermal cracking, 389
Thermal energy, 485

Thermal stability, 510
Thermochemistry, 474, 485
Thermometers, 155
Thomson, Sir William. See Kelvin, Lord
Thomson's atomic model, 19, 20
Thymol blue indicator, 336
Tires, 146, 159, 166, 185
Titrants, 328, 596, 751
Titrant solutions, standardizing, 330–31
Titration, 312, 328–39, 596, 751, 804
    acid–base indicators for, 336
    of acid–base reactions, 333
    pH changes during, 333
    pH curves, 334–35
Titration analysis, 314, 328–29
Toluene, 382
Torricelli, Evangelista, 149
Toxicologists, 204
Traditional ecological knowledge, 574. See also Indigenous knowledge (IK)
Transfats, 450, 451
Transistors, 127
Transition elements, 16
Transuranic elements, 16
Trial-and-error approach, 219, 228, 276, 318, 536, 615
Tri-esters, 449
Trigonal planar, 92
Triple bonds, 85

U

Ultraviolet (UV) radiation, 530
Uncertainty in science, 9, 44, 292, 795, 814
Universal gas constant, 173
Unsaturated fats, 450
Unsaturated hydrocarbons, 374
Urea, 349, 696–97
Urry, Lewis, 619

V

Valence electrons, 78, 81, 83, 85, 87, 124
Valence orbitals, 80
Valve trays, 386
Vanadium, 128, 586, 608
Van der Waals forces, 105–106, 173
Van't Hoff, Jacobus, 78, 82
Vaporization, 106, 109
Vapour deposition, 650
Vinyl benzene, 447
Vinyl chloride, 447
Viscosity, 386
Vitamin C, 438, 743, 744
Vitamins, water-soluble, 438
Volatile organic compounds (VOCs), 118, 383, 384, 426
Volatility, 113
Volta, Alessandro, 612, 622
Voltage, 613, 628
Voltaic cells, 622–26, 627, 639, 641
Voltaic pile, 612
Voltammetry, 337

Voltmeter, 613, 627, 628, 632
Volts, 613
Volume
    absolute temperature and, 154
    calculation of, 209
    and gas volume changes, 694
    Le Châtelier's principle and, 694
    pressure and, 150–52, 694
    temperature and, 152–56
Volumetric glassware, 215
VSEPR theory, 91–94
    multiple bonds in, 96–100

W

Waage, Peter, 684
Waste disposal, 810
Water. See also Aqueous solutions; Hydrologists
    acid–base reaction with, 249–50
    for fossil fuel extraction and refining, 360
    generalizations about solubility in, 222–23
    geometry of, 94
    hard, 63, 216
    ionization constant for. See Ionization constant for water, K_w
    oil recovery and, 388
    purity, 190, 238, 714
    repellents, 105
    soft, 63
    solubility in, 201
    as solvent, 197
    as universal solvent, 190
    as vapour, 53
Water and waste treatment plant operators, 200
Watson, James D., 113
Watts, 614
Weak acids, 201, 254–55, 718–20, 726, 737, 740–41
Weak bases, 256–57, 744, 745–46
Weather forecasts, 167
Weight, 149
Wöhler, Friedrich, 696
Wood alcohol. See Methanol
Wood pulp, bleaching of, 594
Workplace Hazardous Material Information System (WHMIS), 809
Written response questions, 824–25

X

Xu, Yunjie, 115

Y

Yener, Aslihan, 567

Z

Zero energy, 510

Credits

Quirks & Quarks Audio Clips. Originally broadcast on CBC radio program Quirks & Quarks. Produced by the Canadian Broadcasting Corporation. Use is courtesy of the Canadian Broadcasting Corporation.

REVIEW UNIT: p.2 (Fig 1) courtesy Margaret Au, p.2-3 Dick Hemingway.

CHAPTER 1: p.5 (a-f) Dave Starrett, (g) Shutterstock/Johanna Goodyear. p.7 Richard Megna/Fundamental Photographs. p.12 A. Smith, University of Guelph. p.13 (a) Sheila Terry/SPL, (b) Nelson Photo, (c) Dr. E.R. Degginger. p.15 AIP-Niels Bohr Library. p.18 Edgar Fahs Smith Collection, Special Collections, Dept. Van Pett-Dietrich Library Center, University of Pennsylvania. p.19 (Fig 2) D. Boone/CORBIS, (Fig 3b) Dave Starrett. p. 20 Bettman/CORBIS. p.23 David Spears/CORBIS. p. 25 Science Photo Library. p.26 (Fig 12,13) Richard Megna/Fundamental Photographs. p.28 Richard Siemens. p.29 Richard Siemens. p.31 (Fig 7) Richard Siemens, (Fig 8) Jeremy Jones. p.35 Richard Siemens. p.41 Courtesy of the National Research Council of Canada.

CHAPTER 2: p. 43 Thomas Michael Corcoran. p.44 Jeremy Jones. p.46 (Fig 1a) AFP/CORBIS, (b) CORBIS, (c) European Space Agency/SPL. p.48 Richard Siemens. p.51 Richard Siemens. p.52 Dave Starrett. p.54 Andre Maslennikov/Peter Arnold, Inc. p.55 Richard Siemens. p.58 PhotoDisc/Don Farrell. p.59 (Fig 3) Richard Siemens, (Fig 4) Nelson Photo. p.61 Richard Siemens. p.62 Richard Siemens. p.63 Dave Starrett. p.66 P&T Lee. p. 69 Bill Lai/The Image Works. p.70 Michael Coyne/The Image Bank. p.71 (Fig 3) Andrew Lambert Photography/Photo Researchers, Inc., (Fig 4) Charles D. Winters/Photo Researchers, Inc.

UNIT 1: p.73 Duomo/CORBIS. p.75 (Fig 2) Courtesy of Intel, (Fig 3) José Manuel Sanchis Calvete/CORBIS.

CHAPTER 3: p.77 W. Perry Conway/CORBIS. p.78 Science Pictures Ltd./CORBIS. p.81 Roger Ressmeyer/CORBIS. p.82 (Fig 6) Sidney Moulds/SPL, (Fig 7a) Dave Starrett. p.83 Royal Ontario Museum. p.87 Klaus Guldbransen/SPL. p.90 Richard Siemens. p.91 courtesy Dr. Ronald Gillespie. p.85 Bettmann/CORBIS. p.97 Bryan Peterson/Taxi. p.103 Nelson Photo. p.105 (Fig 1) Nelson Photo, (Fig 2) Thomas Wiewand/MaXx Images Inc. p.107 Duke University Archives. p.108 Courtesy National Research Council. p.112 (Fig 10) Will & Deni McIntyre/Photo Researchers, Inc., (Fig 11) Kenneth G. Libbrecht. p.113 Robert Pickett/CORBIS. p.114 (Fig 14) Sinclair Stammers/SPL, (Fig 15) Courtesy of Dr. Robert J. Le Roy. p.115 courtesy Dr. Yunjie Xu, Chemistry Dept. University of Alberta. p.116 (Fig 18) Chris Collins/CORBIS, (Fig 19) PictureQuest, (Fig 20) Courtesy of Brian M. Goldstein. p.118 D'Arcy Butler, Dawson City, Yukon. p.119 (Fig 1) Dave Starrett, (Fig 2) Warren Morgan/CORBIS. p.120 John Mead/SPL. p.121 Andrew Syred/Tony Stone Images. p.123 (Fig 7) Bruce Iverson Photomicrography, (Fig 8) University of Alberta. p.125 (Fig 12 a&b) José Manuel Sanchis Calvete/CORBIS, (Fig 12 c) Biophotoassociates/SPL/Photo Researchers, Inc., (Fig 13) Photos.com. p.126 Courtesy Look4Ideas.com LED Lighting. p.127 (Fig 18) Charles O'Rear/CORBIS, (Fig 19) NASA/AFP/CORBIS. p.128 Mary M. Martin/Chips Magazine. p.130 (Fig 21 top) Charles O'Rear/CORBIS, (Fig 21 bottom) James L. Amos/CORBIS. p.138 C.R. Tompkins. p.139 C.R. Tompkins. p.140 Phil Degginger/Color-Pic, Inc. p.141 Nelson Photo.

CHAPTER ?: p.143 Roger Ressmeyer/CORBIS. p.144 (Fig 1) Chris ?/CORBIS, (Fig 2) Royalty Free/CORBIS, (Fig 3) Dick Hemingway.

CHAPTER 4: p.147 Richard Olivier/CORBIS. p.149 Bryan & Cherry ?tography. p.151 Bettmann/CORBIS. p.152 (Fig 5 & 6) ?BIS. p.156 Adam Woolfitt/CORBIS. p.158 ?BIS. p.159 (Fig 13) Courtesy of the New York State ?N.Y., (Fig 14) Michael Coyne/The Image Bank. p.160 ?ig 16) Courtesy Moteur Developpement

International, (Fig 17) Lowell Georgia/CORBIS. p.161 CORBIS. p.162 (Fig 19) Boreal. (Fig 20) CORBIS. p.164 Ashley & Crippen/National Archives of Canada. p.165 Charles D. Winters/Photo Researchers, Inc. p.168 Courtesy Shell Canada. p.169 Richard Siemens. p.170 PhotoDisc/John A. Rizzo. p.171 Anne Rippy/The Image Bank. p.173 Photo Researchers, Inc. p.174 Kevin R. Morris/CORBIS. p.176. Todd Powell/Index Stock Imagery, Inc. p.183 (Fig 1) J.A. van Kessel Enterprises Ltd. (Fig 2) Ian Davis-Young/Valan Photos.

UNIT 3: p.187 First Light/Larry McDougal.

CHAPTER 5: p.191 Royalty Free/CORBIS. p.192 A. Smith/University of Guelph. p.193 Courtesy of Humco Holding Group. p.194 Matt Meadows/SPL/Photo Researchers, Inc. p.195 Anne Bradley. p.196 (Fig 5) Anne Bradley, (Fig 6) Shutterstock. p.203 Nelson Photo. p.204 Dave Starrett. p.206 courtesy Dr. David Schindler, University of Alberta. p.207 Anne Bradley. p.208 (Fig 6) Dave Starrett, (Fig 7) Nelson Photo. p.212 Dave Starrett. p.213 Randy Faris/CORBIS. p.215 (Fig 1) Ohaus Corporation, (Fig 2) Richard Siemens. p.216 (Fig 3) Culligan of Canada Ltd., (Fig 4) Richard Siemens. p.217 Richard Siemens. p.220 Anne Bradley. p.221 Richard Megna/Fundamental Photographs. p.224 (Fig 3) Gary Houlder/CORBIS, (Fig 4) Richard Siemens. p.226 P&T Lee. p.231 Richard Siemens. p.232 Nelson Photo. p.233 Tom Myers/Photo Researchers, Inc.

CHAPTER 6: p.235 Ted Spiegel/CORBIS. p.236 Stephen Sharnoff. p.237 Richard Megna/Fundamental Photographs. p.238 Richard Siemens. p.239 Dave Starrett. p.242 Richard Siemens. p.245 (Fig 1) Richard Siemens, (Fig 2) Richard Megna/Fundamental Photographs. p.246 Richard Siemens. p.252 Canapress. p.255 (Fig 1) CORBIS, (Fig 2) Richard Siemens. p.256 Richard Megna/Fundamental Photographs. p.257 Anne Bradley. p.258 P&T Lee. p.266 Nelson Photo. p.267 Corel. p.268 P&T Lee

UNIT 4: p.271 TEK Image/SPL. p.272 (Fig 1, 2 & 4) Dave Starrett, (Fig 3) Charles D. Winters/Photo Researchers, Inc.

CHAPTER 7: p.275 courtesy Agrium Inc. p.276 John Turner/CORBIS. p.277 (Fig 2) Dick Hemingway, (Fig 3) Ron Sanford/Photo Researchers, Inc. p.278 (Fig 4) Canadian Space Agency, (Fig 5) Richard Siemens. p.279 (Fig 6) MSFC/NASA, (Fig 7) Photo courtesy of The University of Kansas, Museum of Anthropology, L.L. Dyche Collection, E2053. p.280 (Fig 8) Richard Siemens, (Fig 9) Dick Tompkins. p.281 (Fig 10) Royalty Free/CORBIS, (Fig 11) Richard Siemens. p.282 Jeremy Jones. p.284 Eva Omes, City of Toronto Works and Emergency Services. p.285 (Fig 15) Raymond Gehman/CORBIS, (Fig 16) Melissa King/Shutterstock. p.286 (a) Adrian Wyld/CP Picture Archive, (b) National Defense Imagery Library. p.289 devries mikkelsen/First Light. p.290 (Fig 3) Dr. E.R. Degginger, (Fig 4) Alcan Smelters & Chemical Ltd. Kitmat, B.C. p.291 (Fig 5a) Science Photo Library, (Fig 5b) Nelson Photo, (Fig 6) Richard Siemens. p.292 Dave Starrett. p.294 Richard Siemens. p.297 Bettmann/CORBIS. p.298 Bruce Logan. p.299 (Fig 5a) Methane Hydrate Library/Texas A&M University-Geochemical & Research Environmental Group, (Fig 5b) Courtesy IFM-Geomar, (Fig 6) GSD-Natural Resources Canada. p.300 Potash & Phosphate Institute. p.303 Henry Birks & Sons. p.304 (Fig 1a) Martin Bond/Photo Researchers, Inc., (Fig 1b) José Manuel Sanchis Calvete/CORBIS. p.309 Al Harvey/The Slide Farm. p.310 (Fig 2 & 4) Nelson Photo, (Fig 3) Jeremy Jones. p.311 (Fig 5) Dick Hemingway, (Fig 6) Husky Oil.

CHAPTER 8: p.313 Jim Varney/Photo Researchers, Inc. p.314 Dave Starrett. p.315 (Fig 2) Dave Starrett, (Fig 3) Richard Siemens, (Fig 4) Scott Bauer/USDA K9286-1, (Fig 5) Jan Curtis. p.317 Richard Siemens. p.318 Dick Tompkins. p.321 Rick Eglinton/Toronto Star. p.323 Nelson Photo. p.325 CORBIS. p.326 (Fig 6) courtesy Agrium Inc., (Fig 7) Potash & Phosphate Institute. p.327 (Fig 9) Charles D. Winters/Photo

Researchers, Inc., (Fig 10) Hudson Bay Diecasting. p.332 (Fig 2) Nelson Photo, (Fig 3) Alan Towse/CORBIS. p.333 Martyn F. Chillmaid/Photo Researchers, Inc. p.337 www.bioanalytical.com. p.338 (Fig 6) Todd Gray, (Fig 7 & 8) Courtesy Jed Harrison. p.340 Dick Tompkins. p.341 Dick Tompkins. p.343 Anne Bradley. p.344 Dave Starrett. p.345 P&T Lee. p.349 Courtesy Agrium Inc. p.350 J & L Weber/Peter Arnold Inc. p.351 (Fig 3) dpa/HO/Landov, (Fig 4) Dave Starrett.

UNIT 5: p.352 (inset) courtesy Jean Cooley, p.352-353 Mike Dobel/Masterfile.

CHAPTER 9: p.357 (top) Al Harvey/Slide Farm, (centre) Daryl Benson/Masterfile, (right) Syncrude Canada Ltd. (bottom) Royce Hopkins/Lone Pine Photo. p.358 (Fig 1) Richard B. Hoover, Elena Pikuta and Asim Bej, NASA/NSSTC, University of Alabama at Huntsville, and the University of Alabama at Birmingham, (Table 1, top to bottom) Alan Sirulnikoff/SPL/Photo Researchers, Inc., Charles D. Winters/Photo Researchers, Inc., Andrew Lambert Photography/SPL/Photo Researchers, Inc., Astrid & Hans Frieder Miehler/SPL/Photo Researchers, Inc., Suncor Energy Inc., PhotoDisc/Getty Images. p.360 CP Picture Archives. p.363 (Fig 2 & 3) Shell Canada Ltd., (Fig 4) Frank Jenkins Enterprises. p.367 Richard Siemens. p.370 P&T Lee. p.381 Dr. E. Keller, Kristallographisches Institut der Universitat Freiburg, Germany, using the computer program SCHAKAL92. p.385 Richard Siemens. p.386 Dick Tompkins. p.390 Courtesy Korite International. p.395 (Fig 9) Suncor Energy Inc., (Fig 10) courtesy Syncrude, (Fig 11) Shell Canada, (Fig 12) Paul Rapson/Photo Researchers, Inc. p.398 Thomas Del Brase/Stone/Getty Images.

CHAPTER 10: p.418 NASA. p.425 Ron Stroud/Masterfile. p.426 Frank Jenkins. p.430 Ron Watts/CORBIS. p.431 Frank Jenkins. p.432 Chiquita Banana. p.433 Nova Chemicals. p.436 (Fig 1) AFP/CORBIS, (Fig 2a) Michael Freeman/CORBIS. p.437 (Fig 3a) Andrew Lambert Photography/Photo Researchers, Inc., (Fig 3b) Catherine Karnow/CORBIS. p.446 Greg Epperson/MaXx Images. p.447 (Fig 4) Wire Images Stock/Masterfile, (Fig 5) istockimages/Bojan Tezek. p.450 Tony Freeman/PhotoEdit. p.458 Dave Starrett. p.469 Dow Chemical Canada Inc.

UNIT 6: p.474-475 Royce Hopkins/Lone Pine Photo, p. 476 (Fig 1a) courtesy Chris Lees/Cochrane High School, Alberta Sustainable Development Project, (Fig 1b) Janet Foster/Masterfile, (Fig 1c) Canadian Hydro-Preston Stuart Communications, (Fig 1d) Bill Donahue/Freshwater Research Ltd.

CHAPTER 11: p.479 (Fig 1) Al Harvey/Slide Farm, (Fig 2) Frank Jenkins. p.480 Sherman Hines/Masterfile. p.481 (Fig 2) Courtesy Drake Landing Solar Community, (Fig 3) Courtesy Ballard Power Systems, (Fig 4) Trans-Canada Pipelines. p.482 (Fig 5) LLC, Fogstock/Index Stock, (Fig 6) IT Stock Free/MaXx Images. p.485 Dave Starrett. p.487 Science Photo Library. p.491 Sergio Dorantes/CORBIS. p.496 Greg Locke/First Light. p.497 Courtesy of Inco. p.498 CORBIS. p.500 Dr. E.R. Degginger. p.501 Sherry Ward Design, Calgary, Alberta. p.504 (Fig 3) Frank Jenkins, (Fig 4) CP Rail. p.507 (Fig 6) Dave Olecko/Bloomberg News/Landov, (Fig 7) Courtesy Canadian Hydro Developers, Inc. p. 510 Photo Researchers, Inc. p.511 Courtesy Graymont Lime Group, Alberta. p.513 P&T Lee. p.514 Imperial Oil Inc. p.515 P&T Lee.

CHAPTER 12: p.523 Alfred Pasieka/SPL/Photo Researchers, Inc. p.527 (Fig 4a) NASA, (Fig 4b) UPI/Bettmann Newsphotos. p.531 Charles D. Winters/Photo Researchers, Inc. p.532 (Fig 1) AP Photo/Mark Humphrey, (Fig 2) Bill Brooks/Masterfile. p.534 Brian Willer/Masterfile. p.535 (Fig 1) Ontario Ministry of Natural Resources, (Fig 2) EyeSquared. p.536 P&T Lee. p.537 Thomas Eisner and Daniel Aneshansley/Cornell University. p.541 P&T Lee. p.546 Shell Canada Ltd.

UNIT 7: p.552 (inset) Don Lawton, p.552-553 Robert Pickett/CORBIS, p.555 Dave Starrett.

CHAPTER 13: p.557 EyeSquared. p.558 (Fig 1, left to right) photo courtesy of P.E. Martin, S.R. Martin and Michigan Technological University Archaeology Lab., Jonathan Blair/CORBIS, Charles Philip/Visuals Unlimited, Angelo Hornak/CORBIS, Jim Amos/SPL/Photo Researchers, Inc. p.559 (Fig 2) Bettmann/CORBIS, (Fig 3) Photo by Cary Smith. p.560 Gianni Dagli Orti/CORBIS. p.561 (Fig 5a) Bettmann/CORBIS, (Fig 5b) The Nobel Foundation, (Fig 6) Dave Starrett. p.562 Richard Siemens. p.564 Health & Safety Laboratory/Science Photo Library. p.567 (Fig 10) Dave Starrett, (Fig 11) Courtesy of Aslinhan Yener, University of Chicago. p.568 Richard Siemens. p.590 Sindo Farina/Science Photo Library. p.571 Dave Starrett. p.575 Tom Brakefield/CORBIS. p.576 Dave Starrett. p.577 Dave Starrett. p.579 NASA p.580 Jim Varney/Science Photo Library. p.585 Sherman Hines/Masterfile. p.591 Nativestock Pictures. p.593 Courtesy CMI, Inc. p.594 Courtesy Abitibi-Price Inc. p.596 Richard Siemens. p.598 Courtesy of Mark Fritz/Epocal Inc. p.599 Richard Seimens. p.603 Richard Siemens. p.606 Al Harvey/Slide Farm. p.607 BrandX Pictures/G.K.& Vikki Hart/Alamy. p.608 (Fig 3) Nelson Photo, (Fig 4) GC Minerals/Alamy.

CHAPTER 14: p.611 Pierre Paul Poulin/MAGMA. p.615 Lester V. Bergman/CORBIS. p.618 Courtesy of UTC Fuel Cells. p.619 Ron Kuntz/Associated Press. p.621 AFP/CORBIS. p.634 Craig Aurness/CORBIS. p.635 Dave Starrett, (Fig 14) EyeUbiquitous/CORBIS. p.637 CORBIS. p.638 Science VU/NASA/Visuals Unlimited. p.639 Maximilian Stock Ltd./SPL. p.646 Mary Evans Picture Library. p.648 Courtesy Dow Chemical. p.650 CORBIS. p.651 CP Photo/Ken Gigliotti. p.653 Science Photo Library. p.655 Kenncott. p.656 Boden/Ledingham/Masterfile. p.662 Sciencephotos/Alamy. p.665 James L. Amos/Peter Arnold Inc. p.668 Richard Megna/Fundamental Photographs.

UNIT 8: p.670-671 Alfred Pasieka/Science Photo Library

CHAPTER 15: p.675 (Fig 1) Ken Davies/Masterfile, (Fig 2 & 3) Dave Starrett. p.676 Richard Siemens. p.677 Dave Starrett. p.681 Visuals Unlimited. p.683 Courtesy Paul Kebarle, University of Alberta. p.684 Reproduced courtesy of the Library and Information Centre, Royal Society of Chemistry. p.688 Richard Siemens. p.690 (Fig 1) The Edgar Fah Smith Collection, (Fig 2) Richard Siemens. p.697 (Fig 9) Courtesy of Agrium, Calgary, Alberta, (Fig 10) Imperial Oil Ltd. p.698 Dave Starrett. p.699 Royalty Free/CORBIS. p.701 Richard Siemens. p.705 Courtesy of Inco. p.707 K. Imamura/zefa/CORBIS. p.709 Richard Siemens.

CHAPTER 16: p.712 Richard Siemens. p.713 Richard Siemens. p.716 P&T Lee. p.719 Richard Siemens. p.721 National Research Council Canada. p.722 Reproduced courtesy of the Library & Information Centre, Royal Society of Chemistry. p.723 Dave Starrett. p.724 Dave Starrett. p.728 Courtesy Farideh Jalilehvand, University of Calgary. p.729 (Fig 8l) P&T Lee, (Fig 8r) Dave Starrett. p.731 Richard Siemens. p.732 Richard Siemens. p.733 (Fig 11) Leonard de Selva/CORBIS, (Fig 12) Michael Nicholson/CORBIS. p.734 (Fig 13 & 16) Bettmann/CORBIS, (Fig 14 & 15) Reproduced courtesy of the Library and Information Centre, Royal Society of Chemistry. p.736 P&T Lee. p.743 (Fig 2) P&T Lee. p.746 Elena Segatini Bloom/CORBIS. p.750 Charles Bach/Photo Researchers, Inc. p.751 Ontario Ministry of the Environment. p.753 Richard Siemens. p.758 Martin Bond/Science Photo Library. p.766 University of Toronto. p.778 Nelson Photo.

APPENDICES: p.780 Phil Schermeister/CORBIS/MAGMA.

SOME TYPICAL COMMERCIAL REAGENT SOLUTIONS

Reagent (• strong acids)	Formula	Concentration (mol/L)	Concentration (mass %)
acetic acid	$CH_3COOH(aq)$	17.4	99.5
ammonia	$NH_3(aq)$	14.8	28
carbonic acid	$H_2CO_3(aq)$	0.039	0.17
• hydrochloric acid	$HCl(aq)$	11.6	36
• nitric acid	$HNO_3(aq)$	15.4	69
phosphoric acid	$H_3PO_4(aq)$	14.6	85
sodium hydroxide	$NaOH(aq)$	19.1	50
sulfurous acid	$H_2SO_3(aq)$	0.73	6
• sulfuric acid	$H_2SO_4(aq)$	17.8	95

SELECTED SI PREFIXES

Prefix	Symbol	Factor
peta	P	10^{15}
tera	T	10^{12}
giga	G	10^9
mega	M	10^6
kilo	k	10^3
milli	m	10^{-3}
micro	μ	10^{-6}
nano	n	10^{-9}
pico	p	10^{-12}
femto	f	10^{-15}

DEFINED (EXACT) QUANTITIES

1 t	=	1000 kg = 1 Mg
STP	=	0 °C and 101.325 kPa (use 0 °C and 101 kPa)
SATP	=	25 °C and 100 kPa
0 °C	=	273.15 K (use 273 K)
1 atm	=	101.325 kPa (use 101 kPa)
1 atm	=	760 mm Hg
1 bar	=	100 kPa

MEASURED (UNCERTAIN) QUANTITIES

N_A	=	6.02×10^{23}/mol
R	=	8.314 kPa·L/(mol·K)
F	=	9.65×10^4 C/mol
K_w	=	1.0×10^{-14}
c (in a vacuum)	=	3.00×10^8 m/s
V_{STP}	=	22.4 L/mol
V_{SATP}	=	24.8 L/mol
d_{H_2O}	=	1.00 g/mL

SOME COMMON ACID–BASE INDICATORS

Common name of indicator	Suggested symbol	Colour of HIn(aq)	pH range of colour change	Colour of In$^-$(aq)	K_a
methyl violet	HMv	yellow	0.0 - 1.6	blue	$\sim 10^{-1}$
cresol red	H_2Cr	red	0.0 - 1.0	yellow	$\sim 10^{-1}$
thymol blue	H_2Tb	red	1.2 - 2.8	yellow	2.2×10^{-2}
orange IV	HOr	red	1.4 - 2.8	yellow	$\sim 10^{-2}$
benzopurpurine-4B	HBp	violet	2.2 - 4.2	red	$\sim 10^{-3}$
congo red	HCo	blue	3.0 - 5.0	red	$\sim 10^{-4}$
methyl orange	HMo	red	3.2 - 4.4	yellow	3.5×10^{-4}
bromocresol green	HBg	yellow	3.8 - 5.4	blue	1.3×10^{-5}
methyl red	HMr	red	4.8 - 6.0	yellow	1.0×10^{-5}
chlorophenol red	HCh	yellow	5.2 - 6.8	red	5.6×10^{-7}
bromothymol blue	HBb	yellow	6.0 - 7.6	blue	5.0×10^{-8}
litmus	HLt	red	6.0 - 8.0	blue	$\sim 10^{-7}$
phenol red	HPr	yellow	6.6 - 8.0	red	1.0×10^{-8}
cresol red	HCr^-	yellow	7.0 - 8.8	red	3.5×10^{-9}
metacresol purple	HMp	yellow	7.4 - 9.0	purple	$\sim 10^{-8}$
thymol blue	HTb^-	yellow	8.0 - 9.6	blue	6.3×10^{-10}
phenolphthalein	HPh	colourless	8.2 - 10.0	red	3.2×10^{-10}
thymolphthalein	HTh	colourless	9.4 - 10.6	blue	1.0×10^{-10}
alizarin yellow R	HAy	yellow	10.1 - 12.0	red	6.9×10^{-12}
indigo carmine	HIc	blue	11.4 - 13.0	yellow	$\sim 10^{-12}$
1,3,5-trinitrobenzene	HNb	colourless	12.0 - 14.0	orange	$\sim 10^{-13}$